Physics of Shock Waves and High-Temperature Hydrodynamic Phenomena

Ya. B. Zel'dovich and **Yu. P. Raizer**
Academy of Sciences, Moscow

Edited by

Wallace D. Hayes and **Ronald F. Probstein**
Princeton University *Massachusetts Institute of Technology*

DOVER PUBLICATIONS
Garden City, New York

EDITOR'S NOTE TO DOVER EDITION, 2002: The translation of this work from Russian to English was done by Scripta Technica, Inc., for the publication of the work in English in two volumes in 1966 (Volume I) and 1967 (Volume II). The basis for this translation was the second, expanded and corrected Russian edition, published in 1966, but the work in English is not identical to that Russian edition, as the U.S. editors, Professors Wallace D. Hayes and Ronald F. Probstein, made further changes and additions to the text, including the addition of explanatory notes and source notes. This Dover edition restores the original one-volume format of the 1963 Russian first edition. The Author Index to Volume I, retained as pages 447–451 of this edition, and the Subject Index to Volume I, retained as pages 452–464 of this edition, contain information that also is included in the complete Author Index and Subject Index to the entire work, found on pages 887–916. Also, the Appendix appears twice in this edition, on pages 441–445 and again on pages 881–885, because it has been retained where it appeared in both Volume I (1966) and Volume II (1967) of the first English-language edition.

Bibliographical Note

This Dover edition, first published in 2002, is a republication in one volume of the two-volume work originally published in English in 1966 and 1967 by Academic Press, Inc., New York. Except for the elimination of duplication of material, such as prefaces, that appeared in each of the original two volumes, and for correction of a few typographical errors, this edition is an unabridged, unaltered republication of the original edition. The Dedication and the Preface to the Dover Edition were added to the 2002 edition.

Library of Congress Cataloging-in-Publication Data

Zel'dovich, IA. B. (IAkov Borisovich)
[Fizika udarnykh voln i vysoktemperaturnykh gidrodinamicheskikh iavlenii. English]
Physics of shock waves and high-temperature hydrodynamic phenomena / Ya. B. Zel'dovich and Yu. P. Raizer.
p. cm.
Originally published (in English, in 2 v.): New York : Academic Press, 1966–1967. With new pref.
Includes bibliographical references and index.
ISBN-13: 978-0-486-42002-8 (pbk.)
ISBN-10: 0-486-42002-7 (pbk.)
1. Shock waves. 2. Gases at high temperatures. I. Raizer, IU. P. (IUrii Petrovich). II. Title.

QC168 .Z3813 2002
533'293—dc21

Manufactured in the United States of America
42002714
www.doverpublications.com

This Dover edition is dedicated to the memories of

Wally Hayes (1918–2001)

and

Yakov Borisovich Zel'dovich (1914–1987)

Preface to the Dover Edition

It has been thirty-five years since the first English-language edition of this book was published. It is a credit to the authors that, despite the many advances that have taken place in this period, the book remains fresh and current in its relevance to the astrophysical, nuclear, aerodynamic, and seismological communities. The need to make the subject material available to younger workers in these fields was recognized by Dmitri Mihalas of the Los Alamos National Laboratory. Together with many of his colleagues at Los Alamos, he brought this to the attention of the editors at Dover Publications, who agreed to bring out the book in an inexpensive edition.

Many thanks are due to John Grafton, who formulated the editorial plan for publishing the two-volume work in one volume. The original edition is unabridged and unaltered, except for the correction of a few typographical errors and the elimination of duplicate front matter that appeared in the second of the original two English-language volumes.

The Dover edition is dedicated to the memory of my close friend, colleague, and co-editor of the English version of the book, Wally Hayes, whose scientific contributions profoundly shaped our understanding of shock waves and high-speed flight. He knew of the forthcoming appearance of this edition but did not live to see it in print. It also is dedicated to the memory of Yakov Borisovich Zel'dovich, a colleague and guiding author of the book–a man of unusual scientific breadth who will be remembered as one of the giants in physics and astrophysics in the 20th century.

Ronald F. Probstein
Cambridge, Massachusetts
June, 2001

Editors' foreword

The lack of a comprehensive book on high-temperature physical gasdynamics has been felt for a long time. Since we wrote the first edition of "Hypersonic Flow Theory" we have particularly felt the need for a complementary text covering this field. A few books in the field have appeared, treating some of the pertinent topics. The brilliant text of Zel'dovich and Raizer first appeared early in 1964, and was not only outstanding but completely unique. The revised and updated second edition of this text is presented here in English, shortly after the Russian version. We hope that these two volumes together with the second edition of our own "Hypersonic Flow Theory" will serve to present a comprehensive picture of high-temperature high-speed flows in both their physical and hydrodynamical aspects. Our second edition will be in two volumes, the first on Inviscid Flows already published, and the second on Viscous and Rarefied Flows planned for 1969.

Zel'dovich and Raizer's text is truly comprehensive in the field of high-temperature gasdynamics, dealing thoroughly with all the essential physical aspects and their influence on the dynamics and thermodynamics of continuous media. The authors bring a deep physical insight to bear in explaining the nature of seemingly most complicated phenomena. Mathematical and formal treatments are kept to a minimum, while the results of such treatments are reported and compared with those of simplified approaches. The actual scope of the text is discussed in the authors' prefaces.

The standard approach of the theoretical physicist for many of the subjects treated here is quite formal, and not readily connected in the student's mind with physical ideas. Relatively rare are approaches which are correct in essentials and which are based upon sound physical reasoning from fundamental concepts. It is here that this text excels and is unduplicated. The authors consistently explain the physical phenomena of interest on the simplest correct physical basis, using classical instead of quantum physics where this is possible.

As an inevitable consequence of the comprehensive nature of the text, it is a large one. In the English version we have been forced to split the work into two volumes. This split is a fairly natural one, however, as the first volume covers primarily the fundamental hydrodynamic, physical, and chemical processes involved; the second volume covers primarily the application and interaction of these processes in a multitude of important problems. However, the two volumes together certainly constitute a fully integrated and interrelated work.

It is truly difficult to qualify the audience to which this work is directed. Our physicist colleagues who have used parts of the book in class have found it exceptionally well adapted to teaching in graduate courses. At the same time, they, like the editors, have learned from the book and have obtained a better appreciation for the subject as a whole. The book is also ideally suited for engineers, presenting for them not only the basis for acquiring the physical understanding they need, but also specific formulas, methods, and experimental results for use in making practical calculations. Thus, without exaggeration, we can say that it is well suited to researchers, engineers, students, and professors. In a word, the authors have succeeded in the aims set forth in their original preface.

The authors are thoroughly acquainted with the world literature, and the references cited are comprehensive. The Soviet journals cited are mostly those now regularly translated into English, so that very few of the references cited in this edition are to be found only in Russian.

The editors are grateful for the close and friendly cooperation of the authors. We have exchanged comments, clarifications, and lists of errata. Some editorial changes have been incorporated with the authors' consent. Where it was thought helpful to indicate a different point of view on a topic, to discuss terminology, or occasionally to amplify a statement, we have added an editors' footnote. Subject and author indexes have been added, covering Volume I at the end of that volume, and covering both volumes at the end of Volume II.

The editors gratefully acknowledge the financial support and assistance of the Advanced Research Projects Agency through a contract technically administered by the Fluid Dynamics Branch of the Office of Naval Research with the Massachusetts Institute of Technology. Without this support the editors would not have been able to carry out this project. For the main part of the translating we are indebted to Scripta Technica, Inc. We wish to thank Miss Margaret Gazan for her skillful handling of the secretarial work and of many editorial details. A number of our colleagues have given us valuable comments and corrections, for which we express our sincere thanks.

We also wish to thank Mezhdunarodnaya Kniga for furnishing us the figures. We express our most sincere appreciation to our publishers for their cooperative attitude in this endeavor.

Our warmest thanks go to the authors for their wholehearted cooperation in this undertaking.

September, 1966

Wallace D. Hayes
Ronald F. Probstein

Preface to the English edition

This book considers a large variety of problems of modern physics and engineering, which deal with shock waves, high temperatures and pressures, plasmas, strong explosions, very strong electrical discharges, interaction of intense laser radiation with a medium, etc. We have attempted not only to present the clearest possible interpretation of the physical bases of the phenomena arising in these fields, but also to give practical guidance to those who work with these subjects in science and modern technology.

The content of this book is determined to a large extent by the tastes of the authors. In particular, we have considered in more detail those phenomena and problems which were investigated by us personally. Naturally, more attention has been given to the work of Soviet authors. However, we have attempted to reflect in a sufficiently complete manner the work of American and British scientists, which has led to important advances in the solutions of the problems considered.

The text prepared by us for the English edition is almost identical with that of the second Russian edition, which is to be published in 1966. It contains important additions (and corrections) not contained in the first Russian edition of 1963.

We are very glad that this book has been translated into English and will become accessible to many foreign scientists and engineers. We are grateful to the translators for their work and value most highly the initiative of Professors Hayes and Probstein, who have undertaken the editorship of the translation, have shown great care and thoroughness, and have made a number of valuable comments.

October, 1965

YA. B. ZEL'DOVICH
YU. P. RAIZER

Preface to the first Russian edition

The requirements of modern technology have made it necessary for scientific research to penetrate into the domain of very high values of the state variables, such as with high concentrations of energy, very high temperatures and pressures, and extreme velocities. In practice, such conditions are encountered in strong shock waves, in explosions, in hypersonic flight of bodies in the atmosphere, in very strong electrical discharges, etc.

A great variety of physical and physical-chemical processes can occur in gases at high temperatures: excitation of molecular vibrations, dissociation, chemical reactions, ionization, and radiation of light. These processes affect the thermodynamic properties of gases, while at high velocities and with sufficiently rapid changes in the state of the fluid the rates of these processes also affect the motion of the fluid. Of special importance at very high temperatures are processes related to the emission and absorption of light and radiative heat transfer. The enumerated processes are of interest not only from the point of view of their energetic effect on the motion of the gas, but also frequently lead to changes in the composition of the gas and its electrical properties, to the emission of radiation from the gas and many optical phenomena, etc. An appreciable portion of this book is devoted to the study of all of these problems, comprising the newly-arisen branch of science termed "physical gasdynamics."

Of great scientific and practical interest is the study of strong shocks in solids. Recent achievements, which have permitted the compression of solid bodies by means of shock waves to pressures of millions of atmospheres, have opened new paths for the investigation of solid media at ultra-high pressures. Considerable attention has, therefore, been devoted to these problems in the present book.

Many scientific disciplines are interwoven here, including gasdynamics, shock wave theory, thermodynamics and statistical physics, molecular physics, physical and chemical kinetics, physical chemistry, spectroscopy, the general theory of radiation, the elements of astrophysics, and solid state physics. Many of the physical processes and phenomena considered are of differing character and are not directly related to each other. The result of such diversity in the material is the absence of obvious continuity in the contents of the book. Certain chapters are quite independent, and deal with completely different areas of physics or mechanics, so that not all chapters are related to each other. Hence, the reader interested in one or another particular topic will find it sufficient to become acquainted only with the corresponding chapters.

In examining the most diverse problems, even those of mathematical character, we endeavored primarily to explain the physical essence of the phenomena using the simplest mathematical tools, frequently resorting to estimates and semiqualitative analysis. At the same time, we attempted to help those physicists, fluid mechanicians, and engineers who work in the corresponding areas of applied physics and engineering, and to supply them with practical tools for independent analysis of many different and complex physical phenomena. For this purpose, the treatment of the majority of the phenomena examined is carried through to specific numerical results, the formulas for the calculation and evaluation of various quantities are presented in a convenient form for practical work, a large amount of useful experimental data and reference material is cited, etc.

The book is of a theoretical character and the description of experimental apparatus and methods is kept to a minimum. However, the presentation and comparison of experimental results with theoretically predicted values has been given an appropriate amount of attention.

The journal literature in "physical gasdynamics" is immense. As far as we know, however, no attempt has been made, either in the Soviet Union or elsewhere in the world, to present a systematic and generalized exposition — in a single book and from a single point of view — of the material in this new area of science. Apparently this book is the first attempt in this direction.

The literature cited in the text reflects the fact that the book was written during 1960 and 1961. However, references to more recent works and brief additions were added in those sections dealing with problems which are being developed at an especially rapid pace. This refers primarily to Chapters V, VI and VII.

The great variety of the phenomena and the large scope of the material forced us to limit the presentation to far from all the problems related to the vast area under consideration. We have not considered the more mathematical aspects of hydrodynamics, nor such problems as that of supersonic flow past bodies. We have only barely touched upon electromagnetic phenomena, and have not dealt at all with thermonuclear fusion, behavior of a plasma in a magnetic field, nor magnetohydrodynamics and magnetogasdynamics, combustion, detonation, etc. A great many books dealing with such problems are already available.

The selection of the material for the book has been to some extent a subjective one. A significant place in the text is devoted to phenomena which were investigated by the authors in their own studies. Thus, the authors' original works have served as a basis for part of the text — almost completely in Chapters VIII and IX, to a considerable extent in Chapters VII, X and XII, and partially in Chapter XI. Chapter I represents a complete revision of an earlier book by one of the authors, "Theory of Shock Waves and Introduction to Gasdynamics," published by the Academy of Sciences of the USSR in 1946.

We should like to express our especial gratitude to A. S. Kompaneets, who is responsible for working out a number of problems dealt with in the book and for many useful criticisms and remarks on the manuscript. We are grateful to L. B. Altshuler and S. B. Kormer for their remarks on the manuscript for Chapter XI, which is based on their work to a large extent. We are also grateful to M. A. El'yashevich who read the manuscript carefully and made valuable comments.

YA. B. ZEL'DOVICH
YU. P. RAIZER

Preface to the second Russian edition

The general structure of the book and the major part of the text in this second edition were retained without change. At the same time, certain chapters were thoroughly revised and a considerable amount of new material was added. Chapter V now contains a part devoted to breakdown (high-intensity ionization) processes and to the heating of gases by a focused laser beam. This is one of the most interesting phenomena connected with the interaction between an intense light beam and a medium. It was discovered experimentally several years ago, shortly after the development of lasers, which produce high pulse intensities measured in tens of megawatts and higher, and immediately attracted the attention of many physicists (including the authors of this book, who have published works on the theory of this phenomenon).

In connection with problems of gas ionization by laser radiation we have added sections to Chapter V in which emission and absorption of light by free electrons on collision with neutral atoms is considered. The lively interest which is now shown toward lasers has induced us to write a special section (in Chapter II) devoted to the semiclassical treatment of induced emission and of the laser effect.

Extensive changes were made in Part 3 of Chapter VI, in which we consider problems of ionization, recombination, and electronic excitation. This part has been virtually rewritten and extensively expanded in order to take into account modern views on these processes. According to these views an important role is played by stepwise ionization of atoms (first excited and then ionized) and electron capture into upper atomic levels through three-body collisions with subsequent deexcitation of the excited atoms through electron impact and radiative transition. Ionization in air has been considered in more detail. The presentation of the closely related problem of ionization of a gas in a shock wave (in Chapter VII) was also changed.

Sections of Chapter VIII, pertaining to the rate of change in the degree of ionization and of the "freezing" accompanying a sudden expansion of an ionized gas into a vacuum have been rewritten. This problem has been recently reexamined with account taken of electron capture into upper atomic levels as a result of recombination through three-body collisions.

On the basis of material which was contained in the first edition and of more recent results we have added in Chapter XII a part dealing with the propagation of shock waves in an inhomogeneous atmosphere with an exponential density distribution. We have added an appendix wherein are collected certain constants, relations between atomic constants, and relations between units and formulas which are frequently encountered in practice when dealing with the subject matter of this book.

We have here mentioned only the principal, but by far not all of the changes and additions which were made (we also note that mistakes and printing errors which were found in the first edition have been corrected).

Topics of physics and mechanics which were touched upon in the book are developing at an extremely rapid rate, with the consequent discovery of more and more new fields of application (an example of this is the phenomenon of breakdown and heating of gases in the focus of a laser beam).

As an evidence of the interest shown toward these scientific disciplines we cite the fact that immediately after publication of this book, an English translation was undertaken in the United States, and a need for a new edition very soon arose. We hope that this second revised and supplemented edition will be of use to specialists already working in the above fields of science and engineering and to those who are about to enter these fields.

YA. B. ZEL'DOVICH
YU. P. RAIZER

Contents

2. *Gases with Coulomb interactions* 215

IV. Shock tubes

V. Absorption and emission of radiation in gases at high temperatures

1. *Continuous spectra* 248

The Cited References, Author Index, and Subject Index listed on this page relate to chapters I–VI only. All three items for the entire work are listed at the end of the Contents.

VIII. Physical and chemical kinetics in hydrodynamic processes

IX. Radiative phenomena in shock waves and in strong explosions in air

I. Elements of gasdynamics and the classical theory of shock waves

1. *Continuous flow of an inviscid nonconducting gas*

§1. The equations of gasdynamics

Extremely high pressures, of the order of thousands of atmospheres, are required to achieve an appreciable compression of liquids (and solids). Therefore, under normal conditions it is possible to regard liquids as incompressible media. With the density changes small, the speed of the flow of a liquid is much smaller than the speed of sound; the sound speed serves as a characteristic velocity scale in describing continuous media. With small density changes and with flow velocities much smaller than the speed of sound, gases may also be considered as incompressible and their motion may be described in terms of the hydrodynamics of incompressible fluids. In contrast to liquids, however, appreciable changes in density and flow velocities close to the speed of sound are relatively easy to achieve in gases. In such cases the pressure change can be of the order of the pressure itself, e.g., when the gas is initially at atmospheric pressure and $\Delta p \sim 1$ atm. Under these conditions it is necessary to take into account the compressibility of the medium. The gasdynamic equations differ from the hydrodynamic equations for incompressible fluids by the fact that they account for the possibility of large density changes.

The state of a moving gas whose thermodynamic properties are known can be defined in terms of its velocity, density, and pressure as functions of position and time. These functions are, in turn, defined by the differential equations that describe the general laws of conservation of mass, momentum, and energy. These equations are given below without proof, and may be found, for example, in the book by Landau and Lifshitz [1].

We shall disregard gravitational effects, viscosity, and thermal conductivity*. A partial derivative with respect to time at a given point in space is denoted by $\partial/\partial t$, and a total derivative, describing the time change in any quantity following a moving fluid particle, by D/Dt. If $\mathbf{u}$ is the vector velocity

* Gasdynamic equations which take into account viscosity and thermal conductivity will be considered in §20.

of the fluid particle whose components are u_x, u_y, and u_z or u_i, where $i = 1$, 2, 3, then

$$\frac{D}{Dt} = \frac{\partial}{\partial t} + \mathbf{u} \cdot \nabla. \tag{1.1}$$

The first equation—the continuity equation—describes the conservation of mass of the fluid, that is, the fact that the density in a given volume element changes as a result of flow of the fluid into or out of this element:

$$\frac{\partial \rho}{\partial t} + \nabla \cdot \rho \mathbf{u} = 0. \tag{1.2}$$

Using (1.1), the continuity equation can be rewritten as

$$\frac{D\rho}{Dt} + \rho \nabla \cdot \mathbf{u} = 0. \tag{1.3}$$

For an incompressible fluid, where $\rho = const$, the continuity equation is

$$\nabla \cdot \mathbf{u} = 0. \tag{1.4}$$

The second equation expresses Newton's law and does not differ from the corresponding equation of motion for an incompressible fluid (p is the pressure)

$$\rho \frac{D\mathbf{u}}{Dt} = -\nabla p \tag{1.5}$$

or, in the form of Euler's equation,

$$\frac{\partial \mathbf{u}}{\partial t} + \mathbf{u} \cdot \nabla \mathbf{u} = -\frac{1}{\rho} \nabla p. \tag{1.6}$$

It is evident that the equations of motion and continuity when combined are equivalent to the law of conservation of momentum expressed in a form similar to (1.2),

$$\frac{\partial}{\partial t} \rho u_i = -\frac{\partial \Pi_{ik}}{\partial x_k}, \tag{1.7}$$

where Π_{ik} is the momentum flux density tensor

$$\Pi_{ik} = \rho u_i u_k + p\delta_{ik}. \tag{1.8}$$

Equation (1.7) expresses the fact that a change in the ith component of momentum at a given point in space is related to the flux of momentum out of (or into) a small volume (first term in (1.8)) plus the force from the pressure field (second term)*.

* The summation on the right-hand side of (1.7) is carried out with respect to the subscript k ($k = 1, 2, 3$); $\delta_{ik} = 1$ for $i = k$ and $\delta_{ik} = 0$ for $i \neq k$.

The third equation is essentially new to the hydrodynamics of incompressible fluids and is equivalent to the first law of thermodynamics, i.e., to the law of conservation of energy. It can be formulated as follows: A change in the specific internal energy ε of a given particle is a result of the work of compression done on the particle by the surrounding medium, and of the energy generated by external sources

$$\frac{D\varepsilon}{Dt} + p\frac{DV}{Dt} = Q. \tag{1.9}$$

Here $V = 1/\rho$ is the specific volume and Q is the energy generated by the external sources per unit mass of the material per unit time (Q can be negative when nonmechanical energy losses, as for example radiation losses, are present).

Using the equations of motion and continuity, the energy equation can be reduced to a form similar to (1.2) and (1.7)

$$\frac{\partial}{\partial t}\left(\rho\varepsilon + \frac{\rho u^2}{2}\right) = -\nabla\cdot\left[\rho\mathbf{u}\left(\varepsilon + \frac{u^2}{2}\right) + p\mathbf{u}\right] + \rho Q. \tag{1.10}$$

In physical terms, this equation states that a change in the total energy per unit volume at a given point in space occurs as a result of energy flux (in or out) during the fluid motion, the work of the pressure forces, and the energy supplied from external sources.

The continuity, motion, and energy equations form a system of five equations (the equation of motion is vectorial and is equivalent to three scalar equations) with five unknown functions of space and time: ρ, u_x, u_y, u_z, and p. It is assumed that the external energy sources Q are known, and that the internal energy ε can be expressed in terms of density and pressure, since the thermodynamic properties of the fluid are also assumed to be known: $\varepsilon = \varepsilon(p, \rho)$.

If the energy, as is frequently the case, is given not as a function of pressure and density, but either as a function of temperature T and density, or of temperature and pressure, then the equation of state $p = f(T, \rho)$ must be added to the system. The equation of state for a perfect gas is

$$pV = RT, \qquad p = R\rho T, \tag{1.11}$$

where R is the gas constant per unit mass*.

The energy equation (1.9) is a general one and is applicable when the fluid is not in a state of thermodynamic equilibrium. In the particular case of practical importance when the fluid is in thermodynamic equilibrium, this equa-

* $R = \mathcal{R}/\mu_0$, where $\mathcal{R}$ is the universal gas constant and μ_0 is the molecular weight.

tion can be written in a different form based on the second law of thermodynamics

$$T\,dS = d\varepsilon + p\,dV, \tag{1.12}$$

where S is the specific entropy. In the absence of external heat sources, the third gasdynamic equation is equivalent to the entropy equation for a particle, which is the same as the adiabatic flow condition

$$\frac{DS}{Dt} = 0. \tag{1.13}$$

The entropy of a perfect gas with constant specific heats can be expressed in a simple form in terms of pressure and density (specific volume)

$$S = c_v \ln pV^{\gamma} + const, \tag{1.14}$$

where γ is the isentropic exponent, equal to the ratio of the specific heats at constant pressure and at constant volume, $\gamma = c_p/c_v = 1 + R/c_v$. The entropy (or energy) equation (1.13) can, in this case, be written as a differential equation relating the pressure and density (volume)

$$\frac{1}{p}\frac{Dp}{Dt} + \gamma\frac{1}{V}\frac{DV}{Dt} = 0. \tag{1.15}$$

To the system of gasdynamic equations must be added the appropriate initial and boundary conditions.

§2. Lagrangian coordinates

The flow equations which consider the gasdynamic variables as functions of the space coordinates and time are called the Euler equations, or the flow equations in Eulerian coordinates.

Lagrangian coordinates are frequently used to describe one-dimensional flow, that is, plane and cylindrically and spherically symmetric flow. In contrast to Eulerian coordinates, Lagrangian coordinates do not determine a given point in space, but a given fluid particle. Gasdynamic flow variables expressed in terms of Lagrangian coordinates express the changes in density, pressure, and velocity of each fluid particle with time. Lagrangian coordinates are particularly convenient when considering internal processes involving individual fluid particles, such as a chemical reaction whose progress with time depends on the changes of both the temperature and the density of each particle. The use of Lagrangian coordinates also occasionally yields a shorter and easier way of obtaining exact solutions to the gasdynamic equations, or provides a more convenient numerical integration of them. The derivative with respect to time in Lagrangian coordinates is simply equivalent to the total derivative D/Dt. The particle can be described either in terms of the mass of fluid separating it from a given reference particle (in one dimension), or in terms of its position at the initial instant of time.

The use of Lagrangian coordinates is especially simple in the case of plane motion, when the flow is a function of only one cartesian coordinate x. Let us denote the Eulerian coordinate of a particular fluid particle by x and the coordinate of a reference particle by x_1 (as a reference particle we can choose a particle near a solid wall or near a gas-vacuum interface). Then the mass of a column of fluid of unit cross section between the reference particle and the particular fluid particle of interest is equal to

$$m = \int_{x_1}^{x} \rho \, dx. \tag{1.16}$$

The increment in mass resulting from the passage from one particle to a neighboring one is

$$dm = \rho \, dx. \tag{1.17}$$

The quantity m may be chosen as the Lagrangian coordinate.

If, as is frequently the case, the gas is initially at rest and its initial density is constant, $\rho(x, 0) = \rho_0$, then it is convenient to take the initial coordinate of the particle (relative to x_1), which we shall denote by a, as the Lagrangian coordinate. Then

$$a = \int_{x_1}^{x} \frac{\rho}{\rho_0} \, dx, \qquad da = \frac{\rho}{\rho_0} \, dx. \tag{1.18}$$

The equations for plane motion of a gas in Lagrangian coordinates take on a simple form. The continuity equation, written in terms of the specific volume $V = 1/\rho$ and the single x component of the velocity u, is

$$\frac{\partial V}{\partial t} = \frac{\partial u}{\partial m} \qquad \text{or} \qquad \frac{1}{V_0} \frac{\partial V}{\partial t} = \frac{\partial u}{\partial a}. \tag{1.19}$$

Here, as in the following equations, the derivative with respect to time is the total derivative D/Dt, though it is better to express it in the form of a partial derivative $\partial/\partial t$, in order to emphasize that it is taken with m and $a = const$, that is, for a given particle with a specified m or a coordinate. The equation of motion in Lagrangian coordinates is

$$\frac{\partial u}{\partial t} = -\frac{\partial p}{\partial m} \qquad \text{or} \qquad \frac{\partial u}{\partial t} = -V_0 \frac{\partial p}{\partial a}. \tag{1.20}$$

The energy equation, written either in the form (1.9) or in the entropy form

(1.13) retains the same form in the absence of external heat sources and dissipative processes (viscosity and heat conduction). Here, it is only necessary to replace D/Dt by $\partial/\partial t$. For a perfect gas with constant specific heats, (1.13) gives

$$pV^{\gamma} = f[S(m)], \tag{1.21}$$

where the function f depends only on the entropy of the given particle m. In so-called isentropic flow, where the entropy of all the particles is identical and does not vary with time, $f = const$, in which case the equation $pV^{\gamma} = const$ is valid in Lagrangian as well as Eulerian coordinates.

The Eulerian coordinate x does not enter the equation explicitly in the one-dimensional (plane) case. After the Lagrangian equations are solved and the function $V(m, t)$ is found, the dependence of the flow variables on the Eulerian coordinate may be obtained by integrating (1.17)

$$dx = V(m, t)\, dm, \qquad x(m, t) = \int_0^m V(m, t)\, dm + x_1(t). \tag{1.22}$$

In the cylindrical and spherical cases, the gasdynamic equations in Lagrangian coordinates are slightly more complicated than in the plane case. Here, the Eulerian coordinate enters the equations explicitly and an additional equation, relating the Lagrangian and Eulerian coordinates, must be added to the system. For example, in the spherical case it is possible to define the Lagrangian coordinate as the mass contained within a spherical volume about the center of symmetry

$$m = \int_0^r 4\pi r^2 \rho\, dr, \qquad dm = 4\pi r^2 \rho\, dr. \tag{1.23}$$

If the gas density is initially constant, then it is possible to take as the Lagrangian coordinate the initial radius r_0 of the "particle", considered here as an elementary spherical shell

$$\frac{4\pi r_0^3}{3}\rho_0 = \int_0^r 4\pi r^2 \rho\, dr, \qquad dr_0 = \frac{r^2}{r_0^2}\frac{\rho}{\rho_0}\, dr. \tag{1.24}$$

The continuity equation in spherical Lagrangian coordinates is

$$\frac{\partial V}{\partial t} = \frac{\partial}{\partial m} 4\pi r^2 u \qquad \text{or} \qquad \frac{1}{V_0}\frac{\partial V}{\partial t} = \frac{1}{r_0^2}\frac{\partial}{\partial r_0} r^2 u. \tag{1.25}$$

The equation of motion is

$$\frac{\partial u}{\partial t} = -4\pi r^2 \frac{\partial p}{\partial m} \qquad \text{or} \qquad \frac{\partial u}{\partial t} = -\frac{1}{\rho_0}\frac{r^2}{r_0^2}\frac{\partial p}{\partial r_0}. \tag{1.26}$$

The energy or entropy equations remain the same as in the plane case. As a supplementary equation, the differential (or integral) relationship (1.23) or (1.24), relating m and r or r_0 and r, must be included in the system.

The equations for the cylindrical case are set up in exactly the same manner as those for the spherical case. It should be noted that in the general case of two- and three-dimensional flows, changing to Lagrangian coordinates is inconvenient as a rule, since the equations become very complex.

§3. Sound waves

The speed of sound enters the gasdynamic equations as the velocity of propagation of small disturbances. In the limiting case, where changes in the density and pressure $\Delta\rho$ and Δp accompanying the fluid motion are very small in comparison with the average values of the density and pressure ρ_0 and p_0, and where the flow velocities are small in comparison with the speed of sound c, the gasdynamic equations become acoustic equations describing the propagation of sound waves.

Let us write the density and the pressure as $\rho = \rho_0 + \Delta\rho$, $p = p_0 + \Delta p$ and consider the quantities $\Delta\rho$, Δp and also the velocity u as small. Neglecting second-order quantities and considering only the plane case of a uniform fluid, we rewrite the Eulerian equations of motion and continuity. The continuity equation yields

$$\frac{\partial\, \Delta\rho}{\partial t} = -\rho_0 \frac{\partial u}{\partial x}. \tag{1.27}$$

The equation of motion takes the form

$$\rho_0 \frac{\partial u}{\partial t} = -\frac{\partial p}{\partial x} = -\left(\frac{\partial p}{\partial \rho}\right)_S \frac{\partial\, \Delta\rho}{\partial x}. \tag{1.28}$$

We have here used the fact that the particle motion in the sound wave is isentropic, whence a small change in pressure is related to a small change in density by the isentropic derivative, $\Delta p = (\partial p/\partial \rho)_S\, \Delta\rho$. As we shall presently see, this derivative represents the square of the sound speed

$$c^2 = \left(\frac{\partial p}{\partial \rho}\right)_S, \tag{1.29}$$

and refers to the undisturbed fluid state.

Differentiating (1.27) with respect to time, and (1.28) with respect to the space coordinate, we eliminate the cross derivative $\partial^2 u/\partial t\, \partial x$ and obtain a wave equation for the density change

$$\frac{\partial^2\, \Delta\rho}{\partial t^2} = c^2 \frac{\partial^2\, \Delta\rho}{\partial x^2}. \tag{1.30}$$

The pressure change $\Delta p = c^2 \Delta\rho$ (proportional to $\Delta\rho$), the velocity u, and all other fluid parameters, such as the temperature, also satisfy a similar equation*. A wave equation of the type (1.30) permits two families of solutions

$$\Delta\rho = \Delta\rho(x - ct), \qquad \Delta p = \Delta p(x - ct), \qquad u = u(x - ct) \tag{1.31}$$

and

$$\Delta\rho = \Delta\rho(x + ct), \qquad \Delta p = \Delta p(x + ct), \qquad u = u(x + ct) \tag{1.32}$$

where c denotes the positive root $c = +\sqrt{(\partial p/\partial\rho)_S}$.

The first solution describes a disturbance that propagates in the positive x direction, and the second describes a similar motion but in the opposite direction. In the first case, for example, the given value of the density corresponds to a particular value of the argument $x - ct$, that is, the disturbance moves with the velocity c in the direction of positive x. Thus, c here denotes the propagation velocity of sound waves.

Noting that $\partial u(x \mp ct)/\partial x = \mp(1/c)\, \partial u(x \mp ct)/\partial t$ and that in the undisturbed gas ahead of the wave $u = 0$ and $\Delta\rho = 0$, we find from (1.27) a relationship between the particle velocity of the gas u and the changes in density or pressure

$$u = \pm\frac{c}{\rho_0}\Delta\rho = \pm\frac{\Delta p}{\rho_0 c}, \qquad \Delta p = c^2\, \Delta\rho = \pm\rho_0 c u. \tag{1.33}$$

The upper sign refers to a wave propagating in the positive x direction and the lower sign to a wave propagating in the negative x direction. In both cases the particle velocity is in the direction of wave propagation where the fluid is compressed, and in the opposite direction at points where the fluid is expanded.

The general solution of the wave equations for $\Delta\rho$ and u is made up from two particular solutions, corresponding to waves propagating in the positive and negative x directions. From (1.31) to (1.33), the solutions for the density and the velocity are

$$\Delta\rho = \frac{\rho_0}{c} f_1(x - ct) + \frac{\rho_0}{c} f_2(x + ct), \tag{1.34}$$

$$u = f_1(x - ct) - f_2(x + ct), \tag{1.35}$$

* To obtain the wave equation for the velocity, we differentiate (1.30) with respect to time and use (1.27) and (1.28)

$$\frac{\partial^3 \Delta\rho}{\partial t^3} = c^2 \frac{\partial^3 \Delta\rho}{\partial x^2\, \partial t} = -\rho_0 \frac{\partial}{\partial x}\frac{\partial^2 u}{\partial t^2} = -c^2\rho_0 \frac{\partial}{\partial x}\frac{\partial^2 u}{\partial x^2},$$

from which $\partial^2 u/\partial t^2 = c^2\, \partial^2 u/\partial x^2 + f(t)$. Noting that $u = 0$ in the undisturbed fluid ahead of the wave, we find that $f(t) = 0$.

where f_1 and f_2 are arbitrary functions of their arguments, determined by the initial distributions of density and velocity

$$f_1 = \frac{1}{2}\left[\frac{c}{\rho_0}\,\Delta\rho(x,0) + u(x,0)\right],$$

$$f_2 = \frac{1}{2}\left[\frac{c}{\rho_0}\,\Delta\rho(x,0) - u(x,0)\right].$$

For example, if the initial density disturbance is rectangular and the gas is everywhere at rest, then rectangular disturbances will propagate to the right and to the left, as shown in Fig. 1.1. If the density and velocity distributions

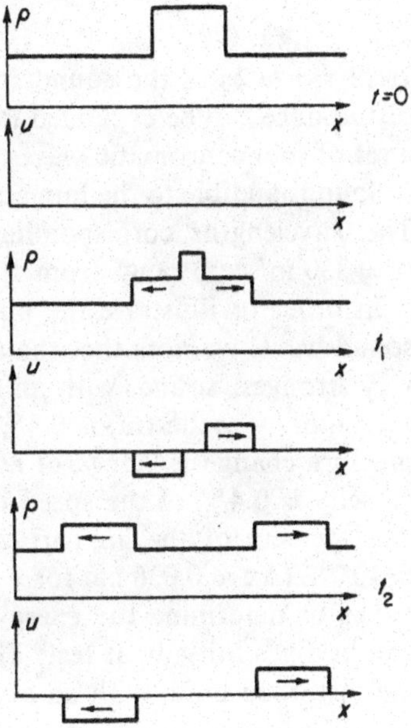

Fig. 1.1. Propagation of a rectangular density and pressure pulse in linear acoustics.

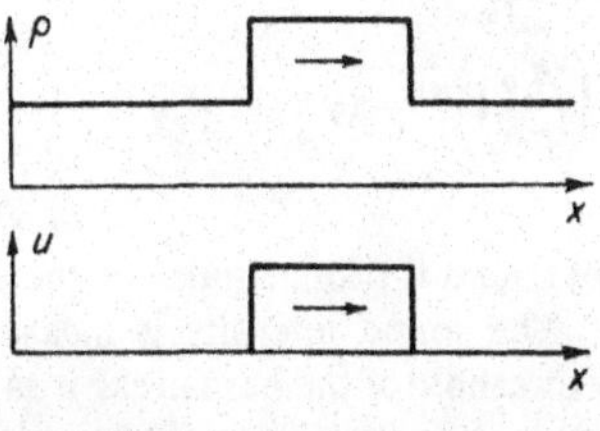

Fig. 1.2. Propagation of a rectangular density and pressure pulse in linear acoustics.

are initially of the form shown in Fig. 1.2, with $u = (c/\rho_0)\,\Delta\rho$ so that $f_2 = 0$, then the rectangular pulses will propagate in one direction only. (Such a disturbance can be created by a piston which at the initial instant of time begins to move into the undisturbed gas with a constant velocity u, and which stops "instantaneously" after a certain time. If the length of the rectangular pulse is L, then, obviously, the time during which the piston acts on the gas is $t_1 = L/c$.)

Of particular importance in acoustics are monochromatic sound waves, in which all quantities are periodic functions of time of the type

$$f = A\cos\left(\frac{\omega}{c}x - \omega t\right),$$

or in complex form

$$f = A\exp\left[-i\omega\left(t - \frac{x}{c}\right)\right].$$

Here $\nu = \omega/2\pi$ is the sound frequency and $\lambda = c/\nu$ is the wavelength. Any disturbance can be expanded in a Fourier integral, i.e., can be represented as a set of monochromatic waves of different frequencies.

Sounds audible to the human ear have a frequency ν from 20 to 20,000 cps. The wavelengths corresponding to the speed of sound in atmospheric air ($c = 330$ m/sec*) range from 15 m to 1.5 cm.

In order to illustrate the numerical values of the various quantities in a sound wave, we note that the amplitude of the density change in air for the very strongest sound, with an intensity† 10^5 times that of the fortissimo of a symphony orchestra, is 0.4% of the normal density; the amplitude of the pressure change is 0.56% of atmospheric pressure and the amplitude of the velocity is 0.4% of the speed of sound, or 1.3 m/sec. The amplitude of the displacement of the air particles Δx is of the order $u/2\pi\nu = (u/c)(\lambda/2\pi) \approx 6 \cdot 10^{-4}\lambda$ ($\Delta x \approx 0.036$ cm for $\nu = 500$ cps).

Let us determine the energy for a small disturbance that is propagated within a gas initially at rest. The increment in the specific internal energy of the disturbed fluid, with an accuracy up to second order with respect to $\Delta\rho$

* The specific heat ratio for air under normal conditions is

$$\gamma = 1.4, \quad \text{and} \quad c = \left(\frac{\partial p}{\partial \rho}\right)_S^{1/2} = \left(\frac{\gamma p_0}{\rho_0}\right)^{1/2} = (\gamma R T_0)^{1/2}$$

(since $p \sim \rho^\gamma$ for constant S).

† As will be shown below, the energy or intensity of sound is proportional to the square of the amplitude of the pressure or density change. The sound intensity is measured in decibels on a logarithmic scale. The average sensitivity threshold of the human ear is taken as zero. An increase in the volume by n decibels corresponds to an increase in the sound energy by a factor of $10^{n/10}$. An increase in the volume from the rustle of leaves or a whisper (~10 decibels) to the fortissimo of a symphony orchestra (~80 decibels) corresponds to an increase in the sound energy by a factor of 10^7.

(or Δp or u), is

$$\varepsilon - \varepsilon_0 = \left(\frac{\partial \varepsilon}{\partial \rho}\right)_0 \Delta\rho + \frac{1}{2}\left(\frac{\partial^2 \varepsilon}{\partial \rho^2}\right)_0 (\Delta\rho)^2.$$

Since the motion is isentropic, the derivatives are taken at constant entropy. They can be evaluated from the thermodynamic relationship $d\varepsilon = T\,dS - p\,dV = (p/\rho^2)\,d\rho$. We obtain

$$\varepsilon - \varepsilon_0 = \frac{p_0}{\rho_0^2}\Delta\rho + \frac{c^2}{2\rho_0^2}(\Delta\rho)^2 - \frac{p_0}{\rho_0^3}(\Delta\rho)^2.$$

The increment in internal energy per unit volume to the same order of accuracy is

$$\rho\varepsilon - \rho_0\varepsilon_0 = (\rho_0 + \Delta\rho)(\varepsilon - \varepsilon_0) + \varepsilon_0\,\Delta\rho$$
$$= \left(\varepsilon_0 + \frac{p_0}{\rho_0}\right)\Delta\rho + \frac{c^2}{2\rho_0}(\Delta\rho)^2 = h_0\,\Delta\rho + \frac{c^2}{2\rho_0}(\Delta\rho)^2,$$

where $h = \varepsilon + p/\rho$ is the specific enthalpy.

The internal energy density associated with the disturbance is, in first approximation, proportional to $\Delta\rho$. The kinetic energy density $\rho u^2/2 \approx \rho_0 u^2/2$ is a second-order quantity. Equations (1.33) which are valid for a traveling plane wave show that the second-order term in the expression for internal energy density and the kinetic energy term are exactly the same; thus the total energy density of the disturbance is

$$E = h_0\,\Delta\rho + \frac{c^2}{2\rho_0}(\Delta\rho)^2 + \frac{\rho_0 u^2}{2} = h_0\,\Delta\rho + \rho_0 u^2. \tag{1.36}$$

The first-order change in the above expression is related to the change in total volume of the gas that occurred as a result of the disturbance. If the disturbance was created in such a manner that the total volume remained unchanged, then the perturbation energy of the entire gas is a quantity of second order in $\Delta\rho$, since the term proportional to $\Delta\rho$ vanishes in the process

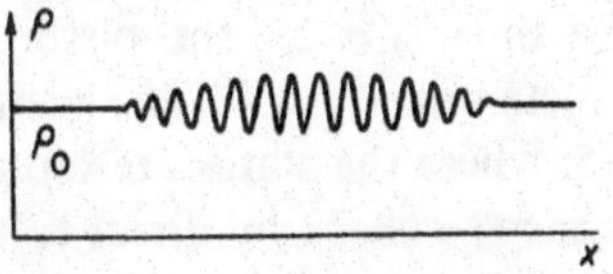

Fig. 1.3. Density distribution in a wave packet.

of integration over the volume. This, for example, is the situation in a wave packet which is propagated within a gas occupying an infinite space and in which the gas at infinity is undisturbed (Fig. 1.3). The density changes in the compression regions are compensated by the changes in the expansion regions, with an accuracy up to terms of second order. The energy of a sound wave is thus a quantity of second order, proportional to the square of the amplitude*.

$$E_{sw} = \rho_0 u^2. \tag{1.37}$$

If the disturbance causes a change in the gas volume, then the perturbation energy will contain a term proportional to the first power of $\Delta\rho$. However, this basic fraction of the energy which is proportional to $\Delta\rho$, can be "returned by the gas", if the source of the disturbance returns to its initial position. The energy remaining in the disturbed gas will be a quantity of second order. Let us explain this situation by means of a simple example. Assume that at the initial instant of time a piston begins to move into the undisturbed gas with a constant velocity u (much smaller than the speed of sound, $u \ll c$). At the time t_1 the piston stops "instantaneously". A compression pulse of length $(c-u)t_1 \approx ct_1$, whose energy is equal to the work done by the external force in moving the piston, $put_1 = (p_0 + \Delta p)ut_1 \approx p_0 ut_1$ will travel through the gas

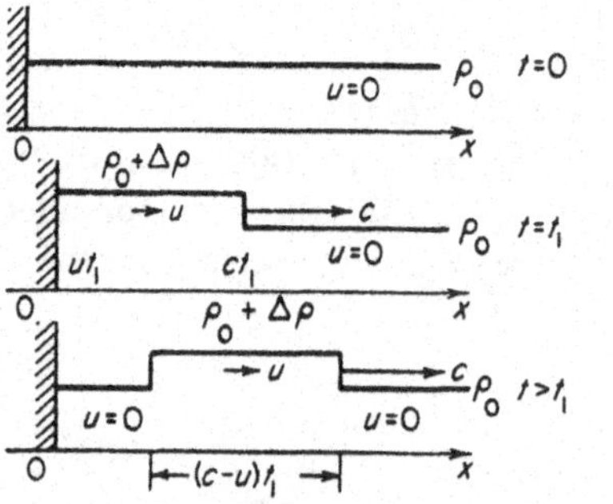

Fig. 1.4. Propagation of a compression pulse created by a piston moving into a gas.

(this case was considered above and is illustrated in Fig. 1.4). The energy, in first approximation, is proportional to the amplitude of the wave u, $\Delta\rho$, Δp, and the compression time (that is, to the length of the disturbance). Suppose now that the piston returns to its initial position in the following manner: its velocity u is "instantaneously" reversed (becomes $-u$) at the time t_1 and at the time $t_2 = 2t_1$ the piston, having returned to its initial position, stops "instantaneously". The disturbance will now have the form shown in Fig. 1.5, where the states are represented at the times $t = 0, t_1, t_2$, and $t > t_2$. It is easy to check by direct calculation that, to first approximation, during the second period from t_1 to t_2 the gas does work on the piston equal to that which the piston does on the gas during the first period from zero to t_1. The

* The expression (1.37) should be either time or space averaged

$$E_{sw} = \rho_0\overline{u^2} \quad (\overline{u} \sim \overline{\Delta\rho} \sim \overline{\Delta p} = 0, \quad \text{while} \quad \overline{u^2} \sim (\overline{\Delta\rho})^2 \sim (\overline{\Delta p})^2 > 0).$$

lengths of the positive and negative regions of the pulse, in first approximation, are also equal to each other and are both equal to $ct_1 = c(t_2 - t_1)$. Thus, if we add the energies in the compression and expansion regions of the pulse, the first-order terms cancel out. However, if higher order terms are retained*,

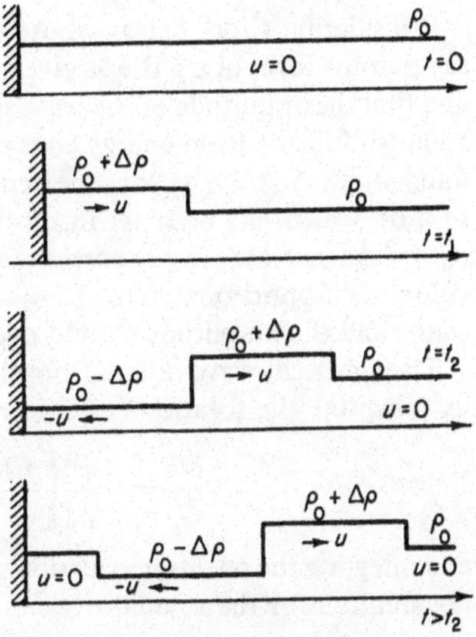

Fig. 1.5. Propagation of compression and rarefaction pulses created by a piston initially pushed into the gas and then withdrawn to the initial position.

the energy will contain a second-order term and the energy density of the perturbation will be given by the general equation (1.37).

§4. Spherical sound waves

In the absence of absorption (that is, without taking into account viscosity and heat conduction; see §22), neither the amplitude nor density of a plane wave decreases with time. For example, the pulses shown in Figs. 1.4 and 1.5 continue to "infinity" without either their form or amplitude changing. This is not the case, however, with a spherical wave. Linearizing the continuity equation in the spherically symmetric case, we obtain

$$\frac{\partial \Delta\rho}{\partial t} = -\frac{\rho_0}{r^2}\frac{\partial}{\partial r} r^2 u.$$

The linearized equation of motion is the same as (1.28)

$$\frac{\partial u}{\partial t} = -\frac{c^2}{\rho_0}\frac{\partial \Delta\rho}{\partial r}.$$

* In particular, the lengths of the compression and rarefaction pulses will differ by an amount $2ut_1$ (for $t_2 - t_1 = t_1$).

Hence, as in the plane case, we obtain a wave equation for $\Delta\rho$, whose solution, describing a wave spreading out from the center, is

$$\Delta\rho = \frac{f(r - ct)}{r}. \tag{1.38}$$

Considering short pulses whose lengths are much shorter than r, we can say that the form of a pulse as given by the function $f(r - ct)$ does not change, and that the amplitude of the wave decreases as $1/r$. This is completely natural and also follows from energy considerations. Let us assume that a pulse of a finite width Δr travels from the center. As the pulse is propagated, the mass of fluid which has been set in motion and which is approximately equal to $\rho_0 4\pi r^2 \, \Delta r$ increases in proportion to r^2. The energy of a sound wave per unit volume is proportional to $(\Delta\rho)^2$. Since the total energy is conserved, $(\Delta\rho)^2 r^2 =$ *const*, and the amplitude should decrease as $\Delta\rho \sim 1/r$.

The spherical wave differs from the plane wave in still another respect. Let us substitute the solution (1.38) into the equation of motion

$$\frac{\partial u}{\partial t} = -\frac{c^2}{\rho_0}\left[\frac{f'(r - ct)}{r} - \frac{f(r - ct)}{r^2}\right],$$

and integrate the resulting equation with respect to time. We obtain the following solution for the velocity:

$$u = \frac{c}{\rho_0}\left[\frac{f(r - ct)}{r} - \frac{\int^{r-ct} f(\xi)\, d\xi}{r^2}\right] = \frac{c}{\rho_0}\left[\Delta\rho - \frac{\varphi(r - ct)}{r^2}\right], \tag{1.39}$$

which differs from the plane wave solution (1.33) by the presence of an additional term. With a plane wave in the situation shown in Fig. 1.4, the fluid can be compressed only in the region of the disturbance. This is impossible in the case of a spherical wave, and a compression region must be followed by an expansion region. Indeed, outside the disturbance region both $\Delta\rho$ and u become zero. In the plane case, by virtue of the proportionality $u \sim \Delta\rho$, this condition is satisfied automatically and independently of the pulse form. In the case of a spherical wave, however, this is possible only when $\varphi(r - ct) = 0$ outside the disturbance region, that is, when the integral over the entire disturbance region is equal to zero

$$\varphi(r - ct) = \int f(\xi)\, d\xi = \int r\, \Delta\rho\, dr = 0.$$

It is thus evident that $\Delta\rho$ changes sign in a spherical wave, and that the compression region is followed by an expansion region.

The additional fluid included in the wave is equal to $\int \Delta\rho\, 4\pi r^2\, dr$. Since

$\Delta\rho \sim 1/r$, the additional mass in the compression wave increases as it spreads out from the center. An increase in the amount of compressed fluid during propagation causes the appearance of the lower density wave which follows the higher density wave.

As in the case of the plane wave, a change in pressure within a spherical wave is proportional to the change in density. The velocity, however, as shown

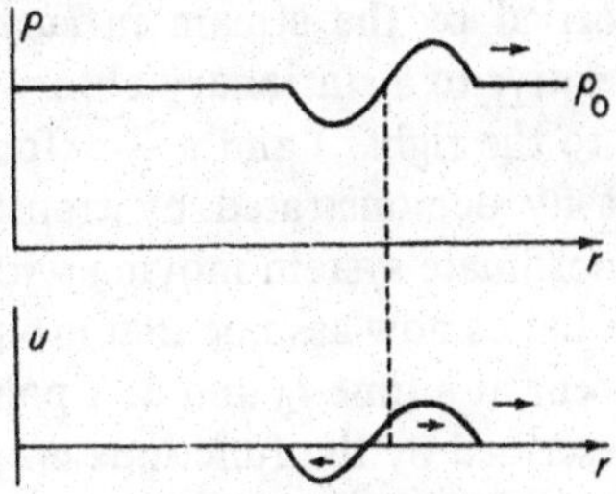

Fig. 1.6. Density and velocity distribution in a spherical sound wave.

by (1.39) is not proportional to either $\Delta\rho$ or Δp. In fact, the velocity and the change in density reverse their signs at different points, so that the density and velocity distributions in a wave traveling from the center assume the shapes shown in Fig. 1.6.

§5. Characteristics

It was shown in §3 that if arbitrarily small disturbances in velocity and pressure (or density*) are created at an initial time t_0 at any point x_0 of a stationary gas with uniform density and pressure, then two waves carrying the disturbances will travel from this point in both directions with the speed of sound. The small changes in the wave variables, which are propagated to the right in the positive x direction, are related by†

$$\Delta_1 u = \frac{\Delta_1 p}{\rho_0 c} = \frac{c}{\rho_0} \Delta_1 \rho = f(x - ct).$$

For a wave propagated to the left, these relationships are

$$\Delta_2 u = -\frac{\Delta_2 p}{\rho_0 c} = -\frac{c}{\rho_0} \Delta_2 \rho = -f_2(x + ct).$$

The arbitrary disturbances Δu and Δp arising at the initial instant of time can always be broken into two components $\Delta u = \Delta_1 u + \Delta_2 u$, $\Delta p = \Delta_1 p + \Delta_2 p$

* Since the flow is isentropic, the changes in density and pressure are not independent, but are related to one another by the thermodynamic relationship $\Delta p = c^2 \Delta\rho$.

† We use here Δu instead of u for consistency of notation.

satisfying the above relationships, because the initial disturbance is propagated in different directions in the form of two waves in general. If the initial disturbances Δu and Δp are not arbitrary, but are connected through one of the above relations, then the disturbance will travel in one direction only (this corresponds to the vanishing of one of the functions f_1 or f_2).

If the gas is not stationary, but moves as a whole with a constant velocity u, then the picture does not change, except for the fact that the waves are now carried by the stream in such a manner that their propagation velocities, relative to a stationary observer, become equal to $u + c$ (for a wave traveling "to the right") and $u - c$ (for a wave traveling "to the left"*). This can be easily demonstrated by transforming the gasdynamic equations to a new coordinate system moving with the gas at a velocity u.

Let us now assume that arbitrary small disturbances in velocity and density occur at a time t_0 and at a point x_0, in an arbitrary isentropic plane gas flow described by the functions $u(x, t)$, $p(x, t)$ (or $\rho(x, t)$, see first footnote in §5). Considering a small region about the point x_0 and a small time interval (a small neighborhood of the point x_0, t_0 in the x, t plane), we may in first approximation disregard the changes in the undisturbed functions $u(x, t)$ and $p(x, t)$. Consequently the changes in $\rho(x, t)$ and $c(x, t)$ in this region can also be neglected and the functions may be considered as constant and equal to their values at the point x_0, t_0. Then, the description given above for the propagation of disturbances is also completely applicable to this case. If the disturbances $\Delta u(x_0, t_0)$ and $\Delta p(x_0, t_0)$ are arbitrary, they can be broken into two components, one of which begins to propagate "to the right" with the velocity $u_0 + c_0$ and the other "to the left" with the velocity $u_0 - c_0$; here u_0 and c_0 denote the local values of these quantities at the point x_0, t_0.

Since u and c vary from point to point, then the paths along which disturbances are propagated in the x, t plane (described by the equations $dx/dt = u + c$ and $dx/dt = u - c$) will curve over a long period of time. These curves in the x, t plane, along which the small disturbances are propagated are called the characteristic curves, or simply characteristics. In the case of a plane isentropic flow there exist two such families of characteristics described by

$$\frac{dx}{dt} = u + c, \qquad \frac{dx}{dt} = u - c.$$

They are termed the C_+ and C_- characteristics, respectively. Two characteristics, one belonging to each of the C_+ and C_- families, can be drawn through

* We include the phrase "to the left" in quotation marks: when $u > c$, the wave will also travel to the right, but, obviously, more slowly than the first one. (*Editors' note.* Analogously, the phrase "to the right" has been included in quotation marks.)

each point in the x, t plane. In general the characteristics are curved, as shown in Fig. 1.7. In the region of undisturbed flow, where u, p, c, and ρ are constant in space and time, the characteristics of both families are straight lines.

If the flow is not isentropic but only adiabatic, that is, the entropy of different particles does not change with time but differs for each particle,

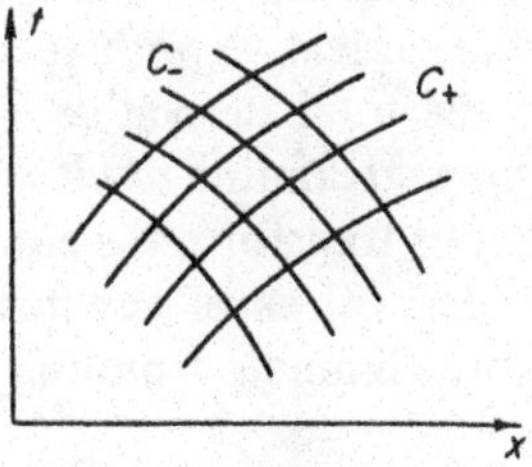

Fig. 1.7. A set of two families of characteristics in the isentropic case.

then disturbances in entropy are also possible. Since the motion is adiabatic $DS/Dt = 0$, that is, each disturbance in entropy not accompanied by a disturbance in the other variables (p, ρ, u), remains localized within the particle and is displaced together with the particle along a streamline. Consequently, in the case of nonisentropic flow these lines are also characteristics. They are described by the equation $dx/dt = u$ and are termed C_0 characteristics. In nonisentropic flow three characteristics pass through each point and the entire x, t plane is covered by a set of three families of characteristics C_+, C_-, and C_0 (Fig. 1.8).

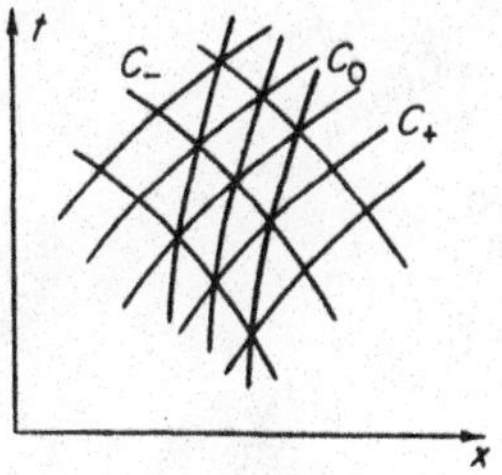

Fig. 1.8. A set of three families of characteristics in the nonisentropic case.

Up to now we have discussed the characteristics as curves in the x, t plane along which small disturbances are propagated. This, however, does not exhaust the significance of characteristics. The gasdynamic equations can be transformed so as to contain derivatives of the flow variables along the characteristics only. As will be shown in the following section, in isentropic flow not only small disturbances but also certain combinations of the flow variables are propagated along characteristics.

It is well known that a function f of the two variables x, t can be differentiated with respect to time along a given curve $x = \varphi(t)$ in the x, t plane. The time derivative of the function $f(x, t)$ along an arbitrary curve $x = \varphi(t)$ is determined by the angle between the tangent to the curve at the given point and the t axis $dx/dt = \varphi'$, and is equal to

$$\left(\frac{df}{dt}\right)_{\varphi} = \frac{\partial f}{\partial t} + \frac{\partial f}{\partial x}\frac{dx}{dt} = \frac{\partial f}{\partial t} + \frac{\partial f}{\partial x}\varphi'.$$

We are already familiar with two cases of differentiation along a curve: the partial derivative with respect to time $\partial/\partial t$ (along a curve $x = const$, $\varphi' = 0$) and the total derivative $D/Dt = (\partial/\partial t) + u(\partial/\partial x)$ (along a particle path or along a streamline $dx/dt = \varphi' = u$).

Let us transform the equations of plane adiabatic motion so that they contain derivatives of the flow variables along the characteristics only. To do this we eliminate from the continuity equation

$$\frac{D\rho}{Dt} + \rho\frac{\partial u}{\partial x} = 0$$

the density derivative, and replace it by the derivative of the pressure. Since density is related to the pressure and entropy by the thermodynamic relationship $\rho = \rho(p, S)$, and since $DS/Dt = 0$, we have

$$\frac{D\rho}{Dt} = \left(\frac{\partial\rho}{\partial p}\right)_S\frac{Dp}{Dt} + \left(\frac{\partial\rho}{\partial S}\right)_p\frac{DS}{Dt} = \frac{1}{c^2}\frac{Dp}{Dt}.$$

Substituting this expression into the continuity equation and multiplying by c/ρ, we find

$$\frac{1}{\rho c}\frac{\partial p}{\partial t} + \frac{u}{\rho c}\frac{\partial p}{\partial x} + c\frac{\partial u}{\partial x} = 0.$$

We add this equation to the equation of motion

$$\frac{\partial u}{\partial t} + u\frac{\partial u}{\partial x} + \frac{1}{\rho}\frac{\partial p}{\partial x} = 0$$

and obtain

$$\left[\frac{\partial u}{\partial t} + (u + c)\frac{\partial u}{\partial x}\right] + \frac{1}{\rho c}\left[\frac{\partial p}{\partial t} + (u + c)\frac{\partial p}{\partial x}\right] = 0.$$

Subtracting one equation from the other, we find, analogously,

$$\left[\frac{\partial u}{\partial t} + (u - c)\frac{\partial u}{\partial x}\right] - \frac{1}{\rho c}\left[\frac{\partial p}{\partial t} + (u - c)\frac{\partial p}{\partial x}\right] = 0.$$

The first of these equations contains derivatives only along the C_+ characteristics, and the second, only along the C_- characteristics. Noting that the adiabatic condition $DS/Dt = 0$ can be regarded as an equation along the C_0

characteristics, we can write the gasdynamic equations as

$$du + \frac{1}{\rho c}\,dp = 0 \qquad \text{along } C_+ : \frac{dx}{dt} = u + c, \tag{1.40}$$

$$du - \frac{1}{\rho c}\,dp = 0 \qquad \text{along } C_- : \frac{dx}{dt} = u - c, \tag{1.41}$$

$$dS = 0 \qquad \text{along } C_0 : \frac{dx}{dt} = u. \tag{1.42}$$

In Lagrangian coordinates the equations for the characteristics become (see (1.18))

$$C_+ : \frac{da}{dt} = \frac{\rho c}{\rho_0}; \qquad C_- : \frac{da}{dt} = -\frac{\rho c}{\rho_0}; \qquad C_0 : \frac{da}{dt} = 0.$$

The equations along the characteristics are the same as (1.40)–(1.42).

In spherically symmetric flow, the equations for the characteristics in Eulerian coordinates are the same as for the plane case (it is only necessary to replace the x coordinate by the radius r). The equations along the $C_\pm$ characteristics, however, contain additional terms that depend on the functions themselves but not on their derivatives,

$$du \pm \frac{1}{\rho c}\,dp = \mp \frac{2uc}{r}\,dt \qquad \text{along } C_\pm : \frac{dr}{dt} = u \pm c.$$

In many cases the gasdynamic equations written along the characteristics are more convenient for numerical integration than in the usual form.

§6. Plane isentropic flow. Riemann invariants

In isentropic flow the entropy, being constant in space and time, disappears completely from the equations. The flow can be described by two functions, the velocity $u(x,t)$ and any one of the thermodynamic variables: $\rho(x,t)$, $p(x,t)$, or $c(x,t)$. The latter variables are uniquely related to each other at every point by the purely thermodynamic relations: $\rho = \rho(p)$, $c = c(\rho)$, or $p = p(\rho)$, $c = c(\rho)$; $c^2 = dp/d\rho$.

The differential expressions $du + dp/\rho c$ and $du - dp/\rho c$ represent total differentials of the quantities

$$\begin{aligned} J_+ &= u + \int \frac{dp}{\rho c} = u + \int c\,\frac{d\rho}{\rho}, \\ J_- &= u - \int \frac{dp}{\rho c} = u - \int c\,\frac{d\rho}{\rho}, \end{aligned} \tag{1.43}$$

which are called the Riemann invariants* (see, e.g., [14]). By means of the thermodynamic relations the integral quantities $\int dp/\rho c = \int c\, d\rho/\rho$ can, in principle, be expressed in terms of the thermodynamic variables, let us say, the speed of sound c. For example, in a perfect gas with constant specific heats, we have

$$p = const\ \rho^{\gamma}, \qquad c^2 = \gamma\ const\ \rho^{\gamma-1},$$

and

$$J_{\pm} = u \pm \frac{2}{\gamma - 1} c. \tag{1.44}$$

The Riemann invariants are determined to within an arbitrary constant, which can always be dropped for convenience, as was done above in (1.44).

Equations (1.40) and (1.41) show that in isentropic flow the Riemann invariants are constant along characteristics

$$\begin{aligned} dJ_{+} &= 0, \quad J_{+} = const \qquad \text{along } C_{+} : \frac{dx}{dt} = u + c\,; \\ dJ_{-} &= 0, \quad J_{-} = const \qquad \text{along } C_{-} : \frac{dx}{dt} = u - c. \end{aligned} \tag{1.45}$$

This statement can be regarded as a generalization of relations which hold for the case of acoustic waves propagating through a gas with constant velocity, density, and pressure. These relations may be obtained from the general expressions for the invariants as a first approximation. Setting $u = u_0 + \Delta u$, $p = p_0 + \Delta p$, we obtain in first approximation

$$J_{\pm} = u_0 + \Delta u \pm \int \frac{d\,\Delta p}{\rho_0 c_0} = \Delta u \pm \frac{\Delta p}{\rho_0 c_0} + const. \tag{1.46}$$

The equations of the characteristics are given in first approximation by

$$\frac{dx}{dt} = u_0 \pm c_0, \qquad x = (u_0 \pm c_0)t + const.$$

Thus, the quantity $\Delta u + \Delta p/\rho_0 c_0$ is conserved along the path $x = (u_0 + c_0)t + const$. This shows that it can be represented as a function of the constant in the equation $x = (u_0 + c_0)t + const$ in the following way:

$$\Delta u + \frac{\Delta p}{\rho_0 c_0} = 2f_1[x - (u_0 + c_0)t].$$

* In nonisentropic flow ρ and c are functions of the two variables p and S, and the expressions $du + dp/\rho c$ are no longer total differentials. In this case the combinations (1.43) do not have a precise physical meaning.

Along the path $x = (u_0 - c_0)t + const$ the quantity

$$\Delta u - \frac{\Delta p}{\rho_0 c_0} = -2f_2[x - (u_0 - c_0)t]$$

is conserved. Changes in velocity and pressure are represented as a superposition of the two waves f_1 and f_2 traveling in opposite directions $\Delta u = f_1 - f_2$, $\Delta p = \rho_0 c_0(f_1 + f_2)$, where the variables in each equation are related by the relations given previously

$$\Delta_1 u = \frac{\Delta_1 p}{\rho_0 c_0} = f_1, \qquad \Delta_2 u = -\frac{\Delta_2 p}{\rho_0 c_0} = -f_2 .$$

The Riemann invariants J_+ and J_- may be used to describe the motion of a gas in place of the old variables—the velocity u and one of the thermodynamic quantities (e.g., the speed of sound c). They are uniquely related to the variables u and c by equations (1.43). By solving these equations for u and c, we can transform back from the functions J_+ and J_- to the functions u and c. For example, for a perfect gas with constant specific heats, we have, from (1.44),

$$u = \frac{J_+ + J_-}{2}; \qquad c = \frac{\gamma - 1}{4}(J_+ - J_-).$$

Considering the invariants as functions of the independent variables x and t, the equations of the characteristics can be expressed as

$$C_+ : \frac{dx}{dt} = F_+(J_+, J_-); \qquad C_- : \frac{dx}{dt} = F_-(J_+, J_-). \tag{1.47}$$

Here F_+ and F_- are known functions, whose form is determined only by the thermodynamic properties of the fluid. For a perfect gas with constant specific heats, we have

$$F_+ = \frac{\gamma + 1}{4} J_+ + \frac{3 - \gamma}{4} J_-; \qquad F_- = \frac{3 - \gamma}{4} J_+ + \frac{\gamma + 1}{4} J_-.$$

Equations (1.45) show that the characteristics have a property that permits them to preserve a constant value of one of the invariants. Since $J_+ = const$ along a specified C_+ characteristic, a change in the slope of a characteristic is determined only by the change in the invariant J_-. Similarly, J_- is constant along a C_- characteristic and the change in slope in going from one point in the x, t plane to another is determined only by the change in the J_+ invariant.

The flow equations when written in characteristic form make the causal connection of phenomena in gasdynamics quite apparent. Let us consider any plane isentropic gas flow in an infinite space. We assume that at the initial time $t = 0$ the distributions of flow variables are specified along the x coordinate: $u(x, 0)$ and $c(x, 0)$ or, equivalently, that the distributions of the invariants $J_+(x, 0)$ and $J_-(x, 0)$ are specified. A set of C_+ and C_- characteristics, originating from different points on the x axis, exists in the x, t plane (Fig. 1.9)*. The values of the flow variables at any point $D(x, t)$

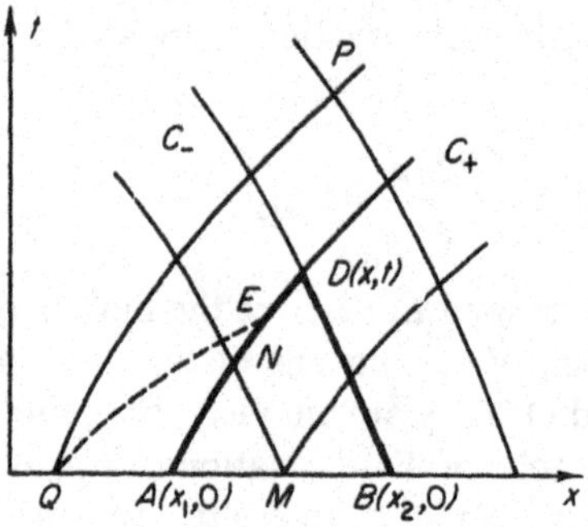

Fig. 1.9. An x, t diagram, illustrating the domain of dependence.

(the coordinate point x at the time t) are determined only by the values of the quantities at the initial points $A(x_1, 0)$ and $B(x_2, 0)$:

$$J_+(x, t) = J_+(x_1, 0)\,; \qquad J_-(x, t) = J_-(x_2, 0).$$

For example, solving these equations for u and c for a perfect gas with constant specific heats, we can write the physical variables at the point D in the explicit form

$$u(x, t) = \frac{u_1 + u_2}{2} + \frac{2}{\gamma - 1}\,\frac{c_1 - c_2}{2},$$
$$c(x, t) = \frac{c_1 + c_2}{2} + \frac{\gamma - 1}{2}\,\frac{u_1 - u_2}{2}, \tag{1.48}$$

where u_1 and c_1 are values at the point $A(x_1, 0)$ and u_2 and c_2 refer to the point $B(x_2, 0)$.

Obviously, we cannot claim that the state of the gas at point D depends on the given initial conditions at points A and B alone, because the position of the point D as the point of intersection of the C_+ and C_- characteristics, originating from the points A and B, depends on the paths of these characteristics. These paths are determined by imposing initial conditions along the entire segment AB of the x axis. For example, the slope of the C_+ characteristic AD at the intermediate point N (see Fig. 1.9) is determined not only by the invariant $J_+(A)$, but also by the value of the invariant $J_-(M)$ which has been propagated to N from the intermediate point M of the segment AB.

* Such a family of curves can be constructed only after the problem is solved.

The state of the gas at the point D is, however, completely determined by the initial conditions on the segment AB of the x axis and is absolutely independent of the initial values outside this segment. A slight change in the initial conditions at the point Q will have no effect on the state of the gas at the point D simply because the disturbance created by this change will not be able to reach the point x at the time t. It will arrive at this point at a later time (at the point P along the C_+ characteristic QP). In the same way, the initial state of the gas along the segment AB of the x axis can influence the state of the gas at subsequent times only at those points inside the region bounded by the C_- characteristic AP and the C_+ characteristic BQ (Fig. 1.10). There is no

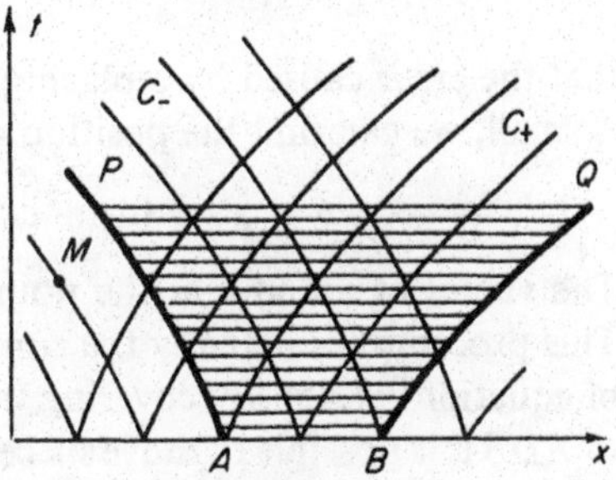

Fig. 1.10. An x, t diagram, illustrating the region of influence.

influence on the state at M, since the "signals" from the initial conditions along the segment AB will not be able to arrive at the point x_M at the time t_M.

Let us emphasize that the above discussion of the causality of phenomena in gasdynamics is valid only when the characteristics of one family do not intersect each other. For example, if the C_+ characteristic originating from Q (see Fig. 1.9) were to trace the path QE, shown by the dashed line, then the state of the gas at Q would affect the state at D. However, in regions of continuous flow characteristics of the same family never intersect, since such intersection would lead to a nonuniqueness in the flow variables. Actually, the invariant J_+ would have two different values at the point of intersection of two C_+ characteristics, corresponding to the values of J_+ associated with both characteristics. On the other hand, to every point in the x, t plane there can be only one value of J_+ and J_-, which derive from the unique values of the velocity and speed of sound at the point. As we shall see below, the intersection of characteristics of the same family leads to the breakdown of the continuity of the flow and to the formation of the discontinuities in the flow variables which are called shock waves.

It is possible to draw characteristic curves over the entire x, t plane only if the solution to the gasdynamic problem is known. If the solution is not known, then it is not possible to give the exact position of the point D in Fig. 1.9 as the point of intersection of two characteristics originating from A and B. It is possible, however, to find the approximate point of intersection by replacing the actual curved paths AD and BD by straight lines whose slopes correspond to the initial values u_1, c_1 and u_2, c_2 at points A and B (or $J_+(A)$, $J_-(B)$) (Fig. 1.11). Having chosen points A and B sufficiently close together

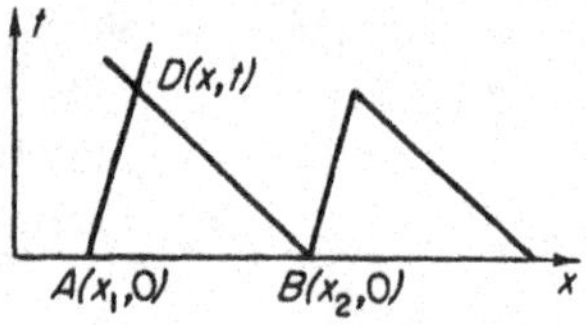

Fig. 1.11. Local approximation of the characteristics by straight lines.

that the error caused by replacing the actual characteristics by straight lines is small, we can find the position of the point of intersection from

$$x - x_1 = (u_1 + c_1)t, \qquad x - x_2 = (u_2 - c_2)t.$$

The values of u and c at the point of intersection are determined by (1.48). This procedure is actually the simplest method for the numerical integration of equations (1.45). By covering the x, t plane with a grid of triangles similar to ADB it is possible to move successively through the solution of the equations with respect to time, starting with the initial conditions $u(x, 0)$, $c(x, 0)$ or $J_+(x, 0)$, $J_-(x, 0)$.

§7. Plane isentropic gas flow in a bounded region

Let us consider any plane isentropic gas flow in a bounded region. Let the gas occupy the space between two plane surfaces—pistons—whose motions are described by $x_1 = \psi_1(t)$ and $x_2 = \psi_2(t)$. The coordinates of the pistons at the initial time $t = 0$ are x_{10} and x_{20}. The distributions of the velocity u and the thermodynamic variable c along the x coordinate in the interval $x_{10} < x < x_{20}$ at the initial time $u(x, 0)$ and $c(x, 0)$ are specified or, equivalently, the distributions of the invariants $J_+(x, 0)$ and $J_-(x, 0)$ are given.

We sketch in Fig. 1.12 the characteristics net and the boundaries of the pistons in the x, t plane. Points such as F, through which pass the C_+ and C_- characteristics originating from points inside the interval O_1O_2 of the x axis, do not differ from points in a gas in an unbounded medium. As previously discussed, the initial values of the invariants J_+ and J_- are transferred to these points.

Let us now consider a point on the boundary of a piston, the point D on the left piston, for example. Only one invariant J_- is transferred to the point D from the "past"; it is transferred along the C_- characteristic originating from the point A of the initial segment O_1O_2 in such a way that $J_-(D) = J_-(A)$.

The second invariant J_+ is not transferred to D because the C_+ characteristic does not reach D (from the "past"). Instead, a C_+ characteristic emerges from D (into the "future") carrying with it the value of the invariant J_+ "generated" at this point. The state of the gas at the point D is determined by the value of the invariant J_- and one other quantity. The other quantity is the velocity u, which by virtue of the boundary conditions coincides at the

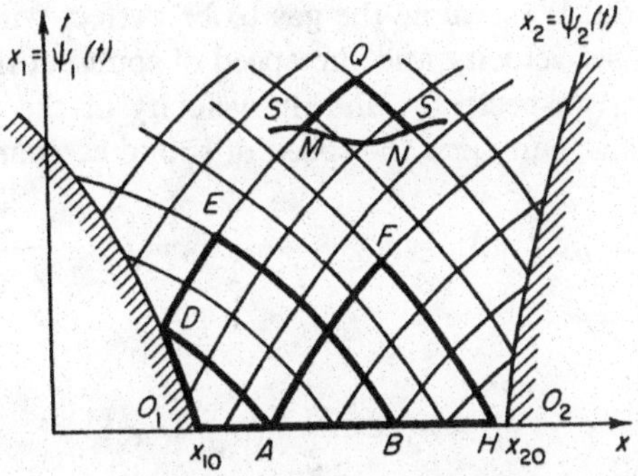

Fig. 1.12. Sketch of characteristics for plane isentropic gas flow between two pistons.

point D with the known velocity of the left piston $u_1(D)$. This pair of quantities $J_-(D) = J_-(A)$ and $u = u_1(D)$ now replaces the pair of quantities J_+ and J_-, which arrives at points in the gas that are not in contact with the pistons. At point D the second invariant J_+ is determined by the quantities $J_-(D)$ and $u_1(D)$ following the relation $J_+(D) = 2u_1(D) - J_-(D)$. This invariant is carried away by the C_+ characteristic. Now, for example, the C_- characteristic arriving at the point E originates at the point B of the initial segment of the x axis and carries with it the invariant $J_-(B)$, so that $J_-(E) = J_-(B)$. On the other hand, the C_+ characteristic arriving from the point D on the boundary of the piston brings with it the invariant J_+ equal to $J_+(D)$, so that $J_+(E) = J_+(D)$. Thus, the state of the gas at E depends only on $J_-(B)$, $u_1(D)$, and $J_-(A)$, hence on conditions at A, B, and D. But the location of E depends on the intervals O_1D, O_2B. We conclude that in the case of a gas flow in a bounded region, the state of the gas at any given point depends not only on the initial conditions, but also on the boundary conditions.

In general, the state of the gas at an arbitrary point of the x, t plane is determined by specifying the values of u and c or of J_+ and J_- on a segment of an arbitrary curve which is cut off by the C_+ and C_- characteristics passing through the point under consideration. For example, the state at the point Q is determined by the state on the segment MN of a curve S (see Fig. 1.12).

In a similar manner the invariant J_+ is transferred from the "past" to the right piston along the C_+ characteristics, and the C_- characteristics begin at the points on the trace of the piston and carry away into the "future" the invariant J_-. The values of the J_- invariant are determined by the values of the arriving J_+ invariant and the values of the piston velocity u (which equals the velocity of the gas layer adjacent to the piston).

The pressure on the piston is uniquely determined by one of the arriving invariants and by the piston velocity. Let us consider, for example, the point D on the left piston. We assume the gas to be perfect with constant specific heats. The initial gas velocity and the speed of sound at the point A are denoted by u_A and c_A, respectively, and the velocity of the piston at the point D by u_{1D}. The gas velocity and the speed of sound at D are obtained from

$$u_D = u_{1D}, \qquad J_- = u_D - \frac{2}{\gamma - 1} c_D = u_A - \frac{2}{\gamma - 1} c_A,$$

whence

$$c_D = c_A + \frac{\gamma - 1}{2}(u_{10} - u_A)$$

or in terms of the invariant

$$c_D = \frac{\gamma - 1}{2}[u_{10} - J_-(A)].$$

The pressure at the piston p_D is related to the speed of sound c_D in a purely thermodynamic manner with $p_D = const\ c_D^{2\gamma/(\gamma-1)}$.

From the above considerations we can give a clear physical meaning to the Riemann invariants. Let us consider the following experiment. We introduce at a particular time t and at a point x a flat plate parallel to the piston surface. We then place a pressure gauge on the left side of the plate sensitive to the changes in gas pressure on that side.

The invariant $J_+ = u + \int dp/\rho c = u + w(p)$ arrives at the left of the gauge at a time t at a point x. Here u and p are the velocity and pressure in the gas undisturbed by the plate ($w(p)$ is a function of the pressure which depends only on the thermodynamic properties of the gas and on its entropy). At the time t the gas slows down near the plate and stops, since the plate is at rest. The new pressure to the left of the plate, corresponding to the gas at rest ($u = 0$), is denoted by p_1. Then $J_+ = u + w(p) = w(p_1)$. The gauge will register the reflected pressure p_1. Since the function w is known, the gauge scale can be calibrated to give directly the value of the invariant J_+. In a similar manner, a pressure gauge placed on the right side of the plate will measure the invariant J_- arriving from the right.

If we place a very thin plate perpendicular to the piston surface and parallel to the direction of the flow velocity so that the gas flows freely past the plate without changing its velocity, the gauge will then register the pressure p of the undisturbed flow. By calibrating the gauge to give directly the value of $w(p)$, the gauge will measure the combination of the invariants

$$w(p) = \tfrac{1}{2}(J_+ - J_-).$$

§8. Simple waves

Equation (1.46) describing the propagation of small disturbances (acoustic waves) in a gas shows that if the wave is propagated in one direction only, then one of the invariants is constant in both space and time. Thus, if the wave travels to the right and $\Delta u(x,t) = \Delta p(x,t)/\rho_0 c_0 = f_1[x - (u_0 + c_0)t]$, then the invariant J_- is constant:

$$J_- = \Delta u - \frac{\Delta p}{\rho_0 c_0} + const = const.$$

If the wave travels to the left, then the J_+ invariant is constant.

We shall prove that the existence of waves propagating in one direction is not limited to those cases in which the amplitude of the waves is small, for which we have seen that one of the Riemann invariants remains constant. First we observe that it is possible to achieve practically the constancy of one of the invariants, for example of J_-. If the gas occupies an unbounded region, it is sufficient to give the initial distributions $u(x, 0)$ and $c(x, 0)$ in such a manner that $J_-(x, 0) = const$ at the initial time. Since this constant value of J_- is carried along the C_- characteristics originating from all points on the x axis, the invariant J_- also remains constant at subsequent times; thus, $J_-(x, t) = const$.

Let the gas occupy a half space bounded on the left by a piston moving according to the equation $x_1 = \psi_1(t)$. If, at the initial time $J_-(x, 0) = const$ in the entire region occupied by the gas, $x > x_{10}$ (x_{10} is the initial coordinate of the piston), then at subsequent times J_- will also remain constant in the entire region bounded by the piston $x > x_1 = \psi_1(t)$. Actually, the left piston, as has been shown in the preceding section, "excites" only the C_+ characteristics; the C_- characteristics arrive at the boundary of the piston from the "past" and they terminate their existence there, because the piston sends into the "future" only the invariant J_+ and not the J_-. The values of the invariant J_- in the entire region of the x, t plane which correspond to the gas (this region is bounded by the piston boundary $x_1 = \psi_1(t)$) are determined by the initial values of J_- on the x axis, that is, they are constant. On the other hand, if the gas occupies a half space bounded from the right by a moving piston (the piston boundary is $x_2 = \psi_2(t)$, $x_{20} = \psi_2(0)$), and at the initial instant $J_+(x, 0) = const$ for $x < x_{20}$, then the J_+ invariant is constant over that part of the x, t plane where $x < x_2 = \psi_2(t)$.

We return now to the original problem, assuming that $J_-(x, t) = const$. It follows from the equation for the characteristics, written in the form (1.47), that the C_+ characteristics constitute a family of straight lines ($F_+ = const$ because $J_+ = const$ along the characteristics and $J_- = const$ in general). Integrating the equations for the C_+ characteristics, we can write

$$x = F_+(J_+, J_-)t + \varphi(J_+), \tag{1.49}$$

where $\varphi(J_+)$ is a constant of integration, regarded as a function of that value of J_+ which is propagated along the characteristic. This constant is determined by the initial and boundary conditions of the problem. For example, if the given characteristic originates from the initial interval of the x axis, then φ is the coordinate of that point on the x axis from which the characteristic starts, and for which the value of J_+ defining the argument of φ has been given. Equation (1.49), together with the imposed condition

$$J_-(x, t) = const, \tag{1.50}$$

represents the general solution of the gasdynamic equations for the case considered. Equation (1.49) determines implicitly the other desired function, $J_+(x, t)$. (We recall that the function F_+ is specified by the thermodynamic properties of the fluid which are assumed known.)

The solutions (1.49) and (1.50) can be written in the form of equations involving the usual gasdynamic variables, the gas velocity and the speed of sound. It follows from (1.50)

$$J_- = u - \int \frac{dp}{\rho c} = const$$

that the speed of sound or any other thermodynamic variable, say the pressure, is a function of the velocity u alone, with $c = c(u)$ and $p = p(u)$, and does not explicitly contain the independent variables x and t. Equation (1.49) is equivalent to

$$x = [u + c(u)]t + \varphi(u), \tag{1.51}$$

where the constant of integration φ is expressed as a function of u. This equation determines u as an implicit function of x and t. It is evident from (1.51) that the given values of u and $c(u)$ are carried through the gas along the x axis with a constant velocity $u + c(u)$. In other words, the solution is a wave, traveling to the right, which is defined by the functional relation

$$u = f\{x - [u + c(u)]t\}, \qquad c = g\{x - [u + c(u)]t\}.$$

Here the form of the functions f and g is determined by the boundary and initial conditions of the problem. However, in contrast to a traveling wave of small amplitude, different values of the gas velocity and thermodynamic variables are propagated with different velocities, so that the initial profiles of $u(x, 0)$ and $c(x, 0)$ become distorted with time. This is a result of the nonlinearity of the gasdynamic equations. The solution we have obtained in the form of a traveling wave is called a simple wave.

A simple wave traveling in the opposite direction can be obtained in a similar manner. It has a constant invariant J_+, and the C_- characteristics are straight lines. The general solution in this case is

$$J_+ = const, \qquad x = F_-(J_+, J_-)t + \varphi_1(J_-)$$

or

$$J_+ = u + \int \frac{dp}{\rho c} = const, \qquad x = [u - c(u)]t + \varphi_1(u),$$

$$u = f_1\{x + [c(u) - u]t\}, \qquad c = g_1\{x + [c(u) - u]t\}.$$

We note that the simple wave solution is a particular solution of the equations of one-dimensional isentropic flow. It is possible to find a general integral of these equations for an arbitrary flow (see [1]). The particular solution is not directly contained in the general solution.

§9. Distortion of the wave form in a traveling wave of finite amplitude. Some properties of simple waves

We shall use the solution obtained for the simple wave to explain what happens with an acoustic-type wave if we do not limit ourselves to the first approximation, as was done in §3, but instead start with the exact gasdynamic equations. We shall not present here an analytic solution, but will explain the qualitative character of the phenomena by graphical means. The gas will be taken to be a perfect one with constant specific heats.

Let the initial profiles of the velocity and of the speed of sound $u(x, 0)$ and $c(x, 0)$ be as shown in Fig. 1.13 and, in addition, let the variables be related to each other in such a way that $J_-(x, 0) = const$ (we consider a wave traveling to the right). According to (1.44) we have $c = \frac{1}{2}(\gamma - 1)u + c_0$, where the constant value of the invariant J_- has been chosen in accordance with the condition that $u = 0$ and $c = c_0$ in the undisturbed gas. Since $p \sim c^{2/(\gamma-1)}$, $\rho \sim c^{2\gamma/(\gamma-1)}$ (for $c = c_0$, $p = p_0$, $\rho = \rho_0$), both the pressure and density profiles are completely similar qualitatively to the profile of the speed of sound.

The invariant $J_-(x, t)$, which is constant initially, is also constant during the entire time following, so that the motion is a simple wave traveling to the right. Characteristics of the C_+ family are the straight lines $dx/dt = u + c = \frac{1}{2}(\gamma + 1)u + c_0$. They are depicted in Fig. 1.13. They emerge parallel to each other from the points A_0, B_0, and D_0, where $u = 0$ so that $dx/dt = c_0$ (and also parallel to the C_+ characteristics, emerging from points on the x axis corresponding to the undisturbed region). In order not to complicate Fig. 1.13

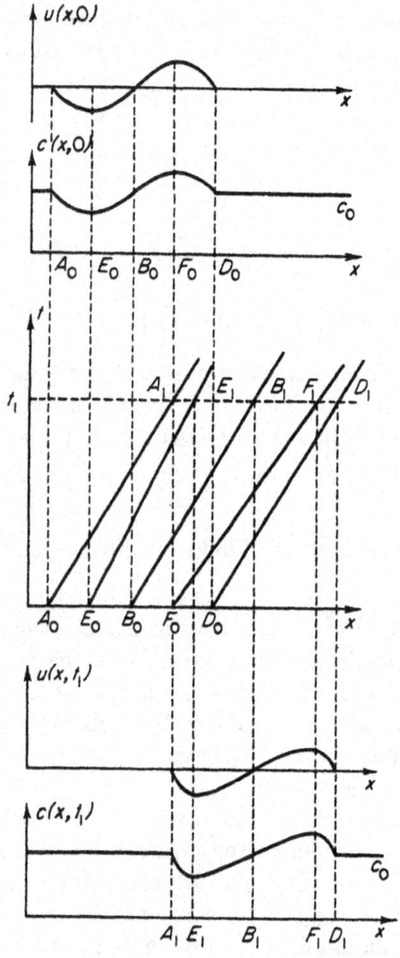

Fig. 1.13. Propagation of a wave traveling to the right. A graphical construction which permits the determination of the wave form distortion. Top, initial velocity and speed of sound profiles. Bottom, distorted profiles at time t_1. Center, sketch of the C_+ characteristics.

we draw only two more C_+ characteristics, those from the points E_0 and F_0 corresponding to the minima and maxima of the initial distributions of $u(x, 0)$ and $c(x, 0)$.

We now construct the velocity and speed of sound profiles $u(x, t_1)$ and $c(x, t_1)$ at a time t_1. Since the values of u and c are constant along the C_+

characteristics, the values of u and c at the points A_1, E_1, etc., are equal to the corresponding values at the points A_0, E_0, etc. Carrying out the graphical construction as shown in Fig. 1.13, we find the profiles of u and c at the time t_1. We see that the "head" (D) and the "tail" (A) of the wave, adjacent to the undisturbed regions where $u = 0$ and $c = c_0$, have been displaced along the x axis through distances equal to $c_0 t_1$ (they have been propagated along the characteristics $D_0 D_1$ and $A_0 A_1$ in the x, t plane). The heights of the maxima and minima of u and c have not changed, but the relative positions of the maxima and minima are different, and the profiles have been distorted.

In acoustics, where the gasdynamic equations are linearized, such distortion does not take place and the profiles are displaced as a "frozen picture". The distortion of the wave forms is a result of the nonlinearity of the gasdynamic equations. The physical reason for the distortion lies in the fact that the wave crests travel relatively faster, due to the higher velocity with which they are propagated through the fluid (higher speed of sound), as well as due to the fact that they are carried forward faster together with the fluid (higher gas velocity). On the other hand, the valleys travel relatively slower, since in this region both velocities are lower.

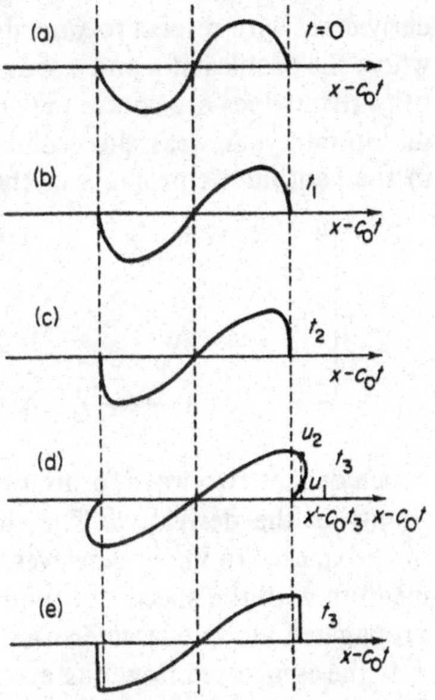

Fig. 1.14. Diagram showing the steepening and "overshooting" of a finite amplitude wave in the nonlinear theory. The figure shows velocity profiles at successive instants of time. To compare these waves at different instants of time the combination $x - c_0 t$ has been plotted along the abscissa. The wave form (d) corresponds to a physically unrealistic condition. Actually, the wave has the form (e) with discontinuities at the time t_3.

Figure 1.14 shows that the wave forms become increasingly distorted with time. If we formally extend the analytic solution over a sufficiently long period, then an "overshooting" of the wave, shown in Fig. 1.14d, will occur.

This picture has no physical meaning, however, since it does not give a unique solution. For example, at the point $x = x'$ three values of the velocity $u = 0$, u_1, and u_2 exist at the same time. This nonuniqueness is mathematically attributable to the intersection of the characteristics of one family (C_+); such a tendency can be seen from an examination of Fig. 1.13. In reality the "overshooting" does not take place, and when the front and rear parts of the wave become extremely steep, discontinuities (i.e., shock waves) are formed as shown in Fig. 1.14e (we shall discuss this below). Thus, the solution in the form of a simple wave is, in this case, valid only for a limited period, up to the time when discontinuities are formed. The solution is always valid whenever the wave has the character of a rarefaction wave throughout the region, that is, whenever there are no segments where the velocity, pressure, and density of the gas decrease in the direction of wave propagation. The segments AE and FD shown in Fig. 1.13 constitute compression waves. The simple rarefaction wave will be considered in the next section.

We note an important property of the simple wave illustrated by the above example. The head of a simple wave is always propagated along a characteristic (in our example along the characteristic D_0D_1). The quantities u and c are continuous at the forward edge of the simple wave at the point D but their derivatives with respect to x are discontinuous (this is apparent from Fig. 1.13, where the profiles of u and c show breaks). A discontinuity in which the quantities themselves are continuous but their derivatives are not is called a weak discontinuity. A weak discontinuity may be imagined as a small disturbance in the continuous progress of the flow variables. This is shown in Fig. 1.15

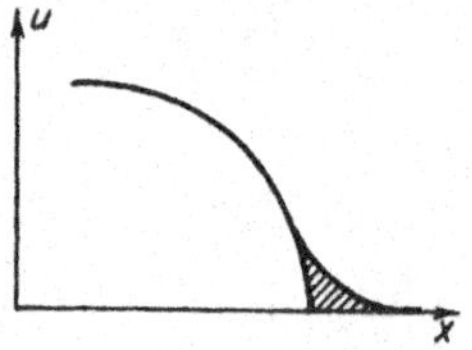

Fig. 1.15. Illustration of a weak discontinuity.

which depicts two wave forms, one smoothed out and the other with a discontinuity in the derivative. The shaded segment may be regarded as a small disturbance. We know, however, that small disturbances are propagated in a medium with the speed of sound. Therefore weak discontinuities are always propagated along characteristics.

If the isentropic flow has a common boundary with a region of uniform flow, then this flow is necessarily a simple wave, and, conversely, only a simple wave can have a mutual boundary with a region of uniform flow. Actually, the C_+ and C_- characteristics in the uniform flow region represent families of parallel straight lines, and the invariants $J_+(x, t)$ and $J_-(x, t)$ are constant. One of the characteristics (let us say a C_+ characteristic) serves as the interface of an isentropic flow region I with a uniform flow region II (Fig. 1.16). Then the C_- characteristics extending from region II into region I carry the constant value of J_-, so that $J_-(x, t) = const$ in region I as well. Consequently, this

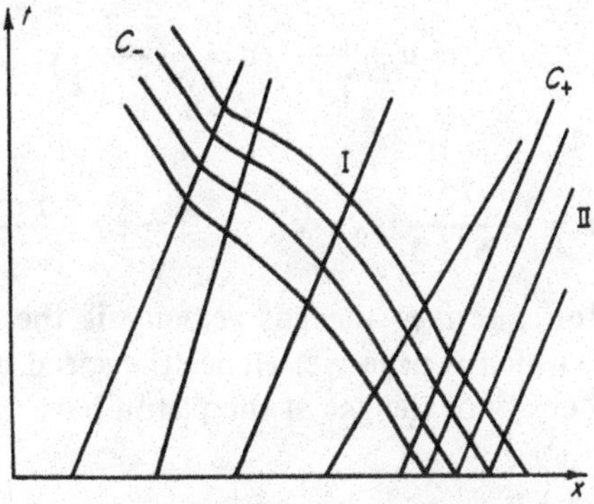

Fig. 1.16. Illustration of the two families of characteristics for the wave shown in Fig. 1.13.

region is a simple wave traveling to the right. Figure 1.16 illustrates the characteristics for the case of an impulse of one "wavelength" used as an example in the above discussions.

§10. The rarefaction wave

Let us consider the motion of a gas caused by the action of a receding piston. Let the gas initially have constant density, pressure, and speed of sound ρ_0, p_0, c_0 and occupy a half space $x > 0$, which is bounded on the left by a stationary piston whose initial position is $x = 0$. At the time $t = 0$ the piston begins to move to the left, gradually accelerating from zero velocity to a certain constant velocity U. The equation of motion of the piston is $x = X(t)$. When the speed of the piston becomes constant, the line $X(t)$ becomes a straight line described by $X(t) = -Ut + const$.

As shown in the preceding section, the motion of a gas for $t > 0$ is a simple wave traveling to the right. The head of the wave, that is, the initial disturbance imparted by the piston, propagates to the right with the speed of sound along the C_+ characteristic OA, whose path is $x = c_0 t$ (Fig. 1.17). Also shown in Fig. 1.17 are the piston path $X(t)$ and the characteristics of the C_+ and C_- families. The gas contained in the region I between the x axis and the C_+ characteristic OA is undisturbed. In this region the characteristics are straight lines whose slopes are $(dx/dt)_+ = c_0$ and $(dx/dt)_- = -c_0$. Intersecting the straight line OA, the C_- characteristics are extended to the piston trace, where they terminate. For the sake of simplicity we shall assume the gas to be a perfect one with constant specific heats. We emphasize, however, that qualitatively the picture remains the same for a gas with other thermodynamic properties. The invariant J_- is constant over the entire physical part of the x, t plane and is equal to

$$J_- = u - \frac{2}{\gamma - 1} c = -\frac{2}{\gamma - 1} c_0 .$$

From this we have

$$u = -\frac{2}{\gamma - 1}(c_0 - c), \qquad c = c_0 + \frac{\gamma - 1}{2} u.$$

At the gas-piston interface the gas velocity is the same as the velocity of the piston $w(t)$, which is negative. Hence the speed of sound, and also the pressure and the density of the gas at the piston have values lower than their

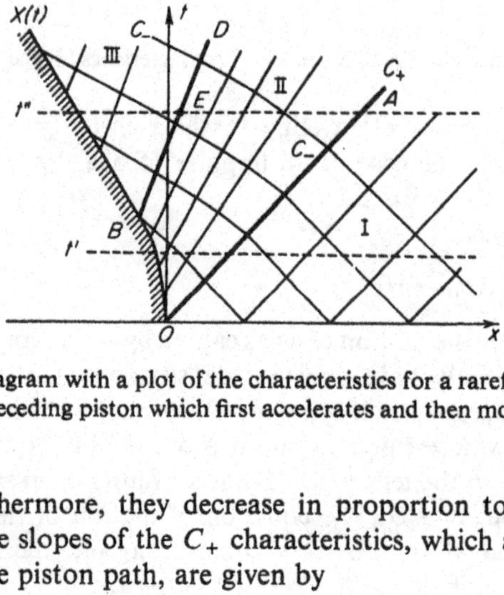

Fig. 1.17. An x, t diagram with a plot of the characteristics for a rarefaction wave arising from the motion of a receding piston which first accelerates and then moves with a constant velocity.

initial values; furthermore, they decrease in proportion to the increase in piston velocity. The slopes of the C_+ characteristics, which are straight lines emanating from the piston path, are given by

$$\left(\frac{dx}{dt}\right)_+ = u + c = c_0 + \frac{\gamma + 1}{2} u = c_0 - \frac{\gamma + 1}{2} |w|.$$

Since the piston only accelerates and does not decelerate, all the C_+ characteristics emanating from the piston path are divergent, as shown in Fig. 1.17. The C_+ characteristics emanating from that section of the piston path where the piston has already reached a constant velocity have the same slopes $(dx/dt)_+ = c_0 - \frac{1}{2}(\gamma + 1)U$, so that they are parallel to each other. Suppose for example, that starting at time t_1 (point B on the piston path) the piston velocity becomes strictly constant and equal to $w = -U$ $(U > 0)$. In region

III of the x, t plane, which is bounded by the piston path and the C_+ characteristic BD, all the flow variables are constant with $u = -U$, $c = c_0 - \frac{1}{2}(\gamma - 1)U = c_1$ *. In this region J_- is constant by virtue of its general constancy and J_+ is constant because the gas velocity on the piston path from which all the C_+ characteristics emerge is the same at all points, with

$$J_+ = u + \frac{2}{\gamma - 1} c = \frac{2}{\gamma - 1} c_0 + 2u = \frac{2}{\gamma - 1} c_0 - 2U.$$

The flow variables in the region II lying between the C_+ characteristics OA and BD and the segment OB of the piston path are the same functions of x and t as obtained for the simple wave. The C_+ characteristics emerging from the segment OB of the piston path carry with increasing time decreasing values of the speed of sound and of the gas velocity (increasing absolute values of the gas velocity). Hence, the distribution of u and c in the gas at any given time $t' < t_1$ (corresponding to the horizontal line $t = const = t'$ in the x, t plane) is that shown in Fig. 1.18a. Because of the direct relationship

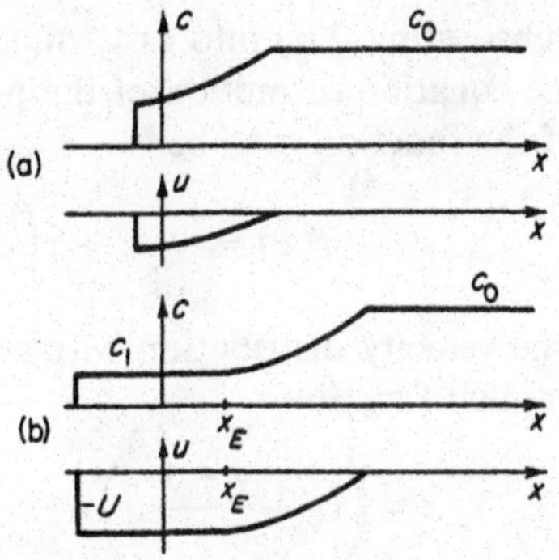

Fig. 1.18. Distribution of the speed of sound and gas velocity in a rarefaction wave arising from the motion of a piston (see Fig. 1.17): (a) up to the time when the piston velocity becomes constant $t' < t_1$; (b) after the piston velocity becomes constant $t'' > t_1$.

between p, ρ, and c the density and pressure distributions are qualitatively similar to the distribution of the speed of sound. The distributions of the flow variables at a still later time $t'' > t_1$ (the line $t = const = t''$ in the x, t plane) are shown in Fig. 1.18b. In this case a uniform flow region $u = -U$, $c = c_1$ is adjacent to the piston. The coordinate x_E of the point separating the constant and variable flow regions III and II corresponds to the point E on the characteristic BD.

Starting with a concrete example of a relationship governing the motion of a piston, we can find the solution to the problem in analytic form. Let us

* These equations are valid only if c_1 is a positive quantity; this imposes a limitation on the final velocity of the piston, that $U < [2/(\gamma - 1)]c_0$. The case where $U > [2/(\gamma - 1)]c_0$ will be considered in §11.

assume that the velocity of the piston varies with time according to

$$w = -U(1 - e^{-t/\tau}), \qquad \tau > 0$$

and approaches the constant value $-U$ asymptotically for $t \to \infty$. The piston path is described by

$$X(t) = \int_0^t w\,dt = -U\tau\left[\frac{t}{\tau} - (1 - e^{-t/\tau})\right],$$

and asymptotically approaches the straight line $X = -U(t - \tau)$. The desired solution is found by applying the boundary condition $u = w(t)$ for $x = X(t)$ to the general solution (1.51). This condition defines the arbitrary function $\varphi(u)$ as

$$\varphi(w) = X(t) - [w + c(w)]t,$$

where

$$c(w) = c_0 + \frac{\gamma - 1}{2} w \qquad \text{and} \qquad w = w(t).$$

Substituting $X(t)$ into this equation and expressing time in terms of w from the equation of motion of the piston $t = -\tau \ln(1 + w/U)$, we find the form of the function φ to be

$$\varphi(w) = -w\tau + \tau\left(c_0 + \frac{\gamma + 1}{2} w + U\right) \ln\left(1 + \frac{w}{U}\right).$$

The velocity distribution with respect to x at different times is given by the implicit function

$$x = \left(c_0 + \frac{\gamma + 1}{2} u\right)t - u\tau + \tau\left(c_0 + \frac{\gamma + 1}{2} u + U\right) \ln\left(1 + \frac{u}{U}\right),$$

which is valid in the interval $X(t) < x < c_0 t$.

Let us again assume that the velocity of the piston becomes strictly constant at a certain time t_1. We fix the constant value of the final velocity of the piston $-U$; we consider that the initial acceleration of the piston becomes greater and greater, while the constant final velocity is approached faster and faster ($t_1 \to 0$). Segment OB of the piston path where the piston velocity is not constant becomes smaller and smaller (see Fig. 1.17). Points B and O, where the C_+ characteristics BD and OA which bound the variable flow region II originate, will therefore approach each other. In the limit $t_1 = 0$ when the points B and O coincide (corresponding to the piston reaching the constant velocity $w = -U$ instantaneously) both the characteristics BD and OA originate from the same point. This point is the origin of the coordinate system $x = 0$ and $t = 0$ in the x, t plane. All the other C_+ characteristics covering the variable flow region II also fan out from the origin O. Thus, in the limiting case when the piston begins to move with the constant velocity $w = -U$ at the time $t = 0$, the picture in the x, t plane takes the form shown in Fig. 1.19.

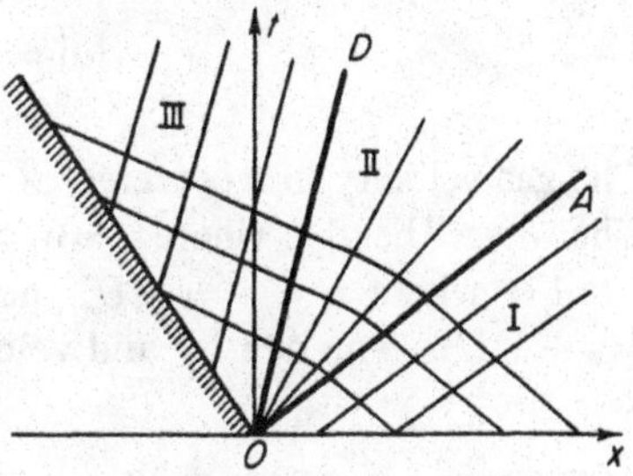

Fig. 1.19. An x, t diagram showing the characteristics for a centered rarefaction wave.

The three special lines, the line of the "head" rarefaction wave OA, the line of the "tail" wave OD beyond which the flow parameters take on their constant and final values, and the piston path, all originate from the "center" O. All the C_+ characteristics which are between the C_+ characteristics OA and OD also originate from this "center". This type of wave is called a centered simple wave. Since all the C_+ characteristics in a centered simple wave, those in the variable flow region II, originate from the point $x = 0$, $t = 0$, the function $\varphi(u)$ will vanish in the solution (1.51); this solution is at the same time the equation for these characteristics. The solution for the centered wave takes the form

$$x = [u + c(u)]t. \tag{1.52}$$

This solution can also be obtained formally by taking the limit $\tau \to 0$ in the example considered above. The function φ is proportional to τ, so that for $\tau \to 0$, $\varphi(u) \to 0$.

Let us write an explicit solution for the centered rarefaction wave for the case of a perfect gas with constant specific heats. The relationship between the thermodynamic variables and the gas velocity u is given by the result derived earlier from the constancy of the invariant J_-

$$c = c_0 - \frac{\gamma - 1}{2}|u|, \qquad u < 0. \tag{1.53}$$

Since $p = p_0(\rho/\rho_0)^\gamma$, $c^2 = \gamma p/\rho = c_0^2(\rho/\rho_0)^{\gamma-1}$, we have

$$\rho = \rho_0\left[1 - \frac{\gamma - 1}{2}\frac{|u|}{c_0}\right]^{2/(\gamma-1)}, \tag{1.54}$$

$$p = p_0\left[1 - \frac{\gamma - 1}{2}\frac{|u|}{c_0}\right]^{2\gamma/(\gamma-1)}. \tag{1.55}$$

In order to express these quantities as functions of x and t, it is necessary to substitute here the value of $|u|$ found by elimination of c between (1.52) and (1.53), or

$$|u| = \frac{2}{\gamma + 1}\left(c_0 - \frac{x}{t}\right). \tag{1.56}$$

The gas velocity in a centered rarefaction wave is thus a linear function of x. The wave "head", where $u = 0$, moves along the line $x = c_0 t$, while the wave "tail", where $u = w = -U$, moves along the line $x = (c_1 - U)t = (c_0 - \frac{1}{2}(\gamma + 1)U)t$. The density and velocity profiles are shown in Fig. 1.20.

§11. The centered rarefaction wave as an example of self-similar gas motion

The one-dimensional plane motion of a gas considered in the preceding section resulting from the influence of a piston receding with a constant velocity exhibits a special property. All the flow variables describing the motion $u(x, t)$, $c(x, t)$, $\rho(x, t)$, and $p(x, t)$ do not depend upon the x coordinate and time independently but are functions only of the combination x/t. For the region II, where these quantities are changing, this is directly evident from (1.53)–(1.56). This also applies to the regions of uniform flow I and III bounded in the x, t plane by the straight lines $x/t = c_0 = const$ (region I) and $x/t = w = const$, $x/t = w + c_1 = const$ (region III), which are also described by equations containing x and t only in the combination x/t. In other words, the distributions of all quantities with respect to the x coordinate, as shown in Fig. 1.20, change with time without changing their form; they remain similar

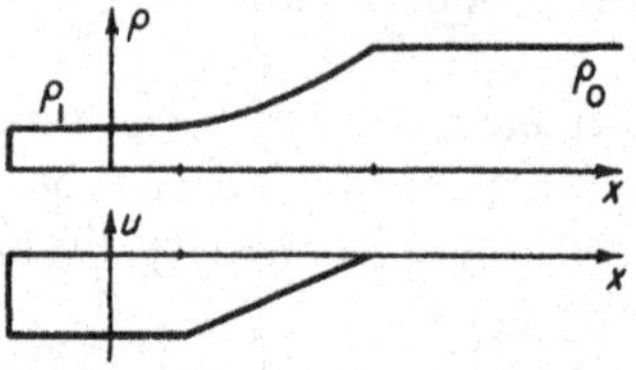

Fig. 1.20. Density and velocity profiles in a centered rarefaction wave.

to themselves. If we were to draw the distributions of u, c, ρ, and p using as the abscissa not x but the ratio x/t (or one of the dimensionless quantities $x/c_0 t$, x/wt), we would obtain a "frozen" picture, one which does not vary with time. This type of motion in which the distributions of the flow variables remain similar to themselves with time and vary only as a result of changes in scale is called self-similar. (In the case being considered the scale is the length $c_0 t$ or wt.) In §25 we shall consider a more complicated example of self-similar motion, where not only the length scale but also the flow variable scales change. In this case the similarity variable has the more general form $\xi = xt^{\alpha}$, where $\alpha = const$. The centered rarefaction wave is an elementary example of a self-similar motion, where $\alpha = -1$, $\xi = x/t$, and the scales of the flow variables remain constant. With time the flow profiles $u(x, t)$ and $c(x, t)$ move in a self-similar fashion with respect to the abscissa but do not change with respect to the ordinate (the scales of u, c, ρ, and p remain unchanged).

The physical reason for the self-similar character of a centered rarefaction wave can be explained by dimensional considerations. If we neglect the dissipative processes associated with viscosity and heat conduction, then the gasdynamic equations and also the relations describing the thermodynamic properties of the fluid do not contain any characteristic length or time scales. The only length and time scales in the gas are the length and time of a mean free path of molecular motion, with which the coefficients of viscosity and thermal conductivity are related. However these scales characterize only the microprocesses, which take place in distances and times of the order of those of the mean free path of the molecules, but not the macroscopic motion. Fluids do possess a dimensional parameter, the speed of sound, which together with the velocity of the fluid enters the description of the fluid flow. Thus, if the boundary and initial conditions of the problem do not contain any characteristic length or time scales, then the flow can depend on the coordinate and time only in the combination x/t, which has the dimension of a velocity. This is precisely the problem of the rarefaction wave which results from the motion of a piston receding with a constant velocity w. The initial and boundary conditions introduce only the velocity scales c_0 and w (and, obviously, also density and pressure scales ρ_0 and p_0, but not length nor time scales*).

Self-similar motion is of great importance in gasdynamics. In this case the flow variables do not depend on the coordinates and time separately, but depend only on particular combinations of them. This decreases by one the number of independent variables in the system of equations. In particular, for one-dimensional motions, only one independent variable ($\xi = x/t$ in our problem) appears instead of the two variables x and t (or r and t in the case of spherical or cylindrical symmetry). Therefore, the flow can be described by

* If the piston velocity is not a constant, but a function of time, then the time or length scales will appear immediately. In this case the problem of the rarefaction wave will no longer be a self-similar one. Mathematically this follows from (1.51): if $\varphi(u) \neq 0$, then u becomes a function of x and t separately. However, if the piston velocity tends to become constant in time, as in the problem considered in the preceding section, then the actual solution will asymptotically approach the self-similar solution. For $t \gg \tau$ ($t/\tau \to \infty$), the function $\varphi(u) \sim \tau$ can be omitted in the solution. Physically, this corresponds to the fact that the parameter τ becomes small in comparison with the characteristic time t of the problem and its role becomes increasingly less. For a more detailed discussion of the asymptotic approach of actual solutions to self-similar solutions, see Chapters X and XII.

ordinary rather than by partial differential equations, and this simplifies the problem considerably from a mathematical point of view.

In view of the importance of self-similar flows, of which the centered simple wave appears as an example, we again solve the piston problem, using the general system of flow equations and taking advantage of the possibility of reducing the number of independent variables. To do this we express the Euler equations in terms of a new independent variable $\xi = x/t$. If $f(x, t)$ is some function of x and t which depends only on the ratio $\xi = x/t$ we obtain

$$\frac{\partial f}{\partial x} = \frac{1}{t}\frac{df}{d\xi},$$

$$\frac{\partial f}{\partial t} = -\frac{x}{t^2}\frac{df}{d\xi} = -\frac{\xi}{t}\frac{df}{d\xi},$$

$$\frac{Df}{Dt} = \frac{\partial f}{\partial t} + u\frac{\partial f}{\partial x} = \frac{u - \xi}{t}\frac{df}{d\xi}.$$

Using these relations we can rewrite the equations of continuity, motion, and entropy for the plane case as

$$\begin{aligned} \frac{D\rho}{Dt} &= -\rho\frac{\partial u}{\partial x} &&\rightarrow\quad (u-\xi)\frac{d\rho}{d\xi} = -\rho\frac{du}{d\xi}, \\ \rho\frac{Du}{Dt} &= -\frac{\partial p}{\partial x} &&\rightarrow\quad (u-\xi)\rho\frac{du}{d\xi} = -\frac{dp}{d\xi}, \\ \frac{DS}{Dt} &= 0 &&\rightarrow\quad (u-\xi)\frac{dS}{d\xi} = 0. \end{aligned} \tag{1.57}$$

As expected, the variables x and t are eliminated from the equations. In this form the equations immediately yield the trivial solution $u = const$, $p = const$, $\rho = const$, and $S = const$, corresponding to a uniform flow of the gas.

To obtain a nontrivial solution, we eliminate $du/d\xi$ from the first pair of equations, and note that the third equation gives $S = const$*; thus, the self-similar flow is isentropic. Replacing in (1.57) the pressure derivative by the density derivative $dp/d\xi = (dp/d\rho)(d\rho/d\xi) = c^2\, d\rho/d\xi$ (since the flow is isentropic $dp/d\rho = (\partial p/\partial\rho)_S = c^2$), we obtain

$$[(u-\xi)^2 - c^2]\frac{d\rho}{d\xi} = 0,$$

* The assumption that $u - \xi = 0$ rather than $dS/d\xi = 0$ contradicts the first equation in (1.57).

whence

$$u - \xi = \pm c, \qquad \xi = \frac{x}{t} = u \mp c. \tag{1.58}$$

Substituting this relationship into equations (1.57), we find

$$du \pm c\,\frac{d\rho}{\rho} = du \pm \frac{dp}{\rho c} = 0$$

or

$$J_\pm = u \pm \int \frac{dp}{\rho c} = const. \tag{1.59}$$

We have thus arrived at the solution found previously for the problem of a centered rarefaction wave. The lower sign in (1.58) and (1.59) refers to waves traveling to the right, while the upper sign refers to waves traveling to the left. Again, the entire flow picture can be constructed from the solutions (1.58) and (1.59) and the trivial solutions $u = const$, $c = const$ which also satisfy the self-similar equations. The solutions must be combined in such a manner so as to satisfy the boundary condition $u = w$ at the piston.

Let us consider certain properties of the rarefaction wave in somewhat greater detail. The character of the solution shows that it remains valid even if the gas does not extend from the piston to infinity, $x \to \infty$. The presence of the boundary does not affect the flow in any way until such time as the head of the rarefaction wave (which travels through the undisturbed gas to the right with the speed of sound c_0) reaches the right-hand boundary of the gas $x = x_1 > 0$, that is, until $t_1 = x_1/c_0$*. Therefore, the solution obtained will always describe the initial stage of the gas motion caused by a receding piston, even when the gas occupies a finite region.

Let us follow a gas particle whose initial coordinate is, say, x_0. The particle is at rest up to the time $t = t_0 = x_0/c_0$, that is, before the arrival of the rarefaction wave. Then the particle begins to move to the left, accelerating and expanding at the same time. After the particle density decreases to its limiting value ρ_1, and the particle velocity becomes equal to the velocity of the piston w, further acceleration and expansion stop and the particle then moves with a constant velocity w. Some particle paths in the x, t plane are shown in Fig. 1.21. The equations of these lines in the expansion region II can be easily obtained by integrating the streamline equation $dx/dt = u = [2/(\gamma - 1)] \times (c_0 - x/t)$ with the initial condition $x = x_0$ at $t = t_0 = x_0/c_0$.

Let us now examine what happens with larger and larger values of the

* We recall here the discussion of the regions of influence in §6.

piston speed $|w|$. It is evident from (1.53)–(1.56) that the higher is $|w|$, the lower are the speed of sound, the density, the pressure, and temperature ($T \sim \sqrt{c}$) of the gas in the final state ($c_1 = c(w)$, $\rho_1 = \rho(w)$, etc.). Finally, at the particular piston velocity $|w|_m = [2/(\gamma - 1)]c_0$, the final values of c_1, ρ_1, and p_1 become zero. If the piston recedes even faster, then the solutions (1.53)–(1.56) lose their meaning, because for $|u| > |w|_m$ the value of c_1 is negative,

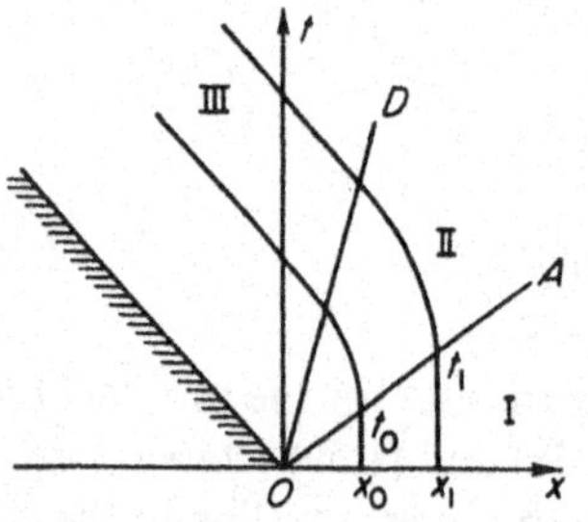

Fig. 1.21. Particle paths in x, t diagram for a centered rarefaction wave; OA is the wave head and OD is the wave tail.

while ρ_1 and p_1 are complex. Physically this means that a vacuum region is created between the piston and the left boundary of the gas when $|w| > |w|_m$. In this case the system behaves as if the piston were altogether " withdrawn " at the initial time $t = 0$, and as if the gas then flows into the vacuum left. The gas expands until its density, pressure, and temperature (speed of sound) reach zero, and its boundary moves to the left with the velocity

$$u = -\frac{2}{\gamma - 1} c_0, \qquad |u|_{\max} = \frac{2}{\gamma - 1} c_0. \tag{1.60}$$

The velocity and density profiles for unsteady flow into a vacuum are shown in Fig. 1.22. For example, for air at ordinary temperatures $\gamma = 7/5$ and

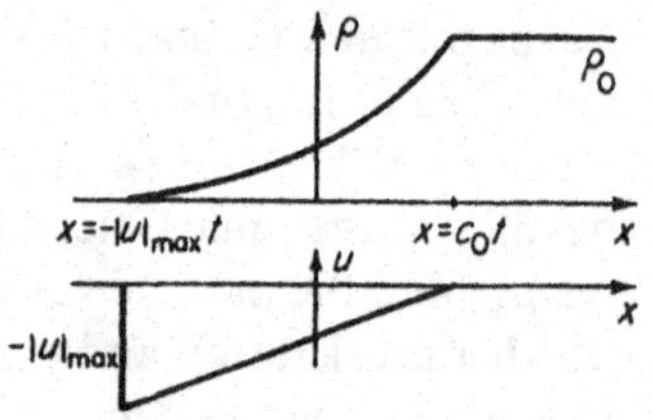

Fig. 1.22. Density and velocity profiles in plane unsteady gas flow into a vacuum.

$|u|_{\max} = 5c_0$. This value is more than twice the limiting velocity for a steady flow expanding from a large reservoir into a vacuum; here the Bernoulli equation $h + u^2/2 = h_0 = c_0^2/(\gamma - 1)$ holds and $u_{\max} = [2/(\gamma - 1)]^{1/2} c_0 \approx 2.2c_0$ for $\gamma = 1.4$ (h denotes the specific enthalpy, $h = \varepsilon + p/\rho$). In steady flow the particle only acquires a kinetic energy per unit mass of $u_{\max}^2/2$ equal to its initial enthalpy h_0. In the case of unsteady flow into a vacuum, the kinetic energy is larger than the initial enthalpy h_0 (by a factor of five for $\gamma = 1.4$). The additional kinetic energy is acquired at the expense of heat energy in the neighboring particles. The total energy, equal to the sum of the kinetic and internal energies in the region occupied by the rarefaction wave, is conserved and is equal to the initial internal energy in that region.

Spherically or cylindrically symmetric rarefaction waves, created when "spherical" or "cylindrical" pistons are suddenly withdrawn from a gas occupying the space $r > r_0$ or $r < r_0$, can be considered in a manner analogous to the plane case. A rarefaction wave, whose head travels through the undisturbed gas with the speed of sound c_0, is also formed in this case. However, in these cases the regions between the piston and the tail of the rarefaction wave are not ones of uniform flow. We note that spherical and cylindrical rarefaction waves, in contrast to their plane counterpart, are not self-similar; in this problem there is a characteristic length scale—the initial radius of the piston r_0.

§12. On the impossibility of the existence of a centered compression wave

It would seem that the solution to the problem of a piston moving with a constant velocity is equally applicable regardless of whether the piston is pushed into or is withdrawn from the gas, whether it causes an expansion or a compression. The motions with expansion are self-similar, and their solutions can be constructed from trivial regions corresponding to uniform flow and a nontrivial region corresponding to a centered simple wave. Let us attempt to construct formally a continuous solution for a self-similar compression wave formed by a piston pushed into the gas with a constant initial velocity $w > 0$ (the gas is to the right of the piston). The "head" of the wave travels through the gas with the speed of sound c_0 along the line $x = c_0 t$ in the x, t plane. The piston is contiguous to the uniform flow region where $u = w$ and $c = c_1$, and both of the uniform flow regions (I and III in accordance with the terminology introduced in the preceding sections) are divided by the centered simple wave region II, where $J_- = u - [2/(\gamma - 1)]c = const = -[2/(\gamma - 1)]c_0$. It follows that $c_1 = c_0 + \frac{1}{2}(\gamma - 1)w$, so that the "tail" of the wave travels along the line $x = (w + c_1)t = [\frac{1}{2}(\gamma + 1)w + c_0]t$. The velocity distribution with respect to the x coordinate in region II is described by an equation similar to (1.56)

$$u = \frac{2}{\gamma - 1}\left(\frac{x}{t} - c_0\right).$$

The result is obtained that the "tail" of the wave propagates at a higher speed than the "head" of the wave, with $\frac{1}{2}(\gamma + 1)w + c_0 > c_0$, and that the velocity and density profiles are of the shapes shown in Fig. 1.23. This result has no physical meaning and the solution is not single-valued in region II. However, this is the only continuous solution that follows from the gasdynamic equations. Consequently, a continuous solution for the given case simply does not exist. Historically speaking, this difficulty was one of the starting points for the development of discontinuous solutions to the equations of gasdynamics, that is, for the development of shock wave theory.

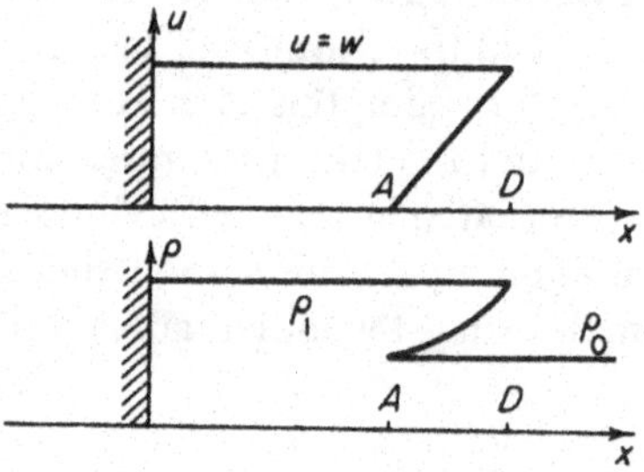

Fig. 1.23. Velocity and density profiles corresponding to a continuous solution for a self-similar (centered) compression wave. *A* is the wave head and *D* is the wave tail. The solution is not single-valued and has no physical meaning.

We note that if the piston, instead of moving into the gas with a constant velocity, gradually accelerates from rest, it is possible to find a continuous solution for a simple (but no longer centered) compression wave which describes the initial stage of the motion. The situation in this case is completely analogous to that for sound waves of finite amplitude (see §7). Characteristics of the C_+ family (if the piston is to the left of the gas) approach each other and tend to intersect. The steepness of the compression wave form increases with time (Fig. 1.24) and at a certain instant of time "overshooting"

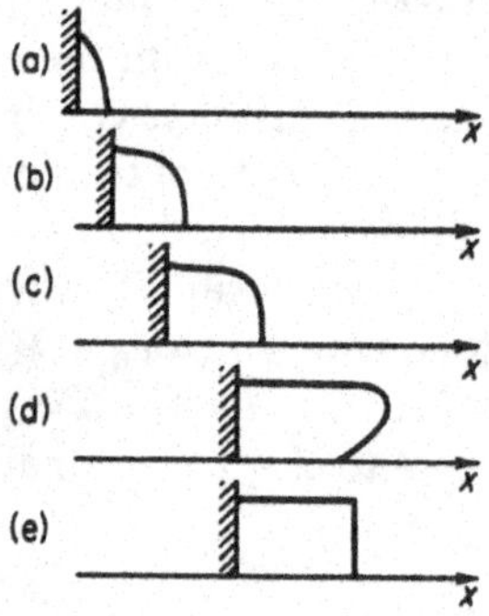

Fig. 1.24. Gradual steepening of the velocity profile in a compression wave generated by an accelerating piston. (d) corresponds to a physically meaningless solution with "overshooting" of the wave; (e) shows the actual profile with a discontinuity after the occurrence of the "overshooting".

takes place and the solution becomes multi-valued in a manner analogous to that described above and in §7. Essentially, this means that a discontinuity, i.e., a shock wave, has been formed.

2. *Shock waves*

§13. Introduction to the gasdynamics of shock waves

Let us examine a gas initially at rest, with constant density and pressure ρ_0, p_0, bounded on the left by a plane piston. Let us further assume that the gas is compressed at an initial time by the piston moving into the gas with a constant velocity, which we shall denote by u. As shown in the preceding section, an attempt to find a continuous solution to this problem leads to a physically meaningless result. Since the problem is self-similar, the only solutions satisfying the gasdynamic equations are the trivial solutions, in which the quantities u, ρ, and p are constant, and the centered simple wave solution. Thus, there remains only one possibility for constructing a solution that would satisfy the boundary conditions of the problem in the undisturbed gas, $u = 0$, $p = p_0$, and $\rho = \rho_0$, while having in the region next to the piston the gas velocity equal to the piston velocity. This solution eliminates the physically unreal region II and brings together the uniform flow regions I and III with the assumption that the flow variables are discontinuous at the boundary between these regions; the solution is shown in Fig. 1.25.

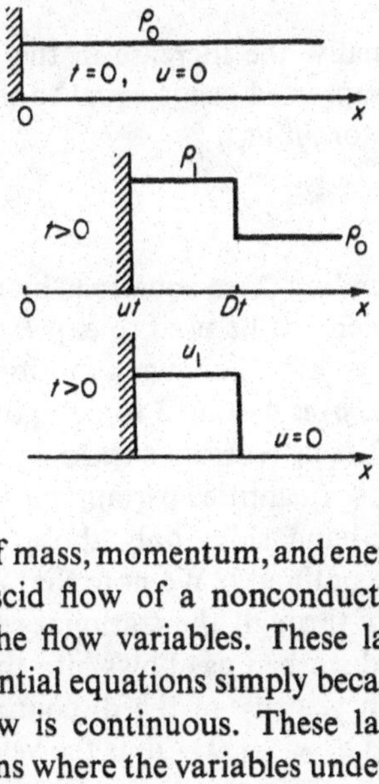

Fig. 1.25. Density and velocity profiles in a shock wave. The wave is produced by a piston which moves into the gas with a constant initial velocity. The top figure shows the initial state.

Generally speaking, the laws of conservation of mass, momentum, and energy that form the basis for the equations of inviscid flow of a nonconducting gas do not necessarily assume continuity of the flow variables. These laws were originally formulated in the form of differential equations simply because it was assumed at the beginning that the flow is continuous. These laws, however, can also be applied to those flow regions where the variables undergo a discontinuous change. From a mathematical point of view, a discontinuity can be regarded as the limiting case of very large but finite gradients in the flow variables across a layer whose thickness tends to zero. Since in the dynamics of an inviscid and nonconducting gas (with molecular structure disregarded) there are no characteristic lengths, the possibility of the existence of arbitrarily thin transition layers is not excluded. In the limit of vanishing thickness these layers reduce to discontinuities. Such discontinuities represent shock waves.

Let us apply the general laws of conservation of mass, momentum, and energy to find the unknown quantities: the density ρ_1, and the pressure p_1 in the compressed region, and the propagation velocity of the discontinuity through the undisturbed fluid D. The parameters of the undisturbed gas ρ_0 and p_0 and the piston velocity u (which is equal to the gas velocity) are assumed to be known. A mass of gas equal to $\rho_0 Dt$ contained in a column of unit cross-sectional area is set in motion at the time t. This mass occupies a volume $(D - u)t$, that is, the density of the compressed gas ρ_1 satisfies the condition

$$\rho_1(D - u)t = \rho_0 Dt.$$

The mass $\rho_0 Dt$ acquires a momentum $\rho_0 Dt \cdot u$ which, according to Newton's law, is equal to the impulse due to the pressure forces. The resultant force acting on the compressed gas is equal to the difference between the pressure on the piston side and on the side of the undisturbed fluid, that is,

$$\rho_0 Dut = (p_1 - p_0)t.$$

Finally, the increase in the sum of the internal and kinetic energies of the compressed gas is equal to the work done by the external force acting on the piston, $p_1 ut$,

$$\rho_0 Dt\left(\varepsilon_1 - \varepsilon_0 + \frac{u^2}{2}\right) = p_1 ut.$$

Dividing these equations by t, we obtain a system of three algebraic equations which can be used to express the three unknown quantities p_1, ρ_1, and D in terms of the known quantities u, ρ_0, and p_0 (the thermodynamic relationship $\varepsilon(p, \rho)$ is assumed to be known).

Let us rearrange these equations so that the right-hand sides contain only those quantities pertaining to the region in front of the discontinuity, and the left-hand sides only those quantities pertaining to the region behind the discontinuity. We note that if D is the propagation velocity of the discontinuity through the stationary gas, then $u_0 = -D$ is the velocity at which the undisturbed gas flows into the discontinuity. Likewise, $D - u$ is the propagation velocity of the discontinuity with respect to the gas moving behind it, and $u_1 = -(D - u)$ is the velocity of the gas flowing out of the discontinuity. With this notation we can rewrite the law of conservation of mass as

$$\rho_1 u_1 = \rho_0 u_0. \tag{1.61}$$

Using (1.61), the law of conservation of momentum takes the form

$$p_1 + \rho_1 u_1^2 = p_0 + \rho_0 u_0^2. \tag{1.62}$$

Using (1.61) and (1.62), the law of conservation of energy becomes

$$\varepsilon_1 + \frac{p_1}{\rho_1} + \frac{u_1^2}{2} = \varepsilon_0 + \frac{p_0}{\rho_0} + \frac{u_0^2}{2}. \tag{1.63}$$

Introducing the specific enthalpy $h = \varepsilon + p/\rho$, we can rewrite this equation as

$$h_1 + \frac{u_1^2}{2} = h_0 + \frac{u_0^2}{2}. \tag{1.64}$$

These equations, expressed above in their simplest form, relate the flow variables at the surface of the discontinuity into which the gas is flowing normal to the surface. It is important to note that these equations do not require any assumptions regarding the properties of the fluid and express only the general laws of conservation of mass, momentum, and energy. Equations (1.61)–(1.63) can also be derived directly by treating the discontinuity in a coordinate system in which it is stationary. Since the discontinuity is infinitesimally thin, no accumulation of mass, momentum, or energy can take place within it. Consequently, the fluxes of these quantities on both sides of the discontinuity are equal. If a gas with density ρ_0 and velocity u_0 flows normally to the surface, then the mass flux or mass flow per unit area per unit time into the discontinuity is equal to $\rho_0 u_0$ and is also equal to the mass flux $\rho_1 u_1$ on the other side of the discontinuity. We thus obtain the result (1.61). The momentum of the mass $\rho_0 u_0$ per unit area per unit time is $\rho_0 u_0 \cdot u_0$. The increase in momentum after passing through the discontinuity $\rho_1 u_1^2 - \rho_0 u_0^2$ is equal to the impulse due to the pressure force per unit time $p_0 - p_1$; or, equivalently, the momentum fluxes $p + \rho u^2$ are equal on each side of the discontinuity (the fact that the quantity $p + \rho u^2$ is the momentum flux density for plane motion is evident from (1.7) and (1.8)). We thus obtain the result (1.62).

The increase in the total (internal and kinetic) energy of the gas flowing per unit area of the discontinuity surface per unit time $\rho_0 u_0[(\varepsilon_1 + \frac{1}{2}u_1^2) - (\varepsilon_0 + \frac{1}{2}u_0^2)]$ is equal to the work done by the pressure forces per unit area of the same surface per unit time. This work is equal to $p_0 u_0 - p_1 u_1$. To clarify the derivation of the latter quantity let us imagine a gas flowing from right to left through a pipe and passing through a discontinuity somewhere in the middle of the

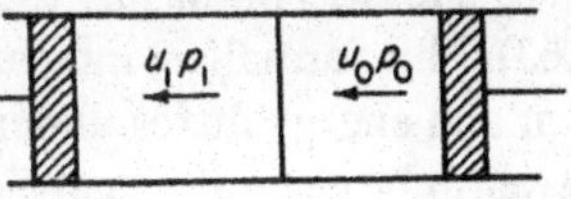

Fig. 1.26. Experiment to clarify the derivation of the work equation.

pipe (Fig. 1.26). Pistons placed at each side of the tube move with velocities u_0 and u_1 in such a manner that the discontinuity surface is at rest. The pressure at the right piston p_0 pushes the gas through the tube, doing the work $p_0 u_0$ per unit time per unit area. The work done by the gas on the left piston is $p_1 u_1$ (the piston "performs" negative work $-p_1 u_1$ on the gas). The total work done on the gas is therefore $p_0 u_0 - p_1 u_1$. Equating the total work to the increase in the gas energy we obtain (1.63). This equation can also be interpreted in a different manner by noting that the total energy fluxes $\rho u(\varepsilon + \frac{1}{2}u^2 + p/\rho)$ are equal on both sides of the discontinuity. (The expression for the energy flux follows from the energy equation written in the form of (1.10).)

Equations (1.61)–(1.63) describing the conservation of mass, momentum, and energy through a discontinuity surface can also be obtained formally from the differential equations (1.2), (1.7), and (1.10). Let us write these equations for the plane case

$$\frac{\partial \rho}{\partial t} = -\frac{\partial}{\partial x}(\rho u),$$

$$\frac{\partial}{\partial t}(\rho u) = -\frac{\partial}{\partial x}(p + \rho u^2), \tag{1.65}$$

$$\frac{\partial}{\partial t}\left(\rho\varepsilon + \frac{\rho u^2}{2}\right) = -\frac{\partial}{\partial x}\left[\rho u\left(\varepsilon + \frac{u^2}{2} + \frac{p}{\rho}\right)\right].$$

First, we shall formally consider the discontinuity as a thin layer containing large gradients of all parameters and integrate the equations from x_0 to x_1. For example,

$$\int_{x_0}^{x_1} \frac{\partial}{\partial t}(\rho u)\, dx = -\int_{x_0}^{x_1} \frac{\partial}{\partial x}(p + \rho u^2)\, dx.$$

We then take the limit, letting the thickness of the layer $x_1 - x_0$ go to zero. The integrals on the left-hand side are proportional to $x_1 - x_0 \to 0$, and vanish (which corresponds to the absence of mass, momentum, and energy accumulation in the discontinuity). The integrals on the right-hand side give the difference between the various fluxes on each side of the discontinuity, that is, we arrive again at (1.61)–(1.63).

We should emphasize the formal nature of the last derivation of (1.61)–(1.63). This formalism indicates only that the expressions for the mass, momentum, and energy fluxes under the divergence signs in the differential equations are entirely general, regardless of whether the flow is continuous or not. If we consider the discontinuity not as a mathematical surface, but as a thin layer of finite thickness where the flow variables change exceedingly sharply, but continuously, we can no longer apply equations (1.65), which do not take either viscosity or heat conduction into account. It will be shown later that the entropy of the gas on each side of the discontinuity is different, while the differential equations (1.65) are subject to the condition that the entropy is constant along streamlines (the flow is adiabatic). We note the outward similarity of the energy equation at the shock discontinuity (1.64) and the Bernoulli equation for steady flow

$$h + \frac{u^2}{2} = const,$$

which holds along streamlines.

§14. Hugoniot curves*

Equations (1.61)–(1.63) relating the flow variables on each side of the discontinuity form a system of three algebraic equations with six variables u_0, ρ_0, p_0, u_1, ρ_1, and p_1. It is assumed that the thermodynamic properties of the fluid (the function $\varepsilon(p, \rho)$ or $h(p, \rho)$) are known. Knowing the thermodynamic properties of the gas ahead of the discontinuity ρ_0, p_0 and assuming that the value of a parameter describing the strength of the shock wave is known (for example, the pressure behind the wave front p_1 or the velocity of the "piston" creating the wave $|u| = u_0 - u_1$), we can calculate all the remaining variables. Let us derive some general relationships which follow from the conservation laws (1.61)–(1.63). In place of density we introduce the specific volumes $V_0 = 1/\rho_0$ and $V_1 = 1/\rho_1$. From (1.61), we obtain

$$\frac{V_0}{V_1} = \frac{u_0}{u_1}. \tag{1.66}$$

Eliminating the velocities u_0 and u_1 from (1.61)–(1.62), we find

$$u_0^2 = V_0^2 \frac{p_1 - p_0}{V_0 - V_1}, \tag{1.67}$$

$$u_1^2 = V_1^2 \frac{p_1 - p_0}{V_0 - V_1}. \tag{1.68}$$

If the shock wave is created in the undisturbed gas by the motion of a piston, we obtain the following equation for the flow velocity of the compressed gas (equal to the "piston" velocity) with respect to the undisturbed gas:

$$|u| = u_0 - u_1 = [(p_1 - p_0)(V_0 - V_1)]^{1/2}. \tag{1.69}$$

* *Editors' note.* Termed shock adiabatics by the authors.

We note here a useful formula for the difference between the kinetic energy of the gas on each side of the discontinuity in a coordinate system in which the shock is at rest

$$\tfrac{1}{2}(u_0^2 - u_1^2) = \tfrac{1}{2}(p_1 - p_0)(V_0 + V_1). \qquad (1.70)$$

Substituting (1.67) and (1.68) into the energy equation (1.63) we obtain a relationship between the pressure and the specific volume on each side of the discontinuity

$$\varepsilon_1(p_1, V_1) - \varepsilon_0(p_0, V_0) = \tfrac{1}{2}(p_1 + p_0)(V_0 - V_1). \qquad (1.71)$$

Replacing the specific internal energies by the specific enthalpies according to the definition $h = \varepsilon + pV$, we can rewrite (1.71) as

$$h_1 - h_0 = \tfrac{1}{2}(p_1 - p_0)(V_0 + V_1). \qquad (1.72)$$

By analogy with the equation relating the initial and final pressures and volumes during adiabatic compression of a fluid, relation (1.71) or its equivalent (1.72) is termed the shock adiabatic or the Hugoniot relation. The Hugoniot curve is represented by the function

$$p_1 = H(V_1, p_0, V_0), \qquad (1.73)$$

which in many practical cases, when the thermodynamic function $\varepsilon = \varepsilon(p, V)$ has a simple form, can be found in an explicit form.

Hugoniot curves differ appreciably from ordinary isentropes* or isentropic adiabatics. While the ordinary isentrope belongs to a one-parameter family of curves $p = P(V, S)$, where the only parameter is the entropy S, the Hugoniot curve is a function of two parameters, the initial pressure p_0 and volume V_0. In order to cover all the curves $p = P(V, S)$ it is sufficient to traverse a one-dimensional series of values of the entropy S. In order to cover all the curves $p = H(V, p_0, V_0)$ it is necessary to construct an "infinity squared" of curves corresponding to all possible values of p_0 and V_0.

§15. Shock waves in a perfect gas with constant specific heats

The shock wave equations for a perfect gas with constant specific heats are particularly simple. It is convenient to use this case to explain all the basic qualitative relationships governing the changes in the variables across a shock wave. Let us substitute the following relationships into the Hugoniot relations (1.71) and (1.72):

$$\varepsilon = c_v T = \frac{1}{\gamma - 1} pV; \qquad h = c_p T = \frac{\gamma}{\gamma - 1} pV. \qquad (1.74)$$

* *Editors' note.* Termed Poisson's adiabatics by the authors.

We can now obtain the equation of the Hugoniot curve in the explicit form

$$\frac{p_1}{p_0} = \frac{(\gamma+1)V_0 - (\gamma-1)V_1}{(\gamma+1)V_1 - (\gamma-1)V_0} \tag{1.75}$$

from which the specific volume ratio is given by

$$\frac{V_1}{V_0} = \frac{(\gamma-1)p_1 + (\gamma+1)p_0}{(\gamma+1)p_1 + (\gamma-1)p_0}. \tag{1.76}$$

The temperature ratio follows from

$$\frac{T_1}{T_0} = \frac{p_1 V_1}{p_0 V_0}. \tag{1.77}$$

Using (1.76) we can express the velocities in the formulas (1.67) and (1.68) as a function of the pressures and the initial volume

$$u_0^2 = \frac{V_0}{2}[(\gamma-1)p_0 + (\gamma+1)p_1], \tag{1.78}$$

$$u_1^2 = \frac{V_0}{2}\frac{[(\gamma+1)p_0 + (\gamma-1)p_1]^2}{[(\gamma-1)p_0 + (\gamma+1)p_1]}. \tag{1.79}$$

Considering a perfect gas with constant specific heats as an example, we can clarify certain relationships governing the behavior of shock waves. The Hugoniot curve is a curve on the p, V diagram passing through the initial state p_0, V_0. This curve is shown in Fig. 1.27. In principle, (1.75) can be also

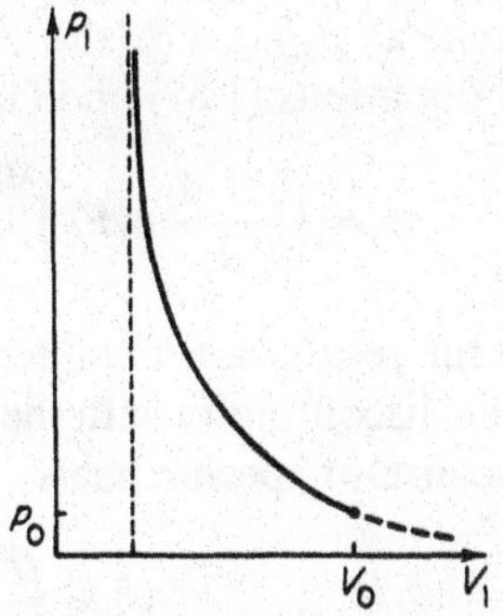

Fig. 1.27. Hugoniot curve.

extended to pressures lower than the initial pressure $p_1 < p_0$. As we shall see in §17, this part of the curve corresponds to physically unattainable states. Hence, it is sketched in Fig. 1.27 by a dashed line. It is evident from (1.76) that the density ratio across a very strong shock wave, where the pressure behind the wave front is much higher than the initial pressure, does not increase with increasing strength indefinitely, but approaches a certain finite value. This limiting density or volume ratio across the shock wave is a function of the specific heat ratio only, and is equal to

$$\frac{\rho_1}{\rho_0} = \frac{V_0}{V_1} = \frac{\gamma+1}{\gamma-1}. \tag{1.80}$$

The limiting density ratio for a monatomic gas with $\gamma = 5/3$ is equal to 4. For a diatomic gas $\gamma = 7/5$ (assuming that the vibrational modes have not been excited) and the limiting density ratio is 6; if, on the other hand, full vibrational excitation is assumed, then $\gamma = 9/7$ and the density ratio is 8. In reality, at high temperatures and pressures, the specific heats and the specific heat ratio are no longer constant because of molecular dissociation and of ionization. Hugoniot curves taking these processes into account will be considered in Chapter III. Even in this case, however, the density ratio remains finite and does not increase without limit; generally it does not exceed 11–13. The increase in density across a shock wave in a perfect gas with a high pressure ratio is greater the higher are the specific heats and the lower is the specific heat ratio.

At high pressures p_1 the density increases very slowly with increasing pressure and consequently the increase in temperature of the compressed gas is proportional to the pressure (see (1.77) for $V_1 \approx const$). In the limit of a strong shock, when $p_1/p_0 \gg 1$, and $V_1/V_0 \approx (\gamma - 1)/(\gamma + 1)$,

$$\frac{T_1}{T_0} = \frac{\gamma - 1}{\gamma + 1} \frac{p_1}{p_0}. \tag{1.81}$$

In the limit, as $p_1/p_0 \to \infty$ the velocities increase as the square root of the pressure. Equations (1.67) and (1.68) show that for $p_1 \gg p_0$

$$u_0 = \left(\frac{\gamma + 1}{2} p_1 V_0\right)^{1/2}, \qquad u_1 = \left(\frac{(\gamma - 1)^2}{2(\gamma + 1)} p_1 V_0\right)^{1/2}. \tag{1.82}$$

Important results are obtained by comparing the gas velocities on both sides of the discontinuity with the corresponding speeds of sound. In a perfect gas with constant specific heats

$$c^2 = \left(\frac{\partial p}{\partial \rho}\right)_S = \gamma \frac{p}{\rho} = \gamma p V.$$

The gas velocities with respect to the discontinuity divided by the speed of sound are then given by

$$\left(\frac{u_0}{c_0}\right)^2 = \frac{(\gamma - 1) + (\gamma + 1)p_1/p_0}{2\gamma}, \tag{1.83}$$

$$\left(\frac{u_1}{c_1}\right)^2 = \frac{(\gamma - 1) + (\gamma + 1)p_0/p_1}{2\gamma}. \tag{1.84}$$

In the limiting case of a weak shock wave the pressures on both sides of the discontinuity are close to each other, so that $p_1 \approx p_0$, $(p_1 - p_0)/p_0 \ll 1$. From (1.76) the density increase is also small, so that $V_1 \approx V_0$, and the sound speeds are almost equal, so that $c_1 \approx c_0$. It becomes obvious from (1.83) and (1.84) that in this case $u_0 \approx c_0 \approx c_1 \approx u_1$. However u_0 is the propagation velocity of the discontinuity through the undisturbed gas. Therefore, a weak shock wave travels through the gas with a velocity which is very close to the speed of sound; thus, the weak shock wave is practically the same as an acoustic compression wave. This is not surprising since the difference between p_1 and p_0 is small; we are dealing with a small disturbance.

Equations (1.83) and (1.84) also show that with a shock wave, across which the gas is compressed ($V_1 < V_0$, $p_1 > p_0$), the gas flows into the discontinuity with a supersonic velocity $u_0 > c_0$ and flows out with a subsonic velocity $u_1 < c_1$. The fact that $V_1 < V_0$, $\rho_1 > \rho_0$ for $p_1 > p_0$ follows from the general relations (1.67) and (1.68). We can formulate this statement in a different manner: The shock wave propagates at a supersonic velocity with respect to the undisturbed gas and at a subsonic velocity with respect to the compressed gas behind it. The stronger is the shock wave (i.e., the larger is the ratio p_1/p_0), the higher is the velocity of the wave front u_0 in comparison with the speed of sound c_0 in the undisturbed gas. On the other hand, in the limiting case of a strong shock ($p_1 \gg p_0$) the ratio u_1/c_1 approaches a constant value, $u_1/c_1 \to [(\gamma - 1)/2\gamma]^{1/2} < 1$.

Let us examine what happens to the entropy of a gas compressed by a shock wave. To within an arbitrary constant the entropy of a perfect gas with constant specific heats is given by $S = c_v \ln pV^\gamma$. The difference between the entropy on each side of the shock front, as derived from (1.76), is

$$S_1 - S_0 = c_v \ln \frac{p_1 V_1^\gamma}{p_0 V_0^\gamma} = c_v \ln \left\{ \frac{p_1}{p_0} \left[\frac{(\gamma - 1)(p_1/p_0) + (\gamma + 1)}{(\gamma + 1)(p_1/p_0) + (\gamma - 1)} \right]^\gamma \right\}. \tag{1.85}$$

In the limiting case of a weak wave ($p_1 \approx p_0$) the expression in braces is close to unity and $S_1 \approx S_0$. As the strength of the wave increases, that is, as the ratio p_1/p_0 increases beyond unity, the expression in braces increases monotonically, and, as can be easily verified, approaches infinity as $p_1/p_0 \to \infty$. Thus, the entropy jump of a gas compressed by a shock increases with the strength of the shock wave. The increase in entropy indicates that irreversible dissipative processes (which can be traced to the presence of viscosity and heat conduction in the fluid) occur in the shock wave. A theory that does not take into account these processes is not capable of describing either the mechanism of shock compression or the structure of the very thin but finite layer where the gas undergoes a transition from the initial to the final state. Indeed, such a theory formally represents a shock discontinuity as a mathematical surface of zero thickness. As previously noted, this theory does not include any characteristic length that could serve as a scale for the thickness of the discontinuity. This scale appears when the molecular structure of the gas, in terms of the viscosity and thermal conductivity, is taken into account. It is the molecular mean free path, which is proportional to the viscosity and the thermal conductivity, which serves as a measure of the width of the discontinuity. It is significant, however, that the entropy increase across a compression shock is entirely independent of the dissipative mechanism and is defined exclusively by the conservation laws of mass, momentum, and energy. Only the thickness of the discontinuity, which depends upon the rate of the irreversible heating of the gas experiencing the shock compression, depends on the dissipative mechanism. Analogously, a glass of hot water will invariably cool to a given room temperature, independently of the mechanism of heat exchange with the surrounding medium, while the mechanism determines only the rate of cooling.

The dissipative mechanism controls only the values of the gradients of the flow variables in the transition layer, but does not affect the jumps in these quantities between the initial and the final states. These changes are determined solely by the conservation laws. For example, if $\Delta p = p_1 - p_0$ is the pressure jump across the shock wave and Δx is the thickness of the transition layer, then Δx and $dp/dx \sim \Delta p/\Delta x$ change with both the viscosity and the thermal conductivity, while the product $\Delta x \, dp/dx \approx \Delta p$ remains unchanged. In the limit, when the coefficients of viscosity and thermal conductivity tend to zero, $\Delta x \to 0$ and $dp/dx \sim 1/\Delta x \to \infty$, so that the gradients become infinite corresponding to a discontinuity.

The gasdynamic equations which do not take into account either viscosity or thermal conductivity admit the existence of discontinuities, but are not capable of describing the continuous transition from the initial to the final state. These equations are based on the assumption that the process is adiabatic, that $DS/Dt = 0$, which is equivalent to the energy equation. The gasdynamic equations describe four conservation laws: conservation of mass, momentum, energy, and entropy. Only the first three of these equations, but not the entropy equation, are satisfied across the discontinuity.

We shall return to the problem of the thickness of a shock front in §23. This problem can be solved only when the molecular structure of the fluid is taken into account, that is, the shock compression process is subjected to a careful "microscopic" examination. At present, we shall continue the "macroscopic" description of the shock compression phenomena based on the conservation laws of mass, momentum, and energy.

§16. Geometric interpretation of the laws governing compression shocks

The p, V diagram is very useful for clarifying the theoretical relations governing shock waves and the properties of Hugoniot curves. A Hugoniot curve HH(Fig. 1.28) is drawn through the point A on the p, V plane. This

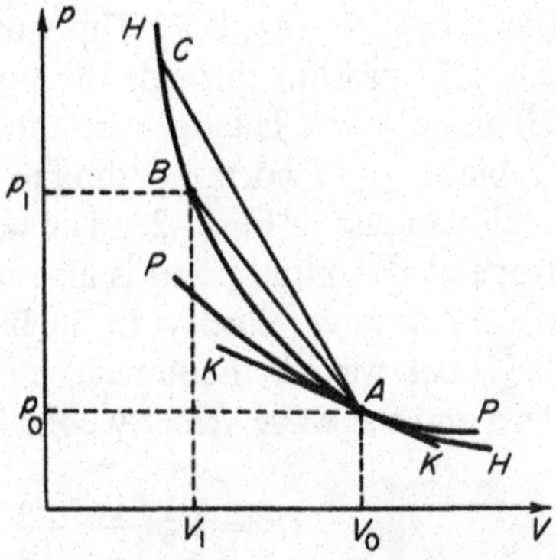

Fig. 1.28. A p, V diagram. HH is a Hugoniot curve, PP is an isentrope, and KK is a tangent to both the Hugoniot and isentrope at the initial state represented by the point $A(V_0, p_0)$.

point denotes the initial state of the fluid p_0, V_0. We assume that the character of this curve is similar to the Hugoniot curve for a perfect gas with constant specific heats, that is, the curve is everywhere convex downward, the second derivative d^2p/dV^2 being positive at all points. For clarity we shall illustrate some statements by specific calculations using a perfect gas with constant specific heats as an example. It can be shown, however, that the nature of the behavior is quite general and applicable to fluids with other thermodynamic properties. The only imposed condition is that the Hugoniot curve must be convex downward.

Let the fluid be transformed by the shock compression from the state $A(p_0, V_0)$ to a state $B(p_1, V_1)$ represented by the point B on the Hugoniot curve. According to (1.67), the propagation velocity of a shock wave through the undisturbed fluid is given by

$$D^2 = u_0^2 = V_0^2 \frac{p_1 - p_0}{V_0 - V_1}.$$

This velocity can be determined graphically from the slope $((p_1 - p_0)/(V_0 - V_1))$ of the straight line AB, drawn between the initial and the final states. It is clear from Fig. 1.28 that the higher the final pressure (the stronger the shock wave), the larger the slope and the higher the wave velocity. Two straight lines AB and AC are shown in Fig. 1.28 to illustrate the above statement.

Let us examine what determines the initial slope of the Hugoniot curve at the point A. We calculate the derivative dp_1/dV_1 from formula (1.75) for a perfect gas with constant specific heats

$$\frac{dp_1}{dV_1} = -\frac{(\gamma - 1)p_0}{(\gamma + 1)V_1 - (\gamma - 1)V_0} - \frac{p_0[(\gamma + 1)V_0 - (\gamma - 1)V_1](\gamma + 1)}{[(\gamma + 1)V_1 - (\gamma - 1)V_0]^2}.$$

Evaluating the derivative at the point A by setting $V_1 = V_0$, we obtain $(dp_1/dV_1)_0 = -\gamma p_0/V_0$. This quantity is simply the slope of the isentrope $p \sim V^{-\gamma}$ passing through the point A, since $(\partial p/\partial V)_S = -\gamma p/V$. Thus, the Hugoniot curve is tangent to the isentrope at the point A. The ordinary isentropic curve PP corresponding to the initial entropy of the gas $S_0 = S(p_0, V_0)$ is also shown in Fig. 1.28. The tangency of the Hugoniot curve and the isentrope at the initial point is also illustrated by the general equation (1.67) for the shock wave velocity. In the limit of a weak shock, when $(p_1 - p_0)/p_0 \to 0$, the shock wave is the same as a sound wave; the entropy change approaches zero and the wave velocity coincides with the speed of sound

$$D^2 = V_0^2 \frac{p_1 - p_0}{V_0 - V_1} \quad \to \quad -V_0^2\left(\frac{\Delta p}{\Delta V}\right)_S \quad \to \quad c_0^2.$$

Generally, however, the slope of the straight line AB is always greater than the slope of the tangent to the Hugoniot curve at the point A, so that we always have $D = u_0 > c_0$.

The initial slope of the Hugoniot curve is determined by the speed of sound in the initial state. This will be proven rigorously for the general case of an arbitrary fluid in §18. Direct calculation using the formulas for a perfect gas with constant specific heats shows that not only the first, but also the second derivatives of the Hugoniot and isentropic curves coincide at the point A, that there is a second-order tangency at this point. This result is also one of a general nature (see §18).

As indicated in Fig. 1.28, the Hugoniot curve always passes above the ordinary isentrope emanating from the initial point. Indeed, the entropy increases during the shock compression from the volume V_0 to the volume $V_1 < V_0$, while, of course, it remains constant during the isentropic compression. With the volume fixed, the pressure increases with increasing entropy*.

Equation (1.71) for the Hugoniot curve shows that the increase in the specific internal energy $\varepsilon_1 - \varepsilon_0$ during the shock compression from the state A to the state B is numerically equal to the area of the trapezoid $MABN$, shaded horizontally in Fig. 1.29. If the gas is compressed isentropically from state A to a volume V_1 (state Q) then the work performed is numerically equal to the area $MAQN$ (shaded vertically) bounded by the ordinary isen-

* *Editors' note.* In applying this reasoning to a fluid with a general equation of state, we must assume the coefficient of thermal expansion at constant pressure $V^{-1}(\partial V/\partial T)_p$ is positive.

trope P on the top and by the V axis on the bottom (Fig. 1.29). This area also gives the increase in the internal energy of the gas $\varepsilon' - \varepsilon_0 = -\int_{V_0}^{V_1} p\,dV$ (the integration is performed for $S = S_0$). In order to bring the gas to the final state B it is necessary to heat it at constant volume V_1, transferring an amount of heat numerically equal to the difference between the horizontally and vertically shaded areas, that is, to the area ABQ. The latter area also determines the entropy increase of the gas due to the shock compression. The area is equal to $\varepsilon_1 - \varepsilon' = \int_{S_0}^{S_1} T\,dS = \bar{T}(S_1 - S_0)$, where $\bar{T}$ is a certain average temperature lying in the segment QB (at $V = V_1 = const$).

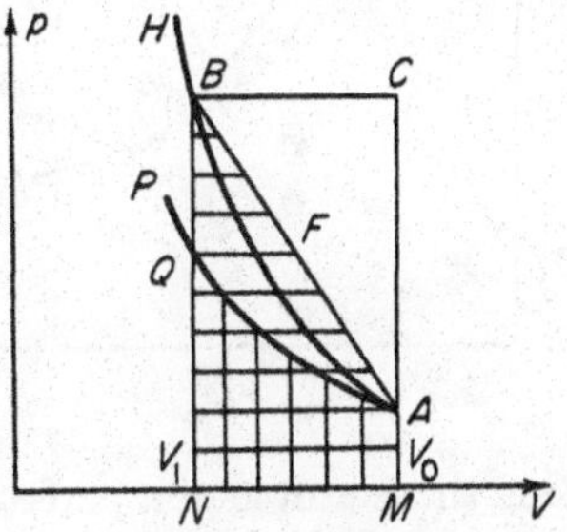

Fig. 1.29. Geometrical interpretation of the energy increase across a shock wave. H is the Hugoniot curve and P is the corresponding isentrope.

In a coordinate system where the gas is initially at rest, the kinetic energy per unit mass acquired by the gas from the compression is equal to

$$\frac{u^2}{2} = \frac{(u_0 - u_1)^2}{2} = \tfrac{1}{2}(p_1 - p_0)(V_0 - V_1).$$

This energy is numerically equal to the area of the triangle ABC in Fig. 1.29, which is the complement of the trapezoid $MABN$ (whose area corresponds to $\varepsilon_1 - \varepsilon_0$) with respect to the rectangle $MCBN$. The area of this rectangle, $p_1(V_0 - V_1)$, represents the total energy transmitted by the "piston" to a unit mass of gas initially at rest. Across a strong shock wave, where $p_1 \gg p_0$, this energy is divided equally between the increase in the internal energy and the increase in the kinetic energy, and the area $MABN$ is approximately equal to the area ABC:

$$\varepsilon_1 - \varepsilon_0 \approx \frac{u^2}{2} \approx \tfrac{1}{2}p_1(V_1 - V_0).$$

Using the p, V diagram let us examine the relationship between the gas velocity and the speed of sound in the final state (Fig. 1.30). On the Hugoniot curve H_A (corresponding to the initial state A) we draw a new Hugoniot curve H_B through point B, where the point B is the initial point of this new curve. The symmetry of the equation for the Hugoniot curve with respect to the permutation of the subscripts "0" and "1" shows that if $p_1 = H(V_1, p_0, V_0)$ then $p_0 = H(V_0, p_1, V_1)$. In other words, the Hugoniot curve H_B formally extended in the direction of decreasing pressure intersects the Hugoniot curve H_A at the point A. The relative location of the Hugoniot curves H_A and H_B is as shown in Fig. 1.30, and this relationship can be easily checked

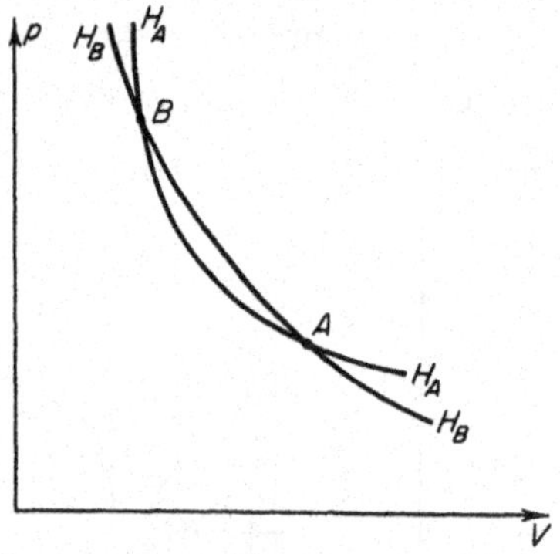

Fig. 1.30. A p, V diagram, illustrating the relationship between the gas velocity and the speed of sound in a shock wave.

using the example of a perfect gas with constant specific heats*. The propagation speed of the wave with respect to the compressed gas is determined by (1.68),

$$u_1^2 = V_1^2 \frac{p_1 - p_0}{V_0 - V_1}.$$

The square of the speed of sound in the compressed gas at the point B is

$$c_1^2 = -V_1^2 \left(\frac{\partial p}{\partial V}\right)_S.$$

The first of these quantities is proportional to the slope of the line BA, and the second to the slope of the Hugoniot curve H_B at the point B (the Hugoniot curve H_B and the isentrope passing through B are tangent). The relative positions of the line BA and the Hugoniot curve H_B correspond to the fact that $u_1 < c_1$.

* The fact that the Hugoniot curve H_B passes to the left of H_A at pressures higher than p_B can be explained as follows: If point B corresponds to compression of the gas from the state A by a very strong shock wave, then the Hugoniot curve H_A becomes almost vertical for $p > p_B$, corresponding to the limiting compression with the volume $[(\gamma - 1)/(\gamma + 1)]V_A$. At the same time, a second shock wave passing through the gas in the state B can cause a compression to the volume

$$\frac{\gamma - 1}{\gamma + 1} V_B = \left(\frac{\gamma - 1}{\gamma + 1}\right)^2 V_A.$$

It was noted at the end of §12 that the Hugoniot curve in contrast to the isentrope is a function of two parameters. By virtue of this fact it is impossible to reach the same final state by compressing the gas by means of several shock waves as is reached by compression with a single shock wave, assuming that both processes start from the same initial state. For example, a strong shock wave propagating through a monatomic gas will yield a density ratio of 4, while two successive strong shock waves will result in a density ratio of 16 with the final pressure remaining the same. At the same time, however, if the final pressure is maintained the same, the same density will be obtained regardless of the number of individual stages of an isentropic process. This situation is illustrated on the p, V diagram of Fig. 1.31, in which are shown an isentrope and several Hugoniot curves corresponding to the compression of a gas by successive shock waves.

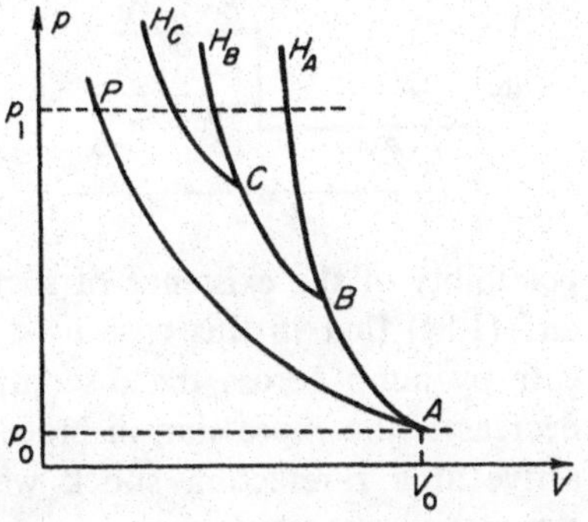

Fig. 1.31. Single and multiple shock and isentropic compression of a gas to the same final pressure p_1. H_A, H_B, and H_C are Hugoniot curves with the initial states A, B, and C, respectively. P is an isentrope.

§17. Impossibility of rarefaction shock waves in a fluid with normal thermodynamic properties

In §15 we presented the equations relating the various flow quantities across a shock wave for the case of a perfect gas with constant specific heats. These equations show that the following inequalities are satisfied across a compression shock wave:

$$p_1 > p_0, \quad \rho_1 > \rho_0, \quad V_1 < V_0, \quad u_0 > c_0, \quad u_1 < c_1, \quad S_1 > S_0. \tag{1.86}$$

The entropy of the fluid increases along with the increase in pressure and density; the wave travels with respect to the undisturbed gas with a supersonic velocity, and with respect to the compressed gas behind it with a subsonic velocity. This situation is represented schematically in Fig. 1.32a.

Let us now extend the relations (1.75) for the Hugoniot curve to include pressures lower than the initial pressure, and let us assume the existence of discontinuities in which an expansion of the gas occurs rather than a compression, with $V_1 > V_0$, $p_1 < p_0$. The conservation laws of mass, momentum, and energy used to derive the equations relating velocities, densities, and pressures on both sides of the discontinuity do not in any way exclude the

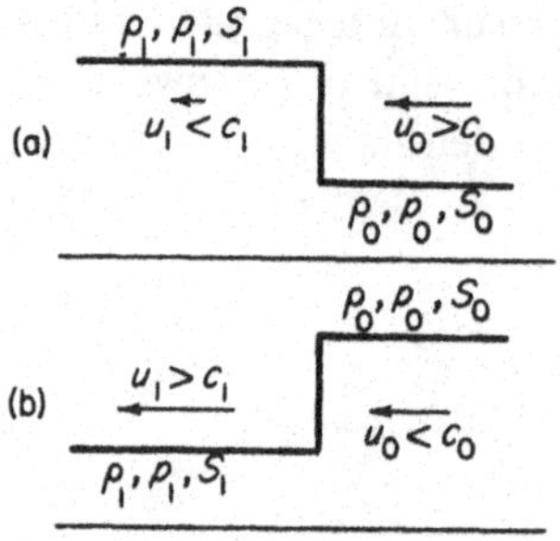

Fig. 1.32. Schematic representation of (a) compression and (b) rarefaction shock waves. The gas flows into the discontinuity from right to left.

possibility of the existence of such discontinuities. It is evident from (1.83) and (1.84) that in this case $u_0 < c_0$, and $u_1 > c_1$. Equation (1.85) for the entropy jump across the discontinuity shows that the entropy of the gas decreases (the expression in braces is less than unity for $p_1 < p_0$). Thus, we arrive at a rarefaction shock wave, where the following inequalities are simultaneously satisfied:

$$p_1 < p_0, \quad \rho_1 < \rho_0, \quad V_1 > V_0, \quad u_0 < c_0, \quad u_1 > c_1, \quad S_1 < S_0. \tag{1.87}$$

This situation is depicted in Fig. 1.32b.

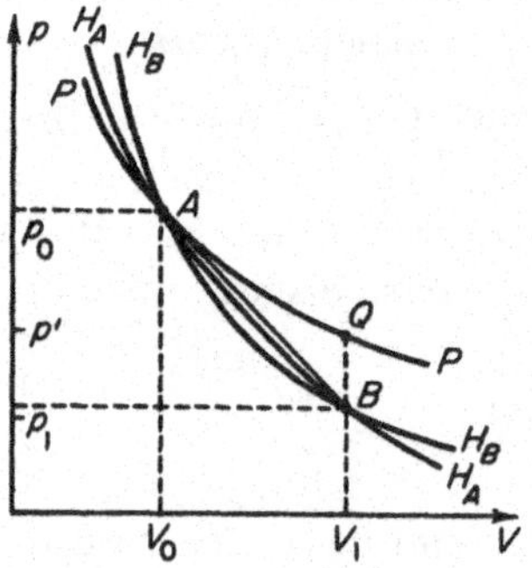

Fig. 1.33. Geometric interpretation of inequalities in a "rarefaction shock wave". H_A is the Hugoniot curve, P is the isentrope passing through the initial state A, and H_B is the Hugoniot curve drawn from the final state B.

The geometric interpretation of these inequalities is similar to that presented in §16 and is shown in Fig. 1.33. The slope of the line AB is less than the slope of the Hugoniot curve H_A at the initial state A ($u_0 < c_0$) and greater than the slope of the second Hugoniot curve H_B in the final state B ($u_1 > c_1$). The isentrope P passing through the point A in the region $p_1 < p_0$ lies above the Hugoniot curve H_A, thus explaining the decrease in entropy across a rarefaction shock. For an isentropic expansion to the same volume V_1, the pressure p' is higher than the final pressure p_1. In order to arrive at B from Q it is necessary to cool the gas at a constant volume, and thus to decrease its entropy.

According to the second law of thermodynamics the entropy of a substance cannot be decreased by internal processes alone, without the transfer of heat to an external medium. This shows that it is impossible for a rarefaction wave to propagate in the form of a discontinuity. Therefore, the requirement that the entropy must increase permits only one of the two possibilities permitted by the conservation laws of mass, momentum, and energy—the compression shock wave. This statement is quite general*. In the next section we shall show that in a weak wave, when the second derivative $(\partial^2 p/\partial V^2)_S > 0$, the inequalities (1.86) or (1.87) are satisfied independently of the thermodynamic properties of the fluid. The validity of this statement for strong waves and for an arbitrary fluid can also be proved. The only condition imposed on the fluid properties is that the Hugoniot curve must be convex downward at all points, that $(d^2p/dV^2)_H > 0$, in the same manner as for a perfect gas with constant specific heats. Indeed, the overwhelming majority of real fluids possess such properties, and therefore the statement regarding the impossibility of the existence of rarefaction shock waves is quite general (certain exceptions will be discussed below).

The impossibility of the existence of a rarefaction shock wave can be explained as follows. Such a wave would propagate through the undisturbed gas with the subsonic velocity $u_0 < c_0$. This means that if a situation similar to that depicted in Fig. 1.32b should arise at any instant of time, any disturbances induced by the density and pressure jumps will begin to travel to the right with the speed of sound c_0, and will outrun the "shock wave". After a certain time the rarefaction region will include the gas in front of the "discontinuity", and the discontinuity will simply disappear. In other words, a rarefaction shock wave is mechanically unstable. Conversely, a compression shock wave propagates through the undisturbed gas at the supersonic speed $u_0 > c_0$; the state behind the wave front can in no way affect the state of the gas ahead of the wave and the discontinuity remains stable.

The compression shock wave is propagated with respect to the compressed gas with a subsonic velocity $u_1 < c_1$, and therefore conditions behind the shock front do affect the strength of the wave. If the gas behind the wave front is either heated or compressed, then the shock wave will be strengthened

* *Editors' note.* Referred to by the authors as Champlain's theorem.

and, conversely, if the gas behind the shock front is either cooled or expanded, then rarefaction disturbances will overtake the shock wave and weaken it. In a rarefaction shock wave the situation would have been exactly opposite: since the rarefaction wave would be propagated through an expanded gas with supersonic velocity, it would not be influenced by any processes occurring behind it, that is, the wave would be "uncontrollable".

It is quite significant that the condition of mechanical stability for a shock wave corresponds with the thermodynamic condition of increasing entropy. Mechanical stability can be present only when the wave is propagated through the undisturbed fluid with supersonic speed, otherwise disturbances induced by the shock wave would penetrate the initial gas at the speed of sound, overtake the shock wave, and thus "wash out" the sharp wave front. The condition of an increase in entropy agrees also with our interpretation of the causality of the phenomena. Namely, with an entropy increase, the compression shock wave propagates through the disturbed gas with a subsonic speed, that is, external factors, such as a piston pushed into the gas, can induce the appearance of a shock wave and subsequently influence its propagation.

Compression shock waves, which correspond to an entropy increase in a fluid with normal thermodynamic properties where $(\partial^2 p/\partial V^2)_S > 0$, turn out to be mechanically stable and affected by external actions. The existence of a rarefaction shock wave is impossible from the points of view of both thermodynamics and stability, and a steep rarefaction front, had such been induced,

Table 1.1

ATTAINABLE STATES BY MEANS OF COMPRESSION AND RAREFACTION WAVES FOR NORMAL FLUIDS

	Compression wave	Rarefaction wave
Discontinuity	Possible; entropy increases; mechanically stable	Impossible; entropy decreases; mechanically unstable
Smooth distribution	Impossible[a]; unlimited increase in the steepness of the wave front with the final result of "overshooting"	Possible; the distribution becomes increasingly smoother with time

[a] *Editors' note.* In the sense of a condition valid for all time.

would disappear with time. In concluding this section we present in Table 1.1 the possibilities of attaining different states through compression and rarefaction waves.

§18. Weak shock waves

Let us consider a weak shock wave, where the jumps in the flow variables can be regarded as small quantities. We shall refrain temporarily from making any assumptions regarding the thermodynamic properties of the fluid and base our discussion upon the conservation laws. Considering the internal energy to be a function of entropy and specific volume, we describe the energy increase in the shock wave as an expansion in terms of small changes in the independent variables with respect to the initial state

$$\varepsilon_1 - \varepsilon_0 = \left(\frac{\partial \varepsilon}{\partial S}\right)_V (S_1 - S_0) + \left(\frac{\partial \varepsilon}{\partial V}\right)_S (V_1 - V_0) + \frac{1}{2}\left(\frac{\partial^2 \varepsilon}{\partial V^2}\right)_S (V_1 - V_0)^2$$
$$+ \frac{1}{6}\left(\frac{\partial^3 \varepsilon}{\partial V^3}\right)_S (V_1 - V_0)^3.$$

All derivatives in this expansion are evaluated at the initial state V_0, S_0. By regarding $V_1 - V_0$ as a small quantity of first order, we shall shortly see that $S_1 - S_0$ is a small quantity of third order. Thus in restricting the expansion for the internal energy to third-order quantities, we can drop terms proportional to $(S_1 - S_0)(V_1 - V_0)$, $(S_1 - S_0)^2$, etc.

According to the thermodynamic identity $d\varepsilon = T\,dS - p\,dV$,

$$\left(\frac{\partial \varepsilon}{\partial S}\right)_V = T, \qquad \left(\frac{\partial \varepsilon}{\partial V}\right)_S = -p.$$

Hence,

$$\varepsilon_1 - \varepsilon_0 = T_0(S_1 - S_0) - p_0(V_1 - V_0) - \frac{1}{2}\left(\frac{\partial p}{\partial V}\right)_S (V_1 - V_0)^2$$
$$- \frac{1}{6}\left(\frac{\partial^2 p}{\partial V^2}\right)_S (V_1 - V_0)^3.$$

We substitute this expression into the Hugoniot equation (1.71) and expand the pressure p_1 on the right-hand side in powers of $(V_1 - V_0)$. Since the left-hand side of the equation is expanded up to quantities of third order, it is sufficient to restrict the expansion of the pressure terms to second-order terms and to neglect the term containing the entropy increase. The latter will yield on the right-hand side a term proportional to $(S_1 - S_0)(V_1 - V_0)$, which is of higher order than $(V_1 - V_0)^3$. We obtain

$$p_1 = p_0 + \left(\frac{\partial p}{\partial V}\right)_S (V_1 - V_0) + \frac{1}{2}\left(\frac{\partial^2 p}{\partial V^2}\right)_S (V_1 - V_0)^2.$$

Substituting into (1.71) and simplifying we obtain for the relationship between the entropy and volume changes

$$T_0(S_1 - S_0) = \frac{1}{12}\left(\frac{\partial^2 p}{\partial V^2}\right)_S (V_0 - V_1)^3. \tag{1.88}$$

If we start with the Hugoniot equation in the form (1.72), where the internal energy has been replaced by the enthalpy, we similarly get

$$T_0(S_1 - S_0) = \frac{1}{12}\left(\frac{\partial^2 V}{\partial p^2}\right)_S (p_1 - p_0)^3. \tag{1.89}$$

The identity of these two equations can be easily shown by substituting the expansion $(p_1 - p_0) = (\partial p/\partial V)_S(V_1 - V_0)$ into (1.89) and noting that

$$\frac{\partial^2 V}{\partial p^2} = \frac{\partial}{\partial p}\frac{\partial V}{\partial p} = \frac{\partial}{\partial p}\left(\frac{1}{\partial p/\partial V}\right) = \frac{1}{\partial p/\partial V}\frac{\partial}{\partial V}\left(\frac{1}{\partial p/\partial V}\right) = -\left(\frac{\partial p}{\partial V}\right)^{-3}\left(\frac{\partial^2 p}{\partial V^2}\right).$$

Equations (1.88) and (1.89) show that the entropy increase in a weak shock wave is a small quantity of third order with respect to the differences $p_1 - p_0$ or $V_0 - V_1$, which characterize the strength of the wave. It is obvious from (1.88) and (1.89) that the sign of the entropy increase in a shock wave is determined by the sign of the second derivatives $(\partial^2 p/\partial V^2)_S$ or $(\partial^2 V/\partial p^2)_S$. If the isentropic compressibility of the fluid $-(\partial V/\partial p)_S$ decreases with pressure, that is, $(\partial^2 V/\partial p^2)_S > 0$ and $(\partial^2 p/\partial V^2)_S > 0$, then the ordinary isentrope is convex downward in the p, V plane (as in the case of a perfect gas with constant specific heats). In this case, when $p_1 > p_0$ and $V_1 < V_0$ the entropy increases ($S_1 > S_0$) in a compression shock wave and decreases in a rarefaction shock wave. If, however, $(\partial^2 V/\partial p^2)_S < 0$ or $(\partial^2 p/\partial V^2)_S < 0$, then the opposite is true, that is, when $p_1 < p_0$ and $V_1 > V_0$ the entropy increases in a rarefaction shock wave and decreases in a compression shock wave. Since for the overwhelming majority of real fluids $(\partial^2 V/\partial p^2)_S > 0$, the impossibility of the existence of rarefaction shock waves follows from the impossibility of an entropy decrease. This postulate has already been formulated above and demonstrated for a perfect gas with constant specific heats.

Let us expand the pressure $p = p(S, V)$ about the initial point S_0, V_0 up to third order in $V_1 - V_0$ and first order in $S_1 - S_0$

$$p_1 - p_0 = \left(\frac{\partial p}{\partial V}\right)_S (V_1 - V_0) + \frac{1}{2}\left(\frac{\partial^2 p}{\partial V^2}\right)_S (V_1 - V_0)^2$$
$$+ \frac{1}{6}\left(\frac{\partial^3 p}{\partial V^3}\right)_S (V_1 - V_0)^3 + \left(\frac{\partial p}{\partial S}\right)_V (S_1 - S_0).$$

We shall use this expansion to describe the initial segments of the Hugoniot and isentropic curves passing through the point S_0, V_0. The terms of first and second order in $V_1 - V_0$ of both curves coincide, that is, the shock and ordinary isentrope have common tangents and common centers of curvature at the initial point (second-order tangency or osculation). The third-order terms are different for both curves. The third term on the right-hand side of the expansion is common to both curves. The fourth and last term disappears for the ordinary isentrope, since $S_1 - S_0 = 0$ ($S = const$), while according to (1.88), for the Hugoniot curve it is equal to

$$\left(\frac{\partial p}{\partial S}\right)_V (S_1 - S_0) = -\frac{1}{12T_0}\left(\frac{\partial p}{\partial S}\right)_V \left(\frac{\partial^2 p}{\partial V^2}\right)_S (V_1 - V_0)^3.$$

The pressure of all normal fluids* increases with entropy at constant volume (during heating at constant volume), that is, $(\partial p/\partial S)_V > 0$ and $(\partial^2 p/\partial V^2)_S$ is also positive. Consequently, the last term is negative when $V_1 > V_0$, and positive when $V_1 < V_0$; when $V_1 > V_0$ the Hugoniot curve passes below the isentrope, and when $V_1 < V_0$ it passes above the isentrope. Thus, at the initial point we have a second-order tangency at the intersection of the two curves. The relative positions of the Hugoniot curve H and the isentrope P are shown in Fig. 1.34. We note that the segment CD is a first-order quantity in $V_0 - V_1$, DE is a second-order quantity, and EF is a third-order quantity.

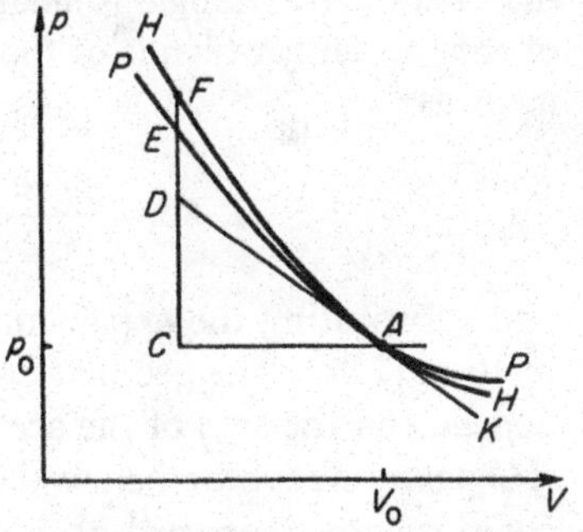

Fig. 1.34. Relative position of the shock H and ordinary isentrope P. DK is the tangent to the curves at the initial state A. In a weak shock wave the segment CD is a quantity of first, DE of second, and EF of third order.

Let us return to the geometric interpretation of the entropy increase across a shock wave (Fig. 1.35). As shown in §16, the quantity $\overline{T}\Delta S$ is described by the area of the figure $AFBCEA$. Let us divide it by the straight line AC into two parts, the segment $ACEA$ and the triangle ABC. The area of the triangle ABC is equal to one half of the product of the base BC and the altitude $V_0 - V_1$. With small changes of all parameters, in a weak wave, the segment

* *Editors' note.* For which the coefficient of thermal expansion at constant pressure $V^{-1}(\partial V/\partial T)_p$ is positive.

BC is equal to $(\partial p/\partial S)_V \, \Delta S$, and

$$T\,\Delta S = F_{\text{segm}} + \frac{1}{2}\left(\frac{\partial p}{\partial S}\right)_V (V_0 - V_1)\,\Delta S,$$

where F_{segm} is the area of the segment $ACEA$. Hence,

$$\Delta S = \frac{F_{\text{segm}}}{T - \alpha}, \qquad \alpha = \frac{1}{2}\left(\frac{\partial p}{\partial S}\right)_V (V_0 - V_1).$$

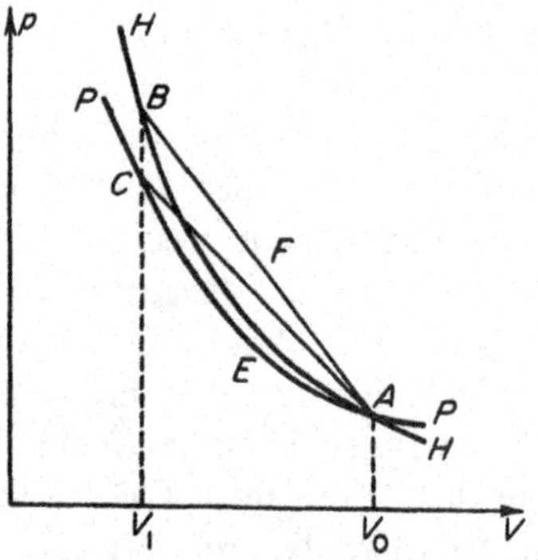

Fig. 1.35. Geometrical interpretation of the entropy increase in a shock wave.

For small volume changes $\alpha \to 0$ and $T\,\Delta S \to F_{\text{segm}}$, that is, the correction for the area of the triangle is small. In fact, it is of a higher order than the area of the segment, which is of the order of $T\,\Delta S$. Expressing the area of the segment as

$$T\,\Delta S = \frac{p_0 + p'}{2}(V_0 - V_1) - \int_{V_1}^{V_0} (p\,dV)_{S=S_0}$$

and substituting the expansions for weak waves we obtain, as expected, the relation (1.88). The geometrical interpretation also shows that the sign of ΔS depends on the sign of the area of the segment, that is, on whether the chord AC passes above or below the isentrope or, what is the same, whether the curve is convex upward or downward.

Let us compare the velocities u_0 and u_1 with the sound speeds c_0 and c_1. As we know, the ratio u_0/c_0 is determined by the ratio of the slope of the line AB (see Fig. 1.28) and of the tangent to the isentrope at the point A. The ratio u_1/c_1 is determined by the ratio of the slope of the line AB and of the tangent to the isentrope at the point B. We write the expressions for the slopes of all three straight lines in the form

$$\frac{p_1 - p_0}{V_1 - V_0} = \left(\frac{\partial p}{\partial V}\right)_{S_0} + \frac{1}{2}\left(\frac{\partial^2 p}{\partial V^2}\right)_{S_0}(V_1 - V_0)$$

for the line AB,

$$\left(\frac{\partial p}{\partial V}\right)_{S_A} = \left(\frac{\partial p}{\partial V}\right)_{S_0}$$

for the tangent to the isentrope at the point A, and

$$\left(\frac{\partial p}{\partial V}\right)_{S_B} = \left(\frac{\partial p}{\partial V}\right)_{S_0} + \left(\frac{\partial^2 p}{\partial V^2}\right)_{S_0} (V_1 - V_0)$$

for the tangent to the isentrope at the point B. The last equation follows from the fact that the isentrope $S_1 = const$ is parallel to the isentrope $S_0 = const$ up to third-order terms in $V_1 - V_0$. Noting that

$$\left(\frac{\partial p}{\partial V}\right)_{S_0} < 0, \qquad \left(\frac{\partial^2 p}{\partial V^2}\right)_{S_0} > 0, \qquad V_1 - V_0 < 0,$$

we see that the straight line AB is steeper than the tangent at the point A, but is less steep than the tangent at point B. Hence, it follows that $u_0 > c_0$ and $u_1 < c_1$. This is directly evident from Fig. 1.30.

Of importance is the inner connection between the condition for entropy increase and that for mechanical stability of the discontinuity $u_0 > c_0$. Both conditions follow directly from the fact that the slope of the isentropic and Hugoniot curves starting at the point A increases with decreasing volume.

Thus, by considering weak shock waves in a fluid with arbitrary thermodynamic properties we have obtained from the conservation laws all the results which were demonstrated previously for a perfect gas with constant specific heats. The only additional necessary condition was that the second derivative $(\partial^2 p/\partial V^2)_S$ should be positive.

§19. Shock waves in a fluid with anomalous thermodynamic properties

Let us now consider a fluid with anomalous thermodynamic properties, where the second derivative $(\partial^2 p/\partial V^2)_S$ is negative at least over some part of the isentropic curve. The ordinary isentrope for such a fluid in the corresponding pressure-volume diagram is convex upward, as shown in Fig. 1.36. It follows from the discussion in the preceding section that if the pressure changes are small, the Hugoniot curve almost coincides with the isentrope (with an accuracy to third order in either $V_1 - V_0$ or $p_1 - p_0$). In this case the area of the figure $APBMNA$ bounded on the top by the isentrope is greater than the area of the trapezoid $AEBMNA$ bounded on the top by the chord AEB, and hence the entropy decreases across the compression shock wave (this is also shown by (1.88)). At the same time, since the slope of the chord is smaller than the slope of the tangent at the point A, the propagation velocity of the shock wave through the undisturbed gas is less than the speed of sound. Also, since the slope of the chord AEB is larger than the slope of the tangent at the point B, the velocity behind the discontinuity is supersonic. Conversely, the entropy of a rarefaction shock increases (see (1.88)). The comparison of the slopes of the chord AC and of the tangents at the points A and C shows that the velocity is supersonic ahead of the discontinuity and subsonic behind it.

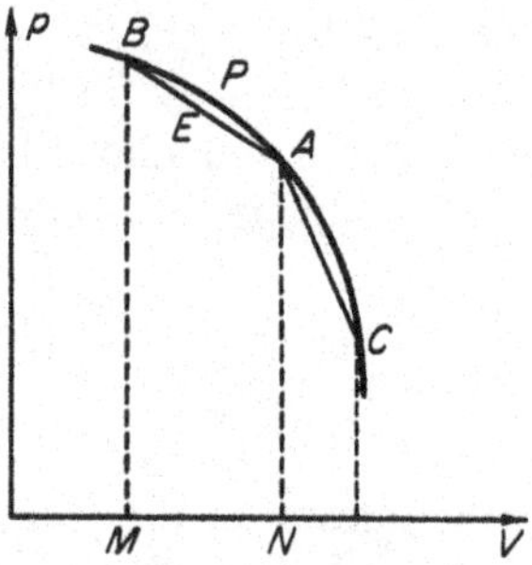

Fig. 1.36. Isentrope for a fluid with anomalous properties and the geometrical interpretation of the relation for compression and rarefaction shock waves.

Thus the entropy increase in a fluid with anomalous properties coincides with the condition of mechanical stability $u_0 > c_0$ and agrees with the condition allowing external factors to influence the wave propagation $u_1 < c_1$. Compression shock waves are impossible, while rarefaction shock waves can occur in such an anomalous fluid. Compression of such a fluid induced by a piston takes place by means of a gradually widening wave similar to rarefaction waves in an ordinary gas. No shock discontinuity is formed and the motion is isentropic. The rarefaction wave, however, is propagated as a steep front, which does not expand with time and whose thickness is determined by the viscosity and thermal conductivity.

Under normal conditions all substances—gaseous, solid, and liquid—have normal properties and their isentropic compressibility decreases with pressure. Anomalous behavior of a fluid can be expected near the gas-liquid critical point. Actually, long before the critical point is reached the gas isotherms show an inflection (at the critical point the inflection becomes horizontal). For a fluid with a sufficiently high specific heat and with the specific heat ratio close to unity, the isentropes and the isotherms do not differ appreciably from each other. We can therefore expect that the isentropes will have an inflection outside the two-phase region, that there will exist a region where the second derivative will have an anomalous sign. This region is shown in Fig. 1.37 (taken from Zel'dovich [2]). Curve I in Fig. 1.37 is the boundary of the two-phase region and curve II is the locus of the points of inflection of the isentropes $(\partial^2 p/\partial V^2)_S = 0$. Curve II bounds a region where $(\partial^2 p/\partial V^2)_S < 0$. An anomalous isentrope is also shown in Fig. 1.37. The curves were calculated using van der Waals' equation of state for the case when the specific heat $c_v = 40$ cal/deg-mole.

The relationship between the sign of the entropy change and the inequalities relating to the gas velocities and the sound speeds, corresponding to the coincidence of the entropy increase with the condition of mechanical stability, can be negated only if for the range of pressures for the particular case considered the sign of $\partial^2 p/\partial V^2$ is both positive and negative, so that the isentrope

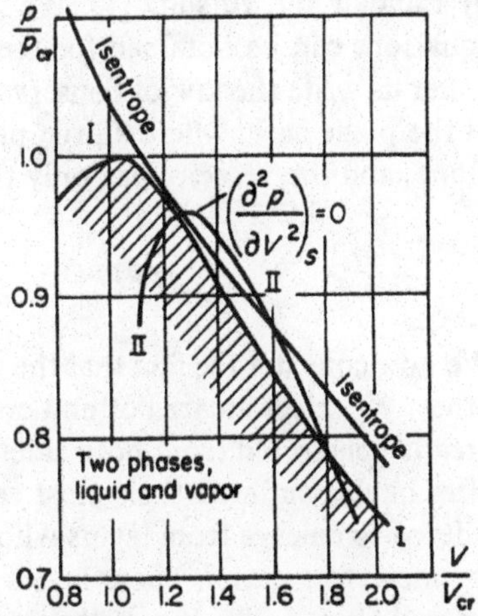

Fig. 1.37. Isentrope with anomalous convexity in a van der Waals' gas with a specific heat $c_v = 40$ cal/deg-mole. The shaded area is a two-phase region. Curve II bounds the region of anomalous convexity of the isentrope. Under curve II $(\partial^2 p/\partial V^2)_S < 0$.

intersects the chord more than twice. This may give rise to more complex conditions with discontinuities and adjoining decaying waves existing together. Another case of anomalous behavior will be considered in Chapter XI; the anomalies in this case are attributable to polymorphous transformations (phase transitions) of solids at the high pressures that are encountered in shock waves. Chapter XI will also deal with the other complex conditions mentioned above.

3. *Viscosity and heat conduction in gasdynamics*

§20. Equations of one-dimensional gas flow

The dissipative processes—viscosity (internal friction) and heat conduction—are attributable to the molecular structure of a fluid. These processes create an additional, nonhydrodynamic transfer of momentum and energy, and result in nonadiabatic flow and in the thermodynamically irreversible transformation of mechanical energy into heat. Viscosity and heat conduction appear only when there are large gradients in the flow variables, which occur, for example, in the boundary layers in flows past solid bodies or within a shock front. In this book we shall consider viscosity and heat conduction principally from the point of view of their effects on the internal structure of shock fronts in gases. In studying this structure we can consider the flow as a function of a single x coordinate only (plane flow), since the thickness of a shock front is always much smaller than the radius of curvature of its surface. We shall therefore not dwell extensively on the derivation of the general flow equations for a viscous fluid or gas, as can be found, for example, in the book by Landau and Lifshitz [1]. We shall only explain the manner in which the equations can be obtained for the one-dimensional, plane case.

Let us write the law of conservation of momentum for an inviscid gas (1.7) in the plane case, where all the parameters are functions of the x coordinate alone, and the velocity has only the x component u

$$\frac{\partial}{\partial t}(\rho u) = -\frac{\partial \Pi_{xx}}{\partial x}, \qquad \Pi_{xx} = p + \rho u^2.$$

We now consider the fact that the gas consists of molecules colliding with each other. We select an area of unit cross section perpendicular to the x axis. This area is penetrated from both sides by molecules moving in definite directions after having suffered their most recent collision. The molecules may be considered to emerge from layers whose thickness is of the order of magnitude of the mean free path l. These layers are adjacent to both sides of the area considered (Fig. 1.38). If n is the number of molecules per unit volume and $\bar{v}$ is

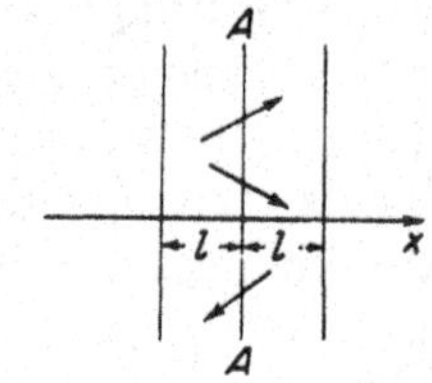

Fig. 1.38. Schematic illustration for the derivation of the equations for the molecular transport of momentum.

their average thermal velocity, then $n\bar{v}$ molecules will pass per unit time through the area from left to right. Each molecule carries through the area the hydrodynamic momentum mu, where m is the mass of the molecule. The total flux density of the hydrodynamic momentum from left to right is, therefore, of the order of $n\bar{v} \cdot mu$. Similarly, the flux density of the hydrodynamic momentum from right to left is approximately equal to $n\bar{v}m(u + \Delta u)$, where Δu is the increase in the hydrodynamic velocity during passage from the left layer to the right one; that is, $\Delta u \approx (\partial u/\partial x)l$. The flux density of the x component of momentum in the x direction caused by the molecular transfer is equal to the difference between the fluxes from left to right and from right to left, or $-n\bar{v}ml(\partial u/\partial x)$. This is the quantity expressing the additional momentum transfer caused by internal friction; it should be added to the momentum flux density $\Pi_{xx} = p + \rho u^2$.

A more rigorous analysis, based on three-dimensional considerations, shows that a numerical coefficient of the order of unity should be introduced into the above expressions. The equation of conservation of momentum for the plane case (for a monatomic gas, *eds.*), including the viscous term, is of the form

$$\frac{\partial}{\partial t}(\rho u) = -\frac{\partial \Pi_{xx}}{\partial x}, \qquad \Pi_{xx} = p + \rho u^2 - \sigma', \qquad \sigma' = \frac{4}{3}\mu\frac{\partial u}{\partial x}, \tag{1.90}$$

where μ is the coefficient of viscosity. This coefficient for gases (in the absence of relaxation processes, see below) is of the order of

$$\mu \sim n\bar{v}ml = \rho\bar{v}l.$$

The quantity σ' is the xx component of the viscous stress tensor. Its appearance in the equation for the momentum flux is equivalent to the appearance of an additional "pressure" created by internal friction forces. With the help of the continuity equation, (1.90) can be easily transformed to the equation of motion

$$\rho\frac{Du}{Dt} = -\frac{\partial}{\partial x}(p - \sigma'), \tag{1.91}$$

where $\partial\sigma'/\partial x$ is the internal friction force per unit volume of gas.

With dissipative processes present, additional terms appear in the energy equation. The additional energy flux is related to the added "viscous" pressure. A quantity $-\sigma' u$, analogous to pu, should be added to the expression for the energy flux density, the divergence of which appears in (1.10). In addition, the energy flux transferred by heat conduction

$$J = -\kappa\frac{\partial T}{\partial x}, \tag{1.92}$$

where κ is the coefficient of thermal conductivity, must be introduced into this expression. Equation (1.92) can be easily obtained by the same approach used in deriving the viscous momentum flux. The result is that for gases the coefficient of thermal conductivity is of the order of $\rho c_p \bar{v} l$.

Taking into account the two dissipation terms, the energy equation (1.10) for the plane case becomes

$$\frac{\partial}{\partial t}\left(\rho\varepsilon + \frac{\rho u^2}{2}\right) = -\frac{\partial}{\partial x}\left[\rho u\left(\varepsilon + \frac{u^2}{2}\right) + pu - \sigma' u + J\right]. \tag{1.93}$$

Rearranging this equation with the aid of the equations of motion and continuity and the thermodynamic identity $T\,dS = d\varepsilon + p\,dV$, we obtain an expression for the rate of change of entropy of the fluid particle

$$\rho T \frac{DS}{Dt} = \sigma' \frac{\partial u}{\partial x} - \frac{\partial J}{\partial x} = \frac{4}{3}\mu\left(\frac{\partial u}{\partial x}\right)^2 + \frac{\partial}{\partial x}\left(\kappa \frac{\partial T}{\partial x}\right). \tag{1.94}$$

The first term on the right-hand side represents the mechanical energy dissipated by viscosity per unit volume per unit time. This term is always positive, since $\mu > 0$ and $(\partial u/\partial x)^2 > 0$; consequently, the internal frictional forces lead to a local increase in the fluid entropy. The second term corresponds to heating or cooling of the fluid by heat conduction. It can be either positive or negative, because conduction results in a heat transfer from the region of higher to the region of lower enthalpy. However, the entropy of the fluid as a whole will always increase as a result of heat conduction. This can be shown by dividing (1.94) by T and integrating over the entire volume. The entropy change caused by heat conduction of a fluid occupying a volume bounded by the surfaces x_1 and x_2 is

$$\int_{x_1}^{x_2} \frac{1}{T}\frac{\partial}{\partial x}\left(\kappa \frac{\partial T}{\partial x}\right) dx = \frac{1}{T}\kappa \frac{\partial T}{\partial x}\bigg|_{x_1}^{x_2} + \int_{x_1}^{x_2} \frac{\kappa}{T^2}\left(\frac{\partial T}{\partial x}\right)^2 dx.$$

If the boundaries x_1 and x_2 are thermally insulated then there is no heat flow through the boundaries (the first term on the right-hand side vanishes), and only the second term, which is always positive ($\kappa > 0$), will remain.

Gasdynamic equations including the viscous and heat conduction terms allow us to determine the conditions under which the dissipative effects become important. To do this we equate the inertial forces in the equation of motion to the viscous forces. If U is the scale of the flow velocity and d is a characteristic dimension of the flow region, then the time scale is of the order of d/U and the inertial term $\rho\, Du/Dt$ is of the order of $\rho U^2/d$. The viscous term in the equation $\partial(\frac{4}{3}\mu\, \partial u/\partial x)/\partial x$ is of the order of $\mu U/d^2$, and its ratio to the inertial term is of the order of

$$\frac{1}{\mathrm{Re}} = \frac{\mu}{\rho U d} = \frac{\nu}{U d} \sim \frac{l}{d}\frac{c}{U}.$$

The reciprocal of this ratio is called the Reynolds number ($\nu = \mu/\rho \sim l\bar{v} \sim lc$ is the kinematic viscosity and $c \sim \bar{v}$ is the speed of sound). Similarly, comparing the heat transferred by conduction with the mechanical energy transfer, we find their ratio to be of the order of

$$\frac{1}{\mathrm{Pe}} = \frac{\kappa}{\rho c_p U d} \sim \frac{\chi}{U d} \sim \frac{l}{d}\frac{c}{U},$$

where Pe is the Peclet number. This number for gases is numerically close to the Reynolds number, because the thermal diffusivity $\chi = \kappa/\rho c_p$ is numerically close to the kinematic viscosity ν. (For example, for air at standard conditions $\nu \approx \chi \approx 0.15\ \text{cm}^2/\text{sec}$.)

Thus, the viscous and heat conduction terms can be neglected for $\text{Re} \approx \text{Pe} \gg 1$. If we consider a flow with velocities smaller than, or close to, the speed of sound, the dimensions of the system must be much greater than the molecular mean free path, $d/l \gg 1$. As we shall see, this condition is not satisfied in a shock front, the thickness of which is comparable to the mean free path. Dissipative processes must appear within the shock front. These processes are responsible for the increase in entropy across a shock wave.

§21. Remarks on the second viscosity coefficient

In writing down the gasdynamic equations and utilizing the thermodynamic relationships between pressure and the other thermodynamic variables of the fluid we have implicitly assumed that the pressure $\bar{p}$ which determines the forces in a moving gas does not differ from the static pressure p_{st} for a gas at rest under the same thermodynamic conditions, for the same composition, density, internal energy, and temperature. The pressure is a scalar quantity independent of the coordinate system, of the direction of flow velocity, and of the velocity gradient. This requirement that the pressure be a scalar quantity invariant with respect to the coordinate system permits us to make an assumption which is much more general than the assumption that the pressure depends only upon the thermodynamic state of the fluid. The pressure can in general depend on the scalar quantity—the divergence of the velocity. With moderate gradients, and with the restriction to the first terms in the appropriate expansion (as in the derivation of the viscous stresses), we can write the general expression

$$p_{st} = \bar{p} + \mu' \nabla \cdot \mathbf{u}, \tag{1.95}$$

where the coefficient μ' characterizes the dependence of the forces acting in the fluid upon the scalar quantity $\nabla \cdot \mathbf{u}$. The coefficient μ' is termed the second viscosity coefficient*. On the other hand, the coefficient μ, the first or ordinary viscosity coefficient, characterizes the forces which depend on the directions and the gradient of the velocity.

The ordinary viscosity coefficient for a gas is related to the translational motion of the molecules. If the time required for establishing the static pressure is of the order of the free time of a molecule l/c, then μ' is of the order of μ. In the plane case when this condition is satisfied, both the ordinary and the second viscosity terms are combined into a single term. In certain cases,

* *Editors' note.* This quantity is generally termed the bulk viscosity coefficient or the dilatational viscosity coefficient. The term second viscosity coefficient is often used for the quantity $\lambda = \mu' - \frac{2}{3}\mu$.

however, μ' takes on an anomalously large value. According to the continuity equation $\nabla \cdot \mathbf{u} = -(1/\rho)D\rho/Dt$, and the coefficient μ' characterizes the dependence of the pressure upon the rate of change of density. In the presence of slowly excited internal degrees of freedom (for example, molecular vibrations) and rapid changes in the state of the fluid, the pressure cannot follow the changes in density and differs from its value for thermodynamic equilibrium. This effect may be described by the second viscosity coefficient (see [1]). The more difficult it is to excite the internal degrees of freedom, the more pronounced is the "inconsistency" between the changes in pressure, the changes in density, and internal state of the fluid, and the larger is the second viscosity. In very fast processes, when this "inconsistency" (deviation from thermodynamic equilibrium) is especially high, the linear relation (1.95) may be inadequate and it may become necessary to introduce an explicit description of the relaxation processes into the gasdynamic equations, to introduce the kinetics of the excitation of the internal degrees of freedom. We shall become more familiar with this phenomenon in Chapters VI, VII, and VIII which consider relaxation processes and their effect on the structure of shock fronts and the absorption of ultrasound.

§22. Remarks on the absorption of sound

As an example of the effect of viscosity and heat conduction in hydrodynamics let us consider the propagation of sound waves. The presence of viscosity and heat conduction leads to the dissipation of sound wave energy by the irreversible conversion of the energy into heat, to absorption of the sound and a decrease in its intensity. The sound absorption coefficient can be found by obtaining the solution to the one-dimensional linearized gasdynamic equations, containing the viscous and heat conduction terms, for plane harmonic waves of the type $\exp[i(kx - \omega t)]$, where k is the wave vector. In this case k has a complex value, whose real part gives the wavelength and whose imaginary part gives the absorption coefficient

$$k = k_1 + ik_2; \qquad \exp[i(kx - \omega t)] = \exp[-k_2 x] \exp[i(k_1 x - \omega t)].$$

The absorption coefficient can also be evaluated from physical considerations. According to (1.94), the energy dissipation per unit volume per unit time is composed of two parts, corresponding to viscous and heat conduction effects. In a sound wave of wavelength λ these quantities are of the order of $\mu u^2/\lambda^2$ and $\kappa(\Delta T)^2/\lambda^2 T$, respectively. Here u is the amplitude of the velocity and ΔT is the amplitude of the temperature change in the wave (the latter is proportional to u). The sound energy per unit volume is $\rho_0 u^2$. The fraction of energy absorbed per unit time is also made up of two parts. The part associated with the viscosity is of the order of $(\mu u^2/\lambda^2)/\rho_0 u^2 \sim \mu/\rho_0 \lambda^2 \sim \mu\omega^2/c^2\rho_0$.

However, the distance traveled by the sound per unit time is c; so that the viscous part of the absorption coefficient per unit length γ_1 is of the order of $\mu\omega^2/c^3\rho_0$. Similarly, the absorption coefficient per unit length associated with heat conduction γ_2 is of the order of $(\kappa/c_p)(\omega^2/c^3\rho_0)$. (This result is easy to understand in the case of gases; we note that $\kappa/c_p \approx \mu$, because of the approximate equality between the kinematic viscosity $\nu = \mu/\rho$ and the thermal diffusivity $\chi = \kappa/\rho c_p$, thus for gases $\gamma_1 \approx \gamma_2$). These expressions are valid for weak sound absorption, where the amplitude decrease over distances of the order of one wavelength is small, i.e., where $\gamma\lambda \ll 1$ ($\gamma = \gamma_1 + \gamma_2$). For gases this condition simply means that

$$\gamma\lambda \sim \frac{\mu\omega^2\lambda}{c^3\rho_0} \sim \frac{\nu}{\lambda^2}\frac{\lambda}{c} \sim \frac{l}{\lambda}\frac{\bar{v}}{c} \sim \frac{l}{\lambda} \ll 1.$$

This means that the expression for the absorption coefficient is valid for wavelengths much larger than the molecular mean free path, which is the case most frequently met with.

Other cases of anomalously large absorption and dispersion of sound (dependence of the speed of sound on frequency) arise in a fluid with slow excitation of the internal degrees of freedom (with a high value of the second viscosity). This problem will be considered in Chapter VIII.

§23. The structure and thickness of a weak shock front

Let us examine the internal structure and thickness of the thin layer representing the shock wave across which the gas undergoes the transition from the initial to the final state; we refer to this layer as the shock front. In this layer occur such phenomena as an increase in density of the fluid, pressure and velocity changes, and, as has been shown by calculations based only on the application of the conservation laws, an increase in entropy. The entropy increase indicates that there is dissipation of mechanical energy and that an irreversible conversion of mechanical energy into heat takes place in the transition layer. Hence, in order to understand the detailed nature of the shock transition it is necessary to account for the dissipative processes, for viscosity and heat conduction.

Let us consider the one-dimensional plane flow of a viscous and heat conducting gas in a coordinate system in which the shock front is at rest. The front thickness is very small in comparison with the characteristic length scales for the gasdynamic process as a whole, for example, in comparison with the distance between the shock front and the piston which acts on the gas and creates the wave. Even if the piston moves with a variable velocity and the strength of the shock wave changes with time, the strength of the wave will remain practically unchanged during the small time interval Δt required to traverse a distance of the order of the front width Δx. Hence, during a certain time interval, which is small in comparison with the overall time scale of the gasdynamic process, but large in comparison with Δt, the entire picture of the distributions of the flow variables across the wave front propagates through the gas as though "frozen". In other words, using a coordinate system in which the gas is at rest, the flow can be considered as steady at any given time.

Let us write the equations of continuity, momentum, and entropy for the steady plane case with the viscous and heat conduction terms included. Since the process is steady, we can drop terms in partial derivatives with respect to time $\partial/\partial t$, and can replace the partial derivative $\partial/\partial x$ by the total derivative d/dx. We obtain*

$$\frac{d}{dx}(\rho u) = 0,$$

$$\frac{d}{dx}\left(p + \rho u^2 - \frac{4}{3}\mu\frac{du}{dx}\right) = 0, \tag{1.96}$$

$$\rho u T\frac{dS}{dx} = \frac{4}{3}\mu\left(\frac{du}{dx}\right)^2 + \frac{d}{dx}\left(\kappa\frac{dT}{dx}\right).$$

Using the second law of thermodynamics $T\,dS = dh - V\,dp$ together with the equations of continuity and momentum, we can rewrite the entropy equation in the form of an energy equation

$$\frac{d}{dx}\left[\rho u\left(h + \frac{u^2}{2}\right) - \frac{4}{3}\mu u\frac{du}{dx} - \kappa\frac{dT}{dx}\right] = 0. \tag{1.97}$$

We now impose appropriate boundary conditions on the solution of these equations, by requiring the gradients of all quantities to vanish ahead of the front at $x = -\infty$ and behind the front at $x = +\infty$. At these limits the variables assume their initial and final values, designated, as before, by the subscripts "0" and "1" (Fig. 1.39).

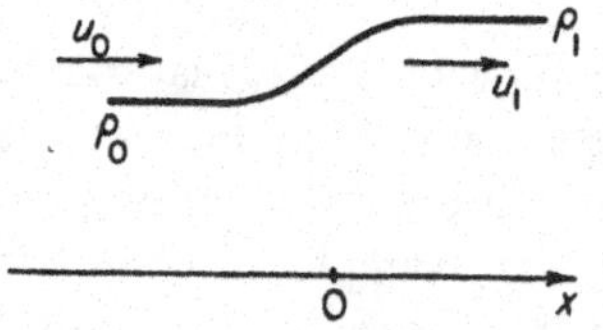

Fig. 1.39. Sketch illustrating the structure of a shock front.

* *Editors' note*. The equations and results of this section remain valid in a gas with a non-negligible second or bulk coefficient of viscosity μ', provided that $\frac{4}{3}\mu$ is replaced by the "longitudinal" coefficient of viscosity $\mu'' = \frac{4}{3}\mu + \mu'$.

First integrals of the system of mass, momentum, and energy equations are readily obtained, and are

$$\rho u = \rho_0 u_0, \tag{1.98}$$

$$p + \rho u^2 - \frac{4}{3}\mu\frac{du}{dx} = p_0 + \rho_0 u_0^2, \tag{1.99}$$

$$\rho u\left(h + \frac{u^2}{2}\right) - \frac{4}{3}\mu u\frac{du}{dx} - \kappa\frac{dT}{dx} = \rho_0 u_0\left(h_0 + \frac{u_0^2}{2}\right). \tag{1.100}$$

The constants of integration are expressed here in terms of the initial values of the variables p, ρ, T, and u, and are considered as functions of the x coordinate*. Equation (1.99) shows that the presence of viscosity, the term containing du/dx, causes the distribution of the flow variables with respect to x in the shock front to be continuous (otherwise, the gradient du/dx would go to infinity, which would contradict the fact that the variables are finite).

In order to understand better the roles of viscosity and heat conduction let us first consider the shock front structure in two particular cases: (1) when viscosity is absent and there is only heat conduction; (2) when there is only viscosity and heat conduction is absent. We shall not seek here exact solutions of the equations (this problem is left to Chapter VII, which is devoted specifically to the structure of shock wave fronts). We shall only explain the qualitative features of the phenomenon and evaluate the front thickness.

(1) Heat conduction is present but there is no viscosity: $\mu = 0$.

In this limiting case the momentum equation (1.99) becomes

$$p + \rho u^2 = p_0 + \rho_0 u_0^2,$$

which is analogous to the equation relating the final and initial values of these quantities. This equation, however, now also describes all the intermediate states in the wave front. Using the equation of continuity (1.98), we obtain

$$p = p_0 + \rho_0 u_0^2\left(1 - \frac{V}{V_0}\right). \tag{1.101}$$

Thus, the point describing the state of the gas within the shock front goes from the initial point A in the p, V plane to the final point B along the straight line AB. The line AB has already been discussed in the description of a Hugoniot curve.

* At $x = +\infty$, $du/dx = 0$, $dT/dx = 0$, $p = p_1$, $\rho = \rho_1$, $u = u_1$ and we arrive at the conservation laws of mass, momentum, and energy across the discontinuity, (1.61), (1.62), and (1.64).

Let us draw the isentropes through the initial and final points in the p, V plane (Fig. 1.40; the Hugoniot curve is not shown). Drawing a series of isentropes, it is evident that one of the curves will be tangent to AB at some point M, as shown in Fig. 1.40. At this point the entropy along the line AB is at its

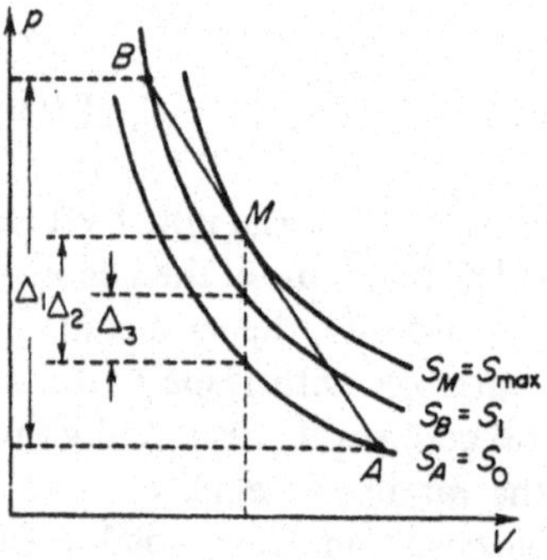

Fig. 1.40. p, V diagram illustrating the structure of shock fronts in the absence of viscosity. The state of the gas in the wave changes along the straight line AB. Segments Δ_1, Δ_2, and Δ_3 are of first, second, and third order, respectively, with respect to the wave strength.

maximum ($S_0 < S_1 < S_M$). It follows from (1.98) and (1.101) that the gas velocity u at the point of tangency M is exactly equal to the local speed of sound ($u = c$ at the point M; we note that $u_0 > c_0$ at point A and $u_1 < c_1$ at point B).

We now wish to determine the maximum value of the entropy S_{max} from the condition of tangency between the isentrope $S = S_{max}$ and the straight line AB. As we shall soon see, the quantity $S_{max} - S_0$ is proportional to either $(V_1 - V_0)^2$ or $(p_1 - p_0)^2$; hence we shall write the equations for the family of isentropes $p(V, S)$ and for the line AB in terms of expansions about the point A, neglecting third-order terms (in this approximation the isentropes S_0 and S_1 coincide; see §18). The equation for the isentrope is

$$p - p_0 = \left(\frac{\partial p}{\partial V}\right)_{S_A}(V - V_0) + \frac{1}{2}\left(\frac{\partial^2 p}{\partial V^2}\right)_{S_A}(V - V_0)^2 + \left(\frac{\partial p}{\partial S}\right)_{V_A}(S - S_0).$$

The equation for the straight line is

$$p - p_0 = \frac{p_1 - p_0}{V_1 - V_0}(V - V_0)$$

$$= \left(\frac{\partial p}{\partial V}\right)_{S_A}(V - V_0) + \frac{1}{2}\left(\frac{\partial^2 p}{\partial V^2}\right)_{S_A}(V_1 - V_0)(V - V_0).$$

The tangency condition is given by the equality $(\partial p/\partial V)_{\text{isent}} = (\partial p/\partial V)_{\text{str. line}}$, which provides an equation for the determination of the volume V_M at the point of tangency M. It can be shown that the point M is situated exactly half way between points A and B, i.e., $V_M - V_0 = \frac{1}{2}(V_1 - V_0)$. Substituting this expression into the equation for the straight line, we find the pressure p_M at the point M. Next, substituting p_M and V_M into the equation for the isentrope and solving for the entropy at the point M, we obtain

$$S_M - S_0 = S_{\max} - S_0 = \frac{1}{8}\frac{(\partial^2 p/\partial V^2)_{S_A}}{(\partial p/\partial S)_{V_A}}(V_1 - V_0)^2.$$

The maximum entropy change within the shock front, taking into account heat conduction alone, is a second-order quantity with respect to the quantity $V_0 - V_1$ or $p_1 - p_0$, in contrast to the total entropy change $S_1 - S_0$ which is third order. This can also be shown on the basis of geometrical arguments: the maximum distance between the straight line AB and the isentropes $S = S_0$ in the p, V plane is proportional to $(V_1 - V_0)^2$ or $(p_1 - p_0)^2$. Thus, the difference between the pressure at the point M and the pressure on the isentrope S_A (or S_B) at the same volume V_M is equal to

$$\begin{aligned} p_M(V_M) - p_{S_A}(V_M) &= \frac{1}{2}\left(\frac{\partial^2 p}{\partial V^2}\right)_{S_A}(V_M - V_0)(V_1 - V_M) \\ &= \frac{1}{8}\left(\frac{\partial^2 p}{\partial V^2}\right)_{S_A}(V_1 - V_0)^2 \end{aligned} \tag{1.102}$$

(the pressure difference between points on the isentropes S_B and S_A at the same volume V_M is a third-order quantity).

The presence of an entropy maximum within the front indicates that the temperature profile $T(x)$ is inflected at the point of maximum entropy; thus, the temperature and entropy distributions in a weak shock wave, considering heat conduction only, can be described by the curves shown in Fig. 1.41. This

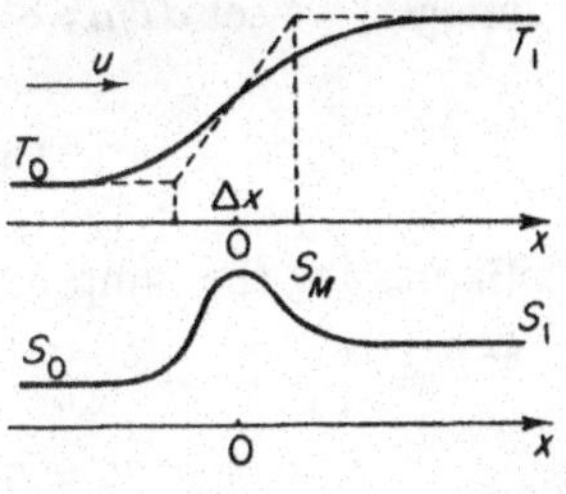

Fig. 1.41. Temperature and entropy distribution in a weak shock front in the absence of viscosity. Δx is the effective front thickness.

conclusion follows from the entropy equation (1.96), which, in the absence of viscosity, takes the form

$$\rho u T\frac{dS}{dx} = \frac{d}{dx}\kappa\frac{dT}{dx} = \kappa\frac{d^2 T}{dx^2} \tag{1.103}$$

(the temperature in a weak wave changes only very slightly and we can assume that the thermal conductivity is constant). The existence of the entropy maximum is attributable to the fact that conductive heat transfer takes place from a high-temperature region to a low-temperature region. Therefore, the gas flowing into the wave is first heated by heat conduction (with an increase in entropy), and is then cooled (with a decrease in entropy). In the end state, the entropy is obviously greater than its initial value. This is illustrated in Fig. 1.41, where flow along the x axis with a velocity $u(x)$ corresponds to the change in state of a given gas particle with time.

Let us now evaluate the thickness of the shock front. We divide (1.103) by T and integrate with respect to x from the initial state A ($x = -\infty$), where $dT/dx = 0$, to any point x in the wave (we also use the fact that $\rho u = \rho_0 u_0 =$ *const*)

$$\rho_0 u_0(S - S_0) = \kappa \int_{-\infty}^{x} \frac{1}{T}\frac{d^2T}{dx^2}\,dx = \kappa\left\{\frac{1}{T}\frac{dT}{dx} + \int_{T_0}^{T}\frac{dT}{dx}\frac{1}{T^2}\,dT\right\}. \tag{1.104}$$

If we apply this equation to the final state B ($x = +\infty$), where $dT/dx = 0$, then the first term in braces disappears and

$$\rho_0 u_0(S_1 - S_0) = \kappa \int_{T_0}^{T_1} \frac{1}{T^2}\frac{dT}{dx}\,dT.$$

We now define the effective thickness of the shock front Δx, in which there is only heat conduction, by

$$\frac{T_1 - T_0}{\Delta x} = \left|\frac{dT}{dx}\right|_{\max},$$

where the geometric meaning is clear from Fig. 1.41. In order to estimate the integral we set $dT/dx \sim (T_1 - T_0)/\Delta x$, from which

$$\rho_0 u_0(S_1 - S_0) \sim \kappa \frac{1}{T_0^2}\frac{(T_1 - T_0)^2}{\Delta x}.$$

Expressing the jump in temperature in terms of the jump in the pressure, we get

$$T_1 - T_0 = \left(\frac{\partial T}{\partial p}\right)_S (p_1 - p_0) = \frac{V_0}{c_p}(p_1 - p_0),$$

where c_p is the specific heat at constant pressure. Using (1.89) for the entropy jump and recalling that approximately $(\partial^2 V/\partial p^2)_S \sim V_0/p_0^2$, $\kappa \sim \rho_0 c_p l c_0$, and that $u_0 \approx c_0$, we obtain from (1.104) for an estimate of the front thickness

$$\Delta x \sim l \frac{p_0}{p_1 - p_0}. \tag{1.105}$$

The front thickness is inversely proportional to the wave strength, and the scale factor, as shown by (1.105), is the molecular mean free path l.

Equation (1.104) can also be used to estimate the maximum entropy increase. At the point of maximum entropy $dS/dx = 0$ and the gradient dT/dx is also a maximum. The dominant term in the expression in braces in (1.104) is the first term which is proportional to $\Delta T/\Delta x \sim \Delta p/\Delta x \sim (\Delta p)^2$, while the second term is proportional to $(\Delta T)^2/\Delta x \sim (\Delta p)^3$. It is therefore evident that $S_{max} - S_0 \sim (\Delta p)^2$, while $S_1 - S_0 \sim (\Delta p)^3$.

Considering the internal structure of the shock front (taking into account heat conduction alone), we can only claim that the temperature in the wave changes continuously, while the other quantities, such as density, velocity, and pressure may, in general, be discontinuous. Indeed, the study of shock wave structure (without considering viscosity) shows that it is impossible to construct a continuous distribution of all wave parameters for a sufficiently strong wave. This difficulty had already been pointed out by Rayleigh (Chapter VII, §3), and indicates the important part played by viscosity in achieving the irreversible compression of a fluid across a shock wave.

Let us now consider the second case.

(2) Viscosity is present but there is no heat conduction: $\kappa = 0$.

In this case we must retain the complete form of the momentum equation (1.99). A point describing the state of the gas in the wave moves in the p, V plane, from point A to point B along some curve (shown in Fig. 1.42 by a

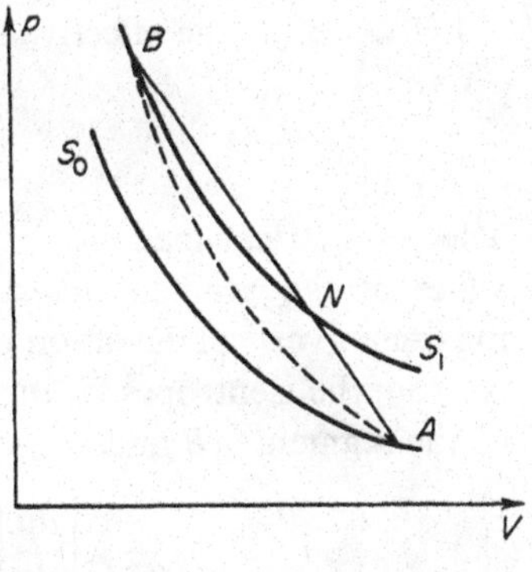

Fig. 1.42. p, V diagram for the structure of a shock front neglecting heat conduction. The state of the gas in the wave changes along the dashed curve AB.

dashed line) rather than along the straight line AB. From the entropy equation without the heat conduction term, we find

$$\rho u T \frac{dS}{dx} = \mu \left(\frac{du}{dx}\right)^2, \qquad (1.106)$$

which shows that the wave entropy increases monotonically from the initial value $S_0 = S_A$ to the final value $S_1 = S_B$; thus the dashed line is entirely contained between the isentropes S_0 and S_1 (see Fig. 1.42). Since the isentropes are convex downward $(\partial^2 p/\partial V^2)_S > 0$, the dashed line lies entirely below the straight line AB*. The equation of the curve describing the transition from point A to point B is given by

$$p = p_0 + \rho_0 u_0^2\left(1 - \frac{V}{V_0}\right) + \frac{4}{3}\mu\frac{du}{dx}. \tag{1.107}$$

Since the curve lies entirely below the straight line, then $du/dx < 0$ at all points within the wave. If the x axis is parallel to the flow then $u > 0$, that is, the gas in the wave only decelerates and, as a result, is compressed monotonically. Thus, consideration of the structure of a shock front including viscosity leads to the conclusion that with $(\partial^2 p/\partial V^2)_S > 0$, the gas in the shock wave can only be compressed. The density and velocity distributions through the wave have the form shown in Fig. 1.43.

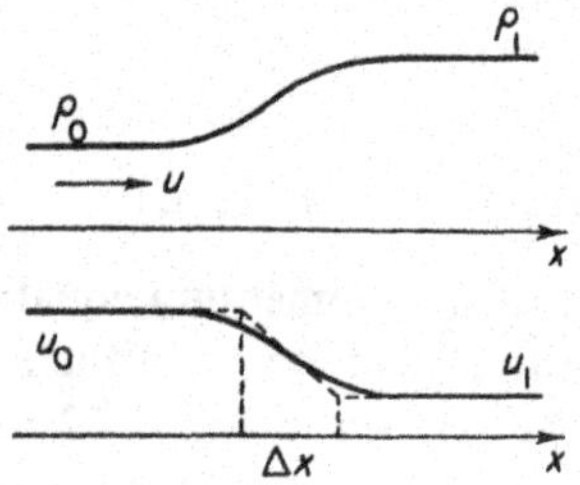

Fig. 1.43. Density and velocity profiles in a shock front. Δx is the effective front thickness.

Let us define the effective front thickness by†

$$\frac{u_0 - u_1}{\Delta x} = \left|\frac{du}{dx}\right|_{\max}. \tag{1.108}$$

The geometrical meaning of the equation is clear. The maximum absolute value of the gradient $|du/dx|_{\max}$ is defined, according to (1.107), by the maximum vertical deviation of the straight line AB from the dashed line, that is, from the isentropes S_0 or S_1. This deviation corresponds to the midpoint of the segment AB and is given by (1.102). Thus,

$$\frac{4}{3}\mu\left|\frac{du}{dx}\right|_{\max} = \frac{1}{8}\left(\frac{\partial^2 p}{\partial V^2}\right)_{S_A}(V_1 - V_0)^2.$$

* Actually, the vertical distance between the isentropes S_1 and S_0 is proportional to $S_1 - S_0 \sim (p_1 - p_0)^3$, while the vertical distance between points A and B is $p_1 = p_0$. Hence, that part of AN, where the dashed line could, in principle, pass above the straight line, is small in comparison with the main part of the straight line AN.

† Δx is sometimes called the Prandtl front thickness.

Substituting this expression for $|du/dx|_{\max}$ into (1.108) and noting that $\mu = \rho_0 \nu \sim \rho_0 l \bar{v} \sim \rho_0 l c_0$ (ν is the kinematic viscosity), and also that

$$u_0 - u_1 = [(p_1 - p_0)(V_0 - V_1)]^{1/2} \sim \left[(p_1 - p_0)^2 \left| \frac{\partial V}{\partial p} \right| \right]^{1/2} \sim \frac{p_1 - p_0}{p_0} c_0,$$

$$\left(\frac{\partial^2 p}{\partial V^2} \right)_S \sim \frac{p_0}{V_0^2},$$

we arrive again at formula (1.105) for the front thickness

$$\Delta x \sim l \frac{V_0}{V_0 - V_1} \approx l \frac{p_0}{p_1 - p_0}.$$

The front thickness can also be estimated from the entropy equation (1.106) by a procedure similar to that used for the first case:

$$\rho_0 u_0 T_0 \frac{S_1 - S_0}{\Delta x} \sim \mu \frac{(u_0 - u_1)^2}{\Delta x^2}.$$

Substituting (1.89) for the entropy jump and rearranging, we arrive at the previous equation for Δx.

The difficulties which arise in constructing a continuous solution with only heat conduction are not encountered in constructing a continuous solution which takes into account only viscosity. This fact, as has been previously mentioned, has a deep physical meaning and indicates the importance of viscosity in a compression shock. Indeed, it is the viscous mechanism that converts a portion of the kinetic energy of the gas flowing into the discontinuity into heat; this conversion is equivalent to the transformation of the energy of ordered motion of gas molecules into the energy of random motion by the dissipation of molecular momentum. In this respect heat conduction has an indirect effect on the conversion process since it only participates in the transfer of the energy of random motion of the molecules from one point to another, but does not directly affect the ordered motion.

If we consider shock waves of moderate strength in an ordinary gas, where the transfer coefficients, the kinematic viscosity ν and the thermal diffusivity χ, are approximately equal to each other and are determined by the same molecular mean free path l ($\nu \approx \chi \sim lc$), we again obtain formula (1.105) for the front thickness. We can easily show that this is true by considering the general entropy equation (1.98) which contains both the viscous and heat conduction terms.

Equation (1.105) shows that if the pressure jump in the wave is of the order of magnitude of the pressure ahead of the front, then the front thickness will be approximately equal to the molecular mean free path. If the strength of the wave increases further, the same equation shows that the thickness becomes smaller than the molecular mean free path. This result, obviously, has no physical significance. If the flow variables change rapidly over distances of the order of the mean free path, then the hydrodynamic treatment of viscosity and heat conduction, which depends upon the assumption that the gradients are small, can no longer apply. The thickness of an arbitrarily strong shock wave cannot, obviously, become smaller than the molecular mean free path, as indicated by studies based on the kinetic theory of gases (see Chapter VII).

Under certain conditions the front of a strong shock wave can become as thick as several mean free paths and it is then possible to divide it into regions of smooth and abrupt changes of the flow variables. In particular, this occurs in a gas with delayed excitation of certain molecular degrees of freedom, or when a reversible chemical reaction takes place in the wave. These problems, as well as many others arising in the detailed study of the internal structure of the front, will be considered in Chapter VII.

4. *Various problems*

§24. Propagation of an arbitrary discontinuity

The flow variables on both sides of a shock front are not independent. They are related by definite equations expressing the conservation laws of mass, momentum, and energy. Furthermore, the discontinuity—the compression shock wave in fluids with ordinary thermodynamic properties—is propagated through the fluid as a stable, nonspreading formation. However, the problem can be formulated to include the existence of a discontinuity surface at the initial time. The flow variables on both sides of such a discontinuity are arbitrary and completely unrelated to each other. Such discontinuities are termed arbitrary discontinuities.

Let us indicate some practical examples which illustrate how arbitrary discontinuities may arise. Imagine a tube filled with gas and divided by a thin partition. Suppose that the density, pressure, and, in general, the gas composition are different on each side of the partition. Let the partition be rapidly removed at some instant of time. At that instant the two regions with entirely arbitrary densities and pressures come into contact at the previous location of the partition. Since the pressures of both gases were different, the pressure jump will cause the gases to move once the partition has been removed. As a second example, let us assume that two shock waves of arbitrary strengths enter from both sides into a tube filled with gas. At the moment of collision of the two waves somewhere in the middle of the tube, a surface is formed separating gases of different pressures, velocities, and temperatures (the possible density differences in this example are somewhat limited; we can say that if the waves are strong, then the densities are the same and equal to the limiting value). After the collision the gas motion will be altered.

Let us consider still a third example. We have approached the theory of shock waves by examining the flow resulting from a piston which is pushed into the gas at a constant velocity. In this case the shock wave was formed immediately at the piston and was propagated through the gas with a constant speed. Actually, however, the piston has a finite mass and cannot instantaneously acquire the final velocity. The piston is, therefore, gradually accelerated by the applied force. As a result, the shock wave will not form immediately but only at some distance away from the piston. We can replace the smooth variation in piston velocity $U(t)$ by some stepped function. This may be done by dividing the time scale into infinitesimal intervals and by assuming that the piston velocity during each of these intervals is constant. We assume further that after each time interval the velocity changes abruptly by a small amount. The piston path in the x, t plane will be a broken curve consisting of very small linear segments. During each small time interval the piston sends a compression disturbance ahead, in the form of a weak shock wave. This wave travels through the gas with a speed slightly exceeding the speed of sound, while the preceding weak shock wave, induced by the preceding abrupt change in piston velocity is propagated relative to the gas moving behind it with a velocity slightly less than the speed of sound (see Fig. 1.44). Hence, each successive

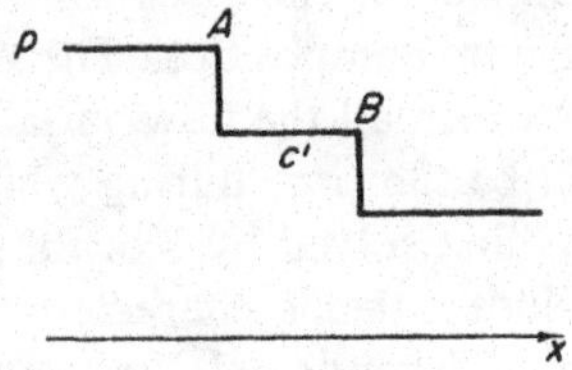

Fig. 1.44. Pressure distribution in a system of two successive weak compression shocks. Wave A travels through the gas ahead of the wave with a speed higher than the speed of sound c' in this gas. Wave B travels through the gas behind it with a subsonic velocity lower than c'. Hence, shock A will eventually overtake shock B.

shock wave overtakes the preceding one and the compressive effects of the individual waves thereby accumulate. If we draw the characteristics in the x, t plane originating from the piston path, they will intersect (Fig. 1.45). It turns out that it is possible to specify the piston acceleration in such a way that all these weak shock waves overtake one another at the same time and at one point. Here all the many small compression pulses coalesce into a single large pressure jump. (All the characteristics intersect at a single point.) The state of the gas across this discontinuity changes from the undisturbed to the final condition almost isentropically. Indeed, if the entire compression of the gas to the pressure p is broken up into n stages, that is, into n weak shock waves with pressure jumps $\Delta p = (p - p_0)/n$, then the entropy increase ΔS at each stage will be proportional to $(\Delta p)^3 \sim 1/n^3$, and the total entropy increase for the sum of the n waves will be proportional to $n\,\Delta S \sim 1/n^2 \to 0$ as $n \to \infty$. Thus, the state of the gas on each side of the discontinuity, which results from the coalescence of n weak shock waves, can be related by the isentropic

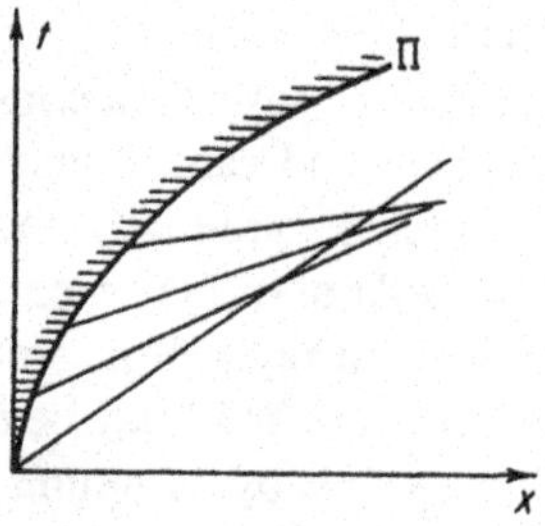

Fig. 1.45. Intersection of characteristics in a gas compressed by an accelerating piston. Π is the piston path.

relation. We know, however, that the states on each side of a shock discontinuity are related not by the isentropic relation but by the Hugoniot equations. It follows that the quantities on both sides of the discontinuity do not satisfy the conservation laws and the discontinuity is an arbitrary one.

Generalizing the cases represented by the above examples we can set up an idealized problem in which we seek a solution for the motion of a gas containing an arbitrary discontinuity. Suppose that at the initial time $t = 0$ in the plane $x = 0$ all the flow variables are discontinuous, the pressure, density, velocity, and the temperature. These quantities are, however, uniform on each side of the discontinuity. In addition, the gas on each side may be of different composition. The greater the distance from the discontinuity surface for which all the flow variables may still be regarded as constant, the longer will be the time during which the desired solution will apply (this problem was first solved by Kochin [3]). Since the conditions of the problem do not include either a characteristic length or time, we should seek for a solution which depends only upon the ratio x/t. It was shown in §11 that self-similar plane flow of a gas can be described in terms of only two types of solutions, those with centered simple rarefaction waves and those with flows in which all the variables are constant. In addition, discontinuities, i.e., shock waves, are also possible. Thus, the desired flow should be constructed from three elements: rarefaction waves, uniform flow regions, and shock discontinuities.

The set of possible flows is limited by the fact that only one wave can move in one direction (regardless of whether the wave is a rarefaction or a shock wave). Indeed, a shock wave propagates with respect to the undisturbed gas at a supersonic speed, and with respect to the compressed gas at a subsonic speed, while the rarefaction wave travels through the gas at the speed of sound. If, for example, a shock wave moves to the right, then a rarefaction wave (and even more so another shock wave) following it in the same direction will always overtake it after a certain time. Since, however, the motion is self-similar, both waves will originate from the same point $x = 0$ and at the same time $t = 0$. Hence, one wave acts as if it had already overtaken the other one at the initial time, and both of them propagate as a single wave. Therefore, it is impossible for a second wave to follow a rarefaction wave. A shock wave would have overtaken the rarefaction wave, while a second rarefaction wave would move at a fixed distance from the first one. Since, however, the motion is self-similar, this distance is equal to zero and the distinction between the two waves disappears.

From the above discussion we see that the desired solution can be constructed only by some combination of two waves, shock and rarefaction, propagating in opposite directions from the initial discontinuity and separated by regions of uniform flow. In general, there are two such regions. They are divided by a plane separating the two gases that were initially located on the opposite sides of the arbitrary discontinuity. Since the hydrodynamics of ideal fluids does not take molecular diffusion into account, the gases do not diffuse into each other and the boundary between them is maintained intact, moving with the gases. The case when the gases are of the same kind is, obviously, not much different (we can imagine that the gas molecules on one side of the initial discontinuity were labeled with a "dye"). This plane boundary between the two gases, which may be called a contact boundary or contact discontinuity, possesses special properties. Obviously, the pressures and velocities are equal on the two sides of the contact discontinuity. Otherwise, a flow in the neighborhood of the discontinuity would take place and the regions on both sides would cease to be uniform flow regions. However, the densities, temperatures, and entropies of the gases on the two sides of the contact discontinuity can have arbitrary values, determined by their arbitrary initial values. Differences in these quantities, at equal pressures and velocities, cannot result in the gases being set into relative motion (assuming, of course, that both diffusion and heat conduction are absent; these effects will be discussed later). The contact discontinuity is stationary with respect to the gas and does not send out disturbances that could affect the waves (shock and rarefaction) traveling in either direction away from it.

Let us list the possible types of flow after the creation of an arbitrary discontinuity, the possible cases of the breaking apart of the discontinuity into different combinations of rarefaction and shock waves. Three typical cases can arise: (1) shock waves are propagated in both directions away from the discontinuity; (2) a shock wave propagates in one direction and a rarefaction wave in the other; and (3) rarefaction waves travel in both directions. Let us look at these cases in more detail using the convenient p, V diagram (Fig. 1.46). First of all, let us fix on the diagram the initial states of the fluid. Point

A represents the gas to the left and point B the gas to the right of the discontinuity. Let the pressure at point A (p_a) be lower than p_b and draw from these points upward the Hugoniot curves characterizing shock compression, and downward the isentropes along which the gas expands in rarefaction waves.

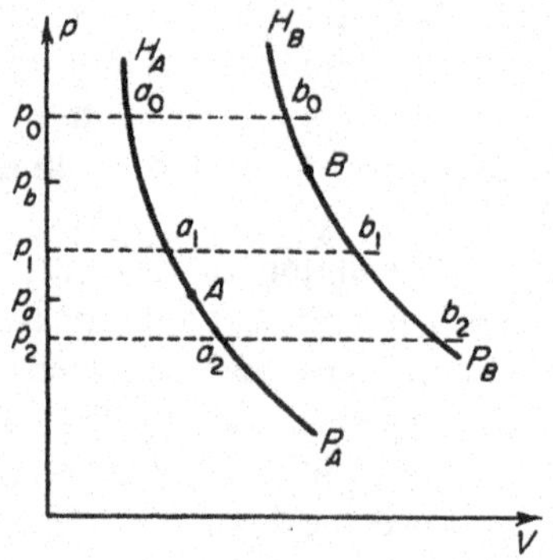

Fig. 1.46. A p, V diagram illustrating different cases of break-up of an arbitrary discontinuity. Points A and B characterize the initial states of the gases A and B. $H_A A$ and $H_B B$ are Hugoniot curves; AP_A and BP_B are isentropes for the gases A and B.

After the discontinuity breaks apart, the pressures in both gases will equalize in the regions affected by the waves.

1. Let this new pressure p_0 be greater than the initial pressures p_a and p_b. In this first case (Fig. 1.47a) compression shocks travel both to the left and to the right of the arbitrary discontinuity (or contact surface). The gases behind these waves are in the states a_0 and b_0 and at the same pressure p_0 and velocity. The gas in the state a_0 moves to the left relative to the gas initially in the state A, and gas b_0 moves to the right relative to the gas initially in the state B. Since the gases a_0 and b_0 move with the same velocity, it is necessary that the gases A and B initially move toward each other. Two shock waves are formed upon collision of two masses of gas moving toward each other at high velocities. The lower the collision velocity, the lower the pressure p_0 behind the shock waves. The second example discussed at the beginning of this section yields this case.

2. At a certain low collision velocity a new condition will arise in which the pressure p_1 is still higher than the pressure p_a but lower than the pressure p_b. In this second case after the discontinuity breaks apart, a shock wave is propagated through the gas A and a rarefaction wave travels through the gas B (Fig. 1.47b). Such conditions will exist when the initial velocities of both gases A and B are equal to zero, that is, when a pressure discontinuity exists initially, as in the example with the partition, and the gas begins to move in the direction of the lower pressure region. This case has important practical applications, since it forms the principle for the operation of shock tubes. Here strong shock waves are obtained under laboratory conditions and can heat the gas A to high temperatures. The shock tube is divided by a thin partition

(diaphragm). On one side of the diaphragm the tube contains at a low pressure the test gas A, while the driver gas B is pumped into the other side—the so-called high pressure chamber. After the diaphragm disintegrates, gas B expands in the direction of the low pressure chamber, sending a strong

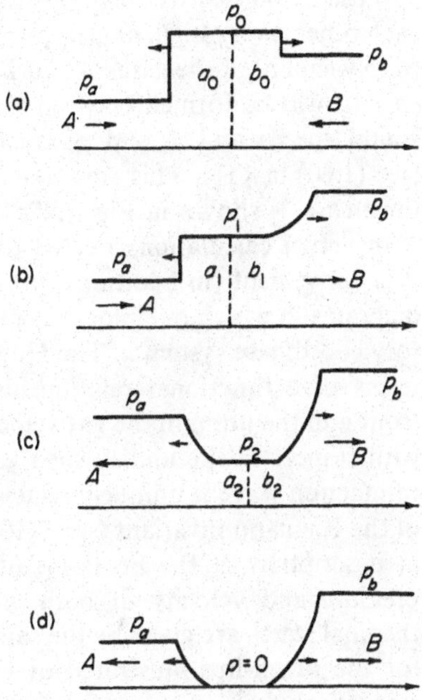

Fig. 1.47. Pressure distributions for different cases of the break-up of a discontinuity. Long arrows with the letters A and B denote the initial velocities of the gases A and B before the discontinuity breaks up. The small arrows show the directions of propagation of the waves through the gas (the direction of propagation in physical space can sometimes be different).

shock wave into gas A. The resulting state, illustrated in Fig. 1.47b, will be considered in detail in Chapter IV, where the operation of shock tubes is examined. An appropriate choice of the gases A and B and of the pressure drop makes it possible to achieve extremely strong shock waves and to heat the test gas to very high temperatures. One of the methods for obtaining even higher temperatures is provided by the collision of two shock waves. A special case of the latter is the reflection of a shock wave from the end wall of a shock tube, which is a means also used for obtaining high temperatures in the laboratory. Reflection of a shock wave by a solid wall is equivalent to a special case of collision of two gas streams. If two completely identical streams collide with one another, the contact discontinuity will remain at rest after the collision, and the situation is the same as in the case where a stationary solid wall replaces the contact discontinuity. The problems of collision of incident and reflected shock waves will be considered in Chapter IV.

3. If after the discontinuity breaks up, the pressure p_2 is lower than p_a and p_b, then we will obtain a rarefaction wave which travels to the right and to the left through each gas. This situation, depicted in Fig. 1.47c, is achieved when both gases A and B are initially moving in opposite directions away from the discontinuity at sufficiently high velocities.

If the initial relative velocity with which the gases A and B move away from each other is very high, namely greater than the sum of the maximum velocities (into vacuum) of the gases A and B, $2c_a/(\gamma_a - 1) + 2c_b/(\gamma_b - 1)$, then a vacuum ($p = 0$) will be formed between the two gases. Here c_a and c_b are the initial sound speeds and γ_a and γ_b are the specific heat ratios of gases A and B (see (1.60) in §11). This situation, which can be regarded as the limit of the third case, is shown in Fig. 1.47d.

In actual calculations related to the break-up of arbitrary discontinuities it is convenient (in addition to the p, V diagrams) to use the so-called p, u diagrams in which one plots the pressure p against the velocity in the laboratory coordinate system u. The Hugoniot relation $p_H(V)$ can be represented in terms of a functional relationship between the pressure behind the wave front and the jump in the gas velocity, i.e., the velocity of the compressed gas with respect to the undisturbed gas. In a similar manner, the pressure in a rarefaction wave is uniquely related to the velocity by virtue of the constancy of the Riemann invariant (see §§10 and 11). The convenience of p, u diagrams in description of the break-up of a discontinuity lies in the fact that the pressure and velocity of both gases are identical in the final state, that the final states are given by the same point on the diagram. The p, u diagrams for the break-ups illustrated in Figs. 1.47a–d are shown in Figs. 1.48a–d, respectively.

Having clarified the types of flow arising in the break-up of an arbitrary discontinuity, we can now verify the initial assumption that such flows depend only on the ratio x/t. In the discussion of the rarefaction wave in §11 this assumption was based on the fact that the width of the rarefaction wave, which is the only scale length in a problem in which dissipative processes are neglected, increases with time as $x \sim ct$. The relative effect of viscosity and heat conduction, which is proportional to l/x, decreases with time and becomes negligibly small in macroscopic flows, when $x \gg l$. Consequently, the only constant scale, the molecular mean free path, will also disappear.

In flows with shock waves, both viscosity and heat conduction, which introduce the length scale l into the equations, are important only in the thin layer of the wave front whose thickness is of the order of l. The thickness of the contact discontinuity is also small. It decays with time due to molecular diffusion and heat conduction. These two mechanisms lead to a discontinuity thickness Δx of the order of $(\chi t)^{1/2} \sim (Dt)^{1/2}$ (where D is the diffusion coefficient, which is approximately equal to the thermal diffusivity $D \sim \chi \sim lc$).

The distance x traveled by the shock and rarefaction waves in a time t is of the order of ct, so that $\Delta x \sim (lx)^{1/2}$. The ratio of the dimensions of the region in which the dissipative forces are not negligible to the dimensions of the entire region occupied by the flow is therefore of the order of l/x for the shock wave and of the order of $(l/x)^{1/2}$ for the contact discontinuity. Both quantities are small in a macroscopic flow where $x \gg l$.

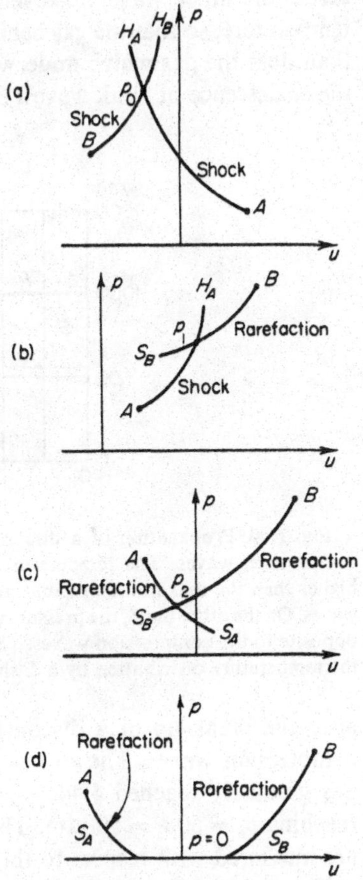

Fig. 1.48. A series of p, u diagrams for the different cases of discontinuity break-up shown in Fig. 1.47. Curves H are Hugoniot curves as a function of p and u; curves S are isentropes as a function of p and u.

Let us return to the third example presented at the beginning of this section and investigate the general conditions which result from the break-up of a discontinuity created by the accumulation of compression waves which were, in turn, produced by an accelerating piston. When two separate waves combine we have on one side of the discontinuity the undisturbed gas A, and on the other, gas in the state B, which has been compressed practically isentropically. We can show that the flow velocity, acquired by the gas as a result of successive compressions by a large number of shock waves, is lower than the velocity acquired as a result of compression to the same pressure by a single shock. It follows, therefore, that the discontinuity breaks up as in case 2. A rarefaction wave travels toward the piston through the compressed gas and a shock wave travels through the undisturbed gas. The pressure p turns out to be lower than the pressure at the piston p_b. Due to the entropy increase across the shock wave, however, this lower pressure corresponds to a higher temperature, so that the gas behind the shock wave undergoes a larger heating than does the gas which underwent the almost isentropic heating induced by the coalescence of weak waves. Figure 1.49 shows the distributions of p and T

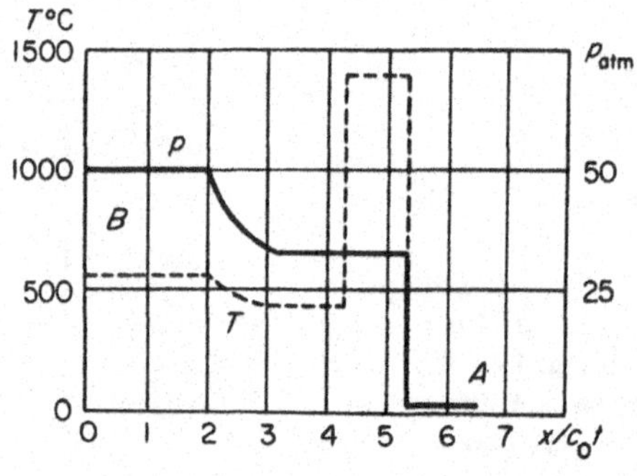

Fig. 1.49. Propagation of a discontinuity resulting from the coalescence of successive compression waves. The temperature behind the generated shock wave is considerably higher than the maximum temperature reached by the superposition of weak compression waves. On the other hand, the pressures are lower, since rarefaction waves move in a direction opposite to the compression waves. The pressure distribution is shown by a solid line, and the temperature distribution by a dashed line.

after the break-up of a discontinuity generated by the coalescence of weak compression waves. The air has been compressed by a piston whose velocity has gradually reached $4.44c_0 = 1500$ m/sec, with the pressure at the piston reaching $p_b = 50p_a = 50$ atm. The distance coordinate and time in the figure are measured with respect to the point and time of coalescence.

The case considered above is of considerable interest in detonation theory. The results obtained explain how a flame, acting on a gas in a manner similar to a piston, can cause the creation of a shock wave at a large distance from the flame by means of successive compressions. By compressing the gas gradually to a moderate temperature (630°C in Fig. 1.49), we can obtain a sharp heating to 1450°C a considerable distance away, and can achieve a "remote ignition" at the time of coalescence. Apparently, this is often the mechanism of detonation in gases.

§25. Strong explosion in a homogeneous atmosphere

The idealized problem of a strong explosion in a homogeneous atmosphere represents a typical example of a self-similar flow, in which the flow variables change with time in such a manner that their distributions with respect to the coordinate variable always remain similar in time. The self-similar problem of a strong explosion was formulated and solved by Sedov [4, 5]. Using a brilliant method, which employed the energy integral, Sedov succeeded in finding an exact analytic solution to the equations of self-similar motion. The same problem was also considered by Stanyukovich (in his dissertation, see [15]) and by Taylor [6], both of whom formulated the equations for the problem and obtained numerical but not analytic solutions. We shall discuss the solution to this problem and some of its consequences, since they will be useful in Chapters VIII and IX where we shall examine certain physical-chemical and optical phenomena accompanying strong explosions in air.

We consider a perfect gas with constant specific heats and density ρ_0, in which a large amount of energy E is liberated in a small volume during a short time interval. A shock wave will propagate through the gas starting from the point where the energy is released. We shall consider the process at the stage when the shock wave has moved through a distance which is extremely large in comparison with the dimensions of the region in which the energy was originally released, when the mass of gas that has been set in motion by the explosion is large in comparison with the mass of the explosion products. In this case, to a very high degree of accuracy, the energy release can be assumed to be both instantaneous and occurring at a point. We shall also assume that this stage of the process is sufficiently early that the shock wave has not moved too far away from the source, so that its strength is still sufficiently large that it is possible to neglect the initial gas pressure or counter-pressure p_0 in comparison with the pressure behind the shock wave. This is equivalent to neglecting the initial internal energy of the gas which has been set in motion in comparison with the explosion energy E and to disregarding the initial speed of sound c_0 in comparison with the velocities of both the gas and the wave front.

The gas motion is determined by two dimensional parameters, the energy of the explosion E, and the initial density ρ_0. These parameters cannot be combined to yield scales with dimensions of either length or time. Hence the motion will be self-similar, that is, will be a function of a particular combination of the coordinate r (distance from the center of the explosion) and the time t. In contrast to the self-similar motion considered in §11, this problem does not contain a characteristic velocity. The initial speed of sound c_0 cannot be used to characterize the process; whenever we set $p_0 = 0$ in a particular approximation, c_0 will also be equal to zero to the same degree of approximation*. Hence, the quantity r/t cannot serve as the similarity variable, as was the case in the self-similar rarefaction wave in §11. In this case the only dimensional combination which contains only length and time is the ratio of E to ρ_0, with the dimensions $[E/\rho_0] = [\text{cm}^5 \cdot \text{sec}^{-2}]$. Hence, the dimensionless quantity

$$\xi = r\left(\frac{\rho_0}{Et^2}\right)^{1/5} \tag{1.109}$$

can serve as the similarity variable.

The shock front is defined by a given value of the independent variable ξ_0. The motion of the wave front $R(t)$ is governed by the relationship

$$R = \xi_0\left(\frac{E}{\rho_0}\right)^{1/5} t^{2/5}. \tag{1.110}$$

The propagation velocity of the shock wave is

$$D = \frac{dR}{dt} = \frac{2}{5}\frac{R}{t} = \xi_0 \frac{2}{5}\left(\frac{E}{\rho_0}\right)^{1/5} t^{-3/5} = \frac{2}{5}\xi_0^{5/2}\left(\frac{E}{\rho_0}\right)^{1/2} R^{-3/2}.$$

We can express the parameters behind the front in terms of its velocity using the limiting formulas for a strong shock wave

$$\rho_1 = \rho_0 \frac{\gamma+1}{\gamma-1}, \qquad p_1 = \frac{2}{\gamma+1}\rho_0 D^2, \qquad u_1 = \frac{2}{\gamma+1} D. \tag{1.111}$$

The density behind the wave front stays constant and equal to its limiting value. The pressure decreases with time according to the relation

$$p_1 \sim \rho_0 D^2 \sim \rho_0\left(\frac{E}{\rho_0}\right)^{2/5} t^{-6/5} \sim \frac{E}{R^3}. \tag{1.112}$$

The physical meaning of the equations governing the propagation of a strong explosion is easily understood. At a time t the wave reaches a radius R and encompasses a volume $\frac{4}{3}\pi R^3$, whose mass is $M = \frac{4}{3}\pi R^3 \rho_0$. The pressure is proportional to the average energy per unit volume, that is, $p \sim E/R^3$. The gas and shock front velocities are proportional and $D \sim u \sim (p/\rho)^{1/2} \sim (E/\rho_0 R^3)^{1/2}$. Integrating $dR/dt = D$, we find the radius of the front as a func-

* This condition actually determines the limits of applicability of the solution. We impose specific requirements on the accuracy of the solution by comparing the pressure p_1 behind the wave front and the propagation velocity of the wave D, with p_0 and c_0, and find the time when the approximation $p_1 \gg p_0$ is no longer satisfied. It should be noted, however, that the condition for neglecting the counterpressure is stricter, namely: $p_1 \gg [(\gamma+1)/(\gamma-1)]p_0$. This is evident from (1.76); with this condition the density ratio across a shock wave is equal to its limiting value $(\gamma+1)/(\gamma-1)$.

tion of time $R \sim (E/\rho_0)^{1/5} t^{2/5}$ (to within a factor of the order of the numerical coefficient ξ_0). Equation (1.112) demonstrates the similarity law for different explosive energies. The pressure behind the front has a fixed value at distances proportional to $E^{1/3}$, or at times proportional to $E^{1/3}$.

The distributions of pressure, density, and gas velocity with respect to the radius are determined as functions of the one dimensionless variable ξ, which can be written as $\xi = \xi_0 r/R$. Since the motion is self-similar, the shapes of the distributions do not change with time, while the scales for the variables p, ρ, and u are exactly the same functions of time as at the shock front. In other words, the solution can be expressed in the form

$$p = p_1(t)\,\tilde{p}(\xi), \qquad u = u_1(t)\,\tilde{u}(\xi), \qquad \rho = \rho_1\,\tilde{\rho}(\xi).$$

Here $p_1(t)$, $u_1(t)$, and ρ_1 are the pressure, velocity, and density behind the shock front, and are functions of time as given by (1.111) and (1.112), while $\tilde{p}(\xi)$, $\tilde{u}(\xi)$, $\tilde{\rho}(\xi)$ are new, dimensionless functions. Substituting these expressions into the gasdynamic equations for the spherically symmetric case and transforming from differentiation with respect to r and t to differentiation with respect to ξ (by using (1.109) in a manner similar to that of §11) we obtain a system of three ordinary first-order differential equations for the three unknown functions $\tilde{p}$, $\tilde{u}$, and $\tilde{\rho}$. The solution of this system must satisfy the conditions at the wave front $\tilde{p} = \tilde{u} = \tilde{\rho} = 1$, for $\xi = \xi_0$.

We shall not present here the detailed solution and the resulting formulas which may be found in the book by Sedov [5] or the one by Landau and Lifshitz [1]. We note, however, that the only dimensionless parameter ξ_0 included in the solution is determined from the condition of conservation of energy

$$E = \int_0^R 4\pi r^2 \left(\varepsilon + \frac{u^2}{2}\right) \rho \, dr, \tag{1.113}$$

evaluated with the solution obtained. This parameter also depends, as does the entire solution, on the specific heat ratio γ.

The specific heat ratio for atmospheric air is not constant and is a function of both temperature and density, as a result of the dissociation and ionization which take place at high temperatures (see Chapter III). However, it is always possible to select some approximate effective value of the ratio and assume it to be constant. In this way the solution to the idealized strong explosion problem can be applied to describe the actual process. For air we can assume a value of γ to be approximately equal to 1.2 to 1.3.

In Fig. 1.50 are plotted the distributions of the functions p/p_1, ρ/ρ_1, u/u_1, T/T_1 with respect to r/R for $\gamma = 1.23$; the parameter ξ_0 is equal here to 0.930. It is characteristic of a strong explosion that the gas density decreases extremely rapidly behind the front of the shock wave to the center. Practically the entire mass of the gas that was at first uniformly distributed inside a sphere of radius R is now contained within a thin layer near the front surface. In the vicinity of the front the pressure decreases as we move toward the center by a factor of 2 to 3, and then remains approximately constant over almost all the sphere. The temperature increases as we move away from the front to the center. The increase is rather smooth at first, while the pressure is still decreasing; then, in the constant pressure region, the temperature increases very rapidly. This increase in temperature as we move toward the center is due to the fact that the particles which are near the center were heated by a very strong shock wave and have, as a result, a very high entropy. For an isentropic expansion to the same pressure, the temperature is higher the greater is the entropy of the particles, that is, the closer they are to the center. The abrupt density decrease toward the center is, in turn, due to the increase in temperature (the pressure is constant).

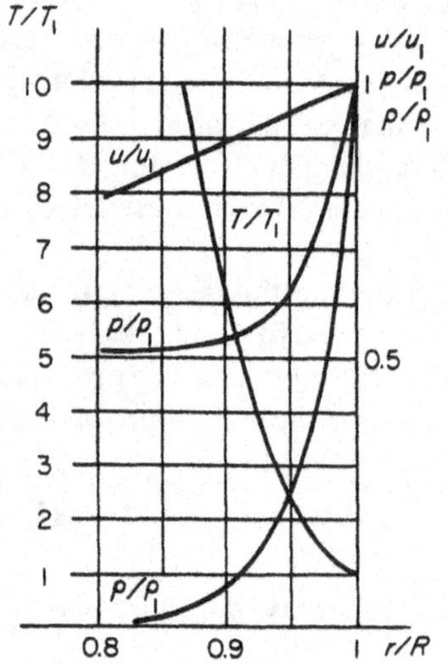

Fig. 1.50. Pressure, density, velocity, and temperature distributions for a point explosion in a gas with $\gamma = 1.23$.

Since the pressure is constant with respect to the radius everywhere except in the region close to the front, we can find the asymptotic distributions of the flow variables as $r \to 0$. It follows from the equation of motion with $p(r) = const$, $\partial p/\partial r = 0$, that $\partial u/\partial t + u\,\partial u/\partial r = 0$, i.e., that $u = r/t$*. To find the asymptotic behavior of the density we carry out a transformation to Lagrangian coordinates (see §2) and characterize the given gas particle by its initial radius r_0 (by "particle" we mean an elementary spherical layer of volume $4\pi r_0^2\,dr_0$). At the time when the shock front passes through the particle, the pressure at the particle p_1 is proportional to $R^{-3} = r_0^{-3}$. Starting at this time, the

* *Editors' note.* This argument is not complete, as $\rho \to 0$ in the limit. A more detailed calculation gives $u = (2/5\gamma)\,r/t$.

particle r_0 expands isentropically so that its density at a later time t is

$$\rho(r_0 t) = \rho_1 \left[\frac{p(r_0 t)}{p_1(r_0)}\right]^{1/\gamma}$$

However, at the given time t the pressure of all the particles in the "cavity" near the center are the same and $p_1(t)$ is proportional to $t^{-6/5}$. Hence, in Lagrangian coordinates the asymptotic result for the density is $\rho \sim r_0^{3/\gamma} t^{-6/5\gamma}$. Let us make a transformation to Eulerian coordinates by means of equation (1.24), $\rho r^2\, dr = \rho_0 r_0^2\, dr_0$. Substituting for the density and integrating, we obtain the Eulerian radius as a function of time $r \sim r_0^{(\gamma-1)/\gamma} t^{2/5\gamma}$. Eliminating r_0 from this expression by means of the function $\rho(r_0 t)$, we obtain the desired asymptotic behavior for the density

$$\rho \sim r^{3/(\gamma-1)} t^{-6/5(\gamma-1)} \qquad \text{as} \quad r \to 0.$$

The asymptotic behavior for the temperature is

$$T \sim \frac{p_c}{\rho} \sim r^{-3/(\gamma-1)} t^{(6/5)(2-\gamma)/(\gamma-1)} \qquad \text{as} \quad r \to 0.$$

§26. Approximate treatment of a strong explosion

The basic relationships governing a strong explosion can be obtained by a simple approximate method proposed by Chernyi [7]. Let us assume that the entire mass of gas encompassed by the explosion wave is concentrated in a thin layer behind the front surface; the density inside the layer is constant and equal to the density behind the front $\rho_1 = (\gamma+1)/(\gamma-1)\rho_0$. The thickness of the layer Δr is determined from conservation of mass

$$4\pi R^2\, \Delta r\, \rho_1 = \frac{4\pi R^3}{3}\rho_0; \qquad \Delta r = \frac{R}{3}\frac{\rho_0}{\rho_1} = \frac{R}{3}\frac{\gamma-1}{\gamma+1}.$$

For example, for $\gamma = 1.3$, $\Delta r/R = 0.0435$.

Since the layer is very thin, the gas velocity inside it remains approximately constant and coincides with the gas velocity behind the front u_1. Let us assume that the density in the layer is infinitely large and that the thickness is, correspondingly, infinitesimally small, while the mass is finite and equal to the mass M contained initially in the sphere of radius R: $M = \frac{4}{3}\pi R^3 \rho_0$. We denote the pressure at the inner side of the layer p_c and let it be the fraction α of the pressure behind the wave front $p_c = \alpha p_1$. Newton's law for the mass M becomes

$$\frac{d}{dt} M u_1 = 4\pi R^2 p_c = 4\pi R^2 \alpha p_1.$$

The mass $M = \frac{4}{3}\pi R^3 \rho_0$ is a function of time so that we differentiate the momentum Mu_1 with respect to time, rather than the velocity. An external force $4\pi R^2 p_c$ acts on the mass from the inside, since p_c is the force per unit area of the surface; the external force is equal to zero since the initial gas pressure is neglected. Expressing u_1 and p_1 in terms of the front velocity $D = dR/dt$, and using (1.111), we get

$$\frac{1}{3}\frac{d}{dt} R^3 D = \alpha D^2 R^2.$$

Noting that

$$\frac{d}{dt} = \frac{dR}{dt}\frac{d}{dR} = D\frac{d}{dR}$$

and integrating the equation, we find

$$D = aR^{-3(1-\alpha)},$$

where a is a constant of integration.

To determine the quantities a and α we use the principle of conservation of energy. The kinetic energy of the gas is equal to $E_k = Mu_1^2/2$. The internal energy is concentrated in the "cavity" bounded by our infinitesimally thin layer, and the pressure inside this "cavity" is equal to p_c (this means that although most of the mass is contained in the layer, the "cavity" also contains a small amount of the fluid). The internal energy is $E_T = [1/(\gamma - 1)] \times \frac{4}{3}\pi R^3 p_c$. Therefore,

$$E = \frac{1}{\gamma - 1}\frac{4\pi R^3}{3} p_c + M\frac{u_1^2}{2}.$$

Writing again $p_c = \alpha p_1$, expressing u_1 in terms of D, and then substituting the expression for D, we obtain

$$E = \frac{4\pi}{3}\rho_0 a^2\left[\frac{2\alpha}{\gamma^2 - 1} + \frac{2}{(\gamma + 1)^2}\right] R^{3-6(1-\alpha)}.$$

Since the explosion energy E is constant, the exponent of the variable R must vanish. This gives $\alpha = \frac{1}{2}$. The equation thus obtained defines the constant a

$$a = \left[\frac{3}{4\pi}\frac{(\gamma - 1)(\gamma + 1)^2}{(3\gamma - 1)}\right]^{1/2}\left(\frac{E}{\rho_0}\right)^{1/2}.$$

From the equation $D \sim R^{-3(1-\alpha)}$ with $\alpha = \frac{1}{2}$ and from (1.111) we obtain the following familiar relationships:

$$D \sim R^{-3/2}, \qquad p_1 \sim R^{-3}, \qquad u_1 \sim R^{-3/2}, \qquad R \sim t^{2/5}.$$

Using the expression for a we can find the coefficient of proportionality in the relation $R \sim t^{2/5}$ from

$$R = \left(\frac{5}{2}a\right)^{2/5} t^{2/5} = \left[\frac{75}{16\pi}\frac{(\gamma-1)(\gamma+1)^2}{(3\gamma-1)}\right]^{1/5}\left(\frac{E}{\rho_0}\right)^{1/5} t^{2/5} = \xi_0\left(\frac{E}{\rho_0}\right)^{1/5} t^{2/5}.$$

Let us compare the approximate solution with the exact solution. The pressure at the center obtained by the approximate method is one-half of the pressure behind the front, regardless of the specific heat ratio. In the exact solution $p_c = 0.35p_1$ for $\gamma = 1.4$ and $p_c = 0.41p_1$ for $\gamma = 1.2$. The approximate numerical values of the coefficient ξ_0 in the equation for the shock radius (1.110) are: $\xi_0 = 1.014$ for $\gamma = 1.4$ and $\xi_0 = 0.89$ for $\gamma = 1.2$. In the exact solution and for the same values of γ, $\xi_0 = 1.033$ and 0.89, respectively. It is thus evident that the approximate solution yields quite accurate results.

Concepts close in spirit to those described above were used by Kompaneets [13] in an approximate treatment of the problem of a strong explosion in an inhomogeneous atmosphere. This problem is treated in Part 5 of Chapter XII.

§27. Remarks on the point explosion with counterpressure

At a later stage of propagation of an explosion wave, when the pressure behind the wave front becomes comparable to the pressure ahead of it (to be more precise, when p_1 becomes of the order of $[(\gamma+1)/(\gamma-1)]p_0$; see footnote in §25), the self-similar solution to the problem of a strong explosion no longer holds. At this stage the process is no longer self-similar, because the problem now contains characteristic length and time scales, which can be constructed from the total explosion energy E and the parameters characterizing the undisturbed gas. The length scale can be represented by the radius of a sphere $r_0 = (E/p_0)^{1/3}$, whose initial energy is comparable to the explosion energy. The time scale is given by the time for a sound wave to travel through this distance $t_0 = r_0/c_0$, where $c_0 = (\gamma p_0/\rho_0)^{1/2}$. Thus, for example, the length and time scales for an explosion in atmospheric air ($\rho_0 = 1.25 \cdot 10^{-3}$ g/cm^3, $p_0 = 1$ atm, and $c_0 = 330$ m/sec) with a liberation of energy $E = 10^{21}$ ergs (approximately equivalent to the energy liberated in exploding 20,000 tons of TNT) are $r_0 = 1$ km and $t_0 = 3$ sec.

The solution to the problem of the propagation of a point explosion with counterpressure has been obtained in a number of papers [8–10] by numerically integrating the appropriate partial differential equations. Numerical results, tables, and graphs of the distributions of the flow variables at different times can be found in these references and in the book of Sedov [5]. We shall limit our discussion here to some qualitative aspects of the process.

The shock wave becomes increasingly weaker with time and the pressure behind the front asymptotically approaches the initial atmospheric pressure.

There is a corresponding decrease in the density behind the wave front and in its propagation velocity, which asymptotically approaches the speed of sound c_0. The relationship governing the shock propagation $R \sim t^{2/5}$ is gradually transformed into the relationship $R = c_0 t$. When the pressure in the central region becomes close to atmospheric, the expansion of the gas in this region stops and the gas stops moving. The region in which the gas does move is found closer to the shock front, which gradually transforms into a spherical wave of acoustic type. The compression zone across such a wave is followed by a rarefaction zone, after which the air arrives at its final state. The final state of the layers that are far from the center and through which only a weak shock wave has passed differs little from the initial undisturbed state. The pressure, velocity, and density distributions with respect to the radius are shown in Fig. 1.51. If we follow the pressure change with time at a

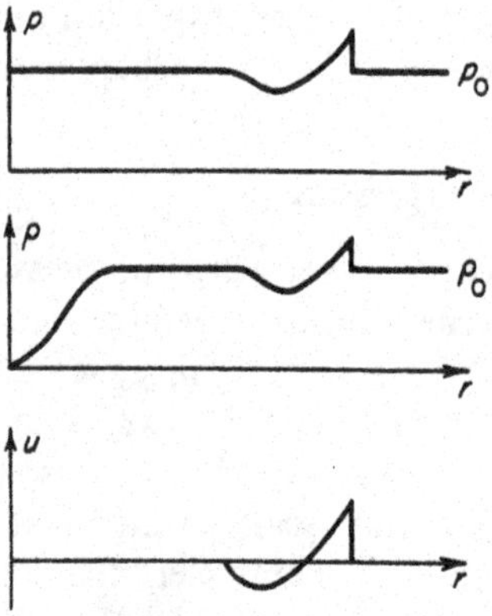

Fig. 1.51. Pressure, density, and velocity distributions at a later stage of an explosion, when the shock wave has become weak.

fixed large distance from the explosion center, we get the picture shown in Fig. 1.52. At the time t_1, when the shock front approaches a given point, the pressure undergoes a discontinuous jump above atmospheric, then decreases below atmospheric (the positive and negative pressure phases), and finally returns to its initial value.

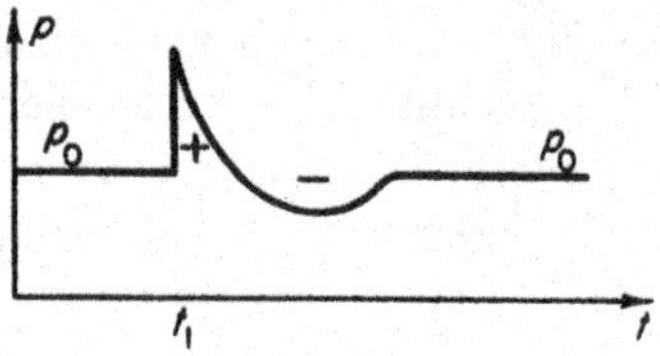

Fig. 1.52. The time dependence of the pressure at a fixed point far from the explosion source.

As previously noted, the final state of the gas far from the explosion source is almost the same as the undisturbed state. At small distances, however, the gas in its final state is very rarefied and heated to a high temperature. This is due to the fact that a strong shock wave passed through the particles located near the center and the entropy of these particles has become much higher than the initial entropy. The asymptotic distributions of the final densities and temperatures with respect to the radius in the vicinity of the center can be found by considering the isentropic expansion to atmospheric pressure of the particles heated in the front of a strong shock wave. Repeating the calculations presented at the end of §25, with p_c no longer a function of t but a constant $p_c = p_0$, we find the same distributions with respect to the radius for $r \to 0$ as we did in the strong explosion problem, $\rho \sim r^{3/(\gamma-1)}$, $T \sim r^{-3/(\gamma-1)}$. The final distributions of $\rho(r)$ and $T(r)$ are shown in Fig. 1.53. A considerable

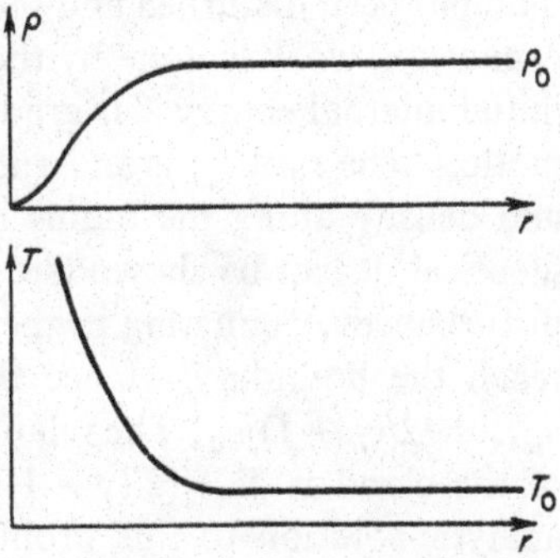

Fig. 1.53. Final density and temperature distributions ($t \to \infty$) for a strong explosion (assuming that the process is adiabatic).

fraction of the explosion energy (dependent on γ, but of the order of several times 10%) is concentrated in the heated region. This energy was spent in irreversibly heating the gas during the shock compression. The remaining energy is carried away by the shock wave and is dissipated in space. In Chapter IX we shall discuss the fate of the energy left behind in the central region (the air in this region is cooled by radiative emission).

The later stage of propagation of an explosion wave has been studied theoretically and experimentally by many authors. Limiting relations governing the wave propagation at large distances were found by Landau [11]. Of great practical importance is the empirical formula of Sadovskii [12], which expresses the pressure behind the wave front as a function of distance from the explosion center. We note that the similarity relationship $p_1 = f(E^{1/3}/R)$ is also valid at a later stage of shock wave propagation, when $p_1 - p_0$ is of the same order as p_0 or less.

§28. Sudden isentropic expansion of a spherical gas cloud into vacuum

Let us consider another gasdynamic problem, the problem of the sudden expansion of a gas cloud into vacuum. This problem will be dealt with in more detail in Chapter VIII. We imagine a gas occupying initially a spherical volume of radius R_0. We suppose that initially the gas has the uniform density ρ_0 and is at rest (the total gas mass is $M = \frac{4}{3}\pi R_0^3 \rho_0$). The initial pressure in the cloud is also assumed to be constant and equal to p_0, so that the total energy of the gas is $E = [1/(\gamma - 1)]\frac{4}{3}\pi R_0^3 p_0$ (it is assumed that the gas is perfect with constant specific heats). The partition containing the gas is removed at $t = 0$ and the cloud starts expanding into a vacuum.

After the partition has been removed, the discontinuity breaks up and a rarefaction wave propagates through the gas toward the center. The front layers of the cloud expand into the vacuum with the maximum escape velocity $u_{max} = [2/(\gamma - 1)]c_0$. When the rarefaction wave arrives at the center, all the fluid has been disturbed and set in motion. In the process of sudden isentropic expansion, work is done by the expanding gas, the gas is accelerated, and its initial internal energy E is gradually transformed into kinetic energy of radial motion. The motion is an isentropic one, since the initially constant pressure and density along the radius result in the entropy of all the particles being identical. It can be shown (see [15]) that during the isentropic expansion the disturbances originating in the interior regions of the spherical cloud do not reach the boundary. Hence this surface moves with the constant velocity $u_{max} = [2/(\gamma - 1)]c_0$. The relationship governing the motion of the boundary of the cloud is $R = [2/(\gamma - 1)]c_0 t + R_0$. It is not possible to find an exact analytic solution to this problem, since the flow is not self-similar and the solution of a system of partial differential equations is required; this can be done analytically only in a very few exceptional cases. We can convince ourselves that the problem is not self-similar by noting that it contains a characteristic length scale, that is, the initial radius of the cloud R_0.

This problem, however, has the property that the flow asymptotically approaches a self-similar one with time. The role of the initial length parameter R_0 becomes less and less important during the later highly expanded stage where $R \gg R_0$, since the length scale R_0 becomes very small in comparison with the characteristic flow scale, the actual radius of the spherical cloud R. The flow "forgets" about the initial radius R_0 with time. However, not all parts of the flow solution completely "forget" the initial conditions, and this is an evidence of the fact that the flow is not self-similar in at least some essential aspects.

Let us examine the asymptotic behavior of the solution as $t \to \infty$. In this limit the force acting on a unit mass of the gas approaches zero. The force $-(1/\rho)\,\partial p/\partial r$ has the order of magnitude of $-p/\rho R$, where p and ρ are the pressure and density at the time t, averaged over the mass. But the average pressure p is proportional to the ratio of the thermal energy of the entire gas to its volume $p \sim E_{heat}/R^3$ and it is in any case lower than E/R^3. The average density $\rho \sim 1/R^3$, and therefore the force tends to zero at least as fast as $1/R$.

Actually, as $R \to \infty$ the force decreases faster than $1/R$ since the thermal part of the energy decreases during the isentropic expansion: $E_{heat} \sim M\varepsilon \sim Mp/\rho \sim \rho^{\gamma-1} \sim R^{-3(\gamma-1)}$. Hence, $p \sim E_{heat}/R^3 \sim R^{-3\gamma}$, and the force decreases as $R^{-3\gamma+2} = R^{-1-3(\gamma-1)}$. The equation of motion for the limits $t \to \infty$ and $R \to \infty$ takes the asymptotic form

$$\frac{Du}{Dt} = \frac{\partial u}{\partial t} + u\frac{\partial u}{\partial r} = -\frac{1}{\rho}\frac{\partial p}{\partial r} \sim \frac{1}{R^{1+3(\gamma-1)}} \to 0, \tag{1.114}$$

that is, the velocities of the fluid particles approach constant values and $u = r/t$.

The sudden expansion becomes inertial as $t \to \infty$. This also follows directly from the condition of conservation of the total energy E, which is made up of the thermal and kinetic energies. The thermal portion of the energy, however, approaches zero asymptotically during the expansion and, consequently, the kinetic energy approaches E. The average (root mean square) velocity of the mass of gas asymptotically approaches the constant limiting value $u_\infty = (2E/M)^{1/2}$, which is a specific numerical fraction of the boundary velocity

$$u_{max} = \frac{2}{\gamma-1}c_0 = \frac{2}{\gamma-1}\left(\gamma\frac{p_0}{\rho_0}\right)^{1/2} = \frac{2}{\gamma-1}[\gamma(\gamma-1)\varepsilon_0]^{1/2}$$
$$= \left(\frac{4\gamma}{\gamma-1}\frac{E}{M}\right)^{1/2} = \left(\frac{2\gamma}{\gamma-1}\right)^{1/2} u_\infty$$

(for example, in a monatomic gas $\gamma = 5/3$ and $u_{max} = 2.9u_\infty$). Substituting the asymptotic solution for the velocity $u = r/t$ into the continuity equation, we see that it is satisfied by the density function

$$\rho = \frac{f(r/t)}{t^3}, \tag{1.115}$$

where f is a completely arbitrary function of r/t. Since the radius of the spherical boundary is $R = u_{max}t$, we can rewrite this equation in the form

$$\rho = \frac{\varphi(r/R)}{R^3}.$$

The asymptotic distribution of the density with respect to the radius does not change with time; it changes in the same ratio as R increases, remaining similar to itself. In the absence of any forces acting on the gas, each particle will indeed move with a constant inertial velocity, there will be no redistribution of mass, and the density profile will remain unchanged. The fact that the problem is not internally self-similar makes it impossible to find the asymptotic density distribution from the equations for the asymptotic flow, as these equations permit any type of distribution. The density distribution is formed in an earlier stage, when the pressure forces are acting on the gas. By the time the gas is strongly expanded, the distribution "freezes". The density distribution depends on the initial conditions and can be found only on the basis of a complete solution of the problem.

As mentioned previously, it is impossible to find an exact analytic solution to the problem with the initial conditions $\rho_0(r) = const$ and $p_0(r) = const$. We can only construct an approximate solution on the basis of the analogous and solvable plane problem of the sudden expansion into vacuum of a gas layer with a finite mass and constant initial distributions. This approximate solution is presented in the book by Stanyukovich [15] and is of the form

$$\rho = \frac{A}{R^3}\left(1 - \frac{r^2}{R^2}\right)^\alpha, \qquad \alpha = \frac{3-\gamma}{2(\gamma-1)}, \qquad R = u_{max} t.$$

This solution is valid only for integral values of $\alpha = 0, 1, 2, 3, \ldots$, which correspond, respectively, to values of the specific heat ratio $\gamma = 3$, $5/3$, $7/5$, $9/7, \ldots$. The constant A can be determined from conservation of mass by integrating the density over the entire volume of the sphere. The appropriate equation is presented in [15].

§29. Conditions for the self-similar sudden expansion of a gas cloud into vacuum

There exists a class of solutions to the problem of the sudden expansion of a spherical gas cloud into vacuum, where the distributions of all flow variables are self-similar, and are from the very beginning functions of r/R alone, where R is the radius of the spherical cloud. The initial distributions of flow variables with respect to the radius, which lead to these solutions, cannot be arbitrarily set but must satisfy particular conditions. The above class of solutions is characterized by a linear velocity distribution with respect to the radius (these solutions were investigated by Sedov [5]),

$$u = rF(t) = \dot{R}\frac{r}{R}, \tag{1.116}$$

where the function of time $F(t)$ is expressed in terms of the velocity of the spherical boundary of the cloud $\dot{R} = dR/dt$ *. Substituting this expression into the equation of motion we get

$$\frac{\partial p}{\partial r} = -\rho r(\dot{F} + F^2), \tag{1.117}$$

* *Editors' note.* The asymptotic solutions of §28 are of this form, with $F = t^{-1}$.

which must be satisfied by the distributions of p and ρ with respect to the radius during the entire process, including the initial time. Only under this condition will the solution belong to the class under consideration.

Let us consider two specific examples of such solutions.

1. Let the density ρ be constant over the entire volume and independent of the radius

$$\rho = f(t) = \frac{M}{\frac{4}{3}\pi R^3}. \tag{1.118}$$

We can easily check that the specification of density and velocity in the form of (1.118) and (1.116) automatically satisfies the continuity equation for an arbitrary function $R(t)$. Substituting (1.118) into (1.117) and integrating, we obtain a pressure distribution parabolic with respect to the radius

$$p = p_0(t)\left(1 - \frac{r^2}{R^2}\right), \tag{1.119}$$

which then must be specified as an initial condition in order to satisfy (1.117). It is evident that the problem is not isentropic, since the density of all the particles is the same while the pressures are different. Substitution of p and ρ into the entropy equation gives a relationship between the unknown functions: pressure at the center $p_0(t)$ and the radius of the spherical cloud $R(t)$

$$p_0(t) = A\rho^\gamma = A\left(\frac{3M}{4\pi}\right)^\gamma \frac{1}{R^{3\gamma}}, \tag{1.120}$$

where A is a constant depending on the initial entropy at the center of the cloud. Finally, substituting (1.118), (1.119), and (1.120) into the equation of motion (1.117) we get a second-order ordinary differential equation governing the motion of the boundary of the cloud $R(t)$. Solving this equation with the initial conditions $t = 0$, $R = R_0$, and $\dot{R} = \dot{R}_0$, we can find the complete solution to the problem. In particular, we can assume that the gas is initially at rest, with $\dot{R}_0 = 0$.

If we are interested in the asymptotic result as $t \to \infty$, we can set $\dot{R} = const = u_1$, where u_1 is the limiting velocity of the spherical boundary (the solution of the differential equation gives, as expected, $\dot{R} \to const$ as $t \to \infty$). The value of u_1 can be calculated from the condition of conservation of energy using the radial distributions of ρ and u and noting that the total energy is converted into kinetic energy as $t \to \infty$. Thus we get

$$u_1 = \left(\frac{5}{3}\right)^{1/2}\left(\frac{2E}{M}\right)^{1/2} = \left(\frac{5}{3}\right)^{1/2} u_\infty, \tag{1.121}$$

where u_∞ is again defined as the square root of the square of the velocity averaged over the entire mass (root mean square velocity)

$$u_\infty = (\overline{u^2})^{1/2} = (2E/M)^{1/2}.$$

2. Let the entropy of all the particles be identical (isentropic flow), so that $S(r, t) = const$, $p/\rho^\gamma = A = const$ (A is the isentropic constant). Substitution of $p = A\rho^\gamma$ into (1.117) gives the pressure and density distributions

$$\rho = \rho_c\left(1 - \frac{r^2}{R^2}\right)^{1/(\gamma-1)}, \tag{1.122}$$

$$p = A\rho_c^\gamma\left(1 - \frac{r^2}{R^2}\right)^{\gamma/(\gamma-1)}, \tag{1.123}$$

which, of course, must be specified as initial conditions.

The density at the center ρ_c can be determined by integrating the density over the volume and equating the resulting integral to the total mass. As usual this gives $\rho_c \sim M/R^3$, with a proportionality coefficient which is a function of γ. Equation (1.117) reduces after substitution of (1.122) and (1.123) to a second-order equation in $R(t)$. The limiting value for the velocity of the boundary u_1 can be obtained from conservation of energy

$$E = \int_0^R \frac{\rho u^2}{2}\, 4\pi r^2\, dr,$$

if the expression for ρ given by (1.122) and $u = u_1 r/R$ are substituted into the above integral. This yields the relationship between u_1 and $u_\infty = (2E/M)^{1/2}$ in terms of a function of γ, as was the proportionality coefficient. Both coefficients are given by definite integrals that are calculated with the aid of gamma functions. Let us present some numerical results. For $\gamma = 5/3$, $\rho_c = 3.4\bar{\rho}$, $u_1 = 1.64u_\infty$; for $\gamma = 4/3$, $\rho_c = 6.6\bar{\rho}$, $u_1 = 1.92u_\infty$, where $\bar{\rho} = M/\frac{4}{3}\pi R^3$ is the density averaged over the volume. In the limit, as $t \to \infty$, $R \approx u_1 t$ *.

We note that Imshennik [16] considers the problem of the sudden isothermal expansion of a gas cloud into vacuum. We also note the paper of Nemchinov [18] which treats the sudden expansion into vacuum of a gas in which a gradual release of energy takes place, and the paper by Nemchinov [19] which considers the expansion into vacuum of a triaxial ellipsoid.

* Reference [17] contains some numerical solutions to the problem of the sudden isentropic expansion of a gas cloud into vacuum for homogeneous initial conditions and $\gamma = 5/3$ (at $t = 0$ the gas in the sphere is at rest with no variation in density with respect to the radius). Unfortunately, the reference does not present the asymptotic density distribution and only gives a plot of $\rho_c(t)$. It is evident that the function approaches $\rho_c \sim 1/t^3$ with time, and the coefficient in this limiting relation turns out to be higher than in the self-similar solution described above by a factor of only 1.22.

II. Thermal radiation and radiant heat exchange in a medium

§1. Introduction and basic concepts

Until recently high temperatures of the order of tens and hundreds of thousands or even millions of degrees had been mainly of interest in the study of astrophysics. The theory of radiative transfer and radiant heat exchange was created and developed in order to understand processes which take place in stellar media and to explain the observed luminosity of stars. To a large extent, this theory can be also applied to other high-temperature systems considered in modern physics and engineering. In this chapter we shall be concerned with the principles of thermal radiation, the theory of radiant energy transfer, and the theory of luminosity of heated bodies; we shall also derive the equations describing the motion of a fluid in the presence of a strong radiation field. In presenting this material we shall be mainly concerned with terrestrial applications and shall emphasize certain topics that are of little or no importance in astrophysics*.

Let us recall the basic concepts and definitions of thermal radiation. Radiation is characterized by the frequency ν of oscillation of an electromagnetic field or by the wavelength λ related to the frequency and the speed of light c by the relation $\lambda = c/\nu$. In what follows we shall always be dealing with media whose indices of refraction are very close to unity, and we can therefore assume that c is the speed of light in vacuum, equal to $3 \cdot 10^{10}$ cm/sec. From the quantum mechanical point of view radiation can be considered as a collection of particles, photons, or light quanta, whose energy is related to the frequency of the equivalent field by the Planck constant $h = 6.62 \cdot 10^{27}$ erg·sec. It is customary to express a quantum of energy† $h\nu$ in electron volts. One electron volt is the energy acquired by an electron moving through a potential difference of one volt; 1 electron volt (1 ev) is equal to $1.6 \cdot 10^{-12}$ erg. Temperature is also frequently expressed in electron volts. The energy of $kT = 1.6 \cdot 10^{-12}$

* More detailed discussions of the theory of radiative transfer and its applications to astrophysics are given by Ambartsumian *et al.* [1], Unsöld [2], and Mustel' [3].

† In quantum theory it is customary to use the "angular" frequency $\omega = 2\pi\nu$ instead of the frequency ν and, consequently, to use the modified Planck constant $\hbar = h/2\pi$. In this book we shall follow the practice of radiative transfer theory and astrophysics and use the quantities ν and h.

erg, where $k = 1.38 \cdot 10^{-16}$ erg/deg is the Boltzmann constant, corresponds to the temperature T of one electron volt. In general, we define

$$T_{ev} = \frac{kT^{\circ}}{1.6 \cdot 10^{-12}} = \frac{T^{\circ}}{11{,}600};$$

thus, 1 ev of temperature is equal to 11,600°K.

The electromagnetic frequency (or wavelength) scale, also termed the radiation spectrum, is conventionally divided into several not too sharply defined bands that have been assigned particular names: radio wave, infrared, visible, ultraviolet, x-ray, and gamma-ray. This division is entirely historical in nature and has no rational physical basis. Some intermediate frequencies are even difficult to classify as belonging to any particular part of the spectrum. An exception is the more or less defined visible part of the spectrum: $\lambda \sim 7500$–4000 Å, $h\nu \sim 1.7$–3.13 ev. It has been shown from the theory of thermal radiation that for thermodynamic equilibrium between the radiation and the medium the frequency distribution of energies has a clearly defined maximum, corresponding to a particular ν which is related to the temperature by the relation $h\nu = 2.82\,kT$. We can say that the most characteristic frequency for a body at a temperature $T = h\nu/2.82\,k$ is ν; hence the frequency range immediately provides some idea about the temperatures corresponding to any given spectral region. Visible radiation is characteristic of bodies whose temperatures are of the order of 7000–13,000°K.

The electromagnetic field or light quanta are characterized not only by their energy but also by their momentum. The absolute value of the momentum of a quantum $h\nu$ is equal to $h\nu/c$. The direction of motion of a quantum is given by the vector representing the energy flux in the electromagnetic field—the Poynting vector.

A radiation field in space is described by the distribution of the intensity of radiation with respect to frequency, to space, and to the direction of the radiant energy transfer. If we consider radiation as a collection of particles—photons—then the field can be described by a photon distribution function which is, to a large extent, analogous to any other particle distribution function. Let $f(\nu, \mathbf{r}, \mathbf{\Omega}, t)\, d\nu\, d\mathbf{r}\, d\mathbf{\Omega}$ be the number of photons in the frequency interval ν to $\nu + d\nu$, contained at the time t in the volume element* $d\mathbf{r}$ about the point $\mathbf{r}$, and having a direction of motion within an element of solid angle $d\mathbf{\Omega}$ (scalar) about a unit vector $\mathbf{\Omega}$. The function f is called the distribution function.

Each photon possesses an energy $h\nu$ and moves with the speed c. Hence the quantity

$$I_{\nu}(\mathbf{r}, \mathbf{\Omega}, t)\, d\nu\, d\mathbf{\Omega} = h\nu c f(\nu, \mathbf{r}, \mathbf{\Omega}, t)\, d\nu\, d\mathbf{\Omega}$$

* The linear dimensions of the volume element $d\mathbf{r}$ are assumed to be much greater than the wavelength λ.

represents a radiant energy in the spectral interval $d\nu$, passing per unit time through a unit area, with directions of energy propagation contained within the element of solid angle $d\mathbf{\Omega}$ about the vector $\mathbf{\Omega}$. The area is located at the point $\mathbf{r}$ and is oriented perpendicular to $\mathbf{\Omega}$. The quantity I_ν is called the spectral radiation intensity. The radiation field is essentially fully defined by specifying either the function I_ν or f. The radiant energy of frequency ν included in a unit spectral interval of frequency and contained in a unit volume at the point $\mathbf{r}$ at the time t is termed the spectral radiant energy density and is given by

$$U_\nu(\mathbf{r}, t) = h\nu \int_{(4\pi)} f\, d\mathbf{\Omega} = \frac{1}{c}\int_{(4\pi)} I_\nu\, d\mathbf{\Omega}. \tag{2.1}$$

Let us imagine a unit area with the normal unit vector $\mathbf{n}$. The photons are passing through this area from left to right and from right to left. The amount of radiant energy in the interval $d\nu$, passing from left to right per unit time, is equal to $h\nu c \int_{2\pi} f \cos\vartheta\, d\mathbf{\Omega}$, where ϑ is the angle between the directions of motion of the photons $\mathbf{\Omega}$ and the normal $\mathbf{n}$; the integral is taken over the right hemisphere, having this area as a base (Fig. 2.1). The integral over the left hemisphere is equal to the energy flowing from right to left. The difference between the individual fluxes from left to right and from right to left gives the

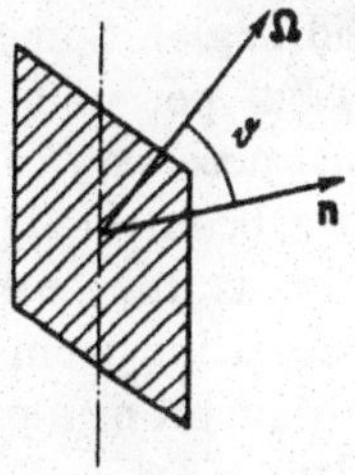

Fig. 2.1. Illustration for the derivation of the relation for radiant energy flux.

net spectral radiant energy flux through the given area. Since $\cos\vartheta$ has different signs in the right and left hemispheres, the spectral radiant energy flux through an area with normal $\mathbf{n}$ is

$$S_\nu(\mathbf{r}, t, \mathbf{n}) = h\nu c \int_{(4\pi)} f \cos\vartheta\, d\mathbf{\Omega} = \int_{(4\pi)} I_\nu \cos\vartheta\, d\mathbf{\Omega}, \tag{2.2}$$

where the integral is taken over the entire solid angle. The energy flux is a vector quantity. Equation (2.2) represents the component of the flux vector in the direction $\mathbf{n}$. The spectral energy flux vector is given by

$$\mathbf{S}_\nu = \int I_\nu \mathbf{\Omega}\, d\mathbf{\Omega}, \tag{2.3}$$

where $\mathbf{\Omega}$ is the unit vector in the direction of motion of the photons.

For an isotropic distribution of radiation, where the distribution function f and the intensity function I_ν are independent of the direction $\mathbf{\Omega}$, the radiant energy density is equal to

$$U_\nu = 4\pi h\nu f = \frac{4\pi}{c} I_\nu. \tag{2.4}$$

In this case there is no flux and $\mathbf{S}_\nu = 0$; the components in any direction are equal to zero (since equal amounts of energy are transferred in any two opposite directions).

The integrated intensity, density, and flux of radiation are obtained from their spectral counterparts by integrating over the entire frequency spectrum

$$I = \int_0^\infty I_\nu\, d\nu, \qquad U = \int_0^\infty U_\nu\, d\nu, \qquad \mathbf{S} = \int_0^\infty \mathbf{S}_\nu\, d\nu. \tag{2.5}$$

Let us now introduce the concept of the optical characteristics of a material*. The amount of energy of frequency ν which is spontaneously emitted by a unit volume of matter per unit time in a unit spectral interval of frequency is called the emission coefficient J_ν. Usually, because of the random orientation and chaotic motion of atoms, molecules, etc., gases radiate uniformly in all directions (isotropically). Therefore, the amount of energy radiated into a solid angle $d\mathbf{\Omega}$ in any direction is simply equal to $j_\nu\, d\mathbf{\Omega} = J_\nu\, d\mathbf{\Omega}/4\pi$ (j_ν is per unit solid angle). Sometimes the emission coefficient is not defined per unit volume, but per unit mass. To obtain the corresponding quantities it is, obviously, necessary to divide J_ν or j_ν by the density of the matter ρ.

When a beam of light rays passes through matter it is attenuated. This attenuation is due to the absorption of photons as well as to their scattering, that is, deviation from their original direction. The relative weakening of a parallel beam along a path element dx is proportional to this element, following the law

$$dI_\nu = -\mu_\nu I_\nu\, dx. \tag{2.6}$$

On traversing the distance x, from the point $x = 0$ to the point x, the beam intensity is decreased exponentially according to the relation

$$I_\nu = I_{\nu 0} \exp\left[-\int_0^x \mu_\nu\, dx\right]. \tag{2.7}$$

The attenuation (total absorption) coefficient μ_ν is composed of the absorption coefficient $\kappa_{\nu a}$ † and the scattering coefficient $\kappa_{\nu s}$. The reciprocals of these

* Here and in what follows the terms "light", "light quanta" or "photons", and "optical" properties will be applied not only to the visible part of the spectrum, as is usually done, but to any frequency.

† We do not consider here the process of stimulated emission, which will be discussed later, and denote the true absorption coefficient by $\kappa_{\nu a}$.

quantities are the mean free paths of light. Thus, $l_\nu = 1/\mu_\nu$ is the total mean free path, $l_{\nu a} = 1/\kappa_{\nu a}$ is the absorption mean free path, and $l_{\nu s} = 1/\kappa_{\nu s}$ is the scattering mean free path ($l_\nu = (l_{\nu a}^{-1} + l_{\nu s}^{-1})^{-1}$). These mean free paths characterize the attenuation per unit length of a beam of light rays relative to the corresponding process. The coefficients defined per unit of mass rather than per unit of path are termed mass coefficients. The mass coefficients are equal to μ_ν/ρ, $\kappa_{\nu a}/\rho$, and $\kappa_{\nu s}/\rho$, respectively. The mean free path represents an average distance traversed by a photon before it is absorbed, scattered, etc. However, the photon travels with the speed c, and hence the average "lifetime" of a photon in a given process is equal to the length of the path divided by the speed of light l/c. For example, if a fraction of photons $dx/l_{\nu a}$ is absorbed along a path element dx, then $c\,dt/l_{\nu a}$ photons are absorbed in a time dt.

The attenuation of a light beam is described by the product of the attenuation coefficient and the path length. The dimensionless quantity

$$\tau_\nu = \int_0^x \mu_\nu\, dx, \qquad d\tau_\nu = \mu_\nu\, dx \tag{2.8}$$

is called the optical thickness of a layer x with respect to light of frequency ν. A light beam passing through a unit optical thickness is attenuated by a factor of e. In the case when photon scattering can be neglected, the optical thickness becomes

$$\tau_\nu = \int_0^x \kappa_{\nu a}\, dx, \qquad d\tau_\nu = \kappa_{\nu a}\, dx. \tag{2.9}$$

§2. Mechanisms of emission, absorption, and scattering of light in gases

Photons are emitted and absorbed during electronic transitions from one energy state to another in atomic systems (atoms, molecules, ions, electron-ion plasmas). The absorption of a photon is accompanied by the excitation of an atom, molecule, etc. In order to emit a photon the atom must first be excited; the atom loses its excitation energy in transferring it to the emitted photon. The emission coefficient is higher the greater is the number of excited atoms, and thus the higher is the temperature.

Figure 2.2 shows an energy level diagram of the most elementary atomic system consisting of a proton and electron, which in the bound state constitutes the hydrogen atom. As usual, the zero energy level separates the free and bound states of the electron with the energy in the bound states negative. In the bound states the electron energy can assume only certain discrete values. The energy of the ground state of the proton-electron system is $E_1 = -13.5$ ev, and its absolute value equals the ionization potential of the hydrogen atom. In the free state with positive energy (hydrogen ion) the electron energy can assume any value and the energy spectrum is continuous.

The energy spectrum of complex atomic systems does not differ qualitatively from the spectrum of elementary systems. All electronic transitions can be divided (as is done in astrophysics) into three groups using the continuity criterion or the discreteness of the energy spectrum of the initial and final states of the atomic system. These groups are bound-bound, bound-free, and free-free (all allowed transitions are shown by arrows in Fig. 2.2).

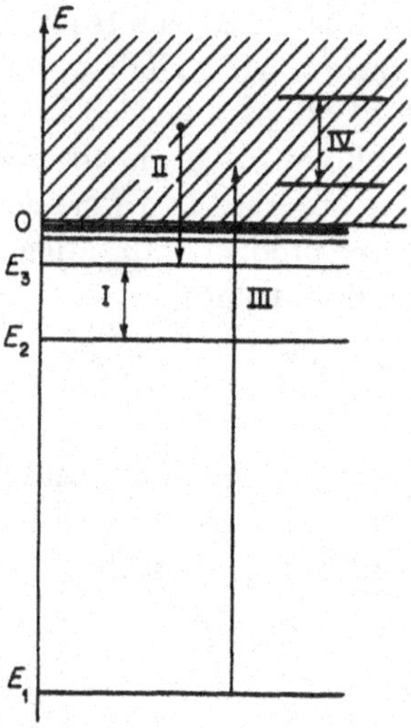

Fig. 2.2. Energy level diagram for proton-electron system. $E_1 = -13.5$ ev is the ground state of the hydrogen atom, E_2 and E_3 are the levels with principal quantum numbers $n = 2$ and 3. $E = 0$ corresponds to the boundary between discrete and continuous spectra. The arrows show the allowed types of transitions: I—bound-bound, II—electron capture by a proton, III—ionization of the atom, and IV—free-free transition.

The bound-bound transitions correspond to electronic transitions in atoms, molecules, and ions from one discrete level to another. By virtue of the discreteness of the energy levels of the bound electronic states, these transitions result in the emission or absorption of line spectra. In molecules where the electronic transitions are accompanied by changes in the vibrational and rotational modes, band spectra are obtained*.

During a bound-free transition resulting from photon absorption, the electron acquires an energy exceeding its binding energy to the atom, molecule, or ion and becomes free—photoionization takes place. The excess of the photon energy over the binding energy is transformed into kinetic energy of the free electron. Reverse transitions, that is, the capture of free electrons by ions in an ionized gas (photorecombination), result in the emission of photons. Since the energy of a free electron can assume any positive value the bound-free transitio n have continuous absorption and emission spectra. We should

* Transitions in molecules are sometimes associated with changes in the vibrational and rotational states only, without any change in the electronic states. In this case the emitted or absorbed photons have very low energy, lying in the infrared part of the spectrum; at temperatures of the order of several thousand degrees and above their role is insignificant.

note that not any arbitrary photon can produce a photoelectric effect in an atom in a given state. The energy of the photon must exceed the binding energy of the electron in this state. However, even a photon of very low energy can remove the electron from a sufficiently strongly excited atom, because the binding energy of the electron becomes increasingly less as the degree of excitation increases.

A free electron traveling through the electric field of an ion in an ionized gas (plasma) can either emit a photon without losing all its kinetic energy and remain free, or it can absorb a photon and acquire additional kinetic energy. These free-free transitions are usually called *bremsstrahlung* (from the German: *Bremse*—brake, and *Strahlung*—radiation, *eds.*), since the electron is slowed down in the field of the ion and loses a part of its energy in the radiation process. Bremsstrahlung has a continuous emission and absorption spectrum. Bremsstrahlung can also occur when an electron passes through the field of a neutral atom. In contrast to the field of an ion, the field of a neutral atom decreases rapidly with distance, and therefore the electron must pass very close to the atom to ensure the emission or absorption of light. The probability of bremsstrahlung with the participation of a neutral atom is much smaller than the probability of this process with an ion.

The coefficients of bound-bound and bound-free absorption are proportional to the number N of absorbing atoms per unit volume of gas. The value of the coefficient per absorbing atom depends only on the internal properties of the atom, its degree of excitation, and the frequency of the photon; it is a characteristic of the atom itself. This quantity $\kappa_{\nu a}/N = \sigma_\nu$ has the dimensions of a length squared ($\kappa_{\nu a}^{-1}$ has the dimensions of length and N^{-1} of length cubed) and is called the *absorption cross section*. Its physical meaning can be easily understood from the following considerations. Consider a parallel beam of light rays of frequency ν and unit cross-sectional area traveling through an absorbing gas. Let us replace each atom by a small opaque disc placed perpendicular to the direction of the beam. We can now visualize absorption as a process of the capture of the photons which strike the discs. If the area of each disc is equal to σ_ν and the number of discs (atoms) per unit volume is N, then the total area of all the discs in a cylinder of gas of thickness dx and unit base area will be equal to $N\sigma_\nu\, dx$. We choose dx small enough that the discs in the gas layer do not overlap. Then, obviously, the fraction of photons "captured" during the passage of light through this layer is equal to the ratio of the opaque area $N\sigma_\nu\, dx$ to the total area; we have $dI_\nu = -I_\nu N\sigma_\nu\, dx$. Recalling the definition of the absorption coefficient (equation (2.6)) we see that $\kappa_\nu = N\sigma_\nu$, and that we can identify the cross section σ_ν as the area of the opaque (to frequency ν) disc which corresponds to one absorbing atom. In the same manner we can speak of the cross section of an atom or any other particle for the scattering of photons.

Bound-bound transitions are caused by photons with a strictly defined energy $h\nu$, lying within extremely narrow limits. This energy corresponds to the difference between two energy levels in the atom. This absorption is therefore termed selective absorption. The absorption cross sections of "isolated" atoms for these "selected" photons are extremely large. For visible light they are of the order of 10^{-9} cm^2 at the center of the spectral line (in the middle of a narrow band of selective absorption)*. These cross sections correspond to very short photon mean free paths. For example, for a density $N \sim 10^{19}$ cm^{-3} (the order of the density of atmospheric air) the mean free path of a photon would be of the order of $l = 1/\kappa = 1/N\sigma \sim 10^{-10}$ cm.

Cross sections for bound-free absorption, i.e., for the photoelectric effect, are much smaller, of the order of 10^{-17} to 10^{-20} cm^2 ($l \sim 10^{-2}$ to 10 cm for $N \sim 10^{19}$ cm^{-3}). These values apply, of course, only to those photons capable of removing an electron from the atom, photons whose energies are higher than the binding energy of the electron.

In the free-free transitions a photon can be absorbed only if the electron passes very close to the ion at the instant of absorption, i.e., it must "collide" with the ion (a free electron is not capable of absorbing a photon, it can only serve as a scattering center). Hence, the coefficient of bremsstrahlung absorption is proportional to the number of ions as well as to the number of free electrons contained per unit volume; $\kappa_{\text{brems}} \sim N_+N_e$. We may speak about the cross section of an ion $\sigma_{\text{brems}} = \kappa_{\text{brems}}/N_+$ (with $N_+ \sim N_e$) only in a restricted sense, since this cross section is proportional to the density of the free electrons. It turns out, however, that the bremsstrahlung absorption coefficient in the case of partial ionization is proportional only to the first power of the gas density, since the product N_+N_e is also proportional to the density. For the photons most frequently occurring at a given temperature, the coefficient of bremsstrahlung absorption is approximately an order of magnitude smaller than the coefficient of bound-free absorption. For complete ionization, however, when only nuclei and electrons are present in the gas (and bound-free absorption is totally absent), the coefficient of bremsstrahlung absorption is proportional to the square of the density.

The photons are scattered mainly by free electrons† (if the photon energy

* The absorption cross section at the center of a line with a certain natural line width is of the order of λ^2, where λ is the photon wavelength. On the wavelength scale, the natural line width in the visible part of the spectrum is of the order of 10^{-4} Å $= 10^{-12}$ cm (1 angstrom (Å) is equal to 10^{-8} cm). (*Editors' note.* Independent of the wavelength.) The width of the spectral lines in gases is usually greater than the natural line width, so that the cross section at the line center based on the natural line width is, correspondingly, smaller than λ^2. For a more detailed discussion see Chapter V, §9.

† We note the existence of resonant scattering in which the absorption of a photon by a bound electron is accompanied by its transition to a bound excited state with the subsequent emission of a photon in a random direction. The resonant scattering cross section at the center of a line, like the absorption cross section, is of the order of λ^2.

is large in comparison with the binding energy of an electron in the atom, then such a bound electron can also be considered as "free"). Photons of intermediate energy, that is, with energies much smaller than the rest mass energy of the electron $m_e c^2 = 500$ ev (these are the usual photon energies in the temperature range under consideration) are scattered without any energy change. The scattering cross section is defined here by the classical electron radius r_0 and is equal to $\sigma_s = \frac{8}{3}\pi r_0^2 = 6.65 \cdot 10^{-25}$ cm^2 (this is the so-called Thomson scattering cross section). This cross section is quite small and corresponds to a scattering mean free path $l_s \sim 10^5$ cm for an electron density $N_e \sim 10^{19}$ cm^{-3}. In estimating the scattering mean free path of high-energy photons, for which conditions all the electrons in the atoms and molecules can be considered as free, it is understood that N_e denotes the total number of electrons in the atoms. For example, in air at standard density $N_{molec} = 2.67 \cdot 10^{19}$ cm^{-3}, while the total number of electrons is approximately 14.4 times greater. The scattering mean free path is equal to 37 meters. It should be noted that the cross section for very high energy photons (energies measured in Mev) is different from the Thomson cross section.

The scattering mean free path of a photon belonging to a continuous spectrum in a partially ionized gas is always much longer than the corresponding absorption mean free path. The scattering becomes considerable only in a very rarefied and fully ionized gas, when the bremsstrahlung absorption, which is proportional to N^2, becomes small. The scattering of light under terrestrial conditions can always be neglected in comparison with absorption*. Therefore, in our subsequent development we shall drop the subscript "a" on the quantities κ_ν, l_ν, and it is to be understood that these quantities denote the absorption coefficient and the absorption mean free path, respectively.

This concludes our general survey of the mechanisms of interaction between radiation and matter. Explicit expressions for the absorption coefficients are not required now. Chapter V is devoted to a detailed consideration of these problems.

§3. Equilibrium radiation and the concept of a perfect black body

Let us imagine an infinite medium in thermodynamic equilibrium at a constant temperature T. Under steady state conditions the radiation field will also be in equilibrium. Thermal radiative equilibrium is characterized by the fact that the number of photons or amount of radiant energy emitted by the medium per unit time per unit volume in a given frequency interval $d\nu$ and in a given differential solid angle $d\Omega$ is exactly equal to the number of absorbed photons or to the radiant energy absorbed by the medium in the same intervals $d\nu$ and $d\Omega$. The equilibrium radiation field is isotropic, is

* Under conditions encountered in astrophysics, scattering is sometimes even greater than absorption.

independent of both the direction and specific properties of the medium; the field is a universal function of the frequency and temperature.

The spectral energy density function for radiative equilibrium $U_{\nu p}$ was derived by Planck early in the development of the quantum theory. The most natural way to obtain this function is to apply the laws of quantum statistics which govern the behavior of a "photon gas" (see, for example, [4]). The

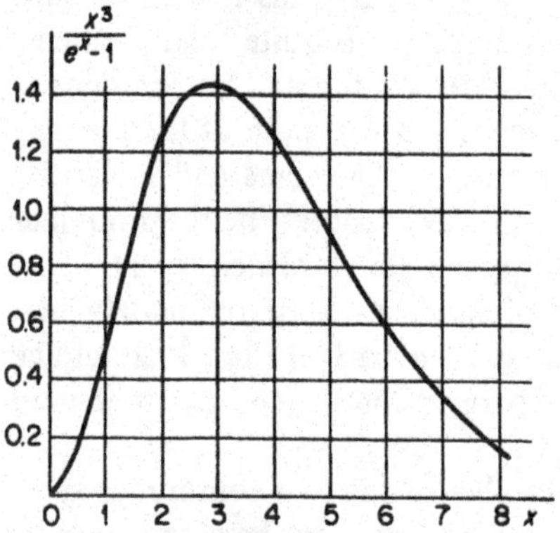

Fig. 2.3. The Planck function $x^3(e^x - 1)^{-1}$, where $x = h\nu/kT$.

amount of energy per unit volume radiated at equilibrium with a frequency ν per unit frequency interval is

$$U_{\nu p} = \frac{8\pi h\nu^3}{c^3} \frac{1}{e^{h\nu/kT} - 1}. \tag{2.10}$$

As a result of the isotropy, the spectral intensity for radiative equilibrium is*

$$I_{\nu p} = \frac{cU_{\nu p}}{4\pi} = \frac{2h\nu^3}{c^2} \frac{1}{e^{h\nu/kT} - 1}. \tag{2.11}$$

The energy distribution for radiative equilibrium as a function of frequency is given by the Planck function (2.10), which is plotted in Fig. 2.3. The maximum of this distribution occurs at a photon energy $h\nu_{max} = 2.822\ kT$. As the temperature increases the maximum is displaced toward the higher frequencies. In the low frequency region $h\nu \ll kT$, the Planck formula reduces to the classical Rayleigh–Jeans law

$$U_{\nu p} = \frac{8\pi kT}{c^3} \nu^2, \qquad h\nu \ll kT. \tag{2.12}$$

In the high frequency region $h\nu \gg kT$ we get Wien's displacement law

$$U_{\nu p} = \frac{8\pi h\nu^3}{c^3} e^{-h\nu/kT}, \qquad h\nu \gg kT. \tag{2.13}$$

* In astrophysics the symbol B_ν is ordinarily used in place of $I_{\nu p}$.

The integrated equilibrium radiant energy density is obtained by integrating the spectral energy density (2.10) over all frequencies from 0 to ∞. This yields the well-known expression

$$U_p = \int_0^\infty U_{\nu p}\, d\nu = \frac{4\sigma T^4}{c}, \tag{2.14}$$

where $\sigma = 2\pi^5 k^4/15h^3c^2 = 5.67 \cdot 10^{-5}$ erg/cm^2·sec·deg^4 is the Stefan–Boltzmann constant ($U_p = 7.57 \cdot 10^{-15}\, T^{\circ 4}$ erg/cm^3).

The fact that the integrated equilibrium radiant energy density is proportional to the fourth power of the temperature follows directly from the second law of thermodynamics and from the well-known result of classical electrodynamics that the pressure of an isotropic radiation field is equal to one-third of the energy density, i.e., $p_\nu = U_p/3$. Substituting this expression into the general thermodynamic relationship $T\, dS = d\varepsilon + p\, dV$*, where the internal energy is understood to denote the product of the radiant energy density and the volume ($\varepsilon = U_p V$), and noting that dS is a total differential, we obtain $U_p = const\; T^4$. We note that the relation $p_\nu = U_p/3$ is an evidence of the fact that equilibrium radiation can be considered from a thermodynamic point of view as a perfect gas with a specific heat ratio $\gamma = 4/3$.

Since an equilibrium radiation field is isotropic the net radiant energy flux at any point inside a body is zero. This means that if we visualize a plane surface drawn inside a body, then the one-sided radiant energy fluxes through the surface from right to left and from left to right are exactly equal in magnitude and opposite in direction. The one-sided flux itself, that is, the amount of radiant energy passing (for example) from left to right per unit time per unit area, can be obtained by substituting (2.11) for the equilibrium intensity into (2.2) and integrating over a hemisphere (rather than over the entire solid angle). The result is the one-sided spectral radiant energy flux

$$S_{\nu p} = \frac{cU_{\nu p}}{4} = \frac{2\pi h\nu^3}{c^2}\frac{1}{e^{h\nu/kT} - 1}. \tag{2.15}$$

The one-sided flux integrated over all frequencies is

$$S_p = \int_0^\infty S_{\nu p}\, d\nu = \frac{cU_p}{4} = \sigma T^4. \tag{2.16}$$

Let us imagine a body of material at a constant temperature T containing a cavity filled with equilibrium radiation. The radiant energy flux delivered by the cavity to a unit area of surface material per unit time is $S_{\nu p}$. This flux is, in general, partially reflected from the walls of the cavity and partially transmitted into the body and finally absorbed by the body (we assume that the

* Here S is the entropy of radiation.

body is of infinite extent and that the flux does not emerge from the body). Let us denote the reflectivity and absorptivity by R_ν and A_ν, respectively, with $A_\nu = 1 - R_\nu$. The radiation transmitted from the cavity into the body and absorbed by it is equal to $S_{\nu p} \cdot A_\nu$. As a consequence of equilibrium the same amount of radiation $J_\nu' = S_{\nu p} \cdot A_\nu$ is emitted by the body per unit time per unit area of surface. The absorptivity, the reflectivity, and the amount of radiation emitted by the surface are characteristics of the material and of its thermodynamic state. Nevertheless, the relationship

$$\frac{J_\nu'}{A_\nu} = S_{\nu p} = \frac{2\pi h\nu^3}{c^2} \frac{1}{e^{h\nu/kT} - 1} \tag{2.17}$$

is independent of the specific properties of the body and is a universal function of the frequency and temperature. This statement is called Kirchhoff's law.

A body which completely absorbs the entire radiation incident upon it is called a perfect black body. By definition, for a perfect black body $R_\nu = 0$ and $A_\nu = 1$. It follows from (2.17) that a spectral energy flux equal to $S_{\nu p}$ emanates from its surface and that the flux integrated over the spectrum is equal to $S_p = \sigma T^4$.

Let us consider an unbounded continuous medium at a constant temperature T in which the radiation is in equilibrium with the matter, and let us again divide the medium by an imaginary plane. One-sided energy fluxes through the surface are equal to $S_{\nu p}$. Photons passing through the surface from left to right are "born" to the left of the surface, while those traveling from right to left are "born" to the right of the surface. We imagine that the matter is removed from one side of the surface, let us say from the right side, and assume that the temperature of the matter to the left remains unchanged. In addition, let us assume that the index of refraction of the medium is equal to unity (the same as the empty space which is created to the right), and thus that the interface boundary does not reflect light. Then after the matter has been "removed" from the right side, there will be no photon flux from that side, while the photon flux from the left side will obviously remain the same and equal to $S_{\nu p}$. Therefore, a half-space filled with matter with an index of refraction equal to unity and at a constant temperature T emits from the surface a radiant energy flux equal to $S_{\nu p}$, that is, it radiates as a perfect black body at a temperature T.

§4. Induced emission

Let us consider the balance between the absorption and emission of light in material placed in a radiation field I_ν. The radiant energy in the frequency interval $d\nu$ and in the element of solid angle $d\mathbf{\Omega}$ absorbed per unit volume per unit time is

$$\kappa_\nu I_\nu \, d\nu \, d\mathbf{\Omega} = \textit{absorption per unit volume per unit time.} \tag{2.18}$$

The amount of energy spontaneously emitted per unit volume of the material per unit time in the same interval $d\nu \, d\mathbf{\Omega}$ is

$$j_\nu \, d\nu \, d\mathbf{\Omega} = \textit{spontaneous emission per unit volume per unit time.}$$

The magnitude of the spontaneous emission (the emission coefficient j_ν) is determined only by the properties and the state of the material, i.e., the type of atoms, the temperature (which affects the degree of excitation of the atoms), and so forth, and is absolutely independent of any radiation which may or may not be present in space. This, however, does not exhaust the total amount of radiation emitted by the material.

There also exists the so-called stimulated or induced emission. The probability of an induced emission of a photon with a given frequency and direction is proportional to the radiation intensity of the same frequency and direction present at the given point in space. The existing photons "facilitate" the transitions of excited atomic systems, which are accompanied by the emission of the same types of photons. The quantum theory shows that the total probability of emission of a photon is proportional to $1 + n$, where n is the number of photons with a definite direction of polarization and located in the same phase cell which admits the emitted photon. This number is equal to $n = c^2 I_\nu / 2h\nu^3$ *. Therefore the total radiation emitted per unit time per unit volume in the interval $d\nu \, d\mathbf{\Omega}$ is

$$j_\nu \, d\nu \, d\mathbf{\Omega} \left(1 + \frac{c^2}{2h\nu^3} I_\nu\right) = \textit{total emission per unit volume per unit time.} \tag{2.19}$$

The first term in parentheses corresponds to the spontaneous emission and the second to the induced emission†.

In a state of thermodynamic equilibrium the emission and absorption of photons of given frequencies and directions are exactly compensated by each

* The phase volume corresponding to an element $d\nu \, d\mathbf{\Omega} \, d\mathbf{r}$, containing $f \, d\nu \, d\mathbf{\Omega} \, d\mathbf{r}$ photons, is equal to $d\mathbf{p} \, d\mathbf{r}$, where $d\mathbf{p}$ is a volume element in the momentum space. Since the momentum of a photon is equal to $\mathbf{p} = h\nu\mathbf{\Omega}/c$, $d\mathbf{p} = p^2 \, dp \, d\mathbf{\Omega} = h^3\nu^2 \, d\nu \, d\mathbf{\Omega}/c^3$. The number of phase cells in an element of phase space $d\mathbf{p} \, d\mathbf{r}$ is equal to $d\mathbf{p} \, d\mathbf{r}/h^3$ and, consequently, the number of photons in one cell is $f \, d\nu \, d\mathbf{\Omega} \, d\mathbf{r} \, h^3/d\mathbf{p} \, d\mathbf{r} = c^3 f/\nu^2 = c^2 I_\nu / h\nu^3$. The number of photons with a specific polarization direction is equal to one half of this number, or $c^2 I_\nu / 2h\nu^3$.

† *Editors' note.* This rule for induced emission depends upon an assumption as to the independence of the emitting atoms in responding to the stimulating radiation. The rule may fail to apply in certain cases as, for example, that of a laser with a very high emission rate.

other, so that (2.18) and (2.19) should be set equal, and the radiation intensity I_ν should be replaced by the equilibrium value $I_{\nu p}$. Using (2.11) for the equilibrium intensity, we find that the ratio of the emission coefficient of any substance to its absorption coefficient is a universal function of frequency and temperature

$$\frac{j_\nu}{\kappa_\nu} = \frac{I_{\nu p}}{1 + (c^2/2h\nu^3)I_{\nu p}} = \frac{2h\nu^3}{c^2}\, e^{-h\nu/kT}. \tag{2.20}$$

This relation is one form of Kirchhoff's law. It is convenient to rewrite (2.20) as

$$j_\nu = I_{\nu p}\kappa_\nu(1 - e^{-h\nu/kT}). \tag{2.21}$$

The total emission coefficient in all directions is equal to

$$J_\nu = 4\pi j_\nu = cU_{\nu p}\kappa_\nu(1 - e^{-h\nu/kT}). \tag{2.22}$$

Kirchhoff's law expresses the general principle of detailed balancing applied to the emission and absorption of light. It allows us to calculate the emission coefficient of a substance if the absorption coefficient is known (and vice versa)*.

* *Editors' note.* The principle of detailed balancing underlies much of the physics presented in this text. It is important to understand the nature of this principle. As used by the authors and by most other authors the principle refers to a balance between a particular process and its reverse process in a system which is in thermodynamic equilibrium. When results of the principle (such as Kirchhoff's law above) are applied to a system which is not in strict thermodynamic equilibrium, it is essential that an appropriately defined thermodynamic quasi or restricted equilibrium apply, so that an appropriate temperature is definable.

A quasi equilibrium in a thermodynamic system appears when certain energy modes in the system equilibrate rapidly, where the interaction between these modes and other energy modes is very weak. The restricted system in which only these rapidly equilibrating modes are taken into account can be approximately in thermodynamic equilibrium and possess a definable temperature. This concept is essential in the theory of radiative transfer, and in the theory of relaxation processes treated in Chapter VI.

In the theory of radiative transfer treated in this chapter from §5 on, Kirchhoff's law is fundamental. The quantities j_ν, κ_ν, and $I_{\nu p}$ are functions of the temperature of the medium. With the radiation not in equilibrium, with $I_\nu \neq I_{\nu p}$, this temperature is definable only if the medium is in a quasi equilibrium; this restricted equilibrium must be maintained, for example, through collisions. Kirchhoff's law does not apply where such a quasi equilibrium does not hold, as with the lasers discussed in §4a.

The principle of detailed balancing rests upon the fundamental physical principle of microscopic reversibility, and some authors use the term for a microscopic statement of balance which does not involve the concept of equilibrium (cf. Landau and Lifshitz [15]). Some results classically derived using the principle of detailed balancing, in particular the Einstein relation of Chapter V, (5.71), rest directly upon microscopic reversibility and do not involve the concepts of equilibrium or temperature. For a discussion of the microscopic basis of detailed balancing see Reif [16].

The existence of induced emission (atomic transitions whose probability depends on the number of "particles"—photons—already present in the final state of the atom-plus-photon system) is characteristic of processes in which "particles" (the photons) obeying Bose quantum statistics take part. Indeed, it is due to the presence of these processes that the distribution function for the photon gas differs from the distribution function for a gas governed by classical Boltzmann statistics. With Boltzmann statistics the number of particles with energy ε is proportional to $e^{-\varepsilon/kT}$ and not to $(e^{\varepsilon/kT} - 1)^{-1}$, as for the photons where $\varepsilon = h\nu$. In order to clarify this statement let us consider a simple case where the atom has only two energy levels ε_1 and ε_2 ($\varepsilon_2 > \varepsilon_1$), and where the transition from the upper to the lower energy level is accompanied by the emission of a photon $h\nu = \varepsilon_2 - \varepsilon_1$ while the transition from the lower to the upper level is accompanied by an absorption of a photon $h\nu$. The probability of absorption κ_ν is proportional to the number of atoms in the lower energy state which, according to Boltzmann statistics, is proportional to $\exp(-\varepsilon_1/kT)$. The probability of spontaneous emission j_ν is proportional to the number of atoms in the upper energy state, that is, to $\exp(-\varepsilon_2/kT)$. Let us assume that there is no induced emission. Then, at equilibrium the total number of emitted photons will be equal to the total number of absorbed photons, that is, relation (2.20) or (2.21) can be replaced by

$$\frac{j_\nu}{\kappa_\nu} = I_{\nu p}, \qquad j_\nu = I_{\nu p}\kappa_\nu, \tag{2.23}$$

but $j_\nu \sim \exp(-\varepsilon_2/kT)$, $\kappa_\nu \sim \exp(-\varepsilon_1/kT)$, so that

$$\frac{j_\nu}{\kappa_\nu} = I_{\nu p} = const \cdot e^{-\frac{\varepsilon_2 - \varepsilon_1}{kT}} = const \cdot e^{-\frac{h\nu}{kT}}.$$

Thus we have obtained Boltzmann's law for the equilibrium radiation intensity, or in other words, the photon distribution function turns out to be the same as for "ordinary" particles. Actually, Boltzmann's law applies only to the high-energy photons $h\nu \gg kT$ in the Wien region.

The correct description of the equilibrium between the emission and absorption, leading to the Planck distribution function, is obtained only by taking into account the phenomenon of induced emission. In our example of an atomic system with two energy levels we obtain

$$\frac{j_\nu}{\kappa_\nu} = \frac{I_{\nu p}}{1 + (c^2 I_{\nu p}/2h\nu^3)} = const \cdot e^{-\frac{\varepsilon_2 - \varepsilon_1}{kT}} = const \cdot e^{-\frac{h\nu}{kT}},$$

which gives us the Planck equation for the intensity $I_{\nu p}$ (with the *const* above equal to $2h\nu^3/c^2$).

The above discussion shows that the induced emission becomes negligible in comparison with the spontaneous emission under equilibrium conditions as $h\nu/kT \to \infty$, that is, in the Wien region of the spectrum. This is directly evident from (2.19), if we note that at equilibrium and in the limit as $h\nu/kT \to \infty$,

$$I_\nu = I_{\nu p} \sim e^{-h\nu/kT} \to 0.$$

On the other hand, in the Rayleigh–Jeans region of the spectrum, where $h\nu \ll kT$, the relative role of the induced emission becomes dominant; in (2.19) we find

$$1 + \frac{c^2}{2h\nu^3} I_{\nu p} = 1 + \frac{1}{e^{h\nu/kT} - 1} \approx \frac{kT}{h\nu} + \frac{1}{2},$$

so that the ratio of probabilities of induced and spontaneous emissions is equal to $kT/h\nu \gg 1$.

It should be noted that in the case when the radiation field is not in equilibrium, the above considerations about the relative roles of spontaneous and induced emissions are generally invalid. The induced emission is proportional to the actual radiation intensity, which in the absence of equilibrium can be arbitrary.

§4a. Induced emission of radiation in the classical and quantum theories and the laser effect

In recent years there has been a great deal of attention paid to the phenomenon of induced emission of radiation, because of the fact that it serves as the basis of laser and maser operation. In order to explain this phenomenon in physical terms we shall first describe it from the classical point of view. As is well known, in the classical model of a radiating atom the atom is represented with an elastically bound electron, that is, as a harmonic oscillator. Let the electric field of a light wave act on this oscillator as an inducing force, with the wave frequency the same as the natural frequency of the oscillator. If the oscillator was initially at rest at $t = 0$, then under the action of the field it will go into resonant oscillations, with the amplitude increasing as t and the energy as t^2. However, if initially the oscillator already possessed a certain amount of energy, then the force which acts with resonance frequency can excite the oscillator even more or, conversely, can damp the oscillations so that the oscillator will lose energy. Which process occurs depends upon the relationships between the phases of the oscillator and the alternating force. Let us emphasize that in this case the inducing force must be in proper resonance with the oscillator frequency, not only in order to reinforce but also in order to damp the vibrations. This classical concept of resonance energy transfer serves as the basis for the concept of induced emission of radiation.

An oscillator which possesses energy and which is in the required phase relationship with the light wave in which it is situated will give up its energy to reinforce the passing light wave. Such an oscillator will increase the energy of a coherent wave. In classical language we can say that E_s, the electric field of the oscillator, is added to the wave field E_0. The oscillator field has a suitable angular distribution. But the energy flux is proportional to the square of the field. For this reason the energy flux of the oscillator in the direction of the passing wave is proportional to $E_s E_0$ and is greater if E_0 is greater; this corresponds to the fact that the intensity of the induced radiation increases with an increase in the intensity of the wave which produces it.

However, this classical picture has a flaw which leads to an incorrect form of the equations. In the classical theory a combination of oscillators with arbitrary phases always absorbs more energy on the average than it emits by induced radiation. In order to obtain correct results this problem must be considered quantum-mechanically. We turn to the quantum treatment and its effect on the harmonic oscillator. It is of extreme importance here that the energy levels of the harmonic oscillator are equidistant from one another (the interval is $h\nu$, where ν is the natural frequency). An oscillator situated at the nth level can, under the action of the resonance force, move to either the $(n+1)$th or the $(n-1)$th level. In this case the transition to the upper $(n+1)$th state, with absorption of energy, is more probable than the transition to the lower $(n-1)$th state, with emission of energy. As is known from quantum mechanics, the ratio between the probabilities of these transitions is $(n+1)/n$. This is precisely the fact which shows that a combination of harmonic oscillators absorbs (rather than emits) light on the average.

The decisive prerequisite for obtaining the laser effect, for the predominance of induced emission of radiation, is that the oscillator be anharmonic, and that consequently the energy intervals between neighboring levels should be unequal. If the levels are situated at unequal intervals, then we can have a frequency which is resonant for the transition $n \to n-1$, but is not resonant for the transition $n \to n+1$. Then it is clear that an oscillator in the nth state will only give up energy under the action of light with frequency ν. This is precisely the situation when, with the levels inversely populated (the nth level is full, while the $(n-1)$th level is not), energy is given up and the wave goes into negative absorption; these are the conditions for laser generation*.

* It is of interest that we encounter the effect of anharmonicity in the case of the radiation from electrons in a magnetic field. In the nonrelativistic approximation the electron gyrates with constant angular frequency. A result of this is that the quantum levels of the energy of transverse electron motion in the magnetic field are equidistant. No such distribution of energy in a magnetic field in this approximation can produce a negative absorption coefficient or generate coherent light. Taking into account relativistic corrections, it may be shown that the levels are actually not equidistant and thus at the same time that it is possible to have electron energy distributions which can produce laser generation.

A population consisting of N atoms with two levels each (and which do not interact with one another, such as the chromium atoms in the lattice of ruby) can also be regarded as a single system with equidistant levels; its energy is $E_n = nh\nu$, where n is the number of excited atoms. Unlike an oscillator, where the spectrum of energy states is unbounded above (and the statistical weight of all the states is the same), here the spectrum is bounded not only below ($n = 0$, $E_n = 0$), but also above ($E_{max} = Nh\nu$). The statistical weight for different n is different, and the maximum statistical weight is obtained at $n = N/2$. This is the reason that in this system absorption predominates at $n < N/2$, while induced emission predominates at $n > N/2$*.

As we stated in the beginning of this section, the concept of induced emission of radiation can be clearly expressed in classical terms. As expected, induced radiation is completely contained in Newton's and Maxwell's equations; it is not without reason that the Rayleigh–Jeans formula (2.12) for the spectral density in the low-frequency range does not contain the Planck constant—after all, induced emission must be taken into account in the quantum-mechanical derivation of the Rayleigh–Jeans formula.

After the discovery of lasers, which have directed the attention of physicists to the phenomenon of induced emission, it has been shown that a large number of processes are most conveniently described in terms of absorption and induced emission. As a remarkable example we can quote the phenomenon of electron scattering from a standing light wave. This phenomenon was predicted by Kapitsa and Dirac [11] more than 30 years ago. A monochromatic light wave, after being reflected by a mirror, creates in space periodically situated regions in which the electric field is small (nodes) and in which it is large (antinodes). The electrons should be scattered at specific "Bragg" angles in the same manner in which electrons are scattered by the periodic field of a crystal lattice. This effect was observed [12] only very recently by using a powerful monochromatic light pulse produced by a laser.

The scattering of electrons by the field of a standing wave may be described quantitatively in new terms as an "induced Compton effect". In fact, the Compton effect is a process of interaction of a γ-photon with an electron

$$\gamma + e = \gamma' + e'.$$

For a given energy and momentum of a photon and an electron before scattering the conservation laws allow any direction (in the center-of-inertia system) for the photon γ' and electron e' after scattering. Thus, under ordinary conditions, the Compton effect produces a statistical distribution in direction of the electrons. Let us return to electron scattering from a standing wave.

* We point out that in [9] the laser effect is considered on the basis of semiclassical concepts, and that in [10] the quantum-mechanical and semiclassical treatments are compared.

A standing wave represents a superposition of an incident wave γ_1 and of a reflected wave γ_2. This means that the Compton scattering can take place as

$$\gamma_1 + e = e' + \gamma_3, \qquad \gamma_2 + e = e'' + \gamma_4.$$

But when the light flux is very powerful, then the law of induced emission predicts the likelihood of the emission of a photon which is the same as the photons already existing, that is, the likelihood of the processes

$$\gamma_1 + e = e_{12} + \gamma_2, \qquad \gamma_2 + e = e_{21} + \gamma_1,$$

which it is natural to term the "induced Compton effect". Since the direction and frequency of both photons (hence their energy and momentum) both before and after scattering have been specified, then the change in the electron direction is fully defined. The electron energy evidently does not change. Moreover, it follows from the conservation laws that this process is possible only when the incident electron has a particular momentum. All the relationships governing diffraction scattering are thus very explicitly obtained from the induced Compton effect concept. A very important general principle is realized in this example, that any action exerted on a system by an external periodic force must, in quantum-mechanical terms, be considered as a combination of absorption and induced emission of appropriate energy quanta by the system. This principle also pertains to problems in which the external force can be regarded as being classical, and implies that we can in this case ignore the reaction of the system on the force, as in the example presented above, the action of an electron on the field of a standing wave. However, even if we consider this force from the classical point of view, its action on the system of photons should, or can*, be considered as absorption and emission of photons. The consideration of a force as classical implies that the induced emission of photons with a frequency determined by the external force is greater by far than the spontaneous emission of any other photons which differ in their frequency, direction, or polarity.

When a system is subjected to a periodic external action, its Hamiltonian can no longer be considered as being independent of time. However, the Hamiltonian is then periodic, $H(t) = H(t + nT)$, which means that there should exist solutions which, after a single period of the acting force, revert to their previous form multiplied by a phase factor

$$\psi_k(t + T) = a_k \psi(t), \qquad |a_k| = 1,$$

$$a_k = e^{i\alpha_k}.$$

Let us introduce the concept of quasi-energy ε_k, where

$$\alpha_k = 2\pi\varepsilon_k T/h, \qquad \psi_k(t + nT) = \psi_k(t)\exp(2\pi i\varepsilon_k nT/h).$$

* We note the paper by Keldysh [13], in which use is made of exact wave functions of an electron in the field of a classical light wave.

This entity is related to the energy in the same way as is the quasi-momentum of an electron in a space-periodic lattice to the proper electron momentum. In this way it is possible to develop an ordered theory of systems subjected to the action of periodic external forces. According to what we have seen this will, at the same time, be the theory of the phenomena of absorption and induced emission.

A second example of the application of the induced emission concept is the theory of the harmonic oscillator. It is well known that the width of a spectral line radiated on transition from state A to state B is the sum of the widths Γ_A and Γ_B of these states. As applied to an oscillator for which the matrix element of the transition from the nth to the $(n-1)$th state is proportional to $n^{1/2}$, we will find that $\Gamma_n = kn$ and $\Gamma_{n-1} = k(n-1)$. It follows from here that the width of the line emitted on transition from the nth to the $(n-1)$th state is proportional to $(2n-1)$. On the other hand, in the classical theory the width of a line emitted by a harmonic oscillator is independent of its amplitude. The radiation intensity is proportional to the square of the amplitude*; consequently, the oscillator energy decreases exponentially $E \sim e^{-\gamma t}$, and the amplitude of its oscillations also decreases exponentially $x = a_0 e^{-\gamma t/2} \cos 2\pi\nu t$. The expansion of this expression in a Fourier integral gives the Lorentz form for the line (see Chapter V)

$$|b(\nu)|^2 = \frac{1}{(\nu - \nu_0)^2 + (\gamma_0/4\pi)^2},$$

with a width which is independent of the amplitude. The paradox of this disagreement between the classical and quantum-mechanical theories for high quantum numbers was noted already by Weisskopf and Wigner [14], and also by Pauli (cited in [14]) at the beginning of the thirties. Reference [14] shows that this paradox is related to a specific property of the harmonic oscillator, the fact that its levels are equidistant, so that the frequencies emitted by transitions between any pair of neighboring levels are strictly identical. Reference [14] considers three states A, B, and C, with cascade emission of two photons $A^{h\nu_1} \rightarrow B^{h\nu_2} \rightarrow C$. If the levels are not equidistant, then $E_A - E_B \neq E_B - E_C$, $\nu_1 \neq \nu_2$, and we get the standard answer for the line ν_1—its width is proportional to $\Gamma_A + \Gamma_B$.

However, if the levels are equidistant (for example, if C is the ground or zero level, B is the first level, and A is the second level of a harmonic oscillator) the situation becomes more complicated, because the emission of two photons of the same frequency can take place also through a second process, $A^{h\nu_2} \rightarrow B^{h\nu_1} \rightarrow C$. To calculate the probability of the entire process of the emission of

* Consequently, the time needed for the emission of one energy quantum is inversely proportional to the energy of the oscillator itself; this conclusion, based on the calculation of the value of the matrix element giving the probability of emission, is beyond doubt.

two photons we must add the amplitudes along both processes, with a resultant change in the width and shape of the line. Using current terminology we may say, without referring to the numbers (1st, 2nd) of the photons, that a photon emitted by the transition $A \rightarrow B$ produces induced emission from the transition $B \rightarrow C$ *. The harmonic oscillator in a state of large quantum number radiates a cascade of photons; in this case the role of induced emission, in comparison with spontaneous emission, is greater the higher is the average number of the state n. The reference here is to emission induced not by external radiation but by the radiation from the oscillator itself. Only by considering the entire cascade process including induced emission will we obtain results which are in agreement with the classical theory.

At any given time the state of the oscillator consists of a superposition of many excited levels. It is also of importance that with induced transitions specific phase relationships between successive levels are obtained, and the state of the system cannot be specified by the same probabilities for the different levels. This situation is also natural: In the classical picture the electron is localized. It is obvious from the uncertainty principle that the description of a localized electron requires taking the superposition of several eigenfunctions (their number being greater the more precise the localization) which must have specific phase relationships. We may ask if the reasoning given applies only to the idealized case of a strictly harmonic oscillator, for which the energy states are equidistant, or also to any anharmonic oscillator? What is the criterion for the application of customary concepts about discrete quantum jumps? Discrete levels themselves may be considered only to the extent that the width of the level Γ_n connected with spontaneous emission is small compared with the distance between the levels, i.e., with the transition frequency,

$$\Gamma_n \ll \nu_{n,n-1} = \frac{E_n - E_{n-1}}{h}.$$

The criterion for the applicability of the customary quantum concepts of the line width (without considering induced emission) has a different form, and is that the width must be small compared with the difference of frequencies

$$\Gamma_n \ll \nu_{n,n-1} - \nu_{n-1,n-2} = \frac{E_n - 2E_{n-1} + E_{n-2}}{h}.$$

It would be very interesting to observe experimentally the effects of "internal" induced emission at high vibrational levels of molecules.

* This point of view is also substantiated by the fact that, according to [14], the spectrum changes when both photons are emitted in the same direction.

The examples presented above show that the concept of induced emission of radiation makes it possible to interpret anew many facts and paradoxes; this concept is an integral part of the interpretation of quantum mechanics.

§5. The radiative transfer equation

Let us set up the kinetic equation for the distribution function of photons of a given frequency. Since this function, to within the constant factor $h\nu c$, is identical with the radiation intensity, we can write down the equation for the intensity directly. The kinetic equation written in this form is usually called the radiative transfer equation.

We shall be interested in radiation of frequency ν in a unit frequency interval propagated within a unit solid angle in the direction $\mathbf{\Omega}$. Let us consider the

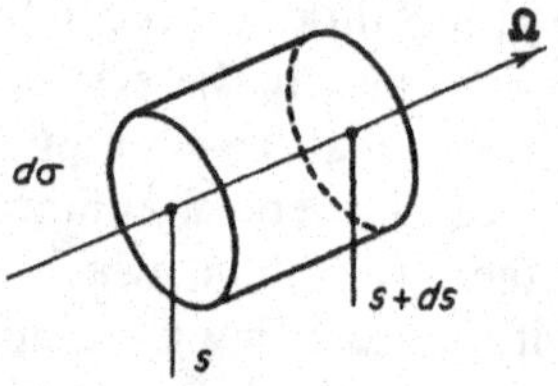

Fig. 2.4. Illustration for the derivation of the radiative transfer equation.

balance of radiation in an elementary cylinder of base area $d\sigma$ and height ds, located at a point in space in such a way that the direction $\mathbf{\Omega}$ coincides with the axis of the cylinder and is therefore perpendicular to its bases (Fig. 2.4). An amount of radiation $I_\nu(\mathbf{\Omega}, \mathbf{r}, t)\, d\sigma\, dt$ flows into the left base during the time dt. An amount of radiation $(I_\nu + dI_\nu)\, d\sigma\, dt$ flows out from the right base during the same time interval dt. The intensity I_ν is a function of both position and time. The change in intensity of the light beam as a result of its passage from the left base to the right one is composed of the local change during the time of passage through the distance ds plus the increment arising from the change in position coordinate from s to $s + ds$ at fixed time,

$$dI_\nu = \frac{\partial I_\nu}{\partial t}\frac{ds}{c} + \frac{\partial I_\nu}{\partial s}\, ds.$$

But within the cylinder the intensity of the beam must change as a result of the emission and absorption of the light of the specified characteristics. (We recall here the remark made at the end of §2 about disregarding scattering of light.) The amount of radiation emitted in the cylinder during the time dt is, according to (2.19),

$$j_\nu\left(1 + \frac{c^2}{2h\nu^3} I_\nu\right) d\sigma\, ds\, dt.$$

On the other hand, the radiation absorbed during the same time interval is $\kappa_\nu I_\nu \, d\sigma \, ds \, dt$. Setting up a balance for the cylinder and dividing the resulting expression through by the product of the differentials $d\sigma \, ds \, dt$, we get

$$\frac{1}{c}\left(\frac{\partial I_\nu}{\partial t} + c\mathbf{\Omega} \cdot \nabla I_\nu\right) = j_\nu\left(1 + \frac{c^2}{2h\nu^3} I_\nu\right) - \kappa_\nu I_\nu . \tag{2.24}$$

We have here replaced on the left-hand side the directional partial derivative $\partial I_\nu/\partial s$ by the equivalent vector expression $\mathbf{\Omega} \cdot \nabla I_\nu$. The term in parentheses on the left-hand side of (2.24) simply represents the total derivative of the intensity with respect to time, i.e., the time derivative of the intensity of a given photon packet (cf. the hydrodynamic equation of motion (1.16)).

Let us rearrange the right-hand side of (2.24) by collecting together the absorption and induced emission terms, since they are both proportional to an unknown function of position and time—the radiation intensity. Let us also express the emission coefficient j_ν in the induced emission term in terms of the absorption coefficient as defined by (2.21) and substitute into the latter the equilibrium intensity from (2.11). The right-hand side of (2.24) then becomes

$$j_\nu - \kappa_\nu(1 - e^{-h\nu/kT})I_\nu . \tag{2.25}$$

It now becomes evident that the induced emission can be treated as a decrease in the absorption: some of the photons act as if they were absorbed and immediately re-emitted again with the same frequency and in the same direction; the probability of this "re-radiation" is equal to $e^{-h\nu/kT}$. Physically such "re-radiation" does not occur and it can be excluded from our considerations by considering the absorption coefficient to have the slightly lower value

$$\kappa_\nu' = \kappa_\nu(1 - e^{-h\nu/kT}). \tag{2.26}$$

The interaction between radiation and matter can therefore be represented as if there existed only spontaneous emission and an effective absorption described by the coefficient κ_ν', which has been corrected for the induced emission in the system. Using this modified representation, Kirchhoff's law (2.21) becomes

$$j_\nu = \kappa_\nu' I_{\nu p}, \qquad \kappa_\nu' = \kappa_\nu(1 - e^{-h\nu/kT}). \tag{2.27}$$

Introducing this expression into the right-hand side of the transfer equation (2.24), we obtain the final form

$$\frac{1}{c}\frac{\partial I_\nu}{\partial t} + \mathbf{\Omega} \cdot \nabla I_\nu = \kappa_\nu'(I_{\nu p} - I_\nu). \tag{2.28}$$

Integrating equation (2.28) with respect to all directions $\mathbf{\Omega}$ (over the entire solid angle) and recalling the definitions of radiation density and flux (2.1) and (2.2), respectively, we obtain

$$\frac{\partial U_\nu}{\partial t} + \nabla \cdot \mathbf{S}_\nu = c\kappa_\nu'(U_{\nu p} - U_\nu). \tag{2.29}$$

This equation can be considered as the equation of continuity for radiation of a given frequency. It is an expression of the law of conservation of radiant energy which is completely analogous to the hydrodynamic energy equation written in the "divergence" form (1.10).

The radiative transfer equation (2.28) is a partial differential equation for the intensity as a function of position, time, and direction $I_\nu(\mathbf{r}, t, \mathbf{\Omega})$, and describes a nonequilibrium radiation field. Thermodynamic equilibrium in matter is usually established very rapidly, and it is therefore possible to consider the material to be in a state of thermodynamic equilibrium at each point of space and at each instant of time. The state of the material can then be described by two parameters, such as, temperature and density. The radiative transfer equation includes quantities dependent on the nature and state of the material, in particular the absorption coefficient κ_ν', which depends on the nature and state of the material, and the equilibrium intensity $I_{\nu p}$, which is a function of the temperature only.

Equation (2.28) also describes, in particular, the history of establishment of equilibrium between the radiation and the matter. Let us imagine an infinite constant density medium. The medium is initially cold, so that radiation is absent. Assume that at the initial time $t = 0$ the matter is "instantaneously" heated and then maintained at a constant temperature T. Let us now examine how the radiation intensity changes with time. Obviously, the spatial gradients in this case are all equal to zero, $\kappa_\nu' = const$ and $I_{\nu p} = const$. The solution of equation (2.28) in this case takes the form

$$I_\nu(t) = I_{\nu p}(1 - e^{-c\kappa_\nu' t}). \tag{2.30}$$

The radiation intensity asymptotically approaches the equilibrium intensity, and the relaxation time required for establishing equilibrium between the radiation and the material is $t_p = 1/c\kappa_\nu' = l_\nu'/c = l_\nu/(1 - e^{-h\nu/kT})c$. For example, with $l_\nu = 1$ cm, in the maximum part of the Planck spectrum $h\nu = 2.8\ kT$ we would have $t_p = 3 \cdot 10^{-11}$ sec.

§6. Integral expressions for the radiation intensity

Let us now obtain a formal solution of the radiative transfer equation by considering those quantities which depend only upon the state of the material $I_{\nu p}(T)$, $\kappa_\nu'(T, \rho)$ to be known functions of position and time. For simplicity, we consider first the steady state case, where the temperature and density distributions and also the radiation field are independent of time. We shall be concerned with the radiation at a point **r** of a body of matter or fluid, propagating in the direction **Ω** (Fig. 2.5). We draw a ray through the given point in the given direction. The coordinate along the ray we denote as s. Noting that the differential expression on the left-hand side of the transfer equation (2.28) represents the total derivative of the intensity of a photon packet along the direction of propagation, we can rewrite this equation as

$$\frac{dI_\nu}{ds} + \kappa'_\nu I_\nu = \kappa'_\nu I_{\nu p}. \tag{2.31}$$

This equation may be considered as an ordinary linear equation for the intensity along the direction of propagation. Its solution is

$$I_\nu(s) = \int_{s_0}^{s} \kappa'_\nu I_{\nu p} \exp\left[-\int_{s'}^{s} \kappa'_\nu\, ds''\right] ds' + I_{\nu 0} \exp\left[-\int_{s_0}^{s} \kappa'_\nu\, ds''\right]. \tag{2.32}$$

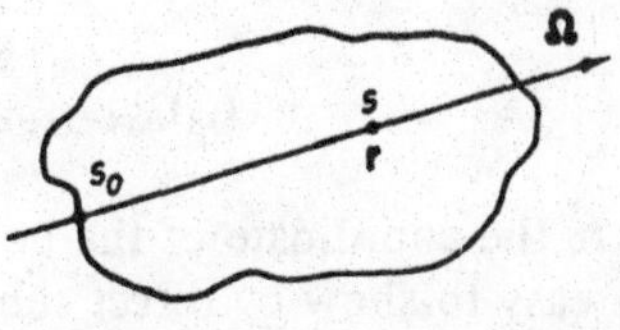

Fig. 2.5. Diagram showing the limits of integration in (2.32).

Here $I_\nu(s)$ is the intensity $I_\nu(\mathbf{r}, \mathbf{\Omega})$, considered as a function of the coordinate s along the ray. The integration along the ray is in general carried out from "$-\infty$", although actually it is from the boundary of the body of matter s_0 (as shown in Fig. 2.5). The constant of integration is denoted by $I_{\nu 0}$.

Let us clarify the physical meaning of the above solution. Radiation passing per unit time through a unit cross-sectional area at the point s (per unit solid angle) is made up of all photons "born" in a cylinder of unit cross section along the ray. An amount of radiation $j_\nu\, ds' = \kappa'_\nu I_{\nu p}\, ds'$ is "born" at the point s' on the ray segment ds' and is propagated along the ray **Ω** within a unit solid angle. Only the fraction $\exp[-\int_{s'}^{s} \kappa'_\nu\, ds'']$ of this radiation arrives from the point s' to the point s while the rest is absorbed along the way. The total intensity is made up of photons "born" in all elementary segments ds', that is, it is equal to the integral along the ray. If the radiating body of matter has finite dimensions, then the integration must be carried out from the boundary of the body s_0 to the point s. In this manner we obtain the first term in (2.32). The second term represents the radiation that entered the body at the boundary s_0 from external sources. The constant of integration $I_{\nu 0}$ is the intensity of this external radiation entering the body. The factor $\exp[-\int_{s_0}^{s} \kappa'_\nu\, ds'']$ accounts for the decrease in intensity along the interval from s_0 to s caused by absorption. The absorption coefficient κ'_ν and the equilibrium intensity $I_{\nu p}$ are functions of position along the ray, because of their dependence on temperature and density which are distributed in some manner along the ray. If these functions are known, then the determination of the intensity at any point of the body of matter reduces, as is easily seen from (2.32), simply to a quadrature, to an integration along the ray.

Let us generalize the solution (2.32) to the unsteady case, where the temperature and density and, consequently $I_{\nu p}$, κ'_ν as well as the intensity I_ν are functions of time. It is obvious that at the time t and the point s photons arrive from the point s' which were born at an earlier time $t - (s - s')/c$. As before, the photons are absorbed by the material at a point s'' along the ray. This absorption is determined by the magnitude of the absorption coefficient at the time $t - (s - s'')/c$. It follows that the unsteady solution of the transfer equation can be written in the form

$$I_\nu(s, t) = \int_{s_0}^{s} (\kappa'_\nu I_{\nu p})_{s',\, t-(s-s')/c} \exp\left[- \int_{s'}^{s} (\kappa'_\nu)_{s'',\, t-(s-s'')/c}\, ds'' \right] ds'$$
$$+ (I_{\nu 0})_{s_0,\, t-(s-s_0)/c} \exp\left[- \int_{s_0}^{s} (\kappa'_\nu)_{s'',\, t-(s-s'')/c}\, ds'' \right], \qquad (2.33)$$

where the coordinate of the boundary s_0 is taken at the time $t - (s - s_0)/c$. It is easy to show by direct substitution that the solution (2.33) satisfies the unsteady transfer equation. It is evident from (2.32) or (2.33) that the contribution of the far sources in a highly absorptive medium to the total intensity at a given point decreases exponentially with distance. Photons arriving at point s are "born" in the immediate vicinity at distances not exceeding several radiation mean free paths or, more precisely, at distances not exceeding several optical thicknesses. This situation is particularly easy to see in the case when the absorption coefficient is constant along the ray. Then the exponential factors become

$$\exp\left[- \int_{s'}^{s} \kappa'_\nu\, ds'' \right] = \exp\left[-\kappa'_\nu(s - s')\right] = \exp\left[- \frac{s - s'}{l'_\nu} \right]; \qquad l'_\nu = \frac{1}{\kappa'_\nu}.$$

The case of sharp temperature gradients constitutes an exception. Here, the effect of the increase in the emission coefficient $j_\nu = \kappa'_\nu I_{\nu p}$ with distance from a point may be greater than the effect of the absorption over the same path. This, however, almost never occurs in practice, and the main contribution to the integrals in (2.32) and (2.33) is provided by the segment of the ray in the vicinity of the point considered. The length of this segment is of the order of several (two or three) radiation mean free paths. Light, however, traverses this distance within a very short time interval l'_ν/c which, as a rule, is much shorter than the characteristic times required to produce any significant change in the state (temperature and density) of the material. For example, for a mean free path $l'_\nu = 3$ cm this time is $l'_\nu/c \approx 10^{-10}$ sec, which is much shorter than the characteristic times encountered in ordinary flows. This is simply a consequence of the fact that the fluid velocity is ordinarily much less than the speed of light.

The fact noted above is quite important. It indicates that the radiation field in practically all cases can be regarded as quasi-steady at any instant of time. This condition corresponds to an instantaneous distribution of emission and absorption sources, that is, a distribution of temperature and density in the fluid. As a result, we can neglect the derivative of the intensity with respect to time in the radiative transfer equation and consider time as a parameter upon which the temperature and density (and therefore, $I_{\nu p}$ and κ'_ν) will depend. Henceforth we shall always use the simplified transfer equation

$$\mathbf{\Omega} \cdot \nabla I_\nu = \kappa'_\nu(I_{\nu p} - I_\nu) \tag{2.34}$$

or its solution in the form (2.32).

§7. Radiation from a plane layer

In general, radiative transfer and radiant heat exchange have an influence on the state of the fluid, on its motion, or on the steady state temperature distribution. This influence is caused by the fact that the fluid by emitting or absorbing light either loses or gains energy, so that it is either cooled or heated. In general the state of the fluid can be described by the hydrodynamic equations which, in the presence of radiant heat exchange, must be generalized to include the interaction between the radiation and the fluid. Since the radiative transfer is also a function of the state of the fluid, of the temperature, and of density, then a system of equations describing both the fluid and the radiation generally consists of a properly generalized set of hydrodynamic equations together with a generalized radiative transfer equation.

In many cases, however, the "reciprocal" effect of the radiation on the fluid is either not too great, or may be accounted for by some approximate method. For example, at sufficiently low temperatures the radiant heat exchange and the energy losses suffered by the fluid from radiation are unimportant. In this case, the state of the fluid is practically independent of the radiation and the problem of determining the radiation field and the state of the fluid may be treated separately. The state of the fluid is described, for example, by the hydrodynamic equations, while the radiation field at any time may then be found from the known temperature and density distributions and the known absorption coefficients.

As a rule, in this case we are not interested in determining the entire radiation field in the medium (since it does not affect the state of the medium), but we are interested rather in determining the radiation emitted from the surface of the body of fluid, that is, the luminosity of the heated body, the surface brightness, the radiation spectrum, the angular distribution of flux, and so forth. If the optical properties of the fluid are known, that is, the absorption coefficient κ_ν' * is given as a function of frequency and of the temperature and density distribution in the body of fluid, then the answer to all of these problems is given by the integral formula for the intensity (2.32).

Considering the radiation emitted from the surface of a body of fluid, we can, without loss of generality, measure the distance s along the ray from the surface into the body of fluid (changing the sign of s) and extend the integration along the ray to infinity,

$$I_\nu(\Omega) = \int_0^\infty I_{\nu p}[T(s)] \exp\left(-\int_0^s \kappa_\nu'\, ds'\right) \kappa_\nu'(s)\, ds. \tag{2.35}$$

If the body has finite dimensions then the coefficient of absorption outside its boundary is equal to zero and the corresponding segment of integration will vanish. If "external" radiation penetrates a body of finite dimensions from "behind" (through its "back" surface) then by extending the integration along the ray to infinity we will also include the effect of these "external" sources.

Let us consider some simple examples that are of practical interest. Let the body of matter occupy an infinite half-space $x > 0$ and be bounded by a plane surface. The temperature of the body is constant while the absorption coefficient can vary arbitrarily from point to point (but in such a manner that the optical thickness of the body $\int_0^\infty \kappa_\nu'\, dx$ remains infinite). The radiation intensity at the surface is, in this case, simply $I_{\nu p}(T)$, since

$$I_\nu(\Omega) = \int_0^\infty I_{\nu p} e^{-z}\, dz = I_{\nu p}; \qquad dz = \kappa_\nu'\, ds, \qquad z = \int_0^s \kappa_\nu'\, ds.$$

The body radiates as a perfect black body at the temperature T. The intensity I_ν is the amount of radiant energy passing per unit time through a unit solid angle and through a unit area perpendicular to the direction of the photon beam†. For a radiating black body this intensity is independent of angle. The radiant energy passing per unit time through a unit area of the surface per unit solid angle and at an angle ϑ to the normal (let us call this quantity the emittance of the body i_ν ‡) is

$$i_\nu = I_\nu(\vartheta) \cos\vartheta. \tag{2.36}$$

* We recall that we are considering here media with indices of refraction equal to unity, such as gases.

† The dimensions of I_ν are energy/cm$^2\cdot$sec$\cdot$sterad$\cdot$frequency = erg/cm$^2\cdot$sterad.

‡ This should not be confused with the emission coefficient of a medium j_ν or J_ν.

For a radiating black body

$$i_\nu = I_{\nu p} \cos \vartheta. \tag{2.37}$$

Let us consider the radiation of a plane layer of finite thickness d at a constant temperature T and with an absorption coefficient κ'_ν. The radiation intensity at the surface in a direction which makes an angle ϑ with the normal (Fig. 2.6) is

$$I_\nu(\vartheta) = \int_0^{d/\cos\vartheta} I_{\nu p} e^{-\frac{\kappa'_\nu x}{\cos\vartheta}} \kappa'_\nu \frac{dx}{\cos\vartheta} = \int_0^{\kappa'_\nu d/\cos\vartheta} I_{\nu p} e^{-\frac{\tau'_\nu}{\cos\vartheta}} \frac{d\tau'_\nu}{\cos\vartheta}$$
$$= I_{\nu p}\left(1 - e^{-\frac{\kappa'_\nu d}{\cos\vartheta}}\right) = I_{\nu p}\left(1 - e^{-\frac{\tau_\nu}{\cos\vartheta}}\right), \tag{2.38}$$

where $\tau_\nu = \int_0^d \kappa'_\nu \, dx$ is the optical thickness of the layer in the direction normal to the surface.

Equation (2.38) shows that the radiation intensity for a layer of finite thickness is always lower than the equilibrium intensity. The spectrum differs from the Planck spectrum $I_{\nu p}(T)$ by the factor $[1 - \exp(-\tau_\nu/\cos\vartheta)]$. This factor is a function of the frequency because of the frequency dependence of the absorption coefficient and it approaches unity only as $d \to \infty$. The difference between this intensity and the Planck intensity is strongest in the direction normal to the surface, where the ray segment containing the sources has a minimum length (equal to d). The intensity spectrum approaches the Planck spectrum at large angles to the normal, when $\vartheta \to \pi/2$ and $\cos\vartheta \to 0$.

The greatest difference between this spectrum and the Planck spectrum as a function of the layer thickness d is observed in the limit of an optically thin layer, at such angles that $\kappa'_\nu d/\cos\vartheta \ll 1$. Expanding the exponential factor we find that to second order

$$I_\nu = I_{\nu p} \frac{\kappa'_\nu d}{\cos\vartheta} \ll I_{\nu p}. \tag{2.39}$$

The intensity at the surface is proportional to $1/\cos\vartheta$, and the emittance of the layer is independent of the angle

$$i_\nu = I_\nu \cos\vartheta = I_{\nu p}\kappa'_\nu d \qquad \text{for} \quad \cos\vartheta \gg \tau_\nu. \tag{2.40}$$

It should be noted that the applicability of the concept of the optical thickness of a layer depends upon the angle; that is, it is always possible to find large enough angles ($\vartheta \approx \pi/2$, $\cos\vartheta \ll 1$), for which the layer will be "optically thick". Thus at an angle $\vartheta \approx \pi/2$, even a layer with $\tau_\nu \ll 1$ will radiate as a black body. For small angles, where $\tau_\nu/\cos\vartheta \ll 1$ and the layer is optically thin, the body will radiate as a volume radiator, and photons "born" at any point emerge from the layer without any absorption in transit. There is no self-absorption in the layer and each volume element contributes equally to the total radiation emanating from the surface. It is for this reason that we use the term "volume radiator". An optically thick body radiates "from the surface", because the photons born deep within the body are absorbed during transit and are not emitted.

In many cases we are not interested in the radiation intensity at a given angle but in the radiant energy flux from the surface of the body, that is, in the amount of energy passing per unit time per unit area of surface in all directions. This quantity is called the surface brightness (spectral or integrated). The spectral surface brightness is, obviously, equal to

$$S_\nu = \int_{\text{over a hemisphere}} \cos\vartheta \, I_\nu(\Omega)\, d\Omega, \tag{2.41}$$

where $I_\nu(\Omega)$ is given by (2.35), and ϑ is the angle between the direction of propagation of the radiation and the normal to the surface.

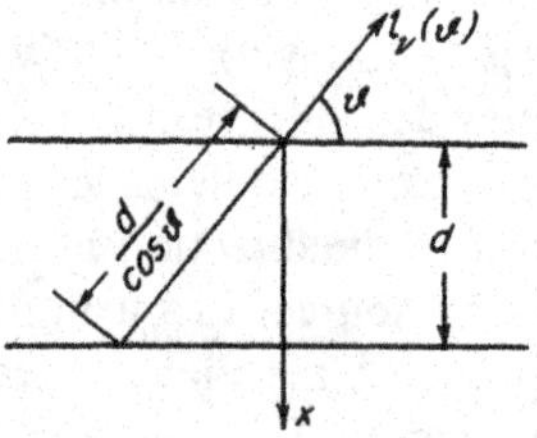

Fig. 2.6. Diagram illustrating the problem of radiation by a plane layer.

Let us find the surface brightness of a plane layer; we shall assume that the temperature and the absorption coefficient are variable but are functions of the x coordinate only (Fig. 2.6). We replace ds in (2.35) by $dx/\cos\vartheta$, and introduce the optical thickness

$$d\tau_\nu' = \kappa_\nu' \, dx, \qquad \tau_\nu' = \int_0^x \kappa_\nu' \, dx. \tag{2.42}$$

Then

$$I_\nu(\vartheta) = \int_0^\infty I_{\nu p} e^{-\frac{\tau_\nu'}{\cos\vartheta}} \frac{d\tau_\nu'}{\cos\vartheta}, \qquad \frac{\pi}{2} > \vartheta > 0. \tag{2.43}$$

Substituting this expression into (2.41) and integrating over all angles ($d\Omega = 2\pi \sin\vartheta \, d\vartheta$)

$$S_\nu = 2\pi \int_0^{\pi/2} \cos\vartheta \sin\vartheta \, d\vartheta \int_0^\infty I_{\nu p} e^{-\frac{\tau_\nu'}{\cos\vartheta}} \frac{d\tau_\nu'}{\cos\vartheta}$$

$$= 2\pi \int_0^\infty I_{\nu p} \, d\tau_\nu' \int_0^1 d(\cos\vartheta) e^{-\frac{\tau_\nu'}{\cos\vartheta}}.$$

We next introduce the variable $w = 1/\cos\vartheta$ and take into account the definition of the tabulated exponential integral,

$$E_n(z) = \int_1^\infty e^{-zw}\frac{dw}{w^n}, \qquad n = 1, 2, \ldots. \tag{2.44}$$

On replacing the equilibrium intensity by the equilibrium energy density through the relation $I_{\nu p} = cU_{\nu p}/4\pi$, we obtain

$$S_\nu = \frac{c}{2}\int_0^\infty U_{\nu p}[T(\tau'_\nu)]E_2(\tau'_\nu)\,d\tau'_\nu \tag{2.45}$$

or, for a layer of finite optical thickness $\tau_\nu = \int_0^d \kappa'_\nu\,dx$,

$$S_\nu = \frac{c}{2}\int_0^{\tau_\nu} U_{\nu p}E_2(\tau'_\nu)\,d\tau'_\nu. \tag{2.46}$$

Using the known property of the exponential integral

$$\int_0^\infty E_2(z)\,dz = \tfrac{1}{2},$$

we get, for a semi-infinite body at constant temperature,

$$S_\nu = \frac{cU_{\nu p}}{4} = S_{\nu p}. \tag{2.47}$$

As should have been expected, the spectral surface brightness is equal to the brightness of a perfect black body. The brightness of a finite thickness layer at a constant temperature is given by

$$S_\nu = \frac{cU_{\nu p}}{2}\int_0^{\tau_\nu} E_2(\tau'_\nu)\,d\tau'_\nu = \frac{cU_{\nu p}}{4}[1 - 2E_3(\tau_\nu)] = S_{\nu p}[1 - 2E_3(\tau_\nu)]. \tag{2.48}$$

This brightness is always lower than the brightness of a perfect black body at the same temperature, and approaches the latter as $\tau_\nu \to \infty$. For an optically thin layer

$$\begin{gathered}\tau_\nu \ll 1, \qquad E_2(\tau'_\nu) \approx E_2(0) = 1,\\ 2E_3(\tau_\nu) \approx 1 - 2\tau_\nu\end{gathered} \tag{2.49}$$

and

$$S_\nu = \frac{cU_{\nu p}}{2}\tau_\nu = S_{\nu p}\cdot 2\tau_\nu, \qquad 2\tau_\nu \ll 1. \tag{2.50}$$

§8. The brightness temperature of the surface of a nonuniformly heated body

The spectral surface luminosity of a nonuniformly heated body is most conveniently characterized by the brightness temperature* $T_{\nu\,\mathrm{br}}$. The latter denotes the temperature of a perfect black body emitting from its surface the same amount of radiation in a given frequency range as the actual body under consideration. Equating (2.46) and (2.47) we obtain an expression for the brightness temperature in the plane case

$$U_{\nu p}(T_{\nu\,\mathrm{br}}) = 2\int_0^{\tau_\nu} U_{\nu p}[T(\tau'_\nu)]E_2(\tau'_\nu)\,d\tau'_\nu, \tag{2.51}$$

or, substituting the Planck function for $U_{\nu p}$,

$$\frac{1}{e^{h\nu/kT_{\nu\,\mathrm{br}}}-1} = 2\int_0^{\tau_\nu} \frac{1}{e^{h\nu/kT}-1}E_2(\tau'_\nu)\,d\tau'_\nu. \tag{2.52}$$

The brightness temperature is a function of the frequency. Only in the case of a perfect black body is it the same for all frequencies, and is then equal to the temperature of the material.

One can also introduce a brightness temperature for the integrated radiation by means of the definition

$$S = \sigma T_{\mathrm{br}}^4 \tag{2.53}$$

where S is the integrated energy flux emerging from the surface of the body. Obviously, the brightness temperature for the integrated radiation is some mean value with respect to the spectral brightness temperatures.

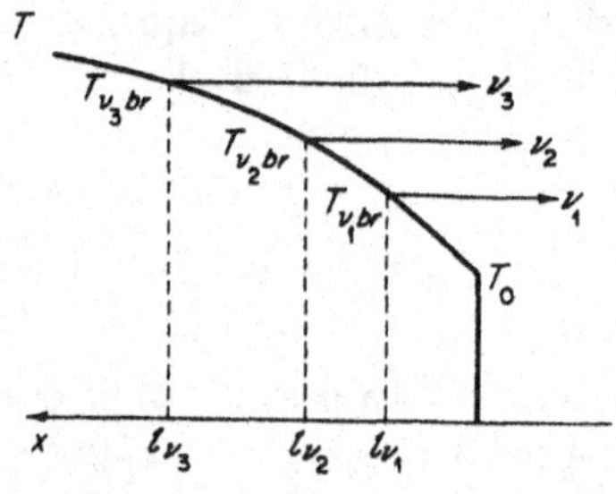

Fig. 2.7. Temperature profile for a radiating body whose temperature decreases toward the surface.

Let us examine the relationship between the radiation spectrum of a body and the frequency dependence of the absorption coefficient. Consider an optically thick body and let the radius of curvature of the surface be large in comparison with the radiation mean free path, so that the body can be regarded as a plane. Let the temperature decrease toward the surface as shown in Fig. 2.7.

* *Editors' note.* Termed effective temperature by the authors.

The radiant energy flux of frequency ν emerging from the surface is determined by the integral over all sources (see (2.45)). Since the exponential integral decreases rapidly with τ'_ν for self-absorption, the main contribution to the integral will come from a layer of the order of the radiation mean free path l'_ν in the neighborhood of the surface (with an optical thickness τ'_ν of the order of unity). In other words, photons, emitted from the surface of the body, are born mainly in a layer near the surface with an optical thickness of the order of unity (to be more precise, of two or three). This layer can be called the radiating layer. Photons born in layers farther removed from the surface are almost totally absorbed before reaching the surface. The brightness temperature, as follows from equation (2.52), is therefore equal to some average temperature of the radiating layer.

Photons emerging from the surface at frequencies for which the absorption is stronger and the radiation mean free path is shorter, are radiated in the less heated layers closer to the surface. Conversely, frequencies at which the absorption is weaker emerge from the deeper and more heated layers. Thus, if the temperature of the fluid decreases toward the surface (as is usually the case), the brightness temperature for the strongly absorbed frequencies is less than that for the weakly absorbed frequencies. This is represented schematically in Fig. 2.7 by the arrows indicating the "spots" from which the photons of different frequencies may be considered to be emitted. These "spots" are located relative to the surface at distances of the order of the photon mean free paths.

The radiation spectrum for a nonuniformly heated body is different from the Planck spectrum; the difference is the more pronounced the stronger is the dependence of the absorption coefficient upon frequency and temperature, and the steeper is the temperature gradient near the surface (at distances of the order of the photon mean free paths). Figure 2.8 is a schematic

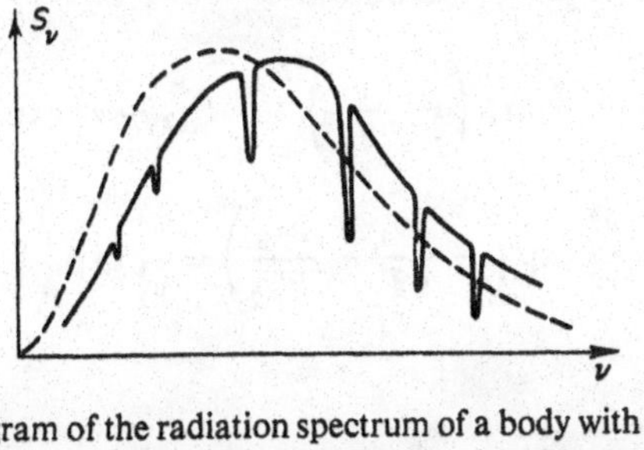

Fig. 2.8. Schematic diagram of the radiation spectrum of a body with a temperature which decreases toward the surface. Low frequencies are absorbed more strongly than high frequencies. The dashed curve shows the Planck spectrum corresponding to the mean brightness temperature of the radiation. Selective absorption lines are "cut" into the spectrum. The flux at the line centers is practically equal to the Planck flux which corresponds to the surface temperature of the body.

representation of the radiation spectrum of a body with a temperature decreasing toward the surface and with an absorption coefficient that varies inversely with the frequency, i.e., with the lower frequencies absorbed more strongly than the higher ones. Discrete lines, corresponding to the bound-bound transitions in atoms or ions, have been superimposed on the continuous spectrum. The absorption coefficients of these lines are always very high, much higher than in the continuous spectrum. Hence, the brightness temperature of these lines almost exactly coincides with the temperature close to the surface of the body (the lines are "cut" into the radiation spectrum of the body). For comparison, the dashed curve in Fig. 2.8 shows the Planck spectrum corresponding to the integrated brightness temperature, which is an average of the spectral temperatures. By virtue of the definition of the integrated brightness temperature, the areas bounded by the solid and dashed curves are exactly equal.

We shall show in Chapter V that the continuum absorption coefficients are not smooth functions of frequency but that they undergo discontinuous jumps at high temperatures. This results in corresponding jumps in the radiation spectrum of the body. (This is not shown in Fig. 2.8, which assumes a smooth dependence of κ_ν' on ν.)

The concept of color temperature is frequently employed in optical measurements of the luminosity of heated bodies. The term color temperature is used to describe the temperature of a perfect black body, which would yield a ratio of brightness in two different spectral regions (for example, in the red and violet regions of the spectrum) equal to the experimentally measured ratio. Using the definitions of brightness and color temperatures, we can easily find their mutual relationship. Let the brightness temperatures at frequencies ν_1 and ν_2 be T_1 and T_2 and let the color temperature be T_{12}. Assuming, for simplicity, that both lines ν_1 and ν_2 lie in the Wien spectral region, that is, $h\nu_1/kT_1 \gg 1$, $h\nu_2/kT_2 \gg 1$, we obtain

$$S_{\nu 1} \sim \nu_1^3 \exp\left(-\frac{h\nu_1}{kT_1}\right), \qquad S_{\nu 2} \sim \nu_2^3 \exp\left(-\frac{h\nu_2}{kT_2}\right);$$

$$\frac{S_{\nu 1}}{S_{\nu 2}} = \frac{\nu_1^3}{\nu_2^3} \exp\left(-\frac{h\nu_1}{kT_1} - \frac{h\nu_2}{kT_2}\right) = \frac{\nu_1^3}{\nu_2^3} \exp\left(-\frac{h\nu_1 - h\nu_2}{kT_{12}}\right),$$

from which

$$\frac{\nu_1 - \nu_2}{T_{12}} = \frac{\nu_1}{T_1} - \frac{\nu_2}{T_2}. \tag{2.54}$$

If the temperature of the body is approximately constant in the radiating layers near the surface throughout the entire basic spectrum, then the color temperature will frequently be closer to the true temperature of the body than is the brightness temperature. This condition is utilized in pyrometry for optical measurement of temperatures. We may note that in the case of a nonuniformly heated "gray" body, one whose absorption coefficient κ'_ν is independent of frequency or $\kappa'_\nu \equiv \kappa'$, the brightness temperature for different frequencies remains a function of the frequency. Only for low-energy photons (in the Rayleigh-Jeans spectral region $h\nu/kT \ll 1$) does the frequency drop out from (2.52) and the brightness temperature remain the same for all frequencies.

§9. Motion of a fluid taking into account radiant heat exchange

We have given above the methods of determining the radiation field in a body of fluid, or the radiation emerging from the surface, when the state of the fluid, i.e., the temperature and density distributions in the medium, is known. Let us now formulate the problem for the simultaneous determination of the state and motion of the fluid and of the radiation field. We shall consider the case when the radiative transfer and the interaction between the radiation and the fluid have a substantial effect on both the state and the motion of the medium (gas). The motion will always be considered as nonrelativistic, that is, we shall assume that the flow velocities are much less than the speed of light.

If the temperature is not too high, and the gas density is not too low, then the radiant energy density and the radiation pressure are negligibly small in comparison with the energy and pressure of the fluid. Let us estimate and compare the equilibrium radiant energy density $U_p = 4\sigma T^4/c$ and the thermal energy contained in a unit volume of monatomic gas $E = \frac{3}{2}nkT$ (n is the number of particles per unit volume). For example, for $n = 2.67 \cdot 10^{19}$ per cm^3 (which corresponds to the number of molecules in air at standard density), both energies are equal at a temperature of 900,000°K. Actually the radiant energy becomes comparable to the energy of the fluid only at even higher temperatures. The heating of the fluid results in ionization of the atoms, which leads first to an increase in the number of particles per unit volume and second to the addition of the ionization energy to the thermal energy*. Thus, the radiant energy in real air at standard density conditions becomes comparable with the internal energy of the air only at temperatures of approximately 2,700,000°K. In a highly rarefied gas the equilibrium radiant energy becomes comparable with the energy of the fluid at much lower temperatures (roughly speaking, the temperature at which both energies are equal is proportional to $n^{1/3}$). However, one must be very careful in comparing the

* Thermodynamic properties of gases at high temperatures are discussed in Chapter III.

energies in this case, since the radiation mean free path in a highly rarefied gas is very long and if the dimensions of the gaseous body are not sufficiently large, then the radiant energy density may turn out to be much lower than the equilibrium energy density (see discussion below).

The radiation pressure and the fluid pressure are related in approximately the same manner as the energies. Actually the radiation pressure (for an isotropic radiation field) is $p_\nu = U/3$ and the fluid pressure is $p = (\gamma - 1)E$. At high temperatures the value of the specific heat ratio lies between 5/3 and ~ 1.15, depending on the composition of the gas, its temperature, and its density.

Thus, at not too high temperatures and not too low densities, the radiant energy density and the radiation pressure have no effect on either the energy balance or the motion of the fluid. The effect of the radiation on the energy balance and motion of the gas manifests itself in another way: The radiant energy lost by the heated body of fluid and the radiant heat transfer in the medium can, generally, turn out to be quite appreciable. These effects frequently play an important role even at much lower temperatures, when both the energy and the radiation pressure are very small.

The cause of the above phenomenon under ordinary conditions is the marked difference between the fluid velocity u and the speed of light c; $u \ll c$. As a result, the energy flux in the fluid and the radiant energy flux can become comparable, even if the radiant energy density is much less than the energy density of the fluid. For example, in the limiting case when all the photons are moving in one direction, the radiant energy flux is $S = Uc$, while the flux of energy of the fluid is of the order of Eu, that is, Uc can be of the order of and even greater than Eu even with $U \ll E$, because $c \gg u$. The radiant energy flux and the flux of energy of the fluid are frequently comparable even in actual cases where the radiation field is relatively isotropic and the resultant net radiant energy flux S (equal to the difference of one-sided fluxes) is considerably smaller than its maximum limiting value Uc (which corresponds to an extreme anisotropy of the radiation field). As we shall presently prove, the amount of energy lost or, conversely, the energy released in the fluid as a result of the interaction with the radiation is given by the divergence of the radiation flux. Thus, a comparison of the radiant energy flux and the flux of energy of the fluid can be used to characterize the importance of the radiant heat transfer in the medium.

Let us find the energy q lost by radiation in a unit volume of the fluid per unit time. This energy represents the difference between the energy emitted by the fluid and the radiation energy absorbed by it. The difference between the emission and absorption of radiation of frequency ν (per unit frequency interval) and direction $\mathbf{\Omega}$ (per unit solid angle) per unit time per unit volume is given by the right-hand side of the radiative transfer equation (2.28). The total energy q lost per unit volume of fluid per unit time can be found by integrating over the entire solid angle and over all frequencies,

$$q = \int_0^\infty d\nu \int \kappa'_\nu (I_{\nu p} - I_\nu)\, d\Omega = c \int_0^\infty \kappa'_\nu (U_{\nu p} - U_\nu)\, d\nu. \tag{2.55}$$

The first term in parentheses corresponds to spontaneous emission, and the second to absorption corrected for "re-radiation". Using the equation of continuity for radiation (2.29) in which, according to our earlier statements on the quasi-steady character of the radiative transfer, we can neglect the derivative with respect to time, we find that the resultant energy loss is equal to the divergence of the integrated radiation flux:

$$q = \int_0^\infty \nabla \cdot \mathbf{S}_\nu \, d\nu = \nabla \cdot \mathbf{S}. \tag{2.56}$$

If the fluid radiates more energy than it absorbs, then the energy is lost in radiation (radiation cooling), and $q > 0$; if, however, more energy is absorbed than emitted, then the fluid is heated by radiation and the energy loss is negative, $q < 0$.

Let us now write the gasdynamic equations including radiant heat transfer, but disregarding the radiation energy and pressure. The first of these equations —the continuity equation—remains unchanged. The equation of motion also remains unchanged, because we are neglecting the radiation pressure. Only the energy equation requires the introduction of a term describing the energy losses by radiation (the radiant energy density and the work done by the radiation pressure forces are neglected). The energy equation (1.10) then becomes*

$$\frac{\partial}{\partial t}\left(\rho\varepsilon + \frac{\rho u^2}{2}\right) = -\nabla \cdot \left[\rho \mathbf{u}\left(\varepsilon + \frac{p}{\rho} + \frac{u^2}{2}\right)\right] - q, \tag{2.57}$$

or replacing q by the divergence of the flux $\mathbf{S}$

$$\frac{\partial}{\partial t}\left(\rho\varepsilon + \frac{\rho u^2}{2}\right) = -\nabla \cdot \left[\rho \mathbf{u}\left(\varepsilon + \frac{p}{\rho} + \frac{u^2}{2}\right) + \mathbf{S}\right]. \tag{2.58}$$

Thus, the radiant energy flux is added to the total hydrodynamic energy flux. By writing the gasdynamic energy equation in the form of an entropy equation (see §1, Chapter I), we obtain

$$\rho T \frac{d\Sigma}{dt} = -q = -\nabla \cdot \mathbf{S}, \tag{2.59}$$

where Σ is the specific entropy of the fluid.

* It is assumed that no other energy sources and no other irreversible processes besides the radiant heat transfer are present.

Serious mathematical difficulties are encountered in finding the radiation field and the temperature distribution in a medium when radiant heat transfer has a large effect on the energy balance. The spatial derivative in the transfer equation (2.34) which describes the radiation field is formulated for a spectral radiation intensity propagated in a given direction. The energy balance equation (2.57) contains, however, the quantities q or $\mathbf{S}$ which are obtained from integrals over frequency and over the entire solid angle. Therefore, together, the transfer and energy equations have an integro-differential character requiring a double integration with respect to frequency and angle. Mathematical simplifications of this integro-differential system are based on approximate descriptions of the spectral and angular distributions, in order to eliminate the "integro" part.

The influence of the spectral distribution on the energy balance arises as a result of the frequency dependence of the absorption coefficient. The spectral characteristics can be excluded from consideration only when the absorption coefficient κ_ν' is independent of frequency, that is, $\kappa_\nu' \equiv \kappa'$. In the case of a "gray" body, the transfer equation (2.34), upon integration over all frequencies, can be written directly for the integrated intensity $I = \int_0^\infty I_\nu \, d\nu$,

$$\mathbf{\Omega} \cdot \nabla I = \kappa'(I_p - I). \tag{2.60}$$

It also becomes possible to find the energy loss by integrating over the spectrum in (2.55)

$$q = \kappa' \int (I_p - I) \, d\mathbf{\Omega} = c\kappa'(U_p - U). \tag{2.61}$$

In general, the absorption coefficients in gases at high temperatures are strongly dependent on frequency, and the concept of a "gray body" represents a considerable idealization. This idealization is very useful for clarifying the relationships governing those phenomena that are not related to the spectral distribution of the radiation. However, in certain important limiting cases, which will be discussed later, we shall introduce an absorption coefficient κ' averaged in the proper manner with respect to frequency; in this approach the spectral characteristics of the radiation can be neglected and equations (2.60) and (2.61) can be used.

The next two sections deal with approximate descriptions of the angular distribution of the radiation field.

§10. The diffusion approximation

It is evident from (2.55) and (2.56) that the radiant energy loss q is not an explicit function of the angular distribution of the radiation, but is defined only by integrals over direction, by the radiant energy density or flux. If it were possible to replace the radiative transfer equation, which depends upon direction, by some other equations directly governing the spectral density and flux, then in general the problem of the angular distribution of the radiation would not arise in considering the effect of radiation on the state and the motion of the fluid. Such an equation exists in the continuity equation (2.29), which for the quasi-steady case is

$$\nabla \cdot \mathbf{S}_\nu = c\kappa_\nu'(U_{\nu p} - U_\nu). \tag{2.62}$$

A second equation relating the radiant energy flux and density to complete the system of equations can be obtained in an approximate form only. Equation (2.62) was found by integrating the transfer equation (2.34) over all angles. Let us now multiply the transfer equation (2.34) by a unit direction vector $\mathbf{\Omega}$ and integrate again over all angles. Noting that the integral of the $\kappa_\nu' I_{\nu p}$ term, which is independent of direction, vanishes, and recalling the definition of flux (2.3), we obtain

$$\int \mathbf{\Omega}(\mathbf{\Omega} \cdot \nabla I_\nu)\, d\mathbf{\Omega} = -\kappa_\nu' \mathbf{S}_\nu. \tag{2.63}$$

The flux $\mathbf{S}_\nu = \int \mathbf{\Omega} I_\nu \, d\mathbf{\Omega}$ vanishes in an isotropic radiation field. The integral on the left-hand side can be easily evaluated if the intensity I_ν is independent of angle*

$$\int \mathbf{\Omega}(\mathbf{\Omega} \cdot \nabla I_\nu)\, d\mathbf{\Omega} = \frac{1}{3}\int \nabla I_\nu \, d\mathbf{\Omega} = \frac{c}{3} \nabla U_\nu. \tag{2.64}$$

The vanishing of this expression means that the isotropy of the radiation field requires the constancy in space of the radiant energy density. If the radiation field is anisotropic, then the flux and the integral (2.63) are both different from zero. However, in the case of weak anisotropy the integral can be again represented to a first approximation in the form of (2.64), by assuming the intensity to be very weakly dependent on angle so that it may be taken independent of angle in the integral. This yields an approximate relationship between the flux and the radiation density

$$\mathbf{S}_\nu = -\frac{l_\nu' c}{3} \nabla U_\nu, \tag{2.65}$$

* Let us find the ith component of the vector integral, replacing the vector operator $\mathbf{\Omega} \cdot \nabla$ by $\Omega_k\, \partial/\partial x_k$ and bearing in mind the summation convention for repeated subscripts

$$\int \Omega_i \Omega_k \frac{\partial I_\nu}{\partial x_k} d\mathbf{\Omega} = \frac{\partial I_\nu}{\partial x_k} \int \Omega_i \Omega_k \, d\mathbf{\Omega} = \frac{\partial I_\nu}{\partial x_k} \frac{4\pi}{3} \delta_{ik} = \frac{4\pi}{3} \frac{\partial I_\nu}{\partial x_i} = \frac{c}{3} \frac{\partial U_\nu}{\partial x_i},$$

since $\int I_\nu \, d\mathbf{\Omega} = 4\pi I_\nu = cU_\nu$; (2.64) follows.

where $l'_\nu = 1/\kappa'_\nu$ is the absorption mean free path (corrected for induced emission).

Dividing both sides of (2.65) by the energy of a photon $h\nu$ we obtain the usual particle diffusion relation between the photon flux $\mathbf{J}_\nu$ and the photon density N_ν

$$\mathbf{J}_\nu = -D_\nu \nabla N_\nu, \qquad D_\nu = \frac{l'_\nu c}{3}.$$

The "diffusion" coefficient for the photons D_ν is analogous to the diffusion coefficient for atoms or molecules; c is the speed of the photons and l'_ν is their mean free path. An important difference, however, exists between the diffusion of atoms and the "diffusion" of photons. The atom does not vanish upon a collision, it only changes the direction of its motion (in a random manner in the case of isotropic scattering); the mean free path appearing in the equation for the diffusion coefficient is the collisional mean free path. The photon, however, after traversing an average distance l'_ν, is absorbed by the fluid and, if the fluid is in a state of thermodynamic equilibrium, the photon's energy, as a result of collisions between atoms, electrons, and so on, is distributed in accordance with the laws of statistical equilibrium. At the place of absorption new photons are emitted with different frequencies and in random directions. In considering the "diffusion" of photons of a given frequency, among all newly born photons we look only at those which are at the same frequency. The process progresses as if the photon traveled, became absorbed, and then is "born" again. After being "reborn", the photon has an equal probability of traveling in any direction, and this corresponds to the isotropic scattering of atoms following a collision*.

As with the diffusion of atoms, the diffusion approximation is applicable only to small gradients of the radiation density. The latter should change little over a distance of the order of the radiation mean free path l'_ν. The radiation field for small gradients is almost isotropic, and it was this condition which was used as the basis of the derivation of the diffusion equation (2.65). In point of fact, the photons arriving at a given point originate primarily in a region with dimensions of the order of the mean free path. If the radiation density in this region is almost constant, then the photons arrive at the given point uniformly from all directions and this leads to the conclusion that the radiation field at this point is isotropic.

* If photon scattering is taken into account in the radiative transfer, then for weak anisotropy we again obtain a diffusion relation of the type (2.65). This relation contains the mean free path corresponding to a total decay coefficient which is equal to the sum of the absorption and scattering coefficients. If the scattering is anisotropic, then, as in the case of diffusion of atoms, the scattering coefficient is replaced by the transport coefficient $\kappa_s(1-\overline{\cos\vartheta})$, where $\overline{\cos\vartheta}$ is the average of the cosine of the scattering angle.

At a vacuum-fluid interface the density changes very rapidly over a distance of the order of the mean free path and the angular distribution of the photons is very anisotropic, that is, the photons leave the body of fluid in the direction of the vacuum, while none arrive from the vacuum. Hence, the diffusion approximation can lead to appreciable errors when applied near a vacuum interface.

In the case of optically thick bodies the density gradients are small and the diffusion approximation is valid. If x is the characteristic scale along which the radiation density has an appreciable change (x is of the order of the dimensions of the body), then the order of the diffusion flux is given by

$$\mathbf{S}_\nu = -\frac{l'_\nu c}{3}\nabla U_\nu \sim \frac{l'_\nu}{x} cU_\nu .$$

The greater is the optical thickness of the body x/l'_ν, the smaller is the change in the radiation density over a mean free path (this change is of the order of $l'_\nu \nabla U_\nu \sim (l'_\nu/x)U_\nu$), the smaller is the flux S_ν in comparison with $U_\nu c$, and the more exact is the diffusion approximation. If the optical thickness of the body is of the order of unity, then $l'_\nu/x \sim 1$, and $S_\nu \sim cU_\nu$. In the case of an optically thin body we have $l'_\nu/x > 1$, and the flux, estimated according to the diffusion formula, becomes greater than cU_ν. This is physically impossible and simply indicates the inapplicability of the diffusion approximation to optically thin bodies.

The flux S_ν can never be higher than cU_ν. The equality $S_\nu = cU_\nu$ corresponds to the case when all the photons travel in exactly the same direction, the case of maximum anisotropy. The quantity cU_ν is sometimes termed the kinetic flux. The ratio of the flux to the kinetic flux S_ν/cU_ν is a measure of the anisotropy of the radiation field and within the framework of the diffusion approximation is of the order of the reciprocal of the optical thickness of the body l'_ν/x. For complete isotropy $S_\nu/cU_\nu = 0$, while when all the photons travel in the same direction then $S_\nu/cU_\nu = 1$. Thus, this flux ratio is always contained within the limits $0 \leq S_\nu/cU_\nu \leq 1$. The dependence of the flux upon the degree of anisotropy of the angular distribution of radiation, at a given radiant energy density, is schematically illustrated in the polar diagrams of the intensity sketched in Fig. 2.9. The areas of all the diagrams are the same and characterize the radiant energy density, while the lengths of the arrows characterize the fluxes. Radiation fields of different densities are capable of giving the same energy flux. The higher the density for a given flux, the lower is the ratio S_ν/cU_ν, and the more isotropic is the radiation field.

The equations (2.62) and (2.65) for the diffusion approximation represent a system of two differential equations for two unknown functions of position, the energy density and flux of the radiation. Boundary conditions at interfaces of media with different optical properties (with different "diffusion coefficients") must also be added to the system. It follows from the condition of continuity of the radiation intensity that both the density and flux are continuous at such interfaces. A density discontinuity in the diffusion approximation (2.65) would entail an infinite flux and a discontinuity in flux would lead to an accumulation of radiant energy, to an unsteadiness in the solution (see (2.29)).

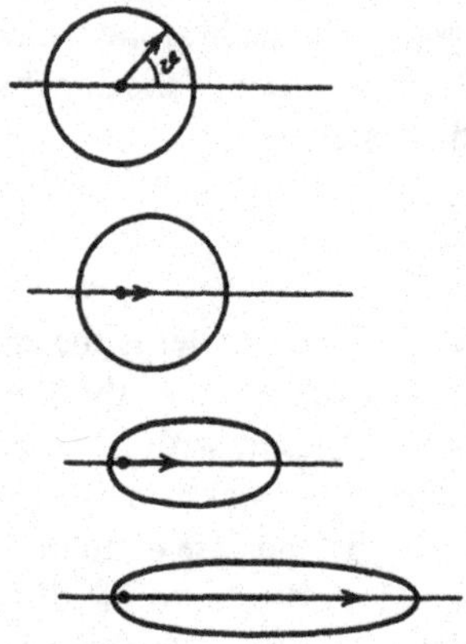

Fig. 2.9. Polar diagrams for the distribution of the radiation intensity with respect to angle for different degrees of anisotropy. The intensity at a given angle ϑ is characterized by the length of the radius vector drawn from the center. The length of the arrow characterizes the magnitude of the flux. The equality of the radiant energy densities for all cases is schematically indicated by the equal areas of the diagrams.

The interface between a fluid and a vacuum requires special examination. Since the vacuum does not supply any photons, the radiation field at the interface is strongly anisotropic (all photons travel only in the direction toward the vacuum) and, strictly speaking, the diffusion approximation is inapplicable. We can obtain an approximate boundary condition from the following considerations. Let us assume that the radiation emerging from the surface of a body into a hemispherical volume, in the direction of the vacuum, has an isotropic angular distribution (which, for optically thick bodies, is close to true). The intensity in the complementary hemisphere is zero, that is, no photons arrive from the vacuum side (Fig. 2.10). Then at the vacuum interface we have

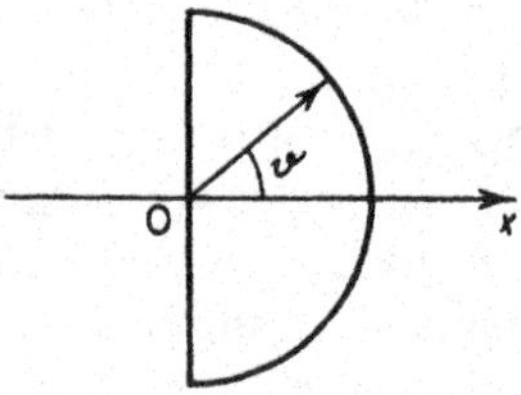

Fig. 2.10. Polar diagram for the intensity distribution at the interface $x = 0$, between the body and the vacuum. The vacuum is on the right, the medium on the left.

$$S_\nu = \frac{cU_\nu}{2}, \tag{2.66}$$

where the flux is in the direction of the outward normal to the surface. The factor of $\frac{1}{2}$ represents the average of the cosine of the direction angle of the photon motion for an isotropic distribution in the hemisphere. We have

$$\mathbf{S}_\nu = \int_{\text{hemisphere}} \mathbf{\Omega} I_\nu \, d\mathbf{\Omega}; \qquad S_\nu = \int_0^{\pi/2} \cos\vartheta \, I_\nu(\vartheta) \, 2\pi \sin\vartheta \, d\vartheta = 2\pi I_\nu \cdot \tfrac{1}{2} = \pi I_\nu,$$

but

$$cU_\nu = \int_{\text{hemisphere}} I_\nu \, d\mathbf{\Omega} = \int_0^{\pi/2} 2\pi \sin\vartheta \, d\vartheta \, I_\nu = 2\pi I_\nu .$$

Equation (2.66) follows.

Equation (2.66) can also be formally obtained from the equations for the diffusion approximation. It is easy to verify that the following expression for the intensity

$$I_\nu(\mathbf{\Omega}) = \frac{cU_\nu}{4\pi}\left[1 + \frac{3\mathbf{\Omega} \cdot \mathbf{S}_\nu}{cU_\nu}\right] = \frac{cU_\nu}{4\pi}\left[1 + 3\cos\vartheta \frac{S_\nu}{cU_\nu}\right],$$

where ϑ is the angle between the direction $\mathbf{\Omega}$ and the direction of the flux S_ν, leads to the diffusion approximation equations (2.62) and (2.65). With the x axis in the direction of the flux we can calculate the one-sided fluxes in the positive and negative x directions. We obtain

$$S_{\nu+} = \frac{cU_\nu}{4} + \frac{S_\nu}{2}, \qquad S_{\nu-} = -\frac{cU_\nu}{4} + \frac{S_\nu}{2} \tag{2.67}$$

(as expected, $S_\nu = S_{\nu+} + S_{\nu-}$). Applying these relations to the interface between the body and the vacuum (the x axis is directed towards the vacuum) and assuming that the one-sided flux from the vacuum is $S_{\nu-} = 0$, we get $S_\nu = S_{\nu+} = cU_\nu/2$, namely, relation (2.66). Equations (2.67) are more general than the expression for the intensity. This is easily verified by applying the relation for the intensity at a point on the interface. In the negative x direction, for example, $\cos\pi = -1$ and $I_\nu = -cU_\nu/8\pi < 0$, which has no physical significance. The point is that the diffusion equation for the intensity is valid for only weak anisotropy, when the second term in the brackets above is much smaller than unity.

§11. The "forward-reverse" approximation

We now consider still another method of approximating the angular distribution of radiation, one which is sometimes used in plane problems of radiative transfer. This method is known as the Schwarzschild approximation or as the "forward–reverse" approximation. In this method we combine into one group all photons moving in the positive x direction at angles ϑ ranging from 0 to $\pi/2$ ("forward"), and into another group all those moving in the opposite direction at angles ϑ from $\pi/2$ to π ("reverse") (Fig. 2.11). We shall assume that the angular distribution in each hemisphere is approximately isotropic and denote the intensity in the "forward" and "reverse"

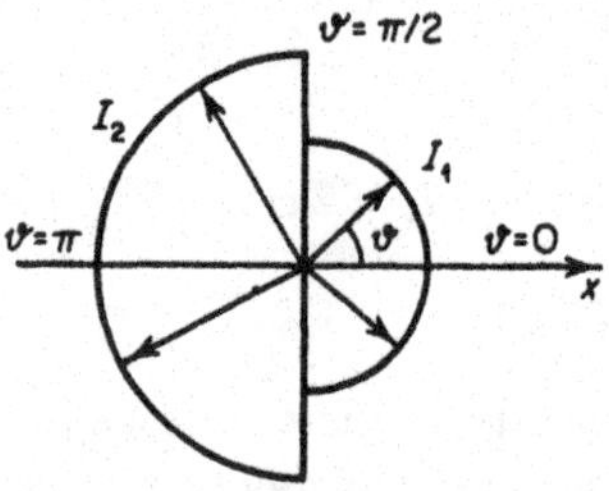

Fig. 2.11. Polar diagram for the radiation intensity distribution in the "forward-reverse" approximation. The flux is directed to the left for the case shown.

directions by I_1 and I_2, respectively (the frequency subscript is dropped to simplify the notation). The radiant energy density and flux are

$$U = \frac{1}{c}\int I\, d\Omega = \frac{2\pi}{c}\int_0^{\pi/2} I_1 \sin\vartheta\, d\vartheta + \frac{2\pi}{c}\int_{\pi/2}^{\pi} I_2 \sin\vartheta\, d\vartheta$$
$$= \frac{2\pi}{c}(I_1 + I_2), \tag{2.68}$$

$$S = \int I \cos\vartheta\, d\Omega = 2\pi \int_0^{\pi/2} I_1 \cos\vartheta \sin\vartheta\, d\vartheta$$
$$+ 2\pi \int_{\pi/2}^{\pi} I_2 \cos\vartheta \sin\vartheta\, d\vartheta = \pi(I_1 - I_2). \tag{2.69}$$

The degree of anisotropy is then given by

$$\frac{S}{cU} = \frac{I_1 - I_2}{2(I_1 + I_2)} \to 0 \qquad \text{for} \quad I_1 \approx I_2 .$$

At the interface between the medium and a vacuum, when the x axis is in the direction of the outward normal to the surface, we have $I_2 = 0$ and $S/cU = \frac{1}{2}$, in agreement with condition (2.66).

The equation for the average "one-sided" intensities I_1 and I_2 is obtained by averaging the transfer equation for the plane case

$$\cos\vartheta \frac{dI}{dx} = \kappa'(I_p - I) \tag{2.70}$$

over both hemispheres. We then obtain (with the average $\overline{\cos\vartheta} = \pm\frac{1}{2}$)

$$\frac{1}{2}\frac{dI_1}{dx} = \kappa'(I_p - I_1); \qquad -\frac{1}{2}\frac{dI_2}{dx} = \kappa'(I_p - I_2). \tag{2.71}$$

This pair of equations serves to determine the average intensities in both hemispheres. Adding and subtracting, we easily arrive at the density and flux equations ($I_p = cU_p/4\pi$):

$$\frac{dS}{dx} = \kappa' c(U_p - U); \qquad S = -\frac{l'c}{4}\frac{dU}{dx}. \tag{2.72}$$

The first equation is the exact continuity equation (2.62) and the second one is almost the same as equation (2.65) for the diffusion approximation; the only difference is that here the "diffusion coefficient" is equal to $l'c/4$ instead of $l'c/3$.

Considering (2.71) as a pair of linear differential equations in I_1 and I_2, we can write their solution in the integral form

$$I_1 = \int_0^\tau I_p e^{-2(\tau-\tau')}2\, d\tau'; \qquad I_2 = \int_\tau^\infty I_p e^{-2(\tau'-\tau)}2\, d\tau'.$$

Here, the x coordinate has been replaced by the optical thickness by means of the relation

$$d\tau = \kappa'\, dx, \qquad \tau = \int_0^x \kappa'\, dx.$$

Adding and subtracting the expressions for I_1 and I_2 and substituting $I_p = cU_p/4\pi$, we obtain the following two approximate integral expressions for the density and flux in the "forward–reverse" approximation:

$$\begin{aligned} U &= \frac{1}{2}\int_\tau^\infty U_p e^{-2(\tau'-\tau)}2\, d\tau' + \frac{1}{2}\int_0^\tau U_p e^{-2(\tau-\tau')}2\, d\tau', \\ S &= -\frac{c}{4}\int_\tau^\infty U_p e^{-2(\tau'-\tau)}2\, d\tau' + \frac{c}{4}\int_0^\tau U_p e^{-2(\tau-\tau')}2\, d\tau'. \end{aligned} \tag{2.73}$$

Generally speaking, the diffusion approximation in the case of weak anisotropy appears to be more justified, although it is not much different from the "forward–reverse" approximation.

§12. Local equilibrium and the approximation of radiation heat conduction

Steady state radiation in an infinite medium at constant temperature will be in thermodynamic equilibrium. The intensity is independent of direction and is determined by the Planck formula. Photons arriving at any point in space are born in the vicinity of that point at distances of not more than several mean free paths; photons born farther away are absorbed in transit. Consequently, only the immediate vicinity of the point "participates" in establishing the equilibrium intensity. Even if the temperature at a farther distance is different from the temperature of this region, there is no practical effect on the radiation intensity at the point under consideration. This means that if the temperature of a sufficiently extended and optically thick medium is not constant but changes sufficiently slowly with distance, so that changes over distances of the order of a radiation mean free path are small, then the intensity will be very close to the equilibrium value corresponding to the temperature in the medium at the given point. The less pronounced is the temperature change over distances of the order of a radiation mean free path, the closer will be the intensity to the equilibrium value. In particular, the radiation will be closer to equilibrium for those frequencies that are more strongly absorbed and for which the mean free path l'_ν is smallest. If the temperature gradient is so small that the temperature changes over distances of the order of the longest mean free path l'_ν are small for all those frequencies which have a significant effect on the equilibrium radiation at the given temperature, then the radiation will be in equilibrium for practically the entire spectral interval, at the temperature corresponding to that of the point. The radiation intensity will then be described as a function of frequency by the Planck function at the temperature of the point. When the radiation at each point of a medium with a nonuniform temperature is close to equilibrium, then the medium is spoken of as being in a state of local thermodynamic equilibrium between the radiation and the fluid.

The necessary condition for the existence of local equilibrium—small gradients in an extended, optically thick medium—serves simultaneously as a justification for the use of the diffusion approximation when considering radiative transfer. In the diffusion approximation the radiation flux is proportional to the gradient of radiation density. However, if the radiation density is close to its equilibrium value, then it is possible to approximate the true density in the flux equation by the equilibrium density at the given point. Thus, for local equilibrium conditions, the spectral flux is given approximately by

$$\mathbf{S}_\nu = -\frac{l'_\nu c}{3} \nabla U_{\nu p}. \tag{2.74}$$

The total flux is

$$\mathbf{S} = \int_0^\infty \mathbf{S}_\nu \, d\nu = -\frac{c}{3} \int_0^\infty l'_\nu \nabla U_{\nu p} \, d\nu. \tag{2.75}$$

We factor out from the integrand an average value of the radiation mean free path, which we denote by l. Noting that $\int_0^\infty U_{\nu p} \, d\nu = U_p = 4\sigma T^4/c$, equation (2.75) gives

$$\mathbf{S} = -\frac{lc}{3} \nabla U_p = -\frac{16\sigma l T^3}{3} \nabla T. \tag{2.76}$$

The radiant energy flux in local equilibrium is proportional to the temperature gradient, and the radiative transfer is similar to heat conduction, and is termed radiation heat conduction. The coefficient of thermal conductivity is equal here to $16\sigma l T^3/3$, and is a function of temperature.

As in the case of ordinary molecular heat conduction, the energy loss q by the medium from radiation is equal to the divergence of the flux of radiation heat conduction (see (2.56)). These losses are determined by the temperature of the fluid at the given point, the average radiation mean free path (which for a given fluid is a function of the temperature and density), and the spatial derivatives of these quantities.

Combining equations (2.75) and (2.76), we can obtain the law for the average of the mean free path with respect to frequency which gives the correct value for the radiant energy flux in the case when the radiant heat exchange has the same character as heat conduction. Noting that both $U_{\nu p}$ and U_p depend upon position only through the temperature dependence, we obtain

$$l = \frac{\int_0^\infty l'_\nu (dU_{\nu p}/dT)\, d\nu}{dU_p/dT} = \frac{\int_0^\infty l'_\nu (dU_{\nu p}/dT)\, d\nu}{\int_0^\infty (dU_{\nu p}/dT)\, d\nu}. \tag{2.77}$$

Differentiating the equilibrium radiation density given by the Planck formula with respect to temperature, and introducing the dimensionless variable of integration $u = h\nu/kT$, we obtain the relation for the average mean free path

$$l = \int_0^\infty l'_\nu G(u)\, du, \tag{2.78}$$

where the weighting factor $G(u)$ is given by

$$G(u) = \frac{15}{4\pi^4} \frac{u^4 e^{-u}}{(1 - e^{-u})^2}. \tag{2.79}$$

The value of l obtained by averaging the radiation mean free path l'_ν using the weighting factor $G(u)$ is called the Rosseland mean free path. If the mean free path l'_ν, corrected for induced emission, is expressed in terms of the actual absorption coefficient $l'_\nu = 1/\kappa'_\nu = 1/\kappa(1 - e^{-u})$, then (2.78) and (2.79) may be rewritten as

$$l = \int_0^\infty \frac{1}{\kappa_\nu} G'(u)\, du, \qquad G'(u) = \frac{15}{4\pi^4} \frac{u^4 e^{-u}}{(1 - e^{-u})^3}. \tag{2.80}$$

The Rosseland weighting factor has a maximum at $h\nu \approx 4kT$, and this means that it is the high-energy photons (with energies several times greater than kT) which are dominant in the energy transfer process.

According to (2.76) the radiation flux is greater, the higher is the coefficient of thermal conductivity, that is, the longer the mean free path. We should not forget, however, that this relationship is valid only if the mean free path is not excessively large; otherwise local equilibrium and formula (2.76) will no longer apply. As we shall see later, in the opposite limiting case, when the radiation mean free path is larger than the characteristic dimensions of the body, the radiation flux decreases with increasing radiation mean free path.

§13. Relationship between the diffusion approximation and the radiation heat conduction approximation

It is customarily assumed in astrophysics that the concepts of the diffusion approximation and of radiation heat conduction are identical. This is due to the fact that stars and stellar photospheres, which are optically thick bodies with small gradients, always simultaneously satisfy the conditions which lead to a weak anistropy of the radiation field, i.e., to a diffusional relationship between the flux and the gradient of the radiant energy density, and to the existence of local equilibrium, i.e., to the replacement of U_ν by $U_{\nu p}$. It has been estimated that generally the deviation from local equilibrium is even less pronounced in the case of small gradients in optically thick bodies than is the degree of anisotropy, that if the diffusion approximation is valid (over the spectrum), the existence of local equilibrium is even more justified. To show this, let the dimensions of the body be of the order of x, which is a characteristic scale for temperature, density, and radiation flux gradients. It follows from the diffusion approximation (2.62) and (2.65) that

$$\frac{S_\nu}{x} \sim \frac{c(U_{\nu p} - U_\nu)}{l'_\nu}, \qquad S_\nu \sim \frac{l'_\nu}{x} cU_\nu,$$

from which

$$\frac{U_{\nu p} - U_\nu}{U_\nu} \sim \left(\frac{l'_\nu}{x}\right)^2.$$

If the degree of anisotropy, which is characterized by the ratio of the diffusion flux to the kinetic flux, $S_\nu/cU_\nu \approx l'_\nu/x$ is small and $l'_\nu/x \ll 1$, then the relative change in the radiation density from its equilibrium value is of second order in the small parameter.

However, in considering problems with more complex conditions than those prevailing in stellar photospheres, it is convenient to draw a clear distinction between the diffusion and radiation heat conduction approximations. Here by the diffusion approximation is meant the method of approximately describing the angular distribution of radiation in which the radiation flux is assumed to be proportional to the gradient of the actual energy density, even if it differs appreciably from the equilibrium value. This may be regarded as a convenient procedure for explaining the characteristics of transfer phenomena of highly nonequilibrium radiation that are not dependent on the nature of the angular distribution of the photons. Rigorous theoretical analysis of the angular distribution of the photons would entail great mathematical difficulty. The diffusion approximation, while resulting in appreciable errors in some cases, does not as a rule alter the qualitative picture of the radiative transfer phenomena even if the angular distribution is strongly anisotropic. This characteristic of the diffusion approximation makes it suitable for the approximate solution of various problems involving nonequilibrium radiation, where the use of the radiation heat conduction approximation would impose certain requirements on the temperature of the fluid which are frequently physically meaningless.

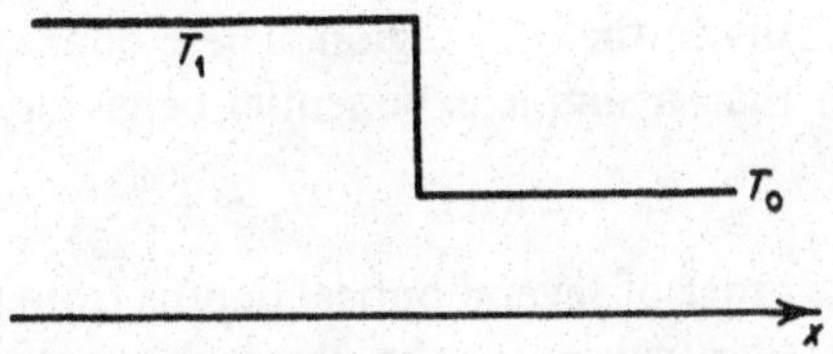

Fig. 2.12. Temperature distribution across a shock wave.

Let us illustrate these remarks by an example. Consider a radiation field in a body with a sharp temperature jump at the surface separating the hot and cold regions, with $T_1 \gg T_0$, as shown in Fig. 2.12 (a case typical of a strong shock wave). The flux density U_1 in the high-temperature region is high and of the order of its equilibrium value $U_{p1} = 4\sigma T_1^4/c$. Practically no photons are emitted in the low-temperature region and its radiation density is determined by the flux emerging from the surface of the heated region. In other words, the radiation density is also proportional to U_1, and since $T_1 \gg T_0$, it is much higher than its equilibrium value $U_{p0} = 4\sigma T_0^4/c$. This case, as we shall see, is completely unlike those for which we have local equilibrium and radiation heat conduction. However, the diffusion approximation for the angular distribution gives a qualitatively correct picture, in stating that if the cold medium absorbs light the radiant energy density and flux decrease with the distance from the hot surface into the cold medium. The length scale for an appreciable decrease in these quantities is the mean free path for photon absorption in the cold medium. Thus, in the given example, the diffusion equations for the cold nonradiating medium are

$$\frac{dS_\nu}{dx} = -\frac{cU_\nu}{l'_\nu}, \qquad S_\nu = -\frac{l'_\nu c}{3}\frac{dU_\nu}{dx}.$$

In terms of the optical thickness, measured from the temperature jump, $\tau_\nu = \int_0^x \kappa'_\nu \, dx$,

$$\frac{dS_\nu}{d\tau_\nu} = -cU_\nu, \qquad S_\nu = -\frac{c}{3}\frac{dU_\nu}{d\tau_\nu}.$$

These equations yield the following solution for the radiation density and flux

$$S_\nu = \frac{cU_\nu}{\sqrt{3}} \sim e^{-\sqrt{3}\tau_\nu},$$

which gives a qualitatively correct description of the attenuation of these quantities.

A more rigorous treatment of the angular distribution, which is possible in this simple case, gives a somewhat different solution for the attenuation of the flux and density in the cold region. These quantities have an exponential integral rather than a simple exponential behavior (see [5]), and

$$S_\nu \sim E_3(\tau_\nu), \qquad U_\nu \sim E_2(\tau_\nu).$$

At distances of the order of several optical depths from the temperature jump the exact solution gives values for the physical quantities of the same order as does the diffusion approximation. If we used the radiation heat conduction approximation, the sharp temperature jump in the fluid would have to spread out, since a discontinuity in temperature would result in the flux $S \sim dT/dx$ becoming infinite.

In general, the diffusion approximation will always give a qualitatively reasonable result. For example, in the limiting case when the angular distribution of photons has a complete anisotropy and all the photons in the cold medium move in one direction, the flux is $S_\nu = cU_\nu$. From the exact continuity equation (2.62) we then see that the flux, as in the case of diffusion, is proportional to the energy density gradient $S_\nu = -l'_\nu c \, dU_\nu/dx$ (the x axis is directed along the light ray) but with a proportionality coefficient three times larger than the ordinary diffusion coefficient. This case of pure absorption of a parallel beam of light in a nonradiating medium has the exact solution

$$S_\nu = cU_\nu \sim e^{-\tau_\nu}, \qquad \tau_\nu = \int_0^x \kappa'_\nu \, dx,$$

which differs from the diffusion approximation solution only by a factor of $\sqrt{3}$ in the exponent and by a factor of $1/\sqrt{3}$ in the relation between the flux and the density.

The quantitative difference for $\tau_\nu \gg 1$ is obviously very high, but qualitatively the diffusion approximation gives a physically correct result; for $\tau_\nu \sim 1$, even the numerical error is moderate.

§14. Radiative equilibrium in stellar photospheres

The study of the temperature and radiation fields in the peripheral layers (photospheres) of stationary stars for the purpose of determining star brightness is a classical problem that served as the basis for the development of radiative transfer theory and of methods of solving the transfer equation*. This problem is of interest here not only as a classical example of the application of radiative transfer theory, but also as a model to which, as will be shown in Chapter IX, we can reduce the problem of the cooling by radiation of a large volume of heated air. Stationary stars comprise tremendous gaseous masses heated to high temperatures, varying from tens of thousands of degrees at the surface to millions and tens of millions of degrees at the center. The gas is maintained in mechanical equilibrium as a result of a balance between the pressure forces tending to burst the gas sphere, and the gravitational forces that prevent such bursting.

The hot gas sphere—a star—radiates from its surface. The energy loss is compensated by the energy released in the nuclear reactions occurring in the central regions of the star. In stationary stars the fluid is at rest and no hydrodynamic motion takes place. The energy released at the center is transferred to the periphery of the star only by radiation and is emitted into space also by radiation. Since no nuclear reaction or energy release occurs in the peripheral layers, the steady state is achieved only by the complete compensation of the emission by the absorption of light in each volume element; the energy loss by radiation q is equal to zero and the temperature at each point does not change with time†.

The equilibrium between the emission and absorption of light in the absence of radiation losses is referred to as radiative equilibrium of a star. It follows from the radiative equilibrium condition $q = 0$ that the divergence of the radiation flux $\nabla \cdot \mathbf{S}$ is also equal to zero. The total radiation flux through a spherical surface of any radius r, $4\pi r^2 S$, is constant and equal to the energy release at the center per unit time ($S \sim 1/r^2$). The temperature and density distributions in the gas as a function of the star radius are determined by simultaneously considering the mechanical equilibrium and radiative transfer. However, in determining the temperature and density distributions in the photosphere

* Detailed presentations of these problems along with bibliographies may be found in the books of Ambartsumian [1] and Unsöld [2].

† The fact that a star is in a steady state and that the distribution of temperature and other quantities with respect to the radius are invariant with time does not mean that such a star does not evolve. When a star is called stationary with reference to a radiative transfer problem, it only means that its state is constant during a time interval comparable to that necessary for the transfer of heat from the center of the star to its surface.

We note that the condition of radiative equilibrium $q = 0$ replaces in this case the hydrodynamic energy equation (2.57).

the problem is essentially broken down into two stages. The temperature distribution as a function of an optical coordinate can be found by considering the radiative transfer alone without any knowledge of the density distribution. Then, if desired, the temperature distribution as a function of radius can be found by introducing the mechanical equilibrium condition and the dependence of the coefficient of light absorption as a function of temperature and density.

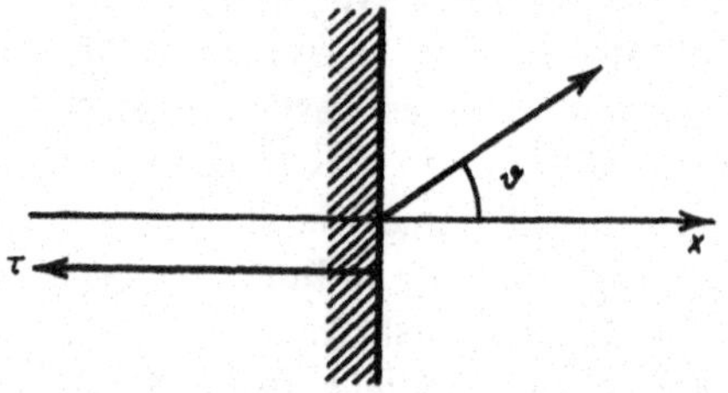

Fig. 2.13. Coordinates for the problem of radiative transfer in stellar photospheres.

Let us formulate the problem of determining the temperature distribution and radiative transfer in the photosphere of a star. Since we are interested in a surface layer whose thickness is much smaller than the radius of the star, we can disregard its curvature and consider the photosphere to be plane. We take the x axis in the direction of the outward normal to the star surface (Fig. 2.13) and write the radiative transfer equation for the plane case

$$\cos\vartheta \frac{dI_\nu}{dx} = \kappa_\nu'(I_{\nu p} - I_\nu), \tag{2.81}$$

where ϑ is the angle between the direction of propagation of the radiation and the x axis. To this equation is added the condition of radiative equilibrium

$$q = \int_0^\infty d\nu \int \kappa_\nu'(I_{\nu p} - I_\nu)\, d\Omega = c \int_0^\infty \kappa_\nu'(U_{\nu p} - U_\nu)\, d\nu = 0, \tag{2.82}$$

and the boundary condition at the surface $x = 0$, which states that no photons arrive from the vacuum side

$$I_\nu(x = 0, \vartheta) = 0 \qquad \text{for } \frac{\pi}{2} < \vartheta < \pi. \tag{2.83}$$

It is usually the case that the absorption coefficient $\kappa_\nu(T, \rho)$ has the same functional dependence on the gas density at different frequencies, so that it may be expressed in the form $\kappa_\nu(T, \rho) = \varphi(\nu, T)f(\rho)$. In this case, by replacing x by the new optical coordinate y defined by $dy = f(\rho)\, dx$, we can eliminate the need to determine directly the gas density distribution as a function of x. Instead we may first determine the temperature and radiation density distributions as a function of this new optical coordinate y. Equations (2.81)–(2.83) describe these distributions completely. The problem has only one arbitrary parameter, the radiation flux S, which in the plane case is constant ($q = \nabla \cdot \mathbf{S} = dS/dx = 0$). The flux S is equal to the energy flux from $x = -\infty$ (from within the star) and is in fact determined by the energy release at the star center. At the same time the flux S represents the total radiant energy flux emanating from the surface of the star, that is, the integrated surface brightness.

The problem as formulated generally presents considerable mathematical difficulties. The principal difficulty arises from the fact that the transfer equation is written for the spectral intensity I_ν while the condition of radiative equilibrium involves quantities integrated over the entire spectrum. To simplify the problem, we introduce an absorption coefficient averaged over the spectrum κ' (which is equivalent to assuming that we are dealing with a "gray" medium) and integrate the transfer equation (2.81) over the spectrum. We then obtain for the integrated intensity $I = \int_0^\infty I_\nu \, d\nu$ the relation

$$\cos\vartheta \frac{dI}{dx} = \kappa'(I_p - I), \qquad I_p = \int_0^\infty I_{\nu p} \, d\nu = \frac{cU_p}{4\pi} = \frac{\sigma T^4}{\pi}. \tag{2.84}$$

Transforming to an optical coordinate, measured from the surface into the photosphere, $d\tau = -\kappa' \, dx$, $\tau = -\int_0^x \kappa' \, dx$, we obtain

$$\cos\vartheta \frac{dI}{d\tau} = I - I_p(T). \tag{2.85}$$

The boundary condition (2.83) takes the form

$$I(\tau = 0, \vartheta) = 0 \qquad \text{for} \quad \frac{\pi}{2} < \vartheta < \pi, \tag{2.86}$$

and the radiative equilibrium condition (2.82) is

$$\int I \, d\Omega = \int I_p \, d\Omega, \qquad U = U_p = \frac{4\sigma T^4}{c} \tag{2.87}$$

(the constant flux S is $S = \int \cos\vartheta \, I \, d\Omega$).

Although the radiation is anisotropic, the integral with respect to angle of the integrated intensity, i.e., the integrated radiation density at each point, is equal to its equilibrium value U_p. Or, more exactly, the temperature of the fluid at each point, controlled by the radiative transfer, is determined by the radiation density at the point $U = U_p$. Even in its simplified form the solution of the system (2.85)–(2.87) (the so-called Milne problem) is very complex from a mathematical point of view. An approximate solution of this system will be presented in the following section. However, we shall now derive an integral equation equivalent to this system, which can serve as the basis for finding an exact solution.

We employ an integral expression for the intensity of the type of (2.32), which in the plane case can be written in a form that follows directly from the differential equation (2.85) for I:

$$I(\vartheta, \tau) = \int_{\tau}^{\infty} I_p[T(\tau')] e^{-\frac{\tau'-\tau}{\cos\vartheta}} \frac{d\tau'}{\cos\vartheta}, \qquad \frac{\pi}{2} > \vartheta > 0. \tag{2.88}$$

$$I(\vartheta, \tau) = -\int_{0}^{\tau} I_p[T(\tau')] e^{-\frac{\tau'-\tau}{\cos\vartheta}} \frac{d\tau'}{\cos\vartheta}, \qquad \pi > \vartheta > \frac{\pi}{2}. \tag{2.89}$$

The first equation gives the radiation intensity propagated toward the surface. The integration is carried out from $\tau = \infty$, since the photosphere is assumed to be semi-infinite. The second equation corresponds to the radiation propagated into the star, taking into account that no photons arrive from the vacuum.

Let us calculate the radiation density $U = (1/c)\int I \, d\Omega$, using the first equation for the integration with respect to ϑ from 0 to $\pi/2$, and the second equation for the interval $\pi/2 < \vartheta < \pi$:

$$cU = 2\pi \int_0^{\pi/2} \sin\vartheta \, d\vartheta \int_{\tau}^{\infty} I_p e^{-\frac{\tau'-\tau}{\cos\vartheta}} \frac{d\tau'}{\cos\vartheta}$$
$$- 2\pi \int_{\pi/2}^{\pi} \sin\vartheta \, d\vartheta \int_0^{\tau} I_p e^{-\frac{\tau'-\tau}{\cos\vartheta}} \frac{d\tau'}{\cos\vartheta}.$$

Reversing the order of integration, introducing in the first integral the variable $w = 1/\cos\vartheta$ and in the second $w = -1/\cos\vartheta$, noting the definition of the exponential integral (2.44), and setting $I_p = cU_p/4\pi$, we get

$$U = \frac{1}{2}\int_{\tau}^{\infty} U_p E_1(\tau' - \tau) \, d\tau' + \frac{1}{2}\int_0^{\tau} U_p E_1(\tau - \tau') \, d\tau'. \tag{2.90}$$

Recalling the radiative equilibrium condition $U = U_p \sim T^4$, we obtain finally an integral equation for the integrated equilibrium density U_p, or equivalently for T^4,

$$U_p(\tau) = \frac{1}{2}\int_0^{\infty} U_p(\tau') E_1(|\tau' - \tau|) \, d\tau'. \tag{2.91}$$

We also write for future reference the integral expression for the flux in the plane case, which is calculated in a manner similar to the density*:

$$S = \frac{c}{2}\int_{\tau}^{\infty} U_p E_2(\tau' - \tau)\,d\tau' - \frac{c}{2}\int_{0}^{\tau} U_p E_2(\tau - \tau')\,d\tau'. \tag{2.92}$$

From (2.91) it is apparent that the solution $U_p(\tau)$ is determined to within a constant factor. This factor corresponds to the arbitrary value of the flux S.

§15. Solution to the plane photosphere problem

We now seek a solution to the problem formulated in the preceding section by use of the diffusion approximation. Averaging the diffusion approximation equations over the spectrum and introducing the average absorption coefficient κ' and the average mean free path $l' = 1/\kappa'$, we write these equations as

$$\frac{dS}{dx} = c\kappa'(U_p - U), \tag{2.93}$$

$$S = -\frac{l'c}{3}\frac{dU}{dx}. \tag{2.94}$$

On replacing the x coordinate by the optical thickness τ ($d\tau = -\kappa'\,dx$), we obtain

$$\frac{dS}{d\tau} = c(U - U_p), \tag{2.95}$$

$$S = \frac{c}{3}\frac{dU}{d\tau}. \tag{2.96}$$

Equation (2.95) demonstrates the equivalence of the radiative equilibrium condition $U = U_p$ and the condition of constancy of flux $S = const$. In the case considered the radiative equilibrium condition also leads to a rigorous equivalence between the diffusion approximation and the radiation heat conduction approximation, since by virtue of the equality $U = U_p$,

$$S = \frac{c}{3}\frac{dU_p}{d\tau} = \frac{4}{3}\sigma\frac{dT^4}{d\tau}. \tag{2.97}$$

Solving equation (2.97) and applying the boundary condition (2.66)

$$S = 2\sigma T_0^4, \tag{2.98}$$

* For the point $\tau = 0$ this equation has already been obtained in §7, equation (2.45). It is interesting to compare the exact equations for the density (2.90) and the flux in the plane case (2.92) with those obtained in the "forward-reverse" approximation (2.73). The latter differ from the former in the numerical coefficients and in the replacement of the ordinary exponentials by exponential integrals.

for the flux S in terms of the surface temperature T_0 we obtain the following temperature and radiation density distribution in terms of the optical thickness:

$$U = U_p = \frac{4\sigma T^4}{c} = \frac{4\sigma T_0^4}{c}(1 + \tfrac{3}{2}\tau). \tag{2.99}$$

The brightness temperature of the surface (from the definition $S = \sigma T_{\text{br}}^4$) is given by $T_{\text{br}} = 2^{1/4}T_0 \approx 1.2T_0$.

The brightness temperature is somewhat higher than the true surface temperature T_0. This is obvious, since the photons emerging from the surface are born in a radiating layer near the surface with a thickness of the order of a mean free path (the optical thickness is of the order of unity). The temperature of the radiating layer is slightly higher than the surface temperature (Fig. 2.14) and therefore the "temperature" of the emerging radiation is also

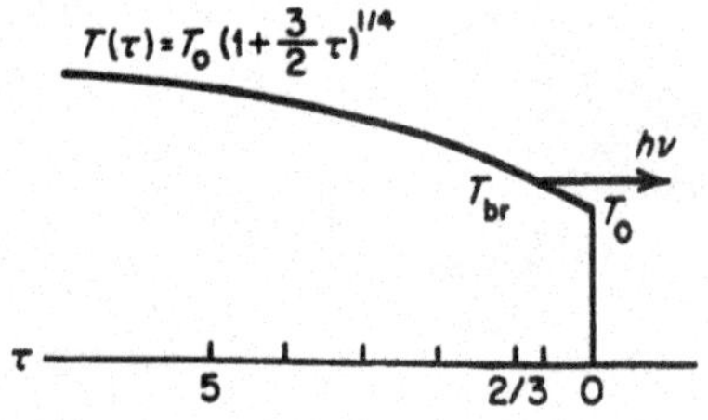

Fig. 2.14. Temperature distribution along the optical coordinate in a plane photosphere according to the diffusion approximation.

slightly higher. The brightness temperature is the same as the temperature of the medium at an optical depth $\tau = 2/3$. We can say that this depth corresponds approximately to the center of the radiating layer.

An exact analytic solution has been found for the problem of radiative equilibrium in a photosphere considered as "gray matter". This solution satisfies the exact integral equation (2.91). The problem has also been solved by various approximate methods, ones that are more exact than the diffusion approximation. (This problem, one of the few in the radiative transfer theory which can be solved exactly, usually serves as a standard for checking different approximate methods.)

The surface temperature T_0 obtained from the exact solution for the same flux S, that is, for the same brightness temperature T_{br}, turns out to be slightly lower than that given by the diffusion approximation. Namely, in the exact solution $T_0^4 = (\sqrt{3}/4)T_{\text{br}}^4$, $T_0 = 0.811T_{\text{br}}$, while in the diffusion approximation $T_0^4 = \frac{1}{2}T_{\text{br}}^4$, $T_0 = 0.841T_{\text{br}}$. The temperature distributions as functions of optical thickness are quite close to each other in the exact and the diffusion solutions (Fig. 2.15). This indicates that good accuracy can be obtained from the diffusion approximation. As expected, the error with the diffusion approximation is smaller, the greater is the optical depth, the farther away we are from the boundary. As $\tau \to \infty$ the exact solution $U_p(\tau)$ asymptotically approaches the diffusion solution (2.99). This can be verified directly from the integral expressions for the density and the flux (2.91) and (2.92). The usefulness of such an asymptotic analysis lies in the fact that it shows the manner in which the diffusion and the radiation heat conduction approximations

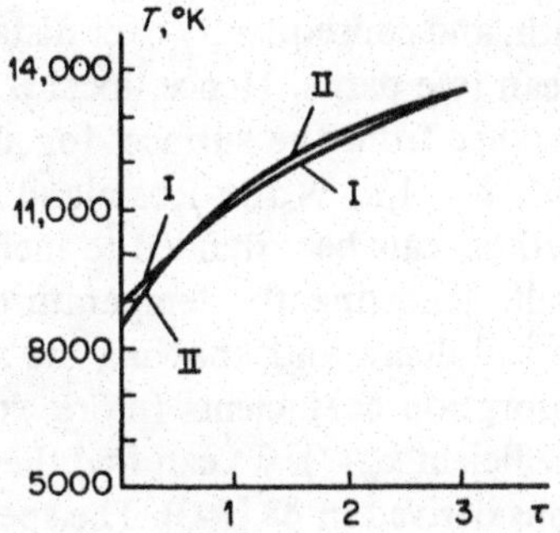

Fig. 2.15. Comparison of temperature distribution in a plane photosphere, calculated by the diffusion approximation (I) and by the exact solution (II). The brightness temperature was taken to be 10,500°K. The figure is taken from [1].

follow from the asymptotic behavior of the exact solution. The diffusion solution (2.99) shows that the relative change in the equilibrium density U_p over a mean free path decreases with the distance from the surface, as $\tau \to \infty$

$$\frac{\Delta U_p}{U_p} \approx \frac{l}{U_p}\frac{dU_p}{dx} = \frac{1}{U_p}\frac{dU_p}{d\tau} = \frac{1}{\frac{2}{3} + \tau} \approx \frac{1}{\tau}.$$

The exponential integrals E_1 and E_2 decrease rapidly with an increase in the argument, so that only the region $|\tau' - \tau| \sim 1$ about the point τ gives any significant contribution to the integrals (2.91) and (2.92). Therefore, the integration with respect to τ' from 0 to τ in the second integral of both (2.90) and (2.92) can for $\tau \gg 1$ be extended to $-\infty$ or, equivalently, the integration with respect to $\tau - \tau'$ from 0 to $\tau \gg 1$ can be extended to the range 0 to ∞. The larger the value of τ, the smaller is the error resulting from this change in the limits of integration.

Let us expand $U_p(\tau')$ about the point τ:

$$U_p(\tau') = U_p(\tau) + \frac{dU_p}{d\tau}(\tau' - \tau) + \frac{1}{2}\frac{d^2U_p}{d\tau^2}(\tau' - \tau)^2 + \cdots.$$

Since $U_p(\tau)$ is a slowly varying function as $\tau \to \infty$, the higher derivatives become smaller and smaller. Substituting this expansion into (2.90) and (2.91) and evaluating the integrals, we obtain (with an accuracy proportional to the retained highest order derivatives of U_p with respect to τ) $S = (c/3)\, dU_p/d\tau$ from (2.92) and $d^2U_p/d\tau^2 = 0$ from (2.91). These results correspond to the diffusion and radiation heat conduction approximations.

The fact that the integrated radiation density at each point is equal to its equilibrium value corresponding to the temperature of the fluid does not mean that the same is true for the spectral densities*. However, the further away from the surface and into the photosphere we are, the less are the relative temperature changes over distances of the order of the average mean free path, and consequently over distances of the order of the frequency dependent mean free paths. Hence, local equilibrium exists for all points at a sufficient distance from the surface, for all frequencies. Here the averaged mean free path $l' = 1/\kappa'$ is the Rosseland mean free path. The Rosseland averaging method can be extended to include the entire photosphere up to the surface itself. Knowing the temperature distribution as a function of the average optical depth and knowing the relations governing the frequency dependent absorption coefficients (more exactly, their ratio to the average absorption coefficient κ'_ν/κ'), we can find the radiation spectrum of the star from the relations derived in §8 [1–3]. The spectrum, in general, does not coincide with the Planck spectrum corresponding to T_{br}, but in a number of cases it is quite close to it.

§16. Radiation energy losses of a heated body

Let us consider the energy losses resulting from the radiation from a heated body. We shall consider ordinary bodies of finite dimensions, heated nonuniformly in general. The total energy Q lost by the body per unit time is, obviously, equal to the volume integral of the energy q lost per unit volume per unit time. Noting that $q = \nabla \cdot \mathbf{S}$, we can write†

$$Q = \int q\, dV = \int S_0\, d\Sigma, \tag{2.100}$$

where dV is an element of volume of the body, $d\Sigma$ is an element of surface, and S_0 is the normal component of the radiation flux through the surface. We may also write $S_0 = \sigma T_{br}^4$ where T_{br} is the brightness temperature of the surface.

* In the same way that it does not follow from $S = const$ that $S_\nu = const$;

$$\frac{dS_\nu}{dx} = \kappa'_\nu(U_{\nu p} - U_\nu) \neq 0.$$

† The quantity q can change sign over the body, so that some parts of the body are cooled while others are heated by radiation.

The brightness temperature does not necessarily have to be close to the average temperature T of the nonuniformly heated body. In the case of an optically thick body, whose dimension x is much greater than the mean free path l (say, corresponding to some average temperature), the order of magnitude of the flux is

$$S_0 \sim lc\frac{U_p}{x} \sim \frac{l}{x}\sigma T^4 \ll \sigma T^4; \qquad T_{br} \sim \left(\frac{l}{x}\right)^{1/4} T.$$

For $(l/x)^{1/4} \ll 1$, then, $T_{br} \ll T$. The brightness temperature is rather close to the surface temperature. Only in the case of bodies which are optically thin enough can the brightness temperature be close to the average temperature of the body (the temperatures T_{br} and T can also be close to each other when the body is maintained by some means at a constant temperature).

Let us now consider an optically thin body, whose dimensions are quite small in comparison with an appropriate average photon mean free path*. If the optical thickness of the body x/l is small, then almost all the photons born at any point in the body emerge at the surface. Only a fraction of the photons, of the order of $x/l \ll 1$, are absorbed during their transit. The radiation density in the body is of the order of x/l of its equilibrium value, and is thus considerably lower than the equilibrium radiation density (the radiation is strongly out of equilibrium). In fact, according to (2.32) the radiation intensity at any point is equal to the integral with respect to the density of the sources along the ray within the limits of the body. Since the body is optically thin, $\int_{s'}^{s} \kappa'_\nu \, ds \sim x''/l_\nu \ll 1$, and the exponential factor characterizing the photon absorption is close to unity. Therefore, the intensity $I_\nu \sim (x/l_\nu)I_{\nu p}$, and the radiation density obtained from integrating over all angles is $U_\nu \sim (x/l_\nu)U_{\nu p}$. Integrating U_ν over the spectrum and introducing a frequency averaged mean free path l_1, we get $U \sim (x/l_1)U_p \ll U_p$. The amount of energy absorbed per unit volume per unit time is a fraction of the order of x/l_1 of the energy emitted per unit volume per unit time. This is true because the ratio of these quantities is U/U_p, as may be seen from the relation (2.61) for q.

Thus, in the case of an optically thin body the energy lost by the fluid per unit volume per unit time turns out to be equal (with an accuracy of the order of x/l_1) to the emitted energy, that is, to the integrated emission coefficient

$$J = \int_0^\infty J_\nu \, d\nu = c \int_0^\infty \kappa'_\nu U_{\nu p} \, d\nu. \tag{2.101}$$

* We denote the frequency averaged mean free path for the case of an optically thin body by l_1 in order to avoid confusion with the Rosseland mean free path l which characterizes an optically thick body. As we shall show below, the law for averaging the absorption over the entire spectrum in the case of an optically thin body differs from the Rosseland averaging law.

We take outside the integral sign the average value of the absorption coefficient, denoted by κ_1 and by definition equal to the reciprocal of the mean free path l_1. Equation (2.101) becomes

$$J = \kappa_1 U_p c = \frac{4\sigma T^4}{l_1}. \qquad (2.102)$$

Equating (2.102) and (2.101) gives us the law for averaging the mean free path for an optically thin body

$$\kappa_1 = \frac{1}{l_1} = \frac{\int_0^\infty \kappa'_\nu U_{\nu p}\, d\nu}{\int_0^\infty U_{\nu p}\, d\nu} = \int_0^\infty \kappa'_\nu G_1(u)\, du. \qquad (2.103)$$

The weighting function is

$$G_1(u) = \frac{15}{\pi^4} \frac{u^3}{e^u - 1}, \qquad u = \frac{h\nu}{kT}. \qquad (2.104)$$

In terms of the actual absorption coefficient we have

$$\kappa_1 = \frac{1}{l_1} = \int_0^\infty \kappa_\nu G'_1(u)\, du, \qquad (2.105)$$

$$G'_1(u) = (1 - e^{-u}) G_1(u) = \frac{15}{\pi^4} e^{-u} u^3. \qquad (2.106)$$

This averaging method is different from the Rosseland method (2.77) in that in the Rosseland method it is the mean free path or reciprocal of the absorption coefficient which is averaged with a weighting function which is proportional to the derivative with respect to temperature of the Planck function. On the other hand, the integrated emission coefficient (2.102) is characterized by a mean free path obtained by averaging the absorption coefficient itself with a weighting function proportional to the Planck function*.

The total energy lost by a heated optically thin body is given by the emission coefficient integrated over the volume

$$Q = \int q\, dV = \int J\, dV. \qquad (2.107)$$

We note that an optically thick body is mainly cooled by radiation "from the surface", whereas in the case of an optically thin body the entire volume participates in the cooling process. Of course, it is possible to introduce the concept

* *Editors' note.* This inverse of a frequency averaged absorption coefficient is usually termed the Planck mean free path.

of radiation flux from the surface even in this case and to write (2.107) in the form of a surface integral, since the relation $q = \nabla \cdot \mathbf{S}$ is always valid. However, in the case of volume cooling this interpretation of energy losses is purely formal, while in the case of an optically thick body the emitted photons are actually born in the surface layer. Accordingly, the radiation spectrum of an optically thick body is to a certain degree close to the Planck spectrum corresponding to the brightness temperature T_{br} or the surface temperature. The radiation spectrum of an optically thin body can be essentially different from the Planck spectrum corresponding to the temperature of the body, if the absorption coefficient of the fluid is strongly frequency dependent. In this case the spectrum is characterized by the frequency function $\kappa_\nu' U_{\nu p}$.

Let us compare the radiant energy losses per unit volume of the body (the cooling rate per unit volume) and those per unit surface area (flux from the surface) for cases of optically thick and optically thin bodies. If the linear dimensions of the body are of the order of x, then its surface area is of the order of x^2 and its volume is of the order of x^3. The cooling rate per unit surface area for an optically thick body is of the order of

$$\frac{Q}{x^2} \sim S \sim \frac{l}{x}\,\sigma T^4 \ll \sigma T^4, \qquad \frac{l}{x} \ll 1, \tag{2.108}$$

and the cooling rate per unit volume is

$$\frac{Q}{x^3} \sim \frac{S}{x} \sim \frac{1}{x}\frac{l}{x}\,\sigma T^4 \sim \left(\frac{l}{x}\right)^2 \frac{\sigma T^4}{l} \ll \frac{\sigma T^4}{l}. \tag{2.109}$$

In the case of an optically thin body however,

$$\frac{Q}{x^2} \sim \frac{Jx^3}{x^2} \sim \frac{x}{l_1}\,\sigma T^4 \ll \sigma T^4; \qquad \frac{x}{l_1} \ll 1, \tag{2.110}$$

$$\frac{Q}{x^3} \sim J \sim \frac{\sigma T^4}{l_1}, \qquad \frac{x}{l_1} \ll 1. \tag{2.111}$$

Let us compare the relative energy losses of two bodies at approximately the same average temperature; one of the bodies has large dimensions and is optically thick, while the other has small dimensions and is optically thin. We assume that for both bodies the densities and the temperatures are nearly equal so that the mean free paths l and l_1 (which are functions of the temperature and density of the fluid only) are of the same order. As a rule, the different methods in averaging over the spectrum do not lead to large numerical differences between the mean free paths; l and l_1 usually do not differ by more than a factor of two to four. From equations (2.108) and (2.110) it is apparent that the losses in both cases per unit surface area are less than σT^4. Only a body whose dimensions are of the order of a mean free path (an optical thickness of the order of unity, $x \sim l \sim l_1$) emits from its surface a radiant energy flux corresponding to a perfect black body at a temperature of the order of the average temperature of the body. Let us consider the losses per unit volume (or per unit mass) of an optically thick body. Here, the mass cooling rate is much lower than in the case of an optically thin body. In the latter case the rate is of the order of the integrated emission coefficient $J \sim \sigma T^4/l_1$ averaged over the volume and independent of the dimensions (by virtue of the volume character of the radiation). The physical reason for this is clear: the photons emitted within the optically thick layer are "locked up" within the body and cannot reach the outside, since they are absorbed during their transit inside the body.

§17. Hydrodynamic equations accounting for radiation energy and pressure and radiant heat exchange

In §9 we have indicated a method of accounting for the interaction between the radiation and the fluid, which simply reduces to the determination of light emission and absorption. The radiation pressure and energy were assumed to be small in comparison with the energy and pressure of the fluid. At very high temperatures or in a very rarefied gas, the radiation energy and pressure cannot be neglected as long as the dimensions of the gaseous body are large compared to the radiation mean free path. It is clear in the case of local equilibrium between the radiation and the fluid, when $U = U_p = 4\sigma T^4/c$ and the radiation pressure is $p_\nu = U_p/3 = \frac{4}{3}\sigma T^4/c$, that the radiation energy and pressure must be added to the internal energy and pressure of the fluid, and that a radiant heat transfer term should be included in the hydrodynamic equations. Let us show how this statement follows from the general equations describing the fluid-radiation system.

In order to write in a complete form the laws of conservation of energy and momentum for the fluid-radiation system (which, in general, is not in equilibrium), it is convenient to start with the divergence form of the equations, which are equivalent to the "continuity" equations for the quantities being conserved. These equations were formulated in Chapter I for the flow of a perfect gas in the absence of radiation (equations (1.7) and (1.10)). The equations for the fluid-radiation system can be easily obtained by a straightforward generalization of (1.7) and (1.10) (we recall that we are considering nonrelativistic motions only). For example, we add to the momentum density $\rho\mathbf{u}$ the momentum density $\mathbf{G}$ of the radiation, and to the tensor momentum flux density of the fluid Π_{ik} we add the tensor momentum flux density of the radiation T_{ik}. The latter is, of course, equivalent to the Maxwell stress tensor for an electromagnetic field. In exactly the same way, we add the integrated radiant energy density U to the energy density of the fluid and to the energy flux density of the fluid we add the integrated radiant energy flux $\mathbf{S}$, i.e., the Poynting vector. We remind the reader that the radiation momentum is related to the Poynting vector by the relation $\mathbf{G} = \mathbf{S}/c^2$. In this manner we obtain the momentum and energy equations for the fluid-radiation system

$$\frac{\partial}{\partial t}(\rho u_i + G_i) + \frac{\partial}{\partial x_k}(\Pi_{ik} + T_{ik}) = 0, \tag{2.112}$$

$$\frac{\partial}{\partial t}\left(\rho\varepsilon + \frac{\rho u^2}{2} + U\right) + \frac{\partial}{\partial x_k}\left\{\rho u_k\left(\varepsilon + \frac{p}{\rho} + \frac{u^2}{2}\right) + S_k\right\} = 0. \tag{2.113}$$

The continuity equation obviously remains unchanged, since the radiation "has no mass"*

$$\frac{\partial \rho}{\partial t} + \frac{\partial}{\partial x_k}(\rho u_k) = 0.$$

Equations (2.112) and (2.113), formulated above by simply generalizing the hydrodynamic equations, have a clear and precise physical meaning. They can be also obtained in a strictly formal manner by starting with the conservation equations written for the four-dimensional energy-momentum tensor for the fluid-radiation system, and then applying the nonrelativistic approximation to the tensor component associated with the fluid (we shall not present here this fairly elementary derivation).

The quantities characterizing the radiation which are included in equations (2.112) and (2.113) may be interpreted in two ways. When treated as components of an electromagnetic field they are expressed in terms of the electric and magnetic field intensities $\mathbf{E}$ and $\mathbf{H}$

$$U = \frac{E^2 + H^2}{8\pi},$$

$$\mathbf{S} = \frac{c}{4\pi}\mathbf{E} \times \mathbf{H} = c^2\mathbf{G}, \tag{2.114}$$

$$T_{ik} = \frac{1}{4\pi}\left\{-E_iE_k - H_iH_k + \tfrac{1}{2}\,\delta_{ik}(E^2 + H^2)\right\}.$$

It is only necessary to keep in mind that radiation is a rapidly varying electromagnetic field; the period of the electromagnetic oscillations is negligibly small in comparison with macroscopic flow times, and hence it is understood that the above equations contain time-averaged quantities averaged over a period large in comparison with the period of oscillation.

* When $U \sim \varepsilon \ll \rho c^2$.

In the quantum-mechanical treatment the macroscopic quantities U, $\mathbf{S}$, and T_{ik} are expressed in terms of the photon distribution function. If $f(\nu, \mathbf{\Omega}, \mathbf{r}, t)$ is the distribution function at the point $\mathbf{r}$ at the time t for a particular frequency ν and a particular direction of motion of the photons $\mathbf{\Omega}$, then, as shown in §1 of the present chapter,

$$U = \int h\nu f \, d\mathbf{\Omega} \, d\nu,$$

$$\mathbf{S} = \int h\nu c \mathbf{\Omega} f \, d\mathbf{\Omega} \, d\nu, \tag{2.115}$$

$$T_{ik} = \int \Omega_i \Omega_k h\nu f \, d\mathbf{\Omega} \, d\nu \, *.$$

Expanding the rapidly varying electromagnetic fields in Fourier integrals, we can represent these fields as a superposition of harmonic oscillations of different frequencies. In averaging over time the terms contained in the relations for U, S_i, and T_{ik} which are quadratic with respect to the field components and which are products of quantities involving different frequencies vanish. Only quadratic terms containing products of the Fourier components corresponding to the same frequency remain. Therefore, the energy, momentum, and the radiation energy and momentum fluxes are represented as a linear superposition of terms corresponding to different frequencies. This allows us to introduce the concept of a radiation intensity at a given frequency $I_\nu(\mathbf{\Omega}, \mathbf{r}, t)$ and to express the macroscopic quantities in terms of integrals of the intensity over the spectrum and over the direction of propagation of the radiation

$$U = \frac{1}{c} \int I_\nu \, d\mathbf{\Omega} \, d\nu,$$

$$\mathbf{S} = \int \mathbf{\Omega} I_\nu \, d\mathbf{\Omega} \, d\nu, \tag{2.116}$$

$$T_{ik} = \frac{1}{c} \int \Omega_i \Omega_k I_\nu \, d\mathbf{\Omega} \, d\nu.$$

This linear superposition also allows us to go over to the quantum-mechanical treatment of intensity as the product of the photon energy and the distribution function ($I_\nu = h\nu c f$).

* The energy of a photon is $h\nu$, its momentum is $\mathbf{\Omega} h\nu/c$, and the flux of the ith component of the momentum in the kth direction is $\Omega_i \Omega_k (h\nu/c)c$, from which we obtain the expression for the momentum flux tensor T_{ik}.

It is well known that electromagnetic fields, frequencies, and directions of propagation of electromagnetic waves and, consequently, the integral quantities U, $\mathbf{S}$, and T_{ik} depend on the coordinate system in which they are measured. The integral quantities contained in equations (2.112) and (2.113) pertain to a stationary, "laboratory" coordinate system, where the given fluid particle moves with the velocity $\mathbf{u}$. It is more convenient, however, to employ the radiation parameters measured in the particle's "own" coordinate system, the system in which the particle is at rest. Indeed, it is the radiant energy density in a fluid at rest which is equal to the equilibrium value $U_p = 4\sigma T^4/c$ in the state of complete thermodynamic equilibrium; and it is the radiation flux with respect to a fluid at rest which has the character of a diffusion. The radiation is "carried away" together with the moving fluid and the total flux includes this "carried away" radiation.

Let us transform the quantities U, $\mathbf{S}$, and T_{ik} in (2.112) and (2.113) to corresponding primed quantities which are taken in a coordinate system attached to the moving particles of the fluid. If the medium moves with nonrelativistic velocities $u/c \ll 1$, so that it is possible to neglect those terms proportional to u/c, the transformation to the moving coordinate system gives (see [6])

$$\begin{aligned} U &= U', \\ S_i &= S_i' + u_i U' + u_k T_{ik}', \\ T_{ik} &= T_{ik}'. \end{aligned} \tag{2.117}$$

Let us introduce the transformed quantities into equations (2.112) and (2.113). We note that the radiation momentum G_i is extremely small in comparison with the momentum of the fluid ρu_i, and hence can be neglected*. Expanding the momentum flux tensor of the fluid as $\Pi_{ik} = \rho u_i u_k + p\,\delta_{ik}$, we get

$$\begin{gathered} \frac{\partial}{\partial t}(\rho u_i) + \frac{\partial}{\partial x_k}(\rho u_i u_k) + \frac{\partial p}{\partial x_i} + \frac{\partial T_{ik}'}{\partial x_k} = 0, \\ \frac{\partial}{\partial t}\left(\rho\varepsilon + \frac{\rho u^2}{2} + U'\right) + \frac{\partial}{\partial x_k}\left\{\rho u_k\left(\varepsilon + \frac{p}{\rho} + \frac{u^2}{2}\right) + S_k' + u_k U' + u_i T_{ik}'\right\} = 0 \end{gathered} \tag{2.118}$$

(these equations were derived by Belen'kii [7]).

Let us consider the case of local thermodynamic equilibrium between the radiation and the fluid. The radiant energy density is then equal to its equilibrium value $U' = 4\sigma T^4/c$. The radiant energy flux with respect to the fluid

* Actually, if the radiant energy is comparable to the energy of the fluid, that is, $U \sim \rho u^2$, then the radiation momentum which is of the order of $G \sim U/c$ is smaller by a factor of u/c than the momentum of the fluid ρu:

$$G \sim \frac{U}{c} \sim \frac{u}{c}\rho u.$$

S'_k is approximately proportional to the gradient of the equilibrium radiation density. According to the relation (2.76) for radiation heat conduction

$$S'_k = -\frac{lc}{3}\frac{\partial}{\partial x_k}\left(\frac{4\sigma T^4}{c}\right) = -\frac{16\sigma l T^3}{3}\frac{\partial T}{\partial x_k}.$$

The momentum flux tensor is most simply obtained from (2.116), by noting that for local equilibrium the radiation field is almost isotropic and that the intensity is only weakly dependent upon angle. We find

$$T'_{ik} = \frac{U'_p}{3}\,\delta_{ik} = p_\nu\,\delta_{ik},$$

where $p_\nu = \frac{1}{3}U'_p = \frac{4}{3}\sigma T^4/c$ is the radiation pressure. Substituting these quantities into equations (2.118) we find for the case of local equilibrium that

$$\frac{\partial}{\partial t}(\rho u_i) + \frac{\partial}{\partial x_k}(\rho u_i u_k) + \frac{\partial}{\partial x_i}(p + p_\nu) = 0, \tag{2.119}$$

$$\frac{\partial}{\partial t}\left(\rho\varepsilon + \frac{\rho u^2}{2} + U_p\right) + \frac{\partial}{\partial x_k}\left\{u_k\left(\rho\varepsilon + U_p + \frac{\rho u^2}{2} + p + p_\nu\right) - \frac{lc}{3}\frac{\partial U_p}{\partial x_k}\right\} = 0, \tag{2.120}$$

where $U_p = 3p_\nu = 4\sigma T^4/c$. The momentum and energy equations of the system are now in closed form, since all the quantities describing the radiation are expressed in terms of temperature and of the optical properties of the fluid.

If the radiation is not in local thermodynamic equilibrium with the fluid, then the radiative transfer equation must be added to equations (2.118). The radiative transfer equation for a moving medium taking into account terms of order u/c is discussed in [8].

§18. The number of photons as an invariant of the classical electromagnetic field

The radiation field is characterized by the density and flux of energy (the momentum density being proportional to the latter); these quantities are determined by equations (2.114). It is remarkable that in the classical theory it is possible to introduce still another characteristic quantity which (to within an arbitrary multiplier) coincides with the number of photons in the field. At first sight the number of photons is, by definition, a quantum concept, has no place, and cannot be defined in the classical theory. Let us recall, however, that in classical mechanics the integral $\oint p\,dq = h(n + \frac{1}{2})$ which yields the number of the quantum state of an oscillator appears as an adiabatic invariant. Similarly, the number of photons in a given electromagnetic field should be an adiabatic invariant. In addition, it should be a time-invariant quantity and a relativistic invariant.

The quantity which is proportional to the number of photons is given by

$$N \sim \iint \frac{\mathbf{E}(\mathbf{r}) \cdot \mathbf{E}(\mathbf{r}') + \mathbf{H}(\mathbf{r}) \cdot \mathbf{H}(\mathbf{r}')}{|\mathbf{r}' - \mathbf{r}|^2} \, d\mathbf{r} \, d\mathbf{r}', \tag{2.121}$$

where the integration is carried out over the spaces $\mathbf{r}$ and $\mathbf{r}'$ ($d\mathbf{r}$ and $d\mathbf{r}'$ denote volume elements). The structure of this expression can be easily understood if we change to the integration variables $\mathbf{r}$ and $\boldsymbol{\rho} = \mathbf{r}' - \mathbf{r}$. We fix the point $\mathbf{r}$ and write an integral over the volume $d\boldsymbol{\rho}$ in polar coordinates with center at the point $\mathbf{r}$

$$\mathbf{E}(\mathbf{r}) \cdot \int \mathbf{E}(\mathbf{r} + \rho) \frac{\rho^2 \, d\rho \, d\Omega_\rho}{\rho^2},$$

where $d\Omega_\rho$ is an element of solid angle about the point $\mathbf{r}$.

The above expression is of the same order of magnitude as $E^2(\mathbf{r})\bar{\lambda}$, where $\bar{\lambda}$ is an average distance at which the correlation between the field at the given point $\mathbf{E}(\mathbf{r})$ and the field in the vicinity of that point is retained. Setting up an analogous expression using $\mathbf{H}$, we find that, to the correct order of magnitude,

$$N \sim \int (E^2 + H^2)\bar{\lambda} \, d\mathbf{r} \sim \int U(\mathbf{r})\bar{\lambda} \, d\mathbf{r}.$$

Let us now derive (2.121) for the number of photons*. We consider a field which contains neither free nor bound charges. Then $\nabla \cdot \mathbf{E} = 0$ and $\nabla \cdot \mathbf{H} = 0$. The number of photons in the given frequency interval $d\nu$ is equal to the energy of the field which is contained in this interval divided by $h\nu$. In order to determine the spectral energy we expand the field in a Fourier integral over the space

$$\mathbf{E}(\mathbf{r}) = \int \mathbf{E}_\mathbf{k} e^{i\mathbf{k} \cdot \mathbf{r}} \, d\mathbf{k}.$$

By virtue of the field equations under the specified conditions, the Fourier components $\mathbf{E}_\mathbf{k}$ are functions of time proportional to $\exp(-i\omega_k t)$, where $\omega_k = ck$ ($\omega_k = 2\pi\nu_k$). We introduce for each direction of the wave vector $\mathbf{k}$ the directions of light polarization $\mathbf{n}_1$ and $\mathbf{n}_2$, which are perpendicular to $\mathbf{k}$. The Fourier components of a field with the given polarity are

$$E_{k_1} = \int \mathbf{E} \cdot \mathbf{n}_1 e^{-i\mathbf{k} \cdot \mathbf{r}} \, d\mathbf{r}.$$

The quantities E_{k_2}, H_{k_1}, and H_{k_2} are defined similarly.

The energy of waves (with the given polarity) propagating in the direction $\mathbf{k}$ is proportional to

$$\varepsilon_{k_1} \sim |E_{k_1} + H_{k_2}|^2, \qquad \varepsilon_{k_2} \sim |E_{k_2} - H_{k_1}|^2.$$

* We are indebted to A. S. Kompaneets for discussion of this problem.

The corresponding number of photons with the given frequency is proportional to

$$f_k = f_{k_1} + f_{k_2} = \frac{\varepsilon_{k_1} + \varepsilon_{k_2}}{h\nu_k} \sim \frac{\varepsilon_{k_1} + \varepsilon_{k_2}}{k}.$$

The total number of photons in the field is proportional to

$$N \sim \int f_k \, d\mathbf{k}.$$

Substituting the expressions for E_{k_1}, etc., and carrying out some simple transformations, we find that

$$N \sim \iiint [\mathbf{E}(\mathbf{r}) \cdot \mathbf{n}_1 e^{i\mathbf{k}\cdot\mathbf{r}} \, d\mathbf{r}][\mathbf{E}(\mathbf{r}') \cdot \mathbf{n}_1 e^{-i\mathbf{k}\cdot\mathbf{r}'} \, d\mathbf{r}'] \frac{d\mathbf{k}}{k} + \cdots$$

to which integral are to be added analogous terms in $\mathbf{E} \cdot \mathbf{n}_2$, $\mathbf{H} \cdot \mathbf{n}_1$, and $\mathbf{H} \cdot \mathbf{n}_2$. This expression can be written as a single integral

$$N \sim \iiint \{[\mathbf{E}(\mathbf{r}) \cdot \mathbf{n}_1 \mathbf{E}(\mathbf{r}') \cdot \mathbf{n}_1 + \mathbf{E}(\mathbf{r}) \cdot \mathbf{n}_2 \mathbf{E}(\mathbf{r}') \cdot \mathbf{n}_2]$$

$$+ [\text{same terms in } \mathbf{H}]\} e^{i\mathbf{k}\cdot(\mathbf{r}-\mathbf{r}')} \, d\mathbf{r} \, d\mathbf{r}' \frac{d\mathbf{k}}{k}. \tag{2.122}$$

It follows from the condition that $\nabla \cdot \mathbf{E} = 0$, $\nabla \cdot \mathbf{H} = 0$, that the waves are transverse, and that the quantities

$$E_{k_3} = \int \mathbf{E} \cdot \mathbf{n}_3 e^{-i\mathbf{k}\cdot\mathbf{r}} \, d\mathbf{r}$$

and H_{k_3} are equal to zero ($\mathbf{n}_3 = \mathbf{k}/k$ is a unit vector in the direction $\mathbf{k}$). This means that (2.122) will not change if we add to the integrand the quantity

$$\mathbf{E}(\mathbf{r}) \cdot \mathbf{n}_3 \mathbf{E}(\mathbf{r}') \cdot \mathbf{n}_3 e^{i\mathbf{k}(\mathbf{r}-\mathbf{r}')} \, d\mathbf{r} \, d\mathbf{r}' \frac{d\mathbf{k}}{k},$$

which is equal to zero. But, after this quantity is added, the expression in brackets in (2.122) will take on the form of a scalar product $\mathbf{E}(\mathbf{r}) \cdot \mathbf{E}(\mathbf{r}')$, since $\mathbf{n}_1$, $\mathbf{n}_2$, and $\mathbf{n}_3$ form three perpendicular directions. As a result we get

$$N \sim \iiint [\mathbf{E}(\mathbf{r}) \cdot \mathbf{E}(\mathbf{r}') + \mathbf{H}(\mathbf{r}) \cdot \mathbf{H}(\mathbf{r}')] e^{i\mathbf{k}\cdot(\mathbf{r}-\mathbf{r}')} \, d\mathbf{r} \, d\mathbf{r}' \frac{d\mathbf{k}}{k}.$$

Evaluating the integral $\int e^{i\mathbf{k}\cdot(\mathbf{r}-\mathbf{r}')} d\mathbf{k}/k$, we obtain (2.121), which proves our earlier statement that the quantity defined in (2.121) is proportional to the number of photons in the field.

Most interesting is the fact that N is written as a double integral over a volume for a given $t = const$. In the Lorentz transformation **E** and **H** are changed, in addition to which it is necessary to make a transition to another volume with $t' = const$ (to another hypersurface in the four-dimensional Minkowski space). A direct proof of the fact that the expression N is a Lorentz invariant* is quite difficult. Since t and t' cannot coincide, we must use Maxwell's equations to make a transition from **E** and **H** in the volume $t = const$ to **E**′ and **H**′ in the volume $t' = const$. However, the relativistic invariance of (2.122) actually stems from the fact that N—the total number of photons in this volume—is, obviously, a relativistic invariant. It is here essential that we consider a field free of charges in which photons are neither born nor absorbed.

* *Editors' note.* That N in an unbounded empty space is constant with time is clear from the fact that ε_k is independent of time.

III. Thermodynamic properties of gases at high temperatures

1. *Gas of noninteracting particles*

§1. Perfect gas with constant specific heats and invariant number of particles

In many real processes the macroscopic parameters characterizing the state of the gas, such as the density ρ, specific internal energy ε, or temperature T, change slowly in comparison with the rates of the relaxation processes leading to thermodynamic equilibrium. Under such conditions a gas particle is at any instant of time in a state that is very close to the thermodynamic equilibrium state corresponding to the instantaneous values of the macroscopic parameters. Exceptions to this are very rapid processes, such as the flow of gas through a shock front. In this chapter we shall only consider gases in thermodynamic equilibrium.

In order to describe the adiabatic motion of a fluid it is necessary to specify either the entropy $S(\rho, p)$ or the specific internal energy $\varepsilon(\rho, p)$ as a function of density and pressure. In the nonadiabatic case, the energy equation usually contains the temperature explicitly (for example, when considering heat conduction or radiation), which must be related to the density and pressure through an equation of state $p = p(\rho, T)$.

As is well known, all thermodynamic properties of a fluid can be obtained from one of the generalized thermodynamic potentials expressed as a function of the appropriate variables, for example, $\varepsilon(S, \rho)$, $h(S, p)$, $F(T, \rho)$, or $\Phi(T, p)$, where F is the free energy, h the enthalpy, and Φ the Gibbs potential*. In calculating the thermodynamic properties of gases it is customary to determine directly the internal energy as a function of temperature and density $\varepsilon(T, \rho)$ or of temperature and pressure $\varepsilon(T, p)$. Then it is necessary to introduce independently the equation of state, which can be derived from the function $\varepsilon(S, \rho)$, but not from the functions $\varepsilon(T, \rho)$ or $\varepsilon(T, p)$.

We shall consider only perfect gases (unless specified otherwise) where, by

* *Editors' note.* The function $\Phi = h - TS$ is referred to by the authors as thermodynamic potential. This function is also variously called Gibbs function or Gibbs free energy in distinction to the Helmholtz free energy F, here referred to simply as free energy. The equation $p = p(\rho, T)$ is often termed the *engineering* equation of state to distinguish it from any equation (of state) which connects state variables.

definition, the interaction between the particles can be disregarded. In many cases of practical importance the perfect gas approximation is quite accurate (the imperfect nature manifests itself only at very high densities; this is discussed in §§11–14). The equation of state for a perfect gas can be written in one of the equivalent forms

$$p = nkT = N\rho kT = \frac{NkT}{V} = \frac{\mathcal{R}}{\mu_0}\rho T = R\rho T, \tag{3.1}$$

where n is the number of particles per unit volume, N the number of particles per unit mass, $\mathcal{R}$ the universal gas constant*, R the gas constant per unit mass, μ_0 the average molecular weight, and V the specific volume. The number of particles per unit mass N, or the average molecular weight μ_0 may depend on the temperature and density as a result of dissociation, chemical reactions, or ionization.

The internal energy of a gas and hence also the specific heat at constant volume are in general made up of a number of contributions which correspond to the different degrees of freedom of the gas, as for example, translational motion, rotation and vibrations of the molecules, or electronic excitation of the atoms and molecules, and also a number of contributions which correspond to molecular dissociation, chemical reactions, and ionization. In the subsequent discussion we shall, for the sake of brevity, include the latter factors as well in the general concept of "degrees of freedom". As with the internal energy, all the remaining thermodynamic potentials and the entropy are sums over the various degrees of freedom. Each of the different degrees of freedom, with the exception of the translational motion of the particles, provides a contribution to the thermodynamic functions only above some more or less definite temperature. For the degrees of freedom associated with a change in the number of particles (dissociation, chemical reactions, ionization) these temperatures depend on the gas density.

At very low temperatures the atoms and molecules are neither ionized nor excited and the chemical composition corresponds to the most energetically probable state (that of minimum energy); the thermal motion is limited to the translational displacements of the particles. The specific internal energy measured from zero temperature is given by $\varepsilon_{\text{trans}} = \frac{3}{2}NkT$, and the specific heat at constant volume is $c_{v\,\text{trans}} = \frac{3}{2}Nk$. For a monatomic gas, the temperature region where the thermodynamic functions are determined by the purely translational motion of the atoms extends to quite high values of the temperature, of the order of several and even tens of thousands of degrees; at still higher temperatures, ionization and electronic excitation occur.

* $\mathcal{R} = 8.31 \cdot 10^7$ erg/deg·mole $= 1.99$ cal/deg·mole, $k = 1.38 \cdot 10^{-16}$ erg/deg $= 8.31$ joule/deg·mole.

In a molecular gas, the rotational degrees of freedom are excited at very low temperatures. This usually takes place at temperatures up to the order of 10°K. Energies of rotational quanta expressed in degrees (that is, divided by the Boltzmann constant k) are very small; for example, 2.1°K for oxygen, 2.9°K for nitrogen, and 2.4°K for nitric oxide. The only exception is the hydrogen molecule, for which this quantity is equal to 85.4°K. Even at 300°K (room temperature), and even more so at higher temperatures, quantum effects are not very significant. The rotational part of the specific heat is equal to its classical value. The specific heat $c_{v\,\mathrm{rot}} = Nk$ for diatomic and linear polyatomic molecules and $c_{v\,\mathrm{rot}} = \frac{3}{2}Nk$ for nonlinear polyatomic molecules. The corresponding internal energy contributions are $\varepsilon_{\mathrm{rot}} = NkT$ and $\frac{3}{2}NkT$, respectively.

Molecular vibrations are excited at much higher temperatures, of the order of several hundred or even thousands of degrees; therefore, a temperature region exists in which the thermal motion of a molecular gas consists of translational and rotational motion only. The specific heat in this region is constant, and for a diatomic gas (for example, air) is equal to $c_v = c_{v\,\mathrm{trans}} + c_{v\,\mathrm{rot}} = \frac{5}{2}Nk$. The corresponding internal energy $\varepsilon = \frac{5}{2}NkT$.

The energies of vibrational quanta expressed in degrees are usually of the order of a few thousand degrees in diatomic molecules. For example, $h\nu/k =$ 2230°K for O_2, 3340°K for N_2, and 2690°K for NO; the lowest vibrational frequency for triatomic molecules is usually smaller, for example, $h\nu/k =$ 916°K, 1960°K, 2310°K for NO_2. At temperatures less than or of the order of $h\nu/k$ the vibrational contribution to the specific heat depends on the temperature and must be calculated using quantum-mechanical formulas. For temperatures appreciably greater than $h\nu/k$, however, the vibrational contribution to the specific heat is constant and equal to its classical value of k per vibrational degree of freedom. A diatomic molecule has one vibrational degree of freedom, a nonlinear m-atomic molecule has $3m - 6$, and a linear molecule has $3m - 5$ degrees of freedom. Thus, for temperatures appreciably greater than the maximum value of $h\nu/k$, the total classical specific heat per molecule $c_v = c_{v\,\mathrm{trans}} + c_{v\,\mathrm{rot}} + c_{v\,\mathrm{vib}}$ is $c_v = \frac{3}{2}Nk + Nk + (3m - 5)Nk = (3m - \frac{5}{2})Nk$ for linear m-atomic molecules, and $c_v = \frac{3}{2}Nk + \frac{3}{2}Nk + (3m - 6)Nk = (3m - 2)Nk$ for nonlinear molecules. For diatomic molecules $c_v = \frac{7}{2}Nk$. The isentropic equation for a perfect gas with constant specific heats and an invariant number of particles is determined from the general thermodynamic relationship

$$T\,dS = d\varepsilon + p\,dV = c_v\,dT + NkT\frac{dV}{V} = 0.$$

Integrating, we obtain

$$T \sim V^{-(\gamma-1)} \sim \rho^{\gamma-1}; \qquad p \sim V^{-\gamma} \sim \rho^{\gamma}, \tag{3.2}$$

where the proportionality coefficients are functions of the entropy alone. Here $\gamma = c_p/c_v$ is the specific heat ratio and $c_p = c_v + Nk$ is the specific heat at constant pressure. For example, for a monatomic gas $\gamma = 5/3$, for a diatomic gas with the vibrational degrees of freedom not excited $\gamma = 7/5$, while with fully excited vibrations $\gamma = 9/7$.

It should be noted, however, that there is not a very wide temperature range in which the molecular vibrations are fully excited while the specific heats and the specific heat ratio remain constant. This is a result of the fact that molecular dissociation and chemical reactions frequently begin at temperatures for which the vibrational contribution to the specific heats has just reached its classical limiting value.

§2. Calculation of thermodynamic functions using partition functions

The most rigorous and consistent method for finding all of the thermodynamic functions is the so-called method of partition functions. We shall briefly present the fundamentals of this method* and use it to obtain an expression for the entropy and the quantum-mechanical formula for the vibrational energy of a molecule. In subsequent sections we shall apply this method to a gas in which the number of particles is variable.

According to statistical mechanics the probability of the nth state of a system consisting of N particles and having an energy equal to E_n is proportional to $\exp(-E_n/kT)$. The sum of these probabilities over all possible states of the system, determined to within a constant multiplicative factor, is given by

$$Q = \sum_n e^{-\frac{E_n}{kT}}. \tag{3.3}$$

This expression is called the partition function of the system.

For an ideal Boltzmann gas consisting of several kinds of molecules, the number of which is $N_A, N_B, \ldots$, the partition function may be factored into a product of co-factors, each corresponding to the particles of one kind

$$Q = \frac{Z_A^{N_A}}{N_A!} \cdot \frac{Z_B^{N_B}}{N_B!} \cdots. \tag{3.4}$$

Here $Z_A, Z_B, \ldots$ are the partition functions of each type for one molecule, and are expressed by equations of a form similar to (3.3)

$$Z = \sum_k e^{-\frac{\varepsilon_k}{kT}}. \tag{3.5}$$

* Detailed derivations can be found in most texts on statistical physics as, for example, in the book by Landau and Lifshitz [1].

Here ε_k is the energy of a molecule in the kth state, and the summation is carried out over all possible states of one molecule.

The general formula for the free energy of a system is

$$F = -kT \ln Q. \tag{3.6}$$

If we replace the factorials in (3.4) by means of Stirling's formula $N! \approx (N/e)^N$ and substitute the result into (3.6), we get

$$F = -N_A kT \ln \frac{Z_A e}{N_A} - N_B kT \ln \frac{Z_B e}{N_B} - \cdots. \tag{3.7}$$

Since the free energy is a thermodynamic potential with respect to the variables, temperature and density (or volume), we can derive all the thermodynamic properties from (3.7) if the partition functions of the molecules are known as a function of the temperature T and the volume V. The general thermodynamic equations define the entropy, internal energy, and pressure, respectively, as

$$S = -\left(\frac{\partial F}{\partial T}\right)_{V,N}, \tag{3.8}$$

$$\varepsilon = F + TS = -T^2 \frac{\partial}{\partial T}\left(\frac{F}{T}\right)_{V,N}^{*}, \tag{3.9}$$

$$p = -\left(\frac{\partial F}{\partial V}\right)_{T,N}. \tag{3.10}$$

Neglecting any interactions between electronic, vibrational, and rotational states, and considering the molecule as a rigid rotator and the vibrations as harmonic, the energy of the molecule may be represented as the sum of the energy contributions of the various degrees of freedom. In this case, as is evident from (3.5), the partition function of one molecule also may be factored into the product

$$Z = Z_{\text{trans}} \cdot Z_{\text{rot}} \cdot Z_{\text{vib}} \cdot Z_{\text{el}}. \tag{3.11}$$

We present here several formulas for the partition functions without proof. The translational partition function for any particle is

$$Z_{\text{trans}} = \left(\frac{2\pi MkT}{h^2}\right)^{3/2} V, \tag{3.12}$$

where M is the mass of the particle and V is the volume occupied by the gas (if N is understood to be the number of particles per unit mass, then V is the specific volume).

* As may be easily checked by direct substitution of (3.6) and (3.3) into (3.9), $\varepsilon = \Sigma E_n \exp(-E_n/kT)/\Sigma \exp(-E_n/kT)$ and the internal energy is simply equal to the energy of the system averaged over all possible states.

The rotational partition function (for temperatures much higher than the energy of a rotational quantum divided by k) is

$$Z_{\text{rot}} = \frac{8\pi^2 IkT}{h^2}\frac{1}{\sigma} \tag{3.13}$$

for a diatomic or a linear polyatomic molecule*, and

$$Z_{\text{rot}} = \frac{8\pi^2}{\sigma}\left(\frac{2\pi IkT}{h^2}\right)^{3/2} \tag{3.14}$$

for a nonlinear polyatomic molecule. Here I represents the moment of inertia of a linear molecule in (3.13), and in (3.14) it denotes the geometric mean of the three moments of inertia of a nonlinear polyatomic molecule $I = (I_1 I_2 I_3)^{1/3}$; σ is the so-called symmetry factor, equal to 1 plus the number of transpositions of identical atoms in a molecule, where the transpositions are equivalent to the rotation of the molecule as a whole†.

The quantum-mechanical expression for the partition function of a harmonic oscillator vibrating at a frequency ν is

$$Z_{\text{vib}} = (1 - e^{-h\nu/kT})^{-1}. \tag{3.15}$$

The vibrational energy in this equation is measured with respect to the lowest quantum vibrational level. It is assumed that the zero-point energy for vibrations $h\nu/2$ is included in the energy of the ground state of the molecule. If the molecule has several vibrational degrees of freedom, then the total vibrational partition function is represented as the product of factors corresponding to all the normal modes.

Finally, the electronic partition function retains its original form

$$Z_{\text{el}} = \sum_n e^{-\frac{\varepsilon_n}{kT}}, \tag{3.16}$$

where ε_n is the energy of the nth electronic quantum state of the atom or molecule. If the energy levels are degenerate, then each contribution enters the partition function as an independent component, so that the number of identical components for each level is equal to the statistical weight of the level.

The different atomic and molecular constants required for calculating the thermodynamic properties of gases are usually obtained from spectroscopic data. The rotational and vibrational energy of various molecules has been discussed in the preceding section. Energies of the first excited electronic states of atoms and molecules ε_1 are usually of the order of several ev, with ε_1/k of the order of tens of thousands of degrees. For example, for the ^1D

* The rotational energy of a quantum $h\nu_{\text{rot}} = h^2/8\pi^2 I$, so that $Z_{\text{rot}} = kT/h\nu_{\text{rot}} \cdot \sigma$.

† For example, in a diatomic molecule composed of identical atoms $\sigma = 2$; and in one composed of different atoms $\sigma = 1$.

term* for O atoms $\varepsilon_1 = 1.96$ ev, $\varepsilon_1/k = 22{,}800°$K, while for the $^2D^0$ term for N, $\varepsilon_1 = 2.37$ ev, $\varepsilon_1/k = 27{,}500°$K. In the case of molecules we have for the $A\ ^3\Sigma_u^+$ term for N_2, $\varepsilon_1 = 6.1$ ev, $\varepsilon_1/k = 71{,}000°$K, while for the $A\ ^2\Sigma^+$ term for NO, $\varepsilon_1 = 5.29$ ev, $\varepsilon_1/k = 61{,}400°$K. There are, however, some exceptions. For example, the first excitation levels for an O_2 molecule lie rather low, so that for the $^1\Delta_g$ term $\varepsilon_1 = 0.98$ ev, $\varepsilon_1/k = 11{,}300°$K, while for the $^1\Sigma_g^+$ term, $\varepsilon_2 = 1.62$ ev, $\varepsilon_2/k = 18{,}800°$K.

For temperatures which are not too high, where $T \ll \varepsilon_1/k$, the electronic partition function reduces essentially to the contributions corresponding to the ground state of the electron. If the energy spacing between the fine structure levels (if such exist) of the ground state is appreciably smaller than kT†, then the corresponding contributions to Z_{el} can be considered as approximately identical. Measuring the energy ε_n from the ground state ($\varepsilon_0 = 0$), we can set Z_{el} equal to the statistical weight of the ground state g_0 (for example, for O atoms the ^3P term has $g_0 = 9$, while for N(^4S) $g_0 = 4$; for molecules: $O_2(^3\Sigma)$ $g_0 = 3$; $N_2(^1\Sigma)$ $g_0 = 1$; NO($^2\Pi$) $g_0 = 4$). The calculation of Z_{el} at high temperatures will be discussed in §6.

Since the partition function Z of a molecule is equal to the product of the individual factors which correspond to the different degrees of freedom, the free energy and other thermodynamic functions of a gas are expressible in terms of the partition functions of the respective contributions. Substituting the expressions for the factors of Z into (3.7), we obtain an explicit expression for the free energy in terms of temperature and density. The density dependence arises because the translational partition function Z_{trans} contains the volume V. The quantities N_A/V, N_B/V, ... appearing in the logarithmic terms in (3.7) correspond to the numbers of particles per unit volume n_A, n_B, ..., expressed in terms of the gas density and the fractions of the various kinds of particles present (which in the case being considered are invariant).

The partition function for a monatomic gas is composed of translational and electronic contributions only; substituting these contributions into (3.7) we find for the free energy of N identical atoms (we assume that $Z_{el} = g_0$)

$$F = -NkT \ln \left(\frac{2\pi MkT}{h^2}\right)^{3/2} \frac{eVg_0}{N}. \tag{3.17}$$

The specific entropy of a monatomic gas in the absence of ionization and electronic excitation is given by (3.8) as

$$S = Nk \ln \frac{e^{5/2} g_0}{n} \left(\frac{2\pi MkT}{h^2}\right)^{3/2}. \tag{3.18}$$

* Spectroscopic notation is discussed in §14 of Chapter V.

† For example, in the case of the O atom the energy spacings for the components of the ground triplet state $^3D_2$ are $\Delta\varepsilon/k = 230°$ and 320°K; for NO the doublet splitting for the $^2\Pi$ ground state is $\Delta\varepsilon/k = 178°$K.

The energy and pressure are given by the familiar expressions

$$\varepsilon = \tfrac{3}{2}NkT, \qquad p = nkT.$$

Proceeding in a similar manner we can easily obtain the rotational and vibrational contributions to the thermodynamic functions. The internal rotational energy is essentially that given by the equations presented in §1, while the internal vibrational energy is expressed by the Planck function. The energy of N identical oscillators (diatomic molecules) is

$$\varepsilon_{\text{vib}} = N \frac{h\nu}{e^{h\nu/kT} - 1}. \tag{3.19}$$

In the limit $kT \gg h\nu$ the energy approaches its classical value $\varepsilon_{\text{vib}} = NkT$ and the specific heat $c_{v\,\text{vib}} = \partial\varepsilon_{\text{vib}}/\partial T \to Nk$. Actually, both the energy and the specific heat are already close to their limiting values when $kT \approx h\nu$. For example, at $kT/h\nu = 0.5$ we find $c_v/Nk = 0.724$, at $kT/h\nu = 1$ we find $c_v/Nk = 0.928$, while at $kT/h\nu = 2$ we find $c_v/Nk = 0.979$. The rotational and vibrational degrees of freedom of the molecules do not affect the pressure; formally this is attributable to the fact that the corresponding partition functions, internal energies, and specific heats are independent of the volume. The pressure of a perfect gas is to be attributed exclusively to the translational motion of the particles.

At high temperatures of the order of several thousand degrees, when the amplitudes of the molecular vibrations become appreciable in comparison with the interatomic distances, the vibrations become anharmonic and coupling appears between the vibrational and rotational degrees of freedom. The anharmonicity results in a slight decrease in the vibrational contribution to the specific heat. The corresponding corrections for this effect are, in first approximation, proportional to the temperature. These corrections are usually not too large (dissociation of the molecules begins before the corrections become appreciable). For the calculation of these corrections see [2], for example.

§3. Dissociation of diatomic molecules

At temperatures of the order of several thousand degrees diatomic molecules usually dissociate into atoms. Polyatomic molecules, in which the bonds are weaker, begin to dissociate at even lower temperatures. The dissociation of a molecule requires a large amount of energy and hence this process has an appreciable effect on the thermodynamic properties of gases.

Let us consider the simplest but at the same time practically important case of a diatomic gas consisting of the same kind of molecules A_2, composed of identical atoms A. Suppose that at a temperature T and gas density ρ a fraction α of the original number of molecules is dissociated into atoms, following the reaction $A_2 \rightleftarrows 2A$. If N is the initial number of molecules per unit mass, then there will be $N \cdot 2\alpha$ atoms and $N(1-\alpha)$ molecules per unit mass of gas. The total number of particles is $N(1+\alpha)$, so that the gas pressure is

$$p = N(1+\alpha)\rho kT. \tag{3.20}$$

For complete dissociation ($\alpha = 1$) the pressure is twice as high as the pressure at the same T and ρ would be if the gas were undissociated.

For small degrees of dissociation ($\alpha \ll 1$) the change in pressure is small, although the change in internal energy and specific heat of the gas may be appreciable. Let ε_{A_2} be the energy of a single molecule at the temperature T, and ε_A be the energy of a single atom. Let us denote the energy required for the dissociation of an unexcited molecule by U (i.e., in the absence of rotational and vibrational excitation, at $T = 0$). The energy U represents the binding energy or dissociation energy of a molecule; for example, for O_2 the energy $U = 5.11$ ev $= 118$ kcal/mole *, and $U/k = 59{,}400°K$; for N_2 the energy $U = 9.74$ ev $= 225$ kcal/mole, and $U/k = 113{,}000°K$; for NO the energy $U = 6.5$ ev $= 150$ kcal/mole, and $U/k = 75{,}500°K$. The specific internal energy of the gas taken with respect to the molecular state at zero temperature is

$$\varepsilon = N\varepsilon_{A_2}(1-\alpha) + N\varepsilon_A\, 2\alpha + NU\alpha. \tag{3.21}$$

Dissociation usually begins at temperatures much lower than U/k, with these temperatures lower the more rarefied is the gas. At standard atmospheric density ($n = 2.67 \cdot 10^{19}$ molecules/cm^3) dissociation in air is already noticeable at $kT/U \sim 1/20$. This phenomenon is due to the high statistical weight of the state in which the molecule is dissociated into atoms. In fact, for $kT \ll U$ the molecules are dissociated by collisions with the very energetic particles corresponding to the far tail of the Boltzmann energy distribution function. In the absence of ionization and electronic excitation $\varepsilon_A = \frac{3}{2}kT$. If kT is larger than the energy of a vibrational quantum $h\nu$, the vibrational energy of a molecule according to (3.19) is approximately equal to kT, and $\varepsilon_{A_2} \approx \frac{7}{2}kT$. The energy of the dissociated gas (3.21) appreciably exceeds the energy in the absence of dissociation $\varepsilon = N\varepsilon_{A_2}$, even for small degrees of dissociation ($\alpha \sim 0.1$ and less), as a result of the importance of the last term corresponding to dissociation energy. Correspondingly, the specific heat $c_v = (\partial\varepsilon/\partial T)_V$ of the dissociated gas also increases appreciably.

We should note that (3.20) and (3.21) are also valid under conditions of nonequilibrium dissociation, where the degree of dissociation differs from the equilibrium value corresponding to the "temperature" and density of the

* 1 ev/molecule is equivalent to 23.05 kcal/mole.

gas. By "temperature" we understand here the temperature corresponding to the translational and rotational degrees of freedom of the particles, which are always taken to be in thermodynamic equilibrium*.

The equilibrium degree of dissociation is uniquely determined by the temperature and density (or pressure) of the gas. The relationship between the degree of dissociation and the temperature and density can be derived from the general expression (3.7) for the free energy of a gas composed of different types of particles. We employ the fact that the equilibrium composition of a mixture which undergoes a chemical transformation, of which dissociation is a particular case, corresponds to a minimum of the free energy.

Let us consider the free energy F as a function of the number of particles N_{A_2} and N_A at a given temperature and volume, with the original number of molecules $N^0_{A_2}$,

$$F = -N_{A_2}kT \ln \frac{Z_{A_2}e}{N_{A_2}} - N_A kT \ln \frac{Z_A e}{N_A}.$$

The variation δF is given by

$$\delta F = -\delta N_{A_2}\left(kT \ln \frac{Z_{A_2}e}{N_{A_2}} - kT\right) - \delta N_A\left(kT \ln \frac{Z_A e}{N_A} - kT\right)^{\dagger}.$$

The variations δN_{A_2} and δN_A are related by the condition of conservation of the number of atoms

$$N_{A_2} + \frac{N_A}{2} = N^0_{A_2} = const\,; \qquad \delta N_{A_2} = -\tfrac{1}{2}\,\delta N_A.$$

Setting δF equal to zero (corresponding to a minimum in the free energy) with the condition of conservation of the number of atoms, we get

$$\frac{N_A^2}{N_{A_2}} = \frac{Z_A^2}{Z_{A_2}}. \tag{3.22}$$

Since the partition functions Z_A and Z_{A_2} are proportional to the volume (which enters in the translational partition functions) in addition to being functions of the temperature, we can replace (3.22) by

$$\frac{n_A^2}{n_{A_2}} = f(T), \tag{3.23}$$

* Equilibrium for the vibrational degrees of freedom is established more slowly than for the rotational and translational ones but usually faster than for dissociation. For details see Chapter VI.

† The quantities in parentheses represent the chemical potentials of the molecules and atoms, respectively, with the signs reversed:

$$\mu_{A_2} = \frac{\partial F}{\partial N_{A_2}}, \qquad \mu_A = \frac{\partial F}{\partial N_A}.$$

or, for the partial pressures $p_i = n_i kT$

$$\frac{p_A^2}{p_{A_2}} = f(T) \cdot kT = K_p(T). \tag{3.24}$$

Equation (3.22), (3.23), or (3.24) represents a particular case of the so-called law of mass action for chemical equilibrium, and the quantity $K_p(T)$ is termed the dissociative equilibrium constant. This constant depends only upon the temperature and the molecular (or atomic) constants. We now assume, for simplicity, that the molecular vibrations are fully excited so that $Z_{vib} \approx kT/h\nu$ (see (3.15)), and that the electronic partition functions contain only the terms corresponding to the ground molecular and atomic states. Substituting the expressions for the partition functions of the molecules A_2 and atoms A into (3.22), we obtain

$$\frac{p_A^2}{p_{A_2}} = K_p(T) = \frac{M_A^{3/2} \nu (kT)^{1/2}}{4\pi^{1/2} I_{A_2}} \frac{g_{0A}^2}{g_{0A_2}} e^{-U/kT}. \tag{3.25}$$

The last two factors in (3.25) enter from the quotient of the electronic partition functions

$$\frac{Z_{el\,A}^2}{Z_{el\,A_2}} \approx \frac{g_{0A}^2}{g_{0A_2}} e^{-(2\varepsilon_{0A} - \varepsilon_{0A_2})/kT} = \frac{g_{0A}^2}{g_{0A_2}} e^{-U/kT}.$$

Here the difference between the zero-point energies $2\varepsilon_{0A} - \varepsilon_{0A_2}$ is, by definition, equal to the dissociation energy U.

Expressing the partial pressures in (3.25) in terms of the degree of dissociation

$$\alpha = \frac{N_{A_2}^0 - N_{A_2}}{N_{A_2}^0} = \frac{N_A}{2N_{A_2}^0},$$

we get

$$\frac{\alpha^2}{1-\alpha} = \frac{1}{4n_{A_2}^0} \frac{K_p(T)}{kT} = \frac{M_A^{3/2} \nu}{16\pi^{1/2} I_{A_2} (kT)^{1/2}} \frac{g_{0A}^2}{g_{0A_2}} \frac{1}{n_{A_2}^0} e^{-U/kT}, \tag{3.26}$$

where $n_{A_2}^0 = \rho / M_{A_2}$ is the original number of molecules per unit volume of gas.

For small degrees of dissociation $\alpha \ll 1$ (when $U/kT \gg 1$), (3.26) shows that $\alpha \sim \rho^{-1/2} e^{-U/2kT}$. Thus, α increases sharply with temperature and slowly with decreasing gas density. The sharp dependence of the degree of dissociation on temperature also results in a rapid increase in the specific heat. At high temperatures, when most of the molecules are dissociated, $\alpha \approx 1$, and the concentration of the molecules is proportional to the density, $1 - \alpha \sim \rho e^{U/kT}$. Since in this case U/kT is not a very large number the change in the concentration with temperature is much slower.

It would appear that at high temperatures, when the dissociation is complete, the specific heat of the gas (which is now monatomic) should decrease and become equal to $\frac{3}{2}k$ per atom or $3k$ per original molecule. The specific heat should become even less than it was before dissociation took place ($\frac{7}{2}k$ per molecule). This situation does not ordinarily occur, since after the dissociation is complete (and sometimes even before) the temperature increase results in ionization of the atoms (and molecules). The resulting contribution of the ionization to the specific heat is appreciable.

The dependence of the degree of dissociation upon the temperature and density of a gas and the effect of dissociation on its thermodynamic properties are illustrated in Tables 3.1 and 3.2, which are based on data for air (79% N_2 + 21% O_2) taken from [3]*. The formation of nitric oxide $N_2 + O_2 \rightleftarrows 2NO$ in air (see the following section) does not strongly influence either the dissociation of N_2 and O_2 molecules or the thermodynamic properties of air.

Table 3.1

EQUILIBRIUM COMPOSITION OF DISSOCIATED AND SLIGHTLY IONIZED AIR

Standard density $\rho_0 = 1.29 \cdot 10^{-3}$ g/cm³

T°K	N_2	N	O_2	O	NO	N^+	O^+	NO^+
2000	0.788	—	0.205	—†	0.007	—	—	—
4000	0.749	0.0004	0.100	0.134	0.084	—	—	—
6000	0.744	0.044	0.006	0.356	0.050	—	—	—
8000	0.571	0.416	0.007	0.393	0.024	—	—	—
10,000	0.222	1.124	—	0.407	0.009	0.0034	—	0.0015
12,000	0.050	1.458	—	0.411	0.003	0.020	0.0034	0.001
15,000	0.006	—	—	—	—	0.096	0.015	—

Density $\rho = 10^{-2}\,\rho_0$

T°K	N_2	N	O_2	O	NO	N^+	O^+	NO^+
2000	0.788	—	0.205†	0.002	0.007	—	—	—
4000	0.777	0.004	0.008	0.378	0.024	—	—	—
6000	0.592	0.394	—	0.413	0.005	—	—	—
8000	0.068	1.440	—	0.416	0.001	0.004	0.001	0.0001
10,000	0.004	1.528	—	0.410	—	0.046	0.008	0.0002
12,000	—	1.380	—	0.384	—	0.202	0.034	—
15,000	—	0.858	—	0.282	—	0.724	0.136	—

Concentrations of all particles c_i are defined here as the ratio of the number of particles of the given species to the original number of molecules. At room temperature $c_{N_2} = 0.791$, $c_{O_2} = 0.209$. The data for argon are not shown, since its effect is small.

* The data given in Table 3.2 were taken from [3] only for those temperatures below 20,000°K. The data for the higher temperatures were taken from [4]. (This reference is discussed in §5.)

† *Editors' note*. These values have been changed by the editors for consistency.

Table 3.2

THERMODYNAMIC PROPERTIES OF AIR

Standard density $\rho_0 = 1.29 \cdot 10^{-3}$ g/cm³					Density $\rho = 10^{-2}\rho_0$			
T°K	ε, ev/molecule	p, atm	γ	$\frac{7}{2}kT$, ev/molecule	T°K	ε, ev/molecule	p, atm	γ
2000	0.515	7.42	1.335	0.604	2000	0.520	0.074	1.330
4000	1.52	15.8	1.240	1.21	4000	2.09	0.177	1.195
8000	5.38	41.7	1.180	2.42	8000	10.6	0.575	1.125
12,000	12.7	88	1.160	3.92	12,000	16.6	0.994	1.140
20,000	24	183	1.175	6.04	20,000	45.3	2.8	1.145
50,000	95	870	1.215	15.1	50,000	158	11.6	1.170
100,000	276	2690	1.225	30.2	100,000	499	37.3	1.175
250,000	922	10,870	1.275	75.4	250,000	1080	125	1.270
500,000	1450	23,150	1.370	151	500,000	3310	412	1.290

The internal energies are given in electron volts per original molecule; for air 1 ev/molecule = 0.8 kcal/g. The effective adiabatic exponent γ is defined as $\gamma = 1 + p/\rho\varepsilon$. The last column gives for comparison the values of the energy $\varepsilon = \frac{7}{2}kT$ in ev/molecule for air in the absence of either dissociation or ionization, but with classical excitation of molecular vibrations.

The latter are basically determined by the dissociation of N_2 and O_2, so that the effects of dissociation and all of the other functional dependence are evident in Table 3.2. For comparison, the table presents values of the energy corresponding to the given temperatures assuming that there is no dissociation (in this case the specific energy is independent of the density). Since ionization begins before the dissociation of nitrogen is complete, the table also presents the ion concentrations (for a discussion of ionization see §5).

It should be noted that for accurate calculations of dissociation and of the thermodynamic functions, in place of the simple equations of the type of (3.26), one must use more exact relations which take into account excitation of the higher electronic states, anharmonic oscillations, etc. In this case the exact expressions (3.22) are used as the starting point, and the partition functions are calculated on the basis of spectroscopic data for the atoms and molecules. A description of the methods used in such computations can be found in [5].

§4. Chemical reactions

The chemical composition of a mixture of gases under normal conditions, i.e., at room temperature, frequently differs from the thermodynamic equilibrium composition. This is due to the fact that even exothermic reactions, resulting in the transition of the gas to an energetically preferred state, generally require an activation energy E. The rate of chemical reaction, proportional to the Boltzmann factor $e^{-E/kT}$, is very slow at low temperatures and large activation energies, with $E/kT \gg 1$, so that for all practical purposes the reaction does not proceed. The system of the mixture of gases is in an equilibrium, but not in thermodynamic equilibrium. This state is termed one of metastable equilibrium*. A typical example is provided by a hydrogen–oxygen mixture with a composition $H_2 + \frac{1}{2}O_2$, which in a condition of thermodynamic equilibrium at low temperatures would be completely transformed into water H_2O (the heat of reaction is 57.1 kcal/mole). However, at normal temperatures and without the influence of external factors, this irreversible reaction does not take place and the mixture remains in a state of metastable equilibrium.

At high temperatures of the order of several thousand degrees (and lower temperatures for some reactions) the chemical reaction rates are very high and chemical equilibrium is established in the gas mixture. The presence of reversible reactions (that is, reactions that can proceed in either direction depending on the conditions for chemical equilibrium at the given temperature and density) affects both the chemical composition and the thermodynamic properties of gases. As an example we can take atmospheric air where, at high temperatures of the order of several thousand degrees, oxidation of a portion of the nitrogen takes place to form nitric oxide following the reaction

$$\tfrac{1}{2}N_2 + \tfrac{1}{2}O_2 + 21.4\ \text{kcal/mole} \rightleftarrows NO. \tag{3.27}$$

The oxidation of nitrogen requires a high activation energy, so that for all practical purposes it does not proceed at temperatures below about 1500°K (a very long time is required to reach equilibrium). However, at temperatures of the order of 3000°K and above, equilibrium is established very rapidly (at standard atmospheric density in 10^{-4} seconds or less) and we can refer to the equilibrium composition of air in which the formation of nitric oxide is taken into account†.

Let us consider chemical equilibrium and its effect on the thermodynamic properties of a gas mixture. As an example, we take a reaction of a type representative of the oxidation of nitrogen,

$$A_2 + B_2 \rightleftarrows 2AB. \tag{3.28}$$

We assume for simplicity that the degree of dissociation of the molecules is small. This assumption is valid at moderate temperatures, for example, in air

* *Editors' note.* Termed "conditional" equilibrium by the authors.

† The reaction rates for the oxidation of nitrogen will be discussed in Chapter VI, §8, and the reaction kinetics in a shock wave will be discussed in Chapter VIII, §5.

at $T \sim 2000°$–$3000°$K the degree of dissociation of the N_2 and O_2 molecules is very small but the equilibrium concentration of nitric oxide is appreciable.

Let a unit mass of the original mixture contain $N^0_{A_2}$ and $N^0_{B_2}$ molecules of A_2 and B_2; their concentrations are $m^0_{A_2} = N^0_{A_2}/N$ and $m^0_{B_2} = N^0_{B_2}/N$, where $N = N^0_{A_2} + N^0_{B_2}$ is the total number of molecules per unit mass of the original gas. The equilibrium number of molecules of each species per unit mass at the temperature T and gas density ρ are denoted by N_{A_2}, N_{B_2}, N_{AB}, and their concentrations $m_i = N_i/N$ are given by m_{A_2}, m_{B_2}, m_{AB}. The number of molecules and concentrations are related through the condition of conservation of the number of atoms

$$N_{A_2} + \tfrac{1}{2}N_{AB} = N^0_{A_2}, \qquad N_{B_2} + \tfrac{1}{2}N_{AB} = N^0_{B_2}, \tag{3.29}$$

$$m_{A_2} + \tfrac{1}{2}m_{AB} = m^0_{A_2}, \qquad m_{B_2} + \tfrac{1}{2}m_{AB} = m^0_{B_2}. \tag{3.30}$$

We denote the energy of one molecule by ε_{A_2}, ε_{B_2}, ε_{AB} and the heat of reaction by $2U'$, that is, the energy generated when the two molecules A_2 and B_2 are transformed into two molecules of AB (if the reaction is endothermic, then $U' < 0$). Assuming the energy of the original mixture $A_2 + B_2$ at $T = 0$ to be zero, the specific internal energy of the gas is then given by

$$\varepsilon = Nm_{A_2}\varepsilon_{A_2} + Nm_{B_2}\varepsilon_{B_2} + Nm_{AB}\varepsilon_{AB} - Nm_{AB}U'. \tag{3.31}$$

The total number of particles in the gas in the above reaction does not change, so that if T and ρ remain the same the reaction does not influence the pressure*. The number of particles participating in the reaction is governed at equilibrium by the law of mass action, which can be derived from the general expression for the free energy, by a procedure analogous to that used in the derivation for molecular dissociation. To do this we determine the minimum in the free energy for constant T, ρ and original number of molecules $N^0_{A_2}$, $N^0_{B_2}$, with N_{A_2}, N_{B_2}, N_{AB} variable. The result is

$$\frac{N^2_{AB}}{N_{A_2}N_{B_2}} = \frac{Z^2_{AB}}{Z_{A_2}Z_{B_2}}, \tag{3.32}$$

which is completely analogous to the corresponding equation (3.22) for the dissociation case.

Factoring out the volumes in the translational partition functions, we get for the number densities and partial pressures

$$\frac{n^2_{AB}}{n_{A_2}n_{B_2}} = \frac{p^2_{AB}}{p_{A_2}p_{B_2}} = K'_p(T), \tag{3.33}$$

* As in the case of dissociation, (3.31) is also valid in the absence of chemical equilibrium, with nonequilibrium concentrations.

where $K'_p(T)$ is the equilibrium constant for the reaction (3.28). As in the case of dissociation, we substitute the expressions for the partition functions and obtain

$$K'_p(T) = 4\left(\frac{M_{\mathrm{AB}}}{M_{\mathrm{A}_2}M_{\mathrm{B}_2}}\right)^{3/2} \frac{I^2_{\mathrm{AB}}}{I_{\mathrm{A}_2}I_{\mathrm{B}_2}} \frac{\nu_{\mathrm{A}_2}\nu_{\mathrm{B}_2}}{\nu^2_{\mathrm{AB}}} \frac{g^2_{0\mathrm{AB}}}{g_{0\mathrm{A}_2}g_{0\mathrm{B}_2}} e^{2U'/kT}\,*. \tag{3.34}$$

For example, the reaction rate constant for the formation of nitric oxide is given quite accurately by

$$\frac{p^2_{\mathrm{NO}}}{p_{\mathrm{N}_2}p_{\mathrm{O}_2}} = K'_p(T) = \frac{64}{3} e^{-43,000/\mathscr{R}T},$$

where $\mathscr{R} = 2$ cal/mole-deg. We have here taken the masses, frequencies, and the moments of inertia of all three molecules to be approximately the same, $U' = -21.4$ kcal/mole, and the ratio of statistical weights equal to 16/3 (see §2).

If the gas is a mixture with several reactions taking place simultaneously, the law of mass action can be derived for each reaction by a similar procedure to that above. This law relates, in a manner similar to (3.32), the number of particles participating in the reaction and their partition functions. Substitution of the expressions for the partition functions gives the equilibrium constants. The number of particles participating in the different reactions is related by the principle of conservation of the number of atoms of each species (similar to (3.29)). The laws of mass action for the various reactions and the conditions of conservation of the number of atoms for the various molecular species constitute a system of nonlinear algebraic equations. This system determines the chemical composition in terms of the number of different particles N_i as a function of the temperature and density (or pressure) of the gas and of the original atomic composition of the mixture. As was pointed out by one of the authors [6], this system has a unique solution, and the equilibrium chemical composition of the mixture is uniquely determined.

Setting up an expression for the energy of the form (3.31), we can calculate the internal energy of the mixture. Using the general expression for the free energy $F(T, V, N_i)$ and the thermodynamic relations (3.9) and (3.10) we can also obtain an expression for the energy and pressure, and with the aid of (3.8) an expression for the entropy of the mixture. An example of such a calculation is provided by the determination of the composition and thermodynamic properties of air taking into account the dissociation of N_2 and O_2 and the formation of nitric oxide (see, e.g., [3] and [5]). Other reactions,

* The factor 4 arises from the ratio of the symmetry factors $\sigma_{\mathrm{A}_2}\sigma_{\mathrm{B}_2}/\sigma^2_{\mathrm{AB}}$, where $\sigma_{\mathrm{AB}} = 1$, $\sigma_{\mathrm{A}_2} = \sigma_{\mathrm{B}_2} = 2$; see footnote following (3.14). As in the case of dissociation we have assumed that $Z_{\mathrm{vib}} = kT/h\nu$, $Z_{\mathrm{el}} = g_0$.

involving the formation of NO_2, O_3, etc., have little effect on these calculations, since the concentrations of these components are extremely small. The chemical composition and thermodynamic properties of air are illustrated in Tables 3.1 and 3.2.

§5. Ionization and electronic excitation

Ionization of atoms (or molecules), as with the dissociation of molecules, begins at values of kT much lower than the ionization potential I. The reason here is the same as for the case of dissociation; the statistical weight of the free electron state is very large. The first ionization potentials of the majority of atoms and molecules vary between 7 and 15 ev ($I/k \sim 80{,}000$–$170{,}000°K$)*. The principal exceptions are the alkali metals, which have very low ionization potentials. Ionization usually begins at temperatures of the order of a few to ten thousand degrees. Ionization begins sooner the lower is the ionization potential and the more rarefied is the gas.

The degree of ionization increases with temperature. When the temperature is of the order of several tens of thousands of degrees practically all of the atoms are singly ionized. In the case of hydrogen this terminates the ionization process; subsequent heating does not change the fully ionized state, in which the gas is made up of protons and electrons. Each particle undergoes translational motion only and the specific heat is equal to $\frac{3}{2}k$ per particle. In a gas composed of heavier atoms the first ionization is followed by a second, a third, etc. Usually, the next ionization begins before the preceding one ends, so that at temperatures above several tens of thousands of degrees the gas contains multiply ionized atoms. If the gas consists of a mixture of several elements then it contains differently charged ions of each element.

As in the case of molecular dissociation, the internal energy of the ionized gas is made up of the thermal energy of the particles (atoms, ions, and electrons) and the potential energy, which is equal to the work required to remove the electrons from the atoms or the ions. In addition, the excitation energy of the unionized electrons in the atoms and ions can also contribute to the total energy of the gas.

Let us consider a simple gas consisting of atoms of a single element, and let us assume, as is usually the case, that all the molecules (if the gas is not monatomic) are completely dissociated into atoms in the region of appreciable ionization. We assume that there are N atoms per unit mass of gas and we denote the successive ionization potentials by I_m, thus I_1 is the energy required to remove the first electron from a neutral atom, I_2 the energy required to remove an electron from a singly ionized atom, etc. It follows that the energy

* For example, $I_O = 13.6$ ev, $I_N = 14.6$ ev, $I_{O_2} = 12.1$ ev, $I_{N_2} = 15.6$ ev, $I_{NO} = 9.3$ ev.

required to remove m electrons from an atom is

$$Q_m = I_1 + I_2 + \cdots + I_m \qquad (I_0 = 0). \tag{3.35}$$

At a given temperature T and density ρ (or specific volume V) let there be N_0 neutral atoms, N_1 singly ionized atoms, etc., in a unit mass of gas. For conciseness we shall term an ion with a charge equal to m an m-ion; the number of m-ions per unit mass is then denoted by N_m (neutral atoms are a particular case of m-ions). The number of free electrons is denoted by N_e. Assuming that the gas is sufficiently rarefied and that the electrons obey Boltzmann statistics*, we must assign to each gas particle a thermal energy of translational motion equal to $\frac{3}{2}kT$. In addition, an m-ion possesses an electronic excitation energy W_m.

If we take the internal energy of an unionized gas at zero temperature as the reference level, then the specific internal energy per unit mass can be written†

$$\varepsilon = \tfrac{3}{2}N(1 + \alpha_e)kT + N \sum_m Q_m \alpha_m + N \sum_m W_m \alpha_m, \tag{3.36}$$

where α_e is the degree of ionization of the gas, that is, the number of free electrons per original atom ($\alpha_e = N_e/N$), and $\alpha_m = N_m/N$ is the concentration of m-ions. The concentrations α_m are connected by the condition of conservation of the number of atoms

$$\sum N_m = N, \qquad \sum \alpha_m = 1 \tag{3.37}$$

and by the condition of charge conservation

$$\sum mN_m = N_e, \qquad \sum m\alpha_m = \alpha_e. \tag{3.38}$$

The pressure of the ionized gas‡ is

$$p = N\rho(1 + \alpha_e)kT. \tag{3.39}$$

The equilibrium ion concentrations satisfy relations similar to the law of mass action for dissociation. This is quite reasonable because the ionization process can be treated as a chemical reaction or "dissociation" of an atom or an ion; for example, the removal of the $(m + 1)$st electron from an m-ion can be written symbolically as

$$\mathrm{A}_m \rightleftarrows \mathrm{A}_{m+1} + e, \qquad m = 0, 1, 2, \ldots. \tag{3.40}$$

* The degenerate electron gas will be considered in §12.

† If at low temperatures the gas is polyatomic the dissociation energy must be added to ε.

‡ We note that, as in the case of dissociation, equations (3.36) for the energy and (3.39) for pressure are also applicable to nonequilibrium ionization, if T is understood to denote the "translational" temperature of the particles.

The "law of mass action" for this reaction can be easily derived from the general expression for the free energy in the same manner as for a dissociation reaction. We write the free energy of a unit mass of ionized gas as

$$F = -\sum_m N_m kT \ln \frac{Z_m e}{N_m} - N_e kT \ln \frac{Z_e e}{N_e}, \tag{3.41}$$

where Z_m and Z_e are the partition functions of an m-ion and an electron. For thermodynamic equilibrium, at constant T and V, the free energy is a minimum with respect to the number of particles. We take the variation δF with respect to the change in the number of m-ions arising from the ionization following the reaction (3.40), setting $\delta N_m = -\delta N_{m+1} = -\delta N_e$ and requiring that the number of all the remaining particles does not change. We set this variation δF equal to zero and obtain

$$\frac{N_{m+1} N_e}{N_m} = \frac{Z_{m+1} Z_e}{Z_m}. \tag{3.42}$$

The translational partition functions of the $(m + 1)$st and the mth ions cancel, since their masses are practically equal. However, in the electronic contribution to the partition function of an ion (atom), we factor out the part corresponding to the zero-point energy (ground state)

$$Z_{el} = \sum e^{-\frac{\varepsilon_k}{kT}} = e^{-\frac{\varepsilon_0}{kT}} \sum e^{-\frac{\varepsilon_k - \varepsilon_0}{kT}} = e^{-\frac{\varepsilon_0}{kT}} u.$$

Denoting by w_k the energy difference $\varepsilon_k - \varepsilon_0$ (the excitation energy of the ion in the kth state), we can write the transformed electronic partition function in the form

$$u = \sum_k e^{-\frac{w_k}{kT}} = g_0 + g_1 e^{-\frac{w_1}{kT}} + g_2 e^{-\frac{w_2}{kT}} + \cdots, \tag{3.43}$$

where $g_0, g_1, \ldots$ are the statistical weights of the 0, 1, ... energy levels of the ion; if the levels are not degenerate $g = 1$.

The partition function for a free electron consists of the product of the translational partition function and the statistical weight of the free electron; this weight is equal to 2, since there are two possible spin orientations. Noting that the difference of the zero-point energies of the $(m + 1)$st and mth ions is equal to the ionization potential of the m-ion, $\varepsilon_{0\,m+1} - \varepsilon_{0\,m} = I_{m+1}$, and dividing (3.42) by the volume ($n_i = N_i/V$), we get

$$\frac{n_{m+1} n_e}{n_m} = 2 \frac{u_{m+1}}{u_m} \left(\frac{2\pi m_e kT}{h^2} \right)^{3/2} e^{-\frac{I_{m+1}}{kT}} = K_{m+1}(T) \tag{34.4}$$

(m_e is the mass of an electron). This relation is known as the Saha equation. Multiplying (3.44) by kT, we obtain a relation for the partial pressures $p_i = n_i kT$. For numerical calculations it is convenient to rewrite the Saha equation in a form relating the particle concentrations $\alpha_i = N_i/N = n_i V/N = n_i/N\rho$,

$$\frac{\alpha_{m+1}\alpha_e}{\alpha_m} = \frac{1}{\rho N} K_{m+1}(T), \qquad m = 0, 1, 2, \ldots . \tag{3.45}$$

Equations (3.45), (3.37), and (3.38) form a complete system of nonlinear algebraic equations for the ion and electron concentrations as functions of the gas temperature and density.

Ordinarily there exists a temperature range from 8,000°–30,000°K, where the gas is singly ionized and where the second ionization has not yet begun (the second ionization potential is approximately twice as great as the first ionization potential). In this region the system is simplified, since only the one equation (3.45) with $m = 0$ remains. Noting that in the singly ionized region $\alpha_1 = \alpha_e = 1 - \alpha_0$ and dropping the subscripts on α_1 and the ionization potential, we obtain an equation for the degree of ionization $\alpha = \alpha_1 = \alpha_e$:

$$\frac{\alpha^2}{1-\alpha} = 2\frac{u_1}{u_0}\frac{1}{\rho N}\left(\frac{2\pi m_e kT}{h^2}\right)^{3/2} e^{-I/kT}. \tag{3.46}$$

This relation is quite similar to the relation (3.26) for the degree of dissociation. For $I/kT \gg 1$, $\alpha \ll 1$, the degree of ionization α is proportional to $\rho^{-1/2}e^{-I/2kT}$, so that it increases very rapidly with temperature and very slowly with decreasing gas density. Equation (3.46) is always valid for a gas composed of hydrogen atoms.

The excitation energies of atomic and ionic energy levels are usually quite high and close to the ionization potential. In some cases there exist low-lying levels which must, of course, be taken into account in the calculations, but their number is very limited. The calculation of the transformed electronic partition functions u will be discussed in greater detail in the next section. We only note here that as a rule it is sufficient to consider only the first few terms in these partition functions. In most cases the dominant role is played by the first term and the partition function simply reduces to the statistical weight of the ground state $u \approx g_0$. The point is that in a gas which is not too dense, the electron in an atom or an ion "prefers" to be removed rather than to jump to a higher energy level. In the singly ionized region with $T \sim$ 10,000°–20,000°K, the value of I_1/kT is usually of the order of 5–10. As the temperature increases, I_1/kT becomes small, but at the same time the singly ionized atoms also disappear because the second ionization begins, and for the majority of ions the quantity I_{m+1}/kT is still of the order of 5–10. Since the energies of the excited atomic levels are of the same order as the ionization potential, then even the second term in the series for u will be very small, of the order of e^{-5}. This illustrates that for most common ions the dominant role in the electronic partition functions is played by the first term g_0.

Rigorous calculations usually take into account the first 5–10 levels of the ions and atoms, with the energies and statistical weights taken from appropriate tables (see [7]). Tables of successive ionization potentials of different atoms may be found, for example, in [8].

The internal energy of the gas can be calculated from (3.36), which follows from the general expression for the free energy (3.41) combined with the thermodynamic relation (3.9). The energy of electronic excitation W is equal to (where the subscript m corresponding to the ionic charge is dropped)

$$W = \frac{\sum w_k \exp(-w_k/kT)}{\sum \exp(-w_k/kT)} = -kT^2 \frac{\partial \ln u}{\partial T}. \tag{3.47}$$

According to (3.8), the entropy is obtained by differentiation of the free energy with respect to the temperature:

$$S = \sum_m N_m k \ln \frac{e^{5/2} V}{N_m} \left(\frac{2\pi M k T}{h^2} \right)^{3/2} u_m + \sum_m N_m \frac{W_m}{T}$$
$$+ N_e k \ln \frac{e^{5/2} V}{N_e} \left(\frac{2\pi m_e k T}{h^2} \right)^{3/2} 2. \tag{3.48}$$

If the excitation can be neglected, then the second term vanishes and $u_m = g_{0m}$.

The simplest calculations are in the singly ionized region, where the degree of ionization can be simply calculated from (3.46). Ionization makes a significant contribution to the specific heat and the energy of the gas and must be taken into account in calculating the thermodynamic functions. A wide range of temperatures and densities in which the atoms are multiply ionized has been considered in the work of Selivanov and Shlyapintokh [4]. These authors have calculated the composition*, thermodynamic functions, and Hugoniot curves for ionized air at temperatures from 20,000° to 500,000°K and densities from 10 ρ_0 to 10^{-3} ρ_0 (ρ_0 is standard atmospheric density). The nature of the dependence of the composition and degree of ionization upon the temperature and the effect of ionization upon the thermodynamic functions is evident from Tables 3.2 and 3.3 for air. These tables are based on the calculations of Selivanov and Shlyapintokh†. See also Kuznetsov [35].

At very high temperatures‡ (or very low densities) the energy and pressure of thermal radiation may be comparable with the energy and pressure of the fluid. When the radiation is in thermodynamic equilibrium with the fluid (whether this condition is satisfied or not must be checked in each particular

* Generalization of the equations presented to the case when the gas is a mixture of elements does not present any difficulties.

† The data given in Table 3.2 were taken from [4] only for temperatures of 20,000°K and above.

‡ For air at standard density these temperatures are greater than a million degrees.

Table 3.3

COMPOSITION OF IONIZED AIR AT STANDARD DENSITY ($\rho_0 = 1.29 \cdot 10^{-3}$ g/cm³) AND HIGH TEMPERATURES

T°K	Atom	0	1^+	2^+	3^+	4^+	5^+	6^+	e
20,000	N	0.589	0.201						0.24
	O	0.172	0.036						
50,000	N	0.018	0.451	0.321	0.001				1.50
	O	0.0065	0.303	0.048					
100,000	N		0.012	0.275	0.463	0.04			2.65
	O		0.005	0.09	0.113	0.005			
250,000	N				0.005	0.183	0.603		5.0
	O				0.005	0.020	0.114	0.074	
500,000	N					0.017	0.75	0.025	5.2
	O						0.010	0.200	

Concentration is defined as the ratio of the number of particles of the given species to the original number of atoms. 0 denotes neutral atoms, 1^+ singly ionized atoms, etc., e are electrons, N and O are nitrogen and oxygen ions, respectively.

case, see Chapter II), the radiation energy and pressure are simply added to the energy and pressure of the gas. The "specific" energy of equilibrium radiation is equal to the radiant energy density divided by the fluid density

$$\varepsilon_\nu = \frac{U_p}{\rho} = \frac{4\sigma T^4}{c\rho}, \tag{3.49}$$

and the radiation pressure is

$$p_\nu = \frac{U_p}{3} = \frac{4\sigma T^4}{c}. \tag{3.50}$$

The radiation entropy can be obtained with the aid of general thermodynamic relationships

$$S_\nu = -\frac{\partial F_\nu}{\partial T}, \qquad F_\nu = -T\int \frac{\varepsilon_\nu}{T^2}\,dT = -\frac{4\sigma T^4}{3c\rho}, \qquad S_\nu = \frac{16\sigma T^3}{3c\rho}. \tag{3.51}$$

The thermodynamic functions of air calculated in [4] take the equilibrium radiation into account.

§6. The electronic partition function and the role of the excitation energy of atoms

An isolated atom (ion or a molecule) in an infinite space has an infinite number of discrete energy levels which converges to a continuum, corresponding to an ionized state with a completely removed electron. The electronic partition function u formally contains an infinite number of terms and is divergent. The average excitation energy of an atom W, calculated from (3.47) with an infinite number of terms, is equal to the ionization potential, since the excitation energies of the higher states asymptotically approach the ionization potential. This difficulty, arising from a purely formal evaluation of u and W, is to a certain extent deceptive, since in reality the atom is never isolated but is always contained in a gas of finite density. The dimensions of the electron orbit increase rapidly as the higher energy electronic states of an atom are excited. These dimensions finally become comparable with the average distance between the gas particles, and this is approximately given by $r \approx N^{-1/3}$ (here N denotes the particle number density). The trajectories of the electrons moving in these large orbits are distorted by the presence of neighboring particles. Such an electron, which is removed from an atom to a distance comparable with the average distance between gas particles, does not differ essentially from a free electron, and such a highly excited atom does not differ essentially from an ionized atom. The finite value of the gas density imposes, therefore, a limitation on the number of possible excited atomic states, on the number of terms in the electronic partition function, and on the average excitation energy of the atom.

Let us consider a gas consisting of hydrogen atoms. The results obtained from a study of the hydrogen atom are quite general, since the highly excited states of any complex atomic system are very similar to the excited states of the hydrogen atom. If an electron in a complex atom (or ion or molecule) moves in a very large orbit, then the field in which it travels is very close to the Coulomb field of a point charge (representing the rest of the atom). Therefore, the structure of the highly excited states of any atom or ion is close to that of hydrogen. In order to apply these results to multiply ionized atoms, we shall introduce into all equations the charge of the "nucleus". In other words, we do not consider hydrogen in a literal sense but rather hydrogen-like atoms represented by a system consisting of a positive "nucleus" with a charge Z and a single electron.

The energy levels of a hydrogen-like atom are characterized by the principal quantum number n (for an energy level diagram see Fig. 2.2 in Chapter II, §2). The energy of the nth level, measured from the boundary of the continuous spectrum, is, as we know, equal to $\varepsilon_n = -I_H Z^2/n^2$, where $I_H = 13.5$ ev is the ionization potential of hydrogen. Its absolute value $E_n = |\varepsilon_n| = I_H Z^2/n^2$ is the binding energy of an electron in the nth level. The binding energy of the ground state $n = 1$* is equal to the ionization potential

$$E_1 = I_H Z^2 = I.$$

The excitation energy of the nth state is given by $w_n = \varepsilon_n - \varepsilon_1 = E_1 - E_n = I_H Z^2(1 - 1/n^2)$. The transformed electronic partition function of the hydrogen-like atom takes the form

$$u = \sum g_n \exp\left(-\frac{w_n}{kT}\right) = \sum 2n^2 \exp\left[-\frac{I_H Z^2}{kT}\left(1 - \frac{1}{n^2}\right)\right],$$

where $g_n = 2n^2$ is the statistical weight of the nth level.

The binding energy of an electron in the nth state is equal to its Coulomb energy in the field of the nucleus at a distance of the order of the orbital dimension, namely, $E_n = Ze^2/2a$, where a is the semimajor axis of the elliptical orbit. Then, $a = Ze^2/2E_n = e^2n^2/2ZI_H = a_0 n^2/Z$, where $a_0 = 0.53 \cdot 10^{-8}$ cm is the Bohr radius. The summation defining the partition function u should in any case be terminated at a value n^* at which the semimajor axis of the orbit becomes comparable with the average distance between gas particles, at $a = a_0 n^{*2}/Z = r$, where $n^* = (Zr/a_0)^{1/2} \sim N^{-1/6}$ (n^* is smaller the higher is the gas density). As a numerical example, let us consider molecular hydrogen, originally at room temperature and atmospheric pressure, and then heated by a strong shock wave to a temperature of the order of ten thousand degrees. The density ratio across the shock wave is approximately 10, so that for complete dissociation of the molecules the number of atoms per cm^3 N will be approximately $5 \cdot 10^{20}$/cm^3. The average distance between the atoms is $r \approx N^{-1/3} = 1.3 \cdot 10^{-7}$ cm and the limiting value n^* is 5 ($Z = 1$). At a temperature $T = 11{,}600°$K $= 1$ ev, the partition function consisting of five terms has a value $u = 2.00053$, and is practically equal to the statistical weight of the ground state $g_1 = 2$. The average excitation energy of the atom calculated from (3.47) using five terms is $W = 0.003$ ev. At these values of T and N, the degree of ionization of hydrogen is $\alpha = 3 \cdot 10^{-3}$, that is, the ionization energy per atom, $I_H\alpha = 0.04$ ev. The excitation energy W is small in comparison with the ionization energy ($W/I_H\alpha = 0.075$). At the higher temperature $T = 23{,}200°$K $= 2$ ev and the same density, $u = 2.212$ (still not much larger than $g_1 = 2$), and $W = 1.16$ ev. The degree of ionization in this case is $\alpha = 0.22$, the ionization energy per original atom $I_H\alpha = 3$ ev, and the excitation energy, also per original atom, is $W(1 - \alpha) = 0.9$ ev. In this case the excitation energy has an appreciable effect, although it is still less than the ionization energy.

* Here, the subscript "1" (rather than "0") is assigned to the ground state because the principal quantum number n for the ground state is equal to unity.

It should be noted that cutting off the upper excitation levels in a gas of finite density will at the same time lower the ionization potential by an amount equal to the electron binding energy at the cutoff boundary, that is, by $\Delta I = E_{n*} = Ze^2/2r = ZI_{\text{H}}a_0/r = 7 \cdot 10^{-8} ZN^{1/3}$ ev (with N per cm^3). In our example this decrease is $\Delta I = 0.55$ ev, so that the calculated degrees of ionization are somewhat low.

At very high temperatures, of the order of 50,000°K and above, the excitation energy of the remaining hydrogen atoms becomes very large and comparable to the ionization potential. On the other hand, the degree of ionization also increases rapidly and the number of neutral atoms becomes small. Thus, the contribution of the excitation energy to the energy of the gas is in any case smaller than the contribution of the ionization energy. This corresponds to the idea that the electron "finds it more convenient" to be removed from the atom than to occupy a high excitation level*.

Our choice of the number of terms to be included in the electronic partition function most likely overestimates the actual number of levels. In the cutoff of the higher excitation levels in atoms and ions there is a significant effect of the electrostatic field of the nearest neighboring charged particles (the Stark effect). In addition, in a sufficiently rarefied gas, the binding energy of an electron moving in the limiting orbit $a \sim r$ is $E_{n*} = \Delta I = 7 \cdot 10^{-8} ZN^{1/3}$ ev, which is smaller than kT (in our example $\Delta I = 0.55$ ev, while the temperatures were 1 and 2 ev). The kinetic energy of an electron in a hydrogen-like atom is equal to its binding energy at the given level. It makes no sense, however, to consider an electron to be bound when its binding energy and kinetic energy are less than kT. Practically each "collision" with a free electron would knock out such a weakly bound electron from the atom. Some authors, therefore, terminate the partition function summation even sooner, at a level where the electron binding energy is equal to kT.

A large number of papers [9–13, 34] are devoted to the problem of the lowering of ionization potentials in an ionized gas and to the calculation of

* This situation can be made clearer by the following semiqualitative argument, which is most easily visualized in the limiting case of a gas of very low density. The ratio of the probability of free and bound electronic states is proportional to the ratio of the translational and electronic partition functions ($Z_{\text{trans}} \sim V \sim 1/\rho$). The electronic partition function in the limit of low densities contains a large number of terms and can be represented approximately by

$$Z_{\text{el}} = \Sigma\, g_n \exp\left(-\frac{\varepsilon_n}{kT}\right) = \Sigma\, 2n^2 \exp\left(\frac{I_{\text{H}}}{n^2 kT}\right) \sim \int_0^{n*} n^2\, dn \sim n^{*3}.$$

But, $n^* \sim r^{1/2}$, so that $Z_{\text{el}} \sim V^{1/2} \sim \rho^{-1/2}$. Hence, $Z_{\text{trans}}/Z_{\text{el}} \sim V^{1/2} \sim \rho^{-1/2}$. Thus, when the density decreases (in the range of small densities), the probability that an electron will be removed from the atom increases even faster despite the increase in the number of possible bound states.

electronic partition functions. It should be noted that there is no general agreement on the subject, and different authors still recommend different procedures for terminating the electronic partition function summation. Fortunately, calculations show that the varying choices of the number of terms taken into account in the summation have, as a rule, a very small effect on the calculated thermodynamic functions of gases. However, the lowering of ionization potentials as a result of the cutoff of the upper levels sometimes has a significant effect on the calculated composition of an ionized gas (see [14]).

In concluding this section we note that the cutoff phenomenon in the upper excitation levels of atoms, ions, and molecules has been confirmed experimentally. Low pressure arc discharge spectra ordinarily show no more than 5–10 spectral lines of the Balmer hydrogen series arising from the transition of an electron from an upper excited level to a level whose principal quantum number $n = 2$. Even in spectra of the extremely low density in nebulae (of the order of tens of particles per cm^3) not more than 50–60 Balmer lines are observed.

§7. Approximate methods of calculation in the region of multiple ionization

The calculations for ionization equilibrium, which form the basis for determining the thermodynamic properties of gases at high temperatures, are extremely involved and time consuming. For each pair of temperature and density values, a nonlinear system of algebraic equations must be solved to determine the concentrations of the ions of different charge. This calculation is even more complicated if the gas is composed of atoms of several elements. In this regard tables covering a wide range of temperatures and densities have been compiled only for air. Obviously, with modern computers the problem of large numerical calculations becomes significantly less acute, but it is still useful for practical purposes to have a simple approximate method which, with a minimum expenditure of time and effort, permits the calculation of the degrees of ionization and the thermodynamic functions of any gas over a wide range of high temperatures and densities, in the range where the atoms are multiply ionized. In this section we shall consider such a method, proposed by one of the authors [15]. For all its simplicity, the method is sufficiently accurate for the solution of most practical problems.

Let us consider a gas consisting of atoms of a single element. Our approximate method is based on two fundamental steps. The first of these is the assumption that the ion number density n_m and the ionization potentials I_{m+1} are considered to be continuous functions of the ionic charge m, obtained by connecting the discrete values of n_m and I_{m+1} by continuous curves. The function $I(m)$ is constructed by joining the points I_m on the I, m diagram

(Fig. 3.1) by a continuous curve, say, by a series of straight lines. The system of recurrence relations defined by the Saha equations (3.44) can then be transformed into a differential equation for the function $n(m)$, by replacing the finite differences by differentials

$$n(m+1) = n(m) + \frac{dn}{dm}, \qquad \Delta m = 1.$$

The ratio of the ion electronic partition functions u_{m+1}/u_m ordinarily varies in an irregular manner when the charge m changes its value for a given element,

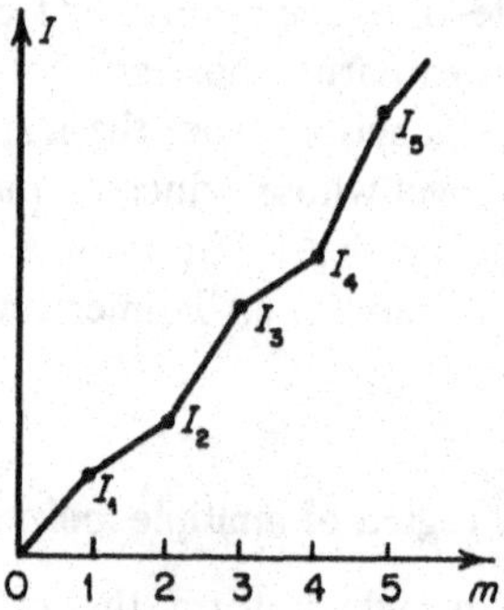

Fig. 3.1. Transition to the continuous curve $I(m)$.

or when the element itself is transformed; this ratio, however, is always of the order of unity. Let us assume that it is approximately equal to unity. We can then replace the system of Saha equations by the differential equation

$$\left(1 + \frac{d \ln n}{dm}\right) n_e = AT^{3/2} \exp\left(-\frac{I(m+1)}{kT}\right), \tag{3.52}$$

where

$$A = 2\left(\frac{2\pi m_e k}{h^2}\right) = 4.8 \cdot 10^{15}\ \text{cm}^{-3} \cdot \text{deg}^{-3/2} = 6 \cdot 10^{21}\ \text{cm}^{-3} \cdot \text{ev}^{-3/2}.$$

We next rewrite the particle and charge conservation conditions (3.37) and (3.38) in the integral form

$$\int n(m)\, dm = n, \tag{3.53}$$

$$\int mn(m)\, dm = n_e. \tag{3.54}$$

The results of exact calculations and results from the system of Saha equations developed below show that a significant number of doubly or at most triply ionized atoms are always present in the gas. In addition, the distribution function $n(m)$ has a very narrow and sharp peak about some value m_{max} which, of course, depends on both the density and temperature of the gas. This leads to the second step, which is the approximation that the average value of the ionic charge, also the average number of free electrons per original atom

$$\overline{m} = \frac{\int mn(m)\,dm}{\int n(m)\,dm} = \frac{n_e}{n}, \tag{3.55}$$

is exactly equal to that value of m_{max} for which the ion distribution function $n(m)$ has a maximum. Obviously, this assumption is more justified the sharper and narrower is the peak of the distribution $n(m)$.

Denoting the ionization potential of ions with an "average" charge $\overline{m}$ by $\bar{I}$, and noting that $dn/dm = 0$ at the peak or maximum point, we get from (3.52), with the aid of (3.55),

$$\overline{m} = \frac{AT^{3/2}}{n} e^{-\bar{I}/kT}. \tag{3.56}$$

In order to convert this expression into a formula for determining the average charge (or the degree of ionization) as a function of the temperature T and density (the number density n of original atoms) it is necessary to determine the relation between $\bar{I}$ and $\overline{m}$. A certain degree of arbitrariness exists here which is connected purely with the formalism (in the exact theory) of assigning subscripts to ionization potentials. If we denote the ionization potential of an m-ion by I_{m+1} (the ionization potential of a neutral atom is I_1), then formally we should have set $\bar{I} = I(\overline{m} + 1)$. Sometimes, however, the ionization potential of an m-ion is denoted by I_m (the ionization potential of a neutral atom is then I_0). In this case in the Saha equation (3.44) I_{m+1} should be replaced by I_m and $\bar{I}$ should be formally set equal to $I(\overline{m})$.

Of course, if we consider heavy elements at very high temperatures, where the degree of ionization is so high that $\overline{m}$ is of the order of several tens, the arbitrariness does not result in any substantial change in the value of $\overline{m}$ (since in this case $I_{m+1} - I_m \ll I_m$). Where the average charge of the ions is not large, however, the arbitrariness has a marked effect on both the values of $\overline{m}$ and of the thermodynamic functions. This results from the approximation of replacing discrete values by continuous functions.

Comparison of the results of approximate and exact calculations shows that best agreement is obtained when, as before, we denote the ionization potential of an m-ion by $I_{m+1} = I(m + 1)$, setting $I_0 = I(0) = 0$, but referring the "average" value of the potential $\bar{I}$ to the point $m + \frac{1}{2}$, i.e., assuming $\bar{I} = \bar{I}(\overline{m} + \frac{1}{2})$. This procedure appears to be quite natural, when we note that the sequence of discrete values of m is separated by finite intervals $\Delta m = 1$.

Taking the logarithm of (3.56) we then obtain a simple transcendental equation for $\bar{m}(T, n)$

$$I(\bar{m} + \tfrac{1}{2}) = kT \ln \frac{AT^{3/2}}{\bar{m}n}. \tag{3.57}$$

Because the right-hand side is a logarithmic function of $\bar{m}$, two or three successive approximations are sufficient to obtain a fairly accurate value of the root of $\bar{m}$ from a plot of the function $I(m)$.

Let us now show that the distribution of ion number density with respect to the charge always has a narrow peak, and let us find the equation governing the peak of the distribution function. Combining successively the Saha equations for the values $m = 1, 2, \ldots$, having set the ratios of the electronic partition functions equal to unity, and using the definition of the "average" potential (3.56), we obtain

$$\frac{n_{m+l}}{n_m} = \exp\left[-\sum_{i=1}^{l} \frac{I_{m+i} - \bar{I}}{kT}\right],$$

$$\frac{n_{m-l}}{n_m} = \exp\left[-\sum_{i=0}^{l} \frac{\bar{I} - I_{m-i}}{kT}\right],$$

where $l = 1, 2, 3, \ldots$. Let us choose m equal to the value for which n_m is maximum. The quantity $\bar{I}$ corresponds approximately to the ionization potential of such ions, so that all the terms in the summation are positive and the ion concentrations decrease on both sides of the maximum. To determine the relation governing the shape and width of the peak, we introduce the continuous functions $n(m)$ and $I(m)$. Expanding, we have approximately

$$I(m) \approx \bar{I} + \left(\frac{dI}{dm}\right)(m - \bar{m}),$$

from which we obtain the Gaussian distribution curve

$$n(m) = n_{\max} \exp\left[-\left(\frac{m - \bar{m}}{\Delta}\right)^2\right] \tag{3.58}$$

with a peak half-width

$$\Delta = [2kT/(\overline{dI/dm})]^{1/2}. \tag{3.59}$$

Noting that on the average, for different elements and different degrees of ionization, the ionization potential increases with ionic charge faster than the charge itself (that $dI/dm > I/m$), we find that

$$\Delta < \left(\frac{2kT\bar{m}}{\bar{I}}\right)^{1/2} = \left(\frac{2\bar{m}}{x_1}\right)^{1/2}, \qquad \bar{x}_1 = \frac{\bar{I}}{kT}. \tag{3.60}$$

Putting in numerical values of $\bar{x}_1$ and $\bar{m}$ determined, for example, from Table 3.3 for air, we see that $\Delta < 1$, that the peak is actually narrow*.

Approximate expressions for the various thermodynamic functions can be obtained from the exact equations by assuming that the ion distribution function $n(m)$ has a very narrow peak, is almost a delta function about m. In other words, we assume that all ions have the same nonintegral "average" charge $\bar{m}$. The specific internal energy (3.36) then becomes

$$\varepsilon = \tfrac{3}{2}N(1 + \bar{m})kT + NQ(\bar{m}) \tag{3.61}$$

(we neglect the electronic excitation energy). Here the continuous function $Q(m)$, as with $I(m)$, is plotted by connecting the discrete values Q_m determined from (3.35) by a continuous curve. We note that here the best agreement with exact calculations is obtained by setting $\bar{Q} = Q(\bar{m})$ in (3.61) in contrast to having set $\bar{I} = I(\bar{m} + \tfrac{1}{2})$. The pressure is given by

$$p = n(1 + \bar{m})kT. \tag{3.62}$$

The specific entropy (3.48) (if we neglect the electronic excitation and assume the statistical weight of all the ions to be the same and equal to g) is found to be

$$S = Nk \ln \left(\frac{2\pi MkT}{h^2}\right)^{3/2} \frac{e^{5/2}g}{n} + Nk\bar{m} \ln \left(\frac{2\pi m_e kT}{h^2}\right)^{3/2} \frac{e^{5/2}g}{n\bar{m}}. \tag{3.63}$$

Setting $S = const$ and using (3.57), we obtain the equation of the isentrope in the parametric form

$$\frac{T^{3/2}}{n} \exp\left\{\bar{m}\left[\frac{I(\bar{m} + \frac{1}{2})}{kT} + \frac{5}{2}\right]\right\} = const. \tag{3.64}$$

The parameter here is $\bar{m}$; the constant on the right-hand side is determined by values of T_0 and n_0 through which the isentrope passes.

The above method for finding the degree of ionization and thermodynamic functions can be easily generalized to include gas mixtures. For example, the "average" ionic charges $\bar{m}_1$ and $\bar{m}_2$ of each of the elements in a two-element mixture are found from the system of two transcendental equations

$$I_1(\bar{m}_1 + \tfrac{1}{2}) = I_2(\bar{m}_2 + \tfrac{1}{2}) = kT \ln \frac{AT^{3/2}}{n(c_1\bar{m}_1 + c_2\bar{m}_2)}, \tag{3.65}$$

* The fact that the peak of the n_m distribution is narrow, with Δ of the order of the "finite" difference $\Delta m = 1$, generally makes the transformation to differentials with respect to m meaningless. However, the method gives results which are better than any rational justification would indicate.

where c_1 and c_2 are the atom concentrations of both elements, I_1 and I_2 their ionization potential curves, and n is the total number density of original atoms. The specific internal energy is

$$\varepsilon = \tfrac{3}{2}N(1 + c_1\bar{m}_1 + c_2\bar{m}_2)kT + Nc_1Q_1(\bar{m}_1) + Nc_2Q_2(\bar{m}), \qquad (3.66)$$

etc. In many cases, however, there is little advantage in complicating the calculations in this manner. If the successive ionization potentials for the different atoms do not differ very much from each other, it is convenient to introduce an "average" potential curve $I(m)$, considering all the atoms as identical and the values of the successive potentials as averaged with respect to all the elements in accordance with their percentage in the mixture.

Table 3.4

COMPARISON OF APPROXIMATE AND EXACT CALCULATIONS OF THE DEGREE OF IONIZATION AND INTERNAL ENERGY OF AIR

T°K	$\rho_0 = 1.29 \cdot 10^{-3}$ g/cm³		$\rho = 10^{-2}\rho_0$	
	$1+\bar{m}$	$\varepsilon, \dfrac{\text{ev}}{\text{atom}}$	$1+\bar{m}$	$\varepsilon, \dfrac{\text{ev}}{\text{atom}}$
30,000	1.68	16.6	2.30	33
	1.77	23	2.21	33
50,000	2.4	40.5	3.35	83
	2.42	47.8	3.26	80
100,000	3.72	126	5.10	243
	3.75	140	5.16	252

The upper numbers in each pair of values were obtained by the approximate method of [15], while the lower ones were taken from the work of Selivanov and Shlyapintokh [4].

Table 3.4 compares for air the approximate values of the degree of ionization and the internal energy with the exact values obtained by Selivanov and Shlyapintokh [4]. It is evident that even at low degrees of ionization, where the error should be particularly large, the approximate method gives fairly good accuracy. The error at high degrees of ionization does not exceed a few percent.

The method correctly reflects all the irregularities in the variation of $\bar{m}$ and ε with temperature and density, corresponding to the sharp jumps in the ionization potentials which take place with transitions from ions with filled electron shells to ions with unfilled shells. Calculations have shown that the method also gives satisfactory accuracy for xenon. Since the ionization potential curves for all elements are similar to each other, we may hope that the approximate method will be accurate enough for other gases as well.

§8. Interpolation formulas and the effective adiabatic exponent

Calculated thermodynamic functions have been tabulated as a function of temperature and density (or pressure). The use of such tables in solving gasdynamic problems is, however, very inconvenient. It is much more convenient to use simple interpolation formulas which approximate the tabulated data. Of particular interest is the approximation of the actual functions for cases in which a suitably defined adiabatic exponent which characterizes hydrodynamic processes turns out to be almost constant. The use of a constant "effective adiabatic exponent" or "effective ratio of specific heats" enables us to apply self-similar and exact solutions of the gasdynamic equations, solutions which as a rule can be obtained only for a gas with constant specific heats.

Isentropic relationships between any two thermodynamic parameters, for example T and ρ or p and ρ, which account for partial vibrational excitation, dissociation, and ionization can no longer be described by the equations for a perfect gas with constant specific heats. An exponent γ can be defined formally at every point in such a way that the actual isentrope in the neighborhood of this point coincides approximately with the perfect gas isentrope. To satisfy this condition we set

$$\left(\frac{\partial \ln T}{\partial \ln \rho}\right)_S = \gamma' - 1 \qquad \text{or} \qquad \left(\frac{\partial \ln p}{\partial \ln \rho}\right)_S = \gamma''*.$$

However, the exponents corresponding to different choices of thermodynamic parameters are different in general. Therefore, in order to introduce an effective adiabatic exponent γ in the range of interest of T and ρ (or p and ρ), it is necessary to define it in a manner closely reflecting the nature of the gasdynamic process.

In the usual scheme the third gasdynamic relation is the energy conservation equation. In order to close the system of hydrodynamic equations for an ideal fluid†, it is sufficient to introduce a relation between the internal energy,

* *Editors' note.* The quantity γ'' is often termed the isentropic exponent. It is the ratio of the square of the speed of sound to p/ρ, and in many problems plays a role more fundamental than that of the effective adiabatic exponent γ. In Hugoniot relations for strong shocks γ is of more fundamental importance than γ''.

† The hydrodynamics of an ideal fluid does not take into account either viscosity or heat conduction.

pressure, and density $\varepsilon(p, \rho)$. This relationship is usually given in terms of the formula

$$\varepsilon = \frac{1}{\gamma - 1}\frac{p}{\rho}.$$

We determine an adiabatic exponent in the range of interest of p and ρ by tabulating the quantity

$$\gamma - 1 = \frac{p}{\rho\varepsilon} \tag{3.67}$$

and choosing a constant value of $\gamma - 1$ which best approximates the different values of $p/\rho\varepsilon$. As a result, the isentropic equation $d\varepsilon + p\,dV = 0$ $(V = 1/\rho)$ takes the constant-γ form $p \sim \rho^\gamma$, $\varepsilon \sim \rho^{\gamma-1}$ with an effective constant value of γ.

The specific internal energy as a function of temperature and density is most conveniently approximated by a power-law relation

$$\varepsilon = aT^\alpha V^\beta \tag{3.68}$$

with constant values of a, α, and β. In the region where vibrational degrees of freedom are fully excited the specific heat is independent of the density and $\beta = 0$. In the dissociation and ionization regions the specific heat always increases with decreasing density, since in this case there is an increase in the degree of dissociation or ionization with a corresponding increase in the energy losses. Therefore, the exponent β is always positive. The exponent α is usually greater than unity, since the specific heat increases with temperature in the region where partial excitation of the vibrational modes takes place, as well as in the dissociation and ionization regions.

When the function $\varepsilon(T, V)$ is approximated by (3.68) with constant exponents α and β, and the functions $p(\varepsilon, \rho)$ or $p(\varepsilon, V)$ are approximated by (3.67) with constant γ, then the three constants α, β, and γ cannot be chosen independently. The functions $p(\varepsilon, V)$ and $\varepsilon(T, V)$ must satisfy the general thermodynamic relation

$$p + \left(\frac{\partial \varepsilon}{\partial V}\right)_T = T\left(\frac{\partial p}{\partial T}\right)_V.$$

Direct substitution shows that the three quantities α, β, and γ are related by

$$\gamma - 1 = \frac{\beta}{\alpha - 1}, \tag{3.69}$$

provided, of course, that these quantities are assumed to be constant. In the above interpolation (as may be easily checked from the isentropic equation $d\varepsilon + p\,dV = 0$) the isentropic relations between T and ρ and ε and ρ are also characterized by a single exponent γ, precisely as in the classical case of a perfect gas of constant specific heats where

$$T \sim \rho^{\gamma-1}, \qquad \varepsilon \sim \rho^{\gamma-1}, \qquad p \sim \rho^{\gamma}, \qquad \gamma = const.$$

This result is obtained in spite of the fact that the specific heats are functions of both temperature and volume.

Table 3.2 illustrates the numerical values of this effective adiabatic exponent by tabulating $1 + p/\rho\varepsilon = \gamma$ for multiply ionized air. It is evident that γ decreases with decreasing density.

In the range of temperatures from 10,000°K to 250,000°K and of densities from $10\rho_0$ to $10^{-3}\rho_0$ (ρ_0 is the standard atmospheric density) the internal energy of air can be roughly approximated by (3.68) with the constants given by the relation

$$\varepsilon = 8.3\left(\frac{T^\circ}{10^4}\right)^{1.5}\left(\frac{\rho_0}{\rho}\right)^{0.12} \quad \text{ev/molecule.} \tag{3.70}$$

The effective adiabatic exponent obtained from (3.69) is $\gamma = 1.24$.

It is important that the quantity γ as determined from (3.67) varies much less than do the exponents α and β in (3.68). This is a favorable situation since the function $\varepsilon(T, V)$ is in fact not necessary for the calculation of isentropic processes; it is sufficient to know $\varepsilon(p, V)$ or $p(\varepsilon, V)$ as given by (3.67). It should be noted that, in approximating over a wide range of temperatures and densities, the effective adiabatic exponent and the exponents α and β in (3.68) differ very little from one gas to another. That this should be so is quite clear, since the ionization potential curves are in general similar to one another and differ only in details affecting the behavior of the energy and pressure within narrow ranges of the temperature and density.

§9. The Hugoniot curve with dissociation and ionization

The changes in the flow variables across a shock wave in a gas with constant specific heats were calculated in Chapter I. In the case of a strong shock wave, where the pressure behind the front is much larger than the initial pressure $p_1 \gg p_0$, the density ratio across the shock approaches its limiting value $K = (\gamma + 1)/(\gamma - 1)$. Thus, in a monatomic gas (inert gases, metal vapors) $c_v = \frac{3}{2}Nk$, $\gamma = \frac{5}{3}$, and $K = 4$, while in a diatomic gas with the vibrational mode unexcited $c_v = \frac{5}{2}NkT$, $\gamma = \frac{7}{5}$, and $K = 6$ *. It is apparent from the equation for K in the case of a gas with constant specific heats that the density ratio across

* Practically, the limiting density ratio of 6 is attained in a diatomic gas with the vibrational mode unexcited only for low initial temperatures T_0. Otherwise, the pressure ratio p_1/p_0 is not large enough to consider the shock as "strong", for those temperatures behind the wave which do not excite the vibrational mode.

the shock is larger, the higher are the specific heats and the closer to unity is the specific heat ratio. The tendency for an increase in the density ratio with an increase in the specific heat is also true in the general case where the specific heat is a function of temperature and density. If a diatomic gas is so dense that upon passage through a strong shock wave the vibrational mode is fully excited even before dissociation begins, then the specific heat behind the wave increases and approaches the value $c_v = \frac{7}{2}NkT$, the specific heat ratio or adiabatic exponent approaches $\gamma = \frac{9}{7}$, and the density ratio across the wave increases to $K = 8$.

Dissociation and ionization lead to a further increase in the density ratio. It is important to note that the density ratio is affected only by that part of the specific heat which is associated with the potential and internal energy of the particles, that is, with the energy of dissociation and ionization, with the rotational and vibrational energy of the molecules, and with the electronic excitation energy of the atoms and ions. The increase in the specific heat as a result of an increase in the number of particles does not affect the density ratio, since the increase in the translational energy of the particles is accompanied by a corresponding increase in the gas pressure. The change in the number of particles does not directly affect the adiabatic exponent γ which determines the density ratio. We can easily prove this statement by writing the internal energy as the sum $\varepsilon = \varepsilon_{\text{trans}} + Q$, where Q includes the potential energy and the energy of the internal degrees of freedom of the particles. Noting that the pressure is $p = \frac{2}{3}\rho\varepsilon_{\text{trans}}$, we substitute these expressions into the Hugoniot relation (1.71). Neglecting the initial energy and pressure, that is, assuming the shock wave to be strong, we find for the limiting value of the density ratio*

$$K = 4 + \frac{3Q}{\varepsilon_{\text{trans}}}. \tag{3.71}$$

The greater is the relative importance of the potential and the internal energies, the greater is the difference from the monatomic gas value of 4.

With dissociation and ionization the potential energy usually turns out to be larger than the translational energy of the particles and the density ratio across the shock is rather large, of the order of 10–12 (or even larger). The density ratio is especially large when the initial density is low, and when the degree of dissociation and ionization is very high at a given temperature†.

* In a paper of the authors [16] the erroneous relation $K = 4/(1 - 3Q/\varepsilon_{\text{trans}})$ was given (equation (2.5)) in place of (3.71).

† Thus, for example, when shock waves are propagated through air at an initial pressure $p_0 = 10^{-4}$ atm with velocities $D \sim 6.5$–12 km/sec (Mach numbers $M \sim 20$–35), the density ratio across the shock is approximately 17.

For heavy gases with ionization the density ratio does not remain constant as the strength of the wave increases. The relative contribution of the potential energy decreases gradually after the density ratio has reached a maximum in the dissociation or first ionization region, because the translational energy increases faster than the potential energy as a result of an increase in the number of particles. The density ratio in this case decreases gradually. This situation prevails until all electrons are removed from any atomic shell. A large jump always exists between the ionization potentials of the last electron in a shell and the first electron of the following closed shell. This jump is especially large between the L and K shells. For example, in nitrogen it is 97 ev and 550 ev, in oxygen it is 137 ev and 735 ev. In air there exists, therefore, a fairly wide range of shock strengths, approximately in the temperature range from 500,000 to 700,000°K, when all electrons of oxygen and nitrogen atoms filling the L shells have already been removed and the ionization of the K shells has not yet begun, so that only helium-like ions exist in the gas. When a further increase in shock strength results in the removal of the K electrons, the ionization energy again increases sharply, the relative contribution of the potential energy (as at the beginning of the first ionization) increases, and the density ratio passes through a second, clearly defined maximum.

From the mass and momentum conservation relations (1.61) and (1.62), the pressure behind a strong shock wave is given by

$$p_1 = \rho_0 D^2\left(1 - \frac{1}{K}\right). \tag{3.72}$$

The pressure is not very sensitive to the value of the density ratio, especially at high density ratios, and is approximately proportional to the square of the propagation velocity of the wave D. For example, with $K \sim 10$ this is to within about 10%. The specific enthalpy behind the shock

$$h_1 = \frac{D^2}{2}\left(1 - \frac{1}{K^2}\right) \tag{3.73}$$

(as derived from (1.61), (1.62), and (1.64)) is even more closely proportional to the square of the velocity. In the example considered this is to within the order of 1%.

The temperature, which in a gas with constant specific heats is also proportional to the square of the velocity, with increasing shock strength does increase with dissociation and ionization present, although much more slowly. In the singly ionized region this slowing-down of the temperature rise comes as a result of the relative increase in the energy lost to ionization, in the increase of the quantity $Q/\varepsilon_{\text{trans}} \sim Q/T$; subsequently, when the contribution of the potential energy to the internal energy decreases in comparison with that of the translational energy, the slower temperature increase can be explained by the increase in the number of particles, to which both ε_{trans} and p are proportional:

$$\varepsilon_{trans} = \tfrac{3}{2}N(1 + \overline{m})kT, \qquad p = n(1 + \overline{m})kT.$$

We note that after complete ionization, when ε_{trans} increases with increasing shock strength and temperature behind the shock, while Q remains unchanged, the density ratio approaches 4 (without considering thermal radiation) as the strength increases. This is evident from (3.71). In hydrogen, for example, in the region of complete dissociation and ionization the potential energy per atom is 15.74 ev (the dissociation energy per H atom is 2.24 ev and the ionization energy is 13.5 ev) and the translational energy per atom (the energy of a proton and electron) is $3kT = 3T$ ev, so that

$$K = 4 + \frac{15.74}{T_{ev}} \rightarrow 4 \quad \text{as} \quad T \rightarrow \infty$$

(effectively complete ionization of hydrogen with atmospheric density ahead of the wave takes place even at $T \sim 100{,}000°\text{K} \sim 10$ ev).

The effect of dissociation and ionization upon the quantities behind a shock wave is illustrated in Table 3.5. The calculations are for air initially at standard density. The low-temperature data with excited vibrational modes are taken from the book of Zel'dovich [17]; the calculations with dissociation and the beginning of the first ionization were made by Davies [18]. The quantities behind a shock over a wide range of temperatures from 20,000 to 500,000°K were calculated by Selivanov and Shlyapintokh [4] (cited previously). The flow quantities behind shock waves in air over a wide range of initial pressures

Table 3.5

FLOW QUANTITIES BEHIND A SHOCK WAVE IN AIR WITH STANDARD CONDITIONS AHEAD OF THE WAVE ($p_0 = 1$ atm, $T_0 = 293$°K)

T°K	D, km/sec	p_1, atm	ρ_1/ρ_0	T°K	D, km/sec	p_1, atm	ρ_1/ρ_0
293	0.33	1	1	14,000	9.31	1,000	11.10
482	0.70	5	2.84	20,000	11.8	1,650	10.10
705	0.98	10	3.88	30,000	15.9	2,980	9.75
2,260	2.15	50	6.04	50,000	23.3	6,380	8.97
4,000	3.35	127	8.58	100,000	40.1	19,200	8.62
6,000	4.54	236	9.75	250,000	81.6	76,500	7.80
8,000	5.64	366	10.30	500,000	114.0	143,900	6.27
10,000	6.97	561	11.00				

(from standard pressure to $p_0 \sim 10^{-5}$ atm) have been calculated by Rozhdestvenskii [19] and Gorban' [20] (for temperatures behind the wave not greater than 12,000°K). A number of authors have calculated quantities behind shock waves for other gases as well: in argon and hydrogen (Prokof'ev [21]), in argon (Resler, Lin, and Kantrowitz [22]), in xenon (Sabol [23]), and in hydrogen and xenon (Kholev [24]). The processes are qualitatively similar for all the gases, and the Hugoniot curves are accordingly all rather similar.

The calculated Hugoniot curves for argon and xenon are in good agreement with experimental results obtained in shock tubes. Satisfactory agreement between calculations and experiment has also been obtained for air. It should be noted that the calculated behavior of the Hugoniot curve for dissociated air is strongly affected by the value assigned to the dissociation energy of nitrogen (the two previously contradictory values were 7.38 ev and 9.74 ev). Experiments by Christian and Yarger [25], who studied shock waves in air using a shock tube, have confirmed that the experimental Hugoniot curve is closer to the calculated one which takes a dissociation energy for nitrogen of 9.74 ev. This value is also supported by the experiments of Model' [26], who measured the wave velocity and (by an optical method) temperature behind the wave.

§10. The Hugoniot relations with equilibrium radiation

At very high temperatures (or very low densities), when the energy and pressure of equilibrium radiation are comparable to the energy and pressure of the fluid, the effect of radiation must be included when calculating the Hugoniot curve. (Obviously, it should be checked first whether equilibrium between the radiation and the fluid is attained under the given conditions of the problem.)

Let us consider a very strong shock wave propagating through a cold gas, and let us assume that the flux of radiation on both sides of the wave is zero. We also assume that the radiation behind the shock front is in equilibrium (the means by which the equilibrium is established is not of interest here). Thus, we consider the problem from a purely thermodynamic point of view, as is usual in deriving the Hugoniot relations*. It is to be emphasized that we are considering the nonrelativistic case, where the shock and fluid velocities are much smaller than the speed of light, and where the energies of the radiation and of the fluid are very much smaller than the rest energy of the fluid. Let us introduce the radiation energy and pressure behind the front $\varepsilon_{\nu 1}$ and $p_{\nu 1}$ into the momentum and energy conservation equations across the shock

* This problem was considered by Sachs [27].

wave (see §13, Chapter I and §17, Chapter II). The conservation relations across the wave are then written

$$\rho_1 u_1 = \rho_0 D,$$
$$p_1 + p_{v1} + \rho_1 u_1^2 = \rho_0 D^2, \tag{3.74}$$
$$\varepsilon_1 + \varepsilon_{v1} + \frac{p_1}{\rho_1} + \frac{p_{v1}}{\rho_1} + \frac{u_1^2}{2} = \frac{D^2}{2}.$$

In order to simplify the problem so that we may clarify the role of the radiation, we shall assume that the gas is one of constant specific heats with ratio (or adiabatic exponent) γ and obeys the usual equation of state

$$p = R\rho T, \qquad R = const\,; \qquad \varepsilon = \frac{1}{\gamma - 1} RT = \frac{1}{\gamma - 1}\frac{p}{\rho}.$$

Substituting ε_{v1} and p_{v1} from (3.49) and (3.50) into (3.74), expressing the pressure p_1 and the energy ε_1 in terms of the temperature T_1, and eliminating u_1 by means of the first of (3.74), we obtain the relations corresponding to (3.72) and (3.73) (in which radiation is not taken into account):

$$R\rho_0 K T_1 + \frac{4\sigma T_1^4}{3c} = \rho_0 D^2\left(1 - \frac{1}{K}\right), \tag{3.75}$$
$$\frac{\gamma}{\gamma - 1} R\rho_0 K T_1 + \frac{16\sigma T_1^4}{3c} = \frac{\rho_0 D^2}{2} K\left(1 - \frac{1}{K^2}\right).$$

Here $K = \rho_1/\rho_0$ is the density ratio across the shock wave. We may eliminate D from these equations and solve the resulting expression in terms of T_1

$$\frac{4\sigma T_1^3}{3Rc\rho_0} = \frac{K(K - K_0)}{(7 - K)}, \tag{3.76}$$

where $K_0 = (\gamma + 1)/(\gamma - 1)$ is the limiting density ratio across a strong shock wave without radiation. This equation can be considered as a defining equation for K in terms of the strength of the shock wave, which is here characterized by the temperature T_1 behind the front.

The left-hand side of (3.76), which is proportional to T_1^3, represents simply K times the ratio of the radiation pressure to the pressure of the fluid behind the shock wave p_{v1}/p_1 *. Equation (3.76) shows that if the radiation pressure is relatively small $p_{v1}/p_1 \ll 1$, then $K \approx K_0$, and the density ratio is almost equal to the usual limiting value $K_0 = (\gamma + 1)/(\gamma - 1)$. In the limit of a very strong

* *Editors' note.* The quantity $(K - K_0)/(7 - K)$ is equal to the ratio p_{v1}/p_1 without the restriction that γ be constant, as long as the shock is strong and the gas opaque. The variation of K with shock strength need not be monotonic in the general case.

shock, where $p_{v1}/p_1 \sim T_1^3 \to \infty$, the density ratio K tends to $K_\infty = 7$. This result might have been expected, since equilibrium radiation from a thermodynamic point of view behaves as a perfect gas with a specific heat ratio $\gamma = 4/3$ (see Chapter II, §3), for which the limiting density ratio across a shock is 7. Between the two limiting cases $p_{v1}/p_1 \to 0$ and $p_{v1}/p_1 \to \infty$, the density ratio K varies monotonically* from $K_0 = (\gamma + 1)/(\gamma - 1)$ to $K_\infty = 7$ as the wave strength increases, independently of whether $K_0 > 7$ or $K_0 < 7$, that is, independently of whether the specific heat ratio γ (without considering radiation) is less than or greater than 4/3.

In the limiting case, when the radiation energy and pressure are much greater than the energy and pressure of the fluid, when the second terms on the left-hand sides of (3.75) are much greater than the first terms, the temperature behind the front $T_1 \sim D^{1/2}$, unlike the usual case without radiation (in a gas with constant specific heats), where $T_1 \sim D^2$. We note that the relative importance of the energy and pressure of equilibrium radiation is greater, the lower is the density of the fluid, that $p_v/p \sim 1/\rho$ (in a gas of constant number of particles). For example, in completely ionized hydrogen, the radiation pressure is equal to the gas pressure at $T = 10^6$ °K, for a number density (protons and electrons) $n = 10^{19}$ cm^{-3}, while for $n = 10^{16}$ cm^{-3} the two pressures are equal at $T = 10^5$ °K.

2. *Gases with Coulomb interactions*

§11. Rarefied ionized gases

Let us consider the departure of an ionized gas from an ideal one as a result of the Coulomb interactions between charged particles. In this section we shall restrict ourselves to the case of "weakly" imperfect gases, where the terms representing the Coulomb interactions in the thermodynamic functions can be considered as small corrections to the terms describing a perfect gas.

In order to consider an ionized gas as a perfect one, it is necessary that the energy of the Coulomb interactions between neighboring particles be small in comparison with the thermal energy of the particles, that the condition $(Ze)^2/r_0 \ll kT$ holds. Here, Z is the average charge of the particles (ions and electrons) and $r_0 \approx n^{-1/3}$ is the average distance between them (n is the number of particles per cm^3). This condition can be rewritten as

$$n \ll \left(\frac{kT}{Z^2e^2}\right)^3 = 2.2 \cdot 10^8 \left(\frac{T^\circ}{Z^2}\right)^3 \text{ cm}^{-3}. \tag{3.77}$$

* *See footnote, p. 214.*

For example, if the degree of ionization is of the order of unity ($Z \sim 1$) and $T \sim 30{,}000°$K, the gas can be considered as a perfect one when $n \ll 6.2 \cdot 10^{21}$ cm^{-3} (for comparison, we recall that the number of molecules in air at standard conditions is $2.67 \cdot 10^{19}$ cm^{-3}).

Coulomb corrections to the thermodynamic functions in the case of weakly imperfect gases can be calculated by the Debye–Hückel method as was done in the text by Landau and Lifshitz [1] (see also the paper of Timan [11]). A spherically symmetric nonuniformly charged cloud of particles of like charge is formed about each ion or electron. The distribution of charge density in each cloud is the Boltzmann distribution corresponding to the electrostatic potential of the central charge and the cloud. The solution of Poisson's equation for the electrostatic potential distribution with respect to the radius r from the central ion of charge $Z_i e$ leads in a first approximation to the equation

$$\varphi_i = Z_i e\, r^{-1} e^{-r/d},$$

where d is the so-called Debye radius characterizing the dimensions of the cloud, with

$$d = \left(\frac{4\pi e^2}{kT} \sum n_i Z_i^2\right)^{-1/2} = 6.90\left(\frac{T^\circ}{n\overline{Z^2}}\right)^{1/2} \text{cm} \tag{3.78}$$

(n_i is the number of ions of charge $Z_i e$ per cm^3; electrons are here included in the concept of "ions" by setting $Z = -1$ for the electrons).

The statistical treatment using the Debye–Hückel method is justified if the cloud contains many particles, that is, if the Debye radius d is much greater than the average distance between the particles $r_0 \approx n^{-1/3}$. The requirement $d \gg r_0$ leads to the condition $n \ll (kT/4\pi e^2\overline{Z^2})^3 = 1.1 \cdot 10^5 (T^\circ/\overline{Z^2})^3$ cm^{-3}, which is even stricter than the requirement (3.77) for a perfect gas. The Debye approach, therefore, assumes a very weakly imperfect gas.

Near the center for $r \ll d$, $\varphi_i = Z_i e/r - Z_i e/d$. The first term is the potential of the central ion itself, and the second one $\varphi_i' = -Z_i e/d$ is the potential due to all the other surrounding charges at the point where the given ion is located. The Coulomb energy of the gas in a volume V, according to the general equation of electrostatics, is given by

$$E_{\text{coul}} = V \cdot \tfrac{1}{2} \sum e Z_i n_i \varphi_i' = -Ve^3\left(\frac{\pi}{kT}\right)^{1/2}\left(\sum n_i Z_i^2\right)^{3/2}. \tag{3.79}$$

The free energy correction can be found by integrating the thermodynamic relation $E/T^2 = -\partial/\partial T\,(F/T)$ to give

$$F_{\text{coul}} = \tfrac{2}{3} E_{\text{coul}} = -\tfrac{2}{3} e^3\left(\frac{\pi}{kTV}\right)^{1/2}\left(\sum N_i Z_i^2\right)^{3/2}, \tag{3.80}$$

where $N_i = n_i V$ is the total number of particles of the ith species in the volume V. The pressure correction is

$$p_{\text{coul}} = -\left(\frac{\partial F_{\text{coul}}}{\partial V}\right)_{T,N_i} = \frac{E_{\text{coul}}}{3V}. \tag{3.81}$$

On the average, the forces between particles are attractive, since each ion surrounds itself mostly with charges of the opposite sign; therefore, both the Coulomb energy and pressure are negative. The Coulomb interaction affects the state of the gas in two ways. First, it decreases the energy and the pressure (and also the entropy, since $S_{\text{coul}} = -\partial F_{\text{coul}}/\partial T = E_{\text{coul}}/3T$). Second, and this effect is the more important, it displaces the ionization equilibrium toward higher degrees of ionization. Indeed, a free electron in an interacting gas has a negative potential energy, and behaves as if it were weakly bound to the ions. Therefore, slightly less work is required to remove an electron from an atom or an ion, and this corresponds to an effective decrease in the ionization potentials.

The equation for ionization equilibrium taking into account Coulomb interactions is derived in the same manner as in §5. The total free energy of the system is expressed as

$$F = F_{pg} + F_{\text{coul}},$$

where F_{pg} is given by (3.41) and F_{coul} by (3.80). We next calculate the variation δF with respect to the variation in the number of m-ions due to ionization. Using the condition $\delta N_m = -\delta N_{m+1} = -\delta N_e$ and setting the variation δF equal to zero, we obtain a corrected expression for the law of mass action in place of (3.42). In order not to confuse the partition function with the charge, we denote the partition function with a tilde ($\tilde{Z}$). We then have

$$\frac{N_{m+1}N_e}{N_m} = \frac{\tilde{Z}_{m+1}\tilde{Z}_e}{\tilde{Z}_m}\exp\left(\frac{\Delta I_{m+1}}{kT}\right), \tag{3.82}$$

where the quantity ΔI_{m+1}, equal to the change in the Coulomb part of the chemical potentials

$$\Delta I_{m+1} = \mu_{m,\text{coul}} - \mu_{m+1,\text{coul}} - \mu_{e,\text{coul}}\,; \qquad \mu_{i,\text{coul}} = \left(\frac{\partial F_{\text{coul}}}{\partial N_i}\right)_{T,V},$$

can be treated as a decrease in the ionization potential of the m-ions (we recall that $\tilde{Z}_{m+1}\tilde{Z}_e/\tilde{Z}_m \sim \exp(-I_{m+1}/kT)$).

Calculation gives for the correction to the ionization potential

$$\Delta I_{m+1} = 2(Z_m + 1)e^3\left(\frac{\pi}{kT}\right)^{1/2}\left(\sum n_i Z_i^2\right)^{1/2}, \tag{3.83}$$

where Z_m is the charge of an m-ion; actually $Z_m = m$. Recalling the definition of the Debye radius (3.78), we can rewrite (3.83) as

$$\Delta I_{m+1} = \frac{(Z_m + 1)e^2}{d} = \frac{Z_{m+1}e^2}{d}. \tag{3.84}$$

The decrease in the ionization potential of an m-ion is exactly equal to the Coulomb interaction energy of an $(m + 1)$-ion, which is obtained as a result of ionization of an m-ion, with the removed electron located at a distance equal to the Debye radius. In agreement with the conditions for the validity of the Debye–Hückel method and the condition for a weakly imperfect gas, (3.84) is valid for $d \gg r_0$, or equivalently for $\Delta I \ll kT$.

In the singly ionized case (3.84) becomes ($i = 0, 1, e$; $Z_0 = 0, Z_1 = 1, Z_e = -1$)

$$\Delta I_1 = 2e^3\left(\frac{2\pi n\alpha}{kT}\right)^{1/2}, \tag{3.85}$$

where $\alpha = n_e/n = n_1/n$ is the degree of ionization. For multiple ionization, replacing, as in §7, all the ions by ions with an "average" charge $\bar{m} = n_e/n$ (n is the number of original ions per cm^3) and setting $\overline{Z_i^2} = \bar{m}^2$, we obtain for the change in the "average" ionization potential

$$\overline{\Delta I} = 2(\bar{m} + 1)e^3\left[\frac{\pi\bar{m}(\bar{m} + 1)n}{kT}\right]^{1/2}. \tag{3.86}$$

As an example, we consider air at a temperature $T = 100{,}000°$K and at standard density $n = 5.34 \cdot 10^{19}$ cm^{-3}. The degree of ionization without considering Coulomb interactions is $\bar{m} = 2.72$, and the "average" ionization potential $\bar{I} = 60$ ev ($\bar{I}/kT = 6.9$). The correction to the "average" ionization potential corresponding to this value of $\bar{m}$ is $\overline{\Delta I} = 5.4$ ev ($\overline{\Delta I}/kT = 0.63$). Thus the Coulomb interactions decrease the "average" ionization potential by almost 10% which would in turn correspond to an increase in the degree of ionization by approximately 14%*. The effect of Coulomb corrections on the shift in ionization equilibrium in argon at $T = 45{,}000°$K and $p \sim 10^{-3} - 10^2$ atm has been considered in [14]. This effect was found to be quite noticeable, even though the corrections to the thermodynamic functions did not exceed 1%.

§12. Dense gases. Elements of Fermi–Dirac statistics for an electron gas

In our discussion of ionized gases we have always assumed that free electrons obey classical Boltzmann statistics. Strictly speaking, an electron gas is described by Fermi–Dirac quantum statistics, which reduces to Boltzmann

* Formally, under these conditions we are almost at the limit of applicability of the method, since $\overline{\Delta I} = 5.4$ ev is only slightly less than $kT = 8.6$ ev.

statistics only in the limit of sufficiently high temperatures or sufficiently low densities. The transition from Fermi–Dirac to Boltzmann statistics occurs if the temperature of the electron gas is much greater than the so-called degeneracy temperature T_0, which is determined by the number density of the electrons n (per cm^3) through the relation

$$T_0 = \frac{1}{8}\left(\frac{3}{\pi}\right)^{2/3} \frac{h^2}{m_e k} n^{2/3} = 4.35 \cdot 10^{-11} n^{2/3} \text{ deg.} \tag{3.87}$$

At ordinary gas densities and temperatures, where free electrons are present as a result of ionization, the condition $T \gg T_0$ is always satisfied. For example, in air at atmospheric density and with atoms approximately singly ionized, $n = 5.34 \cdot 10^{19}$ cm^{-3}, and the degeneracy temperature $T_0 = 610$°K; in this example the corresponding gas temperature is $T \sim 35{,}000$°K, so that $T/T_0 \approx 60$. The condition for the applicability of Boltzmann statistics is violated only at very low temperatures or very high densities of the electron gas. The first case usually does not arise since gases are not ionized at low temperatures. The second case is, however, of considerable importance. There are many processes that are accompanied by the formation of a very dense, highly heated gas containing electrons. Usually such a situation arises when a solid is rapidly heated to very high temperatures of the order of tens or hundreds of thousands of degrees*; under such conditions the material actually becomes a dense gas, since the energy of thermal motion at these temperatures frequently exceeds the binding energy of the atoms in the solid or liquid.

When the density is of the order of the density of solid matter and the number of free electrons per atom is of the order of unity, the degeneracy temperature is of the order of several tens of thousands of degrees (for example, for $n = 5 \cdot 10^{22}$ cm^{-3}, $T_0 = 59{,}000$°K). Then, even at a temperature of 100,000°K the electrons cannot be described by Boltzmann statistics. It should be noted that at densities close to the density of solids and temperatures of the order of tens or hundreds of thousands of degrees the Coulomb energy of interaction between the charged particles (electrons and ions) is comparable to their kinetic energy, and the electron-ion gas is actually imperfect†.

* For example, rapid heating occurs during the impact of meteorites traveling at high velocities (on the order of several tens of km/sec) with the surface of a planet; in exploding conductors by electric currents; upon heating of the anode needle in pulse-type x-ray tubes by electron impact (see Tsukerman and Manakova [28]); upon the heating of solids by a very strong shock wave, and so forth. We shall not discuss here such classical examples as the application of quantum statistics to free electrons in metals under normal conditions.

† For example, for $n = 5 \cdot 10^{22}$ cm^{-3} and $Z = 1$, the Coulomb energy $e^2/r \approx e^2 n^{1/3}$ is equal to kT at $T = 60{,}000$°K. The kinetic energy of the free electrons, which is determined not simply by the temperature but also by the degeneracy temperature T_0, is also comparable to the Coulomb energy, since T_0 is in this case equal to 59,000°K.

Thermodynamic properties of a gas under these conditions can be determined approximately by a method which is a generalization of the Thomas–Fermi statistical method to include the case of nonzero temperature. To present the essence of this method let us recall the basic concepts of the Fermi–Dirac statistics (for a more detailed discussion see, for example, [1]). Consider a free electron* gas at zero temperature (the so-called completely degenerate gas). The number of quantum states in an element of volume dV with absolute values of electron momenta from p to $p + dp$, the number of cells in the phase space of coordinates and momenta, is equal to $4\pi p^2\, dp\, dV/h^3$. Each cell may contain two electrons with opposite spins, so that the total number of quantum states in the element $dp\, dV$ is $8\pi p^2\, dp\, dV/h^3$. According to the Pauli exclusion principle, no more than one electron may be in any quantum state with a given direction of spin. The N electrons contained in the volume V ($n = N/V$ is the number of electrons per unit volume) fill all the lowest energy states with momenta ranging from 0 to p_0, so that

$$N = V \int_0^{p_0} \frac{8\pi p^2\, dp}{h^3} = \frac{8\pi p_0^3}{3h^3} V.$$

This equation leads to the expression for the maximum kinetic energy $\varepsilon_0 = p_0^2/2m_e$ of electrons at zero temperature, the so-called Fermi limiting energy

$$\varepsilon_0 = \frac{1}{8}\left(\frac{3}{\pi}\right)^{2/3} \frac{h^2}{m_e}\left(\frac{N}{V}\right)^{2/3} = \frac{1}{8}\left(\frac{3}{\pi}\right)^{2/3} \frac{h^2}{m_e} n^{2/3}. \tag{3.88}$$

The degeneracy temperature (3.87) is defined as $T_0 = \varepsilon_0/k$. The kinetic energy of N electrons in the volume V is

$$E_k = V \int_0^{p_0} \frac{p^2}{2m_e} \frac{8\pi p^2\, dp}{h^3} = \tfrac{3}{5}\varepsilon_0 N \sim N^{5/3} V^{-2/3}. \tag{3.89}$$

The average kinetic energy of an electron is $\frac{3}{5}\varepsilon_0$. Since it is assumed that the electrons are free, the kinetic energy is equal to the total energy $E_k = E$ and, by virtue of the thermodynamic relation $T\, dS = dE + P\, dV$ applied at zero temperature, the pressure of a degenerate free electron gas is

$$P = -\frac{dE}{dV} = \frac{2}{3}\frac{E_k}{V} = \frac{2}{5}\varepsilon_0 n = \frac{1}{20}\left(\frac{3}{\pi}\right)^{2/3} \frac{h^2}{m_e} n^{5/3}. \tag{3.90}$$

The pressure is proportional to the 5/3 power of the density. The relation between the pressure and the kinetic energy, $P = \frac{2}{3}E_k/V$, is the same as for a

* The gas is free in the sense that no forces act on the electrons. At the same time it is assumed that the electron gas does not diffuse. This can be thought of as an electrically neutral mixture of ions and electrons, in which the average self-consistent field is assumed to be zero (everywhere, except at the boundary).

monatomic Boltzmann gas. That this should be so is evident, since the "kinetic" pressure is determined by the momentum transferred by the particles, and its relation to the kinetic energy of the particles is a purely mechanical one and is independent of the type of statistics obeyed by the particles.

As the temperature of the electrons increases, the electrons which have earlier filled the lowest energy levels begin to occupy the higher quantum states. It is shown in Fermi–Dirac statistics that the particle distribution function over the quantum states, i.e., the average number of electrons in a quantum state of energy ε, is

$$f = \frac{1}{e^{(-\mu+\varepsilon)/kT} + 1}. \tag{3.91}$$

Here μ is a constant which depends upon the temperature and density of the electrons, and which represents the chemical potential of the electron gas. In a free electron gas, the energy ε is equal to the kinetic energy, $\varepsilon = p^2/2m_e$.

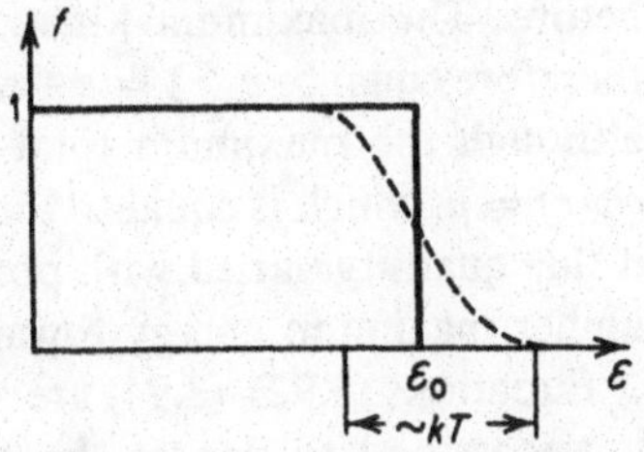

Fig. 3.2. Distribution function for an electron gas according to Fermi–Dirac statistics.

At zero temperature the distribution function is equal to 1 if $\varepsilon < \mu$ $((-\mu + \varepsilon)/kT = -\infty)$, and is equal to 0 if $\varepsilon > \mu((-\mu + \varepsilon)/kT = +\infty)$. Thus we obtain the same distribution already found above, with the additional result that the chemical potential of the free electron gas is equal to the Fermi limiting energy ε_0. For nonzero temperatures the distribution function is spread out, as shown in Fig. 3.2.

The number of electrons per unit volume with momenta between p and $p + dp$ is

$$\rho(p)\,dp = \frac{8\pi p^2\,dp}{h^3} f = \frac{8\pi}{h^3} \frac{p^2\,dp}{e^{(-\mu+\varepsilon)/kT} + 1}, \tag{3.92}$$

and the total number of electrons per unit volume is

$$n = \int_0^\infty \rho(p)\,dp = \frac{8\pi}{h^3} \int_0^\infty \frac{p^2\,dp}{e^{(-\mu+\varepsilon)/kT} + 1}. \tag{3.93}$$

This equation defines implicitly the chemical potential μ as a function of temperature and density. The kinetic energy of the electrons per unit volume is

$$\mathscr{E}_k = \int_0^\infty \frac{p^2}{2m_e} \rho(p)\,dp = \frac{8\pi}{h^3} \int_0^\infty \frac{p^2}{2m_e} \frac{p^2\,dp}{e^{(-\mu+\varepsilon)/kT} + 1}. \tag{3.94}$$

Statistics can be also applied to an electron gas in a potential field. Of course, the spatial variation of the potential must be slow enough that a sufficient number of particles are contained in an elementary volume dV in which the field can be taken to be constant. Otherwise the application of Fermi–Dirac statistics to the particles is meaningless*. If we denote the electrostatic potential at a point r by $\varphi(r)$, then the energy of an electron ε can be written

$$\varepsilon = \frac{p^2}{2m_e} - e\varphi(r). \tag{3.95}$$

Statistical mechanics shows that if a gas is in a potential field, then at equilibrium its chemical potential μ must be the same at all points. If this is not the case the gas will move.

If we consider an electron gas at zero temperature in a potential field then, according to (3.91) and (3.95), the distribution function f is equal to 1 for $\varepsilon = p^2/2m_e - e\varphi(r) < \mu$, and is equal to 0 for $\varepsilon = p^2/2m_e - e\varphi(r) > \mu$, as before. The maximum kinetic energy of an electron at a given point r is therefore equal to $\varepsilon_0(r) = \mu + e\varphi(r)$. This energy is now a function of position, although the maximum total energy of the electron $p_0^2/2m_e - e\varphi(r) = \varepsilon_0 - e\varphi(r) = \mu$, which is equal to the chemical potential, is independent of position; if this quantity varied with position the electrons would move from points of higher maximum energy toward points of lower maximum energy.

Equations (3.92)–(3.94) are also valid for gas placed in a potential field, if ε is understood to denote the quantity given by (3.95). Equation (3.93) then gives an implicit relation connecting the gas density at point r, $n(r)$ with the quantity $\varepsilon_0(r) = \mu + e\varphi(r)$, with the potential at the given point and the temperature T. At $T = 0$ this relation is again expressed by (3.88).

§13. The Thomas–Fermi model of an atom and highly compressed cold materials

When a dense gas is described by the Thomas–Fermi method, no distinction is made between "free" and "bound" electrons. The point of view is taken that the gas is not composed of ions and electrons as is normally supposed at low densities, but rather of nuclei and electrons. The nuclei obey Boltzmann statistics and contribute separately to the total pressure and specific internal thermal energy. At high temperatures this contribution corresponds to that of an ordinary monatomic gas

$$P_a = n_a kT, \qquad \varepsilon_a = \frac{3}{2}\frac{n_a}{\rho} kT$$

* The field should also vary but very little over a distance of the order of the de Broglie wavelength of an electron.

(n_a is the number of nuclei per unit volume and ρ is the density of the medium). The total interaction energy of the particles is associated with the electrons only. To calculate the electron contributions to the energy and pressure, the gas is divided into atomic cells, each of which contains a nucleus with a charge Ze and Z electrons. For simplicity the cell is taken to be spherical and its volume V is taken equal to the average volume of the material per nucleus, thus, $V = 1/n_a$ and the radius $r_0 = (3V/4\pi)^{1/3} = (3/4\pi n_a)^{1/3}$.

There are no coupling forces between atomic cells taken into account in the Thomas–Fermi model, and therefore this model does not describe the coupling of atoms in solids. The cells exert a positive pressure on each other, equal to the pressure of the electron gas; the model describes only the repulsive forces and the "thermal" pressure. Hence, this model yields sensible results either at high densities, for strongly compressed solids in which the repulsive forces predominate over the forces of attraction between the atoms, or at high temperatures, for which the coupling forces can be neglected. Then it follows that the "ionization", "excitation", and "thermal motion" energies of the electrons are no longer calculated separately in the Thomas–Fermi model, as they were in the case of rarefied gases. These energies are automatically included in the total electron energy of the atomic cell. In order to separate from the total energy the "thermal" portion of the energy, one specifically related to the existence of a finite temperature, it is necessary to subtract out the energy of a cell of the same volume but corresponding to zero temperature. The same is also true for the pressure.

Let us first consider an atomic cell at zero temperature, according to the classical Thomas–Fermi statistical model of an atom*. The basic assumption of the model is that in atoms with large numbers of electrons the majority of the electrons have large principal quantum numbers, and that consequently their motion is quasi-classical. The electrons in the atom are considered as a gas placed in a self-consistent electrostatic field† $\varphi(r)$ which varies sufficiently slowly with respect to the radius; the field is determined by the charges of the nucleus and the electrons, thereby accounting for the fact that the electron gas is not a perfect one. Fermi–Dirac statistics apply to this gas.

The maximum kinetic energy of an electron at a given distance r from the nucleus $\varepsilon_0(r) = \mu + e\varphi(r)$ is related to the electron density at this point through (3.88), so that the density can be expressed in terms of the potential by

$$n(r) = \frac{8\pi}{3} 2^{3/2} \frac{m_e^{3/2}}{h^3} [e\varphi(r) + \mu]^{3/2}. \tag{3.96}$$

* A detailed treatment of this subject can be found in the book of Gombàs [29]. Landau and Lifshitz [30] also give a brief but clear presentation.

† The possibility of the statistical description of an electron gas in a potential field was discussed in the preceding section.

The electrostatic potential $\varphi(r)$ satisfies Poisson's equation

$$\Delta\varphi = \frac{1}{r}\frac{d^2}{dr^2}[r\varphi(r)] = 4\pi e n(r), \tag{3.97}$$

which, after substituting (3.96) and introducing a new "potential" $\psi = \varphi + \mu/e$ (the potential is determined to within an additive constant), becomes

$$\frac{1}{r}\frac{d^2}{dr^2}(r\psi) = \frac{32\cdot 2^{2/3}\pi^2}{3}\frac{e^{5/2}m_e^{3/2}}{h^3}\psi^{3/2}. \tag{3.98}$$

Boundary conditions must also be added to (3.98). At the center, as $r\to 0$, the field approaches the Coulomb field of the nucleus, and

$$\varphi(r) = \frac{Ze}{r} \quad \text{as} \quad r\to 0. \tag{3.99}$$

Since the cell is electrically neutral, the electric field at its boundary is zero (the potential outside the cell is constant), so that

$$\frac{d\varphi}{dr} = 0 \quad \text{at} \quad r = r_0. \tag{3.100}$$

This condition is equivalent to the obvious relation

$$Z = \int_0^{r_0} n(r) 4\pi r^2\, dr. \tag{3.101}$$

Introducing the dimensionless variables

$$x = \frac{r}{a}, \qquad a = \frac{1}{4}\left(\frac{9\pi^2}{2}\right)^{1/3}\frac{a_0}{Z^{1/3}} = \frac{0.885a_0}{Z^{1/3}}, \tag{3.102}$$

where $a_0 = h^2/4\pi^2 m_e e^2 = 0.529\cdot 10^{-8}$ cm is the Bohr radius, and

$$\chi = \frac{r}{Ze}\left(\varphi + \frac{\mu}{e}\right) = \frac{r}{Ze}\psi, \tag{3.103}$$

equation (3.98) reduces to the universal form

$$x^{1/2}\frac{d^2\chi}{dx^2} = \chi^{3/2}. \tag{3.104}$$

The boundary conditions (3.99) and (3.100) become

$$\chi(0) = 1; \qquad \chi(x_0) = x_0\left(\frac{d\chi}{dx}\right)_{x_0}.$$

The dimensionless form of these equations demonstrates the similarity of the problem with respect to the number of electrons Z. In particular, the density distribution with respect to radius can be written, according to (3.96), (3.102), and (3.103), as

$$n(r) = Z^2 f\left(\frac{rZ^{1/3}}{b}\right), \qquad b = 0.885a_0, \tag{3.105}$$

where the function f is proportional to $(\chi/x)^{3/2}$.

The solution of (3.104) under appropriate boundary conditions (this is carried out by means of a numerical integration) yields the electron potential and density distributions as a function of radius. With these two quantities known we can calculate all other quantities of interest. The solutions show that the electron density in a free neutral atom, one which is not compressed by any external forces, extends to infinity; we find that $\chi \to 0$ and $n \to 0$ as $x \to \infty$* (Fig. 3.3). If we take zero potential energy for the state when all the charges are at infinite separation, then it follows that the potential φ at infinity should be set equal to zero. The chemical potential in this case is zero. The pressure at the boundary of such a free atom, and as a consequence also the pressure† in the entire space, is equal to zero. According to the virial theorem for a Coulomb field of infinite extent, the total kinetic and potential energies

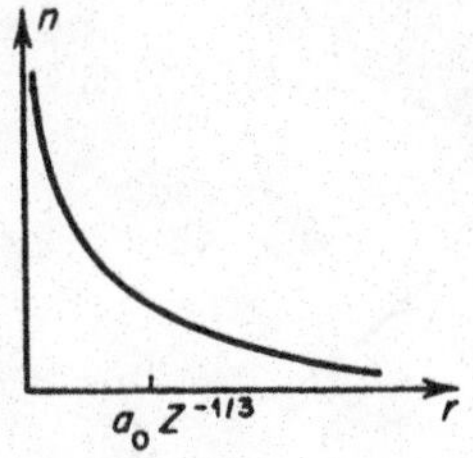

Fig. 3.3. Electron density distribution, (schematic) in a free atom.

of the particles are connected through the relation $2E_k^\infty = -E_p^\infty$. The total energy of an atom is $E^\infty = E_k^\infty + E_p^\infty = -E_k^\infty = \frac{1}{2}E_p^\infty$. The virial theorem in this case expresses the fact that the kinetic repulsion of the electrons is exactly balanced by their Coulomb attraction to the nucleus and hence the total pressure (which is equal to the sum of the "kinetic" and "potential" pressures) is equal to zero everywhere. Although the electron density extends to

* Since the field of an electrically neutral atom must decrease at infinity faster than r^{-2}, the "potential" ψ decreases faster than r^{-1}; in this case the outer boundary condition takes the form $r\psi \sim \chi \to 0$ as $r \to \infty$.

† The pressure in a system consisting of interacting particles is composed of two parts: the "kinetic" pressure, related to the motion of the particles and their kinetic energy by the usual relation $P_k = 2n\varepsilon_k/3$, where n is the number density and ε_k their average kinetic energy; and the "potential" pressure, equivalent to the forces acting on the particles (Coulomb forces in the given case). Formally, this separation follows from the relation (at zero temperature) $P = -\partial E/\partial V = -\partial E_k/\partial V - \partial E_p/\partial V = P_k + P_p$. The kinetic pressure is always positive, while the potential pressure $P_p > 0$, if the particles are repulsive, and $P_p < 0$, if they are attractive.

infinity in principle, the main charge is concentrated in a finite volume V_{ef}. According to (3.105) the Bohr radius a_0 serves as a characteristic scale for this region and $V_{ef} \sim Z^{-1}$ (see Fig. 3.3). This result also follows from the virial theorem. The order of magnitude of the potential energy of an atom is $E_p^\infty \sim -e^2Z^2/V_{ef}^{1/3}$. The kinetic energy, according to (3.88) and (3.89), is given in order of magnitude by

$$E_k^\infty \sim \varepsilon_0 Z \sim \frac{h^2}{m_e} Z\left(\frac{Z}{V_{ef}}\right)^{2/3}.$$

From the condition of mechanical equilibrium or from the virial theorem we find $V_{ef} \sim (m_e e^2/h^2)^3 Z^{-1} \sim a_0^3 Z^{-1}$. The total energy of the atom $E^\infty = E_p^\infty/2$ is of the order of $-e^2Z^{7/3}/a_0 \sim -I_H Z^{7/3}$. The exact value is $E^\infty = -20.8\, Z^{7/3}$ ev; this is the absolute value of the energy required to remove all the charges to infinity (the total ionization energy of the atom).

Let us now consider a "compressed" atom, in an atomic cell with a finite volume V. Now, the pressure (which is equal to the "external" force acting per unit area of the cell surface) differs from zero and is positive. Consequently, the electron density at the boundary of the cell is also finite (Fig. 3.4).

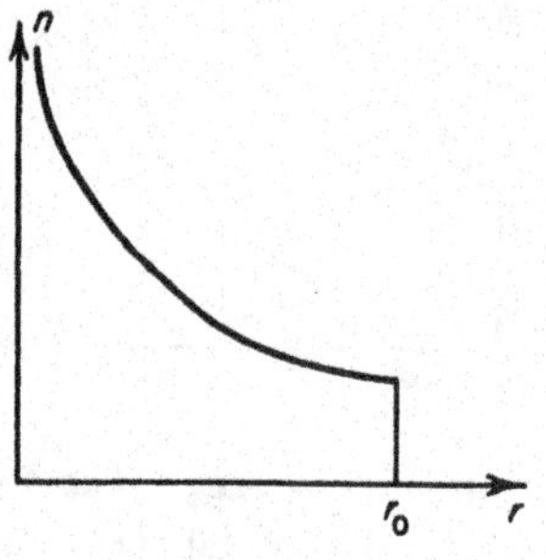

Fig. 3.4. Electron density distribution (schematic) in a "compressed" atom, one that is in an atomic cell of radius r_0.

There is no field at the boundary of the cell. The electrons at the boundary behave as free electrons, so that the entire pressure at the boundary is of "kinetic" origin. By definition, the "kinetic" pressure is equal to the normal component of momentum transferred per unit area of cell surface per unit time. Since the electron distribution is isotropic with respect to direction of motion

$$P = \int_0^\infty \rho(p, r_0) p \frac{v}{3}\, dp, \tag{3.106}$$

where $\rho(p, r_0)$ is the momentum distribution function at the cell boundary r_0, and $v = p/m_e$ is the velocity of the electrons. The pressure, as we would expect, is

$$P = \tfrac{2}{3} n(r_0)\, \varepsilon_k(r_0) = \tfrac{2}{5} n(r_0)\, \varepsilon_0(r_0), \tag{3.107}$$

where $\varepsilon_k(r_0) = \frac{3}{5}\varepsilon_0(r_0)$ is the average kinetic energy of an electron at the boundary of the cell. The pressure is the same everywhere, i.e., $P = P_k + P_p =$ *const*, although the "kinetic" and "potential" components change from point to point. The "kinetic" pressure P_k at any point is given in terms of the kinetic energy by an equation similar to (3.107).

Expressing the total kinetic and potential energies of the entire cell E_k and E_p, in terms of energy density integrals (which are proportional to the electron density) taken over the volume of the cell, direct calculation shows that

$$PV = \tfrac{2}{3}E_k + \tfrac{1}{3}E_p. \tag{3.108}$$

In calculating the potential energy the potential should be broken up into two parts, one corresponding to the potential of the nucleus and the other to the potential of the electrons $\varphi = \varphi_a + \varphi_e$; $\varphi_a = Ze/r$:

$$\begin{aligned} E_p = E_{pe} + E_{pa} &= -\tfrac{1}{2}4\pi e \int_0^{r_0} r^2\, dr\, n(r)\varphi_e(r) - 4\pi e \int_0^{r_0} r^2\, dr\, n(r)\varphi_a(r) \\ &= -\tfrac{1}{2}4\pi e \int_0^{r_0} r^2\, dr\, n(r)\left[\varphi(r) + \frac{Ze}{r}\right]. \end{aligned} \tag{3.109}$$

The factor $\frac{1}{2}$ in E_{pe} is required because the interaction energy of each pair of electrons appears twice in the integral. In order to measure the potential energy from a reference value corresponding to the removal of all the electrons to infinity, the potential $\varphi(r_0)$ at the boundary of a neutral cell should be set equal to zero. Since the density at the boundary of a compressed atom is different from zero (it is proportional to the pressure), (3.96) with the property $\varphi(r_0) = 0$ shows that the chemical potential is nonzero and positive.

Equation (3.108) can be derived from the virial theorem applied to a system of particles occupying a finite volume in a Coulomb field. The virial theorem for the motion of a system of particles in a Coulomb field states: $2E_k = I = -\overline{\Sigma_i r_i F_i}$, where r_i is the radius vector of the ith electron, and F_i is the force acting on it. The averaging is performed with respect to all electron positions (or time). Breaking up the virial I into the three parts corresponding to the forces exerted on the electron by the other electrons I_{ee}, by the nucleus I_{ea}, and by the boundary I_0, respectively, and rearranging terms (see [31]), we get

$$I_0 = r_0 \overline{\sum_i |F_{\text{bound}}|} = 4\pi r_0^3 P = 3PV,$$

$$I_{ea} = \overline{\sum_i \frac{Ze^2}{r_i}} = -E_{pa},$$

$$I_{ee} = -e^2 \overline{\sum_i \sum_j \frac{r_i(r_i - r_j)}{|r_i - r_j|^3}} = -\frac{e^2}{2}\overline{\sum_i \sum_j \frac{1}{|r_i - r_j|}} = -E_{pe}.$$

Substituting all these terms into the virial theorem, we obtain (3.108). In applying (3.108) to the case of a free atom, we set $P = 0$ and find $2E_k^\infty = -E_p^\infty$, as already noted above.

Upon compression of an atom the pressure and density at the boundary increase. By virtue of the relation $dE = -P\,dV\ (P > 0)$, the energy of the cell also increases. This is evident physically from the observation that the electron cloud in the absence of external forces tends to occupy a state corresponding to the minimum energy of the system, and therefore diffuses out to infinity. If the compression energy of the cell is of interest then it should be measured relative to the energy of a free atom, that is, the energy of the free atom E^∞ should be subtracted from the total energy of the cell $E(V)$. Since the pressure in a free atom is equal to zero it is not necessary to subtract anything from the pressure.

We should emphasize here that the Thomas–Fermi model describes essentially only the repulsive forces acting between the atoms (atomic cells), equivalent to a positive pressure; the model does not describe the attractive forces, which appear only when the exchange energy is taken into account. Hence, this model cannot account for the binding of atoms in a solid. In order to compress the atomic cell of the Thomas–Fermi model to its dimensions in a solid body, work must be done against the pressure forces; therefore, the energy of such a cell is greater than the energy of a free atom, while in fact the pressure in a solid at zero temperature is zero and the energy of an atom in the bound state is less than the energy of a free atom.

When the free atom in the Thomas–Fermi model is only slightly compressed, so that the volume $V \gg V_{ef}$, the electron density is redistributed only near the boundary (Fig. 3.5) and the pressure and energy $\Delta E = E - E^\infty$

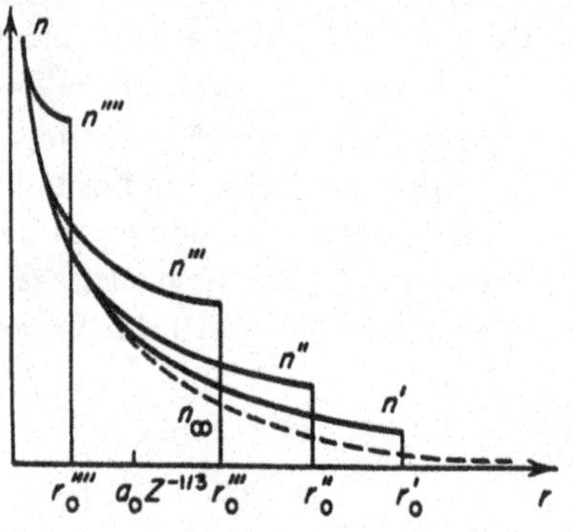

Fig. 3.5. Redistribution of the electron density upon the compression of an atom. n', n'', n''', and n'''' represent the distribution in the cells of radii r_0', r_0'', r_0''', and r_0''''; n_∞ is the distribution in a free atom ($r_0 = \infty$).

are not too large. An approximate relation between the pressure and cell volume can be obtained by assuming that the density at the boundary r_0 is, in first approximation, the same as the density at the point $r = r_0$ in the free atom. As can be easily checked, the asymptotic solution as $x \to \infty$ of (3.104) for a free atom is given by $\chi = 144x^{-3}$. According to (3.105) and (3.102) the density at the boundary is

$$n(r_0) \sim Z^2\left(\frac{\chi}{x}\right)^{3/2} \sim Z^2x^{-6} \sim Z^2r_0^{-6}Z^{-2} \sim r_0^{-6} \sim V^{-2},$$

and the pressure according to (3.107) is $P \sim n\varepsilon_0 \sim n^{5/3} \sim V^{-10/3}$ and is independent of Z.

A substantial increase in pressure and energy takes place at large densities, where the volume of the atomic cell becomes of the order or less than the effective volume V_{ef} occupied at low densities by most of the electrons in the atom. Under these conditions the electrons occupy the entire volume of the cell (see Fig. 3.5), and the average distance $\bar{r}$ between the particles is of the order of $V^{1/3}$; the average density is $\bar{n} \sim Z/V$. In this case, $E_k \sim Z\bar{n}^{2/3} \sim Z^{5/3}V^{-2/3}$, and $E_p \sim -Z^2/\bar{r} \sim -Z^2V^{-1/3}$. These estimates show that the kinetic energy increases with compression more rapidly than does the potential energy and in the limit of small volumes, i.e., large densities of the medium, $E_k \gg E_p$, $E \sim E_k$, $P \sim E_k/V$. The entire pressure becomes "kinetic", and the limiting relation has the form

$$P \sim Z^{5/3}V^{-5/3} \sim \bar{n}^{5/3}. \tag{3.110}$$

The pressure of a highly compressed cold medium is proportional to $\rho^{5/3}$, just as in the case of a free degenerate electron gas. Here ρ is the density of the medium, proportional to the average electron density $\bar{n}$. Accordingly, the specific energy ε is proportional to $\rho^{2/3}$. It is to be emphasized that these limiting relationships are applicable only at very high densities, exceeding the densities of ordinary solids at least by a factor of 10. The actual dependence of the pressure and energy of cold compressed solids upon density will be discussed in Chapter XI.

§14. Calculation of thermodynamic functions of a hot dense gas by the Thomas–Fermi method

The general procedure used to describe a dense gas at high temperatures within the framework of the Thomas–Fermi model was presented at the beginning of the preceding section. The equations obtained for the model of a cold atomic cell can be easily generalized to the case of temperatures different from zero. The basis for this generalization is the Poisson equation (3.97) for the electrostatic potential $\varphi(r)$ in the cell*. As before, the potential satisfies

* We note that the Poisson equation can also be used in the Debye–Hückel method, where it forms the basis for calculation of the Coulomb interaction between a given ion and the electron-ion cloud created around it. In contrast with that method, however, the Coulomb energy here is not assumed to be small compared to the kinetic energy, and an exact expression is used for the charge density. Moreover, the Fermi–Dirac, rather than the Boltzmann, distribution function is used to describe the electrons.

the boundary conditions (3.99) and (3.100), and is set equal to zero at the boundary of the cell for a convenient reference level for the potential energy. However, the simple expression (3.96) which relates the electron density $n(r)$ with the potential is now replaced by the integral relation (3.93) which contains the distribution function $f(p)$. The distribution function has the temperature dependence given by (3.91), in which the electron energy is again given by (3.95).

The normalization condition (3.101) is also valid in this case. The total kinetic energy of the cell is determined by integrating the kinetic energy density (3.94) over the volume of the cell, and the potential energy is given in terms of the electron density and of the potential by means of (3.109). Equation (3.106) gives the pressure; with $\rho(p, r_0)$ understood to denote the momentum distribution function the temperature dependence of this distribution function is given by (3.92). As before the virial theorem holds here and leads to (3.108), which can also be obtained directly from the expressions for P, E_k, and E_p.

Certain difficulties arise in calculating the entropy of the cells. Brachman [32] determined the entropy directly using thermodynamic relations and the expressions for the energy E and pressure P of a cell. Latter [31] found the entropy by a less rigorous procedure through an approximate evaluation of the partition function. The entropy of a cell is

$$S = \frac{1}{T}\left[\tfrac{5}{3}E_k + 2E_{pe} + E_{pa} - Z\mu\right], \tag{3.111}$$

where E_{pe} and E_{pa} are the interaction potential energies between the electrons themselves and between the electrons and the nucleus (see (3.109)). The normalization condition (3.101) determines the chemical potential μ as a function of T and V. It can be shown that as $T \to 0$, the bracketed expression in (3.111) tends to zero more rapidly than T, so that $S \to 0$ in agreement with the Nernst theorem.

The system of equations for the functions $\varphi(r)$ and $n(r)$ and also the expressions for the energy, pressure, and entropy can be expressed in dimensionless variables (the cell radius r_0 is used as a length scale). Here, as in the case of zero temperature, the model permits a similarity transformation with respect to Z. At zero temperature the density distribution was given by (3.105), from which it follows that the density at the boundary of a cell can be expressed as $n(r_0) = Z^2F(VZ)$ (note that $r_0Z^{1/3} \to VZ$), the pressure from (3.107) as $P = Z^{10/3}F_1(VZ)$, and the energy from (3.108) as $E = Z^{7/3}F_2(VZ)$. At temperatures different from zero these similarity relations can be generalized in such a way that the temperature always appears in the equations in the combination $TZ^{-4/3}$, so that

$$PZ^{-10/3} = f(VZ, TZ^{-4/3}), \qquad EZ^{-7/3} = f_1(VZ, TZ^{-4/3}).$$

The entropy and the chemical potential always appear in the combination SZ^{-1} and $\mu Z^{-4/3}$, respectively.

The equations for the Thomas–Fermi model were solved numerically by means of an electronic computer and the thermodynamic functions plotted (see Latter [31]*) over a wide range of the variables VZ and $TZ^{-4/3}$ (reduced density and temperature). In Latter's article the energy of a cold free atom E^∞ was subtracted from the energy E (E_k^∞ and E_p^∞ were subtracted from E_k and E_p, respectively).

The calculations show that as the temperature in a cell of a given volume increases, the kinetic energy, total energy, and pressure increase monotonically. The potential energy changes very slowly and only as a result of the redistribution of the electron density; the electron density approaches a uniform distribution throughout the cell as the temperature increases. In the limit of very high temperatures, when the degeneracy of the electron gas disappears (at $kT \gg (h^2/m_e)(Z/V)^{2/3}$; see (3.87)), the energy and pressure approach their classical values

$$E \approx E_k \approx \tfrac{3}{2}ZkT\,; \qquad P \approx \frac{Z}{V}kT = \bar{n}kT.$$

If the atomic cell is compressed isothermally, then its pressure will increase monotonically at a rate slower than in the case of zero temperature; this is clear from the fact that in the limit of high temperatures $P \sim 1/V$, while at $T = 0$ and $V \to 0$, $P \sim 1/V^{5/3}$. At moderate temperatures the energy as a function of volume has a flat minimum; the energy increase during expansion is caused by the fact that with increasing dimensions of the cell the electrons attempt to occupy a slightly larger volume than in the case of the cold cell (as a result of the temperature and "thermal" pressure). This in turn, results in some increase in the potential energy.

As an example of the temperature dependence of the energy we note that the energy of a cell which includes one atom of iron (at the normal density of solid iron) can be approximated in the temperature range of 20,000°K to 30,000°K by the interpolation formula

$$E = 0.865\ T_{\text{ev}}^{1.8}\ \text{ev/atom}$$

(the energy E^∞ has been subtracted from the cell energy; the temperature is measured in ev).

At densities lower than those of the solid state, the energy is only weakly volume-dependent, and roughly speaking $E \sim V^{0.15}$. The total energy and

* Even before Latter's work, several authors [33] attempted to consider the correction to the zero temperature solution by means of a perturbation method. Such a procedure, however, involves numerical computations that are almost as complicated as for the solution of the exact equations, and is valid only for a much narrower temperature range.

pressure of the material can be obtained by adding the nuclear contributions to the electronic contributions E and P corresponding to an atomic cell (see the beginning of §13); we set

$$P_{\text{tot}} = P_a + P = n_a kT + P_e(V, T), \qquad P_e \equiv P,$$

$$E_{\text{tot}} = E_a + E = \tfrac{3}{2}kT + E_e(V, T) \text{ per atom.} \qquad E_e \equiv E.$$

These results can be slightly improved when the density of the material is equal to the density in the unstressed solid state. To do this we subtract from the pressure and the energy the corresponding quantities associated with a cold cell with the same volume. (This is permissible since actually the pressure in a solid at zero temperature is zero.) Finally, we add to the energy the binding energy of the atoms in the solid (the heat of vaporization U)

$$P_{\text{tot}} = n_a kT + P_e(V, T) - P_e(V, 0),$$

$$E_{\text{tot}} = \tfrac{3}{2}kT + E_e(V, T) - E_e(V, 0) + U \text{ per atom.}$$

In this case the energy is measured from the normal state of the solid.

IV. Shock tubes

§1. The use of shock tubes for studying kinetics in chemical physics

In the preceding chapters we have discussed various chemical-physical processes which take place in gases at temperatures of the order of a thousand or several thousand degrees and higher, such as the excitation of molecular vibrations, the dissociation of molecules, chemical reactions, ionization, and the emission of radiation. We have considered the effects of these processes on the thermodynamic properties of gases, without being concerned with their kinetics, reaction rates, or the time required for the establishment of thermodynamic equilibrium. In reality, however, the kinetics have a large and often decisive effect when the rate of the gasdynamic process is so high that insufficient time is available for the establishment of thermodynamic equilibrium and the gas particles remain essentially in a state of nonequilibrium. These questions are especially timely in connection with problems of the entry of missiles and artificial satellites into the atmosphere, supersonic flows in high power jet engines, strong explosions, strong electric discharges, and so forth.

In contrast to the thermodynamic properties of gases, which can be calculated comparatively easily by theoretical means, our knowledge of the cross sections for elementary processes and of the rates of various reactions of chemical physics is obtained primarily by experiment. At the present time the shock tube is the most convenient and widely used tool for obtaining high temperatures in the laboratory and for studying the chemical physics of gases. The shock tube is used to create a shock wave in a gas which heats the gas to the required temperature*. As we know, an initially cold gas is heated practically instantaneously by a shock to a high temperature†, which can be controlled by varying the strength of the shock wave. Following this rapid heating, various processes take place in the heated gas, including excitation of molecular vibrations, dissociation, ionization, etc., whose relative importance and rates depend upon the temperature and density. Gradually these relaxation processes lead to the establishment of thermodynamic equilibrium corresponding to the given shock strength. The nonequilibrium layer behind

* Shock waves can also be obtained by other methods such as explosions, strong electric discharges, etc.

† Here, the term "temperature" denotes the temperature of the translational degrees of freedom of the atoms and molecules.

the compression shock where the relaxation processes take place (and which can be conceptually included within the shock front) is the layer studied experimentally. The density and temperature distributions in the relaxation layer can be theoretically related to the reaction rates. Hence, the experimental measurement of these distributions makes it possible to determine the rates of the relaxation processes. In some cases it is possible to record the reaction kinetics directly.

The structure of the relaxation layer behind a shock front will be discussed in detail in Chapter VII, while Chapter VI will be devoted to various chemical-physical processes which take place in heated gases along with estimates of their rates. Since very many rate measurements have been obtained in shock tubes it is important to acquaint the reader with the design and methods of operation of this important device. We wish to stress, however, that our presentation is purely supplemental and will, therefore, be extremely brief. It in no way reflects the actual amount of experimental work, which is in fact extremely large. More detailed presentations of the design and operation of shock tubes as well as the associated methods for the measurement of various quantities may be found in the reviews [1, 2] and in the books [3, 4, 18, 19]. In these cited references the reader will also find many original references, in contrast to here, where the references to original work are few and somewhat random.

While we shall not dwell on other methods of obtaining high temperatures (see [16]), we should like to note the very interesting work of Ryabinin [17] on the adiabatic compression of gases. In this method gas contained in a tube is compressed by a fast moving free piston by a factor of several hundred, up to pressures of the order of 10,000 atm, and thereby adiabatically heated to temperatures of the order of 9000°K. Using experimental apparatus designed by himself, Ryabinin studied the thermodynamic and radiative properties and the electrical conductivity of gases at very high temperatures.

§2. Principle of operation

A shock tube consists of a long tube, usually of circular or rectangular cross section, which is separated by a thin diaphragm into two parts. One of them, the low pressure chamber, is filled with the test gas. The compressed driver gas is fed into the second part, the high pressure chamber. The dimensions of the tube can vary. Usually, however, the tube is several meters long and the inner diameter is of the order of several centimeters. The low pressure chamber is several times longer than the high pressure chamber. As a rule, the pressure of the test gas does not exceed atmospheric pressure, and usually it is lower, of the order of several centimeters of mercury. In the high pressure chamber is created as high a pressure as possible, one of the order of tens or hundreds of atmospheres.

At a given time the diaphragm is rapidly burst by means of a special device, and the highly compressed driver gas flows into the low pressure chamber. A shock wave is then propagated through the test gas, while a rarefaction wave travels through the driver gas. The pressure distributions before and after bursting of the diaphragm and also the temperature distribution after bursting of the diaphragm are shown schematically in Fig. 4.1.

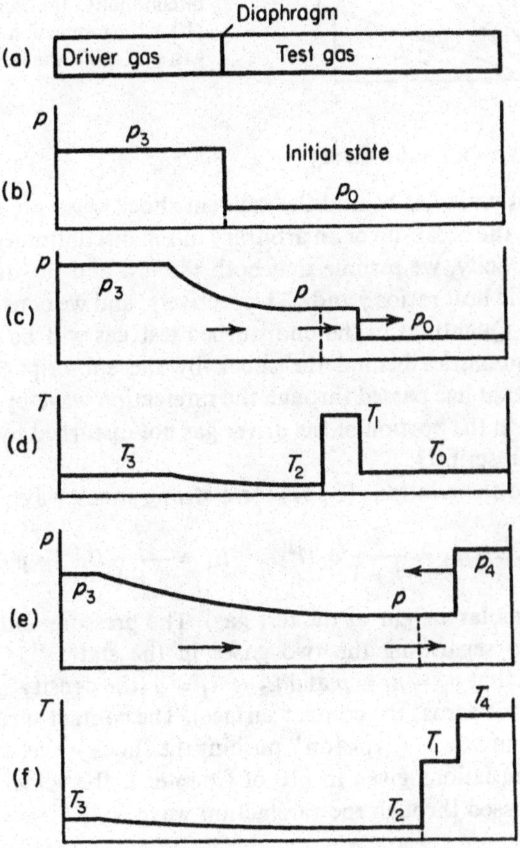

Fig. 4.1. Operation of a shock tube.

Not shown on this figure are the details of the distribution of the various physical parameters through the shock front, which is simply represented as the "classical" jump. After the shock wave reaches the end of the tube (which is usually closed with a fixed cap) it is reflected and travels toward the driver gas. The pressure and temperature behind the reflected shock increase sharply over the values behind the incident wave. The gas behind the reflected shock wave is at rest with respect to the tube walls. An x, t diagram for the process is shown in Fig. 4.2.

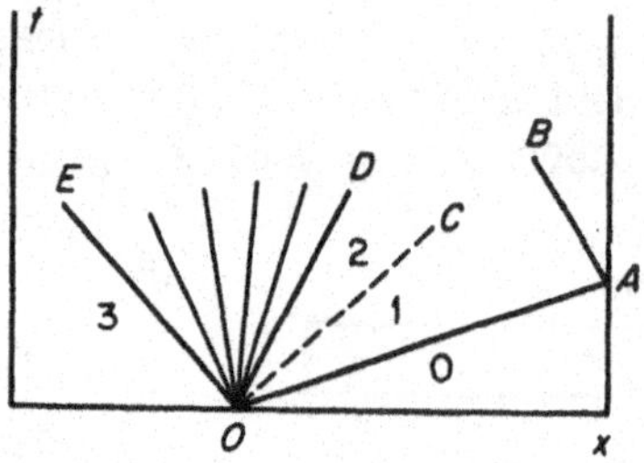

Fig. 4.2. An x, t diagram for the flow in the shock tube shown in Fig. 4.1. OA is the shock wave, OC is the contact discontinuity, the fan between OE and OD is the rarefaction wave in the driver gas, and AB is the reflected shock wave.

§3. Elementary shock tube theory

The physical variables behind the incident shock wave are easily estimated by considering the break-up of an arbitrary initial discontinuity (see Chapter I, §24). For simplicity, we assume that both the test and the driver gases have constant specific heat ratios γ and γ', respectively, and we consider only strong shock waves*. Quantities in the undisturbed test gas will be denoted by the subscript 0, quantities behind the shock by the subscript 1, quantities in the driver gas that has passed through the rarefaction wave by the subscript 2, and quantities in the portion of the driver gas not disturbed by the rarefaction wave by the subscript 3.

According to the relations (1.111) for a strong shock wave, we have

$$\rho_1 = \frac{\gamma+1}{\gamma-1}\rho_0, \qquad p_1 = \frac{2}{\gamma+1}\rho_0 D^2, \qquad u_1 = \frac{2}{\gamma+1}D, \qquad p = \frac{\mathcal{R}}{\mu_0}\rho T \tag{4.1}$$

(μ_0 is the molecular weight of the test gas). The pressure and velocity at the contact surface separating the two gases in the states "1" and "2" are continuous, so that $p_2 = p_1 = p$ and $u_2 = u_1 = u$ (the density and temperature are discontinuous across the contact surface). The contact surface moves with the velocity u and acts as a "piston" pushing the shock wave. According to the characteristic equations given in §10 of Chapter I, the velocity of the driver gas that has passed through the rarefaction wave is

$$u = \frac{2}{\gamma'-1}(c_3 - c_2) = \frac{2}{\gamma'-1}c_3\left(1 - \frac{c_2}{c_3}\right), \tag{4.2}$$

where the sound speeds c_2 and c_3 are connected by the isentropic relation

$$\frac{c_2}{c_3} = \left(\frac{p}{p_3}\right)^{(\gamma'-1)/2\gamma'}$$

* We also assume that the mass of the diaphragm can be neglected, so that we consider only times long enough that the shock wave has traversed a sufficiently large mass of the test gas.

Expressing $p = p_1$ in terms of $u = u_1$ by means of (4.1), we obtain an implicit equation for the speed of the "piston" in terms of the known initial values of the variables

$$\frac{u}{c_3} = \frac{2}{\gamma' - 1}\left\{1 - \left[\frac{\gamma'(\gamma + 1)}{2}\frac{\rho_0}{\rho_3}\frac{u^2}{c_3^2}\right]^{(\gamma'-1)/2\gamma'}\right\}. \tag{4.3}$$

The sound speed c_3 is given by $c_3 = [\gamma'(\mathcal{R}/\mu_0')T_3]^{1/2}$, where μ_0' is the molecular weight of the driver gas. The strength of the shock wave is completely determined by the "piston" speed u. In particular, the temperature behind the front is $T_1 = \frac{1}{2}(\gamma - 1)\mu_0 u^2/\mathcal{R}$.

All other conditions being the same, the strongest possible shock is developed when the ratio of the initial densities ρ_0/ρ_3 is negligibly small, so that after bursting of the diaphragm the driver gas flows essentially into vacuum at the maximum exhaust velocity

$$u_{max} = \frac{2}{\gamma' - 1} c_3 = \frac{2}{\gamma' - 1}\left(\gamma' \frac{\mathcal{R}}{\mu_0'} T_3\right)^{1/2}. \tag{4.4}$$

The corresponding upper limit on the temperature behind the shock is

$$T_{1max} = \frac{2\gamma'(\gamma - 1)}{(\gamma' - 1)^2}\frac{\mu_0}{\mu_0'} T_3. \tag{4.5}$$

This equation shows that a light driver gas should be used for producing high temperatures, and that higher temperatures are attainable in heavy monatomic test gases (the lower the specific heat, the higher is the quantity $(\gamma - 1)\mu_0 = \mathcal{R}/c_v$ appearing in the numerator). It is most convenient to use hydrogen ($\mu_0' = 2$, $\gamma' = 7/5$, $T_{1max} = 8.75(\gamma - 1)\mu_0 T_3$) as the driver gas, although helium ($\mu_0' = 4$, $\gamma' = 5/3$, $T_{1max} = 1.87(\gamma - 1)\mu_0 T_3$) is also used.

In order to obtain the maximum possible velocity (4.4) it is required that the ratio of the initial gas densities ρ_0/ρ_3 be extremely small (the pressure drop p_3/p_0 be extremely large). For values of the pressure drop attainable in practice, the test gas offers considerable "resistance" to the flow of the driver gas, and the velocity u calculated from (4.3) is found to be several times smaller than the exhaust velocity into vacuum. The temperature behind the shock wave is decreased even more sharply. Let us consider an actual example. Let hydrogen be the driver gas and argon ($\mu_0 = 40$, $\gamma = 5/3$) be the test gas. Both gases are assumed to be initially at room temperature $T_0 = T_3 = 300°K$. The initial pressure ratio $p_3/p_0 = 7600$, say, $p_0 = 5$ mm Hg and $p_3 = 50$ atm. Calculation then gives $u = 2.78$ km/sec, $D = 3.7$ km/sec, the Mach number $M = D/c_0 = 11.5$, $T_1 = 41T_3 = 12{,}300°K$, and $p_1 = 164p_0 = 2.1$ atm. The maximum exhaust velocity $u_{max} = 6.65$ km/sec*. Actually, the temperature behind the shock wave will be somewhat less than 12,300°K, because the

* The calculated value of T_{1max} from (4.5) is meaningless, since it gives 70,000°K with a specific heat ratio of $\gamma = 5/3$. At such high temperatures ionization is very important; hence the actual temperature is much lower.

energy absorbed in ionizing the argon is appreciable even at this temperature, and the effective adiabatic exponent γ of the argon is somewhat lowered. For more exact calculations it is necessary to use the actual Hugoniot curves for the test gas, in which ionization is taken into account. The gas velocities u calculated from (4.3), the values of the shock velocity, the pressure, and the internal energy behind the shock depend very little on the assumed values of the thermodynamic properties of the test gas. However, the temperature calculated without taking into account the energy absorbed in ionization, dissociation, etc., can be much too high.

When air is used as the test gas in a shock tube with hydrogen as the driver gas, shock velocities up to 4 km/sec (Mach numbers of the order of 12) and temperatures behind the shock of the order of 4000°K are obtainable. Various methods are available for increasing the effectiveness of a shock tube, i.e., for increasing the shock strength. In particular, a convenient method is to raise the initial temperature of the driver gas T_3 (see (4.5)). To do this an explosive mixture of hydrogen and oxygen (which is usually diluted by a light inert gas such as helium) is often used as the driver gas. At the given time the mixture is ignited and the driver gas heats up as a result of the reaction. In this manner shock velocities D of the order of 5 km/sec (Mach numbers of the order of 15) and temperatures of the order of 6000°K are attainable in air. Shock tubes of varying cross-sectional area and of other designs have also been developed (see [4]).

Let us now calculate the physical variables behind the reflected shock wave, assuming again that the specific heats are constant. Quantities behind the reflected wave will be denoted by the subscript 4, while quantities behind the incident wave will again be denoted by the subscript 1. We apply (1.69) relating the differences in pressure, specific volume, and velocity across the shock wave. The difference in velocity, which represents the relative velocity of the gas behind the shock with respect to the gas ahead of it, is the same for the incident and the reflected wave. Assuming that the incident wave is strong, we then obtain

$$u^2 = (p_4 - p_1)(V_1 - V_4) = p_1(V_0 - V_1).$$

The Hugoniot equation (1.76) for the reflected wave (not necessarily strong) is

$$\frac{V_4}{V_1} = \frac{(\gamma+1)p_1 + (\gamma-1)p_4}{(\gamma-1)p_1 + (\gamma+1)p_4}.$$

Noting that $V_0/V_1 = (\gamma+1)/(\gamma-1)$ and eliminating V_4/V_1 from the preceding two equations, we can find the pressure ratio p_4/p_1 behind the reflected wave, after which the corresponding density and temperature ratios may be obtained. The results are

$$\frac{p_4}{p_1} = \frac{3\gamma-1}{\gamma-1}, \qquad \frac{\rho_4}{\rho_1} = \frac{\gamma}{\gamma-1}, \qquad \frac{T_4}{T_1} = \frac{3\gamma-1}{\gamma}. \tag{4.6}$$

Caution should be exercised in using the above relations for making numerical estimates. The temperatures behind the reflected wave are usually so high that the specific heats are not constant, because of dissociation, ionization, etc. Strictly speaking, quantities behind the reflected wave should be calculated using the actual thermodynamic functions of the gas. For a rough estimate, however, one may use (4.6) with some effective value for the specific heat ratio. For example, for a dissociated or ionized low density gas one can take $\gamma = 1.20$ for an estimate. This gives $p_4/p_1 \approx 13$, $\rho_4/\rho_1 \approx 6$, and $T_4/T_1 \approx 2.17$. For heavy monatomic gases we can obtain temperatures of the order of tens of thousands of degrees behind the reflected shock wave. In air, for an initial pressure $p_0 = 10$ mm Hg and an incident wave velocity $D \approx 5$ km/sec, with $T_1 \approx 5800°$K and $\rho_1/\rho_0 \approx 10$, we find (using actual equilibrium properties) that behind the reflected wave $T_4 \approx 8600°$K and $\rho_4/\rho_1 \approx 7$. The actual processes taking place in a shock tube are much more complex than is indicated in the above idealized scheme. The shock wave approaches a constant velocity not immediately after the bursting of the diaphragm, but only after a finite time. Wall friction, boundary layer interactions (particularly with the reflected shock wave), nonuniform heating across the tube cross section, energy losses through the walls and from radiation (at very high temperatures), mixing of the gases at the contact surface, and many other effects can play an important role (see [2, 4, 5, 19] and the many original references cited therein).

§4. Electromagnetic shock tubes

Shock tubes in which the shock wave in the test gas results from the sudden expansion of a compressed driver gas are widely used in the study of various high-temperature processes. However, the maximum shock velocities (Mach numbers) and thus temperatures obtainable even in highly efficient devices operating on this principle are quite limited.

Newer types of shock tubes, based on different operating principles, have been introduced more recently. In these devices, frequently referred to as electromagnetic or magnetic shock tubes, strong shock waves are developed by heating a gas by means of an electrical discharge and accelerating it by magnetic forces. The earliest such device, built by Fowler and his co-workers [6], consists of the T-tube shown schematically in Fig. 4.3. The tube is filled

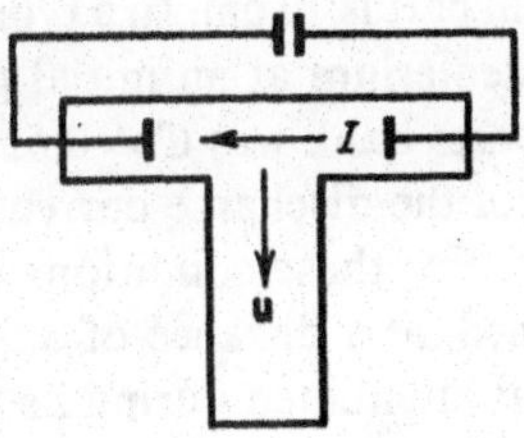

Fig. 4.3. Sketch of the Fowler T-tube.

with the test gas at a low pressure, of the order of a millimeter of mercury. Electrodes are placed in the horizontal part of the "T" and a condenser bank is discharged through the gas between the electrodes. The gas in the discharge region is rapidly heated to a high temperature, and is expelled at a high velocity by the large pressure into the "vertical" part of the T-tube, pushing a shock wave ahead of it.

In contrast to the Fowler tube, where the electrical discharge is used to heat the gas rapidly, a T-tube built by Kolb [7] utilizes the electromagnetic interaction between currents for accelerating a gas plasma. The lead carrying the return current in the discharge circuit is extremely close to the discharge part of the tube, as shown in Fig. 4.4. As is known, a repulsive force exists between

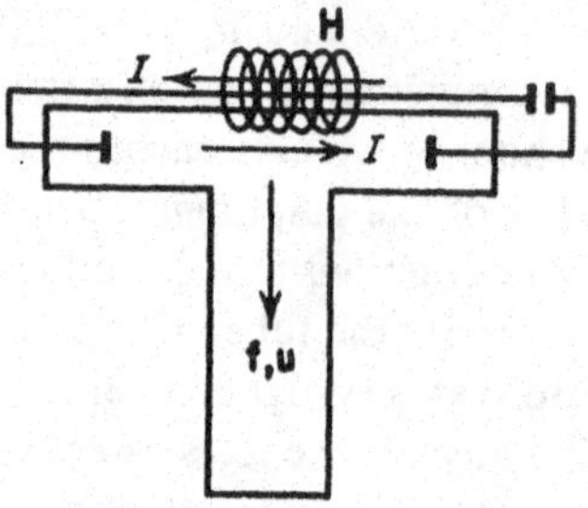

Fig. 4.4. Sketch of Kolb's electromagnetic shock tube.

parallel conductors carrying oppositely directed currents. This force can be regarded as the result of the action of the magnetic field of one current upon the other current-carrying conductor. The force acting per unit volume of current-carrying conductor is determined by the vector product of the current density $\mathbf{j}$ and the magnetic field intensity $\mathbf{H}$, with $\mathbf{f} = (1/c)\mathbf{j} \times \mathbf{H}$ (the magnetic permeability of the plasma is very close to unity). This force is perpendicular to the direction of the current and the magnetic field. In this case the magnetic field of the current flowing in the return lead repels the plasma carrying the discharge current in the direction of the vertical part of the tube, thus imparting an additional acceleration to the plasma. The plasma is thus subjected to the action of a so-called "magnetic piston", and a shock wave is produced in the vertical part of the tube which is much stronger than that produced in the absence of a magnetic field. The dimensions of this magnetic shock tube are not large; its radius is approximately 1.5 cm, while the length of the vertical part is 12 cm. In a typical one of Kolb's experiments the tube was filled with deuterium at an initial pressure of 0.7 mm Hg. The capacitance of the condenser bank was $C = 0.52\ \mu$f and it was charged to $V = 50$ kv. An oscillogram of the discharge current showed the discharge frequency to be $\nu = 700$ kc/sec. For these conditions a maximum shock velocity $D = 90$ km/sec was obtained, at a distance of 3.5 cm from the discharge region of the tube. The wave is attenuated during its travel, so that for example, its velocity dropped to 75 km/sec at a distance of 9 cm. The temperature behind the shock at $D = 90$ km/sec is approximately 120,000°K*.

By means of a simple estimate, we can show that for the given conditions the magnetic force is capable of accelerating the plasma to such a high speed. Neglecting the attenuation (which is not too large) the discharge current from the moment of breakdown varies sinusoidally $I = I_{max} \sin \omega t$, where $\omega = 2\pi\nu$ and $I_{max} = V(C/L)^{1/2} = VC\omega$ (L is the self-inductance of the configuration, in this case equal to 0.1 mh). The maximum current $I_{max} = 115{,}000$ amp $= 1.15 \cdot 10^5\, c/10$ in esu units. The current I, flowing through the return lead, induces a magnetic field $H = 2I/cr$, with r distance from the lead. The tube radius may be used as the average distance between the lead and the plasma. The magnetic field acts on the plasma like a piston exerting a pressure $H^2/8\pi$. The velocity u acquired by the plasma under the action of this pressure is determined by the obvious relation $H^2/8\pi \approx \rho u^2$, where ρ is the density; hence $u = H/(8\pi\rho)^{1/2} = I/cr(2\pi\rho)^{1/2}$. The average current is taken as $I = (\overline{I^2})^{1/2} = I_{max}/\sqrt{2}$. Substituting $r = 1.5$ cm, $\rho = 0.74 \cdot 10^{-7}$ g/cm³ (the density of deuterium at a pressure $p_0 = 0.7$ mm Hg and room temperature), and the current into the equation for the velocity, we obtain $u = 80$ km/sec. The magnetic piston, therefore, should accelerate the plasma to a velocity which is of the order of that observed experimentally ($D_{max} \approx 90$ km/sec). We note that the time over which the magnetic piston acts is of the order of $t \approx r/u \approx 1.9 \cdot 10^{-7}$ sec, which is less than a quarter of the discharge period $T/4 \approx 1/4\nu = 3.6 \cdot 10^{-7}$ sec. The entire process of accelerating the plasma takes place during the first quarter of the discharge period, before the current increases to its maximum value. In our calculations we have neglected the acceleration caused by the pure thermal expansion of the plasma heated by the discharge current. The estimates show that the principal factor causing the acceleration of the plasma is the magnetic rather than the thermal pressure. In order to increase the magnetic pressure acting on the plasma, an additional external field (~15,000 oersteds) was applied in some experiments. The external field had the same direction as the magnetic field of the return current, which with $\bar{I} = I_{max}/\sqrt{2} = 80{,}000$ amp and $r = 1.5$ cm was approximately equal to 11,000 oersteds. In the Kolb T-tube it is very important to obtain a large rate of increase of the current and a large current amplitude (high discharge frequency), and it is necessary to use special techniques to reduce the self-inductance of the circuit to a minimum†.

* This temperature is calculated from the shock velocity using the Hugoniot equation with dissociation and ionization effects taken into account, but neglecting the radiation flux from the front; this flux is small because of the transparency of the gas.

† We point out that in [8] is described the production of very strong shock waves in a T-tube filled with hydrogen or helium, with intensity measurements of different spectral lines in the heated plasma.

The principle of the "magnetic piston" is also used in another shock tube developed by Kholev and Poltavchenko [9] and shown schematically in Fig. 4.5. The discharge current flows radially between the electrodes, one of which is a rod placed along the tube axis and the other a cylinder near the tube surface. The radial discharge current interacts with the concentric magnetic

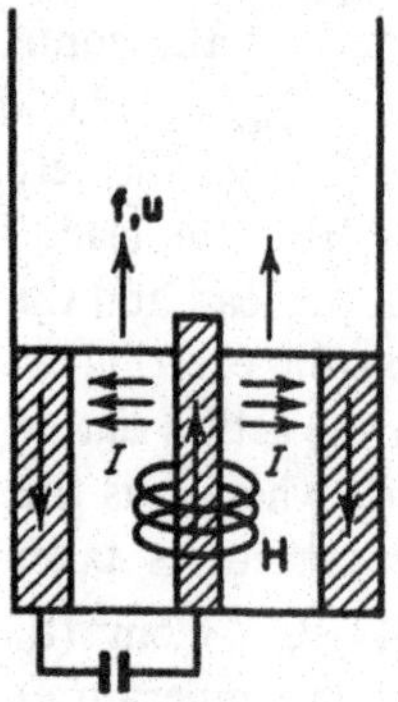

Fig. 4.5. Sketch of the Kholev and Poltavchenko shock tube. The electrodes are cross-hatched.

field induced by the current flowing in the central electrode. The pondermotive force acts along the tube axis and accelerates the plasma in that direction. A shock wave is propagated down the tube. A characteristic of this process is the expulsion of the plasma from the region between the electrodes and its separation from the "bottom" of the tube by the action of the magnetic field, which in this case acts like a piston.

The Kholev and Poltavchenko experiments were carried out with air. The strongest shock had a Mach number $M \approx 250$, $D \approx 80$ km/sec, and $T_1 \approx$ 130,000°K; it was obtained with $C = 2400$ μf, $V = 5$ kv, and $I \approx 560{,}000$ amp (the tubes were made from Plexiglas and had diameters of from 2 to 5 cm and lengths of from 50 to 90 cm). The shock was rapidly attenuated as it traveled down the tube. Weakly attenuated shock waves of smaller amplitudes ($D < 10$ km/sec) were obtained in a tube designed by Kholev and Krestnikova [10]. The principle of operation of the Kholev and Poltavchenko shock tube is closely related to that of the magnetic annular shock tube designed by Patrick [11].

Josephson [12] has described a shock tube, referred to as a "taper tube", which consists of a conical section joined to a cylindrical tube (Fig. 4.6). The center electrode is at the small end of the conical section. A ring at the junction of the cylindrical tube and large end of the conical part serves as the second electrode. The return current flows through leads placed along the cone. The discharge is accompanied by a magnetic compression of the plasma toward the axis, the so-called "pinch effect". The radial pinching starts at the small end first and involves adjacent gas layers in succession. Finally, the hot compressed plasma is ejected into the cylindrical tube where it creates a shock wave. In [13] is reported the use of such a tube to accelerate very low density air to velocities of the order of 12 km/sec ($M = 40$, $T_1 = 12{,}000$°K), and to study the flow past nose cone models.

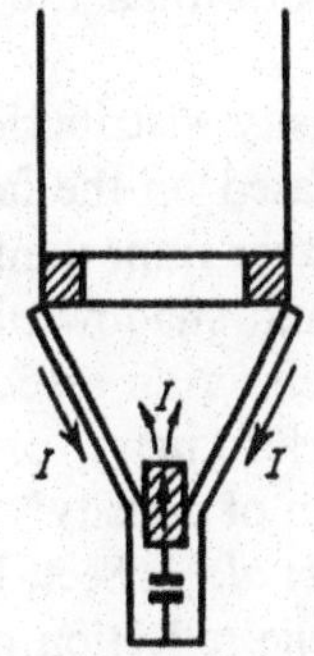

Fig. 4.6. Sketch of a "taper tube". The electrodes are cross-hatched.

Additional details on the design and operation of electromagnetic shock tubes may be found in the collection [14].

§5. Methods of measurement for various quantities

Various methods have been developed for the observation of the rapid processes which take place in shock tubes and for the measurement of different quantities, such as shock speed, density, temperature, etc. Descriptions of these methods and discussion of the results obtained are to be found in a rather extensive literature. One can, however, become familiar with many of the problems through the review articles [1, 2] and the books [3, 4, 18, 19], all of which contain numerous references to the original journal articles.

We shall not consider the experimental methods in detail and shall give only a brief listing of the more important ones. Our presentation will follow the classification of methods given in [2].

(1) *High-speed photography*. The gasdynamic process may be photographed using either luminosity resulting from the heating of the gas to a very high temperature or illumination from an external source. Cameras for filming high-speed processes with framing speeds of up to a million frames per second have been developed*. Another widely applied photographic method uses a rotating mirror camera. The mirror reflects a continuous light source which passes over the film, and rotates at such a rate that the moving luminous object (such as a shock front) describes a continuous inclined line on the film. From the slope of this line the speed of the object can be determined.

* References to the works of Soviet scientists and designers who have developed unique high-speed cameras will be found in Chapter XI.

(2) *Measurement of density.* The measurement of the density distribution in the nonequilibrium layer behind a compression shock is of particular importance, since the density distribution is closely related to the various relaxation rates (see Chapter VII). This was in fact the basis used to determine the rates of vibrational excitation and of molecular dissociation at high temperatures.

The density distribution is measured most frequently by the interference method, based on the fact that the refractive index of a gas varies with the density. Other important optical methods for observing a flow field, such as schlieren and shadow photography, are also based on the variation of the index of refraction associated with the motion of a compressible gas. However, interferometry provides the best quantitative information on the distribution of density*.

Hornig *et al.* [3, Sect. E] determined the density distribution in weak shock fronts by the reflection of light from the front surface. The initial density of the gas was chosen in such a manner that the thickness of the shock front was close to the wavelength of light. Under this condition the reflection coefficient depends on the thickness of the transition layer and the density distribution (that is, the index of refraction). This was the basis for the measurement of front thicknesses and of the rotational excitation rates for molecules in weak shock waves.

The density distribution in a gas has also been measured by the scattering of an electron beam and by the absorption of x-rays.

(3) *Measurement of concentration.* In a number of cases, where molecular dissociation or chemical reactions take place in the nonequilibrium layer behind the compression shock, the change in concentration of some species can be followed directly. This is usually possible when the light absorption of some particles differs sharply from the light absorption of others. This, for example, was the method used for studying the dissociation of molecules of bromine, iodine, and oxygen, etc., in a shock wave. Bromine and iodine molecules strongly absorb visible light, while their atoms do not; oxygen molecules have a characteristic system of absorption bands in the ultraviolet region (see Chapter V).

(4) *Measurement of the emission and absorption of light.* Many authors have made spectral measurements of the intensity of light emitted by a gas heated by a shock wave. Knowing the gas density and the temperature it is then possible to determine the emission coefficient at different temperatures and in different parts of the spectrum. The light is usually recorded photographically or by means of photomultipliers. Kirchhoff's law can be used with the emis-

* The above optical methods utilizing illumination from an external source are usually used at moderate temperatures, when the self-luminosity of the heated gas is low.

sion coefficient to determine the light absorption coefficient in a heated gas (Chapter V). Absorption coefficients are sometimes also determined directly by the attenuation in the gas of a beam of light from an external source.

(5) *Measurement of temperature.* Optical methods are most frequently used for measuring high temperatures. There exists a vast amount of literature describing optical pyrometry methods. In particular we recommend the collection [15] and the review article [16].

(6) *Measurement of electron concentration and electrical conductivity.* The Langmuir probe usually used in the study of gas discharges is frequently employed for measuring the degree of ionization and the electron concentration behind a shock wave. The technique of absorption and reflection of microwaves is also used. Electron concentration is also measured by the luminosity of the gas (for example, the intensity of recombination luminosity is proportional to the square of the electron concentration). Electromagnetic methods based, in particular, on the displacement of an external magnetic field by a moving plasma are also widely used, the displacement being a function of the electrical conductivity. Knowing the electrical conductivity the electron concentration can then be determined.

(7) *Measurement of pressure.* Pressure is most frequently measured by piezoelectric transducers with the sensing element made of barium titanate.

(8) *Measurement of shock speed.* The simplest method for measuring this velocity is to record by one or another means the time at which the shock wave passes definite points in the shock tube which are separated by known distances. The time is recorded by piezoelectric pressure gauges, ionization probes, various electrical contact probes, etc. The very high velocities obtained in electromagnetic shock tubes are usually measured by photographic means (see method 1).

V. Absorption and emission of radiation in gases at high temperatures

§1. Introduction. Types of electronic transitions

It was shown in Chapter II that the light* absorption coefficient is the fundamental optical characteristic of a gas which determines the degree of blackness of a heated body, the spectral radiation intensity, and the energy balance in a fluid undergoing radiant heat exchange. When the absorption coefficient is known, Kirchhoff's law, which is an expression of the general principle of detailed balancing, may be used to determine the emission coefficient of the fluid.

In §2 of Chapter II we have presented a short review and classification of the various mechanisms of absorption and emission. In accordance with the general scheme of allowed energy states of atomic systems (the simplest of which consists of one proton and one electron and constitutes the hydrogen atom in the bound state), all allowed electronic transitions accompanying the absorption and emission of light are subdivided into three types. These are:

(1) free-free transitions (bremsstrahlung emission and absorption);
(2) bound-free transitions (photoelectric absorption);
(3) bound-bound (discrete) transitions.

Free-free and bound-free transitions result in continuous absorption and emission spectra. Bound-bound transitions in atoms result in line spectra, while in molecules they result in the formation of band spectra. Band spectra consist of a great number of spectral lines which are closely spaced with respect to frequency. Under certain conditions the individual lines are so close to one another that they even partially overlap and the resulting spectrum is almost continuous (quasi-continuous).

From an energy point of view continuous (quasi-continuous) spectra are of primary interest. Let us imagine, for example, a body heated to a uniform temperature T. If the body is perfectly black, then the radiation flux emitted from its surface will have the Planck spectral distribution. The spectral flux as a function of the frequency ν is shown by the dashed curve in Fig. 5.1. The

* We recall that the terms "light", "light quanta", "photons", and "optical" properties refer to radiation at all frequencies and not just to those frequencies which lie in the visible part of the spectrum.

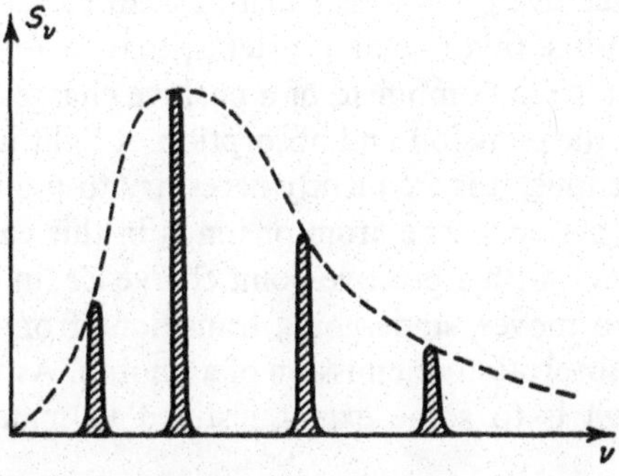

Fig. 5.1. Emission spectrum of a heated body which is perfectly transparent to the continuous spectrum, but is opaque to a line spectrum. The dashed curve corresponds to the Planck spectrum at the given temperature.

area bounded by this curve gives the integrated radiant energy emitted per unit area of surface per unit time σT^4. Let us now assume that the medium is perfectly transparent to the continuous spectrum and that it absorbs and emits only a line spectrum, where the line radiation for the given frequencies is in thermodynamic equilibrium with the medium. The spectral radiation flux from the body surface is now described by a system of individual narrow lines whose height corresponds to the Planck function, as shown in Fig. 5.1 by the solid curves. The integrated radiant energy emitted per unit area of body surface per unit time is numerically equal to the cross-hatched area of these lines. Since the line widths are very narrow, this energy flux is considerably lower than the integrated Planck flux σT^4. The radiant energy losses and also the surface brightness are, in this case, considerably smaller than in the case of a continuous spectrum. Similarly, for radiant energy transfer within the body the line spectra are frequently of little importance in comparison with that of continuous spectra. Therefore, most of our attention in this chapter will be devoted to continuous and quasi-continuous molecular spectra, rather than to line spectra.

At high temperatures, when the molecules are dissociated and the gas consists of atoms or (at even higher temperatures) of ions and electrons, the continuous absorption and emission spectra arise as a result of bound-free and free-free transitions. The calculation of the electronic transition probabilities, the results of which could then be used to find the absorption (and emission) coefficient for the case of multi-electron atoms (complex atomic systems), is a quantum-mechanical problem of considerable difficulty. This problem requires a separate analysis for each particular case, for each atom or ion, and also for each quantum state of the system. Such calculations have been carried out only for a few particular cases.

Complete and relatively simple calculations can be carried out only for the simplest hydrogen-like (hydrogenic) systems, that is, for the transitions of a single electron in the Coulomb field of a positive charge Ze. In practice, even when considering the emission and absorption of light in gases composed of complex atoms or ions, it is frequently necessary to use the relations derived for hydrogen-like systems. The atom or ion is in this case represented as an "atomic remainder" with a positive point charge Ze, in the field of which an "optical" electron moves, undergoing transitions from one energy level to another with the absorption or emission of a photon. As will be shown below, this approximation is to some extent justified in many cases of practical importance.

In calculating molecular absorption coefficients, the coefficient is usually determined as a function of frequency and temperature to within a factor termed the oscillator strength for the electronic transition considered; this factor is determined experimentally, as a rule.

In the following sections of this chapter we consider in detail the different mechanisms of light absorption and emission in gases at high temperatures, and the calculations for the corresponding absorption coefficients. We shall be primarily interested in the fundamental physical aspects of the problem and shall not dwell in detail on the various approximate methods for improving the formulas for calculating absorption coefficients.

Very frequently several different mechanisms which are independent of each other participate in the absorption and emission of light in gases under different conditions. The total absorption and emission coefficients in each spectral region are composed of quantities corresponding to these different mechanisms. Therefore, it is quite proper to examine independently the effect of each individual mechanism. At the end of the chapter we shall consider the radiative properties of high-temperature air as the most important practical example illustrating the combined effect of many mechanisms.

1. Continuous spectra

§2. Bremsstrahlung emission from an electron in the Coulomb field of an ion

As is well known from classical electrodynamics, radiation is emitted from a free electron moving in an external electric field, let us say, in the Coulomb field of an ion of positive charge Ze. In the process the electron loses a part of its kinetic energy and slows down. Hence such radiation is called bremsstrahlung.

The radiant energy S emitted by an electron per unit time is determined by its acceleration $\mathbf{w}$

$$S = \frac{2}{3}\frac{e^2}{c^3}\mathbf{w}^2. \tag{5.1}$$

The total radiation emitted during the entire time of travel past an ion is equal to the time integral of this expression

$$\Delta E = \int_{-\infty}^{\infty} S\, dt = \frac{2}{3}\frac{e^2}{c^3}\int_{-\infty}^{\infty} \mathbf{w}^2\, dt. \tag{5.2}$$

The spectral composition of the radiation may be found by expanding the acceleration vector $\mathbf{w}$ in a Fourier integral and substituting the expansion into (5.2). This yields

$$\Delta E = \frac{16\pi^2}{3}\frac{e^2}{c^3}\int_0^{\infty} \mathbf{w}_\nu^2\, d\nu = \int_0^{\infty} S_\nu\, d\nu, \tag{5.3}$$

where

$$\mathbf{w}_\nu = \frac{1}{2\pi}\int_{-\infty}^{\infty} \mathbf{w}(t)\, e^{-i2\pi\nu t}\, dt$$

is the Fourier component of the acceleration vector $\mathbf{w}(t)$. The quantity

$$S_\nu = \frac{16\pi^2}{3}\frac{e^2}{c^3}\mathbf{w}_\nu^2 \tag{5.4}$$

represents the radiant energy per unit frequency interval* emitted with a frequency ν by an electron passing an ion.

According to classical mechanics, when energy losses by radiation are

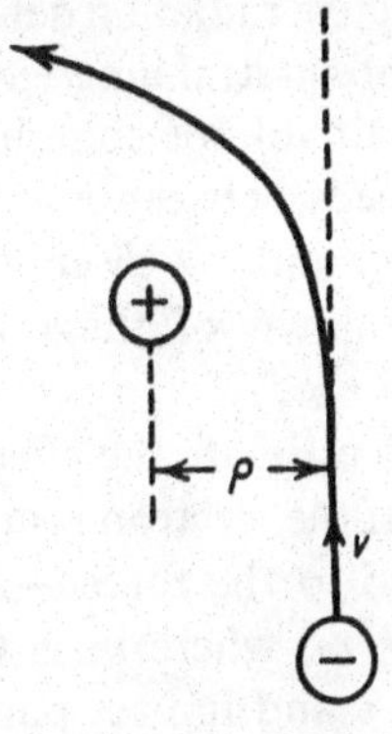

Fig. 5.2. Trajectory of an electron passing a positive ion.

* Following astrophysical practice, we shall always use the ordinary frequency ν rather than the angular frequency $\omega = 2\pi\nu$.

absent, a free electron (the sum of whose kinetic and potential energy is positive) passes the ion along a well-defined hyperbolic orbit characterized by the impact parameter ρ, the meaning of which is clear from Fig. 5.2. The total radiant energy and spectral composition of the radiation can be calculated approximately from equations (5.2)–(5.4), by taking for the acceleration $\mathbf{w}(t)$ the value corresponding to the motion of the electron in the absence of radiation. This is equivalent to assuming that the radiation is weak.

Let a parallel beam of electrons with initial velocity v at infinity and constant number density N_e (the electron flux is $N_e v$) be incident on the ion from infinity. Through an elementary ring of area $2\pi\rho\, d\rho$ about the ion $N_e v \cdot 2\pi\rho\, d\rho$ electrons pass per unit time. Each electron emits ΔE ergs of energy. The radiant energy emitted by these electrons per unit time is $\Delta E\, N_e v \cdot 2\pi\rho\, d\rho$ erg/sec. The energy emitted per unit time by the electrons passing the ion along all possible orbits can be obtained by integrating this expression with respect to ρ from 0 to ∞. The total emitted energy per ion for a unit electron flux $N_e v = 1\ \text{cm}^{-2}\ \text{sec}^{-1}$ is

$$q = \int_0^\infty \Delta E\, 2\pi\rho\, d\rho\ (\text{erg} \cdot \text{cm}^2). \tag{5.5}$$

We can also speak of the energy radiated in the frequency interval ν to $\nu + d\nu$, the so-called effective radiation dq_ν $\left(\int_{\nu=0}^{\nu=\infty} dq_\nu = q\right)$. In accordance with the definition given by (5.3) the effective radiation, the energy emitted in the frequency interval $d\nu$ per ion and per unit electron flux, is

$$dq_\nu = d\nu \int_0^\infty S_\nu 2\pi\rho\, d\rho\ (\text{erg} \cdot \text{cm}^2). \tag{5.6}$$

The effective radiation determines the spectral emission coefficient of a medium due to bremsstrahlung emission.

If a unit volume contains N_+ ions of a particular species and dN_e electrons with speeds between v and $v + dv$, then the energy in the frequency interval between ν and $\nu + d\nu$ emitted per unit time by a unit volume as a result of the slowing down of these electrons in the field of the ions is $N_+ v\, dN_e\, dq_\nu$ (erg/cm^3 · sec).

Let us estimate the effective radiation of electrons in the Coulomb field of an ion. If the electron is at a distance r from the ion (radius vector $\mathbf{r}$), then it is subjected to the force $-Ze^2\mathbf{r}/r^3$. The acceleration due to this force is $\mathbf{w} = -Ze^2\mathbf{r}/r^3 m$, where m is the electron mass. Let the electron have an initial velocity v and impact parameter ρ with respect to the ion. The time during which the force acts t is of the order of ρ/v, and the maximum acceleration interval of the electron during this time w is of the order of $Ze^2/\rho^2 m$. The principal role in the expansion of the acceleration vector in the Fourier integral is played by frequencies ν of the order of $1/2\pi t \sim v/2\pi\rho$ *. We can say that the frequency ν is radiated mainly by the electrons passing the ion at the distance $\rho \sim v/2\pi\nu$, and that frequencies in the interval from ν to $\nu + d\nu$ are mainly emitted by electrons with impact parameters in the interval $d\rho \sim (v/2\pi\nu^2)\, d\nu \sim 2\pi(\rho^2/v)\, d\nu$. The energy emitted by each of these electrons is given in order of magnitude by

$$\Delta E \sim \frac{2}{3}\frac{e^2}{c^3} w^2 t \sim \frac{2}{3}\frac{Z^2 e^6}{m^2 c^3 \rho^3 v}.$$

The effective radiation at the frequency ν corresponds to the radiation emitted by electrons with impact parameters from ρ to $\rho + d\rho$ and from the above relation is given by

$$dq_\nu \sim \Delta E\, 2\pi\rho\, d\rho \sim \frac{4\pi}{3}\frac{Z^2 e^6\, d\rho}{m^2 c^3 \rho^2 v} \sim \frac{8\pi^2}{3}\frac{Z^2 e^6}{m^2 c^3 v^2}\, d\nu. \tag{5.7}$$

The exact calculation of the effective radiation using (5.6) and (5.4), using the acceleration vector found by solving the mechanical problem of the motion of an electron in a hyperbolic orbit about an ion, is given in the book of Landau and Lifshitz [1]. The result is

$$dq_\nu = \frac{32\pi^2}{3\sqrt{3}}\frac{Z^2 e^6}{m^2 c^3 v^2}\, d\nu \qquad \text{for} \quad \nu \gg \frac{mv^3}{2\pi Z e^2}, \tag{5.8}$$

$$dq_\nu = \frac{32\pi}{3}\frac{Z^2 e^6}{m^2 c^3 v^2} \ln\frac{mv^3}{1.78\pi\nu Z e^2}\, d\nu \qquad \text{for} \quad \nu \ll \frac{mv^3}{2\pi Z e^2}. \tag{5.9}$$

It is evident that the exact result at high frequencies differs from the simple estimate (5.7) only by the numerical factor $4/\sqrt{3} = 2.3$. At low frequencies the exact result differs from the approximate one by a logarithmic factor which is a function of the frequency, as well as by a numerical factor. To explain this we note that the low frequencies radiated come from distant collisions with large impact parameters ρ; as $\nu \to 0$ and $\rho \to \infty$, the collisions with impact parameters $\rho > v/2\pi\nu$ give a relatively larger and larger contribution to the radiation at the frequency ν in comparison with that from the collisions with impact parameters $\rho \sim v/2\pi\nu$. The latter collisions are the only ones accounted for in deriving the simple formula (5.7).

The divergence of the effective radiation in the low frequency range is characteristic of a Coulomb field, which decreases slowly with distance; the result is that distant collisions become of considerable importance. This divergence can be eliminated by taking into account the screening effect, which is always present in an actual ionized gas. Actually, the integration

* For greater accuracy we shall retain the numerical coefficient 2π. (The fundamental role in the expansion is played by the "angular" frequencies such that $\omega t \sim 1$.)

with respect to ρ in (5.6) should be taken not to infinity but, say, to the Debye radius d; the radiation from the low frequency region is then cut off at $\nu_{min} \sim v/2\pi d$.

It should be noted, however, that the radiation integrated over the spectrum $q = \int dq_\nu$ converges in the low frequency region, since the divergence in dq_ν is only logarithmic and the contribution of the peak value of dq_ν in the integral with respect to ν is not large as $\nu \to 0$. Therefore, if we are interested in the integrated radiation only, then the question of cutting off the impact parameters ρ from above and the frequencies from below is not too important.

In the classical theory, the high frequency radiation is independent of frequency and the effective radiation per unit frequency interval $dq_\nu/d\nu$ remains finite even as $\nu \to \infty$ *. Formally, the integrated radiation $q = \int dq_\nu$ diverges in the high frequency region. This contradiction in the theory is a result of the imperfection of the classical concepts about the motion of an electron and is eliminated in the quantum theory. High frequencies, as we have seen, are radiated when an electron with a small impact parameter passes an ion. But, according to quantum-mechanical concepts, an electron having an initial momentum $p = mv$ cannot be located with greater precision than that allowed by the uncertainty principle $\Delta r\ \Delta p \sim h/2\pi$. Since the uncertainty in the momentum cannot exceed the momentum itself, there is no point in discussing impact parameters smaller than $\rho_{min} \sim h/2\pi mv$. The maximum frequency radiated for such minimum impact parameters is of the order of $\nu_{max} \sim v/2\pi\rho_{min} \sim mv^2/h$. This upper limit to the emitted frequency has a clear physical meaning. The quantum theory represents the bremsstrahlung as follows. A free electron with an initial energy $E = \frac{1}{2}mv^2$ passing near an ion can emit a photon $h\nu$. If the electron remains free after the emission, that is, has a positive energy E' upon moving away from the ion, then, obviously, the electron cannot emit a photon whose energy exceeds the initial energy E. Thus, $\nu_{max} = E/h = \frac{1}{2}mv^2/h$, which coincides with the frequency limit allowed by the uncertainty principle to within a factor of $\frac{1}{2}$.

In quantum mechanics a free electron is represented by a plane wave and the concept of the impact parameter does not have a precise meaning. We can speak of the probability of emission of a photon of a particular frequency, or, more precisely, about the cross section for the emission of photons with energies between $h\nu$ and $h\nu + d(h\nu)$. The energy emitted in the frequency interval $d\nu$ per unit flux of electrons interacting with a single ion is equal to the product of the photon energy $h\nu$ and the emission cross section $d\sigma_\nu$. This quantity corresponds to the effective radiation of the classical theory

$$dq_\nu = h\nu \cdot d\sigma_\nu\ (\text{erg} \cdot \text{cm}^2). \tag{5.10}$$

* This is true only when the colliding particles are oppositely charged (an electron and a positive ion). For the interaction between similarly charged particles $dq_\nu/d\nu \to 0$ as $\nu \to \infty$.

In the light of the correspondence principle, the effective radiation of frequency ν is related to the transition of an electron from one "stationary hyperbolic orbit", corresponding to an electron energy E, to another, corresponding to the energy $E' = E - h\nu$. The cross section $d\sigma_\nu$ and, consequently, the effective radiation dq_ν are calculated in quantum mechanics by the usual methods, using the matrix elements of the interaction energy between the electron and the ion.

Before discussing the results of quantum-mechanical calculations of bremsstrahlung emission, let us examine the limits of applicability of the classical relations (5.8) and (5.9) and the conditions under which it is necessary to replace these relations by quantum-mechanical ones. According to the classical derivation, (5.8) is valid for high frequencies where $\nu \gg mv^3/2\pi Ze^2$. Of course, there is no point in extending the inequality beyond those frequencies given by the upper limit $\nu_{max} = E/h = \frac{1}{2}mv^2/h$ dictated by quantum-mechanical energy considerations. Let us rewrite the limits imposed upon the frequency in (5.8) in the form

$$1 = \frac{h\nu_{max}}{E} > \frac{h\nu}{E} \gg \frac{h}{E}\frac{mv^3}{2\pi Ze^2} = \frac{hv}{\pi Ze^2}. \tag{5.11}$$

The inequality $hv/\pi Ze^2 \ll 1$, except for a factor of 2, represents nothing else but the condition for quasi-classical motion of an electron in a Coulomb field (see [2], for example)

$$\frac{hv}{2\pi Ze^2} \ll 1. \tag{5.12}$$

Therefore, the classical formula (5.8) for the effective radiation at a frequency ν, limited from above and below by the inequalities (5.11), can be used as an approximation for all electron velocities satisfying the inequality (5.12). If the quasi-classical condition (5.12) is satisfied, then the region of applicability of (5.8) extends down to very low frequencies, those for which $h\nu/E \sim hv/\pi Ze^2 \ll 1$. Since the energies of the photons which are ordinarily of interest are not too small in comparison with kT, in comparison with the energies of the electrons, and since the contribution of the peak value to the integrated radiation as $\nu \to 0$ is not too large, (5.8) can be extended to $\nu = 0$ by replacing it by (5.9). The divergence in dq_ν as $\nu \to 0$ is thus formally eliminated.

Let us transform the quasi-classical condition (5.12), which is also the condition of applicability of (5.8), so as to obtain the condition imposed upon the energy of an electron

$$E = \frac{mv^2}{2} \ll \frac{m}{2}\left(\frac{2\pi Ze^2}{h}\right)^2 = \frac{Z^2e^2}{2a_0} = I_H Z^2 = 13.5\,Z^2 \text{ ev}, \tag{5.13}$$

where $a_0 = h^2/4\pi^2 me^2$ is the Bohr radius, and $I_H = 13.5$ ev is the ionization potential of a hydrogen atom. Thus the quasi-classical condition for the motion of an electron in a Coulomb field is equivalent to the condition of the smallness of the electron energy in comparison with its energy in the first Bohr orbit. In the case of a hydrogen plasma, for example, (5.8) is applicable up to temperatures of the order of 10 ev $\sim$ 100,000°K; in a gas composed of heavier elements it is applicable to even higher temperatures, since the ionic charge Z increases as a result of multiple ionization. Thus, for air at standard density and $T = 10^6$ °K we have $Z \approx 6$, and the average energy of the electrons is still four times less than the "quasi-classical" limit.

At very high temperatures, when conditions opposite to the quasi-classical conditions (5.12) and (5.13) are satisfied, the Born approximation* of quantum mechanics is valid. For nonrelativistic energies ($E \ll mc^2 = 500$ kev) the effective radiation in the Born approximation is given by the expression (see [3])

$$dq_\nu = h\nu\, d\sigma_\nu = \frac{32\pi}{3}\frac{Z^2 e^6}{m^2 c^3 v^2} \ln \frac{[E^{1/2} + (E - h\nu)^{1/2}]^2}{h\nu}\, d\nu,$$

where dq_ν automatically vanishes for $h\nu = E$ and has a weak logarithmic dependence on the frequency over the entire frequency interval from 0 to ν_{max}.

It is remarkable that the quantum-mechanical result yields values of the effective radiation which are quite close to those given by the classical equation (5.8) (with the obvious exception of very low frequencies and frequencies close to the maximum). This is evident from Table 5.1, which gives values of the ratio

$$g_1 = \left(\frac{dq_\nu}{d\nu}\right)_{\text{quant}} \bigg/ \left(\frac{dq_\nu}{d\nu}\right)_{\text{class}} = \frac{\sqrt{3}}{\pi} \ln \frac{[E^{1/2} + (E - h\nu)^{1/2}]^2}{h\nu}$$

$$= \frac{\sqrt{3}}{\pi} \ln \frac{[1 + (1 - x)^{1/2}]^2}{x}$$

as a function of the dimensionless quantity $x = h\nu/E = \nu/\nu_{max}$. The quantity g_1 which distinguishes the quantum expression for bremsstrahlung from the classical expression is sometimes called the Gaunt factor. The integrated radiation calculated using the quantum-mechanical result is usually written in the form

$$q_{\text{quant}} = \int_0^{\nu_{max}} \left(\frac{dq_\nu}{d\nu}\right)_{\text{quant}} d\nu = \left(\frac{dq_\nu}{d\nu}\right)_{\text{class}} \nu_{max} \int_0^1 g_1\, dx = 1.05\, q_{\text{class}}.$$

* The Born approximation requires that both the initial and final electron velocities satisfy the conditions (5.12) and (5.13); otherwise one must use the exact wave functions of an electron in a Coulomb field; this introduces the well-known Coulomb factor into the resulting equations (see [2, 3]).

Table 5.1

x	0	0.1	0.2	0.3	0.4	0.5	0.6	0.7	0.8	0.9	1
g_1	∞	2.01	1.61	1.34	1.13	0.97	0.81	0.68	0.53	0.36	0

Thus, the classical relation (5.8) gives satisfactory approximate results at practically any nonrelativistic temperature.

§2a. Bremsstrahlung emission from an electron scattered by a neutral atom

We now calculate the effective bremsstrahlung emission from an electron in collision with a scattering center, without as yet specifying the interaction law of the electron with the scatterer. The scattering body may be an atom, molecule, or ion. Let us assume that the interaction time τ_0 is small in comparison with the period of electromagnetic oscillations which are radiated, or more precisely that in this case we have $\omega\tau_b \ll 1$, where $\omega = 2\pi\nu$. This assumption may be regarded as valid for visible light frequencies, electron energies of the order of several electron volts, and scattering by a neutral atom*.

If $\omega\tau_b \ll 1$, the scattering takes place "instantaneously", and it is natural to set the acceleration vector $\mathbf{w}(t) = \Delta\mathbf{v}\ \delta(t)$, where $\Delta\mathbf{v}$ is the change in the electron vector velocity on scattering and δ is the delta function. Then the Fourier component of the acceleration vector is $\mathbf{w}_\nu = \Delta\mathbf{v}/2\pi$. Substituting this expression into (5.4) we find that the energy radiated in the frequency interval from ν to $\nu + d\nu$ upon scattering is

$$S_\nu\, d\nu = \frac{4}{3}\frac{e^2}{c^3}(\Delta\mathbf{v})^2\, d\nu.$$

This expression should be averaged over the scattering angle ϑ. Assuming approximately that the absolute electron velocity v does not change appreciably on scattering, which corresponds to the emission of photons $h\nu$ with energies low in comparison with the electron energy $mv^2/2$, we obtain $\overline{(\Delta\mathbf{v})^2} = 2v^2(1 - \overline{\cos\vartheta})$, where $\overline{\cos\vartheta}$ is the average of the cosine of the scattering angle.

In order to find the effective emission we must multiply the quantity $\bar{S}_\nu\, d\nu$, averaged over the scattering angle, by the scattering cross section σ (cf. (5.6)). This yields

$$dq_\nu = \bar{S}_\nu\, d\nu\, \sigma = \frac{8}{3}\frac{e^2 v^2 \sigma_{tr}}{c^3}\, d\nu, \tag{5.13a}$$

* For example, for red light $\lambda = 7000$ Å, $h\nu = 1.8$ ev, and $\omega = 2.7 \cdot 10^{15}$ sec^{-1}. If the radius of the atom is 10^{-8} cm and the electron velocity is 10^8 cm/sec (the energy is 3 ev), then $\tau_b = 10^{-16}$ sec and $\omega\tau_b = 0.27$.

where $\sigma_{tr} = \overline{\sigma(1 - \cos \vartheta)}$ is the so-called transport scattering cross section. This equation describes, in particular, bremsstrahlung emission from electrons in collision with neutral atoms; the cross section σ for elastic collisions between an electron and an atom is usually of the same order of or slightly smaller than the gaskinetic cross sections*. Using (5.10) and (5.13a) we find the relationship between the differential cross sections for the emission of a photon $h\nu$ and the cross section for elastic scattering of an electron

$$\frac{d\sigma_\nu}{d\nu} = \frac{8}{3} \frac{e^2 v^2}{c^3 h\nu} \sigma_{tr}. \tag{5.13b}$$

The derivation given above again clearly illustrates the physical nature of the light emission process in classical electrodynamics. The electron is accelerated upon colliding with a scattering center, and the emission is as if "superimposed" on the scattering; here the probability (cross section) of emission is determined only by the mechanical probability (cross section) of scattering. This relationship between photon processes and the processes associated with electron collisions holds also in quantum mechanics†.

We now apply (5.13a) to the scattering of an electron by a Coulomb center, by an ion. Coulomb forces are long range forces. Scattering as a result of collisions between charged particles, with an appreciable change in the electron momentum vector, takes place when the particles approach each other to within a distance r_0 for which the kinetic energy of the electron $mv^2/2$ is comparable with the potential energy Ze^2/r_0; this distance can then be set $r_0 = 2Ze^2/mv^2$. The cross section for Coulomb "collisions" is of the order of $\pi r_0^2 = 4\pi Z^2 e^4/(mv^2)^2$ (for additional details see Part 3 of Chapter VI). If we substitute this cross section into (5.13a), we will obtain a value for dq_ν which is smaller than the exact value given by (5.8) only by a factor of $\sqrt{3}/\pi$. Thus the cross sections for bremsstrahlung emission for the scattering of an electron by an ion and by a neutral atom are in the same ratio to each other as the corresponding elastic scattering cross sections

$$\frac{(d\sigma_\nu)_{ion}}{(d\sigma_\nu)_{neut}} = \frac{\pi r_0^2}{\sigma_{tr}} = \frac{\pi Z^2 e^4}{\sigma_{tr} E^2} = \frac{\pi a_0^2}{\sigma_{tr}} \left(\frac{2I_H}{E}\right)^2 Z^2.$$

Usually $\sigma_{tr}/\pi a_0^2 \sim 1 - 10$, and for an energy E of the order of several electron volts the effectiveness of neutral atoms with respect to bremsstrahlung emission (and absorption) is one or two orders less than the effectiveness of

* A great deal of experimental data on the value of the cross section σ has been collected in [60].

† In particular, a relationship exists between the ionization cross sections by atom-electron impact and by photoionization [53].

ions. Thus electron-neutral collisions are of importance only in a very weakly ionized gas.

Above we have calculated bremsstrahlung emission from an electron which is scattered by an isolated atom. If the electron-atom collisions take place sufficiently infrequently (in comparison with the frequency of the radiated electromagnetic wave), then the successive collisions can be regarded as independent and the energy radiated in many collisions is simply the sum of the energies radiated in each individual collision. In this case, according to (5.13a) the electron emits the following amount of energy per unit time in the frequency interval $d\nu$:

$$dQ_\nu = \bar{S}_\nu \, d\nu \cdot Nv\sigma = \frac{8}{3}\frac{e^2v^2}{c^3}\,\nu_{\text{eff}}\,d\nu \;(\text{erg}\cdot\text{sec}^{-1}),$$

where $\nu_{\text{eff}} = Nv\sigma_{\text{tr}}$ is the (effective) collision frequency (N is the atom number density). However, if the collision frequency is comparable with the frequency of the emitted light, then the collisions can no longer be treated as independent. A correlation exists between the changes in the vector velocity of the electrons for successive collisions, which produces interference between the partial waves radiated in the individual collisions. The amplitude of two successive electromagnetic waves is found, on the average, to be directed in opposite directions, which decreases the total energy radiated.

In order to calculate the bremsstrahlung emission by an electron subjected to a large number of collisions n ($n \to \infty$), we can represent its acceleration $\mathbf{w}(t)$ in the form

$$\mathbf{w}(t) = \sum_{k=1}^{n} \Delta\mathbf{v}_k\,\delta(t - t_k),$$

where t_k is the instant of the kth collision and $\Delta\mathbf{v}_k$ is the corresponding change in the vector velocity. The square of the modulus of the Fourier component of the acceleration is then

$$|\mathbf{w}_\nu|^2 = \frac{1}{4\pi}\sum_{j=1}^{n}\sum_{k=1}^{n}\Delta\mathbf{v}_j\cdot\Delta\mathbf{v}_k\,e^{i2\pi\nu(t_j - t_k)}.$$

This expression should be averaged over the electron velocity directions and over the collision times. Substituting the expression thus obtained in (5.4) and dividing by the time during which the electron experiences n collisions, $n/Nv\sigma$, we will obtain the quantity dQ_ν corrected for the correlation effect. This calculation was carried out by one of the authors in [61], and yields

$$dQ_\nu = \frac{8}{3}\frac{e^2v^2}{c^3}\,\nu_{\text{eff}}\,\frac{\nu^2}{\nu^2 + (\nu_{\text{eff}}/2\pi)^2}\,d\nu = (dQ_\nu)_{\text{uncorr}}\,\frac{\nu^2}{\nu^2 + (\nu_{\text{eff}}/2\pi)^2}\,.$$

In the limiting case of $\nu_{\text{eff}}/2\pi \ll \nu$ the correction factor becomes one, corresponding to the vanishing of the correlation effect.

Practically the correlation effect shows up only for radiation of very low frequencies (in the radio wave range). For example, for $N = 10^{19}\ \text{cm}^{-3}$, $v = 10^8$ cm/sec, $\sigma_{\text{tr}} = 10^{-15}\ \text{cm}^2$, $\nu_{\text{eff}} = 10^{12}\ \text{sec}^{-1}$, while for red light $\omega = 2\pi\nu = 2.7 \cdot 10^{15}\ \text{sec}^{-1}$.

§3. Free-free transitions in a high-temperature ionized gas

Let us find the emission coefficient of an ionized gas due to bremsstrahlung emission. Let a unit volume of gas contain N_+ positive ions with a charge Ze and N_e electrons with a Maxwell velocity distribution $f(v)\,dv = 4\pi(m/2\pi kT)^{3/2} \exp(-mv^2/2kT)v^2\,dv \left(\int_0^\infty f(v)\,dv = 1\right)$. The temperature of the electron gas is denoted by T. The energy emitted by electrons having velocities from v' to $v' + dv'$ per unit volume per unit time in the frequency interval from ν to $\nu + d\nu$ is

$$N_+ N_e f(v')\,dv'\,v'\,dq_\nu(v'). \tag{5.14}$$

It is assumed that the velocities of the ions are very small in comparison with the electron velocities. The spontaneously emitted energy as a result of free-free transitions in the interval $d\nu$ per unit volume per unit time is obtained by integrating (5.14) with respect to the electron velocities from v_{min} to ∞; here v_{min} is the minimum velocity of electrons capable of emitting a photon $h\nu : \frac{1}{2}mv_{\text{min}}^2 = h\nu$. Using (5.8) for the effective radiation and integrating, we find the spectral emission coefficient due to bremsstrahlung emission

$$J_\nu\,d\nu = \frac{32\pi}{3}\left(\frac{2\pi}{3kTm}\right)^{1/2} \frac{Z^2 e^6}{mc^3} N_+ N_e e^{-h\nu/kT}\,d\nu. \tag{5.15}$$

The emission of high energy photons $h\nu \gg kT$ is exponentially small. This comes from the fact that the high energy photons are emitted by the most energetic electrons, concentrated at the tail of the Maxwell velocity distribution.

The integrated emission coefficient for bremsstrahlung emission is

$$J = \int_0^\infty J_\nu\,d\nu = \frac{32\pi}{3}\left(\frac{2\pi kT}{3m}\right)^{1/2} \frac{Z^2 e^6}{mc^3 h} N_+ N_e$$
$$= 1.42 \cdot 10^{-27} Z^2 T^{\circ 1/2} N_+ N_e\ \text{erg/cm}^3 \cdot \text{sec}. \tag{5.16}$$

(T° is the absolute temperature in degrees Kelvin.) The integrated bremsstrahlung emission is only weakly dependent on the temperature (it is proportional to $T^{1/2}$). If the gas contains ions with different charges Z, then the expressions in (5.15) and (5.16) should be summed over all the species of ions.

We shall now find the coefficient of bremsstrahlung absorption using the principle of detailed balancing. If $U_{\nu p}$ is the spectral equilibrium radiation density, defined by the Planck equation (2.10),

$$U_{\nu p} = \frac{8\pi h\nu^3}{c^3} \frac{1}{e^{h\nu/kT} - 1}, \tag{5.17}$$

and a_ν is the spectral coefficient of the actual bremsstrahlung absorption per ion and per electron moving with velocity v, then the amount of radiation in the frequency interval between ν and $\nu + d\nu$, absorbed under conditions of thermodynamic equilibrium per unit time per unit volume, by electrons with velocities from v to $v + dv$, is

$$N_+ N_e U_{\nu p}\, d\nu \cdot c f(v)\, dv \cdot a_\nu (1 - e^{-h\nu/kT}). \tag{5.18}$$

The factor $(1 - e^{-h\nu/kT})$ accounts for the effective decrease in absorption caused by induced emission (re-emission, see §4 of Chapter II). In thermodynamic equilibrium the emission is exactly canceled out by the absorption, i.e., the expressions (5.18) and (5.14) are equal to each other. Here the velocities of electrons emitting and absorbing photons $h\nu$ are related by the conservation of energy relation

$$\frac{mv'^2}{2} = \frac{mv^2}{2} + h\nu.$$

Noting that $v\, dv = v'\, dv'$ and $dq_\nu = h\nu\, d\sigma_\nu$, we find the general relation between the unit absorption coefficient a_ν and the radiation cross section $d\sigma_\nu$

$$a_\nu = \frac{c^2 v'^2}{8\pi\nu^2 v} \frac{d\sigma_\nu(v')}{d\nu}. \tag{5.19}$$

Using (5.8) for dq_ν, we find from (5.14) and (5.18)

$$a_\nu = \frac{4\pi}{3\sqrt{3}} \frac{Z^2 e^6}{hcm^2 v\nu^3} = 1.8 \cdot 10^{14} \frac{Z^2}{v\nu^3} \text{ cm}^5. \tag{5.20}$$

This equation was derived by Kramers in 1923. Multiplying a_ν by $N_+ N_e$ and averaging with respect to the electron velocities using the Maxwell distribution function, we obtain the spectral coefficient of the actual bremsstrahlung absorption in the gas at the electron temperature T:

$$\kappa_\nu = \frac{4}{3}\left(\frac{2\pi}{3mkT}\right)^{1/2} \frac{Z^2 e^6}{hcm\nu^3} N_+ N_e = 3.69 \cdot 10^8 \frac{Z^2}{T^{\circ 1/2}\nu^3} N_+ N_e \text{ cm}^{-1}$$

$$= 4.1 \cdot 10^{-23} Z^2 \frac{N_+ N_e}{T^{\circ 7/2} x^3} \text{ cm}^{-1}, \qquad x = \frac{h\nu}{kT}. \tag{5.21}$$

In a more exact theory the formulas for the absorption coefficient (5.20) and (5.21) (and in all the other corresponding relations) contain a Gaunt factor g, which takes into account the deviations from Kramers' theory: $\kappa_\nu = (\kappa_\nu)_{\text{Kramers}} \cdot g$. An expression for the Gaunt factor may be found in Spitzer [87].

Recalling the definition of the frequency-averaged absorption coefficient characterizing the emission coefficient (2.102), let us calculate this quantity for the bremsstrahlung mechanism:

$$\kappa_1 = \frac{J}{cU_p} = \frac{J}{4\sigma T^4} = 6.52 \cdot 10^{-24} Z^2 \frac{N_+ N_e}{T^{\circ 7/2}} \text{ cm}^{-1}. \tag{5.22}$$

The corresponding mean free path is

$$l_1 = \frac{1}{\kappa_1} = 1.53 \cdot 10^{23} \frac{T^{\circ 7/2}}{Z^2 N_+ N_e} \text{ cm}. \tag{5.23}$$

Let us also calculate the Rosseland mean free path (2.80) for the case of a fully ionized gas with absorption only by bremsstrahlung (and all the ions having the same charge Z)

$$l = 4.8 \cdot 10^{24} \frac{T^{\circ 7/2}}{Z^2 N_+ N_e} \text{ cm}. \tag{5.24}$$

The Rosseland mean free path l for bremsstrahlung is equal to the spectral mean free path when the energy of the photons $h\nu = 5.8\,kT$. It is evident that when the radiant energy transfer occurs by heat conduction, the main role in the bremsstrahlung absorption is played by very high energy photons in the Wien region of the spectrum. On the other hand, for volume radiation the main role is played by the low energy photons. The mean coefficient κ_1 is equal to the spectral coefficient corresponding to $h\nu = 1.73\,kT$, corrected for induced emission ($\kappa_\nu(1 - e^{-h\nu/kT})$).

In order to give some idea of the order of magnitude of the optical characteristics of a plasma under conditions of bremsstrahlung emission we present a specific example. Let us consider hydrogen at a density $\rho = 1.17 \cdot 10^{-6}$ g/cm^3, which corresponds to a pressure of 10 mm Hg at room temperature, at a temperature $T = 100{,}000$°K. Under these conditions the hydrogen is completely dissociated and ionized, so that $N_+ = N_e = 7 \cdot 10^{17}$ cm^{-3}. The absorption coefficient for red light for which $\lambda = 6500$ Å is in this case equal to $\kappa_\nu = 5.7 \cdot 10^{-3}$ cm^{-1} and the mean free path $l_\nu = 1/\kappa_\nu = 175$ cm. The Rosseland mean free path $l = 3.1 \cdot 10^6$ cm. The mean free path characterizing the emission coefficient $l_1 = 0.98 \cdot 10^5$ cm.

If the dimensions of the body are much smaller than l_1, then the body emits as a volume radiator (see §16 of Chapter II) and the rate of energy loss by radiation is

$$\frac{d(\rho\varepsilon)}{dt} = -J;$$

ε is the specific internal energy. In our example $J = 2.2 \cdot 10^{11}$ erg/cm^3 · sec. When the dissociation and ionization energies are taken into account, $\varepsilon = 41.6$ ev/atom, $\rho\varepsilon = 4.66 \cdot 10^7$ erg/cm^3. The initial time scale for radiative cooling $\tau = \rho\varepsilon/J = 2.12 \cdot 10^{-4}$ sec.

§4. Cross section for the capture of an electron by an ion with the emission of a photon

Let us consider the capture of a free electron by a hydrogen-like "ion" accompanied by the emission of a photon and the formation of a hydrogen-like "atom". We shall, as in §2, base our considerations on semiclassical concepts. In classical mechanics, without taking radiation into account, the transition from free to bound electronic states is continuous. The state or orbit of an electron is characterized by the total energy E of the electron-ion system and (in general) instead of the "impact parameter" ρ, the angular momentum is used. The angular momentum with the energy determines the geometric parameters of the trajectory. If the energy decreases with the momentum remaining constant, the hyperbolic orbits (which correspond to positive energy $E > 0$) make a continuous transition to a parabolic one ($E = 0$) and then, in the bound state of the system (characterized by negative energy $E < 0$), to elliptic orbits (Fig. 5.3). In view of the correspondence principle, the capture of a free electron, accompanied by the emission of a photon whose energy exceeds the initial kinetic energy of the electron E, is related to the transition of the electron from a hyperbolic to an elliptic orbit.

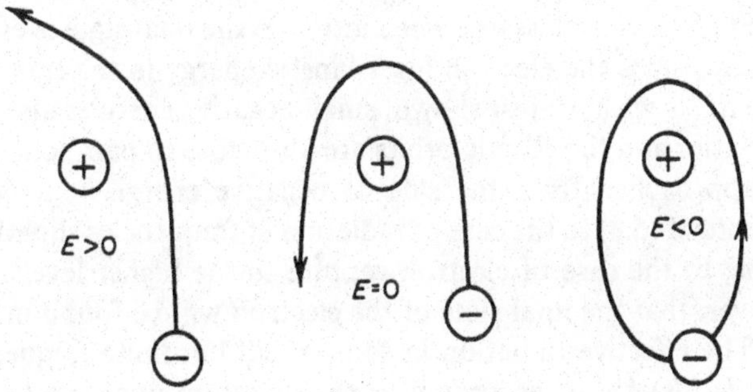

Fig. 5.3. Hyperbolic, parabolic, and elliptic electron orbits.

In classical mechanics the energy of the electron-ion system can be arbitrary. In quantum mechanics the energy spectrum of the system is continuous only if the electron is free and $E > 0$. In the bound state, when $E < 0$, the energy can assume discrete values only. The energy levels of a hydrogen-like atom E_n are described by the principal quantum number n, which assumes values from 1 to ∞,

$$E_n = -\frac{I_H Z^2}{n^2} = -\frac{I}{n^2}, \qquad I_H = \frac{e^2}{2a_0} = \frac{2\pi^2 me^4}{h^2}. \tag{5.25}$$

$I = I_H Z^2 = |E_1|$ is the absolute value of the ground-state energy, that is, the ionization potential. The binding energy of an electron in the nth quantum state is $E_n = |E_n| = I/n^2$. The energy level diagram for hydrogen was given in Fig. 2.2. It is well known that the kinetic energy averaged with respect to time of a bound electron moving in the Coulomb field of an ion is equal to one half of the average potential energy with the sign reversed and is equal to the total energy, also with the sign reversed: $E_{kin} = -E_{pot}/2 = -E$ $(E = E_{kin} + E_{pot})$. Consequently, averaging with respect to time gives

$$E_{kin} = \frac{m\overline{v^2}}{2} = \frac{I}{n^2} = \frac{I_H Z^2}{n^2}.$$

Taking into account the inequality (5.13), this equation shows that the motion of an electron in strongly excited quantum states with a large quantum number n is quasi-classical.

In considering bremsstrahlung emission in §2 we used the classical equation (5.8) for effective radiation to describe "transitions" of an electron from one hyperbolic orbit to another one at a lower energy. And we have extended the equation to include transitions to orbits of infinitesimal positive energy, that is, almost parabolic orbits, corresponding to radiation close to the maximum frequency $\nu_{max} = E/h$. Here, of course, the initial energy E was assumed to be sufficiently small, $E \ll I_H Z^2$, $v \ll 2\pi Ze^2/h$, in order for the motion in the initial state to be quasi-classical. The motion in the final state is even closer to quasi-classical, since the electron loses kinetic energy in the transition and is slowed down. As we have just shown, small negative energies also correspond to low velocities and the elliptic orbits are also close to parabolic (except that these are approached from the side of negative energies); it is, therefore, natural to extend (5.8) to the case of radiation at frequencies slightly exceeding ν_{max}, that is, to the case of electron capture to the higher levels. We should note, however, that the final state of the electron will be found in the discrete spectrum. The effective radiation in some small but finite frequency interval $\Delta\nu$, $\Delta q_\nu = (dq_\nu/d\nu)\,\Delta\nu$, is, according to the quantum interpretation, equal to $h\nu\,\Delta\sigma_\nu$, where $\Delta\sigma_\nu$ is the cross section for photon emission in the small interval $\Delta\nu$. But now the emission of photons with energies in the range from $h\nu$ to $h\nu + \Delta(h\nu)$ corresponds to capture into a well-defined and finite number of levels Δn; and the capture cross section $\Delta\sigma_\nu$ can be represented as the product $\sigma_{cn}\,\Delta n$, where σ_{cn} is the average capture cross section into any level within this interval. This cross section depends upon the average number n in the small interval Δn. Thus,

$$\sigma_{cn} = \frac{\Delta\sigma_\nu}{\Delta n} = \frac{1}{h\nu}\frac{\Delta q_\nu}{\Delta n} = \frac{1}{h\nu}\left(\frac{dq_\nu}{d\nu}\right)\frac{\Delta\nu}{\Delta n}. \tag{5.26}$$

Using (5.25) to determine the energy spacing between the levels for large n, $|dE_n/dn| = h\,\Delta\nu/\Delta n = 2I_{\mathrm{H}}Z^2/n^3$, and (5.8) for the effective radiation, we obtain the capture cross section into the level n of a free electron with an initial energy $E = mv^2/2$:

$$\sigma_{cn} = \frac{128\pi^4}{3\sqrt{3}}\frac{Z^4e^{10}}{mc^3h^4v^2\nu}\frac{1}{n^3} = \frac{2.1\cdot 10^{-22}}{n^3}\frac{I_{\mathrm{H}}Z^2}{E}\frac{I_{\mathrm{H}}Z^2}{h\nu}\ \mathrm{cm}^2. \tag{5.27}$$

The energy of a photon emitted during this capture is

$$h\nu = E + |E_n| = \frac{mv^2}{2} + \frac{I_{\mathrm{H}}Z^2}{n^2}. \tag{5.28}$$

As shown by quantum-mechanical calculations (see the following section), the semiclassical equation (5.27) also gives good results for capture into the lower levels, including the ground level ($n = 1$), despite the fact that the motion of an electron in the ground state is no longer quasi-classical ($E_{\mathrm{kin}} = \frac{1}{2}m\bar{v}^2 = I_{\mathrm{H}}Z^2$). It is assumed, however, that the initial motion of the free electron is quasi-classical, that is, its initial energy $E \ll I_{\mathrm{H}}Z^2$.

Let us use (5.27) to calculate the total cross section for the radiative capture of an electron with a given energy $E = mv^2/2$ into any level of a hydrogen-like ion. For this purpose we must take a sum of the cross section σ_{cn} given by (5.27), over all values of n from 1 to ∞, noting that the photons are emitted at different energies as given by (5.28)

$$\sigma_c = 2.1\cdot 10^{-22}\frac{I_{\mathrm{H}}Z^2}{E}\sum_{n=1}^{\infty}\frac{1}{n^3}\frac{1}{(E/I_{\mathrm{H}}Z^2 + 1/n^2)} = \frac{2.8\cdot 10^{-21}Z^2}{E_{\mathrm{ev}}}\varphi\left(\frac{I_{\mathrm{H}}Z^2}{E}\right). \tag{5.29}$$

Here φ denotes the summation over n. Roughly, for small electron energies $E \ll I_{\mathrm{H}}Z^2$, $\varphi \approx (\sum_{n=1}^{n^*} 1/n) + \frac{1}{2}$, where $n^* \approx (I_{\mathrm{H}}Z^2/E)^{1/2}$. For not too large (but also not too small) electron energies, when E is less than but still comparable with $I_{\mathrm{H}}Z^2$, the sum φ is of the order of unity and the capture cross section is approximately $\sigma_c \approx 3\cdot 10^{-21}Z^2/E_{\mathrm{ev}}\ \mathrm{cm}^2$.

It is interesting to compare the integrated effective radiation of a free electron with a given energy E which is slowed down in the field of a hydrogen-like ion, with the integrated radiation in the process of radiative capture, that is, the quantities $q_{\text{brems}} = \int dq_\nu = \int h\nu \, d\sigma_{\text{brems}}$ and $q_{\text{cap}} = \sum_n h\nu\sigma_{cn}$. The first quantity, according to (5.8), is equal to $q_{\text{brems}} = (dq_\nu/d\nu)E/h$, and the second, by virtue of the derivation of the cross section σ_{cn} (see (5.26)), is given by $q_{\text{cap}} = (dq_\nu/d\nu)I_{\text{H}}Z^2/h$, with the constant $(dq_\nu/d\nu)$ determined from (5.8). Both q_{brems} and q_{cap} are proportional to the energy intervals of the possible final

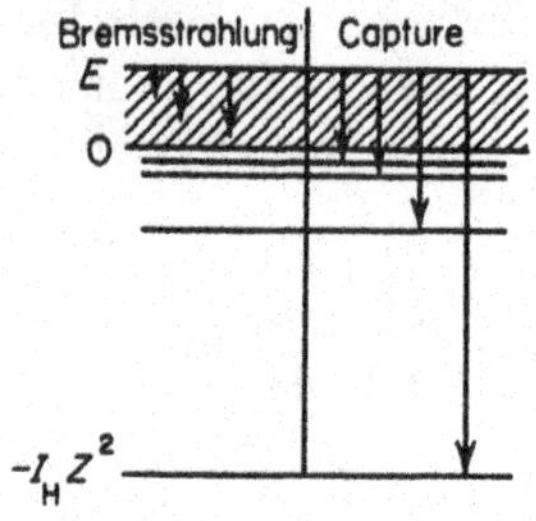

Fig. 5.4. Diagram showing the relationship between the energy intervals of the final states of an electron for bremsstrahlung and capture by an ion.

electronic states (Fig. 5.4) and their ratio is directly proportional to the ratio of these intervals

$$q_{\text{cap}}/q_{\text{brems}} = I_{\text{H}}Z^2/E.$$

§5. Cross section for the bound-free absorption of light by atoms and ions

Let us consider a process which is the reverse of radiative capture, namely the photoionization of a hydrogen-like atom, in other words the absorption of a photon accompanied by the transition of an electron into the continuous energy spectrum. As in the bremsstrahlung calculations, we shall use the principle of detailed balancing. The number of radiative captures of electrons with speeds from v to $v + dv$ into the nth ionic level per unit volume per unit time is given by

$$N_+ N_e f(v) \, dv \cdot v \cdot \sigma_{cn}. \tag{5.30}$$

This results in the emission of photons with the frequencies between ν and $\nu + d\nu$ related to the electron speeds by (5.28). The number of reverse processes, that is, photoionization of "atoms" in the nth quantum state by photons with frequencies between ν and $\nu + d\nu$, is

$$N_n \frac{U_{\nu p}}{h\nu} \, d\nu \cdot c \cdot \sigma_{\nu n}(1 - e^{-h\nu/kT}) \tag{5.31}$$

per unit volume per unit time, where $\sigma_{\nu n}$ is the cross section for the absorption of a photon $h\nu$ by an atom in the nth state, N_n is the number of such atoms per unit volume, while the factor $(1 - e^{-h\nu/kT})$ again accounts for induced emission. Under the conditions of complete thermodynamic equilibrium, $f(v)$ is the Maxwell distribution function for the electrons and $U_{\nu p}$ is the Planck function; the number of excited atoms N_n is given by the Boltzmann equation

$$N_n = N_1 \frac{g_n}{g_1} e^{-\frac{(E_n - E_1)}{kT}} = N_1 \frac{g_n}{g_1} e^{-\frac{I}{kT}\left(1 - \frac{1}{n^2}\right)}, \tag{5.32}$$

where $g_n = 2n^2$ is the statistical weight of the nth level of the hydrogen-like atom, N_1 is the atom number density in the ground state, $g_1 = 2$, and $E_n - E_1 = I_H Z^2(1 - 1/n^2) = I(1 - 1/n^2)$ is the excitation energy of the nth state.

The number of free electrons, ions, and "neutral" atoms (if $Z > 1$, then a "neutral" hydrogen-like atom represents an ion with a charge $Z - 1$) are related by the Saha equation (see (3.44) in §5 of Chapter III):

$$\frac{N_+ N_e}{N} = 2\left(\frac{2\pi mkT}{h^2}\right)^{3/2} \frac{u_+}{u} e^{-I/kT}, \tag{5.33}$$

where the electronic partition function of the ion $u_+ = 1$. The "neutral" atom number density is $N = uN_1/g_1$, where u is the electronic partition function of the atom.

Equating the rates for the forward and reverse processes (5.30) and (5.31), and recalling all of our statements pertaining to the terms in these equations, we find the relation between the photoionization and radiative capture cross section

$$\sigma_{\nu n} = \frac{u_+}{g_n}\left(\frac{mvc}{h\nu}\right)^2 \sigma_{cn}.$$

Substituting into this relation σ_{cn} from (5.27), we obtain the cross section for the absorption of a photon $h\nu$ by a hydrogen-like atom whose "atomic" remainder charge is Z and which is in the nth quantum state

$$\sigma_{\nu n} = \frac{64\pi^4 e^{10} m Z^4}{3\sqrt{3}\, h^6 c \nu^3 n^5} = 7.9 \cdot 10^{-18} \frac{n}{Z^2}\left(\frac{\nu_n}{\nu}\right)^3 \text{cm}^2. \tag{5.34}$$

Here ν_n denotes the minimum frequency of a photon still capable of removing an electron from the nth level: $h\nu_n = I_H Z^2/n^2$ (see (5.28)). The basic characteristic of this cross section is that it is inversely proportional to the cube of the frequency $\sigma_{\nu n} \sim (\nu_n/\nu)^3$. The cross section has a maximum at the absorption threshold, when $\nu = \nu_n$. Equation (5.34) is known in the literature as Kramers' formula.

A somewhat more rigorous, quantum-mechanical derivation of the photoionization of hydrogen-like atoms from high levels leads to an equation differing from (5.34) by the following correction factor (see [4]):

$$g' = 1 - 0.173\left(\frac{h\nu}{I_H Z^2}\right)^{1/3}\left[\frac{2}{n^2}\left(\frac{I_H Z^2}{h\nu}\right) - 1\right].$$

In the majority of cases of practical interest this factor is very close to unity so that it can, as a rule, be neglected.

The semiclassical relation (5.34), which by its derivation is valid only for strongly excited states $n \gg 1$, nevertheless gives good results even when applied to photoionization from the ground state $n = 1$. Quantum-mechanical calculations of the cross section for the photoelectric effect for the K atomic shell, i.e., for the ground state of the hydrogen-like atom, carried out using the exact wave functions for a free electron in a Coulomb field, give the following results (evaluated per electron, as in (5.33); see [5]):

$$\sigma_{\nu 1} \approx \frac{6.34 \cdot 10^{-18}}{Z^2}\left(\frac{\nu_1}{\nu}\right)^{8/3}, \qquad \nu - \nu_1 \ll \nu_1; \tag{5.35}$$

$$\sigma_{\nu 1} \approx \frac{8.32 \cdot 10^{-18}}{Z^2}\left(\frac{\nu_1}{\nu}\right)^{3}, \qquad \nu - \nu_1 \sim \nu_1; \tag{5.36}$$

$$\sigma_{\nu 1} \approx \frac{5.42 \cdot 10^{-17}}{Z^2}\left(\frac{\nu_1}{\nu}\right)^{3.5}, \qquad \nu \gg \nu_1. \tag{5.37}$$

The first of these equations corresponds to the region near the absorption boundary, while the last one applies when the energy of the electron being detached is appreciably larger than the binding energy $h\nu_1 = I_H Z^2$, corresponding to conditions for which the Born approximation is applicable.

Comparison of (5.34) in which $n = 1$ has been set with (5.35) and (5.36) shows that at the absorption boundary, for $\nu = \nu_1$, the "semiclassical cross section" (5.34) is equal to $7.9 \cdot 10^{-18}/Z^2$ cm^2 and exceeds the quantum-mechanical cross section (5.35) by only 25%. For $\nu - \nu_1 \sim \nu_1$, when the energy of the detached electron is of the order of its binding energy in the ground state, (5.34) and (5.35) agree to within 5% and also predict the same dependence on frequency. A pronounced difference occurs only for $h\nu \gg I_H Z^2$, when the energy of the detached free electron is large $E \gg I_H Z^2$, i.e., in the Born region where the situation is the opposite of the quasi-classical situation. We shall see later that such high energy photons are always in the far Wien region of the spectrum and that they have practically no importance under conditions close to thermal equilibrium. The semiclassical equation (5.34) can therefore be used with good approximation to calculate the photoionization from all levels for hydrogen-like atoms. Similarly, the radiative capture equation (5.27) is applicable to the capture of an electron into all levels, including the ground state; this fact was used in the preceding section for calculating the total capture cross section.

Let us briefly consider what one may expect from the application to complex atomic systems of the equations derived for hydrogen-like atoms. The low energy photons whose energies are considerably smaller than the ionization potential I of the atom or ion are absorbed (accompanied by the removal of an electron) only by strongly excited atoms (ions), whose excitation energy is not less than $I - h\nu$. However, in strongly excited states the optical electron moves in a very large orbit, where the field of the "atomic remainder" is very close to the Coulomb field produced by a charge equal to the charge of the "remainder." Hence, we may expect that the "hydrogen-like" approximation will be justified in this case. Unfortunately, no rigorous quantum-mechanical calculations of absorption by strongly excited atoms and ions are available to verify this assumption (though there is little reason to doubt its validity, *eds.*).

The limited available numerical calculations pertain primarily to the photoelectric effect from the ground state of atoms (calculations for ions are even scarcer). In this case the field in which the absorbing electron is moving is produced by a complex system of nuclear charges and charges of the remaining electrons; the dimensions of this system are comparable to the electron "orbit" and, of course, the field differs markedly from a Coulomb field. Correspondingly, the wave function of the electron differs greatly from the wave function of the S state of a hydrogen-like atom. For many atoms, the cross section for photoionization from the ground state differs markedly from the corresponding cross section for the hydrogen atom which, according to (5.34), is equal to $\sigma_{\nu 1} = 7.9 \cdot 10^{-18} (\nu_1/\nu)^3$ cm^2 (at the absorption boundary $\sigma^* = 7.9 \cdot 10^{-18}$ cm^2). For other states the cross sections are quite close to each other at the absorption boundary, but show a different frequency dependence. Thus, for example, the cross sections for oxygen and fluorine at the absorption boundary are approximately equal to $2.5 \cdot 10^{-18}$ cm^2, and are almost independent of frequency up to $\nu \approx 2\nu_1$. For nitrogen at the absorption boundary $\sigma^* = 7.5 \cdot 10^{-18}$ cm^2, and for carbon $\sigma^* = 10 \cdot 10^{-18}$ cm^2, but the cross sections decrease with increasing ν slower than ν^{-3}, as with hydrogen-like atoms†. For lithium $\sigma^* = 3.7 \cdot 10^{-18}$ cm^2 and for calcium $\sigma^* = 25 \cdot 10^{-18}$ cm^2. The divergence from "hydrogen-likeness" is especially pronounced in alkali metals. For sodium $\sigma^* = 0.31 \cdot 10^{-18}$ cm^2. The experimental values for rubidium are $\sigma^* = 0.1 \cdot 10^{-18}$ cm^2 and for cesium $\sigma^* = 0.6 \cdot 10^{-18}$ cm^2. A more detailed review of the available data can be found

† Plots of photoionization cross sections from the ground states of O, N, F, and C are given in [6].

in the article by Bates [7]. Fortunately, as we shall show below, in sufficiently rarefied gases close to thermodynamic equilibrium the role of the high-energy photons whose energy exceeds the ionization potentials of the atoms and ions is relatively small. Thus, the strong divergence from "hydrogen-likeness" in this case does not invalidate the use of the "hydrogen-like" approximation.

Recently Ivanova [84, 85] has carried out rigorous quantum-mechanical calculations of the photoionization cross sections from the ground state and from many excited levels of lithium-like atomic systems: lithium atoms, quadruply ionized nitrogen, and quintuply ionized oxygen. The wave functions were calculated by the Hartree–Fock method. On the basis of cross sections thus found she has calculated the absorption coefficients for N^{+4} and O^{+5} ions from 20 levels over a wide range of frequencies and temperatures.

In some cases the radiative capture of electrons by neutral atoms accompanied by the formation of negative ions and the reverse process of the photoelectric absorption of photons by negative ions* is of great importance. Examples may be found in the absorption of light in stellar atmospheres, in which negative hydrogen ions may play an important role, and the absorption of light in air under certain conditions in which negative oxygen ions are of considerable importance. The binding energies or ionization potentials of negative ions, which define the lower absorption limit $h\nu_{min}$, are equal to 0.75 ev for the atomic hydrogen ion H^-, 1.45 ev for the atomic oxygen ion O^-, and 0.46 ev for the molecular ion O_2^- see [53]. The relationship between the frequency and the absorption cross section does not follow the inverse cube law. Figure 5.5 presents the results of quantum-mechanical calculations of the effective photoionization (photodetachment) cross section for O^-.

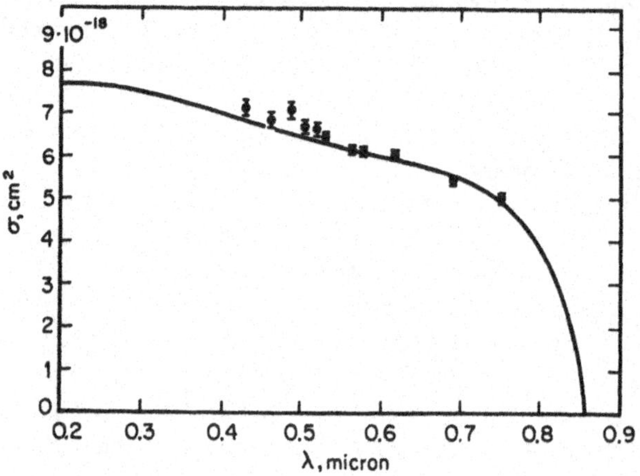

Fig. 5.5. Cross section for light absorption by negative oxygen ions O^-.

* *Editors' note.* Often referred to as photodetachment.

The curve is taken from [8]. Experimental points, taken from [9], fall near the theoretical curve. Data on absorption by H^- ions may be found in [6].

§6. Continuous absorption coefficient in a gas of hydrogen-like atoms

Let us calculate the coefficient of bound-free absorption of photons $h\nu$ by hydrogen-like atoms with a "nuclear" charge equal to Z. At a given temperature the gas contains atoms occupying all possible excitation levels. If N_n is the atom number density in the nth quantum state, and $\sigma_{\nu n}$ is the cross section for the absorption of a photon $h\nu$ by these atoms, then the absorption coefficient* is

$$\kappa_\nu' = \sum_{n^*}^{\infty} N_n \sigma_{\nu n}. \tag{5.38}$$

The lower limit in this summation is determined from the condition that the photon energy is greater than the binding energy of the electron in the atom, $h\nu > |E_n|$. Otherwise, the photon cannot remove an electron and, as a result, atoms excited to a state $n < n^*$ for which $|E_n| > h\nu$ cannot participate in the absorption of photons with energy $h\nu$. In particular, if the energy of the photon exceeds the binding energy of an electron in the ground state of the atom, i.e., the ionization potential $I = I_H Z^2$, then all the atoms will participate in the absorption ($n^* = 1$). Only strongly excited atoms ($n^* \gg 1$) participate in the absorption of low energy photons $h\nu \ll I_H Z^2$.

The absorption curve as a function of frequency is of the "sawtooth" shape, as shown in Fig. 5.6. As soon as the energy $h\nu$ reaches the value of the binding energy of an electron in any state $|E_n|$, the atoms excited into this level participate in the absorption and the absorption coefficient jumps to a higher

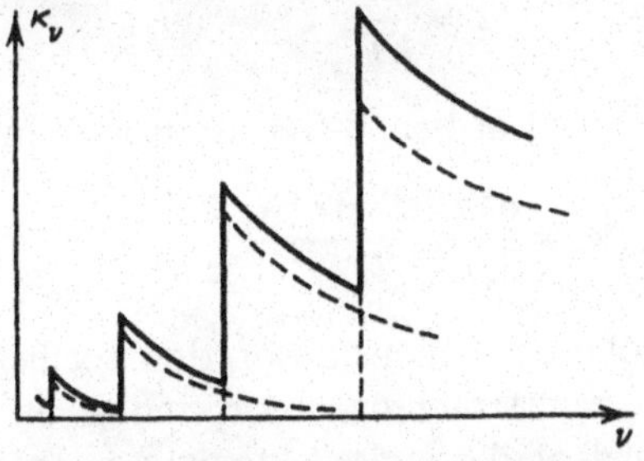

Fig. 5.6. Absorption "sawtooth fence". The dashed lines correspond to absorption by atoms in a given quantum state. The solid curve is the total absorption coefficient. The diagram is schematic.

* We denote here the bound-free absorption by a prime in order to distinguish it from the free-free absorption coefficient, which will be denoted by a double prime.

value. Then, until the next state is excited, κ_ν' decreases as ν^{-3}, in accordance with the relationship $\sigma_{\nu n} \sim \nu^{-3}$. Each level contributes its "tooth" to the "sawtooth fence" $N_n\sigma_{\nu n}$ (the dashed lines in Fig. 5.6), and the total absorption coefficient κ_ν' is obtained by summing all the "teeth" (the solid line in Fig. 5.6).

If the gas is in thermodynamic equilibrium, then N_n, the number of atoms in the nth state, is determined from the Boltzmann equation (5.32). For large values of n, where the exponent is practically independent of n, the number of atoms N_n is simply proportional to n^2 ($g_n = 2n^2$). Since the cross section $\sigma_{\nu n} \sim n^{-5}$, the terms in the summation (5.38) decrease as $1/n^3$ for $n \to \infty$, so that the contribution of the successive levels to the absorption of light at a given frequency decreases very rapidly and the infinite sum converges*.

We shall consider temperatures for which the degree of ionization is not very large. As shown in §5 of Chapter III, appreciable ionization in a not too dense gas begins when kT is still much less than the ionization potential I. The number of excited atoms in this case is very small, since the excitation of the lowest state $n = 2$ requires an energy close to the ionization potential, equal to $\frac{3}{4}I$. For $kT \ll I$ the number of atoms N_1 in the ground state is thus quite close to the total number of atoms $N = \sum N_n$, so that we can approximately set $N_1 \approx N$ in the Boltzmann equation (5.32). Substituting N_n and $\sigma_{\nu n}$ as given by (5.32) and (5.34) into (5.38), and introducing the notation

$$x_n = \frac{|E_n|}{kT} = \frac{|E_1|}{kT}\frac{1}{n^2} = \frac{x_1}{n^2},$$

$$x_1 = \frac{|E_1|}{kT} = \frac{I_{\mathrm{H}}Z^2}{kT} = \frac{I}{kT}, \tag{5.39}$$

$$x = \frac{h\nu}{kT},$$

we obtain the coefficient of bound-free absorption in the form

$$\kappa_\nu' = \frac{64\pi^4}{3\sqrt{3}}\frac{e^{10}mZ^4N}{h^6c\nu^3}\sum_{n^*}^{\infty}\frac{1}{n^3}e^{-(x_1-x_n)}. \tag{5.40}$$

To obtain the total continuous absorption coefficient we must add to κ_ν' the coefficient of bremsstrahlung absorption given by (5.21) for the free electrons in the field of the ionized atoms, i.e., the "hydrogen-like ions". Expressing the product N_+N_e in (5.21) in terms of the number of "neutral"

* In an actual ionized gas, due to the interaction between the atoms and ions, the upper excitation levels are cut off (see §6 of Chapter III), so that the number of terms in the summation (5.38) is finite. In our case it is not necessary to cut off the summation with respect to n, since the sum converges very rapidly.

atoms given by the Saha equation (5.33) and noting that $u \approx g_1 = 2$ and $N \approx N_1$, we rewrite the bremsstrahlung absorption coefficient as

$$\kappa_\nu'' = \frac{16\pi^2}{3\sqrt{3}} \frac{Z^2 e^6 kTN}{h^4 c\nu^3} e^{-I/kT} = \frac{64\pi^4}{3\sqrt{3}} \frac{Z^4 e^{10} mN}{h^6 c\nu^3} \frac{e^{-x_1}}{2x_1}. \tag{5.41}$$

The total coefficient $\kappa_\nu = \kappa_\nu' + \kappa_\nu''$ is then

$$\kappa_\nu = \frac{64\pi^4}{3\sqrt{3}} \frac{e^{10} mZ^4 N}{h^6 c\nu^3} \left\{ \sum_{n^*}^{\infty} \frac{1}{n^3} e^{-(x_1 - x_n)} + \frac{e^{-x_1}}{2x_1} \right\}. \tag{5.42}$$

The above equation can be considerably simplified if the photon energy is small in comparison with the ionization potential, so that the photon is absorbed by strongly excited atoms only (n^* is large). Since the density of the levels increases rapidly with increasing n, the summation for large values of n can be replaced by integration (the "differential" corresponds to $\Delta n = 1$). Integration with respect to n is equivalent to integration of the energy states over the spectrum with the transformation from the discrete to a continuous spectrum, in accordance with the relationship $dn/n^3 = -\frac{1}{2}\, dx_n/x_1$. The lower limit of the integral with respect to x_n should, obviously, be the dimensionless photon energy $x = h\nu/kT$. Thus,

$$\sum_{n^*}^{\infty} \frac{1}{n^3} e^{-(x_1 - x_n)} \approx -\frac{e^{-x_1}}{2x_1} \int_x^0 e^{x_n}\, dx_n = \frac{e^{-x_1}}{2x_1} (e^x - 1). \tag{5.43}$$

If we formally extend the summation or integration to "negative binding energies", or, equivalently, to "excitation" energies $x_1 - x_n$ exceeding the ionization potential, then the integral with respect to x_n from 0 to $-\infty$ gives $e^{-x_1}/2x_1$, in complete correspondence with the free-free transition. This should have been expected, since the states of an atom with an "excitation" exceeding the ionization potential represent the states with a detached electron, and the cross section of bound-free absorption was derived by assuming that the transition from bound to free electron states is continuous. Substituting (5.43) into (5.42), factoring out $I/kT = I_H Z^2/kT$ from the coefficient in front of the braces, and canceling it with x_1 in the denominator of (5.43), we obtain the final equation for the absorption coefficient for low energy photons $h\nu \ll I$*:

$$\kappa_\nu = \frac{16\pi^2}{3\sqrt{3}} \frac{e^6 Z^2 kTN}{h^4 c\nu^3} e^{-\frac{I - h\nu}{kT}} = 0.96 \cdot 10^{-7} \frac{NZ^2}{T^{\circ 2}} \frac{e^{-(x_1 - x)}}{x^3} \text{ cm}^{-1}. \tag{5.44}$$

The absorption coefficient κ_ν is not proportional to Z^4, as might appear by glancing at (5.40), but only to Z^2. In order to clarify this, we recall how the

* This equation is frequently referred to as the Kramers–Unsöld formula.

factor Z^4 arose in (5.40). One factor Z^2 enters in the coefficient because the absorption cross section is proportional to the square of the "acceleration" of an electron in a Coulomb field (according to the classical theory) or to the square of the matrix element of the interaction energy between an electron and the "nucleus" (according to the quantum-mechanical treatment). The second factor Z^2 appears because the absorption cross section $\sigma_{\nu n}$ is proportional to the spacing between the levels, which, in turn, is proportional to the total energy interval of the bound states, $I = I_H Z^2$. The cross section $\sigma_{\nu n}$ is actually proportional to the energy spacing between the levels, since the radiative capture cross section, which is related to the absorption cross section by the principle of detailed balancing, is itself proportional to this separation (see (5.21)). In summing the partial coefficient $\kappa'_{\nu n} = N_n \sigma_{\nu n}$ over the levels or, equivalently, integrating over the energy interval of bound states participating in the absorption of the given photon, the latter dependence on Z^2 disappears. The preceding remarks about the relationship between κ_ν and Z become significant when we consider multiply charged ions (see below).

It is evident from (5.42) and (5.43) that the contributions of bound-free transitions and free-free transitions to the total continuous absorption coefficient κ_ν are in the ratio

$$(e^x - 1) : 1 = (e^{h\nu/kT} - 1) : 1.$$

Hence, it follows that the fundamental role in the absorption of high energy photons $h\nu \gg kT$ is played by the bound-free transitions, while in the case of absorption of low energy photons $h\nu \ll kT$ the main contribution is given by the free-free transitions.

§7. Continuous absorption of light in a monatomic gas in the singly ionized region

Let us consider the continuous absorption of light in inert monatomic gases such as argon, xenon, and others and in monatomic metallic vapors, in the singly ionized region. The reason for assuming that the gas is monatomic is to exclude from consideration the quasi-continuous molecular spectra (if the molecules are almost fully dissociated then, obviously, any gas appears as monatomic). The singly ionized region lies in a range of temperatures of the order of 6000–30,000°K (depending on the ionization potential of the atoms and on the gas density) and is therefore of great interest for many laboratory studies and practical applications. At still higher temperatures, double and triple ionizations become important; they must be taken into account in considering the absorption process and will be discussed in the next section.

We consider a strongly excited atom as a hydrogen-like system, so that the "optical" electron (which is one of the outer valence electrons) moves in a large orbit in the field produced by the nucleus and by the remaining electrons. If the dimensions of the system of charges forming the "atomic remainder" are not very large in comparison with the orbit of the "optical" electron (which is the case in a strongly excited atom) then the whole system can be represented as a point charge $Z = 1$ producing a Coulomb field (if we are not dealing with a neutral atom but with an ion, then Z is one greater than the ionic charge; see the following section).

Extending the results obtained for hydrogen-like atoms to multi-electron atoms, it is natural to replace the ionization potential in all equations by the actual potential of the given atom. Indeed, the basic factor which determines the temperature dependence of the absorption coefficient for photons whose energy is considerably smaller than the ionization potential is the Boltzmann factor $\exp[-(I - h\nu)/kT]$; the number of atoms sufficiently excited for the removal of an electron by a photon is proportional to this factor. Of course, the Boltzmann factor is completely independent of whether the atom is hydrogen-like or more complex. One of the terms in the Boltzmann factor $\exp(-I/kT)$ describes the degree of ionization or, more precisely, the product N_+N_e, to which the bremsstrahlung absorption coefficient is proportional, again independently of the type of atom.

In complex atoms each of the "hydrogen-like" levels with a given principal quantum number n splits into several levels in accordance with its statistical weight. This comes from the fact that the field of complex atoms differs from a Coulomb field, and as a result there is an absence of l-degeneracy. Therefore, the energies of the levels with the same principal quantum number n, but with different orbital numbers l, do not coincide (in contrast to the case of hydrogen-like atoms). If we take into account this "multiplicity" of levels in complex atoms, which leads to the appearance of a large number of more closely spaced "teeth" in the "sawtooth fence" curve for $\kappa(\nu)$, then the replacement of the summation over levels by an integration or, equivalently, the replacement of the "sawtooth fence" by an averaged smooth curve, becomes even more applicable than in the case of hydrogen-like atoms (see Unsöld [10]).

The absorption of the low energy photons whose energies are considerably smaller than the ionization potential should, apparently, be fairly well described by the relation (5.44) derived for hydrogen-like atoms; here, Z for neutral atoms should be taken equal to unity. Actually, the upper levels, that is, the only levels that are accessible for the removal of electrons by low energy photons, are in complex atoms very similar to "hydrogen-like" levels, because the field at large distances from the atomic "remainder" behaves very much like a Coulomb field. The use of (5.44) for high energy photons, which are absorbed by atoms in the ground or low excitation states, can lead to considerable error*. Equation (5.44) becomes completely invalid for photon energies exceeding the ionization potential $h\nu > I$, $x > x_1$. In this case all levels from $n = 1$ to ∞ are included in the summation with respect to n, and (5.43) with a variable lower limit of integration becomes meaningless. The summation over the levels is in this case simply a constant and is independent of ν (of x). A dominant role in the absorption of these photons $h\nu > I$ is played by atoms in the ground state, so that the sum can be taken to be given by the first term only, that is, unity. This gives the approximate equation

$$\kappa_\nu = \frac{32\pi^2}{3\sqrt{3}} \frac{e^6 Z^2 N}{h^4 c \nu^3} I = 0.96 \cdot 10^{-7} \frac{NZ^2}{T^{\circ 2}} \frac{2x_1}{x^3} \text{ cm}^{-1} \quad \text{for} \quad x > x_1, \; h\nu > I, \tag{5.45}$$

which should replace (5.44) for $h\nu > I$.

Let us find the Rosseland mean free path for a monatomic gas in the singly ionized region. The Rosseland mean free path is determined by the reciprocal of the absorption coefficient, which we can call the transmissivity. The spectral mean free path $l_\nu = 1/\kappa_\nu$ which characterizes the transmission is shown schematically in Fig. 5.7. The transmission frequency regions are the opposite of

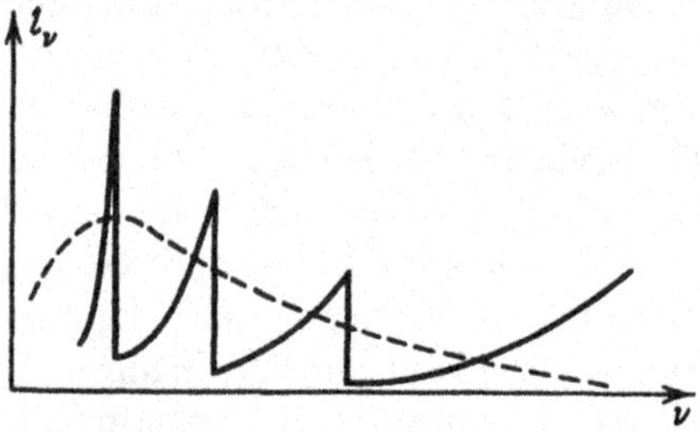

Fig. 5.7. "Sawtooth" transmission curve. The solid line shows the spectral mean free path as a function of frequency. The dashed line is the spectral mean free path with the teeth "smoothed". The diagram is schematic.

the absorption regions and are situated near the jumps, on the lower frequency side. There is practically no transmission in the region of photon energies $h\nu$ exceeding the ionization potential, since such photons are very strongly absorbed by atoms in the ground state. The main contribution to the radiant energy transfer is made by the photons corresponding to the maximum of the weighting funtion in the Rosseland integral (2.80), at $x = h\nu/kT \approx 4$. If the temperature (times k) is much less than the ionization potential, as is usually the case with not too dense a gas in the singly ionized region, then the absorption

* See §5, p. 265, where calculated photon absorption cross sections are given for some atoms in their ground states.

of these photons can be approximately described by (5.44). The smaller is $h\nu$ the more exact is this equation, which can also be used to calculate the mean free path (the transmission "sawtooth" curve is, according to this equation, replaced by the smooth curve x^3e^{-x} shown in Fig. 5.7 by the dashed line). Since the high frequencies $x > x_1$ make practically no contribution to the Rosseland integral (2.80), we can evaluate the latter by extending (5.44), which applies only for $x < x_1$, to $x > x_1$. In fact, (5.44) formally ensures rapid damping of transmission for $x > x_1$, $x \to \infty$. Substituting (5.44) into the Rosseland integral (2.80), we get for the Rosseland mean free path*

$$l = 0.9 \cdot 10^7 \frac{T^{\circ 2}}{NZ^2} e^{I/kT} \text{ cm.} \tag{5.46}$$

It should be noted that if the Rosseland mean free path is calculated using an absorption coefficient obtained not from (5.44) but from the "exact" equation (5.42) for hydrogen-like atoms, that is, without replacing the transmission "sawtooth" curve by a smooth curve, the resulting mean free paths will be approximately five times larger than those given by (5.46) (for $x_1 = I/kT \sim 10$). Let us give an example of calculating the Rosseland mean free path using (5.46). For hydrogen at $T = 11{,}600°\text{K} = 1$ ev, $N = 10^{19}$ cm^{-3}, we obtain $l = 100$ cm (the degree of ionization under these conditions is 0.02).

If the mean absorption coefficient κ_1 characterizing the integrated emission from the gas is formally calculated from (2.105) using (5.44) and (5.45), then the corresponding mean free path will be given by

$$l_1 = \frac{1}{\kappa_1} = 2.3 \cdot 10^7 \frac{T^{\circ 2}}{NZ^2} e^{I/kT} \frac{kT}{I} \text{ cm.} \tag{5.47}$$

It should be noted, however, that this equation may yield a considerable error, since the main contribution to the integral (2.105) comes from the high-frequency region $x > x_1$, where the hydrogen-like approximation is not satisfactory (the emission of high energy photons comes from the capture of electrons into the ground atomic levels).

The temperature dependence of the continuous absorption coefficient for moderate energy photons, those with energies much smaller than the ionization potential $kT \ll I$ in the singly ionized region, is essentially given by the relationship $\kappa_\nu \sim \exp(-I/kT)$, and is thus very sharp. Accordingly, the mean free path l is proportional to $\exp(I/kT)$. The Boltzmann temperature dependence of the absorption is characteristic of both bound-free transitions and bremsstrahlung absorption, of both the components κ'_ν and κ''_ν in κ_ν (since $\kappa''_\nu \sim N_+N_e \sim e^{I/kT}$).

* The resulting integral $\int_0^\infty x^3e^{-x}G'(x)\,dx$ is equal to 0.87.

A number of authors have proposed various methods for the improvement of the Kramers and Kramers–Unsöld formulas for application to more complex atoms than the hydrogen-like atoms for which they were derived. Unsöld [11] introduced an effective charge Z^* in place of the "atomic remainder" charge Z. The effective charge is defined in such a way that the quantity $E_{n,l} = -I_H Z^{*2}/n^2$ corresponds to the actual energy of the level of the complex atom with the given principal and orbital quantum numbers n and l. In addition, the Kramers formula is multiplied by γ/Σ_0, where γ is equal to the ratio of the number of sublevels of the complex atom for the given n and l to the analogous quantity for the hydrogen atom, and Σ_0 is the partition function of the atom. Unsöld [11] and others [12] recommend $Z^{*2} \approx 4$–7 for all levels which correspond to the energy of the ground atomic state.

Burgess and Seaton [13], using one-electron semiempirical wave functions found by the quantum defect method*, obtained a general expression for the photoionization cross section of an arbitrary atom or ion. Biberman and Norman [14], starting with the Burgess and Seaton formula, developed an approximate method for calculating the continuous absorption coefficient for nonhydrogen-like atoms. They represent the absorption coefficient in the form of an equation of the Kramers–Unsöld type, where the factor Z^2 is replaced by some function $\xi(\nu, T)$ of frequency and, in general, of temperature. They calculated this function for O, N, and C atoms (where it is approximately independent of temperature). The upper atomic levels are always "hydrogen-like", hence the low energy photons are absorbed in the same manner as in hydrogen; thus, as $h\nu \to 0$, $\xi \to 1$ (for $Z = 1$). As the photon energy increases from zero to $h\nu \sim 4$ ev, the coefficient ξ for these atoms decreases monotonically to a value of about 1/5. Functions for several other atoms (Li, Al, Hg, Kr, Xe, and Ar) are calculated in [15]. For example, for argon in the visible region of the spectrum $h\nu \sim 2$–3 ev, $\xi \sim 1.5$–2. The parameter ξ varies irregularly from atom to atom.

A survey of the results obtained by the quantum defect method is given in [55]. The results agree rather well with the available experimental data. There exist experimental indications that the Kramers–Unsöld theory gives fairly accurate results for inert gases. Dronov, Sviridov, and Sobolev [42] studied the continuous luminous spectrum of krypton and xenon in a shock tube. The measured intensities agree satisfactorily with the values calculated by the Kramers–Unsöld theory.

* The term quantum defect refers to the quantity $\Delta n(E_{n,l}) = n - n_l^*$, where n is the principal quantum number for the $E_{n,l}$ atomic level and n_l^* is an effective number such that $E_{n,l} = -I_H Z^2/n_l^{*2}$. The quantum defect characterizes the deviation of the energy for some particular level of a complex atom or ion from the energy of the corresponding level of a hydrogen-like atom.

§8. Radiation mean free paths for multiply ionized gas atoms

At very high temperatures, of the order of several tens of thousands of degrees and above, the atoms of a gas are multiply ionized. At these temperatures the molecules are completely dissociated, so that all gases are "monatomic" and behave in the same manner with respect to the absorption of light. We shall determine the radiation mean free paths in a multiply ionized gas. (The results presented below were obtained by one of the present authors [18].) For simplicity we shall consider the gas to consist of atoms of the same element.

Ionization equilibrium calculations show that for most temperatures and densities the gas contains only doubly or triply charged ions in any significant amount (see §7, Chapter III). Each of these ions contributes to the continuous absorption and participates in the bound-free as well as in the free-free transitions. The same calculations show that for a gas of not too high density the ionization potentials of the ions present in large numbers are always much greater than kT. For example, in air at a density 1/100 of atmospheric, the "average" ionization potential of the ions $\bar{I}$ (corresponding to ions with an "average" charge at the given density and temperature) is approximately 11 times greater than kT. Consequently, photons with energies $h\nu$ exceeding kT by a factor of 3–5, which play the principal role in radiant energy transfer, are absorbed not from the ground state, but from the excited ionic levels. As in the case of neutral atoms, this fact can be used as a basis for extending the equations derived for hydrogen-like atoms to multiply charged ions. Moreover, the hydrogen-like approximation for multiply charged ions is even more justified than for neutral atoms, since the field produced by the "atomic remainder" of a multiply charged ion is closer to a Coulomb field the larger is the charge of the "remainder".

We shall regard continuous absorption by multiply charged ions as an absorption by hydrogen-like atoms with an appropriate charge. Let a gas containing N nuclei per unit volume at a temperature T have N_m atoms ionized m times per unit volume (for brevity we shall refer to them as m ions). The total coefficient of the bound-free absorption by m ions and of the free-free absorption in the field of $m+1$ ions will be described by (5.44) and (5.45), where the charge Z is set equal to the charge of the "atomic remainder" of m ions, $Z = m + 1$, and in place of the ionization potential we introduce the actual potential of the m ion, I_{m+1}. The combining of the bound-free and free-free coefficients for multiply charged ions corresponds in all respects to the same procedure in the singly ionized region. Actually, the free-free absorption coefficient in the field of $m+1$ ions is proportional to the product $N_{m+1}N_e$, which as before can be expressed in terms of the number of m ions

N_m from the Saha equation (3.44). Let us write down the total absorption coefficient in the form

$$\kappa_{\nu m} = \frac{aN_m(m+1)^2}{T^2} e^{-x_{1m}} F_m(x), \tag{5.48}$$

where

$$a = \frac{16\pi^2}{3\sqrt{3}} \frac{e^6}{hck^2} = 0.96 \cdot 10^{-7} \text{ cm}^2 \text{ deg}^2,$$

$$x_{1m} = I_{m+1}/kT, \qquad x = h\nu/kT,$$

and $F_m(x)$ depends on the frequency according to

$$F_m(x) = \frac{e^x}{x^3} \qquad \text{for} \quad x < x_{1m}. \tag{5.49}$$

For photons whose energy exceeds the ionization potential, we shall set, according to (5.45),

$$F_m(x) = 2x_{1m} \frac{e^{x_{1m}}}{x^3} \qquad \text{for} \quad x > x_{1m}. \tag{5.50}$$

To obtain the total absorption coefficient at a frequency ν we must sum the partial coefficients $\kappa_{\nu m}$ over all types of ions, i.e., over the charge m

$$\kappa_\nu = \sum_m \kappa_{\nu m}. \tag{5.51}$$

Let us first determine the mean absorption coefficient κ_1 which characterizes the integrated emission coefficient. Substituting the spectral coefficient κ_ν into (2.105) and evaluating the integral over the spectrum, we get

$$\kappa_1 = \frac{1}{l_1} = \frac{45a}{\pi^4 T^2} \sum_m N_m(m+1)^2 x_{1m} e^{-x_{1m}}. \tag{5.52}$$

Now let us find the Rosseland mean free path by substituting κ_ν into (2.80)

$$l = \frac{T^2}{a} \int_0^\infty \frac{G'(x)\,dx}{\sum_m N_m(m+1)^2 e^{-x_{1m}} F_m(x)}. \tag{5.53}$$

Here $G'(x)$ is the Rosseland weighting factor. The integration over the spectrum in this expression cannot be carried out, as was done for κ_1, without the help of additional considerations. This comes from the fact that here we do not average the absorption coefficient, which is additive, but its reciprocal. It is still possible, however, to carry out an approximate integration. According to (5.49), all ions in their transmission region, with $x < x_{1m}$ ($h\nu < I_{m+1}$), absorb light in exactly the same manner with respect to frequency variation. In fact, the upper limit of the integral (5.53) is the minimum transmission limit of the ions present in the gas in numbers large enough to yield a significant contribution to the absorption.

As noted above, for most temperatures and densities only doubly or triply charged ions are found in appreciable numbers. Since the average ionization potential $\bar{I}$ is much greater than kT, the transmission limits for these ions x_{1m} lie outside the limits of that spectral region which makes an appreciable contribution to the integral (5.53). Therefore, approximately, we may neglect the dependence of the function $F_m(x)$ on m and take out the function from the summation with respect to m. In addition, we may extend (5.49) for $F_m(x)$ to values of $x > x_{1m}$ in a manner similar to that in the preceding section. After these simplifications the integral becomes exactly the same as for neutral atoms (see footnote to (5.46)). We obtain

$$l = \frac{0.87T^2}{a} \frac{1}{\sum_m N_m(m+1)^2 e^{-x_{1m}}}. \tag{5.54}$$

For an approximate evaluation of the summations over m in (5.52) and (5.54) we shall use the method applied in §7, Chapter III, in the calculation of the thermodynamic functions of multiply ionized gases. We shall regard the distribution of N_m ions as a δ-function about an "average" charge $\bar{m}$ defined by (3.57). As shown in §7 of Chapter III, the ion distribution function has a sharp peak described by the Gaussian curve $N_m \sim \exp[-(m-\bar{m})^2/\Delta^2]$ (see (3.58)). If we expand the factor $e^{-x_{1m}}$ contained in the summations (5.52) and (5.54) about an average value $\bar{x}_{1m}$, we find that this factor depends on $m - \bar{m}$ according to $e^{-x_{1m}} \approx \exp[-\bar{x}_{1m} - const(m-\bar{m})]$; thus it is a weaker function of $m - \bar{m}$ than is N_m. Therefore, in this case the application of the approximate method for the calculation of the summation over m is, as in §7 of Chapter III, permissible. Taking the average coefficients of N_m in the summation terms outside the summation sign and noting that $\sum N_m = N$, we obtain

$$l = \frac{0.87T^2}{a} \frac{e^{\bar{x}_1}}{N(\bar{m}+1)^2},$$

$$l_1 = \frac{\pi^4 T^2}{45a} \frac{e^{\bar{x}_1}}{N(\bar{m}+1)^2 \bar{x}_1},$$

where $\bar{x}_1 \equiv \bar{x}_{1m} = \bar{I}/kT$. Using (3.56) to replace the exponential, and substituting the numerical value of a, we obtain finally

$$l = \frac{4.4 \cdot 10^{22} T^{\circ 7/2}}{N^2 \bar{m}(\bar{m}+1)^2} \text{ cm}, \tag{5.55}$$

$$l_1 = \frac{1.1 \cdot 10^{23} T^{\circ 7/2}}{N^2 \bar{m}(\bar{m}+1)^2 \bar{x}_1} \text{ cm}. \tag{5.56}$$

The average charge m and the average relative ionization potential $\bar{x}_1 = \bar{I}/kT$ are determined as functions of temperature and density by solving (3.57).

We can verify that the error introduced by the approximate evaluation of the summation over m is very small, and in all cases is certainly smaller than the possible errors arising from the use of the hydrogen-like approximation for complex ions. It is, however, reasonable to expect that (5.55) and (5.56) give the correct order for the mean free paths and correctly describe their dependence on the gas temperature and density. For illustration, numerical values of the mean free paths for air are shown in Table 5.2*. Unfortunately, it is not possible to give a good power law interpolation formula—so convenient for practical purposes—which would describe the dependence of $l(T, N)$ and $l_1(T, N)$ over a wide range of the variables. Roughly, the exponents in the law $l \sim T^{\alpha}N^{-\beta}$ are: $\alpha \sim 1.5$–3, and $\beta \sim 1.6$–1.9.

Table 5.2

RADIATION MEAN FREE PATHS IN AIR FOR CONDITIONS OF MULTIPLE IONIZATION

		N/N_{stand}; $N_{\text{stand}} = 5.34 \cdot 10^{19}$ cm^{-3}		
T, °K		1	10^{-1}	10^{-2}
50,000	$\overline{m}$	1.4	1.85	2.35
	l, cm	0.053	2.8	170
	l_1, cm	0.02	0.8	39
100,000	$\overline{m}$	2.72	3.47	4.1
	l, cm	0.13	7	470
	l_1, cm	0.05	2	110
250,000	$\overline{m}$	4.85	5.15	5.2
	l, cm	0.72	61.5	6,000
	l_1, cm	0.24	15.6	1,200
500,000	$\overline{m}$	5.2	5.4	5.85
	l, cm	6.8	610	50,000
	l_1, cm	2.0	140	9,500

If we consider the dependence of the mean free path on temperature at low temperatures, we find that the function $l(T)$ has a minimum. In the singly ionized region with $kT \ll I_1$, then $l \sim e^{I_1/kT}$ (see (5.46)), so that it decreases very rapidly with temperature. The mean free path has a minimum in the region where the double ionization begins (in air at $T \sim 20{,}000$–$40{,}000$°K). Thereafter, the mean free path increases with temperature, initially slower than

* The table presented in [18] contains an error. All values of the mean free paths l and l_1 were erroneously decreased by a factor of 10.

$T^{7/2}$ and then, after complete ionization when only the bremsstrahlung mechanism remains important, in proportion to $T^{7/2}$ (see (5.24)). It should be noted that the Rosseland mean free path does not increase indefinitely; light scattering becomes significant for very small absorptions (see §2 of Chapter II), and this has not been taken into account in our calculations. The mean free path for Compton scattering of photons of energy $h\nu \ll mc^2 = 500$ kev in air at standard density is 37 m. This is the upper limit of the Rosseland mean free path in standard density air. We would emphasize that the character of the relationships $l(T, N)$ and $l_1(T, N)$, as well as the orders of magnitude of the mean free paths under conditions of multiple ionization, are approximately the same for all gases, because the potentials for each successive ionization are more or less similar for all elements.

As an example let us estimate the emission coefficient and rate of radiative cooling for a transparent parcel of air with dimensions $R \ll l_1$. At $T = 50{,}000°\text{K}$ and $N = 10^{-2}N_{\text{stand}}$, $l_1 = 39$ cm and $J = 4\sigma T^4/l_1 = 3.6 \cdot 10^{13}$ erg/cm$^3 \cdot$ sec. The internal energy of air under these conditions is $\varepsilon = 83$ ev/atom. The initial time scale for cooling $\tau = N\varepsilon/J$ is $\tau = 1.9 \cdot 10^{-6}$ sec ($d(N\varepsilon)/dt = -J$). The method given above for calculating radiation mean free paths has been improved upon somewhat in [50]. Tables of l and l_1 for air, calculated from (5.52) and (5.53), are given in Kuznetsov [88].

§8a. Absorption of light in a weakly ionized gas

In weakly ionized gases the absorption coefficient corresponding to free-free transitions in the fields of ions and to bound-free transitions is proportional to the square of the electron density $\kappa_\nu = \kappa'_\nu + \kappa''_\nu \sim N_e \times e^{-(I-h\nu)/kT} \sim N_e^2$. Therefore at moderate temperatures in very weakly ionized gases we encounter primarily free-free transitions in the field of neutral atoms, whose absorption coefficient is proportional to the first power of the electron density $\kappa_{\nu\,\text{neut}} \sim NN_e \sim N^{3/2}e^{-I/2kT}$. Let us find this coefficient approximately. For this calculation we use (5.13b) for the emission cross section, and the principle of detailed balancing with (5.19). The true absorption coefficient per electron and per atom thus obtained can be written in the following form:

$$a_{\nu\,\text{neut}}(E) = a_{\nu\,\text{class}}\left[\frac{2}{3}\frac{E+h\nu}{E}\frac{E+h\nu}{h\nu}\frac{\sigma_{\text{tr}}(E+h\nu)}{\sigma_{\text{tr}}(E)}\right],$$

$$a_{\nu\,\text{class}} = \frac{e^2 v\sigma_{\text{tr}}(E)}{\pi m c\nu^2}\ \text{cm}^5, \qquad (5.57)$$

where $E = \frac{1}{2}mv^2$ is the electron energy before absorption of the photon. This formula was derived in a paper by the authors [62], treating the breakdown in gases under the action of a laser beam (see §§22 and 23). The factor

$a_{\nu\,\text{class}}$ is the absorption coefficient of electromagnetic waves in a weakly ionized gas according to the purely classical theory [63]. In this theory a solution in the form of a traveling wave with a complex wave vector is substituted into Maxwell's field equations. The imaginary part of the vector which describes the absorption of the wave is then expressed in terms of the electrical conductivity of the medium. Then the electrical conductivity of a weakly ionized gas is determined through the Boltzmann equation for the electrons in a field, which leads to the appearance of the absorption coefficient of the wave energy $a_{\nu\,\text{class}}$.

Evidently, in the limit $h\nu/E \to 0$, the quantum theory must yield the classical results. In this limiting process, however, we must take into account the fact that the concept of a "true" absorption, which is a quantum-mechanical concept, does not exist in classical mechanics. In classical theory there is an "effective" absorption, which is defined as the average over all collisions of the difference between the energy acquired and given up by an electron under the action of an electromagnetic wave. Whether energy is acquired or given up in any given collision depends on the relationship between the electron velocity directions before and after scattering and the vector of the wave electric field at the time of scattering.

Effective absorption in quantum theory corresponds to the differences between the true absorption and induced emission, and it is particularly this quantity with which we should operate when making a transition to the classical limit. According to the Einstein relation for the continuous spectrum, the coefficient of reradiation (induced emission) by an electron with an energy $E' = E + h\nu$ is

$$b_{\nu\,\text{neut}}(E') = \frac{v}{v'}\,a_{\nu\,\text{neut}}(E) = \left(\frac{E}{E + h\nu}\right)^{1/2} a_{\nu\,\text{neut}}(E). \tag{5.57a}$$

The effective absorption coefficient of low energy photons $h\nu \ll E$ is*

$$a'_{\nu\,\text{neut}}(E) = a_{\nu\,\text{neut}}(E) - b_{\nu\,\text{neut}}(E) = a_{\nu\,\text{neut}}(E) - \left(\frac{E - h\nu}{E}\right)^{1/2} a_{\nu\,\text{neut}}(E - h\nu).$$

Substituting here $a_{\nu\,\text{neut}}$ from (5.57) and taking the limit $h\nu/E \to 0$, we obtain

$$(a'_{\nu\,\text{neut}})_{h\nu/E \to 0} = \tfrac{1}{3} a_{\nu\,\text{class}} \left[1 + 2\,\frac{d(\ln a_{\nu\,\text{class}} E)}{d \ln E}\right].$$

We see from this that the limiting value, to within a numerical factor of the order of one, agrees with the classical value $a_{\nu\,\text{class}}$, and when $a_{\nu\,\text{class}}(E) =$ *const* ($v\sigma_{tr}(E) =$ *const*) the agreement is exact.

* If we average this quantity over a Maxwell distribution for the electrons, we will obtain the ordinary absorption coefficient corrected for induced emission $a'_{\nu\,\text{neut}} = a_{\nu\,\text{neut}}(1 - e^{-h\nu/kT})$.

Calculations show that the quantity $a'_{\nu\,\text{neut}}(E)$ is quite close to the classical value $a_{\nu\,\text{class}}$ also in the case when $h\nu$ is not very small in comparison with E. Thus, in approximate calculations of the absorption coefficient in a gas with the induced emission correction we can use the classical value

$$\kappa_{\nu\,\text{neut}} = N_a N_e a_{\nu\,\text{class}} = \frac{e^2 N_e}{\pi m c \nu^2} \nu_{\text{eff}}, \tag{5.57b}$$

where $\nu_{\text{eff}} = N_a v \sigma_{\text{tr}}$ is the effective frequency of electron-atom collisions. We note that if the collision frequency is not small in comparison with the angular frequency of light, an additional factor $\nu^2/[\nu^2 + (\nu_{\text{eff}}/2\pi)^2]$ appears in the coefficients $a_{\nu\,\text{class}}$ and $\kappa_{\nu\,\text{neut}}$. This has already been discussed in §2a.

Up to recently (5.57b) for the absorption coefficient of electromagnetic waves in weakly ionized gases was used primarily for radio wave and microwave frequencies (centimeter waves) [63]. In fact, however, its range of applicability, and in particular, the range of applicability of (5.57) and (5.57a) for coefficients of true absorption and induced emission is wider. For electron energies of the order of several electron volts these equations can be used for estimating the absorption at optical frequencies of photons whose energies are of the order of electron volts. In particular, these equations can be used for investigating the absorption of laser emission in gases during the breakdown stage (see §§22 and 23).

The values of the absorption coefficient given by the semiclassical equation (5.57) are in satisfactory agreement with results of quantum-mechanical calculations for hydrogen carried out by Chandrasekhar and Breen [16]. The quantum-mechanical approach is also presented in [17].

Let us compare the coefficients of light absorption for scattering of an electron by an ion and by a neutral atom. According to (5.20) and (5.57)

$$\frac{a_{\nu\,\text{ion}}(E)}{a_{\nu\,\text{neut}}(E)} = \frac{\pi}{\sqrt{3}} \frac{\pi a_0^2}{\sigma_{\text{tr}}(E')} \left(\frac{2I_{\text{H}}}{E'}\right)^2, \qquad E' = E + h\nu.$$

For example, for hydrogen $E = 1$ ev, $h\nu = 2$ ev, $\sigma_{\text{tr}} \approx 15\pi a_0^2$ [53], and $a_{\nu\,\text{ion}}/a_{\nu\,\text{neut}} \approx 10$. According to [16], for approximately the same conditions $\lambda = 5965$ Å, $h\nu = 2.08$ ev, $T = 7200$°K, $\bar{a}_{\nu\,\text{neut}} = 2.5 \cdot 10^{-39}$ cm^5, and $a_{\nu\,\text{ion}}/a_{\nu\,\text{neut}} = 12$; $a_{\nu\,\text{ion}} = 3 \cdot 10^{-38}$ cm^5.

2. *Atomic line spectra*

§9. Classical theory of spectral lines

Line spectra are emitted and absorbed as a result of bound-bound transitions in atoms (ions), in transitions of the atom from one energy state to another. The classical theoretical model of a radiating atom is an elastically bound electron, which oscillates about some equilibrium position. In the zeroth approximation, without accounting for energy losses by radiation, such a system constitutes a harmonic oscillator. Since the oscillating electron is accelerated, it radiates light. If the energy loss per period of oscillation is very small in comparison with the energy of oscillation W, then the rate of radiation can be calculated from the general equation (5.1) by substituting the acceleration of the harmonic oscillator. Let us denote the natural frequency of the oscillator by ν_0. If $\mathbf{r}$ is the coordinate of the electron measured from the equilibrium position, then its acceleration is $\mathbf{w} = 4\pi^2\nu_0^2\mathbf{r}$. The time-averaged rate of radiation energy loss by the electron, according to (5.1), is

$$\frac{dW}{dt} = -S = -\frac{32\pi^4}{3}\frac{e^2}{c^3}\nu_0^4\langle\mathbf{r}^2\rangle = -\frac{32\pi^4}{3}\frac{\nu_0^4}{c^3}\langle\mathbf{d}^2\rangle, \tag{5.58}$$

where $\mathbf{d} = e\mathbf{r}$ is the dipole moment. The symbol $\langle\ \rangle$ denotes an average with respect to time. Expressing the mean square of the electron's deflection $\langle\mathbf{r}^2\rangle$ in terms of the oscillator energy W, we obtain the energy radiated per unit time

$$S = -\frac{dW}{dt} = \frac{8\pi^2e^2}{3mc^3}\nu_0^2W = \gamma W. \tag{5.59}$$

The combination

$$\gamma = \frac{8\pi^2e^2\nu_0^2}{3mc^3} = 2.47\cdot 10^{-22}\nu_0^2\ \text{sec}^{-1} \tag{5.60}$$

represents the reciprocal of the time during which the oscillator energy decreases by a factor of e (if the initial energy of the oscillator is W_0, then $W = W_0e^{-\gamma t}$). The quantity γ is called the damping constant. The condition of weak damping $\gamma \ll \nu_0$, which serves as the basis for the derivation of (5.58), is always satisfied to a high degree of accuracy*. Thus, for example, for ultraviolet light $\lambda = 4000$ Å, $\nu = 7.5\cdot 10^{14}$ sec^{-1} ($h\nu = 3.1$ ev) and $\gamma = 1.4\cdot 10^8$ sec^{-1}; $\tau = 1/\gamma = 0.7\cdot 10^{-8}$ sec.

If we take into account the energy losses by radiation in a successive approximation, the oscillator performs not harmonic but damped oscillations of amplitude proportional to $W^{1/2} = W_0^{1/2}e^{-\gamma t/2}$. As a result, the emitted frequency is no longer the natural frequency ν_0, but encompasses the entire frequency spectrum. The spectral composition of the radiation can be found by expanding the oscillator acceleration in a Fourier integral (assuming that for $t < 0$ there is no motion and that $\mathbf{r} = 0$ and $\mathbf{w} = 0$). The energy $S_\nu\, d\nu$ radiated over the total period in the spectral interval $d\nu$ is given by the Fourier

* Using quantum-mechanical concepts, we can rewrite this condition in the form $8\pi^2e^2\nu_0^2/3mc^3 \ll \nu_0$; $h\nu_0 \ll (3/8\pi^2)(hmc^3/e^2) = (3/4\pi)(hc/2\pi e^2)mc^2 = (3/4\pi)$"137"$mc^2 =$ 163 Mev.

component of the acceleration according to (5.4). Calculations, which can be found in [19], give for $\nu - \nu_0 \ll \nu_0$:

$$S_\nu \, d\nu = \frac{2e^2\nu_0^2}{3mc^3} \frac{W_0}{(\nu - \nu_0)^2 + (\gamma/4\pi)^2} \, d\nu. \tag{5.61}$$

It can be easily verified by integrating this expression over the entire spectrum from $\nu = 0$ to $\nu = \infty$ that the total radiated energy is equal to the initial energy of the oscillator

$$\int_0^\infty S_\nu \, d\nu = \int_0^\infty S \, dt = \int_0^\infty \gamma W_0 e^{-\gamma t} \, dt = W_0 .$$

We can refer to the energy radiated by the oscillator in the frequency interval $d\nu$ per unit time. This quantity is equal to $\gamma S_\nu \, d\nu$, in which case W_0 in (5.61) should be replaced by W—the oscillator energy at the given instant of time.

The spectral energy distribution of a damped oscillator, given by (5.61), is shown in Fig. 5.8. The half width of the peak, the so-called natural line width, the meaning of which is clear from Fig. 5.8, is equal to $\Delta\nu = \gamma/2\pi$. In the wavelength scale the natural line width is independent of the wavelength and is equal to $\Delta\lambda = c\,\Delta\nu/\nu_0^2 = \frac{4}{3}\pi e^2/mc^2 = \frac{4}{3}\pi r_0 = 1.2 \cdot 10^{-4}$ Å ($r_0 = e^2/mc^2 = 2.8 \cdot 10^{-13}$ cm is the "electron radius").

We have considered above the spontaneous emission of light by an initially excited oscillator. Let us now assume that a monochromatic light wave of frequency ν and with an amplitude independent of time is incident on the

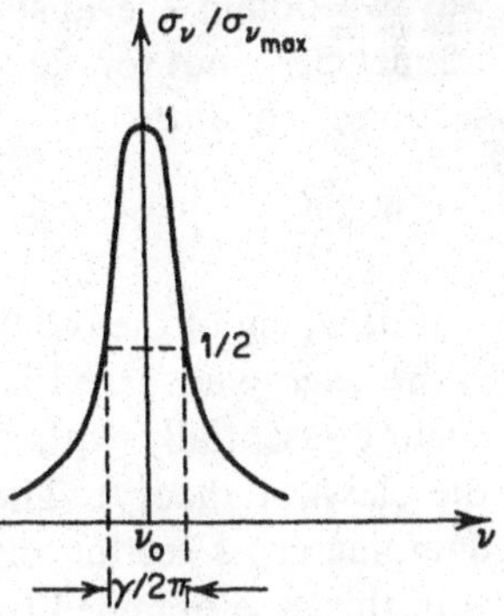

Fig. 5.8. Shape of the absorption curve.

oscillator. Under the action of the electric field of the wave, the elastically bound electron will undergo forced oscillations. If there were no damping, the light wave, shortly after it had been "switched on", would excite the oscillator transferring to it a certain amount of energy, and after that (on the average with respect to time) would no longer perform any work. However, if damping is present, the forced oscillations are accompanied by a continuous radiation of energy by the oscillator. This energy is provided by the work performed by the external field.

Let us find the work done on the oscillator by the periodic field of a light wave. In order to do this we solve the equation of motion for the oscillator

$$m\ddot{\mathbf{r}} + m(2\pi\nu_0)^2\mathbf{r} + m\gamma\dot{\mathbf{r}} = e\mathbf{E}_0 e^{i2\pi\nu t}.$$

Here $\mathbf{E}_0$ is the amplitude of the electric field intensity. The term $m\gamma\dot{\mathbf{r}}$ accounts for the "frictional force" due to damping. The solution of this equation is

$$\mathbf{r} = \mathbf{r}_0 e^{i2\pi\nu t}, \qquad \mathbf{r}_0 = \frac{1}{4\pi^2}\frac{e}{m}\mathbf{E}_0 \frac{1}{\nu_0^2 - \nu^2 + i\nu(\gamma/2\pi)}. \tag{5.62}$$

The work done by the external force per unit time is equal to the scalar product of the force and the velocity $\dot{\mathbf{r}}$. Multiplying the equation of motion by $\dot{\mathbf{r}}$ and averaging with respect to time (as a result of which the terms $\langle\ddot{\mathbf{r}}\cdot\dot{\mathbf{r}}\rangle$ and $\langle\mathbf{r}\cdot\dot{\mathbf{r}}\rangle$ disappear) we obtain for the work done per unit time

$$\langle e\mathbf{E}_0 e^{i2\pi\nu t}\cdot\dot{\mathbf{r}}\rangle = 2\pi^2 m\gamma\nu^2|\mathbf{r}_0^2|. \tag{5.63}$$

As may be seen, this quantity is determined by the modulus of the complex quantity $\mathbf{r}^2$. This work is equal to the energy received by the oscillator from the light wave per unit time, the energy absorbed by the oscillator.

Leaving aside for the time being the question of the future fate of the absorbed energy, let us calculate the absorption cross section. It is, by definition, equal to the energy absorbed per unit time divided by the time-averaged energy flux of the light wave. The average flux is equal to $(c/8\pi)\mathbf{E}_0^2$. In this way we obtain the absorption cross section for light of frequency ν. For frequencies ν not too far from the resonant frequency $|\nu - \nu_0| \ll \nu_0$, the cross section is equal to

$$\sigma_\nu = \frac{e^2}{mc}\frac{\gamma}{4\pi}\frac{1}{(\nu - \nu_0)^2 + (\gamma/4\pi)^2}. \tag{5.64}$$

If damping of the oscillations is attributable to the radiation only, then the entire energy absorbed is used for the emission of light. In this case we are dealing essentially with the scattering of light and not with absorption (in the classical theory). The damping constant is then given by (5.60)*. The damping cross section of an incident light wave attenuated by an oscillator is in this case obtained from (5.64)

$$\sigma_\nu = \frac{4.23\cdot 10^{20}}{\nu_0^2}\frac{1}{1+\xi^2}\,\mathrm{cm}^2 = \frac{7.2\cdot 10^{-9}}{(h\nu_{\mathrm{ev}})^2}\frac{1}{1+\xi^2}\,\mathrm{cm}^2, \qquad \xi = \frac{\nu - \nu_0}{\gamma/4\pi}. \tag{5.65}$$

* Substituting the solution (5.62) into (5.58), we obtain $S = \frac{2}{3}(e^2/c^3)(2\pi\nu)^4|\mathbf{r}_0^2|/2$. Equating this expression to (5.63) we obtain (5.60) for the damping constant.

At the line center the cross section is given by $\sigma_{\nu_{max}} = (3/2\pi)\lambda^2$ or $\sigma_{\nu_{max}} = 7.2 \cdot 10^{-9}/(h\nu_{ev})^2$ cm^2 ($\lambda = c/\nu$ is the wavelength of the incident light). This cross section is very large. For visible light $h\nu \sim 2$–3 ev and $\sigma_{\nu_{max}} \sim 10^{-9}$ cm^2 which corresponds to a light mean free path $l \sim 10^{-10}$ cm for an atmospheric atomic density of $N \sim 10^{19}$ cm^{-3}.

The excited oscillator can also lose its energy by interatomic collisions. In this case the absorbed energy of the light wave is partially transformed into heat. It can be shown (see [19]) that the motion of the oscillator is in this case also described by (5.62), except that γ is now understood to denote not the natural line width (5.60) but the sum of the natural line width and the quantity $2/\tau_{col}$, where τ_{col} is the average time between collisions resulting in damping of the oscillator. Similarly, the form of (5.64) for the absorption cross section will remain unaltered if γ is understood to denote the total width of the line, broadened as a result of collisions.

The fate of the absorbed light energy is determined by the relationship between the natural line width γ and the reciprocal of the time between collisions $2/\tau_{col}$. If $\gamma \gg 2/\tau_{col}$, which occurs in a very rarefied gas, then the absorbed energy is re-emitted (the light is scattered); on the other hand, if $\gamma \ll 2/\tau_{col}$ the energy is basically transformed into heat (absorption in the literal sense of the word). There also exist other mechanisms for broadening spectral lines in gases (see [10, 53,54]) . Let the "atom" contain f_k oscillators, each with a natural frequency ν_{0k}, and let the atom number density be equal to N. Then, the total absorption coefficient for light of frequency ν is

$$\kappa_\nu = N \sum_k f_k \sigma_{\nu k}. \tag{5.66}$$

Usually the different lines ν_{0k} are separated from each other by a distance much greater than a line width. The major role in the absorption of light of a given frequency is played by oscillators with a natural frequency ν_0 which is close to the absorbed frequency; then, in fact, only one term will remain in the summation (5.66). Since the lines are extremely narrow, only the frequencies that are very close to the natural frequencies of the oscillators are absorbed, and the absorption is selective. Let us assume that a continuous radiation spectrum with energy density U_ν is incident upon the atoms, in which, as is usually the case, the changes in U_ν are small over a spectral interval of the order of a line width. The total energy absorbed per unit time per unit volume by oscillators with a frequency ν_0 is equal to $\int_0^\infty U_\nu \, d\nu \, cN\sigma_\nu f = U_\nu cNf \int_0^\infty \sigma_\nu \, d\nu$ (the subscript k has been dropped). The absorption per atom is characterized by a quantity obtained by integrating the cross section (5.64). The integral of the cross section with respect to frequency for a single line, i.e., the area of the spectral line, is equal to

$$f \int_0^\infty \sigma_\nu \, d\nu = \frac{\pi e^2}{mc} f = 2.64 \cdot 10^{-2} f \text{ cm}^2 \cdot \text{sec}^{-1}. \tag{5.67}$$

This is a constant depending only on the number of oscillators f and independent of the line width. Hence, if the line is broadened by collisions, for example, then the cross section will be smaller than that for a line with a natural width.

The absorption of light by an oscillator is frequency-dependent in the same manner as is the radiation (see (5.61) and (5.64)). This is in agreement with the principle of detailed balancing, which is easily shown to be satisfied in this case*.

§10. Quantum theory of spectral lines. Oscillator strength

Let us consider radiation and absorption of light from the quantum-mechanical point of view. A strong parallelism exists between the results of the quantum and classical theories. In the zeroth approximation of the quantum theory for an atom, that corresponding to stationary states, only strictly defined energy levels are possible (analogous to the constancy of the energy for the undamped vibrations of a classical oscillator). In the next approximation the possibility of transitions between energy states of the atom appears. By virtue of the fact that the states are nonstationary, the uncertainty principle requires that the value of the energy corresponding to each energy level (with the exception of the ground level) is uncertain by an amount $\Delta E \sim h/\Delta t$, where Δt is the "lifetime" of the atom in the state under consideration, equal to the reciprocal of the probability of a spontaneous transition to lower levels. However, the broadening of energy levels also results in the broadening of the spectral lines by an amount of the order of $\Delta\nu \sim \Delta E/h \sim 1/\Delta t$, i.e., of the order of the "damping" constant $1/\Delta t$, just as in the classical theory. The width of the nth energy level, according to the above statements, is equal to the sum of the probabilities of transitions to all lower levels

$$\Gamma_n = \sum_{n'} A_{nn'}, \tag{5.68}$$

where $A_{nn'}$ sec^{-1} is the probability of the spontaneous transition $n \to n'$, the so-called Einstein coefficient for emission.

The radiation rate given by quantum mechanics is

$$S = h\nu_{nn'}A_{nn'} = \frac{64\pi^4}{3}\frac{\nu_{nn'}^4}{c^3}|\mathbf{d}|^2, \tag{5.69}$$

where $|\mathbf{d}|$ is the matrix element of the dipole moment. Equation (5.69) is very similar to the classical expression (5.58), the difference lying only in the

* $\gamma S_\nu\, d\nu = U_\nu c\, d\nu\, \sigma_\nu$; this relationship is satisfied by substituting the thermodynamic equilibrium values of the energy and the radiation density of the oscillator (three-dimensional) W and U_ν, calculated either from classical theory ($W = 3kT$, $U_\nu = 8\pi\nu^2kT/c^3$) or according to quantum theory.

replacement of the mean square of the dipole moment by twice the square of the matrix element of the same dipole moment. The numerical values of the emission probability $A_{nn'}$, are of the same order of magnitude as the classical "probability", as the damping constant γ.

The matrix element $|\mathbf{d}|$ is calculated for a transition between two completely defined quantum states. In the case of degeneracy of the energy levels we can turn our attention to the average probability of transition from one level to the other. For this we must sum $|\mathbf{d}|^2$ over the end states and average over the initial states. If α and α' are the quantum numbers (or sets of numbers), which correspond to levels n and n', then in general the probability of transition between the levels is

$$A_{nn'} = \frac{64\pi^4}{3}\frac{\nu_{nn'}^3}{hc^3}\left\{\frac{1}{g_n}\sum_{\alpha\alpha'}(n', \alpha'|\mathbf{d}|n, \alpha)^2\right\}, \tag{5.69'}$$

where g_n is the statistical weight of the nth level. We can also consider transitions between particular groups of states belonging to the given levels. For example, in the case of the hydrogen atom this will refer to the probabilities of transitions $n, l \to n', l'$ (l' is the orbital quantum number). In this case the summation in (5.69′) should be carried out only over the magnetic quantum numbers m and m' (correspondingly, g_n is replaced by g_{nl}).

Table 5.3

TRANSITION PROBABILITIES IN A HYDROGEN ATOM IN UNITS OF 10^8 SEC^{-1}

Initial state	Final state	$n = 1$	$n = 2$	Total	Lifetime, 10^{-8} sec
2s	np	—		0	
2p	ns	6.25		6.25	0.16
2	Average	4.69		4.69	0.21
3s	np	—	0.063	0.063	16
3p	ns	1.64	0.22	1.86	0.54
3d	np	—	0.64	0.64	1.56
3	Average	0.55	0.43	0.98	1.02

Table 5.3 presents the probabilities for some transitions in the hydrogen atom* (for an energy level diagram see Figs. 2.2 and 5.9). Knowing the probability coefficients $A_{nn'}$ it is easy to calculate the intensities of the corresponding emission lines. Namely, if N_n is the atom number density in the nth

* These data are taken from the book by Bethe and Salpeter [5].

excited state (which can be calculated from the Boltzmann equation), then the energy emitted in the line $\nu_{nn'}$ per unit volume per unit time is equal to $N_n A_{nn'} h\nu_{nn'}$.

The principle of detailed balancing establishes a relationship between the probabilities of light emission and absorption for the given transition $n \leftrightarrows n'$. The energy absorbed per unit time per unit volume by atoms in the n' state upon their transition to the nth state is

$$\int U_\nu c\, d\nu\, \sigma_{\nu n'n} N_{n'} = N_{n'} U_\nu c \int \sigma_{\nu n'n}\, d\nu = N_{n'} U_\nu c h\nu_{nn'} B_{n'n},$$

where $\sigma_{\nu n'n}$ is the absorption cross section for the frequency ν within the limits of the given transition $n' \to n$, and $B_{n'n}$ is a coefficient characterizing the total absorption in the given line (the so-called Einstein coefficient for absorption). This coefficient is proportional to the "area" of the line

$$B_{n'n} = \frac{1}{h\nu_{nn'}} \int \sigma_{\nu n'n}\, d\nu. \tag{5.70}$$

Multiplying the absorption rate by $(1 - e^{-h\nu/kT})$ in order to account for the induced emission (see §4 of Chapter II), equating the resulting expression to the emission rate, substituting the radiation density U_ν given by the Planck formula, and the number of atoms N_n given by the Boltzmann equation, we obtain the relationship between the Einstein coefficients*

$$B_{n'n} = \frac{c^2}{8\pi h\nu_{nn'}^3} \frac{g_n}{g_{n'}} A_{nn'}. \tag{5.71}$$

It is conventional to characterize the emissivity of an atom in a given line $\nu_{nn'}$ determined by the area of the line $\int \sigma\nu_{n'n}\, d\nu$ by a number which is equal to that number of classical oscillators with natural frequency $\nu_{nn'}$ that would yield the same effect as the atom under consideration. This number $f_{\text{abs}_{n'n}}$ is called the oscillator strength for absorption and is no longer an integer. Equating the line areas given by (5.70) and (5.67) and noting the relation

*Editors' note. It should be noted that the relation (5.71) connecting the Einstein coefficients is completely independent of any temperature, even though the derivation has been based upon equilibrium concepts. In fact, induced emission can be considered as the process of absorption with the time reversed, and spontaneous emission can be considered as induced emission induced by the zero-point quantum fluctuations in the electric and magnetic fields, so that all three processes are related microscopically in a fundamental way. The Einstein relation (5.71) can be derived on the basis of microscopic reversibility, from the invariance of a quantum matrix element under time reversal. The derivation given here is essentially the classical one of Einstein, and has the advantage that it is convenient and that it sheds light on the role of induced emission. The fact that (5.71) does not depend upon any concepts of equilibrium should be kept in mind. See also the footnote following (2.22) in Chapter II.

(5.60), we find the relationship between the oscillator strength and the Einstein coefficient

$$f_{\mathrm{abs}_{n'n}} = \frac{mc}{\pi e^2} h\nu_{nn'} B_{n'n}. \tag{5.72}$$

This relation essentially defines the concept of the oscillator strength.

Together with $f_{\mathrm{abs}_{n'n}}$ there is introduced a negative oscillator strength for emission (the transition $n \to n'$) defined by*

$$f_{\mathrm{emiss}_{nn'}} = -\frac{g_{n'}}{g_n} f_{\mathrm{abs}_{n'n}} = -\frac{A_{nn'}}{3\gamma} \tag{5.73}$$

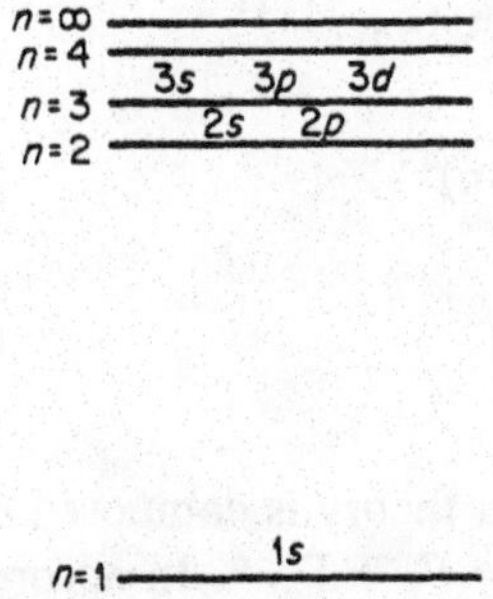

Fig. 5.9. Energy level diagram for the hydrogen atom.

(the second equality is obtained from (5.72) with the use of (5.71)). The oscillator strengths can be expressed directly in terms of matrix elements of the dipole moment. Using (5.73), (5.69′), and (5.60) we find

$$g_{n'} f_{\mathrm{abs}_{n'n}} = -g_n f_{\mathrm{emiss}_{nn'}} = \frac{8\pi^2 m\nu_{nn'}}{3e^2 h} \sum_{\alpha\alpha'} (n', \alpha'|\mathbf{d}|n, \alpha)^2. \tag{5.73′}$$

As can be seen, the quantities $g_n f_{nn'}$, where the first subscript on f corresponds to the initial level, are the same for emission and absorption and, consequently, are symmetric with respect to transposition of numbers which characterize the initial and final levels. In what follows we shall always use the positive oscillator strength for absorption, dropping the subscript "abs" and denoting the initial state by the first subscript with $f_{nn'}$, $n \to n'$, $E_{n'} > E_n$.

For the frequency distribution of the absorption within the limits of the line, the quantum theory predicts the same dependence of the probability of absorption on frequency as does the classical equation for the cross section

* We note that the number f represents the average strength of an oscillator per degree of freedom of an electron. The total oscillator strength is three times greater because the electron in an atom has three degrees of freedom. The total oscillator strength for emission $|3f_{\mathrm{emiss}}| = A_{nn'}/\gamma$ represents the ratio of the quantum "damping constant" (i.e., the probability of transition) to the classical one γ.

σ_ν. By appropriately normalizing the probability, we can write the quantum-mechanical equation for the absorption cross section in a form analogous to the classical equation (5.64) (we interchange the subscripts n and n', and we denote the lower state, from which a transition accompanied by the absorption of photon takes place, by n)

$$\sigma_{\nu nn'} = \frac{e^2}{mc} \frac{\Gamma_{nn'}}{4\pi} f_{nn'} \frac{1}{(\nu - \nu_{nn'})^2 + (\Gamma_{nn'}/4\pi)^2}. \tag{5.74}$$

If the value of $f_{nn'}$ given by (5.73) is substituted into (5.74) using the expression for γ and noting that $c/\nu = \lambda$, then the cross section can be written as

$$\sigma_{\nu nn'} = \frac{\lambda^2}{8\pi^2} \frac{g_n}{g_{n'}} A_{n'n} \frac{\Gamma_{nn'}}{4\pi} \frac{1}{(\nu - \nu_0)^2 + (\Gamma_{nn'}/4\pi)^2}$$

$$= \sigma_{\nu nn' \max} \frac{(\Gamma_{nn'}/4\pi)^2}{(\nu - \nu_0)^2 + (\Gamma_{nn'}/4\pi)^2},$$

where the cross section at the line center is

$$\sigma_{\nu nn' \max} = \frac{\lambda^2}{2\pi} \frac{g_n}{g_{n'}} \frac{A_{n'n}}{\Gamma_{nn'}}.$$

The (natural) line width given by the quantum theory is composed of the sum of the transition probabilities (5.68): $\Gamma_{nn'} = \Gamma_n + \Gamma_{n'}$ *. In accordance with the definitions of oscillator strength and the Einstein coefficient $B_{nn'}$ in (5.72) and (5.70) the area of the line is

$$\int \sigma_{\nu nn'}\, d\nu = \frac{\pi e^2}{mc} f_{nn'} = 2.65 \cdot 10^{-2} f_{nn'} \text{ cm}^2 \cdot \text{sec}^{-1}.$$

The area is entirely independent of the line width $\Gamma_{nn'}$, which in the presence of collisions also includes the term $2/\tau_{\text{col}}$. This result is quite natural, since the principle of detailed balancing uniquely relates the area of a line to the probability of spontaneous emission which is determined solely by the structure of the atom itself and which obviously cannot depend on such external factors as atomic collisions.

In a real gas there are usually several factors responsible for the broadening of the spectral lines; they include particle collisions, the Doppler effect, and the Stark effect. Thus, the broadening due to collisions increases the natural width γ by an amount equal to twice the collision probability, $\gamma_{\text{col}} = 2/\tau_{\text{col}}$. The Doppler broadening is approximately equal to $\Delta\nu = \nu\bar{v}/c$, where $\bar{v}$ is the thermal speed. We shall not consider the details of this problem here; see [10, 53, 54].

* The widths of the levels Γ_n and $\Gamma_{n'}$ also include the probabilities corresponding to induced emission. These terms are proportional to the emission density and are important only at sufficiently high densities.

§11. The absorption spectrum of hydrogen-like atoms. Remarks on the effect of spectral lines on the Rosseland mean free path

Let us assume that light with a continuous spectrum in which all frequencies are present is incident upon a gas made up of hydrogen-like atoms (in particular, atomic hydrogen). We shall find those frequencies that will be absorbed by the atoms in a definite nth state, and the absorption intensity. The atoms absorb selectively the frequencies $\nu_{nn'}$, corresponding to electronic transitions from the nth level to a more highly excited level $n' > n$. Noting formula (5.25) for the energy of a level, we find the relationship between these frequencies and the quantum numbers n and n', the so-called Balmer series formula

$$\nu_{nn'} = \frac{I_H Z^2}{h}\left(\frac{1}{n^2} - \frac{1}{n'^2}\right) = \nu_1\left(\frac{1}{n^2} - \frac{1}{n'^2}\right), \tag{5.75}$$

where $\nu_1 = I_H Z^2/h = \nu_R Z^2$. The frequency $\nu_R = I_H/h = 3.29 \cdot 10^{15}$ sec^{-1} corresponds to the ionization potential of a hydrogen atom. It is frequently used as a unit of frequency and is called a "rydberg". As n' increases, the spacing between the levels and correspondingly also between the lines $\nu_{nn'}$ becomes closer and, as $n' \to \infty$, a continuum (continuous spectrum) is formed, because the absorption of frequencies exceeding the upper limit of the series $\nu_n = \nu_{n,\infty} = \nu_1/n^2$ results in ionization, and the final state of the electron is then found in the continuous part of the energy spectrum. The absorption spectrum from the given energy level n is shown in Fig. 5.10 (which, for comparison, also contains a schematic diagram of the energy levels). To be specific,

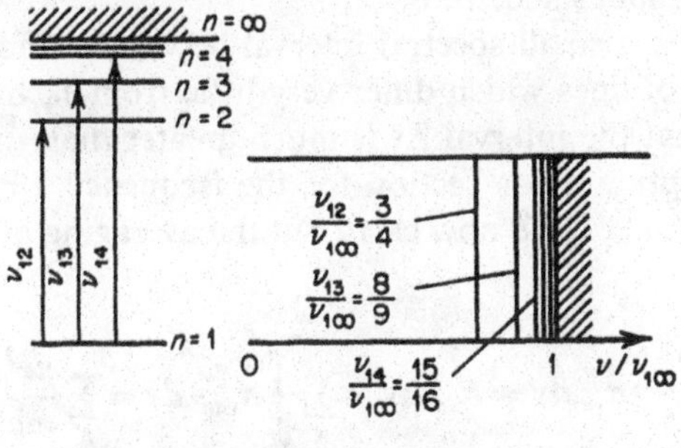

Fig. 5.10. Absorption spectrum for hydrogen atoms in their ground state. The diagram at the left illustrates the transitions.

we have assumed $n = 1$, that is, Fig. 5.10 represents the absorption spectrum of a cold gas consisting of hydrogen-like atoms, in which all the atoms are in the ground state. In a heated gas there are excited levels and the absorption spectrum represents a set of series corresponding to the absorption by atoms in different states.

Near the upper limit of the series, where the lines become strongly crowded, overlapping of the individual lines begins. This occurs when the frequency distance between the lines, which decreases very rapidly as $n' \to \infty$, becomes comparable to the line width. Overlapping of lines is promoted by their broadening due to collisions, Doppler effect, etc. Usually overlapping of atomic lines begins at such large quantum numbers n' and so close to the upper limit of the series $\nu_n = \nu_{n\infty}$ that the entire frequency region of overlapping levels is very narrow and is practically of no importance. In a real atomic gas the overlapping region also does not exist because the upper levels are cut off on account of the interaction between atoms and the effective decrease in the ionization potential. Actual overlapping of the individual lines arises only for the absorption of light by molecules, where the number of lines is much greater than in atoms and where they are spaced much closer to each other (this is discussed later).

Let us consider transitions $n \to n'$ with the absorption of light between high levels with large quantum numbers. The motion of an electron in these levels is quasi-classical and the absorption of light, accompanied by transitions $n \to n'$, with both n and $n' \gg 1$, can be studied by means of the semi-classical concepts. In the spectral region corresponding to transitions with both n and $n' \gg 1$, where the lines are very close to each other and almost overlap, it seems natural to smooth out the dependence of the absorption cross section on frequency by introducing an average cross section. The averaging should be carried out in such a manner as to leave unchanged the total area of the lines which characterizes the attenuation of the external radiation flux with a continuous spectrum.

Let us consider a small spectral interval between ν and $\nu + \Delta\nu$ containing a large number of lines which differ very little from each other. In addition, let us assume that the interval $\Delta\nu$ is much greater than the width of a single line. The absorption cross section for the frequency ν by atoms in the nth state is $\sigma_{\nu n} = \sum_{n'} \sigma_{\nu n n'}$. We now carry out the averaging of the cross section in the interval $\Delta\nu$:

$$\int_{\nu}^{\nu+\Delta\nu} \sigma_{\nu n}\, d\nu = \bar{\sigma}_{\nu n}\, \Delta\nu = \sum_{n'} \int \sigma_{\nu n n'}\, d\nu = \sum_{n'} \frac{\pi e^2}{mc} f_{nn'}.$$

Let us also find the average oscillator strength by determining the average value of $\bar{f}_{nn'} = f_{n\bar{n}'} = f_n(\nu)$ for the given interval $\Delta\nu$. If the frequency interval between ν and $\nu + \Delta\nu$ contains $\Delta n'$ lines corresponding to the final states from n' to $n' + \Delta n'$, then the average cross section can be written

$$\bar{\sigma}_{\nu n} = \frac{\pi e^2}{mc} f_n(\nu) \frac{\Delta n'}{\Delta\nu}. \tag{5.76}$$

The number of lines per unit spectral interval can be obtained by differentiating Balmer's formula (5.75)

$$\frac{\Delta n'}{\Delta \nu} = \left(\frac{d\nu_{nn'}}{dn'}\right)^{-1} = \left(\frac{2\nu_1}{n'^3}\right)^{-1}. \tag{5.77}$$

In §4 we found the cross section for bound-free transitions by extending the classical expression for the effective radiation for free-free transitions to the case when one of the states is in the discrete spectrum. The justification for this was that the motion of an electron in states with a high quantum number n is quasi-classical, and that the motion in an "elliptical" orbit (corresponding to large n and to small negative energy) is very close to the motion in a "hyperbolic" orbit (corresponding to small positive energy). Let us take one further step and consider in the same approximation the case when both states are in the discrete spectrum with large quantum numbers. We consider transitions from the nth level occurring with the absorption of a photon, on the basis of semiclassical concepts. As the frequency increases the electron in its final state will lie in "elliptical" orbits approaching a parabolic shape; for $\nu = \nu_n$ it will lie in a parabolic orbit, and for a frequency ν only slightly exceeding ν_n it will lie in "hyperbolic" orbits which are close to parabolic. Since the motion of an electron in its final state changes continuously, it should also be expected that the average absorption cross section by atoms in the nth state $\bar{\sigma}_{\nu n}$ will also be continuous upon transition from the discrete to continuous spectrum (see Fig. 5.11).

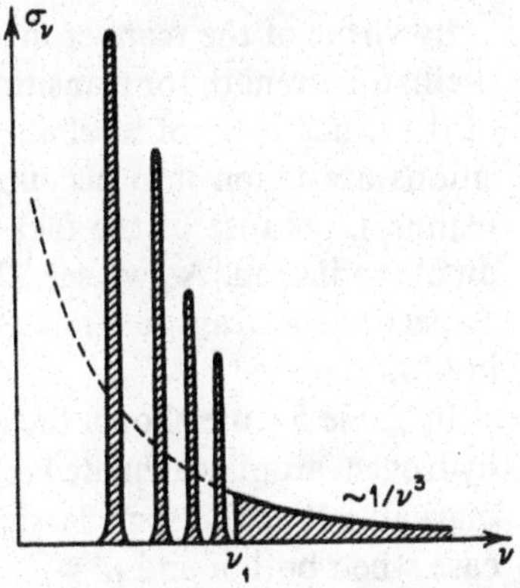

Fig. 5.11. Cross section for the absorption of light by hydrogen atoms from the ground state. Transition of the discrete spectrum into a continuum. The dashed line shows the cross section in the region of the discrete spectrum averaged with respect to the lines. The diagram is schematic.

Let us extend (5.34) for the photoionization cross section from the nth level to include the absorption of frequencies slightly lower than the limiting frequency for the photoionization* ν_n and let us equate the cross section

* In a manner similar to that in §4 in which we extended the expression for the effective bremsstrahlung emission to frequencies slightly exceeding the maximum possible frequency for free-free transitions; this allowed us to describe radiative capture.

(5.34) to the expression for the average cross section for bound-bound transitions given by (5.76). Recalling the definition of the ionization potential for a hydrogen atom I_H given by (5.25) and the expression for the limiting frequency of the series $\nu_n = \nu_1/n^2$ (see (5.75)), we find the average oscillator strength $f_n(\nu)$ for the transition from the nth level to one of the n' levels included in the narrow interval $\Delta n'$, $\Delta\nu$. Denoting the oscillator strength by $f_{nn'}$ and the frequency ν by $\nu_{nn'}$, we get

$$f_{nn'} = \frac{16}{3\pi\sqrt{3}} \frac{1}{n^5} \left(\frac{\nu_1}{\nu_{nn'}}\right)^2 \frac{1}{\nu_{nn'}} \frac{\Delta\nu}{\Delta n'}.$$

Substituting here the average spacing between the levels $\Delta\nu/\Delta n'$ calculated from (5.77), and replacing the transition frequency $\nu_{nn'}$ through Balmer's formula (5.75), we finally obtain the oscillator strength $f_{nn'}$ for the transition $n \to n'$

$$f_{nn'} = \frac{32}{3\pi\sqrt{3}} \frac{1}{n^5} \frac{1}{n'^3} \frac{1}{(1/n^2 - 1/n'^2)^3}. \tag{5.78}$$

For transitions from levels $n' \gg n$ we find the asymptotic equation

$$f_{nn'} = \frac{32}{3\pi\sqrt{3}} \frac{n}{n'^3} = \frac{1.96n}{n'^3}, \qquad n' \gg n. \tag{5.79}$$

By virtue of the manner in which it was derived, $f_{nn'}$ represents the average oscillator strength for transition from any given state l, m for a given n to any of the states l', m' of level n'. In this case the selection rules for dipole transitions are taken into account automatically (of course, in an approximate manner), because of the fact that we started from the classical formula for dipole radiation. As we see, the quantity $g_n f_{nn'} = 2n^2 f_{nn'}$ is symmetrical with respect to the transposition of n and n', in accordance with what was said in §10.

In Table 5.4 are shown the oscillator strengths for several transitions in the hydrogen atom, calculated by quantum-mechanical methods [5]. It is remarkable that the semiclassical equations (5.78) and (5.79) derived for the case when both n and $n' \gg 1$, give a fairly good estimate even for transitions between levels with quantum numbers which are not large, including transitions from the ground state. For example, the semiclassical values are $f_{12} = 0.585$, $f_{13} = 0.104$ and the asymptotic value is $f_{1n'} = 1.96n'^{-3}$, while the table gives $f_{12} = 0.416$, $f_{13} = 0.079$ and for the asymptotic case $f_{1n'} = 1.6n'^{-3}$. Here we encounter the same situation as in the case when we compared the semiclassical and the quantum-mechanical photoionization cross sections from the ground level of the hydrogen-like atom.

Table 5.4

OSCILLATOR STRENGTHS FOR A HYDROGEN ATOM

Initial state	1s	2s	2p	
Final state	*np*	*np*	*ns*	*nd*
$n=1$	—	—	−0.139	—
2	0.4162	—	—	—
3	0.0791	0.425	0.014	0.694
4	0.0290	0.102	0.0031	0.122
5	0.0139	0.042	0.0012	0.044
6	0.0078	0.022	0.0006	0.022
7	0.0048	0.013	0.0003	0.012
8	0.0032	0.008	0.0002	0.008
from $n=9$ to ∞, Σ	0.0101	0.026	0.0007	0.053
Asymptotic formula	$1.6 \cdot n^{-3}$	$3.7 \cdot n^{-3}$	$0.1 \cdot n^{-3}$	$3.3 \cdot n^{-3}$
Line spectrum	0.5641	0.638	−0.119	0.923
Continuous spectrum	0.4359	0.362	0.008	0.188
Total	1.000	1.000	−0.111	1.111

Negative oscillator strengths correspond to transitions accompanied by the emission of a photon.

Under some conditions the absorption lines of atoms can have an appreciable effect on the Rosseland mean free path. The main contribution to the mean free path comes from spectral intervals with a small continuous absorption coefficient which are located in the region of the maximum of the weighting function (see §7, Fig. 5.7). These are intervals which come before the boundaries of the series, i.e., before the beginnings of the corresponding continua. Spectral lines appear in these intervals. Since the absorption at the line centers is usually quite strong, the corresponding frequency intervals are practically cut out from the integral over the spectrum, as is shown in Fig. 5.12. If the lines are narrow, then the width of intervals cut out is small. However, in a sufficiently dense gas, where the lines are extremely wide, the intervals cut out and the reduction in the Rosseland mean free path can be rather significant. According to Biberman and Lagar'kov [51], in hydrogen at densities of 10^{17}–10^{19} atoms/cm^3 and temperatures of 12,000–20,000°K the line absorption can reduce the Rosseland mean free path by a factor of 2–4 in comparison with the mean free path calculated without taking the lines into account.

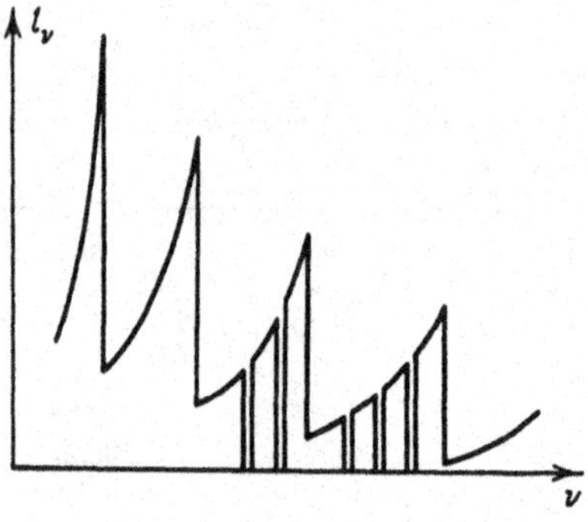

Fig. 5.12. The effect of lines on the mean free path.

§12. Oscillator strengths for the continuum. The sum rule

We have seen in the preceding sections that the probability of transitions between discrete atomic levels accompanied by the absorption of photons is described in terms of the oscillator strengths. The oscillator strength determines the area of the absorption line, i.e., the integral with respect to frequency of the cross section for the absorption of light of frequency ν in the given line. Analogously, we can introduce the concept of an oscillator strength for bound-free transitions. In this case the quantity f_n characterizes the integral with respect to frequency of the cross section for the absorption of light accompanied by the transition of an electron from the nth level of the atom to the continuous spectrum. If $\sigma_{\nu n}$ is the cross section for bound-free absorption at the frequency ν, then for such a transition

$$\int_{\nu_n}^{\infty} \sigma_{\nu n}\, d\nu = \frac{\pi e^2}{mc} f_n, \tag{5.80}$$

where the integration with respect to frequencies is carried out starting from the lowest frequency ν_n at which a transition to the continuous spectrum is possible.

Let us calculate the oscillator strength f_n for bound-free absorption by hydrogen-like atoms. Using the semiclassical equation (5.34) for $\sigma_{\nu n}$ and noting that $\nu_n = I_H Z^2/hn^2$, we obtain after integration

$$f_n = \frac{8}{3\pi\sqrt{3}}\frac{1}{n} = \frac{0.49}{n}. \tag{5.81}$$

The results of quantum-mechanical calculations for the hydrogen atom are given in Table 5.4. For example, for $n = 1$ the exact value is $f_1 = 0.436$, while (5.81) gives $f_1 = 0.49$.

In the classical theory each electron participating in the emission and absorption of light is replaced by an oscillator. The total number of oscillators is, consequently, simply equal to the number of electrons in the atom. The quantum analog of this situation is found in the sum rule for oscillator strengths, according to which the sum $\sum_{n'} f_{nn'}$ over all possible transitions in the atom from a given state n is equal to the number of electrons. If we limit ourselves to transitions with the participation of outer optical electrons only, then the above sum is equal to the number of electrons. In particular, in the case of the hydrogen-like atom the sum is equal to unity. The summation over the final states also includes the transitions to the continuous spectrum, included in the term f_n. This term, as we shall see below, can be represented as an integral over the final states of the continuous spectrum. In addition, the sum also includes negative terms corresponding to the transitions to still lower levels $n' < n$, that is, to transitions accompanied by the emission of light (see [5]). The data in Table 5.4 satisfy, of course, the sum rule, which can be verified by direct calculation.

In describing bound-free transitions (the continuum) and also bound-bound transitions between closely spaced levels in the molecular band spectra (quasi-continuum) one frequently uses the concept of a differential oscillator strength or oscillator strength per unit frequency interval. The differential oscillator strength $df/d\nu$ is defined as follows: If σ_ν is the absorption cross section of the frequency ν for the transition from the nth level, then

$$\sigma_{\nu n} = \frac{\pi e^2}{mc}\left(\frac{df}{d\nu}\right)_n = 2.64 \cdot 10^{-2}\left(\frac{df}{d\nu}\right)_n \text{cm}^2 = 8 \cdot 10^{-18}\left[\frac{df}{d(\nu/\nu_R)}\right]_n \text{cm}^2 \quad (5.82)$$

(ν/ν_R is the frequency expressed in rydbergs). Hence, the total oscillator strength for the entire continuum can be defined in accordance with (5.80) as

$$\int_{\nu_n}^{\infty} \sigma_\nu \, d\nu = \frac{\pi e^2}{mc}\int_{\nu_n}^{\infty}\left(\frac{df}{d\nu}\right)_n d\nu = \frac{\pi e^2}{mc} f_n. \quad (5.83)$$

Let us calculate the differential oscillator strength for bound-free transition from the nth level of a hydrogen-like atom. Comparing the formula (5.34) and the definition (5.82), we find

$$\left(\frac{df}{d\nu}\right)_n = \frac{16}{3\pi\sqrt{3}}\frac{1}{n}\frac{\nu_n^2}{\nu^3} = \frac{0.98}{n}\frac{\nu_n^2}{\nu^3}, \qquad \nu_n = \frac{I_H Z^2}{hn^2}. \quad (5.84)$$

Integrating this expression with respect to ν from ν_n to ∞, naturally, we obtain (5.81).

If the absorption spectrum is an aggregate of many lines, then the cross section $\sigma_{\nu n}$ is understood to denote the average value $\bar{\sigma}_{\nu n}$ (see (5.76)), and the differential oscillator strength is equal to the average oscillator strength for one transition multiplied by the number of lines per unit frequency interval

$$\left(\frac{df}{d\nu}\right)_n = f_n(\nu)\frac{\Delta n'}{\Delta \nu} = f_n(\nu)\frac{dn'}{d\nu} = f_{nn'}\frac{dn'}{d\nu}. \quad (5.85)$$

Table 5.5, taken from Unsöld's book [10], shows the oscillator strengths for hydrogen and alkali metal atoms for the continuous spectrum corresponding to absorption from the ground level. It also includes the values of the differential oscillator strengths at the absorption edge $(df/d\nu)_n$ for $\nu = \nu_n$ (ν is expressed in rydbergs). These data were obtained from quantum-mechanical calculations. They show the degree of "nonhydrogen-likeness" of the alkali metal atoms.

Table 5.5

OSCILLATOR STRENGTHS f FOR THE CONTINUOUS SPECTRUM AND THE DIFFERENTIAL STRENGTHS $df/d\nu$ AT THE EDGES OF THE PRINCIPAL SERIES (ν IS IN RYDBERGS)

Atom	λ_{edge}, Å	f	$df/d\nu$	I, ev
H	912	0.436	0.78	13.5
Li	2281	0.24	0.46	5.4
Na	2442	0.0021	0.038	5.05
K	2857		0.0024	4.32

§13. Radiative emission in spectral lines

Let us consider spontaneous radiative transitions in a hydrogen atom and let us calculate approximately the average transition probabilities from the level n to a lower-lying level n'. We start from the general expression of probability (5.73) in terms of the oscillator strength for the absorption $n' \to n$

$$A_{nn'} = 3\gamma \frac{g_{n'}}{g_n} f_{nn'} = \frac{8\pi^2 e^2}{mc^3} \nu_{nn'}^2 \frac{n'^2}{n^2} f_{n'n}.$$

We substitute here $f_{n'n}$ from (5.78), having transposed n and n' in accordance with the fact that n now denotes the upper level. Substituting also the transition frequency $\nu_{nn'} = \nu_1(1/n'^2 - 1/n^2)$, where $\nu_1 = I_H/h$, we obtain

$$A_{nn'} = \frac{8\pi^2 e^2 \nu_1^2}{mc^3} \frac{32}{3\pi\sqrt{3}\, n^5 n'^3} \frac{1}{(1/n'^2 - 1/n^2)} = \frac{1.6 \cdot 10^{10}}{n^3 n'(n^2 - n'^2)} \sec^{-1}.$$

This equation describes with good accuracy not only the transitions for large n and n' but also those between relatively low-lying levels and even to the ground state. We can satisfy ourselves of this by comparing the approximate results with calculations using exact quantum-mechanical values as given in [5], some of which are presented in Table 5.3. ($A_{nn'}$ should be compared with the average probabilities presented in the table.) For example, $A_{51\,\text{appr}} = 5.3 \cdot 10^6$ sec^{-1}, while $A_{51\,\text{exact}} = 4 \cdot 10^6$ sec^{-1}, $A_{21\,\text{appr}} = 6.7 \cdot 10^8$ sec^{-1}, $A_{21\,\text{exact}} = 4.7 \cdot 10^8$ sec^{-1}.

Let us consider the dependence of the transition probability from a given high level $n \gg 1$ to a final level with quantum number n'. For transitions to lower-lying levels with $n' \ll n$, we have approximately

$$A_{n,n' \ll n} \approx \frac{1.6 \cdot 10^{10}}{n^5 n'} \text{ sec}^{-1}.$$

In particular, the probability of a direct transition to the ground state is approximately*

$$A_{n,1} \approx \frac{1.6 \cdot 10^{10}}{n^5} \text{ sec}^{-1}.$$

For transitions to nearby levels with $n' = n - \Delta n$, $\Delta n \ll n$,

$$A_{n,n-\Delta n} \approx \frac{0.8 \cdot 10^{10}}{n^5 \, \Delta n} \text{ sec}^{-1}.$$

As a function of n', $A_{nn'}$ passes through a minimum when $n' = n/\sqrt{3}$, with a value equal to

$$A_{\min} = A_{n,n/\sqrt{3}} = \frac{4.15 \cdot 10^{10}}{n^5 n} \text{ sec}^{-1}.$$

Thus the most probable transition, on the average, for an atom in the nth level ($n \gg 1$), is a direct transition to the ground state $n' = 1$ with complete deexcitation. Transitions to the first excited state ($n' = 2$) and the nearest state ($n' = n - 1$) are also probable:

$$A_{n,2} = A_{n,n-1} = \tfrac{1}{2} A_{n,1}.$$

* This result is quite close to that given by the exact quantum-mechanical calculations. Only atoms which are in p states can make a transition to the ground state $1s$. For large n the probability of the transition $A_{np,1s}$ is (see [5])

$$A_{np,1s} = \frac{8 \cdot 10^9 \cdot 2^8 n(n-1)^{2n-2}}{9(n+1)^{2n+2}}.$$

It is evident that in the limit $n \gg 1$ this quantity is approximately equal to $A_{np,1s} = 8 \cdot 10^9 \cdot 2^8/9n^3$. The average probability of the transition $n \to 1$ is equal to the product of the probabilities of the transition $np \to 1s$ times the probability that an atom with energy E_n is found in the p state ($l = 1$), that is,

$$A_{n,1} = \frac{2l+1}{n^2} A_{np,1s} = \frac{3}{n^2} A_{np,1s}.$$

This gives

$$A_{n,1} = \frac{1.29 \cdot 10^{10}}{n^5} \text{ sec}^{-1},$$

which is quite close to the quasi-classical value.

Transitions to states intermediate between the lower and the neighboring states are less probable. Of course, transitions to lower and neighboring states with close probabilities do not have the same effect. The transition to a neighboring level is accompanied by the emission of a very low energy photon and the energy of the atom changes very little, while the transition to the ground state is accompanied by the emission of a high energy photon and the energy of the atom changes by a large amount.

It might seem that the conversion into radiation of the energy of an atom at a very high excitation level, when the motion of the optical electron is quasi-classical, should be describable on the basis of classical electrodynamics. If we calculate the rate of radiation by an electron traveling in a circular orbit about an ion by means of (5.1), we obtain

$$S = \left(\frac{dE}{dt}\right)_{\text{class}} = \frac{32}{3}\frac{E^4}{m^2c^3e^2},$$

where E is the binding energy of the electron in the atom (the change in the binding energy is equal to the change in the total energy of the electron). This value of dE/dt is found to be $\pi\sqrt{3}/4 = 1.35$ times higher than the quantum-mechanical rate of change of the electron energy corresponding to a radiative transition to the neighboring level only, $(dE/dt)_{n,n-1} = h\nu_{n,n-1} \cdot A_{n,n-1}$, where $n = (I_H/E)^{1/2}$. However, the overwhelming contribution to the actual radiation rate dE/dt is made by the transition to the ground state $(dE/dt)_{n,1} = h\nu_{n,1}A_{n,1}$, which cannot be described by classical electrodynamics*.

Knowing the transition probabilities (and the distribution of atoms in the excited states), we can calculate the emission coefficient of a gas associated with line radiation

$$J = \sum_{n=1}^{n^*} N_n \sum_{n'<n} h\nu_{nn'}A_{nn'},$$

* Nevertheless, classical electrodynamics does provide a reasonable estimate for the lifetime of an atom with respect to the complete radiative deexcitation, which, within the framework of the classical theory, is understood to mean "the radiative falling of an electron toward the center". Namely,

$$\tau_{\text{class}} = \int_{E_0}^{\infty} \frac{dE}{(dE/dt)_{\text{class}}} = 1.6 \cdot 10^{-11}(I_H/E_0)^3 \text{ sec},$$

where E_0 is the binding energy in the "initial" state. The quantum-mechanical lifetime with respect to deexcitation (without taking into account cascade transitions)

$$\tau_{\text{quant}} \approx \frac{1}{A_{n,1}} = 6.2 \cdot 10^{-11} n_0^5 = 6.2 \cdot 10^{-11}(I_H/E_0)^{2.5} \text{ sec}.$$

Ordinarily E_0 is not less than kT, and I_H/kT is not an exceedingly large quantity, so that both times are of the same order.

where N_n is the atom number density in the nth state and n^* is the quantum number of the highest state attainable, which is determined by the cutoff of the upper atomic levels in the gas.

Line radiation plays an important role in energy losses from an optically thin body. This is indicated by the fact that the area of the absorption lines is comparable with the absorption area in the continuous spectrum. For example, for absorption from the ground level of a hydrogen atom approximately half of the oscillator strength comes from the continuous spectrum, while half comes from the discrete spectrum (see Table 5.4).

If the medium is opaque to the lines, then the relative role of energy losses by radiation in the discrete spectrum decreases because of self-absorption. However, in a gas of sufficiently high density, where the lines are strongly broadened, the energy losses from the discrete spectrum can nevertheless be significant, and can even exceed the losses from the continuous spectrum (if the radiation in the continuous spectrum does not have a Planck distribution). In a gas which is rarefied but is optically thick to the lines, the energy role of the lines is determined by their small total width and is usually not large, so that the main role is played by the continuous spectrum. Calculations which show the relative role of line and continuum radiation for different densities, temperatures, and optical thicknesses were carried out in [49] for hydrogen and in [52] for nitrogen.

3. *Molecular band spectra*

§14. Energy levels of diatomic molecules

Absorption of light by molecules can be meaningfully considered at temperatures below 12,000–8,000°K, since at higher temperatures the molecules dissociate completely into atoms. The energy of an atom is determined by its electronic states only. The energy of a molecule, in addition to the electronic state, depends also on the intensity of excitation of the vibrational and rotational modes. Hence, the number of energy levels and the number of possible transitions between them is, in the case of molecules, much greater than for the atoms and, consequently, molecular spectra are much more complex than atomic spectra. Sometimes the individual lines of the spectrum are so close to each other and so numerous that in certain regions they form an almost continuous spectrum. At high gas temperatures and densities, strong broadening can even cause the overlapping of lines. Therefore, the molecular absorption and emission band spectra can, under certain conditions, exert a considerable energetic effect, analogous to that of the continuous spectra. Of great importance are molecular absorption and emission spectra for radiation in air at temperatures of the order of several thousands and tens of thousands of degrees.

We shall consider the simplest, but practically important, case of diatomic molecules. In first approximation the electronic, vibrational, and rotational modes of motion in the molecule occur independently of each other and the total energy of a molecule can be represented as a sum of the corresponding contributions. For vibrations which are small enough that they may be considered to be close to harmonic the energy of the oscillator is

$$E_{\text{vib}} = hc\omega_e(v + \tfrac{1}{2}), \tag{5.86}$$

where $\omega_e = \nu_{\text{vib}}/c$ is the wave number in cm^{-1} (in spectroscopy it is customary to use wave numbers $1/\lambda = \nu/c \text{ cm}^{-1}$ instead of the frequencies $\nu \text{ sec}^{-1}$ *) and $v = 0, 1, 2, \ldots$ is the vibrational quantum number. The energy of the rotational motion is characterized by the rotational quantum number $J = 0, 1, 2, \ldots$ and by the moment of inertia of the molecule I

$$E_{\text{rot}} = \frac{h^2 J(J+1)}{8\pi^2 I} = hcB_e J(J+1), \tag{5.87}$$

where $B_e = h/8\pi^2 cI$ is the rotational constant in cm^{-1}.

Thus, if U_e is the electron energy for the given state, then in first approximation the total energy of the molecule is†

$$E = U_e + hc\omega_e(v + \tfrac{1}{2}) + hcB_e J(J+1). \tag{5.88}$$

In subsequent approximations, other terms are added to (5.88). These additional terms account for anharmonicity of the oscillations, interaction between the vibrational and rotational modes, etc. (see [20, 41]). These terms will not be considered in the present discussion.

The wave numbers of emitted or absorbed radiation $1/\lambda = \nu/c$ (in spectroscopy they are sometimes referred to as "frequencies", though measured in cm^{-1}) are determined by the difference between the energies of the initial and final states. In what follows we shall always denote the upper state by a prime, and the lower state by a double prime. Thus the wave number is given by

$$\frac{1}{\lambda} = \frac{E' - E''}{hc} = \left[\frac{U_e' - U_e''}{hc}\right] + [\omega_e'(v' + \tfrac{1}{2}) - \omega_e''(v'' + \tfrac{1}{2})]$$
$$+ [B_e' J'(J'+1) - B_e'' J''(J''+1)]. \tag{5.89}$$

Between the differences in the electronic, vibrational, and rotational energies (the quantities $hc\omega_e$ and hcB_e are, respectively, the scales for the last two) the

* The wave number 1 cm^{-1} corresponds to: wavelength $\lambda = 10^8$ Å, frequency $\nu = 3 \cdot 10^{10}$ sec^{-1}, and photon energy $h\nu = 1/8067$ ev, $h\nu/k = 1.44$°K.

† The rotational energy in (5.87) and (5.88) is determined to within a constant, which depends on the type of coupling between the rotational and electronic states; the exact meaning of the rotational quantum number also depends on the type of coupling. The constant is of the order of hcB_e, and it can be included in U_e by expressing the energy in the form of (5.88); for a discussion of this point see [20].

following inequalities always hold:

$$\Delta E_{el} \gg \Delta E_{vib} \gg \Delta E_{rot}; \qquad \frac{1}{\lambda_{00}} \gg \omega_e \gg B_e, \tag{5.90}$$

where $1/\lambda_{00} = (U'_e + \omega'_e/2 - U''_e - \omega''_e/2)/hc$ is the wave number corresponding to the electronic transition in the absence of the vibrational and rotational modes. The validity of the inequalities (5.90) can be checked by examining Table 5.6, which shows the spectroscopic constants of the more important states and transitions for O_2, N_2, and NO molecules and the N_2^+ ions*.

Table 5.6

SPECTROSCOPIC CONSTANTS FOR THE MOST IMPORTANT MOLECULES

Molecule	State	Electron energy U_e, ev	$h\nu_{\infty} = \frac{hc}{\lambda_{00}}$, ev	Transition $\frac{1}{\lambda_{00}}$, cm^{-1}	ω_e, cm^{-1}	B_e, cm^{-1}	Transition and name of band system
O_2	$B\,^3\Sigma_u^-$	6.11	6.11	49,363	700.4	0.819	$B \to X$ Schumann–Runge
	$X\,^3\Sigma_g^-$	0		0	1580	1.446	
N_2	$C\,^3\Pi_u$	11.1	3.69	29,670	2035	1.826	$C \to B$ 2nd positive
	$B\,^3\Pi_g$	7.4	1.18	9,557	1734	1.638	$B \to A$ 1st positive
	$A\,^3\Sigma_u^+$	6.17	6.17	49,757	1460	1.440	$A \to X$ Vegard–Kaplan forbidden band
	$X\,^1\Sigma_g^+$			0	2360	2.010	
NO	$B\,^2\Pi$	5.63	5.63	45,440	1038	1.127	$B \to X$ β-bands
	$A\,^2\Sigma^+$	5.48	5.47	44,138	2371	1.995	$A \to X$ γ-bands
	$X\,^2\Pi$			0	1904	1.705	
N_2^+	$B\,^2\Sigma_u^+$		3.16	25,566	2420	2.083	$B \to X$ 1st negative
	$X\,^2\Sigma_g^+$				2207	1.932	

* The various electronic states of a molecule (or ion) differ in the shapes of the potential curves describing the interaction of the atoms as a function of the internuclear distance, and also in the average internuclear distances (hence the transition from one electronic state to another results in a change in the frequency of vibrations, moment of inertia, and the rotational constant). The table is taken from [8].

A molecular energy level diagram has the form shown in Fig. 5.13. The electron energies at the levels A and B are shown by dashed lines. The first actual levels of the molecule, corresponding to the absence of vibrational excitation ($v = 0$), lie slightly above, because of the zero-point vibrational energy. In each electronic state there are a large number of vibrational levels, and in turn in each vibrational level there are a large number of rotational levels. As the excitation increases, the vibrational levels become somewhat crowded as a result of anharmonicity, and, in the limit $v \to \infty$, they pass into the continuum corresponding to dissociation. On the other hand, the rotational levels spread apart with increasing J (provided J is not too large and the approximation (5.87) applies*).

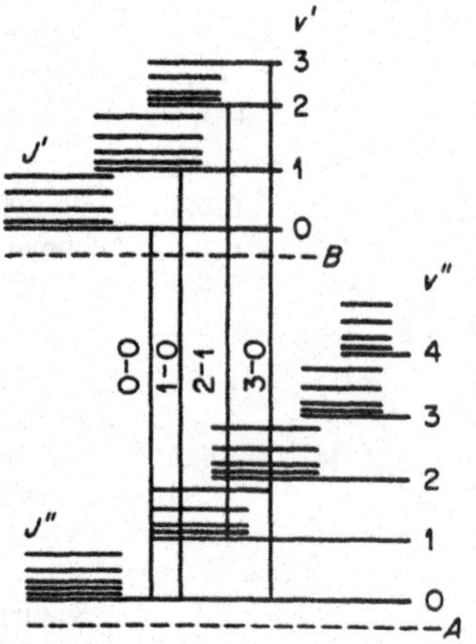

Fig. 5.13. Energy level diagram showing transitions in a diatomic molecule. The vertical lines denote different bands.

The energy level diagram for a nitrogen molecule, indicating all terms and their energies as well as the vibrational states, is shown in Fig. 5.14. For oxygen and nitric oxide molecules we shall present in §16 diagrams of the potential energy curves, on which are also shown the terms and their energies. In what follows we shall frequently make use of spectroscopic notation to define the electronic states of a molecule; therefore, let us briefly recall the basic principles of spectroscopic notation.

An electronic state is characterized by the component of the orbital angular momentum of the electrons in the direction of the molecular axis in terms of the quantum number Λ, the total electron spin S, and the symmetry properties of the state. States with $\Lambda = 0, 1, 2, \ldots$ are denoted by the

* For very intense rotations (extremely large J) the change in the potential energy curve of the molecule from the centrifugal forces becomes appreciable. In the limit $J \to \infty$, the rotational as well as the vibrational levels begin to crowd together and pass into a continuum.

Greek letters Σ, Π, Δ, ..., respectively. The component of the spin in the direction of the molecular axis can assume $2S + 1$ values, with a corresponding splitting of each term or energy level. The multiplicity $2S + 1$ of the term is shown by a superscript at the left, as for example $^3\Sigma$, $^2\Pi$ ($S = 1$, $S = 1/2$, respectively).

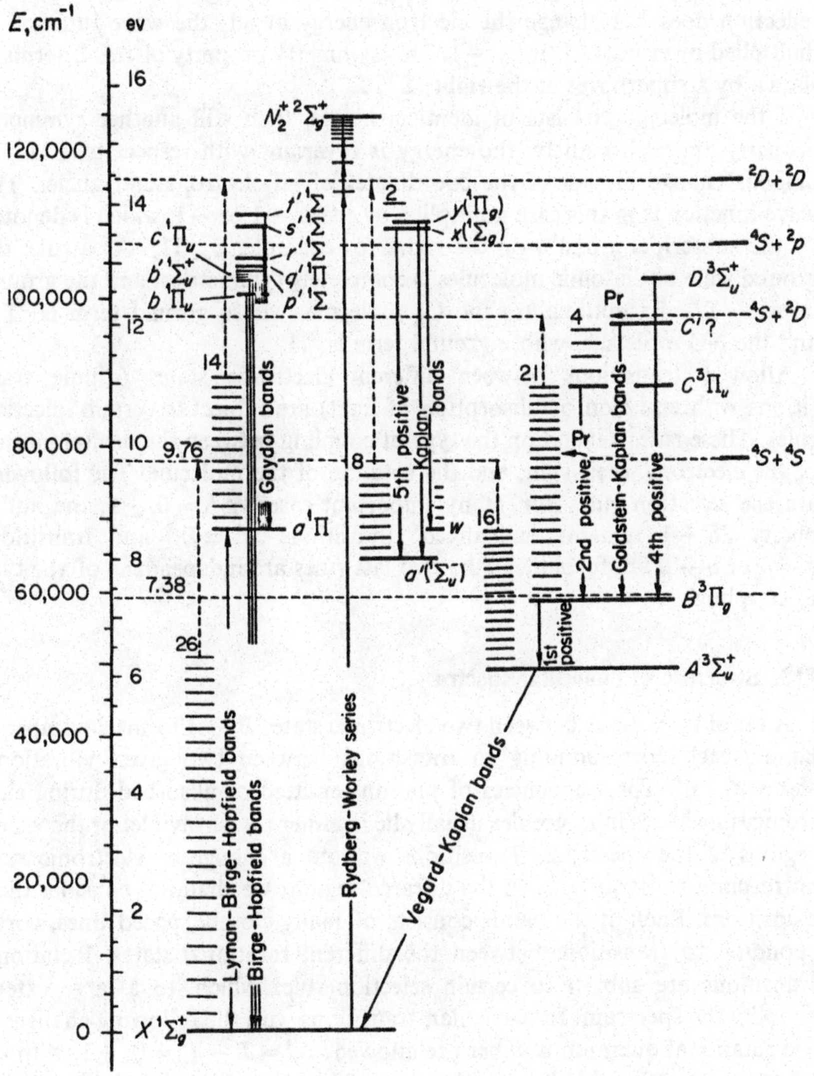

Fig. 5.14. Energy level diagram of the nitrogen molecule.

Upon reflection in a plane in which the molecular axis lies, the axial component of the orbital angular momentum of the electron changes sign (because it is a pseudo- or polar vector); corresponding to this fact, the terms with nonzero orbital angular momentum are doubly degenerate. More precisely, these terms are split into two as a result of the interaction between the rotation of the molecule and the motion of the electrons. This phenomenon is called Λ-type doubling ("lambda"-type doubling). If, however, $\Lambda = 0$, the reflection does not change the electron energy at all; the wave function is multiplied by either $+1$ or by -1. This symmetry property of the Σ terms is shown by a superscript at the right: Σ^+, Σ^-.

If the molecule consists of identical atoms, then still another symmetry property appears, namely, the energy is invariant with respect to a simultaneous change in sign of the coordinates of all electrons and nuclei. The wave function is in this case multiplied by either $+1$ or -1, which is denoted by the subscripts g and u on the right, for example, Σ_g, Π_u. As a rule, the ground state of diatomic molecules is completely symmetrical and the ground term is ${}^1\Sigma_g^+$. Exceptions are the O_2 molecule, whose ground term is ${}^3\Sigma_g^-$, and the NO molecule, whose ground term is ${}^2\Pi$.

Allowed transitions between different electronic states (dipole transitions with emission or absorption of light) are subject to certain selection rules. These rules depend on the type of coupling between the orbital motion of the electrons, their spin, and the rotation of the molecule. The following are the selection rules for many important cases: $\Delta\Lambda = 0, \pm 1$; the multiplicity $2S + 1$ remains unchanged; transitions $\Sigma^+ \rightleftarrows \Sigma^-$ and transitions $g \to g$ or $u \to u$ are forbidden (the two last rules are independent of the type of coupling).

§15. Structure of molecular spectra

A set of transitions between two electronic states $B - A$ forms a number of bands, each corresponding to transitions between two given vibrational states $v' - v''$. The frequencies of photons emitted or absorbed during electronic transitions in molecules usually lie in either the ultraviolet or the visible regions of the spectrum. Transitions without a change of electronic state correspond to frequencies in the infrared region; we shall not consider these transitions. Each of the bands consists of many closely spaced lines, corresponding to transitions between the different rotational states. Rotational transitions are subject to certain selection rules which, to a large extent, simplify the spectrum. In particular, transitions with the following changes in the rotational quantum number are allowed: $\Delta J = J' - J'' = 0, \pm 1$, with the transition $0 - 0$ forbidden; in the case of $\Sigma \to \Sigma$ transitions, all the $\Delta J = 0$ transitions are forbidden. The vertical lines in Fig. 5.13 denote transitions between different vibrational states of two electronic levels (bands $v' - v''$: $0 - 0$, $1 - 0$, etc.). In Fig. 5.15 we have purposely separated one band $v' - v''$ to show its rotational structure. It was assumed here that $\Lambda \neq 0$ in at least one of the states B or A, so that $\Delta J = 0$ transitions exist. The series of lines with $\Delta J = 0, +1, -1$ are called Q-, R-, and P-branches, respectively.

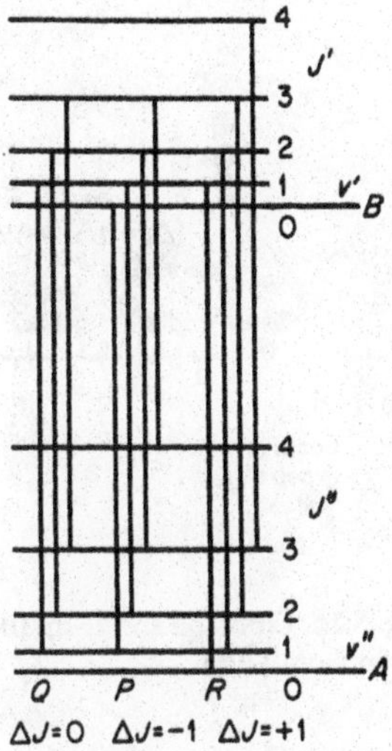

Fig. 5.15. The rotational structure of bands. Diagram of transitions corresponding to the Q-, P-, and R-branches.

If the vibrational levels in different electronic states of a diatomic molecule had identical spacings, i.e., if the frequencies ω_e' and ω_e'' were the same and if the crowding due to anharmonicity had taken place in the same manner, then, as may be seen from Fig. 5.13, bands with the same value of the difference $\Delta v = v' - v''$ would be exactly superimposed on each other. Actually, the distributions of vibrational levels in different electronic states differ only slightly from each other and the difference of vibrational frequencies $\omega_e' - \omega_e''$ is usually much smaller than the frequencies themselves. Hence, bands with the same difference Δv are close to each other and form so-called sequences (or diagonal groups) of bands, while bands with different Δv are separated by large frequency intervals. This situation is illustrated by a photograph of the emission spectrum of the so-called second positive system for nitrogen* (transition $C\,^3\Pi_u \rightarrow B\,^3\Pi_g$; see energy level diagram, Fig. 5.14). On the photograph (Fig. 5.16) is superposed a wavelength scale along with the numbers of the vibrational transitions (the first number corresponds to the upper electronic state). As may be seen from the photograph the distances between two successive bands of the same series, for example, $\Delta v = -2$, are

* Band systems corresponding to various electronic transitions are usually designated by some name. The more important systems are shown on the energy level diagrams.

approximately equal to 50 Å; while the distance between the closest bands of neighboring series is greater, for $\Delta v = -2$ and $\Delta v = -1$ it is approximately equal to 230 Å. As the frequency increases, the bands become more crowded due to the crowding of the vibrational levels as $v \to \infty$. Finally, the bands merge into a continuum associated with dissociation of the molecule.

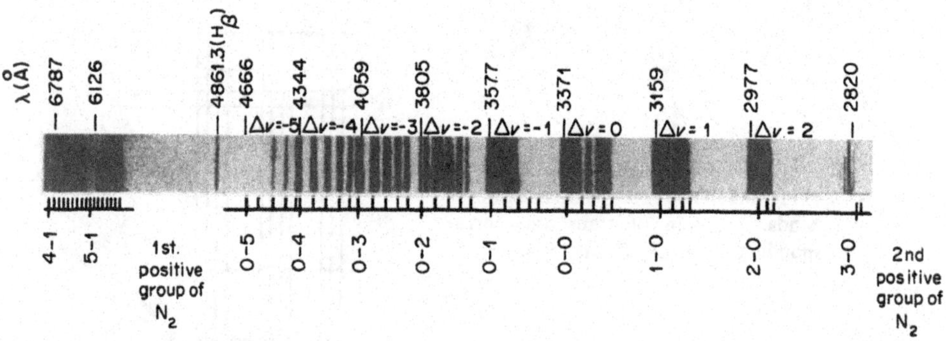

Fig. 5.16. Emission spectrum for the second positive system of nitrogen. The photograph is taken from [20a].

The distribution of lines in the rotational structure of the band can be easily established from (5.89) and the selection rules: $J' - J'' = 0, +1, -1$. The following relationships are then obtained for the three branches:

$$P:\quad J' = J'' - 1, \qquad \frac{1}{\lambda} = \frac{1}{\lambda_{v'v''}} + (B_e' - B_e'')J''^2 - (B_e' + B_e'')J'', \qquad J'' \geqslant 1; \quad (5.90')$$

$$Q:\quad J' = J'', \qquad \frac{1}{\lambda} = \frac{1}{\lambda_{v'v''}} + (B_e' - B_e'')J''^2 + (B_e' - B_e'')J'', \qquad J'' \geqslant 1; \quad (5.91)$$

$$R:\quad J' = J'' + 1, \qquad \frac{1}{\lambda} = \frac{1}{\lambda_{v'v''}} + (B_e' - B_e'')J''^2 + (3B_e' - B_e'')J'' + 2B_e', \qquad J'' \geqslant 0. \quad (5.92)$$

Here $1/\lambda_{v'v''}$ is a constant representing the wave number which corresponds to the electronic-vibrational transition in the absence of any rotational structure (without the third term in (5.89)). The rotational structure depends on which of the rotational constants B_e' or B_e'' is greater. The dependence of the wave numbers $1/\lambda$ on the quantum number J'' and the spectrum are shown schematically for the two cases in Figs. 5.17 and 5.18 (the so-called Fortrat diagrams). It is evident from Fig. 5.17 that when $B_e' > B_e''$ the spectrum has a low-frequency limit, near which the lines are crowded ("red" edge), the lines extend in the direction of higher frequencies, and the spacing between them

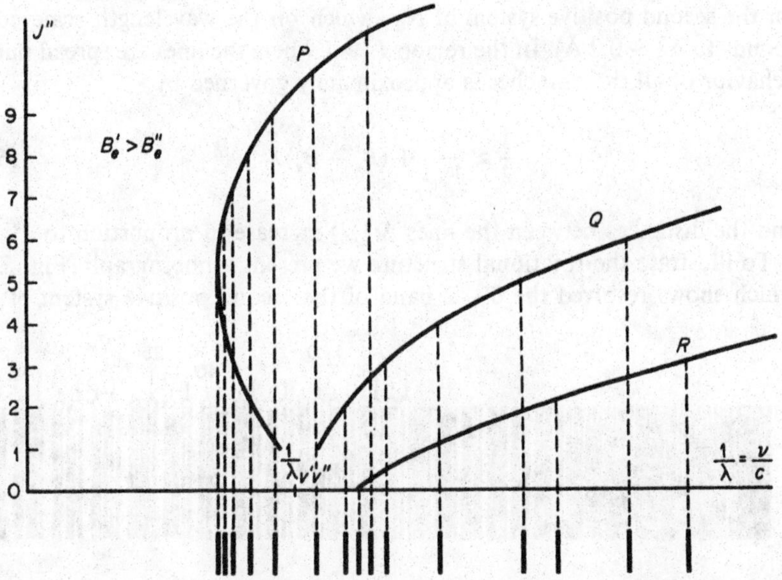

Fig. 5.17. The wave number in the *P*-, *Q*-, and *R*-branches of a band as a function of the rotational quantum number J'' for the case $B_e' > B_e''$ (red edge).

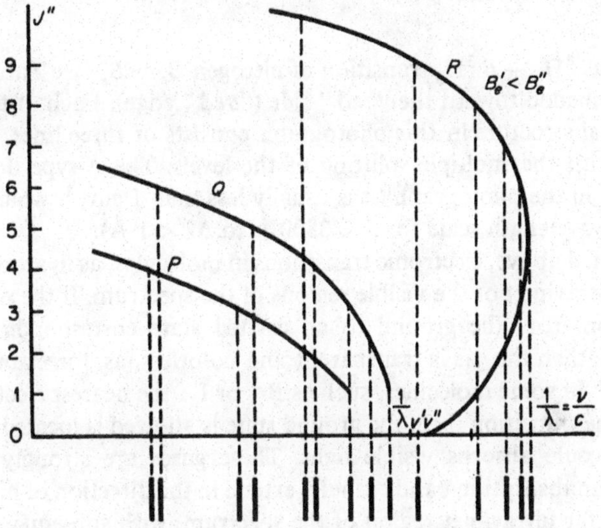

Fig. 5.18. The wave number in the *P*-, *Q*-, and *R*-branches of a band as a function of the rotational quantum number J'' for the case $B_e' < B_e''$ (violet edge).

increases. Conversely, for $B_e' < B_e''$ we have a "violet" edge and the lines extend in the direction of low frequencies. In the region around the edge the "frequency" spacing between the lines is of the order of $B_e' - B_e''$ ($\approx 0.2\ \text{cm}^{-1}$ for the second positive system of N_2, which on the wavelength scale corresponds to $\Delta\lambda \sim 0.2$ Å). In the region $J'' \gg 1$ where the lines are spread out the behavior of all the branches is approximately governed by

$$\frac{1}{\lambda} \approx \frac{1}{\lambda_{v'v''}} + (B_e' - B_e'')J''^2, \tag{5.93}$$

and the distances between the lines $\Delta(1/\lambda)$ increase in proportion to J''.

To illustrate the rotational structure we present a photograph (Fig. 5.19), which shows resolved the $0-2$ band of the second positive system of N_2.

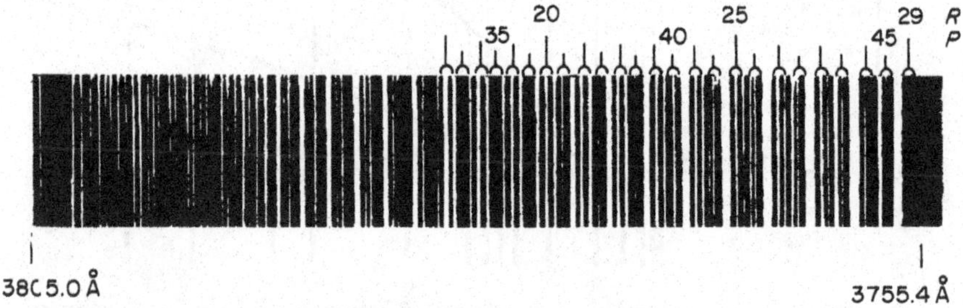

Fig. 5.19. Spectrum in the 0–2 band of the second positive system of nitrogen. The photograph is taken from [20a].

For the $C\,^3\Pi_u \rightarrow B\,^3\Pi_g$ transition of nitrogen $B_e' > B_e''$ (see Table 5.6) and the band is shaded toward the "red" side ("red" edge). Each of the lines of the rotational structure in this photograph consists of three lines, in correspondence with the multiple splitting of the levels. The Λ-type doubling is not resolved in the photograph (it is usually less than $1\ \text{cm}^{-1}$, which corresponds on the wavelength scale for $\lambda \approx 3800$ Å to $\Delta\lambda < 1$ Å).

As noted above, electronic transitions in molecules, as in atoms, correspond to the ultraviolet or the visible regions of the spectrum. If the nearest allowed transition from the ground to an excited state corresponds to ultraviolet photons, then the gas is transparent and colorless, as, for example, N_2, O_2, and NO. In some molecules, such as Br_2 or I_2, the nearest electronic level for which the transition from the ground state is allowed is located quite low and the molecule absorbs visible light. These gases are strongly colored. The molecular absorption bands usually extend in the direction of high frequencies into the far ultraviolet region of the spectrum, with subsequent transition to the continuum.

§16. The Frank–Condon principle

Electronic transitions in a molecule are related to the simultaneous change of all three characteristics of the particular state. The tremendous number of all possible combinations of initial and final states is limited by the selection rules. The selection rules, however, apply only to the changes in the electronic and rotational parameters of the molecule and say nothing about the possible changes of vibrational parameters. In order to establish which of the combinations of vibrational quantum numbers represent the most probable transitions, we return to the potential energy curves of a molecule, with rotation neglected.

The potential energy of a molecule depends on the internuclear distance. Repulsive forces dominate when the nuclei are close together, while attractive forces become dominant when the nuclei are far apart. At a certain distance r_e the repulsive and attractive forces cancel each other, and the potential energy exhibits a minimum at this point. The absolute value of this minimum corresponds to the energy of the electronic state U_e. The difference between the energy for infinite separation of the nuclei and U_e gives the dissociation energy (to within the accuracy of the zero-point vibrational energy). The shape and position of the potential curve depend on the electronic state, so that several potential curves are associated with each molecule. Figures 5.20 and

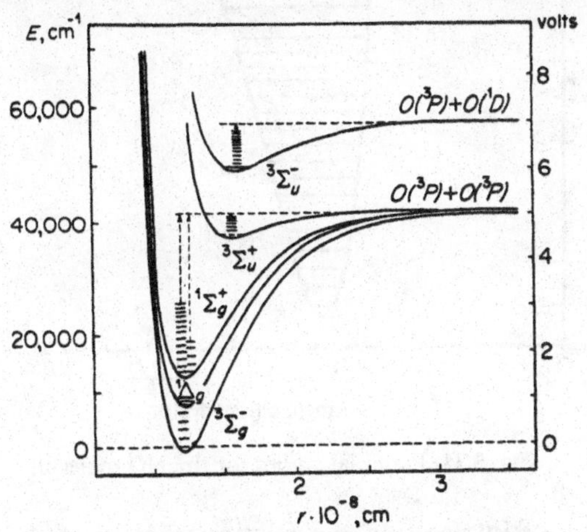

Fig. 5.20. Potential curves for the O_2 molecule.

5.21 give the potential energy curves for the O_2 and NO molecules, respectively, based on spectroscopic data*. The horizontal lines on these figures correspond to the vibrational energy levels for each of the electronic states.

* The diagrams are taken from [20, 21].

From the classical point of view the internuclear distance for a given vibrational energy varies periodically about the equilibrium position r_e. The variation takes place over the interval between the points of intersection of the horizontal line denoting the vibrational energy level with the potential curve.

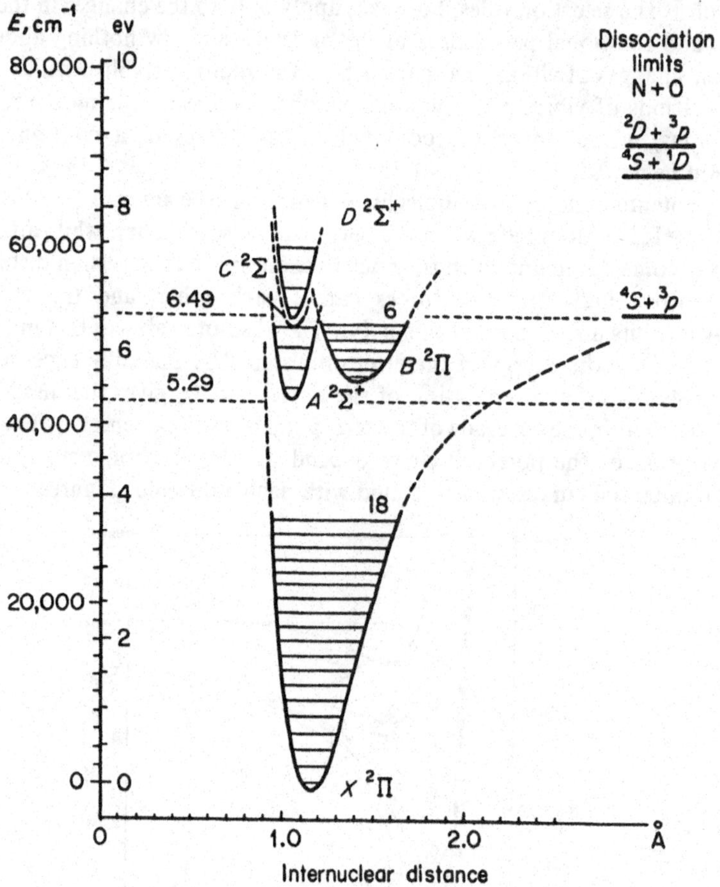

Fig. 5.21. Potential curves for the NO molecule.

At these points of intersection the relative velocity of the nuclei vanishes, since the direction of motion is reversed and at these positions (turning points) the molecule spends the longest time. On the other hand, the molecule passes the equilibrium position very rapidly since its velocity there is a maximum. Therefore, the spontaneous transition from the upper to the lower electronic state is most likely when the nuclei occupy the extreme positions. The re-arrangement of the electron shell during the transition with the accompanying photon emission takes place so rapidly that neither the positions of the nuclei nor their kinetic energies have sufficient time to change their values. In this regard, the duration of the rearrangement is determined by the time during which an electron travels through a distance of the order of a molecular dimension, i.e., $\sim 10^{-16}$ sec (for an electron velocity of $\sim 10^{8}$ cm/sec and a molecular dimension of $\sim 10^{-8}$ cm). However, the distance between the nuclei changes appreciably only over a time of the order of the period of oscillation, i.e., over a time $\sim 1/\omega_e c \sim 10^{-14}$ sec (for an $\omega_e \sim 1000$ cm^{-1}, which is appropriate for light molecules; ω_e for heavy molecules is even smaller, and the period of oscillation correspondingly greater)*.

Electronic transition to a lower state takes place at a constant internuclear distance, i.e., principally along vertical lines drawn from the turning points on the potential curves (Fig. 5.22). The molecule arrives at the final state with zero velocity, that is, it starts its vibrational motion, also at the turning points, with a new vibrational energy. Thus, the most probable transitions are those to such lower vibrational states for which one of the turning points is located at the same internuclear distance as one of the turning points in the upper state. This principle is known as the Frank–Condon principle and is illustrated

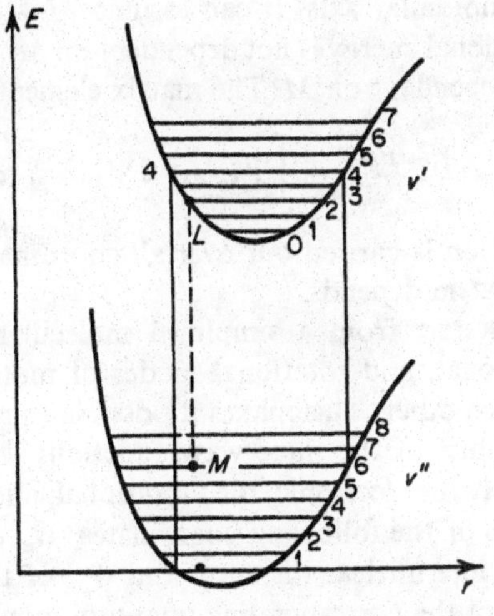

Fig. 5.22. Diagram of the potential curves and transitions illustrating the Frank–Condon principle.

* The probability of allowed electronic transitions from the upper to the lower states in atoms and molecules is of the order of 10^{8} sec^{-1}. An excited molecule, therefore, remains in the upper excited state for a time of the order of 10^{-8} sec (during which the atoms undergo a large number of oscillations $\sim 10^{6}$), and then, in a time $\sim 10^{-16}$ sec, the molecule makes the transition to a lower state, emitting a photon.

in Fig. 5.22, which shows the vertical lines for the most probable transitions from the upper state $v' = 4$ to the lower states $v'' = 0$ and $v'' = 6$. On the other hand, the transitions for which the vertical lines emanating from the upper turning points are found either in the middle of the lower level (as, for example, the $2 - 6$ transition, shown in Fig. 5.22 by a dashed line) or entirely outside the interval bounded by the potential energy curve are highly improbable.

§17. Probability of molecular transitions with the emission of light

Let us consider the transition of a molecule from an upper to a lower state from a quantum-mechanical point of view. The probability of a spontaneous dipole transition with the emission of a photon is proportional to the square of the matrix element **d** of the dipole moment of the system and is described by the general equation (5.69). Let us consider a transition from an upper state $Bv'J'M'$ to a lower state $Av''J''M''$. The indices B and A denote the electronic states of the molecule; v' and v'' denote its vibrational states and J' and J'' are the rotational quantum numbers. M is the "magnetic" quantum number, defined as the component of the rotational angular momentum in the direction of the molecular axis. It can assume $2J + 1$ values: $M = J, J - 1, \ldots, -J$. The rotational energy is not dependent on M, while the wave function of the system is dependent on M. The matrix element is equal to

$$\bar{D}^{Bv'J'M'}_{Av''J''M''} = \int \Psi^*_{Bv'J'M'} \, \mathbf{d} \, \Psi_{Av''J''M''} \, d\tau, \tag{5.94}$$

where the integration is carried out over all coordinates on which the wave function of the system depends.

We shall again start from a simplified molecular model, in which the electronic, vibrational, and rotational modes of motion are assumed to be independent of each other. This makes it possible to represent the total wave function as a product of the three wave functions ψ_{el}, ψ_{vib}, and ψ_{rot}, describing respectively the electronic, the vibrational, and the rotational modes. They are functions of the following coordinates: ψ_{el} of the electron coordinates, ψ_{vib} of the internuclear distance, and ψ_{rot} of the angles of molecular rotation and also of the corresponding quantum numbers. For example, for the upper state, we have

$$\Psi_{Bv'J'M'} = \psi_{el\,B} \psi_{vib\,v',B} \psi_{rot\,J',M'}. \tag{5.95}$$

The function ψ_{vib} depends on the electronic state, since the frequencies of oscillations are dependent on the particular state of the electron.

Let us represent the dipole moment of the system $\mathbf{d} = \sum e_i \mathbf{r}_i$ (the summation is carried out over all particles) as the sum of the electronic and nuclear moments

$\mathbf{d} = \mathbf{d}_e + \mathbf{d}_a$. The electronic wave functions are, by definition, orthogonal to one another in the different electronic states (the nuclear coordinates enter only as parameters). Substituting $\mathbf{d}$ and Ψ into the integral (5.94) we can factor out from the term containing the nuclear moment the expression $\int \psi^*_{\mathrm{el}\,B} \psi_{\mathrm{el}\,A}\, d\tau_e$ which vanishes for $B \neq A$, so that the matrix element of the nuclear moment will also vanish. Since ψ_{vib} and ψ_{rot} are independent of the electronic coordinates, we can represent the remaining matrix element of the electronic moment as the product

$$D = D_e = \int \psi^*_{\mathrm{el}} |\mathbf{d}_e| \psi_{\mathrm{el}} \int \psi^*_{\mathrm{vib}} \psi_{\mathrm{vib}} \cdot \int \psi^*_{\mathrm{rot}} \mathbf{n} \psi_{\mathrm{rot}} = D_{\mathrm{el}} \cdot D_{\mathrm{vib}} \cdot D_{\mathrm{rot}}. \quad (5.96)$$

Here the rotational matrix element includes only the direction of the averaged electronic dipole moment, that is, of the unit vector $\mathbf{n}$, which is averaged over the "rotations" of the molecule. (For simplicity we have here dropped the indices, i.e., the quantum numbers on the wave functions and the differentials.) The condition for D_{rot} to be different from zero gives the selection rule for the change of the rotational quantum numbers for an allowed transition.

In our approximation, the energy of a molecule is independent of the direction of the rotational angular momentum; therefore, to obtain the probability of transition from one energy state, $Bv'J'$, to another, $Av''J''$, the probability should be averaged over all possible directions of the rotational angular momentum in the initial state and summed over all possible directions in the final state. By this means, the transition probability (in units of $\sec^{-1}$) according to (5.69) is given by*

$$A^{Bv'J'}_{Av''J''} = \frac{64\pi^4}{3hc^3} \nu^3_{Bv'J',\,Av''J''} D^2_{\mathrm{el}\,BA} q_{v'v''} p_{J'J''}, \quad (5.97)$$

where

$$q_{v'v''} = D^2_{\mathrm{vib}\,v'v''} = \left| \int \psi^*_{\mathrm{vib}\,v'}(r) \psi_{\mathrm{vib}\,v''}(r)\, dr \right|^2, \quad (5.98)$$

$$p_{J'J''} = \frac{1}{2J' + 1} \sum_{M'M''} D^2_{\mathrm{rot}\,J'M',\,J''M''}. \quad (5.99)$$

* Strictly speaking, the electronic matrix element D_{el} depends on the internuclear distance r (it is calculated at a particular time for a fixed internuclear distance which enters in the electronic wave functions). The quantity D^2_{el}, which enters in the equation for the transition probability (5.97), should be understood to denote a certain value of D^2_{el} averaged with respect to r, say, corresponding to the equilibrium position r_e in the upper electronic state.

The intensity of the corresponding spectral line in erg/cm^3 · sec is equal to the product of the transition probability A (sec^{-1}), the photon energy $h\nu$ (erg), and the number of molecules in the upper quantum state N (cm^{-3}): $I = h\nu NA$ (the indices are dropped for brevity).

The dimensionless probability $p_{J'J''}$ determines the intensity distribution in the rotational lines within the given band $Bv' \to Av''$. It is proven in the quantum theory of molecules that $p_{J'J''}$ obeys the rule

$$\sum_{J''} p_{J'J''} = \sum_{J''} \sum_{M'M''} \frac{1}{2J'+1} D^2_{\mathrm{rot}\, J'M', J''M''} = 1. \qquad (5.100)$$

The meaning of this relation is that after a transition from an upper electronic-vibrational state to a lower one, the molecule must terminate in one of the allowed rotational levels J'' (these transfers correspond to a total probability of one). The probability of a $Bv' \to Av''$ transition to any of the rotational levels is obtained by summing (5.97) over J''. In accordance with the condition (5.100), the probability is given by

$$A^{Bv'}_{Av''} = \frac{64\pi^4}{3hc^3} \nu^3_{Bv', Av''} D^2_{\mathrm{el}\, BA} q_{v'v''}, \qquad (5.101)$$

where $\nu_{Bv', Av''}$ is some average frequency for the given band. By virtue of the smallness of the rotational energies in comparison to the vibrational energies, the spread of frequencies within the band is not too great and the introduction of an average frequency for the band is justified.

The relative probability of the vibrational transfer $v' \to v''$ during the electronic transition $B \to A$ is characterized by the dimensionless factor $q_{v'v''}$, defined by (5.98). Let us consider the integral in (5.98). The wave functions belong to different electronic states, and they have different frequencies of vibration and different equilibrium positions r_e. Because of this fact the integral is different from zero for the different combinations of $v'v''$, and no selection rules exist for the vibrations (if the electronic state were unchanged and the vibrations were strictly harmonic, then the integral in (5.98), by virtue of the orthogonality condition, would be equal to zero for all $v' \neq v''$).

A schematic diagram of wave functions for various vibrational states is given in Fig. 5.23. In order for the integral of the product of the oscillating factors (5.98) (only $\psi_{\mathrm{vib}, v=0}$ does not oscillate) to have an appreciable value, it is necessary that, first, the factors should not have "opposite phases" and, second, that the largest maxima of both factors overlap. However, the largest maxima of vibrational wave functions lie very near the "turning points", which correspond to the maximum probability of these positions. Hence, those transitions are the most probable ones, where at least one of the turning points is in the upper state at the same internuclear distance as the turning point in the lower state. The above analysis provides the quantum-mechanical basis for the Frank–Condon principle. The quantity $q_{v'v''}$, which is frequently called the Frank–Condon factor, is the probability of a given vibrational

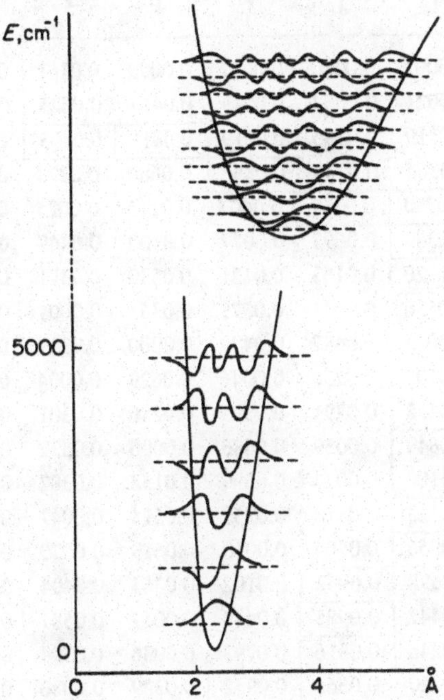

Fig. 5.23. Potential curves and wave functions of several vibrational states for a RbH molecule. (The figure is taken from [20b].) The number of zeros (nodes) of each wave function is equal to the vibrational quantum number v.

transition $v' \to v''$ for a specified electronic transition, since according to the sum rule for matrix elements, the total probability of a transition from the given v' to any v'' is equal to unity,

$$\sum_{v''} q_{v'v''} = \sum_{v''} D^2_{\text{vib}\,v'v''} = 1. \tag{5.102}$$

To illustrate the quantum-mechanical interpretation of the Frank–Condon principle, we present in Table 5.7 for the β-system of NO ($B\,^2\Pi \to X\,^2\Pi$) the values of $|\int \psi^*_{v'} \psi^*_{v''}\, dr|$ the square of which is equal to $q_{v'v''}$. These integrals were taken from [21]. It is instructive to consider Table 5.7 together with the potential curves of Fig. 5.21 to see how the Frank–Condon principle is satisfied.

The Frank–Condon factors must be known in order to calculate the relative probabilities of various $v' \to v''$ transitions, that is, the relative intensities of

Table 5.7

SQUARE ROOT OF THE FRANK–CONDON FACTOR $q_{v'v''}^{1/2}$ FOR THE β-SYSTEM OF BANDS IN A NO MOLECULE (v' REFERS TO THE UPPER STATE AND v'' TO THE LOWER ONE)

v''	Vibrational quantum number of the upper state v'								
	0	1	2	3	4	5	6	7	8
0	0.0000	0.0002	0.0010	0.0032	0.0079	0.0161	0.0280	0.0429	0.0587
1	0.0003	0.0024	0.0087	0.0219	0.0414	0.0625	0.0788	0.0811	0.0707
2	0.0021	0.0119	0.0336	0.0619	0.0819	0.0803	0.0569	0.0257	0.0040
3	0.0086	0.0364	0.0735	0.0896	0.0680	0.0273	0.0016	0.0065	0.0286
4	0.0250	0.0750	0.0967	0.0607	0.0115	0.0025	0.0286	0.0471	0.0362
5	0.0554	0.1069	0.0693	0.0077	0.0100	0.0447	0.0448	0.0146	0.0001
6	0.0972	0.1020	0.0153	0.0121	0.0530	0.0371	0.0027	0.0097	0.0341
7	0.1380	0.0556	0.0041	0.0573	0.0363	0.0001	0.0231	0.0401	0.0170
8	0.1603	0.0075	0.0497	0.0489	0.0000	0.0317	0.0389	0.0055	0.0066
9	0.1522	0.0101	0.0756	0.0046	0.0629	0.0004	0.0567		
10	0.1276	0.0452	0.0391	0.0395	0.0286	0.0301	0.0299		
11	0.0964	0.0849	0.0059	0.0686	0.0006	0.0572	0.0003		
12	0.0657	0.1100	0.0033	0.0599	0.0158	0.0382	0.0198		
13	0.0405	0.1123	0.0318	0.0252	0.0515	0.0047	0.0506		
14	0.0226	0.0956	0.0704	0.0010	0.0619	0.0070	0.0399		
15	0.0113	0.0698	0.0962	0.0102	0.0363	0.0404	0.0070		
16	0.0051	0.0442	0.0985	0.0449	0.0057	0.0587	0.0046		
17	0.0021	0.0246	0.0816	0.0793	0.0036	0.0394	0.0361		
18	0.0007	0.0120	0.0565	0.0932	0.0329	0.0080	0.0557		
19	0.0002	0.0051	0.0334	0.0838	0.0693	0.0022	0.0374		
20	0.0001	0.0019	0.0169	0.0610	0.0881	0.0297	0.0068		
21		0.0006	0.0074	0.0369	0.0821	0.0663	0.0031		
22		0.0002	0.0028	0.0188	0.0603	0.0852	0.0326		
23			0.0009	0.0801	0.0361	0.0784	0.0685		
24				0.0030	0.0179	0.0559	0.0865		
25				0.0009	0.0074	0.0331	0.0727		
26					0.0026	0.0150	0.0486		
27					0.0007	0.0057	0.0258		
28						0.0018	0.0110		
29							0.0038		

$v' =$	9	10	11	12	13	14	15	16	17
$v'' = 0$;	0.0731	0.0837	0.0892	0.0866	0.0841	0.0744	0.0601	0.0445	0.0303

$v' =$	18	19	20	21	22	23	24	25	26
$v'' = 0$;	0.0190	0.0110	0.0059	0.0029	0.0013	0.0006	0.0002	0.0001	0.0000

The rectangles denote the most probable transitions from each upper state.

the different bands within the limits of the given electronic transition. The Frank–Condon factors have been calculated for a number of systems of the more important molecules NO, O_2, N_2, and the ion N_2^+ (see [8, 21–24]).

In order to find the absolute values of the transition probabilities and the intensities of lines or bands it is necessary to know the values of the electronic matrix element D_{el}. Theoretical calculation of the electronic matrix element entails considerable difficulties and it is usually determined experimentally (see §§18 and 21). By analogy with the theory of atomic transitions, the electronic matrix element is usually replaced by the concept of the oscillator strength.

Let us sum the probability of the $A_{Bv' \to Av''}$ transition according to (5.101) over the final vibrational states v'', and average it over the original states v'. We then obtain the probability of the $B \to A$ electronic transition for arbitrary vibrational and rotational transitions

$$A_{BA} = \frac{64\pi^4}{3hc^3} \nu_{BA}^3 D_{el\,BA}^2, \tag{5.103}$$

where ν_{BA} is some average frequency for the electronic transition (as before, the justification for introducing an average frequency is the fact that the differences between the vibrational energies are small in comparison with the differences in the electronic energies). Using (5.69), (5.73), and (5.60), we can determine the oscillator strength for the $B \to A$ electronic transition

$$f_{BA} = \frac{8\pi^2 m}{3he^2} \nu_{BA} D_{el\,BA}^2. \tag{5.104}$$

The ratio of the statistical weights of the upper and lower electronic states is here set equal to unity under the assumption that the multiplicity of both terms is the same. It may be expected that the oscillator strengths for molecular transitions are of the same order of magnitude as for atomic transitions. The numerical values of the oscillator strengths for the more important systems will be given below.

§18. Light absorption coefficient in lines

The light absorption coefficient in a line, corresponding to the $Av''J'' \to Bv'J'$ transition (which is the reverse of the transition with the emission of light considered in the preceding section), will be calculated from the principle of detailed balancing. This principle establishes the relationship between the Einstein coefficients for direct and reverse transitions as given by (5.71)*.

* The ratio of statistical weights in this equation is taken, as before, to be equal to one; see (5.104).

Substituting into (5.71) the emission probability given by (5.97), noting the definition (5.70), and replacing D_{el}^2 in (5.97) by the oscillator strength from (5.104), we obtain the absorption coefficient in the form

$$\kappa_{Av''J'',\,Bv'J'} = \frac{\pi e^2}{mc} f_{BA} q_{v''v'} p_{J''J'} N_{Av''J''} F(v), \tag{5.105}$$

where $N_{Av''J''}$ is the molecule number density in the lower state $Av''J''$, and $F(v)$ is a function describing the absorption distribution within the line; it is normalized to unity, with $\int F(v)\,dv = 1$.

Let us integrate the absorption coefficient of the $A \to B$ electronic transition over the entire spectrum. Obviously, the same result will be obtained by integrating the absorption coefficient of each line (5.105) with respect to frequency and then summing over all the spectral lines*. Summation over lines is equivalent to summation over all initial and final states $v''J''$ and $v'J'$. Summation over the final states is carried out using the sum rules (5.100) and (5.102); summation over the initial states reduces to a summation over the number of molecules: $N_A = \sum_{v''J''} N_{Av''J''}$, where N_A is the molecule number density in the A electronic state. If A is the ground state, then N_A is practically equal to the total molecule number density N.

The integral over the spectrum of the absorption cross section for a molecule in the A state upon its transition to the B state ($\sigma_v = \kappa_v/N_a$) is

$$\int_0^\infty \sigma_v\,dv = \frac{\pi e^2}{mc} f_{BA}. \tag{5.106}$$

This result agrees with the corresponding result for atomic transitions. Thus, as in the case of atoms, the absorption "area" corresponding to the given electronic transition is determined solely by the oscillator strength. The difference consists of the fact that in molecules this "area" is distributed over a large number of lines, as a result of which only a small part of the area is attributed to each line. Consequently, the "height" of the molecular lines is much smaller than the "height" of the atomic lines. Both multiplet splitting of lines and Λ-type doubling lower the "height" of the lines several times without, however, changing the total area.

The oscillator strength for a given band system can be determined from experimental studies of molecular absorption spectra. In these studies the attenuation of light by an optically thin gas layer (still transparent at the peaks of the lines) is measured. In this manner the "area" of the absorption spectrum can be found, and (5.106) can then be used to calculate the oscillator

* We should also include here the continuum that starts with the frequency for which the bands converge toward the dissociation limit.

strength. If the Frank–Condon factor has been calculated, then the absorption curve for an individual line or band may be used directly to estimate the oscillator strength (relatively simple formulas for the rotational transition probabilities $p_{J''J'}$ are available).

Using this method the authors of [25] measured the oscillator strengths for the γ- and β-bands of the NO molecule (γ: $X\,^2\Pi \to A\,^2\Sigma^+$; β: $X\,^2\Pi \to B\,^2\Pi$). They found that $f_\gamma \approx 0.0025$ and $f_\beta \approx 0.008$. In [26] the oscillator strength for the Schumann–Runge bands of an oxygen molecule (the $X\,^3\Sigma_g^- \to B\,^3\Sigma_u^-$ transition) was reported as $f = 0.259$. Here, one part of the total absorption "area" $\Delta f = 0.044$ is due to the bands and the remainder $\Delta f = 0.215$ is due to the continuum corresponding to the dissociation of the O_2 molecules into atoms $O(^3P) + O(^1D)$. The continuum begins at $\lambda = 1760$ Å (the lower edge of the band is at $\lambda = 2030$ Å). The cross section in the continuum has a maximum at $\lambda = 1450$ Å given by $\sigma_\nu = 1.81 \cdot 10^{-17}$ cm^2 and is reduced to one half of this value at $\lambda = 1567$ Å and $\lambda = 1370$ Å. It should be noted that the oscillator strength for the Schumann–Runge bands, obtained from data on light emission at high temperatures, is found to be much smaller (see §20; measurements of the absorption of light by cold oxygen were reported in [26]). Possible reasons for this difference have been discussed in [27].

In general, it should be noted that the oscillator strength for a molecular transition, unlike those for atomic transitions, is not strictly a constant quantity (in particular, it depends on the degree of vibrational excitation and the method of averaging over the internuclear distance). The data on oscillator strengths available in the literature frequently differ appreciably for the same transitions. A summary of known results on oscillator strengths for molecular transitions as well as references to the literature may be found in the review article by Soshnikov [27a].

§19. Molecular absorption at high temperatures

At room temperature practically all molecules are in the ground electronic and vibrational states. For example, only one vibrational quantum is excited per 10^{-5} nitrogen molecules, approximately. The long wave absorption edge for diatomic molecules always lies in the ultraviolet or visible regions of the spectrum, for example, in the far ultraviolet region for O_2, NO, and N_2 molecules*, and in the visible region for Br_2, I_2, and CN molecules. As the temperature increases, excited molecules appear in the gas; these molecules

* Transitions to the low-lying levels $^1\Delta_g$ and $^1\Sigma_g^+$ in the O_2 molecule are forbidden. Also forbidden is the transition to the $^3\Sigma_u^+$ level. The Hertzberg bands which correspond to the latter transition are extremely weak. The transition $X\,^1\Sigma_g^+ \to A\,^3\Sigma_u^+$ in N_2 (the very weak Vegard–Kaplan bands) is also forbidden.

are in upper vibrational states from which a transition into the same upper electronic state requires less energy. Thus, an increase in temperature causes the long wave absorption edge to be displaced in the "red" direction. At temperatures of the order of 10,000°K there also appear molecules which are in the upper electronic states from which new transitions to even higher electronic states are possible. This is the way in which the absorption in the first and second positive systems of nitrogen takes place (the transitions $A\,^3\Sigma_u \rightarrow B\,^3\Pi_g$ and $B\,^3\Pi_g \rightarrow C\,^3\Pi_u$; see Table 5.6 and Fig. 5.14).

Let us consider the molecular absorption of light at high temperatures, using NO molecules as an example. In air at temperatures of the order of 2,000–10,000°K there is an appreciable concentration of nitric oxide, of the order of several percent (see §4 of Chapter III). As we shall show later, the absorption of light by NO molecules under certain conditions plays an important role in determining the optical properties of high-temperature air. Calculations of the absorption by NO molecules have been presented in [21], and we shall essentially follow the approach given there.

There exist three important systems of NO absorption bands from the ground electronic state: γ (the $X\,^2\Pi \rightarrow A\,^2\Sigma^+$ transition), β ($X\,^2\Pi \rightarrow B\,^2\Pi$) and δ ($X\,^2\Pi \rightarrow C\,^3\Sigma$). The long wave edges of the first two systems correspond to $\sim 45{,}000\ \text{cm}^{-1}$ and the long wave edge of the third to $\sim 52{,}000\ \text{cm}^{-1}$ (see Table 5.6). The principal role in the absorption of light at temperatures $T \sim$ 3,000–10,000°K is played by the β-system, since, according to the Frank–Condon principle, the probable transitions in the γ- and δ-systems occur without large changes in the vibrational quantum number; in these two systems, basically only the high frequencies of the order of 45,000–52,000 cm^{-1} are absorbed, and these frequencies lie in the far ultraviolet region and are not very important at these temperatures. On the other hand, in the β-system transitions from the high lower vibrational states with $v'' \sim 12$ to the upper ground state $v' \sim 0$ are probable, and these produce absorption in the near ultraviolet and visible regions with frequencies $\sim 25{,}000\ \text{cm}^{-1}$.

At high densities and temperatures of the gas, the molecular lines broaden strongly and may even overlap, forming an almost continuous spectrum. Let us compare the line widths and the average distance between the lines in the β-system of NO. For our estimate we take the gas temperature T to be 8000°K and the density to be the same as that of atmospheric air at standard conditions. At 8000°K the Doppler width for lines with frequencies $\sim 25{,}000\ \text{cm}^{-1}$ is of the order of 0.3 cm^{-1}. If we assume that each gaskinetic collision changes the state of the vibrational or rotational motion, then the broadening due to collisions is even greater, of the order of

$$\frac{\Delta\nu}{c} = \frac{N\sigma_{\text{gas}}\bar{v}}{2\pi c} \approx \frac{3 \cdot 10^{19} \cdot 5 \cdot 10^{-15} \cdot 1.5 \cdot 10^{6}}{2\pi \cdot 3 \cdot 10^{10}} \approx 1.2\ \text{cm}^{-1}.$$

Let us estimate the average distance between the lines for the β-system of NO in the frequency interval from 15,000 to 45,000 cm^{-1}. In order to absorb even a very low energy photon at 15,000 cm^{-1} the molecule must be excited to an energy of $45{,}000 - 15{,}000 = 30{,}000$ cm^{-1}, that is, to the vibrational level $v'' \approx 20$. Examining the system of potential curves and taking the Frank–Condon principle into account, we can conclude that transitions to approximately five upper states are possible from each lower vibrational level, that is, the interval contains approximately $20 \cdot 5 = 100$ bands. At $T = 8000°K$ the rotational excitation of 2–3 kT, which corresponds to 7500 cm^{-1}, is appreciable, so that approximately $J'' \approx (2.5\, kT/hcB_e)^{1/2} \approx 80$ rotational levels of the lower state participate in the transitions. Each of these yields the two lines $J' = J'' + 1$ and $J'' = J'' - 1$ (the Q-branch $J' = J''$ is usually very weak for $J'' \gtrsim 10$), and there are 160 rotational lines in all per band. Each of them is doubled by the multiplet splitting and is then split again into two lines by Λ-type doubling. Thus in the frequency interval considered of 30,000 cm^{-1} we have approximately $100 \cdot 160 \cdot 2 \cdot 2 = 64{,}000$ lines. The average distance between them is of the order of 0.5 cm^{-1}; since the line width is ~ 1 cm^{-1}, the lines strongly overlap and the spectrum is actually almost continuous.

Let us make a rough estimate of the absorption coefficient. When the average vibrational excitation of a molecule is of the order of kT, that is, of the order of 5000 cm^{-1} at $T = 8000°K$, the most probable transitions are those to the lowest vibrational levels of the upper state. For purposes of our estimate let us assume that light is absorbed mainly during the transitions to the $v' = 0$ level of an upper electronic state. Then, the $h\nu$ photons are absorbed only by molecules excited to the energy $E_0 - h\nu$, where E_0 is the energy of the upper electronic state. According to the Boltzmann relation the number of such molecules is proportional to $\exp[-(E_0 - h\nu)/kT]$. We write the absorption coefficient in the form of (5.82), expressing the absorption cross section in terms of the differential oscillator strength

$$\kappa_\nu = \frac{\pi e^2}{mc} N \frac{df}{d\nu},$$

where N is the number density of NO molecules. Noting that according to the Frank–Condon principle the probability of absorption of photons with energies exceeding E_0 is very small, we can assume that the entire absorption "area" $\int_0^\infty (df/d\nu)\, d\nu$ is concentrated mainly in the frequency interval between 0 and E_0/h, and that the contribution to this integral of the frequency region between E_0/h and ∞ is very small. Recalling that $\kappa_\nu \sim df/d\nu \sim \exp[-(E_0 - h\nu)/kT]$, we find the proportionality coefficient from (5.106) or, equivalently, from the integral $\int_0^{E_0/h} (df/d\nu)\, d\nu$ which is equal to the oscillator strength f. Thus, we obtain

$$\frac{df}{d\nu} = f \frac{h}{kT} e^{-\frac{E_0 - h\nu}{kT}}$$

and

$$\kappa_\nu = \frac{\pi e^2}{mc} f N \frac{h}{kT} e^{-\frac{E_0 - h\nu}{kT}}. \tag{5.107}$$

Let us replace N, the number density of NO molecules, by its concentration in air $c_{NO} = N\rho_0/N_0\rho$, where N_0 is the molecular number density for atmospheric air and ρ/ρ_0 is the ratio of the air density to standard density ($\rho_0 = 1.27 \cdot 10^{-3}$ g/cm^3). We also replace the frequencies by the wave numbers $1/\lambda = \nu/c$. We then obtain

$$\kappa_\lambda = \frac{3.4 \cdot 10^7}{T^\circ} f \cdot c_{NO} \frac{\rho}{\rho_0} \exp\left[-\frac{1.44}{T^\circ}\left(\frac{1}{\lambda_{00}} - \frac{1}{\lambda}\right)\right] \text{cm}^{-1}. \tag{5.108}$$

Here $1/\lambda_{00} = E_0/hc$ (this quantity for the β-system of NO is equal to 45,440 cm^{-1}). In (5.108) $1/\lambda_{00}$ and $1/\lambda$ have the dimensions of cm^{-1}. Knowing the oscillator strength we can now estimate the absorption coefficient. Setting $f_\beta \approx 0.006$ for the β-system of NO (see §20), we find that for red light $\lambda =$ 6500 Å with $\rho/\rho_0 = 1$, $T = 8000°$K ($c_{NO} = 0.036$), that $\kappa_{NO} \approx 4.1 \cdot 10^{-3}$ cm^{-1} (the cross section per molecule is $\sigma_{NO} = \kappa_{NO}/N_{NO} = 4.3 \cdot 10^{-21}$ cm^2).

With respect to the distribution of the potential curves and the satisfaction of the Frank–Condon principle for transitions, the principal absorption system of molecular oxygen (the Schumann–Runge system) is completely analogous to the β-system of NO. Thus, the absorption coefficient for O_2 at high temperatures can be estimated from (5.107) and (5.108) where we must, of course, substitute the constants pertaining to O_2.

§20. More exact calculation of the molecular absorption coefficient at high temperatures

More accurate calculations of the absorption coefficient in a line (see [8, 21, 28]) must be based on the exact equations for the absorption coefficient and the actual probabilities of vibrational transitions. We shall again assume that the lines are broadened so that they almost (or quite appreciably) overlap.

We introduce an average absorption coefficient for the frequency ν and the given electronic transition $A \rightarrow B$. This coefficient is obtained by averaging the actual coefficient in a small spectral interval between ν and $\nu + \Delta\nu$, as was done in §12. To carry out the averaging we integrate the absorption coefficient for an individual line (5.105) with respect to the frequency (this yields the "area" of one line) and sum the integral over all lines contained in the frequency interval between ν and $\nu + \Delta\nu$. The result obtained will be equal to $\bar{\kappa}_\nu \Delta\nu$. Proceeding in the same manner as in the derivation of (5.106), we find the averaged absorption coefficient for the frequency ν and the given electronic transition,

$$\bar{\kappa}_{\nu AB} = \frac{\pi e^2}{mc} \frac{f_{BA}}{\Delta\nu} \sum_{\text{bands}} q_{v''v'} \sum_{J''} N_{Av''J''}. \tag{5.109}$$

The summations over J'' and over the bands extend up to those initial rotational levels and to those bands which give rise to lines and are contained in the spectral interval $\Delta\nu$. The number of molecules in the $A_{v''}J''$ state at a temperature T can be calculated by substituting the energy of the molecule in the given vibrational and rotational state into the Boltzmann equation (see [29])

$$N_{Av''J''} = N_A \frac{\exp\left(-\frac{hc\omega_A v''}{kT}\right)}{Z_{vA}} \frac{(2J''+1)\exp\left(-\frac{hcB_A J''(J''+1)}{kT}\right)}{Z_{rA}}. \tag{5.110}$$

Here N_A is the number of molecules in the electronic state A, and

$$Z_{vA} = \left(1 - e^{-\frac{hc\omega_A}{kT}}\right)^{-1} \approx \frac{kT}{hc\omega_A} \quad \text{and} \quad Z_{rA} = \frac{kT}{hcB_A}$$

are the vibrational and rotational partition functions in the lower electronic state, respectively.

The band is mainly filled by lines with large rotational numbers $J'' \gg 1$ for which, according to (5.90) and (5.91), the wave numbers are approximately given by

$$\frac{1}{\lambda} = \frac{1}{\lambda_{v''v'}} + (B_B - B_A)J''^2. \tag{5.111}$$

The frequency interval $\Delta\nu$ is filled with lines of the $v''v'$ band corresponding to rotational numbers from J'' to $J'' + \Delta J''$, where $\Delta J''$ is determined by differentiating (5.111),

$$\frac{\Delta\nu}{c} = \Delta\left(\frac{1}{\lambda}\right) = (B_B - B_A)2J''\,\Delta J''. \tag{5.112}$$

Assuming that $J'' \gg 1$ and $\Delta J'' \ll J''$ (the interval $\Delta\nu$ is sufficiently small), we can regard all the $\Delta J''$ terms in the summation over J'' in (5.109) to be the same. Neglecting unity in comparison with J'', we substitute J''^2 given by (5.111) into the exponent of (5.110) and replace $\Delta\nu$ in (5.109) through (5.112). Factoring out the exponential factor in (5.109) in order to obtain a final expression analogous to the approximate relation (5.107), we find

$$\bar{\kappa}_{\nu_{AB}} = \frac{\pi e^2}{mc} f_{BA} N_A \frac{h}{kT} e^{-\frac{(E_B - E_A) - h\nu}{kT}} \varphi, \tag{5.113}$$

where the dimensionless factor φ is defined as

$$\varphi = \frac{kT}{hc|B_B - B_A|} \frac{1}{Z_{vA} Z_{rA}} \exp\left[-\frac{hc}{kT}\left(\frac{1}{\lambda} - \frac{1}{\lambda_{00}}\right)\right] \times \sum_{\text{bands}} \exp\left\{-\left[\omega_A v'' + \frac{B_A}{B_B - B_A}\left(\frac{1}{\lambda} - \frac{1}{\lambda_{v''v'}}\right)\right]\right\}. \tag{5.114}$$

Here $1/\lambda_{00} = (1/hc)(E_B - E_A)$ is again the wave number corresponding to an electronic transition in the absence of vibrations and rotations (E_B and E_A are the energies of electronic states including the zero-point energy $E = U_e + hc\omega/2$) and $1/\lambda_{v''v'}$ is the wave number corresponding to the transition $Av'' \to Bv'$ in the absence of rotations. If $B_B > B_A$, then the bands have an edge on the "red" side and extend toward the "violet". If, however, $B_B < B_A$ then the opposite is true (see (5.111)). Hence, the sum over the bands in (5.114) is taken over those bands with $\lambda_{v''v'} > \lambda$ for $B_B > B_A$ (as in the case of the γ-system of NO) and over those bands with $\lambda_{v''v'} < \lambda$ for $B_B < B_A$ (as in the case of the β-system of NO or in the O_2 Schumann–Runge system). An excellent illustration of this situation is given in Figs. 5.24 and 5.25 taken from [21], which shows the value of the summation in the factor φ as a function of wavelength for the γ- and β-systems of NO, respectively. The curves have a "sawtooth" character, with each new tooth appearing when a new band

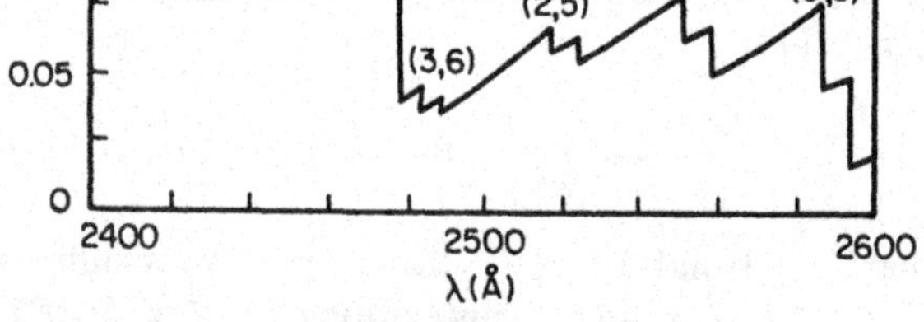

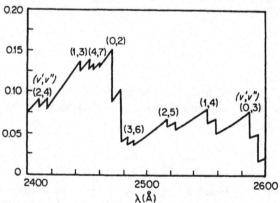

Fig. 5.24. Spectral absorption coefficient in the γ-system of NO, in relative units. $T = 8000°$K. The absorption jump at $\lambda = 2480$ Å corresponds to the inclusion of the 0–2 vibrational transition.

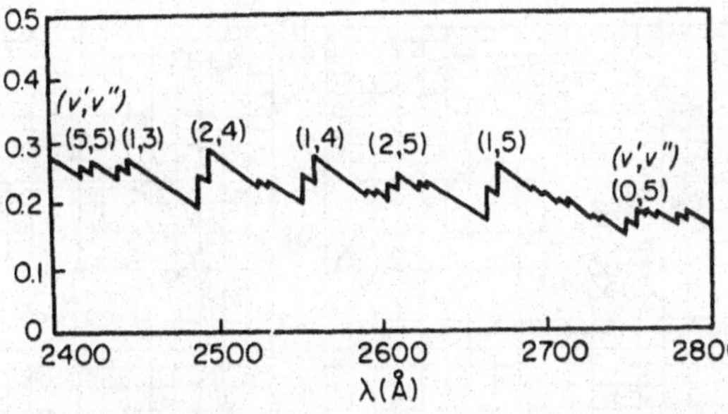

Fig. 5.25. Spectral absorption coefficient in the β-system of NO, in relative units. $T = 8000°K$.

is included in the absorption process. In the case of the γ-system, as λ decreases the absorption increases abruptly (in a jump), and in the case of the β-system the same happens with an increase in λ.

The more exact equation (5.114) is transformed into the approximate equation (5.107) by setting the factor φ (which takes into account the probability of various transitions) equal to unity (since (5.107) pertains to transition from the ground state, for which $E_A = 0$ and $N_A = N$). Calculations show that the coefficient φ is close to one, so that (5.107) may be used for rough estimates.

We have seen that molecular absorption coefficients can be calculated theoretically with the aid of spectroscopic data on molecules, energy level diagrams, vibrational and rotational constants, and potential curves, to within a constant factor–the oscillator strength f—which can be determined experimentally. Figures 5.26 to 5.28 show the results of calculations for the

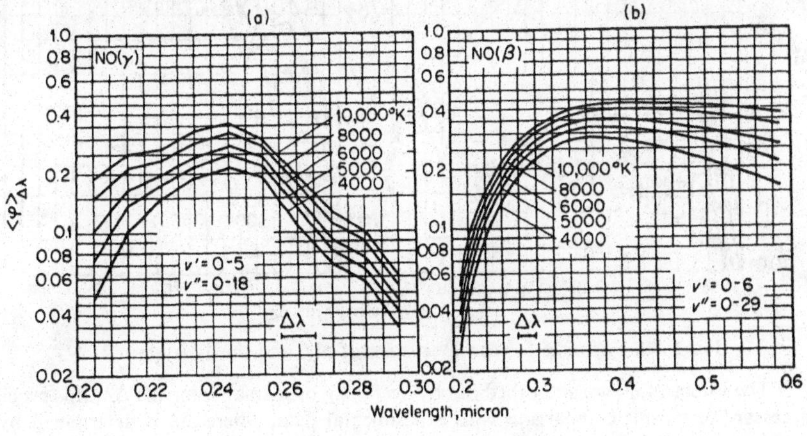

Fig. 5.26. Factor φ for the γ- and β-systems of NO.

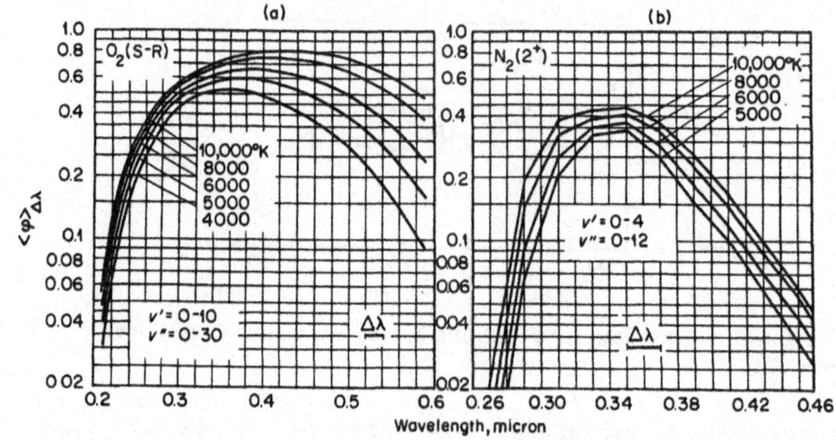

Fig. 5.27. Factor φ for the Schumann–Runge system of O_2 and for the 2^+ system of N_2.

factor φ_λ at several temperatures for the more important absorption systems* which determine the absorption properties of high-temperature air: the γ- and β-systems of NO, the Schumann–Runge system of O_2, the 1^+ and 2^+ systems of N_2, and the 1^- system of N_2^+ (ionized nitrogen molecule); the values of $1/\lambda_{00}$ for these systems are given in Table 5.6. The table of oscillator strengths for these systems is presented in the following section. Figures 5.26 to 5.28 are taken from [8].

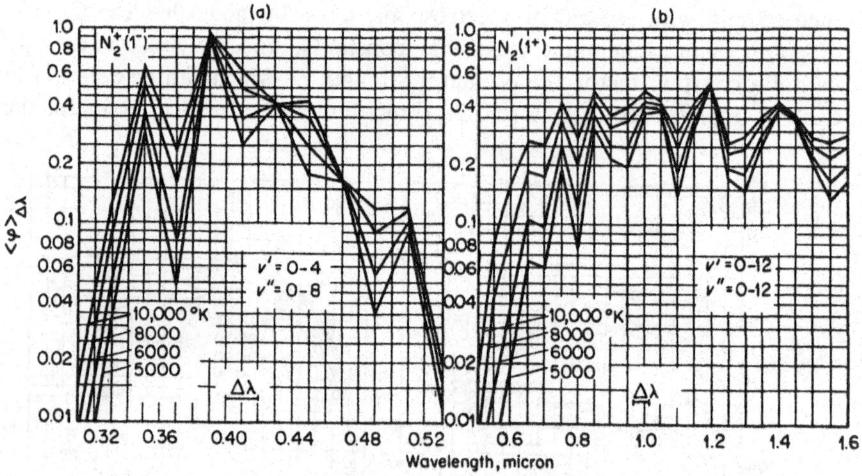

Fig. 5.28. Factor φ for the 1^- system of N_2^+ and the 1^+ system of N_2.

* The values of φ_λ are smoothed out by averaging over small intervals $\Delta\lambda$; this process is necessary to permit comparison with experimental data, where $\Delta\lambda$ is determined by the apparatus (a monochromator).

4. *Air*

§21. Radiative properties of high-temperature air

The absorption and emission of light by high-temperature air is of primary significance for such important practical problems as the study of phenomena occurring in the fire ball of a strong explosion (see Chapter IX) or the calculation of radiative heating of ballistic missiles and artificial satellites during re-entry into the atmosphere. The first problem involves a wide range of temperatures, up to hundreds of thousands and even a million degrees. Of greatest interest for the second problem are temperatures ~5000–20,000°K, which are developed behind the detached shock wave in front of bodies moving in the atmosphere with speeds of the order of up to 10 km/sec. The density range in these phenomena is also wide, from $\sim 10\rho_0$ (behind a shock wave propagating in air at standard density ρ_0) to very low densities $\sim 10^{-3}$–$10^{-4}\rho_0$ and even less in the central regions of a fireball and at high altitudes.

Cold air, as we know, is transparent to visible light. Absorption begins in the ultraviolet region and is attributed to the system of Schumann–Runge bands for oxygen molecules. The absorption reaches a practically measurable value at $\lambda \approx 1860$ Å. The experimental curve for the absorption coefficient of cold air at standard density as a function of wavelength is shown in Fig. 9.3 (§2 of Chapter IX).

At temperatures above 15,000–20,000°K, when the molecules are almost completely dissociated into atoms which are in turn appreciably ionized, the absorption of light in the continuous spectrum is composed of photoelectric absorption by atoms and ions and of bremsstrahlung absorption in the field of the ions. These mechanisms were considered in detail in Part 1 of the present chapter, where approximate formulas were obtained for the calculation of absorption coefficients and radiation mean free paths, based on the hydrogen-like approximation. In Table 5.2 of §8 were shown the results of calculations of the mean free paths in air for conditions of multiple ionization, that is, for temperatures above approximately 50,000°K. At temperatures below ~15,000°K all of the above mechanisms participate in the absorption, with the relative role of the several mechanisms strongly dependent on the light frequency and on the thermodynamic conditions, i.e., on the temperature and density. Associated with the mechanisms of continuous and quasi-continuous absorption are: molecular transitions in the molecules present in high-temperature air N_2, O_2, N_2^+ (ion), NO and NO_2, photoelectric absorption by the species O_2, N_2, NO, O, N, and O^-, free-free transitions in the fields of O^+, N^+, NO^+, O_2^+, and N_2^+ ions and also, possibly, in the fields of neutral atoms and molecules. Actual calculations of the absorption coefficients obviously require a knowledge of the concentrations of all the above components of air and also the concentration of free electrons (see Chapter III).

Radiative properties of high-temperature air have been studied in shock tubes by the AVCO-Everett Research Laboratory in the United States. Experimental and calculated results are presented in [8, 31, 32, 32a, 43–46], and in the reviews [28, 30, 47] (see also [33, 48]). The calculation of absorption and emission coefficients in high-temperature air are contained in a series of papers by Biberman and his associates. A review of these papers is given in [56], which considers the problem of the radiative heating of a body moving in the atmosphere at hypersonic speed. This article contains an extensive bibliography. Reference [64] is also devoted to problems of light absorption in air.

The main result of experimental studies of the radiative properties of air in shock tubes is the determination of oscillator strengths for the more important molecular transitions. In carrying out the experiment one measures the spectral intensity of radiation of a column of heated gas at different temperatures and densities. Behind the incident shock wave temperatures of the order of 3000–5000°K are studied, while behind the reflected wave temperatures of the order of 8000°K are studied. Conversion of the measured intensities into absorption coefficients can be accomplished using the well-known equation for radiative flux from a heated layer of a given thickness d (see §7, Chapter II, (2.38)). We recall that the amount of radiant energy in the wavelength interval $d\lambda$ emerging per unit time per unit area from a surface, per unit solid angle normal to the surface, is

$$I_\lambda \, d\lambda = I_{\lambda p} \, d\lambda \, (1 - e^{-\kappa'_\lambda d}), \tag{5.115}$$

where $I_{\lambda p}$ is the corresponding quantity for a perfect black body, given by

$$I_{\lambda p} = \frac{2hc^2}{\lambda^5} \frac{1}{e^{hc/kT\lambda} - 1},$$

and $\kappa'_\lambda = \kappa_\lambda(1 - e^{-hc/kT\lambda})$ is the absorption coefficient corrected for induced emission. If the layer is optically thin, then the self-absorption can be neglected (even in the line centers), i.e., if $\kappa'_\lambda d \ll 1$. The intensity of the radiation is determined in this case by the emission coefficient

$$I_\lambda = \frac{J_\lambda d}{4\pi} = I_{\lambda p} \kappa'_\lambda \, d.$$

The ratio of the measured intensity of radiation per unit layer thickness to the intensity of a perfect black body yields directly the corrected absorption coefficient κ'_λ.

The oscillator strength for the Schumann–Runge band system was determined by studying the radiation intensity behind an incident shock in pure oxygen at relatively low temperatures, of the order of 3000–4000°K. The degree of ionization is very small and the number of negative oxygen ions at these temperatures is low, and thus practically all the absorption can be attributed to molecular transitions. These data were used with (5.113) and (5.114) and with the previously calculated coefficients φ_λ to derive the oscillator strength $f_{\text{Sch-R}} = 0.028 \pm 0.008$. In the wavelength range from 3300 to 4700 Å this oscillator strength was found to be independent of λ, T, and ρ.

Data for oscillator strengths of NO and N_2 were obtained by analyzing the radiation spectra in air at different temperatures and densities. These quantities were extracted from a study of those spectral, temperature, and density regions where all of the unknown mechanisms, except one, play a minor role; the absorption due to the known mechanisms was then subtracted out from the measured quantities. In this manner the oscillator strengths for all the more important systems were found*; they are presented in Table 5.8. Figure 5.29 presents the experimentally and theoretically determined intensities of radiation at $T = 8000°K$ and $\rho = 0.83\rho_0$.

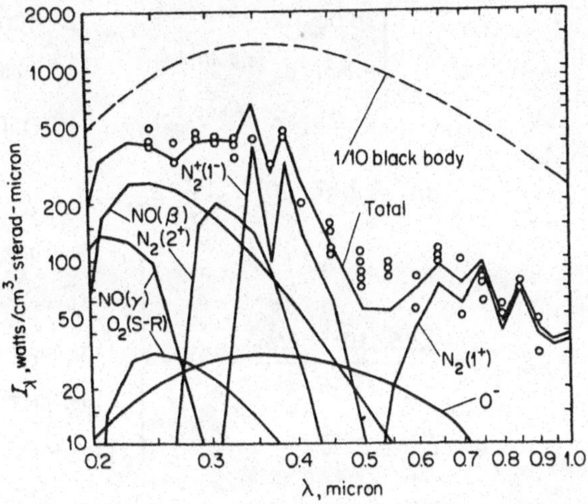

Fig. 5.29. Spectral intensity of radiation I_λ (watts/cm³-sterad-micron) of an air layer with a thickness $d \approx 1$ cm. $T = 8000°K$, $\rho = 0.83\ \rho_0$ (ρ_0 is standard density). The experimental points and the calculated curves correspond to the various emission mechanisms. The dashed curve gives the value $0.1\ I_{\lambda p}$ (1/10 of the radiation intensity of a black body). Since $d \approx 1$ cm, the ratio $I_\lambda / I_{\lambda p}$ gives directly the value of κ'_λ in cm^{-1}. (The diagram is taken from [8].)

* The value of f for the $N_2(1^+)$ system is a strong function of λ, because of the sharp changes in the internuclear distance with λ.

Table. 5.8

OSCILLATOR STRENGTHS FOR THE MORE IMPORTANT BAND SYSTEMS

System	$O_{2(\text{Sch-R})}$	NO_β	NO_γ	$N_2(2^+)$	$N_2^+(1^-)$	$N_2(1^+)$
f	0.028	0.006	0.001	0.09	0.18	0.025
Error	±0.008	±0.002	±0.0005	±0.03	±0.07	±0.008
Interval λ, Å	3300–4700	3500–5000	2500–2700	2900–3300	3300–4500	10,460

For convenience in the calculation of the absorption coefficient for air we present numerical formulas for the individual components in the region of molecular absorption and single ionization, for $T \lesssim 20{,}000°$K.

$$\kappa_{i\,\text{molec}} = \frac{10^5 c_i \rho/\rho_0}{T°} e^{h\nu/kT} \times \begin{cases} 9.5\varphi_{\text{Sch-R}} e^{-71{,}000/T} & \text{Schumann–Runge, } O_2 \\ 2.04\varphi_\beta e^{-65{,}300/T} & \beta\text{-system of NO,} \\ 0.34\varphi_\gamma e^{-63{,}500/T} & \gamma\text{-system of NO,} \\ 30.6\varphi_{2^+} e^{-127{,}500/T} & 2^+ \text{ system of } N_2, \\ 8.5\varphi_{1^+} e^{-84{,}900/T} & 1^+ \text{ system of } N_2, \\ 61.2\varphi_{1^-} e^{-36{,}800/T} & 1^- \text{ system of } N_2^+, \end{cases}$$

(column heading: System)

$$\kappa_{O^-} = 2.67 \cdot 10^{19} c_{O^-} \frac{\rho}{\rho_0} \sigma_{O^-},$$

$$\kappa_{i\,(\text{Kramers})} = \frac{2.56 \cdot 10^{12} c_i \rho/\rho_0}{T°^2 x^3} e^{h\nu/kT} \times \begin{cases} e^{-140{,}000/T} & O_2, \\ e^{-181{,}000/T} & N_2, \\ e^{-158{,}000/T} & O, \\ e^{-169{,}000/T} & N, \\ e^{-108{,}000/T} & NO, \end{cases}$$

$$x = \frac{h\nu}{kT} = \frac{1.44}{\lambda T}$$

$$\kappa_{NO_2} = 2.67 \cdot 10^{19} c_{NO_2} \frac{\rho}{\rho_0} \sigma_{NO_2}$$

(T in °K, λ in cm, κ in cm^{-1}).

In these formulas we define the concentrations of all particles c_i as the ratio of the number of particles to the original number of molecules in cold air. The oscillator strengths in the equations for the absorption coefficients were taken from values given in Table 5.8. The effective charge in the equations for the Kramers absorption was taken equal to one*.

Concentrations of the negative oxygen ions can be obtained from the Saha equation when the concentrations both of the oxygen atoms and of the free electrons are known. The absorption cross section for negative O^- ions is given in Fig. 5.5 in §5. There are experimental data on the existence of negative nitrogen ions N^- [57]. The absorption of light by these ions was discussed in [58]. When all the components of the absorption coefficient are known, the total coefficient and emission coefficient can be calculated at any temperature and density. In Figs. 5.30 and 5.31 are shown the reconstructions of the radiation for several values of temperature and density, obtained in this manner (the data were taken from [28]). The contributions of the individual absorption components are shown on the graphs.

Reference [32] considered the problem of free-free absorption by electrons in a field of neutral atoms. The emission coefficient was measured at $T = 8000°K$ and $\rho/\rho_0 = 0.85$ in the infrared spectral region with $\lambda \sim 20{,}000$ Å to 40,000 Å, where according to calculations all other mechanisms should play an

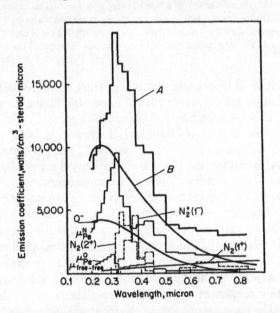

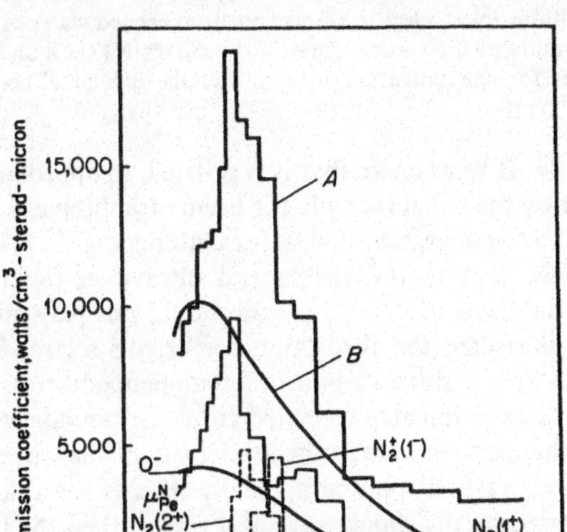

Fig. 5.30. Radiation intensity of air at $T = 12{,}000°K$, $\rho = \rho_0$ (standard density). The contributions of the various mechanisms are shown. μ_{Pe} are free-bound transitions (photoelectric absorption from excited states); A is the total radiation; B is 1/10 of the black body radiation intensity.

* The Biberman–Norman correction factor (see §7) has not been taken into account.

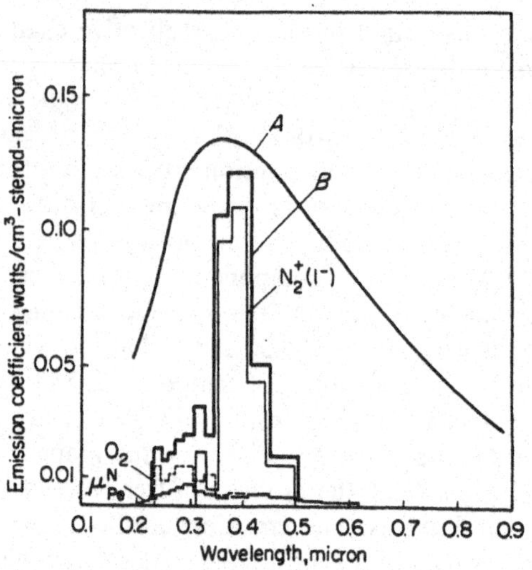

Fig. 5.31. Intensity of radiation for air at $T = 8000°K$ and $\rho = 10^{-3}\rho_0$. The curve A is 10^{-5} times the radiation intensity of a black body, B is the total radiation, and μ^{N}_{Pe} denotes free-bound transitions (photoelectric absorption from excited states of N). (*Editors' note.* The lateral separation of the vertical parts of the curves $N_2^+(1^-)$, and O_2 from the total radiation curve is for illustrative purposes so that the individual contributions may be distinguished.)

insignificant role. It was shown that it is possible to approximate the absorption coefficient by the usual formula for bremsstrahlung absorption, with the square of the effective charge $Z^2 = 0.04$ for O atoms and $Z^2 = 0.02$ for N atoms. These data show that in the visible and ultraviolet regions, the free-free absorption in the fields of neutral atoms should not play an important role.

Model' [34] measured the absorption coefficients for red light $\lambda = 6500$ Å in air behind a shock wave at two different temperatures. The detonation wave in Model's experiments originated from the boundary of the explosive with air. The change with time in the luminosity of the wave front normal to the surface was measured photographically. If d is the thickness of the air layer encompassed by the shock wave at a time t, then the luminosity of the front surface can be determined by (5.115). When the heated air layer becomes optically thick, $\kappa_\nu' d \gg 1$, the shock front radiates as a black body, and $I_\nu \approx I_{\nu p}$. The absorption coefficient was obtained from the curves of increasing luminosity $I_\nu(d)$. The temperature behind the front was determined independently of the luminosity of the front at the stage when $\kappa_\nu' d \gg 1$ and the front radiates as a black body. Model' obtained values of the absorption coefficient for two temperatures: $T = 10{,}900°K$, $\kappa_\lambda = 3.7\ \text{cm}^{-1}$, and $T = 7480°K$, $\kappa_\lambda = 1.66\ \text{cm}^{-1}$ ($\lambda = 6500$ Å, $\rho/\rho_0 \approx 10$). The first value agrees satisfactorily with the value calculated from the formulas listed above. The primary role is played by absorption in the 1^+ system of N_2 and by the Kramers mechanism. The experimental value of the second point is much higher than that obtained theoretically*.

A characteristic feature of all the absorption components considered above (see the summary of formulas following Table 5.8) is the sharp, Boltzmann-type dependence on temperature with appreciable activation energies. At moderate temperatures, of the order of 3000–4000°K, all of the coefficients in the visible region become very small; for example, at $T = 4000$°K and $\rho/\rho_0 = 1$, $\kappa \sim 10^{-6}$ cm^{-1}. At such temperatures the main role in absorption is played by the molecular absorption of the nitrogen dioxide present in air in very small amounts (see Table 5.9)†, but a strong absorber of light in both the visible and ultraviolet regions. Molecular NO_2 bands form a very complex system with practically all lines overlapping. Figure 5.32 shows the absorption cross sections of cold NO_2 molecules plotted from the data of [35]. The cross section decreases monotonically from $\sigma = 6.5 \cdot 10^{-19}$ cm^2 to $\sigma \approx 10^{-20}$ cm^2 in the wavelength interval from $\lambda = 4000$ Å to $\lambda = 7000$ Å. The measurements of [36] show that the absorption cross sections in the infrared region are very small; for $\lambda = 10{,}000$–$20{,}000$ Å, $\sigma < 4.5 \cdot 10^{-23}$ cm^2. In the near ultraviolet region at $\lambda = 3020$ Å, the cross section passes through a minimum [37]; this fact and the curve in Fig. 5.32 indicate that the absorption maximum lies in the blue part of the spectrum at $\lambda \sim 4000$ Å. It may be expected, therefore, that the absorption spectrum at temperatures of the order of 2000–4000°K is strongly displaced in the red direction and that the cross section of NO_2 in the entire visible region becomes of the order of several

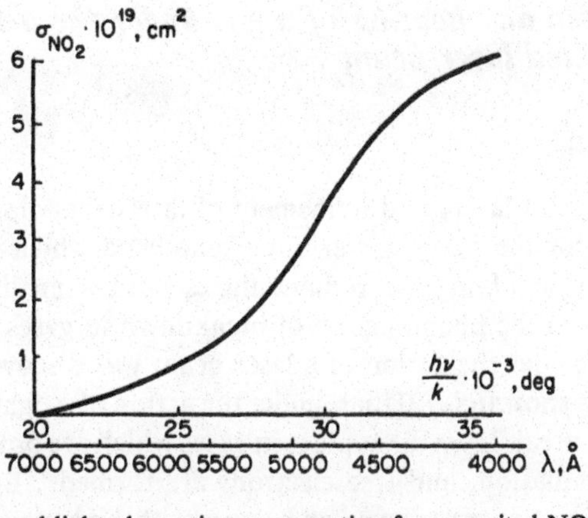

Fig. 5.32. Averaged light absorption cross section for unexcited NO_2 molecules.

* No definite explanation for this disagreement can be offered.

† Concentrations of NO_2 were calculated in [39].

Table 5.9

EQUILIBRIUM CONCENTRATIONS OF NITROGEN DIOXIDE IN HIGH-TEMPERATURE AIR, $c_{NO_2} \cdot 10^4$

	ρ/ρ_0				ρ/ρ_0		
T, °K	10	5	1	T, °K	10	5	1
2000	1.11	0.79	0.35	3500	2.91	1.92	0.79
2600	2.02	1.42	0.63	4000	2.86	1.90	0.67
3000	2.24	1.58	0.69	5000	2.11	1.29	0.25

Note: c_{NO_2} = number of NO_2 molecules per original number of molecules in air.

times 10^{-19} cm^2 (additional details are given in [38]; see also §7 of Chapter IX). For example, for a concentration of NO_2 molecules in air of the order of 10^{-4}, the absorption coefficient at standard density is of the order of 10^{-3} cm^{-1}. In [59] the intensity of radiation of nitrogen dioxide was studied in the temperature range 1400–2100°K (the dioxide was mixed with argon and heated in a shock tube). Absolute intensities in the visible part of the spectrum were obtained.

In conclusion, we should note that the problem of the radiative properties of high-temperature air still does not appear to be fully solved and research in this area is continuing.

5. *Breakdown and heating of a gas under the action of a concentrated laser beam*

§22. Breakdown

The invention of lasers and refinement of laser technology have opened up wide avenues for the study of various phenomena which occur as the result of the interaction of intense radiant fluxes with a medium. In particular, several years ago the phenomenon of breakdown in gases and the formation of a "spark" under the action of a laser beam was discovered.

Experiments show [65–72] that, under the action of a light flux of sufficiently high intensity, breakdown takes place in gases which are ordinarily transparent to the given radiation, and free electrons are formed*. Breakdown requires very large radiant energy fluxes and, even with high power modern optical

* The phenomenon of high frequency breakdown under the action of radiation in the microwave frequency range is well known and has been thoroughly studied [60].

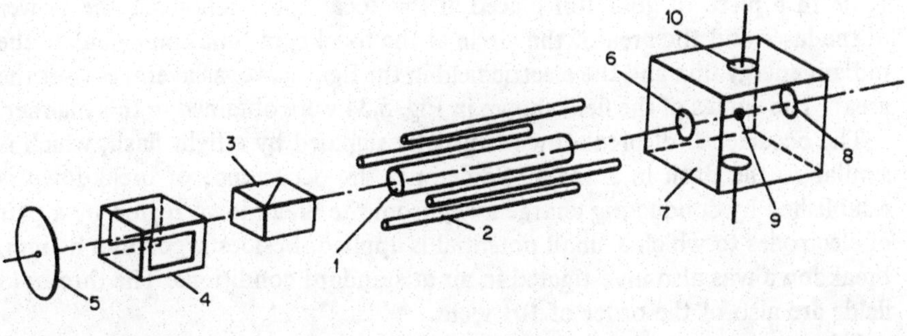

Fig. 5.33. Schematic view of breakdown experiment. (1) Ruby, (2) xenon flash lamps, (3) polarizer, (4) Kerr cell, (5) mirror, (6) lens, (7) vessel with test gas, (8) outlet aperture, (9) focus, (10) collection electrode.

generators (with Q-switching) such fluxes can be obtained only by focusing the laser beam by a lens (see Fig. 5.33). The breakdown threshold, which is usually extremely sharp, is conventionally characterized by the intensity of the electric field of the light wave. Figure 5.34 shows, by way of example, measurements from [65] of the threshold fields for breakdown in argon and helium at different pressures. In these experiments a Q-switched ruby laser gave pulses with a duration of $3 \cdot 10^{-8}$ sec and with a maximum (peak) power of up to 30 megawatts (an energy in the pulse up to 1 joule). The beam was focused by a lens into a circle with a radius of approximately $r_0 = 10^{-2}$ cm. The radius of the focal spot was estimated, first, from the angular spread of

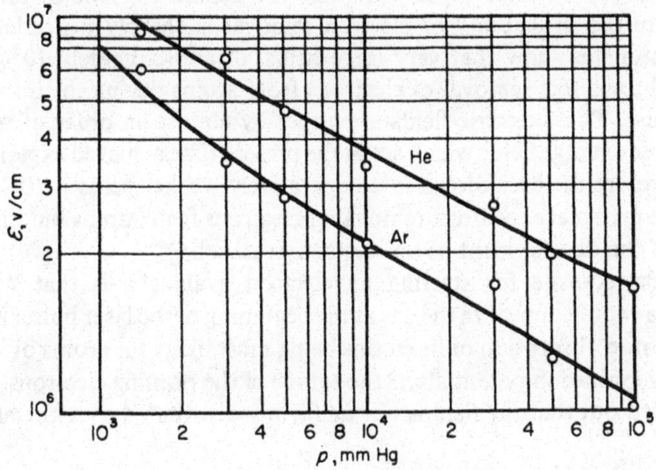

Fig. 5.34. Breakdown fields in argon and helium as a function of gas pressure.

the unfocused laser beam and second, from the size of the hole burned by the beam in a piece of gold foil placed at the focal spot. Knowing the power of the laser and the area of the circle at the focal spot, one can calculate the radiant energy flux and the electric field in the light wave, as averages over the area*. The values of the field shown in Fig. 5.34 were obtained in this manner.

The occurrence of breakdown is usually signaled by a light flash, which is similar to the flash in a spark. Sometimes the occurrence of breakdown is established by conducting charge away from the breakdown region by a pair of electrodes to which a small potential is applied. Besides argon and helium, breakdown was also investigated in air at standard conditions. The threshold fields are also of the order of 10^7 v/cm.

It is necessary to assume that for sufficiently high intensity radiant energy fluxes or electric fields in the light wave, it is possible to have direct removal of electrons from atoms due to radiation. The corresponding quantum-mechanical problem was solved by Keldysh [73], who has found a general expression for the probability of electron removal. In the limit of low frequencies it reduces to the known formula for the probability of the tunnel effect under the action of a static field, and in the case of sufficiently high (and, in particular, optical) frequencies, it describes the multiphoton effect, when ionization proceeds as a result of simultaneous absorption of n photons, whose energy $nh\nu$ exceeds the ionization potential I. The probability of the multiphoton photoelectric effect is proportional to the nth power of the radiant energy flux or to $\mathscr{E}^{2n}$, where $\mathscr{E}$ is the electric field intensity. For atoms and molecules with high ionization potentials, such as Ar, He, N_2, and O_2, a large number of photons is required; for example, in a ruby laser $h\nu = 1.8$ ev and for argon $I = 15.8$ ev, $n = 9$, while for helium $I = 24.6$ ev and $n = 14$. Therefore the probability of electron removal is sharply dependent on the field. Estimates show that very large fields, of approximately 10^8 v/cm, are required for direct removal of electrons from atoms during the duration of a laser pulse. These electric fields are larger by almost an order of magnitude than those average fields which are at the present time achieved experimentally, and therefore the breakdown in comparatively weaker fields $\sim 10^6$–10^7 v/cm does not take place by direct removal of electrons from atoms but instead as a result of the development of an electron avalanche.

The prerequisite for starting an electron avalanche is that "priming" electrons should appear in the gas at the beginning of the laser pulse. Evidently, multiphoton absorption of light, this being most likely for atoms of impurities with low ionization potentials, is the source of the priming electrons. It should be pointed out that the field is not uniformly distributed over the area of the

* For example, for a power of 30 megawatts and a radius of 10^{-2} cm, the energy flux is approximately 10^{18} erg/cm^2 · sec and the field is approximately $0.6 \cdot 10^7$ v/cm.

focal circle. There exist very small regions with local fields which appreciably exceed the average field over the circle. These are the regions in which the first electrons, which start up the avalanche, are born. The nonuniform distribution of the field at the focal spot is a result of the nonuniformity and divergence of the unfocused laser beam and is also a result of the lens aberration. The latter effect has been studied in detail in the paper of Zel'dovich and Pilipetskii [83].

Let us consider the manner in which the electron avalanche develops. The electrons absorb photons on collisions with neutral atoms (see §8a), and thus acquire sufficient energy for ionization. As a result of the ionization there appear in place of the one "fast" electron two "slow" ones, which again acquire energy from the radiation, ionize atoms, and so forth. Together with absorption under the action of the high-intensity light flux there appears also induced emission of photons with the same energies and directions; however, the resulting effect is positive, and an electron, on the average, acquires energy from the radiation and is thus accelerated.

The rate of increase of the electron energy can be estimated by using (5.57b) for the effective absorption, which also describes exactly the resulting effect of the true absorption and induced emission of photons. We denote the radiant energy flux by

$$G = \frac{c}{8\pi}\overline{(\mathscr{E}^2 + H^2)} = \frac{c\overline{\mathscr{E}^2}}{4\pi} \text{ erg/cm}^2\cdot\text{sec},$$

where $\overline{\mathscr{E}^2}$ is the mean square of the electric field in the light wave. The radiant energy absorbed per unit time per unit volume is $G\kappa_{\nu\,\text{neut}}$, and the energy absorbed per electron is $G\kappa_{\nu\,\text{neut}}/N_e$. This quantity also represents the mean rate of increase of the electron energy:

$$\frac{dE}{dt} = \frac{e^2 G}{\pi m c \nu^2}\nu_{\text{eff}} = \frac{e^2\overline{\mathscr{E}^2}}{m\omega^2}\nu_{\text{eff}}\,^*. \tag{5.116}$$

In order to acquire the energy of one photon $h\nu$, the electron must, on the average, undergo $h\nu\pi mc\nu^2/e^2G$ collisions with atoms. For example, for a flux $G = 10^{18}$ erg/cm²·sec, the field $(\overline{\mathscr{E}^2})^{1/2} = 0.6\cdot 10^{17}$ v/cm, and $h\nu = 1.8$ ev,

* This relation can be interpreted in the following manner. Under the action of the alternating electric field of the light wave, oscillations are superposed on the rectilinear motion of the electron. The momentum of the oscillations is $p = e\mathscr{E}/\omega$, and the average energy $\overline{p^2}/2m = e^2\overline{\mathscr{E}^2}/2m\omega^2$. The electron velocity direction changes sharply on colliding with an atom, and a quantity of the order of the energy of the oscillations is added to the energy of random motion. The electron experiences ν_{eff} collisions per unit time and the rate of increase of the additional energy of random motion is equal to (5.116).

this number is 200. In order to acquire an energy equal to the ionization potential I, disregarding energy losses, the number of collisions needed is increased by a factor of $I/h\nu$; for example, in helium this number is 2700. Under atmospheric pressure the frequency of collisions of an electron with an energy $E \approx$ 10 ev in helium is $\nu_{eff} \approx 2 \cdot 10^{12}$ sec^{-1}, so that the time required is approximately $2700/\nu_{eff} = 1.3 \cdot 10^{-9}$ sec. Thus during the time of a laser pulse of $3 \cdot 10^{-8}$ sec, $3 \cdot 10^{-8}/1.3 \cdot 10^{-9} = 23$ generations of electrons would be born, so that on the average $2^{23} \approx 10^7$ electrons would be born for each "priming" electron*. The avalanche development is extremely sensitive to the intensity of the light flux and to the gas density. For example, when the flux or the gas pressure is increased by a factor of 2 (the field is increased by 40%) the rate of energy acquisition and the number of generations in the avalanche would have doubled, so that toward the end of a pulse of the same duration 10^{14} rather than 10^7 new electrons would have been born for each "priming" electron. This extreme sensitivity also explains the experimental discovery of the existence of an abrupt threshold for breakdown in a gas, both with respect to the laser pulse intensity and with respect to the gas pressure.

The above simple considerations on energy acquisition by an electron under the action of a light wave only give the general outline of the process. In fact, the development of the electron avalanche is significantly more complex. An appreciable role is played by the excitation of atoms, by electrons with energies insufficient for ionization but sufficient for excitation. In not too strong fields the excitation decelerates the development of the avalanche, since during excitation the electron discards the acquired energy and must start acquiring it anew. For strong fields with strengths of one to several times 10^7 v/cm, the situation is reversed, and excitation of the atoms promotes the avalanche development, since the excited atoms are rapidly ionized under the action of the radiation (as a result of either single or successive absorption of a small number of photons). In light gases, such as helium, a significant role is also played by the electron energy loss due to elastic collisions with atoms†. In certain cases (small focusing volumes and low gas density) it is possible for electrons to leave the volume of focus by diffusion.

* The number of electrons in the avalanche increases with time according to the relation $N_e = N_{e_0} \exp t/\theta$, where $\theta \ln 2$ is the doubling time for the electron number density.

† In each collision an electron, on the average, loses $2m/M$ of its energy (where M is the mass of the atom). Therefore, when we take into account elastic losses, (5.116) takes the form

$$\frac{dE}{dt} = \nu_{eff}\left[\frac{e^2\overline{\mathscr{E}^2}}{m\omega^2} - \frac{2mE}{M}\right].$$

The electron cannot, on the average, acquire energy which exceeds $E_m = (M/2m)e^2\mathscr{E}^2/m\omega^2$. For example, for helium in a field whose strength is $6 \cdot 10^6$ v/cm, $E_m = 33$ ev. Actually, we note that in the expression for the elastic losses, in place of E the difference of average energies of the electrons and atoms (ions) should appear; however, in the given case the atom gas is cold.

Under conditions when the avalanche development is retarded as a result of losses in electron energies due to excitation of the atoms, a simple formula for energy acquisition by an electron, of the type of (5.116) but with a negative term to take into account energy losses, is insufficient for describing this complex process. It becomes necessary to consider a kinetic equation for the electron distribution function with respect to energy. This was done by the authors in [62], where under several assumptions threshold fields for breakdown were calculated and where satisfactory agreement was obtained with the experimental results for argon and helium from [65]. Problems of breakdown in gases at the focal spot of a laser beam have also been treated theoretically [73–76].

§23. Absorption of a laser beam and heating of a gas after initial breakdown

If the radiant energy flux in the focus appreciably exceeds the threshold value for breakdown, the gas becomes highly ionized and the plasma thus produced will practically completely absorb the beam, as a result of free-free electron transitions in the ion field. In this case the gas is heated to high temperatures. Thus, for example, measurements of the intensity of x-ray radiation from the region of the focal spot in the experiments reported in [71] show that the radiation brightness (and color) temperatures, which characterize the electron temperatures, are approximately 600,000°K. In these experiments the breakdown in atmospheric air was studied using a ruby laser with an energy pulse of 2.5 joules and a duration of $4 \cdot 10^{-8}$ sec, and with the radius of the focal circle $r_0 = 10^{-2}$ cm (the breakdown threshold energy for the given duration and radius is about 1 joule).

We now consider the manner in which the beam is absorbed, and we also estimate the temperature to which the gas is heated (this was carried out by one of the authors [77]). We assume that breakdown takes place at the focal spot, in the narrowest part of the light column (see Fig. 5.35), where the radiant energy flux is a maximum, and that a high degree of ionization and a high temperature have already been established. The light is absorbed in a layer of the order of a photon mean free path l_ν and it heats the gas. The mean free path for photon absorption can be estimated from (5.21), multiplying

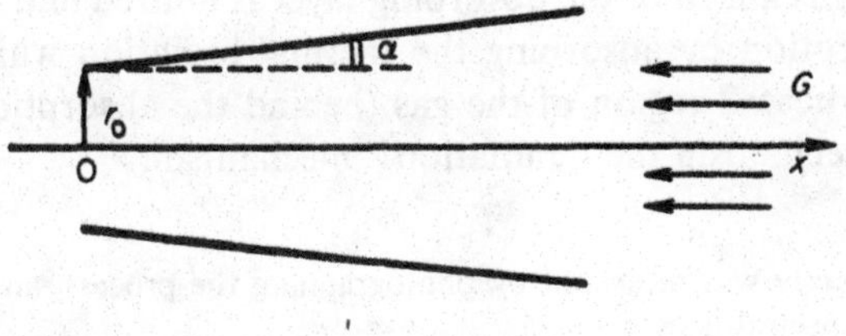

Fig. 5.35. Schematic diagram of light column in the region of the focal spot.

the coefficient κ_ν by the quantity $1 - e^{-h\nu/kT} \approx h\nu/kT$ ($h\nu = 1.8$ ev $\ll kT$) to take into account the induced emission. The mean free path in standard density air at temperatures of 10^5–10^6 °K is found to be equal to (2–7) · 10^{-3} cm (the degree of ionization or the ionic charge is $Z = 2.7$–6.6).

One of the most remarkable features of the process, which is rather obvious physically and which was discovered experimentally [69, 71], consists of the displacement of the beam absorption zone toward the light flux. Velocities measured in these experiments were approximately 100 km/sec*. It is easy to understand the reasons for the displacement of the absorption zone. The photons are strongly absorbed in the highly ionized medium. As soon as the degree of ionization ahead of the gas layer which is absorbing at the given time becomes sufficiently high, the new layer becomes opaque and it begins to absorb the beam strongly. Thus an "absorption and heating wave" is propagated along the light column toward the beam. This effect prevents the release of the entire energy of the pulse in the very small volume of the focal spot in which the breakdown started, and it limits the heating of the gas.

We can point out three different and independent mechanisms which lead to the appearance of the absorption wave:

(1) If the radiant energy flux at the focal spot exceeds appreciably the breakdown threshold, then it also exceeds the threshold value over a certain length of the light column which expands in the direction of the lens. Breakdown also occurs in these parts of the column, but with a time lag with respect to the narrowest point, the lag being the greater the wider is the column cross section, the smaller is the flux. Thus a "breakdown wave" moves toward the beam.

(2) The heated gas in the absorbing layer expands and sends out a shock wave in all directions, including the direction along the light column toward the beam. Across the shock wave the gas is heated and ionized, so that the zone of light absorption and energy release in the gas is displaced behind the shock front. This hydrodynamic mechanism is similar in many respects to the detonation of explosives. A "detonation mechanism" was noted by Ramsden and Savic [78], which is in agreement with the experimental values of wave speed. However, this article contains an incorrect assumption regarding the heating temperature. For a critique of this article see [77].

(3) The gas ahead of the absorbing layer is ionized and becomes capable of light absorption by absorbing the thermal radiation which is emitted from the highly heated region of the gas (behind the absorption wave front). We call this mechanism the "radiation" mechanism.

* The velocity was measured by photographing the process, and also by the Doppler shift of the spectral lines.

The effectiveness of each of the mechanisms is characterized by the rate of displacement of the absorption wave which it produces; the actual wave, naturally, moves with the highest of the possible speeds*.

In a certain sense we can regard the absorption wave as a gasdynamic discontinuity (see Chapter I). In a coordinate system moving with the wave the process is quasi-steady. In fact, in the time Δt during which the wave travels through a distance of the order of its width Δx the light flux and the wave velocity $D = \Delta x/\Delta t$ cannot change appreciably ($\Delta x \approx l_\nu \lesssim 10^{-2}$ cm, $D \approx 100$ km/sec, and $\Delta t \lesssim 10^{-9}$ sec).

Let us set up the energy balance without at the moment considering the fact that it is the heating which sets the gas into motion. An amount of energy $G\,dt$, where G is the radiant energy flux, falls on a unit area of the wave surface during the time dt. It is expended in heating up the mass $\rho_0 D\,dt$ which is swept out by the wave in this time (ρ_0 is the initial density of the gas). Consequently, the specific internal energy $\varepsilon(T)$ which the gas acquires after complete absorption of the light flux is determined by

$$\rho_0 D\varepsilon(t) = G. \tag{5.117}$$

This equation is a direct expression of the law of energy conservation, and is independent of the particular mechanism of wave propagation.

A more detailed consideration must be based on the general conservation laws for mass, momentum, and energy, applied to the flow of the gas through the shock front in exactly the same manner as was done in deriving the relations across a shock wave (see Chapter I). From such a calculation we obtain the equation of a "Hugoniot curve" for the absorption wave, which relates the pressure and density of the gas behind the front with the initial density and with the energy flux G incident on the wave†. A Hugoniot curve for an absorption wave is shown schematically in Fig. 5.36.

An energy balance equation of the type (5.117) does not change much when it is modified to take into account the work of compression and changes in the kinetic energy of the gas. The changes reduce to the fact that G in the equation is replaced by a slightly different quantity $G\beta$, where the coefficient β is bounded within the rather narrow limits $1 < \beta \leqslant 2\gamma/(\gamma + 1)$, where γ is the adiabatic exponent or ratio of specific heats of the gas. For air at temperatures of 10^5–10^6 °K, $\gamma \approx 1.33$ and $1 < \beta < 1.14$, and the elementary energy

* It has been estimated that the heating and ionization ahead of the absorbing layer due to the electron thermal conductivity and electron diffusion do not play an important role.

† Despite common features, it differs from the Hugoniot curve for an explosive in which energy is released, since for the explosive medium the energy per unit mass is a constant quantity, while the energy release per unit mass accompanying light absorption $G/\rho_0 D$ depends on the wave speed.

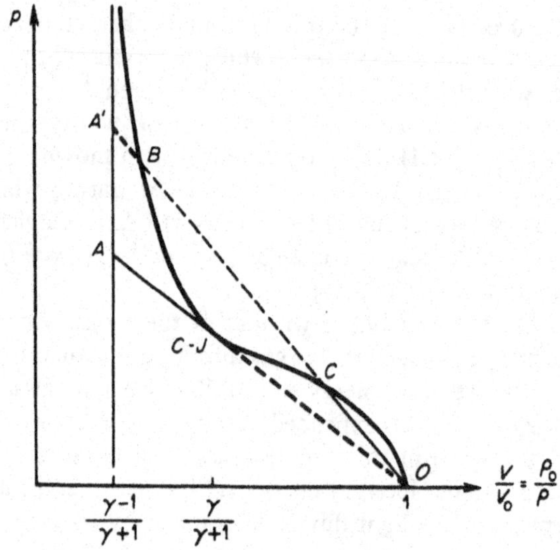

Fig. 5.36. Hugoniot curve for absorption wave.

equation (5.117) is found to be always valid with a sufficient degree of accuracy. This equation relates the speed of the absorption wave to the heating temperature and makes it possible to estimate the temperature on the basis of the experimentally obtained velocity, even if the mechanism of the absorption wave propagation is not known.

As in the case of a shock wave, the wave speed D is determined by the slope of a line drawn on the p, V diagram ($V = 1/\rho$) of Fig. 5.36 from the initial state point 0 to the point representing the final state of the gas behind the wave. It can be seen from Fig. 5.36 that for a given radiant energy flux G there exists a minimum possible propagation speed corresponding to the final state point C–J. This is the so-called Chapman–Jouguet point, at which the speed of the wave relative to the heated fluid behind it is identical with the local sound speed. The physical meaning of the Chapman–Jouguet point and the reasons why the corresponding state is achieved in the detonation of an explosive were explained by one of the authors [81], and also independently by J. von Neumann and by W. Döring. Problems in detonation theory are treated in the book [82].

When the effectiveness of other ionization mechanisms (sparking) is less than that of ionization by a shock wave, it is the hydrodynamic (or detonation) mode which results. In this case the gas is compressed and heated by the shock wave to the state A and then having received additional energy due to absorption of light, it expands along the line A C–J, reaching the Chapman–Jouguet point at the instant when the energy release ceases. The minimum speed of the absorption wave is

$$D = \left[2(\gamma^2 - 1) \frac{G}{\rho_0} \right]^{1/3} \tag{5.118}$$

The heating in this state has the maximum possible value

$$\varepsilon = \frac{\gamma}{(\gamma^2 - 1)(\gamma + 1)} D^2 = \frac{2^{2/3}\gamma}{(\gamma^2 - 1)^{1/3}(\gamma + 1)} \left(\frac{G}{\rho_0} \right)^{2/3} \tag{5.119}$$

If any of the ionization mechanisms (for example, the breakdown mechanism) gives for a given flux G a wave propagation speed higher than the "normal detonation" velocity (5.118), then no shock wave is formed in the light column. The gas, absorbing the light flux, goes from the initial state O to the final state C by continuous heating and compression along the straight line OC; here the higher pressure and density are not caused by but rather are the result of the appearance of the wave. In this case the wave propagates with a supersonic velocity relative to the gas behind it*. Let us give a numerical example. When $G = 2 \cdot 10^{18}$ erg/cm$^2 \cdot$ sec, $\rho_0 = 1.3 \cdot 10^{-3}$ g/cm^3, and $\gamma = 1.33$, corresponding to the conditions of the experiments of [71], we get from (5.118) and (5.119) that $D = 133$ km/sec, $\varepsilon = 1.35 \cdot 10^{14}$ erg/g, which at equilibrium corresponds to a temperature $T = 910{,}000$°K (the experimental values are $D = 110$ km/sec and $T = 600{,}000$°K). If we take into account energy losses due to lateral expansion of the gas, as a result of which the "acting" flux G is decreased by approximately a factor of 2, then we find that $D =$ 105 km/sec, $\varepsilon = 8.5 \cdot 10^{13}$ erg/g, and $T = 720{,}000$°K are quite close to the values obtained experimentally.

As was shown in [77], the speed of the "breakdown" wave can be estimated from the relation

$$D = \frac{r_0}{t_c \tan \alpha},$$

where r_0 is the minimum radius of the light column at the focal spot, α is the half-angle of the light column (see Fig. 5.35), and t_c is the time for the initial breakdown at the focal spot, measured from some effective start of the laser pulse. For example, for the experiments of [77], $r_0 = 10^{-2}$ cm, $\tan \alpha = 0.1$, $t_c \approx 10^{-8}$ sec, and the velocity of the "breakdown" wave is $D \approx 100$ km/sec, close to the detonation speed.

In the experiments described in [69] use was made of a short-focal-length lens which gave $r_0 = 4 \cdot 10^{-3}$ cm, $\tan \alpha = 1$, and $t_c = 7 \cdot 10^{-9}$ sec. Under

* Hydrodynamic states involving ionization by a shock wave, but with a speed exceeding that of a detonation ($O \to A' \to B$), are not achieved. The motion behind the wave in this case would be subsonic; the expansion of the heated gas behind the wave would immediately weaken the wave, bringing it to the state of "normal detonation".

these conditions the breakdown velocity is 6 km/sec, while the detonation speed is more than 100 km/sec, so that in these experiments there was definitely no role played by the breakdown mechanism. The breakdown mechanism is the principal mechanism and determines the motion of the absorption wave for very powerful short laser pulses with the use of long-focal-length lenses (small angles α).

Treatment of the radiation mode by any exact analysis is extremely difficult. The approximate calculation of the speed in this mode made in [77] leads to rather cumbersome formulas, which we shall not write out here. We only wish to point out that the speed in the radiation mode is found to be quite close to the detonation speed and, in addition, both speeds depend in approximately the same manner on the laser intensity. Thus, within the limits of accuracy of the theory, it is not possible to give preference to one of these two mechanisms. Practically this means that in those cases when the effectiveness of the breakdown mechanism is low, the real wave moves with a speed which is approximately the same as the speed of the two remaining modes.

After termination of the laser pulse there remains in the gas a highly heated column (when the wave speed is 100 km/sec and the pulse duration is $3 \cdot 10^{-8}$ sec, its length is 3 mm). The gas expands and the subsequent process is similar to that of a strong explosion.

We note the papers of Basov and Krokhin [79] and also [80], in which preliminary estimates are given of the laser intensities needed to heat hydrogen to thermonuclear temperatures. The problems of breakdown and heating of gases under the action of a laser beam are considered in more detail in a review article by one of the authors [86]. This reference also contains an extensive bibliography of experimental and theoretical papers on this topic.

VI. Rates of relaxation processes in gases

1. Molecular gases

§1. Establishment of thermodynamic equilibrium

The state of a gas depends on the concentrations of the various components, such as atoms, molecules, ions, and electrons, and on the distribution of the internal energy among the various degrees of freedom. Generally, the internal energy of a gas consists of the energy of translational motion of the particles, rotational and vibrational energies of the molecules, chemical energy, ionization energy, and the electronic excitation energy of the atoms, molecules, and ions. For complete thermodynamic equilibrium the state of the gas is completely determined by the concentration of each element in the gas mixture and any two macroscopic thermodynamic parameters as, for example, density and specific internal energy.

Excitation of any of the degrees of freedom* and the establishment of thermodynamic equilibrium require a certain time, whose scale is the so-called relaxation time. Relaxation times for exciting the various degrees of freedom frequently differ appreciably from each other. Therefore, under certain conditions it is possible to achieve thermodynamic equilibrium for some but not all of the degrees of freedom. Equilibrium is established most quickly for the translational degrees of freedom. If some arbitrary velocity distribution of atoms or molecules exists initially, then even after a small number of elastic collisions of particles with their neighbors, the particle velocity distribution will become Maxwellian. The Maxwell distribution is established as a result of the exchange of momentum and kinetic energy among the particles. It should be noted that both the kinetic energy and momentum transfer during the collision of particles with comparable masses are on the average of the same order as the kinetic energy and momentum of the colliding particles. Therefore, the relaxation time for establishing a Maxwell distribution in particles of the same species, or in particles of different species but with comparable masses, is of the order of the average time between gaskinetic collisions

$$\tau_{\text{trans}} \sim \tau_{\text{gas}} = \frac{l}{\bar{v}} = \frac{1}{n\bar{v}\sigma_{\text{gas}}}. \tag{6.1}$$

* For brevity we shall refer to dissociation, chemical reactions, and ionization as "degrees of freedom".

Here l is the gaskinetic mean free path, $\bar{v}$ is the average particle velocity, n is the particle number density, and σ_{gas} is the gaskinetic cross section. For example, in air at standard conditions $l \approx 6 \cdot 10^{-6}$ cm and $\tau_{trans} \sim 10^{-10}$ sec.

Usually the gaskinetic times are very small in comparison with the flow times over which appreciable changes in the macroscopic parameters of the gas, such as density or energy, take place. Therefore, as a rule, it is possible to ascribe to the gas at every instant of time a "translational" temperature, which characterizes the average kinetic energy of translational motion of the particles*. Under conditions of partial thermodynamic equilibrium, where thermodynamic equilibrium for the individual degrees of freedom is referred to, it is implied that the distribution of energy (and of the concentrations of the respective components of the gas mixture) for these degrees of freedom is in equilibrium with the "translational" temperature of the gas. On the other hand, quantities associated with nonequilibrium degrees of freedom may be arbitrary; they depend upon many factors, including the previous "history" of the process in which the gas takes part. Such conditions are encountered in rapid gasdynamic processes or in regions where the macroscopic parameters change rapidly, as for example, in an ultrasonic wave or across a shock front. In these cases the time scale of the phenomenon† is comparable to or even much smaller than the corresponding relaxation time. The distribution of energy and of concentrations of the respective particles is not determined in this case simply by the temperature, density, and composition of the gas, as in the case of thermodynamic equilibrium, but also by the kinetics of the chemical-physical processes leading to the establishment of equilibrium for the given degrees of freedom.

In certain cases the relaxation time for the establishment of thermodynamic equilibrium in a given degree of freedom is so large that the nonequilibrium state of the system is found to be very stable, and appears essentially steady. Usually such a situation arises in a gas mixture capable of undergoing a chemical reaction, which actually does not proceed because of the high activation energy required for the reaction. As a typical example we may cite the explosive mixture $2H_2 + O_2$, which in a state of strict thermodynamic equilibrium at low temperatures should be completely transformed into water. These cases are referred to as metastable equilibria.

* It should be noted that for an isotropic distribution of particles with respect to the direction of the velocity of translational motion, the gas pressure is determined by the energy E_k of the translational motion of particles contained in a unit volume: $p = 2E_k/3$, and is completely independent of the distribution of particles with respect to the absolute values of the velocity, that is, is independent of the existence of a Maxwell distribution and "temperature".

† In a shock wave this is the time over which there is a rapid compression of the gas.

As previously stated, the relaxation times for the establishment of equilibrium in the different degrees of freedom are often very markedly different. If, at the given temperature and density, there is a range from fast to slower relaxation processes, then it is usually possible to establish the following order: translational degrees of freedom, molecular rotations, molecular vibrations, dissociation and chemical reactions, ionization, and electronic excitation. Thanks to the very pronounced difference in the various relaxation times, it is possible to study each relaxation process separately, isolating it from the remaining ones; we assume that in the easily excited degrees of freedom equilibrium exists at all times, while slower relaxation processes generally do not take place during the time period considered.

All relaxation processes exhibit certain common features, independent of the nature of the process. Namely, the given degree of freedom approaches thermodynamic equilibrium asymptotically following an exponential law. If we were to characterize the "state" of the given degree of freedom by some parameter, say a number of particles N (for example, by the number of molecules with vibrational modes excited, or by the number of molecules of a given species in the case of chemical reactions), then at a given temperature and density (and composition) of the gas, we may write

$$\frac{dN}{dt} = \frac{N_{eq} - N}{\tau}, \tag{6.2}$$

where N_{eq} is the number of particles at equilibrium and τ is a quantity with the dimensions of time which characterizes the rate of approach to equilibrium. It is evident from the solution of (6.2)

$$N = N_{init} e^{-t/\tau} + N_{eq}(1 - e^{-t/\tau}) \tag{6.3}$$

that τ is the relaxation time for the given process. In general, the rates of chemical-physical processes are by far not always describable by linear equations of the type (6.2). However, as equilibrium is approached, with $N_{eq} - N \ll N_{eq}$, (6.2) is valid as a first approximation. To justify this, the right-hand side of the general rate equation

$$\frac{dN}{dt} = f(N, T, \rho, \ldots) \tag{6.4}$$

is represented as an expansion in the small departure from equilibrium measured in terms of $(N_{eq} - N)/N_{eq}$. It should be noted that the time τ defined by (6.2), as a rule, also characterizes the time scale for the establishment of equilibrium in the case of the general rate equation (6.4) (we shall convince ourselves of this through a number of specific examples in the subsequent sections).

The study of the kinetics of chemical-physical relaxation processes has two aspects. The first one deals with the rates of the elementary processes leading to the excitation of one or another degree of freedom, i.e., the problem of determining cross sections for the inelastic collisions which cause excitation. Usually, these cross sections serve to determine the characteristic relaxation time τ. The second aspect deals with the kinetics of the relaxation process itself under specific conditions. This problem takes into account both the changes with time of the macroscopic parameters of the system and the effect of the reverse process on the change in these parameters. In this chapter we shall be concerned with only the first of the two aspects (the second one will be considered in Chapters VII and VIII). We shall always assume here that the temperature, density, and concentration of particles that do not participate in the process are kept constant.

§2. Excitation of molecular rotations

The excitation energies of molecular rotations are usually very small. When divided by the Boltzmann constant, they are of the order of several degrees, for example, 2.1°K for oxygen and 2.9°K for nitrogen. Therefore, even at room temperature ($T \approx 300$°K) and even more so at higher temperatures, the quantum effects of molecular rotations are not pronounced. Some exceptions to this rule are the very light hydrogen and deuterium molecules, with very small moments of inertia and comparatively large rotational energies; these are 85.4 and 43°K, respectively. Since the rotations are "classical", there is a rather strong exchange of translational and rotational energies during the collisions of the molecules. Indeed, the collision time, the time during which interaction between the colliding molecules takes place, is of the order of $a/\bar{v}$, where a is the dimension of the molecule and $\bar{v}$ is the mean thermal speed. If the rotational energy is of the order of kT, then the collision time is comparable to the period of the rotational motion*. Consequently, molecular collisions may be thought of as the collision of two slowly rotating "dumbbells" with a moderate asymmetry so that when the particles approach each other they do acquire an appreciable angular momentum.

Experimental data confirm the fact that the rotational modes are easily excited. With the exception of H_2 and D_2, the rotational energy of a molecule reaches its classical equilibrium value of kT (in the case of diatomic molecules) after approximately 10 gaskinetic collisions. Most of the experimentally

* The rotational energy is of the order of $kT \sim M\omega^2a^2$, where ω is the angular frequency of rotation and M is the mass of the molecule. The period of rotation is

$$t = \frac{2\pi}{\omega} \sim a\left(\frac{M}{kT}\right)^{1/2} \sim \frac{a}{\bar{v}}.$$

determined rotational relaxation times have been obtained from studies of ultrasonic dispersion and absorption of sound (for further details see §§3 and 4 of Chapter VIII). These relaxation times are in qualitative agreement with the data of Hornig, Greene, and Cowan [1–3] who measured the thickness of weak shock fronts by the reflection of light (for further details see §5 of Chapter IV). Some data on the rotational relaxation times and on the number of collisions required to establish thermodynamic equilibrium among the rotational degrees of freedom are presented in Table 6.1. More detailed information with numerous references to original articles may be found in the surveys by Leskov and Savin [4] and Losev and Osipov [5], and in the book of Stupochenko, Losev, and Osipov [77].

Table 6.1
ROTATIONAL RELAXATION OF MOLECULES

Molecule	Temperature, °K	Relaxation time at atmospheric pressure, sec	Number of collisions	Method	References
H_2	300	$2.1 \cdot 10^{-8}$	300	Ultrasonic	[6]
H_2	300	$2.1 \cdot 10^{-8}$	300	Shock wave	[1]
D_2	288	$1.5 \cdot 10^{-8}$	160	Ultrasonic	[7]
N_2	300	$1.2 \cdot 10^{-9}$	9	Ultrasonic	[8]
N_2	300		20	Shock wave	[2, 3]
O_2	314	$2.2 \cdot 10^{-9}$	12	Ultrasonic	[9]
O_2	300		20	Shock wave	[3]
NH_3	293	$8.1 \cdot 10^{-10}$	10	Ultrasonic	[10]
CO_2	305	$2.3 \cdot 10^{-9}$	16	Ultrasonic	[11]

With the exception of hydrogen at not too high temperatures, it is practically always possible to assume that equilibrium in the rotational degrees of freedom is established as quickly as in the translational degrees of freedom, that is, the rotations always have the "translational" temperature*.

§3. Rate equations for the relaxation of molecular vibrational energy

Vibrational energies of the more important diatomic molecules, expressed in °K by dividing by the Boltzmann constant, are of the order of a thousand to several thousand degrees; for example, for oxygen $h\nu/k = 2230°K$, and

* *Editors' note.* Although not generally important as a relaxation process, rotational nonequilibrium does contribute an effective bulk viscosity, which affects viscous shock front structure and sound absorption.

for nitrogen 3340°K. From (3.19) for the vibrational energy of a gas, it can be seen that the vibrational degrees of freedom make an appreciable contribution to the specific heat, starting at temperatures for which $h\nu$ is several times larger than kT. Thus, when $h\nu/kT = 4$, the energy per vibration is 7.25% of its classical value kT; at $h\nu/kT = 3$, it comprises 15%; for air, this corresponds to a temperature of about 1000°K. Thus, in contrast to molecular rotations, the problem of vibrational relaxation becomes of practical importance under these conditions, when the vibrations exhibit an essentially quantum character. On the other hand, in the "far" classical region $kT \gg h\nu$, at temperatures of the order of 10,000–20,000°K, the problem loses its meaning to a considerable extent, since almost all the molecules are dissociated into atoms. In the "far" classical region where $kT \gg h\nu$ the excitation of molecular vibrations and also of the rotational modes does not require many collisions. However, at temperatures of the order of a thousand to several thousand degrees, when vibrational relaxation is of practical interest, the relaxation times are quite large: as shown both by theory and experiment, thousands and hundreds of thousands of collisions are required for the excitation of the vibrational modes.

Let us formulate a rate equation for the excitation of the vibrational modes, where for simplicity we consider a diatomic gas of a single species. Let $T < h\nu/k$, so that only the first vibrational level of the molecules is appreciably excited* (for air these temperatures are of the order of 1000–2000°K). If n_0, n_1, and $n = n_0 + n_1$ denote the number density of unexcited, excited, and all molecules, respectively, τ_{col} is the average time between gaskinetic collisions as defined by (6.1), and p_{01} and p_{10} are, respectively, the probabilities of vibrational excitation and deexcitation of a molecule as a result of a collision, then the rate equation may be written in the form

$$\frac{dn_1}{dt} = \frac{1}{\tau_{col}}(p_{01}n_0 - p_{10}n_1). \tag{6.5}$$

From the principle of detailed balancing and in accordance with the Boltzmann relation

$$\frac{p_{01}}{p_{10}} = \frac{n_{1\,eq}}{n_{0\,eq}} = e^{-h\nu/kT} \tag{6.6}$$

(the subscript eq is used here to denote equilibrium values). With $kT \ll h\nu$, $n_{1\,eq} \ll n_{0\,eq} \approx n$, we obtain, approximately

$$\frac{dn_1}{dt} = \frac{n_{1\,eq} - n_1}{\tau}, \tag{6.7}$$

* If the molecules are polyatomic, then we consider vibrations of the lowest frequency only.

where the relaxation time

$$\tau = \frac{\tau_{col}}{p_{10}} \tag{6.8}$$

is proportional to the number of collisions necessary to deexcite the molecule, $1/p_{10}$. Multiplying (6.7) by $h\nu$, we obtain an equation for the relaxation of the vibrational energy per unit volume $E = h\nu n_1$ (with $E_{eq}(T) = h\nu n_{1\,eq}(T)$):

$$\frac{dE}{dt} = \frac{E_{eq}(T) - E}{\tau}. \tag{6.9}$$

As we see, for the vibrational excitation process considered the rate equations (6.7) and (6.9) have the form of (6.2) for any departure from equilibrium.

Let us now consider moderate temperatures with $kT \gtrsim h\nu$, at which the gas contains molecules in different vibrational states. In this general case we must write a system of rate equations for n_l molecules with l vibrational quanta ($l = 0, 1, 2, \ldots$). However, an equation of the form of (6.9) for the relaxation of the total vibrational energy will still retain its validity, but the relaxation time will now be given by an equation of a form which is slightly altered from that of (6.8).

We know from quantum mechanics* that the oscillator energy in the case of harmonic vibrations can change only by the value of one vibrational quantum. Here, the probabilities of a transition from a state with $l - 1$ quanta to a state with l quanta, $p_{l-1,l}$, and of a transition from the l level to the $l - 1$ level, $p_{l,l-1}$, are proportional to l. Thus, if we regard the molecule as a harmonic oscillator, which is valid as long as we deal with vibrational states which are not too high, i.e., with temperatures not too large in comparison with $h\nu/k$, then we may write

$$p_{l-1,l} = l p_{01}; \qquad p_{l,l-1} = l p_{10}; \qquad l = 1, 2, 3, \ldots. \tag{6.10}$$

The rate equation for the number of molecules with l quanta, taking into account transitions to the lth state from the $(l - 1)$st as well as the $(l + 1)$st state, is

$$\frac{dn_l}{dt} = \frac{1}{\tau_{col}} (p_{l-1,l}\, n_{l-1} + p_{l+1,l}\, n_{l+1} - p_{l,l-1}\, n_l - p_{l,l+1}\, n_l). \tag{6.11}$$

From the principle of detailed balancing, in analogy with (6.6), we obtain

$$\frac{p_{l-1,l}}{p_{l,l-1}} = \frac{n_{l,\,eq}}{n_{l-1,\,eq}} = e^{h\nu/kT}. \tag{6.12}$$

* See, for example, the book by Landau and Lifshitz [12].

Multiplying (6.11) by hvl, substituting (6.10) into (6.11), summing over l, and noting that $E = \sum hvln_l$ is the total vibrational energy per unit volume, we obtain

$$\frac{dE}{dt} = \frac{1}{\tau_{col}} [p_{01} hvn - (p_{10} - p_{01})E], \tag{6.13}$$

where $n = \sum n_l$ is the total molecular number density. Noting (6.6) and the fact that $E_{eq} = hvn(e^{hv/kT} - 1)^{-1}$ represents the vibrational energy per unit volume at thermodynamic equilibrium (see (3.19)), we arrive at (6.9) with the relaxation time given by

$$\tau = \frac{\tau_{col}}{p_{10}(1 - e^{-hv/kT})}. \tag{6.14}$$

The average number of collisions necessary for establishing equilibrium in the vibrational degrees of freedom is

$$Z = \frac{1}{p_{10}(1 - e^{-hv/kT})} = \frac{Z_1}{1 - e^{-hv/kT}}, \tag{6.15}$$

where $Z_1 = 1/p_{10}$ is the number of collisions required to deexcite a molecule with a single vibrational quantum. For $hv \gg kT$, $Z = Z_1$ and (6.14) reduces to (6.8). At high temperatures, where the average number of vibrational quanta in the molecule is large, $\bar{l} = kT/hv \gg 1$,

$$Z = \bar{l} Z_1, \qquad \tau = \frac{\tau_{col} \bar{l}}{p_{10}} = \frac{\tau_{col}}{p_{\bar{l}, \bar{l}-1}} \bar{l}^2.$$

Equation (6.11), expressing the change in the number of molecules in the lth quantum state, accounts only for those transitions that are accompanied by an exchange in energy between the translational and vibrational degrees of freedom. Actually, molecular collisions can also be accompanied by the exchange of vibrational quanta, and it may be shown that the probability of such an exchange is much greater than the probability of an exchange between the translational and vibrational energies [13]. Therefore, a Boltzmann distribution of the molecules with respect to the vibrational levels, consistent with the total vibrational energy in the gas, is established rapidly. We can say that in a nonequilibrium system a "vibrational" temperature is established initially, followed by the equilibration of the "vibrational" and "translational" temperatures [14].

§4. Probability of vibrational excitation and the relaxation time

Let us consider the simplest case when an atom A collides with a diatomic molecule BC along the direction of the molecular axis, as shown in Fig. 6.1. If no chemical affinity exists between the colliding particles A and BC, then, as they approach each other, the resulting repulsive force will first slow down the atom A and then repel it from the molecule BC. Atom C is then subjected to a force which will initially tend to displace it from the equilibrium position in the direction of atom B. If atom A approaches atom C very slowly, then atom C will move slowly away from its position. When atom A and molecule BC repel each other and begin to move apart, atom C will also slowly return to its initial position. The collision is said to be "adiabatic" and no vibrations

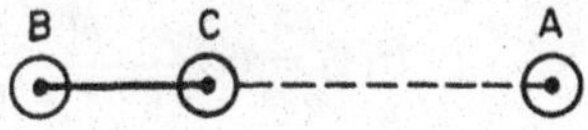

Fig. 6.1. The problem of vibrational excitation in a molecule during a collision with an atom.

will arise. The condition for adiabaticity is that the time of interaction between the atom and molecule, which is of the order of a/v, where a is the range of the forces and v is the relative velocity of the particles for infinite separation, is large in comparison with the vibrational period: $a\nu/v \gg 1$. The above condition may also be presented as follows: In order to get the molecule strongly "swinging" it is necessary that the expansion of the driving force in the Fourier integral contain large resonant components with frequencies close to the natural frequency ν; this, however, requires that the collision time a/v be of the order of $1/\nu$, or more precisely, that the condition $a\omega/v \sim 1$, where $\omega = 2\pi\nu$, be satisfied.

Landau and Teller [15] estimated the dependence of the probability of exciting the vibrational modes on the collisional speed and, in the final analysis, on the temperature, by using the correspondence principle. In order for the quasi-classical approximation to remain valid, it is necessary that the wavelength of the particles be small in comparison with the scale of the force of the field, i.e., that $aMv/h \gg 1$, where M is the reduced mass of the colliding particles. It is easy to check that this condition is automatically satisfied if together with the adiabatic condition $a\nu/v \gg 1$ the kinetic energy of the relative motion is very much larger than the energy of a quantum $Mv^2 \gg h\nu$. Thus the quasi-classical case corresponds to an adiabatic collision, with a very small probability of vibrational excitation.

The probability of vibrational excitation as a result of a collision is proportional to the square of the matrix element of the interaction energy of particles A and BC as a function of the distance between them $U(x)$. In the quasi-classical approximation, the matrix element reduces to the Fourier component of the interaction energy

$$\int_{-\infty}^{\infty} U[x(t)]e^{i2\pi\nu t}\,dt. \tag{6.16}$$

For simplicity we set $\mathscr{E} = Mv^2/2 = const$, and we assume a repulsive law of the form $U = const\ e^{-x/a}$, so that the atom can come up to the molecule very closely. Integrating the equation of motion

$$\frac{dx}{dt} = \left(\frac{2}{M}[\mathscr{E} - U]\right)^{1/2}, \qquad t = \int \frac{dx}{\left(\frac{2}{M}[\mathscr{E} - U(x)]\right)^{1/2}},$$

we find the functions $t(x)$, $x(t)$, and thereby the dependence of $U(t)$

$$U(t) = 4\mathscr{E}\frac{e^{vt/a}}{(e^{vt/a}+1)^2} = 4\mathscr{E}\frac{e^{-vt/a}}{(1+e^{-vt/a})^2} = \frac{\mathscr{E}}{\cosh^2(vt/2a)}$$

($t = 0$ corresponds to the time of closest approach of the particle).

The integral (6.16) can be evaluated as an integral in the complex t plane, with the path changed to the path $\mathscr{I}m\, t = 2\pi a/v$. This path is the straight line at a distance above the real axis equal to twice the distance to the nearest pole $t_1 = i\pi a/v$. By the method of residues, the integral is found to be

$$\int_{-\infty}^{\infty} Ue^{i2\pi\nu t}\, dt = \frac{4\pi^2 Ma^2\nu}{\sinh(2\pi^2 a\nu/v)} \approx 8\pi^2 Ma^2\nu \exp(-2\pi^2 a\nu/v),$$

with the exponential law valid if the adiabatic condition holds.

The physical meaning of the Landau–Teller derivation becomes especially clear if we solve the problem of the excitation of an oscillator by particle impact on the basis of classical mechanics. Such a derivation is given in the book by Stupochenko, Losev, and Osipov [77]. Here we shall consider only one particular case, which is particularly simple. We assume that atom B is much heavier than atom C ($m_B \gg m_C$) and that the arriving atom interacts with atom C only, so that in the center of mass system for the particles BC and A it is only the light atom C which is set into motion. We denote the displacement of the atom C about the equilibrium position (along the line of collision x) by y, and we write the equation of motion for the oscillator

$$m_C(\ddot{y} + \omega^2 y) = F(t), \qquad \omega = 2\pi\nu.$$

The force in this case is simply $F = -\partial U/\partial x$, that is, when $U = \mathscr{E}e^{-x/a}$, $F = U/a$.

We now calculate the oscillator energy $\varepsilon(t) = \frac{1}{2}m_C(\dot{y}^2 + \omega^2 y^2)$. To do this we multiply the equation of motion by $e^{i\omega t}$ and integrate with respect to t from $-\infty$ to t. Integrating the left-hand side by parts and applying the initial condition $y(-\infty) = 0$, $\dot{y}(-\infty) = 0$, we obtain

$$m_C(\dot{y} - i\omega y)e^{i\omega t} = \int_{-\infty}^{t} F(t')e^{i\omega t'}\, dt'.$$

The square of the modulus of this quantity divided by $2m_C$ gives the energy $\varepsilon(t)$. Noting that $F = U/a$, we find the energy acquired by the oscillator as a result of the collision:

$$\varepsilon = \varepsilon(\infty) = \frac{1}{2m_C a^2}\left|\int_{-\infty}^{\infty} U[x(t)]e^{i\omega t}\,dt\right|^2.$$

If we now pass over to quantum-mechanical concepts, then we must set $\varepsilon = h\nu p_{01}(v)$, where p_{01} is the probability of excitation of the oscillator by collision. Substituting the value of the integral above, we obtain the probability

$$p_{01}(v) = \frac{32\pi^4 M^2 a^2 \nu}{m_C h}\exp(-4\pi^2 a\nu/v).$$

It decreases exponentially as the adiabatic factor $a\nu/v$ increases. When the mass ratio of the atoms in the molecule is arbitrary, the problem becomes somewhat more difficult; however, the only change in the results of the calculations is in the formula for p_{01} where the factor $m_B/(m_B + m_C)$ appears (see [77]).

The probability as a function of the relative particle velocity v should be averaged using the Maxwell distribution in terms of the relative velocity, i.e., using a function proportional to $\exp(-Mv^2/2kT)$ .This yields an integral with respect to velocity whose integrand contains the factor $\exp(-4\pi^2 a\nu/v - Mv^2/2kT)$. The principal role in this integral is played by the velocities $v^* = (4\pi^2 a\nu kT/M)^{1/3}$ for which the exponent has a minimum absolute value. Collisions occurring at these velocities are mainly responsible for the excitation and deexcitation of the vibrational modes. The integral and the transition probabilities p_{01} and p_{10} are proportional to the maximum value of the exponential*

$$p_{10} \sim p_{01} \sim \exp\left(-\frac{4\pi^2 a\nu}{v^*} - \frac{Mv^{*2}}{2kT}\right) = \exp\left(-\frac{3}{2}\frac{Mv^{*2}}{kT}\right)$$
$$= \exp\left[-\left(\frac{54\pi^4 a^2 \nu^2 M}{kT}\right)^{1/3}\right]. \qquad (6.17)$$

Substitution of the numerical constants into the exponent in (6.17) as well as experimental evidence show that at not too high temperatures the exponent

* It is interesting to note that the rate of thermonuclear reactions has the same temperature dependence $\exp(-const\ T^{-1/3})$. This is due to the fact that the probability of the nuclei approaching each other with repulsive Coulomb forces acting between them also depends on the relative velocity of approach as $\exp(-const \cdot v^{-1})$; this relationship is then averaged using the Maxwell distribution with respect to the velocities of the nuclei.

is much greater than unity*. This means that the adiabatic condition is satisfied for the collisions which provide the main contribution to the excitation and deexcitation of the vibrational modes and that the kinetic energy of the colliding particles is much greater than kT.

Quantum-mechanical calculations of the deexcitation probability p_{10} which determine the relaxation time (see Zener [17], Schwartz and Herzfeld [18]) also lead, in the adiabatic limit, to an equation containing the exponential factor (6.17). Reference [18] considers a very general case of collisions and derives the following equation for the number of collisions prior to deexcitation

$$Z_1 = \frac{1}{p_{10}} = \pi^2 \left(\frac{3}{2\pi}\right)^{1/2} \left(\frac{h\nu}{\varepsilon_0}\right)^2 \left(\frac{kT}{\varepsilon_0}\right)^{1/6} \exp\left(-\frac{h\nu}{2kT} - \frac{\varepsilon_1}{kT}\right) \exp\left[\frac{3}{2}\left(\frac{\varepsilon_0}{kT}\right)^{1/3}\right], \tag{6.18}$$

where $\varepsilon_0 = 16\pi^4 a^2 \nu^2 M$. The last exponential corresponds precisely to the exponential in (6.17); for temperatures which are not too high, due to the large value of the exponent, this term essentially characterizes the temperature dependence of the number of collisions. The factor $\exp(-\varepsilon_1/kT)$ takes into account a certain facilitation of the transitions due to the acceleration of the particles when they approach each other; this acceleration is caused, in turn, by the long range attractive forces that are described by a "potential well" with an energy ε_1. This energy is usually of the order of a few tenths of an electron volt. Equation (6.18) has been somewhat refined in a later paper of Herzfeld [18a].

The theoretical considerations presented above show that the dependence of the vibrational relaxation time on temperature obeys the relation

$$\tau = Z\tau_{\text{col}} = \tau_{\text{col}} A\, e^{bT^{-1/3}}, \tag{6.19}$$

where b is a constant and A is a slowly varying function of temperature. Thus, a plot of log τ as a function of $T^{-1/3}$ should yield an almost straight line.

Vibrational relaxation times are measured experimentally at room temperature and with small heating by ultrasonic absorption and dispersion. Over a wider temperature range vibrational relaxation times are measured from the establishment of equilibrium behind a shock front generated in a shock tube. A careful study of relaxation in oxygen and nitrogen was carried out by Blackman [16] using a shock tube. His results are presented in Table 6.2. This table also presents the theoretical values for oxygen calculated by Schwartz and Herzfeld [18]. As is evident the theoretical and experimental data show satisfactory agreement. Experimental values measured for many

* For example, in oxygen at $T = 1000°$K the exponent is approximately equal to 10 (according to the data of [16]; see below).

Table 6.2

VIBRATIONAL RELAXATION IN OXYGEN AND NITROGEN MEASURED BY BLACKMAN [16]. THEORETICAL VALUES TAKEN FROM SCHWARTZ AND HERZFELD [18]

T, °K	p_{10} (experimental)[a]	p_{10} (theoretical)	Number of collisions Z (experimental)	τ in seconds normalized to the density $n = 2.67 \cdot 10^{19}$ cm^{-3}
		Oxygen		
288	$4.0 \cdot 10^{-8}$ [b]	$1.0 \cdot 10^{-8}$	$2.5 \cdot 10^{7}$	
900	$1.1 \cdot 10^{-5}$	$3.0 \cdot 10^{-6}$	$1.0 \cdot 10^{5}$	$96 \cdot 10^{-7}$
1200	$2.4 \cdot 10^{-5}$	$1.3 \cdot 10^{-5}$	$5.0 \cdot 10^{4}$	$41 \cdot 10^{-7}$
1800	$9.8 \cdot 10^{-5}$	$8.6 \cdot 10^{-5}$	$1.4 \cdot 10^{4}$	$9.5 \cdot 10^{-7}$
2400	$3.7 \cdot 10^{-4}$	$5.5 \cdot 10^{-4}$	$4.5 \cdot 10^{3}$	$2.7 \cdot 10^{-7}$
3000	$1.2 \cdot 10^{-3}$	$1.5 \cdot 10^{-3}$	$1.6 \cdot 10^{3}$	$0.83 \cdot 10^{-7}$
		Nitrogen		
600	$3.0 \cdot 10^{-8}$ [c]	$3.3 \cdot 10^{7}$		
3000	$3.1 \cdot 10^{-5}$	$4.6 \cdot 10^{4}$		$2.1 \cdot 10^{-6}$
4000	$9.7 \cdot 10^{-5}$	$1.8 \cdot 10^{4}$		$0.67 \cdot 10^{-6}$
5000	$2.5 \cdot 10^{-4}$	$0.8 \cdot 10^{4}$		$0.27 \cdot 10^{-6}$

[a] In calculating p_{10} on the basis of the experimentally determined values of the time τ, the gaskinetic cross sections $\sigma_{O_2} = 3.6 \cdot 10^{-15}$ cm^2 and $\sigma_{N_2} = 4.1 \cdot 10^{-15}$ cm^2 were used.

[b] This point was obtained by the ultrasonic method [19].

[c] This point was obtained by Huber and Kantrowitz [20] from studies of the discharge from a nozzle.

different gases more or less follow the straight line dependence of ln τ or ln Z on $T^{-1/3}$. This is evident from Fig. 6.2 which is taken from [5]. The departures from a straight line can be partially explained by the temperature dependence of the pre-exponential factor A in (6.19).

The vibrational relaxation time for oxygen is, at a given temperature, smaller than that for nitrogen, since the natural frequency in nitrogen is one and a half times higher than in oxygen. This fact makes the excitation of the vibrational modes in nitrogen more difficult. Therefore, vibrational relaxation in air takes place in two stages: first oxygen and then nitrogen reach equilibrium. It should be noted that the collisions of N_2 with O_2 molecules are only 2.5 times less effective with respect to the excitation of vibrational modes in O_2 than are the O_2—O_2 collisions. Generally, some molecules are very active in exciting vibrational modes; for example, H_2O molecules excite vibrations in O_2 50–100 times faster than do the O_2 molecules themselves. Hence, high purity of the gas is essential for measurement of vibrational relaxation.

A detailed review of the literature on vibrational relaxation times in different gases, as well as references to many experimental and theoretical papers, may be found in the reviews [4, 5]. We would also cite several articles concerned with the study of vibrational excitation: in O_2 [58, 59], in NO [60], in CO [61], and in CO_2 [62, 63]. We also note the review [64] and the

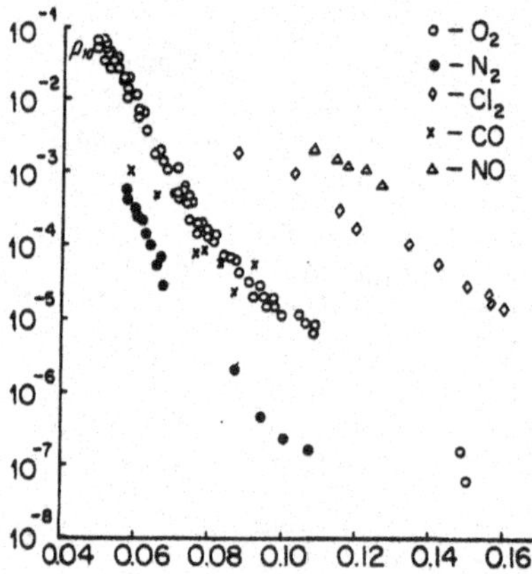

Fig. 6.2. Experimental data on the probability of deexcitation of molecules with excited vibrations.

papers [65, 66] dealing with vibrational relaxation in gas mixtures. The theory of vibrational relaxation is presented in detail in the book [77], in which may be found a detailed bibliography, including references to recent theoretical and experimental studies.

§5. Rate equation for dissociation of diatomic molecules and the relaxation time

Dissociation of diatomic molecules as a result of collisions of sufficiently energetic particles usually corresponds to the reaction

$$A_2 + M \rightleftarrows A + A + M, \tag{6.20}$$

where M is any particle*. In a homogeneous diatomic gas the particle M can be either the molecule A_2 or the atom A. The reverse process results in the recombination of atoms as a result of three-body collisions, where the third body M absorbs a portion of the binding energy released in the process.

The rate equation for the reaction (6.20), taking into account the fact that a molecule as well as an atom can serve as the particle M, takes the form

$$\frac{dA_2}{dt} = -\frac{1}{2}\frac{dA}{dt} = -k_d A_2^2 + k_r A^2 \cdot A_2 - k_d' A_2 \cdot A + k_r' A^3. \tag{6.21}$$

* There is a very small probability that a sufficiently strongly excited molecule will decompose directly into atoms $A_2 \rightleftarrows A + A$, and that these atoms will recombine into a molecule without the participation of a third body which could absorb a portion of the energy liberated during the reaction. The probabilities of photodissociation and of recombination with the emission of a photon are also very small.

For brevity we denote the particle number densities by their symbols. The reaction rate constants depend only on the temperature and are related to each other by the principle of detailed balancing

$$\frac{k_d}{k_r} = \frac{k_d'}{k_r'} = \frac{(A)^2}{(A_2)} = K(T), \tag{6.22}$$

where the parentheses denote the equilibrium values of the number of particles at the given temperature and density; $K(T)$ is the equilibrium constant defined in terms of concentrations; it differs from the equilibrium constant defined in terms of partial pressures $K_p(T)$ by the factor $(kT)^{-1}$. Thus we have $K(T) = K_p(T)/kT$. The equilibrium constant determines the equilibrium degree of dissociation α at a given temperature and density. From (3.26) we have

$$\frac{\alpha^2}{1-\alpha} = \frac{K(T)}{4N} = \frac{1}{4N}\frac{M_A \nu}{4I_{A_2}}\left(\frac{M_A}{\pi kT}\right)^{1/2}\frac{g_A^2}{g_{A_2}}e^{-U/kT}, \tag{6.23}$$

where N is the initial molecular number density and M_A is the mass of an atom (the remaining symbols are defined in §3 of Chapter III).

In contrast to vibrational relaxation, the rate equation for molecular dissociation is in general nonlinear. However, in accordance with the general procedure outlined in §1 for small departures from equilibrium, it can be reduced to the linearized form (6.2) for the number of particles A or A_2. In this case the relaxation time τ is determined by

$$\frac{1}{\tau} = 4\alpha(2-\alpha)N^2\left(k_r + k_r'\frac{2\alpha}{1-\alpha}\right). \tag{6.24}$$

Calculations show that the time τ characterizes not only the final stage of approach to equilibrium, but the entire dissociation kinetics even during the stage when only the nonlinear equation (6.21) is valid. Therefore, τ is of the same order as the time necessary to establish equilibrium dissociation for the general case of arbitrary initial conditions.

In the limiting cases of weak and strong equilibrium dissociation, (6.24) is simplified. When $\alpha \ll 1$, very few atoms are present and molecular dissociation occurs mainly as a result of collisions between molecules. In this case from (6.22) and (6.23),

$$\frac{1}{\tau} = 8\alpha N^2 k_r = \frac{2}{\alpha} N k_d. \tag{6.25}$$

For $1-\alpha \ll 1$, even if no atoms are present initially, the most important part is played by the later stage, when only a few molecules are present and the remaining molecules are broken up by collisions with atoms. In this case

$$\frac{1}{\tau} = \frac{8}{1-\alpha}N^2 k_r' = 2Nk_d'. \tag{6.26}$$

Thus, the problem of the time necessary to establish equilibrium reduces to the problem of the dissociation or recombination reaction rates. Since both rates are related by the principle of detailed balancing (6.22), it is sufficient to know only one of them, either from theory or from experiment.

§6. Atom recombination rates and dissociation rates for diatomic molecules

A rough estimate of the rate of recombination of atoms in diatomic molecules can be obtained from very elementary considerations, assuming that each gaskinetic collision of atoms in the presence of a third body results in recombination. The number of collisions among A atoms per unit volume per unit time is equal to $A \cdot \bar{v} \cdot \sigma \cdot A$, where $\bar{v} = (8kT/M_A\pi)^{1/2}$ is the mean thermal speed and σ the gaskinetic cross section. The probability that at the time of collision a third body will be found "in the neighborhood", i.e., at a distance of the order of the molecular radius r, is approximately equal to the average number of particles in a volume equal to the volume of one molecule: $\frac{4}{3}\pi r^3/N$, where N is the particle number density. The number of three-body collisions per unit volume per unit time is thus equal to $A \cdot \bar{v} \cdot \sigma \cdot A \cdot \frac{4}{3}\pi r^3/N$. Introducing, for generality, the numerical coefficient β equal to the probability that a recombination will follow a three-body collision, we obtain the following expression for the recombination rate constant:

$$k_r = \beta \bar{v} \sigma \tfrac{4}{3} \pi r^3 . \tag{6.27}$$

For example, for nitrogen atoms $\bar{v} = 3.9 \cdot 10^3 T^{\circ 1/2}$ cm/sec and $\sigma \approx 10^{-15}$ cm^2. Setting $r = 3.4 \cdot 10^{-8}$ cm and $\beta = 1$, we find $k_r = 2.2 \cdot 10^{14} T^{\circ 1/2}$ cm^6/mole^2 · sec (one mole contains $6 \cdot 10^{23}$ atoms). At $T = 300$°K, $k_r = 3.8 \cdot 10^{15}$ cm^6/mole^2 · sec.

The recombination of nitrogen atoms is usually studied by measuring the change in the number of nitrogen molecules with time during afterglow*. In this manner the recombination rate constant, with nitrogen molecules acting as the third bodies, has been determined. In the temperature range between 297 and 442°K the constant was found to be almost independent of temperature and given by $k_r = 5.8 \cdot 10^{15}$ cm^6/mole^2 · sec [21]. This value is in good agreement with the estimate given above. Similar results were also obtained by other authors [22, 23]. In [70] is reported an examination of the

* The phenomenon of nitrogen afterglow can be described as follows. During the recombination of nitrogen atoms, the N_2 molecules are found to be in the ${}^5\Sigma_g^+$ excited state. Subsequent collisions with other molecules or atoms partially deexcite the molecules which then undergo a transition to the lower state $B\,{}^3\Pi_g$. Following this transition, photons of the first positive system of N_2 ($B\,{}^3\Pi_g \rightarrow A\,{}^3\Sigma_g^+$) are emitted and are measured in the experiment. Information about the rate of recombination is obtained on the basis of the changes in the afterglow intensity.

dissociation and recombination of nitrogen in a shock tube by measuring the nonequilibrium radiation. It was found that the recombination rate constant at $T = 6400°K$ is equal to $k_{rN} = 6.5 \cdot 10^{15}$ cm^6/mole$^2 \cdot$ sec with a nitrogen atom acting as the third body and 13 times smaller when a nitrogen molecule acted as the third body.

In general, the recombination rate constants at not too high temperatures ($T \sim 300$–$1000°K$), are usually of the order of 10^{14}–10^{16} cm^6/mole$^2 \cdot$ sec, which indicates rather high recombination probabilities β during three-body collisions. The recombination rate exhibits a relatively weak dependence on temperature and usually shows a certain tendency to decrease with increasing temperature. This can be understood if we recall that the recombination probability during a three-body collision is greater, the greater the time of interaction between the colliding particles, that is, the lower their speed or the lower the temperature. Thus, the probability β has an inverse dependence on temperature. For example, if $\beta \sim 1/T$, then, $k_r \sim \bar{v}\beta \sim 1/T^{1/2}$, in agreement with Wigner's theoretical calculations [24].

The rate of recombination of atoms depends on the type of third body. For example, in the recombination of nitrogen atoms, the nitrogen atoms acting as third bodies are 13 times more effective than nitrogen molecules (at $T = 6400°K$). A study of the dissociation rate of iodine in a shock tube described in [25] (the concentration of I_2 molecules was measured by the absorption of light) showed that the iodine molecules at $T = 1300°K$ were 35 times more effective as third bodies than argon atoms in the recombination of iodine atoms. The recombination rate for iodine during three-body collisions with argon at $T = 1300°K$ is $k_r = 4.5 \cdot 10^{14}$ cm^6/mole$^2 \cdot$ sec [25]; at $T = 298°K$, $k_r = 2.9 \cdot 10^{15}$ cm^6/mole$^2 \cdot$ sec [26].

Dissociation of a molecule following collision with another particle takes place only if the energy of the colliding particles exceeds the dissociation energy. The total number of collisions per unit time of the given molecule with other particles (the number density of which is N) is $\nu = N\overline{v'}\sigma$, where $\overline{v'}$ is the average velocity of the relative motion of the particles $\overline{v'} = (8kT/\mu\pi)^{1/2}$ with μ the reduced mass*. For a Maxwell velocity distribution, the number of molecular collisions with the kinetic energy of relative motion exceeding the dissociation energy U is the fraction $(U/kT + 1)e^{-U/kT}$ of the total number of collisions (usually $U/kT \gg 1$, so that $(U/kT) + 1 \approx U/kT$). It is usually assumed that only that part of the kinetic energy of the particles which corresponds to the relative velocity component directed along the line of centers of the colliding particles (if the latter are considered to be solid spheres) is effective with respect to dissociation. On the basis of this assumption the fraction

* In estimating the recombination rate, $\overline{v'}$ was replaced by $\bar{v}$ for simplicity, that is, μ was replaced by the atomic mass M_A.

of "sufficiently energetic" collisions is simply equal to $e^{-U/kT}$, rather than to $(U/kT)\,e^{-U/kT}$.

It is natural to assume that not only the kinetic energy of translational motion of the colliding particles is expended in breaking the molecular bond, but also the energy in their internal degrees of freedom, that is, the vibrational and rotational energy. It can be shown (see [27]) that the fraction of collisions in which the total energy of the colliding particles (taking into account the energy of the internal degrees of freedom) exceeds the dissociation energy is*

$$\frac{1}{s!}\left(\frac{U}{kT}\right)^{s} e^{-U/kT},$$

where each vibrational degree of freedom contributes 1 to the exponent s, and each rotational degree of freedom contributes 1/2 (in the case of half integral s, the factorial s is replaced by the gamma function $\Gamma(s+1)$). At the present time the theory of molecular dissociation by collision with other particles is still far from complete. Therefore, experimental values of the dissociation rate constant are often compared with a relation of the type

$$k_d = P\overline{v'}\sigma\,\frac{1}{s!}\left(\frac{U}{kT}\right)^{s} e^{-U/kT}. \qquad (6.28)$$

The number s which characterizes the degree of participation of the internal degrees of freedom in the dissociation, and the factor P representing the probability of dissociation upon the collision[1] of particles with sufficient energies, are parameters which must be determined experimentally.

According to present ideas, the main role in the dissociation process is played by the vibrational energy of the molecule. Stupochenko and Osipov [28] have shown that the probability of dissociation of an unexcited molecule is extremely small, even if the translational energy of the colliding particles exceeds the binding energy U. It is mainly the molecules in very high vibrational levels, whose energy is close to the dissociation energy, which can be dissociated. In this case, the translational energy of the particles cannot differ very much from the mean thermal energy. If we assume that the distribution of molecules over the vibrational states is a Boltzmann distribution, then the dissociation rate can still be described by an equation of the type (6.28) with an appropriate value of the exponent s. Stupochenko and Osipov [29] have shown that this assumption is not always justified. "Withdrawal" of molecules from the highest vibrational levels caused by dissociation can sometimes seriously distort the Boltzmann distribution for the molecules with respect

* It is assumed in the derivation of this equation that the distribution of molecules over the energy states of all the internal degrees of freedom is a Boltzmann distribution corresponding to the translational temperature T.

to the highest vibrational states. In this case the dissociation kinetics must be considered together with the kinetics of excitation of the highest vibrational levels. The process proceeds in such a manner that the collisions "deliver" the molecules to the upper levels from which they make a transition to the dissociated state. During the process of recombination of atoms in the presence of a third body, the dissociation energy is primarily converted into vibrational energy of the formed molecule. The theory underlying these processes is presented in the review [76] and in the book [77].

Experimental studies have been carried out mainly for the dissociation of oxygen behind a shock front in a shock tube (articles by Matthews [30], Byron [31], Generalov and Losev [32], Camac and Vaughan [67], Rink *et al.* [68]; a summary of bibliographical data and references to other works are given in [4, 5, 77]. A very thorough study was carried out by Matthews, who used an interferometer to determine the changes in density in the nonequilibrium zone behind a shock. These changes were compared with the theoretical values obtained from a dissociation rate equation similar to (6.28) (see Chapters IV and VII). Equilibrium in the vibrational degrees of freedom was established within a time which was at least an order of magnitude less than the dissociation time*. Thus, the effect of vibrational relaxation did not interfere with the study of the dissociation rate. The investigation covered the temperature range from 2000 to 4000°K. The degree of dissociation in Matthews' experiments was not large, $\alpha \sim 0.05$–0.1, so that the main role in dissociation was played by the O_2—O_2 collisions†. In the calculations it was assumed that $s = 3$, for which case the effectiveness of collisions was found to be $P_{O_2-O_2} = 0.073$, while the dissociation rate constant was given by

$$k_{dO_2-O_2} = 5.4 \cdot 10^{10} T^{\circ 1/2} \left(\frac{59{,}380}{T^\circ}\right)^3 e^{-59{,}380/T^\circ} \text{ cm}^3/\text{mole} \cdot \text{sec}. \quad (6.29)$$

Calculations for $s = 0$ yielded an improbably high value of P (greater than unity). This shows that the energy of the internal degrees of freedom of the molecule plays an important role in the dissociation process. Knowing the equilibrium constant for the dissociation of O_2

$$K(T) = 1.85 \cdot 10^3 T^{\circ -1/2} e^{-59{,}380/T^\circ} \text{ mole/cm}^3, \quad (6.30)$$

* For the not too high vibrational states which are occupied by the majority of molecules.

† The reaction $O_2 + O_2 = 2O + O_2$ has a "competitor", the two-stage dissociation reaction of O_2 with the intermediate formation of ozone $O_2 + O_2 = O + O_3$; $O_3 + M = O + O_2 + M$ (the reverse process, i.e., recombination, can also take place). This process plays an insignificant role at high temperatures (in particular, in Matthews' experiments). However, at low temperatures and with a low degree of dissociation, oxygen recombines primarily through the formation of ozone, since the collisions $O + O + M$ occur much less frequently than the collisions $O + O_2 + M \rightarrow O_3 + M$. The reaction rate constants for processes involving the formation of ozone are given in [33].

we may find the recombination rate with O_2 molecules acting as the third bodies:

$$k_r = 6.1 \cdot 10^{21} T^{\circ -2}\ \mathrm{cm}^6/\mathrm{mole}^2 \cdot \mathrm{sec}^*. \tag{6.31}$$

The relaxation time at $T = 3500°K$ and standard density is $\tau = 0.95 \cdot 10^{-6}$ sec ($\alpha = 0.084$). This result is very close to the data obtained by Glick and Wurster [34], who measured the relaxation time for the dissociation of oxygen in a shock tube. Their times, reduced to standard density, are

T, °K	3100	3300	3400	3850
$\tau \cdot 10^6$ sec	2	0.8	0.5	0.06

References [25, 35, 36] report the use of shock tubes to study the dissociation rates of bromine and iodine (the concentration of Br_2 and I_2 molecules was measured by the absorption of light from an external source). In [35] is reported the following data for the dissociation rate of bromine molecules during collision with argon atoms in a temperature range up to 2000°K: $s = 2$ and $P_{Br_2\text{—}Ar} = 0.12$. Satisfactory agreement with this result was obtained in the theoretical work of Nikitin [37]. Reference [5] presents a review of articles on molecular dissociation. We would also note [69], in which a study of the dissociation of hydrogen in a shock tube is reported. A discussion of the dissociation rates for O_2 and N_2 and other relaxation processes in air may also be found in [53]. Additional details on the theory and kinetics of molecular dissociation as well as experimental data and references to numerous original articles may be found in the book [77] which has been often cited.

§7. Chemical reactions and the activated complex method

Chemical reactions are usually divided into two types from the point of view of the energy involved: endothermic reactions, requiring a definite amount of energy, and exothermic reactions, accompanied by the liberation of heat. Examples of both types of reactions are, respectively, the dissociation of molecules and the recombination of atoms into molecules considered above. It is evident that an endothermic reaction can occur only when the colliding molecules possess a definite minimum amount of energy, the so-called activation energy E. The rate of such a reaction, therefore, is proportional to the Boltzmann factor $e^{-E/kT}$ and increases rapidly with temperature. In the dissociation process, the binding energy of the molecule U serves as the activation energy. However, it has been shown experimentally that a majority of exothermic reactions also require an activation energy and that the corresponding reaction rates increase exponentially with temperature as

* This equation is applicable only in the temperature range investigated $T \approx 2000$–4000°K. Extrapolation to room temperature gives significantly too high a recombination rate.

$e^{-E/kT}$. This relation is known as Arrhenius' law. Recombination of atoms into molecules is in this respect atypical, since it proceeds without activation and hence can easily occur at low temperatures, as can many other reactions in which free atoms participate.

An elementary chemical reaction, such as the exchange of atoms during a collision of molecule XY with a molecule WZ

$$\mathrm{XY} + \mathrm{WZ} \rightarrow \mathrm{XW} + \mathrm{YZ}, \tag{6.32}$$

can occur only when both molecules approach each other very closely. Regardless of whether this process is or is not energetically favorable, that is, whether it results in the release or absorption of energy, a very small separation between the particles will, as a rule, give rise to repulsive forces. A definite amount of energy is required to overcome such forces. It is usually said that a potential barrier must be overcome before a chemical reaction can take place. This situation is illustrated schematically in Fig. 6.3, which shows the potential energy of a system of four atoms XYWZ as a function of a "decomposition coordinate" which characterizes the relative configuration of the atoms in space. For definiteness let us assume that the forward reaction (6.32) is exothermic. The energy difference between the initial and final states of the system is equal to the energy Q released by the reaction. It is clear from Fig. 6.3 that the activation energy of the reverse reaction E_2 exceeds the activation energy of the forward reaction E_1 by the reaction energy Q. Therefore, the rate of the reverse endothermic reaction is more strongly temperature dependent than the rate of the forward exothermic reaction.

The rate equation for the reaction (6.32), considering both the forward and the reverse processes, can be written in the form

$$\frac{d\,\mathrm{XY}}{dt} = k_1 \mathrm{XY} \cdot \mathrm{WZ} - k_2 \mathrm{XW} \cdot \mathrm{YZ}\,*. \tag{6.33}$$

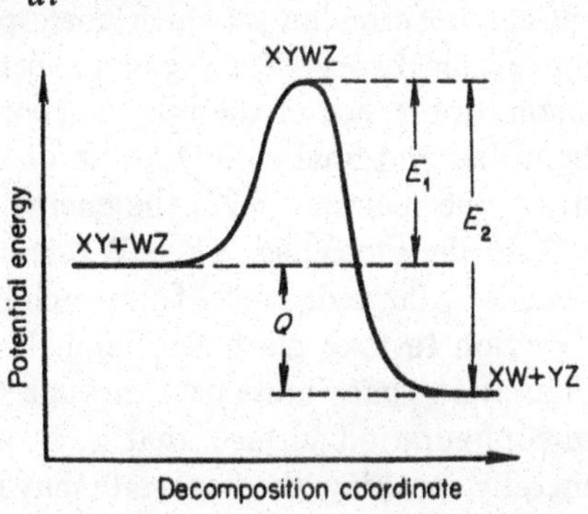

Fig. 6.3. The potential barrier in chemical reactions.

* Reactions with the participation of two molecules (atoms) are called bimolecular, while unimolecular reactions are those where the molecule is dissociated into simpler molecules (or atoms), for example, $\mathrm{XY} \rightarrow \mathrm{X} + \mathrm{Y}$.

The reaction rate constants, which depend only on temperature, are related as usual by the principle of detailed balancing

$$\frac{k_2}{k_1} = \frac{(XY)(WZ)}{(XW)(YZ)} = K(T). \tag{6.34}$$

Using the concepts of collision theory, we can express the reaction rate constants similarly to the dissociation rate constants. Thus, assuming for simplicity that the only effective contribution to overcoming the potential barrier E is made by that portion of the translational energy of the colliding particles along their line of centers and that other portions, including those for the internal molecular degrees of freedom, are ineffective in this respect, we obtain

$$k_1 = P\sigma\overline{v'}e^{-E/kT}, \tag{6.35}$$

Here P, as before, is the probability of the occurrence of a chemical reaction as a result of a sufficiently energetic collision (P is sometimes called the steric factor). It has been experimentally demonstrated that many reactions, especially those where complex molecules are involved, proceed much slower than would have been expected on the basis of the number of sufficiently energetic collisions; the probability P is frequently found to be very small, even of the order of 10^{-8}.

In a number of cases a more definite estimate of the reaction rate can be obtained by using the so-called activated or intermediate complex method*. The potential energy of a system of atoms participating in an elementary reaction depends on their relative configuration. If the positions of the atoms change sufficiently slowly (and this is practically always true), then the electronic state of the system changes continuously and the potential energy depends on the nuclear position only (which corresponds to the adiabatic approximation in molecular theory). The potential energy forms a continuous surface in the configuration space of the nuclear coordinates. The potential energy for both the initial and final configuration of the atoms is at a minimum. For example, in the reaction (6.32) the energy is minimum when the atoms combine to form the molecules XY + WZ and also XW + YZ, with the molecules separated by large distances from each other in each case.

In order for a reaction to take place the point describing the motion of the system in configuration space must pass through the maximum dividing the minima on the configuration surface, that is, it must cross over the potential barrier. Generally speaking, the final state may be reached along many paths from the initial state. The path actually taken is the one which is most

* A detailed presentation of this method and its applications in calculating the rates of a number of reactions can be found in [38]; see also [27].

convenient for the given reaction, and corresponds to the lowest value of the energy maximum. The energy surface around this path resembles a "trough". Figure 6.3 illustrates schematically a section of the energy surface along the "bottom of the trough"; here, the reaction path corresponds to the decomposition coordinate.

The peak of the potential barrier corresponds to the closest approach between the reacting particles. In the neighborhood of the peak, in a region with linear dimensions δ of the order of the molecular dimensions, the atoms are formed into something not unlike a molecule. This state is called an activated complex. However, the principal difference between the activated complex and a molecule is that a molecule is in a stable state with a minimum potential energy, while the complex is in a state of unstable equilibrium with the potential energy a maximum as a function of the decomposition coordinate. The point describing the state of the system moves along the reaction path with a speed of the order of the relative velocities of the atoms, i.e., with a mean speed $\bar{v}$ which is of the order of the thermal speed. The time spent in the vicinity of the peak, that is, the lifetime of the activated complex, is of the order of $\tau = \delta/\bar{v}$. For $\delta \approx 10^{-8}$ cm and $\bar{v} \approx 10^4$ cm/sec, $\tau \approx 10^{-12}$ sec. The lifetime of a complex is very short compared to the characteristic reaction time (the time required to reach chemical equilibrium in a gas mixture). These considerations provide the basis for the theoretical assumption that the complexes may be regarded as certain types of molecules possessing the usual thermodynamic properties and being in a chemical equilibrium with the reactants, with the concentration of the complexes following the change in the reactant concentrations*.

If we assume that each complex formed decomposes in the direction of the reaction products, then the number of reactions per unit volume per unit time is equal to the number of complex decompositions, to the number of complexes per unit volume divided by their lifetime. Denoting by the chemical symbols A, B, M the amount per unit volume of the reactants A and B and complexes M (for example, for reaction (6.32) A and B are XY and WZ and M = XYWZ), we find that the number of forward reactions per unit volume per unit time is given by $k_1 \cdot \mathrm{A} \cdot \mathrm{B} = \mathrm{M}/\tau$. Thus, the forward reaction rate constant is $k_1 = (\mathrm{M}/\mathrm{AB})(1/\tau)$.

According to the law of mass action (see §3 of Chapter III), the ratio of the number of particles participating in the reaction $\mathrm{A} + \mathrm{B} \rightarrow \mathrm{M}$ is equal at equilibrium to the ratio of the partition functions of these particles. (Since A, B, and M denote the number of particles per unit volume, the volumes V which enter the translational partition functions should be set equal to unity.) Factoring out from the partition functions terms of the type $e^{-\mathcal{E}/kT}$

* Actually the relaxation time for establishing such an equilibrium is of the order of the lifetime of the complex, and thus very short.

which correspond to the zero-point energy of the particles, and noting that $\varepsilon_M - (\varepsilon_A + \varepsilon_B) = E$ is equal to the activation energy, we obtain

$$\frac{M}{AB} = \left(\frac{Z_M}{Z_A Z_B}\right) e^{-E/kT}.$$

The partition functions Z_A and Z_B are calculated by the usual methods, while the partition function of the complex should be treated as follows. As with a normal molecule, the complex is stable with respect to all changes in the atomic configuration, except for the direction along the reaction path. Hence, if we consider normal vibrations of the complex, the normal vibrational frequency with respect to the decomposition coordinate is imaginary. If we assume that the peak of the potential barrier is sufficiently flat, then the motion along the decomposition coordinate may be considered as translational with an average speed $\bar{v}_x = (kT/2\pi m^*)^{1/2}$, where m^* is the effective mass of the complex. The partition function for one-dimensional translational motion of particles with a mass m^* along an interval δ, equivalent to the "volume" occupied by the complexes along the decomposition coordinate, is equal to $Z_{\text{trans } 1-d} = (2\pi m^* kT/h^2)^{1/2}\delta$ (cf. (3.12)). In calculating the partition function for the complex Z_M, we must replace the partition function for one of the normal vibrational modes by this translational partition function. The reaction rate constant is thus given by

$$k_1 = \frac{M}{AB}\frac{1}{\tau} = \frac{Z_M}{Z_A Z_B} e^{-E/kT} \frac{\bar{v}_x}{\delta} = \frac{Z_M^*}{Z_A Z_B} e^{-E/kT} \left(\frac{2\pi m^* kT}{h}\right)^{1/2} \delta \left(\frac{kT}{2\pi m^*}\right)^{1/2} \frac{1}{\delta},$$

where Z_M^* denotes the partition function of the complex excluding a factor corresponding to one normal vibrational mode*. This equation shows that the undefined quantities δ and m^* cancel out. Introducing the so-called transmission coefficient κ characterizing the probability of decomposition of the complex in the direction of the reaction products (but not in the direction of the initial particles; κ is usually of the order of unity) we finally obtain the reaction rate constant

$$k_1 = \kappa \frac{kT}{h} \frac{Z_M^*}{Z_A Z_B} e^{-E/kT}. \tag{6.36}$$

The source of the factor kT/h in (6.36), which has the dimensions of frequency and is universal for all reactions, may be visualized in the following manner. We consider a degree of freedom of the complex along the reaction path as a normal vibration with frequency ν. Its partition function is equal to $kT/h\nu$ (for $h\nu < kT$), so that $Z_M = Z_M^* kT/h\nu$. But each vibration actually

* This partition function can be calculated in the usual manner if the "molecular" constants of the complex are known.

results in the decomposition of the complex, so that the lifetime τ is equal to the period of vibrations $\tau = 1/\nu$. From this we get $(kT/h\nu)1/\tau = kT/h$, and thus (6.36).

Substituting into (6.36) the actual expressions for the partition functions and equating the resulting equation to (6.35), we can obtain an explicit value for the steric factor P. Let us first of all formally consider the imaginary reaction in which two atoms combine into a molecule without the participation of a third body. Then Z_A and Z_B are purely translational partition functions and Z_M^* is made up of the translational and rotational partition functions (vibration of the diatomic complex has been excluded). Substituting $Z_{A,B} = (2\pi m_{A,B}kT/h^2)^{3/2}$ and $Z_M^* = [2\pi(m_A + m_B)kT/h^2]^{3/2}(8\pi^2 IkT/h^2)$ into (6.36), and noting that the moment of inertia of the complex is $I = d_{12}^2 m_A m_B/(m_A + m_B)$, where d_{12} is the average diameter of the atoms $d_{12} = (d_A + d_B)/2$, we obtain precisely the relation (6.35), which was derived on collisional arguments, if we identify the steric factor P with the transmission coefficient κ (the collision cross section is $\sigma = \pi d_{12}^2$).

It is generally convenient for purposes of estimation to write the partition functions of the reactants and of the complex as a product of partition functions, each of which corresponds to a single degree of freedom. It is also convenient not to distinguish between the partition functions pertaining to the same degree of freedom but to different particles. For example, in the case when A and B are diatomic molecules, $Z_A \sim Z_B \sim Z_{\text{trans}}^3 Z_{\text{rot}}^2 Z_{\text{vib}}$. Assuming that the complex is nonlinear, we write $Z_M^* \sim Z_{\text{trans}}^3 Z_{\text{rot}}^3 Z_{\text{vib}}^5$ (the complex contains 4 atoms and 6 vibrational degrees of freedom, one of which is excluded). Thus, to within an order of magnitude

$$k_1 \sim \kappa \frac{kT}{h} \frac{Z_{\text{trans}}^3 Z_{\text{rot}}^3 Z_{\text{vib}}^5}{Z_{\text{trans}}^6 Z_{\text{rot}}^4 Z_{\text{vib}}^2} e^{-E/kT}.$$

Similarly, for the reaction in which two atoms combine into a molecule, the factor

$$\frac{kT}{h} \frac{Z_{\text{trans}}^3 Z_{\text{rot}}^2}{Z_{\text{trans}}^6} \approx \frac{kT}{h} \frac{Z_{\text{rot}}^2}{Z_{\text{trans}}^3} \approx \pi d_{12} \overline{v'}$$

gives the approximate number of collisions entering into (6.35), so that the order of magnitude of the steric factor is

$$P \sim \kappa \frac{Z_{\text{vib}}^3}{Z_{\text{rot}}^3}.$$

At room temperatures Z_{vib} is of the order of unity. Z_{rot} is of the order of 10–100, and is smaller, the lighter is the molecule. It is evident, therefore, that the steric factor can be a very small quantity of the order of 10^{-3}–10^{-6}.

In §10 we shall employ the activated complex method for estimating the rate of formation of nitrogen dioxide in high-temperature air. This is important for the understanding of certain optical phenomena observed in a strong explosion.

§8. Oxidation of nitrogen

When air is heated to a temperature of several thousand degrees, the chemical reaction

$$\frac{1}{2}N_2 + \frac{1}{2}O_2 + 21.4\,\frac{\text{kcal}}{\text{mole}} = NO \tag{6.37}$$

takes place. This reaction results in the formation of an appreciable amount of nitric oxide, NO. Equilibrium concentrations of nitric oxide reach several percent at temperatures of 3000–10,000°K and air densities of the order of atmospheric (see Table 3.1, Chapter III). Some of the nitric oxide is oxidized to form nitrogen dioxide NO_2. Under the above conditions, the equilibrium concentrations of the latter are of the order of $10^{-4} = 10^{-2}\%$. Nitrogen oxides play an important role in the radiation and absorption of light by high-temperature air. Especially important in this respect is the role of nitrogen dioxide in the temperature range of 2000–4000°K, where the optical properties of air in the visible part of the spectrum are almost entirely determined by the NO_2 molecules. When air is heated by a strong shock wave, for example, in an explosion, both the temperature and the density of air are subjected to very rapid changes, hence the kinetics of formation and decomposition of nitrogen oxides are of considerable importance in calculating their concentrations. As will be shown in Chapters VIII and IX, features of the kinetics determine some of the characteristic optical effects observed in a strong explosion. In this section we shall consider the kinetics of the formation of nitric oxide and in the following section the kinetics of the oxidation of nitric oxide into the corresponding dioxide.

The oxidation of nitrogen requires a large activation energy, and thus it occurs only at sufficiently high temperatures, of the order of 2000°K and above. This reaction has been studied in detail, experimentally as well as theoretically, by Zel'dovich, Sadovnikov, and Frank-Kamenetskii [39]. The experiment dealing with the formation and decomposition of nitric oxide was carried out with the aid of bombs in which a mixture of hydrogen and oxygen was burned. In this manner temperatures of the order of 2000°K were obtained. To the mixture of H_2 and O_2 nitrogen and different concentrations of nitric oxide were added. When the amounts added were small, the oxide formation resulted from the combination of oxygen with the nitrogen; however, when the additions were large the oxide initially introduced decomposed. The remaining amount of the oxide was determined after the explosion, and the formation and decomposition rates were found by comparing the experimental results with the theory. The combination of oxygen with hydrogen exerted almost no effect on the formation and decomposition of the oxide and served only as a means for obtaining high temperatures.

If we assume that the reaction takes place according to the bimolecular mechanism, that is, a collision between an N_2 and O_2 molecule results in the formation of two NO molecules, then we can write the simple expression for the reaction rate constant which follows from the collision theory (see (6.35)): $k' = P\bar{v}\sigma e^{-E/kT}$. The experimentally determined value of the pre-exponential factor is $1.1 \cdot 10^3 O_2^{-1/2}$, where O_2 denotes the number density of oxygen molecules. Thus, if we substitute $O_2 = 10^{18}$ molecules/cm^3, we obtain a value of the pre-exponential factor equal to $1.1 \cdot 10^{-6}$ cm^3/sec. At $T = 2500°$K, $\bar{v} \approx 2 \cdot 10^5$ cm/sec and $\sigma \approx 10^{-15}$ cm^2 and the transformation probability P has the improbably high value $P \approx 5000$. Thus the assumption that the reaction occurs according to the bimolecular mechanism leads to a physically unrealistic result; it has been shown experimentally that the true reaction rate is much higher.

N. N. Semenov suggested that the oxidation of nitrogen proceeds according to a chain mechanism, in which active roles are played by the free O and N atoms

$$O + N_2 \underset{k_3}{\overset{k_1}{\rightleftarrows}} NO + N - 75.5 \text{ kcal/mole}, \tag{6.38}$$

$$N + O_2 \underset{k_4}{\overset{k_2}{\rightleftarrows}} NO + O + 32.5 \text{ kcal/mole}. \tag{6.39}$$

The heats of reaction correspond to the dissociation energies of the N_2 and NO molecules which are equal to 9.74 ev = 225 kcal/mole and 6.5 ev = 150 kcal/mole, respectively*. The overall rate of the process is determined by the first, endothermic reaction, which requires an activation energy of not less than 75.5 kcal/mole. As soon as an N atom is liberated by the exchange $O + N_2 \rightarrow NO + N$, it immediately reacts with the molecular oxygen, restoring the oxygen atom O that had previously disappeared. Therefore, the concentration of O atoms during the reaction remains constant and corresponds to equilibrium with the O_2 molecules; this equilibrium is established faster than that for the oxidation of nitrogen†.

* In [39] the old values of the dissociation energies of N_2 and NO were used, 7.38 ev and 5.3 ev; however, as shown by the calculations presented below (and also by later experiments [40]), the new values of the dissociation energies do not contradict the assumption of a chain mechanism. All the numerical values of the constants in the subsequent presentation correspond to the later dissociation energies.

† Since the oxygen atoms are in equilibrium with O_2 molecules, the mechanism of dissociation of O_2 does not affect the progress of the oxidation of nitrogen.

Denoting the rate constants as in (6.38) and (6.39), we can write the general rate equations

$$\frac{d\,\mathrm{NO}}{dt} = k_1 \cdot \mathrm{O} \cdot \mathrm{N_2} + k_2 \cdot \mathrm{N} \cdot \mathrm{O_2} - k_3 \cdot \mathrm{N} \cdot \mathrm{NO} - k_4 \cdot \mathrm{O} \cdot \mathrm{NO}, \qquad (6.40)$$

$$\frac{d\,\mathrm{O}}{dt} = -\frac{d\,\mathrm{N}}{dt} = -k_1 \cdot \mathrm{O} \cdot \mathrm{N_2} + k_2 \cdot \mathrm{N} \cdot \mathrm{O_2} + k_3 \cdot \mathrm{N} \cdot \mathrm{NO} - k_4 \cdot \mathrm{O} \cdot \mathrm{NO}. \qquad (6.41)$$

By virtue of the fact that the concentration of O is constant, we can equate the right-hand side of (6.41) to zero, express the concentration of N in terms of O, and substitute the resulting expression into (6.40). This yields

$$\frac{d\,\mathrm{NO}}{dt} = 2\frac{\mathrm{O}}{k_2 \cdot \mathrm{O_2} + k_3 \cdot \mathrm{NO}}(k_1 \cdot k_2 \cdot \mathrm{N_2} \cdot \mathrm{O_2} - k_3 \cdot k_4 \cdot \mathrm{NO}^2). \qquad (6.42)$$

Let us rearrange some of the equations. The constants k_3 and k_2 determine the rates of the exothermic reactions between an atom and a molecule, and are, most probably, of the same order. Since the concentration of NO is much smaller than that of O_2, we can neglect the term $k_3 \cdot \mathrm{NO}$ in the denominator of (6.42). The concentration of O atoms can be expressed in terms of the equilibrium constant $O_2 \rightleftarrows 2O$, which will be denoted by C_0:

$$\mathrm{O} = C_0 \mathrm{O}_2^{1/2} = 6.6 \cdot 10^{12} e^{-61{,}000/\mathscr{R}T} \mathrm{O}_2^{1/2}. \qquad (6.43)$$

Here, as well as in the subsequent discussion, all the numerical values of the equilibrium constants and of the reaction rate constants correspond to concentrations expressed in molecules/cm^3. The energies are given in cal/mole. The gas constant $\mathscr{R} = 2$ cal/mole · deg. The rate constants are connected by the principle of detailed balancing, and we have*

$$C_1 = \frac{(\mathrm{NO})(\mathrm{N})}{(\mathrm{N_2})(\mathrm{O})} = \frac{k_1}{k_3} = \frac{32}{9} e^{-75{,}500/\mathscr{R}T},$$

$$C_2 = \frac{(\mathrm{NO})(\mathrm{O})}{(\mathrm{O_2})(\mathrm{N})} = \frac{k_2}{k_4} = 6\, e^{32{,}500/\mathscr{R}T}.$$

From this we get the following identity:

$$C_1 C_2 = \frac{k_1 k_2}{k_3 k_4} = \frac{(\mathrm{NO})^2}{(\mathrm{N_2})(\mathrm{O_2})} = C^2 = \frac{64}{3} e^{-43{,}000/\mathscr{R}T}. \qquad (6.44)$$

* The pre-exponential factors in the equilibrium constants C_1, C_2, and C are calculated by assuming that the masses of N and O atoms, their moments of inertia, and the frequencies of N_2, O_2, and NO are approximately equal, taking into account the different symmetries and multiplicities of the terms. This approximation is sufficiently accurate.

Factoring out k_3k_4 in (6.42) and using (6.44), we finally obtain the rate equation for the oxidation of nitrogen

$$\frac{d\,\mathrm{NO}}{dt} = k' \cdot \mathrm{N_2} \cdot \mathrm{O_2} - k \cdot \mathrm{NO}^2 = k\{(\mathrm{NO})^2 - \mathrm{NO}^2\}, \tag{6.45}$$

where the rate constants are

$$k' = \frac{2C_0k_1}{\mathrm{O}_2^{1/2}}, \qquad k = \frac{k'}{C^2} = \frac{2C_0k_1}{C^2\mathrm{O}_2^{1/2}}. \tag{6.46}$$

In this notation (NO) is the equilibrium value of NO corresponding to the given concentrations O_2 and N_2. Equation (6.45) differs from the usual bimolecular reaction equation in the fact that the rate constants are dependent on the concentration of one of the reactants—molecular oxygen.

The physical meaning of the expression for the rate of oxide formation $k' \cdot \mathrm{N_2} \cdot \mathrm{O_2} = 2C_0k_1\mathrm{N_2} \cdot \mathrm{O}_2^{1/2}$ is very simple: $C_0\mathrm{O}_2^{1/2}$ is the concentration of atomic oxygen and $k_1C_0\mathrm{O}_2^{1/2} \cdot \mathrm{N_2}$ is the rate of the first reaction in the chain; however, since the second reaction is exothermic it follows the first one "instantaneously", so that each event in the first reaction which "leads" the process results in the formation of two NO molecules. Factoring out $e^{-E_1/\mathscr{R}T}$ in the rate constant k_1 and noting that according to (6.43) $C_0 \sim e^{-61{,}000/\mathscr{R}T}$, it may be seen that the activation energy E' for the formation of nitric oxide ($k' \sim e^{-E'/\mathscr{R}T}$) is composed of the energy necessary for the formation of one oxygen atom -61 kcal/mole*, and the activation energy for the reaction between an oxygen atom and a nitrogen molecule $-E_1$.

In the experiments described in [39], nitric oxide was obtained by exploding a combustible mixture containing oxygen and nitrogen. The amount of oxide formed was measured after the explosion products cooled. The activation energy for the formation of the oxide $E' = 125 \pm 10$ kcal/mole and the absolute value of the rate constant in the investigated temperature range of 2000–3000°K were derived from the theory of the reaction kinetics of the cooling process. It was also noted that the higher value of the activation energy, that is, $E' = 125 + 10$, is the more probable one. This yielded an activation energy for the first reaction in the chain, $\mathrm{O} + \mathrm{N_2} \rightarrow \mathrm{NO} + \mathrm{N}$ of $E_1 = 135 - 61 = 74$ kcal/mole, which agrees with the heat of the endothermic reaction. This means that the reverse reaction $\mathrm{N} + \mathrm{NO} \rightarrow \mathrm{O} + \mathrm{N_2}$ proceeds practically without activation (or with a very low activation energy), which is typical for an exothermic reaction between a free atom and a molecule. The asbolute values of the rate constants derived from the experimental data are

$$k' = \frac{1.1 \cdot 10^3}{\mathrm{O}_2^{1/2}}\, e^{-135{,}000/\mathscr{R}T}, \qquad k = \frac{53}{\mathrm{O}_2^{1/2}}\, e^{-92{,}000/\mathscr{R}T}\, \frac{\mathrm{cm}^3}{\mathrm{sec}} \quad (\mathrm{O_2} \text{ is in cm}^{-3}). \tag{6.47}$$

* This value is valid for temperatures of the order of 2000–5000°K; it differs slightly from the formation energy at absolute zero, which is -58.5 kcal/mole.

The rate constant for the first reaction in the chain is $k_1 = 8.3 \cdot 10^{-11} \times e^{-74{,}000/\mathscr{R}T}$. Comparison of this value with (6.35) of the collision theory gives a steric factor $P = 0.086$ (if we take the effective diameter $d_{12} = 3.75 \cdot 10^{-8}$ cm to be equal to the diameter of the N_2 molecule, as determined from data on viscosity). This value of P appears to be entirely reasonable.

A later investigation of the kinetics of formation of nitric oxide was carried out by Glick *et al.* [40] with the aid of a shock tube, in which a gas mixture containing nitrogen and oxygen was heated by a shock wave to a temperature of 2000–3000°K. These authors found the activation energy to be $E' = 135 \pm 5$ kcal/mole ($E_1 = 74 \pm 5$ kcal/mole), which is in good agreement with the data of [39] and confirms the agreement of the activation energy E_1 of the first reaction in the chain with the heat of reaction. Absolute values of the rate constant were also found to be close to the values given in the first reference.

It is evident from (6.47) that the activation energy for the decomposition of nitric oxide is also quite large, $E = E' - 43 = 92$ kcal/mole, and that the oxide decomposes very slowly at low temperatures. As a result, nitric oxide which formed in air at high temperatures is retained for a long time after the air is cooled, so that its concentration considerably exceeds the equilibrium value, which is very low at low temperatures (this effect is called "freezing"; we shall return to it in §5 of Chapter VIII). As is apparent from the rate equation (6.45), and (6.44) and (6.47), the relaxation time required to establish an equilibrium concentration of the oxide is*

$$\tau = \frac{1}{2k(\mathrm{NO})} = \frac{0.95 \cdot 10^{-2}\mathrm{O}_2^{1/2}}{(\mathrm{NO})} e^{92{,}000/\mathscr{R}T} = \frac{2.06 \cdot 10^{-3}}{\mathrm{N}_2^{1/2}} e^{113{,}500/\mathscr{R}T} \text{ sec}, \tag{6.48}$$

with (NO) the equilibrium value of NO. The relaxation time decreases rapidly with increasing temperature. Given below are several values for air at standard density ($N_2 = 2.1 \cdot 10^{19}$ molecules/cm³):

T, °K	1000	1700	2000	2300	2600	3000	4000
τ, sec	$2.2 \cdot 10^{12}$	140	1	$5.3 \cdot 10^{-3}$	$1.4 \cdot 10^{-3}$	$7.8 \cdot 10^{-5}$	$7.2 \cdot 10^{-7}$

§9. Rate of formation of nitrogen dioxide at high temperatures

Since the formation of nitrogen dioxide from nitric oxide

$$2\,\mathrm{NO} + \mathrm{O}_2 = 2\,\mathrm{NO}_2 + 25.6 \text{ kcal/mole} \tag{6.49}$$

* According to the definition of relaxation time given by (6.2), when NO differs only slightly from (NO) we have $\{(\mathrm{NO})^2 - \mathrm{NO}^2\} \approx 2(\mathrm{NO}) \cdot \{(\mathrm{NO}) - \mathrm{NO}\}$, from which (6.48) follows. The time τ characterizes not only the approach to equilibrium, but the process of reaching equilibrium in general, even if no oxide was present initially.

is exothermic, the equilibrium shifts in the direction of oxidation of the oxide with a decrease in temperature. This reaction has wide industrial application and has been well studied experimentally at temperatures below 1000°K. The reaction has a very low, practically undetectable, activation energy and therefore takes place easily at ordinary temperatures. The rate equation for the reaction is

$$\frac{d\,NO_2}{dt} = 2\{k_1'NO^2 \cdot O_2 - k_2'NO_2^2\} = 2k_2'\{(NO_2)^2 - NO_2^2\}. \quad (6.50)$$

Here again (NO_2) is the equilibrium value of NO_2. The reaction rate constants describe a number of reaction events; the factor of 2 takes into account the fact that two NO_2 molecules are either formed or disappear during each reaction event. The relaxation time for establishing chemical equilibrium of the nitrogen dioxide with the oxide and the oxygen is

$$\tau' = \frac{1}{4k_2'(NO_2)} = \frac{C^2}{4k_1'(NO_2)}, \quad (6.51)$$

where $C^2 = (NO_2)^2/NO^2 \cdot O_2$ is the equilibrium constant between the dioxide and the true amounts of the oxide and the oxygen, which may not be at their equilibrium values*. The equilibrium constants can be calculated by the statistical method. After substitution of all the appropriate parameters it is found to be†

$$C = \frac{(NO_2)}{NO \cdot O_2^{1/2}} = \frac{1.25 \cdot 10^{-11}}{T^{\circ\,3/4}}$$

$$\times \frac{(1 - e^{-2740/T})(1 - e^{-2270/T})^{1/2} e^{6460/T}}{(1 + e^{-174/T})(1 - e^{-916/T})(1 - e^{-1960/T})(1 - e^{-2310/T})}, \quad (6.52)$$

where the temperature is expressed everywhere in degrees and the units of C correspond to the concentrations expressed in number of particles per cm^3.

The rate constant k_1' was calculated in [38] by the activated complex method; it was found to be in good agreement with the experimental values of M. Bodenstein and his associates [44], who investigated the reaction rate in the temperature range between 353 and 845°K. Comparison shows that the reaction takes place without activation energy. The equation for the rate constant k_1' derived in [38] can also be used for estimating both the rate and the relaxation time at high temperatures that have not been investigated experimentally. The

* We note that the time τ' characterizes the relaxation only under the condition $(NO_2) \ll NO$, i.e., at sufficiently high temperatures. In the opposite case it is also necessary to consider simultaneously the change in the concentration of NO.

† It was noted by one of the authors [41] that the equilibrium constant quoted in the widely used handbook [42] was taken from [43], in which an incorrect value 2.42 times larger than the true value was given.

calculated relaxation times for the formation of nitrogen dioxide in high-temperature air at several temperatures and several values of density are presented in Table 6.3 (the equilibrium concentrations of (NO_2) have been calculated on the basis of equilibrium values of the oxide (NO) and oxygen (O_2) concentrations).

Table 6.3

RELAXATION TIMES FOR ESTABLISHING EQUILIBRIUM CONCENTRATION OF NITROGEN DIOXIDE IN AIR, IN SEC (τ' DENOTES A TERMOLECULAR REACTION AND τ'' A BIMOLECULAR ONE)

T, °K	ρ/ρ_{stand}					
	10		5		1	
	τ'	τ''	τ'	τ''	τ'	τ''
1600			$8.0 \cdot 10^{-3}$	$3.5 \cdot 10^{-1}$	$0.9 \cdot 10^{-1}$	$6.9 \cdot 10^{-1}$
1800			$3.5 \cdot 10^{-3}$	$3.9 \cdot 10^{-2}$	$0.4 \cdot 10^{-1}$	$0.9 \cdot 10^{-1}$
2000	$6.75 \cdot 10^{-4}$	$3.1 \cdot 10^{-3}$	$1.95 \cdot 10^{-3}$	$4.5 \cdot 10^{-3}$	$2.2 \cdot 10^{-2}$	$0.1 \cdot 10^{-1}$
2300	$1.42 \cdot 10^{-4}$	$2.7 \cdot 10^{-4}$	$4.0 \cdot 10^{-4}$	$4.0 \cdot 10^{-4}$	$4.5 \cdot 10^{-3}$	$0.9 . 10^{-3}$
2600	$4.75 \cdot 10^{-5}$	$4.4 \cdot 10^{-5}$	$1.35 \cdot 10^{-4}$	$6.3 \cdot 10^{-5}$	$1.5 \cdot 10^{-3}$	$1.4 \cdot 10^{-4}$
3000	$1.75 \cdot 10^{-5}$	$6.6 \cdot 10^{-6}$	$4.75 \cdot 10^{-5}$	$9.4 \cdot 10^{-6}$	$5.5 \cdot 10^{-4}$	$2.1 \cdot 10^{-5}$
4000	$2.50 \cdot 10^{-8}$	$2.8 \cdot 10^{-7}$	$7.5 \cdot 10^{-8}$	$4.0 \cdot 10^{-7}$	$1.05 \cdot 10^{-6}$	$1.0 \cdot 10^{-6}$

At high temperatures, and especially at low densities, another mechanism for the formation of nitrogen dioxide, namely

$$NO + O_2 + 45\,\frac{\text{kcal}}{\text{mole}} = NO_2 + O, \tag{6.53}$$

competes with the termolecular reaction (6.49). In spite of the fact that this reaction is endothermic, it has an advantage over the reaction given by (6.49) through the fact it occurs by binary, rather than three-body molecular collisions. This advantage should manifest itself at high temperatures where the activation conditions are favorable. Reaction (6.53) has not been subjected to any experimental studies; a theoretical estimate of its rate was given by one of the present authors in [41].

The rate equation for the reaction (6.53) can be written as

$$\frac{d\,NO_2}{dt} = k_1'' \cdot NO \cdot O_2 - k_2'' NO_2 \cdot O = k_2'' \cdot O\{(NO_2) - NO_2\}. \tag{6.54}$$

The relaxation time is

$$\tau'' = \frac{1}{k_2'' \cdot O}. \tag{6.55}$$

Let us estimate the rate constant by the activated complex method, the basic details of which were presented in §7. In particular, this estimate can serve as an illustration of a concrete application of the method. For convenience we shall consider the reverse reaction $NO_2 + O \rightarrow NO_3^* \rightarrow NO + O_2$, where the asterisk denotes the complex. From the general relation (6.36) the rate constant k_2'' is

$$k_2'' = \kappa \frac{kT}{h} \frac{Z^*_{NO_3}}{Z_{NO_2} \cdot Z_O}.$$

The partition functions for the O atoms and NO_2 molecules can be calculated without difficulty, since the spectroscopic constants for the NO_2 molecules are known. The complex, however, contains a large number of unknown quantities that must be chosen in a reasonable manner to carry out the estimate.

The mass of the NO_3^* complex is 1.39 times greater than the mass of the NO_2 molecule. Assuming that its dimensions somewhat exceed the dimensions of the NO_2 molecule, we assume that the average moment of inertia of the complex is 1.5 times greater than the average moment of inertia of the NO_2 molecule. The natural frequencies of the NO_3 molecule, which would be used to estimate the frequencies of the complex, are not known. We can only expect that the three highest frequencies are lower than the frequencies of the NO_2 molecule: $h\nu_{NO_2}/k = 960, 1960, 2310°K$, since the bonds in the complex are weaker. It can be easily checked that the rate constant at temperatures of 2000–4000°K is not very sensitive to the frequencies selected for the complex as long as they are taken to be within a reasonable range. For purposes of calculation let us consider the following five frequencies: $h\nu/k = 600$, 800, 900, 1500, and 2000°K (the sixth frequency is excluded from Z^*). The complex is not symmetric, so that the symmetry factor $\sigma = 1$. The statistical weight of the electronic state is $g^* \geqslant 2$, since the complex contains one unpaired electron. Let us set $g^* = 2$. The activation energy for the exothermic reaction $NO_2 + O \rightarrow NO + O_2$ is, apparently, very small, as is usual when one of the reactants is a free atom. As an estimate take $E = 10$ kcal/mole, which, at worst, can lower the estimate of the reaction rate by a factor of 2–3 at temperatures of 2000–4000°K. Substituting these as well as the other known constants into the expressions for the partition functions, and taking the transmission coefficient κ equal to unity, we obtain the rate constant

$$k_2'' = \frac{1.16 \cdot 10^{-12}}{T^{\circ 1/2}} \frac{\prod_{i=1}^{5} Z^*_{\mathrm{vib}\, i}}{\prod_{i=1}^{3} Z_{\mathrm{vib\, NO_2}\, i}} e^{-5030/T} \ \mathrm{cm^3/sec}, \tag{6.56}$$

where the vibrational partition function is given by $Z_{\text{vib}} = (1 - e^{h\nu/kT})$. In order to obtain a rate constant from the collision theory (see (6.35)) of the same order as that given by (6.56), we would have to assume that the steric factor P is of the order of $2 \cdot 10^{-4}$. To choose such a low value without any apparent justification would be quite difficult, so that in this case the collision theory is practically useless and the reaction rate can be estimated only by using the activated complex method.

The relaxation times for air calculated from (6.55) and (6.56) are also presented in Table 6.3. Comparison of these relaxation times shows that at air densities of the order of or less than standard density, and at temperatures of ~2000–3000°K, the second reaction proceeds more rapidly and can thus be considered to be the principal reaction.

2. *Ionization and recombination. Electronic excitation and deexcitation*

§10. Basic mechanisms

Excitation of the higher electronic states in atoms (molecules, ions) and ionization have very much in common. Essentially, ionization is a limiting case of electronic excitation, when a bound electron in an atom acquires sufficient energy to leave the atom and pass into the continuous spectrum. If sufficient energy is available, each of the elementary processes resulting in the excitation of electrons in atoms can also result in ionization.

All the elementary excitation and ionization processes can be divided into two categories: excitation and ionization of atoms (molecules, ions) caused by collisions with particles, and photoprocesses, in which the role of one of the "particles" is played by a photon. In the first category we must distinguish between ionization and excitation by electron impact and by inelastic collisions of heavy particles; this distinction is necessary since the probability of each type of collision is very different. According to this classification we can denote the basic ionization reactions in the following symbolic form (A and B are the heavy particles, e are the electrons, and $h\nu$ are the photons)

$$\mathrm{A} + e = \mathrm{A}^+ + e + e, \tag{6.57}$$

$$\mathrm{A} + \mathrm{B} = \mathrm{A}^+ + \mathrm{B} + e, \tag{6.58}$$

$$\mathrm{A} + h\nu = \mathrm{A}^+ + e. \tag{6.59}$$

The reverse processes, proceeding from right to left, result in recombination of electrons with ions: the first two represent recombinations by three-body collisions with the participation of an electron or a heavy particle as the third body, and the last reaction is photorecombination or radiative capture of electrons.

Each of the processes (6.57)–(6.59) corresponds to an excitation process (an excited atom is denoted by an asterisk and the negative charge on e is omitted)

$$A + e = A^* + e, \tag{6.60}$$

$$A + B = A^* + B, \tag{6.61}$$

$$A + h\nu = A^*. \tag{6.62}$$

The first two reverse processes represent deexcitation of excited atoms by so-called collisions of the second kind, and the third represents spontaneous emission of an excited atom.

The atoms which are ionized may be not only those in the ground state, but also excited atoms, so that to the list of reactions (6.57)–(6.59) we must also add reactions of the type

$$A^* + e = A^+ + e + e, \tag{6.63}$$

$$A^* + B = A^+ + B + e, \tag{6.64}$$

$$A^* + h\nu = A^+ + e. \tag{6.65}$$

The same is true with respect to the excitation processes; to the list of reactions (6.60)–(6.62) we must add the reactions resulting in an increase in the degree of excitation

$$A^* + e = A^{**} + e, \tag{6.66}$$

$$A^* + B = A^{**} + B, \tag{6.67}$$

$$A^* + h\nu = A^{**}. \tag{6.68}$$

In spite of the fact that the number of excited atoms is usually appreciably smaller than the number of atoms in the ground state, the role of ionization of excited atoms is not insignificant, since the ionization of excited atoms is caused by collisions with particles of lower energies. Indeed, the number of particles capable of ionizing an unexcited atom is proportional to $e^{-I/kT}$ where I is the ionization potential. However, the number of ionization events for atoms excited to the level E^* is also proportional to $e^{-E^*/kT}e^{-(I-E^*)/kT} = e^{-I/kT}$, since the number of excited atoms is proportional to the first factor in this expression, and the number of particles capable of ionizing the excited atoms is proportional to the second factor. The relative importance of ionization of excited and unexcited atoms under conditions of equilibrium excitation is determined principally by the ionization cross sections for the two types of atoms in collisions with particles whose energy is higher than the threshold energy.

Generally speaking, processes of all three types occur simultaneously in a gas. Quite frequently, however, one of the processes is found to be dominant. For energies of the order of excitation or ionization potentials of the atom, that is, of the order of several or 10 electron volts, the cross sections for inelastic collisions of heavy particles are several orders of magnitude smaller than the cross sections for inelastic electron impacts. In addition, the velocities of the heavy particles with comparable energies are approximately a hundred times smaller than the electron velocities (in the ratio of the square roots of their masses). Therefore, processes of the type (6.58) and (6.61) in a high-temperature gas are important only if free electrons are practically absent. When the degree of ionization is of the order of 10^{-5}–10^{-4} or higher, the rates of processes of the first type (6.57) and (6.60) are greater than the rates of processes with the participation of heavy particles, and the role of the latter is negligibly small. If the gas is "instantaneously" heated, as sometimes occurs, for example, during the passage of a strong shock wave, ionization by atom or molecular collisions is of importance only for the formation of a small number of initially free "priming" electrons. In some cases the initial ionization in an "instantaneously" heated gas is created by a sufficiently intense radiation flux or by rapidly moving electrons which are supplied from the outside, from previously heated regions; these mechanisms eliminate even the "priming" role of the second process.

The comparative role of the first and third processes depends on the macroscopic conditions in a more complicated manner. The number of ionization events caused by electron impacts per unit time per unit volume is proportional to the electron density, while the number of photoionization events is proportional to the radiation density. If the dimensions of the region occupied by the heated gas are sufficiently large in comparison with the photon mean free paths, so that the radiation density is appreciable (of the order of the equilibrium density), then the radiation will be independent of the gas density and will depend only on the temperature. In this case the rate of ionization by electron impact in a sufficiently rarefied gas is found to be small and the dominant role is played by photoionization. The same is true of the excitation processes and also of the reverse processes, such as recombination and deexcitation: photorecombination dominates over recombination by three-body collisions, and spontaneous emission of excited atoms dominates over the deexcitation of atoms excited by collisions of the second kind. This is precisely the situation observed, for example, in stellar photospheres.

If the region occupied by the heated gas is bounded and transparent ("optically thin"), then the photons radiated in the gas are not captured but leave the heated volume, and the radiation density in the gas is less than its equilibrium value. Under these conditions, even if the electron density is low, the rate of ionization by electron impact can turn out to be higher than the photoionization rate; the ratio of the rates of the inverse recombination processes can remain the same as before, so that photorecombination can still dominate.

In a sufficiently dense ionized gas, photoionization and photorecombination are of secondary importance in comparison with processes (6.57) and (6.63). Actually, the ionization and recombination processes most often take place in a more complicated manner than indicated by the simple reactions (6.57)–(6.59). So-called step-wise ionization takes place, in which the atom is first excited, let us say, by electron impact, and then it is either immediately ionized by succeeding electron impacts, or first passes through several stages during which its degree of excitation increases. The reverse process of recombination by three-body collisions also frequently takes place in a complex manner. The electron is captured by an ion into an excited atomic level, and then the atom is deexcited in stages; there are competing deexcitation mechanisms, electron collisions of the second kind and spontaneous radiative transitions between the levels.

In a molecular gas in which the atoms and molecules have ionization potentials not much greater than the dissociation energies, ionization begins long before dissociation ends, so that there is a temperature range in which there are appreciable concentrations of both electrons and molecules simultaneously. As an example we can cite air at temperatures of the order of 7000–15,000°K, which are in a range important for practical applications. In this case, the above ionization processes are accompanied by more complex processes, the most important of which is the recombination of atoms into a molecule with simultaneous ionization (associative ionization). From an energy point of view this process has an advantage over the others in the fact that it requires the expenditure of less energy than the ionization potential by an amount equal to the dissociation energy. At comparatively low temperatures and small degrees of ionization, the most important role in the ionization of air can be attributed to the reaction

$$\mathrm{N} + \mathrm{O} + 2.8\ \mathrm{ev} = \mathrm{NO}^+ + e, \tag{6.69}$$

which proceeds faster by several orders of magnitude than the simple ionization of NO caused by atomic or molecular collisions*. An important role in the recombination of electrons with ions in molecular gases is played by the so-called dissociative recombination. In particular, the following processes take place in air:

$$\begin{aligned} e + \mathrm{O}_2^+ &\to \mathrm{O}^* + \mathrm{O}^* + 6.9\ \mathrm{ev}, \\ e + \mathrm{N}_2^+ &\to \mathrm{N}^* + \mathrm{N}^* + 5.8\ \mathrm{ev}, \\ e + \mathrm{NO}^+ &\to \mathrm{N}^* + \mathrm{O}^* + 2.8\ \mathrm{ev}. \end{aligned} \tag{6.70}$$

* At comparatively low air temperatures the main supply of free electrons is provided by the NO molecules with an ionization potential $I_{\mathrm{NO}} = 9.25$ ev, which is lower than those of all the other constituents of air ($I_{\mathrm{O}_2} = 12.15$ ev, $I_{\mathrm{N}_2} = 15.56$ ev, $I_{\mathrm{O}} = 13.57$ ev, $I_{\mathrm{N}} = 14.6$ ev, and $I_{\mathrm{Ar}} = 15.8$ ev).

Dissociative recombination results in the formation of excited atoms. The liberated binding energy of the electron is partly used up in the dissociation of the molecule, while the remainder goes into exciting the atoms and into kinetic energy.

If the gas contains atoms or molecules with electron affinity (for example, H, O, O_2, Cl, Br, I, and others), then negative ions are formed at comparatively low temperatures. The formation of these negative ions has an appreciable effect on the rates of formation and disappearance of free electrons. In addition to reactions of the types (6.57)–(6.59), in which A and A^+ are replaced by A^- and A, respectively, there can also take place more complicated but energetically more advantageous reactions of the type (6.69). For example, in air there are the exothermic reactions

$$N + O^- = NO + e + 4\ \text{ev},$$
$$O + O^- = O_2 + e + 3.6\ \text{ev}.$$

A list of the reactions taking place in air which lead to the formation and disappearance of free electrons and also to charge exchange, along with the respective energy yields, is given in [73].

§11. Ionization of unexcited atoms by electron impact

Let us examine the single ionization process in a gas consisting of identical atoms. We assume that all atoms are ionized from the ground state and also that in the process of recombination (the reaction (6.57)) an electron is captured in the ground level. The ionization cross section for the collisions depends on the relative velocities of the colliding particles. Since the velocities of the atoms at comparable atom and electron temperatures are always considerably lower than the electron velocities, the relative velocities are approximately the same as those of the electrons; the reduced mass, which characterizes the kinetic energy of the relative motion, then is approximately the same as the electron mass.

If N_a and N_e are the atom and electron number densities, respectively, $f_e(v)\,dv$ is the Maxwell velocity distribution function for the electrons which corresponds to the electron temperature* T_e $\left(\int_0^\infty f_e(v)\,dv = 1\right)$, and $\sigma_e(v)$ is the ionization cross section for electron impact, then the number of ionization events per unit volume per unit time is

$$Z^e_{\text{ion}} = N_a N_e \int_{v_k}^{\infty} \sigma_e(v) v f_e(v)\,dv = N_a N_e \alpha_e, \tag{6.71}$$

* As a result of the great difference between the masses of the electrons and the atoms, energy exchange between electrons and heavy particles during elastic collisions takes place quite slowly. Therefore, the electron temperature can in general differ from the translational temperature of the heavy particles (see Part 3 of this chapter).

where the integration extends over electron velocities whose energy exceeds the ionization potential: $m_e v_k^2/2 = I$. Denoting the recombination rate constant by β_e, we can write down the rate equation for the reaction (6.57)

$$\frac{dN_e}{dt} = \alpha_e N_a N_e - \beta_e N_+ N_e^2, \tag{6.72}$$

where the number of ions N_+ is equal to the number of electrons N_e. The rate constants α_e and β_e are related by the principle of detailed balancing

$$\beta_e = \frac{\alpha_e}{K(T_e)}, \tag{6.73}$$

where the equilibrium constant is determined by the Saha equation (3.44):

$$K(T_e) = \frac{(N_e)(N_+)}{(N_a)} = \frac{g_+}{g_a} \frac{2(2\pi m_e kT_e)^{3/2}}{h^3} e^{-I/kT_e}$$

$$= 4.85 \cdot 10^{15} \frac{g_+}{g_a} T^{\circ\, 3/2} e^{-I/kT} \text{ cm}^{-3}. \tag{6.73'}$$

The number of recombinations per unit volume per unit time is sometimes expressed in the form $Z_{rec} = b_e N_+ N_e$. The quantity $b_e = \beta_e N_e$ is called the recombination coefficient. The dimensions of b_e are cm^3/sec, which are the same as the dimensions of the ionization rate constant α_e.

If the electron (and ion) concentration is much smaller than its equilibrium value, then recombination is unimportant and the development of ionization by electron impact has the character of an electron avalanche: if we assume that the electron temperature is independent of time, then the electron concentration increases exponentially with time $N_e = N_e^0\, e^{t/\tau_e}$. Here N_e^0 is the initial electron number density, and the time scale for the avalanche is approximately (for $N_a \approx const$) given by*

$$\tau_e = \frac{1}{\alpha_e N_a}. \tag{6.74}$$

It can easily be shown that the quantity τ_e also characterizes the relaxation time for the approach to ionization equilibrium by means of the mechanism

* We would emphasize the fact that the simple exponential dependence of the electron avalanche with the time scale τ_e is valid only for $T_e = const$. Under actual conditions the electron temperature itself may be time-dependent. The point is that for $kT_e \ll I$ the ionization process requires a very large fraction of the thermal energy of the electrons; roughly speaking, the birth of each new electron requires the thermal energy from I/kT_e electrons. If there is no source to compensate for the energy lost by the electron gas in ionization, the electron temperature decreases with time, and the factor $\alpha_e \sim e^{-I/kT_e}$ decreases sharply, with the result that the progress of the avalanche is damped out. The loss of electron energy across a shock front is compensated for by the flow of energy from the atoms (ions) to the electrons. For further details, see §10 of Chapter VII.

(6.57). More precisely, for $|(N_e) - N_e| \ll (N_e)$, the relaxation time according to the general definition (6.2) is smaller than τ_e by a factor of one half.

A typical curve of the dependence of the ionization cross section σ_e on the electron velocity or energy is shown in Fig. 6.4. The cross section increases from the ionization threshold $\varepsilon_e = I$, reaches a maximum at an electron energy several times the threshold energy, and then slowly decreases. As a rule, the

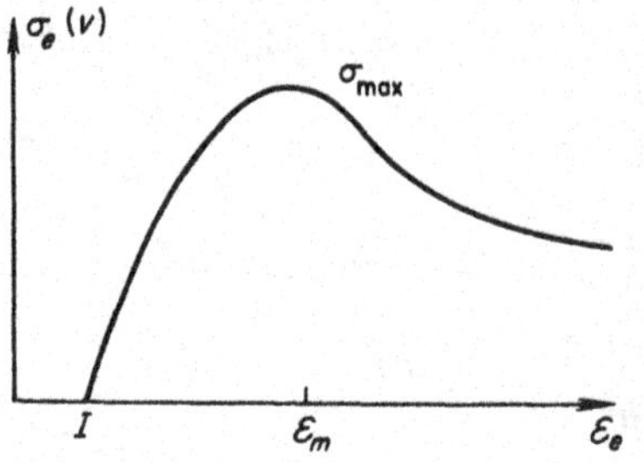

Fig. 6.4. Dependence on the electron energy of the ionization cross section for electron impact.

maximum cross section is of the order of 10^{-16} cm^2. In not too dense a gas ionization usually begins at much lower temperatures than that corresponding to the ionization potential: $I/kT_e \gg 1$. Thus, for example, in atomic hydrogen with $N_a = 10^{19}$ cm^{-3} (which corresponds to a pressure of 135 mm Hg for undissociated molecular hydrogen at room temperature) and $T = 10{,}000$°K, the equilibrium degree of ionization is $6.25 \cdot 10^{-3}$; here $I/kT = 15.7$.

Only those electrons which correspond to the tail of the Maxwell distribution function possess sufficient energy for ionization; the number of such electrons is exponentially small (proportional to $\exp(-m_e v^2/2kT_e) \ll 1$). Therefore, the dominant role in the integral of (6.71) is played by those electrons whose energies only slightly, of the order of kT_e ($kT_e \ll I$), exceed the ionization potential. It has been shown both theoretically and experimentally that the cross section near the threshold depends linearly on the electron energy ε_e, with

$$\sigma_e(v) \approx C(\varepsilon_e - I), \qquad C = const. \tag{6.75}$$

Substituting this quantity into (6.71) and integrating, we find the rate constant for ionization from the ground level of the atoms to be given by

$$\alpha_e = \int_{v_k}^{\infty} \sigma_e(v) v f_e(v)\, dv = \sigma_e \bar{v}_e \left(\frac{I}{kT_e} + 2\right) e^{-I/kT_e}. \tag{6.76}$$

Here

$$\bar{v}_e = \left(\frac{8kT_e}{\pi m_e}\right)^{1/2} = 6.21 \cdot 10^5 (T_e^{\circ})^{1/2} = 6.7 \cdot 10^7 (T_{ev})^{1/2}\ \text{cm/sec}$$

is the mean thermal speed of the electrons, and σ_e is an average value of the cross section $\sigma_e(v)$, the value which corresponds precisely to the electron energy $\varepsilon_e = I + kT_e$; thus $\sigma_e = CkT_e$.

According to (6.73) and (6.76) the recombination rate constant with capture of an electron into the ground atomic level is

$$\beta_e = \frac{g_a}{g_+}\left(\frac{I}{kT_e} + 2\right)\frac{h^3\sigma_e}{2\pi^2 m_e^2 kT} = 1.1 \cdot 10^{-14} C \frac{g_a}{g_+}\left(\frac{I}{kT_e} + 2\right)\frac{\text{cm}^6}{\text{sec}}. \quad (6.77)$$

In not too dense a gas, when $I/kT_e \gg 1$ and $\beta_e \sim T_e^{-1}$, the characteristic time τ_e for small degrees of ionization has a temperature dependence given by $\tau_e \sim e^{I/kT_e}$.

Table 6.4

IONIZATION BY ELECTRON IMPACT

Atom, Molecule	I, ev	$C \cdot 10^{17}$, cm²/ev	Region of applicability, ev	ε_{max}, ev	$\sigma_{e\,max} \cdot 10^{16}$, cm²	Reference
H_2	15.4	0.59	16—25	70	1.1	[46]
He	24.5	0.13	24.5—35	100	0.34	[46]
N	14.6	0.59	15—30	~100	~2.1	[47]
N_2	15.6	0.85	16—30	110	3.1	[46]
O	13.6	0.60	14—25	~ 80	~1.5	[48]
O_2	12.1	0.68	13—40	110		[46]
NO	9.3	0.82	10—20	~100	3.25	[49]
Ar	15.8	2.0	15—25	100	3.7	[46]
		1.7	15—18			
Ne	21.5	0.16	21.5—40	~160	0.85	[46]
Hg	10.4	7.9	10.5—13	42	5.4	[46]
Hg		2.7	10.5—28			

In Table 6.4 we have presented experimental data on ionization cross sections by electron impact for some atoms and molecules (for notation, see Fig. 6.4)*. The numerical value of the constant C coincides with the cross section (in cm²) for electron energies exceeding the ionization potential by 1 ev, that is, with an average cross section σ_e at a temperature $T_e = 1$ ev = 11,600°K, which is characteristic of the singly ionized region. As may be seen from the table, the cross section σ_e is of the order of 10^{-17} cm².

In order to get some idea of the orders of magnitude of the quantities involved, let us consider the concrete example of argon at $T_e = 13{,}000$°K

* A detailed survey and analysis of data available in the literature is given in the books by Massey and Burhop [45] and Brown [78]. We also recommend Granovskii's book [46].

and $N_a = 1.7 \cdot 10^{18}$ cm^{-3} (this density corresponds to a pressure of 50 mm Hg at standard temperature). The equilibrium degree of ionization under these conditions is 0.14, $\sigma_e = 2.24 \cdot 10^{-17}$ cm^2, $\bar{v}_e = 7.1 \cdot 10^7$ cm/sec; the rate constants are: $\alpha_e = 2 \cdot 10^{-14}$ cm^3/sec, $\beta_e = 5.9 \cdot 10^{-31}$ cm^6/sec and the characteristic time $\tau_e = 2.9 \cdot 10^{-5}$ sec. At $T_e = 16{,}000$°K and the same density, the time τ_e is approximately 1/15 as large, and $\tau_e = 2 \cdot 10^{-6}$ sec.

§12. Excitation of atoms from the ground state by electron impact. Deexcitation

By analogy with our previous discussion we assume for simplicity that the atom has only one excited level E^*, so that the atom can be excited only by transition from the ground state. Let us write the rate equation for excitation

$$\frac{dN^*}{dt} = \alpha_e^* N_a N_e - \beta_e^* N^* N_e. \tag{6.78}$$

Here, α_e^* is the excitation rate constant and β_e^* is the deexcitation rate constant given by $\beta_e^* = \bar{v}_e \sigma_{e2}$, where σ_{e2} is the cross section for electron collisions of the second kind, averaged with respect to the Maxwell distribution. The excitation rate constant is expressed in terms of the excitation cross section $\sigma_e^*(v)$ through exactly the same integral as for α_e (see (6.76)), with the only difference that the lower limit is replaced by the velocity v^* which corresponds to the excitation threshold $m_e v^{*2}/2 = E^*$. The dependence of the cross section $\sigma_e^*(v)$ on velocity or energy (the so-called excitation function) has the same character as that in the ionization curve shown in Fig. 6.4. In exactly the same manner it can be approximated near the threshold by the straight line $\sigma_e^*(v) = C^*(\varepsilon - E^*)$ *. Therefore,

$$\alpha_e^* = \int_{v^*}^{\infty} \sigma_e^*(v) v f_e(v)\, dv = \sigma_e^* \bar{v}_e \left(\frac{E^*}{kT_e} + 2\right) e^{-E^*/kT}. \tag{6.79}$$

where σ_e^* corresponds to an electron energy of $E^* + kT_e$.

Using the principle of detailed balancing and noting that

$$\frac{(N^*)}{(N_a)} = \frac{g^*}{g_a} e^{-E^*/kT_e} \tag{6.80}$$

(g^* and g_a are the statistical weights of the excited and ground states, respectively), we can relate the rate constants α_e^* and β_e^*, or the excitation and deexcitation cross sections:

$$\sigma_{e2} = \sigma_e^* \frac{g_a}{g^*}\left(\frac{E^*}{kT_e} + 2\right); \qquad \beta_e^* = \bar{v}_e \sigma_{e2}. \tag{6.81}$$

* This is possible for many but not for all atoms; in any case the approximation results in only a small error. We note that the excitation cross section for positive ions close to the threshold value as $\varepsilon \to E^*$ is different from zero. See article by M. J. Seaton in [83] (Fig. 4, p. 389).

The characteristic time corresponding to excitation by electron impact is the same as the relaxation time required to establish the Boltzmann distribution (6.80) under the condition that $T_e = const$ (see footnote to (6.74)), and is

$$\tau_e^* = \frac{1}{\beta_e^* N_e} = \frac{1}{\bar{v}_e \sigma_{e2} N_e}. \tag{6.82}$$

Data on excitation cross sections have been collected in the books [45, 46, 78]. Some results are presented in Table 6.5. The average excitation cross sections σ_e^* are of the order of 10^{-17} cm^2. This also is the order of magnitude of the cross section for collisions of the second kind σ_{e2} (the factor in parentheses (6.81) is of the order of 10, but the ratio of the statistical weights g_a/g^* is usually ~ 1–10^{-1}).

Table 6.5

IONIZATION BY ELECTRON IMPACT

Atom	Levels	Potential E^*, ev	Interpolation of total cross section σ_e^*, cm^2	Source[a]
H	$2p$	10.1	$25 \cdot 10^{-18}(\varepsilon_{ev} - 10)$	[83]
He	$2s\,^3S$[b]	19.7	$4.6 \cdot 10^{-18}(\varepsilon_{ev} - 20)$	[46]
	$2s\,^1S$	20.6		
Ne	$3s\,^3P_2$	16.6	$1.5 \cdot 10^{-18}(\varepsilon_{ev} - 16)$	[46]
		18.5		
Ar	$4s\,^3P_2$	11.5	$7 \cdot 10^{-18}(\varepsilon_{ev} - 11.5)$	[50]
Hg	$6p\,^3P_1$	4.87	Max. cross section for $\varepsilon = 6.5$ev $\sigma^*_{max} = 1.7 \cdot 10^{-16}$	[46]
H_2		8.7	$7.6 \cdot 10^{-18}(\varepsilon_{ev} - 8.7)$	[78]

[a] Data with the source [46] were taken from tables given in Granovskii's book. References to the original papers can be found in this book.

[b] *Editors' note.* This level has been changed from $2p\,^3P$ to $2s\,^3S$ for consistency with the excitation potential. The error noted appears in the original source [46].

As an example, let us estimate the relaxation time for argon at $T_e = 13{,}000°$K, with $N_a = 1.71 \cdot 10^{18}$ cm^{-3}. The cross sections are $\sigma_e^* = 10^{-17}$ cm^2 and $\sigma_{e2} \sim 10^{-17}$ cm^2. If we take the equilibrium concentration of electrons $(N_e) = 2.4 \cdot 10^{17}$ cm^{-3}, we get $\tau_e^* \approx 6 \cdot 10^{-9}$ sec. This time is considerably less than the ionization time τ_e. It is interesting to compare the characteristic times for ionization and excitation by electron impact. From (6.74) and (6.82) we obtain

$$\frac{\tau_e}{\tau_e^*} = \frac{\bar{v}_e \sigma_{e2} N_e e^{I/kT}}{\bar{v}_e \sigma_e (I/kT + 2) N_a} \approx \frac{1}{10} \frac{N_e}{N_a} e^{I/kT} \quad \text{when} \quad \frac{I}{kT} \sim 10.$$

Close to equilibrium and with moderate degrees of ionization $N_e/N_a \approx (6 \cdot 10^{21}/N_a)^{1/2} T_{ev}^{3/4} e^{-I/2kT_e}$; the Boltzmann distribution with respect to excitation at ordinary gas densities is always established faster than is ionization equilibrium. In our example with argon for $N_a = 1.7 \cdot 10^{18}$ cm^{-3} and $T_e = 13{,}000°$K, $\tau_e/\tau_e^* \approx 5000$. The times may turn out to be comparable only at the beginning of the ionization process, when the number of electrons is very much smaller than the equilibrium value.

§13. Ionization of excited atoms by electron impact

Let us consider the ionization of an atom on the basis of classical mechanics, assuming that the collision of a free electron with an atom, or rather a collision with the optical electron of an atom, takes place in a time short in comparison with the period of rotation of a bound electron in its orbit, so that the latter acquires energy from the impact in the same manner as would a free electron. This was first done by Thomson in 1912 (classical formulas for the ionization cross section were used in the study of elementary ionization and recombination processes in a hydrogen plasma in [79, 80], while the Born approximation was used in [81]).

As is known from basic mechanics [82], if an electron with kinetic energy ε passes near another electron, the differential cross section of energy transfer to the "target" electron in the range $\Delta\varepsilon$ to $\Delta\varepsilon + d\,\Delta\varepsilon$ is $d\sigma = (\pi e^4/\varepsilon) \times [d\,\Delta\varepsilon/(\Delta\varepsilon)^2]$. The cross section for the transfer of energy exceeding E ($E \leqslant \Delta\varepsilon \leqslant \varepsilon$) is

$$\sigma = \frac{\pi e^4}{\varepsilon}\left(\frac{1}{E} - \frac{1}{\varepsilon}\right).$$

We assume that ionization takes place each time that the impact imparts to the bound electron an amount of energy exceeding the binding energy. Then, if E is the binding energy of the electron in the atom, σ is the ionization cross section. Noting that $e^2 = 2I_H a_0$, where I_H is the ionization potential of a hydrogen atom and a_0 is the Bohr radius, we write the ionization cross section in the form

$$\sigma = 4\pi a_0^2 \frac{I_H^2(\varepsilon - E)}{E\varepsilon^2}. \tag{6.83}$$

This equation gives a qualitatively correct description of the dependence of the ionization cross section of an unexcited atom on the energy $\sigma(\varepsilon)$, as depicted in Fig. 6.4, and gives the correct order of magnitude of the cross sections (in the given case $E \equiv I$). Near the threshold, when $\varepsilon \approx E \equiv I$, (6.83) reduces to the linear dependence (6.75) $\sigma \approx C(\varepsilon - I)$, where the slope constant is found to be

$$C = \frac{4\pi^2 a_0^2}{I_H}\left(\frac{I_H}{I}\right)^3 = 2.6 \cdot 10^{-17}\left(\frac{I_H}{I}\right)^3 \frac{\text{cm}^2}{\text{ev}}.$$

For the majority of atoms and molecules the constants C and cross sections near the threshold as calculated on the basis of classical formulas exceed by several-fold the experimental values (see Table 6.4); correct values are obtained in the case of argon.

Equation (6.83) leads to a similitude relationship with respect to n for ionization from the nth level of a hydrogen atom. For a hydrogen atom $E_n = I_H/n^2$ and

$$\sigma_n(\varepsilon) = 4\pi a_0^2 I_H n^4 \frac{(n^2\varepsilon - I_H)}{(n^2\varepsilon)^2} = n^4\sigma_1(n^2\varepsilon).$$

It is interesting that the quantum-mechanical formulas based on the Born approximation, in the case of the ionization of excited hydrogen atoms, lead to a similar, but slightly different similitude relationship $\sigma_n(\varepsilon) = n^3\sigma_1(n^2\varepsilon)$ [81]. However, we are most justified in considering the ionization of excited atoms on the basis of the classical theory.

We now calculate the ionization rate of the excited atoms. If N_n is the number of atoms per unit volume which are in the nth quantum state, and $f(\varepsilon)\,d\varepsilon$ is the Maxwell distribution function for the electrons normalized to unity, then the number of ionizations of these atoms per unit volume per unit time is

$$Z_{\text{ion},\,n} = N_n N_e \int_{E_n}^{\infty} \sigma(\varepsilon) v_e f(\varepsilon)\, d\varepsilon = \alpha_n N_n N_e. \tag{6.84}$$

Substituting the cross section (6.83), we find the ionization rate constant

$$\alpha_n = 4\pi a_0^2 \bar{v}_e \left(\frac{I_H}{kT}\right)^2 \psi_n, \qquad \bar{v}_e = \left(\frac{8kT}{\pi m}\right)^{1/2};$$

$$\psi_n = \int_{x_n}^{\infty} \left(\frac{1}{x_n} - \frac{1}{x'}\right) e^{-x'}\, dx' = \frac{e^{-x_n}}{x_n} - E_1(x_n); \qquad x_n = E_n/kT \tag{6.85}$$

(we note that the number "n" in this equation is for the present still only a subscript, so that (6.85) can be applied to any atom in any state). When $x_n \gg 1$, we have approximately

$$E_1(x) \approx \frac{e^{-x}}{x}\left(1 - \frac{1}{x}\right), \qquad \psi_n \approx \frac{e^{-x_n}}{x_n^2};$$

$$\alpha_n = 4\pi a_0^2 \bar{v}_e \left(\frac{I_H}{E_n}\right)^2 e^{-E_n/kT} = 2.2 \cdot 10^{-10} T^{\circ 1/2} \left(\frac{I_H}{E_n}\right)^2 e^{-E_n/kT}\, \frac{\text{cm}^3}{\text{sec}}. \tag{6.86}$$

(If $E_n = I$, we obtain (6.76), with the factor 2 dropped in comparison with $I/kT \gg 1$.)

In order to compare the rates of ionization of excited and unexcited atoms, we must make some assumption with respect to the number of excited atoms. We assume that the gas is close to thermodynamic equilibrium and that the excited states have a Boltzmann distribution. In addition, we consider hydrogen atoms. Then,

$$E_n = I_{\mathrm{H}}/n^2, \qquad N_n = n^2 e^{-(x_1 - x_n)} N_1,$$

$$\frac{Z_{\mathrm{ion},n}}{Z_{\mathrm{ion},1}} = \frac{\alpha_n N_n}{\alpha_1 N_1} = n^2 e^{-(x_1 - x_n)} \frac{\psi_n}{\psi_1}.$$

We can always take $x_1 \gg 1$. If x_n is also large (low-lying levels and low temperatures), then

$$\frac{Z_{\mathrm{ion},n}}{Z_{\mathrm{ion},1}} \approx n^2 \left(\frac{x_1}{x_n}\right)^2 \approx n^6 \gg 1.$$

For the upper levels with binding energies $E_n \sim kT$ and $x_n \sim 1$, we can set approximately $\psi_n = \frac{2}{5} e^{-x_n}/x_n$ [80]. In this case

$$\alpha_n \approx 4\pi a_0^2 \bar{v}_e \frac{2}{5}\left(\frac{I_{\mathrm{H}}}{kT}\right) n^2 e^{-E_n/kT}$$

$$\approx 1.4 \cdot 10^{-4} n^2 T^{\circ\,-1/2}\, e^{-E_n/kT} \frac{\mathrm{cm}^3}{\mathrm{sec}} \qquad (E_n \sim kT). \tag{6.86'}$$

Then the ratio of ionization rates is

$$\frac{Z_{\mathrm{ion},n}}{Z_{\mathrm{ion},1}} \approx \frac{2}{5} \frac{x_1^2}{x_n} n^2 \approx \frac{2}{5}\left(\frac{I_{\mathrm{H}}}{kT}\right) n^4 \gg 1.$$

Thus, when the levels have a Boltzmann distribution, the atoms are ionized by electron impacts primarily from the upper levels, in which case the role of the excited states is more important, the higher is the degree of excitation. In drawing conclusions about relative roles of the ionization of unexcited and excited atoms under highly nonequilibrium conditions we must be very careful, since the upper levels may be found to be quite depleted in comparison with the equilibrium distribution. We shall return to the problem of electron-impact ionization in Chapter VII, where we consider the ionization of a gas in a shock wave.

Let us find the rate of electron capture by ions into excited atomic levels by three-body collisions, with the electron playing the role of the third body. The number of captures into the nth level per unit of volume per unit time is $Z_{\mathrm{rec},n} = \beta_n N_e^2 N_+$. According to the principle of detailed balancing, for thermodynamic equilibrium $Z_{\mathrm{rec},n} = Z_{\mathrm{ion},n}$, so that

$$\beta_n = \alpha_n \frac{(N_n)}{(N_e)(N_+)}.$$

The Saha equation together with the Boltzmann law gives

$$\frac{(N_e)(N_+)}{(N_n)} = \frac{2g_+}{g_n}\left(\frac{2\pi mkT}{h^2}\right)e^{-E_n/kT}. \tag{6.87}$$

Substituting α_n from (6.85), we obtain the capture rate constant. For a hydrogen-like atom ($g_+ = 1$, $g_n = 2n^2$)

$$\beta_n = \frac{4}{\pi}\frac{a_0^2 h^3}{m^2 kT}\left(\frac{I_H}{kT}\right)^2 n^2 \psi_n e^{x_n}$$

$$= 0.91 \cdot 10^{-25}\frac{n^2}{T^\circ}\left(\frac{I_H}{kT}\right)^2 \psi_n e^{x_n}\frac{\text{cm}^6}{\text{sec}}. \tag{6.88}$$

If $x_n \gg 1$ (capture into low levels and low temperatures)

$$\psi_n e^{x_n} \approx x_n^{-2}, \qquad \beta_n \approx 0.91 \cdot 10^{-25} n^6/T^\circ \ \text{cm}^6/\text{sec}.$$

For captures into high levels with $E_n \sim kT$,

$$\psi_n e^{x_n} \approx \frac{2}{5}x_n^{-1},$$

$$\beta_n \approx \frac{0.36 \cdot 10^{-25} n^4}{T^\circ}\frac{I_H}{kT} = \frac{5.65 \cdot 10^{-21} n^4}{T^{\circ 2}}\frac{\text{cm}^6}{\text{sec}}. \tag{6.88'}$$

It can be seen from these equations that captures into the upper levels are much more frequent than into the lower levels; the probability of capture increases very rapidly with an increase in the quantum number n. Physically this is simply a result of increasing the radius and area of the orbit of the bound electron with the increase in n (the orbit area is proportional to n^4). We here use the term "capture" rather than "recombination" on purpose, since capture into an upper level, in general, is not always equivalent to recombination—it is easy to remove an electron from the upper levels. We shall return to the problem of recombination by three-body collisions in §17.

Let us discuss briefly the problem of ionization of ions. The cross section relation (6.83) applied to the ionization of hydrogen-like ions from the ground level yields a similarity relationship with respect to Z ($E \equiv I_Z = I_H Z^2$)

$$\sigma_Z(\varepsilon) = \frac{4\pi a_0^2 I_H}{Z^4}\frac{(Z^{-2}\varepsilon - I_H)}{(Z^{-2}\varepsilon)^2} = \frac{1}{Z^4}\sigma_1\left(\frac{\varepsilon}{Z^2}\right).$$

In this case the Born approximation leads to a slight departure from the similarity [83].

Let us estimate the dependence of the rate of ionization of ions on their charge ($Z = 1$ corresponds to a neutral atom). It is obvious that there is no sense in making this comparison at the same temperatures, since the second ionization takes place at temperatures higher than the first, the third at temperatures higher than the second, etc. As was explained in Chapter III, atoms and ions are usually ionized at temperatures satisfying the condition $I_Z/kT \sim$ 5–10. Therefore it is reasonable to compare the ionization rates for constant values of the ratio I_Z/kT. Remembering that in this case $\bar{v}_e \sim T^{1/2} \sim Z$, we obtain from (6.86) ($E_n \to I_Z = I_H Z^2$) the rate constant of the Zth ionization from the ground level $\alpha_Z = \alpha_1/Z^3$. The ionization of ions thus takes place at a relatively slower rate than the ionization of neutral atoms. Physically, this is primarily due to the smallness of the geometric dimensions of ions.

§14. Impact transitions between excited states of an atom

Experimental data on cross sections for electron-impact excitation pertain to transitions from the ground state (see §12). In the study of a heated plasma it is sometimes necessary to estimate the probability of impact transitions between higher levels. The transition cross sections can be estimated on the basis of results given by the quantum-mechanical method of distorted waves [83]. A simplified formula for the cross section of electron impact excitation of an atom for allowed transitions is given in [86]. It is convenient to represent it in a form similar to the classical expression for the ionization cross section

$$\sigma_{nn'} = 4\pi a_0^2 \frac{I_H(\varepsilon - E_{nn'})}{E_{nn'}\varepsilon^2} 3f_{nn'}. \tag{6.89}$$

Here $E_{nn'}$ is the energy of transition $n \to n'$ $(n' > n)$; $f_{nn'}$ is the corresponding oscillator strength for absorption. The quantum-mechanical formula for the transition cross section contains a product of matrix elements, one of which describes the bremsstrahlung emission of a photon $h\nu = E_{nn'}$ from scattering of an electron in the field of an atom, and the other the absorption of this photon by an atom. This latter matrix element is the one expressed in terms of the oscillator strength. The excitation of an atom by electron impact can thus be interpreted as though the electron first emits a bremsstrahlung photon, while the subsequent absorption of this photon brings about the excitation.

If we use (6.89) to describe the excitation of atoms from the ground state due to electron impacts with energies slightly exceeding the threshold energy, it is found that this equation gives values somewhat higher than the experimentally determined cross sections. For hydrogen the cross section given by it is higher than the experimental values by a factor of 3–3.5, while for some other atoms (helium, sodium) the overestimate of the formula is less. It should be noted that in [84] the transitions of a bound electron in an atom under the action of electron impacts were treated on the basis of classical mechanics (the orbital motion of the bound electron was taken into account), and results were obtained which are in agreement within order of magnitude with those given by quantum-mechanical calculations*.

Let us now estimate by means of (6.89) the transition probabilities in a hydrogen plasma (this was done in [79]). The number of excitation events by electron impacts $n \to n'$ per unit volume per unit time is

$$Z_{\text{exc},\, nn'} = N_n N_e \int_{E_{nn'}}^{\infty} \sigma_{nn'}(\varepsilon) v_e f(\varepsilon)\, d\varepsilon = \sigma_{nn'} N_n N_e. \tag{6.90}$$

For the rate constant we obtain an expression which is completely analogous to (6.85),

$$\alpha_{nn'} = 4\pi a_0^2 \bar{v}_e \left(\frac{I_{\mathrm{H}}}{kT}\right)^2 \psi_{nn'} 3 f_{nn'};$$

$$\psi_{nn'} = e^{-x_{nn'}} x_{nn'}^{-1} - E_1(x_{nn'}); \qquad x_{nn'} = \frac{E_{nn'}}{kT}. \tag{6.91}$$

The number of reverse processes of deexcitation $n' \to n$ per unit volume per unit time is $Z_{\text{deex},n'n} = \beta_{n'n} N_{n'} N_e$. According to the principle of detailed balancing

$$\beta_{n'n} = \alpha_{nn'} \frac{n^2}{n'^2} e^{x_{nn'}} = 4\pi a_0^2 \bar{v}_e \left(\frac{I_{\mathrm{H}}}{kT}\right)^2 3 f_{nn'} \frac{n^2}{n'^2} x_{nn'} \psi_{nn'}. \tag{6.92}$$

If we substitute into these equations the oscillator strength given by (5.78), the transition energy $E_{nn'} = I_{\mathrm{H}}(1/n^2 - 1/n'^2)$, and also the values of the constants, we obtain the numerical formulas

$$\alpha_{nn'} = \frac{2 \cdot 10^{-4}}{T^{\circ 1/2}} \frac{1}{n^5 n'^3 (n^{-2} - n'^{-2})^4} [e^{-x_{nn'}} - x_{nn'} E_1(x_{nn'})] \frac{\text{cm}^3}{\text{sec}}, \tag{6.91'}$$

$$\beta_{n'n} = \frac{2 \cdot 10^{-4}}{T^{\circ 1/2}} \frac{1}{n^3 n'^5 (n^{-2} - n'^{-2})^4} [1 - x_{nn'} e^{x_{nn'}} E_1(x_{nn'})] \frac{\text{cm}^3}{\text{sec}}. \tag{6.92'}$$

It is readily seen that the probabilities of deexcitation transitions from the state n' into the states $n' - 1$, $n' - 2$, etc., decrease faster than the sequence $1, 2^{-4}, 3^{-4}$. The ratio of the total probability of transitions to all the low-lying levels to the probability of transition to the closest level is bounded between 1 and $1 + 2^{-4} + 3^{-4} + \cdots = \pi^4/90 = 1.08$. The same is true with respect to the excitation probabilities $n \to n + 1$, $n \to n + 2, \ldots$. Thus, for discrete transitions in atoms under the action of electron impacts the most probable transitions are those between neighboring levels, while "jumps" over levels have a very low probability.

* Collisions between free and bound electrons on the basis of the classical theory are also treated in [85].

Let us now consider transitions between neighboring levels in the region of the upper states with $n \gg 1$. If the distances between levels ΔE are less than kT ($\Delta E \approx 2I_{\rm H}/n^3 = 2E_n/n$, which is the case when $E_n < nkT/2$), the expressions in brackets in (6.91′) and (6.92′) do not differ appreciably from one. In the limit $x_{nn'} \to 0$, they reduce exactly to one. It follows from this that in this region of states the probability of transitions between neighboring levels with excitation and that with deexcitation of an atom ("upward" and "downward") are rather close to each other and are approximately

$$N_e\alpha_{n,n+1} \approx N_e\beta_{n,n-1} \approx \frac{8}{\sqrt{3}} N_e\bar{v}_e a_0^2 n^4 I_{\rm H}/kT$$

$$= 1.25 \cdot 10^{-5} n^4 T^{\circ\,-1/2} N_e \ \sec^{-1} \qquad (6.93)$$

(the average transition cross section is larger by a factor of approximately $I_{\rm H}/kT \gg 1$ than the area of the circular orbit which corresponds to the nth level, $\pi a_0^2 n^4$).

Let us now compare the probabilities of excitation and ionization of an atom by electron impact. If the atom is in the ground state and the temperature is not too high, so that $I/kT \gg 1$, then it is clear that ionization events are less frequent than excitation events simply because a smaller number of electrons possess the energy required for ionization. However, even in those cases when the majority of electrons do possess energy sufficient for ionization, in collisions with atoms whose binding energy is $E_n \sim kT$, ionization takes place less frequently than discrete transitions. The probability of discrete transitions to neighboring states is relatively higher, the smaller is the binding energy (the greater is n). This can be seen by comparing (6.86) and (6.93). Of course, the absolute value of the probability of ionization increases as n increases (but at a slower rate than the probability of discrete transitions).

§15. Ionization and excitation by heavy particle collisions

The formal description of these processes is completely analogous to the previously considered cases of ionization and excitation by electron impact. Thus the rate equation for ionization of unexcited atoms has the form

$$\frac{dN_e}{dt} = \alpha_a N_a^2 - \beta_a N_+ N_e N_a\,^*,$$

where, according to the principle of detailed balancing, $\beta_a = \alpha_a/K(T)$. The characteristic time is given by

$$\tau_a = \frac{(N_e)}{2\alpha_a N_a^2} = \frac{1}{2\beta_a N_a (N_e)}.$$

* By definition, the recombination coefficient is $b_a = \beta_a N_a$.

The ionization rate constant α_a is expressed by the same equation as that for α_e. It should be noted, however, that the ionization cross section $\sigma_a(v')$ depends on the relative velocity of the colliding atoms and that the Maxwell distribution function (with respect to the relative velocities) contains the reduced mass μ; in the case of identical atoms the reduced mass is $\mu = m_a/2$. If we again approximate the cross section near the threshold by a linear dependence on the kinetic energy of the relative motion $\varepsilon' = \mu v'^2/2$, we then obtain for α_a an expression analogous to (6.76). If, however, we simply take outside the integral sign some average value of the cross section σ_a, then the factor $(I/kT) + 2$ will be replaced by approximately the same quantity, namely, $(I/kT) + 1$. Thus,

$$\alpha_a = \sigma_a \overline{v'} \left(\frac{I}{kT} + 1 \right) e^{-I/kT},$$

$$\overline{v'} = \left(\frac{8kT}{\pi\mu} \right)^{1/2},$$

where σ_a corresponds to the energy $\varepsilon' \approx I + kT$. The kinetics of the excitation process can be described in a similar manner.

Unfortunately, unlike the case of electron impact, it is very difficult to make any quantitative estimates of the rates of the processes. Comparison of the rate constants for ionization by electrons and atoms shows that $\alpha_e/\alpha_a = (\overline{v_e}/\overline{v'})(\sigma_e/\sigma_a)$. At comparable temperatures $\overline{v_e}/\overline{v'} \approx (m_a/m_e)^{1/2} \sim 100$. It would appear that the cross section σ_a is several orders of magnitude smaller than σ_e. No experimental data are available on the cross sections for ionization or excitation by atoms for energies of the order of tens of electron volts. Apparently these cross sections are of the order of 10^{-20} to 10^{-22} cm^2.

In order for a collision to be inelastic the impact must be sufficiently sharp, in other words, the velocity with which the particles approach each other must be of the order of velocities of the outer electrons in the atom. In the case of an electron impact with an energy of the order of the ionization potential or of the excitation energy, that is, of the order of several up to 10 ev, this condition is satisfied and the inelastic cross section is large. The collision velocities of heavy particles are comparable only when the energies are greater than the above values by a factor of approximately $(m_a/m_e)^{1/2} \sim 100$, that is, when the energies are of the order of kev's. Indeed, in this case the ionization or excitation cross sections are close to the analogous electron impact cross sections. However, when the energies are of the order of 10 ev the particles approach each other with a very low velocity and the impact is "adiabatic". This situation is completely analogous to the case of vibrational excitation in molecules which was considered in §4. In precisely the same manner, in order for the probability of inelastic energy transfer during such a collision to be high, the adiabatic factor $2\pi a\nu/v$ should not be too great, i.e., of the order of unity. Here $2\pi\nu$ does not denote the angular frequency of molecular vibrations, but that of the rotation of the electron in its orbit ($2\pi a\nu$ is of the order of the electron velocity in an atom, since a is of the order of atomic dimensions). The lowest energies of relative motion* ε', for which it has been still possible to carry out experimental measurements of ionization, were of the order of 30–40 ev. It was found that the cross section for ionization of argon by atoms and ions of argon at $\varepsilon' = 35$ ev was $\sigma_a \sim 3 \cdot 10^{-18}$ cm^2 [51], for ionization of helium by helium atoms $\sigma_a \sim 2 \cdot 10^{-19}$ cm^2 at $\varepsilon' = 35$ ev [52], for ionization of argon by potassium ions $\sigma_a \sim 2 \cdot 10^{-19}$ cm^2 at $\varepsilon' \sim 45$ ev †.

The quantum-mechanical analogue of the adiabatic condition $2\pi a\nu/v \gg 1$ is

$$2\pi a\nu/v \rightarrow \frac{ah\nu}{\hbar v} = \frac{a\,\Delta E}{\hbar v} \gg 1\ ‡,$$

where ΔE is the inelastic energy converted on a collision. The origin of this condition is as follows: The probability of the process is determined by the interaction matrix element, which contains the product of wave functions of the initial and final particle states. Wave functions of translational motion are described by plane waves $e^{i\mathbf{p}\cdot\mathbf{r}/\hbar}$; the product of plane waves of the initial and final states gives the oscillating factor $e^{i\Delta\mathbf{p}\cdot\mathbf{r}/\hbar}$ in the integrand of the matrix element, where $\Delta\mathbf{p}$ is the change in momentum of the arriving particle during a collision. The integral has a considerable magnitude if this factor does not oscillate in the region where the interaction energy is high, that is, at a distance r which is of the order of the atomic dimensions a. The condition for a high probability of the process is thus given by the condition that $\Delta p \cdot a/\hbar \lesssim 1$. The change in the momentum Δp is of the order of $\Delta E/v$, where ΔE is the change in the kinetic energy of the particle in an inelastic energy transfer. From this we obtain the condition for a high probability to be $a\Delta E/\hbar v \lesssim 1$, while the condition for a low probability is $a\Delta E/\hbar v \gg 1$.

In particular, it follows from this condition that the cross sections of processes in which the inelastic conversion of energy ΔE is very small (the so-called resonance case) must be large. In fact, the cross sections for ionization of atoms by excited atoms or molecules are large, when the removal of an

* The usual experimental procedure employs a beam of fast particles which passes through a gas consisting of "stationary" atoms. The ionization threshold with respect to the energy of the arriving particles is then twice as large as the ionization potential. This corresponds to the fact that the reduced mass is half as small as the atomic mass, and for a given relative velocity $\varepsilon' = \varepsilon/2$; and $\varepsilon_{\text{thr}} = 2\varepsilon'_{\text{thr}} = 2I$.

† In [54] theoretical calculations were carried out of the cross sections for the inelastic collisions Ar–Ar and He–He, and the results compared with the experimental data of [51, 52]. Data on ionization cross sections for collisions between ions and atoms with energies of the order of several hundred ev and above are given in the surveys [75].

‡ $\hbar = h/2\pi$.

electron is accomplished by the expenditure of the energy of the internal degrees of freedom, rather than the kinetic energy of translational motion. Thus, the cross sections for processes such as

$$A + B^* \rightarrow A^+ + e + B,$$

where the excitation energy E^* of particle B is close to the ionization potential of particle A, and in order of magnitude close to the gaskinetic cross sections. Therefore, the process of ionization by heavy particles, particularly by molecules, is most likely to proceed in two or more stages: First, one of the particles is excited, and then ionization by collision with the excited particle takes place (so-called ionization by a collision of the second kind) or, conversely, the removal of an electron from the excited particle takes place. Some data on these processes may be found in the books [45, 46].

The problem of estimating the rate constants for ionization or excitation by heavy particles can arise only in considering the earliest stage of ionization of an "instantaneously" heated gas, while the electron concentration remains very small (less than 10^{-4}–10^{-5}), i.e., before an electron avalanche develops.

The estimation of the lower limit of time required for the generation of the "priming" electrons and of an electron avalanche will be considered by the following hypothetical process. Let the gas be "instantaneously" heated to a high temperature T and let the freed electrons instantaneously acquire the same temperature T as the atoms. At the beginning of the process, while the ionization is considerably below its equilibrium value, we may neglect the recombination process. Initially, $dN_e/dt = \alpha_a N_a^2$ and $N_e = \alpha_a N_a^2 t$. The number of electrons increases linearly with time until the rate of ionization by electron impact becomes comparable with the rate of ionization by atom collisions and an avalanche appears. This time t_1 is determined by the condition $\alpha_e N_a N_e = \alpha_a N_a^2$. Substituting here $N_e = \alpha_a N_a^2 t_1$ and noting that, according to (6.74), $\alpha_e N_a = \tau_e^{-1}$, we find $t_1 = \tau_e$. In other words, the minimum required time t_1 is equal to the characteristic time for the development of an avalanche.

The actual "induction" time for the development of an avalanche can be appreciable. This time is determined not by the generation of a sufficient number of free electrons, but by the heating of the electron gas to a sufficiently high temperature to produce appreciable ionization. This time is limited by the slowing-down effect of the energy exchange between atoms (ions) and electrons, which lose an appreciable amount of energy in inelastic collisions, i.e., through ionization and excitation. The topic of energy exchange between ions and electrons will be considered in §20. The conditions under which atoms are "instantaneously" heated to high temperatures with ensuing ionization are attainable in shock tubes. Chapter VII will deal with the kinetics of ionization in a shock front and with the establishment of ionization equilibrium behind the front.

§16. Photoionization and photorecombination

Photoionization and photorecombination processes have already been considered in Chapter V in calculating the absorption and emission coefficients for light; therefore, we recall here some of the considerations and conclusions of that chapter. Let us assume for simplicity that all the atoms are in the ground state and that during recombination the electrons are captured into the ground level. If N_a is the number density of atoms, $U_\nu\, d\nu$ is the radiant energy per unit volume in the spectral interval from ν to $\nu + d\nu$, and $\sigma_{\nu 1}(\nu)$ is the cross section for photoionization from the ground atomic level, then the number of photoionization events per unit volume per unit time is

$$Z_{\text{ion}}^{\nu} = N_a \int_{\nu_1}^{\infty} \frac{U_\nu}{h\nu}\, d\nu \cdot c \cdot \sigma_{\nu 1}(\nu) = \alpha_\nu N_a. \tag{6.94}$$

Here only the photons with $h\nu > h\nu_1 = I$ participate in the absorption; α_ν is the photoionization rate constant. Let $\sigma_{c1}(v)$ denote the cross section for the radiative capture of electrons with a velocity v into the ground atomic level. Then the number of photorecombination events per unit volume per unit time is

$$Z_{\text{rec}}^{\nu} = b_\nu N_+ N_e = N_+ N_e \int_0^\infty f_e(v)\, dv \cdot v \cdot \sigma_{c1}(v)\left(1 + \frac{c^3 U_\nu}{8\pi h\nu^3}\right). \tag{6.95}$$

The term $c^3 U_\nu / 8\pi h\nu^3$ takes into account the induced recombination, corresponding to induced photon emission. The energy of the emitted photon is related to the electron velocity by the photoelectric equation

$$h\nu = \frac{m_e v^2}{2} + I.$$

The integral in (6.95) represents the photorecombination coefficient b_ν.

According to the principle of detailed balancing, the differentials in the integral expressions for Z_{rec}^{ν} and Z_{ion}^{ν} are equal to each other under conditions of complete thermodynamic equilibrium. Substituting for $f_e(v)$ the Maxwell distribution function for the electrons and for U_ν the Planck function, using the Saha equation (6.73) and the photoelectric equation, we obtain a relationship between the cross sections for photoionization and photorecombination

$$\sigma_{c1}(v) = \frac{g_1}{g_+} \frac{h^2 \nu^2}{m_e^2 v^2 c^2} \sigma_{\nu 1}(\nu).$$

The cross sections for photoionization from the nth excited atomic level and for radiative capture into the nth level are similarly related

$$\sigma_{cn}(v) = \frac{g_n}{g_+} \frac{h^2 \nu^2}{m_e^2 v^2 c^2} \sigma_{\nu n}(\nu). \tag{6.95a}$$

Here g_n is the statistical weight of the nth atomic state. The frequency ν and the electron velocity v are also related by the photoelectric equation

$$h\nu = \frac{m_e v^2}{2} + \varepsilon_n = \frac{m_e v^2}{2} + I - w_n,$$

where ε_n is the binding energy of an electron in the nth state and w_n is the excitation energy of the nth atomic level.

The rate equation for the photoprocesses is

$$\frac{dN_e}{dt} = Z_{\text{ion}}^{\nu} - Z_{\text{rec}}^{\nu} = \alpha_\nu N_a - b_\nu N_+ N_e .$$

The relaxation time for the photoprocesses is

$$\tau_\nu = \frac{1}{b_\nu (N_e)} = \frac{(N_e)}{\alpha_\nu N_a}.$$

Let us estimate the photoionization rate constant by assuming that the radiation density is close to its equilibrium value. Unlike the collision ionization cross sections, which are equal to zero at the ionization threshold, the photoionization cross section is different from zero at the threshold and, on the contrary, in many cases is a maximum when $h\nu = I = h\nu_1$. Thus, for hydrogen-like atoms $\sigma_{\nu 1} = \sigma_{\nu 1}^0 (\nu_1/\nu)^3$, where $\sigma_{\nu 1}^0 = 7.9 \cdot 10^{-18}$ cm^2, if the charge of the "nucleus" is equal to unity (see (5.34)). If, as is usually the case, $I/kT \gg 1$, then the ionizing photons are in the Wien region of the spectrum, where $U_\nu \sim e^{-h\nu/kT}$. Taking the average value of the cross section (which can be set equal to the cross section at the ionization threshold to a high degree of accuracy) outside the integral sign in (6.94), we obtain after carrying out the integration

$$\alpha_\nu = \frac{8\pi}{c^2} \frac{I^2}{h^2} \frac{kT}{h} \sigma_{\nu 1}^0 e^{-I/kT} = 3.95 \cdot 10^{23} T_{\text{ev}} I_{\text{ev}}^2 \sigma_{\nu 1}^0 e^{-I/kT} \text{ sec}^{-1}. \quad (6.96)$$

The recombination coefficient b_ν can be found either from the principle of detailed balancing $b_\nu = \alpha_\nu (N_a)/(N_+)(N_e)$, or directly by evaluating the integral (6.95). We note that for $I \gg kT$, the role of the induced recombinations is very small; the factor in square brackets in (6.95) $1 + e^{-h\nu/kT} \approx 1$, since $h\nu > I \gg kT$. The recombination coefficient becomes (for $I/kT \gg 1$)

$$b_\nu = \overline{v_e \sigma_{c1}(v)} = \bar{v}_e \bar{\sigma}_{c1},$$

$$\bar{\sigma}_{c1} = \sigma_{c1}^0 / T_{\text{ev}} = \frac{g_1}{2g_+} \frac{I^2}{m_e c^2 kT} \sigma_{\nu 1}^0 = \frac{g_1}{g_+} \frac{I_{\text{ev}}^2}{T_{\text{ev}}} \sigma_{\nu 1}^0 \cdot 10^{-6} \text{ cm}^2, \quad (6.97)$$

where $\bar{\sigma}_{c1}$ is the average cross section for radiative capture into the ground level ($\bar{v}_e$ is the mean electron thermal speed). The average radiative capture cross section is inversely proportional to the electron temperature. The photoionization and radiative capture cross sections for some atoms at a temperature corresponding to 1 ev ($\sigma_{\nu 1}^0$; $\bar{\sigma}_{c1} = \sigma_{c1}^0/T_{\text{ev}}$) are given in Table 6.6. Regarding the ion cross sections, we can state that if they are considered as hydrogen-like systems, then $\sigma_{\nu 1}^0 \sim Z^{-2}$ and $\sigma_{c1}^0 \sim I_z^2 Z^{-2}$. Usually ionization potentials of ions increase with the charge as $I_z \sim Z$ to Z^2, from which $\sigma_{c1}^0 \sim Z^0$ to Z^2.

Table 6.6

CROSS SECTIONS FOR PHOTOIONIZATION FROM THE GROUND LEVEL OF VARIOUS ATOMS AND RADIATIVE CAPTURE OF ELECTRONS INTO THE GROUND LEVEL

Atom	I, ev	g_1	g_+	$\sigma_{\nu 1}^0 \cdot 10^{18}$, cm^2	Behavior of cross section $\sigma_{\nu 1}$ past threshold	$\sigma_{c1}^0 \cdot 10^{21}$, cm^2 ev
H	13.54	2	1	7.9	Decreases as ν^{-3}	2.9
Li	5.37	2	1	3.7		0.21
C	11.24	9	6	10	Decreases by one half when $h\nu = I + 10$ ev	1.9
N	14.6	4	9	7.5	Decreases slowly	0.7
O	13.57	9	4	3	Almost constant until $h\nu \sim I + 15$ ev	1.24
F	17.46	6	9	2	Almost constant until $h\nu \sim I + 15$ ev	0.41
Na	5.09	2	1	0.31	Decreases more rapidly than ν^{-3}	0.016
Ca	6.25	1	2	25	Decreases as ν^{-3}	0.51

The cross sections $\sigma_{\nu 1}^0$ and the data on the behavior of the cross sections past threshold are taken from [55]. The quantities σ_{c1}^0 were calculated using (6.95).

Let us clarify the role of recombinations with the capture of an electron into an excited level. The recombination coefficient in the general case is (compare with (6.97))

$$b_r = \sum_n \overline{v\sigma_{cn}(v)}, \tag{6.98}$$

where the summation is taken over all the levels n, and the averaging is with respect to a Maxwell distribution of electrons. Here $\sigma_{cn}(v)$ is given by (6.95a). For hydrogen-like atoms $\sigma_{\nu n} \sim 1/n^5$ and $g_n = 2n^2$, so that $\sigma_{cn} \sim 1/n^3$, while ν and v are related by the photoelectric equation, with $\varepsilon_n = I/n^2$. Generally speaking, in summing with respect to n we are faced with the problem of the actual number of levels in the atom (see §6, Chapter III). In the given case, however, the summation over n converges rapidly and it can be extended approximately to $n = \infty$.

Calculating the recombination coefficient by (6.98) for a hydrogen-like atom, we find that it can be expressed in the form

$$b_\nu = b_{\nu 1}\varphi(I/kT); \qquad b_{\nu 1} = 2.07 \cdot 10^{-11} Z^2 T^{\circ\,-1/2}$$
$$= 2 \cdot 10^{-13} Z^2 T_{\text{ev}}^{-1/2}\ \text{cm}^3/\text{sec}, \qquad (6.99)$$

where $b_{\nu 1}$ is a coefficient corresponding to capture into the ground level for $I/kT \gg 1$ (it is determined from (6.97)); $\varphi(I/kT)$ is a very slowly varying function, which is obtained by a summation over n. This function is tabulated in Spitzer's book [56]. For example, for $I/kT = 5$, $\varphi = 1.69$, for $I/kT = 10$, $\varphi = 2.02$, while for $I/kT = 100$, $\varphi = 3.2$. Approximately (see [86]),

$$b_\nu = 2.7 \cdot 10^{-13} Z^2 T_{\text{ev}}^{-3/4}\ \text{cm}^3/\text{sec}. \qquad (6.100)$$

Thus under conditions usually encountered in the singly ionized region, where $\beta = I/kT \sim 10$, capture into all excited levels makes approximately the same contribution to recombination as does capture into the ground level. By virtue of the principle of detailed balancing, if the atoms obey a Boltzmann distribution with respect to their excited states and if radiative equilibrium exists, the same result applies for photoionization. The role of ionization of excited atoms during photoionization is thus comparable to the role of ionization of unexcited atoms, so that our estimates of photoionization and photorecombination rates are low by approximately a factor of 2.

Let us compare the ionization rate of unexcited atoms by electron impact and by photons (assuming that the radiation density is that for equilibrium). Using equations (6.94), (6.96), (6.71), and (6.76), we find

$$\frac{Z^e_{\text{ion}}}{Z^\nu_{\text{ion}}} = \frac{\alpha_e N_e}{\alpha_\nu} = \frac{1.7 \cdot 10^{-16} N_e}{I_{\text{ev}} T_{\text{ev}}^{1/2}} \frac{C(\text{cm}^2/\text{ev})}{\sigma^0_{\nu 1}(\text{cm}^2)}.$$

Evidently, the rates of the reverse processes are in the same ratio. The numerical values of C and $\sigma^0_{\nu 1}$ are of the same order of magnitude ($\sim 5 \cdot 10^{-18}$, see Tables 6.4 and 6.6), ionization potentials I are of the order of 10 ev, and a typical temperature with single ionization is of the order of 1 ev = 11,600°K. These values give

$$\frac{Z^e_{\text{ion}}}{Z^\nu_{\text{ion}}} = \frac{Z^e_{\text{rec}}}{Z^\nu_{\text{rec}}} \sim 10^{-17} N_e,$$

so that when $N_e > 10^{17}$ cm^{-3} it is the electron processes which predominate, while when $N_e < 10^{17}$ cm^{-3} the main role is played by photoprocesses (we wish to emphasize that this pertains only to the ionization of atoms from the ground state and to the capture of an electron into the ground level of an atom, with the radiation and electron densities of the order of those at equilibrium).

Now a few words on the cause of excitation and deexcitation of the first excited states by radiation. The lifetimes of atoms in first excited states, with respect to spontaneous emission, are of the order of $\tau_v^* \sim 10^{-8}$ sec. The lifetime of an atom in these levels, with respect to deexcitation by electron impact at an electron temperature $T_e \sim 1$ ev, is, according to §12, of the order of $\tau_e^* \sim 10^9/N_e$ sec, so that the "quenching" of radiation by electron impact takes place at electron densities $N_e > 10^{17}$ cm^{-3} and, conversely with $N_e < 10^{17}$ cm^{-3} the main role is played by photoprocesses (in the same manner as for ionization of atoms from the ground level and for the capture of an electron into the ground level).

Under conditions close to thermodynamic equilibrium the same is true with respect to the excitation rates by electrons and photons of atomic levels lying close to the ground level. Let us note that the absorption cross sections for resonant radiation which is capable of exciting atoms are very large, and the resonant radiation is most often in equilibrium (the medium is opaque to resonant radiation). For this reason, the time $\tau^* \sim 10^{-18}$ sec characterizes the relaxation time for the establishment of a Boltzmann distribution in the first excited atomic levels due to photoprocesses. The probability of spontaneous radiative transitions from the upper excited states is discussed in §13, Chapter V.

§17. Electron-ion recombination by three-body collisions (elementary theory)

In a highly rarefied plasma the recombination of electrons with ions takes place primarily by binary collisions with the emission of a photon. In a dense plasma this process takes place primarily by three-body collisions with an electron acting as the third body (a neutral atom can also serve as the third body, but this process is of importance only for extremely low degrees of ionization, less than 10^{-7}–10^{-10}). The simplest way to estimate the recombination rate when the electron serves as the third body is to generalize to this case the old Thomson theory [45], which concerns recombination with the participation of a neutral atom. The considerations here are completely analogous to those which were used in §6 of this chapter to estimate the rate of recombination of atoms into a molecule by three-body collisions.

Let us assume that the electron is captured by an ion (with a charge Z) into a closed orbit and recombines if it passes past the ion with an impact parameter (aiming distance) r such that the potential energy of Coulomb attraction to the ion Ze^2/r is greater than the average kinetic energy of the electron $\frac{3}{2}kT$. The impact parameter, consequently, cannot exceed Zr_0, where $r_0 = e^2/\frac{3}{2}kT$ is the effective radius of Coulomb interaction of particles with charges $Z = 1$.

The number of such collisions per unit volume per unit time is $N_e \bar{v}_e \pi r_0^2 Z^2 N_+$. But, in order for capture to take place when the electron passes past an ion (on a path for which the impact parameter is of the order of Zr_0), the electron must interact with another electron which can receive the potential energy which is released on capture. The probability of such an event is approximately $Zr_0\pi r_0^2 N_e$. The number of recombinations per unit volume per unit time is thus

$$Z_{rec} = N_e\bar{v}_e\pi r_0^2 Z^2 N_+ \cdot Zr_0^3\pi N_e = \beta N_e^2 N_+ = bN_e N_+ . \qquad (6.101)$$

From this we obtain for the recombination coefficient

$$b = \bar{v}_e\pi^2 r_0^5 Z^3 N_e = \frac{2^6\pi(2\pi)^{1/2}}{3^5}\frac{e^{10}Z^3}{m^{1/2}(kT)^{9/2}} N_e . \qquad (6.102)$$

The radius of Coulomb interaction between an electron and a proton $r_0 = e^2/\frac{3}{2}kT$ (we shall consider hydrogen), is very close to the radius r_{n*} of the circular orbit of an electron with binding energy $E_{n*} \approx kT$

$$r_{n*} = a_0 n^{*2} = a_0 I_H/kT = e^2/2kT.$$

It is therefore evident that this same recombination coefficient should be obtained by summing the electron capture coefficients $\beta_n N_e$ derived in §14 over all n from 1 to $n^* = (I_H/E_{n*})^{1/2} = (I_H/kT)^{1/2}$ (for $I_H/kT \gg 1$, $n^* \gg 1$, the summation can be replaced by integration). And in fact, this procedure yields an equation which practically coincides with (6.102). This method of calculating the recombination coefficient in terms of the probability of capture into discrete levels was used by Hinnov and Hirschberg [80]. It has the advantage of being physically clear, although it is not rigorous.

It is remarkable that the rigorous (within the framework of certain assumptions) theory which has been developed by Pitaevskii [87] and by Gurevich and Pitaevskii [88] (the latter will be discussed in the next section) leads to the formula

$$b = \frac{4\pi(2\pi)^{1/2}}{9}\frac{e^{10}Z^3}{m^{1/2}(kT)^{9/2}} \ln\Lambda_1 \cdot N_e , \qquad (6.103)$$

which differs from the elementary relation (6.102) only by the numerical coefficient $\frac{27}{16}\ln\Lambda_1$, of the order of unity. Here $\ln\Lambda_1$ is a Coulomb logarithm of a particular kind, which can approximately be set equal to one. Numerically, from (6.103) with $\ln\Lambda_1 = 1$, the recombination coefficient is

$$b = \frac{8.75\cdot 10^{-27}Z^3}{T_{ev}^{9/2}} N_e = \frac{5.2\cdot 10^{-23}Z^3}{T_{1000\,deg}^{9/2}} N_e \frac{\text{cm}^3}{\text{sec}} . \qquad (6.104)$$

The range of applicability of this equation (for hydrogen $Z = 1$) is limited to quite low temperatures $kT \ll I_H$, for which capture takes place into very high levels $n^* = (I_H/kT)^{1/2} \gg 1$. Evidently, these temperatures are below 3000°K ($n^* \geqslant 7$).

Let us compare the coefficients b of three-body recombination as given by (6.104) and those of photorecombination with capture into all levels b_ν from (6.100). The latter predominates when

$$N_e < \frac{3.1 \cdot 10^{13} T_{\text{ev}}^{3.75}}{Z} = \frac{3.2 \cdot 10^{9} T_{1000\ \text{deg}}^{3.75}}{Z}\ \text{cm}^{-3}. \tag{6.105}$$

As may be seen, at low temperatures and not extremely low densities the main role is always played by three-body recombinations.

Reference [80] contains approximate formulas for estimating three-body recombination coefficients at higher temperatures (from ≈ 3000 to $\approx 10{,}000$°K), when capture takes place into low-lying levels. We note that in this range (6.104) gives values of the recombination coefficient which are on the high side, but by not more than a factor of 5–10 even at $T = 10{,}000$°K. Also given in [80] is a convenient graph of the total recombination coefficient $b + b_\nu$ as a function of N_e and T. There is also a description of the results of an experimental study of recombination in helium, which shows good agreement of the theory with experiment.

§18. A more rigorous theory of recombination by three-body collisions

Let us examine in more detail the process of three-body recombination in a low-temperature hydrogen plasma. We assume that the gas is appreciably out of equilibrium with the degree of ionization higher than for equilibrium, or, what is the same thing, that the temperature is lower than that which corresponds to the given degree of ionization, so that it is predominantly recombination which takes place in the plasma. We have pointed out above that the probability of an electron capture in three-body collisions increases rapidly with an increasing orbit radius and a decreased binding energy of the level, so that electrons are primarily captured into the upper levels. As was shown in §13 of Chapter V, the probability of spontaneous radiative transitions from the upper levels is reduced sharply when the quantum number n increases and the binding energy E_n becomes lower (as $1/n^5 \sim E_n^{5/2}$).

Let us assume that the electron density is sufficiently high that the probabilities of radiative transitions from the upper levels are much less than the probabilities of impact transitions. Practically, the opposite situation is not realized at low temperatures: If the densities are so low that radiative transitions from upper levels take place at a higher rate than impact transitions, then in general it is photorecombination which predominates and the electrons are captured primarily into the lower rather than into the upper levels. Thus, the state of a highly excited atom which is formed as a result of electron capture changes under the action of electron impacts. In this case the transitions which are most probable are those into the nearest neighboring states. In the region where the distances between the levels ΔE_n are smaller than kT, transitions with excitation and deexcitation ("upward" and "downward") are, according to the principle of detailed balancing, almost equally probable ($\Delta E_n < kT$ in the region of binding energies $E_n < \frac{1}{2}nkT$ or $E_n < I_H(kT/2I_H)^{2/3}$; this follows from $\Delta E_n \approx |dE_n/dn| = 2I_H/n^3 = 2E_n/n = 2I_H(E_n/I_H)^{3/2}$). Thus, in this region of the energy spectrum, the energy of an atom changes by small amounts in each impact, and "steps" in either direction have approximately equal probability. This is a typical picture of "diffusion". We can say that what is taking place is "diffusion of a bound electron in an atom along the energy axis".

The ratio of probabilities of "upward" and "downward" changes in the region of low-lying levels $E_n \gg kT$, $\Delta E > kT$, where the probability of deexcitation is higher than that of excitation. In addition, in the region of low-lying levels radiative transitions also promote deexcitation and are quite probable. Within the framework of the "diffusion" model this means that a "drain" exists in the region of the low-lying levels and, consequently, the "diffusive flux" is directed "downward" and the highly excited atom which was formed tends to move into the ground state. This process, in fact, is what recombination consists of. We wish to emphasize that the direction of the diffusive flux is determined by the state of the gas. Had the conditions been such that the ionization was lower than that at equilibrium, then excitation would have predominated and the flux would have been directed "upward". It is important that the probability of ionizing an atom on collision at not too high levels is not large and is less than the probability of discrete transistions, so that we can neglect ionization in this region. The probability of ionization, which increases with the level number n, is high in the region of very large n where the probability of capture is also high. The result of this is that in the region of very small binding energies (of the order or less than kT) the Saha-Boltzmann equilibrium of (6.87) is established between the population of the levels and the density of the free electrons. Within the framework of the diffusion model this means that the "source" of particles in the region of low binding energies is such that the given particle density is automatically maintained in this region. The flux along the energy axis, which here arises due to the presence of the "drain" in the region of high binding energies, apparently also determines the rate of draining of excited atoms "downward" and the rate of formation of atoms in the ground state, i.e., essentially the recombination rate. This picture of the recombination process has been formulated mathematically by Pitaevskii [87], who considered the diffusion along the energy axis on the basis of the purely classical Fokker–Planck equation, on the assumption that the principal role is played by the upper states, where the discreteness of energy levels is weak.

For greater clarity and in order to more clearly demonstrate the physical meaning of the approximations made here and of the "diffusion coefficient", we turn here to a classical approach, starting from a system of quantum-mechanical rate equations. Let N_n be the number of atoms per unit volume in the nth levels (the population of the levels). We set up rate equations for these number densities which take into account only discrete transitions between neighboring levels. For convenience we group the direct and reverse processes in pairs, and write

$$\frac{dN_n}{dt} = N_e\{[\beta_{n+1,n}N_{n+1} - \alpha_{n,n+1}N_n] - [\beta_{n,n-1}N_n - \alpha_{n-1,n}N_{n-1}]\}. \quad (6.106)$$

Let us consider the region of large quantum numbers $n \gg 1$ (it is assumed that $I_H/kT \gg 1$). In this region of discrete numbers n and N_n we can treat as continuous and differentiate functions of the number n (recalling that the "differential" $dn = 1$). Expanding (6.106) up to terms of second order, we obtain

$$\frac{\partial N_n}{\partial t} = -\frac{\partial j_n}{\partial n}; \qquad j_n = -\alpha_{n-1,n}N_e\left[\frac{\partial N_n}{\partial n} + \left(\frac{\beta_{n,n-1}}{\alpha_{n-1,n}} - 1\right)N_n\right]. \quad (6.107)$$

We now transform the ratio of probabilities using the principle of detailed balancing, and we restrict ourselves to the consideration of that region of levels where the distance between levels is less than kT

$$\frac{\beta_{n,n-1}}{\alpha_{n-1,n}} = \frac{(N_{n-1})}{(N_n)} = \frac{(n-1)^2}{n^2}e^{\Delta E/kT} \sim 1 - \frac{2}{n} + \frac{\Delta E}{kT}.$$

Now

$$\frac{\partial N_n}{\partial t} = -\frac{\partial j_n}{\partial n}; \qquad j_n = -\alpha_{n-1,n}N_e\left[\frac{\partial N_n}{\partial n} + \left(\frac{\Delta E}{kT} - \frac{2}{n}\right)N_n\right]. \quad (6.108)$$

This equation has the typical form of a diffusion equation in the presence of a volume force. The quantity $D_n = \alpha_{n-1,n}N_e$ plays the role of the diffusion coefficient, and its units are $\sec^{-1}$ (since the "coordinate" n is dimensionless). From the distribution function N_n with respect to n of the excited atoms, it is easy to make a transition to the distribution function with respect to the binding energy $\varphi(E)$. Obviously,

$$N_n = \varphi(E)\left|\frac{dE}{dn}\right| = \varphi(E)\,\Delta E$$

and

$$\frac{\partial}{\partial n} = \frac{\partial}{\partial E}\cdot\left|\frac{dE}{dn}\right| = -\Delta E\frac{\partial}{\partial E}.$$

Transforming (6.108) and dropping higher-order terms, we obtain

$$\frac{\partial\varphi}{\partial t} = -\frac{\partial j}{\partial E}; \quad j = -D\frac{\partial\varphi}{\partial E} + D\left(\frac{1}{kT} - \frac{5}{2E}\right)\varphi;$$
$$D = \alpha N_e(\Delta E)^2; \qquad \alpha = \alpha_{n-1,n}. \tag{6.109}$$

The coefficient of diffusion D along the energy axis has the units of erg^2/sec; ΔE corresponds to the "mean free path", and $1/2\alpha N_e$ is the average time between collisions, which are accompanied by "upward" and "downward" transitions. If we take still another step and transform (6.109) from the distribution function of the atoms with respect to the binding energy $\varphi(E)$ to the distribution function of bound electrons in the phase space of position and momenta, we obtain the Fokker–Planck equation in the form which served as the starting point in [87, 88].

The diffusion coefficient D has been calculated in [85, 88] on the basis of the purely classical theory as the mean square of the change in energy of a bound electron per unit time as a result of collisions with free electrons

$$D_{\text{class}} = \frac{1}{2}\left\langle\frac{\partial(\delta E)^2}{\partial t}\right\rangle = \frac{2(2\pi)^{1/2}}{3}\frac{e^4E\ln\Lambda_1\cdot N_e}{(mkT)^{1/2}}. \tag{6.110}$$

When we calculate the diffusion coefficient in terms of the quantum-mechanical probability of the transitions $\alpha_{n-1,n}N_e$, with $D = \alpha N_e(\Delta E)^2$, where $\alpha_{n-1,n}$ is given by the equations of §14 under the condition $\Delta E \ll kT$, we arrive at exactly the same form for the coefficient as that given by the classical theory; we find that D exceeds D_{class} by a factor of $(8/\sqrt{3})/(\pi\ln\Lambda_1) \approx 4.4$ (for $\ln\Lambda_1 = 1$). It is necessary to conclude that the classical value in the region of closely spaced levels is more accurate than the approximate quantum-mechanical value*.

Let us assume that the highly excited atom which is formed as a result of capture of an electron by an ion is deexcited and returns to the ground state at a rate which is fast in comparison with the rate of change of electron density and temperature, that is, essentially, in comparison with the recombination rate. Then a quasi-steady energy distribution is established among the excited atoms, which "follows" the slow electron density and temperature changes. In other words, in considering recombination with the given N_e and T we can seek a steady state solution of (6.108) or (6.109) in which the flux is constant. In this case it is also unnecessary to solve for the distribution function in an explicit form. It is sufficient to determine the flux, since the quantity j_n in $\text{cm}^{-3}\,\text{sec}^{-1}$ represents the rate of formation of atoms in the ground state,

* It is appropriate to note here that the theoretical quantum-mechanical transition probabilities for the lower levels are on the high side in comparison with experimental results.

i.e., the number of recombinations $Z_{rec} = bN_eN_+$ per unit volume per unit time.

Setting in (6.108) $j_n = const$, integrating the linear equation thus obtained for the function $N(n) \equiv N_n$, and taking into account the fact that $\Delta E = 2I_H/n^3$ and $E_n = I_H/n^2$, we obtain

$$N_n = -\frac{j_n}{N_e} n^2 e^{E_n/kT} \int_{n_0}^{n} \frac{e^{-E_n/kT}}{n^2 \alpha_{n-1,n}} dn ;$$

here n_0 is a constant of integration. As was noted above, a "drain" exists in the region of low-lying levels, so that the level populations are small there and $N_n \approx 0$. Consequently, we take for n_0 a number of the order of unity. Calculations show that the integral converges rapidly in the region where n is not large and is practically independent of the specific value selected for n_0.

Let us now use the second boundary condition in the region of large n, which is a statement of the fact that the Saha–Boltzmann equilibrium (6.87) is established in this region. Consequently,

$$\text{as } n \to \infty, \qquad N_n \to n^2 e^{E_n/kT} \left(\frac{h^2}{2\pi mkT}\right)^{3/2} N_e N_+. \tag{6.111}$$

From this we get an expression for the flux

$$-j_n = \left(\frac{h^2}{2\pi mkT}\right)^{3/2} \left[\int_{n_0 \sim 1}^{n} \frac{e^{-E_n/kT}}{n^2 \alpha_{n-1,n}} dn\right]^{-1} N_e^2 N_+ = bN_eN_+, \tag{6.112}$$

and for the recombination coefficient b.

If we transform from $\alpha_{n-1,n}$ to the diffusion coefficient with $\alpha_{n-1,n} = D/(\Delta E)^2$, make the substitution $\Delta E = 2I_H/n^3$, and evaluate the integral (6.112) using the classical value of D from (6.110), we obtain formula (6.104) for the recombination coefficient*.

Bates, Kingston, and McWhirter [89], in a paper published almost simultaneously with the first article of Pitaevskii [87], have considered recombination with rather general assumptions. In fact, the paper [89] is based on the same ideas on the progress of the recombination as those in [87], but the authors used a system of rate equations for the numbers of discrete state populations (of the type (6.106)) which takes into account all the processes, comprising electron capture, electron impact ionization, discrete transitions to distant levels, radiative capture, and spontaneous radiative transitions. The

* This is precisely what was done in [87, 88], but there the authors operated from the very beginning with the energy of the atom $E = -E_n$, and not with the quantum numbers n. We note that the above method was used in [87] to calculate the recombination coefficient with a neutral atom participating as the third body. The value thus obtained exceeds the result of Thomson's theory [45] by a factor of approximately 6.

condition that the distribution in the upper levels is an equilibrium one was used to limit the number of equations. The system of equations, including that for the steady state case $dN_n/dt = 0$ for $n \neq 1$, were solved numerically over a wide range of densities and temperatures, including also the high temperatures for which there is capture into low-lying levels. The recombination rate was defined as $-dN_+/dt = dN_1/dt$. The calculated results for the recombination coefficients were tabulated*. The authors propose that the general complex process be termed "impact-radiative recombination". In the limit of low densities it becomes photorecombination, at higher densities it becomes what we have called above three-body recombination. Results of numerical calculations for this limiting case are in satisfactory agreement with those given by (6.104).

In [83, 89] are explained the conditions under which the use of the approximation assuming that the energy distribution of excited atoms is quasi-steady is permissible. Actually, for this assumption to be valid the numbers of excited atoms must be much smaller than the numbers of unexcited atoms and free electrons. See [93] for a discussion of the recombination coefficient.

In some cases it is necessary to know not only the recombination rate, but also the subsequent fate of the potential energy which is released on recombination. As follows from what was said above, a part of this energy is imparted to the electron gas on impact deexcitation of the excited atoms which are formed on recombination, and is then transformed into heat. Another part is emitted as a result of spontaneous radiative transitions, and in the case when the gas is transparent to radiation it is essentially lost by the gas. However, if the gas is not completely transparent, then this part of the energy participates in the subsequent, quite complex, reactions connected with the absorption and emission of light (in particular, with the diffusion of resonance radiation); finally, as a result of impact deexcitation of the atoms this energy in part is also transformed into heat and in part leaves the gas volume in the form of radiation. The problem of the rate of deexcitation of highly excited atoms which are formed on recombination, and that of the transformations of the potential energy was considered in a paper by Kuznetsov and one of the authors of this book [90], in connection with a study of radiation recombination kinetics in a gas expanding into a vacuum. The recombination kinetics for the sudden expansion into vacuum will be discussed in §9 of Chapter VIII, where some results obtained in the above reference will also be presented.

§19. Ionization and recombination in air

Ionization and recombination in molecular gases dissociated as a result of very high temperatures takes place approximately in the same manner as

* They are also given in the book [83].

in atomic gases. At lower temperatures, when the molecular concentration is appreciable, the situation is essentially different, and the primary processes are those involving molecular participation. Thus, in air at temperatures below approximately 10,000°K the main ionization mechanism is the reaction (6.69), $N + O + 2.8 \text{ ev} = NO^+ + e$, which requires the minimum activation energy. In reactions of this type the electron removal is accomplished by the binding energy which is released upon recombination of the atoms into a molecule, and hence only a small amount of additional energy is needed, taken from the supply of thermal energy. A reaction of the type $A + B \rightleftarrows AB^+ + e$ can take place if the potential curves of the two systems, denoted by the right- and left-hand sides of the equation, respectively, intersect, as is shown schematically in Fig. 6.5. Let the atoms possess an energy supply (let us say of

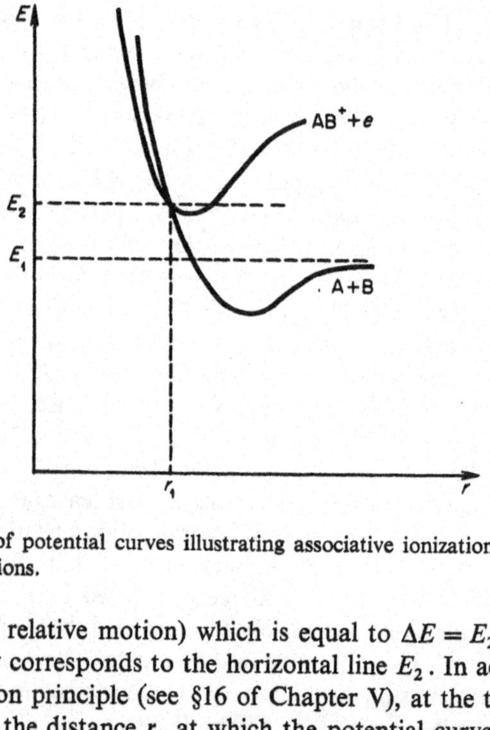

Fig. 6.5. Sketch of potential curves illustrating associative ionization and dissociative recombination reactions.

kinetic energy of relative motion) which is equal to $\Delta E = E_2 - E_1$, that is, their total energy corresponds to the horizontal line E_2. In accordance with the Frank–Condon principle (see §16 of Chapter V), at the time when they are separated by the distance r_1 at which the potential curves intersect, the system $A + B$ can, without changing its total energy, pass into the other state $AB^+ + e$ corresponding to the second potential curve. The cross section for such a process (when the required energy, which exceeds the activation energy, is available) can be presented in the very graphic form [83],

$$\sigma = \gamma \pi r_1^2 [1 - e^{-r_1/v\tau}],$$

where τ is the lifetime of the molecular complex which is formed at the time when the atoms are at a distance of r_1 from each other with respect to a transition to the state with a removed electron, v is the rate at which the atoms approach each other, i.e., r_1/v is the characteristic time for the interaction of the atoms or for the existence of the complex, and the ratio $r_1/v\tau$ characterizes the probability of transition in this time; γ is the probability that the atomic system A + B be in the particular state which corresponds to the intersecting potential curves; it is defined as the ratio of the statistical weight of this state to the sum of statistical weights of all possible states. If $r_1/v\tau \gg 1$, then the cross section is $\sigma \approx \gamma\pi r_1^2$; if, on the other hand, the probability of the transformation is low $r_1/v\tau \ll 1$, then $\sigma = \gamma\pi r_1^2(r_1/v\tau)$. These considerations were used by Lin and Teare [73] in order to estimate the cross section and rate of the reaction (6.69); in this case they used the experimental data of [72], which were obtained by measuring the ionization rate of air in a shock tube, in order to determine the unknown quantity τ. As follows from the potential curves for the NO molecule, $r_1 \approx 10^{-8}$ cm, and γ is estimated as 0.1. The time was found to be given by $\tau \approx 6 \cdot 10^{-13}$ sec. The corresponding cross section of the process is $\sigma = 1.5 \cdot 10^{-16}$ cm^2 (for the speed $v = 3 \cdot 10^5$ cm/sec and $T = 5000$°K). The value of the rate constant for the reaction (6.69) obtained as a result is

$$k_{\text{ion N, O}} = \frac{5 \cdot 10^{-11}}{T^{\circ 1/2}} e^{-32{,}500/T^\circ} \frac{\text{cm}^3}{\text{sec}}. \tag{6.113}$$

Using the equilibrium constant for the reaction which, over a wide range of temperatures, is given approximately by the relation

$$K = (1.4 \cdot 10^{-8}\, T + 1.2 \cdot 10^{-12} T^2 + 1.4 \cdot 10^{-16} T^3)\, e^{-32{,}500/T},$$

we can estimate the rate constant for the reverse reaction, i.e., dissociative recombination*. For low temperatures

$$b_{\text{diss rec NO}^+} = \frac{k_{\text{ion N, O}}}{K} = 3 \cdot 10^{-3} T^{\circ\,-3/2} \frac{\text{cm}^3}{\text{sec}}. \tag{6.114}$$

This value is in agreement with experimental data on dissociative recombination of NO^+ at room temperature and at 2000°K.

At high temperatures together with the reaction (6.69) the analogous reactions

$$\text{N} + \text{N} + 5.8\ \text{ev} \rightarrow \text{N}_2^+ + e, \qquad \text{O} + \text{O} + 6.5\ \text{ev} \rightarrow \text{O}_2^+ + e$$

take place. Rate constants for these reactions are also derived in [73], but not directly as for the reaction (6.69), but by using the equilibrium constants and the

* For the theory of this process see [83].

rates of the corresponding dissociative recombination reactions. The latter are defined by expressions close to those for b_{NO^+}.

The paper of Lin and Teare [73] is a fundamental investigation and one can find in it a large amount of information on the rates of ionization processes in air, and also a review of experimental works and a bibliography. A detailed list of reactions in air in which charged particles participate and the corresponding rate constants selected from various published sources is presented in [92].

Dissociative recombination processes play a most important role in the E and F layers of the ionosphere (at altitudes over 100 km above sea level). A detailed summary of experimental values of the dissociative recombination coefficient $b_{\text{diss rec}}$ is given in the review by Ivanov-Kholodnyi [71]. The value of $b_{\text{diss rec}}$ decreases with increasing temperatures approximately as $T^{-1/2}$ to $T^{-3/2}$ (according to various data). For N_2^+ ions at $T = 300°K$, $b_{\text{diss rec } N_2^+} \approx 10^{-6}$ cm^3/sec, which corresponds to the very large cross section of $\sigma \approx 10^{-13}$ cm^2. The value of $b_{\text{diss rec}}$ for O_2^+ and NO^+ ions is slightly smaller.

For recombination in cold air (in the ionosphere) an important role is played by charge exchange reactions of the type

$$O^+ + O_2 \rightarrow O + O_2^+ + 1.6 \text{ ev}.$$

Atomic ions O^+ which are formed in the upper layers of the atmosphere under the action of ultraviolet radiation from the sun recombine at a slow rate. On the other hand, charge exchange with the subsequent dissociative recombination of O_2^+ is a much faster process. The rate constants for charge exchange reactions which take place with a release of energy are estimated [92] as

$$k_{\text{ch ex}} \approx 1.3 \cdot 10^{-12} T^{\circ 1/2} \text{ cm}^3/\text{sec}.$$

Recombination in cold air at comparatively high densities (in the D layer of the ionosphere below ~80 km) proceeds primarily through the formation of negative oxygen ions. Electrons become attached to oxygen molecules predominantly by three-body collisions $O_2 + e + M \rightarrow O_2^- + M$, and then the O_2^- ions exchange charge with N_2^+ or O_2^+ ions in binary or three-body collisions. The most recent data on recombination in cold air and also on many other inelastic processes which take place in the ionosphere are given in the surveys by Dalgarno [74] and by Danilov and Ivanov-Kholodnyi [91].

3. *Plasma*

§20. Relaxation in a plasma

In an atomic or molecular gas the relaxation time for establishing a Maxwell velocity distribution is characterized by the time between particle collisions, or by gaskinetic cross sections which are of the order of 10^{-15} cm^2. These gaskinetic cross sections are approximately $\sigma \approx \pi a^2$, where the radius a is of the order of the range of interatomic and intermolecular interaction forces, of the order of the dimensions of the particles. The forces acting between the charged particles of a plasma, of electrons and ions, are of a different character. The Coulomb forces decrease very slowly with distance, as $1/r^2$, and do not have a characteristic length scale. Therefore, the problem of "collisions" between the charged particles together with the problem of the corresponding relaxation times must be considered separately.

A plasma can be visualized as a mixture of two gases, an electron gas and an ion gas, with very different particle masses m_e and m, respectively. Owing to this large mass difference, energy transfer becomes difficult, since the energy transferred during a "collision" between an electron and an ion is only a fraction of the order of $m_e/m \ll 1$ of their kinetic energy. Therefore, the average kinetic energy of the electrons and ions, that is, the electron and ion temperatures, can differ greatly from one another over a relatively long time period. These two factors, the long range character of the Coulomb forces and the pronounced difference in the electron and ion masses, determine the specific properties of a plasma.

Let us first consider the interaction of charged particles with masses of the same order. By collisions the particles can transfer energy comparable to their initial energies, hence the Maxwell velocity distribution, and thus a temperature, is established after only a few collisions. If "collision" is understood to denote an interaction between particles involving a significant change in velocity and energy, with a deflection by an appreciable angle (of the order of 90°), then in the case of charged particles "collisions" will occur when the particles approach each other to a distance at which the kinetic and potential (Coulomb) energies become comparable. This characteristic distance r_0 is obviously determined from the condition $Z^2e^2/r_0 \approx \frac{3}{2}kT$, where Z is the charge of the particles. Thus we can take as a measure of the cross section for such "collisions" the quantity

$$\sigma \approx \pi r_0^2 \approx \frac{4}{9}\pi \frac{Z^4e^4}{(kT)^2}. \tag{6.115}$$

Actually, the problem is somewhat more complicated, since an important role in the velocity changes of particles obeying a Coulomb interaction law is played by "distant" collisions, corresponding to large impact parameters (Fig. 6.6). "Distant" collisions occur more frequently than "close" collisions. For a Coulomb law with the force decreasing slowly with distance, the total effect of the "distant" collisions is very large despite the fact that the change in the velocity in each such collision is small. Let us estimate this effect. The force F acting on a particle that passes another particle at a distance r is of

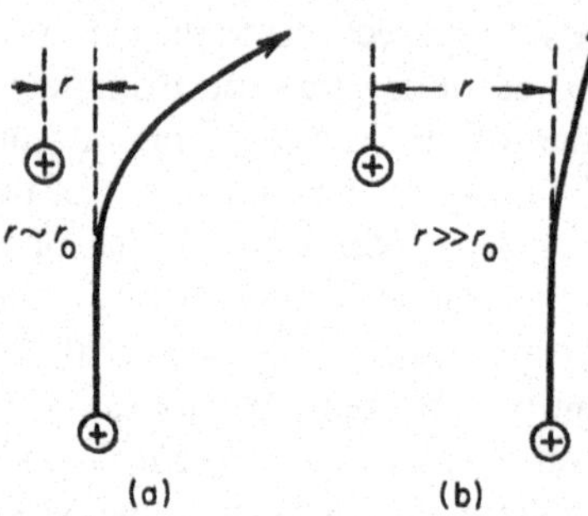

Fig. 6.6. Trajectory of an ion passing another ion with the same type of charge: (a) $r \sim r_0$, strong interaction; (b) $r \gg r_0$, weak interaction.

the order of magnitude of Z^2e^2/r^2. The time during which the force is acting is $t \sim r/v$, where v is the particle velocity. The velocity change during passage Δv is of the order of $Ft/m \sim Z^2e^2/mvr$ *. Since Δv can be both positive and negative, it is only natural to characterize the interaction by the square of the velocity change $(\Delta v)^2 \sim Z^4e^4/m^2v^2r^2$. The probability of such a change is proportional to the area of a ring $2\pi r\, dr$. Thus the rate of change of $(\Delta v)^2$ for a flux of particles Nv is the order of

$$\frac{d(\Delta v)^2}{dt} \sim Nv \int (\Delta v)^2 \cdot 2\pi r\, dr \sim \frac{NvZ^4e^4 2\pi}{m^2v^2} \int \frac{dr}{r},$$

where N is the particle number density. The lower limit on the integral is the minimum distance to which the particles can approach each other $r_0 = 2Z^2e^2/mv^2 = \frac{2}{3}Z^2e^2/kT$. At the upper limit, as $r \to \infty$, the integral diverges logarithmically. However, very "distant" interactions in an electrically neutral gas are screened off by the simultaneous action of the positive and negative charges. The screening radius, which may be taken to be the upper limit, is, clearly the Debye radius d (see §11, Chapter III). Using (3.78) for this quantity, we find

$$\int \frac{dr}{r} \approx \int_{r_0}^{d} \frac{dr}{r} = \ln \frac{d}{r_0} = \ln \Lambda, \qquad \Lambda = \frac{3(kT)^{3/2}}{2(4\pi)^{1/2}Z^3e^3N^{1/2}}. \tag{6.116}$$

If we define the relaxation time τ as the time during which $(\Delta v)^2$ changes by an amount of the order of v^2, and replace approximately v by $\bar{v}$ and mv^2 by $3kT$, we obtain

$$\frac{1}{\tau} = \frac{1}{v^2}\frac{d(\Delta v)^2}{dt} = N\bar{v}\frac{2\pi}{9}\frac{Z^4e^4}{(kT)^2} \ln \Lambda.$$

* *Editors' note.* The velocity v here denotes relative velocity and m the reduced mass of the two particles $m_1m_2/(m_1+m_2)$, approximately the mass of either particle if their masses are of the same order.

More exact considerations [56] lead to the appearance in this equation of an additional numerical factor of the order of one, namely, if $\bar{v} = (8kT/\pi m)^{1/2}$ is the mean thermal speed, then

$$\frac{1}{\tau} = N\bar{v} \cdot 1.1\pi \frac{Z^4 e^4}{(kT)^2} \ln \Lambda = \frac{8.8 \cdot 10^{-2} N Z^4}{A^{1/2} T^{\circ 3/2}} \ln \Lambda \ \sec^{-1}, \quad (6.117)$$

where A is the atomic weight of the particles. In particular, for electron-electron collisions

$$\frac{1}{\tau_{ee}} = \frac{3.8 N_e \ln \Lambda}{T^{\circ 3/2}} \sec^{-1}. \quad (6.118)$$

Using the ordinary relationship between the collision frequency and the gas-kinetic cross section $1/\tau = N\bar{v}\sigma$, one can also introduce the concept of a cross section for "Coulomb collisions" of particles

$$\sigma = 0.69\pi \frac{Z^4 e^4}{(kT)^2} \ln \Lambda = \frac{6 \cdot 10^{-6} Z^4}{T^{\circ 2}} \ln \Lambda = \frac{4.4 \cdot 10^{-14} Z^4}{T_{\text{ev}}^2} \ln \Lambda \ \text{cm}^2. \quad (6.119)$$

As can be seen, it differs from the elementary formula (6.115), which does not take into account "far collisions", by a factor which is approximately equal to $\ln \Lambda$. As follows from Table 6.7*, $\ln \Lambda$ is of the order of 10. The cross

Table 6.7
$\ln \Lambda$ FOR $Z = 1$

T, °K	N, cm^{-3}: 10^{12}	10^{15}	10^{18}	10^{21}
10^3	5.97			
10^4	9.43	5.97		
10^5	12.8	9.43	5.97	
10^6	15.9	12.4	8.96	5.97

section has a weak logarithmic dependence on the density and is inversely proportional to the square of the temperature. It becomes comparable to ordinary gaskinetic cross sections $\sigma \sim 10^{-15}$ cm^2 at a temperature $T \sim$ 250,000°K†.

* The data in the table were taken from [56]. They are somewhat more accurate than those obtained from (6.116).

† It should be noted that when the temperature and energy are so large that the radius r_0 is smaller than the ion (complex) radius and the cross section is smaller than the gas-kinetic cross section, the frequency of collisions and the mean free path of the ions are determined by the gaskinetic rather than by the Coulomb cross section. In this case it should be kept in mind that the gaskinetic ion radius decreases with increasing charge.

The cross section σ and the mean free path $l = 1/N\sigma$ for charged particles are independent of mass, and so are the same for electrons and ions with the same temperatures (for $Z = 1$). The relaxation time, due to its dependence on the velocity, is proportional to the square root of the mass $\tau \sim 1/\bar{v} \sim m^{1/2}$, and the relaxation time for electrons is 100 times smaller than for ions (at the same temperatures). For example, in an electron gas with $T = 20{,}000°\text{K}$ and $N_e = 10^{18}\ \text{cm}^{-3}$, $\sigma \approx 6 \cdot 10^{-14}\ \text{cm}^2$ and $\tau = 2 \cdot 10^{-13}$ sec. In a gas consisting of hydrogen nuclei (protons) at the same temperature and density, the time is greater by a factor of 43, $\tau \approx 8.6 \cdot 10^{-12}$ sec. These estimates show that the temperature in each of the gases is established very rapidly,so that the problem of the relaxation time required to establish a translational temperature practically never arises.

The situation is different with respect to the establishment of thermodynamic equilibrium between an electron and an ion gas, that is, the equalization of the electron and ion temperatures. In a number of physical processes a difference in the temperatures of the ion and electron gases arises; this difference disappears with time as the system approaches thermodynamic equilibrium. Thus, for example, in a shock wave propagating through a plasma the shock heats only the ions, while the electrons remain cold. The gradual energy transfer from the ions to the electrons and the equilibration of their temperatures take place downstream of the shock, over a comparatively long time (see §12, Chapter VII). Let us estimate the relaxation time required for the exchange of energy between the ions and electrons, for the equilibration of their temperatures. The "cross section" (6.119) is independent of the mass of the charged particles and actually characterizes the probability of strong deflection of the particle from its original direction during an interaction. The effect of energy exchange appears, so to speak, to be a result of the deflection. When the masses of the particles are comparable, this strong deflection is simultaneously related to an appreciable exchange of energy, as a result of which the cross section σ does determine the rate of energy transfer during the collision of identical particles. However, when the interacting particles have sharply differing masses (electrons and ions) the energy transfer from a collision, according to the conservation laws, cannot exceed a fraction of the order of m_e/m of their kinetic energy. Therefore, appreciable energy transfer can occur only when the particles are subjected to very many collisions (approximately m/m_e).

In repeating the derivation of the "cross section" for collisions between electrons and ions, we note that the kinetic energy of the colliding particles is understood to denote the kinetic energy of their relative motion. If the electron temperature is not much lower than that of the ions, then the relative velocity is almost equal to the electron velocity. The reduced mass is also equal to the mass of the electron, so that the average energy of relative motion is characterized by the electron temperature. In addition, one of the factors

Z^2 in the expression for the cross section pertains to one of the particles, while the second factor Z^2 pertains to the other one. Since for an electron $Z = 1$, the factor Z^4 in the expression for the cross section is now replaced by Z^2. Thus, the "cross section" for "collisions" between electrons and ions is of the order of $\sigma' \approx \pi Z^2 e^4 \ln \Lambda/(kT_e)^2$, the time between "collisions" is $\tau' \sim 1/N\bar{v}_e\sigma'$, and the characteristic time for energy transfer is

$$\tau_{ei} \sim \frac{m}{m_e}\tau' \sim \frac{m}{m_e}\frac{1}{N\bar{v}_e\sigma'}\,^*.$$

A more rigorous analysis [56] leads to the appearance of a numerical coefficient of the order of unity. After substituting the values of the constants, the expression for the time of energy transfer becomes

$$\tau_{ei} = \frac{252\,A \cdot T_e^{\circ 3/2}}{NZ^2 \ln \Lambda} = \frac{3.5 \cdot 10^8 A T_{\text{ev}}^{3/2}}{NZ^2 \ln \Lambda} \text{ sec}, \tag{6.120}$$

where A is the atomic weight of the ions and N is the ion number density. For example, if $N = 10^{18}$ cm^{-3}, $T_e = 20{,}000$°K, $Z = 1$, and $A = 16$ (oxygen atoms), $\tau_{ei} \approx 2.8 \cdot 10^{-9}$ sec.

When the difference between the ion and electron temperatures is not too great, it is natural to represent the change in temperature of one of the gases as an ordinary relaxation equation such as (6.2)

$$\frac{dT_e}{dt} = \frac{T - T_e}{\tau_{ei}}. \tag{6.121}$$

It turns out, however, that the rate equation for the equilibration of temperatures (6.121) is also valid at large temperature differences. Equation (6.121) with the exchange time given by (6.120) (only differing by a very small amount through a numerical coefficient of the order of unity) was originally derived by Landau [57] in 1937, by means of a rigorous analysis of the kinetic equation for a gas consisting of charged particles interacting in accordance with the Coulomb law.

We note that in the case of a weakly ionized gas the equation (6.121) describing energy exchange is still valid if we take into account exchange with neutral atoms by setting $1/\tau_{ei} = 1/\tau_{e,\text{ion}} + 1/\tau_{e,\text{neut}}$; here $\tau_{e,\text{ion}}$ is given by (6.120) above. As to $\tau_{e,\text{neut}}$, we may write from elementary considerations

$$\frac{1}{\tau_{e,\text{neut}}} \approx N_{\text{neut}}\bar{v}_e\sigma_{e,\text{elast}}\frac{2m_e}{m}, \tag{6.122}$$

where $\sigma_{e,\text{elast}}$ is the average cross section for elastic electron-atom collisions. Precise calculations [50] yield results close to the above. Ordinarily, for $T_e \sim 1$ ev, energy exchange with neutral atoms plays the principal role only when the degree of ionization is less than about 10^{-3}.

* In the case of interaction between electrons and complex ions at sufficiently large energy, the remark made in the second footnote following (6.119) still holds.

Cited references

Editors' note:

In the references cited we have used, as far as possible, the abbreviations for journals and reports used by *Chemical Abstracts*. A list of these abbreviations may be found in the List of Periodicals of the Chemical Abstracts Service published by the American Chemical Society.

Transliteration of Russian names has essentially followed the system adopted by the Library of Congress, but with no distinction between е and ё or between и and й, and with yu used for ю and ya for я. In the case of books translated from Russian into English the transliterated author names are those appearing on the translation. Russian titles have been translated into English, but where a translation is indicated the title given is that appearing on the translated version. A source of an English translation for all cited Russian references has been given whenever known to the editors.

Chapter I

1. Landau, L. D., and Lifshitz, E. M.
   Fluid Mechanics. Addison-Wesley, Reading, Mass., 1959.
2. Zel'dovich, Ya. B.
   Theory of Shock Waves and Introduction to Gasdynamics. Izdat. Akad. Nauk SSSR, Moscow, 1946.
3. Kochin (Kotchine), N. E.
   Sur la théorie des ondes de choc dans un fluide, *Rend. Circolo Mat. Palermo* **50**, 305–344 (1926).
   Roze, A. V., Kibel', N. A., and Kochin, N. E.
   Theoretical Hydromechanics, Part 2. ONTI, Moscow, 1937.
4. Sedov, L. I.
   Le mouvement d'air en cas d'une forte explosion, *Compt. Rend. (Doklady) Acad. Sci. URSS* **52**, 17–20 (1946).
   Propagation of strong blast waves, *Prikl. Mat. i Mekh.* **10**, 241–250 (1946).
5. Sedov, L. I.
   Similarity and Dimensional Methods in Mechanics. Gostekhizdat, Moscow, 4th edition, 1957. English transl. (M. Holt, ed.), Academic Press, New York, 1959.
6. Taylor, G. I.
   The formation of a blast wave by a very intense explosion. II. The atomic explosion of 1945, *Proc. Roy. Soc. (London), Ser. A* **201**, 175–186 (1950).
7. Chernyi, G. G.
   The problem of a point explosion, *Dokl. Akad. Nauk SSSR* **112**, 213–216 (1957). Transl. as *Rep.* no. MOA TIL/T.4871, Ministry of Aviation (Gt. Brit.), 1959.

8. Goldstine, H. H., and von Neumann, J.
Blast wave calculation, *Commun. Pure Appl. Math.* **8**, 327–353 (1955).
9. Brode, H. L.
Numerical solutions of spherical blast waves, *J. Appl. Phys.* **26**, 766–775 (1955).
10. Okhotsimskii, D. E., Kondrasheva, I. L., Vlasova, Z. P., and Kazakova, R. K.
Calculation of a point explosion with counterpressure, *Tr. Mat. Inst. Akad. Nauk SSSR* **50** (1957).
11. Landau, L. D.
On shock waves at far distances from their place of generation, *Prikl. Mat. i Mekh.* **9**, 286–292 (1945).
12. Sadovskii, M. A.
The mechanical effect of blast waves in air with respect to data from experimental studies, *Physics of Explosions, Collection No. 1*, pp. 20–111. Izdat. Akad. Nauk SSSR, Moscow, 1952.
13. Kompaneets, A. S.
A point explosion in an inhomogeneous atmosphere, *Soviet Phys. "Doklady" (English Transl.)* **5**, 46–48 (1960).
14. Courant, R., and Friedrichs, K. O.
Supersonic Flow and Shock Waves. Wiley (Interscience), New York, 1957.
15. Stanyukovich, K. P.
Unsteady Motion of Continuous Media. Gostekhizdat, Moscow, 1955. English transl. (M. Holt, ed.), Academic Press, New York, 1960.
16. Imshennik, V. S.
The isothermal scattering of a gas cloud, *Soviet Phys. "Doklady" (English Transl.)* **5**, 253–256 (1960).
17. Molmud, P.
Expansion of a rarefied gas cloud into a vacuum, *Phys. Fluids* **3**, 362–366 (1960).
18. Nemchinov, I. V.
The sudden expansion of a plane gas layer with a gradual energy release, *Zh. Prikl. Mekhan. i Tekhn. Fiz.*, 1961, No. 1, 17–26.
The sudden expansion of a heated gas mass in the regular regime, *Zh. Prikl. Mekhan. i Tekhn. Fiz.*, 1964, No. 5, 18–29.
19. Nemchinov, I. V.
Expansion of a tri-axial gas ellipsoid in a regular behavior, *Appl. Math. Mech., PMM (English Transl.)* **29**, 143–150 (1965).

Chapter II

1. Ambartsumian, V. A. (ed.)
Theoretical Astrophysics. Gostekhizdat, Moscow, 1952. English transl., Pergamon Press, New York, 1958.
2. Unsöld, A.
Physik der Sternatmosphären, mit Besonderer Berücksichtigung der Sonne. Springer, Berlin, 2nd edition, 1955.
3. Mustel', E. P.
Stellar Atmospheres. Fizmatgiz, Moscow, 1960.
4. Landau, L. D., and Lifshitz, E. M.
Statistical Physics. Addison–Wesley, Reading, Mass., 1958.
5. Raizer, Yu. P.
On the structure of the front of strong shock waves in gases, *Soviet Phys. JETP (English Transl.)* **5**, 1242–1248 (1957).

6. Landau, L. D., and Lifshitz, E. M.
The Classical Theory of Fields. Gostekhizdat, Moscow, 3rd edition, 1960. English transl. (revised 2nd edition), Addison–Wesley, Reading, Mass., 1962.
7. Belen'kii, S. Z.
On the equations of hydrodynamics taking into account radiation, *Tr. Fiz. Inst. Akad. Nauk SSSR* **10**, 15–22 (1958).
8. Imshennik, V. S., and Morozov, Yu. I.
The energy-momentum radiation tensor in a moving medium for conditions close to equilibrium, *Zh. Prikl. Mekhan. i Tekhn. Fiz.*, 1963, No. 3, 3–10.
9. Davis, L. W.
Semiclassical treatment of the optical maser, *Proc. Inst. Elec. Electronics Engrs.* **51**, 76–79 (1963).
10. Jaynes, E. T., and Cummings, F. W.
Comparison of quantum and semiclassical radiation theories with application to the beam maser, *Proc. Inst. Elec. Electronics Engrs.* **51**, 89–109 (1963).
11. Kapitza, P. L., and Dirac, P. A. M.
The reflection of electrons from standing light waves, *Proc. Cambridge Phil. Soc.* **29**, 297–300 (1933).
12. Bartell, L. S., Thompson, H. B., and Roskos, R. R.
Observation of stimulated Compton scattering of electrons by laser beam, *Phys. Rev. Letters* **14**, 851–852 (1964).
13. Keldysh, L. V.
Ionization in the field of a strong electromagnetic wave, *Soviet Phys. JETP (English Transl.)* **20**, 1307–1314 (1965).
14. Weisskopf, V., and Wigner, E.
Über die natürliche Linienbreite in der Strahlung des harmonischen Oszillators, *Z. Physik* **65**, 18–29 (1930).
15. Landau, L. D., and Lifshitz, E. M.
Quantum Mechanics, Non-Relativistic Theory. Addison–Wesley, Reading, Mass., 1958.
16. Reif, T.
Fundamentals of Statistical and Thermal Physics. McGraw-Hill, New York, 1965.

Chapter III

1. Landau, L. D., and Lifshitz, E. M.
Statistical Physics. Addison–Wesley, Reading, Mass., 1958.
2. Godnev, I. N.
The Calculation of Thermodynamic Properties from Molecular Data. Gostekhizdat, Moscow, 1956.
3. Predvoditelev, A. S., *et al.*
Tables of Thermodynamic Properties of Air for Temperatures from 6000°K to 12,000°K and Pressures from .001 to 1000 atm. Izdat. Akad. Nauk SSSR, Moscow, 1957.
Thermodynamic Functions of Air (for Temperatures of 12,000 to 20,000°K and Pressures of 0.001 to 1000 atm). Izdat. Akad. Nauk SSSR, Moscow, 1959. English transl. Assoc. Technical Services, Inc., Glen Ridge, N. J., 1962.
4. Selivanov, V. V., and Shlyapintokh, I. Ya.
The thermodynamic properties of air for thermal ionization and the shock wave, *Zh. Fiz. Khim.* **32**, 670–678 (1958).

5. Stupochenko, E. V., Stakhanov, I. P., Samuilov, E. V., Pleshanov, A. S., and Rozhdestvenskii, I. B.
Thermodynamic properties of air in the temperature range from 1000 to 12,000°K and pressure range from 0.001 to 1000 atm., *Physical Gasdynamics*, pp. 3–38. Izdat. Akad. Nauk SSSR, Moscow, 1959.
6. Zel'dovich, Ya. B.
The proof of singularity of the solution of mass law equations, *Zh. Fiz. Khim.* **11**, 685–687 (1938).
7. Moore, C. E.
Atomic Energy Levels as Derived from the Analyses of Optical Spectra. Vols. I, II, III, Circular 467, National Bureau of Standards, Washington, 1949–1958.
8. Kaye, G. W. C., and Laby, T. H.
Tables of Physical and Chemical Constants and Some Mathematical Functions. Longmans Green, New York, 9th edition, 1941.
9. Fermi, E.
Sopra lo spostamento per pressione delle righe elevate delle serie spettrali, *Nuovo Cimento* **11**, 157–166 (1934).
10. Ecker, G., and Weizel, W.
Zustandsumme und effektive Ionisierungsspannung eines Atoms im Inneren des Plasmas, *Ann. Physik* **17**, 126–140 (1956).
Seaton, M. J.
A comparison of theory and experiment for photo-ionization cross-sections. I. Neon and the elements from boron to neon, *Proc. Roy. Soc. (London), Ser. A* **208**, 408–430 (1951).
Ehler, A. W., and Weissler, G. L.
Ultraviolet absorption of atomic nitrogen in its ionization continuum, *J. Opt. Soc. Am.* **45**, 1035–1043 (1955).
Vitense, E.
Der Aufbau der Sternatmosphären. IV. Kontinuierliche Absorption und Streuung als Funktion von Druck und Temperatur, *Z. Astrophys.* **28**, 81–112 (1951).
11. Timan, B. L.
The effect of the interaction of ions on their equilibrium concentration in the case of a thermally multiply ionized gas, *Zh. Eksperim. i Teor. Fiz.* **27**, 708–711 (1954).
12. Margenau, H., and Lewis, M.
Structure of spectral lines from plasmas, *Rev. Mod. Phys.* **31**, 569–615 (1959).
13. Kudrin, L. P.
The equation of state of partially ionized hydrogen, *Soviet Phys. JETP (English Transl.)* **13**, 798–801 (1961).
14. Benson, S. W., Buss, J. H., and Myers, H.
Thermodynamic properties of ionized gases, *IAS Paper* No. 59–95, Inst. of Aero. Sci., New York, N. Y., 1959.
15. Raizer, Yu. P.
A simple method of calculating the degree of ionization and thermodynamic functions of a multiply ionized gas, *Soviet Phys. JETP (English Transl.)* **9**, 1124–1126 (1959).
16. Zel'dovich, Ya. B., and Raizer, Yu. P.
Strong shock waves in gases, *Usp. Fiz. Nauk* **63**, 613–641 (1957).
17. Zel'dovich, Ya. B.
Theory of Shock Waves and Introduction to Gasdynamics. Izdat. Akad. Nauk SSSR, Moscow, 1946.

18. Davies, D. R.
Shock waves in air at very high pressures, *Proc. Phys. Soc. (London), Sect. A* **61**, 105–118 (1948).
19. Rozhdestvenskii, I. B.
Thermodynamic and gasdynamic properties of air behind a normal shock wave taking into account dissociation and ionization, *Physical Gasdynamics*, pp. 70–82. Izdat. Akad. Nauk SSSR, Moscow, 1959.
20. Gorban', N. F.
The determination of the flow variables behind a normal shock wave taking into account variable specific heat and dissociation, *Physical Gasdynamics*, pp. 83–93. Izdat. Akad. Nauk SSSR, Moscow, 1959.
21. Prokof'ev, V. A.
On the problem of the calculation of radiation in one-dimensional steady gas motion, *Uch. Zap., Mosk. Gos. Univ.*, 1954, No. 172, Mekhanika, 79–125.
22. Resler, E. L., Lin, S. C., and Kantrowitz, A.
The production of high temperature gases in shock tubes, *J. Appl. Phys.* **23**, 1390–1399 (1952).
23. Sabol, A. P.
Flow properties of strong shock waves in xenon gas as determined for thermal equilibrium conditions, *NACA Tech. Note* No. 3091 (1953).
24. Kholev, S. R.
Equilibrium parameters for very strong shock waves in monatomic gases and hydrogen, *Izv. Vysshikh. Uchebn. Zavedenii, Fiz.* 1959, No. 4, 28–37.
25. Christian, R. H., and Yarger, F. L.
Equation of state of gases by shock wave measurements. I. Experimental method and the Hugoniot of argon, *J. Chem. Phys.* **23**, 2042–2044 (1955).
26. Model', I. Sh.
Measurement of high temperatures in strong shock waves in gases, *Soviet Phys. JETP (English Transl.)* **5**, 589–601 (1957).
27. Sachs, R. G.
Some properties of very intense shock waves, *Phys. Rev.* **69**, 514–522 (1946).
28. Tsukerman, V. A., and Manakova, M. A.
Sources of short x-ray pulses for investigating fast processes, *Soviet Phys.—Tech. Phys. (English Transl.)* **2**, 353–363 (1957).
29. Gombàs, P.
Die Statistische Theorie des Atoms und ihre Anwendungen. Springer, Wien, 1949.
30. Landau, L. D., and Lifshitz, E. M.
Quantum Mechanics, Non-Relativistic Theory. Addison-Wesley, Reading, Mass., 1958.
31. Latter, R.
Temperature behavior of the Thomas-Fermi statistical model for atoms, *Phys. Rev.* **99**, 1854–1870 (1955).
32. Brachman, M. K.
Thermodynamic functions on the generalized Fermi-Thomas theory, *Phys. Rev.* **84**, 1263 (1951).
33. Bethe, H. A., and Marshak, R. E.
The generalized Thomas-Fermi method as applied to stars, *Astrophys. J.* **91**, 239–243 (1940).
Feynman, R. P., Metropolis, N., and Teller, E.
Equations of state of elements based on the generalized Fermi-Thomas theory, *Phys. Rev.* **75**, 1561–1573 (1949).

Gilvarry, J. J.
Thermodynamics of the Thomas-Fermi atom at low temperature, *Phys. Rev.* **96**, 934–943 (1954).
Solution of the temperature-perturbed Thomas-Fermi equation, *Phys. Rev.* **96**, 944–948 (1954).
Gilvarry, J. J., and Peebles, G. H.
Solutions of the temperature-perturbed Thomas-Fermi equation, *Phys. Rev.* **99**, 550–552 (1955).

34. Larkin, A. I.
Thermodynamic functions of a low temperature plasma, *Soviet Phys. JETP* (*English Transl.*) **11**, 1363–1364 (1960).
Vedenov, A. A., and Larkin, A. I.
Equation of state of a plasma, *Soviet Phys. JETP* (*English Transl.*) **9**, 806–811 (1959).

35. Kuznetsov, N. M.
Thermodynamic Functions and Hugoniot Curves for Air at High Temperatures. Mashinostroenie, Moscow, 1965.

Chapter IV

1. Penner, S. S., Harshbarger, F., and Vali, V.
An introduction to the use of the shock tube for the determination of physio-chemical parameters, *in Combustion Researches and Reviews 1957*, pp. 134–172. Butterworths, London, 1957.

2. Losev, S. A., and Osipov, A. I.
The study of nonequilibrium phenomena in shock waves, *Soviet Phys.—Usp.* (*English Transl.*) **4**, 525–552 (1962).

3. Ladenburg, R. W., Lewis, B., Pease, R. N., and Taylor, H. S. (eds.)
Physical Measurements in Gas Dynamics and Combustion (Vol. IX of *High Speed Aerodynamics and Jet Propulsion*). Princeton Univ. Press, Princeton, N.J., 1954.

4. Rakhmatullin, Kh. A., and Semenov, S. S. (eds.)
Shock Tubes, A Collection of Papers Translated into Russian. Inostrannoi Lit., Moscow, 1962.

5. Strehlow, R. A., and Cohen, A.
Limitations of the reflected shock technique for studying fast chemical reactions and its application to the observation of relaxation in nitrogen and oxygen, *J. Chem. Phys.* **30**, 257–265 (1959).

6. Fowler, R. G., Atkinson, W. R., Compton, W. D., and Lee, R. J.
Shock waves in low pressure spark discharges, *Phys. Rev.* **88**, 137–138 (1952).
Fowler, R. G., Atkinson, W. R., and Marks, L. W.
Ion concentrations and recombination in expanding low pressure sparks, *Phys. Rev.* **87**, 966–970 (1952).

7. Kolb, A. C.
Production of high-energy plasmas by magnetically driven shock waves, *Phys. Rev.* **107**, 345–350 (1957).
Propagation of strong shock waves in pulsed longitudinal magnetic fields, *Phys. Rev.* **107**, 1197–1198 (1957).

8. Wiese, W., Berg, H. F., and Griem, H. R.
Measurements of temperatures and densities in shock-heated hydrogen and helium plasmas, *Phys. Rev.* **120**, 1079–1085 (1960).

9. Kholev, S. R., and Poltavchenko, D. S.
Acceleration of the plasma of a discharge and production of strong shock waves in a

camera with coaxial electrodes, *Soviet Phys. "Doklady"* (*English Transl.*) **5**, 356–360 (1960).
10. Kholev, S. R., and Krestnikova, L. I.
Experimental investigation of a directed gas flow from an impulsive discharge, *Izv. Vysshikh. Uchebn. Zavedenii, Fiz.* 1960, No. 1, 29–37.
11. Patrick, R. M.
High-speed shock waves in a magnetic annular shock tube, *Phys. Fluids* **2**, 589–598 (1959).
12. Josephson, V.
Production of high-velocity shocks, *J. Appl. Phys.* **29**, 30–32 (1958).
13. Ziemer, R. W.
Experimental investigation in magneto-aerodynamics, *ARS J.* **29**, 642–647 (1959).
14. Kudryavtseva, E. V., and Ionova, V. P. (eds.)
Moving Plasmas, A Collection of Papers Translated into Russian. Inostrannoi Lit., Moscow, 1961.
15. *Optical Pyrometry in Plasmas, A Collection of Papers Translated into Russian.* Inostrannoi Lit., Moscow, 1960.
16. Lochte-Holtgreven, W.
Production and measurement of high temperatures, *Rept. Progr. Phys.* **21**, 312–383 (1958).
17. Ryabinin, Yu. N.
Gases at High Densities and Temperatures. Fizmatgiz, Moscow, 1959.
18. Ferri, A. (ed.)
Fundamental Data Obtained from Shock Tube Experiments. AGARDograph No. 41, Pergamon Press, New York, 1961.
19. Stupochenko, E. V., Losev, S. A., and Osipov, A. I.
Relaxation Processes in Shock Waves. Nauka, Moscow, 1965.

Chapter V

1. Landau, L. D., and Lifshitz, E. M.
The Classical Theory of Fields. Gostekhizdat, Moscow, 3rd edition, 1960. English transl. (revised 2nd edition), Addison-Wesley, Reading, Mass., 1962.
2. Landau, L. D., and Lifshitz, E. M.
Quantum Mechanics, Non-Relativistic Theory. Addison-Wesley, Reading, Mass., 1958.
3. Heitler, W.
The Quantum Theory of Radiation. Oxford Univ. Press, London and New York, 2nd edition, 1944.
4. Menzel, D. H., and Pekeris, C. L.
Absorption coefficients and hydrogen line intensities, *Monthly Notices Roy. Astron. Soc.* **96**, 77–111 (1935).
5. Bethe, H. A., and Salpeter, E. E.
Quantum Mechanics of One- and Two-Electron Atoms. Academic Press, New York, 1957.
6. Ambartsumian, V. A. (ed.).
Theoretical Astrophysics, Gostekhizdat, Moscow, 1952. English transl., Pergamon Press, New York, 1958.
7. Bates, D. R.
Calculation of the cross-section of neutral atoms and positive and negative ions towards the absorption of radiation in the continuum, *Monthly Notices Roy. Astron. Soc.* **106**, 432–445 (1946).

8. Keck, J. C., Camm, J. C., Kivel, B., and Wentink, T., Jr.
Radiation from hot air. Part II. Shock tube study of absolute intensities, *Ann. Phys. (N.Y.)* **7**, 1–38 (1959).
9. Branscomb, L. M., Burch, D. S., Smith, S. J., and Geltman, S.
Photodetachment cross section of the electron affinity of atomic oxygen, *Phys. Rev.* **111**, 504–513 (1958).
10. Unsöld, A.
Physik der Sternatmosphären, mit Besonderer Berücksichtigung der Sonne. Springer, Berlin, 2nd edition, 1955.
11. Unsöld, A.
Über das kontinuierliche Spektrum der Hg-Hochdrucklampe, des Unterwasserfunkens und ähnlicher Gasentladungen, *Ann. Physik* **33**, 607–616 (1938).
12. Vitense, E.
Der Aufbau der Sternatmosphären. IV. Kontinuierliche Absorption und Streuung als Funktion von Druck und Temperatur, *Z. Astrophys.* **28**, 81–112 (1951).
Schirmer, H.
Die Bestimmung der Temperatur der zylindreschen Saüle einer Xenon-Hochdruckentladung, *Z. Angew. Phys.* **6**, 3–9 (1954).
13. Burgess, A., and Seaton, M. J.
Cross sections for photoionization from valence-electron states, *Rev. Mod. Phys.* **30**, 992–993 (1958).
A general formula for the calculation of atomic photo-ionization cross sections, *Monthly Notices Roy. Astron. Soc.* **120**, 121–151 (1960).
Seaton, M. J.
The quantum defect method, *Monthly Notices Roy. Astron. Soc.* **118**, 504–518 (1958).
14. Biberman, L. M., and Norman, G. E.
On the calculation of photoionization absorption, *Opt. Spectr. (USSR) (English Transl.)* **8**, 230–232 (1960).
15. Biberman, L. M., Norman, G. E., and Ul'yanov, K. N.
On the calculation of photoionization absorption in atomic gases, *Opt. Spectr. (USSR) (English Transl.)* **10**, 297–299 (1961).
16. Chandrasekhar, S., and Breen, F. H.
The motion of an electron in the Hartree field of an atom, *Astrophys. J.* **103**, 41–70 (1946).
On the continuous absorption coefficient of the negative hydrogen ion. III, *Astrophys. J.* **104**, 430–445 (1946).
17. Biberman, L. M., and Romanov, V. E.
On the mechanism of formation of continuous background in the emission spectrum of hot gases, *Opt. i Spektroskopiya* **3**, 646–648 (1957).
18. Raizer, Yu. P.
Simple method for computing the mean range of radiation in ionized gases at high temperatures, *Soviet Phys. JETP (English Transl.)* **10**, 769–771 (1960).
19. Becker, R.
Theorie der Elektrizität. Vol. II. Elektronentheorie. B. G. Teubner, Leipzig and Berlin, 1933.
20a. Herzberg, G.
Molecular Spectra and Molecular Structure. I. Spectra of Diatomic Molecules. Van Nostrand, Princeton, N. J., 2nd edition, 1950.
20b. Gaydon, A. G.
Dissociation Energies and Spectra of Diatomic Molecules. Wiley, New York, 1947.

20c. Kondrat'ev, V. N.
The Structure of Atoms and Molecules. Fizmatgiz, Moscow, 1959.

21. Kivel, B., Mayer, H., and Bethe, H.
Radiation from hot air. Part I. Theory of nitric oxide absorption, *Ann. Phys.* (*N.Y.*) **2**, 57–80 (1957).

22. Jarmain, W. R., Fraser, P. A., and Nicholls, R. W.
Vibrational transition probabilities of diatomic molecules: collected results N_2, N_2^+, NO, O_2^+, *Astrophys. J.* **118**, 228–233 (1953).
Fraser, P. A., Jarmain, W. R., and Nicholls, R. W.
Vibrational transition probabilities of diatomic molecules; collected results II, N_2^+, CN, C_2, O_2, TiO, *Astrophys. J.* **119**, 286–290 (1954).
Jarmain, W. R., Fraser, P. A., and Nicholls, R. W.
Vibrational transition probabilities of diatomic molecules; collected results. III. N_2, NO, O_2, O_2^+, OH, CO, CO^+, *Astrophys. J.* **122**, 55–61 (1955).
Jarmain, W. R., and Nicholls, R. W.
Vibrational transition probabilities to high quantum numbers for the nitrogen first and second positive band systems, *Can. J. Phys.* **32**, 201–204 (1954).
Turner, R. G., and Nicholls, R. W.
An experimental study of band intensities in the first positive system of N_2. I. Vibrational transition probabilities, *Can. J. Phys.* **32**, 468–474 (1954).
Fraser, P. A.
A method of determining the electronic transition moment for diatomic molecules, *Can. J. Phys.* **32**, 515–521 (1954).
Nicholls, R. W.
An experimental study of band intensities in the first positive system of N_2. III. Quantitative treatment of eye estimates, *Can. J. Phys.* **32**, 722–725 (1954).
Nicholls, R. W., and Jarmain, W.
r-Centroids: average internuclear separations associated with molecular bands, *Proc. Phys. Soc.* (*London*) *Sect. A* **69**, 253–264 (1956).

23. Biberman, L. M., and Yakubov, I. T.
An approximate method for calculating Franck–Condon factors, *Opt. Spectr.* (*USSR*) (*English Transl.*) **8**, 155–158 (1960).

24. Losev, S. A.
On the absorption of ultraviolet radiation by oxygen heated to temperatures of several thousand degrees. *Nauchn. Dokl. Vysshei Shkoly, Fiz.-Mat. Nauki*, 1958, No. 5, 197–200.

25. Weber, D., and Penner, S. S.
Absolute intensities for the ultraviolet γ bands of NO, *J. Chem. Phys.* **26**, 860–861 (1957).

26. Ditchburn, R. W., and Heddle, D. W. O.
Absorption cross-sections in the vacuum ultra-violet. I. Continuous absorption of oxygen (1800 to 1300 Å), *Proc. Roy. Soc.* (*London*), *Ser. A* **220**, 61–70 (1953).
Absorption cross-sections in the vacuum ultra-violet. II. The Schumann–Runge bands of oxygen (2000 to 1750 Å), *Proc. Roy. Soc.* (*London*), *Ser. A* **226**, 509–521 (1954).

27. Biberman, L. M., Erkovich, S. P., and Soshnikov, V. N.
The transition probability in the Schumann–Runge band system of the O_2 molecule, *Opt. Spectr.* (*USSR*) (*English Transl.*) **7**, 346–347 (1959).

27a. Soshnikov, V. N.
Absolute intensities of electronic transitions in diatomic molecules, *Soviet Phys.—Usp.* (*English Transl.*) **4**, 425–440 (1961).

28. Meyerott, R. E.
Radiation heat transfer to hypersonic vehicles, *Combustion and Propulsion, Third AGARD Colloquium* (M. W. Thring *et al.*, eds.), pp. 431–450. Pergamon, New York, 1958.

29. Landau, L. D., and Lifshitz, E. M.
Statistical Physics. Addison-Wesley, Reading, Mass., 1958.

30. Logan, J. G., Jr.
Recent advances in determination of radiative properties of gases at high temperatures, *Jet Propulsion* **28**, 795–798 (1958).

31. Keck, J., Camm, J., and Kivel, B.
Absolute emission intensity of Schumann–Runge radiation from shock heated oxygen, *J. Chem. Phys.* **28**, 723–724 (1958).

32. Wentink, T., Jr., Planet, W., Hammerling, P., and Kivel, B.
Infrared continuum radiation from high-temperature air, *J. Appl. Phys.* **29**, 742–743 (1958).

32a. Armstrong, B. H., and Meyerott, R. E.
Absorption coefficients for high-temperature nitrogen, oxygen, and air, *Phys. Fluids* **3**, 138–140 (1960).

33. Losev, S. A., Generalov, N. A., and Terebenina, L. B.
On the absorption of ultraviolet radiation behind a shock wave in air, *Opt. Spectr. (USSR) (English Transl.)* **8**, 300–301 (1960).

34. Model', I. Sh.
Measurement of high temperatures in strong shock waves in gases, *Soviet Phys. JETP (English Transl.)* **5**, 589–601 (1957).

35. Dixon, J. K.
The absorption coefficient of nitrogen dioxide in the visible spectrum, *J. Chem. Phys.* **8**, 157–160 (1940).

36. Harris, L., and King, G. W.
The infrared absorption spectra of nitrogen dioxide and tetroxide, *J. Chem. Phys.* **2**, 51–57 (1934).

37. Lambrey, M.
Sur le spectre d'absorption ultraviolet du peroxyde d'azote, *Compt. Rend.* **188**, 251–252 (1929).

38. Raizer, Yu. P.
Glow of air during a strong explosion, and the minimum brightness of a fireball, *Soviet Phys. JETP (English Transl.)* **7**, 331–339 (1958).

39. Raizer, Yu. P.
The formation of nitrogen oxides in the shock wave generated by a strong explosion in air, *Zh. Fiz. Khim.* **33**, 700–709 (1959).

40. Penner, S. S.
Quantitative Molecular Spectroscopy and Gas Emissivities. Addison-Wesley, Reading, Mass., 1959.

41. El'yashevich, M. A.
Atomic and Molecular Spectroscopy. Fizmatgiz, Moscow, 1962.

42. Dronov, A. P., Sviridov, A. G., and Sobolev, N. N.
The continuous emission spectra of krypton and xenon behind a shock wave, *Opt. Spectr. (USSR) (English Transl.)* **12**, 383–389 (1962).

43. Kivel, B., Hammerling, P., and Teare, J. D.
Radiation from the non-equilibrium region of normal shocks in oxygen, nitrogen and air, *Planetary Space Sci.* **3**, 132–137 (1961).

44. Kivel, B.
Radiation from hot air and its effect on stagnation-point heating, *J. Aerospace Sci.* **28**, 96–102 (1961).
45. Allen, R. A., Camm, J. C., and Keck, J. C.
Radiation from hot nitrogen, *J. Quant. Spectr. and Radiative Transfer* **1**, 269–277 (1961).
46. Olfe, D. B.
Mean beam length calculations for radiation from non-transparent gases, *J. Quant. Spectr. and Radiative Transfer* **1**, 169–176 (1961).
47. Penner, S. S.
The determination of absolute intensities and *f*-numbers from shock-tube studies, *Fundamental Data Obtained from Shock Tube Experiments* (A. Ferri, ed.), pp. 261–290. AGARDograph No. 41, Pergamon Press, New York, 1961.
48. Faizullov, F. S., Sobolev, N. N., and Kudryavtsev, E. M.
The temperature of nitrogen and air behind shock waves, *Soviet Phys. "Doklady" (English Transl.)* **4**, 833–836 (1960).
49. Biberman, L. M., Vorob'ev, V. S., and Norman, G. E.
Energy emitted in spectral lines by a plasma at equilibrium, *Opt. Spectr. (USSR) (English Transl.)* **14**, 176–179 (1963).
50. Penner, S. S., and Thomas, M.
Approximate theoretical calculation of continuum opacities, *AIAA J.* **2**, 1572–1575 (1964).
51. Biberman, L. M., and Lagar'kov, A. N.
Effect of spectral lines on the coefficient of radiant heat conduction, *Opt. Spectr. (USSR) (English Transl.)* **16**, 173–175 (1964).
52. Vorob'ev, V. S., and Norman, G. E.
Energy radiated in spectral lines by an equilibrium plasma. II, *Opt. Spectr. (USSR) (English Transl.)* **17**, 96–101 (1964).
53. Bates, D. R. (ed.)
Atomic and Molecular Processes. Academic Press, New York, 1962.
54. Sobel'man, I. N.
Introduction to the Theory of Atomic Spectra. Moscow, 1963.
55. Biberman, L. M., and Norman, G. E.
Recombination and bremsstrahlung in plasmas (free-bound and free-free transitions of electrons in the fields of positive ions), *J. Quant. Spectr. and Radiative Transfer* **3**, 221–245 (1963).
56. Biberman, L. M., Vorob'ev, V. S., Norman, G. E., and Yakubov, I. T.
Radiation heating in hypersonic flow, *Kosmicheskie Issledovaniya* **2**, No. 3, 441–454 (1964).
57. Allen, R. A., and Textoris, A.
Evidence for the existence of N^- from the continuum radiation from shock waves, *J. Chem. Phys.* **40**, 3445–3446 (1964).
58. Norman, G. E.
The role of the negative nitrogen ion N^- in the production of the continuous spectrum of nitrogen and air plasmas, *Opt. Spectr. (USSR) (English Transl.)* **17**, 94–96 (1964).
59. Levitt, B. P.
Thermal emission from nitrogen dioxide, *Trans. Faraday Soc.* **58**, 1789–1800 (1962).
60. Brown, S. C.
Basic Data of Plasma Physics. Wiley, New York, 1959.
61. Raizer, Yu. P.
Bremsstrahlung radiation from an electron due to scattering by neutral atoms with

correlation of collisions taken into account, *Zh. Prikl. Mekhan. i Tekhn. Fiz.*, 1964, No. 5, 149–151.

62. Zel'dovich, Ya. B., and Raizer, Yu. P.
Cascade ionization of a gas by a light pulse, *Soviet Phys. JETP* (*English Transl.*) **20**, 772–780 (1965).
63. Ginzburg, V. L.
Propagation of Electromagnetic Waves in Plasma. Fizmatgiz, Moscow, 1960. English transl. (W. L. Sadowski and D. M. Gallik, eds.), Gordon and Breach, New York, 1962.
64. Huebner, W. F., *et al.* (eds.)
Opacities. Proc. Second International Conf., 1964 *in J. Quant. Spectr. and Radiative Transfer* **5**, 1–280 (1965).
65. Meyerand, R. G., Jr., and Haught, A. F.
Gas breakdown at optical frequencies, *Phys. Rev. Letters* **11**, 401–402 (1963).
66. Damon, E. K., and Tomlinson, R. G.
Observation of ionization of gases by a ruby laser, *Appl. Opt.* **2**, 546–547 (1963).
67. Minck, R. W.
Optical frequency electrical discharges in gases, *J. Appl. Phys.* **35**, 252–254 (1964).
68. Meyerand, R. G., Jr., and Haught, A. F.
Optical-energy absorption and high-density plasma production, *Phys. Rev. Letters* **13**, 7–9 (1964).
69. Ramsden, S. A., and Davies, W. E. R.
Radiation scattered from the plasma produced by a focused ruby laser beam, *Phys. Rev. Letters* **13**, 227–229 (1964).
70. Mandel'shtam, S. L., Pashinin, P. P., Prokhindeev, A. V., and Sukhodrev, N. K.
Study of the "spark" produced in air by focused laser radiation, *Soviet Phys. JETP* (*English Transl.*) **20**, 1344–1346 (1965).
71. Mandel'shtam, S. L., Pashinin, P. P., Prokhorov, A. M., Raizer, Yu. P., and Sukhodrev, N. K.
Investigation of the spark discharge produced in air by focusing laser radiation. II, *Soviet Phys. JETP* (*English Transl.*) **22**, 91–96 (1966).
72. Ambartsumyan, R. V., Basov, N. G., Boiko, V. A., Zuev, V. S., Krokhin, O. N., Kryukov, P. G., Senatskii, Yu. V., and Stoilov, Yu. Yu.
Heating of matter by focused laser radiation, *Soviet Phys. JETP* (*English Transl.*) **21**, 1061–1064 (1965).
73. Keldysh, L. V.
Ionization in the field of a strong electromagnetic wave, *Soviet Phys. JETP* (*English Transl.*) **20**, 1307–1314 (1965).
74. Askar'yan, G. A., and Rabinovich, M. S.
Cascade ionization in a medium by an intense light flash, *Soviet Phys. JETP* (*English Transl.*) **21**, 190–192 (1965).
75. Wright, J. K.
Theory of the electrical breakdown of gases by intense pulses of light, *Proc. Phys. Soc.* (*London*) **84**, 41–46 (1964).
76. Ryutov, D. D.
Theory of breakdown of noble gases at optical frequencies, *Soviet Phys. JETP* (*English Transl.*) **20**, 1472–1479 (1965).
77. Raizer, Yu. P.
Heating of a gas by a powerful light pulse, *Soviet Phys. JETP* (*English Transl.*) **21**, 1009–1017 (1965).

78. Ramsden, S. A., and Savic, P.
A radiative detonation model for the development of a laser-induced spark in air, *Nature* **203**, 1217–1219 (1964).
79. Basov, N. G., and Krokhin, O. N.
Conditions for heating up of a plasma by the radiation from an optical generator, *Soviet Phys. JETP* (*English Transl.*) **19**, 123–125 (1964).
80. Dawson, J. M.
On the production of plasma by giant pulse lasers, *Phys. Fluids* **7**, 981–987 (1964).
81. Zel'dovich, Ya. B.
On the theory of propagating detonations in gas-like systems, *Zh. Eksperim. i Teor. Fiz.* **10**, 542–568 (1940).
82. Zel'dovich, Ya. B., and Kompaneets, A. S.
Theory of Detonation. Academic Press, New York, 1960.
83. Zel'dovich, Ya. B., and Pilipetskii, N. F.
Laser radiation field focused by real systems, *Izv. Vysshikh Uchebn. Zavedenii, Radiofiz.* **9**, 95–101 (1966).
84. Ivanova, A. V.
Photoionization of optical electrons in the ions N V and O VI, *Opt. Spectr.* (*USSR*) (*English Transl.*) **16**, 502–504 (1964).
85. Ivanova, A. V., and Solodchenkova, S. A.
Quantum-mechanical calculation of continuous absorption coefficients for some components of intensely heated air, *Opt. Spectr.* (*USSR*) (*English Transl.*) **20**, 220–223 (1966).
86. Raizer, Yu. P.
Breakdown and heating of a gas under the influence of a laser beam, *Soviet Phys.—Usp.* (*English Transl.*) **8**, 650–673 (1966).
87. Spitzer, L., Jr.
Physics of Fully Ionized Gases. Wiley (Interscience). New York, 2nd edition, 1962.
88. Kuznetsov, N. M.
Thermodynamic Functions and Hugoniot Curves for Air at High Temperatures. Mashinostroenie, Moscow, 1965.

Chapter VI

1. Hornig, D. F.
Energy exchange in shock waves, *J. Phys. Chem.* **61**, 856–860 (1957).
2. Greene, E. F., Cowan, G. R., and Hornig, D. F.
The thickness of shock fronts in argon and nitrogen and rotational heat capacity lags, *J. Chem. Phys.* **19**, 427–434 (1951).
3. Greene, E. F., and Hornig, D. F.
The shape and thickness of shock fronts in argon, hydrogen, nitrogen, and oxygen, *J. Chem. Phys.* **21**, 617–624 (1953).
4. Leskov, L. V., and Savin, F. A.
On the relaxation of nonequilibrium gas systems, *Soviet Phys.—Usp.* (*English Transl.*) **3**, 912–927 (1961).
5. Losev, S. A., and Osipov, A. I.
The study of nonequilibrium phenomena in shock waves, *Soviet Phys.—Usp.* (*English Transl.*) **4**, 525–552 (1962).
6. Zartmann, I. F.
Ultrasonic velocities and absorption in gases at low pressures, *J. Acoust. Soc. Am.* **21**, 171–174 (1949).

7. Van Itterbeek, A., and Vermaelen, R.
Mesures sur l'absorption et la vitesse de propagation du son dans l'hydrogène léger et l'hydrogène lourd entre 300°K et 60°K, *Physica* **9**, 345–355 (1942).
8. Zmuda, A. J.
Dispersion of velocity and anomalous absorption of ultrasonics in nitrogen, *J. Acoust. Soc. Am.* **23**, 472–477 (1951).
9. Connor, J. V.
Ultrasonic dispersion in oxygen, *J. Acoust. Soc. Am.* **30**, 297–300 (1958).
10. Petralia, S.
Interferometria ultrasonora nei gas (IV). Assorbimento di ultrasuoni nell'ammoniaca, *Nuovo Cimento, Ser.* 9 **10**, 817–826 (1953).
11. Ener, C., Gabrysh, A. F., and Hubbard, J. C.
Ultrasonic velocity, dispersion, and absorption in dry, CO_2-free air, *J. Acoust. Soc. Am.* **24**, 474–477 (1952).
12. Landau, L. D., and Lifshitz, E. M.
Quantum Mechanics, Non-Relativistic Theory. Addison-Wesley, Reading, Mass., 1958.
13. Schwartz, R. N., Slawsky, Z. I., and Herzfeld, K. F.
Calculation of vibrational relaxation time in gases, *J. Chem. Phys.* **20**, 1591–1599 (1952).
14. Osipov, A. I.
The relaxation of the vibrational motion in an isolated system of harmonic oscillators, *Soviet Phys. " Doklady" (English Transl.)* **5**, 102–104 (1960).
15. Landau, L. D., and Teller, E.
Zur Theorie der Schalldispersion, *Physik. Z. Sowjetunion* **10**, 34–43 (1936).
16. Blackman, V.
Vibrational relaxation in oxygen and nitrogen, *J. Fluid Mech.* **1**, 61–85 (1956).
17. Zener, C.
Interchange of translational, rotational and vibrational energy in molecular collisions, *Phys. Rev.* **37**, 556–569 (1931).
Low velocity inelastic collisions, *Phys. Rev.* **38**, 277–281 (1931).
18. Schwartz, R. N., and Herzfeld, K, F.
Vibrational relaxation times in gases (three-dimensional treatment), *J. Chem. Phys.* **22**, 767–773 (1954).
18a. Herzfeld, K. F.
Calculations of absolute relaxation times in gases, *Proc. Third International Cong. on Acoustics Stuttgart, 1959. Vol. I. Principles* (L. Cremer, ed.), pp. 503–504. Elsevier (Van Nostrand), New York, 1961.
19. Knötzel, H., and Knötzel, L.
Schallabsorption und Dispersion in Sauerstoff, *Ann. Physik* **2**, 393–403 (1948).
20. Huber, A. W., and Kantrowitz, A.
Heat-capacity lag measurements in various gases, *J. Chem. Phys.* **15**, 275–284 (1947).
21. Herron, J. T., Franklin, J. L., Bradt, P., and Dibeler, V. H.
Kinetics of nitrogen atom recombination, *J. Chem. Phys.* **29**, 230–231 (1958).
22. Harteck, P., Reeves, R. R., and Mannella, G.
Rate of recombination of nitrogen atoms, *J. Chem. Phys.* **29**, 608–610 (1958).
23. Wentink, T., Jr., Sullivan, J. O., and Wray, K. L.
Nitrogen atomic recombination at room temperature, *J. Chem. Phys.* **29**, 231–232 (1958).
24. Wigner, E.
Calculation of the rate of elementary association reactions, *J. Chem. Phys.* **5**, 720–725 (1937).
Some remarks on the theory of reaction rates, *J. Chem. Phys.* **7**, 646–652 (1939).

25. Britton, D., Davidson, N., Gehman, W., and Schott, G.
Shock waves in chemical kinetics: further studies on the rate of dissociation of molecular iodine, *J. Chem. Phys.* **25**, 804–809 (1956).
26. Bunker, D. L., and Davidson, N.
A further study of the flash photolysis of iodine, *J. Am. Chem. Soc.* **80**, 5085–5090 (1958).
27. Fowler, R. H., and Guggenheim, E. A.
Statistical Thermodynamics. Cambridge Univ. Press, London and New York, 1939.
28. Stupochenko, E. V., and Osipov, A. I.
On the mechanism of thermal dissociation of diatomic molecules, *Zh. Fiz. Khim.* **32**, 1673–1674 (1958).
29. Stupochenko, E. V., and Osipov, A. I.
Kinetics of thermal dissociation of diatomic molecules, *Russ. J. Phys. Chem.* (*English Transl.*) **33**, 36–39 (1959).
30. Matthews, D. L.
Interferometric measurement in the shock tube of the dissociation rate of oxygen, *Phys. Fluids* **2**, 170–178 (1959).
31. Byron, S. R.
Measurement of the rate of dissociation of oxygen, *J. Chem. Phys.* **30**, 1380–1392 (1959).
32. Generalov, N. A., and Losev, S. A.
On an investigation of nonequilibrium phenomena behind the front of a shock wave in air. Dissociation of oxygen, *Zh. Prikl. Mekhan. i Tekhn. Fiz.*, 1960, No. 2, 64–73.
33. Zinman, W. G.
Recent advances in chemical kinetics of homogeneous reactions in dissociated air, *ARS J.* **30**, 233–238 (1960).
34. Glick, H. S., and Wurster, W. H.
Shock tube study of dissociation relaxation in oxygen, *J. Chem. Phys.* **27**, 1224–1226 (1957).
35. Palmer, H. B., and Hornig, D. F.
Rate of dissociation of bromine in shock waves, *J. Chem. Phys.* **26**, 98–105 (1957).
36. Britton, D., and Davidson, N.
Shock waves in chemical kinetics. Rate of dissociation of molecular bromine, *J. Chem. Phys.* **25**, 810–813 (1956).
37. Nikitin, E. E.
On the calculation of the velocity distribution of diatomic molecules, *Dokl. Akad. Nauk SSSR* **119**, 526–529 (1958).
38. Glasstone, S., Laidler, K. J., and Eyring, H.
The Theory of Rate Processes; The Kinetics of Chemical Reactions, Viscosity, Diffusion and Electrochemical Phenomena. McGraw-Hill, New York, 1941.
39. Zel'dovich, Ya. B., Sadovnikov, P. Ya., and Frank-Kamenetskii, D. A.
The Oxidation of Nitrogen by Combustion. Izdat. Akad. Nauk SSSR, Moscow, 1947.
40. Glick, H. S., Klein, J. J., and Squire, W.
Single-pulse shock tube studies of the kinetics of the reaction $N_2+O_2 \rightleftarrows 2NO$ between 2000–3000°K, *J. Chem. Phys.* **27**, 850–857 (1957).
41. Raizer, Yu. P.
The formation of nitrogen oxides in the shock wave generated by a strong explosion in air, *Zh. Fiz. Khim.* **33**, 700–709 (1959).
42. Kaye, G. W. C., and Laby, T. H.
Tables of Physical and Chemical Constants and Some Mathematical Functions. Long-

mans Green, New York, 9th edition, 1941.

43. Zeise, H.
Chemische Konstante und thermodynamisches Potential von NO_2-Gas und das Gasgleichgewicht $2NO + O_2 \rightleftarrows 2NO_2$, aus Spektroskopischen daten Berechnet, *Z. Elektrochem.* **42**, 785–789 (1936).

44. Ramstetter, Dr.-Ing.
Messungen der Geschwindigkeit $2NO_2 \leftrightarrows 2NO + O_2$, pp. 106–118, *in* Bodenstein, M., Bildung und Zersetzung der Höheren Stickoxyde, *Z. Physik. Chem.* (*Leipzig*) **100**, 68–123 (1922).
Lindner, Fraulein Dr.
Messungen der Gleichgewichts $2NO_2 \rightleftarrows 2NO + O_2$, pp. 82–87, *in* Bodenstein, M., Bildung und Zersetzung der Höheren Stickoxyde, *Z. Physik. Chem.* (*Leipzig*) **100**, 68–123 (1922).

45. Massey, H. S. W., and Burhop, E. H. S.
Electronic and Ionic Impact Phenomena. Oxford Univ. Press, London and New York, 1952.

46. Granovskii, V. L.
Electric Currents in a Gas. Vol. I. Gostekhizdat, Moscow, 1952.

47. Seaton, M. J.
Electron impact ionization of Ne, O, and N, *Phys. Rev.* **113**, 814 (1959).

48. Fite, W. L., and Brackmann, R. T.
Ionization of atomic oxygen on electron impact, *Phys. Rev.* **113**, 815–816 (1959).

49. Tate, J. T., and Smith, P. T.
The efficiencies of ionization and ionization potentials of various gases under electron impact, *Phys. Rev.* **39**, 270–277 (1932).

50. Petschek, H., and Byron, S.
Approach to equilibrium ionization behind strong shock waves in argon, *Ann. Phys.* (*N.Y.*) **1**, 270–315 (1957).

51. Rostagni, A.
Ricerche sui raggi positivi e neutrali. V. Ionizzazione per urto di ioni e di atomi, *Nuovo Cimento* **13**, 389–406 (1936).
Wayland, H.
The ionization of neon, krypton and xenon by bombardment with accelerated neutral argon atoms, *Phys. Rev.* **52**, 31–37 (1937).

52. Rostagni, A.
Ricerche sui raggi positivi e neutrali. III. Ionizzazione per urto di atomi, *Nuovo Cimento* **11**, 621–634 (1934).

53. Wray, K. L., Teare, J. D., Kivel, B., and Hammerling, P.
Relaxation processes and reaction rates behind shock fronts in air and component gases, *Eighth Symposium* (*International*) *on Combustion*, pp. 328–339. Williams & Wilkins, Baltimore, 1962.

54. Rosen, P.
Low-energy inelastic atomic collisions, *Phys. Rev.* **109**, 348–350 (1958).
Ionization cross section of argon–argon collisions near threshold, *Phys. Rev.* **109**, 351–355 (1958).

55. Ambartsumian, V. A. (ed.)
Theoretical Astrophysics. Gostekhizdat, Moscow, 1952. English transl., Pergamon Press, New York, 1958.

56. Spitzer, L., Jr.
Physics of Fully Ionized Gases. Wiley (Interscience), New York, 2nd edition, 1962.

57. Landau, L. D.
Kinetic equation in the case of a Coulomb interaction, *Zh. Eksperim. i Teor. Fiz.* **7**, 203–209 (1937).

58. Generalov, N. A.
Vibrational relaxation in oxygen at high temperatures. I, *Vestn. Mosk. Univ., Ser. III: Fiz., Astron.*, 1962, No. 3, 51–59.

59. Camac, M.
O_2 vibrational relaxation in oxygen–argon mixtures, *J. Chem. Phys.* **34**, 448–459 (1961).

60. Roth, W.
Shock tube study of vibrational relaxation in the $A\,^2\Sigma^+$ state of NO, *J. Chem. Phys.* **34**, 999–1003 (1961).

61. Matthews, D. L.
Vibrational relaxation of carbon monoxide in the shock tube, *J. Chem. Phys.* **34**, 639–642 (1961).

62. Johannesen, N. H., Zienkiewicz, H. K., Blythe, P. A., and Gerrard, J. H.
Experimental and theoretical analysis of vibrational relaxation regions in carbon dioxide, *J. Fluid Mech.* **13**, 213–224 (1962).

63. Hurle, I. R., and Gaydon, A. G.
Vibrational relaxation and dissociation of carbon dioxide behind shock waves, *Nature* **184**, 1858–1859 (1959).

64. Griffith, W. C.
Vibrational relaxation times, *Fundamental Data Obtained from Shock Tube Experiments* (A. Ferri, ed.), pp. 242–260. AGARDograph No. 41, Pergamon Press, New York, 1961.

65. Valley, L. M., and Levgold, S.
Vibrational relaxation times for gas mixtures, *Phys. Fluids* **3**, 831 (1960).

66. Osipov, A. I.
Vibrational relaxation in binary gas mixtures, *Vestnik. Mosk. Univ., Ser. III: Fiz., Astron.*, 1960, No. 4, 96–99.

67. Camac, M., and Vaughan, A.
O_2 dissociation rates in O_2–Ar mixtures, *J. Chem. Phys.* **34**, 460–470 (1961).

68. Rink, J. P., Knight, H. T., and Duff, R. E.
Shock tube determination of dissociation rates of oxygen, *J. Chem. Phys.* **34**, 1942–1947 (1961).

69. Patch, R. W.
Shock-tube measurement of dissociation rates of hydrogen, *J. Chem. Phys.* **36**, 1919–1924 (1962).

70. Allen, R. A., Keck, J. C., and Camm, J. C.
Nonequilibrium radiation and the recombination rate of shock-heated nitrogen, *Phys. Fluids* **5**, 284–291 (1962).

71. Ivanov-Kholodnyi, G. S.
Intensity of short-wave solar radiation and the rate of ionization and recombination processes in the ionosphere. Review, *Geomagnetism and Aeronomy* **2**, 315–336 (1962).

72. Lin, S. C., Neal, R. A., and Fyfe, W. I.
Rate of ionization behind shock waves in air. I. Experimental results, *Phys. Fluids* **5**, 1633–1648 (1962).

73. Lin, S. C., and Teare, J. D.
Rate of ionization behind shock waves in air. II. Theoretical interpretations, *Phys. Fluids* **6**, 355–375 (1963).

74. Dalgarno, A.
Charged particles in the upper atmosphere, *Ann. Geophys.* **17**, 16–49 (1961).

75. Fedorenko, N. V.
Ionization in collisions between ions and atoms, *Soviet Phys.—Usp.* (*English Transl.*) **2**, 526–546 (1959).
Fedorenko, N. V., Flaks, I. P., and Filippenko, L. G.
Ionization of inert gases by multiply charged ions, *Soviet Phys. JETP* (*English Transl.*) **11**, 519–523 (1960).

76. Osipov, A. I., and Stupochenko, E. V.
Non-equilibrium energy distributions over the vibrational degrees of freedom in gases, *Soviet Phys.—Usp.* (*English Transl.*) **6**, 47–66 (1963).

77. Stupochenko, E. V., Losev, S. A., and Osipov, A. I.
Relaxation Processes in Shock Waves. Nauka, Moscow, 1965.

78. Brown, S. C.
Basic Data of Plasma Physics. Wiley, New York, 1959.

79. Ivanov-Kholodnyi, G. S., Nikol'skii, G. M., and Gulyaev, R. A.
Ionization and excitation of hydrogen. I. Elementary processes for the upper levels, *Soviet Astron.—AJ* (*English Transl.*) **4**, 754–765 (1961).

80. Hinnov, E., and Hirschberg, J. G.
Electron-ion recombination in dense plasmas, *Phys. Rev.* **125**, 795–801 (1962).

81. Biberman, L. M., Toropkin, Yu. N., and Ul'yanov, K. N.
The theory of multistage ionization and recombination, *Soviet Phys.—Tech. Phys.* (*English Transl.*) **7**, 605–609 (1963).

82. Landau, L. D., and Lifshitz, E. M.
Mechanics. Addison-Wesley, Reading, Mass., 1960.

83. Bates, D. R. (ed.)
Atomic and Molecular Processes. Academic Press, New York, 1962.

84. Gryzinski, M.
Classical theory of electronic and ionic inelastic collisions, *Phys. Rev.* **115**, 374–383 (1959).

85. Gurevich, A. V.
Structure of the disturbed zone in the vicinity of a charged body in plasma, *Geomagnetism and Aeronomy* **4**, 1–11 (1964).

86. Allen, C. W.
Astrophysical Quantities. (Athlone Press) Oxford Univ. Press, New York, 1st edition, 1955.

87. Pitaevskii, L. P.
Electron recombination in a monatomic gas, *Soviet Phys. JETP* (*English Transl.*) **15**, 919–921 (1962).

88. Gurevich, A. V., and Pitaevskii, L. P.
Recombination coefficient in a dense low-temperature plasma, *Soviet Phys. JETP* (*English Transl.*) **19**, 870–871 (1964).

89. Bates, D. R., Kingston, A. E., and McWhirter, R. W. P.
Recombination between electrons and atomic ions. I. Optically thin plasmas, *Proc. Roy. Soc.* (*London*), *Ser. A* **267**, 297–312 (1962).
Recombination between electrons and atomic ions. II. Optically thick plasmas, *Proc. Roy. Soc.* (*London*), *Ser. A* **270**, 155–167 (1962).

90. Kuznetsov, N. M., and Raizer, Yu. P.
On the recombination of electrons in a plasma expanding into vacuum, *Zh. Prikl. Mekhan. i Tekhn. Fiz.*, 1965, No. 4, 10–20.

91. Danilov, A. D., and Ivanov-Kholodnyi, G. S.
Research on ion-molecule reactions and dissociative recombination in the upper atmosphere and in the laboratory, *Soviet Phys.—Usp.* (*English Transl.*) **8**, 92–116 (1965).

92. Eschenroeder, A. Q., Daiber, J. W., Golian, T. C., and Hertzberg, A.
Shock tunnel studies of high-enthalpy ionized airflows, *The High Temperature Aspects of Hypersonic Flow* (W. C. Nelson, ed.), pp. 217–254. Pergamon Press, New York, 1964.

93. Makin, B., and Keck, J. C.
Variational theory of three-body electron-ion recombination rates, *Phys. Rev. Letters* **11**, 281–283 (1963).

Appendix

Some often used constants, relations between units, and formulas*

Fundamental constants

Speed of light	$c = 2.998 \cdot 10^{10}$ cm/sec
Planck constant	$h = 6.625 \cdot 10^{-27}$ erg · sec
	$\hbar = h/2\pi = 1.054 \cdot 10^{-27}$ erg · sec
Electron charge	$e = 4.803 \cdot 10^{-10}$ esu
Electron mass	$m = 9.108 \cdot 10^{-28}$ g
Proton mass	$m_p = 1.673 \cdot 10^{-24}$ g
Mass of unit atomic weight	$M_0 = 1.660 \cdot 10^{-24}$ g
Boltzmann constant	$k = 1.380 \cdot 10^{-16}$ erg/deg
Universal gas constant	$\mathcal{R} = 8.317 \cdot 10^{7}$ erg/deg · mole
	$= 1.987$ cal/deg · mole
Avogadro number	$N_0 = 6.023 \cdot 10^{23}$ mole^{-1}
Loschmidt number	$n_0 = 2.687 \cdot 10^{19}$ cm^{-3}

Relations between units

Energy $E_0 = 1$ ev $= 1.602 \cdot 10^{-12}$ erg corresponds to:

temperature	$E_0/k = 11{,}605°$K
frequency	$E_0/h = 2.418 \cdot 10^{14}$ sec^{-1}
wavelength	$hc/E_0 = 1.240 \cdot 10^{-4}$ cm $= 12{,}400$ Å
wave number	$E_0/hc = 8066$ cm^{-1}

* Values of the constants are taken from C. W. Allen, *Astrophysical Quantities.* (Athlone Press) Oxford Univ. Press, New York, 2nd edition, 1963. *Editors' note.* Some of the constants and numerical values in this appendix may differ slightly in the last significant figure from those used in the body of the text, which were based on constants given in the first edition (1955) of Allen's book.

In spectroscopy wave number is often used in place of frequency.

Wave number $1/\lambda = 1\ \text{cm}^{-1}$ corresponds to:

wavelength	$\lambda = 10^8$ Å
frequency	$\nu = 2.998 \cdot 10^{10}\ \text{sec}^{-1}$
temperature	$T = h\nu/k = 1.439°\text{K}$
photon energy	$h\nu = 1.240 \cdot 10^{-4}\ \text{ev} = 1.986 \cdot 10^{-16}\ \text{erg}$

$1\ \text{cal} = 4.185 \cdot 10^7\ \text{erg}$, $1\ \text{kcal} = 10^3\ \text{cal}$

1 ev per molecule corresponds to 23.05 kcal/mole

1 volt = 1/300 units of potential in esu

Constants and relations between them

Bohr radius $$a_0 = \frac{h^2}{4\pi^2 me^2} = \frac{\hbar^2}{me^2} = 0.529 \cdot 10^{-8}\ \text{cm}$$

Ionization potential of hydrogen atom $$I_H = \frac{e^2}{2a_0} = \frac{2\pi^2 e^4 m}{h^2} = \frac{e^4 m}{2\hbar^2} = 13.60\ \text{ev}$$

Rydberg constant $$Ry = \frac{I_H}{h} = \frac{2\pi^2 e^4 m}{h^3} = 3.290 \cdot 10^{15}\ \text{sec}^{-1}$$

Electron speed in first Bohr orbit $$v_0 = \frac{2\pi e^2}{h} = \frac{e^2}{\hbar} = 2.188 \cdot 10^8\ \text{cm/sec}$$

Classical electron radius $$r_0 = \frac{e^2}{mc^2} = 2.818 \cdot 10^{-13}\ \text{cm}$$

Compton wavelength $$\lambda_0 = \frac{h}{mc} = 2.426 \cdot 10^{-10}\ \text{cm}$$

$$\bar{\lambda}_0 = \frac{\lambda_0}{2\pi} = \frac{\hbar}{mc} = 3.862 \cdot 10^{-11}\ \text{cm}$$

Rest mass energy of electron $$mc^2 = 511\ \text{kev} = 8.186 \cdot 10^{-7}\ \text{erg}$$

The number "137" = (fine structure constant)$^{-1}$ $\quad \dfrac{\hbar c}{e^2} = \dfrac{hc}{2\pi e^2} = 137.0$

Length ratio $\quad a_0 = \text{“137”}\, \lambda\!\!\!^{-}_0 = \text{“137”}^2 r_0$

Energy ratio $\quad mc^2 = 2I_{\mathrm{H}} \cdot \text{“137”}^2$

Thomson cross section $\quad \varphi_0 = \dfrac{8}{3}\pi r_0^2 = 6.65 \cdot 10^{-25}\ \mathrm{cm}^2$

Mass ratio proton/electron $\quad m_p/m = 1836$

Electric field of a proton at a distance of the first Bohr radius $\quad \dfrac{e^2}{a_0^2} = 1.715 \cdot 10^7\ \mathrm{esu} = 5.145 \cdot 10^9\ \mathrm{volt/cm}$

Area of spectral line with unit oscillator strength $\quad \dfrac{\pi e^2}{mc} = 0.0265\ \mathrm{cm^2/sec}$

Atomic unit cross section $\quad \pi a_0^2 = 0.880 \cdot 10^{-16}\ \mathrm{cm}^2$

Formulas

Radiant energy flux from surface of a perfect black body

$$S = \sigma T^4 = 5.67 \cdot 10^{-5} T_{\mathrm{deg}}^4$$
$$= 1.03 \cdot 10^{12} T_{\mathrm{ev}}^4\ \mathrm{erg/cm^2 \cdot sec}$$

(σ = Stefan-Boltzmann constant)

Equilibrium radiant energy density

$$U_p = \frac{4\sigma T^4}{c} = 7.56 \cdot 10^{-15} T_{\mathrm{deg}}^4$$
$$= 1.37 \cdot 10^2 T_{\mathrm{ev}}^4\ \mathrm{erg/cm^3}$$

Spectral equilibrium radiant energy density

$$U_{\nu p}\,d\nu = \frac{8\pi h\nu^3}{c^3}\frac{1}{e^{h\nu/kT}-1}\,d\nu \text{ erg/cm}^3$$

(maximum at the frequency $h\nu = 2.822\,kT$)

Spectral equilibrium radiation intensity

$$I_{\nu p}\,d\nu = \frac{cU_{\nu p}\,d\nu}{4}$$

$$= \frac{2h\nu^3}{c^2}\frac{d\nu}{e^{h\nu/kT}-1} \text{ erg/cm}^2\cdot\text{sec}\cdot\text{sterad}$$

Saha equation

$$\frac{N_e N_+}{N_a} = A\frac{g_+}{g_a}T^{3/2}e^{-I/kT}$$

where

$$A = 2\left(\frac{2\pi mk}{h^2}\right)^{3/2} = 4.83\cdot 10^{15}\text{ cm}^{-3}\cdot\text{deg}^{-3/2}$$

$$= 6.04\cdot 10^{21}\text{ cm}^{-3}\cdot\text{ev}^{-3/2}$$

Maxwell distribution function normalized to unity

$$f(v)\,dv = 4\pi\left(\frac{m}{2\pi kT}\right)^{3/2}\exp\left(-\frac{mv^2}{2kT}\right)v^2\,dv$$

$$f(\varepsilon)\,d\varepsilon = \frac{2}{\pi^{1/2}}\frac{\varepsilon^{1/2}}{(kT)^{3/2}}e^{-\varepsilon/kT}\,d\varepsilon$$

Electron speed

$$v_e = 5.93\cdot 10^7\varepsilon_{\text{ev}}^{1/2}\text{ cm/sec}$$

Speed of particle with atomic weight A

$$v = 1.38\cdot 10^6(\varepsilon_{\text{ev}}/A)^{1/2}\text{ cm/sec}$$

Electron mean thermal speed

$$\bar{v}_e = \left(\frac{8kT}{\pi m}\right)^{1/2} = 6.21\cdot 10^5 T_{\text{deg}}^{1/2}$$

$$= 6.69\cdot 10^7 T_{\text{ev}}^{1/2}\text{ cm/sec}$$

Particle mean thermal speed

$$\bar{v} = 1.45 \cdot 10^4 \left(\frac{T_{\text{deg}}}{A}\right)^{1/2}$$

$$= 1.56 \cdot 10^6 \left(\frac{T_{\text{ev}}}{A}\right)^{1/2} \text{ cm/sec}$$

Classical damping constant

$$\gamma = \frac{8\pi^2 e^2 \nu^2}{3mc^3} = 2.47 \cdot 10^{-22} \nu^2_{\text{sec}^{-1}}$$

$$= \frac{0.222 \cdot 10^{16}}{\lambda^2_{\text{Å}}} \text{ sec}^{-1}$$

Cross section σ in terms of P_c = average number of collisions per cm at 1 mm Hg and 0°C $\quad \sigma = 2.83 \cdot 10^{-17} P_c \text{ cm}^2$

Specific energy

$$1 \text{ ev/molecule} = \frac{9.65 \cdot 10^{11}}{M} \text{ erg/g}$$

where

M = molecular weight

Author Index

Volume I

Numbers in brackets are reference numbers and are included to assist in locating references in which the authors' names are not mentioned in the text. Italicized numbers indicate the pages on which the references are listed, and those in parentheses the number of citations for the given page.

Subject Index

Volume I

VII. Shock wave structure in gases

§1. Introduction

The basic ideas on shock waves have been presented in Chapter I. It was shown that the hydrodynamic equations for an ideal fluid admit the existence of discontinuous solutions describing shock waves. The flow parameters, i.e., the density, pressure, and velocity on each side of the discontinuity surface, are connected by finite difference equations which correspond to the differential equations describing the continuous flow regions. Both the hydrodynamic and the difference equations are expressions of the general laws of conservation of mass, momentum, and energy. It follows from the conservation laws that the entropy of the fluid also undergoes a jump (increases) at the discontinuity surface. The increase in entropy across a shock wave is determined only by the conditions of conservation of mass, momentum, and energy and by the thermodynamic properties of the fluid, and is entirely independent of the dissipative mechanisms causing this increase.

It is somewhat paradoxical that the adiabatic equations of motion for a fluid admit the existence of surfaces where the entropy undergoes a jump. The irreversibility of a shock compression indicates the presence of dissipative processes, such as viscosity and heat conduction, which lead to the increase in the entropy. It is precisely the viscosity which results in the irreversible conversion into heat of a major part of the kinetic energy of the stream crossing the discontinuity, in a coordinate system in which the discontinuity is at rest. Thus, if we are interested in the mechanism of shock compression, in the internal structure and thickness of the transition layer in which the fluid is transformed from the initial to the final state and which, within the framework of the hydrodynamics of an ideal fluid, is replaced by a mathematical surface, we must turn to a theory which includes a description of the dissipative processes. In Chapter I this problem was considered for the case of weak shock waves. In this chapter we shall not impose any limitations on the strength of the shock.

Usually in hydrodynamic processes changes in the macroscopic parameters in regions of continuous flow occur very slowly in comparison with the rates of the relaxation processes which lead to the establishment of thermodynamic equilibrium. Each gas particle at any instant of time is in the state of thermodynamic equilibrium which corresponds to the slowly changing macroscopic variables, as though the particle "follows" the changes in the variables.

Therefore, when considering shock discontinuities within the framework of the hydrodynamics of an ideal fluid, it is entirely permissible to assume the state of the gas on both sides of the discontinuity to be in equilibrium.

The density, pressure, etc. change very rapidly in the thin transition layer, through which the gas passes from its initial state of thermodynamic equilibrium into its final, also equilibrium, state. The thermodynamic equilibrium inside this region, called here the shock front, can be appreciably disturbed. Therefore, in studying the internal structure of a shock front it is necessary to consider the kinetics of relaxation processes and to investigate in detail the mechanism of the establishment of the final state of thermodynamic equilibrium in the fluid which is attained behind the shock front.

The study of the internal structure of shock fronts is of interest for many reasons. At first this problem attracted attention as purely a theoretical one, the solution of which aided in understanding the physical mechanism of shock compression, as a truly remarkable phenomenon in gasdynamics. Later shock waves were employed in laboratories with the aim of obtaining high temperatures and of studying various processes which take place in gases at high temperatures, as for example, vibrational excitation in molecules, molecular dissociation, chemical reactions, ionization, and radiation (see Chapter IV). Theoretical considerations of the shock front structure enable one to deduce from the experimental data a good deal of valuable information about the rates of these processes. Finally, the study of the structure of very strong shock fronts in which radiation plays an important role helps to clarify the problem of such an important characteristic as the luminosity of the shock front and makes it possible to explain some interesting optical effects observed in strong explosions in air (see Chapter IX).

The mathematical theory of shock front structure is based on the assumption that the structure is steady. The time it takes the fluid in a shock wave to go from the initial to the final state is very short, much shorter than the characteristic times over which the flow variables change in the continuous flow region behind the shock front. In exactly the same way, the front thickness is much less than the characteristic length scale over which the state of the gas behind the front changes significantly, for example, the distance from the shock front to the piston "pushing" the wave (the piston moves with a nonuniform speed, in general). In the short time during which the shock wave traverses a distance of the order of the front thickness, its propagation velocity, pressure, and the other flow variables behind the front remain practically unchanged. However, the kinetics of the internal processes which take place within a shock front propagating through a gas with given initial conditions depend only on the wave strength. Therefore over some relatively long period, each of the gas particles flowing into the shock discontinuity passes through the same sequence of states as the preceding ones. In other words, the distribution of the various variables across the shock front forms a "frozen" picture which moves during this period as an entity together with the front (Fig. 7.1).

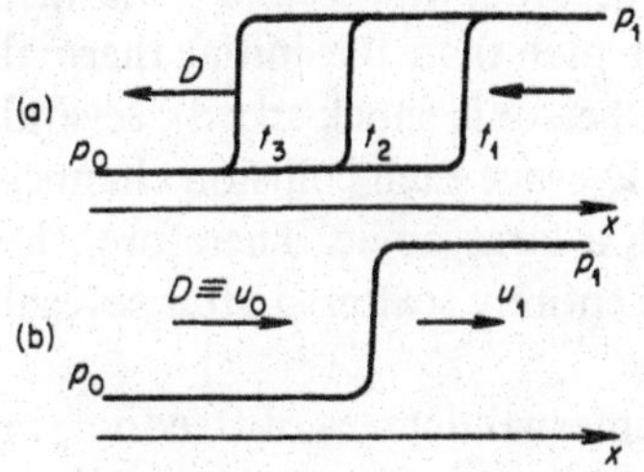

Fig. 7.1. Pressure distributions across a shock wave: (a) propagation of the shock in a laboratory coordinate system; (b) the shock in a coordinate system moving with the front.

If we denote the front velocity by D ($D = |D| > 0$), and the coordinate normal to the front surface at a given point by x, then we can say that all the state variables of the gas inside the wave depend on position and time only in the combination $x + Dt$. In a coordinate system moving with the front the process is steady and is independent of time. This fact (which has already been used in deriving the relationships across the discontinuity) greatly simplifies the study of the problem from a mathematical point of view, since all the flow variables in the coordinate system moving with the wave are functions not of the two variables x and t, but only of a single coordinate, and the process can be described by ordinary differential equations.

In considering the thickness of weak shock fronts in §23 of Chapter I, we have shown that the molecular mean free path serves as a characteristic scale of the shock thickness. As the wave strength increases, the thickness decreases, and when the increase in the pressure behind the front over the initial pressure becomes comparable with the initial pressure, the front thickness is of the order of a molecular mean free path. It is physically clear that the thickness of a strong shock wave, where the compression takes place in the presence of "viscous" forces, is always of the order of a molecular mean free path*. This can be most simply explained by considering a shock wave in a coordinate system in which the gas behind the front is at rest (in a coordinate system moving with the piston) or, equivalently, by considering the bringing

* We wish to emphasize the particular way in which we use the concept of "viscosity" in the case being considered. When referring to viscosity, it is usually understood that the velocity gradients are small and that the velocity changes significantly only over distances much greater than a molecular mean free path. In other words, viscosity in hydrodynamics is a "macroscopic" concept. If a sharp change in gas velocity and density takes place over a distance of the order of a mean free path, then this "microscopic" phenomenon may not be considered from the point of view of hydrodynamics but must be treated on the basis of the kinetic theory of gases. As applied to the case of very large gradients, the term "viscosity" in the shock wave front denotes the mechanism by which the directed velocity of the molecules is changed to a random velocity by molecular collisions.

to rest of a high-velocity gas stream incident on a stationary wall. The kinetic energy of the directed molecular motion (the kinetic energy of hydrodynamic motion) is converted into kinetic energy of random motion, i.e., into heat when the fluid is brought to rest. In order to "brake" the fast molecules, whose directed velocity is much greater than the initial thermal velocity (in the case of a strong wave, with a hypersonic shock speed), several gaskinetic collisions are sufficient, since on the average each collision changes the direction of motion of a molecule through a large angle. Therefore, the directed momentum of the molecules is almost entirely scattered after several collisions and the velocities become random.

The distribution of energy over the various internal degrees of freedom, in particular vibrational excitation of the molecules, dissociation, and ionization, usually requires many collisions. The thickness of the relaxation layer in which the final thermodynamic equilibrium is established is much greater than the thickness of the initial shock wave. Hence the entire transition layer of the shock front can be divided into two regions with appreciably different thicknesses, a very thin "viscous" shock front and an extended relaxation layer.

In a sufficiently strong shock front in which the gas is heated to high temperatures, an important role is played by radiation and radiative heat transfer. The structure of the front in this case becomes even more complex. The total front thickness is determined by the largest scale characterizing the transition process associated with radiant heat exchange, namely the radiation mean free path, which ordinarily is many times greater than the gaskinetic mean free path.

In the following sections we shall consider in detail the characteristic properties of the structure of shock fronts. We start by considering relatively weak shock waves, after which we shall consider increasingly stronger waves.

1. The shock front

§2. Viscous shock front

Since the compression shock in a shock front takes place over distances comparable with the gaskinetic mean free path, we should actually begin our study of shock front structure on the basis of the kinetic theory of gases. As a first step in this direction, however, it is natural to consider the problem in the framework of the hydrodynamics of a real fluid, in which dissipative processes are taken into account, i.e., with viscosity and heat conduction. Here, in contrast to §23 of Chapter I, we do not impose any limitations on the strength of the shock wave. To provide continuity of presentation we repeat here some conclusions and calculations of that section. In order not to complicate the presentation by unnecessary detail connected with the slow excitation of nontranslational degrees of freedom, we regard the gas as monatomic and neglect ionization.

The one-dimensional flow equations for a viscous and heat conducting gas flow which is steady in a coordinate system moving with the wave front are

$$\frac{d}{dx}\rho u = 0,$$

$$\rho u \frac{du}{dx} + \frac{dp}{dx} - \frac{d}{dx}\frac{4}{3}\mu\frac{du}{dx} = 0, \tag{7.1}$$

$$\rho u T \frac{d\Sigma}{dx} = \frac{4}{3}\mu\left(\frac{du}{dx}\right)^2 - \frac{dS}{dx}.$$

Here Σ is the specific entropy*, μ the coefficient of viscosity†, and S is a

* *Editors' note.* The reader is cautioned to note that the symbol S used elsewhere in the book to denote specific entropy is going to be consistently used for energy flux, particularly for radiant energy flux, and therefore a new symbol had to be used for entropy in this chapter.

† For the case considered the concepts of first and second viscosities are indistinguishable.

Editors' note. More specifically, their effects are combined into a single term. As in the analysis of §23, Chapter I, we may include second or bulk viscosity by replacing $\frac{4}{3}\mu$ by the longitudinal coefficient $\mu'' = \frac{4}{3}\mu + \mu'$.

Dilute monatomic gases have $\mu' = 0$, and the usual physical origin of bulk viscosity is in rotational relaxation. As discussed in §2 of Chapter VI, the rotational relaxation time τ_{rot} is extremely short, though it may be appreciably larger than the characteristic translational or gaskinetic collision time of (6.1), Chapter VI.

On the assumptions that τ_{rot} is large compared with τ_{gas} but small compared with a characteristic macroscopic time scale, and that vibrational modes are unexcited, a relation between μ' and τ_{rot} may be established. The quantity $\bar{p}$ of (1.95), Chapter I, is given by $\bar{p} = \rho R T_{trans}$, while p_{st} is given by $p_{st} = \rho(\gamma - 1)(c_{trans}T_{trans} + c_{rot}T_{rot})$, with $c_{trans} + c_{rot} = c$ and $(\gamma - 1)c_v = R$. Thus we identify

$$\bar{p} - p_{st} = \mu' \frac{D\rho}{\rho Dt} = \rho R(T_{trans} - T_{rot})\frac{c_{rot}}{c_v}.$$

With $\tau_{rot} \gg \tau_{trans}$ and the rotational mode governed by a relaxation law, we have

$$\frac{DT_{rot}}{Dt} = \frac{T_{trans} - T_{rot}}{\tau_{rot}}.$$

Finally, with τ_{rot} small in comparison with the macroscopic time scale, we may approximate

$$\frac{DT_{rot}}{Dt} \approx \frac{DT_{trans}}{Dt} \approx (\gamma - 1)T_{trans}\frac{D\rho}{\rho Dt} \approx \frac{(\gamma - 1)^2\varepsilon}{R}\frac{D\rho}{\rho Dt}.$$

nonhydrodynamic energy flux, equal according to the ordinary heat conduction law to

$$S = -\kappa \frac{dT}{dx}, \tag{7.2}$$

where κ is the coefficient of thermal conductivity. To the system of equations (7.1) must be added the boundary conditions expressing the absence of any gradients ahead of and behind the wave front, and also the fact that the flow variables must approach their initial (for $x = -\infty$) and final (for $x = +\infty$) values.

Rewriting the third equation in (7.1) by means of the relation

$$T\, d\Sigma = d\varepsilon + p\, dV = dh - \frac{1}{\rho} dp$$

and integrating each equation in (7.1), we obtain the first integrals of the system

$$\begin{gathered} \rho u = \rho_0 D, \\ p + \rho u^2 - \frac{4}{3}\mu \frac{du}{dx} = p_0 + \rho_0 D^2, \\ h + \frac{u^2}{2} + \frac{1}{\rho_0 D}\left(S - \frac{4}{3}\mu u \frac{du}{dx}\right) = h_0 + \frac{D^2}{2}. \end{gathered} \tag{7.3}$$

The constants of integration are expressed here in terms of the initial values of the flow variables, distinguished by the subscript "0" and by the front velocity $D \equiv u_0$.

If we refer the system (7.3) to the final state (denoted by the subscript "1")

Eliminating $T_{\text{trans}} - T_{\text{rot}}$ and $D\rho/Dt$ yields the result

$$\mu' = (\gamma - 1)^2 \rho\varepsilon \frac{c_{\text{rot}}}{c_v} \tau_{\text{rot}}.$$

See Kohler [100].

If τ_{rot} is not much larger than τ_{gas}, the relaxation analysis above does not apply. The relation above then may be considered to indicate correct orders of magnitude. For either diatomic or polyatomic molecules,

$$\frac{\mu'}{\mu} \approx 0.16 \frac{\tau_{\text{rot}}}{\tau_{\text{gas}}}.$$

The quantity τ_{rot} is the relaxation time for a process in which T_{trans} is kept fixed. For a process in which $\varepsilon = c_{\text{trans}} T_{\text{trans}} + c_{\text{rot}} T_{\text{rot}}$ is kept fixed, the relaxation time is $c_{\text{trans}} \tau_{\text{rot}} / c_v$. The distinction between these two definitions of the relaxation time should be kept in mind.

we obtain the already well-known relations across a discontinuity, which we repeat here for convenience

$$\begin{gathered} \rho_1 u_1 = \rho_0 D, \\ p_1 + \rho_1 u_1^2 = p_0 + \rho_0 D^2, \\ h_1 + \frac{u_1^2}{2} = h_0 + \frac{D^2}{2}. \end{gathered} \tag{7.4}$$

It follows from these equations that the entropy jump across a shock wave $\Sigma_1 - \Sigma_0 = \Sigma(p_1, \rho_1) - \Sigma(p_0, \rho_0)$ is entirely independent of both the dissipative mechanisms involved, or in this case of the values of the coefficients of viscosity and thermal conductivity μ and κ. The latter determine only the internal structure of the wave front and its thickness δ. The thickness δ of the viscous shock front is proportional to the coefficients μ and κ, which in turn are proportional to the molecular mean free path l. In the limit $l \to 0$ the hydrodynamics of a real fluid becomes, in the continuous flow regions, the hydrodynamics of an ideal fluid. The shock front in the limit $l \to 0$ becomes a mathematical surface, since $\delta \sim l \to 0$. In this case, the gradients of all the flow variables across the front tend to infinity as $1/l$ but their jumps remain finite.

Specifying the coefficients of viscosity and thermal conductivity and also the thermodynamic relation $h(p, \rho)$ (in a monatomic gas $h = c_p T = \frac{5}{2} p/\rho$), we can numerically integrate the system (7.3) and (7.2) with the given boundary conditions. However, it is much more convenient to have an analytic solution, since it illustrates graphically all the relationships governing the phenomenon. Unfortunately, it is not possible in general to find an analytic solution to the system. The equations can be integrated analytically if we limit ourselves to weak waves and expand the solution in a series with respect to the small change in one of the flow variables. This method was used in §23 of Chapter I for estimating the front thickness (the complete solution is given in the book by Landau and Lifshitz [1]). An exact analytic solution for a wave of arbitrary strength can be found in one special case. This solution, first obtained by Becker [2] and subsequently investigated by Morduchow and Libby [3], describes all of the physical laws governing the structure of a shock front, and is both simple and graphic. Let us describe this solution in more detail.

Usually the transport coefficients in gases, that is, the values of the kinematic viscosity $\nu = \mu/\rho$ and of the thermal diffusivity $\chi = \kappa/c_p\rho$, are close to each other and to the diffusion coefficient $l\bar{v}/3$. Let us set the dimensionless group $Pr = \mu c_p/\kappa = \nu/\chi$, called the Prandtl number, equal to 3/4. In this case the expression in parentheses in the last of equations (7.3) becomes a total differential of the quantity $h + u^2/2$, and the equation becomes

$$\left(h+\frac{u^2}{2}\right)-\frac{4}{3}\frac{\mu}{\rho_0 D}\frac{d}{dx}\left(h+\frac{u^2}{2}\right)=h_0+\frac{D^2}{2}.$$

In writing the integral of this linear equation, it is evident that $h + u^2/2$ can be finite at $x = +\infty$ only if it is independent of x,

$$h+\frac{u^2}{2}=h_0+\frac{D^2}{2}^{*}. \tag{7.5}$$

Thus, for a Prandtl number equal to 3/4, (7.5) is satisfied not only behind the shock front (see (7.4)), but also at any intermediate point x.

Equation (7.5) gives a curve in the p, V plane along which the gas changes from the initial to the final state. Noting that for the monatomic gas under consideration $h = \frac{5}{2}pV$, and introducing the dimensionless velocity or specific volume

$$\eta=\frac{u}{D}=\frac{V}{V_0}=\frac{\rho_0}{\rho},$$

we find the equation of this curve is

$$\frac{p}{p_0}=\frac{1+\frac{1}{3}M^2(1-\eta^2)}{\eta}=\frac{4\eta_1-\eta^2}{(4\eta_1-1)\eta}. \tag{7.6}$$

Here η_1 refers to the final state behind the shock front

$$\eta_1=\frac{1}{4}+\frac{5}{4}\frac{p_0}{\rho_0 D^2}=\frac{1}{4}+\frac{3}{4}\frac{1}{M^2}, \tag{7.7}$$

and M is the Mach number $M = D/c_0$, where c_0 is the speed of sound in the initial state ($c_0^2 = \frac{5}{3}p_0 V_0$). In deriving (7.6) and (7.7) we have made use of the equations relating the variables on both sides of the shock front. The Hugoniot equation in terms of the variables p_1/p_0 and η_1 is

$$\frac{p_1}{p_0}=\frac{4-\eta_1}{4\eta_1-1}.$$

Figure 7.2 shows the Hugoniot curve and the curve along which the state of a particle changes through the wave (and also the characteristic straight line connecting the initial and the final states).

Using (7.6) and the first two equations of (7.3), let us write a differential equation describing the velocity and specific volume distributions through

* This equation is analogous to the Bernoulli equation in steady flow theory.

the shock front $\eta(x)$

$$\frac{5}{3}\frac{\mu}{\rho_0 D}\eta\frac{d\eta}{dx} = -(1-\eta)(\eta-\eta_1). \tag{7.8}$$

For simplicity, the coefficient of viscosity is assumed to be independent of temperature and equal to $\mu = \rho_0 l_0 \bar{v}_0/3$ (it is independent of density, since

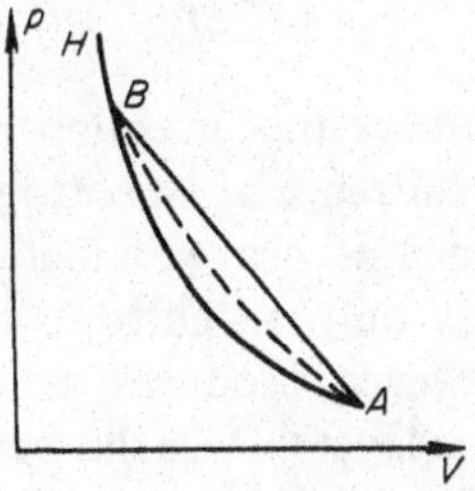

Fig. 7.2. Shock transition $A \to B$ on a p, V diagram. H is the Hugoniot curve. The point describing the state inside the wave front passes from A to B along the dashed line.

$\mu \sim \rho l$, and $l \sim 1/\rho$). The integral of (7.8) contains an additive constant consistent with the arbitrariness in the choice of the coordinate origin. Locating the origin at the point of inflection of the velocity profile (in the "center" of the wave) and using (7.7), we find for $\eta(x)$ the expression

$$\frac{1-\eta}{(\eta-\eta_1)^{\eta_1}} = \frac{1-\sqrt{\eta_1}}{(\sqrt{\eta_1}-\eta_1)^{\eta_1}}\exp\left[a\frac{M^2-1}{M}\frac{x}{l_0}\right] \qquad \left(a = \frac{27}{40}\sqrt{\frac{5\pi}{6}} = 1.1\right). \tag{7.9}$$

Knowing the velocity profile $u = D\eta$, it is easy to determine the distributions of the other variables. Thus, for the temperature we have from (7.5) $T/T_0 = 1 + \frac{1}{3}M^2(1-\eta^2)$; the pressure is defined in terms of η by (7.6) and the entropy is

$$\Sigma - \Sigma_0 = c_p \ln\frac{T}{T_0} - R\ln\frac{p}{p_0} \qquad \left(R = \frac{p}{\rho T}\right).$$

It is evident from (7.9) that as $x \to -\infty$, $\eta \to 1$ and as $x \to +\infty$, $\eta \to \eta_1$, with the initial and final values being approached asymptotically in an exponential manner. All the flow variables velocity, density, pressure, and temperature change monotonically across the wave from their initial to the final values, which are approached asymptotically as $x \to \mp\infty$*. The entropy, on the other hand, does not change monotonically, and has a maximum within the wave (this has already been shown in §23, Chapter I). We can easily satisfy

* The inflection points for the various variables in the wave front do not coincide.

ourselves that this is so by rewriting the entropy equation (the third of equations (7.1)) with the aid of the second law, the "Bernoulli equation" (7.5), and the second of equations (7.1). We find

$$\rho u T \frac{d\Sigma}{dx} = \rho u\left(\frac{dh}{dx} - V\frac{dp}{dx}\right) = \rho u\left(-u\frac{du}{dx} - V\frac{d}{dx}\frac{4}{3}\mu\frac{du}{dx} + V\rho u\frac{du}{dx}\right)$$

$$= -u\frac{d}{dx}\frac{4}{3}\mu\frac{du}{dx} = -\frac{4}{3}\mu u\frac{d^2u}{dx^2}.$$

It is evident that the entropy has an extremum at the inflection point in the velocity, that is, at the wave "center". The existence of an entropy maximum in the wave is connected with the presence of heat conduction. One of the dissipative processes, viscosity, produces only an entropy increase, proportional to $(du/dx)^2$. Heat conduction, however, produces an irreversible transfer of heat from the hotter to colder gas layers. Here the increase in entropy of fluid particles through heat conduction in the colder layers (where $dS/dx \sim -d^2T/dx^2 < 0$) is positive, and in the hotter layers (where $dS/dx \sim -d^2T/dx^2 > 0$) is negative.

The entropy decrease in the more heated layers does not in any way contradict the second law. The entropy of the gas as a whole or of an individual particle increases across the whole shock discontinuity as a result of the process of shock compression. However, an individual layer of gas passing through the wave is no longer an isolated system. Its entropy increases at the beginning, when it is supplied heat through heat conduction and the work of the viscous forces, and then decreases when the heat loss due to heat conduction in the direction of the colder gas layers behind it exceeds the heat supplied by the work of the viscous forces.

The front thickness, as in §23 of Chapter I, is given by

$$\delta = \frac{D - u_1}{(du/dx)_{\max}}.$$

It is evident from (7.9) that the order of magnitude of the front thickness is

$$\delta \sim l_0 \frac{M}{M^2 - 1}.$$

In a weak shock wave when $M - 1 \ll 1$, $\delta \sim l_0/(M - 1)$, in agreement with the results presented in §23 of Chapter I. In this case, the front thickness can be equal to many molecular free paths. In the case when $M = 2$, shown in Fig. 7.3, the front thickness is approximately equal to three molecular free paths l_0. In the limiting case of a strong wave, when $M \to \infty$, $\delta \sim l_0/M \to 0$. The statement that the front thickness vanishes as the wave strength increases should, of course, not be taken literally*. The fact is simply that when the front thickness becomes of the order of a mean free path, the hydrodynamic theory loses its meaning, since it is based on the assumption that the gradients are small, that the mean free path is small in comparison with the distance over which appreciable changes in the flow variables take place. Hence the theory

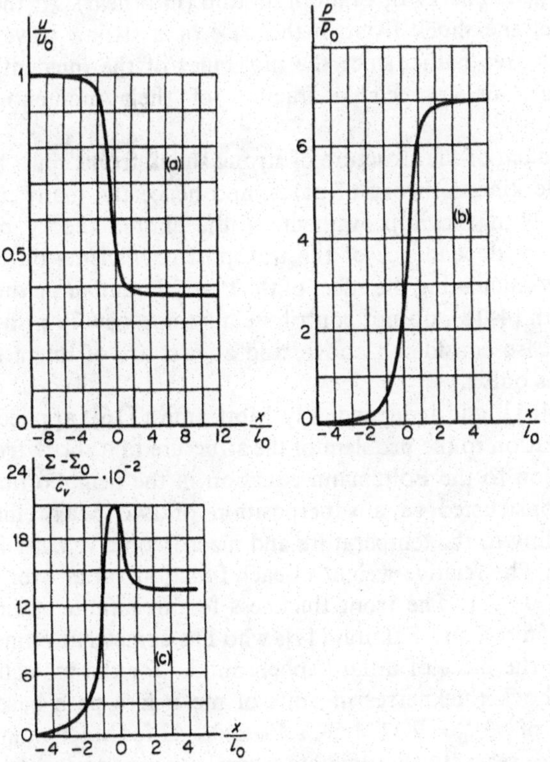

Fig. 7.3. Distributions of (a) velocity, (b) pressure, and (c) entropy through a viscous shock front with Mach number $M = 2$ in a gas with a specific heat ratio $\gamma = 7/5$ and a temperature-independent viscosity coefficient. The abscissa is in units of molecular mean free path in the undisturbed gas (the graphs were taken from [3]).

* *Editors' note*. The result that $\delta \sim l_0/M \to 0$ as $M \to \infty$ is somewhat artificial. It is based on the assumption that μ is constant through the wave. Since $\mu \sim \rho \bar{v} l$ and $\rho_0/\rho_1 \sim 1$, $\bar{v}l$ remains of the same order through the wave. But $\bar{v}_1 \sim M\bar{v}_0$, and $l_1 \sim l_0/M \to 0$ as $M \to \infty$ if l_0 is kept fixed. But $\delta \sim l_1$ in the same limit, and δ remains finite if l_1 does.

The conclusion $\delta \sim l_1$, that the shock front thickness is of the order of a few mean free paths in the flow *behind* the shock, is a general one for strong shocks. Even though the Navier–Stokes equations are invalid for strong shocks, they nevertheless give not only correct order-of-magnitude results, but also quite faithful numerical results for the macroscopic flow variables except in the upstream part of the shock front.

is simply inapplicable for sufficiently strong waves. It is evident physically that the thickness of the shock front for a wave of any strength cannot become smaller than the mean free path, since the gas molecules flowing into the discontinuity must make at least several collisions in order to scatter the directed momentum and to convert the kinetic energy of the directed motion into the kinetic energy of random motion (into heat). At the same time, the thickness of the shock front in the case of a strong wave cannot include many mean free paths, since the molecules of the incident stream lose, on the average, an appreciable fraction of their momentum during each collision.

The problem of the structure of strong shock fronts must be treated on the basis of the kinetic theory of gases, and hence the many numerical studies concerned with the improvement of the simple theory presented above, by taking the dependence of the transport coefficients on temperature into account, by calculating the effect of the Prandtl number on the front structure, and so forth [4–13], do not contribute anything new in principle beyond the particular case considered above, and at best are of interest for the case of weak waves only*.

Tamm [101], and independently Mott-Smith [16], applied the Boltzmann kinetic equation to the problem of the structure of a shock front. An approximate solution to the Boltzmann equation in the neighborhood of the shock front is constructed as a superposition of two Maxwellian distributions corresponding to the temperature and macroscopic velocity in the initial and final states. The relative weight of each function varies over the width of the wave from 0 to 1. The front thickness for an infinite strength shock wave tends to a finite limit. Sakurai [17], who has somewhat refined Mott-Smith's method on the basis of a hard-sphere model for the molecular interactions, obtained shock thicknesses in units of mean free path based on the initial conditions, of† $\delta/l_0 = 2.11, 1.68, 1.46$, and 1.42 for Mach numbers of $M = 2.5$, 4, 10, and ∞, respectively. We note several other papers which have developed

* An attempt to refine the hydrodynamic approximation by taking into account second derivatives in the expressions for the transfer terms (the so-called Burnett approximation), undertaken by Zoller [14], somewhat improves the results for weak waves and, essentially, only points out the limits of applicability of the hydrodynamic theory. For a wave strength $p_1/p_0 = 1.5$, the thickness of the front, according to Zoller, is equal to 17 mean free paths, and for $p_1/p_0 = 4$ is equal to 6 mean free paths. The front thickness of weak shock waves in monatomic gases was measured by Greene, Cowan, and Hornig [15] using a method based on the reflection of light (see §5 of Chapter IV). The thickness was found equal to 30, 19, and 13 mean free paths for Mach numbers $M = 1.1$, 1.5, and 2.5, respectively. Zoller's calculations are in not too bad agreement with these results. See also [56].

† The front thickness δ is defined in the following manner. If f_α and f_β are the molecular distribution functions in the initial and final states, then the distribution function at an intermediate point x in the wave is, according to the theory, $f = \nu(-x)f_\alpha + \nu(x)f_\beta$, where $\nu(x) = \frac{1}{2}(1 + \tanh(2x/\delta))$.

Mott-Smith's method and treated the shock front on the basis of the Boltzmann equation [52–55].

§3. The role of viscosity and heat conduction in the formation of a shock front

Despite the fact that the values of the transport coefficients—kinematic viscosity and thermal diffusivity, or the corresponding dissipative terms in the energy equation—are close to each other, the contribution of each of these dissipative processes to the formation of a shock are far from equal. Physically, it is clear that the principal role in the mechanism of shock compression is played by viscosity rather than heat conduction, since it is the viscous mechanism which causes the scattering of the directed momentum of the incident gas and the conversion of the kinetic energy of the directed molecular motion into the kinetic energy of random motion, i.e., the conversion of mechanical energy into heat. Heat conduction, on the other hand, only indirectly affects the conversion of mechanical energy as a result of the redistribution of pressure. In order to satisfy ourselves of this, it is useful to consider the problem of a one-dimensional steady flow of a gas through a shock front under the assumption that viscosity is completely absent and that the dissipation is caused by heat conduction only. An investigation of this problem (first carried out by Rayleigh [18]) is of considerable interest, since it illustrates the features of the structure of a shock front in the presence of other mechanisms of heat exchange, e.g., radiative heat transfer or electron heat conduction (in a plasma).

By neglecting viscosity, the first integrals in the hydrodynamic equations for one-dimensional steady flow (7.3) take the form

$$\begin{gathered} \rho u = \rho_0 D, \\ p + \rho u^2 = p_0 + \rho_0 D^2, \\ h + \frac{u^2}{2} + \frac{S}{\rho_0 D} = h_0 + \frac{D^2}{2}. \end{gathered} \tag{7.10}$$

It follows from the first two equations of (7.10) that in the process of shock compression, in the absence of viscosity, the state of a gas particle must change continuously along a straight line in the pressure-specific volume diagram

$$p = p_0 + \rho_0 D^2(1 - \eta), \qquad \eta = \frac{V}{V_0}. \tag{7.11}$$

This important property of inviscid gas flow is illustrated in Fig. 7.4, which shows a Hugoniot curve and a straight line connecting the initial and final states of the gas.

We shall attempt to solve the system of equations (7.10) where, as before, we shall eliminate all variables except the dimensionless velocity or relative specific volume η. For generality, we shall not restrict our problem to a

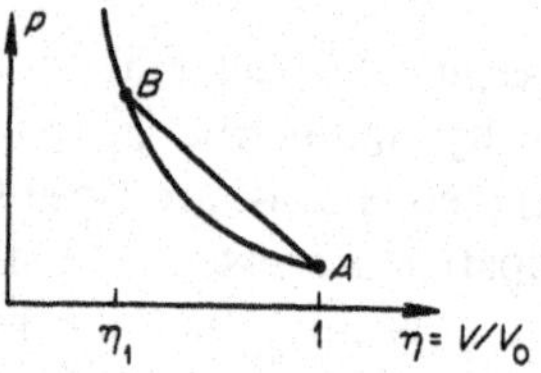

Fig. 7.4. Straight line shock transition for an inviscid gas.

monatomic gas and shall retain an arbitrary constant specific heat ratio. Noting the equation of state

$$p = \frac{\mathscr{R}}{\mu_0}\rho T = R\rho T, \qquad R = \frac{\mathscr{R}}{\mu_0} \tag{7.12}$$

(μ_0 is the molecular weight), and the thermodynamic relation $h = [\gamma/(\gamma-1)]p/\rho$, we express by means of the third equation of (7.10) and (7.11) the nonhydrodynamic energy flux and temperature in terms of η,

$$\frac{T}{T_0} = 1 + \gamma M^2(1-\eta)\left(\eta - \frac{1}{\gamma M^2}\right), \tag{7.13}$$

$$S = -\frac{\rho_0 D^3}{2}\frac{\gamma+1}{\gamma-1}(1-\eta)(\eta-\eta_1). \tag{7.14}$$

Here, as before, the quantity

$$\eta_1 = \frac{\gamma-1}{\gamma+1} + \frac{2}{\gamma+1}\frac{1}{M^2}$$

is the dimensionless velocity in the final state and $M = D/c_0$ is the Mach number.

The function $T(\eta)$ passes through a maximum at the point

$$\eta = \eta_{max} = \frac{1}{2} + \frac{1}{2\gamma M^2}.$$

Two cases can be encountered in considering shock waves of different strengths. If the shock is sufficiently weak, then $\eta_1 > \eta_{max}$. Indeed, for a Mach number close to unity ($M - 1 \ll 1$), we find $\eta_1 \approx 1 - [4/(\gamma+1)](M-1)$, so that it is also close to unity, while $\eta_{max} \approx (\gamma+1)/2\gamma < 1$. In this case, a monotonic compression of the gas from the initial to the final volume (from $\eta = 1$ to $\eta = \eta_1$) results in a monotonic increase in the temperature from the initial value T_0 to the final value T_1, given (under the assumed conditions) by

$$\frac{T_1}{T_0} = 1 + \frac{2\gamma(\gamma - 1)}{(\gamma + 1)^2}(M^2 - 1)\left(1 + \frac{1}{\gamma M^2}\right).$$

The curves of $T(\eta)$ and $S(\eta)$ in this case have the shape illustrated in Fig. 7.5.

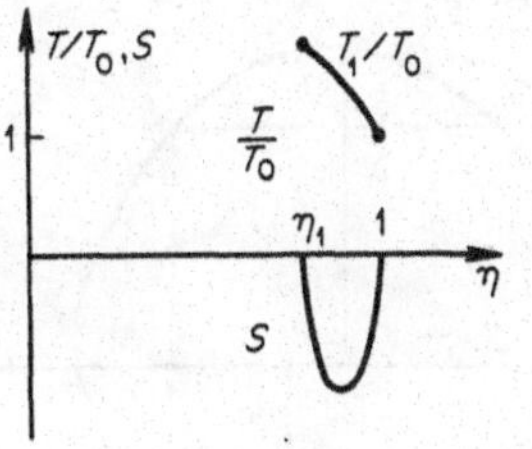

Fig. 7.5. T,η and S,η diagrams for the case when a continuous shock transition with heat conduction only is possible, without considering viscosity.

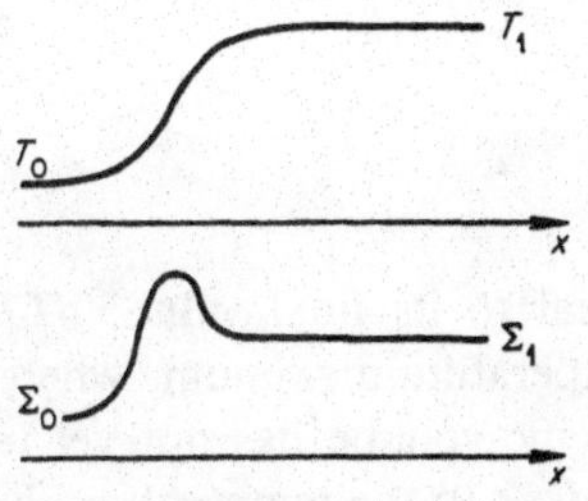

Fig. 7.6. Temperature and entropy distributions in a shock wave with heat conduction only (without viscosity) in the case when a continuous transition is possible.

By eliminating η from (7.13) and (7.14) and substituting (7.2) for S, we obtain a differential equation of the type $dT/dx = f(T)$ which has a continuous solution. The temperature and entropy distributions in such a wave are represented schematically in Fig. 7.6. They are quite similar to the distributions found in the preceding section. It is evident from the entropy equation (7.1) with $\mu = 0$, that the entropy has a maximum at the point where $d(\kappa\, dT/dx)/dx = 0$, or in the case of $\kappa = const$ where the temperature in the wave $T(x)$ has an inflection point, i.e., where $d^2T/dx^2 = 0$. Therefore, a weak shock wave with a continuous distribution of flow variables across the front can also exist in the absence of viscosity, with only heat conduction present.

Let us now consider a sufficiently strong shock wave. In this case the specific volume for which the temperature assumes its maximum value lies between the initial and final values: $\eta_1 < \eta_{max} < 1$. Actually, for $M \gg 1$, $\eta_{max} \approx 1/2$, and $\eta_1 = (\gamma - 1)/(\gamma + 1) < 1/4$, since the specific heat ratio cannot exceed 5/3. Thus, for a continuous monotonic compression of the gas from its initial to its final volume, the temperature across the wave front must necessarily pass through a maximum. Plots of the functions $T(\eta)$ and $S(\eta)$ for this case are given in Fig. 7.7. Let us investigate the possibility of the existence of a continuous solution of (7.13) and (7.14) in this case. It can be seen from (7.14) and Fig. 7.7 that the heat flux S caused by heat conduction does not change sign over the entire range in which the relative volume changes from $\eta = 1$ to $\eta = \eta_1$ and it is opposite to the direction of gas flow, i.e., $S < 0$. In accordance with the definition of the flux $S = -\kappa\, dT/dx$, the temperature can only increase as the volume changes from its

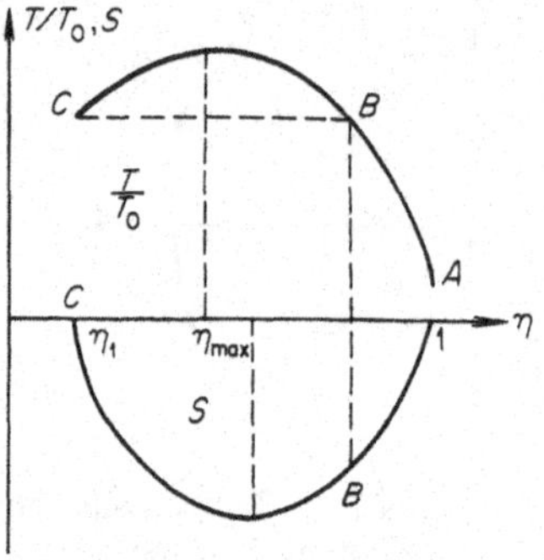

Fig. 7.7. T,η and S,η diagrams for the case of an isothermal jump, taking into account heat conduction alone but without taking viscosity into account.

initial to its final value: $dT/dx > 0$. Consequently, the region behind the temperature maximum, where $dT/d\eta > 0$, is not realized. In this region the specific volume has not yet reached its final value and must decrease, with $d\eta/dx < 0$; the temperature, however, also decreases with decreasing volume, with

$$\frac{dT}{dx} = \frac{dT}{d\eta}\frac{d\eta}{dx} < 0,$$

and the heat flux would have to be directed in the opposite direction ($S > 0$), in contradiction to (7.14).

Thus, a continuous distribution of temperature and density with respect to position is impossible in the case of a strong wave in which only heat con-

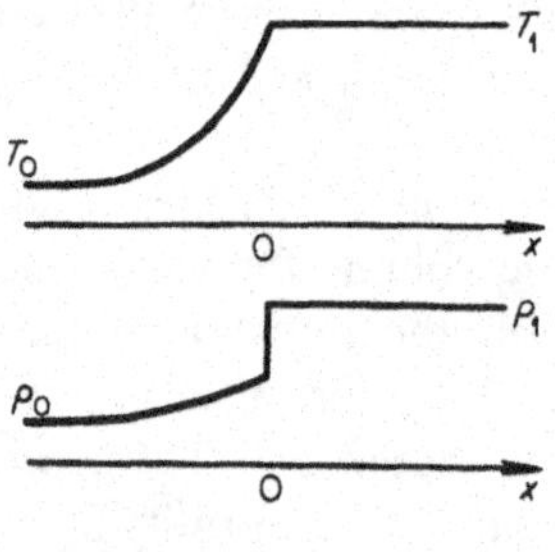

Fig. 7.8. Temperature and density profiles in a shock wave with an isothermal jump.

duction is taken into account. Starting from the initial state, the final state can be reached without passing through the temperature drop region caused by the increased compression only if a discontinuity is allowed in the solution.

In particular, there is a continuous change from the initial state (point A on Fig. 7.7) to point B, and then there is a jump to the final point C. The formation of a density jump indicates that viscous forces must be present, and that a strong discontinuity can disappear only through the action of viscosity but not of heat conduction. The temperature in the jump remains constant, and only its derivative, i.e., the flux, changes. The temperature and density profiles in such a wave, which is called an "isothermal" shock, are shown in Fig. 7.8*.

It is easy to find the largest strength for which a continuous solution in the absence of viscosity is still possible. This strength corresponds to the case in which the maximum of the function $T(\eta)$ coincides with the final state, when $\eta_{max} = \eta_1$. In this case, the Mach number and the pressure ratio across the front are given by

$$M' = \left(\frac{3\gamma - 1}{\gamma(3-\gamma)}\right)^{1/2}, \qquad \frac{p_1'}{p_0} = \frac{\gamma + 1}{3 - \gamma}.$$

For example, for $\gamma = 5/3$, $M' = 1.35$ and $p_1'/p_0 = 2$, while for $\gamma = 7/5$, $M' = 1.2$ and $p_1'/p_0 = 1.5$.

If we consider the other limiting case, when there is only viscosity and heat conduction is absent, we obtain a continuous solution for the flow variables across the shock wave which does not differ basically from the solution of the preceding section, with the single exception that the entropy in this case also increases monotonically (see the third equation of (7.1) without the term dS/dx). The behavior of the entropy in both limiting cases can be clarified by considering a p, V or a p, η diagram (Fig. 7.9). In the absence of viscosity

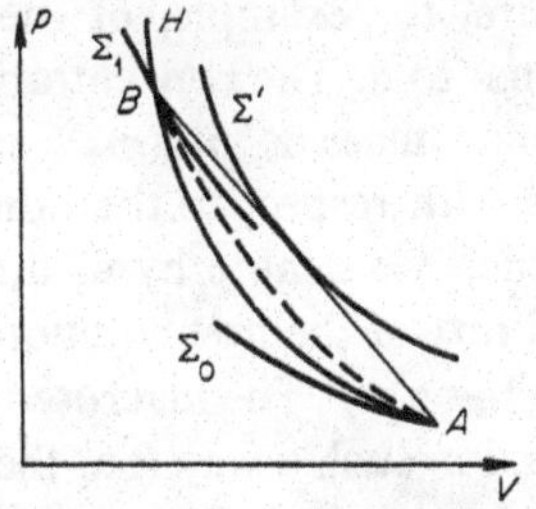

Fig. 7.9. p, V diagram for a shock wave, taking viscosity into account. H is the Hugoniot curve; Σ_0, Σ_1, and Σ' are isentropes; the transition from the initial to the final state takes place along the dashed curve.

the state in the wave changes along the straight line AB and the entropy (as is evident from a comparison of the Hugoniot curve with the isentropes) first

* We note that the "isothermal" character of the shock, i.e., the continuity of temperature across the shock front, is a consequence of the assumption that the heat flux is proportional to the temperature gradient. In the third part of this chapter, in considering radiant heat exchange in a shock front, we shall see that without this assumption the temperature will also have a discontinuity.

increases, reaching a maximum at the point of tangency of the straight line and the isentrope Σ', and then decreases. In the absence of heat conduction the state changes along the dashed curve which passes below the straight line AB (the equation of this curve is $p = p_0 + \rho_0 D^2(1 - \eta) + \frac{4}{3}\mu\, du/dx$, where $du/dx < 0$), and it is nowhere tangent to the isentropes. The situation here is completely analogous to that which exists in the weak waves considered in §23 of Chapter I.

§4. Diffusion in a binary gas mixture

The presence of gradients in the thermodynamic quantities of a gas mixture gives rise to diffusional fluxes in the components of the mixture, which results in a redistribution of their concentrations. In general, diffusion tends to equalize the spatial concentrations of the components. However, the existence of pressure and temperature gradients or of an external force field such as gravity, centrifugal forces in rotating mixtures, or the presence of an acceleration results in component separation in an initially homogeneous mixture. In particular, such a situation arises in a shock wave propagating through a gas mixture. The concentrations of the components behind and ahead of the front are uniform and constant in space. In the region of the front, where gradients are present, these concentrations change. As with viscosity and heat conduction, diffusion represents an irreversible molecular mass transfer of a specific component (viscosity is responsible for momentum transfer and heat conduction for internal energy transfer) and is one of the sources for the dissipation of mechanical energy.

The diffusional flux is defined as follows. Let the mass concentration of one of the components in a binary gas mixture, for example, of the light component whose molecular mass is m_1, be equal to α. The concentration of the second, heavy component whose molecular mass is $m_2 (m_2 > m_1)$ is $1 - \alpha$*. As a result of the diffusion of one gas with respect to the other the gases have different macroscopic velocities, which we denote by $\mathbf{u}_1$ and $\mathbf{u}_2$. If ρ is the density of the mixture, then the total flux of the first component is $\rho\alpha\mathbf{u}_1$, and the flux of the second component is $\rho(1 - \alpha)\mathbf{u}_2$. The macroscopic or hydrodynamic velocity of the mixture $\mathbf{u}$ is defined in such a way that the total mass flux of the gas is equal to $\rho\mathbf{u}$ ($\mathbf{u}$ is the momentum per unit mass). Thus, $\rho\mathbf{u} = \rho\alpha\mathbf{u}_1 + \rho(1 - \alpha)\mathbf{u}_2$ or $\mathbf{u} = \alpha\mathbf{u}_1 + (1 - \alpha)\mathbf{u}_2$. Within the framework of the

* The mass concentration α is equal to the mass of the first, light component, per unit mass of the mixture. If the number of molecules per unit mass of mixture is N_1 and N_2 ($N_1 + N_2 = N$), then $\alpha = N_1 m_1$ and $1 - \alpha = N_2 m_2$. The molar concentrations are

$$\frac{N_1}{N} = \frac{\alpha}{Nm_1}, \qquad \frac{N_2}{N} = \frac{1-\alpha}{Nm_2}.$$

hydrodynamics of an ideal fluid (without diffusion), the velocities of both components of the mixture are the same and equal to $\mathbf{u}$. The fluxes of the components are equal to $\rho\alpha\mathbf{u}$ and $\rho(1-\alpha)\mathbf{u}$.

In the next approximation there appear in the hydrodynamic theory viscosity, heat conduction, and diffusion (in the mixture). The diffusional flux $\mathbf{i}$ refers to the difference between the total and the hydrodynamic fluxes of one component, say the first component, $\mathbf{i} = \rho\alpha\mathbf{u}_1 - \rho\alpha\mathbf{u} = \rho\alpha(\mathbf{u}_1 - \mathbf{u})$. The total flux of the first component is equal to the sum of the hydrodynamic and diffusional fluxes $\rho\alpha\mathbf{u} + \mathbf{i}$. The total flux of the second component is, obviously, $\rho(1-\alpha)\mathbf{u}_2 = \rho(1-\alpha)\mathbf{u} + \rho(1-\alpha)(\mathbf{u}_2 - \mathbf{u}) = \rho(1-\alpha)\mathbf{u} - \mathbf{i}$. The diffusional fluxes of the two components in a binary mixture are of the same magnitude but of opposite direction.

As pointed out above, diffusion arises as a result of the presence of concentration, pressure, and temperature gradients in a gas*. In the one-dimensional case the gradients are simply equal to derivatives with respect to x, and the vector $\mathbf{i}$ has only an x component which will be denoted simply by i. The diffusional flux is given by (see [1])

$$i = -\rho D\left(\frac{d\alpha}{dx} + \frac{k_p}{p}\frac{dp}{dx} + \frac{k_T}{T}\frac{dT}{dx}\right). \tag{7.15}$$

Here D is the diffusion coefficient, k_pD is the pressure diffusion coefficient, and k_TD is the thermal diffusion coefficient. The dimensionless quantity k_p is determined only by the thermodynamic properties of the mixture and is given by (see [1]†)

$$k_p = (m_2 - m_1)\alpha(1-\alpha)\left(\frac{1-\alpha}{m^2} + \frac{\alpha}{m_1}\right)^{\ddagger}. \tag{7.16}$$

For $m_2 > m_1$, $k_p > 0$ and the pressure diffusional flux of the light component is in the direction of decreasing concentration. The flux caused by the concentration gradient is also in the direction of decreasing concentration (for either component). The thermal diffusional flux of the light component for

* The state of a binary mixture is characterized by three thermodynamic variables: the concentration and any two of the usual variables such as temperature, pressure, and density. In studying diffusion, it is convenient to choose pressure and temperature as the independent variables.

† In the absence of viscous momentum transfer (see below).

‡ The quantity k_p is most simply derived by considering an equilibrium binary mixture in a gravity field at constant temperature. In equilibrium the molecular number densities n_1 and n_2, from the Boltzmann equation, are expressed in proportional form as $n \sim \exp(-m_1gx/kT)$, and $n_2 \sim \exp(-m_2gx/kT)$, where g is the acceleration of gravity and x is the altitude. Since the equilibrium diffusional flux is equal to zero, $d\alpha/dx + (k_p/p)\,dp/dx = 0$. Using the relationship between the concentration α and the particle number densities n_1 and n_2 and noting that $p = (n_1 + n_2)kT$, we find the above equation for k_p.

most mixtures is in the direction of increasing temperature (for $m_2 > m_1$, $k_T < 0$).

In contrast to k_p, the thermal diffusion ratio k_T depends not only on the component concentrations (for $\alpha = 0$ or 1, $k_T = 0$) and the molecular masses, but also on the law governing the molecular interactions. The quantity k_p is determined purely by the thermodynamic properties of the gas, since thermodynamic equilibrium is possible even in the presence of a pressure gradient in an external force field. If a temperature gradient is also present, then the gas is no longer in a state of equilibrium. If only repulsive forces varying as $1/r^n$ are acting between the molecules, then for $n > 5$, which is usually the case, $k_T < 0$: the light gas will diffuse in the direction of increasing temperature. For $n < 5$, which is rarely encountered, the light gas diffuses in the direction of decreasing temperature (the case of $n < 5$ includes the Coulomb law for the interaction of charged particles, for which $n = 2$). When $n = 5$ thermal diffusion is absent and $k_T = 0$. Usually, when the relative gradients $\nabla p/p$ and $\nabla T/T$ are comparable, the importance of thermal diffusion is small in comparison with pressure diffusion. For further details on thermal diffusion see [19]. With the diffusional flux is connected an additional irreversible energy flux $\mathbf{q}$, which is proportional to the diffusional flux $\mathbf{i}$ (see [1]).

Zhdanov, Kagan, and Sazykin [19a] have introduced an important correction to the classical diffusion concepts presented above. These authors derived an expression for the diffusional flux from the kinetic equation, using the Grad "thirteen moment" method. This method of approximation has a number of advantages over the Chapman–Enskog method used to obtain (7.15), whenever it is necessary to take into account higher approximations in the expansion of the distribution function. It is found that equation (7.15) for the diffusional flux is valid only in the absence of viscous momentum transfer in the gas. Under conditions when viscous momentum transfer is present (that is, when there is a velocity gradient) equation (7.15) must also include terms proportional to the viscous forces. In spite of the fact that these forces are determined by second-order derivatives of macroscopic quantities (such as velocity), they can be of the same order as terms proportional to first-order derivatives such as the term with pressure gradient. For example, in the case of a purely viscous steady flow without acceleration, the pressure gradient is simply balanced by the viscous forces. In the case of unsteady flow, the inclusion of viscous forces in the expression for the diffusional flux actually introduces into this expression terms proportional to the accelerations of the gas.

In the case of purely viscous flow, the replacement of the viscous force by the pressure gradient which balances it results in a change in the pressure diffusion constant k_p in comparison with its purely thermodynamic value (7.16). The pressure diffusion constant is no longer a thermodynamic quantity in a viscous flow; it depends on the character of the molecular interaction. The pressure diffusion constant under some conditions can become negative (in the case when the molecular weights of the components differ very slightly from each other and when the molecular cross sections are appreciably different). When viscous momentum transfer is taken into account, the thermal diffusion ratio k_T also changes.

§5. Diffusion in a shock wave propagating through a binary mixture

Let us consider what happens when a shock wave is propagated through a binary gas mixture. Large gradients in the thermodynamic quantities are present in the shock front and, as a result, conditions for diffusion are favorable. Physically, it is clear that the light component will concentrate in the shock front. Indeed, in the heated gas behind the shock front the molecules of the light component have a higher thermal velocity than the molecules of the heavy component ($\bar{v} \sim (T/m)^{1/2}$). Therefore, the light molecules "pull ahead" and leave the heavy molecules slightly behind (in the laboratory coordinate system, where the original mixture is at rest).

Let the heavy gas have a small admixture of the light gas. Then the density distributions of both the heavy and light gases (ρ_2 and ρ_1) in a strong shock wave have the form shown in Fig. 7.10. This figure also shows the concentration distribution of the light component $\alpha = \rho_1/(\rho_2 + \rho_1)$. The thickness of the

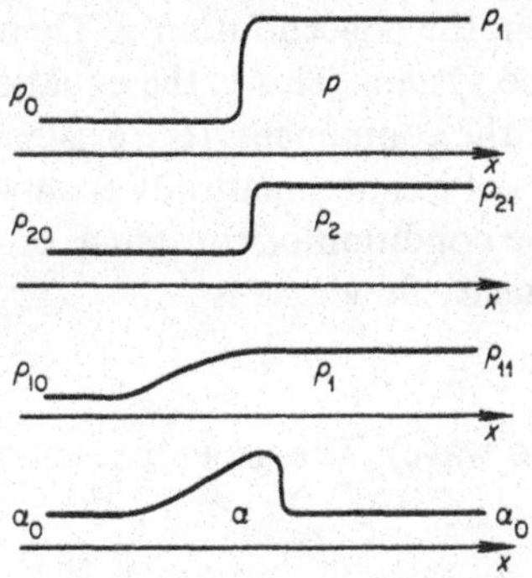

Fig. 7.10. Pressure and density distributions of the heavy (ρ_2) and light (ρ_1) components and concentration of the light component (α) in a shock wave propagating through a binary gas mixture.

region containing the higher concentration of the light component is of the order of $\Delta x \sim D/u_0$, where D is the diffusion coefficient and u_0 here denotes the shock velocity*. The diffusion coefficient D is of the order of $l\bar{v}_1$, where

* This follows from the condition that the total flux of the light component is steady in a coordinate system moving with the front. Approximately $\rho_1 u_0 = D\, d\rho_1/dx$, from which $\rho_1 = \rho_{11} \exp(-u_0|x|/D)$. Here we have used an approximate boundary condition according to which we can assume that the density of the light component at the point $x = 0$, where the viscous shock front is located, is equal to its final value ρ_{11}.

$\bar{v}_1$ is the thermal speed of the light gas heated by the shock wave and l is the molecular mean free path. The velocity of the front u_0 is of the order of the thermal speed of the heated heavy gas, $u_0 \sim \bar{v}_2$. But $\bar{v}_1/\bar{v}_2 \approx (m_2/m_1)^{1/2}$, so that $\Delta x \approx (m_2/m_1)^{1/2} l$. The thickness of the viscous shock front is of the order of l. Consequently, the thickness of the region in which the light component is concentrated is greater by a factor of $(m_2/m_1)^{1/2}$ than the thickness of the shock front. The components are most sharply separated when the particle masses are appreciably different ($m_2/m_1 \gg 1$).

This effect would be most pronounced in the case of a plasma, owing to the very large difference between the electron and ion masses. In a plasma, however, an important role is played by the electrostatic interaction between the electrons and ions, which strongly limits the diffusion process (for a discussion of this see §13).

Together with viscosity and heat conduction, diffusion affects the structure of a shock front. To describe this structure we must set up the equations for the planar steady flow case, in a manner similar to that of §2 for treating the viscous shock front. The equations for the conservation of mass and momentum and the first two equations of (7.3) remain unchanged (μ is now understood to represent the coefficient of viscosity of the mixture). To the equation of conservation of energy (the last equation of (7.3)), we must add the molecular heat flux resulting from diffusion and replace the molecular flux due to heat conduction S by the sum $S + q$. The system of equations will now include the diffusional flux i (to which the heat flux q is proportional), and the system will contain a new unknown function, the concentration α. Therefore an additional equation must be added to the system. This is the equation of continuity (conservation of mass) for one of the components (the existence of an equation of continuity for the entire mass of gas automatically ensures the conservation of the second component). The condition of constancy of mass flux of the light component in the planar steady flow case is

$$\rho\alpha u + i = const = \rho_0 \alpha_0 u_0$$

(the diffusional flux disappears ahead of the wave). The general equation of continuity for one of the components [1] is

$$\frac{\partial \rho\alpha}{\partial t} + \nabla \cdot (\rho\alpha \mathbf{u} + \mathbf{i}) = 0.$$

It is evident then that behind the wave, where the diffusional flux also vanishes, the concentration is equal to its initial value $\alpha_1 = \alpha_0$ (since $\rho_1 u_1 = \rho_0 u_0$).

The system of equations for one-dimensional steady flow in a binary mixture can be solved, in principle, in the same manner as for a single-component gas (see §2). The solution will yield a distribution of all quantities in the wave front. This problem was considered by D'yakov [20] for the case of a weak shock, when it is possible to expand all quantities in powers of a small parameter (see §23, Chapter I)*. As was shown in §§18 and 23 of Chapter I, if we regard the pressure change in a weak shock $\Delta p = p_1 - p_0$ as a first-order quantity, then the volume and temperature changes will also be first-order quantities. The total entropy change in the transition from the initial to the final state $\Sigma_1 - \Sigma_0$ is a third-order quantity and the entropy change inside the wave front, let us say, $\Sigma_{max} - \Sigma_0$ is a second-order quantity. The thickness of the shock wave front is of the order of $\Delta x \approx lp_0/\Delta p$, where l is the molecular mean free path. From the conservation equation for one of the components, which can be rewritten as

$$\alpha - \alpha_0 = -\frac{i}{\rho_0 u_0},$$

and from the expression for the diffusional flux, it is evident that the changes in concentration in the wave $\Delta\alpha$ and the flux i are second-order quantities, with

$$\alpha - \alpha_0 \sim i \sim \frac{dp}{dx} \sim \frac{\Delta p}{\Delta x} \sim (\Delta p)^2.$$

Consequently, the term containing the concentration gradient in the expression for the diffusional flux can be neglected ($d\alpha/dx \sim \Delta\alpha/\Delta x \sim (\Delta p)^3$, while $dp/dx \sim (\Delta p)^2$).

D'yakov [20] obtained an analytic solution for the concentration distribution in the front of a weak shock wave. This solution will not be given here (the distribution has the form shown in Fig. 7.11), but we shall estimate the

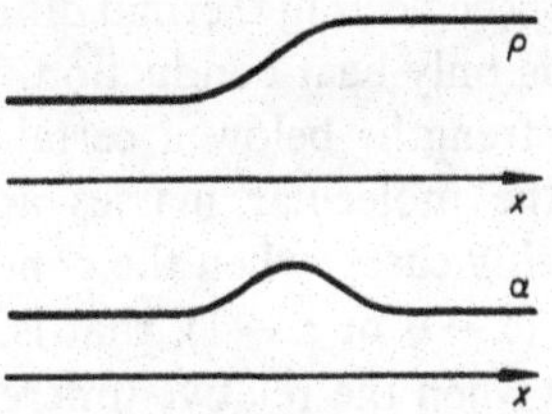

Fig. 7.11. Density and concentration distributions in a weak shock propagating through a binary gas mixture.

order of magnitude of the change in concentration. If we neglect thermal diffusion, which normally plays a less important role than pressure diffusion (since the value of k_T is usually lower than that of k_p), then we may write

$$\Delta\alpha = \alpha - \alpha_0 \sim \frac{|i|}{\rho_0 u_0} \sim \frac{D}{u_0}\frac{k_p}{p}\frac{\Delta p}{\Delta x}.$$

* For another treatment of this problem (including a numerical integration for shocks of arbitrary strength, *eds.*) see also the article by Sherman [21].

The diffusion coefficient $D \sim l\bar{v}$, where the thermal speed of the molecule $\bar{v}$ is of the order of the speed of sound, that is, of the order of u_0. Noting that $\Delta x \sim (p/\Delta p)l$, we find $\Delta\alpha \sim k_p(\Delta p/p)^2$. The excess amount of the light component collected by the shock wave (per unit area of front surface) is of the order of

$$M = \rho \int_{-\infty}^{\infty} (\alpha - \alpha_0)\, dx \sim \rho\, \Delta\alpha \cdot \Delta x \sim \rho k_p \frac{\Delta p}{p} l.$$

In a sufficiently strong shock, where $\Delta p \sim p$, we have $\Delta\alpha \sim k_p$, $M \sim \rho k_p l$. If the difference between the molecular masses is comparatively large ($k_p \sim (m_2 - m_1)$), then the change in concentration in a strong wave is of the order of the concentration itself and the excess mass of one component is of the order of the mass of the component itself in a layer of thickness of a molecular mean free path.

We have noted above that diffusion, like viscosity and heat conduction, results in the dissipation of mechanical energy and in an increase in the entropy of the gas (for a discussion of this point see [1])*. We know that with dissipative processes excluded, in the framework of the hydrodynamics of an ideal fluid a shock wave is represented as a mathematical discontinuity. The discontinuity vanishes and becomes a layer of finite thickness with a continuous distribution of flow variables only if dissipative processes are considered. Here heat conduction by itself can ensure a continuous transition in the shock wave only in the case when the wave strength is not too great (see §3). It is interesting to consider whether a dissipation of diffusional origin, without taking into account viscosity or heat conduction, can ensure a continuous transition in a shock wave propagating through a binary mixture. This question was investigated by Cowling [22] (neglecting thermal diffusion). It was found that, as in the case when there is only heat conduction, a continuous solution is possible only for shock strengths below a certain limit which depends on the difference between the molecular masses and the concentration of the components. In the limiting cases, when the concentration of one of the components tends to zero ($\alpha \to 0$ or $\alpha \to 1$), that is, when the gas becomes a single-component fluid, or when the relative difference of the masses tends to zero, the upper limit of the allowable shock strength also tends to zero. When the difference between the molecular masses is large and the numbers of molecules of both species are comparable, diffusion provides a continuous transition up to quite high shock strengths, being more effective

* Like heat conduction, diffusion can also cause a local entropy decrease (see §2). Diffusion does, however, increase the entropy of the entire system as a whole, and of a particle taken over the transition from the initial to the final state in a shock wave. Unlike heat conduction and diffusion, viscosity can only cause a local increase of entropy, and can only result in an increase in the entropy of a particle.

in this respect than heat conduction. Thus, for example, in a mixture of hydrogen and oxygen ($m_1/m_2 = 1/8$) with a 10% molar concentration of oxygen (N_2/N), a continuous compression of the mixture by a shock is possible up to a density ratio of 4.78 (the limiting density ratio for $\gamma = 7/5$ of 6 was used in the calculations). Heat conduction alone can provide a continuous compression to a density ratio of no more than $(3\gamma - 1)/(\gamma + 1) = 4/3$.

2. *The relaxation layer*

§6. Shock waves in a gas with slow excitation of some degrees of freedom

Often the excitation of certain degrees of freedom of a gas* requires many molecular collisions, with the necessary number of collisions (or the relaxation times) appreciably different for different degrees of freedom. The time required to establish complete thermodynamic equilibrium in a shock front and, consequently, the front thickness, is determined by the slowest of the relaxation processes. Here, of course, we should consider only those processes which result in the excitation of degrees of freedom which make an appreciable contribution to the specific heat for finite values of the flow variables. If τ_{max} is the longest relaxation time, and u_1 is the gas velocity behind the front with respect to the front, then the front thickness is of the order of $\Delta x \sim u_1 \tau_{max} = D(\rho_0/\rho_1)\tau_{max}$†.

The translational degrees of freedom of a particle are those which are "excited" most rapidly. Therefore the mechanical energy of a gas flowing into a discontinuity is primarily converted into thermal energy of translational motion of the gas atoms and molecules. As shown in §2, the thickness of a viscous shock front for strong shocks is of the order of one or several gaskinetic mean free paths. At room temperatures molecular rotations are also rapidly excited as a result of a small number of collisions, while vibrations are ordinarily unimportant at these temperatures. Consequently, the front thickness for weak shocks propagating through a molecular gas heated to room temperature is of the order of several gaskinetic mean free paths‡.

At temperatures of the order of 1000°K, when kT is comparable with the energy $h\nu_{vib}$ of the vibrational quanta of molecules, excitation of the vibrational modes requires many thousands and sometimes tens and hundreds of thousands of collisions. The thickness of a shock front of corresponding

* Let us recall that for the sake of brevity we include within the term "degree of freedom" also the potential energy of dissociation, of chemical reactions, and of ionization.

† In what follows, we shall continue to denote the shock front velocity by D.

‡ Exceptions are molecular hydrogen and deuterium, which require hundreds of gaskinetic collisions for rotational excitation (see §2, Chapter VI).

strength is determined by the relaxation time of the vibrational degrees of freedom.

The rates of relaxation processes always increase rapidly with increasing temperature; thus, for example, at temperatures of the order of 8000°K, when $kT \gg h\nu_{vib}$, several collisions are sufficient for the excitation of the vibrational modes. Those processes which took place rather slowly for certain shock strengths and which determined the front thickness become rapid in a strong wave and are replaced by other processes. For example, in a diatomic gas at temperatures of the order of 4000–8000°K, the establishment of thermodynamic equilibrium is delayed principally by the slow molecular dissociation (the vibrational modes are excited comparatively fast and ionization is not yet important). At temperatures of the order of 20,000°K a small number of collisions is sufficient to dissociate the molecules, and the front thickness is determined by the rate for single ionization (double ionization is still unimportant). At $T \sim 50{,}000$°K single ionization is replaced by double, and so forth.

Obviously, the limit of the temperature region in which one or another relaxation process is slow is not clearly defined. In exactly the same manner, the front thickness at a given temperature is not always determined by only one of the processes. However, as an approximation for a shock wave of a given strength it is always possible to subdivide the excitation processes for the different degrees of freedom which make significant contributions to the heat capacity into rapid and slow processes. Here the term rapid is understood to refer to processes whose relaxation times τ_{rel} are comparable with the gaskinetic relaxation times and for which the characteristic lengths $\Delta x = u_1\tau_{rel}$ are of the order of a moderate number of gaskinetic mean free paths and thus are comparable with the thickness of the viscous shock front. It follows that slow refers to those processes which require a very large number of gaskinetic collisions.

The problem of the structure of a shock front in a gas with slow excitation of part of the specific heat was first analyzed by one of the present authors in 1946 [23, 24] using a reversible chemical reaction and excitation of molecular vibrations as examples.

Let us consider qualitatively the process of shock compression in a gas with slow excitation of some of the degrees of freedom. We shall not as yet specify the kinds of degrees of freedom and we divide them only into two categories: those which are excited rapidly and those which require many gaskinetic collisions. The principal dissipative processes—viscosity and heat conduction—play a role only in the region where there are large gradients of the flow variables, in the region where the rapidly relaxing degrees of freedom are excited. This region coincides to a certain degree with the viscous shock front region. In the slow relaxation region, extending over a distance of many gaskinetic mean free paths, the gradients are small and this dissipation can be neglected.

We shall not consider the structure of the narrow region in which the rapid processes take place. In principle, it does not differ from the structure of the viscous shock front treated in §2. The increase in specific heat due to the rapid excitation of the nontranslational degrees of freedom introduces only some quantitative changes in the structure of the viscous front without changing the basic qualitative relations. Since the thickness of this region is not large, of the order of several mean free paths, we can consider it approximately as infinitesimally thin and relate the quantities on both sides of it by conservation equations which are in all respects similar to equations (7.4). In what follows, for the sake of definiteness, we shall term the region of rapid relaxation the "compression shock" to differentiate it from the concept of the "shock wave front" which includes the entire transitional region from the initial to final thermodynamic equilibrium state.

Denoting the flow variables directly behind the compression shock by a prime, we can write the equations defining these quantities

$$\rho' u' = \rho_0 D; \qquad p' + \rho' u'^2 = p_0 + \rho_0 D^2; \qquad h' + \frac{u'^2}{2} = h_0 + \frac{D^2}{2}.$$

The enthalpy $h' = h'(p', \rho') = h'(T', \rho')$ includes only the rapidly excited degrees of freedom of the gas (with the slowly excited degrees of freedom frozen at the state ahead of the shock, *eds.*). The extended region of slow relaxation is described by integrals of the one-dimensional steady flow equations of the type (7.3), in which the dissipation terms can be neglected. Considering ρ, p, ε, h, and u as functions of the coordinate x, we can write the integrals of the equations in this region as

$$\begin{gathered} \rho u = \rho_0 D = \rho' u', \\ p + \rho u^2 = p_0 + \rho_0 D^2 = p' + \rho' u'^2, \\ h + \frac{u^2}{2} = h_0 + \frac{D^2}{2} = h' + \frac{u'^2}{2}. \end{gathered} \tag{7.17}$$

It is convenient to place the coordinate origin $x = 0$ on the "infinitesimally thin" compression shock. In exactly the same manner, if we follow the change of state of a given particle passing through the shock front with respect to time, then it is convenient to let $t = 0$ be the time of rapid compression in the compression shock. The boundary or initial conditions for the flow variables $\rho(x)$, $u(x)$ have the form $\rho(0) = \rho'$, $u(0) = u'$, and so forth. When $x \to +\infty$ we have as before $\rho(\infty) = \rho_1$, $u(\infty) = u_1$, and so forth.

Shown on the p, V diagram of Fig. 7.12 are two Hugoniot curves originating from a point A corresponding to the initial state of the gas. One of them (II) corresponds to complete thermodynamic equilibrium, that is, it corresponds to the final state of the gas behind the shock front. The other curve (I)

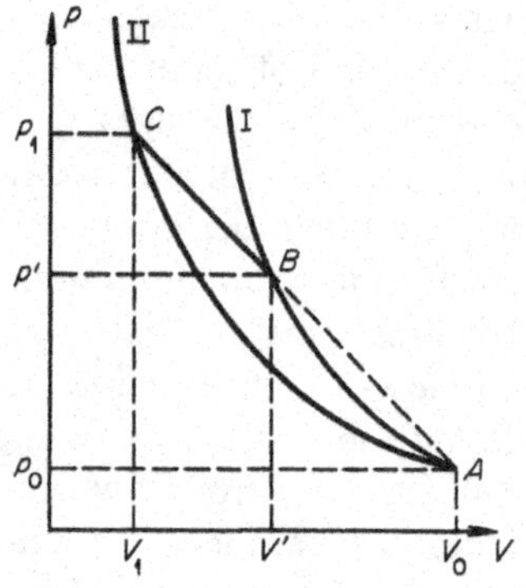

Fig. 7.12. p, V diagram for a shock propagating through a gas with slow excitation of some of the degrees of freedom.

corresponds to the excitation of only the rapidly relaxing degrees of freedom and to the "frozen" state of slowly relaxing degrees of freedom. In calculating curve I the specific internal energy of the slowly excited degrees of freedom is taken to be the same as in the initial state, in spite of the fact that the density and pressure of the gas change. As may be seen from the figure, curve I is steeper than curve II. Indeed, at the same density, the temperature and pressure are greater when some of the degrees of freedom are frozen, since, roughly speaking, the same compression energy is distributed among a smaller number of degrees of freedom*.

Let us draw a straight line AC, connecting the initial and final states of the gas. As is well known, the slope of this line determines the propagation velocity of the shock wave through the undisturbed gas D. It follows from the first two equations of (7.17) that the state of the gas particles in the relaxation region changes along this straight line:

$$p = p_0 + \rho_0 D^2\left(1 - \frac{V}{V_0}\right) = p' + \rho' u'^2\left(1 - \frac{V}{V'}\right). \tag{7.18}$$

Thus, the point describing the successive states of a gas particle for a given front velocity jumps from the initial state $A(p_0, V_0)$ to the intermediate state $B(p', V')$ behind the compression shock and then moves to the final state $C(p_1, V_1)$ along the straight line (7.18). In this case the pressure and density ratio increase as the final state is approached and the gas velocity relative to

* In this case, numerical calculations show that the increase in the number of particles due to dissociation or ionization cannot compensate for the temperature decrease caused by the expenditure of energy in dissociation and ionization at constant volume. Therefore, the pressure in case II is still lower than in case I.

the front decreases. In the case of a wave so weak that its velocity is smaller than the speed of sound corresponding to the frozen degrees of freedom, the straight line AC passes below the tangent to the Hugoniot curve I at point A (Fig. 7.13). In this case, the state changes continuously along the straight line AC from point A to point C, and the gas experiences from the very beginning a gradual excitation of the lagging part of the specific heat.

It is evident from (7.18) that the pressure in the relaxation region of a strong shock increases by only a small amount. Indeed, even if in the rapid compression zone only the translational degrees of freedom are excited, so that

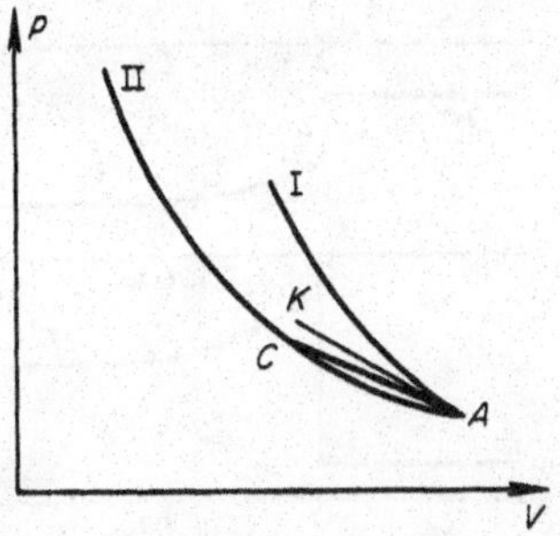

Fig. 7.13. p, V diagram for a weak shock propagating through a gas with slow excitation of some of the degrees of freedom. AK is the tangent to the Hugoniot curve I at the point A.

$V'/V_0 = 1/4$, then the pressure in the relaxation layer can increase by not more than 25% of its final value, since the quantity $1 - V/V_0$ to which the pressure change $p - p_0$ is proportional lies in the range $1 > 1 - V_1/V_0 > 1 - V'/V_0 \geqslant 3/4$. If, however, other degrees of freedom are also rapidly excited, then $V'/V_0 < 1/4$ and the pressure change in the relaxation region is even smaller. The enthalpy increase in the relaxation region is extremely small. It follows from the third and first equations of (7.17) that

$$h = h_0 + \frac{D^2}{2}\left(1 - \frac{V^2}{V_0^2}\right). \tag{7.19}$$

The quantity $(V'/V_0)^2 < 1/16$ for a strong shock, so that the enthalpy increase in the relaxation region does not in any case exceed 5–6%. Since the specific enthalpy is almost constant in the relaxation region and the specific heats increase as the previously frozen degrees of freedom are excited, the temperature in this region will decrease. The temperature decrease can be appreciable if the lagging part of the specific heat is large and makes a large contribution to the final specific heat of the gas. The final temperature T_1 can be one half or one third as much as the temperature T' behind the compression shock. In exactly the same manner, the gas density can also increase appreciably (roughly speaking, $p \sim \rho T$; p changes slightly and T changes appreciably). The profiles of p, ρ, u, and T in a shock front propagating through a gas with slow excitation of part of the specific heat are shown schematically in Fig. 7.14. For specific calculations of the distributions, the rate equations for the corresponding relaxation processes must be used. This will be done for several cases in the following sections.

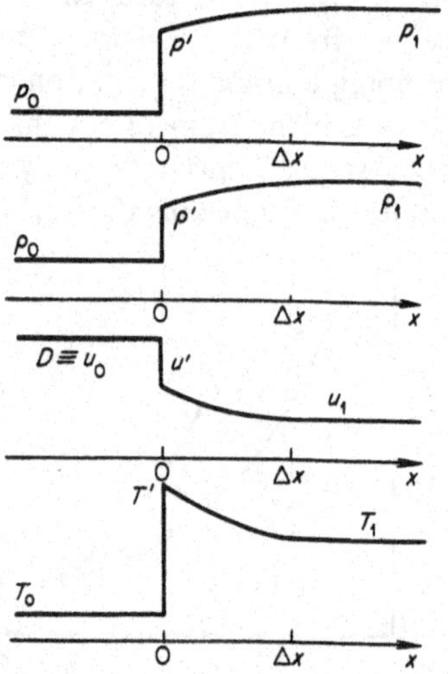

Fig. 7.14. Pressure, density, velocity, and temperature profiles in a shock front propagating through a gas with slow excitation of some of the degrees of freedom; $\Delta x \approx u\tau_{\text{rel}}$ is the front thickness.

We note that if the shock wave is generated by a piston moving with a constant velocity u, then the velocity with which the gas moves behind the compression shock relative to the undisturbed gas $D - u'$ is not the same as the piston velocity (it is lower); only the relative velocity of the gas in the final state behind the wave front, $D - u_1$, is the same as the piston velocity.

§7. Excitation of molecular vibrations

At temperatures of the order of 1000–3000°K (depending on the type of molecule) behind the shock front, molecular dissociation is very small and the contribution made by the chemical energy to the internal energy of the gas may be neglected. In this case the front thickens basically as a result of the slow vibrational excitation of the molecules. Molecular rotations at these temperatures are excited very rapidly, in only a few collisions, so that the rotational energy at each point of the wave front is in equilibrium and corresponds to the "translational" temperature of the gas.

We shall consider a diatomic gas composed of molecules of the same species initially heated to room temperature of the order of $T_0 \approx 300$°K. At this temperature the vibrational energy is extremely small and the specific heat ratio is equal to 7/5. The flow variables behind the compression shock can be calculated from the ordinary equations for a perfect gas with constant specific heats, corresponding to the participation of only the translational and rotational degrees of freedom of the molecules, with a specific heat ratio of $\gamma' = 7/5$. We shall write these equations, characterizing the strength of the shock wave by the Mach number ($M = D/c_0$; $c_0^2 = \frac{7}{5}p_0V_0$), as is conventional in laboratory studies:

$$\frac{\rho'}{\rho_0} = \frac{6}{1 + 5M^{-2}},$$

$$\frac{p'}{p_0} = \frac{7}{6}M^2 - \frac{1}{6},$$

$$\frac{T'}{T_0} = \frac{1}{36}(7 - M^{-2})(M^2 + 5).$$

The flow variables in the final state behind the shock front can be calculated from the general relations at the front by specifying the functions $h_1(T_1)$ or $\varepsilon_1(T_1)$, taking into account the vibrational energy.

In general, the final values of the flow variables are not expressed by simple equations, since the vibrational energy in the region where quantum effects must be considered is a complicated function of temperature (see (3.19)). If we consider sufficiently strong shock waves with temperatures behind the front greater than the energy of vibrational quanta divided by the Boltzmann constant ($T_1 > h\nu/k$), then the vibrational energy per molecule is equal to its classical value kT and $\varepsilon = [1/(\gamma - 1)]p/\rho$, where the specific heat ratio $\gamma = 9/7$. In this limiting case $\varepsilon_1 = \frac{7}{2}p_1V_1$ and the Hugoniot relation takes the simple form*

$$\frac{p_1}{p_0} = \frac{6 - V_1/V_0}{8V_1/V_0 - 1} \quad \text{or} \quad \frac{V_1}{V_0} = \frac{p_1/p_0 + 6}{8p_1/p_0 + 1}. \tag{7.20}$$

From the general relation (1.67) it follows that

$$\frac{7}{5}M^2 = \frac{p_1/p_0 - 1}{1 - V_1/V_0}. \tag{7.21}$$

One can also easily express p_1/p_0, as well as V_1/V_0 and $T_1/T_0 = p_1V_1/p_0V_0$ in terms of the Mach number M. It should be noted that the region of applicability of the above simple equation for the Hugoniot of a diatomic gas is very limited. If $T_1 < h\nu/k$, then the vibrational energy is not equal to kT, while at temperatures appreciably greater than $h\nu/k$, molecular dissociation becomes important.

As an example, let us consider a shock wave at a Mach number $M = 7$ in oxygen with an initial temperature $T = 300°K$. If the initial pressure is

* We emphasize that these equations are not the same as the equations for a gas with the constant specific heat ratio $\gamma = 9/7$, since in the initial state $\gamma = 7/5$ and $\varepsilon_0 = \frac{5}{2}p_0V_0$.

atmospheric and the speed of sound $c_0 = 350$ m/sec, then the shock velocity $D = 2.45$ km/sec. The flow variables behind the compression shock are $\rho'/\rho_0 = 5.45$, $p'/p_0 = 57$, $T'/T_0 = 10.5$, $T' = 3150°K$. The parameters in the final state behind the shock front are $\rho_1/\rho_0 = 7.3$, $p_1/p_0 = 60$, $T_1/T_0 = 8.2$, and $T_1 = 2460°K$. The value of $h\nu/k$ for oxygen is 2230°K; since T_1 is slightly larger than this value it is possible to use the simple equation for calculating T_1 (dissociation of oxygen at this temperature and not too low a density is sufficiently small that it can be neglected).

Let us find the distribution of the flow variables in the relaxation region and estimate its thickness. The specific internal energy of the gas at any point x consists of the energy of the translational and rotational degrees of freedom, equal to $\frac{5}{2}RT$, with T the "translational" temperature at the point x and R the gas constant per gram, and of the nonequilibrium vibrational energy which will be denoted by ε_{vib}: thus, $\varepsilon = \frac{5}{2}RT + \varepsilon_{\text{vib}}$. As was pointed out above, the specific enthalpy remains practically constant in the relaxation region (in our numerical example the change amounts to only 1%), and hence

$$h = \tfrac{7}{2}RT + \varepsilon_{\text{vib}} \approx const \approx h_1 \approx h'.$$

This equation relates the nonequilibrium vibrational energy to the temperature at the point x. Directly behind the compression shock the vibrational modes are not excited (in the initial state at $T = T_0 \approx 300°K$ the vibrational energy is very small), so that at the point $x = 0$ behind the compression shock $\varepsilon_{\text{vib}} = 0$. Behind this point a gradual excitation of the vibrational modes takes place, ε_{vib} increases, and the temperature decreases from T' to the final value T_1, at which value the vibrational energy attains its equilibrium value corresponding to this temperature.

The temperature distribution with respect to x may be found from the rate equation for vibrational excitation (6.9):

$$\frac{D\varepsilon_{\text{vib}}}{Dt} = \frac{\varepsilon_{\text{vib}}(T) - \varepsilon_{\text{vib}}}{\tau_{\text{vib}}}.$$

Here $\varepsilon_{\text{vib}}(T)$ is the equilibrium vibrational energy corresponding to the translational temperature T, and τ_{vib} is the relaxation time. Let us for simplicity consider only strong shocks where the temperature is sufficiently high and the equilibrium vibrational energy is expressed by the classical formula $\varepsilon_{\text{vib}}(T) = RT$. In this case $\varepsilon_{\text{vib}} = h_1 - \frac{7}{2}RT = \frac{9}{2}RT_1 - \frac{7}{2}RT$. Substituting these expressions into the rate equation and replacing the material derivative with respect to time by a derivative with respect to position by taking into account the fact that the process is steady, $D/Dt = \partial/\partial t + u\,\partial/\partial x = u\,d/dx$, we obtain the equation

$$\frac{dT}{dx} = \frac{9}{7}\frac{T_1 - T}{u\tau_{\text{vib}}}.$$

The relaxation time τ_{vib} depends on the temperature and density (or pressure). This dependence can be approximately described by (6.19), which was derived in §4 of Chapter VI, namely

$$\tau_{vib} \approx \frac{const}{\rho} e^{const/T^{1/3}}.$$

In order to make these considerations physically clearer we shall assume that $u\tau_{vib}$ is approximately constant in the relaxation region and corresponds to some average temperature and density between T' and T_1 and ρ' and ρ_1 ($u = D\rho_0/\rho$). This approximation is meaningful since the temperature and density changes are not large. In our numerical example, the temperature changes by a factor of 1.28, $T^{1/3}$ by a factor of 1.08, and the density and the velocity change by a factor of 1.34. Integrating the temperature equation with the initial condition $T = T'$ at $x = 0$, and noting that since $h' = h_1$, $T' = \frac{9}{7}T_1$, we obtain the temperature distribution

$$T = T_1\left(1 + \frac{2}{7} e^{-9x/7u\tau_{vib}}\right) = T'\left(\frac{7}{9} + \frac{2}{9} e^{-9x/7u\tau_{vib}}\right).$$

Recalling that the pressure is almost constant ($p \sim \rho T \approx const$), and that the temperature variation is also not too rapid, we find the approximate density distribution

$$\rho = \rho_1 - (\rho_1 - \rho') e^{-9x/7u\tau_{vib}} = \rho' + (\rho_1 - \rho')(1 - e^{-9x/7u\tau_{vib}}). \quad (7.22)$$

Thus, as $x \to \infty$ the temperature and density asymptotically approach their final values T_1 and ρ_1, and the effective thickness of the relaxation region and of the shock front is approximately given by

$$\Delta x = \tfrac{7}{9} u\tau_{vib}. \quad (7.23)$$

Equations (7.22) and (7.23) can serve for an experimental determination of the vibrational relaxation time. Ordinarily for this purpose, interferometric methods are used to measure the density distribution behind the compression shock and the thickness of the shock front (see Chapter IV). To extract from the experiment better data than can be obtained using the above simple theory one can refine the simple theory by taking into account the quantum dependence of the vibrational energy on temperature, the change in velocity $u = u(x)$, etc. Of course, all of these refinements do not change either the qualitative picture of the distributions or the order of magnitude of the front thickness.

The theory presented above can also be extended to vibrational relaxation in polyatomic molecules if the strength of the shock wave is such that only the lowest-frequency vibrational mode is excited*. Calculations and measurements for CO_2 and N_2O may be found in [25]. Table 7.1 gives several values of shock front thicknesses in oxygen and nitrogen as determined by vibrational relaxation (from the experiments of Blackman [26]). These data are reduced to

Table 7.1

SHOCK FRONT THICKNESS IN OXYGEN AND NITROGEN WITH VIBRATIONAL RELAXATION [26]

M	D, km/sec	T_1°	ρ_1/ρ_0	$\tau_{vib} \cdot 10^6$, sec	Δx, cm
		Oxygen			
5.95	2.08	2000	6.3	5	0.165
8.0	2.8	3300	7.1	0.8	0.031
		Nitrogen			
7.42	2.43	3000	6.55	30	1.11
9.97	3.26	5000	7.14	5	0.23

correspond to a pressure behind the front $p_1 = 1$ atm ($\Delta x \sim \tau_{vib} \sim 1/p_1$), and an initial temperature $T_0 = 296°$K.

The most detailed summary of all theoretical work devoted to the calculation of the structure of the vibrational relaxation region in a shock front is to be found in the review by Blythe [57]. This paper considers a large variety of approximate solutions and also presents the results of exact solutions obtained with the aid of digital computers (see also [58]). We also note several experimental papers which describe studies of the vibrational relaxation in a shock front and determine the corresponding relaxation times and rates of vibrational excitation. Oxygen was studied in [59, 60], nitric oxide in [61], carbon monoxide in [62], and carbon dioxide in [63, 64]. A detailed survey of the experiments relating to this problem along with numerous references is given in the book of Stupochenko, Losev, and Osipov [90].

§8. Dissociation of diatomic molecules

At temperatures of the order of 3000–7000°K behind a shock front in a diatomic gas there is still no ionization, molecular vibrations are excited relatively quickly, and the thickness of the wave front is connected with the slowest relaxation process—molecular dissociation. Estimates show that the

* In the case of nonlinear polyatomic molecules the numerical coefficient 9/7 in (7.21) and (7.22) must be replaced by 5/4, corresponding to the different rotational specific heat ($\frac{3}{2}k$ per molecule instead of k).

vibrational relaxation time in the above temperature range is approximately an order of magnitude less than the time required for establishing dissociative equilibrium. Approximately, therefore, one may take the vibrational, as well as the rotational, energies at each point of the relaxation region to have their equilibrium values. The flow variables behind the compression shock correspond to an intermediate value of the specific heat ratio $\gamma' = 9/7$ (the vibrations at such high temperatures are completely "classical"). They can be calculated from (7.20) and (7.21).

Appreciable dissociation appears only in sufficiently strong shocks, where the density ratio across the compression shock is close to its limiting value of 8 corresponding to the specific heat ratio of $\gamma = 9/7$ (we assume that the shock wave propagates through a gas at room temperature $T \approx 300°\mathrm{K}$). In this case (7.20) and (7.21) simplify to give, approximately,

$$\frac{\rho'}{\rho_0} = 8, \qquad \frac{p'}{p_0} = \frac{49}{40} M^2, \qquad \frac{T'}{T_0} = \frac{49}{320} M^2,$$

where M is the Mach number (defined with $\gamma = 7/5$ ahead of the shock, *eds.*). The flow variables behind the shock front, with dissociation taken into account, cannot be expressed by simple formulas (see §9, Chapter III); they are calculated from the general relationships at the shock front.

Let us find the distribution of the flow variables in the relaxation region. The specific internal energy of the gas taking dissociation into account is (see (3.21))

$$\varepsilon = \frac{7}{2}(1-\alpha)RT + 2\alpha\frac{3}{2}RT + \alpha U = \left(\frac{7}{2} - \frac{\alpha}{2}\right)RT + \alpha U,$$

where U is the dissociation energy per unit mass of gas and α is the degree of dissociation (which may be out of equilibrium). Since the density ratio just behind the compression shock is already very large (almost 8), the pressure change in the relaxation region is small and the enthalpy change is negligibly small. From this it follows that

$$p = R(1+\alpha)\rho T \approx const = p' = R\rho' T', \tag{7.24}$$

$$h = \left(\frac{9}{2} + \frac{\alpha}{2}\right)RT + \alpha U \approx const = h' = \frac{9}{2}RT'. \tag{7.25}$$

These equations allow us to express the density and degree of dissociation at a point x of the wave in terms of the temperature, or the temperature and density in terms of the degree of dissociation. Thus, for example, neglecting α ($\alpha < 1$) in comparison with 9, we find from (7.25) that

$$\alpha = \frac{9}{2}\frac{R}{U}(T' - T) = \frac{9}{2}\frac{T' - T}{T_{\mathrm{dis}}}, \tag{7.26}$$

where $T_{\text{dis}} = U/R$ (for example, for oxygen $T_{\text{dis}} = 59{,}400°\text{K}$). There is still no dissociation at the point $x = 0$ behind the compression shock: $\alpha = 0$ and $T = T'$. Beyond that point dissociation begins; the degree of dissociation increases and the temperature decreases on account of the energy lost in dissociation. This process continues until dissociative equilibrium corresponding to the gas temperature is attained.

To find the distribution of the flow variables with respect to x we shall use the rate equation for dissociation (see §5, Chapter VI). We shall consider here shock waves which are not very strong, in which the degree of dissociation reached behind the front is small, i.e., $\alpha_1 \ll 1$. In this case we can neglect molecular dissociation due to collisions with atoms and retain in the rate equation (6.21) only those terms which correspond to dissociation as a result of collisions with molecules and recombination of atoms by three-body collisions (with a molecule acting as the third body). When replacing in the rate equation (6.21) the atom number density by the degree of dissociation according to the relation $N_A = 2\alpha N_0$ (N_0 is the original molecule number density) only the degree of dissociation and not the gas density (i.e., N_0) should be differentiated with respect to time, since (6.21) does not contain any term describing the density change. If such a term is added to (6.21), it cancels out with the term $2\alpha\, dN_0/dt$, obtained by differentiating N_0 in the expression

$$N_A = 2N_0\alpha.$$

Neglecting α in comparison with unity in all terms, using the definition of the relaxation time τ given by (6.25), and replacing the material derivative with respect to time by a derivative with respect to position, we can rewrite the rate equation in the form

$$\frac{d\alpha}{dx} = \frac{(\alpha)}{2u\tau}\left[1 - \frac{\alpha^2}{(\alpha)^2}\right],$$

where (α) is the equilibrium degree of dissociation corresponding to the temperature and density of the gas at the point x (see (6.23)).

As in the preceding section we shall take both the relaxation time $\tau(T, \rho)$ and the velocity of the gas with respect to the compression shock $u = D\rho_0/\rho$ to be constant, corresponding to some average value of the temperature and density in the relaxation region. If the final degree of dissociation is very small, and the temperature and density changes are not too large, then this approximation may be used for a rough estimate. The equilibrium degree of dissociation (α) which depends on T and ρ will also be taken constant and equal to the degree of dissociation in the final state α_1. Integrating the rate equation under these assumptions and imposing the initial condition $\alpha = 0$ at $x = 0$, we obtain

$$\frac{\alpha_1 - \alpha}{\alpha_1 + \alpha} = e^{-x/u\tau}. \tag{7.27}$$

Substitution of the degree of dissociation α calculated from this expression into (7.26) gives the temperature distribution $T(x)$ (for $\alpha = \alpha_1$, $T = T_1$), from which (7.24) will give the density distribution $\rho(x)$. We shall not write out the equations for the $T(x)$ and $\rho(x)$ distributions. It is obvious from these distributions as well as from (7.27) that T and ρ asymptotically approach their final values behind the wave front T_1 and ρ_1. The effective thickness of the relaxation region and of the shock front, as should have been expected, is approximately equal to

$$\Delta x \approx u\tau,$$

where τ is some average relaxation time in the nonequilibrium region.

Nonequilibrium dissociation in a shock front has been studied experimentally by many authors. A large number of these studies were carried out in oxygen. Matthews [27] used an interferometer to measure the density distribution in the nonequilibrium region behind the compression shock in a shock tube. The experimental data were compared with the theoretical calculations based on a solution of the rate equation for the dissociation. Matthews calculated a number of distributions with different values of the reaction rate constants; the constants were selected to obtain the closest agreement with the experimental data. (The distributions were calculated more exactly than in the method presented above.) The experimentally determined dissociation

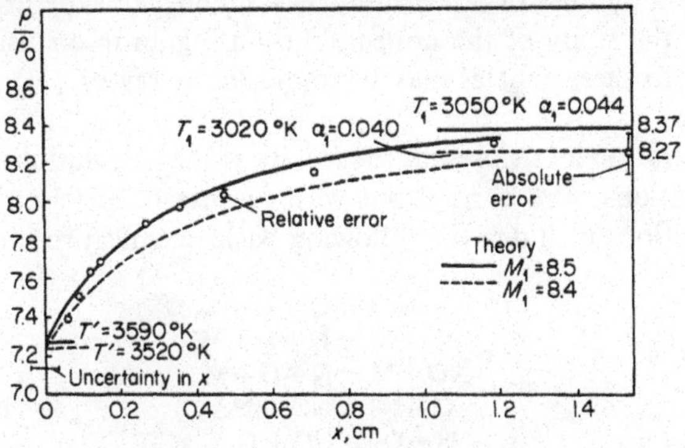

Fig. 7.15. Density distribution across a compression shock in oxygen according to the data of [27]. The initial pressure $p_0 = 19.6$ mm Hg and the initial temperature $T_0 = 300°$K.

rate for oxygen was given in §6, Chapter VI. Figure 7.15 shows the density distribution in the nonequilibrium region of a shock wave in oxygen according to the data of Matthews. It is evident from Fig. 7.15 that the thickness of the shock front under experimental conditions is of the order of $\Delta x \approx 1$ cm. Losev [28] and Generalov and Losev [29] measured the temperature distribution behind a compression shock in the region in which oxygen was out of equilibrium by means of the temperature-dependent absorption of ultraviolet radiation in the Schumann–Runge bands of O_2 molecules. Light absorption was used to study the dissociation rate of bromine and iodine in a shock wave [30].

In the papers by Camac and Vaughan [65], Rink *et al.* [66], and Wray and Freeman [91], the dissociation of oxygen in a shock wave was investigated; in [67] the dissociation of hydrogen was studied and in [68] the dissociation and recombination of nitrogen were studied. For a survey of the literature and a detailed bibliography see the book [90].

§9. Shock waves in air

Air is a mixture of two diatomic gases: nitrogen and oxygen (79 and 21% with respect to the number of molecules). In shock waves with strengths corresponding to final temperatures $T_1 \sim 3000$–$8000°$K a considerable broadening of the shock front is observed as a result of the dissociation of nitrogen and oxygen molecules. Besides the dissociation reactions at high temperature, oxidation of nitrogen also takes place. The determination of the distributions of the flow variables in the wave front and the thickness of the front requires the simultaneous solution of the rate equations for all of these reactions. Such calculations have been carried out by Duff and Davidson [32] and also by a number of other authors. A number of experimental papers are devoted to the study of the nonequilibrium region in air using shock tubes. References to these articles may be found in the review [31] and in the book [90].

For illustration let us present the results of the calculations reported in [32] (the calculations were carried out with the use of an electronic computer). The calculations included the following basic chemical reactions:

$$O_2 + M \rightleftarrows O + O + M,$$
$$N_2 + M \rightleftarrows N + N + M,$$
$$NO + M \rightleftarrows N + O + M,$$
$$O + N_2 \rightleftarrows NO + N,$$
$$N + O_2 \rightleftarrows NO + O.$$

In all of these reactions M denotes any atom or molecule. The following recombination rate constants were taken for the three dissociation reactions:

$3 \cdot 10^{14}$, $3 \cdot 10^{14}$, and $6 \cdot 10^{14}$ $\text{mole}^{-2} \cdot \text{cm}^6 \cdot \text{sec}^{-1}$, respectively. The forward rates of the fourth and fifth reactions were taken in the following form:

$$k_4 = 5 \cdot 10^{13} \exp\left(-\frac{75{,}500}{\mathcal{R}T}\right) \quad \text{mole}^{-1} \cdot \text{cm}^3 \cdot \text{sec}^{-1},$$

$$k_5 = 1 \cdot 10^{11} T^{1/2} \exp\left(-\frac{6200}{\mathcal{R}T}\right) \quad \text{mole}^{-1} \cdot \text{cm}^3 \cdot \text{sec}^{-1}$$

(cf. the data of §8, Chapter VI).

Two assumptions were made in these calculations: (1) the vibrational degrees of freedom at each point of the nonequilibrium region are in equilibrium and (2) the rate of vibrational excitation was calculated simultaneously with the rates of the chemical reactions. The temperature and density distributions behind the compression shock in a shock wave with a Mach number $M = 14.2$ propagating through air with $p_0 = 1$ mm Hg and $T_0 = 300°\text{K}$ are shown in Fig. 7.16. The temperature behind the compression shock T' is

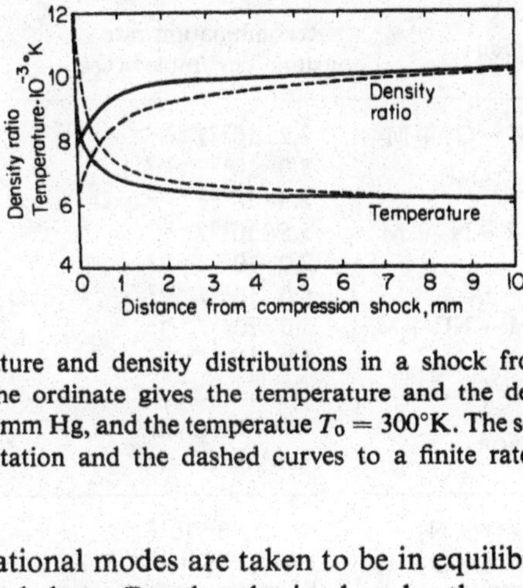

Fig. 7.16. Temperature and density distributions in a shock front in air at a Mach number $M = 14.2$. The ordinate gives the temperature and the density ratio ρ/ρ_0. The initial pressure $p_0 = 1$ mm Hg, and the temperatue $T_0 = 300°\text{K}$. The solid curves correspond to instantaneous excitation and the dashed curves to a finite rate of excitation of the vibrational modes.

9772°K if the vibrational modes are taken to be in equilibrium and 12,000°K if they are unexcited there. Results obtained under the assumption of vibrational equilibrium are shown by the solid lines and those obtained from the assumption of no vibrational excitation behind the compression shock by the dashed lines. The difference between these curves, although not very large, is still significant, since the chemical reaction rates do not strongly exceed the vibrational excitation rate. The front thickness under these conditions, as can be seen from Fig. 7.16, is of the order of 5 mm.

The thickness of the relaxation layer in air in the dissociation region was measured by Losev and Generalov [33]. The temperature change in the relaxation layer was measured from the change in the absorption of light from an external source in the Schumann–Runge bands of oxygen molecules. The pressure behind the shock front was very close to atmospheric. For $D = 3.7$ km/sec, $\Delta x \approx 0.5$ cm (the average temperature in the layer $\bar{T} \approx$ 4500°K); for $D = 2.8$ km/sec, $\Delta x \approx 1.3$ cm ($\bar{T} \approx 3200$°K). Comparison with the Duff and Davidson calculations presented above indicates the correctness of the reaction rate constants chosen for these calculations.

Wray, Teare, Kivel, and Hammerling [69] list rate constants for chemical reactions in high temperature air. The constants were chosen by the authors on the basis of analysis of the available experimental data and are recommended by them for the calculation of nonequilibrium processes in shock waves. Calculations carried out by the authors on the front structure in air are in agreement with the shock tube measurements of Lin [70]. The list of the rate constants is given below. The last two lines give the reaction rate

Table 7.2

RECOMBINATION RATE CONSTANTS [69]

Reaction	Recombination rate constant, $cm^6/mole^2 \cdot sec$	Third body
$O+O+M \to O_2+M$	$2.2 \cdot 10^{20} T^{-3/2}$	O
	$8.0 \cdot 10^{19} T^{-3/2}$	O_2
	$2.5 \cdot 10^{15} T^{-1/2}$	N_2, N, NO, A
$N+N+M \to N_2+M$	$5.5 \cdot 10^{20} T^{-3/2}$	N
	$2.0 \cdot 10^{20} T^{-3/2}$	N_2
	$6.0 \cdot 10^{15} T^{-1/2}$	O_2, O, NO, A
$N+O+M \to NO+M$	$2.0 \cdot 10^{21} T^{-3/2}$	NO, O, N
	$1.0 \cdot 10^{20} T^{-3/2}$	O_2, N_2, A

Reaction	Rate constant, $cm^3/mole \cdot sec$
$NO+N \to O+N_2$	$1.3 \cdot 10^{13}$
$NO+O \to N+O_2$	$1.0 \cdot 10^{12} T^{1/2} e^{-3120/T}$
$N+O \to NO^+ + e$	$3 \cdot 10^{13} T^{-1/2} e^{-32,500/T}$
$NO^+ + e \to N+O$	$1.8 \cdot 10^{21} T^{-3/2}$

constants for ionization and electron recombination which play an important role in establishing ionization equilibrium in air at comparatively low temperatures. The rate constants for the reverse chemical reactions can be expressed in terms of the rate constants for the forward reactions and the corresponding equilibrium constants. A detailed summary of the rates of various reactions which take place in air, including those involving charged particles, is given in Chapter 11 of the collection [92]. We note also [71–75], which have reported studies on the relaxation layer in a shock wave in air and related problems. A more detailed survey with respect to shock waves in air is given in the book [90].

§10. Ionization in a monatomic gas

For temperatures behind the shock front of the order of 15,000–20,000°K the gas is appreciably ionized. The establishment of ionization equilibrium at these temperatures is the slowest of the relaxation processes and therefore determines the thickness of the wave front*.

From the point of view of experimentally studying ionization in a shock tube it is most attractive to deal with monatomic gases. Due to the absence of a number of the degrees of freedom possessed by molecular gases it is easier in monatomic gases to obtain high temperatures ~15,000–20,000°K. Monatomic gases are also very suitable for checking the theory of the phenomenon, since ionization (single) is the only relaxation process responsible for broadening the shock front. The first detailed study of this kind was carried out by Petschek and Byron [35] for argon.

Let us consider a shock wave in a monatomic gas. Appreciable ionization is attained only for rather strong waves, where the limiting density ratio reached across the compression shock is 4, since the specific heat ratio $\gamma' = 5/3$. The flow variables behind the compression shock are expressed in terms of Mach number by the simple relations

$$\frac{\rho'}{\rho_0} = 4, \quad \frac{p'}{p_0} = \frac{5}{4} M^2, \quad \frac{T'}{T_0} = \frac{5}{16} M^2.$$

For example, for $M = 18$ and an initial temperature $T_0 = 300°\text{K}$, which corresponds to a shock velocity $D = 5.75$ km/sec, the temperature behind the compression shock $T' = 30{,}000°\text{K}$. At equilibrium behind the shock front in argon for an initial pressure $p_0 = 10$ mm Hg, the gas is approximately 25% ionized and the temperature $T_1 = 14{,}000°\text{K}$.

The thickness of the compression shock is approximately two to three gaskinetic atomic mean free paths. If the gas is ionized ahead of the shock front and thus also immediately behind the compression shock, it is only weakly so. Ionization occurs after the shock compression and rapid heating of the gas particles. The basic ionization mechanism is ionization by electron impact (see Chapter VI). However, ionization by electron impact with the formation of an electron avalanche requires the presence of some initial "priming" electrons. One of the mechanisms which can lead to this initial ionization is ionization by atom-atom collisions. As was noted in Chapter VI,

* Molecular dissociation at these temperatures proceeds very rapidly, with a small number of collisions.

the cross section for this process is extremely small. Therefore, the formation of the "priming" electrons by atom-atom collisions requires an appreciable time. Correspondingly, the region behind the compression shock where the flow variables correspond to a negligibly small degree of ionization and are simply equal to ρ', p', and T', etc., extends over a rather large distance.

Avalanche ionization begins when the rate of ionization by electron impact becomes greater than the rate of ionization by atom collisions, or of ionization due to other processes. Possible mechanisms for the formation of the priming electrons will be discussed below. Since the rate of ionization by atomic collisions is extremely low, avalanche ionization begins with only very few "priming" electrons, when the degree of ionization $\alpha \sim 10^{-5}$–10^{-3}. We shall not examine here the formation of the "priming" electrons but shall consider the basic process of ionization from the very low to the equilibrium values ($\alpha_1 = 0.25$ in our example).

At a constant electron temperature T_e the avalanche increases exponentially as $n_e \sim \alpha \sim e^{t/\tau}$ (see §11 of Chapter VI), until recombination begins to compensate appreciably for the ionization. Thereafter, the degree of ionization gradually approaches its equilibrium value at which the recombination exactly balances ionization. In actuality, the formation of an avalanche proceeds in a more complex manner. Each act of ionization results in the electron gas losing an amount of energy equal to the ionization potential I (which in argon is equal to 15.8 ev). On the other hand, the temperature of the electron gas is of the order of 10,000°K, that is, the thermal energy of a single electron is of the order of 1.5 ev. Thus, the formation of a single new electron requires an energy equal to the thermal energy of approximately ten electrons. If the thermal energy of the electrons were not replenished, the electron temperature would drop very rapidly. This would also result in a drop in the rate of ionization which depends very strongly on the electron temperature through a Boltzmann factor of the type e^{-I/kT_e} (see §11, Chapter VI). The electron energy loss in ionization is compensated for by energy transfer to the electrons from the atom gas heated by the compression shock. However, the energy exchange between the heavy particles and the electrons, due to the great difference in their masses, progresses extremely slowly and it is just this exchange process which limits the rate of development of the electron avalanche and determines the time required to reach equilibrium ionization.

For small degrees of ionization, very few ions are present and the electrons acquire their energy by collisions with neutral atoms. However, the effectiveness of such collisions for electron temperatures $T_e \sim 1$ ev $\approx 10^4$°K is approximately 10^3 times less effective than the electron-ion collisions. Therefore, the transfer of energy from atoms to electrons is important only at the very beginning of the process; already at a low degree of ionization $\alpha \sim 10^{-3}$, the energy exchange between the ions and electrons is of primary importance. The temperature of the ions is equal to that of the atoms since the energy exchange between them proceeds very rapidly as a result of their equal masses. Thus, a small number of ions serves in this case as an "intermediary" in the energy exchange from the atoms to the electrons. In the electron gas the energy is distributed very quickly, so that we can speak of an electron temperature T_e which is, naturally, different from the temperature T of the heavy particles (atoms and ions).

The electrons not only ionize the atoms but also excite them. The energy of the first excited level of an argon atom is $E^* = 11.5$ ev. In the case of appreciable electron concentrations the excited atoms are deexcited by electron collisions of the second kind. In this case the excitation energy is again returned to the electron gas. However, when the electron temperatures are of the order of and, in particular, higher than 1 ev, electron impact ionization of an excited atom becomes more probable than deexcitation (only a moderate amount of energy $I - E^* = 4.3$ ev is required for ionization). The ionization in this case takes place in two stages, the atom is first excited and then it is ionized. The energy expended for ionization in this two-stage process still remains equal to the ionization potential: $E^* + (I - E^*) = I$. Other multistage processes in which an atom is not immediately ionized by electron impact, but is first subjected to several increases in the degree of excitation, are also possible (see Chapter VI).

If the rate of ionization of excited atoms is high in comparison with those of deexcitation and of the excitation of unexcited atoms, then the ionization rate is essentially determined by the excitation rate only (in accordance with (6.79)). This is precisely the assumption made by Petschek and Byron [35], who assumed that each atom is "instantaneously" ionized following the excitation. The excited atoms emit part of their energy. The photon generated as a result of this emission is absorbed by a neighboring unexcited atom (the absorption cross section for the resonant photons is very large), which, in turn, reemits and so forth*.

Let us set up a system of equations which approximately describes the ionization process and the distributions of flow variables in the shock wave. For simplicity we restrict ourselves to the case of a small degree of ionization

* Resonant photons born in the heated zone behind a shock front may leave the heated region by diffusing through the gas and penetrating the front surface. After that they diffuse through the unexcited gas and leave the propagating shock wave behind. The diffusion of resonance radiation at a large distance ahead of the front results in a significant concentration of excited atoms. This process was considered by Biberman and Veklenko [34]. They have shown that at a distance of 1 m ahead of the wave front in argon with $p_0 = 10$ mm Hg, $M = 18$, and $T_1 = 14{,}000$°K, the concentration of excited atoms reaches $5 \cdot 10^{13}$ cm^{-3}, which corresponds to an excitation "temperature" of ~13,500°K, only slightly lower than the temperature of resonance radiation passing through the front surface and equal to T_1.

$\alpha \ll 1$. For convenience we shall refer to the energy and all other thermodynamic quantities not per unit mass but per original gas atom.

The enthalpy per original atom is equal to

$$h = \tfrac{5}{2}kT + \tfrac{5}{2}\alpha kT_e + \alpha I.$$

From the condition that the enthalpy is approximately constant in the relaxation zone and that $\alpha \ll 1$ we obtain a relationship between the degree of ionization and the atom temperature analogous to that of (7.26)

$$\alpha = \frac{5}{2}\frac{T' - T}{T_{\text{ion}}}, \tag{7.28}$$

where $T_{\text{ion}} = I/k$ (in argon $T_{\text{ion}} = 1.83 \cdot 10^5$°K).

The gas pressure is $p = nkT + n\alpha kT_e \approx nkT$, where $n = n_a + n_i = n_a + n_e$ is the total number of atoms and ions per unit volume. It can be expressed in terms of the atom temperature and the degree of ionization from (7.18). It can be determined less exactly from the condition that the pressure in the relaxation zone is approximately constant: $p \approx nkT \approx n'kT'$. This yields

$$n = 4n_0\left(1 - \frac{2}{5}\alpha\frac{T_{\text{ion}}}{T'}\right)^{-1}, \tag{7.28'}$$

where n_0 is the atom number density ahead of the front.

The rate equation for the degree of ionization $\alpha = n_e/n$ is

$$\frac{D\alpha}{Dt} = u\frac{d\alpha}{dx} = \frac{q}{n}. \tag{7.29}$$

Here q is the algebraic sum of all terms which describe the appearance and disappearance of free electrons per unit volume per unit time. Over the main region q is determined by electron impact ionization. Within the framework of the Petschek–Byron assumptions, for example, q in this region represents the rate of atom excitation $q = \alpha_e^* n_e n_a$, where the rate constant α_e^* is given by (6.79). In the very beginning of the process, immediately behind the compression shock, q is determined by processes which result in the formation of "priming" electrons (atom-atom collisions, etc., see below). In the final stage, in the region where equilibrium is approached, recombination must also be included in q.

The rate of ionization by electron impact depends on the temperature of the electron gas, which is governed by the equation of electron energy balance. We denote the entropy and enthalpy of the free electrons per original atom by Σ_e and $h_e = \tfrac{5}{2}\alpha kT_e$, and the electron pressure by $p_e = n\alpha kT_e$. Remembering that $D/Dt = u\, d/dx$, we may write the energy balance equation

$$uT_e \frac{d\Sigma_e}{dx} = u\left(\frac{dh_e}{dx} - \frac{1}{n}\frac{dp_e}{dx}\right) = \frac{1}{n}(\omega_{ea} - \omega_i), \tag{7.30}$$

where ω_{ea} is the heat inflow per unit volume per unit time due to the energy exchange from the ions and atoms to the electrons, and ω_i is the energy lost by the electrons in ionization (per unit volume per unit time)*. According to (6.121)

$$\omega_{ea} = \frac{3}{2}k\left(\frac{DT_e}{Dt}\right)_{\text{exch}} n_e = \frac{3}{2}kn_e\frac{T - T_e}{\tau_{\text{exch}}}, \tag{7.31}$$

where $1/\tau_{\text{exch}} = 1/\tau_{ei} + 1/\tau_{ea}$; τ_{ei} is the characteristic time for energy exchange between ions and electrons (equation (6.120)) and τ_{ea} is the characteristic time for energy exchange between atoms and electrons (equation (6.122)). The

* Equation (7.30) can also be derived from equations of the type of (1.10) and (1.6) written for an electron gas. In this case, however, we must take into account the fact that a small polarization (relative displacement of positive and negative electric charge, *eds.*) takes place in an ionized gas in the presence of gradients of the macroscopic quantities, as a result of which electric fields arise which prevent an appreciable charge separation (for more details, see §13). The polarization field **E** ensures a "rigid" connection between the electron and atom-ion gases. When this field is taken into account, additional terms appear in the electron equations of motion and energy

$$m_e n_e \frac{D\mathbf{u}_e}{Dt} = -\nabla p_e - en_e\mathbf{E}$$

$$\frac{\partial}{\partial t}\left[n_e\left(\frac{3}{2}kT_e + \frac{m_e u_e^2}{2}\right)\right] + \nabla\cdot\left[n_e\mathbf{u}_e\left(\frac{5}{2}kT_e + \frac{m_e u_e^2}{2}\right)\right] = \omega_{ea} - \omega_i - en_e\mathbf{E}\cdot\mathbf{u}_e.$$

Because the mass of the electrons is extremely small, the inertia term in the equation of motion of the electron gas is extremely small, and the gradient of the electron pressure is balanced by the action of the polarization field; thus, $-en_e\mathbf{E} = \nabla p_e$. We now substitute this quantity into the energy equation, neglecting the small kinetic energy of the electron gas, and consider the one-dimensional steady-state case. Noting also that the electron gas velocity $\mathbf{u}_e$ is practically the same as that of the atom-ion gas $\mathbf{u}$, using the integral of the continuity equation $nu = const$ and the definition $n_e = \alpha n$, we arrive at (7.30). We also write the equations of motion and energy for the atom-ion gas in the form:

$$m_a n \frac{D\mathbf{u}}{Dt} = -\nabla p + en_i\mathbf{E}, \qquad p = p_a + p_i, \qquad n = n_a + n_i,$$

$$\frac{\partial}{\partial t}\left[n\left(\frac{3}{2}kT + \frac{m_a u^2}{2}\right)\right] + \nabla\cdot\left[n\mathbf{u}\left(\frac{5}{2}kT + \frac{m_a u^2}{2}\right)\right] = -\omega_{ea} + en_i\mathbf{E}\cdot\mathbf{u}.$$

energy expended in ionization is equal to

$$\omega_i = Iq = Inu \frac{d\alpha}{dx}. \tag{7.32}$$

The system of differential equations (7.29) and (7.30) for $\alpha(x)$ and $T_e(x)$ and the algebraic equations (7.28) and (7.28′) which give $T(\alpha)$ and $n(\alpha)$, together with the ionization rate q defined in an appropriate manner, can be used to find the distributions of all the quantities in the relaxation zone. In fact, the rates of exchange and of inelastic losses ω_{ea} and ω_i compensate one another to an appreciable extent, $\omega_{ea} - \omega_i \ll \omega_{ea}, \omega_i$. Thus in by far the greatest part of the relaxation region the energy balance equation (7.30) reduces to the algebraic relation $\omega_{ea} \approx \omega_i$, which makes it possible to express α in the form of a function of T_e. This was the approach used by Petschek and Byron for calculating the width of the relaxation region.

The basic difficulty in considering ionization in the relaxation region is presented by the problem of the formation of the priming electrons. The cross sections for ionization by atom collisions are practically unknown (see §15, Chapter VI). The published experimental data for argon [37, 38] pertain to energies of several tens of ev. Therefore some reasonable estimate of the cross section must be assumed for calculation purposes. The structure of the relaxation region in argon was calculated by Bond [36] and Biberman and Yakubov [93]*. For illustration we show in Fig. 7.17 the distributions taken from [93] of the atom and electron temperatures and of the degree of ionization for a Mach number $M = 16$, $D = 5.1$ km/sec, and inital pressure $p_0 =$ 10 mm Hg. These curves were calculated on the assumption that the atoms are ionized by electron impact directly from the ground level, and that the cross section for ionization by atom collisions near the threshold as a function of the collision energy ε is approximated by a straight line with slope $C =$ $1.2 \cdot 10^{-20}$ cm^2/ev, so that when $\varepsilon = I + 1$ ev the cross section $\sigma = 1.2 \cdot 10^{-20}$ cm^2 (see Chapter VI). It can be seen from the figure that initially the ionization is extremely small and that it increases very slowly. The electron temperature rises quite rapidly to a value $T_e \approx 1.3$ ev, and then remains almost constant.

In this region the energy supplied by the ions is compensated by the energy lost in ionization. It should be noted that the time for accumulating priming electrons due to atom-atom collisions should depend only weakly on the cross section chosen. In the case of $T_e = const$ it is completely independent of the cross section of atom-atom ionization and is determined by the time of the succeeding electron avalanche development (this was shown at the end of §13, Chapter VI). As shown by calculations in [93], the taking into account of the step-wise character of electron impact ionization with preliminary

* In [35] only values of the total width of the region were obtained.

excitation of the atoms significantly reduces the ionization time and width of the relaxation region only for shock Mach numbers less than 12–13; for $M > 13$ this reduction is insignificant.

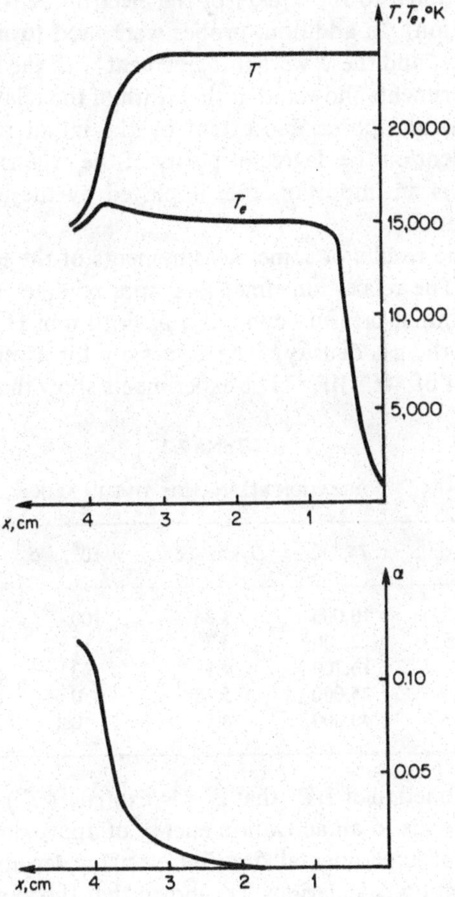

Fig. 7.17. Distribution of electron and atom temperatures and degree of ionization in a shock wave in argon on the assumption that the initial electrons are formed by atom-atom collisions. The Mach number is 16, the pressure ahead of the front 10 mm Hg, and the initial temperature 293°K.

Experiments in a shock tube on ionization relaxation in argon were carried out by Petschek and Byron [35]. In order to expand the nonequilibrium region and to increase the relaxation times so as to make them accessible to measurement, the experiments were carried out at quite low initial pressures of argon. The most reliable measurements were made at $p_0 = 2$ mm Hg. The electron density distribution in the shock wave was determined by recording the continuous luminous spectrum which results from electron-ion recombination. The intensity in this spectrum, at a given cross section x of the shock wave, is proportional to the square of the electron density (the gas is transparent to radiation). In addition, probes were used to measure the electron density gradients, and these were in agreement with the luminosity measurements. The experiments showed that the width of the relaxation region, which is determined to an appreciable extent by the initial rate of ionization, is strongly dependent on the degree of purity of the argon; in the formation of priming electrons an important role is played by the impurities (with low ionization potentials).

We present the results of some measurements of the ionization relaxation time in argon. The relaxation times and approximate front thicknesses are reduced to the initial pressure condition $p_0 = 10$ mm Hg (they are inversely proportional to the gas density). The values are for a very pure gas, with an impurity content of $\sim 5 \cdot 10^{-5}$. The experiments show that, roughly speaking,

Table 7.3

IONIZATION RELAXATION TIME IN ARGON

M	T', °K	D, km/sec	$\tau \cdot 10^6$, sec	Δx, cm
10.3	10,000	3.3	100	~6.5
11.5	12,500	3.7	17	~1
13.4	16,700	4.3	3	~0.2
16.4	25,000	5.25	0.5	~0.032
20.3	40,000	6.5	0.1	~0.006

$\log \tau$ is a linear function of $1/T'$, that is, $\tau \sim \exp(const/T')$. The constant in this relation corresponds to an activation energy of approximately 11.5 ev.

Comparison of experimental data on electron density distributions with avalanche ionization calculations has shown that the avalanche is developed only after the initial ionization reaches a value of the order of 0.1 of the equilibrium ionization, or an absolute value of $\alpha \sim 10^{-2}$. The question of the nature of the initial ionization was not resolved in [35]. Estimates have shown that ionization as a result of atom-atom collisions or photoionization by photons born in the equilibrium region cannot ensure the rapid formation of the large number of priming electrons which are required to explain the experimental data. The evident insufficiency of the mechanism of atom-atom collisions is shown by the fact that calculations which take into account only this mechanism give relaxation times which are tens of times larger than the experimental relaxation times [93].

A number of authors have proposed different mechanisms for the initial ionization: in particular, it has been suggested that a role is played by the diffusion of electrons from a region with a high level of ionization into a region where the degree of ionization is low and even into the gas ahead of the shock front. (Electron diffusion in a shock wave is treated in [39, 76, 77].) The role of atom excitation ahead of the shock front by resonance radiation emerging from the equilibrium region was pointed out in [94].

An analysis of the various ionization mechanisms in a shock wave in argon (and in monatomic gases in general) is contained in the paper by Biberman and Yakubov [93] cited previously. The authors have investigated the effect of variations in the choice of ionization cross sections by electron and atom impacts, the role of step-wise and radiative processes. They have come to the conclusion that the decisive role in the rapid formation of the priming electrons must be played by the excitation of atoms by resonant radiation coming from the equilibrium region. This effect brings about a large increase in the concentration of excited atoms, which are easily ionized by electron impacts. Taking into account the above effect, the authors were able to considerably narrow the gap between calculated and experimental values of the relaxation times and to obtain satisfactory agreement.

We must note that the problem of ionization relaxation and, in particular, of the initial ionization mechanism is still not entirely clear. We also wish to note [95] in which relaxation in xenon is studied, and [96], which is concerned with the effect of radiation.

§11. Ionization in air

Ionization in shock waves in air was studied in the early papers [40, 41, 70, 78–80], and especially thoroughly in the experiments described in [87]. In the latter experiments waves with velocities of 4.5–7 km/sec (Mach numbers 14–20) were studied in shock tubes at initial pressures of 0.02–0.2 mm Hg. The measurements showed that the ionization develops very rapidly and reaches a value of the order of equilibrium ionization at distances from the shock equal to 10–40 gaskinetic mean free paths in the undisturbed air. Here the degree of ionization is of the order of 10^{-4}–10^{-3} at equilibrium, but the degree of ionization in the relaxation region passes through a maximum which can exceed the equilibrium value by several times.

It was already noted in Part 2 of Chaper VI that the ionization mechanism in a molecular gas such as air, for shock waves which are not too strong, differs appreciably from the ionization mechanism in monatomic gases. The free electrons in air are formed primarily by associative ionization, in which two atoms combine into a molecule with the simultaneous removal of an electron and formation of a molecular ion. The principal process requiring the lowest activation energy is the reaction

$$N + O + 2.8 \text{ ev} \rightarrow NO^+ + e.$$

Since the ionization potentials of all the components of air are much higher than the amount of energy expended in this reaction, the latter (for not too high temperatures) takes place at a much faster rate than direct ionization of atoms and molecules by particle impacts. The rate constant of the above principal ionization reaction is given in Table 7.2 of §9. Since an appreciable role in the ionization of air is played by atoms, calculations of ionization rates in air are based on calculations of molecular dissociation (chemical reactions in general). These calculations were carried out by Lin and Teare [86], and they are in good agreement with measurements [87]. Calculations (for shock wave velocities not exceeding 9 km/sec) have shown that ionization occurs rapidly, even faster than the chemical reactions, so that the ionization in the relaxation region, to a certain extent, comes into equilibrium with the chemical composition of the gas and "follows" the change in the degree of molecular dissociation.

Ionization in air in shock waves with velocities somewhat greater than 10 km/sec (9–15 km/sec) was considered by Biberman and Yakubov [97]. They took into account the chemical composition of the air in the relaxation region and the excitation of the atoms and molecules. Unlike the case of low velocities, dissociation takes place more rapidly than does ionization and the ionization occurs mainly in the atom gas. Associative ionization plays the determining role in the creation of the initial electrons; as the electron density increases, the role of step-wise ionization by electron impacts becomes more important, with the energy of the electrons, as is also the case in a monatomic gas, replenished by energy transfer from the ions.

The ionization rate of atoms and molecules by electron impact was calculated in [97] using a method of combining the excited and ionized states into one group. This method, suggested by Biberman and Ul'yanov [99], may also be useful for other problems connected with the disturbance of ionization equilibrium. It consists of the following: It is assumed that impact excitation and ionization of atoms which are in the ground state, and also the inverse processes of deexcitation with the atom moving to the ground level and recombining by three-body collisions with capture of an electron into the ground level, take place at a relatively slow rate. At the same time it is assumed that the increase in the degree of excitation by electron impact and ionization of excited atoms, as well as the corresponding inverse processes, takes place at a relatively rapid rate.

The rates of actual processes behave in this manner to some extent, so that the above assumptions are reasonable. But if these assumptions hold we can assume approximately that a Boltzmann distribution is established among the different excited states, while a Saha distribution is established between the excited and ionized atomic states. In other words, all the excited and ionized states can be combined into a single group, by assigning to this group of states the particular temperature equal to the temperature of the electron gas. On the other hand, the electron density, as well as the relation between the electron (or excited atom) densities and the densities of atoms in the ground state, is no longer described by the Saha equation, and is not an equilibrium relation. It is determined from the rate equation which describes the transition of atoms between the ground state and the excited and ionized states. If desired, this method can be refined by separating out from the excited and ionized states the lowest excited states, and writing separate rate equations for the atom concentrations in these states. Reference [99] has used the above method for considering the effect of radiation leaving a bounded gas volume when the state of the gas is disturbed from thermodynamic equilibrium.

Ionization in air for very high velocity shock waves, of the order of tens of km/sec (applied to the problem of the motion of meteors in the atmosphere) was considered by Bronshten [98]. One of the most characteristic properties of ionized gases is their ability to conduct an electric current. A rather large number of papers have been devoted to calculations and theoretical studies of the electrical conductivity of ionized air (and other gases). See, for example, [70, 81–84].

§12. Shock waves in a plasma

The structure of a shock front propagating through an ionized gas has a number of interesting features. These features were noted by one of the present authors [42]; quantitative calculations of the front structure have been made by Shafranov [43]; see also the papers by Imshennik [51], Jukes [44], Tidman [44a], and Pikel'ner [85]. The basic features of the structure are related to the slow character of the energy exchange between ions and electrons and to the high electron mobility, as a result of which the electron heat conduction greatly exceeds the ion heat conduction. Maxwellian distributions in the electron and ion gases are established quite rapidly, in a time of the order of the time between particle "collisions"*. On the other hand, the equilibration of the temperatures of both gases takes place much more slowly, because of the large difference between the electron and ion masses. This relaxation process determines the shock front thickness in a plasma.

For a qualitative discussion of the consequences of the low rate of energy exchange between the electrons and ions, we shall first assume that the electron heat conduction does not differ from the ion heat conduction. In

* For the concept of "collision" of charged particles interacting according to the Coulomb law see §20, Chapter VI.

addition, we shall assume that ionization does not take place in the shock wave, but that the wave is propagated through a gas which is already ionized. In a coordinate system moving with the wave, a considerable part of the kinetic energy of the gas entering the compression shock is irreversibly converted into heat through the action of ion viscous forces. The increase in ion temperature across the compression shock is of the order $\Delta T_i \sim m_i D^2/k$, where m_i is the mass of an ion and D is the velocity of the incident gas, equal to the shock front velocity. The thickness of the viscous shock is determined by the time between ion collisions τ_i; it is of the order of an ion mean free path $l_i \sim \bar{v}\tau_i$, where $\bar{v} \sim D$ is the thermal velocity of the ions in the compression shock (the definition of τ_i is given in §20, Chapter VI). During the compression time τ_i the ion gas does not succeed in transferring to the electron gas any appreciable thermal energy, since the characteristic exchange time $\tau_{ei} \sim (m_i/m_e)^{1/2} \cdot \tau_i$ is very large. For ions of average mass, τ_{ei} is hundreds of times larger than τ_i; for protons it is only 43 times larger than τ_i. The increase in electron temperature across the compression shock due to the conversion of the kinetic energy of the incident electron gas into heat through the action of electron viscous forces is negligibly small. It is of the order $\Delta T_e \sim m_e D^2/k$, that is, smaller by a factor of m_e/m_i than ΔT_i. The electron gas in the compression shock is heated by another process.

The electrons and ions are bound together by the forces of electric interaction and this bonding is very strong. The smallest separation of the electron and ion gases results in the formation of strong electric fields which prevent further separation. Hence, each small parcel of the plasma remains electrically neutral. The electron density n_e is always the same as the positive charge density Zn_i (Z is the ionic charge, n_e and n_i are the electron and ion number densities, respectively). In the compression shock the electron gas does not behave independently, but is compressed in the same manner as the ion gas. We can say that the electrons are "rigidly coupled" to the ions by the electric forces. These forces are "external" with respect to the electron gas and do not lead to any dissipation. Since the dissipation of energy due to electron viscous forces is negligibly small, the compression and heating of the electron gas in the compression shock is adiabatic. Thus, for example, when a hydrogen plasma is compressed by a strong shock wave the density across the compression shock increases by a factor of 4, corresponding to the specific heat ratio of 5/3. The ion temperature can increase very markedly if the wave is strong, while across the compression shock the electron temperature increases only by a factor of $4^{\gamma-1} = 4^{2/3} = 2.5$. Therefore in a strong shock wave propagating through a plasma with equal ion and electron temperatures, behind the compression shock there is a marked difference in the temperatures of the two gases. After a small parcel has undergone the shock compression it begins to transfer thermal energy from the ions to the electrons, and this leads to an equilibration of the temperatures in a time of the order of the exchange time τ_{ei} (see §20, Chapter VI). The thickness of the relaxation region behind the compression shock in which the plasma approaches equilibrium (with equal temperatures $T_e = T_i = T_1$) is of the order of $\Delta x \sim u_1\tau_{ei}$ ($u_1 = (\rho_0/\rho_1)D$). The final temperature T_1 is determined by the general equations of conservation for the shock front. Thus, in the absence of effects connected with the existence of increased electron heat conduction, the temperature distribution in the wave front would have the form shown in Fig. 7.18.

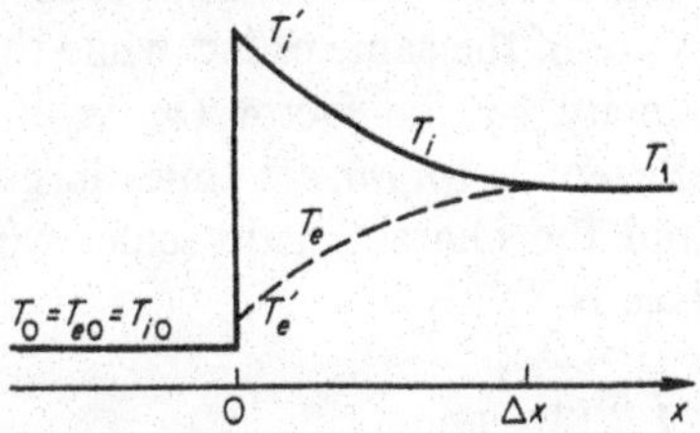

Fig. 7.18. Ion and electron (dashed line) temperature profiles in a shock front in a plasma without taking into account electron heat conduction.

If none of the other degrees of freedom other than the previously "frozen" translational degrees of freedom of the electron are excited (which is the situation in a fully ionized gas), then the density and pressure in the relaxation region remain strictly constant. Indeed, the specific heat ratios for a gas with "frozen" and equilibrium degrees of freedom are the same and equal to $\gamma = 5/3$, so that the compression across the compression shock takes place along a Hugoniot curve which coincides with the Hugoniot curve of the final state. The physical reason is, obviously, the fact that the pressure is only determined by the average translational energy of the particles, which remains constant during the exchange, the value being independent of its distribution among the particles.

Let us now consider the effect of electron heat conduction on the shock front structure. Up to this point we have assumed (with ample justification) that the dissipative processes, viscosity and heat conduction, are important only in the region of large gradients in the compression shock, where the macroscopic quantities change significantly over distances of the order of a gaskinetic mean free path. In the relaxation region, which extends over a distance of many mean free paths, the gradients are small and the dissipation processes can be neglected. Actually, the characteristic scale, which serves as a criterion for the smallness of the gradients, is a length scale based on the transport coefficients and the front velocity. The transport coefficient, for example, for the atom thermal diffusivity, is of the order of $\chi \sim l\bar{v}/3$ and the length scale $\lambda \sim \chi/D \sim l\bar{v}/D \sim l$ is of the order of a gaskinetic mean free path, since the thermal speed of the atoms in the front $\bar{v}$ is of the order of the front velocity D. The coefficient of electron thermal diffusivity χ_e is approximately

$$\chi_e = \frac{l_e \bar{v}_e}{3} \approx \frac{\bar{v}_e^2 \tau_e}{3},$$

where l_e is the electron mean free path, $\bar{v}_e$ is the thermal speed of electrons, and τ_e is the time between the "collisions" of electrons with each other. As was shown in §20 of Chapter VI, the mean free path for charged particles is independent of their mass and depends only on the charge and temperature, i.e., $l \sim T^2/Z^4$.

At comparable temperatures and in light gases, for example, in hydrogen ($Z = 1$), the electron and ion mean free paths are of the same order, while the electron velocity is larger than the ion velocity by the factor $(m_i/m_e)^{1/2}$. Therefore the electron heat conduction coefficient is $(m_i/m_e)^{1/2}$ times larger than the ion heat conduction coefficient, and the characteristic scale over which the electron heat conduction takes place is

$$\lambda_e \sim \frac{\chi_e}{D} \sim \left(\frac{m_i}{m_e}\right)^{1/2} \frac{\chi_i}{D} \sim \left(\frac{m_i}{m_e}\right)^{1/2} l_i.$$

This length scale is of the same order as the thickness of the relaxation region in which the electron and ion temperatures equilibrate:

$$\Delta x \sim D\tau_{ei} \sim D\left(\frac{m_i}{m_e}\right)^{1/2} \tau_i \sim \frac{D}{\bar{v}}\left(\frac{m_i}{m_e}\right)^{1/2} l_i \sim \left(\frac{m_i}{m_e}\right)^{1/2} l_i.$$

Therefore, the gradients in the relaxation region are not small with respect to electron heat conduction and the heat exchange by conduction in this region is comparable to the heat exchange between ions and electrons. The electron heat conduction promotes more rapid equilibration of the temperatures behind the viscous shock, since it transports the heat from the layers which are further removed from the compression shock to the more forward layers where the electron temperature is lower. In addition, and this effect is extremely important, the electron heat conduction results in a preheating of the gas ahead of the viscous compression shock. While the "hot" ions cannot move away too far from behind the compression shock into the region ahead of the shock (their thermal velocity is comparable with the propagation velocity of the shock through the undisturbed gas), the "hot" electrons successfully move ahead and leave the compression shock behind, since their velocity is approximately $(m_i/m_e)^{1/2}$ times greater than the front velocity. A preheating layer is thus formed ahead of the compression shock. In this layer the electron temperature is higher than the ion temperature, since the electron gas is heated first and only then is the heat partially transferred to the ions. A sharp increase in the ion temperature takes place in the compression shock.

The electron temperature remains constant, since its tendency to a discontinuous increase is prevented by the "smoothing out" due to the large heat conduction. The compression shock has an "isoelectron-thermal" character. The temperature profiles in the wave front, taking into account electron heat conduction, are shown in Fig. 7.19.

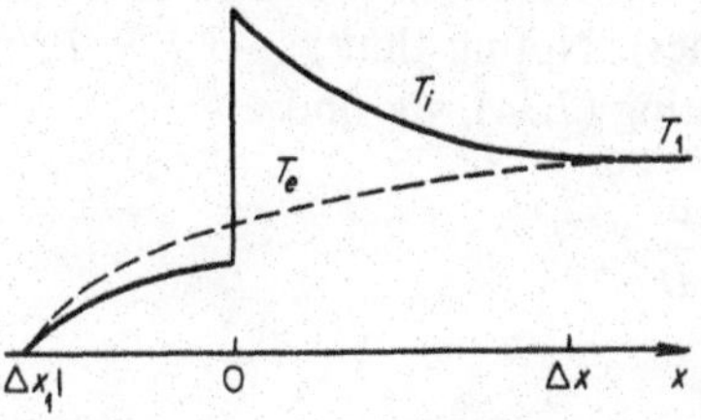

Fig. 7.19. Ion and electron (dashed line) temperature profiles in a shock front propagating through a cold plasma.

Let us estimate the thickness of the preheating layer ahead of the compression shock. We shall assume, for simplicity, that no energy is transferred from the electron gas which is being preheated to the ion gas, and also that the gas ahead of the compression shock is not compressed and not slowed down (in a coordinate system where the front is at rest). Exact calculations justify these simplifying assumptions. The electron heat conduction flux is

$$S = -\kappa_e \frac{dT_e}{dx} = -\chi_e c_e \frac{dT_e}{dx}, \tag{7.33}$$

where $\kappa_e = \chi_e c_e$ is the coefficient of thermal conductivity and c_e is the specific heat at constant volume of the electron gas per unit volume. The effective coefficient of electron thermal conductivity is

$$\kappa_e = \xi \frac{(kT_e)^{5/2}k}{m_e^{1/2}Ze^4 \ln \Lambda} = \xi \cdot 1.93 \cdot 10^{-5} \frac{T_e^{\circ 5/2}}{Z \ln \Lambda} \quad \frac{\text{erg}}{\text{cm} \cdot \text{sec} \cdot \text{deg}},$$

where $\ln \Lambda$ is the Coulomb logarithm (see §20 of Chapter VI), and ξ is a number which depends only weakly on Z; $\xi(1) = 0.95$, $\xi(2) = 1.5$, $\xi(4) = 2.1$*.

* We present, for reference purposes, the formula for the electrical conductivity of a plasma

$$\sigma = 2.63 \cdot 10^{-4}\gamma(Z) \frac{T^{\circ\, 3/2}}{Z \ln \Lambda} \quad (\text{ohm} \cdot \text{cm})^{-1},$$

$$= 2.38 \cdot 10^{8}\gamma(Z) \frac{T^{\circ\, 3/2}}{Z \ln \Lambda} \quad \text{sec}^{-1},$$

$$\gamma(1) = 0.58; \qquad \gamma(2) = 0.68; \qquad \gamma(4) = 0.78.$$

Since the process is steady, the heat conduction flux in the preheating layer is equal to the hydrodynamic flux of electron energy*

$$-S = Dc_eT_e = \chi_e c_e \frac{dT_e}{dx} \tag{7.34}$$

(the initial electron temperature ahead of the front is assumed to be zero; far ahead of the wave the flux S vanishes). Noting that $\chi_e \sim \bar{v}_e l_e \sim T_e^{5/2}$ or $\chi_e = aT_e^{5/2}$, where $a = const$, and integrating (7.34), we find

$$x - x_0 = \frac{2}{5}\frac{a}{D} T_e^{5/2}$$

or

$$T_e = \left[\frac{5}{2}\frac{D}{a}(x - x_0)\right]^{2/5}, \tag{7.35}$$

where x_0 is the coordinate of the leading edge of the preheating region, from which the temperature departs from the value zero. The temperature profile described by this equation is shown schematically in Fig. 7.19. If we place the coordinate origin $x = 0$ at the point of the compression shock and denote the temperature at this point by T_{e0} (the electron temperature does not change across the shock), then the expression for the thickness of the preheating layer can be written

$$|x_0| = \frac{2}{5}\frac{a}{D} T_{e0}^{5/2} = \frac{2}{5}\frac{\chi_e(T_{e0})}{D}. \tag{7.36}$$

When electron heat conduction is taken into account, the electron temperature at the shock is of the same order as behind the entire wave front, so that the thickness of the preheating layer is of the same order as the thickness of the relaxation region behind the shock

$$|x_0| \sim \frac{\chi_e(T_1)}{D} \sim \frac{\bar{v}_e l_e}{D} \sim \left(\frac{m_i}{m_e}\right)^{1/2} l_e \sim \left(\frac{m_i}{m_e}\right)^{1/2} l_i \sim \Delta x_{\text{exch}}.$$

The thickness of the preheating layer ahead of the compression shock increases rather rapidly with an increase in wave strength. If we consider the fact that $\chi_e \sim T_e^{5/2} \sim T_1^{5/2}$, and $D \sim T_1^{1/2}$, then we find from (7.36) that $|x_0| \sim T_1^2 \sim D^4$.

The temperature profile we have found is characteristic of nonlinear heat

* This is a first integral of the energy equation for the given case:

$$c_e \frac{dT_e}{dt} = -\frac{\partial S}{\partial x}; \qquad Dc_e \frac{\partial T_e}{\partial x} = -\frac{\partial S}{\partial x}; \qquad Dc_eT_e = -S.$$

conduction with the coefficient of thermal conductivity decreasing with decreasing temperature*. In the case of ordinary heat conduction with a constant coefficient of thermal conductivity $\kappa = const$, $\chi = const$ we would have found from the energy equation that the preheating extends exponentially to infinity:

$$T = T_1 e^{-|x|/x_1}, \qquad T_1 = T(x = 0),$$

where the characteristic scale is $x_1 = \chi/D$. In the case of ordinary heat conduction, the effective thickness of the preheating region x_1, in contrast to the nonlinear case, decreases with increasing wave strength: $x_1 \sim D^{-1} \sim T_1^{-1/2}$.

When electron heat conduction is taken into account in very strong shocks, the un-ionized gas is strongly preheated and ionized even before the compression shock, so that the qualitative features of the structure of a wave propagating through an ionized gas remain the same also in the case when the wave travels through an un-ionized gas.

For an exact calculation of the structure of a shock front in a fully ionized gas we must add to the hydrodynamic equations which account for electron heat conduction, such as (7.10), the entropy equation for an electron gas similar to (7.30)

$$n_i u T_e \frac{d\Sigma_e}{dx} = -\frac{dS}{dx} + \omega_{ei}, \tag{7.37}$$

where ω_{ei} is the energy transferred per unit volume per unit time from the ion to the electron gas; it is given by (7.31). The velocity and the charge density of both gases at any point are assumed to be equal ($n_e = Zn_i$). The enthalpy and pressure are

$$h = h_i + h_e = \tfrac{5}{2}kT_i + Z\tfrac{5}{2}kT_e,$$

$$p = p_i + p_e = n_i kT_i + n_e kT_e.$$

The entropy of the electron gas per ion is

$$\Sigma_e = Zk \ln \frac{T_e^{3/2}}{n_e} + const = Zk \ln \frac{T_e^{3/2}}{\rho} + const. \tag{7.38}$$

Let us find the condition which determines the electron temperature in the compression shock. To do this we integrate (7.37) over the region of the compression shock as its thickness tends to zero, noting that the electron temperature across the shock is continuous (it is equal to T_{e0}). Denoting

* More details on nonlinear heat conduction are given in Chapter X.

quantities ahead of and behind the shock by the subscripts 01 and 02, we can write the result of integration as

$$\rho_0 D T_{e0} \ln \frac{\rho_{02}}{\rho_{01}} = S_{02} - S_{01}. \tag{7.39}$$

The flux has a jump at the discontinuity: the difference between the fluxes on both sides of the discontinuity corresponds to the work to isothermally compress the electron gas by the "external" forces exerted by the ions.

The system of equations describing the front structure can be solved only by numerical integration. This was done by Shafranov [43] for the limiting case of a strong wave ($p_1/p_0 \gg 1$) in a hydrogen plasma ($Z = 1$) with zero initial temperature. The temperature and density distributions are shown in Fig. 7.20. The temperature T_1 here is arbitrary (it is proportional to the

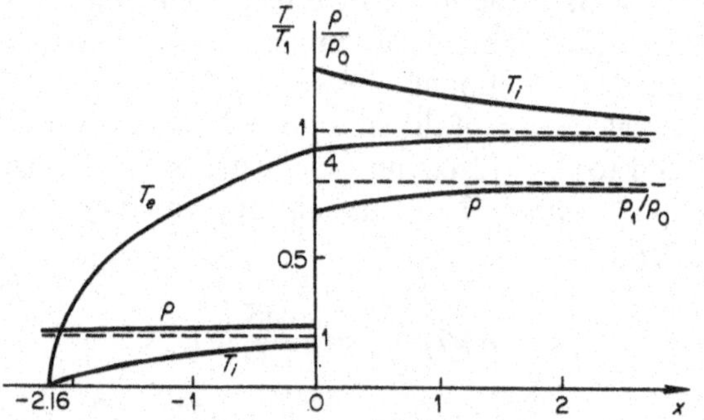

Fig. 7.20. Temperature and density distributions for a strong shock wave in a plasma (the figure is taken from [43]). The electron temperature at the compression shock is $T_{e0} = 0.93T_1$; the ion temperatures ahead and behind the compression shock are $T_{i_1} = 0.16T_1$ and $T_{i_2} = 1.24T_1$, respectively. The densities ahead and behind the compression shock are $\rho_{01}/\rho_0 = 1.13$ and $\rho_{02}/\rho_0 = 3.53$.

square of the wave velocity D). The unit of length that has been used is $0.019 D\tau_{ei_1}$, where τ_{ei_1} is the characteristic exchange time in the final state behind the shock wave front; for example, at an initial density $n_{i0} = n_{e0} = 10^{17}$ cm^{-3} and temperature behind the front $T_1 = 10^{5}$°K, $\tau_{ei_1} = 3.3 \cdot 10^{-9}$ sec, $D = 94$ km/sec, and the unit of length is $5.9 \cdot 10^{-4}$ cm.

§13. Polarization of a plasma and the creation of an electric field in a shock wave

In the preceding section we have assumed that the electrons and ions are rigidly bound to each other by electric forces, and that the plasma is electrically neutral at each point in the shock; the electron density changes from point to point and is exactly proportional to the ion density. Actually, however, this assumption is not strictly fulfilled. Due to the presence of large electron density gradients in the compression shock, and to the high mobility of electrons resulting from their exceedingly small mass, conditions are favorable for the diffusion of the electron gas with respect to the ion gas, for changes in the electron concentration, and for the creation of space charges.

The effects of diffusion on the propagation of a shock wave in a binary gas mixture were considered in §5. Diffusion in a plasma, however, is substantially different from diffusion in a mixture of neutral gases. The point is that the smallest change in the relative concentration of the electrons and ions leads to the creation of space charges, i.e., to polarization of the plasma accompanied by the creation of a strong electric field. This field prevents further polarization and inhibits the diffusion electron current.

Let us estimate the order of magnitude of the polarization of a plasma in the presence of gradients of the macroscopic quantities, and the degree to which, on average, the condition of electrical neutrality is satisfied. For simplicity, let us consider a hydrogen plasma ($Z = 1$). We assume the electron temperature to be of the order of T and the electron and ion number density to be given approximately by $n_e = n_i = n$. Furthermore, let us also assume the presence of gradients in the macroscopic quantities, such as density, pressure, etc., such that the characteristic dimension of the region in which appreciable changes in these quantities take place is of the order of x. As a result of electron diffusion in a region of the order of x there will be a difference between the electron and ion densities, $\delta n = n_i - n_e$, which results in the creation of a space charge $e\delta n$. This gives rise to an electric field $E \sim 4\pi e \cdot \delta n \cdot x$* and to a potential difference across the boundaries of the region $\delta\phi \sim Ex \sim 4\pi e\delta n \cdot x^2$. But in the absence of external fields the separation of ions and electrons and the potential difference are maintained only by the thermal motion of the electrons; consequently, the potential energy of the electrons $e\delta\phi$ cannot exceed a value of the order of kT; $e\delta\phi \sim 4\pi e^2 \cdot \delta n \cdot x^2 \sim kT$. From this, the degree of polarization, that is, the extent of the departure from electrical neutrality in the region being considered, is of the order of

$$\frac{\delta n}{n} \sim \frac{kT}{4\pi e^2 n x^2}.$$

A strong separation of the ions and electrons, for which $\delta n/n \sim 1$, can come

* We recall that the electrostatic equations for the field intensity E and the potential ϕ are

$$\nabla \cdot \mathbf{E} = 4\pi e \cdot \delta n, \qquad \mathbf{E} = -\nabla\phi.$$

about only in a thin layer whose thickness d is determined by the condition $\delta n/n \approx 1 \approx kT/4\pi e^2 nd^2$. From this

$$d \approx \left(\frac{kT}{4\pi e^2 n}\right)^{1/2} = 6.9\left(\frac{T°}{n}\right)^{1/2} \text{ cm.}$$

The length d is simply the Debye radius (see §11, Chapter III)*. It characterizes the distance over which the plasma screens the electric field of any charged body, that is, the thickness of the so-called double layer which forms around a charged body. In particular an individual ion can serve as the "charged body" (this is precisely the manner in which the concept of the Debye radius was introduced in §11, Chapter III).

Using the definition of d, the departure from electrical neutrality may be expressed as $\delta n/n \sim (d/x)^2$. The largest gradients in the plasma appear in the viscous compression shock of a propagating strong shock wave, when the macroscopic variables change appreciably over a distance of the order of the mean free path of the charged particles

$$l \sim \frac{(kT)^2}{ne^4 \ln \Lambda} \sim 3.5 \cdot 10^4 \frac{T^{\circ 2}}{n} \text{ cm†.}$$

The average departure from electrical neutrality in the compression shock region ($x = x_{\min} \sim l$) is

$$\frac{\delta n}{n} \sim \left(\frac{d}{l}\right)^2 \sim \frac{e^6 n(\ln \Lambda)^2}{4\pi(kT)^3} \approx 3.9 \cdot 10^{-8} \frac{n}{T^{\circ 3}}.$$

This quantity is very small for all reasonable values of density and temperature; for example, at $T = 10^{5}$ °K, $n = 10^{18}$ cm^{-3}, $d \approx 0.8 \cdot 10^{-6}$ cm, $l \approx 3.5 \cdot 10^{-4}$ cm, $\delta n/n \sim 4 \cdot 10^{-5}$, $\delta\phi \sim kT/e = 8.6$ volts‡, and $E \sim \delta\phi/l \sim 2.5 \cdot 10^4$ volts/cm.

We note that the compression of the electron gas in a strong compression shock whose thickness is of the order of a mean free path l is due only to the electric forces exerted by the ions (the compression of the ion gas is usually due to viscosity). Consequently, the potential difference in the compression shock is determined by the work done per electron in compressing the electron gas $e(\phi_{02} - \phi_{01}) = kT_{e0} \ln(\rho_{02}/\rho_{01})$. For a several-fold compression the logarithm is of the order of unity and, consequently, $e\delta\phi \sim kT$, as was stated above.

The distribution of charge, electric field, and potential in a shock front in a

* More precisely, the Debye radius multiplied by $\sqrt{2}$, since n in (3.78) is the total number of ions and electrons.

† The logarithmic factor in the mean free path (see §20, Chapter VI) is usually of the order $\ln \Lambda \sim 10$.

‡ The numerical value of $\delta\phi$ is of the order of the temperature in electron volts.

plasma are shown in Fig. 7.21. The significant difference between the distributions of the concentrations of the different components in a plasma and the distributions of the concentrations in a mixture of neutral gases lies in the fact that, together with the region of increased electron concentration in the

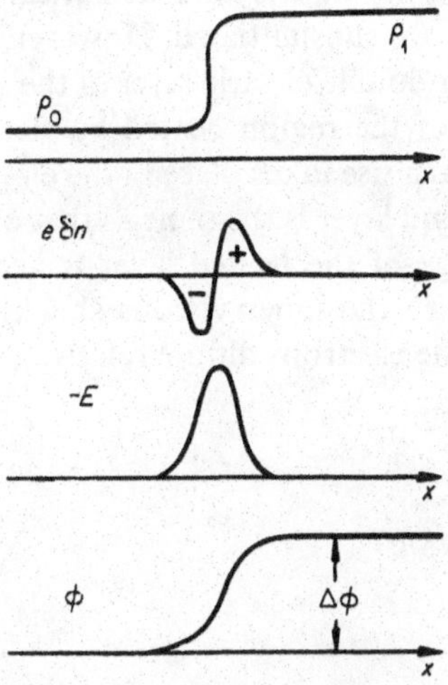

Fig. 7.21. Mass density, space charge density, electric field intensity, and electrostatic potential distributions in a shock front propagating in a plasma with electron diffusion taken into account.

forward part of the shock front, there arises a region of decreased concentration in the rear part of the wave. For a mixture of neutral gases only the light component concentrates in the shock front (the excess mass of the light component arrives from "infinity"). This situation is not possible in the case of a plasma. Concentration of electrons without a simultaneous concentration of positive ions in a neighboring region would lead to the appearance of an electric field at "infinity", that is, it would require the expenditure of an infinite amount of energy.

In [45], the structure of a weak shock front in a plasma was examined, taking into account only the effect of electron diffusion impeded by the electric forces and without taking into account either viscosity or heat conduction, in a manner similar to that of Cowling [22] for a mixture of electrically neutral gases (see §5)*. As in the electrically neutral case diffusion guarantees the spreading out of a shock discontinuity which is not too strong. Because of the restraining effect of the electric field the thickness of the transition layer is smaller than in the neutral gas mixture.

Jaffrin and Probstein [89] have considered the structure of a shock wave in a fully ionized plasma in a general form, simultaneously taking into account the viscosity, the thermal conductivity, and the polarization and charge separation

* See also [44, 46].

of the plasma. They have used as their starting point a system of hydrodynamic equations for a mixture of electron and ion gases and Poisson's equations for the electric field. The mass density, electron, and ion temperature distributions which were obtained in this paper agree with the distributions as described in the preceding section. The qualitative considerations given on the formation of a double electric layer, shown schematically in Fig. 7.21, in the region of the viscous shock are substantiated. However, in a strong shock wave there appears still another double electric layer of the same type, which is located at the leading edge of the region heated by electron thermal conduction, at the point where a sharp rise takes place in the electron temperature. The physical nature of this second layer is the same as that of the basic layer: the temperature rise at the edge of the heated layer is accompanied by a fairly small (in comparison with the inner viscous shock) but very sharp compression, and in this region the electrons diffuse relative to the ions and a charge separation results.

3. *Radiant heat exchange in a shock front*

§14. Qualitative picture

When a shock wave is propagated through a gas occupying a large volume, and the dimensions of the heated region are very large in comparison with the mean free path of a photon so that the gas temperature changes very little over a distance of the order of a mean free path, the thermal radiation in a wave is brought into local thermodynamic equilibrium with the fluid. Radiative equilibrium also exists immmediately behind the shock front.

The energy density and radiation pressure become comparable with the energy density and pressure of the fluid only at extremely high temperatures or extremely low gas densities. For example, in standard density air this occurs at a temperature of $\approx 2.7 \cdot 10^6$ °K. The radiation energy and pressure in shock waves of not too high a strength are much smaller than the energy and pressure of the fluid, and therefore have almost no effect on quantities behind the front. The relationship between the radiation energy flux and the energy flux of the fluid is, however, different, since the shock velocities encountered in practice are much smaller than the speed of light. The ratio of energy fluxes $\sigma T^4/D\rho\varepsilon \sim (U_{\text{rad}}/\rho\varepsilon)(c/D)$, is, roughly speaking, greater by a factor of c/D than the ratio of energy densities. Thus, for $D = 100$ km/sec, $c/D = 3 \cdot 10^3$. In atmospheric air, for example, both fluxes become equal already at a temperature $T \sim 300{,}000$°K, for which the radiation density is still very small.

It would seem that radiative transfer of energy away from the front of a strong shock must play an important role, and that therefore in the third equation of (7.4) with the energy flux of the fluid we should also include the energy flux carried away from the surface of the front by radiation $S = \sigma T_1^4$. This could have an appreciable effect on the final state behind the shock front, and could lead to a high density behind the front similar to the high density obtained with increased specific heats. Actually, however, the energy lost by radiation from the front surface is rather limited and the effect of this loss is usually negligible. The point is that in a continuous spectrum gases are transparent only to the photons of comparatively low energy. Both atoms and molecules strongly absorb photons whose energies exceed the ionization potentials, giving rise to the photoelectric effect, while molecules, as a rule, absorb photons of even lower energies; for example, the boundary of the transparent region for cold air lies at $\lambda \sim 2000$ Å and $h\nu \sim 6$ ev. When the temperature behind the front is high, the energy contained in the low-frequency region comprises only a small fraction of the total energy in the spectrum. Thus, at a temperature behind the front of $T = 50{,}000$°K only 4.5% of the energy of the Planck spectrum is concentrated in the transparent region of air $h\nu < 6$ ev. In this case the low energy photons are in the Rayleigh–Jeans region of the spectrum and their flux (and the corresponding possible energy losses) is in any case proportional not to the fourth but only to the first power of temperature.

The major part of the radiation from the shock front actually escapes to "infinity" only at temperatures for which the maximum of the Planck spectrum lies in the transparent region of the spectrum, at temperatures behind the front of the order of 1–2 ev. At such temperatures, however, the absolute magnitude of the radiation flux σT_1^4 is very small and the additional density increase resulting from radiation losses in air at standard density does not exceed one percent.

Thus, the presence of thermal radiation has only a very small effect on the flow variables behind the front of a not too strong shock. However, this is not the case for the effect of the radiation on the internal structure of the transition layer between the initial and final thermodynamic equilibrium states of the gas, on the structure of the shock front itself. Here the radiation in strong waves (which are of real interest) is found to play a very important role and, moreover, it is precisely the radiant heat exchange which determines the front structure. The problem of the structure of a shock front taking into account radiant heat exchange, to which §§14–17 of the chapter are devoted, has been studied by the present authors in [42, 47–49]. Although the flux of radiation going out from the wave front to "infinity" is very small and exerts no influence in terms of energy on the shock wave flow variables, the fact that it exists is of tremendous importance since it enables the observation of the wave by optical methods. The problem of shock wave luminosity and the brightness of the front surface is closely interwoven with the problem of the front structure. It will be considered in Chapter IX.

Owing to the opaqueness of the cold gas, the radiation emanating from the surface of the shock discontinuity in strong waves is almost entirely absorbed ahead of the discontinuity and heats the layers of gas flowing into the discontinuity. The energy which goes into the heating is produced by emission from the gas layers which have already suffered a shock compression and which as a result are cooled by the radiation. The effect thus reduces to the transfer of energy from one gas layer to the others by radiation. Radiant heat exchange takes place in distances of the order of the absorption mean free path of the photons. Usually the photon mean free path is several orders of magnitude larger than the gaskinetic mean free path of the particles (see Chapter V) and is larger than the thickness of the relaxation layer in which thermodynamic equilibrium is established in the fluid. Thus, in air at standard density the mean free paths for photons with energies $h\nu \sim$ 10–100 ev, corresponding to temperatures behind the front of $T_1 \sim 10^4$–10^5 °K, are of the order of 10^{-2}–10^{-1} cm, while the gaskinetic mean free path is of the order of 10^{-5} cm.

The shock front thickness, in which the radiant heat exchange plays an important role in the energy balance, is determined by the photon mean free path, which is the greatest length scale. In a sense we can speak about the relaxation of radiation in a shock front and about the establishment of equilibrium between the radiation and the fluid behind the front. Let us follow qualitatively the change in the front structure going from weak to strong waves. Here we shall consider the phenomenon on a "large scale", disregarding the "small scale" details related to the relaxation of the various degrees of freedom of the gas; we assume that the fluid at each point of the wave is in thermodynamic equilibrium. The viscous compression shock together with the relaxation region behind it will be considered as a mathematical discontinuity.

In the limiting case of a sufficiently weak wave when the role of the radiation on the energy balance is small, the profiles of all the variables across the

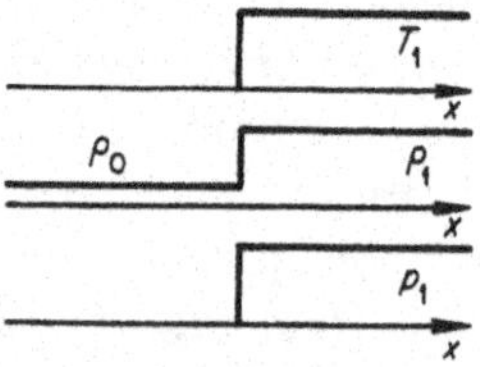

Fig. 7.22. Temperature, density, and pressure profiles in a "classical" shock wave.

shock wave have the "classical" step-like character (Fig. 7.22). As the strength increases, the radiation flux from the front surface σT_1^4 increases very rapidly.

The radiation is absorbed ahead of the discontinuity at a distance of the order of a photon mean free path and heats the gas; the heating drops off with distance from the discontinuity as a result of absorption of the radiation flux. The compression shock is now propagated not through a cold, but through a heated gas and the temperature behind the shock T_+ is higher than without heating, that is, it is higher than in the final state. The temperature behind the compression shock decreases from T_+ to T_1. In other words, a gas particle passing through the shock wave is first heated by radiation and, after being subjected to a shock compression, is then cooled by the emission of part of its energy as radiant flux. Heating of the gas ahead of the discontinuity leads to an increase in its pressure and to some density increase (and also to a slowing down, in a coordinate system in which the front is at rest). In the compression shock the gas is compressed to a density slightly lower than the final one. The cooling of the gas behind the compression shock helps to compress it further to the final density (as in the case of the decrease in temperature resulting from the excitation of additional degrees of freedom). In this process the pressure also increases. The profiles of temperature, density, and pressure for the wave described are shown schematically in Fig. 7.23.

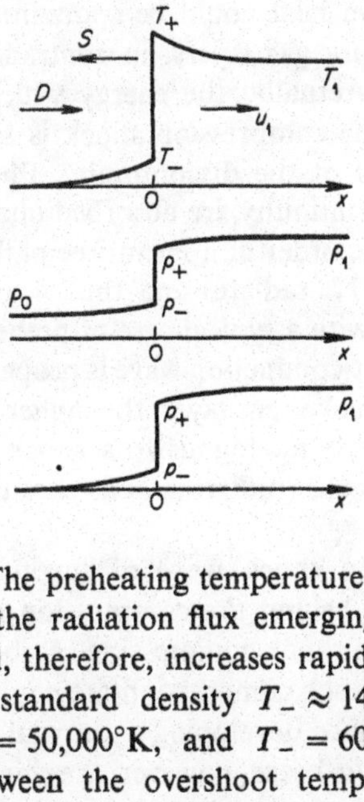

Fig. 7.23. Temperature, density, and pressure profiles in a shock front of not too large a strength, taking into account radiant heat exchange.

The preheating temperature ahead of the discontinuity T_- is proportional to the radiation flux emerging from the discontinuity surface $-S_0 \approx \sigma T_1^4$, and, therefore, increases rapidly with increasing wave strength. Thus, in air at standard density $T_- \approx 1400°\text{K}$ for $T_1 = 25{,}000°\text{K}$, $T_- = 4000°\text{K}$ for $T_1 = 50{,}000°\text{K}$, and $T_- = 60{,}000°\text{K}$ for $T_1 = 150{,}000°\text{K}$. The difference between the overshoot temperature behind the shock T_+ and the final temperature T_1 increases correspondingly (roughly speaking, $T_+ - T_1 \approx T_-$). At some temperature behind the front $T_1 = T_{cr}$, the preheating temperature T_- reaches the value of T_1 and the temperature profile assumes the shape shown in Fig. 7.24. This temperature T_{cr} is approximately equal to 300,000°K for air, and can be called critical since it divides two rather different types of shock front structure.

Fig. 7.24. Temperature profile in a shock wave of "critical" strength.

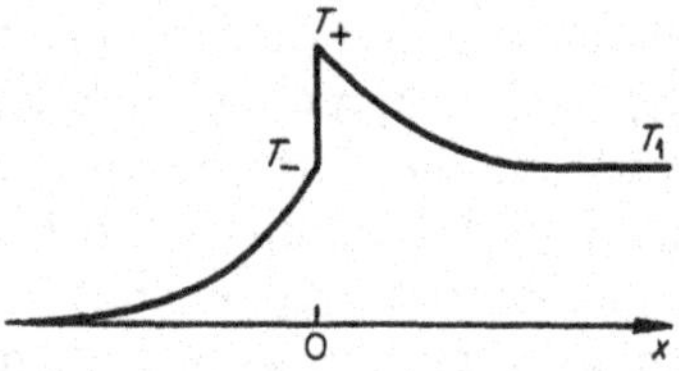

Let us consider a strong wave of supercritical amplitude, with a temperature behind the front $T_1 > T_{cr}$. The photon energy flux emitted by the gas behind the compression shock and emerging from the discontinuity surface toward the cold gas would be sufficient to heat a layer whose thickness is of the order of a photon absorption mean free path to a very high temperature, higher than T_1. Can such an intensive heating be actually achieved? Obviously, it cannot, since otherwise the preheating layer would begin to radiate strongly and cool down very rapidly to the temperature T_1. The formation of a state with $T_- > T_1$ would mean that in a closed system heat could be spontaneously transferred from the low- to high-temperature gas layers, in contradiction with the second law of thermodynamics*. Actually, the energy which the radiation removes from the gas heated in the compression shock is simply used up in heating the thicker layers ahead of the discontinuity. Photons emerging from behind the surface of the discontinuity are absorbed ahead of the discontinuity in a layer of thickness of the order of a mean free path and the fluid, heated to a temperature close to T_1, radiates and thus heats the neighboring layers, etc. We are dealing here with a typical case of preheating of the gas by radiation heat conduction. A heat-conduction wave is propagated ahead of the discontinuity, encompassing a thicker gas layer, the higher is the shock strength. The phenomenon is completely analogous to a shock wave with electron heat conduction considered in §12 (radiation heat conduction is also nonlinear).

The temperature and density profiles in a shock wave of supercritical strength are shown in Fig. 7.25. As before, behind the compression shock there is a temperature peak resulting from the shock compression. As before, the gas particles which have undergone the shock compression are cooled by emitting a part of their energy, which goes into developing a thermal wave ahead of the discontinuity. Unlike the subcritical case, however, the thickness of the peak is now less than a radiation mean free path and decreases with increasing wave strength (see also §17).

* Additional details on the impossibility of a state with $T_- > T_1$ will be presented in §17. A rigorous proof of this statement is given in [42].

In the radiation heat conduction approximation, where the details of the phenomena occurring at distances less than a mean free path are not considered, the peak is "cut off" as shown by the dashed line in Fig. 7.25, and

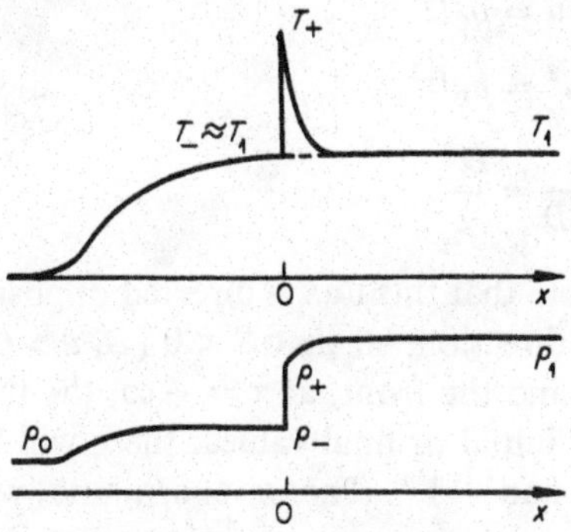

Fig. 7.25. Temperature and density profiles in a very strong shock front taking into account radiant heat exchange. The dashed line corresponds to the radiation heat conduction approximation (isothermal jump).

the shock takes on the character of an "isothermal" shock (see §3 of this chapter). In the following sections the physical picture whose general features have been outlined above will be justified mathematically.

§15. Approximate formulation of the problem of the front structure

As usual, we shall consider a one-dimensional steady flow in a coordinate system in which the front is at rest. A number of simplifications will be introduced to make clear the specific features of the front structure which are related to radiant heat exchange. The gas will be assumed to be a perfect one with constant specific heats, so that its pressure and specific internal energy may be expressed by the simple relations

$$p = R\rho T, \qquad \varepsilon = \frac{1}{\gamma - 1} RT.$$

The viscous compression shock, together with the relaxation layer in which thermodynamic equilibrium in the fluid is established, will be replaced by a mathematical discontinuity. We shall neglect relaxation phenomena, viscosity, heat conduction, and also electron heat conduction in the radiant heat exchange region*. The shock wave is taken to be strong (the initial pressure and energy of the fluid are small in comparison with the final values). We shall not consider, however, extremely strong waves; in our case we may neglect the energy and pressure (but not the flux!) of radiation. We also neglect the small flux of low energy photons escaping from the wave front to

* Estimates in a number of actual cases, including the practically important case of a shock wave in standard density air, show that electron heat conduction plays less of a role than that of the transfer of energy by radiation (see [48]).

"infinity" by assuming that the radiation flux ahead of the front is zero.

Using the above assumptions the system of integrals of the hydrodynamic equations (7.10) takes the following form:

$$\begin{gathered} \rho u = \rho_0 D, \\ p + \rho u^2 = \rho_0 D^2, \\ \varepsilon + \frac{p}{\rho} + \frac{u^2}{2} + \frac{S}{\rho_0 D} = \frac{D^2}{2}. \end{gathered} \tag{7.40}$$

Here S is the radiation energy flux. We note that this flux is directed opposite to the gas flow moving in the positive x direction, so that $S < 0$ $(D, u > 0)$. Ahead of the front, at $x = -\infty$, and behind the front, at $x = +\infty$, the flux $S = 0$, and all the variables take on their initial or final values; these will be denoted again by the subscripts "0" and "1". The x coordinate will be measured from the compression shock.

In order to determine the radiation flux, we must add to the hydrodynamic equation (7.40) the radiative transfer equation. We shall consider the angular distribution of photons within the framework of the diffusion approximation, replacing the rigorous kinetic equation for the intensity by two equations for the density and flux of radiation (see §10, Chapter II). We emphasize that the diffusion approximation does not formally make any assumptions regarding the proximity of the radiation density to its equilibrium value, and that the diffusion approximation is by no means equivalent to the radiation heat conduction approximation. The diffusion approximation can also be used to describe nonequilibrium radiation, by taking into account the angular distribution of photons only approximately (see the discussion of this point in §13, Chapter II). We shall use only the spectrally integrated values of the radiation density and flux U and S, introducing for this purpose a photon mean free path l appropriately averaged with respect to frequency. As noted in Chapter II, this approximation is, strictly speaking, possible only in definite limiting cases. It does not, however, alter the qualitative relations governing the radiative heat transfer and is, therefore, sufficient for our purposes.

The radiation equations in the above approximations (see (2.62) and (2.65)) can be written

$$\frac{dS}{dx} = \frac{c(U_p - U)}{l},$$

$$S = -\frac{lc}{3}\frac{dU}{dx}.$$

Here $U_p = 4\sigma T^4/c$ is the energy density of equilibrium radiation, corresponding to the temperature of the fluid at the given point x.

Since the hydrodynamic and radiative transfer equations do not explicitly contain x, we can transform to a new coordinate—the optical thickness τ, measured from the point $x = 0$ in the positive x direction

$$d\tau = \frac{dx}{l}, \qquad \tau = \int_0^x \frac{dx}{l}. \tag{7.41}$$

If the mean free path l is known as a function of the temperature and density, we can easily transform the various quantities in the final solution from distributions with respect to the optical coordinate to distributions with respect to x, by means of equations (7.41) (for $l = const$ both distributions are obviously the same). In terms of the optical thickness the transfer equations take the form

$$\frac{dS}{d\tau} = c(U_p - U), \tag{7.42}$$

$$S = -\frac{c}{3}\frac{dU}{d\tau}. \tag{7.43}$$

The hydrodynamic equations (7.40) and the radiative transfer equations (7.42) and (7.43), together with the natural boundary conditions expressing the absence of radiation in the cold gas ahead of the wave and the fact that the radiation behind the wave front is in thermodynamic equilibrium*

$$\tau = -\infty, \quad S = 0, \quad U = 0, \qquad T = 0, \tag{7.44}$$

$$\tau = +\infty, \quad S = 0, \quad U = U_{p_1} = \frac{4\sigma T_1^4}{c}, \qquad T = T_1, \tag{7.45}$$

completely describe the structure of the shock front within the present statement of the problem. The system of differential equations is of second order. The order can be decreased by eliminating τ from the system by dividing equations (7.42) and (7.43) by each other

$$\frac{dS}{dU} = \frac{c^2}{3}\frac{U - U_p}{S}. \tag{7.46}$$

The (p, V), (T, V), and (S, V) diagrams considered in §3 are very convenient for clarifying the physical meaning of the relations governing the front structure. Introducing again the relative specific volume $\eta = V/V_0$, equal to the reciprocal of the density ratio or to the velocity ratio

$$\eta = \frac{V}{V_0} = \frac{\rho_0}{\rho} = \frac{u}{D},$$

* Only two of these conditions are independent, the others follow from the equations.

we find from the first two equations of (7.40) that in the regions where the flow variables are continuous the pressure changes along the straight line

$$p = \rho_0 D^2(1 - \eta). \tag{7.47}$$

The dependence of temperature and flux on the density ratio is described by relations similar to (7.13) and (7.14). These relations are obtained from (7.40) for the case of a gas with constant specific heats. Replacing the Mach number in (7.13) and (7.14) by the temperature behind the shock front T_1, we obtain

$$T = \frac{T_1 \eta(1 - \eta)}{\eta_1(1 - \eta_1)}, \tag{7.48}$$

$$S = -\frac{\rho_0 D R T_1 (1 - \eta)(\eta - \eta_1)}{2\eta_1^2(1 - \eta_1)}, \tag{7.49}$$

where $\eta_1 = (\gamma - 1)/(\gamma + 1)$. Radiation in a shock wave plays an important role only at high temperatures, when the gas is strongly ionized. For numerical estimates the effective specific heat ratio in the ionization region can be taken equal to $\gamma = 1.25$. The corresponding density ratio across the wave front is $1/\eta_1 = 9$, $\eta_1 = 0.111$.

The functions $T(\eta)$, $S(\eta)$, and $p(\eta)$ are shown in Figs. 7.26 and 7.27.

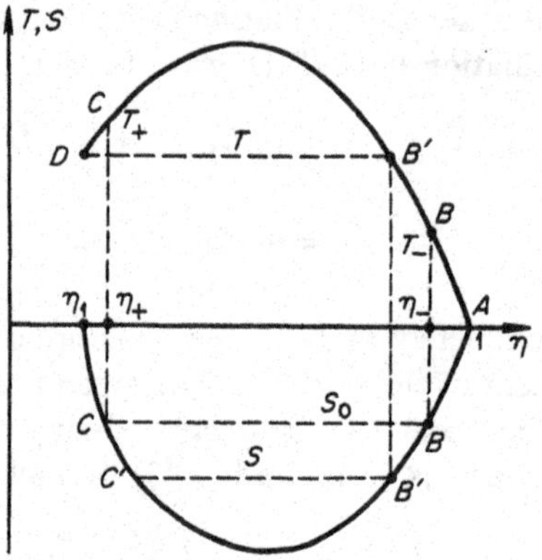

Fig. 7.26. T,η and S,η diagrams for a shock wave taking radiant heat exchange into account.

Figure 7.26 shows that the $T(S)$ curve, which can be obtained from (7.48) and (7.49), has two branches. One of them, which in the limit $S \to 0$ gives $T \to 0$ $(\eta \to 1)$, corresponds to states close to the initial state, and thus to the preheating region ahead of the discontinuity; the other, which in the limit $S \to 0$ gives $T \to T_1$ $(\eta \to \eta_1)$, corresponds to the states close to the final state, and thus to the region behind the discontinuity.

In the following two sections we shall find approximate solutions of the equations for the two limiting cases described in §14, i.e., for shock waves of subcritical and supercritical strengths. It should be noted that the transition from the first to the second case is continuous. It is simply that for intermediate strengths, close to critical, it is not possible to find solutions in analytic form. Numerical integration for these intermediate strengths is not difficult. However, this is not really necessary, since the limiting analytic solutions found are valid for strengths very close to critical from either side.

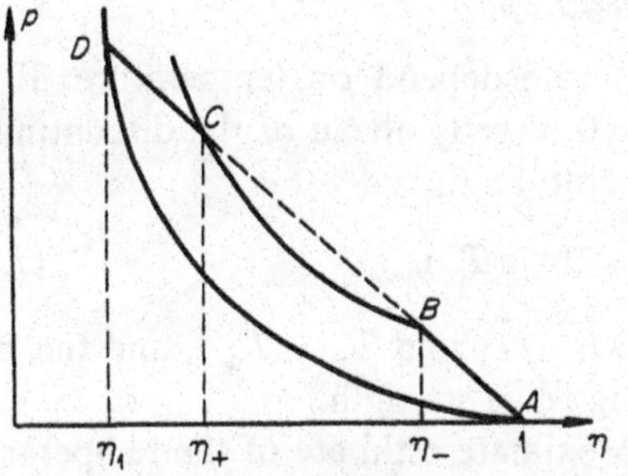

Fig. 7.27. p, η diagram for a shock wave taking radiant heat exchange into account.

§16. The subcritical shock wave

Let us consider a shock of moderate strength in which the radiation-induced effects are small. The temperature behind the compression shock will then be close to the final temperature and a radiation flux of absolute magnitude $|S_0| \approx \sigma T_1^4$ will come out from the surface of the discontinuity. Following the state of a gas particle incident on the wave, we note that the point representing the particle on the (T, η), (S, η), and (p, η) diagrams moves from the initial position A in the direction of increasing compression up to position B at which the flux is equal to S_0. The gas density, temperature, pressure, and radiation flux in the parcel increase monotonically as the discontinuity is approached. It follows from equations (7.48) and (7.49), and this is also evident from Fig. 7.26, that the increase in density on those branches of the curves which originate from the initial point A, in the preheating region, is very small. Even for $T = T_1$, the density ratio on this branch will be only $1/(1 - \eta_1) = 1.13$ (if $\gamma = 1.25$ and $\eta_1 = 0.111$), and for temperatures ahead of the discontinuity T_- smaller than T_1, the density ratio in the preheating region is even smaller. If we approximately eliminate η from (7.48) and (7.49) and express S in terms of T to second order in η_1, we obtain

$$-S = \frac{D\rho_0 RT}{\gamma - 1} = D\rho_0 \varepsilon. \tag{7.50}$$

This equation, which is obtained from the energy integral (7.40) if the terms p/ρ, $D^2/2$, and $u^2/2$ are dropped, has a simple physical meaning. It denotes the fact that the radiant energy absorbed in the preheating region is used up only in raising the gas temperature. Actually, it is easy to show that the compression work p/ρ and the change in the kinetic energy $D^2/2 - u^2/2$, which are essentially proportional to η_1, balance each other to within order η_1^2.

The equation of conservation of energy in the case when the gas is neither slowed down nor compressed, written as

$$-S = D\rho_0 \varepsilon(T, \rho_0), \tag{7.51}$$

is valid in general when the specific heats depend on temperature. If we evaluate this equation at the point $x = 0$ directly ahead of the discontinuity we find the maximum preheating temperature T_-

$$|S_0| \approx \sigma T_1^4 = D\rho_0 \varepsilon(T_-). \tag{7.52}$$

In a gas with constant specific heats $D \sim \sqrt{T_1}$ and $T_- \sim T_1^{3.5}$, and the preheating rapidly increases with increasing wave strength.

We may use (7.52) to obtain an approximate estimate of the temperature behind the front for which the temperature ahead of the compression shock T_- reaches the value T_1 (we call such a wave critical). The approximation lies in the fact that the flux from the discontinuity surface is again assumed to be σT_1^4, although it is actually slightly greater, because the temperature behind the discontinuity is slightly higher than T_1. The approximate equation for the critical condition $T_1 = T_{cr}$ is

$$\sigma T_{cr}^4 = D(T_{cr})\rho_0 \varepsilon(T_{cr}). \tag{7.53}$$

Table 7.4

TEMPERATURE AHEAD OF COMPRESSION SHOCK IN AIR AT STANDARD DENSITY

D, km/sec	T_1, °K	ε_-, ev/mol.	T_-, °K	D, km/sec	T_1, °K	ε_-, ev/mol.	T_-, °K
23.3	50,000	3.7	6,800	56.5	150,000	122	60,000
28.5	65,000	8.4	9,000	81.6	250,000	635	175,000
32.1	75,000	13.1	12,000	86.2	275,000	910	240,000
40.6	100,000	32.7	25,000	88.1	285,000	1020	285,000

In Table 7.4 are presented values of the temperature ahead of the compression shock in air at standard density calculated from (7.52), taking into account the actual variation of $\varepsilon(T)$. It is evident from the table that the critical temperature in air is approximately equal to 300,000°K (285,000°K according to (7.53)). As follows from (7.53), the critical temperature is the temperature for which the energy flux of the radiation and of the fluid become approximately equal (we recall the remark made at the beginning of §14).

Returning to the original equations for a gas with constant specific heats, let us find the approximate solution in the preheating region of a subcritical wave. If the temperature in the preheating region is small in comparison with the temperature behind the front ($T_- < T_1$), then the equilibrium radiation density, which is proportional to the fourth power of the gas temperature ($U_p \sim T^4$), is much less than the actual density U, which is determined by radiation passing through the gas and coming out from behind the surface of discontinuity at a temperature T_1 ($U \sim |S_0| \sim T_1^4$). The radiation born in the preheating region in this case makes only a small contribution to the total flux and density. The radiation density in the preheating region is therefore significantly out of equilibrium. Neglecting in (7.42) and (7.46) U_p in comparison with U, we find for the solution ahead of the discontinuity for $\tau < 0$

$$-S = \frac{cU}{\sqrt{3}} = -S_0 e^{-\sqrt{3}|\tau|}, \tag{7.54}$$

$$T = T_- e^{-\sqrt{3}|\tau|}, \tag{7.55}$$

$$\frac{\rho - \rho_0}{\rho_0} = \frac{\rho_- - \rho_0}{\rho_0} e^{-\sqrt{3}|\tau|}, \tag{7.56}$$

$$p = p_- e^{-\sqrt{3}|\tau|}. \tag{7.57}$$

All the quantities decrease exponentially with optical thickness with increasing distance from the discontinuity (Fig. 7.28). The values of T_-, ρ_-, and

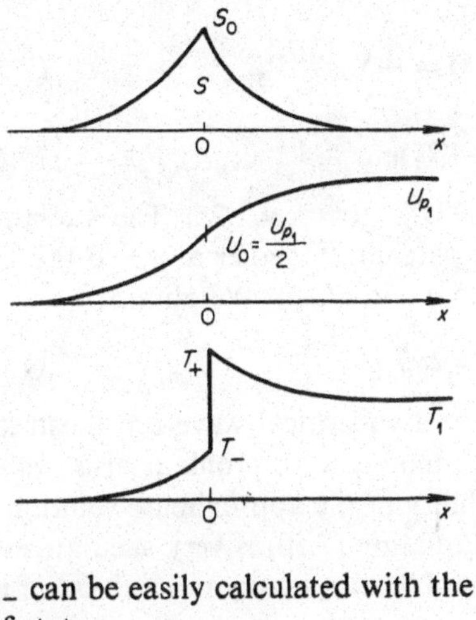

Fig. 7.28. Radiation flux, density of radiation, and temperature profiles in a subcritical shock wave.

p_- can be easily calculated with the aid of (7.52) and (7.48) and the equation of state.

The radiation density and flux at the shock discontinuity remain continuous. Actually, according to (7.43), a discontinuity in the radiation density would correspond to an infinite value of the flux, which from conservation of energy must actually be finite. A discontinuity in flux would lead to an unsteady accumulation or loss of radiant energy at the discontinuity. Consequently, a representative point in the T, η and S, η diagrams on passing through the viscous shock jumps from a point B on one of the branches to position C on another branch, corresponding to the same flux S_0. (The derivative of the flux in this case, of course, is discontinuous.) The same is true on the p, η diagram: the gas is compressed through the compression shock following the Hugoniot curve CB passing through the point B. After the shock compression the state of the particle monotonically approaches its final position at the point D. Here, the temperature and radiation flux decrease and the gas density and pressure increase.

As is evident from the T, η diagram, in a wave of not too large strength the temperature difference between the value behind the discontinuity T_+ and the value behind the front T_1 is, like the volume change behind the discontinuity, not large. Eliminating η as before from (7.48) and (7.49) for the second branch, we find to second order in η_1 the relation between the flux and temperature behind the discontinuity

$$-S = \frac{1}{\eta_1(3-\gamma)} D\rho_0 R(T - T_1). \tag{7.58}$$

For an approximate solution of the radiative transfer equation in the region behind the discontinuity, we note that the temperature here changes very little and we can set $U_p \approx U_{p1} = 4\sigma T_1^4/c = const.$ We obtain for $\tau > 0$

$$-S = \frac{c}{\sqrt{3}}(U_{p1} - U) = -S_0 e^{-\sqrt{3}\tau}, \tag{7.59}$$

$$T - T_1 = (T_+ - T_1)e^{-\sqrt{3}\tau}, \tag{7.60}$$

with $T_+ - T_1 = [(3-\gamma)/(\gamma+1)]T_- \approx 0.78T_-$ for $\gamma = 1.25$. The radiation density at the discontinuity is found by patching together at $\tau = 0$ the two branches of the curves $U(S)$ given by (7.54) and (7.59). We obtain*

$$U_0 = \tfrac{1}{2}U_{p1} = 2\sigma T_1^4. \tag{7.61}$$

The radiation density and flux profiles in a subcritical wave are illustrated in Fig. 7.28, in which for comparison the temperature profile is also shown.

Let us determine the limits of applicability of the approximate solution of the equations in the preheating region. Equation (7.50) is very accurate even in a wave of critical strength, since for $\gamma = 1.25$ the density ratio ahead of the

* We note that here another value $-S_0 = (2/\sqrt{3})\sigma T_1^4$ is obtained for the flux, which differs slightly from the preceding value $-S_0 = \sigma T_1^4$. The small discrepancy is a result of the inaccuracy of the diffusion approximation, and it disappears when the exact radiative transfer equation is used; for additional details see [47].

discontinuity is small: $\rho_-/\rho_0 \leqslant 1.13$. The solution of the radiative transfer equation (7.54), however, was obtained under the approximation $U_p \ll U$ and becomes invalid when the radiation density U is comparable with its equilibrium value. It follows from (7.54) and (7.50) that this occurs at a temperature T_c satisfying the equation

$$\frac{4\sigma T_c^4}{\sqrt{3}} = \frac{D\rho_0 R T_c}{\gamma - 1}. \tag{7.62}$$

Comparing this equation with (7.53) in which the specific heats are assumed constant [$\varepsilon = RT/(\gamma - 1)$], and noting that D has only a weak dependence on temperature ($D \sim \sqrt{T_1}$ with $\gamma = const$), we see that T_c is very close to the critical temperature T_{cr}. It follows from this that the radiation density in the preheating region is always out of equilibrium for temperatures below critical and that our approximate solution is valid for waves with strengths close to the critical.

Essentially, the radiation density becomes of the order of its equilibrium value when the radiant energy and the hydrodynamic fluxes are comparable. As can be seen from (7.54) to (7.57), the optical thickness of the preheating region in a subcritical wave is of the order of unity. The geometric thickness of the region is, consequently, of the order of a radiation mean free path averaged over the spectrum. In air at standard density this thickness is of the order of 10^{-2}–10^{-1} cm. The thickness is the greater, the higher is the temperature behind the front, since the radiation mean free path increases with increasing photon energies. The thickness of the region behind the compression shock where the gas and the radiation approach their final states is also of approximately the same order of magnitude.

§17. The supercritical shock wave

Let us consider a shock wave of large, supercritical strength when the temperature behind the front $T_1 > T_{cr}$. The temperature in the preheating region increases from zero to T_-, which is equal to the final temperature T_1 and thus also higher than T_{cr}. Since the temperature T_c defined by (7.62) is close to T_{cr}, T_- is greater than T_c. The density ratio in the preheating region is not large and (7.50) still applies.

At the leading edge of the region where the temperature is lower than T_c, the radiation is again out of equilibrium, and the solution is of the type of (7.54) and (7.55) in which T, S, and U decrease exponentially with the optical thickness. At the point where the temperature reaches the value T_c, the radiation density is of the order of its equilibrium value and the flux S is of the order of the Stefan–Boltzmann flux σT^4. Moving further toward the discontinuity, the radiation flux increases by virtue of the conservation relation

(7.50) in proportion to the temperature ($S \sim T$), so that it becomes smaller than the Stefan–Boltzmann flux. This means that in the temperature region where $T > T_c$, oppositely directed one-sided fluxes (which are of the order of σT^4) to a large extent balance each other, the generated radiation at each point is comparable to the absorption, and thus the radiation density is close to its thermodynamic equilibrium value. In other words, in this layer of the preheating region the radiation is in local equilibrium with the fluid and the radiative transfer has the character of radiation heat conduction. The flux S is now determined by the temperature gradient, and its smallness in comparison with the Stefan–Boltzmann flux is due to the fact that the temperature changes very little over a distance of the order of a photon mean free path.

A solution for the radiation heat conduction layer can be obtained by replacing the radiation density U in (7.43) by the equilibrium value $U_p \approx U$

$$S = -\frac{c}{3}\frac{dU_p}{d\tau} = -\frac{16\sigma c T^3}{3}\frac{dT}{d\tau}. \tag{7.63}$$

Solving this equation together with the algebraic equation (7.50), we find the temperature, flux, and radiation density profiles in the equilibrium layer of the preheating region. They should be patched together with the solution at the leading edge of the nonequilibrium layer at a point with temperature $T = T_c$, which effectively separates the two layers. The optical coordinate of this point will be denoted by τ_c. After some elementary calculations we obtain in the nonequilibrium layer, for $\tau < \tau_c$, $|\tau| > |\tau_c|$,

$$\frac{T}{T_c} = \frac{cU}{4\sigma T_c^4} = \frac{\sqrt{3}S}{4\sigma T_c^4} = e^{-\sqrt{3}|\tau - \tau_c|}, \tag{7.64}$$

and in the equilibrium layer for $\tau_c < \tau < 0$, $0 < |\tau| < |\tau_c|$,

$$\frac{T}{T_c} = \frac{\sqrt{3}S}{4\sigma T_c^4} = \left(\frac{cU}{4\sigma T_c^4}\right)^{1/4} = \left(1 + \frac{3\sqrt{3}}{4}|\tau - \tau_c|\right)^{1/3}, \tag{7.65}$$

where τ_c is expressed in terms of the temperature ahead of the discontinuity by

$$|\tau_c| = \frac{4}{3\sqrt{3}}\left[\left(\frac{T_-}{T_c}\right)^3 - 1\right]. \tag{7.66}$$

Since the temperature in the nonequilibrium layer decreases exponentially, decreasing by several times over an optical distance equal to unity, the value of $|\tau_c|$ in the case of a very strong wave, with $T_-^3 \gg T_c^3$, is effectively the optical thickness of the preheating region. We note that in the equilibrium layer the photon mean free path is taken to be the Rosseland mean value (see §12, Chapter II). The temperature ahead of the discontinuity in the supercritical wave almost coincides with the temperature T_1 behind the front.

The temperature ahead of the discontinuity T_- can never become higher than the final temperature behind the front T_1. Indeed, if $T_- > T_1$, then the radiation density in the preheating region ahead of the discontinuity, which is approximately equal to $U_- \approx 4\sigma T_-^4/c$, would become higher than the radiation density behind the front $U_{p_1} = 4\sigma T_1^4/c$. Consequently, the radiation density in the region between the discontinuity and the final state ($0 < \tau < +\infty$) would decrease in the direction away from the discontinuity. The flux $S \sim -dU/d\tau$ would be positive and directed in the direction of gas motion. However, this contradicts (7.48) and (7.49), which follow from the conservation laws and which show that the flux in the shock wave is everywhere negative and directed opposite to the gas motion. The temperature T_- is thus bounded above by the value of T_1. The impossibility of heating the gas ahead of the compression shock to a temperature exceeding the temperature behind the front T_1 attests at the same time to the fact that a discontinuity must necessarily arise in the solution (a representative point on the T, η and S, η diagrams, in order to reach the final position D, must "jump over" from a position B' on another branch). The above physical considerations on the impossibility of the temperature T_- exceeding T_1 and the necessity that a discontinuity must arise, are confirmed by a rigorous analysis of the equations for this regime (see [42]).

Since in a supercritical wave $T_- \approx T_1$, the radiation density ahead of the discontinuity is close to equilibrium and even ahead of the discontinuity it almost reaches its final value. Thus,

$$U_0 = U_- \approx \frac{4\sigma T_-^4}{c} \approx \frac{4\sigma T_1^4}{c} = U_{p_1}, \tag{7.67}$$

and in the region behind the discontinuity the radiation density remains practically constant

$$U(\tau) \approx U_{p_1} = \frac{4\sigma T_1^4}{c} \quad \text{for} \quad \tau > 0. \tag{7.68}$$

It follows from (7.65) that the flux at the discontinuity is equal to

$$S_0 = \frac{4\sigma T_c^4}{\sqrt{3}} \frac{T_1}{T_c}. \tag{7.69}$$

On the T, η and S, η diagrams a representative point in a supercritical wave moves along the curves from the point A to the point B' and then jumps to the point C', where the flux is the same. The temperature behind the discontinuity T_+ can be obtained from (7.48) and (7.49). It is

$$T_+ = (3 - \gamma)T_1. \tag{7.70}$$

Using the radiation heat conduction approximation in the region behind the discontinuity together with the fact that the radiation density in this region is constant, we find that the temperature is also constant. The temperature across the compression shock is continuous and equal to T_1. A representative point on the T, η and S, η diagrams moves directly from the position B' ahead of the discontinuity to the final position D. Here the flux will obviously undergo a discontinuity, since it is different from zero and equal to S_0 ahead of the compression shock and in the final state (point D) is equal to zero. We are thus dealing with a typical case of an "isothermal shock" which has already been encountered in §§3 and 12.

The formation of the "isothermal shock" is a result of the mathematical approximation in which the flux is taken to be proportional to the temperature gradient. This excludes the possibility of a temperature jump, since a temperature discontinuity would result in an infinite flux. Actually, however, by virtue of the fact that the process in the compression shock is steady it is the flux which is continuous while the temperature undergoes a jump. This does not present any contradiction. It is simply that the radiation in the region immediately behind the discontinuity is out of equilibrium (the radiation density is lower than at equilibrium, since the density corresponds to the temperature $T_- \approx T_1$ while the gas temperature is $T_+ > T_1$) and the flux, which is determined by the gradient of the actual radiation density, is not expressed in terms of the gas temperature gradient. Again, a temperature peak exists behind the shock discontinuity and the temperature profile in a supercritical wave has the shape shown in Fig. 7.25.

Let us estimate the optical thickness of the temperature peak behind the discontinuity from simple physical considerations. The geometrical thickness of the peak Δx is such that radiation born in this region gives a flux S_0, coming out from the surface of the discontinuity and used up in preheating the gas entering the wave. The energy radiated per unit time in a layer Δx (per unit area of discontinuity surface) is of the order of

$$\frac{\sigma T_+^4}{l}\Delta x \sim \frac{\sigma T_1^4}{l}\Delta x.$$

This quantity is approximately equal to the flux S_0, which according to (7.69) is of the order of $S_0 \sim \sigma T_c^3 T_1$. From this we obtain the thickness of the temperature peak

$$\Delta\tau = \frac{\Delta x}{l} \sim \left(\frac{T_c}{T_1}\right)^3. \tag{7.71}$$

The thickness decreases rapidly with increasing wave strength and the peak in a very strong wave is much narrower than a radiation mean free path. It is for this reason that this peak is "cut off" in the radiation heat conduction approximation, which disregards details related to distances shorter than a radiation mean free path.

In conclusion, let us cite some values of the thickness of the preheating region in a supercritical shock wave propagating through air at standard density. These values were estimated using (7.66) and the values of the Rosseland mean free path for real air calculated by the method presented in §8 of Chapter V. At $T_1 = 500{,}000°\text{K}$, $\tau_c = 3.4$ and the thickness is of the order of 40 cm. At $T_1 = 750{,}000°\text{K}$, $\tau_c = 14$ and the thickness is of the order of 2 m. Since the temperature peak is very narrow, these thicknesses represent at the same time the thickness of the entire shock front.

§18. Shock waves at high energy densities and radiation pressures

It was shown in §3 that in not too weak a shock wave, when heat conduction is present but viscosity is absent, the gas cannot make a continuous transition from the initial to final state. A discontinuity is necessarily formed, which corresponds to a viscous compression shock and which within the framework of the given approximation is infinitesimally thin (since the viscosity of the fluid was excluded from the very beginning). If the heat conduction flux is proportional to the temperature gradient, then all the flow variables, with the exception of temperature, undergo a discontinuous jump at the discontinuity and an "isothermal shock" takes place. In §§12 and 17 we have examined specific examples of "isothermal shocks" which are caused by electron and radiation heat conduction.

However, for extremely strong shock waves, when the energy density and the radiation pressure become sufficiently large in comparison with the energy and pressure of the fluid, the situation changes. The discontinuity disappears and the gas in the shock wave makes a continuous transition from the initial to the final state through radiation heat conduction alone, even if the fluid viscosity is not considered. This problem was treated by S. Z. Belen'kii and (later) by Belokon' [50].

We shall describe the internal structure of the shock front on the basis of equations (7.40), where S is the radiation heat conduction flux. The total pressure and energy are composed of quantities pertaining both to the fluid and to the radiation; here the radiation is taken to be in thermodynamic equilibrium. The point which describes the state in the wave on a p, V diagram moves along the straight line

$$p = \rho_0 D^2(1 - \eta), \qquad \eta = \frac{V}{V_0}, \tag{7.72}$$

where

$$p = R\rho_0 \frac{T}{\eta} + \frac{4}{3}\frac{\sigma T^4}{c}. \tag{7.73}$$

The temperature and the relative volume behind the wave front T_1 and η_1 are related by the Hugoniot equation (with temperature and volume as variables) including both the radiation energy and pressure. This equation was derived in §10, Chapter III (equation (3.76)). Let us rewrite it here in the form

$$\frac{R\rho_0 T_1}{\eta_1}\left(\frac{\eta_1}{\eta_{10}} - 1\right) = \frac{4\sigma T_1^4}{3c}(1 - 7\eta_1), \tag{7.74}$$

where $\eta_{10} = (\gamma - 1)/(\gamma + 1)$ is the final relative volume without taking into account the radiation density and pressure. It is evident from this equation that for $p_{rad} \gg p_{gas}$, $\eta_1 = 1/7$; and for $p_{rad} \ll p_{gas}$, $\eta_1 = \eta_{10}$.

The temperature as a function of the volume for the compression of the gas in the wave front is obtained by substitution of (7.73) into (7.72), and is

$$R\rho_0 \frac{T}{\eta} + \frac{4}{3}\frac{\sigma T^4}{c} = \rho_0 D^2(1 - \eta).$$

It is more convenient to write it in a somewhat different way by replacing $\rho_0 D^2$ by T_1 and η_1

$$R\rho_0 \frac{T}{\eta} + \frac{4}{3}\frac{\sigma T^4}{c} = \left(R\rho_0 \frac{T_1}{\eta_1} + \frac{4}{3}\frac{\sigma T_1^4}{c}\right)\frac{1 - \eta}{1 - \eta_1}. \tag{7.75}$$

The function $T(\eta)$ has a maximum in the interval $0 < \eta < 1$ (the coordinates of the maximum are denoted by T_{max} and η_{max}).

The radiation flux in the wave S, as in the case when both the radiation density and pressure are small, is always unidirectional and opposite to the gas flow. It vanishes only when $x = -\infty$ and $x = +\infty$, ahead and behind the wave front. Therefore, the temperature in the wave must increase monotonically from $T = 0$ to T_1, since otherwise the flux $S \sim -dT/dx$ would change sign within the wave.

If the radiation pressure is low, $p_{rad} \ll p_{gas}$, then as follows from (7.75) $\eta_{max} = 1/2 > \eta_1 \approx \eta_{10} = (\gamma - 1)/(\gamma + 1)$. In this case the point on the T, η diagram jumps from one branch of the $T(\eta)$ curve to the other without passing through the maximum, and this results in an isothermal shock (see §§3 and 17). If the radiation pressure is high, $p_{rad} \gg p_{gas}$, then the point η_{max} at which the function $T(\eta)$ passes through its maximum lies beyond the range of volumes actually encountered, $1/7 \approx \eta_1 < \eta < 1$: the value of η_{max} is close to zero (this is evident from (7.75)). Thus, in this case the gas density in the wave and the temperature change continuously and there is no discontinuity. This case is shown on a diagram of T^4 versus η (Fig. 7.29). Temperature, gas density, and radiation flux profiles in such a wave are shown schematically in Fig. 7.30.

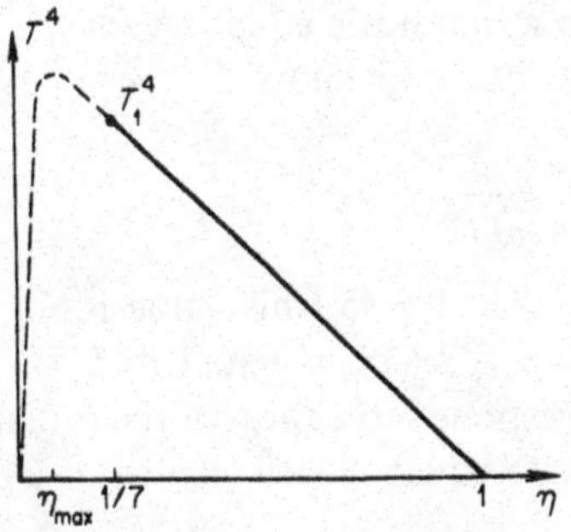

Fig. 7.29. T^4, η diagram for a shock wave with radiation and without a discontinuity.

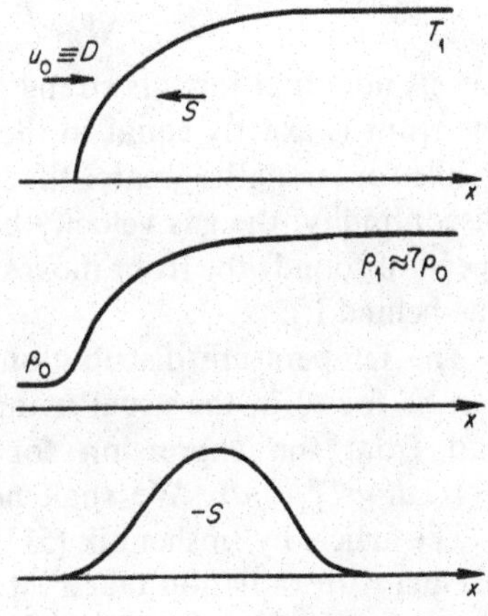

Fig. 7.30. Temperature, gas density, and radiation flux profiles in a shock wave taking into account the radiation energy and pressure in the case when a discontinuity is not present.

Let us find the wave strength at which the discontinuity disappears. Obviously, the strength corresponds to the case when the point of maximum temperature $T_{\max}$, $\eta_{\max}$ coincides with the final point T_1, η_1 (exactly as in §3). Indeed, a discontinuity is present for $\eta_1 < \eta_{\max}$ ("small" strengths), while for $\eta_{\max} < \eta_1$ ("large" strengths) the discontinuity is absent. The front variables corresponding to the transition strength which separates the continuous solution and isothermal jump regions will be denoted by T^*, η^*. Differentiating (7.75) for the function $T(\eta)$ and setting $dT/d\eta = 0$, $T = T^*$, $\eta = \eta^*$, and also setting in this equation and in the Hugoniot equation (7.74) $T_1 = T^*$ and $\eta_1 = \eta^*$, we obtain a system of two equations for the unknowns T^* and η^*

$$\frac{R\rho_0 T^*}{\eta^{*2}} = \left(R\rho_0 \frac{T^*}{\eta^*} + \frac{4}{3}\frac{\sigma T^{*4}}{c}\right)\frac{1}{1-\eta^*},$$

$$\frac{R\rho_0 T^*}{\eta^*}\left(\frac{\eta^*}{\eta_{10}} - 1\right) = \frac{4\sigma T^{*4}}{3c}(1 - 7\eta^*).$$

Eliminating T^* from this system we obtain a quadratic equation for η^*, one of whose roots corresponds to the real state and is given by

$$\eta^* = \frac{1}{4 + (2 + 1/\eta_{10})^{1/2}}. \tag{7.76}$$

Thus, for example, for $\gamma = 5/3$, $\eta_{10} = 1/4$, $\eta^* = 1/6.45$ (this value is slightly higher than the limiting volume for $p_{\text{rad}} \gg p_{\text{gas}}$ which is equal to 1/7). The transition strength, according to (7.74), corresponds to a ratio of the radiation pressure to the fluid pressure in the final state equal to

$$\left(\frac{p_{\text{rad}}}{p_{\text{gas}}}\right)^* = \frac{4\sigma T^{*4}/3c}{R\rho_0 T^*/\eta^*} = 4.45.$$

Let us note that for this strength the gas velocity behind the front relative to the front is exactly equal to the isothermal speed of sound in the final state (while for strengths greater than the transition strength, for which there is no discontinuity, the gas velocity behind the front is higher than the isothermal speed of sound; the front moves with a supersonic velocity with respect to the gas behind it).

The temperature distribution in a shock wave without a discontinuity can be found in the usual manner from the hydrodynamic equations (7.40) and from the expression for the radiation heat conduction flux $S = -\frac{1}{3}cl\, d(4\sigma T^4/c)/dx$. We shall not discuss this further here.

The article by Imshennik [51] considers a shock wave in a two-temperature plasma with radiation taken into account (the electron and ion temperatures are not assumed to be equal). In [88] the structure of a shock front is studied, with radiant energy transfer taken into account on the basis of the equations of radiation hydrodynamics (in the nonrelativistic approximation).

VIII. Physical and chemical kinetics in hydrodynamic processes

1. Dynamics of a nonequilibrium gas

§1. The gasdynamic equations in the absence of thermodynamic equilibrium

In the preceding chapter we have studied the structure of a shock front in a gas with slow excitation of some of the degrees of freedom and have become acquainted with one of the simplest problems of nonequilibrium gasdynamics. The variables behind the shock wave front, in the region where complete thermodynamic equilibrium is established, are independent of the mechanism and the rates of the nonequilibrium processes. The rates of these processes, however, have an appreciable effect on the distribution of the hydrodynamic variables in the nonequilibrium region and on the thickness of this region. Distortions of gasdynamic flows caused by the nonequilibrium processes are attributable mainly to changes in the specific heats and in the effective specific heat ratio of the nonequilibrium gas on which the progress of the gasdynamic process depends. The effect of the specific heat ratio on gasdynamic solutions may be seen from the examples of those problems which were treated in Chapter I. Thus, in the unsteady expansion of a gas initially at rest from a tube into vacuum, the exhaust velocity is equal to $u = 2c_0/(\gamma - 1)$ where $c_0 = (\gamma p_0/\rho_0)^{1/2}$ is the speed of sound in the initial state. Let us assume that a diatomic gas, contained by a diaphragm in a tube, is initially at equilibrium and is then heated to a temperature at which the vibrational modes are "classically" excited. After the diaphragm is broken it is assumed that the gas expands so rapidly that the vibrational modes remain frozen, and that in the expansion the vibrational energy does not have sufficient time to be converted into the kinetic energy of the expansion*. This would mean that the exhaust velocity does not correspond to the equilibrium value of the specific heat ratio $\gamma = 9/7$ but to $\gamma' = 7/5$, and thus is smaller roughly by a factor of $7/5 = 1.4$.

* During the expansion the density decreases, the rates of various processes are decreased, and the conversion of vibrational energy into the translational energy of the molecules, which is necessary for the subsequent conversion into the energy of directed, hydrodynamic motion, takes a long time.

This simple example shows the appreciable effect that nonequilibrium in a gas can exert on the dynamics of the process. The need to consider the rate of establishment of equilibrium arises whenever we are dealing with rapidly changing processes, or with processes whose characteristic scales are comparable with the relaxation "lengths". One of the more important practical problems of this type deals with the problem of a very rarefied gas flowing past a body, in which the relaxation times are comparable with the flow time about the body, in which the relaxation "length" is comparable with the characteristic dimensions of the body. The reentry of ballistic missiles into the atmosphere at hypersonic speeds is accompanied by the formation of a detached shock wave ahead of the body, as shown in Fig. 8.1. The distance

Fig. 8.1. Detached shock wave for supersonic flow past a body.

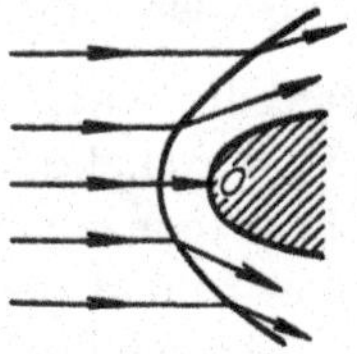

between the shock and the nose of the body is usually of the order of one-tenth the radius of curvature of the nose. If the gas is sufficiently rarefied that there are not a sufficiently large number of gaskinetic mean free paths over the stand-off distance between the shock and the body, then the slowly relaxing degrees of freedom in the gas particles behind the shock front do not have time to become excited, or in other words, not enough time is available to establish chemical equilibrium. As a result, the temperature of the gas compressed by a shock wave is found to be higher than under conditions of thermodynamic equilibrium, and this changes the manner in which the body is heated. In fact, we are dealing here with a case in which the character of the distribution of the flow variables in the nonequilibrium region which forms behind the compression shock is rather important.

In a number of cases it is possible to approximately describe the dynamics of a nonequilibrium gas by using some effective value of the specific heat ratio, corresponding to some degree of freezing of part of the specific heat, as for example when the energy change in some degrees of freedom can be neglected over the characteristic hydrodynamic flow time. In general, however, one must consider the gasdynamic process simultaneously with the kinetics of the nonequilibrium processes, which complicates the system of equations describing the phenomenon.

The dynamics of a nonviscous and nonheat-conducting gas at thermodynamic equilibrium is described by the equations of continuity, momentum, and entropy:

$$\frac{D\rho}{Dt} + \rho \nabla \cdot \mathbf{u} = 0, \tag{8.1}$$

$$\rho \frac{D\mathbf{u}}{Dt} + \nabla p = 0, \tag{8.2}$$

$$\frac{DS}{Dt} = 0, \tag{8.3}$$

to which we add the thermodynamic relationship for entropy as a function of pressure and density, $S(p, \rho)$ (for example, in a gas with constant specific heats $S = c_v \ln(p\rho^{-\gamma}) + const$).

We shall now consider the motion of a gas whose state departs from thermodynamic equilibrium. Here again we shall neglect viscosity and heat conduction and assume that the nonequilibrium state is entirely connected with the delayed progress of internal processes which take place only within a given parcel of the fluid, such as, for example, the delayed excitation of molecular vibrations.

In the case of nonequilibrium we replace the entropy equation (8.3), which no longer applies, by the more general equation of conservation of energy, which is always valid. Assuming the absence of any external energy sources*, we can write in place of (8.3)

$$\frac{D\varepsilon}{Dt} + p \frac{DV}{Dt} = 0. \tag{8.4}$$

By virtue of the thermodynamic identity

$$T\,dS = d\varepsilon + p\,dV \tag{8.5}$$

(8.4) and (8.3) are equivalent under conditions of thermodynamic equilibrium. While in the equilibrium case the internal energy ε is determined by the pressure and density only, $\varepsilon = \varepsilon(p, \rho)$, in the absence of equilibrium it also depends on other variables characterizing the state of the system which are not in equilbrium (for example, on the degree of dissociation). Without specifying these parameters, we shall term them λ. To close the system of gasdynamic equations, we must add to (8.1), (8.2), and (8.4) an equation

* The thermal effect of a reversible chemical reaction is not an external energy source; it is taken into account by introducing an appropriate term into the expression for the internal energy of the gas.

connecting the internal energy with the pressure, density, and the state variables λ,

$$\varepsilon = \varepsilon(p, \rho, \lambda),$$

and also rate equations which describe the changes in the variables λ in the gas with time,

$$\frac{D\lambda}{Dt} = f(\lambda, p, \rho).$$

Usually, the functions $\varepsilon(p, \rho, \lambda)$ and $f(\lambda, p, \rho)$ are not expressed explicitly in terms of density and pressure, but instead in terms of temperature

$$\varepsilon = \varepsilon(\rho, T, \lambda), \qquad \frac{D\lambda}{Dt} = f(\lambda, \rho, T).$$

In this case, we must also add the equation of state

$$p = p(T, \rho, \lambda).$$

The temperature T, unless otherwise noted, will always denote the temperature corresponding to the translational degrees of freedom of the molecules (atoms, ions). These are usually in equilibrium even in the most rapid gasdynamic processes, since the Maxwell distribution of molecular velocities is established extremely rapidly.

As an example of a nonequilibrium system let us consider a diatomic gas without dissociation but with slow excitation of the vibrational modes of the molecule (we consider only not too high temperatures, for which the degree of dissociation is still negligibly small). The role of the variable λ is played here by the nonequilibrium vibrational energy ε_{vib} (per unit mass of the gas). For the given case, the equations which must be added to the system, (8.1), (8.2), and (8.4), may be written in the form

$$\varepsilon = \varepsilon_1 + \varepsilon_{\text{vib}} = \tfrac{5}{2}RT + \varepsilon_{\text{vib}}, \tag{8.6}$$

$$p = R\rho T, \tag{8.7}$$

$$\frac{D\varepsilon_{\text{vib}}}{Dt} = \frac{\varepsilon_{\text{vib}}(T) - \varepsilon_{\text{vib}}}{\tau(T, \rho)}. \tag{8.8}$$

Here ε_1 is the sum of the energies of the translational and rotational degrees of freedom of the molecules. (It is assumed that the rotational energy has its equilibrium value, and corresponds to the translational temperature T.) The quantity $\varepsilon_{\text{vib}}(T)$ is the vibrational energy which the gas would have in thermodynamic equilibrium with the translational degrees of freedom, and $\tau(T, \rho)$ is the relaxation time for establishing vibrational equilibrium.

Similar equations which are, however, of a more complex form, can also be written for all the other cases, where there is nonequilibrium dissociation, chemical reactions, ionization, or where the translational temperatures of the electron and atom (ion) gases differ. All these cases were examined in the preceding chapter when we considered the structure of the nonequilibrium layer in a shock front.

§2. Entropy increase

An extremely important property of nonequilibrium gasdynamic processes is the increase in the entropy of the gas and the dissipation of mechanical energy. As with the internal energy ε, the entropy of a nonequilibrium gas is no longer determined by only the two variables pressure and density or temperature and density, but depends on the other variables which characterize the nonequilibrium state; thus $S = S(p, \rho, \lambda)$ or $S(T, \rho, \lambda)$. The increase in entropy dS is no longer equal to the heat supplied by the external sources divided by temperature, as was true for the equilibrium case, so $dS \neq dQ/T$. The entropy increases with time even without a supply of heat (when $dQ = 0$) as a result of the nonequilibrium internal processes only.

We shall clarify our preceding remarks with the aid of the example of nonequilibrium vibrational excitation. The total specific entropy of a gas S is composed of the entropies corresponding to the translational and rotational degrees of freedom, which because of their equilibrium character can be combined, plus the vibrational entropy*. We denote these two parts of the entropy by S_1 and S_{vib}, respectively, with

$$S = S_1 + S_{\text{vib}}. \tag{8.9}$$

For the entropy of the translational and rotational degrees of freedom we can write the thermodynamic relation

$$T\,dS_1 = d\varepsilon_1 + p\,dV. \tag{8.10}$$

Usually, the exchange of vibrational energy by molecules takes place much faster than the exchange between the vibrational and translational energies. Thus a Boltzmann distribution with respect to the vibrational excitations of the molecules is established quite rapidly, and we can assign a definite temperature T_{vib} to the vibrations. This temperature corresponds to the actual supply of vibrational energy $\varepsilon_{\text{vib}} = \varepsilon_{\text{vib}}(T_{\text{vib}})$. If we denote the vibrational specific heat by c_{vib}, then $d\varepsilon_{\text{vib}} = c_{\text{vib}}\,dT_{\text{vib}}$. Here, of course, the

* For nonequilibrium dissociation or ionization an expression for the entropy is written in terms of the number of different species of particles (molecules and atoms, for example) which are assumed to be out of equilibrium.

vibrational temperature T_{vib} can be appreciably different from the translational temperature of the molecules T; this difference is the manifestation of the nonequilibrium state of the gas*. If we can assign the specific temperature T_{vib} to the vibrational modes, then for the vibrational contribution to the entropy we can also write the thermodynamic relation

$$T_{vib}\, dS_{vib} = d\varepsilon_{vib}. \tag{8.11}$$

The vibrational energy and entropy are independent of the gas volume.

It is easy to see that the entropy of a nonequilibrium system only increases with time, independent of the transformations the gas undergoes. Indeed, by virtue of (8.9), (8.10), (8.4), and (8.6), we have

$$\frac{DS}{Dt} = \frac{DS_1}{Dt} + \frac{DS_{vib}}{Dt} = \frac{1}{T}\left(\frac{D\varepsilon_1}{Dt} + p\frac{DV}{Dt}\right) + \frac{1}{T_{vib}}\frac{D\varepsilon_{vib}}{Dt} = \frac{D\varepsilon_{vib}}{Dt}\left(\frac{1}{T_{vib}} - \frac{1}{T}\right). \tag{8.12}$$

Taking into account the rate equation (8.8) in which

$$\varepsilon_{vib} = \int_0^{T_{vib}} c_{vib}(T')\, dT' \qquad \text{and} \qquad \varepsilon_{vib}(T) = \int_0^{T} c_{vib}(T')\, dT',$$

we see that for $T_{vib} < T$ the vibrational modes take away energy from the translational and rotational degrees of freedom, $D\varepsilon_{vib}/Dt > 0$, and $DS/Dt > 0$. For $T_{vib} > T$ the vibrations give up their energy $D\varepsilon_{vib}/Dt < 0$, but again $DS/Dt > 0$. The above example illustrates the second law of thermodynamics according to which, without the participation of external factors, heat is always transferred from the hotter to the cooler object, and as a result of which the entropy of the entire system increases. In our case the "objects" are not bodies touching one another, but different degrees of freedom of the same body.

If at a given time the gas is in a state of thermodynamic equilibrium, and it then takes part in a rapidly progressing process during which the equilibrium is disturbed and subsequently the state of the gas changes slowly in order to return to equilibrium, then the entropy of the gas will increase. This increase in entropy is accompanied by the dissipation of mechanical energy, namely by its irreversible conversion into heat. If the process proceeds without the participation of external energy sources, satisfying the energy equation (8.4), then the dissipated energy cannot under any conditions be again converted into mechanical energy. We shall study the phenomenon

* We recall that a similar situation occurs in the case of a plasma. The Maxwell distributions and the temperatures in the electron and ion gases are established very rapidly. The electron and ion temperatures, however, differ from each other as a result of the slow exchange of energy between the electron and ion gases.

of dissipation in more detail in the next section, in which we shall consider the absorption of sound in a relaxing medium. Absorption of sound waves is a characteristic example of the dissipation of mechanical energy.

An example of the incomplete utilization of available energy as a consequence of "irreversibility" is provided by the idealized case of expansion of a gas into vacuum with completely frozen vibrational modes. Only the reversible part of the internal energy, i.e., the energy of the translational and rotational degrees of freedom, is converted into the kinetic energy of the sudden expansion; the vibrational energy remains in the molecules, as a result of which the exhaust velocity of the gas is lower. Such irreversible effects in the presence of nonequilibrium processes can result in additional losses in high-speed turbines at high temperatures, in rocket engine nozzles, etc. The effect of the increase in entropy with time served as the basis for an independent method of measuring the vibrational relaxation time τ by Kantrowitz [1] in investigating relaxation in CO_2.

There is an extensive literature devoted to gasdynamic calculations which take nonequilibrium processes into account, relating primarily to flow problems and to the aerodynamic heating of bodies reentering the atmosphere (satellites, ballistic missiles). For examples, see [2, 2a], which also contain references to many other papers. We shall not consider here problems of the effects of physical and chemical kinetics on gasdynamic motions. In this chapter we shall be interested in another problem, that of the kinetics of nonequilibrium processes not from the point of view of its effect on the motion of a gas, but from the point of view of determining the concentration of the various components under conditions significantly out of equilibrium in chemical reactions, ionization, and vapor condensation in various hydrodynamic phenomena. As a rule, the hydrodynamics will be considered only approximately, by means of effective values of the specific heat ratio, and the kinetics of the processes of interest will be "imposed" on the already known hydrodynamic solution. Exceptions to this plan are the following two sections, in which we shall consider the phenomena of the absorption and dispersion of sound in a relaxing medium. Thus we shall be considering the effect of nonequilibrium processes on a gasdynamic process—the propagation of sound waves.

§3. Anomalous dispersion and absorption of ultrasound

Dispersion and absorption of sound in gases, which are connected with viscosity and heat conduction, ordinarily are appreciable only at very short wavelengths of the sound waves, those comparable with the mean free path of the gas particles, and thus at frequencies comparable with the frequency of gaskinetic collisions (see §22 of Chapter I). Propagation of ultrasonic waves in molecular gases is, however, sometimes accompanied by an anomalously strong dispersion and absorption in a region of much greater wavelengths and lower frequencies. These phenomena are related to the relaxation processes for the establishment of equilibrium in slowly excited degrees of freedom of the gas.

In the limiting case of low frequencies, the relaxation times for establishing equilibrium in those degrees of freedom which make an appreciable contribution to the specific heat are small in comparison with the period of sound vibrations. Under these conditions a gas particle is in a state of thermodynamic equilibrium at any instant of time, and "follows" the changes in pressure and density in the sound wave. The speed of sound, defined as the square root of the isentropic derivative of the pressure with respect to density, corresponds to its own thermodynamic equilibrium value

$$a^2 = \left(\frac{\partial p}{\partial \rho}\right)_S = \gamma \frac{p_0}{\rho_0}, \qquad \gamma = \frac{c_p}{c_v} = 1 + \frac{R}{c_v}.^* \tag{8.13}$$

On the other hand, in the limiting case of very high frequencies, the slowly relaxing degrees of freedom in the sound wave do not have time to become excited, and their energy simply corresponds to the temperature of the undisturbed state T_0. These degrees of freedom do not participate in the periodic changes in the state of the gas; they are "frozen", and do not affect the isentropic relationship between the changes in pressure and density. The active part of the specific heat is now less than at equilibrium, and the specific heat ratio and the speed of sound are greater than at low frequencies.

A gradual change of the speed of sound from the equilibrium value a_0 to the value a_∞ corresponding to the frozen part of the specific heat takes place in the intermediate frequency region; thus, there is dispersion of sound. For example, measurements of Kneser [3, 4] show the speed of sound in carbon dioxide at room temperature to vary between $a_0 = 260$ m/sec at a frequency ν of the order of 10^4 sec^{-1} (10 kc) to $a_\infty = 270$ m/sec at $\nu \sim 10^6$ sec^{-1} (1 Mc). The lower speed of sound corresponds to the equilibrium value of the specific heat

$$c_v = c_{\text{trans}} + c_{\text{rot}} + c_{\text{vib}} = \tfrac{3}{2}R + R + 0.8R = 3.3R.$$

The CO_2 molecule is linear, so that $c_{\text{rot}} = R$; only the low-frequency molecular vibrations with $h\nu/k = 954°$K are excited at room temperature, and for these the vibrational specific heat is even smaller than its classical value R. The higher speed of sound corresponds to frozen vibrations, with a specific heat $c_v = c_{\text{trans}} + c_{\text{rot}} = 2.5R$. It follows from these data that the relaxation time for vibrational excitation in a CO_2 molecule (at atmospheric pressure)

* We are here using specific heat capacities; R is the gas constant per unit mass. To avoid confusion, we denote the speed of sound here by a instead of c.

corresponds to some intermediate sound frequency, so that roughly $\tau_{vib} \sim 1/\nu \sim 10^{-5}$ sec. Molecular rotations at room temperature are excited very rapidly and dispersion related to slow rotational excitation can be observed at atmospheric pressure only at extremely high frequencies $\nu \sim 1/\tau_{rot} \sim 10^9$–$10^{10}$ sec^{-1} (the only exception is hydrogen; see §2, Chapter VI).

Dispersion of sound is also observed in gases in which slow chemical reactions take place as a result of the temperature (and density) changes in a sound wave. An example is the polymerization of nitrogen dioxide $2NO_2 \rightleftarrows N_2O_4$, which takes place easily at room temperature since its heat of activation in both directions is very low. It was in connection with systems of this type that the theory of sound dispersion was first developed by Einstein in 1920 [5]. Apparently, analogous phenomena also occur in the propagation of ultrasound in some liquids.

Measurements of ultrasonic dispersion and absorption provide one of the most important methods of studying relaxational processes and of experimentally determining relaxation times. There is an extensive literature devoted to this subject* but we shall not consider it in detail here. We shall examine only the basic physical properties and laws governing this phenomenon.

Dispersion of sound in a relaxing fluid is always accompanied by increased absorption, which considerably exceeds the natural absorption due to ordinary viscosity and heat conduction. A fluid parcel in a sound wave performs successive cyclical transformations, returning to its initial state upon the completion of each cycle. If internal nonequilibrium processes take place in the parcel, they inevitably lead to an increase in entropy and to a dissipation of mechanical energy, and thereby to the absorption of sound. It is to be emphasized that in the presence of dissipation the state of a parcel upon completion of a cycle differs somewhat from its initial state (since its entropy increases). However, this difference, let us say the temperature increase, is proportional to the entropy increase, and is a second-order quantity in comparison with the small amplitude of the sound wave Δp or ΔT. This follows from the fact that the entropy increase ΔS is proportional to the sound energy which, in turn, is proportional to $(\Delta p)^2$ (see §3, Chapter I). Therefore, in first approximation the motion in a sound wave even in the presence of absorption is isentropic and we can regard the cycles to be closed.

The mechanism of dissipation of mechanical energy and of sound absorption can be made clearer by considering the cycle in the gas on a p, V diagram. Figure 8.2 shows two families of isentropes, one of which (I) corresponds to equilibrium changes of state, and the other (II) to the frozen part of the specific heat. The isentropes were drawn near the undisturbed region, denoted

* A survey of the literature and references may be found, for example, in [6].

by the point O. For very slow sound vibrations the point describing the state of the gas, p, V, oscillates about the center O along the equilibrium isentrope, denoted in Fig. 8.2 by I′. In the limiting case of very high frequency, the

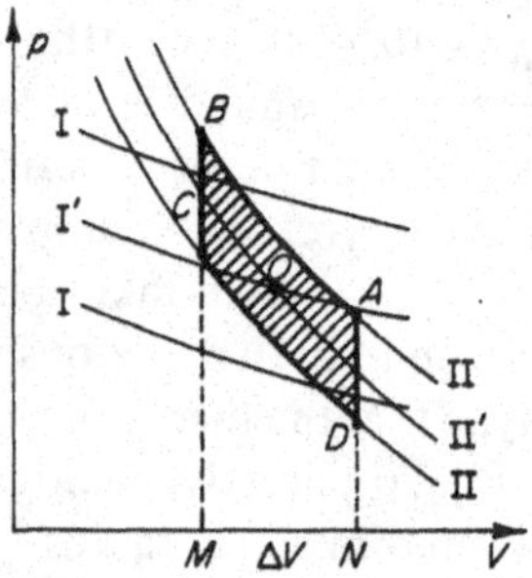

Fig. 8.2. p, V diagram for a cycle in a sound wave with a rectangular profile.

point oscillates about the center along the "frozen" isentrope denoted by II′. In both cases nonequilibrium processes are absent, the entropy of the gas does not change, and there is no sound absorption. The work done on the gas per cycle, which is numerically equal to the area of the figure described by the point on the p, V diagram, is equal to zero, which shows that absorption is absent. Using the example of vibrational relaxation it is easy to see that the entropy of the gas does not change in the second case, as well as in the first case of thermodynamic equilibrium. It is evident from (8.12) that the rate of change of the entropy in a nonequilibrium process is proportional to the rate of change of the vibrational energy. But, for strictly frozen vibrational modes this energy does not change, $\varepsilon_{vib} = const$, and $DS/Dt = 0$.

Let us now consider sound waves of intermediate frequencies, where relaxation processes are important (for definiteness we again consider vibrational relaxation). For simplicity we imagine that the density profile in the sound wave has the square wave shape shown in Fig. 8.3a*. This plot can be considered either as a density distribution with respect to the position coordinate at a given instant of time, or as the relation governing the change in density in a given gas particle with time. The same applied to Fig. 8.3b, which illustrates the corresponding temperature (or pressure) profiles (the temperature and pressure profiles are in this case similar).

We shall follow the change of state of a gas particle in a wave on the p, V diagram of Fig. 8.2, as well as on Figs. 8.3a and b. When the gas is very rapidly compressed from point A to point B, its state changes along a frozen isentrope II. In this case the entropy does not change, and positive work, numerically equal to the area $NABM$, is done on the gas. The gas temperature and pressure increase sharply while the vibrational energy remains unchanged

* This example, which is distinguished by its clarity, has been considered before, for example in the book by Gorelik [7].

and corresponds to the old, low temperature. Then, for a certain period of time, the gas density remains unchanged (the transition $B \rightarrow C$). The vibrational modes are excited, part of the energy is transferred from the transla-

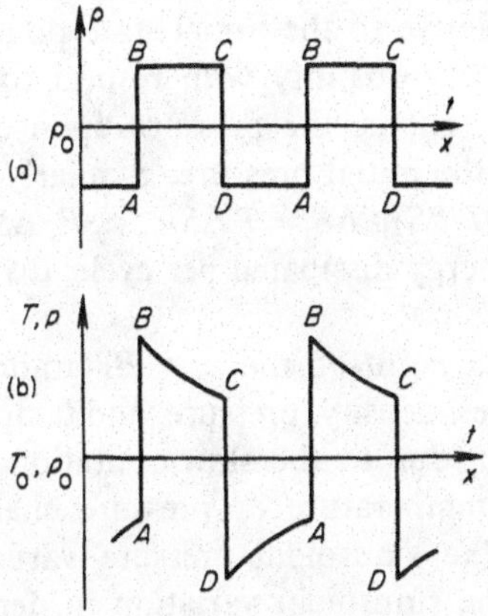

Fig. 8.3. An acoustic wave in a relaxing gas with a square wave density profile: (a) density profile; (b) temperature or pressure profile.

tional and rotational degrees of freedom, the temperature and pressure decrease, and the entropy increases (see (8.12): $T_{vib} < T$, $D\varepsilon_{vib}/Dt > 0$, and $DS/DT > 0$). Since the volume does not change, no work is performed during the transition $B \rightarrow C$.

Following this the gas then expands very rapidly (the transition $C \rightarrow D$) along a frozen isentrope II. The temperature and pressure fall, the entropy remains unchanged, and the vibrational energy also remains unchanged at the value it had at point C. The work done by the gas is numerically equal to the area $MCDN$ (negative work is done on the gas). And, finally, during the slow transition at constant volume $D \rightarrow A$, the vibrational modes are partially deexcited since their energy exceeds the value corresponding to the decreased temperature; the vibrational energy is partially transformed into translational and rotational energy, the temperature and pressure increase, and the entropy also increases ($T_{vib} > T$, $D\varepsilon_{vib}/Dt < 0$, $DS/Dt > 0$). No work is done in this case.

Thus, during the expansion stage $C \rightarrow D$ the gas particle does less work on the surrounding gas than done by the gas on the particle during the compression stage $A \rightarrow B$. The particle does not fully "give back" the work. Part of the energy expended during the compression period remains "forever" in the particle. This energy, numerically equal to the difference in work, to the area of the figure $ABCD$, is the mechanical energy which has been irreversibly converted into heat. As a result of the dissipation of mechanical energy the sound wave is also attenuated (absorbed); the absorption of sound energy per period (or per wavelength) is exactly equal to the area $ABCD$.

On the other hand, irreversible generation of heat is related to the entropy increase per cycle, and is equal to $T_0 \Delta S$. This quantity, as is evident from Fig. 8.2, is proportional to $\Delta V \cdot \Delta p \sim (\Delta p)^2$. It follows therefore that the displacement of the final state point A' relative to the initial state point A, $\delta p = (\partial p/\partial S)_V \cdot \Delta S \sim (\Delta p)^2$, is a second-order quantity with respect to the amplitude Δp. Since $(\partial p/\partial S)_V > 0$, $\delta p > 0$, that is, the pressure upon completion of the cycle is slightly higher than the initial pressure. Similarly, the temperature is also slightly higher, $\delta T = (\partial T/\partial S)_V \, \Delta S = T \, \Delta S/c_v \approx T_0 \, \Delta S/c_v$. The temperature increase is equal to the energy dissipated per cycle, divided by the specific heat at constant volume.

In a sinusoidal (harmonic) sound wave a point on the p, V diagram describes a smooth curve. All the state variables, density, pressure, and temperature vary harmonically with time. However, due to the slow excitation and deexcitation of molecular vibrations, the temperature or pressure changes cannot follow the density changes and the sinusoidal pressure variation undergoes a phase shift with respect to the sinusoidal variation in density (volume). It can be shown that a point on the p, V diagram describes, in this case, an elliptical trajectory, with the axes of the ellipse inclined with respect to the p, V coordinate axes.

At low frequencies ν (or angular frequencies $\omega = 2\pi\nu$) the ellipse is stretched out along an equilibrium isentrope (see curve 1 in Fig. 8.4). The thickness of

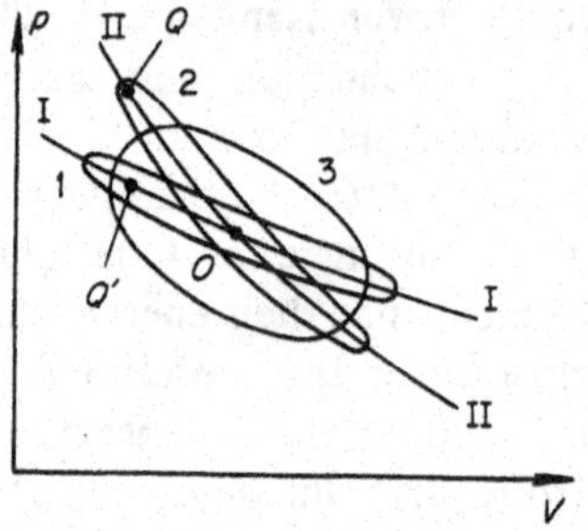

Fig. 8.4. p, V diagram for the cycles in harmonic sound waves of different frequencies.

the ellipse in the limit of small frequencies is proportional to the frequency (to the first term of an expansion in the small quantity ω). The sound energy absorbed per period is proportional to ω and that absorbed per unit time is proportional to the number of cycles, that is, to ω^2. At high frequencies the ellipse is stretched out along a frozen isentrope (curve 2). The thickness of the ellipse is proportional to $1/\omega$ (as can also be seen from an expansion), and the absorption per unit time is proportional to $\omega \cdot 1/\omega$, that is, it is independent of the frequency. The strongest absorption per period occurs in the intermediate case, when the frequency is of the order of the reciprocal of the relaxation time. In this case the thickness of the ellipse is maximum (curve 3); it is of the order of the vertical distance between the equilibrium and frozen isentropes at the maximum pressure change equal to the amplitude of the wave (the distance between the points Q and Q' in Fig. 8.4). If the relative difference between the equilibrium and frozen adiabatic exponents is large (it is precisely this difference which characterizes the angle between the isentropes I and II, and thus the distance QQ'), then the thickness of the ellipse may even become of the order of its length. This corresponds to a phase shift between the pressure and density of the order of $\pi/2$ (when the ellipse degenerates into a circle the phase shift becomes exactly $\pi/2$).

§4. The dispersion law and the absorption coefficient for ultrasound

The qualitative considerations on the dispersion and absorption of sound in the presence of relaxation processes in a fluid presented in the preceding section can be put into an elegant mathematical form. This was done in a general form by Mandel'shtam and Leontovich* [8]; dispersion and absorption relations, in which the relaxation time τ appears, usually serve to determine that time from the experimentally measured dispersion or absorption curves as a function of the ultrasonic frequency.

Let us show how we may derive the dispersion relation and the expression for the absorption coefficient in a relaxing medium. For simplicity and clarity we shall carry out the calculation using the specific example of a gas with nonequilibrium vibrational modes, for which the complete system of gasdynamic equations (8.1), (8.2), (8.4), (8.6), (8.7), and (8.8) has been formulated in §1. All the variables in the sound wave, pressure, density, etc., will be written in the form $f = f_0 + f'$, where f_0 is the average value corresponding to the undisturbed gas and f' is a variable part or perturbation, which will be considered a small quantity (the velocity $u = u_0 + u' = u'$, since the undisturbed gas is at rest and $u_0 = 0$). The true vibrational energy can also be expressed as $\varepsilon_{\text{vib}} = \varepsilon_{\text{vib}_0} + \varepsilon'_{\text{vib}}$, where $\varepsilon_{\text{vib}_0}$ is the vibrational energy in the undisturbed gas at equilibrium. The perturbation of the equilibrium vibrational energy we shall write in the form $\varepsilon'_{\text{vib}}(T) = c_{\text{vib}}T'$, where c_{vib} is the vibrational specific heat corresponding to the average temperature T_0 (if at T_0 the vibrational modes have their classical value, then $c_{\text{vib}} = R$, while if this is not the case, c_{vib} is expressed by a quantum-mechanical relation; see §2, Chapter III).

* A presentation of this theory may be found in the book by Landau and Lifshitz [9].

Let us substitute into the equations all of the quantities in the form indicated above and neglect second-order terms, thus linearizing the equations as is usual in acoustics (see §3 of Chapter I). We then obtain for the one-dimensional plane case the following system of equations for the perturbations

$$\begin{aligned}
&\frac{\partial \rho'}{\partial t} + \rho_0 \frac{\partial u'}{\partial x} = 0, &&\varepsilon' = \tfrac{3}{2}RT' + \varepsilon'_{\text{vib}}, \\
&\frac{\partial u'}{\partial t} + \frac{1}{\rho_0}\frac{\partial p'}{\partial x} = 0, &&\frac{p'}{p_0} = \frac{T'}{T_0} + \frac{\rho'}{\rho_0}, \\
&\frac{\partial \varepsilon'}{\partial t} - \frac{p_0}{\rho_0^2}\frac{\partial \rho'}{\partial t} = 0, &&\frac{\partial \varepsilon'_{\text{vib}}}{\partial t} = \frac{c_{\text{vib}}T' - \varepsilon'_{\text{vib}}}{\tau}.
\end{aligned} \tag{8.14}$$

Here the specific volume in the energy equation (8.4) has been replaced by the density, and both sides of the equation of state have been divided by $p_0 = R\rho_0 T_0$. The relaxation time τ is taken to be constant and given by $\tau = \tau(T_0, \rho_0)$.

A solution of the system (8.14) will be sought in the form of a harmonic plane wave, with all the primed quantities expressed in the form

$$f' = f'^* e^{-i(\omega t - kx)}. \tag{8.15}$$

The wave number k is in general complex: $k = k_1 + ik_2$. The real part k_1 is proportional to the reciprocal of the wavelength $k_1 = 2\pi/\lambda$ and determines the actual speed of sound—the phase velocity of wave propagation $a_1 = \omega/k_1$; the imaginary part k_2 gives the sound absorption coefficient

$$f' = f'^* e^{-i\omega(t - x/a_1)} e^{-k_2 x}. \tag{8.16}$$

The quantity $a = \omega/k$ may be called the complex speed of sound. The amplitudes f'^* are in general also complex, with $f'^* = |f'^*|e^{i\varphi}$. The complex character of the amplitudes testifies to the phase shifts of some of the quantities with respect to the others (through the differences of the angles φ).

Substituting into (8.14) all of the quantities in the form defined by (8.15) and noting that $\partial f'/\partial t = -i\omega f'$, $\partial f'/\partial x = ikf'$ we obtain a system of algebraic equations for the primed quantities (or for the amplitudes, if we cancel out the exponential factor)

$$\begin{aligned}
&-i\omega\rho' + \rho_0 iku' = 0, &&\varepsilon' = \tfrac{3}{2}RT' + \varepsilon'_{\text{vib}}, \\
&-i\omega u' + \frac{1}{\rho_0} ikp' = 0, &&\frac{p'}{p_0} = \frac{T'}{T_0} + \frac{\rho'}{\rho_0}, \\
&-i\omega\varepsilon' + \frac{p_0}{\rho_0^2} i\omega\rho' = 0, &&-i\omega\varepsilon'_{\text{vib}} = \frac{c_{\text{vib}}T' - \varepsilon'_{\text{vib}}}{\tau}.
\end{aligned} \tag{8.17}$$

Solving the last equation for $\varepsilon'_{\text{vib}}$ we obtain

$$\varepsilon'_{\text{vib}} = \frac{c_{\text{vib}} T'}{1 - i\omega\tau}. \tag{8.18}$$

It is precisely this complex relation between the perturbations of the true vibrational energy and of the temperature that causes dispersion and absorption to arise. It can already be seen from this relation that in the limiting cases $\omega\tau \to 0$ and $\omega\tau \to \infty$, for which $\varepsilon'_{\text{vib}} = c_{\text{vib}} T'$ and $\varepsilon'_{\text{vib}} = 0$, the imaginary unit i drops out completely from the system of equations (8.17) and all the quantities are real (if p', ρ', etc., are understood to denote the amplitudes p'^*, ρ'^*, etc.). Neither absorptions nor phase shifts occur in this case.

The first two equations of the system (8.17), which were obtained from the equations of continuity and motion by eliminating the velocity, yield the usual relationship

$$p' = \frac{\omega^2}{k^2}\,\rho' = a^2 \rho', \tag{8.19}$$

where a is now the complex speed of sound. Eliminating ε', $\varepsilon'_{\text{vib}}$, and T' from the remaining four equations, we obtain still another relationship between p' and ρ',

$$p' = \gamma \frac{p_0}{\rho_0}\,\rho', \qquad \gamma = \frac{\frac{7}{2}R + c_{\text{vib}}/(1 - i\omega\tau)}{\frac{5}{2}R + c_{\text{vib}}/(1 - i\omega\tau)}. \tag{8.20}$$

The quantity γ may be termed the complex specific heat ratio. Introducing the notation where $c_{v0} = \frac{5}{2}R + c_{\text{vib}}$ and $c_{p0} = \frac{7}{2}R + c_{\text{vib}}$ are the equilibrium specific heats at constant volume and pressure, respectively, and $c_{v_\infty} = \frac{5}{2}R$ and $c_{p_\infty} = \frac{7}{2}R$ are the specific heats with the vibrational modes completely frozen, we can write the complex specific heat ratio and the expression for the complex speed of sound, from (8.19) and (8.20), as

$$a^2 = \gamma \frac{p_0}{\rho_0}, \qquad \gamma = \frac{c_{p0} - i\omega\tau c_{p_\infty}}{c_{v0} - i\omega\tau c_{v_\infty}}^{*}. \tag{8.21}$$

* Landau and Lifshitz derive a slightly different equation in their book [9] (Chapter VIII, §78, equations (78.3)),

$$\gamma = \frac{1}{1 - i\omega\tau}\left[\frac{c_{p0}}{c_{v0}} - i\omega\tau\frac{c_{p\infty}}{c_{v\infty}}\right].$$

This difference arises from a difference in the definitions of the relaxation time τ which enters in the rate equation. The quantity $\varepsilon_{\text{vib}}(T)$ in our equation (8.8) represents the equilibrium vibrational energy which corresponds to the translational temperature T. Let us denote the relaxation time in our rate equation by the subscript "T". If the gas volume is constant, and the translational temperature is also maintained constant: $T = const$, then

In the limiting case of very low frequencies $\omega\tau \ll 1$, $\gamma = c_{p0}/c_{v0} = \gamma_0$, $a^2 = \gamma_0 p_0/\rho_0 = a_0^2$, we obtain the equilibrium specific heat ratio and speed of sound. In the limit of high frequencies $\omega\tau \gg 1$,

$$\gamma = \frac{c_{p\infty}}{c_{v\infty}} = \gamma_\infty, \qquad a^2 = \gamma_\infty \frac{p_0}{\rho_0} = a_\infty^2,$$

we obtain a specific heat ratio and a speed of sound corresponding to frozen vibrational modes. In both limiting cases the speed of sound and consequently the wave number, $k = \omega/a$, are real, and there is no absorption.

(8.8) gives an exponential relationship with a characteristic time τ_T for the approach to equilibrium

$$\varepsilon_{vib} = \varepsilon_{vib}(T) + [(\varepsilon_{vib})_{t=0} - \varepsilon_{vib}(T)]\, e^{-t/\tau_T}$$

The energy of the gas $\varepsilon = c_{v\infty}T + \varepsilon_{vib}$ is not constant in this case. However, if we assume that the total energy (and, of course, the volume) is constant and use (8.8), then instead of a simple exponential law we obtain a more complex law governing the approach to equilibrium.

Landau and Lifshitz [9] write a rate equation of the type (8.8), but defined so that the equilibrium energy term denotes the vibrational energy at an equilibrium temperature T_{eq} which corresponds to both the translational and vibrational degrees of freedom, and which thus depends on the given volume V and total energy ε of the gas. Let us denote the relaxation time which enters in this rate equation (according to [9]) by τ_S. The equation yields the exponential relationship for the approach to equilibrium

$$\varepsilon_{vib} = \varepsilon_{vib}(T_{eq}) + [(\varepsilon_{vib})_{t=0} - \varepsilon_{vib}(T_{eq})]\, e^{-t/\tau_S}$$

if the gas volume, the energy (i.e., the equilibrium temperature T_{eq}), and the time τ_S are constant. Actually, τ_S depends on the translational temperature, but it is assumed that the departure from equilibrium is slight, so that at $T_{eq} = const$ the translational temperature changes only very little. For a slight departure from equilibrium, we can consider the condition $V = const$, $\varepsilon = const$ as a condition of approximate constancy of the entropy, $S \approx const$.

Let us consider small changes of all quantities in a sound wave about their average values. Using the definition

$$\varepsilon' = c_{v\infty}T' + \varepsilon'_{vib} = c_{v\infty}T'_{eq} + \varepsilon'_{vib}(T_{eq}) = c_{v0}T'_{eq},$$

we can transform one rate equation to the other. In so doing, we find that $\tau_S = (c_{v\infty}/c_{v0})\tau_T$. In the equation for γ given at the beginning of this footnote, τ should be understood to denote τ_S, and in our equation (8.21) it should denote τ_T. By using the relationship between τ_S and τ_T we can easily verify that these equations are identical.

The "τ_S" method, used by Mandel'shtam and Leontovich [8], makes it possible to obtain the general equation for γ given above, independently of the actual relaxation mechanism. When considering the particular case of vibrational relaxation it is more convenient to use the "τ_T" method, as was done in this text. Let us note that this is the same manner in which vibrational relaxation in ultrasound was considered in older works (Kneser [3], Landau and Teller [10]).

In the intermediate frequency region, the sound speed a and the wave number $k = \omega/a$ are complex. If we set up an equation for $k = \omega/a$ using (8.21) and separate the real and imaginary parts, we obtain the dispersion relation $a_1(\omega) = \omega/k_1(\omega)$ and the absorption coefficient $k_2(\omega)$*. In general, this leads to rather cumbersome expressions. In the limit of low frequencies $\omega\tau \ll 1$, we obtain approximately

$$k = k_1 + ik_2 = \frac{\omega}{a_0} + i\frac{\omega^2\tau}{2a_0}\frac{c_{v\infty}}{c_{v0}}\left(\frac{\gamma_\infty}{\gamma_0} - 1\right). \qquad (8.22)$$

The absorption coefficient is of the order $k_2 \sim \omega^2$; the absorption over a distance of one wavelength is of the order $k_2\lambda \sim \omega$. In the limit of high frequencies $\omega\tau \gg 1$, we have

$$k = k_1 + ik_2 = \frac{\omega}{a_\infty} + i\frac{1}{2a_\infty\tau}\frac{c_{p0}}{c_{p\infty}}\left(\frac{\gamma_\infty}{\gamma_0} - 1\right). \qquad (8.23)$$

The absorption coefficient $k_2 \approx const$ is independent of the frequency, and the absorption over a wavelength is of the order $k_2\lambda \sim 1/\omega$. The dispersion curve $a_1(\omega)$ and the frequency dependence of the absorption over a wavelength $k_2\lambda = k_2 a_1 2\pi/\omega = 2\pi k_2/k_1$ are shown schematically in Fig. 8.5. It is

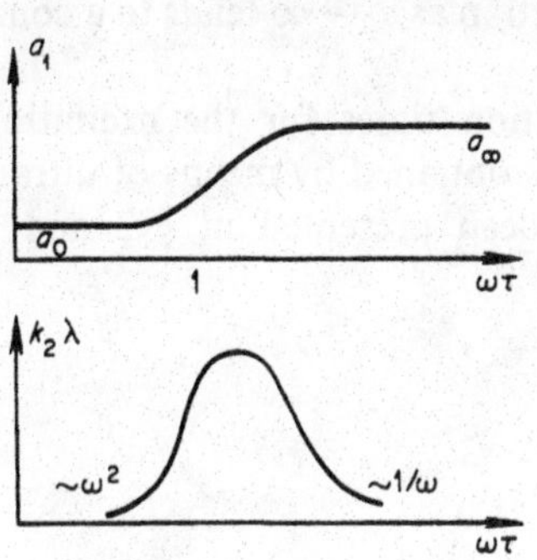

Fig. 8.5. Dependence of the propagation speed a_1 and absorption coefficient $k_2\lambda$ on the ultrasonic frequency in the relaxation region.

not difficult to show that the quantity k_2/k_1 has a maximum at $\omega\tau = (c_{v0}c_{p0}/c_{v\infty}c_{p\infty})^{1/2} \sim 1$. For a value of $\omega\tau$ which is close to but different from the above value, the dispersion curve has an inflection point.

It follows from (8.19) that the pressure in the sound wave is shifted in phase relative to the density. In fact, if the speed of sound is a complex quantity, then $p' = a^2\rho' = |a^2|e^{i\varphi}\rho'$. In the limiting cases $\omega\tau \ll 1$ and $\omega\tau \gg 1$, where the imaginary part of the speed of sound tends to zero, the

* Let us note that between the functions $k_1(\omega)$ and $k_2(\omega)$ there exists a perfectly general relationship, independent of the dispersion and absorption mechanism. This relationship was first derived by Ginzburg [11].

phase shift φ vanishes. For $\omega\tau \sim 1$, for which the real and imaginary parts are comparable, the phase shift φ is appreciable.

If several nonequilibrium processes with appreciably different relaxation times take place in the fluid, strong absorption and dispersion occurs whenever $\omega\tau \sim 1$ and these frequency regions are clearly separated. However, in the case of relaxation times that are only slightly different from each other the regions merge, and it is difficult to separate them experimentally to obtain the relaxation times from experimental data.

Dispersion and absorption of sound related to nonequilibrium processes are determined by fluctuations in the fluid density, and by virtue of the continuity equation $D\rho/Dt + \rho\nabla \cdot \mathbf{u} = 0$ are related to the divergence of the velocity. They can be formally described by the second coefficient of viscosity μ', which characterizes the dissipation term in the equation of motion which is proportional to the divergence of the velocity (see §§20 and 21 of Chapter I). The second coefficient of viscosity can be formally related to the quantity τ and to the limiting speeds of sound a_0 and a_∞ (see, for example, [9]). The use of the second coefficient of viscosity in describing anomalous absorption is only possible at not too high frequencies ($\omega\tau \ll 1$). Due to viscosity the absorption coefficient increases in proportion to $k_2 \sim \omega^2$ (see §22, Chapter I). Therefore, as $\omega \to \infty$ viscous absorption increases without limit, while actually the coefficient of anomalous absorption as $\omega \to \infty$ tends to a constant value $k_2 = const$ (see equation (8.23)).

Some experimental data on the relaxation times for the excitation of molecular vibrational and rotational modes obtained by means of ultrasonic dispersion and absorption have already been presented in §§2 and 4 of Chapter VI.

2. *Chemical reactions*

§5. Oxidation of nitrogen in strong explosions in air

Atmospheric air consists of nitrogen and oxygen molecules; chemically it is in equilibrium and very stable. Dissociation of the molecules into atoms or their partial transformation into molecules of nitric oxide NO requires heating of the air to temperatures of several thousand degrees. The reaction of nitrogen oxidation requires a large activation energy. The activation energy required for breaking up the oxide molecules into oxygen and nitrogen is somewhat lower, but it is still large. Therefore, despite the ease from the point of view of energy considerations of transforming nitric oxide into oxygen and nitrogen at low temperatures, NO molecules are extremely stable.

It was shown in §8 of Chapter VI that if the time required for establishing the equilibrium concentration of nitric oxide in air of standard density at 4000°K is $\sim 10^{-6}$ sec, then at 2000°K it is approximately equal to 1 sec, and at 1000°K it has the colossal value of the order of 10^{12} sec, approximately 30 thousand years! Once the nitric oxide has been formed and cooled to normal temperature it remains in air for an indefinitely long time. Actually, the oxidized nitrogen remains in the form of the dioxide NO_2 (or of N_2O_4 complexes, the preferred form for NO_2 molecules), since nitric oxide reacts rapidly with atmospheric oxygen and is oxidized to form the dioxide. This exothermic reaction requires a very small activation energy and takes place easily even at room temperature (see §9, Chapter VI). Thus, the chemical processes in heated and subsequently cooled air lead to a state substantially out of equilibrium. This result is in sharp contradiction with the laws of chemical equilibrium, according to which the oxides of nitrogen at low temperatures should be completely transformed into nitrogen and oxygen. This effect, well known in laboratory practice, is called the effect of "freezing" of the oxides of nitrogen.

A large amount of nitrogen oxides is formed in a strong explosion in air. The atmospheric nitrogen is oxidized at that stage of the process when the air in the explosion wave is heated to a temperature of several thousand degrees, at which stage a few percent of the nitrogen is oxidized. As the explosion wave is propagated, the air initially heated by the shock front is cooled very rapidly. The nitric oxide which was formed does not have sufficient time to decompose on cooling, and remains in the air "forever". The total weight of oxides of nitrogen which is formed in air during an explosion with an energy of 10^{21} erg, equivalent to approximately 20,000 tons of TNT, is 100 tons. Several tens of seconds or a minute after the detonation all of the oxide has been transformed into the dioxide.

In its ordinary state nitrogen dioxide is a strongly colored brownish-red gas; this is caused by the preferential absorption of green and blue light by the NO_2 molecules. It imparts the reddish hue to the cloud which rises after the explosion*. This effect has been observed experimentally and is described in [12]; see also §5 of Chapter IX. The presence of the oxides and in particular of small amounts of nitrogen dioxide in the high-temperature air encompassed by an explosion wave has a strong effect on the optical properties of the air behind the wave, since, unlike oxygen and nitrogen molecules, the molecules of the dioxide intensely absorb and emit light in the visible part of the spectrum (the NO molecule also does not absorb visible light).

The specific features of the chemical kinetics for the formation and decom-

* The molecular complexes N_2O_4 do not absorb visible light, so that the N_2O_4 gas is colorless. However, the dioxide disappears only after the explosion cloud dissipates in the atmosphere, since the reaction $2NO_2 \rightarrow N_2O_4$ does not proceed too rapidly.

position of nitrogen oxides in an explosion wave give rise to interesting optical phenomena observed in a strong explosion, a description of which may also be found in [12]. These phenomena are: luminosity of the shock wave at comparatively low temperatures, of the order of 4000—2000°K behind the front, at which the gas, consisting of oxygen and nitrogen atoms only, should not be incandescent; a rather sudden cut-off of the shock wave luminosity at a temperature of about 2000°K, and the separation of the shock wave from the boundary of a glowing body, the so-called "fireball"; the distinctive effect of a minimum in the brightness of the fireball at the instant of separation, when the glow starts to die down and after which the ball again flares up. These effects will be considered in §§5–7 of Chapter IX. Here, we shall consider in more detail the kinetics of the nitrogen oxidation reaction in an explosion wave; these considerations are necessary to explain the above optical phenomena. This problem was considered by one of the present authors [13]. It should be pointed out that the study of the kinetics is in itself very interesting, since it provides a characteristic example of a chemical process substantially out of equilibrium within the gasdynamic phemomenon of a strong explosion.

The gasdynamics of a strong explosion was described in §25 of Chapter I. The process is self-similar, the shock front is propagated from the explosion center as $R_f \sim t^{2/5}$. The distributions of all the flow variables with respect to radius are given in Fig. 1.50. These distributions do not change with time, because the process is self-similar; only the scales are time dependent.

Of interest to us here is the course of the chemical reaction in specific parcels of air. For this it is first necessary to know how the thermodynamic state of a given parcel changes with time. The r, t diagram of Fig. 8.6 shows

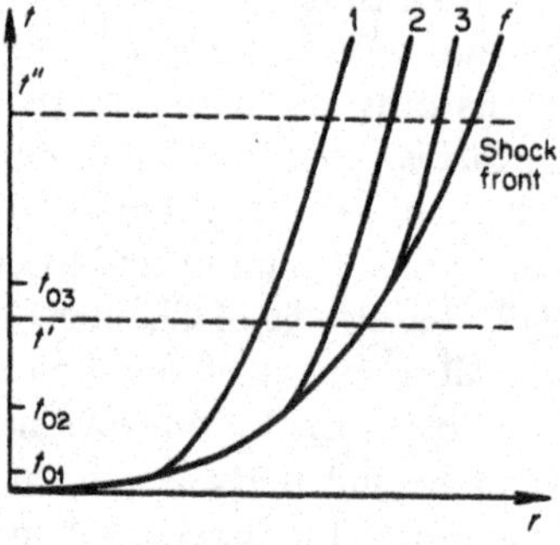

Fig. 8.6. An r, t diagram for a strong explosion in air. f is the trace of the shock front; 1, 2, and 3 are the traces of three parcels over which the front passes at the times t_{01}, t_{02}, and t_{03}.

schematically the trace of the shock front path, and of several of the particle paths behind the front, denoted by the numbers 1, 2, and 3. The parcels which are heated and compressed at the times t_{01}, t_{02}, and t_{03} when the wave front passes them are carried away suddenly by the explosion wave from the center and in the process are isentropically expanded and cooled until the pressure decreases to atmospheric and the parcels stop moving. The expansion and cooling curves for the air parcels as a function of time are shown schematically in Figs. 8.7 and 8.8.

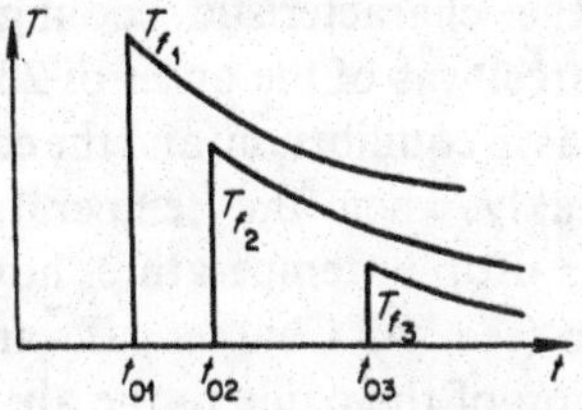

Fig. 8.7. Schematic time dependence of the temperature in three parcels heated by an explosion wave.

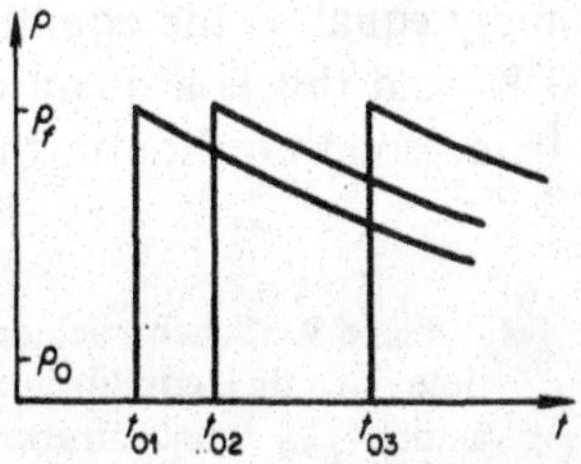

Fig. 8.8. Schematic time dependence of the density in three parcels compressed by an explosion wave.

Calculations carried out with the formulas of §25 of Chapter I show that in an explosion with an energy $E = 10^{21}$ erg, (this value will be used in all our numerical examples), the temperature at the shock front decreases to $T_f = 2000°K$ in a time of the order of 10^{-2} sec from the moment of energy release. The times for cooling the air parcels from a temperature of, let us say, 5000°K to 2000–1500°K are of the same order. The time $t \sim 10^{-2}$ sec is thus a time scale for the gasdynamic process in an explosion with an energy $E = 10^{21}$ erg, to which the characteristic times for the chemical reactions should be compared.

Let us first follow the reaction kinetics in a particular parcel of air. Let, for example, parcel 1 be heated by the shock front to a temperature $T_{f_1} = 3000°K$. The rate of nitrogen oxidation at this temperature is very high and the equilibrium concentration is reached in a time of the order of 10^{-6} sec. Approximately 5% of the nitrogen in the parcel of air is oxidized "instantaneously", after which the concentration of the oxide slowly changes (decreases) in accordance with the laws of chemical equilibrium, following the cooling and expansion. The decomposition of the oxide molecules begins to lag behind the cooling only when the parcel is cooled down to a temperature of the order of 2300°K, at which the relaxation time τ has increased from the initially small value of $\sim 10^{-6}$ sec to a value comparable to the gasdynamic time scale for cooling, i.e., 10^{-2} sec. On further cooling the decomposition stops abruptly, since the decomposition rate drops very rapidly with decreasing temperature. Thus, already at a temperature of 2000°K the decomposition rate is characterized by a relaxation time $\tau \sim 1$ sec. The residual frozen amount of the oxide in the given parcel corresponds approximately to that concentration which existed at the time when the relaxation time τ was comparable with the characteristic cooling time $\tau \sim 10^{-2}$ sec, when the temperature in the parcel was of the order of 2300°K. However, slightly earlier the concentration was in equilibrium, and the equilibrium concentration changes only very weakly when the temperature is decreased by several hundred degrees; such a drop in temperature, however, appreciably changes the decomposition rate (see §4, Chapter III, and §8, Chapter VI). Hence, the residual concentration of the oxide in the air parcel is simply equal to the equilibrium concentration at a temperature of about 2300°K, and this is a quantity of the order of 1%. The time dependence of the oxide concentration in the parcel is shown schematically in Fig. 8.9. Of

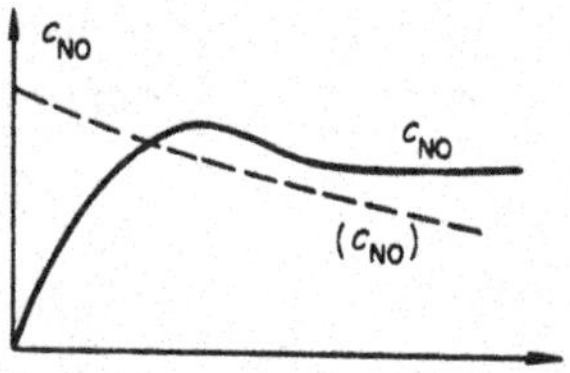

Fig. 8.9. Schematic dependence on time of the equilibrium (c_{NO}) and actual c_{NO} concentrations of nitric oxide in a given air parcel in an explosion wave.

course, the exact value of the residual concentration depends on the particular parcel, in particular on the density with which it arrived at the critical reaction temperature (for this reaction $\approx$2300°K) at which $\tau \sim t$, as well as on the cooling time. These details, however, do not affect the order of magnitude of the residual concentration. The oxidation of the oxide to the dioxide at temperatures of ~2000°K proceeds quite rapidly (see §9, Chapter VI). Therefore, the concentration of the dioxide corresponds to equilibrium, but the dioxide is in this case in equilibrium with the actual frozen amount of the oxide and not with the equilibrium amount of the oxide. At temperatures of the order of 2000°K the concentration of the dioxide is approximately 0.01% (see Table 5.9 in §21 of Chapter V). Subsequently, the entire oxide is gradually oxidized to the dioxide; here the process at first follows the cooling, and then, at temperatures of ~1500°K and below, it lags behind the cooling. Total oxidation of the oxide is completed when the parcel is quite cold, some tens of seconds after the explosion.

In air parcels which were heated by the shock front to temperatures below ~2200–2000°K, in general, no nitric oxide is formed at all, since the oxidation rate at such temperatures is very low, and the parcel quickly passes through the temperature region of approximately 2000°K in which the reaction rate is still appreciable. Thus, a spherical air layer, heated by a shock front to a temperature of ~2200–2000°K, limits, in general, the mass of air in which the oxide and then the dioxide have appeared (the motion of this layer is described on the r, t diagram of Fig. 8.6, say, by the trace 3). From this, we also get the estimate of the total amount of oxides of nitrogen formed in a strong explosion. This amount is determined by the mass of air heated by the shock front to a temperature above ~2200–2000°K, and by the equilibrium concentration of the oxide at this temperature (actually, at a slightly higher temperature—2300°K), since the freezing takes place particularly at these temperatures*. In an explosion with an energy of 10^{21} erg the radius of the shock front at a front temperature $T_f = 2000°$K is approximately 100 m. The mass of air in a spherical volume of this radius is approximately 5000 tons, and at a concentration of ~1% the mass of the oxide is ~50 tons. The mass of the dioxide, upon addition of still another oxygen atom to each NO molecule, is ~75 tons, that is, about 100 tons as was stated previously.

Let us now consider the distribution of the concentration of oxides with respect to the radius at a given instant of time. Two typical cases are possible. If at the time t' (Fig. 8.10) the temperature at the wave front is above

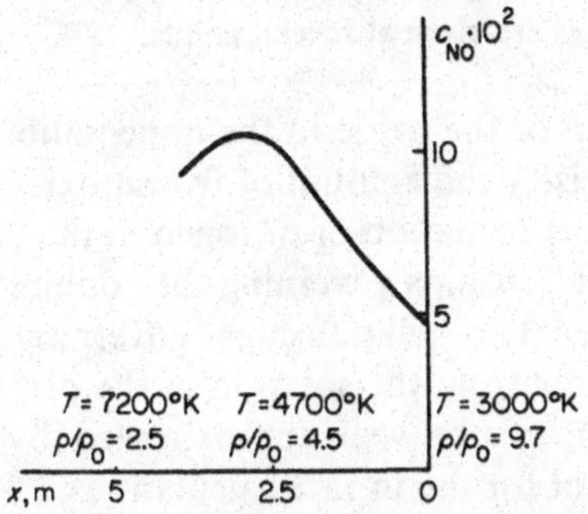

Fig. 8.10. Distribution of nitric oxide concentration behind a shock front in an explosion with an energy $E = 10^{21}$ erg. The temperature at the front $T_f = 3000°$K. The concentration is at equilibrium almost everywhere. Values of temperatures and densities are given at several points. x is the distance behind the shock.

~2300°K, then for practically all the parcels behind the front the concentrations of the oxide and dioxide are equal to their equilibrium values and the distributions of the concentrations are simply determined by the temperature and density distributions behind the front. The only exception is a very thin layer of air immediately behind the front, in which the oxides at this time have not yet formed (Fig. 8.10).

If we consider now the time t'', when the temperature behind the front is less than ~2000°K, for example 1600°K, then we find in the vicinity of the front parcels which have been heated by the front to temperatures below

* We recall that the equilibrium concentration of nitric oxide in air depends only on the temperature, and not on the density (see §4, Chapter III, and §8, Chapter VI).

2000°K; in general, this gas does not contain any oxide. Far from the front, at temperatures above ~2500°K, the concentration is at its equilibrium value, while in an intermediate layer the oxide is present, but not in its equilibrium concentration. Closer to the front it is lower than its equilibrium value, and slightly farther away, in those parcels in which freezing has already begun, it is higher than at equilibrium (Fig. 8.11).

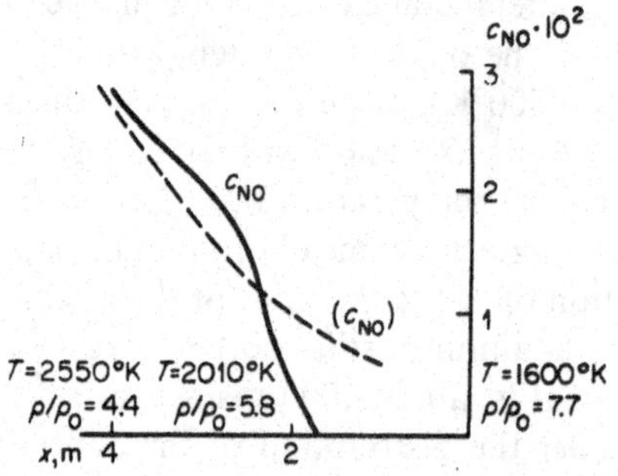

Fig. 8.11. Distribution of the nitric oxide concentration behind a shock wave front in an explosion with $E = 10^{21}$ erg. The temperature at the front $T_f = 1600$°K. The solid curve is the actual concentration, the dashed curve the equilibrium concentration. For $x > 4$ m, $c_{NO} \approx (c_{NO})$. Values of temperatures and densities are given at several points.

In order to calculate the concentration of the oxide in the nonequilibrium region, and also to determine more precisely the amount of frozen oxide, we must solve the rate equation (6.45) for the oxidation of nitrogen in the given parcel of air, taking into account the relationships governing the cooling and expansion behind the explosion wave. The relationships governing the expansion and cooling of air, which follow from the solution of the problem of a strong explosion (§25, Chapter I), can be well approximated by the following equations, which are convenient for use in rate calculations

$$\rho = \rho_0 \left(\frac{t_0}{t}\right)^{2b},$$

$$\frac{1}{T} = \frac{1}{T_0} + \frac{a}{T_0} \ln \frac{t}{t_0}.$$

Here T_0 and ρ_0 are the temperature and density in the parcel at the initial time t_0, when the shock front passed through it, and a and b are numerical constants, depending only on the effective specific heat ratio in the gas-dynamic solution. For $\gamma = 1.30$, $a = 0.44$ and $b = 0.75$.

It has been shown (see [13]) that by an appropriate choice of new variables in the rate equation (6.45) this equation, together with the known relations for the cooling and expansion, can be reduced to the universal dimensionless form

$$\frac{dy}{dx} = x^{3-\delta}(y^2 - x^2). \tag{8.24}$$

Here the quantity x is related to the time variable, y is proportional to the oxide concentration, and δ is a numerical constant less than unity. The initial condition corresponding to the absence of oxide at the initial time $t = t_0$ reduces to the condition $y = 0$, with x equal to some value x_0 which depends only on the time t_0, the initial values of the state variables, and on the constants entering the rate equation (6.45). Zel'dovich, Sadovnikov, and Frank-Kamenetskii [14] have studied the kinetics of the nitrogen oxidation reaction under laboratory conditions with a cooling law of the type $1/T = (1/T_0) + (a'/T_0)t$, at constant density. With new variables introduced, the rate equation (6.45) was also reduced to an equation of the type (8.24) with the initial condition $y = 0$ when $x = x_0$. Reference [14] gives a tabulation of the solution of the equation $y = y(x, x_0)$*.

Knowing the initial state variables for an air parcel from the gasdynamic solution of the problem of a strong explosion, we can thus obtain a complete solution of the dependence of the oxide concentration c_{NO} on time. This solution is in complete agreement with the qualitative considerations presented above. The curve presented in Fig. 8.11 was calculated in this manner.

3. *Disturbance of thermodynamic equilibrium in the sudden expansion of a gas into vacuum*

§6. Sudden expansion of a gas cloud

The phenomenon of sudden expansion of a gas cloud into vacuum is encountered in various natural, laboratory, and industrial processes. Meteorite impacts on planet surfaces result in the sudden braking of the meteorites and the conversion of their kinetic energy into heat. If the impact velocity is high, of the order of several tens of kilometers per second, very high temperatures result, of the order of tens or hundreds of thousands of degrees. The meteorite and a part of the planet soil are vaporized during the impact. This phenomenon resembles a strong explosion on the planet's surface†. If the

* As was noted by Kompaneets, the equation (8.24) with $\delta = 0$ can be solved exactly in terms of Bessel functions.

† The hydrodynamics of this process will be considered in Chapter XII.

planet is without an atmosphere (as for example in the case of the moon) the vapor cloud generated, with tremendous expansion velocities, overcomes the force of gravity and freely expands into vacuum. A hypothesis has been advanced that the lunar craters were formed as a result of such "explosions" from the impacts of extremely large meteorites. Similar phenomena also take place in the much more frequent collisions between small bodies in the solar system—the asteroids.

Sudden expansion into vacuum of tremendously large gas clouds is observed during nova outbursts, in which a disturbance in the energy balance of the star leads to the release of a large amount of energy and a shock wave is propagated from the central layers to the periphery. This shock wave separates from the surface of the star and emits a gas cloud into space. To some extent, similar phenomena (but on very much smaller scale) are also encountered under laboratory conditions. An example is the vaporization of the anode needle of a pulsed x-ray tube caused by a strong electron discharge (Tsukerman and Manakova [15]); another is the explosion of wires by electric currents in vacuum systems. Of course, the expansion under laboratory conditions is not unbounded since it is bounded by the walls of the vacuum chamber; however, at the stage when the gas has not yet reached the walls, the expansion into the vacuum takes place in the same manner as if the vacuum were "infinite".

Experiments in which a gas cloud suddenly expands into a vacuum were also carried out in connection with rocket probe studies of the upper layers of the atmosphere when sodium vapor and nitric oxide were released into space. The same phenomenon also occurred when an artificial comet was created during the moon flight of the Soviet cosmic rocket.

The dynamics of the sudden expansion of a gas cloud into vacuum is very simple; an idealized problem of the sudden isentropic expansion of a gas sphere into vacuum, for a gas with constant specific heats, was considered in §§28 and 29 of Chapter I. Here we shall be interested in certain fine points concerning the state of the gas during the later stage of the expansion to infinity, when the expansion can be treated on the basis of a very simple scheme. In this scheme we shall consider the behavior of only the mass-averaged flow variables. It is clear that the variables describing any particular gas particle change with time in exactly the same manner as do the averaged quantities and differ from the average values only by numerical factors which are of the order of unity and which are not of great importance for our purpose.

Let us consider a gas sphere of mass M and energy E*. Almost the entire initial energy has been transformed into kinetic energy during the earlier

* For convenience, we shall repeat here some of the results of §28, Chapter I.

phase of the expansion, and the fluid expands by inertia with the average velocity

$$u = \left(\frac{2E}{M}\right)^{1/2}.$$

The sphere radius is of the order of $R = ut$ and the gas density decreases with time according to the relation

$$\rho = \frac{M}{4\pi R^3/3} = \rho_0 \left(\frac{t_0}{t}\right)^3, \tag{8.25}$$

where the characteristic time scale is approximately expressed in terms of the initial radius of the sphere R_0 and the initial fluid density ρ_0 by

$$t_0 = \left(\frac{M}{\rho_0 4\pi u^3/3}\right)^{1/3} = \frac{R_0}{u}. \tag{8.26}$$

If we are interested in the gas temperature during the later stage of the expansion, we must consider the small amount of internal energy that still remains in the gas, which we neglected in calculating the expansion velocity. We take into account the fact that the specific entropy of the gas S remains constant during the expansion. Assuming for simplicity that the fluid behaves as a gas with some constant effective value of the specific heat ratio we obtain for the cooling of the gas the relation

$$T = A(S)\rho^{\gamma-1} \sim t^{-3(\gamma-1)}, \tag{8.27}$$

where $A(S)$ is a constant which depends on the entropy and which can be calculated from the well-known formulas of statistical mechanics and thermodynamics. If we consider relatively high temperatures then, taking into account the processes of ionization, dissociation, etc., we can take as an approximate value for the specific heat ratio $\gamma \approx 1.2$–1.3. In any case, the ratio is not greater than $5/3 = 1.66$, which corresponds to a complete freezing of all the internal degrees of freedom in the gas.

§7. Freezing effect

Let us consider the physical and chemical processes taking place in a gas expanding according to the cubic relation $\rho \sim t^{-3}$ and cooling as $T \sim t^{-3(\gamma-1)}$. We assume that the initial temperature was high, so that the molecules were dissociated and the atoms strongly ionized. We further assume that the initial density of the gas was also high, as is usually the case when a gas cloud is formed by the rapid release of energy in an initially solid substance. Then, during the early stage of the sudden expansion, at high density and temperature, all relaxation processes proceed very rapidly and the gas remains in thermodynamic equilibrium; the characteristic state variables of the gas, as for example the degrees of ionization or dissociation, follow the expansion and cooling. If during the entire expansion the gas were to remain in thermodynamic equilibrium, then in the process of expansion and cooling all the electrons would rather quickly combine with the ions into neutral atoms, and all the atoms with a chemical affinity would combine into molecules.

Actually, the equilibrium degrees of ionization and dissociation have an exponential dependence on temperature but only a power-law dependence on the density: $\alpha \sim \rho^{-1/2} \exp(-I/2kT)$, where I is the ionization potential or dissociation energy. For an expansion with cooling to low temperatures, the equilibrium degrees of ionization and dissociation rapidly approach zero, since as $T \sim \rho^{\gamma-1} \to 0$ the exponential term decreases extremely rapidly, much more rapidly than the preexponential factor increases. It can be easily seen, however, that no matter how high the initial rate of establishment of thermodynamic equilibrium in comparison with the cooling and expansion rates, there is a time at which the ratio of these rates will be reversed, thermodynamic equilibrium will no longer be established, and the degrees of ionization and dissociation will begin to depart increasingly from their equilibrium values.

In fact, the equilibrium degrees of ionization and dissociation are established as a result of the mutual compensation of the direct and reverse processes. But at low temperatures the ionization and dissociation, which require large expenditures of energy, drop very sharply. The rates of these processes depend on temperature as $\exp(-I/kT)$, and for $kT \ll I$ they depend extremely strongly on temperature, and thus on time. On the other hand the rates of the reverse recombination processes have only a power-law dependence on density and temperature and consequently on time. Thus, ionization and dissociation will essentially stop at a certain instant, after which the degrees of ionization and dissociation will decrease with time following a power law, while the equilibrium values drop exponentially.

The rates of recombination processes decrease as a result of an expansion, and recombination may cease entirely. We can convince ourselves of this by taking as an example the recombination of atoms into molecules (recombination of electrons with ions will be considered in the following sections). Recombination at high densities takes place by three-body collisions, while two-body collisions are responsible for recombination at low densities, so that it is sufficient to consider the latter in the later stage of the expansion. Let N be the number density, $\bar{v} \sim \sqrt{T}$ the thermal speed, and σ the recombination cross section, which is no larger than the gaskinetic cross section. The recombination rate $dN/dt = -N\bar{v}\sigma$ and the characteristic time during which appreciable changes take place in the degree of dissociation is $\tau \approx 1/N\bar{v}\sigma$. Even if we do not take into account the decrease in the number of atoms as a result of recombination, the atom number density N will decrease in proportion to $1/t^3$ as a result of the expansion of the gas, $\bar{v} \sim \sqrt{T} \sim t^{-\frac{3}{2}(\gamma-1)}$ so that $\tau \sim t^{3+\frac{3}{2}(\gamma-1)} = t^{\frac{3}{2}(\gamma+1)}$, and the characteristic time τ increases faster than t; consequently, at some instant of time it will become greater than t. On the other hand, the time scale characterizing the changes in density and temperature is the time t itself, measured from the beginning of the sudden expansion, since $dT/dt \sim -T/t$ and $d\rho/dt \sim -\rho/t$. Thus, at some time recombination begins to lag ever increasingly behind the cooling. Moreover, beginning approximately at this time, the probability of recombination of the given atom with all other atoms during the remaining expansion to infinity is less than unity; in other words, the recombination does not proceed to completion. In fact, this probability is equal to

$$w = \int_{t_1}^{\infty} N\bar{v}\sigma \, dt = \int_{t_1}^{\infty} \frac{dt}{\tau} \lesssim \frac{1}{\tau_1} \int_{t_1}^{\infty} \left(\frac{t_1}{t}\right)^{\frac{3}{2}(\gamma+1)} dt \approx \frac{t_1}{\tau_1} = const. \qquad (8.28)$$

Starting at the time t_1 at which $\tau_1 > t_1$, the recombination probability of a given atom $w < 1$. Thus, the gas expands to infinity in the dissociated state. This phenomenon is called "freezing" of the atoms.

Starting at some time, the gaskinetic collisions in the gas also cease almost entirely. Deexcitation of the vibrational and rotational modes of molecular excitation by particle collisions stops. This follows from the convergence of the same collision integral (8.28). However, freezing of the molecular vibrational and rotational modes does not take place; the vibrational and rotational energies of the molecules are carried away by the spontaneous emission of photons. The vibrational transitions produce radiation in the infrared region of the spectrum and the rotational transitions result in the emission of radio frequencies.

Owing to the convergence of the collision integral when the gaskinetic cross section (for the neutral atoms) is substituted, the exchange of energy of the random translational motion of the atoms also ceases after a certain time in a spherical sudden expansion. The remainder of the expansion then continues without collisions*. All of the particles move by inertia with the velocity which they have acquired in their last collision. In this case the particles, in general, have a nonradial ("random") velocity component. It would seem that "freezing" of the random velocity, that is, "temperature", should take place. Actually, as was noted by Belokon' [16], this does not happen for purely geometrical reasons. The problem consists in the definition of the concepts of hydrodynamic and internal energy when particle collisions are

* It is interesting to note that for a sudden expansion with $\gamma = 5/3$ the frequency of Coulomb collisions of charged particles does not decrease, since $\sigma \sim T^{-2}$ and $N\bar{v}\sigma \sim NT^{-3/2} \sim N^{5/2-3\gamma/2} = const.$

absent. The internal energy of a unit volume of gas is equal to the difference between the total kinetic energy $\frac{1}{2}Nm\overline{v^2}$ (N is the number of particles per unit volume, m their mass, and $\overline{v^2}$ the mean square velocity) and the kinetic energy of "hydrodynamic" motion $\frac{1}{2}Nm(\bar{v})^2$, where $(\bar{v})^2$ is the square of the mean velocity:

$$E_{\text{int}} = E_{\text{total}} - E_{\text{hydrod}} = \tfrac{1}{2}Nm(\overline{v^2} - (\bar{v})^2).$$

Let us suppose that the collisions stop at a time t_1, when the gas occupies a sphere of radius r_1 (Fig. 8.12). Particles leaving the sphere arrive at points A and B at times t' and t'' with velocities whose directions are included within the cones shown in Fig. 8.12. It is clear that the greater is the distance from

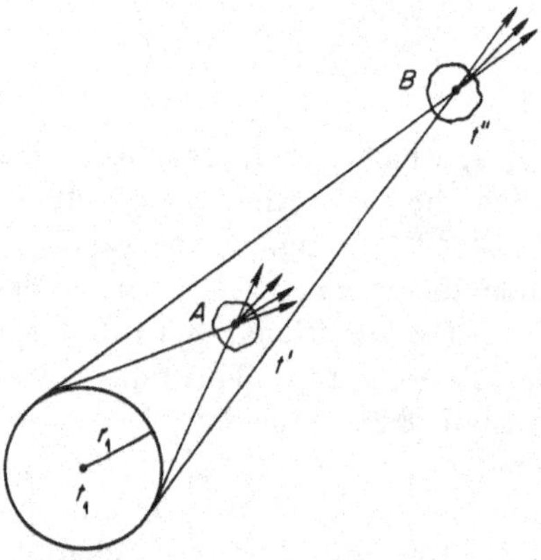

Fig. 8.12. The problem of expansion of a gas into vacuum without collisions.

the center, the narrower is the cone and the closer $\overline{v^2}$ is to $(\bar{v})^2$, the smaller is the difference $\overline{v^2} - (\bar{v})^2$. In the limit $t \to \infty$, $r \to \infty$ all particles move in an exactly radial direction. In this limit $\overline{v^2} = (\bar{v})^2$, and the entire translational internal energy has been transformed into hydrodynamic energy*.

In cylindrical and planar sudden expansions, thermodynamic equilibrium is also definitely disturbed, but of course the changes with time in the degrees of dissociation and ionization follow different laws. It should be noted that, if the mass of the gas is finite, then, when the gas has undergone a sufficiently

* *Editors' note.* Besides the geometrical effect, another effect must be invoked for this conclusion. The particle velocities at the point A are not uniform in magnitude, but lie between $(r' - r_1)/(t' - t_1)$ and $(r' + r_1)/(t' - t_1)$. With t' very large the variation in particle velocity is of the order of r_1/t'. In the limit $t \to \infty$, $r \to \infty$ this variation approaches zero, so that the particle velocities approach uniformity in magnitude as well as in direction. In this limit, then, $\overline{v^2} = (\bar{v})^2$.

large expansion, cases which are initially planar or cylindrical must become spherical cases.

We note a number of other papers [15a, 24–26] devoted to various problems of the free molecular sudden expansion of a gas into a vacuum without collisions (these articles also contain references to other papers). Reference [27] considers the sudden expansion of an ionized gas into a vacuum in which there is a magnetic field.

§8. Disturbance of ionization equilibrium

We now consider in more detail the problem of the disturbance of ionization equilibrium resulting from the expansion of a gas, and show how we can establish approximately the time at which the equilibrium is disturbed (the method presented below was suggested by one of the authors [17]). We assume that initially the gas temperature was high and the atoms were multiply ionized. As the expanding gas cools, electrons are reseated at the appropriate levels in the atoms, and the degree of ionization decreases. Let the ionization equilibrium be disturbed only during the later stage of a sufficiently strong expansion and cooling, when it is the last electrons which are reseated in the atoms, i.e., when the process taking place is the reverse of single ionization of the atoms. The gas at this time is essentially expanding by inertia, with the speed constant and the density changing as $1/t^3$.

The mechanism of recombination of ions and electrons was considered in detail in Chapter VI. The electrons are captured by ions in three-body collisions with an electron acting as the third body; at temperatures which are not too high the electrons, as a rule, are captured into the upper levels of the atoms. Captures by two-body collisions with the emission of a photon are also possible (in this case the electrons are primarily captured into the ground level). Such photorecombination is of importance only at very low electron densities N_e cm^{-3}. The lower the temperature the lower must be the electron density for photorecombination to be important. According to (6.107) this process predominates only under the condition that $N_e < 3.1 \cdot 10^{13} T_{\text{ev}}^{3.75} = 3.2 \cdot 10^9 T_{\text{thous. deg}}^{3.75}$. However, in the majority of cases of sudden expansion which are of interest, at the stage when equilibrium is disturbed at temperatures of several thousand degrees, the electron density is much higher and photorecombination plays no role, neither at the time when the equilibrium is disturbed nor later.

If when the gas is still close to equilibrium the main role is played by recombination from three-body collisions, then ionization takes place as a result of the reverse process, that of the removal of electrons primarily from excited atoms by free electron impact. According to the detailed balancing principle, the ionization rate can be expressed in terms of the recombination coefficient and the equilibrium constant or the equilibrium degree of ionization. In this case the rate equation for the degree of ionization $x = N_e/N$ (N is the number of nuclei, atoms plus ions per unit volume) is expressed in the form

$$\frac{dx}{dt} = bN(x_{eq}^2 - x^2). \tag{8.29}$$

Here b is the recombination coefficient which, at not too high temperatures, not above several thousand degrees, is given by (6.106)

$$b = \frac{ANx}{T^{9/2}}, \tag{8.30}$$

$A = 8.75 \cdot 10^{-27}\ \text{cm}^6 \cdot \text{ev}^{9/2}/\text{sec} = 5.2 \cdot 10^{-23}\ \text{cm}^6 \cdot (\text{thousand deg})^{9/2}/\text{sec}.$

Here x_{eq} is the equilibrium degree of ionization given by the Saha equation. For values of x_{eq} not large in comparison with one we have, approximately,

$$x_{eq} \approx 7 \cdot 10^7 \left(\frac{g_+}{g_a} \frac{T^{\circ\,3/2}}{N} \right)^{1/2} e^{-I/2kT}. \tag{8.31}$$

When the expansion and cooling are governed by the known relations in which $N(t)$ and $T(t)$ are given by (8.25) and (8.27), then (8.29) becomes an ordinary differential equation for the desired function $x(t)$. Since we are primarily interested in the qualitative features of the problem, we shall solve this equation only approximately. The initial ionization and recombination rates, which are proportional to the terms on the right-hand side of (8.29), are large in comparison with the expansion and cooling rates. (For the purpose of comparing rates of the different processes we shall use relative rates expressed in reciprocal seconds, for example, $T^{-1}dT/dt$ and $N^{-1}dN/dt$). Ionization and recombination almost completely cancel each other; the degree of ionization follows the expansion and cooling and remains close to its equilibrium value. Approximately, $x(t) \approx x_{eq}(t) \equiv x_{eq}[T(t), N(t)]$ and the difference $|x_{eq}^2 - x^2| \ll x_{eq}^2$.

The small departure of the degree of ionization from its equilibrium value which inevitably exists (since the temperature and density change with time) can be approximately estimated by setting $dx/dt \approx dx_{eq}/dt$ on the left-hand side of (8.29), replacing x in the expression for the recombination coefficient by x_{eq}, and setting $x_{eq}^2 - x^2 \approx 2x_{eq}(x_{eq} - x)$. It can be easily checked that the relative departure $|x_{eq} - x|/x_{eq}$ increases with time (since the rate of the relaxation process becomes increasingly lower in comparison with the rate of change of the macroscopic parameters temperature and density).

Ionization equilibrium is appreciably disturbed when the difference between the ionization and recombination rates increases to a value of the order of the rates themselves, that is, when the quantity $|x_{eq}^2 - x^2|$ becomes of the order of x_{eq}^2. To estimate the time t_1 when the equilibrium is disturbed and the values of T_1, N_1, and x_1 at this time, we can again set $dx/dt \approx dx_{eq}/dt$ and $x \approx x_{eq}$ in the recombination coefficient, and equate the difference $x^2 - x_{eq}^2$ to the value of x_{eq}^2. Differentiating with respect to time the equilibrium degree of ionization given by (8.31), taking into account the fact that the most rapidly changing factor is the exponential Boltzmann factor, and then using the cooling relation (8.27) (which yields $dT/dt = -3(\gamma - 1)T/t$), we find an equation which determines the time when the equilibrium is disturbed;

$$b_1 N_1 x_{eq_1} t_1 = \tfrac{3}{2}(\gamma - 1)\frac{I}{kT_1}. \tag{8.32}$$

Here $b_1 = b(T_1, N_1, x_{eq_1})$. This equation, together with the expansion and cooling expressions (8.25), (8.27), and the Saha equation (8.31) referred to the time t_1, reduce to a transcendental equation for the temperature T_1. Having found T_1, it is easy to calculate the remaining quantities t_1, N_1, and x_{eq_1}. (Within the approximation used we can take the actual degree of ionization x_1 equal to the equilibrium value x_{eq_1}.)

§9. The kinetics of recombination and cooling of the gas following the disturbance of ionization equilibrium*

After the equilibrium is disturbed the ionization rate, which is proportional to x_{eq}^2, continues to decrease rapidly with time according to the exponential relation $e^{-I/kT(t)}$. The recombination rate, which is proportional to the square of the actual degree of ionization, decreases much more slowly and soon becomes appreciably larger than the ionization rate: $x(t) \gg x_{eq}(t)$. Under these conditions it is possible to neglect ionization and to assume that only recombination takes place. The rate equation (8.29) will then be approximately written as

$$\frac{dx}{dt} = -bNx^2 \qquad \text{for} \quad t > t_1. \tag{8.33}$$

If the recombination coefficient b were entirely independent or only weakly dependent on the temperature, then, as a result of the rapid decrease in density, the recombination rate would have dropped and recombination would soon have stopped entirely. This is precisely the situation that takes place with the recombination of atoms into a molecule (see §7). On the other hand, in our case the recombination coefficient (8.30) is very temperature sensitive ($b \sim T^{-9/2}$), and the decrease in the recombination rate due

* This section is based on a paper by Kuznetsov and one of the authors [28].

to the drop in density in the expanding gas is compensated to an appreciable extent by the increase in the recombination rate coefficient due to cooling. For this reason particular importance is attached to the question of the relation governing the temperature decrease with time. In fact, let us write the formal solution of the differential equation (8.33) using the recombination coefficient (8.30),

$$x = \frac{x_1'}{\left[1 + 2Ax_1'^2 \int_{t_1'}^{t} N^2 T^{-9/2}\, dt\right]^{1/2}}. \tag{8.34}$$

Here the values of t_1' and x_1' are determined from initial conditions (to the order of this approximation we may require that the integral curve $x(t)$ pass through the point where the equilibrium is disturbed, so that $t_1' = t_1$ and $x_1' = x_1$). We shall, as previously, characterize the temperature drop with time by the power-law behavior $T \sim t^{-m}$; the quantity $N \sim t^{-3}$.

According to (8.34) the asymptotic behavior of the degree of ionization depends to a large extent on the cooling rate, that is, on the value of the exponent m. If the gas is cooled "slowly" and $m < 10/9$ (which corresponds to a specific heat ratio $\gamma = 1 + m/3 < 1.37$) the integral in (8.34) converges for $t \to \infty$ and the degree of ionization approaches a constant value different from zero, so that recombination does not go to completion. However, if the gas is cooled "rapidly" and $m > 10/9$, the integral diverges for $t \to \infty$ and the degree of ionization tends to zero, following $x \sim t^{-(9/4)(m-10/9)}$. For $m = 10/9$ it also goes to zero, but logarithmically, following $x \sim (\ln t)^{-1/2}$. Thus for $m \geqslant 10/9$ the electrons and ions must eventually recombine.

But the rate of cooling of the gas itself depends on the mode of recombination, since the recombination releases the potential energy of the free electrons previously removed from the atoms, and this energy is partially transformed into heat. Consequently, in order to solve for the behavior of the degree of ionization with time we must consider together the kinetics of the recombination and the balance of thermal energy in the gas.

In the recombination of an electron in a three-body collision the electron is first captured by the ion into one of the upper levels of the atom with a binding energy E of the order of kT (see Chapter VI). Then, under the action of electron collisions of the second kind but also as a result of spontaneous radiative transitions, the bound electron descends down the energy levels of the atom to the ground level. The process of deexcitation of an excited atom usually takes place rapidly in comparison with the rates of electron capture by ions and of the change in gas temperature. For this reason we may assume approximately that the excited atom which is formed is deexcited immediately following capture, and the potential energy I from the recombination is immediately transformed into other forms of energy. A part E^* of it is transferred directly to the free electrons by the electron collisions of the second kind (and then is distributed over the entire gas as a result of energy transfer between the electrons and ions). The other part of the binding energy $I - E^*$ that is released as a result of radiative transitions is first transformed into line radiation. This radiation in part leaves the gas volume and in part is absorbed by atoms, where the absorption is principally that of the resonance radiation corresponding to the transition of the excited atom directly to the ground state. Through absorption of a resonant photon the atom becomes excited; then it re-emits, the new photon is absorbed by other atoms, etc., this process repeating itself until the photon leaves the gas volume. A so-called diffusion of resonance radiation takes place. During the diffusion of the photon the excited atom can be subjected to a collision of the second kind and give off excitation energy in the form of heat. As a result, some portion of the binding energy $I - E^*$ which was first converted into radiation will with time also be transformed into heat. This part will be less the more transparent is the gas, that is, the shorter the duration of the diffusion of resonance radiation.

We assume for simplicity that the energy $I - E^*$ is completely lost by the gas (corresponding to a transparent gas volume). This assumption underestimates the heat release in the gas and tends to underestimate the temperature; the calculation of the recombination kinetics for this condition will lead to an underestimate of the degree of ionization, and thus yields a lower limit. The above assumption is more justified the more transparent is the gas, and thus the later is the stage of the expansion which we consider. Hence the assumption is valid asymptotically.

In setting up the equation of the energy balance for the gas we shall assume for simplicity that the electron and ion (atom) temperatures are the same. Estimates indicate that in the majority of cases, even after a considerable length of time has passed since the ionization equilibrium was first disturbed, the energy transfer between the electron and ion gases takes place rapidly and the temperatures of these gases are close to each other*. Let us write the energy equation for the gas per heavy particle (per original atom). The thermal energy is $\varepsilon = \frac{3}{2}(1 + x)kT$, the specific volume $V = 1/N$, and the gas pressure $p = N(1 + x)kT$. Then

$$\frac{d\varepsilon}{dt} + p\frac{dV}{dt} = E^*\left(-\frac{dx}{dt}\right). \tag{8.35}$$

From this, using the relation for the expansion $N \sim t^{-3}$, we obtain the equation for the temperature

$$\frac{dT}{dt} + 2\frac{T}{t} = \frac{\frac{2}{3}E^*/k + T}{1 + x}\left(-\frac{dx}{dt}\right). \tag{8.36}$$

* In [28] the equations were set up taking differences between the electron and ion temperatures into account.

To calculate the heat release per recombination E^* we consider the process of the deexcitation of the excited atom which is formed as the result of the capture of an electron by an ion. As noted above, in three-body collisions the electron is, as a rule, captured into one of the very high atomic levels with a binding energy $E \sim kT$. The distances between levels in this region are much less than kT. In collisions with free electrons the bound electron in the excited atom moves to neighboring levels with transitions "upward" and "downward" almost equally probable, so that the change in the energy of an excited atom under the action of electron impacts has the character of a diffusion along the energy axis; on recombination the diffusive flux is directed downward, in the direction of the ground state of the atom (see §18 of Chapter VI). The deexcitation rate, that is the rate of increase in the binding energy of the bound electron dE/dt, can be calculated by multiplying the unsteady diffusion equation (6.109) by the energy E and integrating over the entire spectrum. Since we are considering an unsteady case of the motion of an electron along the energy axis E from a source located in the region of low binding energies, it follows that in the region of high energies we should specify the boundary condition that the distribution function and the diffusive flux are zero. The above procedure gives an approximate expression for the rate of change of the electron binding energy

$$\frac{dE}{dt} \approx \frac{D}{kT},$$

where D is the diffusion coefficient given in §18 of Chapter VI. This equation loses its validity in the region where the distance between the levels is greater than kT, since in this region transitions to the lower levels are significantly more probable than transitions to the upper levels, and the motion has a unidirectional character. In this region

$$\frac{dE}{dt} \approx \beta_{n,n-1} N_e \, \Delta E_{n,n-1},$$

where $\beta_{n,n-1} N_e$ is the transition probability from level n to level $n-1$ (in $\sec^{-1}$), and $\Delta E_{n,n-1}$ is the energy distance between the levels. (The expression for the rate constant of the deexcitation transition $\beta_{n,n-1}$ is given in §15, Chapter VI.)

The transition from a diffusion to a unidirectional motion of the bound electron along the energy axis takes place at a binding energy E' for which the distance between levels $\Delta E_{n,n-1}$ is equal to kT. This energy is (for a hydrogen atom)

$$E' = \frac{1}{2} kT \left(\frac{2I}{kT} \right)^{1/3} = 2.1 \cdot 10^{-4} T^{\circ\, 2/3}.$$

As is known (see §13, Chapter V), the rate of radiative deexcitation, which is very low for low binding energies, increases rapidly as the degree of excitation decreases. After collisional deexcitation of the atom to some level radiative transitions begin to predominate. The radiation rate corresponding to the radiative transitions to neighboring levels, which determines the rate of radiative deexcitation $(dE/dt)_{\text{rad}}$, is given in §13 of Chapter V. The change from impact to radiative deexcitation takes place at a binding energy for which the rates $(dE/dt)_{\text{impact}}$ and $(dE/dt)_{\text{rad}}$ become equal. This binding energy, obviously, is the energy release E^* to be determined. It should be noted that radiative deexcitation is also promoted by radiative transitions in the atom from the upper levels directly to the ground state, after which the atom no longer interacts (see §13, Chapter V). Calculations [28] show that the contribution of this process is comparable with the contribution made by the gradual radiation which takes place with transitions to the neighboring levels.

Calculation of the heat release E^* carried out in [28] gives approximately*

$$E^* = I \times \begin{cases} 4.3 \cdot 10^{-4} N_e^{1/3} T^{\circ\,-1/2}, & \text{if } kT < E^* < E', \\ 3.1 \cdot 10^{-4} N_e^{1/6} T^{\circ\,1/12}, & \text{if } E^* > E'. \end{cases} \tag{8.37}$$

After having determined the heat release E^* we can integrate the system of equations (8.33) and (8.36). As was shown in [28], the order of the system can be reduced, leading to one nonlinear differential equation of first order. Qualitative analysis of the resulting equation and of the numerical solution shows that, depending on the initial conditions, one or the other regime of the recombination process is established.

If the gas cloud expands rapidly (low cloud mass, high speed of the sudden expansion) and the ionization equilibrium is disturbed early, at a high degree of ionization with the store of potential energy of ionization in the gas greater than the thermal energy, then a large amount of heat is generated during recombination. This prevents rapid cooling of the gas and retards the recombination. Under these conditions the recombination soon ceases and the degree of ionization tends to a constant value different from zero. The electrons and ions become frozen. Practically, this happens if at the time the ionization equilibrium is disturbed all the atoms of the gas are at least singly ionized ($x_1 \gtrsim 1$). However, if the gas cloud expands relatively slowly

* If for any value of T and N_e we have $E^* < kT$, this means that radiative deexcitation must start at the beginning of electron capture. This situation is usually not achieved, since in this case photorecombination dominates recombination in three-body collisions, and in photorecombination the electron is ordinarily captured into the lower, rather than into the upper atomic levels. If E^* calculated from (8.37) is found to be larger by an order of magnitude than the ionization potential I, this means that the entire binding energy I is converted into heat, equation (8.37) is inapplicable, and $E^* = I$.

(large mass, low speed of the sudden expansion) and the ionization equilibrium is disturbed late, at a low degree of ionization with the store of potential energy less than the thermal energy, the small amount of heat which is released on recombination is not capable of retarding the rapid cooling of the gas due to expansion, and the recombination rate stays sufficiently high. In this case the recombination proceeds all of the time and the degree of ionization decreases continuously, going to zero. This continues until the energy exchange between the electron and ion gases is disturbed. This happens when the characteristic exchange time τ_{ei} (see §21, Chapter VI) becomes greater than the time t measured from the beginning of the expansion; this time characterizes the relative rates of expansion and cooling $V^{-1}dV/dt$ and $T^{-1}dT/dt$. After the exchange is disturbed, the heat which is released on recombination is no longer distributed uniformly among all the gas particles, but remains only in the electron gas. (The ions and atoms are cooled faster than are the electrons and their temperature decreases relative to the electron temperature.) Under these conditions the energy release in the electron gas is increased comparatively, since the recombination energy is now transferred to only a small number of electrons. As a result the drop in electron temperature takes place more slowly, and the recombination is also retarded. Under these new heat transfer conditions the degree of ionization ceases to approach zero and a residual ionization is retained in the gas (freezing takes place). However, unlike the preceding case, the residual ionization is now very low, since before the time the energy exchange between the electrons and ions is disturbed an appreciable number of the free electrons succeed in recombining.

Figure 8.13 presents curves of $x(t)$ calculated in [28] for two typical cases

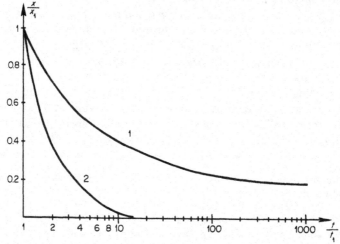

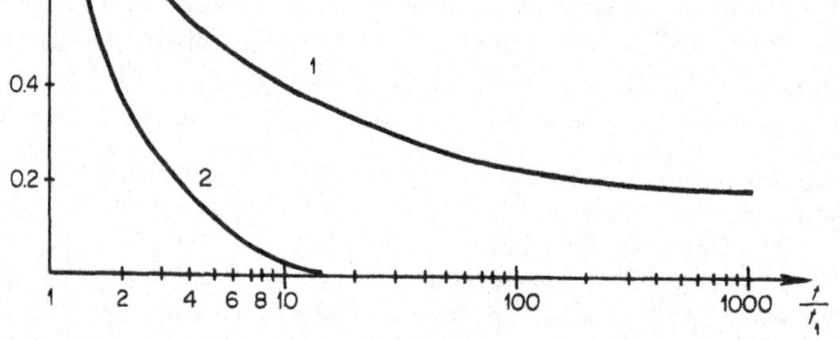

Fig. 8.13. Change in degree of ionization with time for expansion of a gas into vacuum (for explanation see text).

of the sudden expansion of a gas cloud. In the calculations the ionization potential for the atoms was taken to be 13.5 ev and their atomic weight taken to be 14. Curve 1 pertains to the case when the degree of ionization at the time when the ionization equilibrium is disturbed is $x_1 = 0.58$ and the gas temperature and density are $T_1 = 12{,}000°\text{K}$ and $N_1 = 1.7 \cdot 10^{16}\ \text{cm}^{-3}$, respectively. The time when the equilibrium is disturbed is $t_1 = 2 \cdot 10^{-6}$ sec. The radius of the gas cloud at this time is $R_1 = 4.9$ cm, and the expansion velocity $u = 24$ km/sec. The same flow variables given for the time when equilibrium is disturbed can be obtained with different initial conditions. In particular these results can correspond, for example, to an initial temperature, density, and cloud radius $T_0 = 50{,}000°\text{K}$, $N_0 = 2 \cdot 10^{18}\ \text{cm}^{-3}$, and $R_0 = 1$ cm, respectively. As can be seen from the figure, for these conditions the degree of ionization of the gas tends to the rather appreciable constant value of $\approx 0.2x_1 \approx 0.12$, and appreciable freezing takes place. Curve 2 relates to the following values of the parameters at the time when the equilibrium is disturbed: $t_1 = 2.1 \cdot 10^{-6}$ sec, $T_1 = 11{,}700°\text{K}$, $N_1 = 4 \cdot 10^{16}\ \text{cm}^{-3}$, $x_1 = 0.34$, $R_1 = 5$ cm, $u = 24$ km/sec. This case can correspond, for example, to the initial conditions: $T_0 = 50{,}000°\text{K}$, $N_0 = 5 \cdot 10^{18}\ \text{cm}^{-3}$, and $R_0 = 1$ cm. As can be seen, in this case the degree of ionization tends to zero, and it appears that no freezing takes place. Actually, as was noted above, the recombination will also cease at some stage (when the energy transfer from the electrons to the heavy particles slows down), but only when the degree of ionization is very low.

In concluding this section we wish to emphasize again that the diffusion of resonance radiation, which has not been taken into account here, aids in increasing the heat release in the gas, in decreasing the rate of recombination, and in increasing the residual ionization. This process should be considered in relation to the specific problem at hand, since the diffusion of resonance radiation depends on the dimensions and degree of transparency of the gas cloud, the character of the broadening of the spectral lines, the gas composition, etc.

4. *Vapor condensation in an isentropic expansion*

§10. Saturated vapor and the origin of condensation centers

If the vapor of any substance is isentropically expanded and cooled, there is some time t at which the vapor becomes saturated; then it becomes supersaturated, after which condensation begins. It is well known that condensation is greatly facilitated by the presence of ions, dust, and other foreign particles which become condensation centers, about which liquid drops form.

Ions and dust particles only create favorable conditions for the more rapid formation of condensation centers, but their presence is not at all necessary. In a pure supersaturated vapor condensation centers appear as the result of the agglomeration of molecules into molecular complexes. After reaching the so-called critical size the complexes become stable and do not break up, and exhibit a tendency for further growth and transformation into droplets of liquid.

The phenomenon of vapor condensation in an isentropic expansion is met in industry, in the laboratory, and in nature. It serves as the basis of operation of the Wilson cloud chamber, which is widely used in nuclear physics for recording the motion of high-speed charged particles. The Wilson cloud chamber consists of a vessel filled with vapor of water, alcohol, or other liquid. The required supersaturation is created by an isentropic expansion of the vapor by means of a rapidly receding piston. The vapor condenses on the ions which are formed along the trajectory of the rapidly moving particle, and the drops of liquid are recorded by optical means. The condensation of the water vapor in air is frequently observed in the expansion of air in wind tunnels.

The fact that condensation must start at some time during an isentropic expansion of vapor can be easily explained with the aid of a temperature-specific volume diagram. As is known from thermodynamics, the pressure of a saturated vapor which is in equilibrium with the liquid is governed by the Clapeyron–Clausius equation (see, for example, [18]). If the vapor is considered as a perfect gas, then this equation leads to the following relationship between the specific volume of the saturated vapor V_{vap} and the temperature*

$$V_{\text{vap}} = BTe^{U/RT}, \qquad T = \frac{U}{R}\left(\ln \frac{V_{\text{vap}}}{BT}\right)^{-1}, \tag{8.38}$$

where U is the heat of vaporization, R is the gas constant, and B is a coefficient, which can be taken approximately constant. It is evident from this equation that the saturation temperature has only a weak logarithmic dependence on the vapor volume. On the other hand, the isentrope for the vapor is a power-law type curve of the form $T \sim V^{-(\gamma-1)}$, which must intersect the saturation curve (Fig. 8.14). At the point of intersection O the previously unsaturated expanding vapor becomes saturated.

Let us follow the process in time. If the vapor expands continuously, then the specific volume increases monotonically with time. Instead of considering the temperature change with time $T(t)$, we can consider the temperature

* This follows from the relations $p = const\ e^{-U/RT}$, $pV = RT$, where p is the saturated vapor pressure. *Editors' note.* U is assumed constant, and the specific volume of the liquid is assumed negligible.

change with increasing volume $T(V) = T[V(t)]$, using a T, V diagram (see Fig. 8.14). After passing the saturation state, the vapor continues to expand following the vapor isentrope, and becomes supersaturated (supercooled). The rate of formation of condensation centers has an extremely strong dependence on the degree of supersaturation. Therefore a further increase in

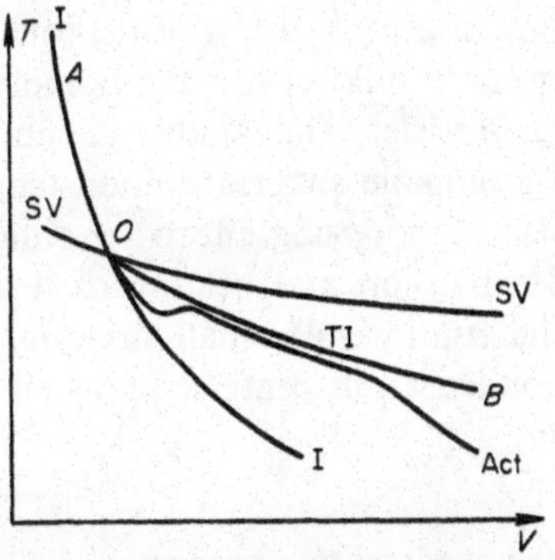

Fig. 8.14. T, V diagram for condensation in an isentropic vapor expansion. I is the isentrope for the vapor, SV is the saturated vapor curve, O is the saturation point, TI is the isentrope for the equilibrium two-phase vapor-liquid system, Act is the actual curve for the vapor-liquid drop systems taking into account the kinetics of condensation.

the degree of saturation results in a rapid increase in the number of nuclei in the liquid phase. Soon after the saturation state is passed, the rate of condensation reaches a value such that the release of the latent heat prevents any further increase in the supersaturation (if, of course, the expansion is not too rapid). The condensation accelerates even if the number of centers remains constant, because of an increase in the surface area of the drops to which the vapor molecules attach. The acceleration of the condensation not only stops the increase in supersaturation, but even leads to a decrease in the degree of supersaturation. The formation of new nuclei, which is highly sensitive to the value of supersaturation, ceases immediately and further condensation proceeds by means of the attachment of molecules to the previously formed drops. Thus all condensation centers, as a rule, are born at the very beginning of the condensation process, as soon as a sufficiently high degree of supersaturation is reached.

In the Wilson cloud chamber the vapor is rapidly expanded to a definite volume, so that a known initial supersaturation is established in the vapor. This supersaturation is chosen to be large enough that all the ions become condensation centers, thus making it possible to determine the number of ions by counting the number of drops*. Therefore, no problem arises as to the number of condensation centers.

It is another matter in gasdynamic processes such as the expansion of gases in wind tunnels, the exhaust from nozzles, or the sudden expansion of a gas cloud which forms as a result of the heating and vaporization of an

* At the same time, practically speaking, nuclei not containing ions do not form.

initially solid substance such as a metal. Here the rate of expansion is determined by the general dynamics of the process, and the number of condensation centers is unknown and depends on the expansion rate. Even if ions are present in the gas (which, of course, does not always happen) by far not all of them become condensation centers if the expansion rate is sufficiently low. As a result of the factors already mentioned, the supersaturation of a system can decrease due to very rapid condensation after only a few of the ions are converted into condensation centers. Even more unknown is the number of pure vapor centers in the absence of foreign particles. The number of condensation centers depends on the maximum attainable supersaturation (supercooling) and is determined by the interplay of opposing effects: cooling of the vapor corresponding to the work of expansion and heating of it as a result of the release of latent heat in condensation. We shall show in §12 how one may calculate the number of condensation centers, knowing the rate of expansion and cooling of the vapor.

§11. The thermodynamics and kinetics of the condensation process

Let us consider the process of condensation in an isentropically expanding fluid from a purely thermodynamic point of view, by assuming that thermodynamic equilibrium exists at any given time. Up to the time of saturation the gas expands along an isentrope. After the saturation state has been reached and condensation begins, the fluid becomes a two-phase vapor-liquid system. The adiabatic equation is complicated because of the conversion of a part of the gas phase into a liquid of different thermodynamic properties and because of the release of the latent heat. The isentrope for this two-phase system satisfies the adiabatic equation

$$[c_1(1-x)+c_2x]\,dT+RT(1-x)\frac{dV}{V}-[U-(c_2-c_1)T]\,dx=0. \quad (8.39)$$

Here c_1 is the specific heat of the vapor at constant volume; c_2 is the specific heat of the liquid; x is the degree of condensation, defined as the ratio of the number of molecules in the liquid phase to the total number of molecules in the given mass of fluid; V is the specific volume of the fluid, which is less by a factor of $(1-x)$ than the specific volume of the vapor: $V=V_{\text{vap}}(1-x)$*. In this equation we have neglected the surface energy of the liquid drops, which is very small in comparison with the latent heat if the drops contain

* The specific volume of the two phase system is $V=V_{\text{liq}}x+V_{\text{vap}}(1-x)$, where V_{liq} is the specific volume of the liquid phase. Since the density of the liquid is much higher than the density of the vapor, for a degree of condensation which is not too close to unity, the first term can be neglected and $V\approx V_{\text{vap}}(1-x)$. The specific heats of the liquid and the vapor c_2 and c_1 in (8.39) are assumed to be constant.

large numbers of molecules. The adiabatic equation (8.39) is also valid in the absence of thermodynamic equilibrium. In the case of nonequilibrium, the degree of condensation x is determined by the condensation kinetics. Under conditions of thermodynamic equilibrium, for an infinitely slow expansion, the vapor is in equilibrium with the liquid at any instant of time, and is always saturated. The state of the fluid changes in this case along the saturation curve (8.38), which, if we replace the specific volume of the vapor by the specific volume of the fluid, takes the form

$$\frac{V}{1-x} = BTe^{U/RT}. \tag{8.40}$$

If we eliminate the degree of condensation x from (8.39) and (8.40), we obtain a differential equation describing the isentropic process in the two-phase system in terms of the variables T and V. The solution of this equation yields the isentrope $T(V)$. The constant of integration in the general solution is determined by the entropy of the fluid. The constant can be expressed in terms of the temperature and volume at the saturation point O, since it is obvious that the isentrope passes through this point. We shall not write out the solution here, but shall instead illustrate the isentrope in Fig. 8.14. The solution lies somewhat below the saturation curve, which can be seen by comparing (8.38) and (8.40) and taking into account the fact that $x > 0$, $1 - x < 1$. For a small degree of condensation, when $x \ll 1$, the isentrope of the two-phase system almost coincides with the saturation curve. The divergence of the two curves determines the degree of condensation x:

$$1 - x = \frac{V(T)}{V_{\text{vap}}(T)}.$$

The degree of condensation increases monotonically along the isentrope with increasing volume.

It is interesting to note that in an infinite isentropic expansion of a fluid, with $V \to \infty$ (and cooling to zero temperature $T \to 0$), the degree of condensation along the isentrope for thermodynamic equilibrium tends to unity, $x \to 1$. In other words, according to the laws of thermodynamics, in the unlimited expansion of a fluid, the vapor should be completely condensed. In an isentropic expansion to a finite volume only a finite amount of vapor is condensed. In reality, of course, the state of the fluid in the condensation process can never follow the equilibrium isentrope exactly, it only more or less approaches the equilibrium state; it is closer to equilibrium, the slower is the change of the external conditions, i.e., the slower is the expansion.

It was already noted above that condensation centers are born primarily immediately after passing the saturation state, at the time when maximum supersaturation is reached. Thereafter, if the expansion does not proceed too rapidly, the accelerating condensation stops the supersaturation process and no new nuclei are formed. The state of the fluid after passing through a point of maximum deviation from the equilibrium isentrope (TI) (Fig. 8.14, maximum supercooling) approaches the equilibrium state. The degree of supercooling, however, does not drop to zero and the curve (Act) never reaches the thermodynamic equilibrium curve (TI), always passing below the latter. Condensation now proceeds by the size of the droplets increasing. Two processes take place simultaneously: the forward process, attachment of vapor molecules to the drops, and the reverse process, evaporation of the drops. The rate of growth of the drops (that is, the condensation rate) is determined by the difference between the rates of the forward and reverse processes and is higher, the higher the degree of supersaturation. In the saturated vapor, for a state on the equilibrium isentrope, the adhesion and evaporation rates are exactly equal and the drops do not increase in size*.

In condensation the degree of supersaturation adjusts to the balance between attachment and evaporation, automatically conforming to the process in such a manner that the attachment exceeds the evaporation and that the condensation rate follows the expansion of the fluid. The system is always in a state close to equilibrium, i.e., close to saturation.

An appreciable departure from thermodynamic equilibrium can take place only in a very strong expansion, when the attachments of vapor molecules become exceedingly rare. Thus, in the sudden expansion of vapor into vacuum the attachment rate, which is proportional to the vapor density and thus to t^{-3}, starting at a certain time is no longer capable of following the expansion; condensation ceases and the remaining vapor expands to infinity, again following the gas isentrope (Fig. 8.14). Freezing takes place, and the fluid expanding to infinity is not completely condensed, as would be required by the equilibrium laws of thermodynamics, but is partially in the form of a gas and partially in the form of condensate drops (for more details, see the following section). For a rapid expansion of a fluid, the condensation cannot "follow" the expansion and the state departs substantially from thermodynamic equilibrium from the very beginning. In a very rapid expansion to a given volume, as takes place in the Wilson cloud chamber, condensation does not take place during the expansion and begins only after the expansion has ceased. For a very rapid expansion into vacuum condensation generally does not occur at all and the fluid flows out to infinity in the gas phase. This corresponds to the maximum departure from thermodynamic equilibrium and to maximum freezing.

* The thermodynamic equilibrium isentrope, strictly speaking, corresponds to a saturated state with respect to a plane surface of the liquid, with respect to drops of infinite radius.

§12. Condensation in a cloud of evaporated fluid suddenly expanding into vacuum

In this section we consider in more detail the condensation process in the expansion of a vapor, setting out the basic scheme for quantitative calculations and presenting some numerical results. We shall examine condensation as it occurs in a cloud of evaporated fluid expanding into vacuum. Here, we have in mind the phenomenon of the explosion of large meteorites on impact with planet surfaces (devoid of atmosphere) or asteroid collisions, which were mentioned at the beginning of §6. We are interested in what the form is of the vaporized material of the planet soil and of the meteorite which expands into interplanetary space: Is the material in the form of a pure gas or in the form of minute particles, and if the latter what are the particle dimensions? A solution to this problem was obtained by one of the present authors [19]*.

All the numerical results will pertain to the condensation of iron vapor as applied to the case of vaporization of iron meteorites. Let us consider when saturation is reached in the expansion of iron vapor. In Table 8.1 we give the calculated vapor temperature T_1 and atom number density n_1 at the time of saturation for several values of vapor entropy S. Assuming that the expansion is isentropic, we can say that the "solid" iron at the time of heating had the same entropy S. The table gives the values of the initial energy of

Table 8.1

PROPERTIES OF IRON VAPOR AT SATURATION

ε_0, ev/atom	T_0, ev	S, $\frac{\text{cal}}{\text{mole} \cdot \text{deg}}$	T_1, °K	n_1, cm^{-3}	u, km/sec
25.6	5	48.3	3100	$8.01 \cdot 10^{19}$	9.2
71.9	10	60.8	2130	$7.15 \cdot 10^{16}$	15.5
138	15	71.5	1700	$2.86 \cdot 10^{14}$	21.4
222	20	81.3	1430	$1.43 \cdot 10^{12}$	27.2

heating ε_0 and temperature T_0 of standard density solid iron, corresponding to these values of the entropy. These quantities were calculated by the method described in §14, Chapter III (both the nuclear and the electronic contributions to the specific heat are taken into account). The last column gives the average velocities of sudden expansion of a gas sphere of iron atoms estimated from the equation $u = (2\varepsilon_0)^{1/2}$ (see §6), that is, by assuming that the vapor is already strongly cooled before the time of condensation and that all the

* Some qualitative remarks on the phenomenon of condensation of vaporized matter were made by the present authors in [20].

initial heating energy has been converted into kinetic energy of the expansion.

Let us now estimate the number of condensation centers, that is, the number of condensate particles in the final state. The theory of the formation of nuclei of the liquid phase in pure supersatured vapor has been developed by a number of authors, M. Volmer, R. Becker and W. Döring, L. Farkas, and Zel'dovich and Frenkel. A detailed presentation of the theory with references to original works may be found in the book by Frenkel [21] (see also [22]). We shall recount here only main the ideas of the theory.

In the vapor phase there occur from time to time fluctations during which the vapor molecules join together forming molecular complexes, nuclei of the liquid phase. In unsaturated vapor, when the gas phase is stable, the complexes are unstable and soon break up (evaporate). In supersaturated vapor only complexes of very small dimensions are unstable. The increase in size of the smallest complexes by the attachment of new molecules is energetically unfavorable, because of the increase in surface energy at the interface between the liquid and the gas phases. On the other hand, the increase in size of sufficiently large complexes is energetically favorable, since the favorable volume energy effect (release of latent heat) becomes greater than the unfavorable surface effect with sufficiently large dimensions. For each degree of supersaturation there is a definite critical complex dimension. Supercritical nuclei (with a radius larger than critical) are stable or "viable", and exhibit a tendency for further growth and transformation into liquid droplets. The rate of formation of these viable nuclei of condensation centers is proportional to the probability of appearance of critical size complexes. The formation of these complexes requires the expenditure of a certain energy $\Delta\Phi_{max}$ necessary to overcome the potential barrier, and hence, according to the Boltzmann relation, the probability of such fluctuations is proportional to $\exp(-\Delta\Phi_{max}/kT)$. The potential barrier $\Delta\Phi_{max}$ or activation energy depends on the critical radius of the complex and is uniquely related to the degree of supersaturation. This degree can be characterized, for example, by the supercooling

$$\theta = \frac{T_{sat} - T}{T_{sat}}.$$

Here T_{sat} is the temperature of saturated vapor at the given density and T is the actual vapor temperature.

The rate of formation of viable nuclei, that is, the number of condensation centers per single vapor molecule which appear per unit time under steady-state conditions, assuming that constant supersaturation (supercooling) is maintained in the system and the supercritical nuclei are removed from the system as formed and replaced by an equivalent amount of vapor, is

$$I = Ce^{-b/\theta^2}, \tag{8.41}$$

where

$$C = n\bar{v}2\omega\left(\frac{\sigma}{kT}\right)^{1/2},$$

$$b = \frac{16\pi\sigma^3\omega^2}{3k^3q^2T}.$$

Here n is the number of vapor molecules per unit volume, $\bar{v}$ their thermal speed, σ the surface tension, ω the volume of liquid per molecule, and $q = U/R$ is the heat of vaporization expressed in degrees. The critical nucleus radius r^* is related to the degree of supersaturation by

$$\theta = \frac{2\sigma\omega}{kqr^*}.$$

The theory can also be generalized to the case of electrically charged nuclei containing an ion (see [19]). The rate of formation of nuclei is again given by (8.41), except that the constant b is now smaller.

Let us set up the rate equation for condensation. We make the basic assumption that the expansion of the vapor proceeds sufficiently slowly that the process of formation of nuclei can be considered as quasi-steady. In this case, the rate of formation always coincides with the steady-state rate (8.41) corresponding to the actual supercooling θ of the system at the given time. If $I(t')$ is the number of condensation centers formed per unit time at the time t' (per vapor molecule), and $g(t, t')$ is the number of molecules at the time t in a liquid drop born from a nucleus which appeared at the time t', then the degree of condensation at the time t, $x(t)$, can be written as

$$x(t) = \int_{t_1}^{t} I(t')g(t, t')\,dt'. \tag{8.42}$$

The integration with respect to time is carried out from the time of saturation, that is, from the time when the nuclei begin to appear.

The rate of increase of a drop of supercritical size is equal to the difference between the rate of attachment of vapor molecules to the drop surface and the rate of evaporation from the drop. It can be written approximately as (see [19, 21])

$$\frac{dg}{dt} = 4\pi r^2 n\bar{v}(1 - e^{-q\theta/T}), \tag{8.43}$$

where $4\pi r^2$ is the surface area of the drop and $n\bar{v}$ is the flux of vapor molecules. The factor in parentheses is proportional to the difference between the attachment and evaporation rates. In the state of saturation, when $\theta = 0$, attachment and evaporation cancel each other and the rate of growth is equal to zero*. In supersaturated vapor $\theta > 0$ and the size of a drop increases on the average, with $dg/dt > 0$; in unsaturated vapor $\theta < 0$ and, on the average, the drop is evaporated, with $dg/dt < 0$.

Equations (8.42), (8.43), and (8.41), together with the adiabatic equation for a two-phase system (8.39), the formula for a saturated vapor (8.38), and the relation for the expansion of the fluid, which in the case of sudden expansion into vacuum is given by (8.25), form a complete system of equations for calculating the condensation kinetics.

In accordance with the qualitative picture presented in the preceding sections, we can break up the solution of this system into two independent stages. The first stage is the analysis of a small time interval immediately after the saturated state has been reached, during which the supercooling first increases, passes through a maximum, and then decreases because of the onset of condensation. The nuclei appear during this short stage. Calculating the number of the nuclei

$$\nu = \int_{t_1}^{\infty} I(t')\, dt',$$

we get the total number of condensate particles (per initial molecule). Actually, the integration with respect to time extends here not to $t = \infty$, but is carried over a rather small time interval, since by virtue of (8.41), the rate I drops very sharply as soon as the supercooling, having passed through the maximum, begins to decrease. The second stage is the analysis of growth of the already known number of drops during all the ensuing states, up to $t \to \infty$.

A rigorous solution of the system of equations is, of course, very difficult. An approximate solution has been given in [19]. The approximate analysis of the first stage is based on the nature of the extremely sharp dependence $I(\theta)$, by virtue of which we can assume that practically all nuclei are formed in a very short time interval close to the time at which the supercooling is maximum (the solution actually gives an extremum for $\theta(t)$).

Referring for details of the solution to [19], we present here the results of calculations of a specific example. We consider a sphere of iron atoms with a mass of 33,000 tons, which is heated and converted into a dense gas, say, as the result of the impact of a very large iron meteorite on the surface of the moon. Let the impact velocity be such that the initial heating of the iron at standard density is $\varepsilon_0 = 72$ ev/atom. The initial temperature in this case is

* The effect of the curvature of the drop is neglected here.

$T_0 = 10$ ev $= 116{,}000°$K. In the stage of strong cooling at the time of vapor saturation the expansion is taking place practically by inertia, with an average velocity of $u = 15.5$ km/sec. The vapor becomes saturated at $t_1 = 6.8 \cdot 10^{-2}$ sec from the start of the expansion, with the corresponding radius of the cloud equal to 1050 m. In this case $T_1 = 2130°$K and $n_1 = 7.15 \cdot 10^{16}$ cm^{-3}.

For the sudden expansion of an initially highly ionized gas into vacuum, even in the stage of strong cooling some residual ionization is retained which is far above the value for thermodynamic equilibrium. The condensation centers in this case will contain ions. As shown by calculations, the number of condensation centers depends very weakly on whether or not they are charged, since the assumption that the condensation takes place on ions does not appear to be important.

The maximum possible supercooling in our example is found to be $\theta_{max} = 0.0765$ $(b/\theta_{max} = 43.1)$. A nucleus of critical size for this supercooling contains 46 atoms. The number of condensation centers is $\nu = 4 \cdot 10^{-11}$ per atom, which is much smaller than the number of ions per atom, in contrast to the process in the Wilson cloud chamber, where all the ions become condensation centers.

Analysis of the second stage, during which the size of the drops increases, shows that during a long period of time the condensation follows the expansion of the fluid and a state close to equilibrium is maintained in the system. Only at the time $t_2 \sim 2.5$ sec, when the sphere has expanded to 40 km, does the density of the fluid become so small that further growth of drops ceases and freezing begins. Up to this time (and thus over the entire process) approximately one half of the iron vapor is condensed. Knowing the degree of condensation x_∞, and the number of condensate particles, we can also find their size (the number of atoms per particle is equal to $x_\infty \nu^{-1}$). In our example the iron particles suddenly expanding to infinity have a radius of $3.1 \cdot 10^{-5}$cm; their total number is $3 \cdot 10^{21}$. Approximately half of the fluid goes out to infinity in gaseous form.

We can establish theoretically approximate similarity relations for transforming to other initial conditions. It appears that for sufficiently slow expansions, when the initial assumptions are valid, the degree of condensation of the given fluid for a sudden expansion to infinity is independent of the initial conditions, and that the size of the condensate particles is proportional to the initial linear dimensions of the vaporizing body (to the cube root of the mass), and decreases rapidly with an increase in initial heating.

§13. On the problem of the mechanism of formation of cosmic dust. Remarks on laboratory investigations of condensation

There are some reasons to believe that the process of condensation of a vaporized substance suddenly expanding into vacuum, which was considered in the preceding section, is one of the mechanisms of formation of cosmic dust in the solar system (this hypothesis was proposed in [19]). Interplanetary space contains small particles of various sizes, which are called cosmic dust. Sometimes these particles fall on the earth in the form of meteor showers. In their motion about the sun the particles experience some slowing down from the aberration component of the light pressure (Poynting–Robertson effect, *eds.*)*. The very small particles with sizes of the order of 10^{-6}–10^{-5} cm fall into the sun and disappear (for further details see [23]). Consequently, the solar system must have a source for replenishing the store of these very small particles of cosmic dust. It was noted (in particular by Stanyukovich), that this source may be the mechanical disintegration of matter in the collisions of small bodies of the solar system (asteroids), or in meteorite impacts on the surfaces of planets devoid of atmosphere. The particles produced have acquired considerable velocity, overcome the gravity field, and, not being slowed down by an atmosphere, fly out into space.

It is reasonable to assume that the phenomenon we have discussed of condensation of vaporized matter of planet soil, meteorites, or asteroids is also a source of these very small particles. In energetic collisions of asteroids, when the kinetic energy of impact is sufficient for complete vaporization of both colliding bodies, mechanical disintegration of the solid is, in general, not present, since the entire mass is completely vaporized. In this case the condensation mechanism is the only one capable of forming the small particles. The liquid droplets, which have grown in the process of condensation, gradually cool down and solidify due to the energy losses from thermal radiation. It can be shown that the process of radiation cooling proceeds much more quickly than the evaporation of the heated particles, which slows down very rapidly as the vapor is cooled. Thus, the condensate particles which were already created will continue to exist in the form of solid dust particles. Since bodies colliding in space have very different sizes and velocities, the condensate particles which are formed are also of various sizes.

The phenomenon of the condensation of a vaporized substance in a gasdynamic expansion can be also used for laboratory investigation of the condensation of vapors of metals and other solids (and liquids) and for the study of optical properties of very small particles. The dimensions of the condensate particles depend on the initial conditions; hence by properly choosing these conditions it is possible to obtain particles of the desired size

* The light pressure acts basically in the radial direction. The force due to this pressure is inversely proportional to the square of the distance between the particle and the sun, and its effect is equivalent to only a small decrease in the gravitational force; the radial component of the light pressure affects only the radius of the orbit. The slowing down, however, is due to the component of the light pressure which is tangent to the orbit and which arises from the aberration of light. For additional details see the book by Fesenkov [23].

in the laboratory. We present here the results of some rough estimates for conditions close to those encountered in the laboratory. If we rapidly vaporize 1 g of iron by transferring to it in some manner an initial energy of $\varepsilon_0 = 13$ ev/atom, corresponding to the initial temperature (at the density of the solid metal) of $T_0 = 35{,}000°K$, then the condensation of the vapor in a sudden expansion into vacuum (in an evacuated vessel) ends at the time $t = 5 \cdot 10^{-5}$ sec, when the cloud has expanded to 30 cm. The dimensions of the condensate particles in this case will be of the order of 10^{-4} cm.

The calculations of condensation kinetics can be easily carried over to other relationships governing the expansion of a fluid which takes place, say, in a wind tunnel or in a nozzle exhaust. These calculations do not involve anything basically new in comparison with the case of sudden expansion into vacuum, and we shall not consider them. We note that if the degree of condensation is not too high, or if the total energy of the vapor is much greater than the heat of vaporization, then the effect of the condensation on the gasdynamics of the process is very small. The kinetics of condensation can then be calculated on the basis of the known gasdynamic solution, found in first approximation without taking condensation into account. This is exactly the procedure which we followed in the preceding section.

IX. Radiative phenomena in shock waves and in strong explosions in air

1. Luminosity of strong shock fronts in gases

§1. Qualitative dependence of the brightness temperature on the true temperature behind the front

Optical measurements are of great importance for determining the temperature of highly heated bodies and for studying high-temperature processes in general. The usual methods are to measure by one or another means the luminosity or brightness of the surface of a luminescent body (by photographic means, with the aid of photoelectric elements, or with electron-optical multipliers [image converters, *eds.*]). The radiation brightness temperature, which is by definition the same as the temperature of a perfect black radiator emitting from its surface exactly the same luminous flux as the object under investigation, can then be found from the luminosity (see §8, Chapter II). Particularly widely used are photographic methods for determining the luminosity and brightness temperature, basically by comparing the degree of darkening on a photographic film produced by the light emitted by the body and that from a calibrated source with a known temperature and spectrum, say, from the sun. For greater accuracy the photographs are ordinarily taken in a narrow region of the spectrum, since the emitting object and the calibrated source have different temperatures and therefore emit different radiation spectra; in addition, the sensitivity of the photographic film depends on the wavelength of the light, which creates some difficulties in converting the degree of darkening into temperature.

Optical (in particular, photographic) methods are also widely used in the study of shock waves. A gas heated to a high temperature by a strong shock wave radiates and the surface of the shock front glows. The luminosity of the emission depends on the strength of the shock and the dimensions of the heated region behind the front. In order to obtain the true temperature of the fluid behind the front from the experimentally measured brightness temperature, we must be sure that the glowing mass radiates as a perfect black body.

If the shock front is a "classical" discontinuity behind which there is a sufficiently extensive, optically thick region with a more or less constant temperature behind the front, then the heated fluid bounded by the front surface radiates from the surface as a perfect black body*. By measuring the luminosity of the front surface we can in this case directly determine the temperature behind the shock front, and hence the wave strength, which not only is important for experimental investigations but also is of great practical significance. It has been shown experimentally that in a certain range of shock strengths (and, of course, for a sufficiently thick heated region behind the front) the shock front actually does radiate as a black body. This is confirmed by the fact that the brightness temperature agrees with the temperature behind the front calculated on the basis of the shock relations and the equation of state, using one of the other experimentally determined front parameters such as the velocity of propagation of the shock wave. We know from experiments and theoretical considerations, however, that this agreement cannot be observed for all wave strengths. The brightness temperature of a sufficiently strong shock becomes lower than the true temperature behind the front and, past a certain strength, it decreases rapidly with increasing strength, reaching a limiting and comparatively low value. Thereafter, the brightness temperature remains almost constant regardless of the strength. A typical dependence of the brightness temperature of a shock front on the true temperature behind the front is shown in Fig. 9.1, which gives a curve of

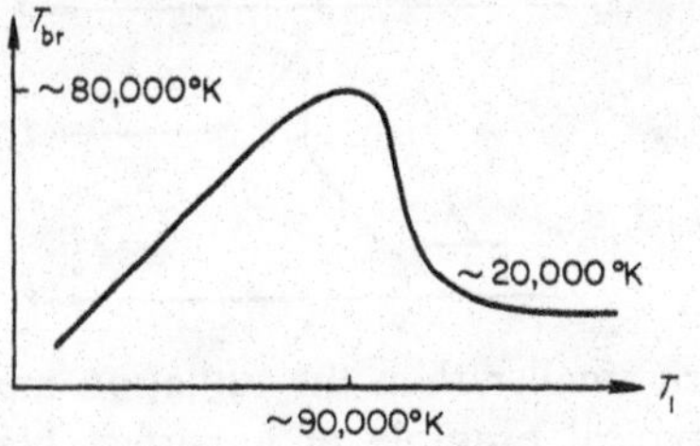

Fig. 9.1. Dependence of the brightness temperature of the surface of a shock front in air on the true temperature behind the front (for red light).

the brightness temperature of red light for a shock wave in standard density air, obtained on the basis of theoretical estimates (carried out in the following sections). Figure 9.1 shows the existence of the luminosity "saturation"

* If the region of heated gas behind the "classical" discontinuity is optically thin (for example, if the shock wave had moved only a small distance from the piston which was pushed into the gas and generated the wave), then the gas radiates as a volume radiator. The spectral flux coming out from the surface of an optically thin plane layer in a direction normal to the surface is equal to $S_\nu = S_{\nu p}[1-\exp(-\kappa_\nu' d)]$, as was shown in §7, Chapter II, where $S_{\nu p}$ is the flux corresponding to a perfect black body at the same temperature, κ_ν' is the absorption coefficient, and d is the thickness of the layer heated by the shock wave. In the case of small optical thickness $\kappa_\nu' d \ll 1$, $S_\nu = S_{\nu p}\kappa_\nu' d \ll S_{\nu p}$. In the limit $\kappa_\nu' d \gg 1$ the flux approaches the Planck value $S_\nu = S_{\nu p}$. As noted in §21, Chapter V, by studying the increase in luminosity S_ν with time $t = d/D$ (D is the wave velocity), Model' [1] measured the absorption coefficient of red light in a shock wave.

effect. No matter how strongly the gas is heated by the shock wave, even to millions of degrees, it is still not possible "to see" temperatures higher than tens of thousands of degrees; there is an upper limit to the temperature behind a shock wave front which can be "seen". This effect can be easily explained on the basis of the structure of a shock front taking radiation into account, as in Part 3 of Chapter VII. The problem of the luminosity of strong shock fronts was considered in [2–4].

We assume that behind the front of a plane shock wave there exists a sufficiently extensive, optically thick region of constant high temperature, and we consider the flux of visible radiation coming out from the front surface and recorded by an instrument situated far from the front, at infinity. We first consider a shock wave which is not too strong, in which the role of radiation is negligible and the gas ahead of the front is not preheated. If we disregard the temperature change in the front related to the relaxation processes in the gas, then the temperature distribution across the shock wave is the classical jump, shown in Fig. 9.2a. The thickness of the jump together

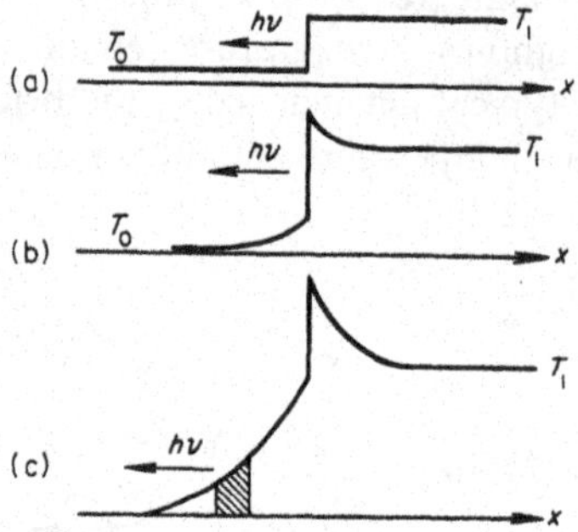

Fig. 9.2. On the problem of the luminosity of a shock wave.

with the relaxation layer is usually much smaller than the radiation mean free path. Therefore we have here a typical example of a perfect black radiator; an optically thick layer of fluid heated to a constant temperature T_1 is bounded by a surface with a very sharp temperature jump. If the cold gas ahead of the front, as is usually the case, is transparent in the visible part of the spectrum (colorless), then the instrument will record a luminous flux corresponding to Planck radiation at the temperature T_1; the radiation brightness temperature will be equal to the true gas temperature behind the front.

Let us now consider a strong shock wave with a temperature of, say, $T_1 = 65{,}000°K$ behind the front. Photons radiated from the surface of the shock discontinuity are for the most part those with energies of the order of ten to thirty ev. (The maximum of the Planck spectrum at $T = 65{,}000°K$ is at $h\nu = 2.8kT = 16$ ev.) These photons exceed the ionization potentials of the atoms and molecules, and are strongly absorbed in the cold gas ahead of the shock discontinuity and preheat it. A heated layer is formed ahead of the shock discontinuity, and the temperature distribution across the shock takes on the form shown in Fig. 9.2b (in air, for example, at $T_1 = 65{,}000°K$ the maximum preheating temperature just ahead of the discontinuity is $T_- = 9000°K$).

Unlike a cold gas, the heated gas always absorbs the low energy visible photons ($h\nu \sim 2$–3 ev). In monatomic gases photons with energies less than the ionization potentials of the atoms I are absorbed by excited atoms, whose excitation energy exceeds $I - h\nu$; in accordance with the Boltzmann relation, the concentration of excited atoms is proportional to $\exp[-(I - h\nu)/kT]$, so that the absorption coefficient also increases sharply with temperature following the Boltzmann relation $\kappa_\nu \sim \exp[-(I - h\nu)/kT]$. In molecular gases, such as air, a number of other mechanisms for absorption of visible light exist, but in any case the absorption coefficient for visible light is always very sensitive to temperature and increases rapidly with heating.

Now the photons of visible light, which are radiated from the surface of the shock discontinuity and whose flux at the discontinuity corresponds, roughly, to the temperature T_1*, must penetrate the preheating layer before arriving at the recording instrument located at infinity. The photons are partially absorbed in this layer. Therefore the brightness temperature of visible radiation from the shock front will be less than the true temperature behind the front. The preheating layer "screens" the highly heated gas behind the shock front. The screening and, consequently, the deviation of T_{br} from T_1 is more pronounced, the greater is the optical thickness of the preheating layer to visible light τ_ν†, that is, the higher is the preheating temperature and the greater is the wave strength. As long as the optical thickness $\tau_\nu \ll 1$, the screening is negligible and the deviation of T_{br} from T_1 is very small; the front glows as a black body at the temperature T_1. By virtue of the fact that the absorption of visible light is strongly temperature dependent, and in turn because the preheating temperature rather strongly depends on the wave strength (see §16, Chapter VII), the onset of strong screening, corresponding to reaching an optical thickness τ_ν of the order of unity, very clearly appears as the wave strength increases. In air, strong screening begins at temperatures behind the front of approximately $T_1 = 90{,}000°K$ (see §3).

* Actually the photon flux is somewhat greater, since the temperature directly behind the discontinuity is higher than the temperature behind the front (see Fig. 9.2b).

† We emphasize that the optical thickness of the preheating layer for visible radiation τ_ν has nothing in common with the optical thickness of the layer averaged over the spectrum, which corresponds to the high energy photons which "lead" the preheating. As was shown in §16 of Chapter VII, the temperature in the preheating layer decreases approximately exponentially with respect to the averaged optical thickness $T = T_- \exp(-\sqrt{3}|\tau|)$ (for $T_- < T_1$), so that the averaged layer thickness is of the order of unity.

In an even stronger shock wave the optical thickness to visible light of the preheating layer is greater than unity and the layer is almost completely opaque to visible light radiated by the highly heated gas behind the wave front; the screening in this region is almost total. Thus, as the wave strength increases, the brightness temperature of the visible light first coincides with the temperature behind the front, then begins to lag behind it, passes through a clearly defined maximum (luminosity "saturation"), and then falls off rapidly. The pronounced screening by the preheating layer does not mean, however, that the luminosity of the front of a very strong shock falls off to zero and that the wave ceases to be luminous. The heated gas ahead of the shock discontinuity not only absorbs, but also radiates visible light. As long as the preheating temperature is not too high and the layer is transparent, this radiation is lost in the background of the passing visible radiation which is emitted by the much more strongly heated gas behind the front. However, when the preheating layer completely stops transmitting the high-temperature light, then its own radiation begins to predominate.

In order to get some idea about the brightness of this natural luminescence of the preheating layer, we note that its temperature increases monotonically, starting from zero, or more exactly, from the temperature of the cold gas ahead of the front. As a result of the sharp temperature dependence of the absorption of visible light, the light is neither absorbed nor radiated in the most forward layers of the preheating region where the temperature is low. In the deeper layers with higher temperature there is a strong emission of photons in the visible region, but they are again absorbed, since they are unable to get through the opaque gas in front. The photons going out from the front surface to infinity are born in some intermediate radiating layer of the preheating region, removed from infinity by an optical distance (corresponding to the frequencies of visible light) of the order of unity. In Fig. 9.2c the radiating layer is shown cross-hatched. Obviously, the radiation brightness temperature agrees with the average temperature of the radiating layer. The position of the layer is determined only by the gas temperature profile $T(x)$ and the temperature dependence of the absorption $\kappa_\nu(T)$ assuming that the layer is located away from the cold gas by an optical distance of the order of unity. As shown in §17 of Chapter VII, the temperature distribution at the leading edge of the preheating region in strong shocks is almost independent of the wave strength. Consequently, the natural luminescence of the preheating layer and the brightness temperature of a very strong shock are also independent of the strength. In air at standard density this limiting brightness temperature for red light is approximately equal to 20,000°K (see §4).

The effect of the screening and the sharp decrease in brightness temperature of the shock front in comparison with the true temperature behind the front were observed experimentally by Model' [1]. He employed photographic means for measuring the brightness temperature of shock fronts in heavy inert gases, xenon, krypton, and argon, in which high temperatures can be produced by shock waves. The front velocity in these experiments was 17 km/sec. The optical thickness of the heated region behind the shock front was known to be large, so that in the absence of screening the front should have radiated as a black body. However, the brightness temperatures observed in the experiments were 30,000–35,000°K, which is several times lower than the temperatures behind the front T_1, calculated on the basis of the front velocity and shock relations (in Xe, $T_1 \approx 110{,}000$°K, in Kr, $T_1 \approx 90{,}000$°K, and in Ar, $T_1 \approx 60{,}000$°K). If we consider the fact that the accuracy with which the brightness temperature of the visible (red) light was determined in this experiment was not worse than $\pm 20\%$, then this disagreement must be attributed to screening by the preheating layer. Unfortunately, only one point with respect to the wave strength was recorded in the experiments of Model', which precludes the possibility of analyzing the character of the entire curve of dependence of the brightness temperature on the true temperature behind the front.

It should be noted that the luminosity "saturation" phenomenon at high temperatures has been observed by many authors in spark discharges*. It is known that starting with a certain rate any further increase in the rate at which energy is delivered into a spark discharge tube does not result in an increase in the brightness temperature above ~45,000°K in air. The brightness temperature is also similarly limited in discharges in argon and xenon (a higher temperature of approximately 90,000°K in air is observed in discharges in capillaries). The saturation effect may be related to the screening of high temperatures in the tube, which is somewhat similar to the screening in a shock wave, although it is possible that the true temperature in the tube is limited by radiation losses, etc.

§2. Photon absorption in cold air

Of great practical interest is the problem, which we shall consider in more detail, of the luminosity of strong shocks in air at standard density. We must determine the upper limit of the shock strength for which the shock front radiates visible light as a perfect black body, and estimate the maximum and limiting brightness temperatures. The problem, obviously, reduces to estimating the optical thickness of the preheating layer to visible radiation, which determines the degree of screening by the highly heated region behind the front, and also to finding the natural luminosity of the preheating layer.

* A bibliography may be found in the review by Vanyukov and Mak [5] on pulsed sources of light of high luminosity as well as in Model's article [1].

In order to do this we must first determine more precisely the geometric thickness of the preheating layer and its temperature distribution with respect to the geometric coordinate. This, in turn, depends on the absorption mechanism in air of the relatively high energy photons with energies of the order of 10–100 ev, which are responsible for preheating the gas ahead of the shock discontinuity. We shall summarize the published data on the absorption of such photons in cold air.

We have already mentioned several times the generally known fact that cold air is completely transparent to visible light. Noticeable absorption begins in the ultraviolet region of the spectrum at a wavelength $\lambda = 1860$ Å ($h\nu = 6.7$ ev)*. The absorption takes place in the Schumann–Runge band system of an oxygen molecule, which at $\lambda = 1760$ Å ($h\nu = 7.05$ ev) becomes a continuum associated with the dissociation of the molecule with the absorption of light. The absorption rapidly increases with increasing photon energy (for $\lambda = 1860$ Å, $\kappa_\nu = 0.0044$ cm^{-1}, and for $h\nu \approx 8$ ev, $\kappa_\nu \approx 100$ cm^{-1}). The experimental curve showing the dependence of the absorption coefficient on wavelength in this region of the spectrum is shown in Fig. 9.3†. Photons with

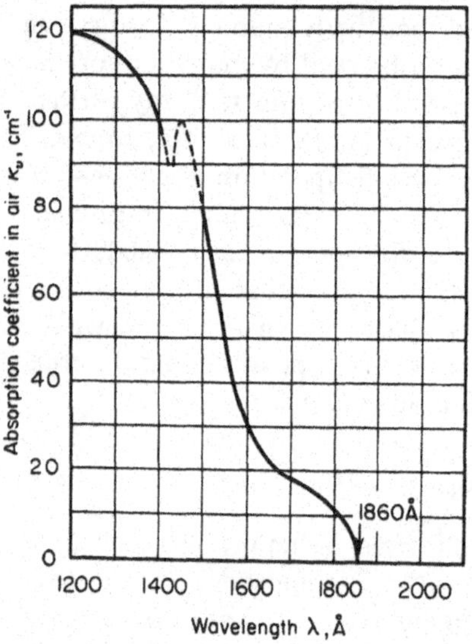

Fig. 9.3. Absorption coefficient for ultraviolet radiation in cold air.

* Strong absorption of the radiation from the sun in the near-ultraviolet region ($\lambda \sim 2000$–3000 Å) is connected with the existence of an ozone layer at an altitude of about 25 km. Oxygen and nitrogen do not absorb in this part of the spectrum; therefore, when speaking about shock waves in air close to the surface of the earth we should take $\lambda \sim 1860$ Å as an upper limit for the transparency of air.

† The curve was taken from the article of Schneider [6].

energies exceeding the ionization potentials of oxygen and nitrogen molecules ($I_{O_2} = 12.1$ ev and $I_{N_2} = 15.6$ ev) experience strong photoelectric absorption. The absorption cross sections from the ground level of the molecules depend only weakly on frequency in the energy range from $h\nu = I$ to $h\nu \sim 25$ ev and are approximately equal to $\sigma_{O_2} = 3 \cdot 10^{-18}$ cm^2 and $\sigma_{N_2} = 5 \cdot 10^{-18}$ cm^2, which gives an absorption coefficient $\kappa_\nu \approx 120$ cm^{-1}. As the frequency increases further, the absorption coefficient should show jumps due to the successive participation in the absorption process of the various electrons filling the L shells in the nitrogen and oxygen atoms. The energy levels in the L shells are, evidently, not spaced very far apart so that the jumps probably lie in the energy region $h\nu$ from 13 to 30–40 ev (as far as we know, there is no experimental evidence on these jumps). Thereafter, the absorption coefficient falls off monotonically with increasing frequency up to an energy $h\nu_K = 410$ ev, equal to the binding energy of a K electron in a nitrogen atom; the K absorption boundary in oxygen is $h\nu'_K = 530$ ev. At an energy $h\nu_K = 410$ ev the absorption coefficient increases sharply, since photons greater than $h\nu_K$ are capable of knocking out K electrons from nitrogen atoms, and then decreases monotonically up to $h\nu'_K = 530$ ev, when the K electrons of oxygen are included in the absorption process. The absorption coefficients for photons $h\nu_K = 410$ ev before and after the K absorption jump in nitrogen, calculated from the data of [7, 8], are 1.6 cm^{-1} and 35 cm^{-1}, respectively. The experimental data with respect to light absorption in air in the intermediate frequency range from tens to hundreds of ev are rather meager; as far as the authors know, measurements have been carried out only for two lines, at $h\nu = 182$ ev [9] and at $h\nu = 280$ ev [10].

On the basis of the available fragmentary data we have compiled a table which gives a picture of the absorption coefficients and mean free paths for

Table 9.1

ABSORPTION COEFFICIENTS AND MEAN FREE PATHS IN COLD AIR AT STANDARD DENSITY FOR DIFFERENT PHOTON ENERGIES

$h\nu$, ev	κ_ν, cm^{-1}	l_ν, cm
8	100	0.01
13–25	~120	0.0083
182	12	0.083
280	5.3	0.19
410 (before the jump)	1.6	0.63
410 (after the jump)	35	0.029

photons with energies in the range of ten to several hundred ev in cold air at standard density.

§3. Maximum brightness temperature for air

It was shown in §16 of Chapter VII that if the shock strength is less than critical (in standard density air the temperature behind the front corresponding to the critical strength is $T_1 \approx 285{,}000°K$), the radiative transfer from the highly heated region behind the front to the layers ahead of the shock discontinuity does not have a diffusive character. The air in these layers is heated to temperatures much lower than those behind the front and the emission of radiation in the preheating region makes practically no contribution to the radiation flux created behind the discontinuity. The air is heated simply by the absorption of photons passing at distances of the order of the absorption mean free path and the thickness of the preheating region Δx is of the order of the mean free path l of those photons that carry the principal energy of the spectrum. This is expressed mathematically by (7.55), which shows the exponential decrease in preheating with respect to the averaged optical thickness τ corresponding to some frequency averaged absorption coefficient $\kappa = 1/l$:

$$\varepsilon = \varepsilon_{-} e^{-\sqrt{3}|\tau|}, \qquad \tau = \int_0^x \kappa \, dx.^* \tag{9.1}$$

This equation shows that the effective optical thickness of the preheating layer is of the order of unity, that is, that the geometrical thickness is of the order of $\Delta x \sim 1/\kappa = l$.

It follows from Table 9.1 in the preceding section that the mean free paths for photons with energies of the order of 10–100 ev in cold air vary between 10^{-2} and 10^{-1} cm. It is evident that the mean free paths of these photons are approximately the same in the not too high temperature air of the preheating region.

Let us, for example, consider a shock wave with a temperature behind the front $T_1 = 65{,}000°K$. The maximum of the Planck spectrum occurs for photons of energy $h\nu = 16$ ev, so that a considerable part of the energy of the spectrum is concentrated in the energy region of photons exceeding the ionization potentials of the atoms and molecules $h\nu > I \approx 13$ ev. The highest preheating temperature, as is shown in Table 7.4, is $T_{-} = 9000°K$. At this temperature the degree of ionization and excitation of the atoms is small, and photons of energy $h\nu \geqslant I$ are absorbed in practically the same manner as in cold air. If we take a stronger shock wave, say with a temperature

* Equation (7.55) is not written for the specific internal energy, but for the temperature. Equation (9.1) is more general; it is valid also in those cases when the specific heats are temperature dependent, which is the case with air.

behind the front $T_1 = 100{,}000°\text{K}$, then photons of energy $h\nu = 24$ ev will correspond to the maximum of the spectrum and the principal energy of the spectrum will be concentrated in a region of higher photon energies of the order of several tens of electron volts. For a preheating temperature $T_- = 25{,}000°\text{K}$, single ionization of the atoms is appreciable, although there is practically no double ionization. Photons with energies of several tens of electron volts knock out from the atoms primarily not the external, optical, but the deeper lying electrons, which at a temperature $\sim 25{,}000°\text{K}$ do not yet undergo thermal ionization and excitation. Thus, in this case also the photons responsible for the preheating are absorbed in approximately the same manner as in cold air.

We can now conclude that the thickness of the preheating region ahead of the shock discontinuity in subcritical strength waves ($T_1 < 285{,}000°\text{K}$) is of the order of the mean free path of photons with energies of 10–100 electron volts in cold air, with $\Delta x \sim 10^{-2}$–10^{-1} cm. In this case, Δx increases within the stated limits for even stronger waves, with temperatures behind the front varying from several tens of thousands of degrees to $T_1 \sim 200{,}000°\text{K}$, and with a corresponding shift in the characteristic photon energies from $h\nu \sim 10$–30 to $h\nu \sim 30$–100 ev.

Let us now consider the extent to which the preheating region screens out the visible light. Table 9.2 gives absorption coefficients and absorption mean free paths for red light $\lambda = 6500$ Å in standard density air at different temperatures. Appreciable screening begins when the mean free path l_ν, which

Table 9.2

ABSORPTION COEFFICIENTS AND MEAN FREE PATHS FOR RED LIGHT $\lambda = 6500$Å IN STANDARD DENSITY AIR AT DIFFERENT TEMPERATURES

$T \cdot 10^{-3}$, °K	κ_ν, cm^{-1}	l_ν, cm
15	4.1	$2.5 \cdot 10^{-1}$
17	13.5	$7.4 \cdot 10^{-2}$
20	60	$1.66 \cdot 10^{-2}$
30	290	$3.45 \cdot 10^{-3}$
50	350	$2.85 \cdot 10^{-3}$
100	2000	$5 \cdot 10^{-4}$

decreases rapidly with increasing temperature, becomes comparable with the thickness of the preheating layer, that is, with the preheating radiation mean free path l (averaged over the spectrum). For convenience, we introduce the concept of a "transparency temperature" T^*, defined by the condition

$$l_\nu(T^*) = l. \tag{9.2}$$

The meaning of this concept is self-evident: the transparency temperature serves as the boundary between two temperature regions in the shock. When $T < T^*$, $l_\nu > \Delta x$ and the air in the preheating region is transparent to visible light. When $T > T^*$, $l_\nu < \Delta x$ and the air is opaque.

Since the absorption of visible light is very strongly temperature dependent, and the mean free path changes comparatively little (no more than by an order of magnitude), the transparency temperature defined by the equality (9.2) lies within quite narrow limits, namely: $T^* \approx 17{,}000$–$20{,}000$°K. We can estimate the optical thickness of the preheating region for visible light by assuming that the temperature dependence of the absorption coefficient for visible light can be described by the Boltzmann relation $\kappa_\nu = const \cdot \exp[-(I - h\nu)/kT]$, and that the mean absorption coefficient κ is constant. We recall that the internal energy of air at standard density and at temperatures of the order of tens of thousands of degrees is, roughly speaking, proportional to $\varepsilon \sim T^{1.4}$. We can then, using (9.1), find the approximate optical thickness in the preheating region from infinity (that is, from the cold air region) to a point at a temperature T. This quantity, $\tau_\nu(T)$ (the total optical thickness of the preheating region is $\tau_\nu(T_-)$), is given by

$$\tau_\nu(T) = \int_{-\infty}^{x} \kappa_\nu \, dx = \int_0^T \kappa_\nu \frac{dx}{dT} dT = \int_0^T \frac{const}{\kappa} \frac{1.4}{\sqrt{3}} e^{-(I-h\nu)/kT} \frac{dT}{T}$$

$$\approx \frac{1}{\kappa} \frac{1.4}{\sqrt{3}} \frac{kT}{I - h\nu} \cdot const \cdot e^{-(I-h\nu)/kT} = \frac{1.4}{\sqrt{3}} \frac{kT}{I - h\nu} \frac{\kappa_\nu(T)}{\kappa}. \tag{9.3}$$

In a shock wave with $T_1 = 90{,}000$°K, where the temperature ahead of the discontinuity is equal to the transparency temperature $T_- = T^* = 20{,}000$°K, the optical thickness of the preheating region, in accordance with the definition of the transparency temperature (9.2), is $\tau_\nu = 0.81kT^*/(I - h\nu) \approx 0.12$ ($I \approx 14$ ev, $h\nu \approx 2$ ev). Consequently, if we look at the surface of a shock front in a direction normal to the surface, then the flux of visible radiation coming out from the surface of the shock discontinuity will be attentuated by the preheating layer by approximately 12%, and the brightness temperature will be approximately 80,000°K instead of 90,000°K (at these temperatures low energy visible photons lie in the Rayleigh–Jeans region of the spectrum and their intensity is proportional to the first power of the temperature; hence the brightness temperature is simply proportional to the luminosity).

As the wave strength increases further, the optical thickness of the layer increases and the luminosity decreases; for example, when the temperature behind the front is increased by only 10,000°K, that is, when $T_1 = 100{,}000$°K $T_- = 25{,}000$°K, $\tau_\nu(T_-) \approx 0.37$, $T_{br} \approx T_1 e^{-0.37} \approx 67{,}000$°K, so that the brightness temperature is already less than 80,000°K. The maximum luminosity corresponds to a temperature behind the front of approximately $T_1 = 90{,}000°K$, and the corresponding maximum brightness temperature is approximately $T_{br\,max} = 80{,}000°K$*. For a temperature behind the front $T_1 = 140{,}000°K$, $T_- \approx 50{,}000°K$, $\tau_\nu(T_-) \approx 1.5$ and the screening is almost total.

§4. Limiting luminosity of very strong waves in air

Let us estimate the natural luminescence of the preheating layer in a strong shock wave, which determines the limiting luminosity of the shock front. We consider a shock wave of supercritical strength with a temperature behind the front much higher than the critical temperature of 285,000°K. It was shown in §17 of Chapter VII that in a wave front the temperature distribution with respect to the average optical thickness τ has the shape shown in Fig. 9.4. The preheating temperature just ahead of the shock discontinuity coincides with the temperature behind the front T_1. The temperature in the preheating layer falls off monotonically until it reaches the

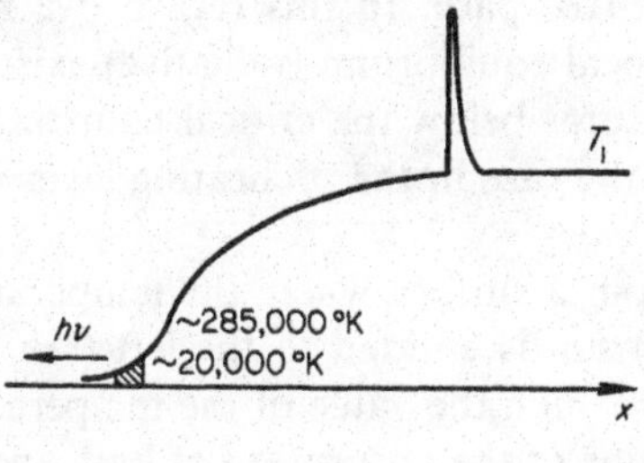

Fig. 9.4. Position of the radiating layer (shaded) in a very strong shock wave.

temperature of the cold air in front. The average optical thickness of the entire preheating region can be very large, and it increases with wave strength. The principal part of the preheating region consists of a region with the temperature ranging from $T_- = T_1$ to a temperature of the order of the critical temperature $T_c \approx 300{,}000°K$. This part of the region, of course, also expands with increasing wave strength (see Fig. 9.4).

At the leading edge of the region, where the temperatures are below 300,000°K, the temperature distribution (as is the case in the subcritical wave) has an exponential character and is almost independent of the strength

$$\varepsilon = \varepsilon_c e^{-\sqrt{3}|\tau-\tau_c|}, \qquad T = T_c e^{-\frac{\sqrt{3}}{1.4}|\tau-\tau_c|}. \tag{9.4}$$

* We emphasize that all these values are only estimates, since the absorption coefficients for visible light in high-temperature air calculated by Kramers' formula cannot be taken as completely reliable.

(See §17, Chapter VII, equation (7.64), and also Fig. 9.4; the optical coordinate τ_c pertains to the point where the temperature is approximately $T \approx T_c \approx 300{,}000°K$.)

It has already been noted at the end of §1 how a body with such a temperature distribution radiates visible light. The air is transparent at low temperatures and does not radiate; at high temperatures the gas is completely opaque and does "not let out" the visible photons. The radiating layer, which basically sends out a flux of visible light to infinity in the cold air, lies somewhere between the transparent and opaque regions (shown shaded in Fig. 9.4). The temperature in the radiating layer is obviously close to the transparency temperature for air defined by the equality (9.2), where l is the frequency averaged mean free path in the region containing the radiating layer. The brightness temperature of the visible radiation also approximately agrees with the transparency temperature. If the mean free path again lies in the range of 10^{-2}–10^{-1} cm, then the limiting brightness temperature should be equal to 17,000–20,000°K (see Table 9.2).

Let us satisfy ourselves that this estimate is valid, that the preheating at the leading edge of the preheating layer in a very strong shock wave is caused by photons which have precisely this mean free path. In this regard, we note that at temperatures above the critical, local equilibrium (radiative) exists in the preheating region, while for temperatures below the critical temperature the radiation is out of equilibrium, as is the case in the preheating layer of a subcritical shock wave.

We can assume approximately that from a surface where the temperature $T_c \approx 300{,}000°K$ a Planck radiation spectrum is emitted to the left (see Fig. 9.4) at this temperature, regardless of how high the value of the temperature is behind this surface. The general behavior of the absorption of high energy photons corresponding to a spectrum with a temperature of 300,000°K (the maximum of the spectrum occurs for photons with $h\nu \approx 70$ ev) is such that the absorption coefficient κ_ν decreases with increasing frequency. As may be seen from Table 9.1, in the region of photon energies of a hundred electron volts, κ_ν decreases monotonically with increasing frequency. Hence, when we move in the direction of decreasing temperatures away from the surface at which $T = 300{,}000°K$, the low energy photons are absorbed first and then the more energetic photons. As we move into the region of low temperatures the spectrum becomes increasingly heavily weighted in the high energies. Calculations presented in [4] show that only very highly energetic photons, with energies $h\nu \approx 200$ ev, can penetrate the region where the temperature is of the order of the transparency temperature, where, as we expect, the radiating layer is located. Table 9.3 gives the energy of the photons carrying out the preheating in the low-temperature region at the leading edge of the preheating region. The table also gives the corresponding mean free paths for these photons, which are approximately equal to the frequency-averaged mean free paths. We see that in the region of temperatures $T \sim 20{,}000°K$ the mean free path $l \sim 10^{-1}$ cm, so that (cf. Table 9.2) the transparency temperature is probably closer to 17,000°K.

Table 9.3

ENERGIES AND CORRESPONDING MEAN FREE PATHS OF PHOTONS WHICH PREHEAT LEADING EDGE OF PREHEATING REGION

T, °K	$h\nu$, ev	l, cm	κ, cm^{-1}
50,000	140		
20,000	200	$0.95 \cdot 10^{-1}$	10.5
15,000	212	$1.02 \cdot 10^{-1}$	9.8
10,000	225	$1.16 \cdot 10^{-1}$	8.6

Thus, the limiting brightness temperature of a very strong shock wave is approximately 17,000°K, regardless of the strength. The general dependence of the brightness temperature (in red light) on the temperature behind the front is shown in Fig. 9.1. We note that the absorption coefficients in the visible region of the spectrum depend only weakly on frequency, and hence the values of brightness temperatures estimated above are approximately applicable not only to the red but in general to the entire visible region of the spectrum.

2. *Optical phenomena observed in strong explosions and the cooling of the air by radiation*

§5. General description of luminous phenomena

An atomic explosion in air produces a very strong shock wave and very high temperatures. The temperature behind the wave front takes on a continuous series of values over a wide range, from hundreds of thousands of degrees down to atmospheric. A number of interesting and rather peculiar optical phenomena are observed in such an explosion. Below we present a general description of the development of an explosion in air near the surface of the earth (that is, in air of standard density). This description is taken entirely from the book "The Effects of Atomic Weapons" [11], published in 1950*.

* We are excerpting paragraphs 2.1, 2.6–2.16, 2.22, 6.2, 6.19, 6.20, 6.22, and 6.23 of the second and sixth chapters of the book and we also present Figs. 2.10, 6.6, 6.18, and 6.20. A second edition of this book [12] appeared in 1957. The second edition has been extensively rewritten. It contains expanded chapters pertaining to the harmful effects of the explosion,

"The fission of the uranium or plutonium in an atomic bomb leads to the liberation of a large amount of energy in a very small period of time within a limited space. In the treatment which follows it will be assumed that the energy released in the atomic bomb is roughly equivalent to that produced by the explosion of 20,000 tons of TNT, namely, about 10^{21} erg (more precisely $8.4 \cdot 10^{20}$ erg). Such a bomb will be referred to as a *nominal atomic bomb*. The resulting extremely high energy density causes the fission products to be raised to a temperature of more than a million degrees. Since this material, at the instant of explosion, is restricted to the region occupied by the original constituents of the bomb, the pressure will also be very considerable, of the order of hundreds of thousands of atmospheres*.

"Because of the extremely high temperature, there is an emission of energy by electromagnetic radiations, covering a wide range of wavelengths, from infrared (thermal) through the visible to the ultraviolet and beyond. Much of this radiation is absorbed by the air immediately surrounding the bomb, with the result that the air itself becomes heated to incandescence. In this condition the detonated bomb begins to appear, after a few millionths of a second, as a luminous sphere called the *fireball* (*ball of fire* in [11], *eds.*). As the energy is radiated into a greater region, raising the temperature of the air through which it passes, the fireball increases in size, but the temperature pressure, and luminosity decrease correspondingly. After about 0.1 msec has elapsed, the radius of the fireball is some 14 m, and the temperature is then in the vicinity of 300,000°K. At this instant, the luminosity (illumination, *eds.*), as observed at a distance of 10,000 m, is approximately 100 times that of the sun as seen at the earth's surface.

"Under the conditions just described, the temperature throughout the fireball is almost uniform; since energy, as radiation, can travel rapidly between any two points within the sphere, there are no appreciable temperature gradients. Because of the uniform temperature the system is referred to as an *isothermal sphere* which, at this stage, is identical with the fireball.

"As the fireball grows, a shock wave develops in the air, and at first the shock front coincides with the surface of the isothermal sphere and of the fireball. Below a temperature of about 300,000°K however, the shock wave advances more rapidly than does the isothermal sphere. In other words, transport of energy by the shock wave is faster than by radiation. Nevertheless, the luminous fireball still grows in size because the great compression

---

but chapters devoted to the description of physical phenomena in the fireball have been abridged. Since we are here interested in precisely the latter problems, we are taking the description of the physics of the explosion from the first edition. All lengths measured in feet, yards, and miles have been converted into meters. *Editors' note*. The paragraphs excerpted are not necessarily complete; phrases in parentheses have been added by the authors.

* *Editors' note*. This paragraph has been combined from paragraphs 2.1 and 2.6 of [11].

of the air due to the passing of the shock wave results in an increase of temperature sufficient to render it incandescent. The isothermal sphere is now a high-temperature region lying inside the larger fireball; and the shock front is coincident with the surface of the latter, which consequently becomes sharply defined. The surface of separation between the very hot inner core and the somewhat cooler shock-heated air is called the *radiation front*.

"The phenomena described above are represented schematically in Fig. 9.5; qualitative temperature gradients are shown at the left, and pressure gradients at the right, of a series of photographs of the fireball taken at various intervals after detonation of an atomic bomb. It can be seen that at first the temperature is uniform throughout the fireball, which is then an isothermal sphere. Later, two distinct temperature levels are apparent, where the fireball has moved ahead of the isothermal sphere in the interior. It may be noted that the luminosity of the outer region of the fireball (brightness of the encompassing shock front) prevents the isothermal sphere from being visible on the photograph. At this time, the rise of the pressure to a peak, followed by a sharp drop at the surface of the fireball, indicates that the latter is identical with the shock wave front.

"The fireball continues to grow rapidly in size for about 15 msec, by which time its radius has increased to approximately 90 m; the surface temperature has then dropped to around 5000°K, although the interior is very much hotter. The temperature and pressure of the shock wave have also decreased to such an extent that the air through which it travels is no longer rendered luminous. The faintly seen shock wave front moves ahead of the fireball, and the onset of this condition is referred to as the *breakaway* (of the shock wave from the luminous sphere). The rate of propagation of the shock wave is then in the vicinity of 4500 m/sec.

"Although the rate of advance of the shock front decreases with time, it continues to move forward more rapidly than the fireball. After the lapse of one second, the fireball has essentially attained its maximum radius of 140 m, and the shock front is then some 180 m further ahead. After 10 sec the fireball has risen about 450 m, the shock wave has traveled about 3700 m and has passed the region of maximum damage.

"An important feature of an atomic explosion in air occurs at about the time of the breakaway of the shock front (from the luminous sphere). The surface temperature (of the luminous region) falls to about 2000°K and then commences to rise again to a second maximum around 7000°K. The minimum is reached approximately 15 msec after the explosion, while the maximum is attained about 0.3 sec later. Subsequently, the temperature of the fireball drops steadily due to expansion and loss of energy.

"It is of interest to note that most of the energy radiated in an atomic explosion appears after the point of minimum luminosity of the fireball.

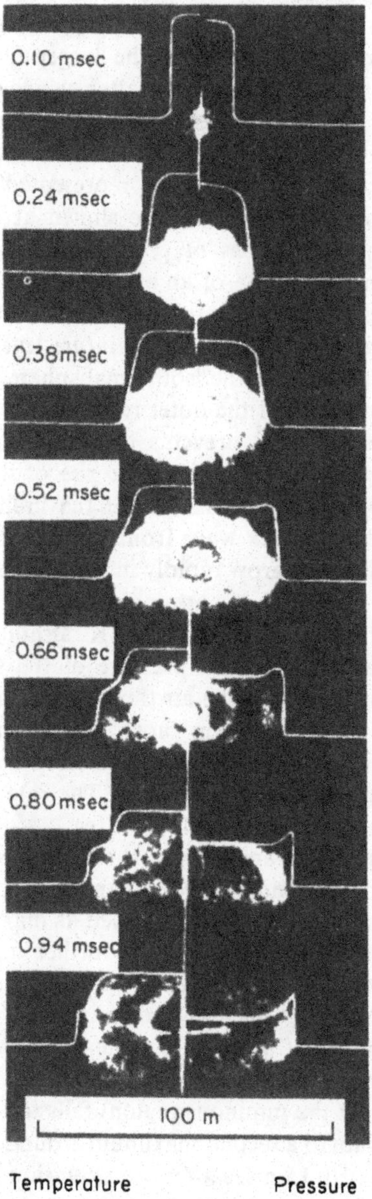

Fig. 9.5. Qualitative representation of the temperature and pressure variations in a fire-ball. The explosion energy is about 10^{21} erg $\approx$ 16 kg of uranium $\approx$ 20,000 tons of TNT.

Only about 1% of the total is lost before this time, in spite of the much higher surface temperature. The explanation of this result lies, of course, in the fact that the duration of the latter period, i.e., about 15 msec, is very short compared with the several seconds during which radiation takes place after the minimum has been passed.

"As stated above, the fireball expands very rapidly to its maximum radius of 140 m, within less than a second from the explosion. Consequently, if the bomb is detonated at a height of less than 140 m, the fireball can actually touch the earth's surface, as it did in the historic "Trinity" test at Alamogordo, New Mexico. Because of its low density, the fireball rises, like a gas balloon, starting at rest and accelerating within a few seconds to its maximum rate of ascent of 90 m/sec.

"After about 10 sec from detonation, when the luminosity of the fireball has almost died out and the excess pressure of the shock wave has decreased to virtually harmless proportions, the immediate effects of the bomb may be regarded as over.

" Because of its high temperature, and consequent low density, the fireball rises, as stated above, and as it rises it is cooled. At temperatures down to about 1800°K the cooling is mainly due to loss of energy by thermal radiation; subsequently, the temperature is lowered as a result of adiabatic expansion of the gases and by mixing with the surrounding air through turbulent convection. When the fireball is no longer luminous, it may be regarded as a large bubble of hot gases rising in the atmosphere, its temperature falling as it ascends.

"An important difference between an atomic and a conventional explosion is that the energy liberated per unit mass is much greater in the former case. As a consequence, the temperature attained is much higher, with the result that a larger proportion of the energy is emitted as thermal radiation at the time of the explosion. An atomic bomb, for example, releases roughly one-third of its total energy in the form of this radiation. For the nominal atomic bomb discussed, the energy emitted in this manner would be about $6.7 \cdot 10^{12}$ cal, which is equivalent to about $2.8 \cdot 10^{20}$ erg.

"The rate at which energy passes through the whole of the spherical surface of the fireball, that is, over a solid angle of 4π, is $\sigma T^4 \cdot 4\pi R^2$, where R is the radius of the ball (and T is the surface temperature; the dependence of R and T on time is shown in Fig. 9.6). Since only the fraction f_0 of this* penetrates the air, the rate at which the radiant energy reaches all points on a spherical area at a moderate distance from the point of detonation is $f_0 \sigma T^4 \cdot 4\pi R^2$.

* It is assumed that the air transmits only those wavelengths exceeding $\lambda_0 = 1680$ Å, so that f_0 is that part of the energy of the Planck spectrum of temperature T, which is included in the interval from $\lambda_0 = 1860$ Å to $\lambda = \infty$. The function f_0 is shown in Fig. 9.7.

The radiant energy flux φ per unit area at a distance D is then obtained upon dividing by the total spherical area $4\pi D^2$, so that

$$\varphi = f_0 \sigma T^4 \left(\frac{R}{D}\right)^2.$$

"From this equation the (radiant energy) flux at a given point, distant D, can be computed for various times after an atomic explosion, using the values of R and T from Fig. 9.6 and of f_0 from Fig. 9.7. In order to avoid plotting values for individual distances, the quantity φD^2, which is equal to $f_0 \sigma T^4 R^2$, is given in Fig. 9.8 as a function of the time; the energy flux is in

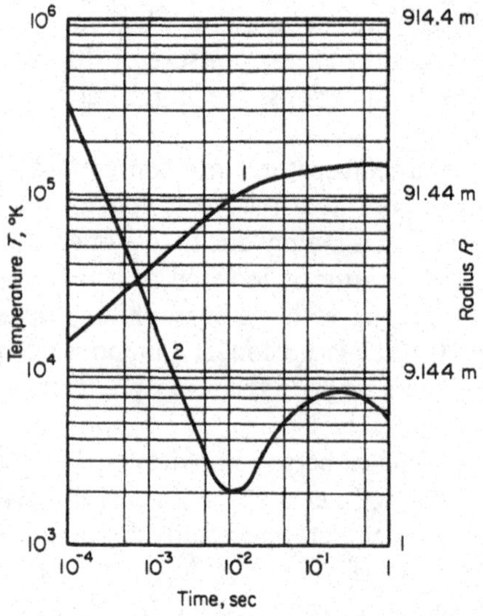

Fig. 9.6. Radius (curve 1) and temperature (curve 2) of fireball as a function of time after explosion.

cal/cm^2 · sec, and the distance is in m. From the curve, the energy flux at any given moderate distance at a specific time can be readily determined.

"In order to obtain some indication of the magnitude of the illumination, it is convenient to introduce a unit called a *sun*; this is defined as a flux of 0.032 cal/cm^2 · sec, and is supposed to be equivalent to the energy received from the sun at the top of the atmosphere. The ordinates at the right of Fig. 9.8 give the value of φD^2, with φ in suns and D in m.

"At the luminosity minimum, the value of φD^2 is about $6.8 \cdot 10^6$ sun-m^2, so that at this point the fireball, as seen at a distance of about 2600 m, should appear about as bright as the sun. Actually, it will be somewhat less bright, to an extent depending on the clearness of the air, because of atmospheric attenuation."

At this point we end the description taken from [11].

As long as the shock wave front has not broken away from the boundary of the luminous body and the latter simply coincides with the shock wave front, the propagation of the fireball is well described by the relation $R \sim t^{2/5}$, which follows from the solution of the strong explosion problem considered

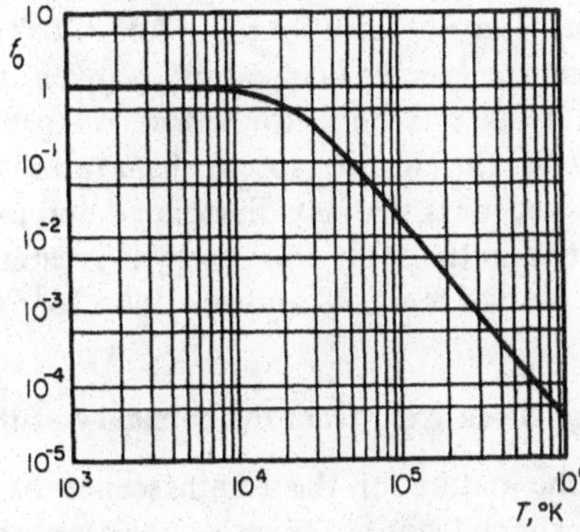

Fig. 9.7. Fraction of equilibrium radiation included in the wavelength interval from $\lambda = 1860$ Å to $\lambda = \infty$ as a function of temperature.

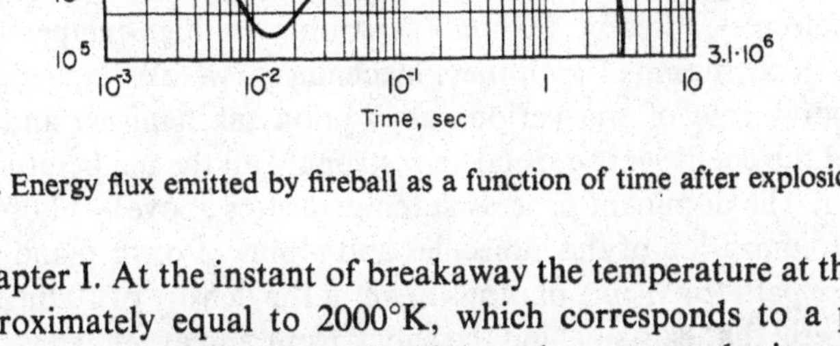

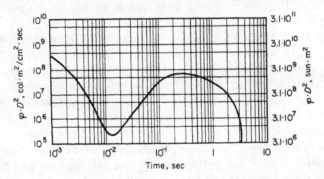

Fig. 9.8. Energy flux emitted by fireball as a function of time after explosion.

in §25 of Chapter I. At the instant of breakaway the temperature at the shock front is approximately equal to 2000°K, which corresponds to a pressure $p_f \approx 50$ atm. This pressure is much higher than atmospheric, and thus the assumptions on which the solution is based ($p_f \gg p_0$) are satisfied.

A comparison of the theoretical relation $R \sim t^{2/5}$ with experiment was presented in the book of Sedov [13]. In the book is given a graph of the straight line $\frac{5}{2}$ log R as a function of log t, on which were plotted the experimental points pertaining to the 1945 New Mexico atom bomb explosion*. The experimental points fit the theoretical curve very well. According to (1.110), the slope of the straight line is related to the explosion energy E by

$$R = \left[\alpha(\gamma)\frac{E}{\rho_0}\right]^{1/5} t^{2/5}; \qquad \frac{5}{2}\log R = \frac{1}{2}\log\left(\alpha\frac{E}{\rho_0}\right) + \log t.$$

Here $\alpha = \xi_0^5$, where ξ_0 is the coefficient in (1.110). For an air density $\rho_0 = 1.25 \cdot 10^{-3}$ g/cm^3 the value obtained for αE is $8.45 \cdot 10^{20}$ erg. The dependence of the coefficient α on the specific heat ratio γ is given in Sedov's book [13]. If we assume $\gamma = 1.4$, as was done by Sedov, we get $\alpha = 1.175$ and $E = 7.19 \cdot 10^{20}$ erg. Actually, the effective specific heat ratio is slightly lower, since at high temperatures the air is strongly dissociated and ionized. Therefore the coefficient α is smaller and the explosion energy is greater. Thus, for example, if we take $\gamma = 1.32$, we find $\alpha = 0.88$ and $E = 9.6 \cdot 10^{20}$ erg.

§6. Breakaway of the shock front from the boundary of the fireball

Let us examine the nature of the luminescence of a fireball for temperatures behind the shock front of the order of several thousand degrees and clarify the causes for the phenomenon of breakaway of the shock front from the boundary of the luminous mass and for the minimum in the luminosity of the fireball. These problems were treated by one of the present authors [14, 15].

Many mechanisms participate in the absorption (and emission) of visible light in high-temperature air. These include photoionization of strongly excited atoms and molecules of oxygen and nitrogen and of nitric oxide molecules, knocking out the additional weakly bound electron from negative oxygen ions, molecular absorption (without the removal of electrons) by excited O_2, N_2, and NO molecules, and, finally, molecular absorption by NO_2 molecules, present in small amounts in high-temperature air. The absorption coefficients for all these mechanisms were estimated in Chapter V. The relative role of the various absorption mechanisms and the absolute values of the coefficients depend very strongly on the temperature and density of the air. The dominant process at temperatures above ~12,000–15,000°K is the photoionization of the molecules and atoms of oxygen and nitrogen. The mean free path for visible photons in air at the density of 10 times higher than atmospheric that exists behind the shock front and at temperatures ~12,000–15,000°K is found to be of the order of millimeters. The mean free path decreases sharply as the temperature increases.

* *Editors' note.* This plot was taken from the paper by Taylor [27] on this subject.

In the lower temperature range of ~6000–8000°K the dominant processes are the photoionization of NO molecules, absorption by negative oxygen ions, and molecular absorption by O_2, N_2, and NO molecules. The mean free paths for visible light in this temperature range are also strongly temperature dependent and are of the order of 10–100 cm (for a density ratio of 10 across the shock wave). At the still lower temperatures below ~5000°K all these mechanisms produce a very weak absorption, which, in addition, decreases rapidly with decreasing temperature. Practically the only absorption mechanism of visible light in air at $T < 5000$°K is absorption by nitrogen dioxide molecules NO_2. Despite its small concentration, nitrogen dioxide absorbs visible light very strongly, leading to mean free paths measured in meters. Thus, the nitrogen dioxide concentration at $T = 3000$°K and at a density five times atmospheric is $c_{NO_2} = 1.6 \cdot 10^{-4}$*, and the mean free path for red light, calculated using an absorption cross section $\sigma_{NO_2} = 2.15 \cdot 10^{-19}$ cm^2, is $l_\nu = 220$ cm.

It is known that the temperature behind the shock front produced by a strong explosion increases from the front to the center (see §25, Chapter I). If we consider that stage of the explosion at which the temperature at the front is equal to several thousand degrees, then for the nominal explosion energy $E = 10^{21}$ erg (corresponding to approximately 20,000 tons of TNT) the explosion wave encompasses a sphere with a radius of the order of hundreds of meters and the temperature behind the front increases appreciably for a distance away from the front toward the center of the order of meters. At this stage the radiant preheating of the air ahead of the shock discontinuity and the screening of the front surface, considered in §§1 and 3, are negligibly small. The thickness of the relaxation layer in the shock front, in which the equilibrium values of dissociation and ionization are established, is much smaller than a photon mean free path. Hence we can state that, as long as the photon mean free path is less than the order of a meter, the shock front is followed by an optically thick region with an almost constant temperature and the front radiates as a perfect black body. The brightness temperature of the fireball as measured by the visible radiation does not in this case differ from the temperature of the shock front. The situation is different when the shock strength falls to a point where the absorption behind the front becomes weak and the photon mean free path becomes comparable to the characteristic distance over which there is an appreciable change of temperature behind the front, that is, where it becomes of the order of a meter or more.

However, before we consider the deviation of the brightness temperature from the temperature of the front, which will be done in the following section, let us determine the point at which the shocked air ceases to be incandescent.

* The concentration c_i is defined as the ratio of the number of ith particles to the original number of molecules in the cold air.

We have noted above that at temperatures below ~5000°K the nitrogen dioxide molecules are responsible for the absorption and emission of the visible light. But, because of the nature of the kinetics of the oxidation of nitrogen in an explosion wave (see §5, Chapter VIII) the nitric oxide molecules, from which the dioxide molecules are subsequently formed, practically do not form in air heated by a shock wave to temperatures below about 2000°K. The reason is the large activation energy of the nitrogen oxidation reaction, which results in the extremely pronounced temperature dependence of the reaction rate. At temperatures below about 2000°K the time required for the formation of any appreciable amount of oxide is extremely large in comparison with the lifetime of the particles in the shock wave, so that the reaction does not take place. Thus nitrogen dioxide, the only absorption agent in the air layers heated by the shock front to temperatures below ~2000°K, is not formed in these layers; they are completely transparent to visible light and are not incandescent.

At the time when the front temperature T_f is below 2000°K, say when $T_f = 1000°K$, a luminous disk with a radius smaller than the radius of the shock front becomes visible from afar. The horizontal cross section of an explosion wave is shown in Fig. 9.9. Rays such as B intersect the air layers

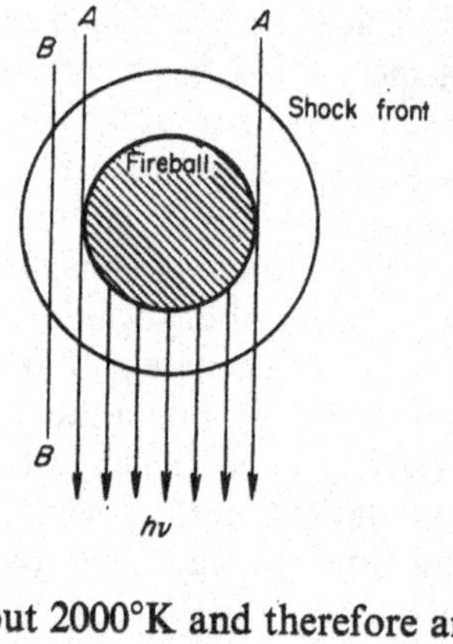

Fig. 9.9. Schematic representation of the luminescence of a fireball after breakaway. The inner circle is the boundary of the luminous mass, the fireball; the outer circle is the shock front.

heated by the shock to a temperature below about 2000°K and therefore are not luminous. The fireball is bounded by the rays A, which are displaced from the center O exactly the distance R_{fb}. This distance is the radius at the given time of those air layers which were previously heated by the front to a temperature of about 2000°K and which still contain a sufficient amount of nitrogen dioxide to produce perceptible luminescence. Since the air particles in the explosion wave expand rapidly away from the center (although slower than the shock wave front), the radius of the fireball R_{fb} increases. The fireball expands until the pressure in the explosion wave drops to atmospheric and the motion ceases*. The nonluminous shock front, after breaking away from the fireball at the time when its temperature was ~2000°K, moves far ahead of it. (The traces of the front and of the fireball are shown schematically in Fig. 9.10.)

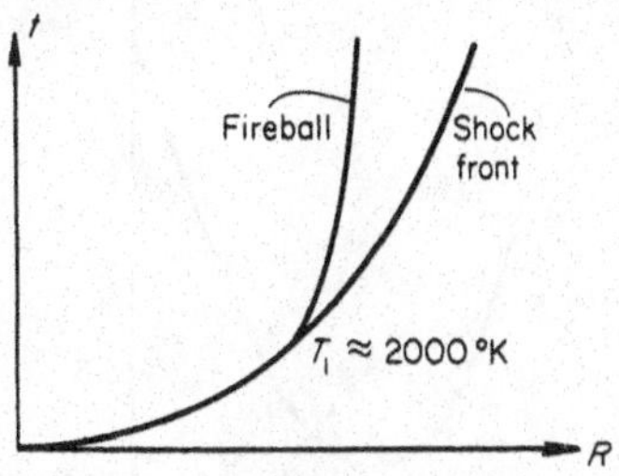

Fig. 9.10. Trace of the shock front and the fireball boundary on an R, t diagram.

§7. Minimum luminosity effect of the fireball

Let us consider how the luminosity and brightness temperature of the fireball surface change with time during the process of the breakaway of the shock wave from the boundary of the luminous mass. When the temperature of the front drops below ~5000°K, the mean free path for visible light increases to a value of the order of a meter and the fireball ceases to radiate as a perfect black body. Under these conditions the brightness temperature should be calculated from the general formula (2.52) in accordance with the distribution of temperature and absorption coefficient with respect to the distance behind the shock front†.

Let us consider as an example the time $t = 1.5 \cdot 10^{-2}$ sec, when the radius of the front $R = 107$ m and the temperature at the front $T_f = 3000$°K (all

* Air particles initially at the temperature $T_f \approx 2000$°K, corresponding to a pressure at the front $p_f \approx 50$ atm, cool to $T \sim 800$°K in an adiabatic expansion to atmospheric pressure. Probably, as time increases the boundary of the luminous region is displaced deeper, into those layers with temperatures closer to 2000°K. The reason is that the emission coefficient, which is proportional to $\exp(-h\nu/kT)$, decreases very rapidly with a decrease in temperature even if the absorption coefficient is unchanged ($h\nu \gg kT$ for $h\nu \approx 2$ ev, $T \sim 2000$–1000°K). To be more precise, the boundary of the fireball is determined by the sensitivity of the recording instrument.

† The radiating layer has a thickness of the order of ten meters, which is considerably less than the sphere radius $R_{fb} \sim 100$ m. Therefore, we can neglect the curvature of the layer and consider it to be planar, and thus we may use equation (2.52). We note that (2.52), in which an exponential-integral enters because the oblique rays are taken into account, gives a brightness temperature averaged over the disk. If we are interested in the luminosity at the center of the disk, then the exponential-integral $E_2(\tau_\nu)$ should be replaced by the ordinary exponential $\exp(-\tau_\nu)$, where τ_ν is the optical thickness measured with respect to the radius from the front surface toward the center of the sphere. We can then calculate the average brightness temperature.

calculations pertain to an explosion with an energy of $E = 10^{21}$ erg). Figure 9.11 shows the distribution of the absorption coefficient for red light $\lambda = 6500$ Å with respect to distance behind the shock front (the x coordinate is measured from the front into the fireball). Also given on the figure are the

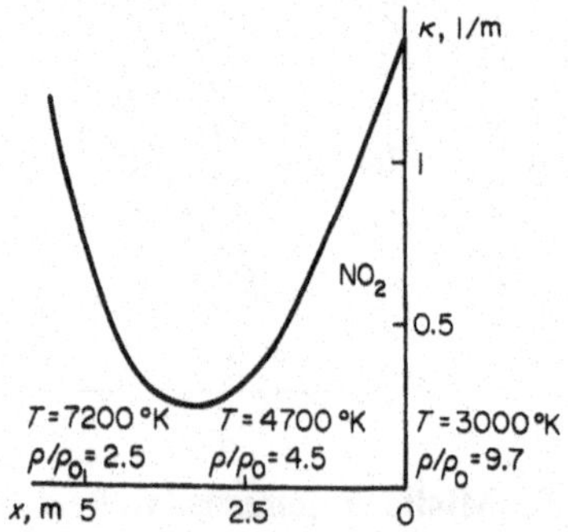

Fig. 9.11. Distribution of the absorption coefficient of red light behind a shock front at a temperature $T_f = 3000°$K for an explosion with $E = 10^{21}$ erg. Values of the temperature and density are shown at several points. The specific heat ratio $\gamma = 1.23$.

temperature and the density ratio ρ/ρ_0 for air at several points. The temperature and density distributions behind the front are taken from the solution of the strong explosion problem; the nitrogen dioxide concentration was calculated by the method of §5, Chapter VIII. Since the exact values of the cross sections for the absorption of red light by excited NO_2 molecules are not known, we have taken for calculational estimates the reasonable values given in Table 9.4 (see §21, Chapter V).

Table 9.4

ESTIMATED CROSS SECTIONS FOR ABSORPTION OF RED LIGHT BY EXCITED NO_2 MOLECULES

T, °K	4000	3000	2600	2000
$\sigma_{NO_2} \cdot 10^{19}$, cm²	3.0	2.15	1.8	0.84

It is evident from Fig. 9.11, that at temperatures above 6000–7000°K, the absorption, due to the many mechanisms listed above, is very strong and that it increases rapidly as we move away from the front where the temperature goes up. In the region of $T \sim 6000°$K the absorption decreases and passes through a minimum, since at this temperature the absorption coefficient for all the other mechanisms is very small and the concentration of the dioxide is still too low (equilibrium for the reaction $NO + \frac{1}{2}O_2 \rightleftarrows NO_2$ at such high temperatures is displaced in the direction of dissociation of the dioxide). The concentration of the dioxide increases at still lower temperatures ~4000–3000°K, which then leads to an increase in the absorption near the wave front.

The air layers at temperatures above ~6000–7000°K are essentially found to be completely opaque, and a Planck radiation flux is emitted through the surface from the internal "hot" sphere at this temperature. The external air layer containing the dioxide plays a dual role. On the one hand, it absorbs this high-temperature radiation emerging from the surface of the "hot" sphere, and on the other hand, it radiates itself. This situation can be described formally by breaking up the integral with respect to τ_ν in (2.51) into two parts: one with respect to the external layer containing the dioxide and with an optical thickness τ_ν^*, and the other with respect to the internal hot region $\tau_\nu^* < \tau_\nu < \infty$:

$$S_\nu(T_{br}) = 2\int_0^\infty S_{\nu p}E_2(\tau_\nu)\,d\tau_\nu = 2\int_0^{\tau_\nu^*} S_{\nu p}E_2(\tau_\nu)\,d\tau_\nu + 2\int_{\tau_\nu^*}^\infty S_{\nu p}E_2(\tau_\nu)\,d\tau_\nu .$$

Here we have replaced the radiation density in (2.51) by the equivalent flux. In the second integral we can factor out some average value of $S_{\nu p}^*$, corresponding to the brightness temperature of the hot sphere T^* ($T^* \sim 7000°K$), and, using the properties of the exponential-integral functions, we can write

$$S_\nu(T_{br}) = 2\int_0^{\tau_\nu^*} S_{\nu p}E_2(\tau_\nu)\,d\tau_\nu + S_{\nu p}(T^*)E_3(\tau_\nu^*).$$

The first term gives the natural radiation of the dioxide layer and the factor $E_3(\tau_\nu^*)$ in the second term takes into account the screening by this layer of the high-temperature radiation from the hot sphere. Calculations show that the relative importance of the second term increases with time, and that the natural luminescence of the dioxide becomes small, that is, the role of the dioxide reduces basically to that of screening the high-temperature radiation. In our example $T_f = 3000°K$, the optical thickness of the dioxide layer $\tau_\nu^* = 2.42$, and the brightness temperature of the fireball is found to be $T_{br} = 4110°K$.

Another typical distribution of the absorption coefficient with respect to the radius is obtained when the temperature at the shock front drops below 2000°K. In this case, the absorption does not begin immediately behind the front, but slightly deeper, since the layers close to the front have been heated only to temperatures below 2000°K and contain no dioxide. These layers do not absorb any light. This situation is shown in Fig. 9.12 ($t = 2.64 \cdot 10^{-2}$ sec, $R = 138$ m, $T_f = 1600°K$).

Let us follow how the radiation brightness temperature changes with time. As long as the temperature at the shock front is higher than ~2000°K, dioxide is formed in the air layers through which the front passes; the total optical thickness of the dioxide layer increases and the luminosity decreases. The brightness temperature for $T_f \lesssim 5000°K$ exceeds the temperature of the front, since the dioxide layer does not completely screen the high-temperature radiation (with $T^* \sim 7000°K$) arriving from inside the sphere. When the

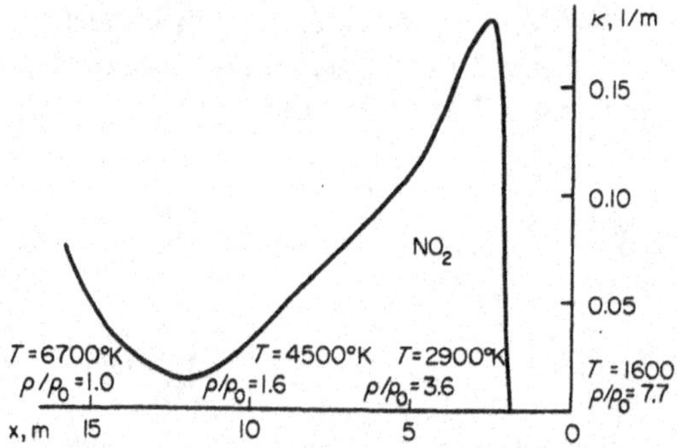

Fig. 9.12. Distribution of the absorption of red light behind a shock front at a front temperature $T_f = 1600°K$, for an explosion of $E = 10^{21}$ erg. Values of the temperature and density are shown at several points. The specific heat ratio $\gamma = 1.30$.

temperature of the front decreases below 2000°K, the dioxide no longer forms in the newer layers over which the front passes. Even if the total number of NO_2 molecules that are present in the air at this time were to remain unchanged, the optical thickness of the screening dioxide layer would still decrease with time, since the expansion of the air would result in a distribution of the same number of NO_2 molecules over a spherical layer of larger and larger radius. It is easy to see that the optical thickness of the dioxide layer

$$\tau_\nu^* = \int_0^R n_{NO_2}\sigma_{NO_2}\,dr,$$

where n_{NO_2} is the number of NO_2 molecules per unit volume, decreases roughly as $R^{-2} \sim t^{-4/5}$ if the total number of NO_2 molecules

$$N_{NO_2} = \int_0^R 4\pi r^2 n_{NO_2}\,dr \approx 4\pi R^2 \int_0^R n_{NO_2}\,dr$$

remains constant. However, the total amount of dioxide, after its formation has stopped, falls off slightly due to the decomposition of the NO_2 molecules (see §5, Chapter VIII). This leads to an even more rapid decrease of the optical thickness τ_ν^*.

Thus, starting at the time when the temperature at the front falls below $\sim 2000°K$, the screening of the dioxide layer decreases and the internal hot region is gradually "revealed". The brightness temperature of the fireball after having passed through a minimum again increases; the fireball thus appears to flare up again—a fact which has been observed experimentally. The above description of the nature of the luminosity minimum is illustrated in Table 9.5, in which are presented the results of calculations of the brightness temperature for an explosion with an energy of $E = 10^{21}$ erg. T_{br} passes through a minimum at 3600°K and τ_ν^* goes through a maximum when the front temperature $T_f = 2600$°K is close to the breakaway temperature $T_f = 2000$°K.

Table 9.5

BRIGHTNESS TEMPERATURE OF A FIREBALL IN RED LIGHT $\lambda = 6500$Å IN THE REGION OF MINIMUM LUMINOSITY, AS CALCULATED IN [15]

$t \cdot 10^2$, sec	R, m	T_f, °K	T_{br}, °K	τ_ν^*
		$E = 10^{21}$ erg		
0.75	82	5000	5930	1.06
1.05	93	4000	4810	1.96
1.50	107	3000	4110	2.42
1.81	109	2600	3600	3.23
1.95	112	2300	4150	2.16
2.25	128	2000	4520	1.80
2.39	132	1800	4810	1.61
2.64	138	1600	5400	1.15
2.94	143	1400	5600	1.11
		$E = 10^{20}$ erg		
0.43	49	5000	6380	0.61
0.61	53	4000	5560	1.16
0.72	58	3000	5060	1.42
0.82	60	2600	4800	1.77
0.95	65	2300	5380	1.18
1.01	66	2000	5850	0.96
1.16	70	1800	6050	0.88
1.38	73	1600	6510	0.71
1.41	75	1400	6980	0.54

It is interesting to observe what happens to the luminosity minimum for different explosion energies. All times and lengths in a strong explosion change in a similar manner, proportionally to $E^{1/3}$ (from the fact that the self-similar solution of the strong explosion problem is approximately valid). Roughly speaking, the optical thicknesses at corresponding times (at the same front temperature), also vary as $E^{1/3}$ (since the dioxide concentration over the main area is in equilibrium and depends primarily on the particle temperature and density, and not on the time it is in the heated state). It follows, therefore, that the screening by the dioxide layer decreases with decreasing explosion energy, and the difference between T_{br} and T_f becomes greater; also, the minimum becomes shallower. As an example, in Table 9.5 are presented the results of calculations of $T_{br}(T_f)$ for an explosion energy $E = 10^{20}$ erg. The position of the minimum remains the same, but the minimum luminosity increases: $T_{br\,min} \approx 4800°K$.

In the limit of very low explosion energies the minimum should disappear entirely. Conversely, in the limit of very high explosion energies all lengths and optical thicknesses become large, the radiation of the fireball approaches closer and closer to black body radiation, and T_{br} approaches T_f up to the time when T_f becomes equal to approximately 2000°K, that is, the minimum becomes lower than $T_{br\,min} \approx 2000°K$. The brightness temperature cannot drop below 2000°K, since even for very high explosion energies and large lifetimes the dioxide does not form for $T < 2000°K$ and the air heated by the shock wave to temperatures $T_f < 2000°K$ is transparent and does not radiate.

§8. Radiation cooling of air

We consider that the gasdynamic process in a strong explosion in air with an energy $E \sim 10^{21}$ erg takes place adiabatically, as described in §25, Chapter I. The expansion of the air encompassed by the shock wave is strongly decelerated up to the time its pressure drops to a value of the order of atmospheric. Thereafter, the shock wave gradually weakens and becomes an acoustic wave which moves far out, carrying with it a large part of the explosion energy. The central regions, however, after atmospheric pressure has been reached and the motion has ceased, contain a large mass of air irreversibly heated by the shock wave. This mass of air contains the "residual" explosion energy, which also constitutes an appreciable fraction (of the order of tens of percent) of the total explosion energy. The air mass is heated to very high temperatures. Thus, for example, the air layers heated to a temperature $T_f = 11{,}000°K$ by the passage of a shock front with a strength such that $p_f = 750$ atm will be at a temperature of the order of 2000°K* after expanding to atmospheric pressure. Layers closer to the center, which were initially heated by the shock front to several hundreds of thousands of degrees (with the pressure at the front of the order of a hundred thousand atmospheres), remain heated to about ten thousand degrees, etc. Thus, the explosion leaves behind a tremendous volume of air with a radius of the

* The residual temperature, roughly speaking, is $T_{res} \approx T_f(1\text{ atm}/p_f\text{ atm})^{(\gamma-1)/\gamma}$. For an estimate we may take an effective value of the specific heat ratio $\gamma \approx 1.3$.

order of hundreds of meters, heated to high temperatures. The temperature in the central regions reaches hundreds of thousands of degrees, while at the periphery it gradually decreases to a thousand degrees and lower, down to standard atmospheric temperature.

We ask the question, what subsequently happens to the residual energy of the air irreversibly heated by the explosion wave, and how does this air cool down? This problem was considered in papers by Kompaneets and the present authors [16, 17]. It is clear that the dissipation of energy by molecular heat conduction plays no role. For a thermal diffusivity coefficient (thermal conductivity) for air of the order of 1 cm^2/sec, a volume with a radius $\sim 10^4$ cm would take about a year to cool down. Convective ascent of the heated sphere due to the difference in density of the hot and cold air at the same atmospheric pressure and the resulting mixing of the hot air with the surrounding mass of cold gas is a more substantial effect. However, the ascent during the first 2–3 seconds after the explosion is not great. It cannot exceed $gt^2/2$ (where g is the acceleration due to gravity), which is 5 m for the first second, 20 m in two seconds, and 45 m during the third second. Hence, convection is also eliminated as a factor in the first few seconds after the explosion.

The basic process which leads to the cooling of the air and to the dissipation of the irreversible thermal energy is the radiation of light. Radiative cooling is made possible by the fact that cold air is transparent in a certain spectral window, in the visible region of the spectrum and in the neighboring ultraviolet and infrared regions. Because of this transparent window, the corresponding photons, radiated by the heated gas, can move away freely to large distances, carrying with them the energy of the heated mass.

A characteristic feature of the process of energy emission from the heated air is the fact that it is unsteady. In this respect it is basically different from the process of stellar radiation (in particular the radiation from the sun, which supplies our planet with energy), which at first glance appears similar. The energy lost by radiation from the surface of a star is compensated for by a flow of energy from within, which is released as the result of nuclear reactions taking place in the central regions (see Chapter II, §14). In the state which results, each volume element receives an amount of radiant energy equal to that it emits, and the temperature distribution with respect to the radius of the star has a steady character (over observable time periods). In our case there are not internal energy sources; the initial temperature distribution is determined by the previous history of the phenomenon, by the gasdynamics of the propagation of the explosion wave, and the air cools gradually by the transfer of energy away by radiation.

Our problem is to clarify how the cooling process proceeds, to determine how the temperature changes at different points of the heated volume, and finally, which is most important, to determine what is the rate of radiative cooling and what is the radiant flux from the surface of the heated mass.

§9. Origin of the temperature drop—the cooling wave

The basic factor which determines the features of this process is the extremely pronounced temperature dependence of the transparency of air, which has already been discussed several times. If we consider the temperature dependence of the mean free path of radiation suitably averaged over a spectrum which is characteristic for a given temperature, say, the mean free path of photons $h\nu$ with energies of 3–5 times kT*, and keep in mind that at a constant pressure close to atmospheric the density of air decreases with increasing temperature, we arrive at the following conclusions: The mean free path of photons changes from kilometers at temperatures of the order of 6000°K to hundreds of meters at $T \sim 8000$°K, tens of meters at $T \sim 10{,}000$°K, and to tens of centimeters at $T \sim 15{,}000$°K.

Obviously, the radiant flux emerging from a heated volume with a smooth temperature distribution is determined by the temperature of that (radiating) layer in which the mean free path is of the order of the characteristic dimension of the problem, that is, of the order of tens of meters. The external, less heated, layers are transparent and practically do not radiate. The deeper layers are opaque and the photons born there cannot move too far away. We have already encountered a similar situation in considering the radiation of the preheating zone of air ahead of the discontinuity in a very strong shock. By analogy, one can also introduce in this problem the concept of a transparency temperature T_2, defined as the temperature at which the mean free path for light is of the order of the characteristic distance over which the temperature changes appreciably. In contrast to the problem of the luminescence of the preheating layer, where the dimensions were 10^{-2}–10^{-1} cm and the transparency temperature was $\sim 20{,}000$°K, the scale here is of the order of 10 m and the transparency temperature $T_2 \sim 10{,}000$°K.

Let us now imagine a spherical volume of stationary air with an initially smooth temperature distribution which varies with radius from $\sim 100{,}000$°K at the center to several thousand degrees at the periphery, and let us consider how this distribution changes with time (here we shall neglect the motion of the air resulting from pressure gradients). In accordance with what was said above, we may expect that radiation and cooling will begin in a layer whose temperature is of the order of the transparency temperature ($T_2 \sim 10{,}000$°K); at the following moment we should observe a "depression" in the

* We recall that the maximum of the Planck frequency spectrum occurs for photon energies $h\nu = 2.8kT$; the maximum of the weighting function appearing in the Rosseland method for averaging the mean free path lies in the region $h\nu \approx 4kT$.

initially smooth temperature distribution, as shown in Fig. 9.13. Subsequently this "depression" acquires the form of a temperature drop which is propagated toward the center of the heated sphere. The air layers are successively

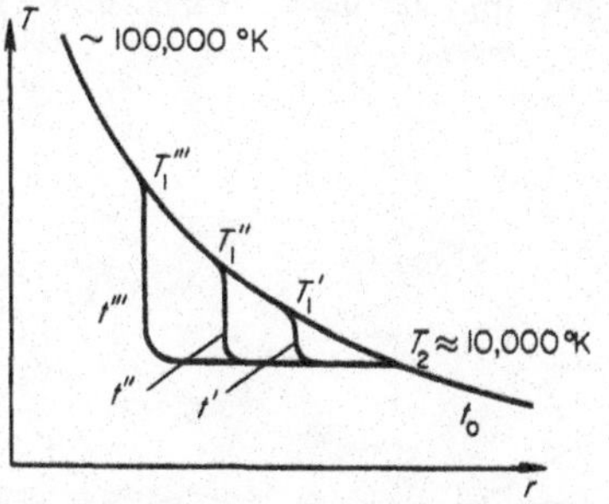

Fig. 9.13. Origin of a temperature drop (cooling wave) in an originally continuous temperature distribution, and its propagation in stationary air; $t_0 < t' < t'' < t'''$.

cooled from the initial temperature to a temperature of the order of 10,000°K, after which they become transparent and practically stop radiating. The internal layers essentially retain their temperature until reached by the drop, since the mean free path for light in these layers is very small and emitted photons are immediately absorbed.

Thus, the air cools as a result of the propagation through it of a certain narrow temperature drop, which can be called a "cooling wave". The temperature in the cooling wave drops sharply (in comparison with the initial smooth distribution) from the initial value T_1, equal to the temperature at the point reached at the given instant by the upper boundary of the wave, to a lower value, the transparency temperature T_2 at which the air practically stops radiating.

In describing the successive changes in the temperature distribution in Fig. 9.13, we have digressed from the consideration of temperature changes due to the gasdynamic motion by taking the air to be stationary. Actually, however, the temperature drop is formed even prior to the time when the air pressure falls to atmospheric and the motion ceases, namely, when the rate of radiative cooling of a layer with temperature $\sim$10,000°K becomes comparable to the rate of adiabatic cooling connected with the expansion of the air in the explosion wave. The adiabatic cooling rate at the earlier stage of the explosion is high and the air has no time to emit its energy, since the temperature region around 10,000°K at which the drop could form is "passed by" very rapidly, and the air becomes transparent without having had a chance to lose any appreciable amount of energy by radiation. Subsequently, however, when the adiabatic cooling slows down as the pressure falls and the expansion becomes slower, the radiant cooling becomes dominant. Estimates show that the drop behind the shock wave front in an explosion with $E = 10^{21}$ erg begins to appear in a layer with $T \sim$ 10,000°K at $t \sim 10^{-2}$ sec, when the temperature at the front is of the order of 2000°K and the pressure is of the order of 50 atm (the pressure in an explosion wave changes by only a small amount from the front to the center; see §25, Chapter I).

The temperature distributions in air through which a cooling wave propagates, including the effect of the adiabatic cooling, are given in Fig. 9.14. The

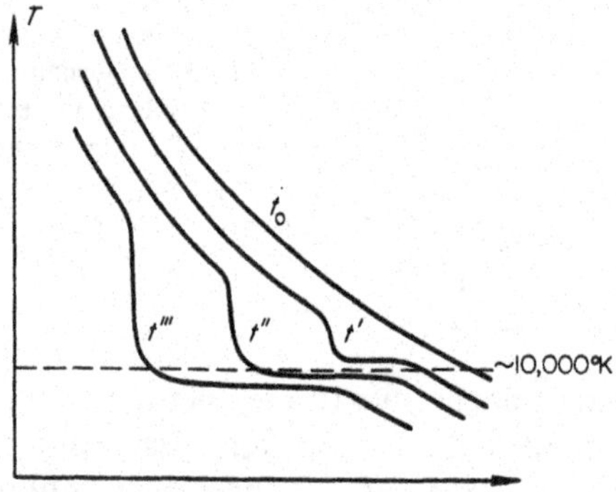

Fig. 9.14. Origin and propagation of a cooling wave in air undergoing an explosion and adiabatic cooling; $t_0 < t' < t'' < t'''$.

abscissa is expressed in terms of a Lagrangian rather than an Eulerian coordinate, and Fig. 9.14 shows the temperature changes of given air particles and the propagation of the cooling wave not through space but with respect to mass in the gas.

§10. Energy balance and propagation velocity of the cooling wave

The cooling wave travels through air that is practically undisturbed by radiative losses. The gas temperature at the time when it is reached by the upper boundary of the temperature drop is determined solely by the previous history of the process and by the hydrodynamic motion (if such exists). This is a result of the fact that at temperatures of the order of tens of thousands of degrees and with temperature gradients of the order of thousands of degrees per meter, which exist initially, the radiative heat transfer is too small to give any appreciable flow of energy in the opaque region that has not yet been reached by the cooling wave. The radiation heat conduction coefficient, which is proportional to the Rosseland mean free path $l(T)$ and to the cube of the temperature*, increases rapidly with increasing temperature and becomes appreciable only in the region of temperatures of the order of hundreds of thousands of degrees, near the center of the explosion. It limits the tem-

* We recall that the energy flux transported by radiation heat conduction is $S = -\kappa\, \partial T/\partial r$, where the radiation heat conduction coefficient is $\kappa = 16\sigma l(T)T^3/3$ (see §12, Chapter II).

perature rise at the center to this order of magnitude and equalizes the temperature near the center.

The radiation conduction coefficient becomes large again in the temperature region below 10,000°K, where the mean free path, increasing sharply with decreasing temperature, becomes very large*. This does not mean, however, that the radiation heat conduction equalizes the temperature also at the low temperatures, since the heated air becomes transparent in this region and the concept of radiation heat conduction, in general, loses its meaning; the character of the radiative transfer changes and, in particular, it leads to the formation of the cooling wave.

Thus, because of the small heat conduction at the upper edge of the cooling wave, the radiant energy flux flowing into the wave from within is close to zero and cannot affect the properties of the wave. The entire radiant flux, carrying away the energy of the air particles being cooled in the wave, is generated within the wave itself. The determination of this flux, which we shall denote by S_2, is a basic theoretical problem (Part 3 of this chapter is devoted to its solution). This problem is nontrivial, since a very steep temperature distribution exists within the wave. It is clear only that the flux is bounded within the limits $\sigma T_1^4 > S_2 > \sigma T_2^4$, since the radiating layer in the wave is at a temperature below the upper temperature T_1 (at which the air is completely opaque) but above the lower temperature T_2 (below which the air is transparent, does not radiate, and is not cooled by energy emission). If the flux S_2 is known, then the velocity u at which the cooling wave propagates through the gas, which in the end determines the cooling time of the heated volume, can be found from an energy balance. According to estimates the cooling wave propagates through air undisturbed by radiative losses with a velocity less than the speed of sound. In this case the pressure along a thin layer—the wave "front"—becomes equalized and practically constant. The gas density automatically adjusts itself to the temperature change, so that an air particle passing through the wave and being cooled is compressed proportionally to $1/T$ (if we assume that the pressure $p \sim \rho T$). This is shown in Fig. 9.15†.

The air is cooled by the wave at a constant pressure. If ρ_1 is the initial density of the air at the time of the wave's approach, then the mass of air per unit area of front surface per unit time is equal to $\rho_1 u$. The change in

* $l(T)$ passes through a minimum at $T \sim 50{,}000$°K, and the coefficient of radiation thermal conductivity, which is proportional to $l(T)T^3$, has a minimum at $T \sim 10{,}000$°K.

† *Editors' note.* The idealized cooling wave described by the authors is an endothermic gasdynamic discontinuity of the detonation type. The Hugoniot curve is simply an isotherm at the transparency temperature. The cooling wave may be considered as the inverse of a weak deflagration. As with most ordinary weak deflagrations (idealized flames), the pressure jump is not strictly zero but is small enough to be negligible.

enthalpy in cooling from T_1 to T_2 is (for constant specific heats) $\rho_1 u c_p(T_1 - T_2)$. This change is, clearly, equal to the energy carried away from the wave front surface by radiation, to the flux S_2. In this manner we obtain the basic

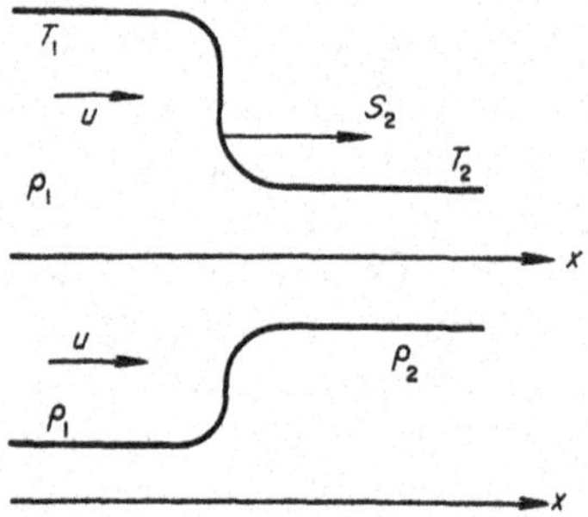

Fig. 9.15. Schematic representation of the gas temperature and density distributions in a cooling wave front. The arrow on u shows the direction of air velocity into the wave.

energy balance equation for the cooling wave, which is here considered as a discontinuity,

$$S_2 = \rho_1 u c_p(T_1 - T_2). \tag{9.5}$$

If we take into account that the specific heat c_p is not constant, then we obtain the more general expression

$$S_2 = \rho_1 u(h_1 - h_2), \tag{9.6}$$

where h is the specific enthalpy of the air.

If γ is the effective adiabatic exponent in front of the wave, then $h_1 = [\gamma/(\gamma - 1)]p/\rho_1$ (see (3.67)) and the wave velocity is

$$u = S_2 \frac{\gamma - 1}{\gamma} \frac{1}{p} \left(1 - \frac{h_2}{h_1}\right)^{-1}. \tag{9.7}$$

It will be shown in the third part of this chapter that the radiation emerging from the surface of the cooling wave is always generated at the lower edge of the temperature drop (at the rear of the wave front), independent of the wave strength characterized by the ratio T_1/T_2 or h_1/h_2; hence, no matter how high the temperature of the hot gas T_1, the temperature of the emerging radiation will be close to T_2.

The flux S_2 is determined basically by the transparency temperature and is approximately equal to

$$S_2 = 2\sigma T_2^4. \tag{9.8}$$

The transparency temperature is not an exactly defined quantity. As was pointed out previously, it serves as a specification of the boundary between the transparent and opaque temperature regions and is found from the condition that the frequency-averaged radiation mean free path at a temperature equal to the transparency temperature is of the order of the characteristic scale d of the problem. This distance is, for example, the distance over which the air temperature drops from T_2 to a sufficiently small value, say, to 2000°K. When the wave travels through expanding air this scale is determined by the hydrodynamics of the process as a whole; it is smaller, the higher is the

Table 9.6

u, KM/SEC AT $p = 1$ ATM

T_1, °K	T_2, °K		
	10,700	9,700	9,300
20,000	2.7	2.1	1.7
50,000	1.8	1.4	1.1
100,000	1.6	1.2	1.0

adiabatic cooling rate. If we describe the absorption coefficient for air approximately by the Boltzmann relation $\kappa \sim \exp(-I/kT)$ with some effective value of the "ionization potential"*, then the transparency temperature is found to have only a weak logarithmic dependence on the scale d, and also on the air density, which is contained only in the preexponential factor;

$$l(T_2) = const\, e^{I/kT_2} = d; \qquad T_2 = \frac{I}{k}\left(\ln \frac{d}{const}\right)^{-1}. \tag{9.9}$$

We have already said that $T_2 \sim 10{,}000$°K for $d \sim 10$ m and atmospheric pressure (when $d \sim 100$ m, $T_2 \sim 8000$°K; when $d \sim 1$ m, $T_2 \sim 12{,}000$°K). Thus, the value of flux $S_2 = 2\sigma T_2^4$ changes within rather narrow limits and, if we consider strong cooling waves with a large temperature drop ($T_1 \gg T_2$, $h_1 \gg h_2$), it will be found that the velocity with which the wave travels through the air in the initial state depends basically only on the gas pressure p and is independent of the upper temperature T_1

$$u \approx S_2 \frac{\gamma - 1}{\gamma} \frac{1}{p} \qquad \text{when} \quad h_1 \gg h_2.$$

To illustrate the numerical values, Table 9.6 lists the velocity u for atmospheric

* Actually, κ is the sum of terms of the type $e^{-I/kT}$, where I is the ionization potential for the components corresponding to photoelectric absorption and the excitation energy for the components of molecular absorption. All values of I are of the order of 5–10 ev; if we consider a small temperature interval, then we can always interpolate $\kappa(T)$ by a relation of the type $\exp(-I/kT)$.

pressure and several values of T_1 and T_2. It is evident from the table that the velocity of the cooling wave is of the order of 1 km/sec.

§11. Contraction of the cooling wave toward the center

The character of the cooling and the dependence of the cooling time on the dimensions of the heated volume in our problem differ appreciably from the values that would have resulted if the heat were dissipated by the mechanism of ordinary heat conduction. The temperature of a body cooled by ordinary heat conduction decreases gradually and uniformly and the cooling time for a body with a radius R is proportional to the square of the radius $t \sim R^2 c_p \rho/\kappa$, where κ is the coefficient of thermal conductivity. In the case of radiative cooling, a wave travels through the gas and the cooling time is proportional to the first power of the radius, $t \sim R/u$.

If the dimensions of the heated body are of the order of $R \sim 100$ m and the pressure is of the order of atmospheric, then a cooling wave traveling at $u \sim 1$ km/sec passes from the periphery to the center in a time $t \sim 0.1$ sec. During this time the air cools from high temperatures of the order of tens and hundreds of thousands of degrees to the transparency temperature $T_2 \sim$ 10,000°K. The trace of the path of the cooling wave, together with traces of the paths of the shock front and of the fireball boundary, are shown schematically on the radius-time diagram of Fig. 9.16.

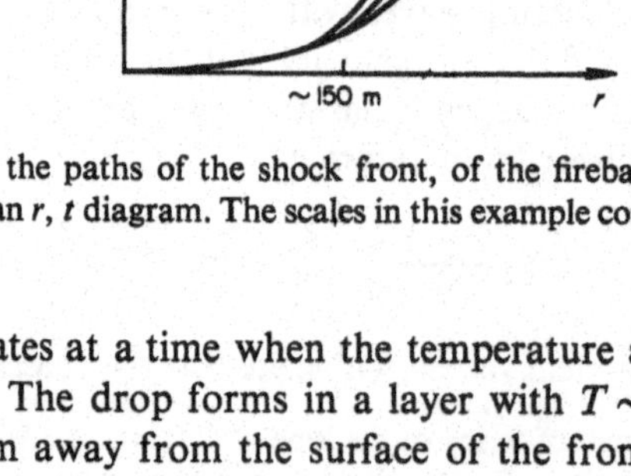

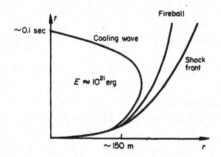

Fig. 9.16. Traces of the paths of the shock front, of the fireball boundary, and of the cooling wave front on an r, t diagram. The scales in this example correspond to an explosion energy $E \approx 10^{21}$ erg.

The wave originates at a time when the temperature at the front is of the order of 2000°K *. The drop forms in a layer with $T \sim 10{,}000$°K, which is approximately 10 m away from the surface of the front. At the beginning,

* Despite the fact that the formation of the cooling wave and the breakaway of the shock front from the boundary of the fireball occur almost simultaneously, no direct physical relationship exists between these two essentially different phenomena.

when the pressure is still high ($p \sim 50$ atm at the time when the drop forms), the wave travels through the gas at a moderate velocity. Thus, despite the fact that with respect to the gas the wave travels toward the center, it actually moves outward in space, being carried away by the rapidly expanding air. Gradually the wave slows down (in space), then changes its direction and "collapses" to the center. The turning point, determining the maximum radius of the cooling wave front, corresponds to zero wave velocity in space, that is, to the point at which the gasdynamic velocity of the expanding air particles and the velocity with which the wave travels with respect to the mass of gas become equal.

After the cooling wave passes through the air heated by the explosion, the air temperature is everywhere below ~10,000°K and the entire volume becomes more or less transparent. The subsequent radiative cooling proceeds much more slowly and has a volume character; each particle emits light corresponding to its emission coefficient, and this light travels almost without absorption far from the point of explosion. Of course, the volume is not completely transparent and some part of the radiation is held back in the outer layers, so that some energy is transferred from the central regions to the periphery. In particular, this process is promoted by the nitrogen dioxide contained in the outer layers at temperatures ~3000–1000°K (which were previously heated by the shock wave to a temperature higher than 2000°K).

A similar energy transport also takes place at the time of passage of the cooling wave, since the radiation flux emerging from the wave surface is partially absorbed in the "transparent" (and actually not completely transparent) peripheral layers. In general, absorption in the ultraviolet region of the spectrum is strong and the ultraviolet photons are absorbed in the neighborhood of the wave front. This, however, does not introduce any substantial changes in the qualitative description of cooling of the air by a wave as presented above. This picture was based on the assumption that a high degree of transparency exists at temperatures lower than T_2, since the strong absorption region with wavelengths $\lambda < 2000$ Å contains less than 4% of the energy of the spectrum corresponding to a temperature of 10,000°K.

It should not be assumed that the cooled air ceases to be luminescent after the cooling wave "collapses" to the center and that the surface of the cooling wave at the stage when it still exists is also the boundary of the fireball. The air that has passed through the cooling wave emits enough radiation to glow brilliantly even when the energy effects of the emission have become small and further cooling ceases. The wave is within the fireball and collapses toward the center, leaving behind it still strongly heated and brightly glowing air. The boundary of the fireball (that is, the incandescent boundary) during the later stage of the explosion is made up of layers with temperatures of the order of 2000–3000°K, which are cooled extremely slowly by radiation. After the pressure becomes equal to atmospheric and the motion practically ceases, these layers are found to be practically stationary. The fireball boundary first moves out from the center together with the expanding air, and then slows down and stops, as shown in Fig. 9.16.

The cooling wave in approaching the center establishes a flow of air from the periphery toward the center, since the wave leaves behind highly cooled particles and cooling at constant pressure is accompanied by compression. For example, if the initial temperature at the center was 100,000°K, and after "collapse" of the wave it dropped to 10,000°K, with the pressure remaining constant at the time of collapse (equal to atmospheric pressure), then the air density at the center is increased by a factor of several tens, and accounts for the flow of air into the center. This inflow, however, does not affect layers that are far from the center and have comparatively low temperatures (of the order of 2000–3000°K), so that the position of the fireball boundary remains unchanged.

At this point we conclude our study of the air cooling process as a whole, of the laws governing the propagation of the cooling wave, and of the luminescence of the fireball, in other words, of the consideration of the "macroscopic" picture*. In the next part we shall study the internal structure of the cooling wave, in a manner similar to that in which general gas flows with shock waves are treated in gasdynamics together with the "microscopic" picture of the internal structure of the shock front. The study of the internal structure of the cooling wave will enable us to find the most important characteristic of the wave—the radiation flux from the wave surface.

§12. The spark discharge in air

Hydrodynamic phenomena which have the character of an explosion also arise in air in connection with spark discharges. These phenomena were investigated by Mandel'shtam and his co-workers [19–24]. We present here the general picture of the process. A thin electrically conducting column is formed in the air discharge gap between the electrodes immediately after the discharge. As a result of Joule heating the air in this column is heated to temperatures of the order of several tens of thousands of degrees and is strongly ionized (at least singly). As a result of the increase in pressure, the column expands and acts on the surrounding air like a piston, sending into it a cylindrical shock wave†.

* We note that a cooling wave also arises in supernovae clouds after emergence of the shock wave at the surface. This was shown by Imshennik and Nadezhin [18], who gave a numerical solution for the problem of the explosion of a massive star as a result of an energy release at the center of the star.

† After the shock wave travels a distance exceeding the column length, it gradually becomes spherical.

In the early stage the air density distributions with respect to radius have a character appropriate to a cylindrical explosion. In [22] the density distributions at successive times were measured by an interferometer. A typical development is shown in Fig. 9.17 (the electrical parameters of the circuit in

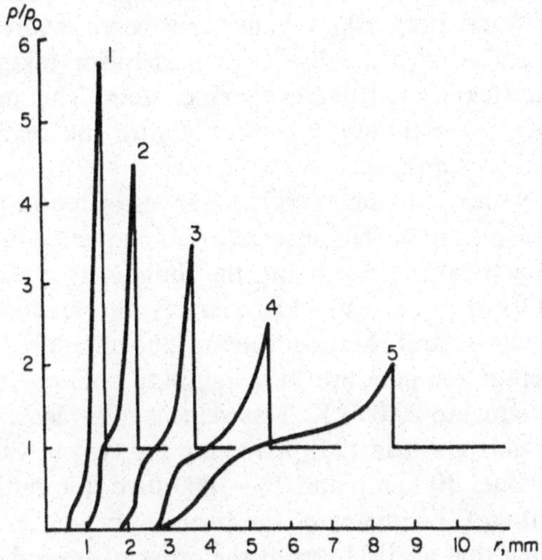

Fig. 9.17. Distribution of air density as a function of spark radius. Curves 1–5 refer to the following times: 1.0, 1.7, 2.9, 5.6, and 9.8 μsec.

this experiment were the following: $C = 0.25$ μf, $L = 2$ μhenries, $V = 10$ kv, and the discharge gap = 5 mm). As can be seen from the figure, at the beginning the shock wave is still rather strong (the velocity is about 2 km/sec) and the density distribution corresponds to that in a strong explosion. At a later stage the wave weakens and the counterpressure of the air ahead of the shock wave begins to have an effect (see §27, Chapter I). In this case the density at the periphery of the blast wave does not differ strongly from standard density, while in the central regions the density is very low. Since the pressure behind the blast wave is equalized spatially, the temperature in the central regions is very high. This central, strongly rarefied and high-temperature region is the electrically conducting column. The average density of air in the column is approximately 10^{-3} of that in the undisturbed air, and the average temperature is approximately 40,000°K. Spectroscopic measurements give consistent results. (Thermodynamic equilibrium in the column is established quite rapidly [23, 24], making it possible to determine the actual gas temperature by spectroscopic means.)

The theory of the explosion wave which results from a spark discharge was developed by Drabkina [20]. It should be noted that the flow differs somewhat from that which takes place with an instantaneous energy release, since in the given case the time for the generation of the Joule heat in the column, which is determined by the half-period of the discharge, is comparable with the time over which the shock wave is observed. This was taken into account in [20]. The rate of energy generation, which enters into the relation governing the motion of the shock wave, was determined experimentally in this case. Braginskii [25] has considered theoretically not only the motion of the air but also the discharge mechanism, taking into account the conductivity and expansion of the spark column. This makes it possible to relate the shock wave parameters directly with the relation governing the increase in discharge current.

Phenomena which are similar to those investigated in the laboratory, but which take place on a much larger scale, are encountered in a storm. Lightning is simply an electrical discharge and the thunder is produced by the shock wave which is formed and which degenerates into an acoustic wave at large distances. Zhivlyuk and Mandel'shtam [26] measured by spectroscopic means the average temperature in a lightning column, and found it to be approximately equal to 20,000°K. This value is in agreement with calculations based on formulas given in [25], assuming for typical values of the current and time the values 30 kamp and 100–1000 μsec (the radius of the lightning column is ~ 10 cm). Estimates of the pressure behind the shock front were found to be such that at distances of the order of several meters the thunder can be quite destructive.

3. *Structure of cooling wave fronts*

§13. Statement of the problem

Up to now, when referring to a cooling wave we have considered it as a discontinuity in which the gas temperature experiences a sharp drop. We have also pointed out the energy balance condition, which is equivalent to the relation describing the conservation of total energy flux for a gas flowing through a shock wave. In constrast to shock waves, however, it was sufficient here to formulate only one energy relation, since the motion across the cooling wave is subsonic and the pressure change across the wave front can be neglected (in this respect the cooling wave is similar to a slow flame front). Such macroscopic considerations do not permit us to draw any conclusions on the most important quantity that determines the wave speed, the radiant flux S_2 which goes out from the wave front to "infinity". The flux S_2 can only be found by considering the internal structure of the transition layer through the wave front, by finding the continuous solution of the equations which describe the radiative transfer in the wave. This was done in the previously cited articles [16, 17].

Leaving aside the specific dimensions and shape of the mass of gas being cooled, we shall seek a solution of the unsteady equations of radiative heat transfer in the form $T(x - ut)$, corresponding to a plane wave which travels with a constant velocity u through a gas with given values of the temperature and density T_1, ρ_1. The velocity u should be found from the equations in a manner similar to the way in which the flame speed in a combustible mixture is determined. Actually the equations do not admit of an exact solution of the form $T(x - ut)$. As the wave travels, the thickness of the cooled gas layer, in which the light absorption is small but different from zero, increases; the transparency temperature, defined by the relation $l(T_2) = d$, where d can denote the thickness of the cooled layer, decreases with time.

In an infinite medium, in which the mean free path has an inverse dependence on temperature, the transparency temperature in general becomes equal to zero, since a gas layer of infinite extent cooled to an arbitrarily low temperature is completely opaque even if the radiation mean free path is extremely large. The radiation flux from the wave front is equal to zero in this case and the cooling wave, in the strict sense, does not exist. To some extent, a similar situation occurs in the theory of steady-state flame propagation. If it is not assumed that the rate of chemical reaction in the unburned mixture is identically equal to zero, despite the fact that the actual rate is finite (although vanishingly small), we could obtain the result that the mixture burns prior to being reached by the flame front. This situation, which is fundamental to the case of an infinite medium, does not lead to any difficulties under actual conditions. The fact is that the region which is heated and then cooled by the cooling wave is always bounded, and the transparency temperature has only a logarithmic dependence on the dimensions of the cooled region, and so changes weakly with the distance traveled by the wave, being, for actual cases of interest, bounded within quite narrow limits. The additional very slow dependence of the solution on time $T(x - ut, t)$ arises only at the lowest, most extended edge of the wave, in a region where the gas is already cooled and almost transparent. The presence of the adiabatic cooling occurring when the wave is propagating through an expanding gas makes this additional dependence even less important, since the air which passed through the wave is cooled by the expansion to lower temperatures and rapidly "skips" the temperature region in which it is still not completely transparent. The additional slow variation with time $T(x - ut, t)$ will exist only in the region of pure adiabatic cooling and will have almost no effect on the temperature distribution in the wave.

In order to find the temperature distribution in the cooling wave front, which in turn also determines the flux S_2, one should consider a planar steady process in a coordinate system moving with the front, as is usually done in the theory of wave structure. This was illustrated in Chapter VII using shock waves as an example. To eliminate the difficulty discussed above and to render the problem steady, that is, to go from the actual solution $T(x - ut, t)$ to an idealized case $T(x - ut)$ (in the laboratory coordinate system), we can use one of two formally artificial methods, which, however, are by virtue of our preceding discussion physically justifiable and correspond to the actual situation. In the first method we can introduce into the energy equation an additional constant term A representing the adiabatic cooling. The quantity A specifies a constant scale d, which determines the transparency temperature T_2 and limits the absorption in the region cooled by radiation, making the optical thickness of this region finite. In the second method, we can disregard the adiabatic cooling, but instead introduce right at the beginning the transparency temperature T_2 obtained from an estimate such as (9.9), and assume formally that for $T < T_2$ the medium is completely transparent (the mean free path $l = \infty$). Then the gas will be cooled only to T_2, after which the emission, which is proportional to $\kappa = 1/l$, and any further cooling will cease.

Since the gas motion across the cooling wave is subsonic (this is shown by estimates), we may neglect the kinetic energy of the gas flow in comparison with the thermal energy. In this case, the energy equation at a point x within the wave is written for the general case by introducing an additional term to describe the adiabatic cooling

$$u\rho_1 c_p \frac{dT}{dx} + \frac{dS}{dx} = -A, \qquad A > 0. \tag{9.10}$$

If the specific heats are not taken constant, it is more convenient to write this equation in terms of the specific enthalpy of the gas

$$u\rho_1 \frac{dh}{dx} + \frac{dS}{dx} = -A. \tag{9.11}$$

Here S is the radiant energy flux at the point x in a wave (from conservation of mass flux $\rho u(x) = \rho_1 u$, where u is the wave speed, equal to the rate at which gas flows into the wave, ρ_1 is the initial density, and ρ and $u(x)$ are quantities at the point x).

The directions of flow, of the x axis, and of the velocity are given in Fig. 9.15, which shows schematically the temperature drop in the wave front. The gas flows into the wave from left to right, while the wave travels through the undisturbed gas from right to left. Ahead of the wave, at $x = -\infty$, the temperature has a known value $T = T_1$ and the flux $S = 0$, in accordance with the remark made in §10 to the effect that the radiative heat transfer in a high-temperature gas is unimportant and that the flux in this region is small. The radiation flux S varies from zero to S_2 as x goes from $-\infty$ to $+\infty$, where S_2 is the flux going out of the wave front to infinity.

If the adiabatic cooling is not taken into account, but it is assumed that the transparency temperature T_2 is given (the second method), then the energy relation (9.11) yields the integral

$$S = u\rho_1(h_1 - h). \tag{9.12}$$

Here the constant of integration is expressed in terms of the enthalpy of the original gas $h_1 = h(T_1)$, in accordance with the boundary condition $x = -\infty$, $T = T_1$, and $S = 0$. If we apply the energy integral (9.12) to the rear of the wave, where $T = T_2$ and the flux is equal to that going out to infinity $S = S_2$, we obtain (as expected) the energy balance equation (9.6), relating the variables on both sides of the wave front if the front is regarded as a discontinuity.

To the energy equation must be added the radiative transfer equation, which determines the flux S. The radiative transfer will be described in the diffusion approximation, as was done when considering the structure of a shock front with radiation taken into account (see Chapter VII, Part 3). In addition, we shall introduce, as before, some frequency-averaged photon mean free path l. The equations for the diffusion approximation are then written as

$$\frac{dS}{dx} = c\,\frac{U_p - U}{l}, \tag{9.13}$$

$$S = -\frac{lc}{3}\frac{dU}{dx}, \tag{9.14}$$

where U is the true radiation density and U_p is its equilibrium value corresponding to the temperature of the medium at the point x: $U_p = 4\sigma T^4/c$.

As will be shown below, the radiation density in a considerable part of the wave will be close to equilibrium. It is well known (see §12, Chapter II), that under these conditions the spectral mean free path l_ν is to be averaged by the Rosseland method. However, local equilibrium no longer exists in the region of highly cooled air, and the Rosseland method of averaging no longer applies. The averaging method, however, cannot introduce any qualitative changes in the results, since the exponential Boltzmann factor of the form $e^{I/kT}$, which effectively describes the basic temperature dependence of the mean free path, is retained regardless of the averaging method; the preexponential factor, which, of course, changes with the method of averaging, has only a very weak logarithmic influence (just as with the transparency temperature T_2) on the effects in the wave. Therefore, for simplicity, we shall always imply by $l(T)$ the Rosseland mean free path.

It is convenient to rewrite (9.13) and (9.14) in terms of the optical co-ordinate τ, which we shall measure from $x = +\infty$, where the gas is transparent and $l = \infty$ (the τ axis is directed opposite to the x axis)

$$d\tau = -\frac{dx}{l}, \qquad \tau = -\int_{+\infty}^{x} \frac{dx}{l} = \int_{x}^{+\infty} \frac{dx}{l}.$$

In this case (9.13) and (9.14) take the form

$$\frac{dS}{d\tau} = -c(U_p - U), \tag{9.15}$$

$$S = \frac{c}{3}\frac{dU}{d\tau}. \tag{9.16}$$

To the radiative transfer equations must be added the boundary conditions. At the front edge of the wave where $\tau = \infty$, as already stated, we have

$$\tau = \infty, \qquad S = 0, \qquad T = T_1. \tag{9.17}$$

At the rear edge, which is the boundary between the absorbing and perfectly transparent media (vacuum), the radiation flux and density should satisfy the known diffusion condition at a vacuum interface (see (2.66))

$$\tau = 0, \qquad S_2 = \frac{cU_2}{2}. \tag{9.18}$$

The radiative transfer equation, together with the energy equation and the boundary conditions completely defines the structure of the wave front, the flux S_2, and the speed u.

§14. Radiation flux from the surface of the wave front

Of practical interest are mainly the strong cooling waves in which the gas, being cooled from the initial temperature T_1 to the transparency temperature T_2, emits a considerable part of its energy: $T_1 \gg T_2$. Weak waves, where the difference between T_1 and T_2 is small*, are of interest mainly from the point of view of methodology, since in this case it is possible to obtain an exact analytic solution of the equations. It is clear that in a weak wave the radiation flux from the front S_2 lies in the range $\sigma T_1^4 > S_2 > \sigma T_2^4$; it is defined quite exactly, since T_1 is close to T_2, with the result that the major problem, that of the flux, does not appear: $S_2 \approx \sigma T_2^4 \approx \sigma T_1^4$. An analytic solution for a weak wave can be found in [16]. We do not consider it here but proceed directly with the consideration of a strong cooling wave.

* Despite the fact that T_1 is close to T_2 we assume, however, that the temperature dependence of the mean free path $l(T)$ is sufficiently sharp that $l(T) \ll l(T_2)$. This inequality is a condition for the existence of the cooling wave.

It was pointed out in the preceding section that the steady solution can be found by one of two methods, either by introducing into the energy equation a constant term representing the adiabatic cooling, or by determining at the very beginning the transparency temperature T_2 and assuming that for $T \leqslant T_2$ the gas is perfectly transparent ($l = \infty$), thus excluding from consideration the region already cooled by radiation and which only very weakly absorbs light. The first approach gives a more complete picture of the temperature distribution, since it permits us to investigate the temperature changes in the cooled air and to take into account its weak absorption. It leads, however, to excessive mathematical complications which arise from considering the temperature distribution within the wave itself (for temperatures above the transparency temperature) and in determining the flux which goes out from the wave front to infinity. In addition, the adiabatic cooling within the wave is small in comparison with the radiative cooling, and therefore the second method is preferable for investigating the internal structure of the wave. In §16 we shall point out certain properties of this regime that are related to the adiabatic cooling.

In the absence of adiabatic cooling the integral of the energy equation is given by (9.12); let us rewrite this equation assuming for simplicity that the specific heats are constant

$$S = u\rho_1 c_p(T_1 - T). \tag{9.19}$$

The problem consists in solving the system of equations (9.15), (9.16), (9.19) together with the boundary conditions (9.17) and (9.18). Before investigating this sytem, we shall attempt to estimate the radiation flux S_2 going out from the front on the basis of the most general physical considerations. These considerations will also tell us what approximations can be made in solving the system of equations and in finding the temperature distribution in the wave.

From the statement of the problem we know that the temperature at any point of the wave cannot be lower than the transparency temperature T_2, since on being cooled to T_2 the gas ceases to absorb and emit radiation, and further cooling stops. Consequently, T_2 is the lowest temperature in the wave, and near the rear edge of the wave the temperature increases as we move away from the interface with the "vacuum", the perfectly transparent region where $T = T_2$ and $l = \infty$, Thus, at the rear edge, where $\tau = 0$, $dT/d\tau \geqslant 0$. It follows from the energy equation (9.19) that as we move away from the rear edge into the wave the flux decreases, and for $\tau = 0$, $dS/d\tau \leqslant 0$. The continuity equation for the radiation (9.15) shows that the radiation density at the rear edge of the wave is not higher than its equilibrium value $U_2 \leqslant U_{p2} = 4\sigma T_2^4/c$ (the divergence of the flux dS/dx is not negative; the fluid is not heated by radiation). In the diffusion approximation the flux at the fluid-vacuum interface is related to the radiation density by the condition (9.18): $S_2 = cU_2/2$.

Noting that $U_2 \leqslant U_{p2}$, we find that the upper limit of S_2 is given by $2\sigma T_2^4$,

$$S_2 = \frac{cU_2}{2} \leqslant \frac{cU_{p2}}{2} = 2\sigma T_2^4.$$

On the other hand, the radiation brightness temperature T_{br}, defined by $S_2 = \sigma T_{\text{br}}^4$, coincides with some average temperature of the radiating layer and thus cannot be less than T_2, since the temperature of the fluid in the radiating layer, as at any other point in the wave, is always higher than T_2. It follows therefore that $S_2 > \sigma T_2^4$, and that the flux S_2 which goes out from the wave front to infinity is found to be bounded within rather narrow limits

$$\sigma T_2^4 < S_2 < 2\sigma T_2^4, \tag{9.20}$$

$$T_2 < T_{\text{br}} < \sqrt[4]{2}T_2. \tag{9.21}$$

Thus, regardless of the wave strength, and for arbitrarily high initial temperatures T_1, the radiation always emerges from the rearmost edge of the wave and the radiation flux from the wave front surface corresponds to a temperature close to T_2. Under no circumstances should one think that the use of the diffusion approximation to describe the radiative transfer, which led to the boundary condition (9.18), was of importance in our estimates. Indeed, the diffusion condition (9.18) corresponds to the assumption that the photons emerging from the medium into vacuum have an isotropic angular distribution, and that no photons arrive in the medium from the vacuum (no light sources exist in the vacuum). Even if we had made the different extreme assumption that the radiation at the vacuum interface is highly anisotropic and all the photons leave the medium normal to its surface, the diffusion condition (9.18) would have been replaced by the condition $S_2 = cU_2$, which would have led to the inequalities $\sigma T_2^4 < S_2 < 4\sigma T_2^4$, and $T_2 < T_{\text{br}} < \sqrt[4]{4}T_2$. These differ from (9.20) and (9.21) by only a small numerical factor. Actually the limitation on the flux $S_2 < 2\sigma T_2^4$ is related to the fact that the solution is steady state, as a result of which the temperature distribution, which completely defines the flux, cannot be arbitrary and is determined completely by the steady state equations.

An important corollary follows from the inequality (9.21) which permits us to solve the whole problem of front structure by very simple means despite the fact that it is described by nonlinear equations. The radiation from a heated body which is bounded by a transparent medium (or vacuum) is essentially generated in a layer near the surface of the body which has an optical thickness of the order of unity (photons born in deeper layers cannot leave the surface, are almost completely absorbed in transit). The radiation brightness temperature coincides with some average temperature of this radiating layer. But, by virtue of the inequality (9.21), the brightness temperature is very close to the temperature of the lower edge of the wave T_2. This means that the temperature of the fluid behind the point $\tau = 0$, where $T = T_2$, changes very little over a distance of the order of several optical thicknesses into the wave. This allows us to draw the following conclusions: In a strong cooling wave, in which $T_1 \gg T_2$, the radiant flux in the radiating layer, at the rear edge of the wave, changes very little and is almost constant. Indeed, when the temperature changes by $\Delta T \lesssim T_2$ the flux changes by

$$|\Delta S| \sim u\rho_1 c_p \, \Delta T \lesssim u\rho_1 c_p T_2 ,$$

and for $T_1 \gg T_2$ the flux at the point $\tau = 0$, $T = T_2$ is approximately equal to $S_2 \approx u\rho_1 c_p T_1$ (see (9.19)). As a result

$$\frac{|\Delta S|}{S_2} \approx \frac{T_2}{T_1} \ll 1.$$

Since the flux near the rear edge of a strong wave is almost constant, the situation is completely analogous to that prevailing in the photospheres of stationary stars, where the radiant flux is strictly constant. Thus, the problem of relating the flux S_2 with the transparency temperature T_2 (the temperature at the vacuum interface) in the limit of a strong wave is equivalent to the well-known Milne problem (see §15, Chapter II). Taking into account the angular distribution of radiation rigorously, it has the exact solution

$$S_2 = \frac{4}{\sqrt{3}} \sigma T_2^4 , \tag{9.22}$$

which differs only slightly from the solution in the diffusion approximation

$$S_2 = 2\sigma T_2^4 . \tag{9.23}$$

It is now evident that the value of the flux S_2, within the framework of the diffusion approximation, coincides with the upper limit of the inequality (9.20) in the limit of a strong wave.

§15. Temperature distribution in the front of a strong wave

The fact that the temperature changes little over the radiating layer with a thickness of the order of several optical units shows that local equilibrium exists between the radiation and the fluid. The relative departure of the radiation density at the rear edge of the wave from its equilibrium value is smaller, the stronger is the wave, the greater is the ratio T_1/T_2. Indeed, it follows from (9.15) that

$$\left(\frac{U_p - U}{U_p}\right)_{\tau=0} = \frac{U_{p2} - U_2}{U_{p2}} = -\frac{1}{cU_{p2}}\left(\frac{dS}{d\tau}\right)_2 .$$

But, by virtue of what was said previously

$$\left|\frac{dS}{d\tau}\right|_2 \sim \frac{|\Delta S|}{\Delta\tau} \lesssim S_2\frac{T_2}{T_1} = 2\sigma T_2^4\frac{T_2}{T_1},$$

since $|\Delta S| \sim S_2T_2/T_1$ is the change in flux over an optical distance $\Delta\tau \sim 1$. Consequently, the relative departure of the radiation density from its equilibrium value in a strong wave is

$$\frac{U_{p2} - U_2}{U_{p2}} \sim \frac{S_2}{cU_{p2}}\frac{T_2}{T_1} \sim \frac{T_2}{T_1} \ll 1.$$

We can show that as we move away from the rear edge into the wave the relative departure of the radiation from equilibrium can only decrease, so that if the wave is sufficiently strong and the departure at the rear edge is small, then the condition of local equilibrium is satisfied throughout the wave. Thus, the equations describing the structure of the front of very strong cooling waves can be solved in the radiation heat conduction approximation by setting

$$S \approx \frac{c}{3}\frac{dU_p}{d\tau} = \frac{16\sigma T^3}{3}\frac{dT}{d\tau}.$$

Combining this equation with (9.19), we obtain an equation for the function $T(\tau)$, which can be integrated by quadratures.

At the rear edge of the wave we obtain an approximate form of the solution which naturally agrees with the diffusional solution of the Milne problem, since the flux $S \approx const$ (see §15, Chapter II)

$$T = T_2(1 + \tfrac{3}{2}\tau)^{1/4}. \tag{9.24}$$

The asymptotic temperature distribution at the front edge of a strong wave has the form

$$T = T_1(1 - e^{-\tau/\tau_{ed}}), \qquad \tau \gg \tau_{ed}, \tag{9.25}$$

where the quantity τ_{ed}, which may be regarded as the effective optical thickness of the wave, depends only on the wave strength

$$\tau_{ed} \approx \frac{8}{3}\left(\frac{T_1}{T_2}\right)^4.$$

The optical thickness increases rapidly with the ratio T_1/T_2. We do not give here the general expression for the distribution $T(\tau)$, which at the rear and front edges takes on the simplified forms given by (9.24) and (9.25) (see [17]), and in Fig. 9.18 only present a curve of the temperature distribution for the case $T_1/T_2 = 5$ and $\tau_{ed} = 1670$.

Knowing the distribution $T(\tau)$ and the mean free path as a function of temperature, it is easy to find the temperature distribution with respect to the geometric coordinate by means of the relation $-x = \int_0^\tau l(T)\,d\tau$*. On Fig. 9.19 is shown the temperature distribution $T(x)$ at the lower edge of the

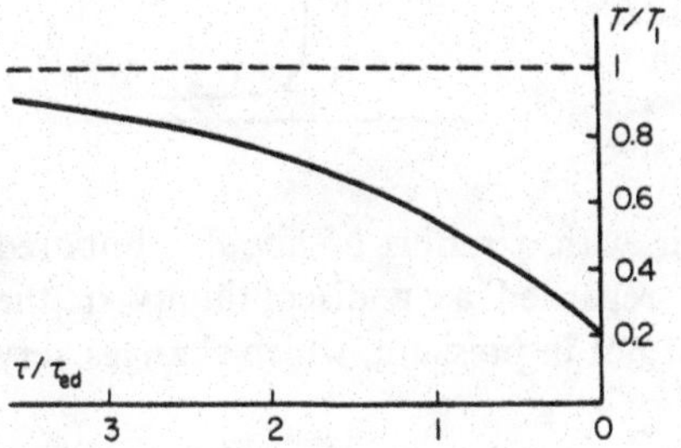

Fig. 9.18. Temperature distribution in a cooling wave as a function of the optical coordinate with $T_2/T_1 = 5$, $\tau_{ed} = 1670$.

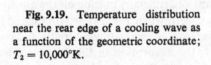

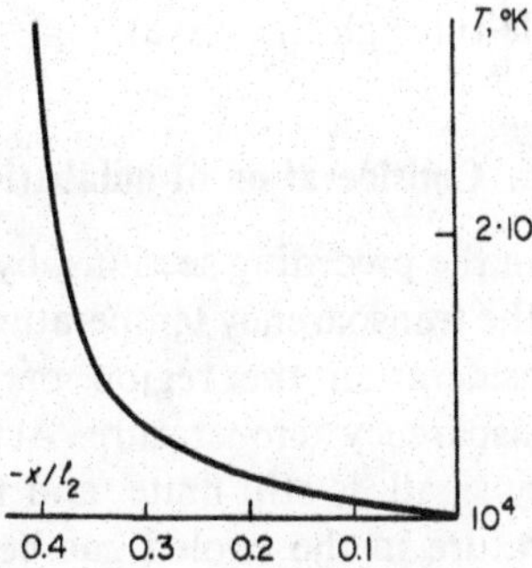

Fig. 9.19. Temperature distribution near the rear edge of a cooling wave as a function of the geometric coordinate; $T_2 = 10{,}000$°K.

wave for the case of a Boltzmann dependence $l(T) = const \exp(I/kT)$. The length scale is taken to be $l_2 = l(T_2)$. The transparency temperature is taken as $T_2 = 10{,}000$°K. Figure 9.19 shows that the wave has the form of a step. In fact, the Boltzmann dependence of $l(T)$, which gives rise to the sharp temperature step in the wave, is valid only to temperatures of about 30,000–40,000°K, with no multiple ionization taking place. At higher temperatures the mean free path passes through a minimum and then begins to increase with temperature. Therefore the front edge of a sufficiently strong wave with $T_1 \sim 50{,}000$–$100{,}000$°K is highly extended (for $l = const$ the distribution of $T(x)$ at the upper edge would be identical with the distribution of $T(\tau)$) in accordance with (9.25). An approximate temperature distribution in a wave with $T_1 = 40{,}000$°K is shown in Fig. 9.20.

If we neglect the extension of the front edge of the wave, which only slightly affects the cooling conditions of the air (since the flux and its divergence, which determines the cooling at the upper edge, are very small), then the

* Actually, since $l(T_2) \neq \infty$, the temperature at the lower edge does not approach T_2 asymptotically, but with a slope different from zero. Therefore the coordinate origin $x = 0$ can be placed at a point where $\tau = 0$ and $T = T_2$.

geometric thickness of the step comprises, as shown in Fig. 9.19, several tenths of the mean free path $l(T_2)$. For $T_2 \sim 10{,}000°K$ and $l_2 \sim 10$ m, the thickness of the wave is of the order of several meters, that is, a cooling wave

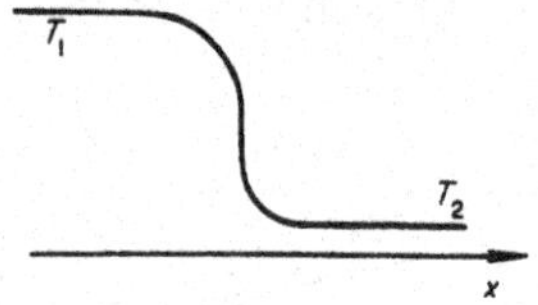

Fig. 9.20. Temperature distribution in a cooling wave.

traveling through a large volume of air with a radius of about a hundred meters is actually narrow and can be regarded as a discontinuity in the temperature and density of the fluid (but not in pressure, which changes very little through the wave).

§16. Consideration of adiabatic cooling

In the preceding sections, by means of artificially cutting off the absorption at the transparency temperature T_2 ($l = \infty$ when $T < T_2$), we eliminated from consideration the region of cooled air whose temperature is below the transparency temperature. Actually, the absorption in this region, although very small, is still finite, and therefore it is natural to inquire how the temperature in the cooled gas behaves under the influence of the radiant flux going out from the wave front. The process in this region is basically unsteady and depends on the particular conditions, on the dimensions, the hydrodynamics of the flow, and the mechanics of light absorption. In this section we shall consider the practically important case in which the cooling wave travels through expanding rather than stationary air, in which the air that is cooled by radiation continues to be cooled adiabatically. The adiabatic cooling rapidly brings the air into the temperature region where it is completely transparent, where it no longer exerts any effect on the behavior of the cooling wave. During the relatively short time interval in which the air undergoing the adiabatic cooling still absorbs an appreciable amount of light, the adiabatic cooling rate changes very little. Hence a process with adiabatic cooling can be approximately regarded as steady and can be described by the energy equation (9.10) with the constant term A. The integral of this equation is

$$u\rho_1 c_p T + S = -Ax + const. \tag{9.26}$$

The constant of integration is here arbitrary, since it is simply determined by the choice of the origin of the x coordinate; it can be set equal to zero.

At the front edge of the wave where $x \to -\infty$, the flux $S \to 0$. It may seem that this artificially imposed condition contradicts the existence of the temperature gradient related to the presence of adiabatic cooling. However, it is assumed that the mean free path $l(T)$ drops so rapidly with increasing temperature that the product $S \sim -l(T)T^3\,dT/dx$ tends to zero as $T \to \infty$; this is physically justified, since the flux of radiation heat conduction from the outside (from the region of hot air) is very small. At the rear edge of the wave, where $x \to +\infty$, the flux tends to the constant value S_0, i.e., the value of the flux going out to infinity*. Therefore, the temperature in the wave, as $x \to \pm\infty$, asymptotically approaches two straight lines

$$u\rho_1 c_p T = -Ax \qquad \text{for} \quad x \to -\infty,$$

$$u\rho_1 c_p T = -Ax - S_0 \qquad \text{for} \quad x \to +\infty.$$

These straight lines are displaced along the ordinate by an amount S_0 (in Fig. 9.21 they are displaced by $S_0/u\rho_1 c_p$). The problem consists in determining the value of S_0. We do not present here the mathematical solution (see [17]), but limit ourselves to a qualitative discussion of the process.

Let us follow the successive changes of state of a gas particle entering the cooling zone, coming from $-\infty$ in the positive x direction (Fig. 9.21). At the

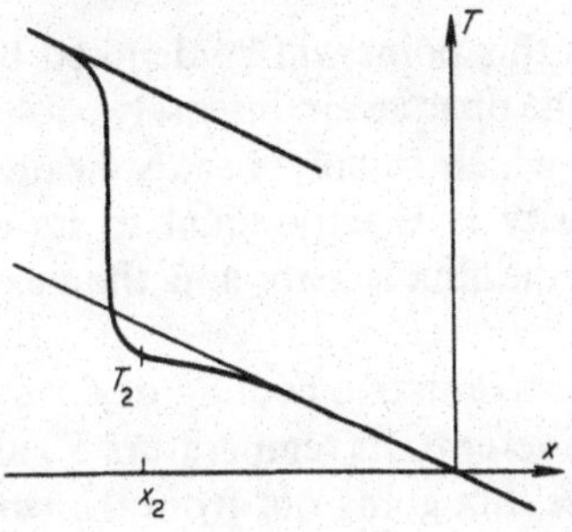

Fig. 9.21. Temperature distribution in a cooling wave taking into account adiabatic cooling.

beginning, when the temperatures are very high, radiation heat conduction is negligible and the particle is cooled entirely adiabatically with its temperature falling off along the upper straight line. Then the particle begins to be increasingly cooled by radiation, and its temperature falls below the upper straight line. The radiation density in the particle in this case is less than its equilibrium value (the particle emits more light than it absorbs), and the radiant flux increases.

The rate of radiative cooling at this stage is considerably higher than the rate of adiabatic cooling and the temperature drops very sharply (the particle passes through the cooling wave). This continues until the time when the

* This flux, as we shall see below, differs slightly from the flux S_2 emerging from an effectively specified surface of the wave front.

particle is cooled to such a low temperature that the rate of radiant heat exchange becomes less than the adiabatic cooling rate. As a result of the extremely sharp decrease in the absorption (and emission) with decreasing temperature, even the small adiabatic cooling that follows makes the particle completely transparent and the radiant heat exchange ceases entirely.

In this case, the radiation density, which is determined by the flux born in hotter layers and passing through the particles, remains almost unchanged. However, the equilibrium radiation density, which is proportional to T^4, rapidly decreases. Therefore the radiation density in the "transparent" region, in contrast to the "opaque" region, is higher than equilibrium, the absorption exceeds emission, and the particle is heated by radiation; the radiant flux decreases as shown in Fig. 9.22*. Consequently, a point $x = x_2$

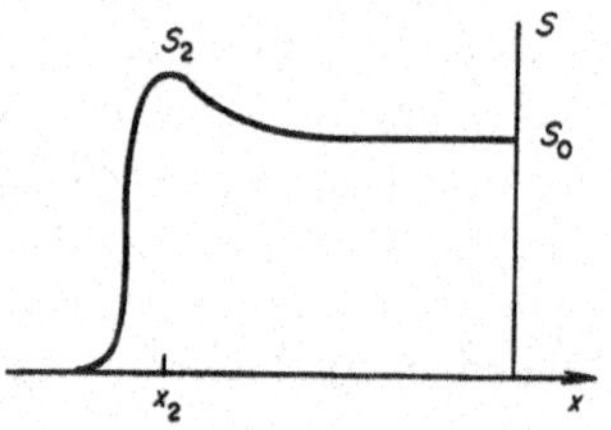

Fig. 9.22. Distribution of radiant flux in a cooling wave taking into account adiabatic cooling.

(the temperature and flux corresponding to this point will be denoted by T_2 and S_2) exists on the x axis which separates the opaque air, intensely cooled by radiation, and the almost transparent air, which is only weakly heated by radiation. At this point the radiation density is exactly equal to its equilibrium value $U_2 = U_{p2}$, the divergence of the flux is zero, and the flux is a maximum $S_{max} = S_2$.

It is natural to take this point at which the radiative cooling of air ceases as the rear boundary of the cooling wave, to refer to its temperature T_2 as the transparency temperature, and to denote the flux going out from the surface of the wave front as S_2. This flux is only slightly absorbed in the almost transparent region, so that the flux S_0 going out to infinity is only slightly smaller than S_2. The temperature and flux distributions $T(x)$ and $S(x)$ corresponding to the situation described are shown in Figs. 9.21 and 9.22. At low temperatures the curve $T(x)$ goes below the lower asymptote, approaching it from below, since the gas is heated by radiation: the maximum flux lies at the point where the temperature has its maximum downward displacement from the asymptote (this follows from (9.26)).

We can show that the flux S_2 is related to the transparency temperature by

* This situation recalls to a certain extent the situation in a shock wave front with radiation: the radiation density behind the shock discontinuity is below equilibrium, the gas is cooled by radiation and emits a flux in the region ahead of the discontinuity, where it is absorbed; the radiation density is higher than at equilibrium and the gas heats up.

the same relation as in a wave without adiabatic cooling, with $S_2 = 2\sigma T_2^4$. The transparency temperature can be estimated from the condition that the rate of radiative cooling at a temperature close to T_2 becomes equal to the adiabatic cooling rate A, which then approximately determines the rear edge of the wave. The temperature T_2 has only a logarithmic dependence on the arbitrarily specified quantity A, because of the exponential dependence of $l(T)$, similar to its earlier logarithmic dependence on the arbitrarily specified characteristic length scale d (based on the condition $l(T_2) = d$). In our case the characteristic scale is the distance over which the temperature drops, as a result of adiabatic cooling, from T_2 to zero. This distance, incidentally, also determines the position of the rear edge of the wave, the coordinate x_2. Actually, the transparency temperature is still determined from the condition $l(T_2) \approx d$, except that now the prescribed quantity is no longer d itself, but the quantity A which defines the scale d.

X. Thermal waves

§1. The thermal conductivity of a fluid

If a fluid body is heated nonuniformly or if energy is released within the body, a thermal flux transported by heat conduction appears. Heat conduction promotes energy diffusion and temperature equalization. In general, with temperature gradients also arise pressure gradients, which set the fluid into motion. In many cases hydrodynamic energy transport dominates over that associated with heat conduction. However, often the motion and hydrodynamic energy transfer are unimportant and heat from any source present is transported by means of thermal conduction alone. For temperatures which are not too high it is ordinary heat conduction which serves as the mechanism of heat transfer.

Ordinary heat conduction transports thermal disturbances comparatively slowly through a medium (we shall prove this subsequently using a gas as our example). Small pressure disturbances propagate with the speed of sound, leaving certain redistributions of density, and the pressure equalizes more rapidly than the temperature. If the temperature changes in the medium are not large, the fluid moves at a speed much less than the speed of sound, and the motion of the fluid can often be neglected when considering heat propagation by thermal conduction, treating the process as one at constant pressure.

The energy balance equation takes the form

$$\rho c_p \frac{\partial T}{\partial t} = -\nabla \cdot \mathbf{S} + W, \tag{10.1}$$

where ρ is the density, which can approximately be taken as constant, c_p is the specific heat at constant pressure, $\mathbf{S}$ is the heat flux vector, and W is the energy release per unit volume per unit time from external sources. The conductive heat flux is in first approximation proportional to the temperature gradient,

$$\mathbf{S} = -\kappa \nabla T, \tag{10.2}$$

where κ is the coefficient of thermal conductivity, which depends on the properties and state of the fluid. Substituting (10.2) into (10.1), we obtain the general heat conduction equation, which describes the temperature of the medium as a function of the coordinates and time

$$\rho c_p \frac{\partial T}{\partial t} = \nabla \cdot (\kappa \nabla T) + W. \tag{10.3}$$

The coefficient of thermal conductivity and specific heat change very little over a moderate range of temperatures and may be considered as practically constant. The heat conduction equation (10.3) then becomes linear (with the exception of those cases where the energy release W is a nonlinear function of the temperature). With $\kappa = const$ we have

$$\rho c_p \frac{\partial T}{\partial t} = \kappa \nabla^2 T + W. \tag{10.4}$$

Dividing the heat conduction equation (10.4) by ρc_p it assumes a form in which the fluid properties are characterized by only one parameter, the coefficient of thermal diffusivity $\chi = \kappa/\rho c_p$:

$$\frac{\partial T}{\partial t} = \chi \nabla^2 T + q, \qquad q = \frac{W}{\rho c_p}. \tag{10.5}$$

The coefficient of thermal diffusivity in gases is approximately equal to the molecular diffusion coefficient,

$$\chi = \frac{l_a \bar{v}}{3},$$

where l_a is the molecular mean free path and $\bar{v}$ is the mean thermal speed of the molecules; for example, in air at standard conditions $\chi = 0.205$ cm^2/sec. The heat conduction mechanism in liquids and solids is more complex and will not be discussed here. We only note that in water at room temperature $\chi = 1.5 \cdot 10^{-3}$ cm^2/sec.

To the heat conduction equation must be added initial and boundary conditions. The initial temperature distribution in the medium is given as

$$T(x, y, z, 0) = T_0(x, y, z). \tag{10.6}$$

The heat flux at the interface between two media 1 and 2 with different properties is continuous,

$$\mathbf{n} \cdot (\kappa \nabla T)_1 = \mathbf{n} \cdot (\kappa \nabla T)_2, \tag{10.7}$$

where $\mathbf{n}$ is the unit vector normal to the interface. The temperature itself is continuous. The temperature or heat flux is given on the boundaries of a body as a function of time or, more generally, a relation between the two.

The mathematical theory of linear heat conduction, which is concerned with the solution of (10.5) in various specific applications is well developed and is extensively used in various fields of physics and engineering.

§2. Nonlinear (radiation) heat conduction

A new heat transfer mechanism comes into play at temperatures of the order of tens and hundreds of thousands of degrees, that of radiation heat conduction. The radiation heat conduction process was discussed in detail in Chapter II, and also in Chapters VII and IX, where we considered the front structure of a very strong shock and the radiative cooling of air. The essential difference between radiation and ordinary heat conduction processes lies in the fact that the coefficient of radiation thermal conductivity is highly temperature dependent, as a result of which the heat conduction equation becomes nonlinear.

The heat flux transported by radiation heat conduction is (see (2.76))

$$\mathbf{S} = -\frac{lc}{3}\nabla U_p = -\frac{lc}{3}\nabla\frac{4\sigma T^4}{c}, \tag{10.8}$$

where $U_p = 4\sigma T^4/c$ is the equilibrium radiation energy density and l is the Rosseland radiation mean free path*. The energy flux (10.8) can be written in terms of the gradient of temperature in the form (10.2), if the coefficient of radiation thermal conductivity is defined as

$$\kappa = \frac{lc}{3}\frac{dU_p}{dT} = \frac{16\sigma T^3 l}{3}. \tag{10.9}$$

The coefficient of radiation thermal conductivity is temperature dependent because the radiation specific heat $c_{\text{rad}} = dU_p/dT \sim T^3$, and because the radiation mean free path l depends on temperature.

The radiation heat conduction mechanism can transfer energy at a speed much faster than the speed of sound in the medium. This is so because the speed of light at nonrelativistic temperatures is very much greater than the speed of sound. If energy is released in a body of fluid and it is heated to a sufficiently high temperature, then this energy will initially be rapidly dissipated by radiation heat conduction. As long as the thermal propagation speed is much higher than the speed of sound, the fluid does not have sufficient time to be set into motion, the pressure does not equalize, and the heat flows through a stationary medium. Later on we shall present an estimate of the conditions under which motion arises. Here, we shall consider the propagation of heat by radiation heat conduction in a stationary medium only, one whose density does not change with time.

* We recall that for radiative transfer to have the character of heat conduction, the radiation energy density at each point of the medium must be close to equilibrium. For this it is necessary that the dimensions of the heated region be appreciably greater than the radiation mean free path. *Editors' note.* This condition is also a sufficient one only within the diffusion approximation. A sufficient condition in general requires that the gradients must be small (see §12, Chapter II).

The energy balance is again given by (10.1) or (10.3) (but not by (10.4) since $\kappa \neq const$), with the only difference that the specific heat at constant pressure c_p is replaced in all the equations by the specific heat at constant volume c_v. It is also assumed that the radiation energy density U_p can be neglected in comparison with the energy density of the medium $\rho\varepsilon(T)$.

If c_v is regarded approximately as being temperature independent, and the heat conduction equation is divided by ρc_v, we get the equation

$$\frac{\partial T}{\partial t} = \nabla \cdot (\chi \nabla T) + q, \tag{10.10}$$

which corresponds to (10.5). The coefficient of radiation thermal diffusivity χ is

$$\chi = \frac{\kappa}{\rho c_v} = \frac{lc}{3}\frac{c_{\text{rad}}}{\rho c_v}. \tag{10.11}$$

There is a close parallel between this quantity and the ordinary coefficient of thermal diffusivity for a gas $\chi = l_a \bar{v}/3$. The latter coincides with the diffusion coefficient of the molecules which transfer the heat. In the case of radiation heat conduction the fluid is heated and cooled and energy is transferred by the radiation, which acts as an "intermediary". For this reason the coefficient of radiation thermal diffusivity is not simply equal to the radiation diffusion coefficient $lc/3$, but is also proportional to the ratio of the specific heats of the radiation and the medium.

In many cases it is possible to consider the radiation mean free path l as proportional to some power of the temperature (it is assumed that the density of the medium is constant)

$$l = AT^m, \qquad m > 0. \tag{10.12}$$

In a fully ionized gas, where the radiation and absorption of light proceeds entirely by bremsstrahlung, $m = 7/2$. In the region of multiply ionized gases $m \sim 1.5$–2.5. For the power law (10.12) the coefficient of radiation thermal conductivity is also proportional to a power of the temperature

$$\kappa = \frac{16\sigma A}{3} T^n = BT^n, \qquad n = m + 3, \tag{10.13}$$

where the exponent $n \sim 4.5$–5.5 in the multiply ionized region. In the approximation in which the specific heat of the gas is assumed constant, we arrive at (10.10) with the radiation thermal diffusivity coefficient equal to

$$\chi = \frac{\kappa}{\rho c_v} = \frac{B}{\rho c_v} T^n = aT^n. \tag{10.14}$$

The nonlinear heat conduction equation then takes the form

$$\frac{\partial T}{\partial t} = a\,\nabla\cdot(T^n\,\nabla T) + q. \tag{10.15}$$

Usually the specific heats and internal energy of a gas at high temperatures in the multiply ionized region can be approximated by power-law functions of temperature

$$\varepsilon = \alpha T^{k+1}, \qquad c_v = \frac{\partial \varepsilon}{\partial T} = (k+1)\alpha T^k,$$

where α is a constant and k is a quantity approximately equal to 0.5 (see §8 of Chapter III). When the specific heat varies as a power of the temperature, the heat conduction equation can also be reduced to the form of (10.15). Let us introduce in place of the temperature another unknown function, namely the internal energy per unit volume

$$E = \rho\alpha T^{k+1}, \qquad T = \left(\frac{E}{\rho\alpha}\right)^{1/(k+1)}.$$

We get

$$\frac{\partial E}{\partial t} = a'\,\nabla\cdot(E^{n'}\,\nabla E) + q', \tag{10.16}$$

where

$$n' = \frac{n-k}{k+1}, \qquad a' = \frac{B}{(k+1)(\rho\alpha)^{(n+1)/(k+1)}}, \qquad q' = W. \tag{10.17}$$

Equation (10.16) is identical with (10.15) and their solutions are the same. The transformation of the solution of (10.15), $T = T(x, y, z, t)$, for any specific problem to the solution of (10.16), $E = E(x, y, z, t)$, for the same problem can be simply carried out by replacing the constants a and n by a' and n', and also by replacing the source function q by $q' = W = q\rho c_v$. We note that when $n = 5$ and $k = 0.5$, $n' = 3$. In what follows, for convenience in comparing the results of the nonlinear and linear heat conduction theories, we shall use the temperature equation (10.15). In so doing, we should bear in mind that the solution which is found for any particular problem can be immediately written also for the case when the specific heat is proportional to some power of the temperature.

In addition to radiation heat conduction, which is of the greatest interest, still another example of nonlinear heat conduction can be given. This is electron heat conduction in a plasma, which was discussed in §12 of Chapter VII. (Ion heat conduction in a plasma is also strongly temperature dependent, but it is considerably less important than electron heat conduction.) The coefficient of electron thermal diffusivity is $\chi_e \sim T_e^{5/2}$.

It is interesting to note that the nonlinear heat conduction equation of the form (10.15) also describes an entirely different process, namely, the motion of a polytropic gas (the pressure and density of which are connected by the equation $p = const\ \rho^n$) in a porous medium. The gas density ρ satisfies the equation

$$\frac{\partial \rho}{\partial t} = b\, \nabla \cdot (\rho^n\, \nabla \rho),$$

where n is the polytropic exponent and b is a constant that is determined by the porosity and permeability of the medium and by the properties of the gas. Problems in filtration theory are also similar to certain problems in nonlinear heat conduction.

Nonlinear heat conduction processes were first considered by Zel'dovich and Kompaneets [1], who, in particular, found an exact solution to the problem of heat propagation from an instantaneous plane source. Analogous problems in filtration theory were studied independently by Barenblatt [2]. He obtained the same solution for the case of an instantaneous concentrated source, and also solved a number of other specific problems.

§3. Characteristic features of heat propagation by linear and nonlinear heat conduction

The basic features of the nonlinear heat conduction process and the characteristic properties distinguishing it from linear heat conduction are best explained by the example of the propagation of heat from an instantaneous plane source in an infinite initially cold medium. Let the energy $\mathscr{E}$ per unit area of surface be the energy released at the initial time $t = 0$ in the plane $x = 0$ ($\mathscr{E}$ is in erg/cm^2). For $t > 0$ the heat propagates in both directions away from the plane $x = 0$.

The heat conduction equation (10.10) for the problem considered takes the form

$$\frac{\partial T}{\partial t} = \frac{\partial}{\partial x} \chi \frac{\partial T}{\partial x}, \tag{10.18}$$

with the spatial temperature distribution satisfying the condition of energy conservation

$$\int_{-\infty}^{\infty} T\, dx = Q. \tag{10.19}$$

The quantity Q is equal to $\mathscr{E}/\rho c_p$ if the process takes place at constant pressure, and to $\mathscr{E}/\rho c_v$, if the specific volume is constant.

In the given case the two equations (10.18) and (10.19) are equivalent to the single heat conduction equation (10.10) with a delta function source (with respect to time as well as with respect to position)

$$q(x, t) = Q\delta(x)\delta(t).$$

Initially at $t = 0$ the temperature of the medium is assumed to be identically equal to zero everywhere, with the exception of the point where the energy release takes place

$$T(x, 0) = Q\delta(x).$$

The solution to the problem posed for the case of linear heat conduction with $\chi = const$ is well known. It is given by the expression

$$T = \frac{Q}{(4\pi\chi t)^{1/2}} e^{-x^2/4\chi t}. \tag{10.20}$$

The characteristic property of linear heat conduction lies in the fact that the heat is concentrated at the point of energy release only at the initial time $t = 0$ (for $x = 0$, $T \to \infty$ as $t^{-1/2}$). For $t > 0$ the heat instantaneously propagates throughout all of space and the temperature tends to zero at infinity, as $x \to \pm\infty$, only asymptotically. The major part of the energy is concentrated in a region whose dimensions are of the order of $x \sim (4\chi t)^{1/2}$, which increases with time proportionally to $\sqrt{t}$. Accordingly, the temperature also decreases as $1/\sqrt{t}$, so that the total amount of heat, which is proportional to $\int T\,dx \sim Tx \sim (1/\sqrt{t})\sqrt{t} \sim 1$, remains constant. The temperature distribution at successive instants of time is shown in Fig. 10.1.

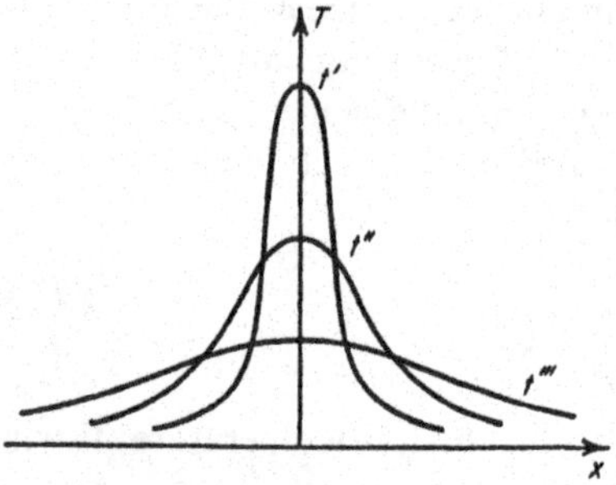

Fig. 10.1. Propagation of heat from an instantaneous plane source by linear heat conduction.

The asymptotic character of the decrease in temperature at infinity and the instantaneous propagation of heat to infinite distance can be explained within the framework of the theory of heat conduction by the fact that the coefficient of thermal conductivity is finite at zero temperature. In practice, of course, only a negligibly small amount of heat reaches a very distant point at a given time; the temperature decrease at infinity is very sharp, following a Gaussian behavior. In principle, however, at any arbitrarily large but finite distance from the source the temperature is finite immediately following the release of energy. It should be noted that the Gaussian law governing the behavior of the temperature at infinity is related to the approximate description of the heat propagation within the framework of the heat conduction theory. Actually the temperature at large distances is not determined by the diffusion of "hot" molecules from the heated region (in a gas), but rather by the direct "unimpeded" molecules that arrive at points distant from the heated region without experiencing any collisions. Therefore, the drop-off in temperature toward infinity is actually not governed by the Gaussian relationship (10.20) but by the exponential relation $T \sim e^{-x/l_a}$, where l_a is the molecular mean free path. It is clear that no matter what the preexponential factor, for a given time the simple exponential $\exp(-x/l_a)$ will eventually become greater than the Gaussian exponential $\exp(-x^2/4\chi t)$, $(\chi = l_a \bar{v}/3)$. However, this region at large distances contains such a negligible amount of heat, that consideration of it is of no interest.

Let us now verify the assumption that the motion of the fluid can be neglected. If we are dealing with a gaseous medium then a compression wave (or shock wave) is propagated from the point of energy release (in our case, from the plane $x = 0$). The shock propagates through the undisturbed medium with a speed of the order of the speed of sound in the heated region, and thus of the order of the thermal speed of the heated molecules $\bar{v}$. The rate of propagation of heat by thermal conduction is

$$\frac{dx}{dt} \sim \frac{d}{dt}(\chi t)^{1/2} \sim \left(\frac{\chi}{t}\right)^{1/2} \sim \frac{\chi}{x} \sim \frac{l_a}{x}\bar{v}.$$

Hence, as soon as the heat propagates a distance greater than the molecular mean free path, the rate of propagation of heat by thermal conduction becomes less than the hydrodynamic speed. Since, in general, there is no reason to consider distances smaller than a molecular mean free path, we can assume that the heat always travels with subsonic speed. If the amount of the energy released is not large, the compression wave is weak and the fluid velocity is small in comparison with the speed of sound. We can assume, as noted at the very beginning, that the role of the hydrodynamics is simply to equalize the pressure, and that the process of heat propagation proceeds at constant pressure. If the amount of energy released is large and the compression wave at a large distance from the energy source is a shock wave, then we are dealing with the purely hydrodynamic process of a strong explosion, considered in §25 of Chapter I. In this case the role of the thermal conductivity of the fluid in the propagation of energy is insignificant.

Let us now assume that the coefficient of thermal conductivity depends on the temperature and, furthermore, let it decrease with decreasing temperature and vanish at zero temperature, as in the case of radiation heat conduction. In this case the heat cannot instantaneously penetrate over an arbitrarily large distance, but propagates from the source with a finite velocity in such a manner that a sharply defined boundary exists between the heated region and the cold region which has not yet been reached by the thermal disturbance. The heat propagates from the source in the form of a wave, with the above boundary surface serving as the wave front. This wave is called a thermal wave. The temperature distribution in a thermal wave at successive times is shown schematically in Fig. 10.2.

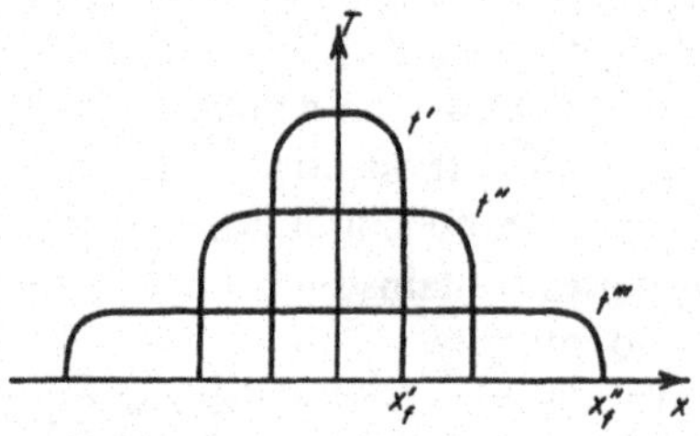

Fig. 10.2. Propagation of a thermal wave from an instantaneous plane source.

The temperature and the heat flux in the cold undisturbed medium are zero, since the coefficient of thermal conductivity goes to zero. From continuity, the flux at the wave front also goes to zero. In the case of linear heat conduction, with $\kappa = const$, the vanishing of the heat flux can be attributed only to the vanishing of the temperature gradient. In the case of nonlinear heat conduction, with the coefficient decreasing to zero as $T \rightarrow 0$, the flux can vanish if the temperature gradient is nonzero because the coefficient of thermal conductivity goes to zero. This condition, in particular, is responsible for the generation of a sharp thermal wave front.

To clarify what has been said, let us consider a layer near the wave front. If we restrict ourselves to small enough time intervals that the wave travels through distances which are small in comparison with the dimensions of the region encompassed by the wave, in comparison with the coordinate of the front x_f (see Fig. 10.2), then the front velocity during this time can be taken to be approximately constant.

The temperature distribution near the front can be sought in the form of a standing wave $T = T(x - vt)$, where v is the front velocity. The temperature distribution near the front is quasi-steady in a coordinate system moving with the front. Substituting a solution in the form $T = T(x - vt)$ into (10.18), we obtain the following equation for the temperature distribution near the front

$$-v \frac{\partial T}{\partial x} = \frac{\partial}{\partial x} \chi \frac{\partial T}{\partial x}. \qquad (10.21)$$

Assuming that $\chi = aT^n$ $(n > 0)$ and integrating this equation twice with the boundary condition $T = 0$ at $x = x_f$, we get for the temperature distribution

$$T = \left[\frac{nv}{a}|x_f - x|\right]^{1/n}. \tag{10.22}$$

This distribution is shown schematically in Fig. 10.2. The front coordinate x_f and its velocity $v = dx_f/dt$ in this equation represent undetermined functions of time. They are found from the complete spatial solution.

The fact that the temperature vanishes in the manner described by (10.22) also justifies the assumption of the existence of a sharp boundary for the heated region, i.e., a thermal wave front. If the exponent $n \leqslant 0$, the coefficient of thermal diffusivity χ does not go to zero for $T = 0$ and (10.21) has no solutions which vanish at a finite distance, and this case corresponds to the instantaneous propagation of heat to arbitrarily large distances. It follows from (10.22) that the temperature gradient near the thermal wave front satisfies the proportionality $dT/dx \sim |x_f - x|^{(1/n)-1}$. If $n > 1$, then the temperature gradient at the front (at $x = x_f$) becomes infinite, i.e., the front is steep. If $n < 1$, $(dT/dx)_{x_f} = 0$. The flux, however, is always zero at $x = x_f$: $S \sim T^n\, dT/dx \sim |x_f - x|^{1/n} \to 0$ for $n > 0$.

In examining the structure of a shock front (§§12 and 17 of Chapter VII) in which electron and radiation heat conduction were considered, it was shown how thermal conductivity produces a preheating "tongue" which travels through the gas ahead of the compression shock. The temperature profile ahead of the discontinuity is given by (10.22) (if the motion of the gas ahead of the discontinuity may be neglected), with the velocity v representing the speed of the shock front. The form of the profile is shown in Fig. 10.3a. The "tongue" lies ahead of the shock by the definite finite distance $\Delta x = x_f - x_1$ (Fig. 10.3a), which depends on the shock front temperature T_1

$$T_1 = \left(\frac{nv}{a}\Delta x\right)^{1/n}, \qquad \Delta x = \frac{aT_1^n}{nv} = \frac{\chi(T_1)}{nv} = \frac{\chi_1}{nv}.$$

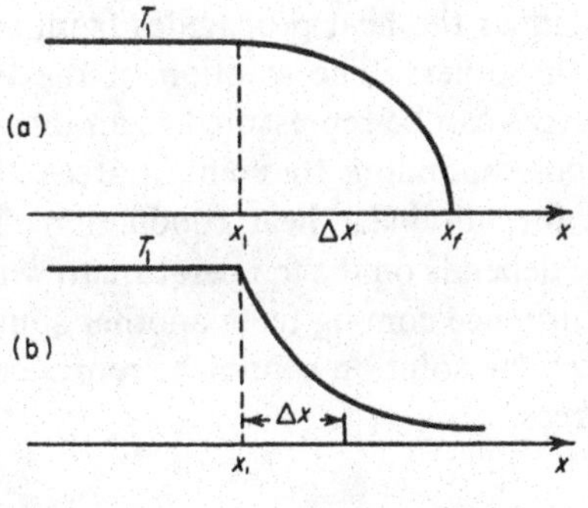

Fig. 10.3. Preheating by heat conduction ahead of a compression shock: (a) for nonlinear heat conduction; (b) for linear heat conduction.

In the case of linear heat conduction $\chi = const$ and the preheating "tongue" extends to infinity, although its effective thickness is finite and constant (for a shock wave moving with a constant velocity). The solution of (10.21) with $\chi = const$ in this case has the form

$$T = T_1 e^{-(x-x_1)/\Delta x'}, \qquad \Delta x' = \frac{\chi}{v}.$$

The temperature profile in the preheating layer is shown in Fig. 10.3b. As previously noted, the temperature vanishes only at infinity.

The relation governing the temperature decrease at infinity due to the "unimpeded" molecules in the case of molecular heat conduction differs from that dictated by the heat conduction theory, which does not consider the motion of individual molecules. Similarly, the thermal wave profile near the boundary, in the case of radiative heat transfer, is given by (10.22) only within the framework of the radiation heat conduction approximation. If we also take into account the presence of "unimpeded" photons, i.e., the fact that the radiation at the leading edge of the wave is out of equilibrium, we find that the temperature at the leading edge of the thermal wave decreases exponentially as $T \sim e^{-x/l}$, where l is the radiation mean free path. This effect was studied in detail in Part 3 of Chapter VII, where we considered the structure of shock fronts including the effect of radiative transfer.

Up to now we have considered the propagation of heat in a medium with zero initial temperature. If $T_0 \neq 0$, then the nonlinear coefficient of thermal conductivity in the undisturbed medium is finite and the relation for the temperature decrease is no longer given by (10.22); however, if T_0 is not too large, the coefficient of radiation thermal conductivity is sufficiently small that this effect can be neglected. Of greater importance is the effect mentioned above of nonequilibrium radiation at the leading edge of the thermal wave, which leads to an exponential temperature drop $T \sim e^{-x/l}$ instead of the power-law relation (10.22).

Let us note one more important difference between nonlinear and linear heat conduction. In the linear case the superposition principle is valid. Therefore, if there are a number of energy sources the heat propagates from each of them completely independently of the others. The solution of the heat conduction equation for distributed sources can be represented as an integral "over the sources" of the solutions corresponding to point sources. The superposition principle does not hold for nonlinear heat conduction. The propagation of heat from a single source depends on the temperature to which the medium is heated by a thermal disturbance coming from another source. In the general case of distributed sources the solution cannot be represented in the form of an integral over the sources.

§4. The law of propagation of thermal waves from an instantaneous plane source

The law of propagation of heat from a source can be easily obtained even without an exact solution by estimating the order of magnitude of the characteristic dimension of the heated region, or from dimensional considerations. Problems on the propagation of heat from an instantaneous concentrated source (plane, point, line) can be solved exactly (see below). However, the semiqualitative estimates describe rather clearly the physical meaning of the governing laws and, in addition, are frequently useful when considering more complex problems for which no exact solutions can be found.

Let us consider the propagation of heat from an instantaneous plane source. The results for the case of linear heat conduction were already presented in the preceding section, where the exact solution to the problem was given. We shall restate these results in order to demonstrate the general procedure for our semiqualitative approach. Let the coefficient of thermal conductivity be constant. Equation (10.18) contains only a single parameter—the coefficient of thermal diffusivity χ measured in cm^2/sec. The other dimensional parameter is the energy per unit area: $\mathscr{E}$ in erg/cm^2 or the quantity Q in deg · cm. If x is the width of the region in which most of the heat is concentrated at the time t, then it is evident from dimensional considerations that $x^2 \sim \chi t$, $x \sim (\chi t)^{1/2}$. The rate of propagation of heat is $dx/dt \sim (\chi/t)^{1/2} \sim \chi/x \sim x/t$. The average temperature in the heated region is of the order of $T \sim Q/x \sim Q/(\chi t)^{1/2}$. These simple results, which agree in order of magnitude with those given by the exact solution (10.20), can be obtained directly from (10.18) by replacing the derivatives $\partial T/\partial t$ and $\partial T/\partial x$ by T/t and T/x, $\partial(\chi\, \partial T/\partial x)/\partial x$ by $\chi T/x^2$. This leads immediately to the same relationships.

We now turn to the case of propagation of a nonlinear thermal wave. We assume that the coefficient of thermal diffusivity has the power-law form $\chi = aT^n$, for which the equation of heat conduction becomes

$$\frac{\partial T}{\partial t} = a \frac{\partial}{\partial x} T^n \frac{\partial T}{\partial x}. \tag{10.23}$$

This equation contains only the single parameter a, in $\text{cm}^2/\text{sec} \cdot \text{deg}^n$. The other dimensional parameter is Q, in deg · cm. We can combine them into a single (independent) dimensional combination containing only the units of length and time, aQ^n in $\text{cm}^{n+2} \cdot \text{sec}^{-1}$. From this the law governing the motion of the thermal wave front follows

$$x_f \sim (aQ^n t)^{1/(n+2)} = (aQ^n)^{1/(n+2)} t^{1/(n+2)}.$$

The speed of propagation of the thermal wave is of the order of

$$\frac{dx_f}{dt} \sim (aQ^n)^{1/(n+2)} t^{1/(n+2)-1} \sim \frac{x_f}{t} \sim \frac{aQ^n}{x_f^{n+1}}.$$

It is evident that when the exponent n is large, the thermal wave is slowed down very rapidly. This is so because, as a result of the thermal propagation, the temperature drops and the coefficient of thermal diffusivity decreases sharply. Recalling that the average temperature in the thermal wave is of the order of Q/x_f and the average coefficient of thermal diffusivity $\chi = aT^n \sim aQ^n/x_f^n$, we can write the law for the propagation of a thermal wave in a form corresponding to the linear theory: $x_f \sim (\chi t)^{1/2}$. It should be noted that the average coefficient of thermal diffusivity in this equation depends on time according to the relation

$$\chi \sim \frac{aQ^n}{x_f^n} \sim \frac{aQ^n}{(aQ^n)^{n/(n+2)} t^{n/(n+2)}} = (aQ^n)^{2/(n+2)} t^{-n/(n+2)}.$$

The law governing the propagation of a thermal wave can be also obtained from the heat conduction equation by replacing approximately the derivatives of all quantities by their values: $\partial T/\partial t \to T/t$; $\partial T/\partial x \to T/x_f$; $\partial(T^n \partial T/\partial x)/\partial x \to T^{n+1}/x_f^2$. We thus obtain $x_f^2 \sim aT^n t \sim \chi t$; using the fact that $T \sim Q/x_f$, we arrive at the relationships previously found.

§5. Self-similar thermal waves from an instantaneous plane source

Let us find the exact solution for the planar problem of a thermal wave propagating in an infinite medium as the result of the instantaneous release of energy at the time $t = 0$ in the plane $x = 0$. The process is described by the nonlinear heat conduction equation (10.23), with the solution satisfying the law of conservation of energy (10.19). It is evident from the dimensional considerations presented in the previous section that the solution to this problem is self-similar*. In fact, the only dimensionless combination that can be obtained from the coordinate x, the time t, and the parameters a and Q of this problem is

$$\xi = \frac{x}{(aQ^n t)^{1/(n+2)}}. \tag{10.24}$$

The quantity $Q/(aQ^n t)^{1/(n+2)} = (Q^2/at)^{1/(n+2)}$ has the dimensions of tem-

* The concept of self-similarity is discussed in §§11 and 25, Chapter I. See also Chapter XII.

perature. Therefore a solution for $T(x, t)$ should be sought in the form

$$T = \left(\frac{Q^2}{at}\right)^{1/(n+2)} f(\xi), \tag{10.25}$$

where $f(\xi)$ is a new unknown function.

Substituting (10.25) into (10.23) and transforming to the similarity variable ξ by means of the relations

$$\frac{\partial f}{\partial t} = -\frac{1}{n+2}\frac{df}{d\xi}\frac{\xi}{t}, \qquad \frac{\partial f}{\partial x} = \frac{1}{(aQ^n t)^{1/(n+2)}}\frac{df}{d\xi},$$

we obtain an ordinary differential equation for the function f

$$(n+2)\frac{d}{d\xi}\left(f^n \frac{df}{d\xi}\right) + \xi\frac{df}{d\xi} + f = 0. \tag{10.26}$$

The solution of this equation must satisfy the following conditions, which follow from the physical conditions of the problem: $T = 0$ at $x = \pm\infty$; or $T = 0$ at $x = +\infty$ and $\partial T/\partial x = 0$ at $x = 0$ (by virtue of the symmetry with respect to the plane $x = 0$). It follows that

$$f(\xi) = 0 \quad \text{for} \quad \xi = \infty; \qquad \frac{df}{d\xi} = 0 \quad \text{for} \quad \xi = 0. \tag{10.27}$$

A solution of (10.25) satisfying the boundary conditions (10.27) was given in [1, 2]. It has the form

$$\begin{aligned} f(\xi) &= \left[\frac{n}{2(n+2)}(\xi_0^2 - \xi^2)\right]^{1/n} \\ &= \left[\frac{n}{2(n+2)}\xi_0^2\right]^{1/n}\left[1 - \left(\frac{\xi}{\xi_0}\right)^2\right]^{1/n} \qquad \text{for} \quad \xi < \xi_0, \\ f(\xi) &= 0 \qquad \text{for} \quad \xi > \xi_0, \end{aligned} \tag{10.28}$$

where ξ_0 is a constant of integration. This constant is to be found from the equation of conservation of energy (10.19), which takes the form

$$\int_{-\infty}^{+\infty} f(\xi)\, d\xi = \int_{-\xi_0}^{+\xi_0} f(\xi)\, d\xi = 1. \tag{10.29}$$

Calculation gives

$$\xi_0 = \left[\frac{(n+2)^{1+n}2^{1-n}}{n\pi^{n/2}}\right]^{1/(n+2)}\left[\frac{\Gamma(1/2 + 1/n)}{\Gamma(1/n)}\right]^{n/(n+2)}, \tag{10.30}$$

where Γ is the Gamma function. The law for the motion of the thermal wave front $\xi = \xi_0$ is

$$x_f = \xi_0(aQ^n t)^{1/(n+2)}. \tag{10.31}$$

As expected, this relation agrees to within the numerical coefficient ξ_0 with the one obtained by semiqualitative considerations in the preceding section.

It is convenient to write the temperature in a plane thermal wave in the form

$$T = T_c\left(1 - \frac{x^2}{x_f^2}\right)^{1/n}, \tag{10.32}$$

where $x_f(t)$ is the front coordinate, whose dependence on time is given by (10.31) and (10.30), and T_c is the temperature in the central plane $x = 0$. This temperature can be expressed in terms of the average temperature in the wave (averaged over the heated volume)

$$T_c = \frac{\overline{T}}{J}, \tag{10.33}$$

where

$$\overline{T} = \frac{Q}{2x_f}; \qquad J = \int_0^1 (1 - z^2)^{1/n}\, dz = \frac{\sqrt{\pi}}{n+2}\frac{\Gamma(1/n)}{\Gamma(1/n + 1/2)}.$$

For example, for $n = 5$, $\xi_0 = 0.77$ and $T_c = 1.12\overline{T}$.

When the specific heat is taken to be variable, the temperature distribution differs very little from (10.32). Indeed, the energy distribution is

$$E = E_c\left(1 - \frac{x^2}{x_f^2}\right)^{1/n'}.$$

But $E \sim T^{1+k}$ and $n' = (n-k)/(k+1)$, from which

$$T = T_c\left(1 - \frac{x^2}{x_f^2}\right)^{1/(n-k)}.$$

Since $n \sim 5$ and $k \sim 0.5$, this expression differs very little from (10.32) (in the first case the exponent $1/n = 1/5$; in the second it is $1/(n-k) = 1/4.5$). The new constant $\xi_0(n')$ in the relation for the propagation equation itself changes more appreciably. For $n = 5$ and $k = 0$ (constant specific heats) $x_f \sim t^{1/7}$, $dx_f/dt \sim x_f^{-6}$; for $n = 5$ and $k = 0.5$ (that is, $c_v \sim T^{0.5}$), $x_f \sim 1/t^{n'+2} \sim t^{1/5}$, $dx_f/dt \sim x_f^{-4}$.

The temperature distribution T/T_c as a function of x/x_f is given in Fig. 10.4a for the case $n = 5$. A thermal wave with a strongly temperature-dependent coefficient of thermal conductivity has a characteristic temperature "plateau". The temperature is almost constant, being equalized by heat conduction within the entire heated region, with the exception of a relatively thin layer near the front where it rapidly drops to zero. This tendency is

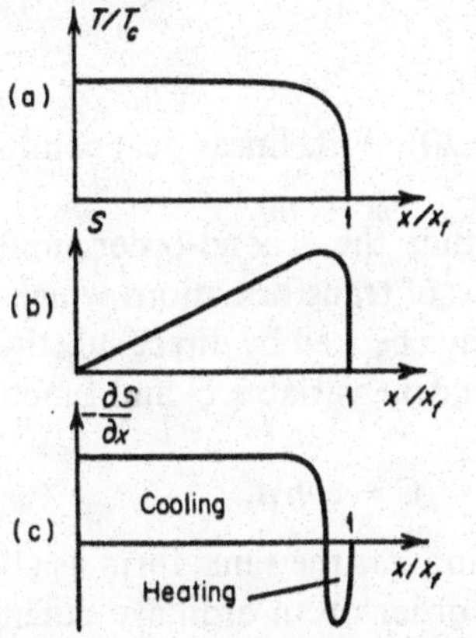

Fig. 10.4. Distributions of temperature, flux, and divergence of flux in a thermal wave.

exhibited more strongly, the larger is the nonlinear exponent n. The distribution of flux with respect to the position coordinate is given by

$$S \sim -T^n \frac{\partial T}{\partial x} \sim \left(1 - \frac{x^2}{x_f^2}\right)^{1/n} x.$$

The flux increases almost linearly from the origin $x = 0$ to the very edge of the wave and drops very rapidly to zero only near the edge, as shown in Fig. 10.4b. The divergence of the flux $\partial S/\partial x$ is almost constant in the entire region of the plateau. The main region of hot gas is cooled almost uniformly, and only near the edge of the wave is the gas heated by energy removed from the main mass of gas (see Fig. 10.4c). The heat propagates in such a manner that the volume of gas is cooled almost uniformly and the energy lost by the gas is absorbed near the wave front, which is the manner by which the wave continually encompasses the new layers of cold gas. The temperature distribution near the front can be approximately given as

$$T \sim \left(1 - \frac{x^2}{x_f^2}\right)^{1/n} \sim \left[\frac{2}{x_f}(x_f - x)\right]^{1/n} \sim (x_f - x)^{1/n},$$

which was already given earlier (see (10.22)).

Let the exponent n in the solution of (10.25), (10.28), and (10.30) go to the limit $n \to 0$, which corresponds to the transition to linear heat conduction (the constant a in (10.23) in the limit $n = 0$ plays the role of a constant coefficient of thermal diffusivity $\chi = const$). For $n \to 0$, $\xi_0 \to 2/\sqrt{n}$

$$T = \frac{Q}{(at)^{1/2}}[f(\xi)]_{n\to 0} = \frac{Q}{(at)^{1/2}}\frac{1}{(4\pi)^{1/2}}\left[\left(1 - \frac{nx^2}{4at}\right)^{1/n}\right]_{n\to 0}$$

$$= \frac{Q}{(4\pi at)^{1/2}} e^{-x^2/4at}, \qquad a = \chi,$$

and we arrive at the well-known solution (10.20) of the linear heat conduction equation.

In concluding this section, let us note that the second-order nonlinear equation (10.26) may be subjected to a group of transformations which leave the equation invariant. Indeed, it can be easily checked by direct substitution, that if ξ and f are replaced by a new independent variable ξ' and function f' given by

$$\xi' = C^n \xi, \qquad f' = C^2 f, \qquad C = const,$$

then in terms of the new variables the equation has the same form as (10.26). According to the theory of Lie groups, the order of an ordinary differential equation which admits a one-parameter group of transformations can be decreased by one. For this purpose it is convenient to introduce the new variables

$$y = \xi^{-2/n} f, \qquad z = \ln \xi.$$

In terms of these variables the new equation contains z only in terms of the differential dz. Therefore, a new variable $p = dy/dz$ can be introduced and z eliminated, yielding a first-order equation in the variables p and y

$$y^n p \frac{dp}{dy} + np^2 y^{n-1} + \frac{4+3n}{n} py^n + \frac{1}{n+2} p + \frac{4+2n}{n^2} y^{n+1} + \frac{y}{n} = 0.$$

Thus, the solution of the second-order equation (10.26) reduces to solving a first-order equation and to quadratures. This situation is characteristic of many self-similar problems of nonlinear heat conduction theory*.

§6. Propagation of heat from an instantaneous point source

Let us consider the spherically symmetric case. Suppose that at the time $t = 0$ an energy of $\mathscr{E}$ erg is released at the point $r = 0$. The heat conduction equation in this case takes the form

$$\frac{\partial T}{\partial t} = \frac{1}{r^2}\frac{\partial}{\partial r}\left(r^2 \chi \frac{\partial T}{\partial r}\right). \tag{10.34}$$

* And also of self-similar problems in gasdynamics. For details, see Chapter XII. *Editors' note.* The Blasius equation of boundary layer theory may be reduced to first order with the help of the same approach.

The law of conservation of energy gives

$$\int_0^\infty T 4\pi r^2\,dr = \frac{\mathscr{E}}{\rho c_v} = Q \quad \text{deg}\cdot\text{cm}^3.$$

The solution of the problem for linear heat conduction $\chi = const$ is well known

$$T = \frac{Q}{(4\pi\chi t)^{3/2}}\, e^{-r^2/4\chi t}. \tag{10.35}$$

The heat flows in such a manner that the main part of the energy is concentrated in a sphere whose radius is of the order of

$$r \sim (4\chi t)^{1/2},$$

in close analogy with the plane case, where we had $x \sim (4\chi t)^{1/2}$. The temperature at the center falls off as $T_c \sim Q/r^3 \sim Q/(\chi t)^{3/2}$. These relationships follow directly from dimensional considerations; they can also be obtained from estimates based on (10.34) and (10.35), by replacing the derivatives by their corresponding values (see §4).

Let us now consider the case of nonlinear heat conduction with $\chi = aT^n$ and $n > 0$. The equation takes the form

$$\frac{\partial T}{\partial t} = \frac{a}{r^2}\frac{\partial}{\partial r}\left(r^2 T^n \frac{\partial T}{\partial r}\right). \tag{10.36}$$

Let us find the equation of motion for the thermal wave front, as was done in the plane case. We have

$$r_f^2 \sim \chi t,$$

where χ is the coefficient of thermal diffusivity, corresponding to the average temperature of the heated region at time t. But

$$T \sim \frac{Q}{r_f^3}, \tag{10.37}$$

so that $r_f^2 \sim aT^n t \sim aQ^n r_f^{-3n} t$, whence

$$r_f \sim (aQ^n)^{1/(3n+2)} t^{1/(3n+2)}. \tag{10.38}$$

The velocity of the thermal wave front is expressed by the proportionality

$$\frac{dr_f}{dt} \sim \frac{r_f}{t} \sim \frac{(aQ^n)^{1/(3n+2)}}{t^{(3n+1)/(3n+2)}} \sim \frac{aQ^n}{r_f^{3n+1}}. \tag{10.39}$$

This velocity decreases extremely rapidly as the wave propagates. For example, with $n = 5$, $dr_f/dt \sim 1/r^{16}$.

We seek an exact solution of the heat conduction equation in the self-similar form

$$T = \left(\frac{Q^{2/3}}{at}\right)^{3/(3n+2)} \varphi(\xi), \tag{10.40}$$

where the similarity variable ξ is defined as

$$\xi = \frac{r}{(aQ^n t)^{1/(3n+2)}}. \tag{10.41}$$

Substituting (10.40) into (10.36) we obtain an ordinary differential equation for the function $\varphi(\xi)$, which differs only slightly from the equation (10.26) for the plane case. This equation was solved by the late S. Z. Belen'kii, and independently by Barenblatt [2]*.

The final solution can be written in a form similar to (10.32)

$$T = T_c\left(1 - \frac{r^2}{r_f^2}\right)^{1/n}, \tag{10.42}$$

where the radius of the front is

$$r_f = \xi_1 (aQ^n t)^{1/(3n+2)}. \tag{10.43}$$

The constant ξ_1 is given by

$$\xi_1 = \left[\frac{3n+2}{2^{n-1} n\pi^n}\right]^{1/(3n+2)} \left[\frac{\Gamma(5/2 + 1/n)}{\Gamma(1 + 1/n)\Gamma(3/2)}\right]^{n/(3n+2)}$$

The temperature at the center T_c is

$$T_c = \frac{4\pi}{3} \xi_1^3 \left[\frac{n\xi_1^2}{2(3n+2)}\right]^{1/n} \bar{T}, \tag{10.44}$$

where

$$\bar{T} = Q \Big/ \frac{4\pi}{3} r_f^3$$

is the temperature averaged over the volume at the time when the radius of the wave front is r_f.

For example, for $n = 5$, $\xi_1 = 0.79$ and $T_c = 1.28\bar{T}$. When the specific heat is variable, with $c_v \sim T^k$,

$$E = E_c\left(1 - \frac{r^2}{r_f^2}\right)^{1/n'}, \quad T = T_c\left(1 - \frac{r^2}{r_f^2}\right)^{1/(n-k)}, \quad n' = \frac{n-k}{1+k},$$

* The propagation of an almost-spherical thermal wave was considered by Andriankin and Ryzhov [3]. Andriankin [4] has also considered a spherical thermal wave with radiant energy taken into account.

as in the plane case. The thermal wave radius, instead of $r_f \sim t^{1/(3n+2)}$, is now given by $r_f \sim t^{1/(3n'+2)}$. For $n = 5$ and $k = 0$, $r_f \sim t^{1/17}$, $dr_f/dt \sim r_f^{-16}$; for $n = 5$ and $k = 0.5$, $n' = 3.0$, $r_f \sim t^{1/11}$, and $dr_f/dt \sim r_f^{-10}$.

The temperature distribution with respect to radius in the spherical case is exactly the same as in the plane case. The flux increases linearly with respect to the radius in almost the entire region from the center to the front and only drops to zero near the front:

$$S \sim -T^n \frac{\partial T}{\partial r} \sim \left(1 - \frac{r^2}{r_f^2}\right)^{1/n} r.$$

The divergence of the flux is almost constant over the entire sphere with the exception of a thin layer near the front. The gas is cooled comparatively uniformly, releasing energy which is absorbed near the front, and in this manner heating new layers of fluid.

Let us imagine a small volume of a gas in which a large amount of energy was rapidly released with the result that the fluid was heated to a very high temperature. A thermal wave will then travel from the place where the energy is released through the surrounding gas. The speed of propagation of a thermal wave, according to (10.39), decreases as the wave propagates and as the temperature of the heated spherical region decreases; the speed is given by $dr_f/dt \sim aQ^n/r_f^{3n+1}$. But $r_f \sim (Q/T)^{1/3}$, so that $dr_f/dt \sim aT^{n+\frac{1}{3}}Q^{-\frac{1}{3}}$. In the case of radiation heat conduction $n = 5$ and $dr_f/dt \sim T^{5.3}$. The speed of sound in a high-temperature gas, roughly speaking, is proportional to $\sqrt{T}$. Consequently, if the initial temperature is very high, the speed of propagation of the thermal wave is necessarily much greater than the speed of sound. When a wave travels through a stationary cold gas of constant density, the pressure of the gas increases. Roughly speaking, the pressure behind the front of a thermal wave is proportional to the temperature $p \sim \rho T$, so that the pressure distribution approximately follows the temperature distribution. The existence of the pressure gradient in the wave results in the acceleration of the gas; it then suddenly expands from the center, and redistributes its mass so that it tends to concentrate at the periphery, near the thermal wave front. The disturbances travel through the gas with the speed of sound. For this reason, when initially the thermal wave travels much faster than the speed of sound, the fluid behind it cannot acquire any appreciable motion. As we have seen, the propagating thermal wave is very rapidly decelerated. After a certain time its speed drops off to a value of the order of the speed of sound and then becomes less than the latter. Starting at this time, the heat conduction wave no longer travels ahead of the sonic disturbances, the fluid is set into motion, and a shock wave is formed which then moves ahead of the thermal wave, traveling with a speed which is of the order of magnitude of the speed of sound in the heated gas behind it. The process gradually comes to resemble that which is described by the solution of the problem of a strong explosion (see §25, Chapter I). The time of formation of a shock wave and of its overtaking of the thermal wave thus approximately coincides with the time at which the speed of the thermal wave drops to the speed of sound in the heated gas.

It has been estimated that for air at standard density this takes place when the temperature in the heated spherical region drops to a value of the order of 300,000°K. If the initial temperature of the air at the time of energy release is much higher than this value, then there is a sharply defined stage at which the energy travels through the stationary air by radiation heat conduction in the form of a thermal wave. When the temperature of the expanding heated spherical region drops to $\sim$300,000°K, a shock wave is formed which then moves ahead, and the role of the radiation heat conduction is reduced exclusively to that of equalizing the temperature in the central region. If, however, the initial energy concentration is such that the air temperature is below 300,000°K, then in general no thermal wave is formed, and the energy from the very beginning is transported by hydrodynamic means, carried by the shock wave.

It was noted at the end of §3 that the temperature profile at the lower edge of a thermal wave is the same as the temperature profile in the preheating zone of a very strong shock wave (in a very strong shock wave a "tongue" preheated by radiation heat conduction moves ahead of the shock front). In particular, the radiation at the very leading edge of the thermal wave is out of equilibrium, due to the presence of "unimpeded" photons, and the temperature drops to zero exponentially in the optical coordinate. This means that the visible luminosity of the surface of a thermal wave front is the same as the luminosity of the surface of a very strong shock front. It was shown in §4 of Chapter IX that this limiting luminosity, in air at standard density, corresponds to a brightness temperature in the visible region of the spectrum of approximately 17,000°K. The brightness temperature at the surface of the front of a thermal wave is also the same. Thus, observing from afar the thermal wave propagating through air, we shall "see" a temperature of the order of 17,000°K, despite the fact that the temperature in the central regions of the wave can be as high as many hundreds of thousands of degrees.

§7. Some self-similar plane problems

Let us consider several self-similar problems. Two of them will be treated by the semiqualitative method presented in §4. We shall obtain an exact solution for one of them.

Constant temperature on the boundary. Let a constant temperature T_0 be maintained at the boundary of the plane half space ($x = 0$) with zero initial temperature. A thermal wave travels from the boundary into the medium, as shown in Fig. 10.5. Since there exists a characteristic temperature T_0, the coefficient of thermal diffusivity is of the order of $\chi \sim aT_0^n$, and the thermal wave front propagates according to the relation

$$x_f \sim (\chi t)^{1/2} \sim (aT_0^n t)^{1/2}.$$

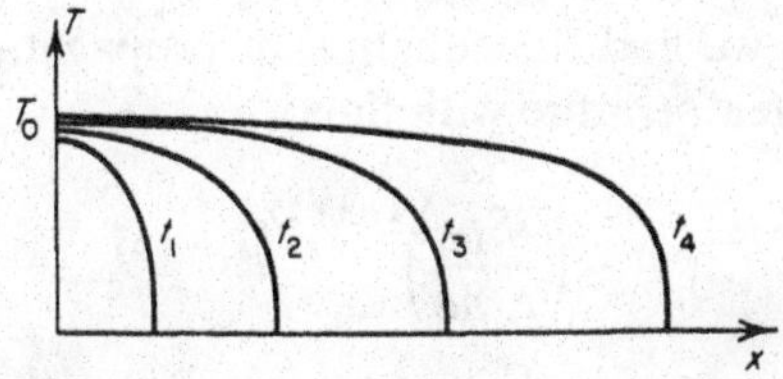

Fig. 10.5. Propagation of a thermal wave for a given temperature on the boundary.

The value of the numerical coefficient in this equation and the temperature distribution (obviously self-similar) can be found by numerical integration of an ordinary differential equation for the dimensionless function $f(\xi)$, where

$$T = T_0 f(\xi), \qquad \xi = \frac{x}{(aT_0^n t)^{1/2}},$$

subject to the boundary conditions $f(0) = 1$ and $f(\infty) = 0$.

The heat flux through the boundary decreases with time as

$$S \sim aT_0^n \frac{\partial T}{\partial x} \sim \frac{aT_0^{n+1}}{x_f} \sim \frac{aT_0^{n+1}}{(aT_0^n t)^{1/2}} \sim \frac{a^{1/2}T_0^{(n+2)/2}}{t^{1/2}}.$$

The change of energy in the thermal wave with time can be estimated by either of two means,

$$\mathscr{E} \sim \int_0^{x_f} T\,dx \sim T_0 x_f \sim t^{1/2}, \quad \text{or} \quad \mathscr{E} \sim \int_0^t S\,dt \sim \int_0^t \frac{dt}{t^{1/2}} \sim t^{1/2}.$$

Constant flux at the boundary. Let us assume that a constant heat flux S_0, externally supplied to the fluid, is specified at the boundary

$$S_0 = -\kappa\left(\frac{\partial T}{\partial x}\right)_0 = -c_v\rho aT^n\left(\frac{\partial T}{\partial x}\right)_0 = const \qquad \text{for} \quad x = 0.$$

The laws of propagation of the thermal wave and of the temperature change in the wave with time may be found by replacing all derivatives by their corresponding ratios. The flux across the thermal wave changes from S_0 to zero. The order of magnitude of the average temperature in the wave is given by

$$S_0 \sim c_v \rho \frac{aT^{n+1}}{x_f}.$$

From the heat conduction equation it follows, however, that in order of magnitude

$$\frac{T}{t} \sim \frac{S_0}{x_f}.$$

From these two approximate relations we find the equation of propagation of the thermal wave and the change of temperature with time:

$$x_f \sim (c_v \rho a S_0^n)^{1/(n+2)} t^{(n+1)/(n+2)}, \qquad T \sim \left(\frac{S_0^2}{c_v \rho a}\right)^{1/(n+2)} t^{1/(n+2)}.$$

For $n = 5$, $x_f \sim t^{6/7}$, $T \sim t^{1/7} \sim x_f^{1/6}$, $dx_f/dt \sim t^{-1/7}$.

The speed of the thermal wave decreases very slowly, and the average temperature slowly increases. The temperature increase appears because as the wave propagates the temperature gradient becomes smaller, and in order to maintain a constant flux the coefficient of thermal conductivity must increase. The propagation of the wave is shown in Fig. 10.6.

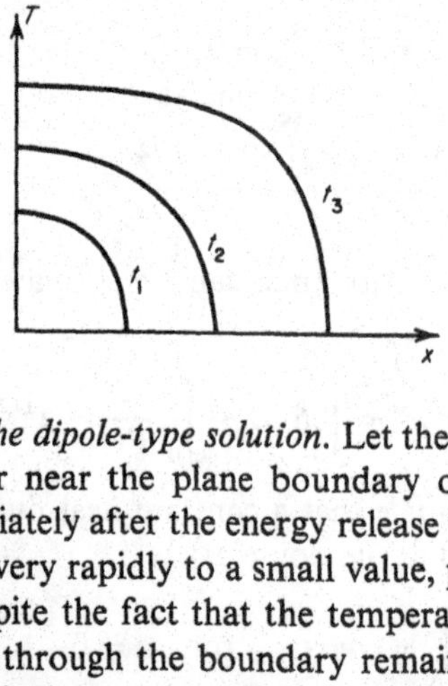

Fig. 10.6. Propagation of a thermal wave for a given flux at the boundary.

The dipole-type solution. Let the energy be released instantaneously in some layer near the plane boundary of the half space. Let us assume that immediately after the energy release the temperature at the boundary is made to fall very rapidly to a small value, practically to zero, and is maintained there. Despite the fact that the temperature at the boundary is very low, the heat flux through the boundary remains finite (correspondingly, the temperature gradient is very large), so that the energy flows out from the fluid. This problem does not admit an energy integral.

Let us idealize the problem in order to eliminate dimensional length parameters (for example, the thickness of the layer where the energy is released, or its distance from the boundary). We shall assume that the energy is released instantaneously in an infinitesimally thin layer at the surface of the body $x = 0$, in such a way that in the limit, when the thickness of the layer in which the energy release took place tends to zero and the layer itself approaches the surface $x = 0$, the first moment of the temperature remains finite

$$\int_0^\infty xT(x, 0)\, dx < \infty.$$

It can be easily shown that in this case, under the condition that the temperature at the boundary is zero, the energy integral in the problem of an instantaneous plane source is replaced by a moment integral, and the first moment of the temperature is conserved in time. This proposition was proved by Barenblatt [5].

Let us multiply the heat conduction equation (10.23) by x and integrate it from 0 to ∞, noting that the flux at infinity vanishes. Integrating by parts, we get

$$\frac{d}{dt}\int_0^\infty xT(x, t)\, dx = +\frac{a}{n+1} T^{n+1}(0, t).$$

If $T(0, t) = 0$ at the boundary, then the first or dipole moment of the temperature is conserved in time,

$$\int_0^\infty xT(x, t)\, dx = P = const. \tag{10.45}$$

This problem is self-similar, since it contains only two dimensional parameters, P in $\text{deg}\cdot\text{cm}^2$ and a in $\text{cm}^2\cdot\text{sec}^{-1}\cdot\text{deg}^{-n}$. It was solved by Barenblatt and Zel'dovich [6], with reference to the process of gas filtration. The thermal wave front propagates as

$$x_f = \xi_0(\alpha P^n t)^{1/2(n+1)}.$$

The temperature can be expressed in the form

$$T = \left(\frac{P}{at}\right)^{1/(n+1)} M\left(\frac{x}{x_f}\right)^{1/(n+1)}\left[1 - \left(\frac{x}{x_f}\right)^{(n+2)/(n+1)}\right]^{1/n}, \tag{10.46}$$

where the numerical constants ξ_0 and M are given by

$$\xi_0 = (n+2)^{1/2}(n+1)^{-n/2(n+1)} n^{-1/2(n+1)} 2^{1/2(n+1)}$$
$$\times\left[B\left(1 + \frac{1}{n}, \frac{n+1}{n+2} + 1\right)\right]^{-n/2(n+1)},$$

$$M = \left[\frac{n}{2(n+2)}\right]^{1/n} \xi_0^{2/n}.$$

Here $B(p, q)$ is the Beta function, which may be found tabulated.

For $n = 5$, the temperature function is

$$T \sim \frac{1}{t^{1/6}}\left(\frac{x}{x_f}\right)^{1/6}\left[1-\left(\frac{x}{x_f}\right)^{7/6}\right]^{1/5},$$

and the front travels as

$$x_f \sim t^{1/12}, \qquad \frac{dx_f}{dt} \sim \frac{1}{t^{11/12}} \sim \frac{1}{x_f^{11}}.$$

The propagation of the thermal wave is shown in Fig. 10.7.

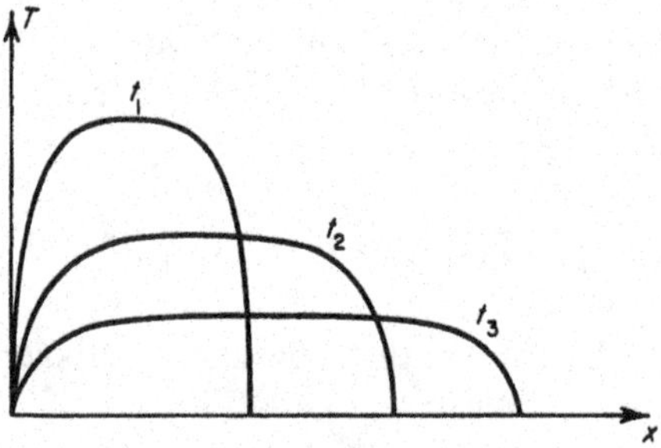

Fig. 10.7. The dipole-type solution.

It is evident that the flux through the boundary $x = 0$ is different from zero, and that energy flows out from the medium. Indeed, for $x/x_f \ll 1$,

$$T \sim \left(\frac{x}{x_f}\right)^{1/(n+1)},$$
$$S \sim T^n \frac{\partial T}{\partial x} \sim \frac{\partial T^{n+1}}{\partial x} \sim \frac{\partial}{\partial x}\left(\frac{x}{x_f}\right) \neq 0. \tag{10.47}$$

(The total energy $\mathscr{E} \sim x_f^{-1} \sim t^{-1/2(n+1)}$, *eds.*)

Barenblatt [2] investigated an entire class of self-similar solutions of plane problems subject to the very general conditions at the half-space boundary

$$T = const\ t^q, \qquad q \geqslant 0,$$

or

$$S = const\ t^q, \qquad q \geqslant 0$$

(the temperature or flux at the boundary increases with time following a power law). He also considered problems with cylindrical and spherical symmetry.

§8. Remarks on the penetration of heat into moving media

It was noted above that it is possible to neglect the motion of the medium when considering thermal waves because in the early stage of propagation of the thermal wave from the source, when the temperature is very high, the propagation speed is much higher than the speed of sound and the fluid is simply unable "to get moving". In some cases, however, the motion of the medium is appreciable from the very beginning.

Let us assume that the temperature at the boundary of the medium increases with time according to the power law $T_0 = const\ t^q$ $(q > 0)$. The distance through which the heat is carried into the medium by the radiation heat conduction mechanism is given by the proportionality

$$x_f \sim (\chi t)^{1/2} \sim T^{n/2} t^{1/2} \sim t^{(nq+1)/2} \qquad (\chi \sim T^n). \tag{10.48}$$

The speed of the thermal wave is expressed by the proportionality

$$\frac{dx_f}{dt} \sim \frac{x_f}{t} \sim t^{(nq-1)/2}.$$

The shock wave travels from the energy source at the boundary of the medium into the medium with a speed of the order of the speed of sound in the heated fluid:

$$D \sim \sqrt{T} \sim t^{q/2}.$$

Let us compare the speeds of the thermal and shock waves, dx_f/dt and D. If $(nq-1)/2 < q/2$, $q < 1/(n-1)$, then at the beginning of the process as $t \to 0$ the speed of the thermal wave is always higher than that of the shock wave and the thermal wave outruns the shock. The motion of the medium at this stage can be neglected, as was done above. It is only starting at a certain time t', when D becomes greater than dx_f/dt, that the shock wave will move in front of the thermal wave and the fluid in the neighborhood of the thermal wave will be set into motion (obviously, there is no clearly defined time t' and the fluid is accelerated gradually; t' represents an effective boundary between these two stages).

If $(nq-1)/2 > q/2$, $q > 1/(n-1)$, the situation is reversed. As $t \to 0$, $D > dx_f/dt$, the shock wave outruns the thermal wave, and the thermal wave from the very beginning travels through a moving medium. Starting at a certain effective time t'', the thermal wave moves in front of the shock wave and then travels through a stationary medium. The mass of the moving fluid, which is proportional to $Dt \sim t^{(q/2)+1}$ (per unit area of surface), then makes up an increasingly smaller fraction of the mass heated by the thermal wave, which is proportional to $x_f \sim t^{(nq+1)/2}$.

In the intermediate case, $(nq-1)/2 = q/2$, $q = 1/(n-1)$, the speeds of the thermal and shock wave increase at the same rate with time. In this case, in general, the stages when the energy penetrates the medium by one method only (either by hydrodynamic means or by heat conduction) are not separable, as they are asymptotically in the cases $q \lessgtr 1/(n-1)$. The fluid is heated by heat conduction and is set into motion almost simultaneously.

It is remarkable that in the particular case $n = 6$ (when the radiation mean free path $l \sim T^3$) the hydrodynamic equations with radiation heat conduction (but not radiation energy and pressure) taken into account admit a self-similar solution. This solution corresponds to the case when the temperature at the boundary of the medium increases as $T_0 \sim t^{1/5}$ (the existence of such a self-similar solution was pointed out by Marshak [7]). The reference density in this case is constant and equal to the initial density of the medium ρ_0, the pressure $p \sim \rho T \sim t^{1/5}$, and the fluid velocity $u \sim (p/\rho)^{1/2} \sim t^{1/10}$. The coordinate of the boundary of the disturbed region (the front of the thermal or shock wave) increases with time according to the relation

$$x \sim ut \sim (\chi t)^{1/2} \sim (T^6 t)^{1/2} \sim t^{11/10}. \tag{10.49}$$

The similarity variable in this case is $\xi = const\, xt^{-11/10}$, so that the solution of the equations is expressed in the form

$$T = const\; t^{1/5} f_1(\xi), \qquad u = const\; t^{1/10} f_2(\xi), \quad \text{etc.}$$

It is significant that a self-similar solution is possible when the thermal conductivity (radiation mean free path) is an arbitrary function of the density, $\chi = f(\rho)T^6$ (since the density scale is not time dependent). That the gasdynamic equations including radiation heat conduction actually admit the above self-similar solution can be easily verified from a direct examination of these equations*.

The character of the self-similar region for the case considered depends on which is larger, the speed of sound $c \sim \sqrt{T}$, or the speed with which the disturbances travel by heat conduction $x/t \sim (\chi/t)^{1/2}$. Both quantities increase with time as $t^{1/10}$, so that the relation between them is determined by the proportionality coefficients. For this reason the character of the process depends on the numerical value of the coefficient in the relation $T_0 \sim t^{1/5}$, determining the temperature increase with time at the boundary of the medium. A state is possible in which a shock wave travels ahead through the undisturbed fluid with a thermal wave following it through the heated and compressed fluid. Another state is possible in which the front of the thermal wave behind which the fluid is set into motion is the boundary between the disturbed and undisturbed regions.

In concluding we note the article by Nemchinov [8], in which some problems of radiative heat transfer including the motion of the medium are considered.

* We recall that the equations of continuity and of motion do not change when radiation heat conduction is included, but that an additional energy flux (10.8) is introduced into the energy equation (see §9, Chapter II).

§9. Self-similar solutions as limiting solutions of nonself-similar problems

Self-similar solutions are of interest not so much as particular solutions of a specific narrow class of problems, but mainly as limits which are asymptotically approached by solutions of more general problems that are not self-similar. This problem was investigated by Zel'dovich and Barenblatt [9], as applied to an initial value problem for the nonlinear heat conduction equation in the one-dimensional planar case (10.23).

The basic physical features of the asymptotic behavior of a solution can best be clarified by using an example of linear heat conduction for which the solution is particularly simple. Let the temperature distribution $T(x, 0) = T_0(x)$ along the x axis at the time $t = 0$ be given, with the temperature differing from zero only on a finite part of the x axis*. As is well known, the solution of the heat conduction equation (10.18) in this case is ($\chi = const$).

$$T(x, t) = \frac{1}{(4\pi\chi t)^{1/2}} \int_{-\infty}^{\infty} T_0(y) e^{-(x-y)^2/4\chi t}\, dy. \tag{10.50}$$

This is a generalization of the solution (10.20) to the case of a distributed source.

Let us consider the behavior of the temperature as $t \to \infty$ at large distances from the place where the heat was initially concentrated, for $x \gg y$. Expanding the integrand in a power series in the small quantity y/x we obtain

$$T(x, t) = \frac{1}{(4\pi\chi t)^{1/2}} e^{-\xi^2} \left[\int_{-\infty}^{\infty} T_0(y)\, dy \right.$$
$$\left. + \frac{\xi}{(\chi t)^{1/2}} \int_{-\infty}^{\infty} T_0(y) y\, dy + \frac{2\xi^2 - 1}{4\chi t} \int_{-\infty}^{\infty} T_0(y) y^2\, dy + \cdots \right], \tag{10.51}$$

where

$$\xi = \frac{x}{(4\chi t)^{1/2}}.$$

The solution is seen to be a sum of self-similar terms, in which the powers of inverse time increase by 1/2 with each successive term, and the coefficients are expressed in terms of successive moments of the initial temperature distribution. In the limit $t \to \infty$ we are left with the first term in the brackets, corresponding to the solution of (10.18) for a concentrated source. The next term

* This initial condition is one applicable for the problem of nonlinear heat conduction. In the linear case the more general condition of a sufficiently rapid temperature decrease at infinity is permissible.

in the expansion, which characterizes the difference between the actual and the limiting solution, is of the order of $1/t^{1/2}$ with respect to the leading term,

$$T = T_{\text{lim}}\left[1 + \frac{\varphi(\xi)}{t^{1/2}} + \cdots\right]. \tag{10.51'}$$

Since the coordinate origin, the time origin, and the temperature scale in (10.18) can be chosen arbitrarily (that is, the group of transformations $x' = x - x_0$, $t' = t + \tau$, $T' = kT$ are admissible), this equation admits of a more general self-similar solution than (10.20), given by

$$T_{\text{sim}}(x - x_0, t + \tau, Q) = \frac{Q}{[4\pi\chi(t+\tau)]^{1/2}} e^{-(x-x_0)^2/4\chi(t+\tau)}. \tag{10.52}$$

This solution corresponds to the instantaneous release of a definite amount of heat $E = c_v \rho Q$ at the point $x = x_0$ and at the time $t = -\tau$.

It can be easily proved that by a proper choice of the parameters x_0, τ, and Q we can obtain a self-similar solution of the type (10.52) which describes the exact solution (10.51) better than does the self-similar solution (10.20), in which $x_0 = 0$ and $\tau = 0$. Indeed, let us expand the function (10.52) in a power series in the small parameters x_0/x and τ/t (in the limit $t \to \infty$, $x \to \infty$). Comparing the expansion with the exact solution (10.51), we see that by choosing the values of Q, x_0, and τ so that

$$Q = \int_{-\infty}^{\infty} T_0(y)\, dy,$$

$$x_0 = \frac{\int_{-\infty}^{\infty} T_0(y) y\, dy}{\int_{-\infty}^{\infty} T_0(y)\, dy}, \tag{10.53}$$

$$\tau = \frac{\int_{-\infty}^{\infty} T_0(y) y^2\, dy}{2\chi \int_{-\infty}^{\infty} T_0(y)\, dy} - \frac{x_0^2}{2\chi},$$

terms of the order of $t^{-1/2}$ and t^{-1} in the expansion disappear, so that

$$T(x, t) = T_{\text{sim}}(x - x_0, t + \tau, Q)\left[1 + \frac{\psi}{t^{3/2}} + \cdots\right]. \tag{10.54}$$

The second term in the brackets in (10.54) is a higher order quantity as $t \to \infty$, than is the second term in (10.51').

The physical reason for the better agreement of the similar solution (10.52) with the exact solution lies in the fact that (10.52) corresponds to an instantaneous release of the same amount of heat at the point x_0, which is the center of gravity $\bar{x}$ of the initial temperature distribution $T_0(x)$. The instant of energy release corresponds to the time required for the heat to acquire the same lateral dispersion (measured by the second moment) relative to the center of gravity as has the initial temperature distribution. The "effective" amount of heat $E = c_v \rho Q$ in the improved similar solution (10.52) is found to be exactly equal to the actual amount of heat $c_v \rho \int_{-\infty}^{\infty} T_0(x)\,dx$. In a similar manner it is possible to find a self-similar solution which best approximates the exact solution with distributed heat sources in the case of nonlinear heat conduction.

The self-similar solution of (10.23) corresponding to an instantaneous heat release at the point $x = 0$ at the time $t = 0$ was given in §5 (equations (10.32), (10.33), (10.31), and (10.30)). It was shown in [9], which also discusses the mathematical features of this problem, that by an appropriate shift in position and time, that is, by properly selecting x_0 and τ, it is possible to obtain a self-similar solution $T(x - x_0, t + \tau, Q)$, which differs from the exact solution $T(x, t, Q)$ by terms of order higher than $t^{-(2n+3)/(n+2)}$*.

§10. Heat transfer by nonequilibrium radiation

Let us imagine that a spherical region of radius R_0 has been formed in low density air at a temperature T high enough that the heated sphere is completely transparent to its own thermal radiation and radiates as a volume radiator. If the radiation mean free path which characterizes the emission

* *Editors' note.* Since the energy and first moment integrals are both conserved in the nonlinear case also, Q and x_0 are still given by the expressions of (10.53). The second moment obeys the law

$$\frac{d}{dt}\int_{-\infty}^{\infty} Tx^2\,dx = \frac{2a}{n+1}\int_{-\infty}^{\infty} T^{n+1}\,dx,$$

and when taken about the center of gravity must approach $(t+\tau)^{2/(n+2)}$ asymptotically. Let us define $\tau'(t)$ by

$$t + \tau'(t) = \frac{\displaystyle\int_{-\infty}^{\infty} Tx^2\,dx - x_0^2\,Q}{\displaystyle\frac{(n+2)a}{n+1}\int_{-\infty}^{\infty} T^{n+1}\,dx},$$

and note that $\tau'(0)$ is given by an analogue of the formula for τ in (10.53). The quantity $\tau' \to \tau$ as $t \to \infty$. We can show (with $n > 0$) that $d\tau'/dt \geqslant 0$, so that τ' approaches its limit τ monotonically from below. Thus $\tau'(0)$, although useless in general as an estimate for τ, does serve as a lower bound.

coefficient of the air is $l_1(T)$, then the condition for transparency is $l_1(T) \gg R_0$. (We recall that the radiation mean free path, as a rule, increases rapidly with increasing temperature.) Photons born in the highly heated region leave almost unimpeded and are absorbed in the surrounding layers of cold air. The air in the central sphere is thus cooled by light emission and the peripheral layers are heated by light absorption. The heated region expands and the temperature in it drops. The process is quite similar to that of propagation of a thermal wave, with the difference, however, that the radiation that transfers the energy is here substantially out of equilibrium. This process of the transfer of heat by nonequilibrium radiation was treated by Kompaneets and Lantsburg [10, 11].

The radiation, having been absorbed first in the opaque peripheral layers of the sphere, heats them to a temperature T^* at which the air becomes transparent. Let the mean free path of photons which carry the major part of the energy of the radiation spectrum of the central sphere with temperature T be l_T in the peripheral region. It depends both on the characteristic frequencies (on T) and on the temperature of the air in which the absorption takes place. It is clear that the approximate condition which defines the transparency temperature T^* is

$$l_T(T^*) = R, \qquad (10.55)$$

where R is the radius of the heated sphere, which increases with time in comparison with the initial radius R_0. The temperature distribution is illustrated in Fig. 10.8.

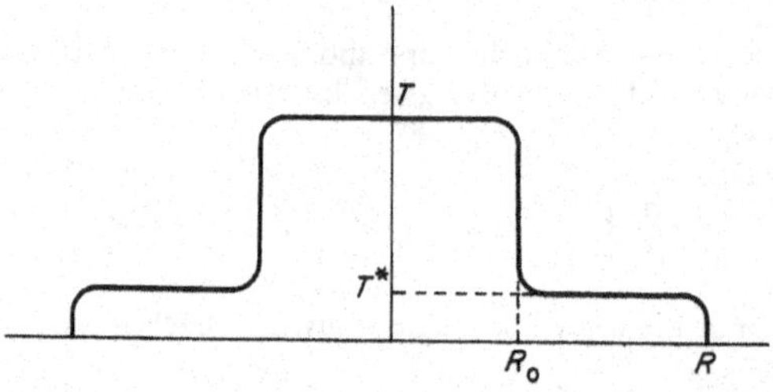

Fig. 10.8. Nonequilibrium radiation from a very hot central region.

If the temperature T in the central sphere is initially very high and the air in it is highly transparent to radiation, then the transparency temperature T^* is found to be appreciably lower than T, and the radiation born in the peripheral layers with temperatures of the order of T^* can be neglected. In this case the rate of expansion of the heated region is determined simply by an energy balance equation. The amount of energy radiated during the time interval dt at a time t in the highly heated central region is of the order of

$$\frac{cU_p(T)}{l_1(T)}\frac{4\pi R_0^3}{3}\,dt,$$

where $U_p = 4\sigma T^4/c$. This energy is absorbed in a peripheral layer of thickness dR and radius R and the air in this layer is heated to a temperature which is of the order of the transparency temperature T^*. It follows that

$$\frac{cU_p(T)}{l_1(T)}\frac{4\pi R_0^3}{3}\,dt = 4\pi R^2\,dR\,\rho\varepsilon(T^*),$$

where ρ is the density and ε is the specific internal energy of the air. From this we obtain the rate of expansion of the sphere

$$v = \frac{dR}{dt} = \frac{cU_p(T)}{\rho\varepsilon(T^*)}\frac{R_0}{3l_1(T)}\left(\frac{R_0}{R}\right)^2. \tag{10.56}$$

Here T, the temperature in the radiating central region, drops in accordance with the cooling equation

$$\rho\frac{d\varepsilon(T)}{dt} = -\frac{cU_p(T)}{l_1(T)}. \tag{10.57}$$

From (10.56) it might appear that the velocity of the sphere boundary v can be as large as we wish, even larger than the speed of light (if the radiation energy density U, which is of the order of $U_p R/l_1$ is higher than the energy density of the medium $\rho\varepsilon$). In fact, however, this is not so for the reason that (10.56) is valid only for the case when $v \ll c$. If the rate of expansion of the sphere is comparable with the speed of light, then this means that the radiated energy is used up not only in heating the medium, but also in "filling" the expanding sphere with radiation. Mathematically this is expressed by the fact that the velocity v is proportional not to $U/\rho\varepsilon$ but to $U/(U + \rho\varepsilon)$, and this automatically sets an upper limit for the velocity v which is less than c.

As the heated sphere expands and the air in the central region is cooled, the latter becomes less and less transparent and the temperatures T and T^* approach each other. When they become equal opaqueness sets in. At the same time the radiation from the peripheral layers, which sends photons not only outward, as from the central region, but also back into the interior of the sphere, becomes appreciable. The radiation density in the sphere then approximates the equilibrium density at the given temperature, and gradually the process takes on the character of the thermal wave considered in the preceding sections.

In order to indicate better how the transition takes place from one regime to the other at the transparency condition for the entire sphere, $T \approx T^*$, $l_1 \approx R$, let us estimate the speed of propagation of the boundary of the heated sphere in this limiting case from the formulas for nonequilibrium and thermal waves, respectively.

In the nonequilibrium regime, in the limit R tending to l_1 (from the side $R \ll l_1$) and T tending to T^*, the radiation comes not only from the central sphere but from the entire heated region, so that in the energy balance equation as expressed by formula (10.56) we should substitute R for R_0. Then, to the right order of magnitude with $T \sim T^*$ and $R \sim l_1$,

$$v \sim c \frac{U_p}{\rho\varepsilon} \frac{R}{l_1} \sim c \frac{U_p}{\rho\varepsilon}. \tag{10.58}$$

On the other hand, in the equilibrium regime the speed of the thermal wave is given approximately by

$$v = \frac{dR}{dt} \sim \frac{\chi}{R} \qquad [R \sim (\chi t)^{1/2}],$$

where the coefficient of radiation thermal conductivity χ, according to the definition (10.11), can be written in the form $\chi \sim lc\, U_p/\rho\varepsilon$. It follows that in order of magnitude the speed of the thermal wave is

$$v \sim c \frac{U_p}{\rho\varepsilon} \frac{l}{R},$$

where in the equilibrium case $R \gg l$. If we now let R tend to l, we obtain at the transparency condition, $R \sim l \sim l_1$, the same value as (10.58), $v \sim c\, U_p/\rho\varepsilon$.

XI. Shock waves in solids

§1. Introduction

The study of the laws governing the propagation of shock waves through condensed media, such as metals, water, etc., is of great theoretical and practical importance. In particular, such studies are necessary for the understanding and calculation of explosion phenomena. The theoretical analysis of data obtained from such studies yields information on the equation of state of solids and liquids subjected to high pressures, which is very important in the solution of a large number of problems in geophysics, astrophysics, and other branches of science.

A knowledge of the thermodynamic properties of the medium is necessary for describing the hydrodynamic processes which take place within it. While no appreciable difficulties are encountered in calculating the thermodynamic properties of gases, a theoretical description of the thermodynamic properties of solids and liquids at the high pressures generated by very strong shocks presents a very complex problem, which at the present time is still very far from understood. Therefore, experimental methods play a major role in the study of condensed media in a compressed state.

Until recently, the physics of high pressures was limited to the study of media compressed under static conditions in piezometers of various design. However, the pressures obtainable by this means can not exceed a hundred thousand atmospheres without the construction of extremely large facilities. Also, and more important, it is impossible to provide conditions for reliable measurements, since under higher pressures the piezometer bomb deforms, preventing the measurement of physical properties to the accuracy desired. Nevertheless, many problems in modern science and engineering are concerned with pressures of hundreds of thousands and millions of atmospheres.

In the postwar years, both in the USSR and in other countries, it was suggested that dynamic methods, based on the utilization of strong shock waves, be used for obtaining high pressures and compressions. Shock waves in metals and other condensed media with pressures of hundreds of thousands and millions of atmospheres were obtained and investigated. In the USSR these new methods were developed in works by Al'tshuler, Kormer, Krupnikov, Ledenev, Bakanova, Sinitsyn, Funtikov, Zhuchikhin, and others [1–5], and in the USA by Walsh, Christian, Mallory, Goranson, Bancroft, McQueen, Marsh, and others [22–26].

Soviet scientists were particularly successful in this direction, having succeeded in obtaining record pressures of five million atmospheres (the American authors investigated weaker shock waves; papers reporting the attainment of pressures of two million atmospheres, which were the highest achieved by them, were published later than those of the Soviet authors*). For the first time in the history of mankind a solid body was compressed by a factor of 2 or more; until then such a dense medium could be "encountered" only in the central regions of the earth and of other cosmic bodies. These outstanding achievements in obtaining high pressures and densities in solids have made it possible to draw a large number of interesting conclusions on the thermodynamic behavior of media under such extraordinary conditions and to determine by semi-empirical means important thermodynamic characteristics of highly compressed metals. The extremely short duration of the impact loads required seeking new measurement techniques that would permit the determination of physical properties under conditions of high-rate processes, and required the design of appropriate instrumentation. A large contribution in this direction was made by the Soviet investigators Tsukerman, Shnirman, Dubovik, Kevlishvili, Zavoiskii, and others [6–12].

The basic feature distinguishing the condensed from the gaseous state and determining the behavior of solids and liquids compressed by shock waves is the strong interaction between the atoms (or molecules) of the medium. The range of interatomic forces is very limited. It is of the order of the dimensions of the atoms and molecules, of the order of 10^{-8} cm. In a sufficiently rarefied gas, where the average distances between particles are very much greater than the particle dimensions, the interaction takes place mainly through collisions, during which the atoms or molecules approach each other closely.

The pressure in a gas is of thermal origin; it is related to the transfer of momentum by particles participating in the thermal motion, and is always proportional to the temperature: $p = nkT$. Relatively small pressures are required to compress a gas strongly. The limiting compression of atmospheric gas across a shock wave, as dictated by the conservation laws, is reached for pressures behind the front of several tens or a hundred atmospheres, so that a shock wave of this strength may be regarded as strong.

The behavior of a condensed medium with respect to a compression is different. The atoms or molecules of solids and liquids are close to each other and interact strongly. This interaction, in particular, is responsible for holding the atoms within the body. The interaction forces have a dual character. On the one hand, particles separated by sufficiently large distances are attracted to each other; on the other hand, when brought close together they

* This refers to the earliest publications (Soviet and American). In subsequent years papers have appeared describing investigations at still higher pressures. See the review [55].

repel each other as a result of the interpenetration of the electronic shells of the atoms. Equilibrium distances of atoms in a solid body in the absence of external pressure correspond to a mutual compensation between the attractive and repulsive forces, to a minimum in the interaction potential energy. In order to separate the atoms by a large distance, it is necessary to overcome the binding forces and to supply energy equal to the binding energy, which for metals is of the order of several tens or hundreds of kcal/mole (of the order of several ev/atom)*. In order to compress a material it is necessary to overcome the repulsive forces, which increase very rapidly as the atoms are brought together. The compressibility of metals is, by definition, $\kappa_0 = -(1/V) \cdot \partial V/\partial p$, and at standard conditions is of the order of 10^{-12} cm^2/dyne $\approx 10^{-6}$ atm^{-1}. In order to compress a cold metal by 10% an external pressure of 10^5 atm must be applied; the compressibility usually decreases with increasing pressure. Compression of metals by a factor of 2 requires pressures of the order of several million atmospheres.

Thus, a strong compression of a condensed medium generates a colossal internal pressure, even in the absence of heating, due only to the repulsive forces between the atoms. The existence of this nonthermal pressure, which is not a property of gases, determines the basic features of the behavior of solids and liquids compressed by shock waves. As we shall see below, the material is also very strongly heated by strong shock waves, and this results in the appearance of a pressure associated with the thermal motion of the atoms (and electrons). This pressure is referred to as "thermal" pressure, in contrast to the elastic or "cold" pressure caused by the repulsive forces. In principle, as the shock strength tends to infinity, the relative importance of the thermal pressure increases and, in the limit, the elastic pressure becomes small in comparison with the thermal pressure; under the action of extremely strong waves the initially solid medium behaves as a gas. However, for shock waves with pressures of the order of millions of atmospheres, as obtained in the laboratory, these two pressures are of comparable magnitude. The elastic pressure is dominant in weaker shock waves, with pressures of the order of hundreds of thousands of atmospheres and below. The thermal energy of the material compressed by the shock wave is also small in this case. Essentially all of the internal energy acquired by the medium from the wave is expended in overcoming the repulsive forces due to the compression and is concentrated in the form of potential elastic energy. The speed of propaga-

* The binding forces in solids are of various types. In accordance with their nature, solids are usually subdivided into five groups: (1) ionic crystals—for example, NaCl, with a binding energy $U = 180$ kcal/mole; (2) crystals with a covalent bond—for example, diamond, $U = 170$ kcal/mole; (3) metals, $U \sim 30$–200 kcal/mole; (4) molecular crystals, bound by van der Waals forces, with a weak bond—for example, for CH_4, $U = 2.4$ kcal/mole; (5) crystals with hydrogen bonds—for example, ice, $U = 12$ kcal/mole. Here we shall be mainly interested in metals.

tion of small disturbances in a condensed medium is, in contrast to gases, not temperature dependent. It is determined by the elastic compressibility of the medium.

The numerical characterization of the "strength" of a shock wave is also different in solids. The strength of a wave in gases is measured by the pressure ratio across the wave front. The limiting density ratio of about 4 to 10 is obtained when this ratio is equal to several tens or a hundred. In this case the shock wave velocity is considerably greater than the speed of sound in the initial gas, and the gas behind the front is accelerated to velocities close to that of the shock wave. If the gas was initially at atmospheric pressure, then a shock wave with a strength of even a hundred atmospheres is regarded as strong. In a solid or liquid, a shock wave with a strength of even a hundred thousand atmospheres is regarded as weak. Such a wave differs little from an acoustic wave: it travels with a speed close to the speed of sound, compresses the material by only a few percent or perhaps of the order of ten percent, and imparts a velocity to the material behind the front which is of the order of a tenth the velocity of the wave itself. If we characterize the strength of the shock wave by the ratio of its speed to the speed of sound in the undisturbed medium or by the closeness of the density ratio to its limiting value, then a strong wave for condensed media is one whose pressure is not less than tens or hundreds of millions of atmospheres.

In this chapter we shall consider in detail the physical behavior of solids at high pressures and densities, we shall familiarize ourselves with the properties of shock compression, we shall describe experimental methods for studying shock waves moving through solids, and we shall discuss the results obtained by these methods. We shall also consider some physical phenomena observed in the passage of shock waves through metals and other media and on unloading of the substance, when a shock wave reaches a free surface.

A great deal of valuable information about these problems may be found in the recently published review by Al'tshuler [55], in which a large amount of experimental data are brought together and analyzed.

1. Thermodynamic properties of solids at high pressures and temperatures

§2. Compression of a cold material

The pressure p and the specific internal energy ε of a solid material can be divided into two parts. The first part, the elastic component p_c or ε_c, is related exclusively to the forces of interaction between the atoms of the medium* and is entirely independent of the temperature. The other part, the thermal component, is related to the heating of the body, that is, with the temperature. The elastic components p_c and ε_c depend only on the density of the material ρ or the specific volume $V = 1/\rho$ and are equal to the total pressure and specific internal energy at absolute zero temperature; that is why they are sometimes called the "cold" pressure and energy. In this section we shall consider only the elastic components of the pressure and energy. Therefore we shall assume that the body is at absolute zero.

The state of mechanical equilibrium of a solid at zero temperature and pressure† is characterized by the mutual compensation of the interatomic forces of attraction and repulsion and by a minimum in the elastic potential energy; this minimum can be taken as the origin for the energy $\varepsilon_c = 0$‡. Let us denote the specific volume of a body in this state ($p = 0$, $T = 0$) by V_{0c}. This volume is slightly smaller than the volume V_0 of the body under standard conditions ($p = 0$ or 1 atm, which are equivalent, and $T_0 \approx 300°\mathrm{K}$), since heating the material from absolute zero to room temperature T_0 results in a thermal expansion, discussed in the following section. The standard volume of metals V_0 is usually 1–2% larger than V_{0c}, which we shall call the zero volume. In many cases the small difference in volume between V_0 and V_{0c} can be neglected. In considering the behavior of a solid whose volume is being changed, we shall be referring to the compression (and expansion) of the body isotropically in all directions, without discussing the effects connected with the anisotropy of the elastic properties, shearing strain, strength, etc., which are present at comparatively low pressures and compressions.

The potential energy curve for a body as a function of its specific volume V has the same qualitative character as the potential energy curve for the interaction of two atoms of a molecule as a function of the internuclear distance. This curve is shown schematically in Fig. 11.1. If the volume V is greater than the zero volume V_{0c}, then the attractive forces predominate. The interaction forces fall off rapidly as the distance between the atoms increases. Hence when the volume increases, when the atoms move farther apart, the potential energy increases asymptotically to a constant value U equal to the binding energy of the atoms in the body. The energy which must be expended to

* We shall be dealing primarily with metals, which are composed of atoms rather than molecules.

† Atmospheric pressure is negligibly small in comparison with the pressures which arise even for very small changes in the volume of the body. Therefore it makes no difference whether the body is in vacuum ($p_c = 0$) or at atmospheric pressure ($p_c = 1$ atm).

‡ At absolute zero the atoms perform so-called zero-point oscillations, with an energy $h\nu/2$ per normal vibration mode of frequency ν. This energy can be accounted for in the potential energy $\varepsilon_c(V)$ in such a way that ε_c is measured from the zero-point vibrational level in the equilibrium state of the body at $p_c = 0$.

remove all atoms of a unit mass of a material to infinity is given by U; it is approximately equal to the heat of vaporization of the body (strictly speaking, it is equal to the heat of vaporization at absolute zero). Heats of vaporization for metals are usually of the order of several tens or hundreds of kcal/mole,

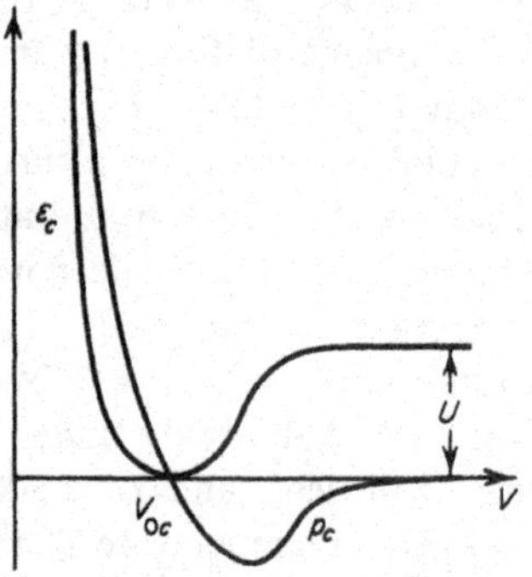

Fig. 11.1. Potential energy and elastic pressure curves of a body as a function of the specific volume.

of several electron volts per atom*. The binding forces weaken at distances of the order of the dimensions of an atomic cell, so that the curve of $\varepsilon_c(V)$ approaches its asymptote $\varepsilon_c(V) = U$ when the body expands by an order of magnitude (say when the distance between the atoms doubles).

The repulsive forces, which increase sharply with decreasing interatomic distance, dominate when a body is compressed, and thus the potential energy $\varepsilon_c(V)$ increases rapidly when the volume is less than its zero value. In order to have some idea of the rate of increase and of the order of magnitude of the energy, we note that according to [1] the energy needed for the cold compression of iron by 7% is $\varepsilon_c = 5.25 \cdot 10^8$ erg/g = 0.03 ev/atom, and for compression by a factor of 1.5 is $\varepsilon_c = 2.42 \cdot 10^{10}$ erg/g = 1.4 ev/atom (here the pressures are equal to $p_c = 1.31 \cdot 10^5$ atm and $p_c = 1.36 \cdot 10^6$ atm, respectively).

The elastic pressure is related to the potential energy by

$$p_c = -\frac{d\varepsilon_c}{dV}, \tag{11.1}$$

which formula has a natural mechanical meaning (increase of energy is equal to the work of compression) and can be regarded as the equation for the isotherm or isentrope of cold compression. Indeed, (11.1) follows from the general thermodynamic relation $T\,dS = d\varepsilon + p\,dV$, in which we set T equal to zero. But if $T = 0$ then the entropy S, according to Nernst's theorem, is

* For example, for iron—94 kcal/mole = 4.1 ev/atom = $6.96 \cdot 10^{10}$ erg/g; for aluminum —55 kcal/mole = 2.4 ev/atom = $8.45 \cdot 10^{10}$ erg/g.

also equal to zero, and hence remains constant. Therefore, the isotherm $T = 0$ is also the isentrope $S = 0$.

The pressure curve $p_c(V)$ is also shown schematically in Fig. 11.1. The elastic pressure at the point $V = V_{0c}$ is zero; the pressure increases rapidly with compression, and, at least formally, becomes negative with expansion. The negative sign on the pressure describes the physical fact that in order to expand the body from the zero volume corresponding to mechanical equilibrium at $T = 0$ and $p = 0$, a tensile force must be applied to the body. This force must overcome the binding forces tending to return the body to the equilibrium volume V_{0c}.

The process of cold expansion $p_c(V)$ for $V > V_{0c}$ cannot be directly followed experimentally, since it is not possible in practice to achieve a strong extension of a metal in all directions. The magnitude of the negative pressures can be estimated from the heat of vaporization of the material. By definition, the area under the curve for cold expansion from zero volume to infinity is

$$\int_{V_{0c}}^{\infty} p_c(V)\, dV = -U. \tag{11.2}$$

If the binding forces weaken considerably when the body expands by a factor of approximately 10 (the interatomic distance is about doubled), then the maximum negative pressure is of the order $p_{max} \sim U/10V_{0c}$, which for iron, for example, is $p_{max} \sim 6 \cdot 10^{10}$ bar $= 6 \cdot 10^4$ atm*.

The slope of the elastic pressure curve at the point of zero pressure corresponds to the definition of the compressibility of the material under ordinary conditions (the isentropic compressibility differs only slightly from the isothermal compressibility; at $T = 0$ they are identical). The compressibility of iron is

$$\kappa_0 = -\frac{1}{V_0}\left(\frac{\partial V}{\partial p}\right)_{T_0} = 5.9 \cdot 10^{-13} \quad \text{bar}^{-1},$$

from which

$$-V_{0c}\left(\frac{dp_c}{dV}\right)_{V=V_{0c}} \approx 1.7 \cdot 10^{12} \quad \text{bar}.$$

The slope of the cold compression curve determines the speed of propagation of elastic waves in the body, the speed of sound. It will be shown later that in a solid there exist several "speeds of sound". For the time being we note that the speed of sound defined in the ordinary manner in terms of the

* This value is appreciably greater than the ultimate tensile strength of iron, which is usually of the order of 10^9 bar $= 10^3$ atm. The low value of the tensile strength is related to the one-sided character of the extension, to the cracks which are present in actual metals, to the polycrystalline structure, etc. We note that the ultimate strength of some types of iron can reach 1 to $2 \cdot 10^4$ atm.

compressibility $c_0 = V|\partial p/\partial V|_S^{1/2}$, is equal to 5.85 km/sec for iron at standard conditions.

Theoretical calculations of cold compression curves $p_c(V)$ or $\varepsilon_c(V)$ in the range of compressions and pressures attainable in practice are based on a quantum-mechanical consideration of the interatomic interaction. In a number of cases it is possible to obtain satisfactory agreement with experimental compressibility data, in particular for alkaline and alkaline-earth metals at low pressures. A detailed presentation of these calculations and a comparison with the experimental data of Bridgman on the static compression of materials up to several tens of thousands of atmospheres can be found in the book of Gombàs [13], in which may also be found references to the literature. Detailed data on cold compression curves for a number of metals (and also for sodium chloride) up to pressures of several million atmospheres and densities approximately twice standard were obtained by Al'tshuler, Krupnikov, Kormer, Bakanova, Trunin, Pavlovskii, Kuleshova, and Urlin [1–5, 14, 15], from theoretical analysis of experimental shock compression results (see [55]). We shall say more about these experiments later on; here for illustration we present the $p_c(V)$ and $\varepsilon_c(V)$ curves for iron (Fig. 11.2).

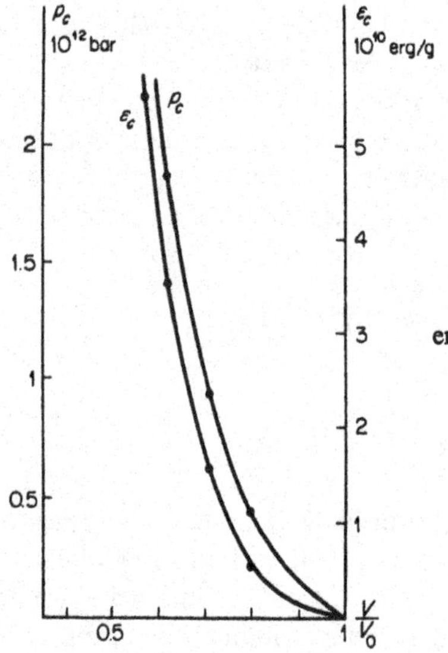

Fig. 11.2. Elastic pressure p_c and energy ε_c of iron (from the data of [1]).

Theoretically it is possible to establish a limiting law for the cold compression of a material at very high pressures and densities. Under conditions of very strong compression the electronic shells of atoms, to some extent, lose their individual structure. The state of the material in this case can be approximately described by the Thomas–Fermi statistical model of an atom, or, somewhat more exactly, by the Thomas–Fermi–Dirac model (the latter model takes exchange energy into account)*. The equation of state for a material in the Thomas–Fermi model was discussed in §13 of Chapter III. In the limit of very high pressures and densities the cold compression pressure is

$$p_c \sim \rho^{5/3} \sim V^{-5/3}. \tag{11.3}$$

This is also only a limiting law for this statistical model of the atom, since for compressions which are not too large the model gives another dependence for $p_c(V)$. In order to compare the actual elastic pressure curves with those obtained from the statistical model, we present a logarithmic plot from [1] which gives the experimental curve for iron and the curves calculated by the Thomas–Fermi and Thomas–Fermi–Dirac methods (Fig. 11.3). It is evident

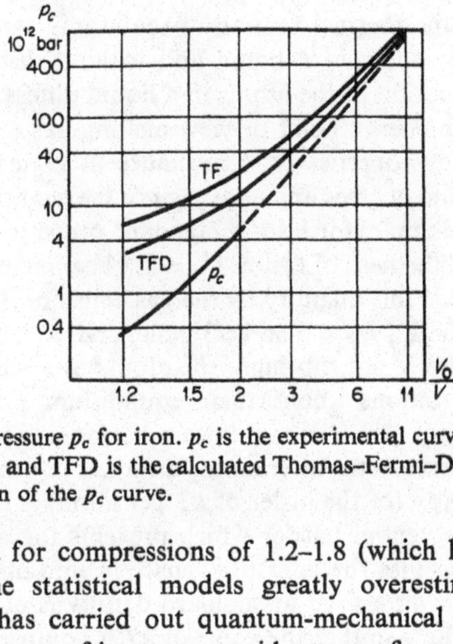

Fig. 11.3. Elastic pressure p_c for iron. p_c is the experimental curve; TF is the calculated Thomas–Fermi curve, and TFD is the calculated Thomas–Fermi–Dirac curve. The dashed line is an extrapolation of the p_c curve.

from the plot that for compressions of 1.2–1.8 (which have been obtained experimentally) the statistical models greatly overestimate the pressure. Gandel'man [37] has carried out quantum-mechanical calculations of the cold compression curve of iron over a wide range of pressures.

§3. Thermal motion of atoms

The atoms of a material are set into motion by heating. A definite energy and pressure are connected with the thermal motion of the atoms. At temperatures of the order of tens of thousands of degrees and above the thermal excitation of the electrons plays an important role. As noted in the preceding

* Calculations by the Thomas–Fermi–Dirac method have a real meaning only in those cases when the exchange correction is small. They essentially indicate the limits of applicability of the Thomas–Fermi method. If the exchange correction is found to be large, it shows that the Thomas–Fermi–Dirac method is no longer valid.

section, the total energy and pressure can be represented as a sum of their elastic and thermal contributions. The thermal contributions, in turn, can be broken up into two parts, one part corresponding to the thermal motion of the atoms (or rather their nuclei) ε_T, p_T, and another part corresponding to the thermal excitation of the electrons ε_e, p_e. The specific internal energy and pressure of a solid can then be written as

$$\varepsilon = \varepsilon_c + \varepsilon_T + \varepsilon_e, \tag{11.4}$$

$$p = p_c + p_T + p_e. \tag{11.5}$$

The electronic terms will be discussed later. At temperatures below approximately ten thousand degrees, these terms are small and can be neglected in (11.4) and (11.5).

Let us consider the thermal motion of the atoms, without making a distinction between a solid and a liquid and without discussing the effect of melting. Thermal motion of the atoms in a liquid differs very little from that in a solid. From an energy point of view melting has a very small effect on the thermodynamic properties of a substance at high temperatures of the order of ten thousand degrees and above, since the heat of fusion is comparatively small. For example, for lead at standard pressure the melting point is $T_{\text{melt}} = 600°\text{K}$, and the heat of fusion $U_f = 1.3$ kcal/mole, which corresponds to 650°K if we divide this quantity by the gas constant $\mathscr{R} = 2$ cal/deg · mole; for iron $T_{\text{melt}} = 1808°\text{K}$, $U_f = 3.86$ kcal/mole, and $U_f/\mathscr{R} = 1940°\text{K}$.

If the temperature is not too high, the atoms of a solid (and of a liquid) undergo small vibrations about their equilibrium positions (the crystal lattice sites in a solid). These vibrations are harmonic as long as their amplitude is much smaller than the interatomic distance, in other words, as long as the vibrational energy (of the order of kT per atom) is appreciably less than the height of the potential barrier which prevents the atoms from jumping from the lattice sites into the interstitial space or into other vacant sites. The height of the barrier in a solid at standard density is of the order of one or several electron volts*, so that the value of kT is comparable with the height of the potential barrier at temperatures of the order of ten or several tens of thousands of degrees. At higher temperatures the atoms are almost completely free to move within the body, and the thermal motion loses its oscillatory character and becomes closer to a random motion, akin to that in a gas. Thus the substance is transformed into a dense gas of strongly interacting atoms.

The situation becomes different, however, when the heating is accompanied

* This quantity is approximately equal to the activation energy for self-diffusion of atoms in the body ΔU. It is usually somewhat less than the binding energy, but of the same order of magnitude, $\Delta U \approx (0.5\text{–}0.7)U$.

by compression. The compression very sharply increases the repulsive forces between neighboring atoms, with the result that the height of the potential barrier which must be overcome by the atom in order to move out of its cell (from its site in the crystal lattice) sharply increases. Free displacement of the atoms in the body becomes very difficult and the motion of the atom remains limited to the space of its cell. This is illustrated in Fig. 11.4.

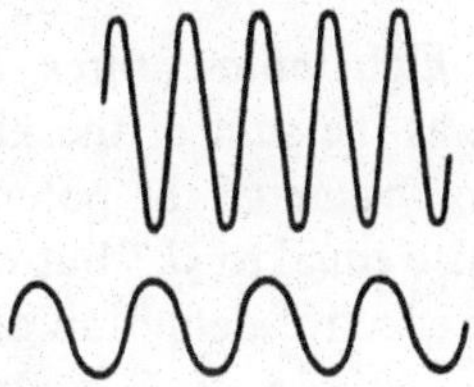

Fig. 11.4. Schematic diagram illustrating the change in height of the potential barrier for atoms in a compressed solid.

Within some rough approximation we can regard the thermal motion of the atoms in a compressed material as small vibrations about their equilibrium positions even at the maximum temperatures of 20,000–30,000°K attainable in the strongest shock waves which have been studied experimentally.

At temperatures above several hundreds of degrees Kelvin quantum effects play no role in the vibrations and the specific heat of a body whose atoms vibrate harmonically is equal to its classical value of $3k$ per atom or $c_v = 3Nk$ per unit mass, where N is the number of atoms per unit mass. To take into account the difference of the specific heat from its value at low temperatures where quantum effects are important, let us express the thermal energy connected with vibrations of the atoms in the form

$$\varepsilon_T = c_v(T - T_0) + \varepsilon_0, \qquad c_v = 3Nk, \tag{11.6}$$

where $\varepsilon_0 = \int_0^{T_0} c_v(T)\, dT$ is the thermal energy at room temperature,obtainable from appropriate tables. For temperatures T much higher than T_0 we can neglect the difference between $c_v T_0$ and ε_0, since both quantities are small in comparison with $c_v T$. In this case

$$\varepsilon_T = c_v T, \qquad c_v = 3Nk. \tag{11.7}$$

The specific heat is equal to $3k$ per atom only when the thermal motion of the atoms has an oscillatory character. At sufficiently high temperatures the atoms move freely through the body; the specific heat then corresponds only to the translational degrees of freedom of the atoms and is equal to $\frac{3}{2}k$ per atom, as in a monatomic gas. The transition from vibrational to translational motion of the atoms and the corresponding decrease in the specific heat occur gradually, in the range of temperatures for which the kinetic energy of an atom $\frac{3}{2}kT$ is of the order of the potential barrier against the motion of the atoms through the body $\Delta U/N$. An effective boundary dividing the regions with the limiting values of specific heat of $3k$ and of $\frac{3}{2}k$ may be defined by the threshold temperature

$$T_k = \frac{2}{3}\frac{\Delta U}{kN}. \tag{11.8}$$

At high temperatures $T \gg T_k$ we can represent the thermal energy per atom as the sum of the kinetic energy of translational motion $\frac{3}{2}kT$ and the average value of the potential energy, which in the case of small vibrations was also equal to $\frac{3}{2}kT$ but which is now of the order of $\Delta U/k$. This corresponds to an effective definition of the specific heat by the discontinuous relation

$$c_v = 3Nk \quad \text{when} \quad T < T_k; \qquad c_v = \tfrac{3}{2}Nk \quad \text{when} \quad T > T_k.$$

For $T > T_k$, the energy is then equal to

$$\varepsilon_T = \int_0^T c_v\,dT = \int_0^{T_k} 3Nk\,dT + \int_{T_k}^T \tfrac{3}{2}Nk\,dT = \tfrac{3}{2}NkT + \Delta U. \tag{11.9}$$

As an example, we note that for iron at standard density

$$\frac{\Delta U}{k} \approx 2.5 \text{ ev} \qquad \text{and} \qquad T_k \approx 20{,}000^\circ\text{K}.$$

As the body is compressed the height of the potential barrier increases and the threshold temperature T_k rises, so that the curves giving the temperature

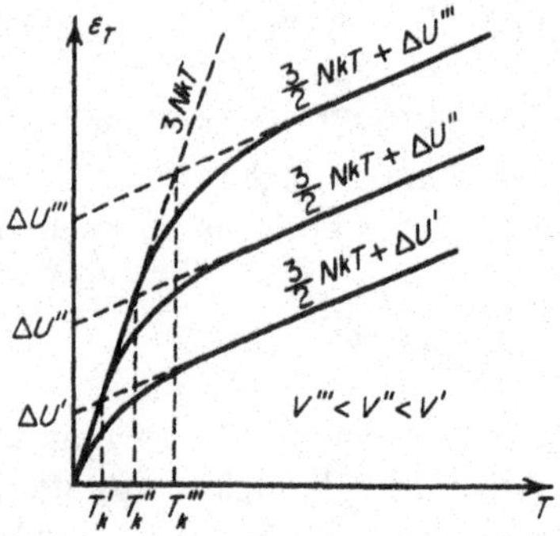

Fig. 11.5. Dependence of thermal energy on temperature for different densities (volumes).

dependence of the thermal energy for different densities (volumes) have the form shown schematically in Fig. 11.5.

In the limiting case $T \gg T_k$, when the thermal motion of the atoms (more precisely, of the nuclei) does not differ from that of a gas, the thermal pressure related to this motion is, as usual,

$$p_T = nkT = \frac{NkT}{V} = \frac{2}{3}\frac{\varepsilon_T}{V}.$$

§4. Equation of state for a material whose atoms undergo small vibrations

We shall assume that the atoms of the body undergo small vibrations about their equilibrium positions, and we shall find the value of the thermal pressure $p_T(V, T)$ corresponding to these vibrations. If the temperature is not too high and we can neglect the electronic excitation, then the equation of state and the internal energy of the body can be expressed in the form

$$p = p_c(V) + p_T(V, T), \tag{11.10}$$

$$\varepsilon = \varepsilon_c(V) + 3NkT. \tag{11.11}$$

The temperature dependence of the thermal pressure can be obtained immediately from the general thermodynamic identity

$$\left(\frac{\partial \varepsilon}{\partial V}\right)_T = T\left(\frac{\partial p}{\partial T}\right)_V - p. \tag{11.12}$$

The elastic terms, in accordance with (11.1), satisfy this equation automatically. Noting that the specific heat $c_v = 3Nk$ is independent of volume, we obtain from (11.12) the result that the thermal pressure is proportional to the temperature: $p_T = \varphi(V)T$, where $\varphi(V)$ is some function of the volume.

Let us rewrite this equation in the form

$$p_T = \Gamma(V)\frac{c_v T}{V} = \Gamma(V)\frac{\varepsilon_T}{V}. \tag{11.13}$$

The quantity Γ characterizing the ratio of the thermal pressure to the thermal energy of the lattice is called the Grüneisen coefficient. The Grüneisen coefficient for a body at standard volume $\Gamma_0 = \Gamma(V_0)$ is related to the other properties of the material through the well-known thermodynamic relation (see, for example, [16])

$$\left(\frac{\partial p}{\partial T}\right)_V\left(\frac{\partial T}{\partial V}\right)_p\left(\frac{\partial V}{\partial p}\right)_T = -1. \tag{11.14}$$

Setting $-(1/V_0)(\partial V/\partial p)_T = \kappa_0$ for the isothermal compressibility of the material at standard conditions, and $(1/V_0)(\partial V/\partial T)_p = \alpha$ for the coefficient of thermal expansion at constant pressure, we obtain

$$\Gamma_0 = \frac{V_0\,\alpha}{c_v\,\kappa_0} = \frac{\alpha}{\rho_0\,c_v\,\kappa_0} = \frac{\alpha c_0^2}{c_v} \tag{11.15}$$

(c_0 is the speed of sound determined by the isothermal compressibility). The properties of several metals at standard conditions are given in Table 11.1, taken from [3]*. The Grüneisen coefficient Γ corresponds to the specific heat ratio for a calorically perfect gas (with constant specific heats) decreased by unity (we recall the equation of state for such a gas is $p = (\gamma - 1)\,\varepsilon/V$).

Table 11.1

SOME CHARACTERISTICS OF METALS AT STANDARD CONDITIONS

	Al	Cu	Pb
ρ_0, g/cm^3	2.71	8.93	11.34
$c_v \cdot 10^{-6}$, erg/g · deg	8.96	3.82	1.29
$\kappa_0 \cdot 10^{12}$, cm^2/dyne	1.37	0.73	2.42
$\alpha \cdot 10^5$, deg^{-1}	2.31	1.65	2.9
Γ_0	2.09	1.98	2.46
c_0, km/sec	5.2	3.95	1.91
$\varepsilon_0 \cdot 10^{-8}$, erg/g	16.1	7.71	3.23
β_0, erg/g·deg^2	500	110	144

By virtue of the condition used in deriving (11.13), that the specific heat c_v is independent of volume, it was found that the Grüneisen coefficient is independent of temperature. However, in reality, in the limit of very high temperatures for which the thermal motion of the atoms (nuclei) becomes random, (11.13) should become the equation of state for a monatomic gas, and $\Gamma \to \frac{2}{3}$ as $T \to \infty$. If we imagine that the atoms are separated and removed to large distances by an external force (the volume increases), then the material becomes a gas even at low temperatures, so that formally, as $V \to \infty$, $\Gamma \to \frac{2}{3}$. As may be seen from Table 11.1, the Grüneisen coefficient for metals at standard conditions is close to 2.

In order to clarify the physical meaning of the Grüneisen coefficient $\Gamma(V)$, which arose formally as an arbitrary function from the integration of (11.12), we turn to a well-known expression from statistical physics for the free energy of a body whose atoms vibrate harmonically. At high temperatures, when kT is much larger than the energy of vibrational quanta $h\nu$, the specific free energy is (see [16])

$$F = \varepsilon_c(V) + 3NkT \ln \frac{h\bar{\nu}}{kT}, \tag{11.16}$$

where $\bar{\nu}$ is a certain average vibrational frequency which is related to the

* The meaning of β_0 will be given in the next section.

Debye temperature θ by $h\bar{\nu} = e^{-1/3}k\theta = 0.715k\theta$ (for example, for iron $\theta = 420°\text{K}$). The first term in (11.16) represents the potential energy of interaction of the atoms and is the same as the energy of the cold body. The second term describes the thermal part of the free energy. From (11.16), using general thermodynamic relationships, we can easily find the specific internal energy and pressure of the body

$$\varepsilon = F - T\frac{\partial F}{\partial T} = \varepsilon_c(V) + 3NkT = \varepsilon_c + \varepsilon_T$$

(we naturally arrived at (11.11)) and

$$p = -\frac{\partial F}{\partial V} = -\frac{d\varepsilon_c}{dV} - 3NkT\frac{\partial \ln \bar{\nu}}{\partial V}.$$

The first term gives the elastic pressure (which we already know), and the second the thermal pressure. Using (11.13), which defines the Grüneisen coefficient, we find

$$\Gamma(V) = -\frac{\partial \ln \bar{\nu}}{\partial \ln V}. \tag{11.17}$$

The Grüneisen coefficient can be related to the function of cold compression by the following simple considerations. The average frequency of the spectrum of elastic vibrations of the lattice $\bar{\nu}$ is, obviously, close to the maximum frequency. The order of magnitude of the maximum frequency is equal to the ratio of the speed of propagation of elastic compression waves c_0 to the minimum wavelength which, in turn, is of the order of the interatomic distance r_0, so that $\bar{\nu} \sim c_0/r_0$. However, the speed of sound is $c_0 = (-V^2\, dp_c/dV)^{1/2}$, and $r_0 \sim V^{1/3}$, whence

$$\bar{\nu} \sim V^{2/3}\left(-\frac{dp_c}{dV}\right)^{1/2}.$$

Taking the logarithmic derivative of this expression, we obtain

$$\Gamma(V) = -\frac{\partial \ln \bar{\nu}}{\partial \ln V} = -\frac{2}{3} - \frac{V}{2}\frac{(d^2p_c/dV^2)}{(dp_c/dV)}. \tag{11.18}$$

This equation was obtained by Slater [17] and by Landau and Stanyukovich [18]. It has been shown experimentally that the Grüneisen coefficient decreases slightly on compression (for a decrease in the specific volume V).

In order to get some idea of the order of magnitude of the thermal pressure (11.13), we note that if, for example, aluminum is heated at constant (standard) volume to a temperature of 1000°K, the pressure will rise to $p_T = 51{,}000$ atm. A solid expands when heated at standard conditions, at constant atmospheric pressure. The reason for this thermal expansion is perfectly clear and may be seen from an examination of equation (11.10) for the pressure. The positive thermal pressure p_T increases on heating. The total pressure can thus remain constant only if the elastic pressure p_c becomes negative, and the body must expand up to that point when the binding forces holding the atoms in the lattice, or the negative pressure, will no longer counterbalance the repulsive effect of the positive thermal pressure. This clarifies the relationship expressed by (11.15) between the Grüneisen coefficient, the coefficient of thermal expansion, and the compressibility. Actually, a small expansion at constant pressure is related to a small amount of heating by the condition

$$dp = dp_c + dp_T \approx \frac{dp_c}{dV}\,dV + \frac{\partial p_T}{\partial T}\,dT = \frac{dp_c}{dV}\,dV + \Gamma_0 \frac{c_v}{V_0}\,dT = 0,$$

from which (11.14) and (11.15) follow*.

As an example let us estimate the expansion of aluminum heated at constant pressure (zero or atmospheric, which is equivalent) from absolute zero to room temperature $T = 300°\mathrm{K}$. Using the constants given in Table 11.1, we find $\Delta V/V \approx \Gamma_0(c_v/V_0)\kappa_0\,\Delta T \approx 2\%$ ($\Delta T = 300°\mathrm{K}$). Moreover, the thermal pressure at $T_0 = 300°\mathrm{K}$ is the same as the absolute value of the elastic pressure, equal to $p_{T_0} = 17{,}000$ atm. It is thus evident that it is always possible to consider atmospheric pressure as being equal to zero, since it is negligibly small in comparison with both pressure components even at room temperature.

If the function $\Gamma(V)$ is known, the entropy of the material is easily found. Considering states with densities differing little from standard, we can regard Γ as constant and equal to its standard value Γ_0. We then get for the entropy the relation

$$dS = \frac{d\varepsilon + p\,dV}{T} = \frac{d\varepsilon_T + p_T\,dV}{T} = c_v\frac{dT}{T} + \Gamma_0 c_v \frac{dV}{V},$$

from which the specific entropy is

$$S = c_v \ln \frac{T}{T_0}\left(\frac{V}{V_0}\right)^{\Gamma_0} + S_0, \tag{11.19}$$

where S_0 is the entropy at standard conditions T_0, V_0, and can usually be found tabulated. The isentropic relation between the temperature and volume is given by†

$$\frac{T}{T_0} = \left(\frac{V_0}{V}\right)^{\Gamma_0}. \tag{11.20}$$

* We only consider materials with normal properties, which expand on heating.

† Compare with the isentropic relation between T and V in a gas with constant specific heats $T \sim V^{-(\gamma-1)}$; Γ corresponds to $\gamma - 1$.

Expressing the temperature in terms of pressure by means of the equation of state

$$p = p_c(V) + \Gamma_0 \frac{c_v T}{V}, \tag{11.21}$$

we find the isentropic relation between the pressure and volume

$$\frac{p - p_c(V)}{p_{T_0}} = \left(\frac{V_0}{V}\right)^{\Gamma_0 + 1}, \tag{11.22}$$

where $p_{T_0} = \Gamma_0 c_v T_0 / V_0$ is the thermal component of the pressure at standard conditions*. In the case of small compressions, which nevertheless are accompanied by a sharp increase in pressure (in comparison with atmospheric pressure, but not with p_{T_0}), the isentrope $p(V)$ passes at an almost constant distance from the cold compression curve $p_c(V)$. For relatively large compressions (by a factor of 1.5 to 2) $p \gg p_{T_0}$ and the relative deviation of the isentrope from the cold compression curve $[p - p_c(V)]/p_c(V)$ becomes small.

§5. Thermal excitation of electrons

In the simplest model of a metal the outer valence electrons of the metal atoms are removed from their places in the atom and, together, form a free electron gas, completely filling the crystalline body whose sites are now filled by ions or atomic remainders†. The electron gas is governed by Fermi–Dirac quantum statistics, the elements of which were presented in §12 of Chapter III.

At absolute zero the electron gas is completely degenerate; in accordance with the Pauli principle the electrons occupy the lowest energy states and their kinetic energy does not exceed the Fermi limiting energy (3.88)

$$E_0 = \frac{h^2}{8\pi^2 m_e} (3\pi^2 n_e)^{2/3}$$

(n_e is the number density of free electrons and m_e is the electron mass). The energy E_0 in metals is usually of the order of several electron volts, and the degeneracy temperature corresponding to it $T^* = E_0/k$ is of the order of several tens of thousands of degrees‡.

The kinetic energy of a completely degenerate electron gas, which is of the order of E_0 per electron, is included in the elastic energy of the body and is not related to the thermal energy. In exactly the same manner, the "kinetic"

* The isotherm is $[p - p_c(V)]/p_{T_0} = V_0/V$. When the volume changes are small the isotherm almost coincides with the isentrope (the pressure change in this case is large).

† We restrict ourselves here to elementary considerations and shall not be concerned with the modern electron theory of metals.

‡ For example, for Na, $T^* = 37{,}000°K$; for K it is 24,000°K, for Ag it is 64,000°K, and for Cu it is 82,000°K.

pressure corresponding to it is included in the elastic pressure, together with the "potential" pressure which arises from the electrostatic interaction between the electrons and ions. The sum of the total pressure of nonthermal origin is equal to zero if the body is in a vacuum at absolute zero.

If the temperature increases, the electrons partially move over to higher energy states, exceeding the Fermi limiting energy, and the energy of the electron gas increases. If the temperature T is much lower than the Fermi temperature T^*, then, roughly speaking, electrons escape from the initial Fermi sphere in momentum space with an energy increase of the order of kT from the Fermi limit. The number of excited electrons is a fraction of the order of kT/E_0 of the total number of electrons. Each of these electrons acquires an additional energy of the order of kT. The order of magnitude of the thermal energy per electron is thus $(kT/E_0)kT$ and is proportional to $V^{2/3}T^2$ (since $E_0 \sim n_e^{2/3} \sim V^{-2/3}$). With the inclusion of a numerical coefficient, the thermal energy of the electrons per unit mass of metal for $T \ll T^*$ is found to be (see, for example, [16])

$$\varepsilon_e = \tfrac{1}{2}\beta T^2, \tag{11.23}$$

where the coefficient β depends on the density of the material and is given by

$$\beta = \beta_0\left(\frac{V}{V_0}\right)^{2/3}, \qquad \beta_0 = \frac{4\pi^4}{(3\pi^2)^{2/3}}\frac{k^2 m_e}{h^2} N_e^{1/3} V_0^{2/3} \tag{11.24}$$

(N_e is the number of free electrons per unit mass of metal and V_0 is the standard specific volume of the metal). The specific heat at constant volume is proportional to the temperature and is equal to

$$c_{v_e} = \beta T. \tag{11.25}$$

Knowing the number of free electrons per atom of metal, we can use (11.24) to calculate the coefficient β_0 and to find the electronic specific heat at a given temperature. Experimentally, the electronic specific heat is determined at very low temperatures, where the specific heat of the lattice is governed by quantum laws and is proportional to T^3. At sufficiently low temperatures the electronic specific heat, which is proportional only to the first power of T, dominates, and thus can be measured. At room temperature, however, the electronic specific heat is usually smaller by a factor of tens and even a hundred than the specific heat of the lattice, which under these conditions is constant and equal to its classical value $c_v = 3Nk$.

Experimental values of the electronic specific heat coefficients β_0 for several metals are given in Table 11.1*. In comparing the values of the electronic and lattice specific heats at different temperatures it becomes apparent that even at temperatures as low as 10,000°K the electronic specific

* They agree in order of magnitude with the values calculated from (11.24).

heat is quite appreciable, and, say, at 50,000°K it becomes even larger than the specific heat of the lattice. It should be noted, however, that (11.25) is valid only as long as the temperature is below the Fermi temperature.

For $T \gg T^*$ a free electron gas with a constant number of electrons is not degenerate, and its specific heat is given by the classical value $c_{v_e} = \frac{3}{2}N_e k$. In reality, however, the actual number of "free" electrons increases at high temperatures and the electronic specific heat of the material can no longer be described by simple equations. The problem of the electronic specific heat of a dense gas at high temperatures was considered in detail in §14 of Chapter III. At temperatures of the order of 10,000–20,000°K, which have been attained in experiments on shock compression of metals, this situation is still far from being reached, and the electronic specific heat can be taken approximately as being proportional to the temperature, as follows from (11.25). We should mention that the degeneracy temperature T^* increases as the metal is compressed ($T^* \sim V^{-2/3}$), so that the temperature range in which the approximation $\varepsilon_e \sim T^2$, $c_{v_e} \sim T$ holds is greater in a compressed material than at standard density.

According to the equation of state for a free electron gas (degenerate, as well as nondegenerate), the thermal part of the electron pressure is

$$p_e = \frac{2}{3}\frac{\varepsilon_e}{V} = \frac{1}{3}\beta\frac{T^2}{V} \sim V^{1/3}T^2. \tag{11.26}$$

If we define the "electronic Grüneisen coefficient" Γ_e by a relationship similar to (11.13),

$$p_e = \Gamma_e\frac{\varepsilon_e}{V}, \tag{11.27}$$

then we find that for a free electron gas it is equal to 2/3.

Kormer (see [3]) carried out a detailed analysis of the thermal behavior of electrons on the basis of the Thomas–Fermi and Thomas–Fermi–Dirac statistical models of an atomic cell (see §§12–14, Chapter III). He used the approximate calculations of Gilvarry [19], who considered the temperature terms as a correction to the Thomas–Fermi model of a cold atom, Latter's calculations [20], which were discussed in §14 of Chapter III, and experimental data. This analysis showed that the electronic specific heat to temperatures of the order of 30,000–50,000°K, as in the free electron model, is proportional to the temperature, with $c_{v_e} \sim T$, $\varepsilon_e \sim T^2$, where with increasing density this relationship remains valid to increasingly higher temperatures.

With respect to thermal pressure, the coefficient Γ_e is equal to 2/3 only in the limiting cases of very high temperatures or very large densities, for which the kinetic energy of the electrons is much greater than the Coulomb energy. The value of Γ_e in the temperature and density ranges obtained in shock compression experiments is slightly lower; it is approximately equal to 0.5–0.6. It was found that it is sufficiently accurate to take $\Gamma_e = const = \frac{1}{2}$. However, for consistency with the thermodynamic identity (11.12), it is necessary to change together with the coefficient Γ_e the exponent related to this coefficient in the relation between energy and volume, in particular to replace $\varepsilon_e \sim V^{2/3}T^2$ by $\varepsilon_e \sim V^{1/2}T^2$ *. Assuming that the coefficient of electronic specific heat at standard volume is equal to its experimental value, we can, according to Kormer, write approximately for $T < 30{,}000$–$50{,}000°K$,

$$\varepsilon_e = \tfrac{1}{2}\beta T^2, \qquad \beta = \beta_0\left(\frac{V}{V_0}\right)^{1/2}, \tag{11.28}$$

$$p_e = \frac{1}{2}\frac{\varepsilon_e}{V}. \tag{11.29}$$

§6. A three-term equation of state

Let us briefly summarize the results of §§2–5. The specific internal energy and pressure of a solid or a liquid can be represented as a sum of three components, which describe the elastic properties of the cold body, the thermal motion of the atoms (nuclei), and the thermal excitation of the electrons. For temperatures which are not too high, not above several tens of thousands of degrees (and large compressions), we can assume approximately that the atoms undergo small vibrations and that their specific heat is $c_v = 3Nk$. The electronic terms at these temperatures are described by the approximate equations (11.28) and (11.29). Thus, the energy and pressure are

$$\varepsilon = \varepsilon_c(V) + \varepsilon_T + \varepsilon_e, \qquad p = p_c(V) + p_T + p_e,$$

where

$$\varepsilon_c(V) = \int_V^{V_{0c}} p_c(V)\,dV,$$

$$\varepsilon_T = 3Nk(T - T_0) + \varepsilon_0,$$

$$\varepsilon_e = \tfrac{1}{2}\beta_0\left(\frac{V}{V_0}\right)^{1/2} T^2, \tag{11.30}$$

$$p_T = \Gamma(V)\frac{\varepsilon_T}{V}, \qquad p_e = \frac{1}{2}\frac{\varepsilon_e}{V}.$$

T_0 is room temperature, and ε_0 is the thermal energy of the atomic lattice at room temperature and is tabulated. The electronic specific heat coefficient at

* For a dependence $\varepsilon_e \sim V^k T^2$ with an equation of state $p = \Gamma_e \varepsilon_e / V$ with $\Gamma_e = const$, it is easy to check that the thermodynamic identity is satisfied only for $k = \Gamma_e$.

standard volume β_0 is obtained from experiments which measure the specific heat at very low temperatures.

The Grüneisen coefficient $\Gamma(V)$ is related to the function $p_c(V)$ by the differential relation (11.18). Only one unknown quantity remains, the elastic pressure as a function of volume $p_c(V)$, and this must be determined experimentally.

2. *The Hugoniot curve*

§7. Hugoniot curve for a condensed substance

The laws of conservation of mass, momentum, and energy across a shock wave (1.61)–(1.63) are entirely general, regardless of the aggregate state of the medium through which the wave propagates. Since the pressures behind even very weak waves are measured in thousands of atmospheres, one may always neglect the initial atmospheric pressure, setting it equal to zero. As usually, we denote by D the propagation speed of the shock wave through the undisturbed medium, and by u the jump in particle velocity across the front, equal to the velocity of the material behind the front (in laboratory coordinates) if the material ahead of the front is at rest. With unsubscripted quantities denoting conditions behind the front, we may write the laws of conservation of mass and momentum in the form

$$\frac{V_0}{V} = \frac{D}{(D-u)}, \tag{11.31}$$

$$p = \frac{Du}{V_0}. \tag{11.32}$$

Eliminating the velocity u from these equations, we get

$$p = \frac{D^2}{V_0}\left(1 - \frac{V}{V_0}\right). \tag{11.33}$$

As the third relation (energy equation) we take the Hugoniot equation (1.71) with $p_0 = 0$

$$\varepsilon - \varepsilon_0 = \tfrac{1}{2}p(V_0 - V). \tag{11.34}$$

The total energy acquired by a unit mass of the substance as a result of shock compression $p(V_0 - V)$ is divided equally between the kinetic energy $u^2/2$ and the internal energy $\varepsilon - \varepsilon_0$ (in a coordinate system in which the undisturbed medium is at rest). The change in the internal energy, in turn, is composed of the changes in the elastic and thermal energies.

Let us first consider a shock wave traveling through a body at zero temperature: $T_0 = 0$, $\varepsilon_0 = 0$, and $V_0 = V_{0c}$. On a p, V diagram (Fig. 11.6) we draw the cold compression curve $p_c(V)$ (which is an isentrope) and the Hugoniot curve $p_H(V)$; the Hugoniot naturally lies above the cold compression curve

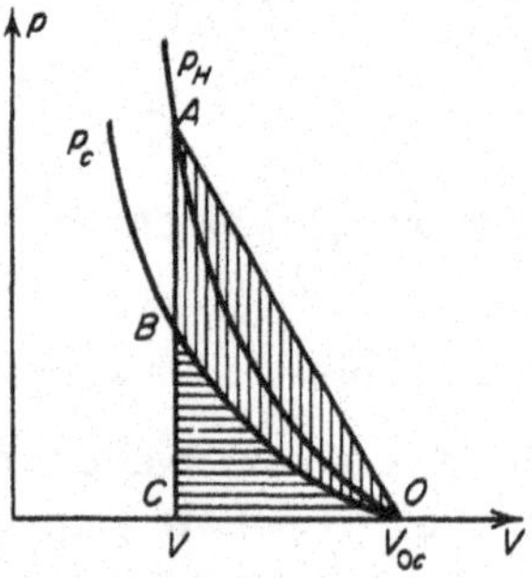

Fig. 11.6. p, V diagram for shock compression of a cold material. p_H is the Hugoniot curve; p_c is the cold compression curve.

since the total pressure behind the front is composed of the elastic and thermal pressure contributions. The elastic energy ε_c acquired by the material is numerically equal to the area of the curved triangle OBC, which is shaded horizontally ($\varepsilon_c = \int_V^{V_{0c}} p_c \, dV$). The total internal energy ε, according to (11.34), is equal to the area of the triangle OAC; the difference between these areas is shaded vertically and comprises the thermal energy of the material subjected to shock compression. As is evident from Fig. 11.6 the area OAC is always greater than the area OBC, as long as the cold compression curve is convex with respect to the volume axis $d^2p_c/dV^2 > 0$, as is ordinarily always the case. Therefore, the material is always heated by a shock wave and its entropy increases. This completely general statement, which was illustrated in Chapter I using as an example a perfect gas with constant specific heats, follows just as obviously in the case of a solid from the elastic properties of the material.

Let us now consider a shock compression of a body initially at standard conditions V_0, T_0. In this case the initial elastic pressure is negative, and the curve of $p_c(V)$ is located as shown in Fig. 11.7. The ordinary isentrope

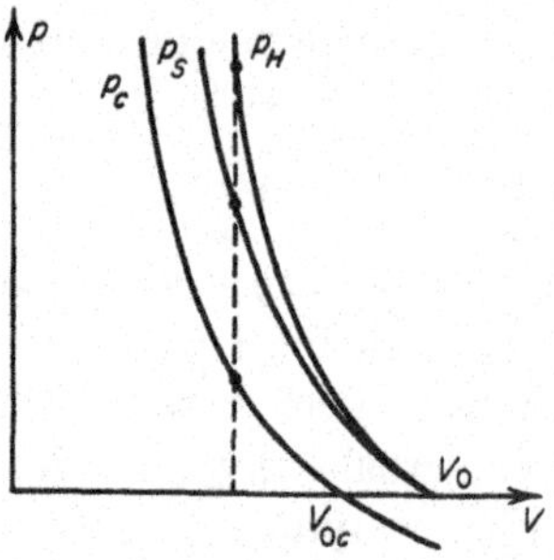

Fig. 11.7. p, V diagram for shock compression of a solid heated to room temperature. p_H is the Hugoniot curve; p_S is the isentrope; p_c is the cold compression curve.

$p_S(V, S_0)$ passing through the initial state lies above the cold compression curve by an amount which increases somewhat with decreasing volume. For small compressions the electron pressure is negligibly small; the Grüneisen coefficient may be taken to be constant and the isentrope $p_S(V, S_0)$ is given by (11.22).

As we know (Chapter I, §18), the Hugoniot curve $p_H(V)$ has a second-order tangency with the isentrope $p_S(V)$ at the initial point, so that the Hugoniot curve is located as shown in Fig. 11.7. Figure 11.7 has been drawn to a scale which makes clear the mutual position of all three curves p_c, p_S, and p_H in a range from relatively small pressures up to values of the order of a hundred thousand atmospheres. If we consider a wider range of pressures, up to millions of atmospheres, then the difference between V_0 and V_{0c} and the difference between the isentrope and the cold compression curve are almost imperceptible, while the deviation of the Hugoniot curve from the isentrope p_S or from the curve p_c becomes appreciable because of the increased effect of the thermal components of energy and pressure, or equivalently, as a result of the significant increase in the entropy. The picture in this case is the same as in Fig. 11.6, where we can assume that $V_{0c} = V_0$ and that the isentrope p_{S_0} coincides with the cold compression curve.

In shock waves with pressures of the order of a million atmospheres the thermal energy which is associated with the increase in the entropy of the material is comparable with the total energy. In exactly the same manner, the thermal pressure is comparable with the total pressure. This is illustrated by Fig. 11.8 taken from [3], on which are shown experimental Hugoniot

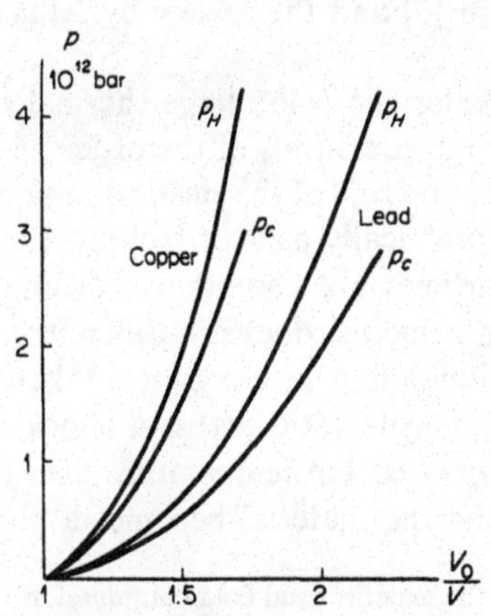

Fig. 11.8. Hugoniot and cold compression curves for copper and lead.

curves for copper and lead up to pressures of the order of $4 \cdot 10^6$ atm, and cold compression curves calculated on the basis of experimental data (the density ratio $\rho/\rho_0 = V_0/V$, rather than the specific volume, is used as the abscissa)*.

Table 11.2 provides an idea of the relative role of the various pressure and energy components for different shock pressures†.

Table 11.2

PARAMETERS BEHIND A SHOCK WAVE IN LEAD

$\frac{\rho}{\rho_0}$	p	p_c	p_T	p_e	$\varepsilon-\varepsilon_0$	ε_c	$c_v(T-T_0)$	ε_e	Γ	T,°K
	in 10^{10} dyne/cm^2 = 10^4 atm				in 10^8 erg/g					
1.3	25.0	21.6	3.35	0.051	25.4	15.3	9.6	0.69	1.9	1,045
1.5	65.5	51.0	13.9	0.63	96.3	46.7	42.3	7.4	1.77	3,580
1.7	133.0	95.3	34.0	3.8	242.0	95.8	107.0	39.4	1.60	8,600
1.9	225.5	156.0	56.0	12.7	471.0	163.2	191.0	118.0	1.35	15,100
2.1	335.5	233.0	73.0	29.0	775.0	248.0	284.0	243.0	1.07	22,300
2.2	401.0	277.0	93.0	41.5	965.0	297.0	337.0	332.0	0.98	26,400

It follows from the table that for the shock compression of lead by a factor of 2.2, the material behind the front is heated to a temperature of 26,400°K; in this case the thermal pressure is 32% of the total pressure and the thermal energy is 69% of the total energy, with half of the thermal energy ascribable to the electrons and the other half to the atomic vibrations. The thermal pressure of the electrons is 34% of the total thermal pressure. Qualitatively the behavior with increasing wave strength is the same for all other metals studied. Quantitative data can be found in [3] and the review by Al'tshuler [55] but will not be given here.

The greater the shock strength, the greater the role of the thermal components of pressure and energy. At very high pressures, of the order of hundreds of millions of atmospheres and above, the role of the elastic components becomes small and the material behaves practically as a perfect gas (perfect in the sense that interactions between particles are absent). Accordingly, a Hugoniot curve under these conditions in principle does not differ from the Hugoniot curve for a perfect gas (when ionization processes are taken into account; see Chapter III), and a limiting density ratio across a shock wave exists also for a solid body. In the limit $p \to \infty$ the temperature also tends to infinity, the atoms are fully ionized, and the material becomes a perfect,

* The experiments and methods for obtaining the experimental cold compression curves are described in §§12 and 13.

† The table is taken from [3]. It has been supplemented with some additional properties for the sake of completeness. These properties were calculated using the constants given in [3].

classical electron-nuclear gas with a specific heat ratio $\gamma = 5/3$, which corresponds to a limiting density ratio of 4 (if effects connected with radiation are disregarded; see Chapter III).

§8. Analytical representation of Hugoniot curves

From the thermodynamic functions $p(V, T)$ and $\varepsilon(T, V)$ it is possible in principle to find an explicit equation for the Hugoniot curve $p_H(V, V_0)$. Practically this cannot be done, since the theoretical dependence of the elastic pressure on the volume $p_c(V)$ is unknown. However, it is useful to write down the equation for the Hugoniot curve in terms of the unknown function $p_c(V)$. We shall consider shock waves that are not too strong, in which the electronic components of pressure and energy can be neglected and the Grüneisen coefficient Γ can be taken to be constant and equal to its value at standard conditions Γ_0. At the same time we assume that the wave is not too weak either, so that the initial energy of the undisturbed medium ε_0 can be neglected. Physically this means that we consider the initial temperature to be equal to zero and make no distinction between the standard volume V_0 and the zero volume V_{0c}.

Let us substitute into the Hugoniot equation (11.34) the energy $\varepsilon = \varepsilon_c + \varepsilon_T$, expressing the thermal energy contribution ε_T in terms of pressure by means of (11.21)

$$p - p_c = p_T = \Gamma_0 \frac{\varepsilon_T}{V}; \qquad \varepsilon = \varepsilon_c + \frac{V(p - p_c)}{\Gamma_0}.$$

Solving the resulting equation for p, we obtain the equation for the Hugoniot curve in the form

$$p_H = \frac{(K-1)p_c(V) - 2\varepsilon_c(V)/V}{K - V_0/V}, \qquad \varepsilon_c = \int_V^{V_{0c}} p_c(V)\, dV, \qquad (11.35)$$

where $K = 2/\Gamma_0 + 1$.

If we formally apply (11.35) to very strong shocks, we find that in the limit $p_H \to \infty$, $V_0/V = K$, so that formally K represents the limiting density ratio across the shock wave. The situation here is completely analogous to that for a perfect gas with constant specific heats. We recall that the Grüneisen exponent Γ corresponds to the specific heat ratio γ minus one. From this, the limiting density ratio K corresponds to the quantity $(\gamma + 1)/(\gamma - 1)$, which is the limiting density ratio for gases. The formal analogy with gases is connected with the fact that in the limit $p_H \to \infty$ the major role is played by the thermal pressure ($p_T = p_H - p_c \to \infty$, while $p_c(V) \to const$), and the equation of state in this case is the same as that for a gas.

It is sometimes convenient to express the Hugoniot curve analytically by means of an interpolation formula. It has been shown experimentally that the relationship between the front velocity and the velocity of the material (relative to the undisturbed medium) behind the front is linear over a wide range of shock strengths:

$$D = A + Bu. \tag{11.36}$$

Thus, for example, for iron $A = 3.8$ km/sec, $B = 1.58$*. Using (11.36) we can easily obtain the equation of the Hugoniot curve from (11.34) and (11.32)

$$p_H = \frac{A^2(V_0 - V)}{(B-1)^2 V^2 \left[\dfrac{B}{B-1} - \dfrac{V_0}{V}\right]^2}. \tag{11.37}$$

The Hugoniot curve $p_H(V, V_0)$ can be interpolated by polynomials of the type

$$p_H = \sum_{k=1}^{m} a_k \left(\frac{V_0}{V} - 1\right)^k,$$

where the constant coefficients are determined partially on the basis of experimental shock compression data and partially from the properties of the material in the standard state.

§9. Weak shock waves

The pressure range of the order of several tens and hundreds of thousands of atmospheres is of great practical importance. These pressures are typical of those generated in detonating explosives, in explosions in water, on the impact of detonation products on metallic obstacles, etc. The following empirical equation of state for a condensed material is frequently used in the isentropic flow region:

$$p = A(S)\left[\left(\frac{V_0}{V}\right)^n - 1\right], \tag{11.38}$$

where the exponent n is assumed to be constant, and where the coefficient A depends on the entropy and in fact is also always taken to be constant. The constants A and n are related by an equation which depends on the compressibility of the material at standard conditions (the speed of sound)

$$c_0^2 = -V_0^2\left(\frac{\partial p}{\partial V}\right)_S = V_0 An. \tag{11.39}$$

* Equation (11.36) cannot be extrapolated to small strengths $p \to 0$ and $u \to 0$, so that the constant A is not the speed of sound in the standard state.

In accord with Jensen's data, Baum, Stanyukovich, and Shekhter [21] took $n = 4$ for metals and calculated the constant A for a number of metals using (11.39) and the experimentally measured values of the compressibility. In a number of cases they obtained good agreement with values of A which they determined experimentally. Thus, for example, for iron, $A_{\text{calc}} = 5 \cdot 10^5$ atm, which is 11% larger than the experimental value; for copper $A_{\text{calc}} = 2.5 \cdot 10^5$ atm, 6% larger than the experimental value; for duraluminum $A_{\text{calc}} = 2.03 \cdot 10^5$ atm, practically the same as the experimental value. For water it is usually assumed that $n \approx 7$–8 and $A \approx 3000$ atm.

In calculating flows with shock waves in the above pressure range we can, to first approximation, neglect the entropy change across the shock wave and use the isentropic equation of state (11.38) with $A = const$ to relate the pressure and the density across the wave front. Here the velocities D and u are found either from (11.30) and (11.32) or from (11.31) and (11.33). The energy equation (11.34) can be used in this case in the next approximation to estimate the increase in internal energy connected with the irreversibility of the shock compression. Indeed, if we consider (11.38) to be the equation of the isentrope, then the internal energy can be determined as a function of V by using the equation $T\,dS = d\varepsilon + p\,dV = 0$,

$$\varepsilon(V) - \varepsilon_0 = -\int_{V_0}^{V} p\,dV = AV_0\left\{\frac{1}{n-1}\left[\left(\frac{V_0}{V}\right)^{n-1} - 1\right] - \left[1 - \frac{V}{V_0}\right]\right\}.$$

This value of the energy and the pressure obtained from (11.38) naturally do not satisfy the energy equation across the shock front (11.34). By definition, the difference

$$\Delta\varepsilon = \tfrac{1}{2}p(V_0 - V) - \int_{V}^{V_0} (p\,dV)_{S=const}$$

is equal to the increase in internal energy caused by the increase in entropy across the shock wave. The smallness of this quantity in comparison with the total increase in energy across the shock wave $\varepsilon - \varepsilon_0$ is a condition for the validity of the isentropic approximation for the shock compression.

Calculating the ratio $\Delta\varepsilon/(\varepsilon - \varepsilon_0)$ with $n = 4$ we find that for $V_0/V = 1.1$ the ratio is equal to 4.5%, and for $V_0/V = 1.2$ it is 17.5% (this ratio is independent of A). A density ratio of 1.1 corresponds to a pressure of the order of 100,000 atm (90,000 atm for aluminum and 210,000 atm for iron). Thus, for pressures of $\sim 10^5$ atm the isentropic approximation for the shock compression yields an error in the energy of not more than 5% (the error in pressure is even less). We can, therefore, consider the shock wave as an acoustic wave in many practical calculations.

§10. Shock compression of porous materials

The process of shock compression of porous bodies exhibits some distinctive features. Experimental studies of the shock compression of the same material at different initial densities make it possible to obtain more complete information on the thermodynamic properties of the material at high pressures and temperatures. Porous media can have quite different forms and structures (powders, bodies with internal voids, fibrous bodies, etc.). All of them are characterized by the presence of more or less large particles or segments of a solid (continuous) material with standard density $\rho_0 = 1/V_0$ and void segments, as a result of which the average specific volume V_{00} is greater than the standard volume V_0 (and the average density ρ_{00} is less than the standard density ρ_0).

Let us imagine that a porous body is subjected to a slow compression on all sides. Initially the work of the external pressure forces is used in closing up the voids, in packing the material and reducing it to standard volume. This work involves overcoming the friction forces between the particles, pulverizing the particles, crumpling the fibers, etc. The completion of this work requires a relatively small pressure, the scale of which is the ultimate strength of the material. For metals this pressure is of the order of a thousand atmospheres, and much less than this for many other materials. If we consider compressions in the range of pressure of hundreds of thousands of atmospheres, then the pressure on the portion of the p, V curve where the material is reduced to standard volume can for practical purposes be taken equal to zero, and the p, V curve starting from the point V_{00} can be taken as a straight line on the axis of V from V_{00} to V_0 $(p = 0)$. For compression above standard density the curve can be taken to be an isentrope for the continuous material (Fig. 11.9).

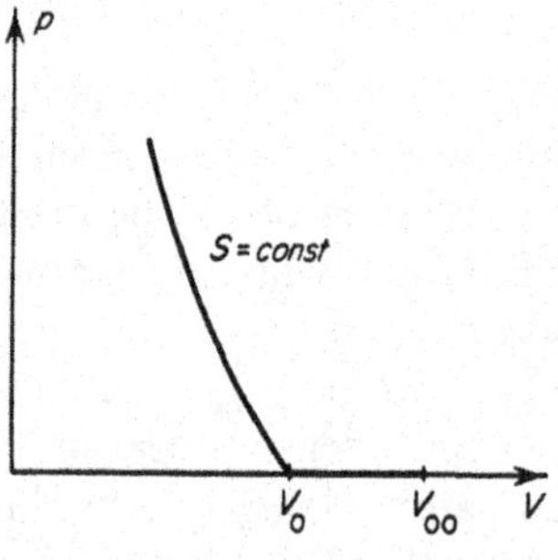

Fig. 11.9. Isentrope for the compression of a porous material.

Let us now consider the shock compression of a porous body. For simplicity we consider shock compression to high pressures, of the order of hundreds of thousands and millions of atmospheres, so that the ordinary isentrope for the continuous material is the same as the cold compression curve. We neglect the effects related to the strength of the material, and the fact that the initial temperature $T_0 \approx 300°K$ is different from zero. We also assume that in the final state behind the front the material is both continuous and homogeneous.

It follows from the conservation laws across the shock wave and from the equation of state of the medium that the Hugoniot curve has the form shown in Fig. 11.10 (this will be clarified below). The point corresponding to the standard volume V_0 and zero pressure $p = 0$ lies on the Hugoniot curve. The internal energy acquired by the medium in the shock wave $\varepsilon = \frac{1}{2}p(V_{00} - V)$ is equal to the area of the horizontally shaded triangle. The elastic part of the internal energy is equal to the area of the curvilinear triangle bounded by the curve $p_c(V)$ and densely cross-hatched in Fig. 11.10. The larger the initial

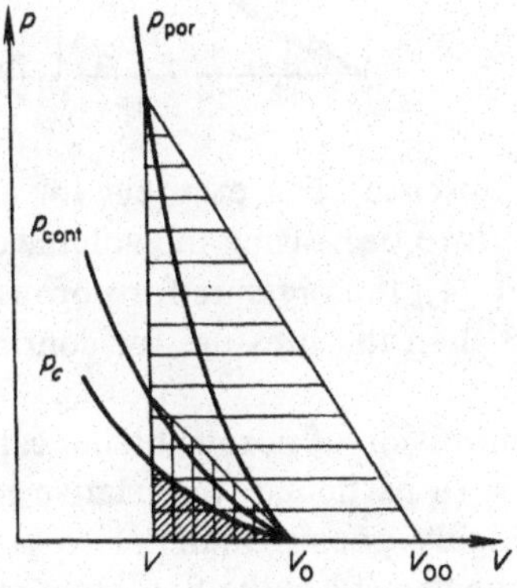

Fig. 11.10. p, V diagram for shock compression of a porous material. p_{por} is the Hugoniot curve for a porous medium; p_{cont} is the Hugoniot curve for a continuous material; p_c is the cold compression curve for a continuous material.

volume V_{00}, i.e., the greater the porosity of the material, for compression of the porous material to the same final volume the greater is the difference between these areas; this difference corresponds to the thermal part of the energy (the elastic energy is the same for a given volume, while the total energy increases). However, the greater the thermal energy, the larger is the thermal pressure. Therefore, the higher the porosity, the steeper is the Hugoniot curve. In particular, the Hugoniot curve for a porous material lies above the Hugoniot curve for a continuous material, as shown in Fig. 11.10. A higher pressure is required to compress a porous material to a given volume than a continuous material. This pressure is higher, the higher is the porosity. The qualitative nature of the picture given will not be altered if we consider the initial temperature (and entropy) to be different from zero.

In order to get some idea of how sharply the thermal components of pressure and energy increase for shock compression of a porous body as compared to the compression of a continuous one, we present experimental Hugoniot curves for iron of standard density and for porous iron with a density lower by a factor of 1.4 ($V_{00} = 1.412V_0$). These curves (Fig. 11.11) were taken from [1] (the density ratio V_0/V with respect to standard density rather than

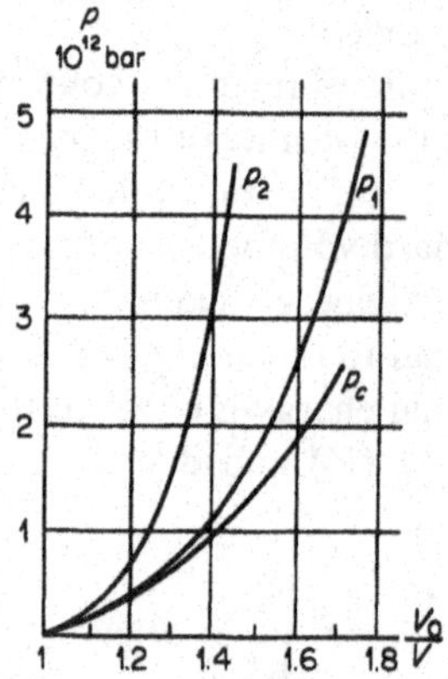

Fig. 11.11. Hugoniot curves for continuous (p_1) and porous (p_2) iron. p_c is the cold compression curve.

the specific volume alone is plotted on the abscissa). For example, for a density ratio of $V_0/V = 1.22$, which corresponds to decreasing the volume of the porous iron by a factor of 1.74 ($V_{00}/V = 1.74$), the pressure for porous iron is found to be greater by a factor of 2.63 than the pressure for continuous iron, while the energy is 8.64 times larger.

The strong heating caused by shock compression of porous bodies can lead to sharp anomalies in the Hugoniot curves. In particular, the relative role of the thermal pressure in compressing a highly porous material to a given pressure can turn out to be so great that the density in the final state at high pressure is below standard density ($V > V_0$). In this case, the volume does not decrease with increasing pressure, as is usually the case, but increases, and the Hugoniot curve has the anomalous shape shown in Fig. 11.12.

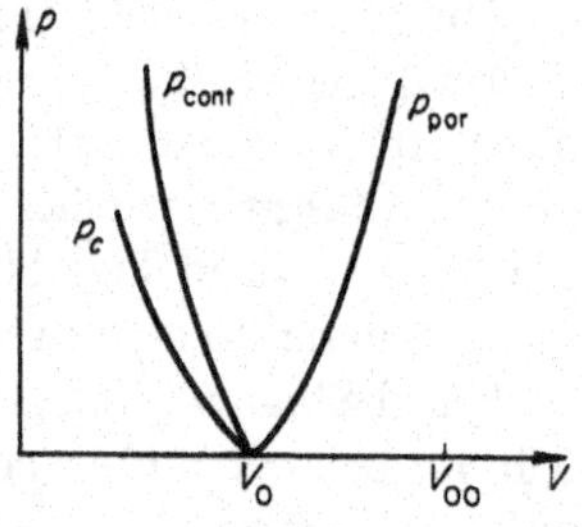

Fig. 11.12. Anomalous Hugoniot curve for a highly porous material.

In order to make clear the origin of this curious effect, we shall use the Hugoniot equation derived under the assumptions that the electron pressure and energy are small, the Grüneisen coefficient is constant, and the initial energy of the medium can be neglected. This is equation (11.35), in which the initial volume V_0 is understood to denote the initial volume of the porous material V_{00} (equation (11.35) was derived without assuming that the material is continuous in its initial state):

$$p_H(V, V_{00}) = \frac{(K-1)p_c(V) - 2\varepsilon_c(V)/V}{K - V_{00}/V}, \qquad K = \frac{2}{\Gamma_0} + 1. \quad (11.40)$$

Equation (11.40) describes a family of Hugoniot curves corresponding to different initial volumes V_{00}, that is, to different degrees of porosity which may be characterized by the coefficient $k = V_{00}/V_0 \geqslant 1$. When $k = 1$, and $V_{00} = V_0$, we have the Hugoniot curve for a continuous material. The point $V = V_0$, $p_H = 0$ satisfies (11.40) for any initial volume V_{00} (since $p_c(V_0) = 0$ and $\varepsilon_c(V_0) = 0$), so that the family of Hugoniot curves represents a bundle of curves emanating from this point. According to (11.40), as $p_H \to \infty$, $V_{00}/V \to K$, and the limiting volume $V_{\lim} = V_{00}/K$. If this value is smaller than V_0, as is the case for small porosity $k < K$, the Hugoniot curves have their normal shape and are higher the larger is the initial volume. If, however, $V_{\lim} > V_0$ (which happens for the case of high porosity, with $k > K$), then the Hugoniot curves are anomalous: the final volume increases with increasing pressure. A family of Hugoniot curves corresponding to different porosities is shown in Fig. 11.13.

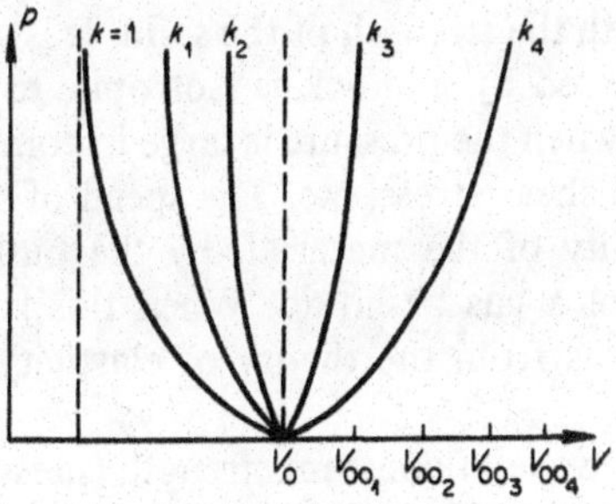

Fig. 11.13. Hugoniot curves for different porosities: $k_4 > k_3 > K > k_2 > k_1 > 1$.

We emphasize again that (11.40) describes only the initial behavior of the Hugoniot curve for low pressures. Actually at large pressures the role of the electronic terms becomes important and the Grüneisen coefficient is not constant. This fact, however, does not invalidate the qualitative conclusion regarding the possibility of having an anomalous Hugoniot curve for a highly porous material.

In the paper by Kormer, Funtikov, Urlin, and Kolesnikova [56] are presented results of studies of the shock compression of porous metals at pressures from 0.7 to 9 million atmospheres. They also give a number of theoretical derivations of an equation of state for porous materials and determine the parameters which enter into the equation of state on the basis of experimental data.

§11. Emergence of weak shock waves from the free surface of a solid

Experimental methods for determining Hugoniot curves for a solid, which will be considered in the following section, make extensive use of the so-called velocity doubling rule in an unloading wave. When a shock wave which propagates through a solid emerges from the free surface, the compressed material expands, or "unloads" practically to zero pressure. The unloading (rarefaction) wave travels backwards into the material with the speed of sound that corresponds to the state behind the shock front, and the material being unloaded acquires an additional velocity in the direction of the initial motion of the shock*.

In this section we shall consider only weak shock waves, which impart to the solid an energy which is not sufficient to melt it and even less so to vaporize it†, so that the final state of the material after unloading may be assumed to be solid. In this case, the final volume of the unloaded material V_1 differs but little from the standard volume of the solid V_0. At the same time we shall also assume that the shock wave is not too weak, so that we can neglect effects associated with the strength of the solid. It is assumed that the pressure in the solid compressed by a shock is isotropic, as in a gas or liquid. This assumption is valid when the pressure is large in comparison with the ultimate strength, the critical shear stress, etc. The speed of sound is then determined by the compressibility of the material, by the bulk compression modulus, just as in the case of a gas or liquid. When this is not so the unloading is described by formulas from the theory of elasticity, which we shall discuss later.

Let a plane shock wave of constant strength (pressure p, particle velocity u, and volume V which is only slightly smaller than the standard volume V_0)

* If the body borders on air rather than on a vacuum, the moving boundary of the unloaded material acts as a piston with respect to the air and pushes ahead an air shock. Therefore, strictly speaking, the material is not unloaded to zero pressure, but to the pressure behind the air shock. However, this pressure, even though it may be large in comparison with atmospheric pressure, is so small in comparison with the initial pressure in the shock-compressed solid that it can always be neglected, and we may consider that unloading into air does not differ from unloading into vacuum. The strength of the shock wave in the air is determined in this case by the equivalent piston velocity, the velocity of the solid after unloading.

† The vaporization of a solid initially compressed by a very strong shock wave will be discussed in §§21 and 22.

propagate through a solid. At a specified time the wave emerges from the free surface, which is assumed to be parallel to the surface of the shock front. A weak shock wave, across which the compression is small, $V_0 - V \ll V_0$, does not differ from an acoustic compression wave and can be described by acoustic equations. The shock travels through the body with the speed of sound c_0. The pressure behind the shock is related to the particle velocity through the relation $p = \rho_0 c_0 u$ ($\rho_0 = 1/V_0$). Starting from the time $t = 0$ at which the shock wave emerges from the free surface, an unloading wave, also an acoustic wave, propagates back into the body. The unloading wave travels through the material with the speed of sound (differing but little from the speed of sound c_0 under standard conditions). The pressure across the wave drops from the initial pressure p to zero and the material acquires a velocity u', related to the pressure change $\Delta p = -p$ by the acoustic formula $u' = -\Delta p/\rho_0 c_0 = p/\rho_0 c_0$ (see Fig. 11.14; the density decreases only slightly, and the final density ρ_1 differs but little from the standard density of the solid: $V_1 - V_0 \ll V_0$). It is evident from comparison of the equations $p = \rho_0 c_0 u$ and $u' = p/\rho_0 c_0$ that u', the additional velocity acquired by the

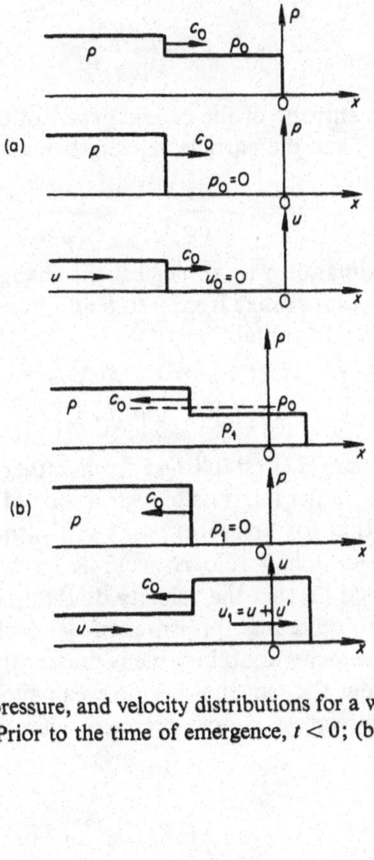

Fig. 11.14. Density, pressure, and velocity distributions for a weak shock wave emerging from a free surface. (a) Prior to the time of emergence, $t < 0$; (b) after emergence, $t > 0$.

material on unloading, is equal to u, the particle velocity behind the original shock. This means that, when a weak shock wave emerges from the free surface, the velocity of the material doubles: $u_1 = u + u' \approx 2u$.

The velocity doubling rule can also be obtained from the general shock and rarefaction wave equations in the limiting case of weak waves. We know from gasdynamics (see §10, Chapter I) that the additional velocity acquired by a material unloaded from an initial pressure p to a final pressure $p_1 = 0$, is

$$u' = \int_0^p \frac{dp}{\rho c} = \int_0^p \left(-\frac{\partial V}{\partial p}\right)_S^{1/2} dp = \int_V^{V_1} \left(-\frac{\partial p}{\partial V}\right)_S^{1/2} dV,$$

where the derivatives, by virtue of the fact that the unloading process is isentropic, are taken at a constant entropy equal to the entropy behind the shock front. The initial particle velocity of the material behind a shock wave, by virtue of the conservation laws (11.31) and (11.32), is

$$u = [p(V_0 - V)]^{1/2}.$$

In a weak shock, where the entropy change is small, the compression is also not large. To first approximation we can express the volume increase in the form

$$V - V_0 = \left(\frac{\partial V}{\partial p}\right)_S p,$$

where S denotes the entropy of the original state of the material prior to the shock compression. Then the particle velocity behind the shock is

$$u \approx \left(-\frac{\partial V}{\partial p}\right)_S^{1/2} p \approx \frac{p}{\rho_0 c_0}.$$

To the same approximation we can neglect the change in the isentropic compressibility in the pressure range from 0 to p and take the derivative constant in the equation for u'. We obtain

$$u' = \int_0^p \left(-\frac{\partial V}{\partial p}\right)_S^{1/2} dp \approx \left(-\frac{\partial V}{\partial p}\right)_S^{1/2} p \approx \frac{p}{\rho_0 c_0} \approx u.$$

Walsh and Christian [22] established from rather general considerations the upper and lower limits for possible variations of the additional velocity u' and have found that for pressures $p \sim 4 \cdot 10^5$ atm for a large number of metals the velocity doubling rule is accurate to within 2%. It has been experimentally verified [3] that the velocity doubling rule for iron is approximately satisfied up to quite high pressures of $\sim 1.5 \cdot 10^6$ atm. In general, the departure from the velocity doubling rule is greater the stronger is the shock.

Let us now consider the fact that even a weak shock wave is not acoustic and that the entropy across it does increase. Here we assume that to first approximation the additional velocity acquired after unloading u' is, as before, equal to u. The density and temperature in the final state will be considered to satisfy the following approximation. When a body is unloaded isentropically to the initial, zero pressure, it is found to be heated and expanded in comparison with the initial state prior to shock compression. The energy of irreversible heating and the final temperature of the unloaded material T_1 can be easily found if the thermodynamic properties and initial state ahead of the shock wave are known. For this we use the equation for isentropic unloading $d\varepsilon + p\,dV = 0$, according to which the final energy ε_1 is equal to

$$\varepsilon_1 = \varepsilon - \int_V^{V_1} (p\,dV)_S. \tag{11.41}$$

Since the energy behind a shock wave is $\varepsilon = \varepsilon_0 + \frac{1}{2}p(V_0 - V)$, the irreversible energy increase after unloading is

$$\varepsilon_1 - \varepsilon_0 = \tfrac{1}{2}p(V_0 - V) - \int_V^{V_1} (p\,dV)_S. \tag{11.42}$$

The magnitude of this energy is given by the difference in the areas of the curvilinear triangle $DBCS$ and the triangle ABC in Fig. 11.15, in which the

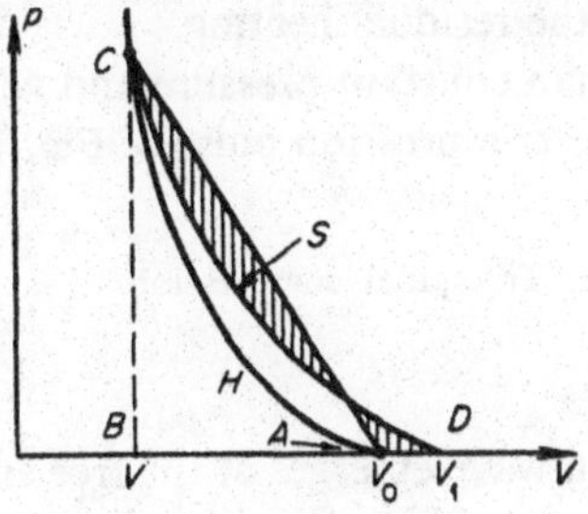

Fig. 11.15. p, V diagram for shock compression and unloading of a solid.

curve H is a Hugoniot curve and the curve S is the unloading isentrope. Numerically this energy is equal to the difference between the upper and lower shaded areas.

Let us assume that the shock wave is weak, so that all three volumes V, V_0, and V_1 differ little from each other and that the Grüneisen coefficient can be assumed to be constant and equal to its standard value Γ_0. In this case the isentropic relation between temperature and volume is given by (11.20), so that the final temperature T_1 is related to the temperature behind the shock wave T by

$$\frac{T_1}{T} = \left(\frac{V}{V_1}\right)^{\Gamma_0}. \tag{11.43}$$

On the other hand, considering the process of thermal expansion of the body at constant, zero pressure from the initial volume V_0 to V_1, we can write

$$V_1 - V_0 = V_0 \alpha (T_1 - T_0), \tag{11.44}$$

where α is the coefficient of thermal expansion at constant pressure. The irreversible increase of energy (11.42) is expressed in terms of the temperature increase by

$$\varepsilon_1 - \varepsilon_0 = c_p(T_1 - T_0),$$

where c_p is the specific heat at constant pressure* of the body. If V and T, the volume and temperature behind the shock, are known, then the volume and temperature in the final state may be calculated from the system of two equations (11.43) and (11.44).

As an example, we present the results for aluminum obtained in [23]. When aluminum is compressed by a shock wave to a pressure $p = 2.5 \cdot 10^5$ atm, the volume decreases to $V = 0.82 V_0$, and the temperature increases by $T - T_0 = 331°\text{K}$ (the initial temperature T_0 was 300°K). After unloading, the residual increase in temperature was $T_1 - T_0 = 134°\text{K}$†. When $p = 3.5 \cdot 10^5$ atm, $V = 0.78 V_0$, $T - T_0 = 522°\text{K}$, then $T_1 - T_0 = 216°\text{K}$. Naturally, the stronger the shock wave, the greater is the entropy increase imparted to the material and the higher is the residual heating.

If a shock wave with decreasing rather than constant pressure and velocity behind the front (for example, the triangular compression pulse of Fig. 11.16)

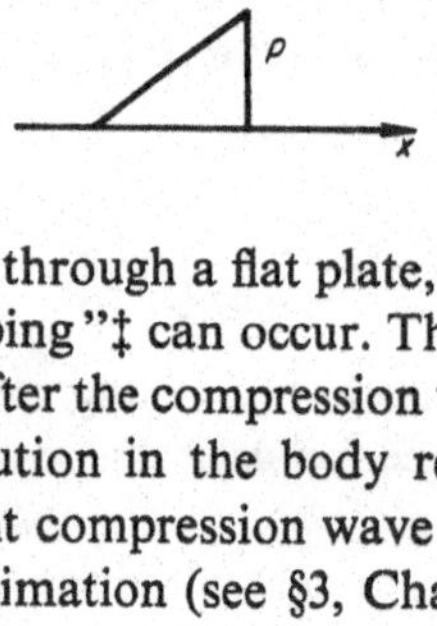

Fig. 11.16. Triangular compression pulse.

travels through a flat plate, then after such a wave emerges at the free surface "scabbing"‡ can occur. The phenomenon of scabbing consists of the following: After the compression wave is reflected from the free surface, the pressure distribution in the body results from the superposition of two waves, the incident compression wave and the reflected unloading wave. In the acoustic approximation (see §3, Chapter I)

$$p = \rho c[f_1(x - ct) + f_2(x + ct)],$$

* In solids over small ranges of temperature changes, it is practically the same as the specific heat at constant volume c_v.

† The residual temperature was calculated in [23] with greater accuracy than that given by (11.43), taking into account the small change in the Grüneisen coefficient with the change in volume; the more exact isentropic relation with a variable $\Gamma(V)$ was integrated for this purpose.

‡ *Editors' note.* Also commonly termed spalling. The authors' term is "break-away", or more literally, "split-off".

where the function f_1 describes the incident wave propagating to the right with the speed of sound, and f_2 describes the reflected wave propagating to the left. In this case the function f_1 corresponds to the triangular pressure distribution shown in Fig. 11.16. The function f_2 can be determined starting from the boundary condition that the pressure at the free surface is zero.

The functions f_1 and f_2 and the resulting pressure distribution in the body at the time of emergence of the shock at the free surface and at two subsequent instants of time are shown in Fig. 11.17. If x_1 is the coordinate of the free surface (Fig. 11.17), then the region $x > x_1$ corresponds to vacuum and the determination of the functions f_1 and f_2 in this region is purely formal and without physical significance. The only physically real values of f_1 and f_2 and of pressure are those for $x < x_1$, within the body. To emphasize this point, the functions f_1 and f_2 for $x > x_1$ in Fig. 11.17 are shown dashed.

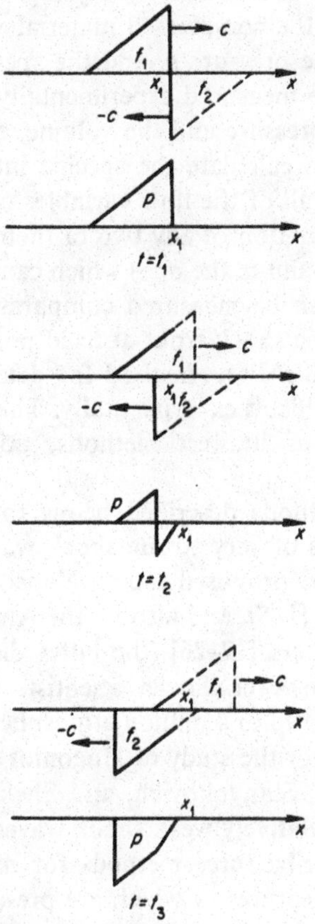

Fig. 11.17. Reflection of an acoustic triangular compression wave from a free surface. (a) $t = t_1$ refers to the time at which the leading edge of the wave emerges at the free surface; (b) $t = t_2$ and (c) $t = t_3$ refer to subsequent instants of time.

It is evident from Fig. 11.17 that after the compression wave is reflected from the free surface, negative pressures arise in the body, i.e., the material experiences a tensile stress. If the tensile stress exceeds the ultimate tensile strength of the material, then a fracture or "scabbing" occurs at this point of the body, and a layer of material (the scab) is split away from the surface and separates from the remaining material, moving away from the surface with a definite speed. Thus, for example, steel breaks down under impulsive loadings when the stresses become of the order of 30,000 kg/cm^2 *.

§12. Experimental methods of determining Hugoniot curves for solids

The conservation laws of mass and momentum (11.31) and (11.32) connect four shock front variables: the propagation speed of the shock wave through the undisturbed material D, the jump in the particle velocity u, equal to the velocity of the compressed material with respect to that of the undisturbed material, the pressure p, and the specific volume V (or density $\rho = 1/V$). If D and u are measured experimentally, then we may use (11.31) and (11.32) to find the pressure and the volume, and then, using the energy equation (11.34), we can calculate the specific internal energy ε. Thus, the problem of determining all of the flow variables of a shock front reduces to an experimental determination of any two of them, in particular, of the two kinematic parameters D and u, the ones which can be most easily measured.

The front speed D can be measured comparatively simply by recording the time of passage of the shock front at fixed points separated by a known distance. The use of such a direct method for determining the jump in particle velocity u is more difficult experimentally. Therefore, the second variable is determined by various indirect methods, utilizing certain mechanical considerations.

The experimental methods described below for studying the compressibility of solids by means of very strong shock waves and for measuring the shock front variables were proposed and developed by Al'tshuler, Krupnikov, Ledenev, and Bakanova [1–5], and also by the American authors Walsh and Christian, as well as others [22–26] (the latter did not use the "collision" method; see below). However, Soviet scientists have investigated a much wider range of pressures, up to 4 million atmospheres. The idea of measuring the kinematic variables for the study of Hugoniot curves was also developed independently by Baum, Stanyukovich, and Shekhter [21], who carried out their experiments with relatively weak shock waves.

References [1–3] describe three methods for measuring the shock wave variables, the principal features of which are presented here (see also [55]).

* *Editors' note*. If the pulse is strong enough, the process may be repeated, producing a second or more additional scabs.

1. *The "free surface" method**. This method is based on measuring the velocity of the free surface of a body which is unloaded after a shock wave emerges from its surface, and on applying the velocity doubling rule, according to which the particle velocity u is approximately equal to one half of the free surface velocity u_1. The applicability of this method is limited, since the departures from the doubling rule begin to be appreciable at very high pressures, resulting in experimental errors in the determination of u. The basic experimental setup consists in the following: A flat plate of the test material is placed against an explosive charge, as shown in Fig. 11.18 (the corresponding diagram of motion in the x, t plane is given in Fig. 11.19).

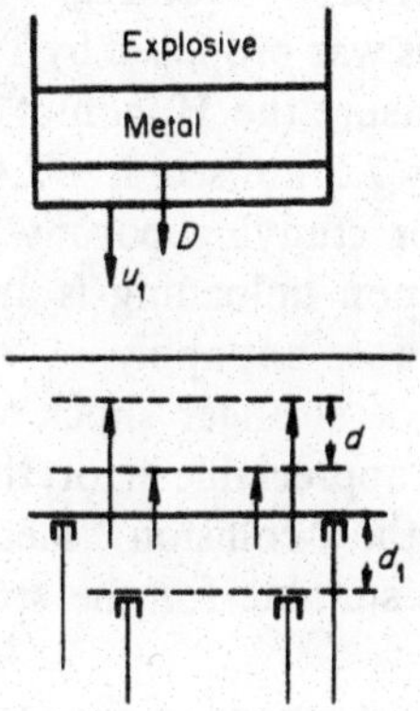

Fig. 11.18. Experimental arrangement for the "free surface" method.

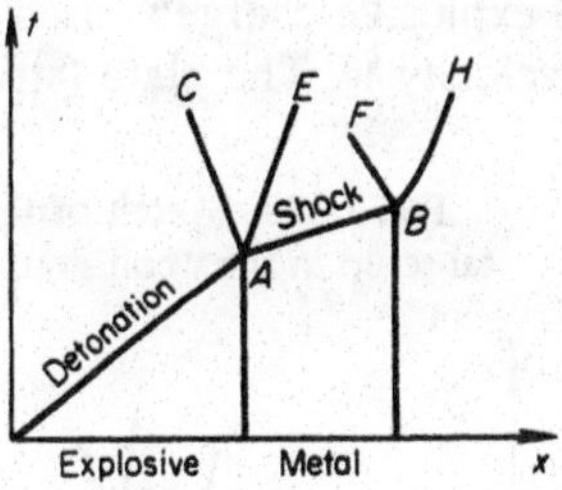

Fig. 11.19. x, t diagram for the "free surface" method.

When the detonation wave emerges from the explosive at the boundary with the metal, the discontinuity breaks up; a shock wave (line AB) travels through the metal with velocity D, the velocity of the contact surface between the explosive and the metal (line AE) is equal to the particle velocity of the metal u (a reflected wave AC propagates through the explosive). When the shock emerges at the free surface (point B) the discontinuity breaks up again, an unloading wave BF travels back into the specimen, and the boundary of

* *Editors' note*. Called the "break-away" or "split-off" method by the authors, which is the term they use for the scabbing phenomenon. The velocity doubling approximation is sometimes referred to as the free surface approximation (see, e.g., [60]).

the metal acquires the doubled velocity $u_1 \approx 2u$ (line BH). The front speed D was measured in experiments described in [1–5] by placing electrical contact pickups at known distances inside the specimen, as shown in Fig. 11.18, which closed at the time the wave passed and sent a pulse which was recorded by means of a special circuit and an oscilloscope.

Dividing the distance d by time the average wave speed across the measuring "base" d could be found (the bases d were of the order of 5–8 mm, the speeds $D \sim 5$–10 km/sec, and the time $\sim 10^{-6}$ sec; this required the development of special methods for recording such short time intervals). By the use of the pickups the time at which the free boundary of the unloaded material passed through fixed points was similarly measured* (see Fig. 11.18). The electrical contact method for measuring velocities was proposed by Tsukerman and Krupnikov. This method was used to measure the Hugoniot curve for iron up to pressures of $p \sim 1.5 \cdot 10^6$ atm ($D \sim 7.5$ km/sec, $u \sim 2.4$ km/sec).

The free surface method is not suitable for studying porous materials, since in this case the additional velocity u' upon unloading is appreciably less than the velocity u, and the doubling rule does not apply.

2. *The "collision" method*†. For the study of stronger shock waves, for which the velocity doubling rule introduces an appreciable error, the authors of [1] used another method, which is termed the "collision" method. This method is perfectly exact in principle and is suitable for the study of all materials, including porous ones.

An explosive charge is used to accelerate a plate made of the test material to a velocity w. The plate (striker) strikes another plate (target), which is at

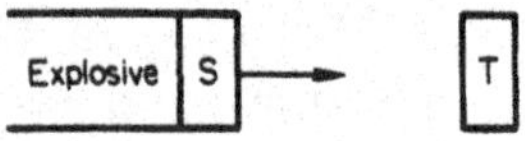

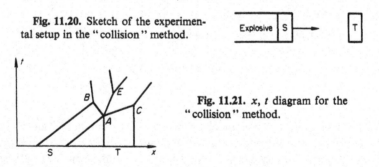

Fig. 11.20. Sketch of the experimental setup in the "collision" method.

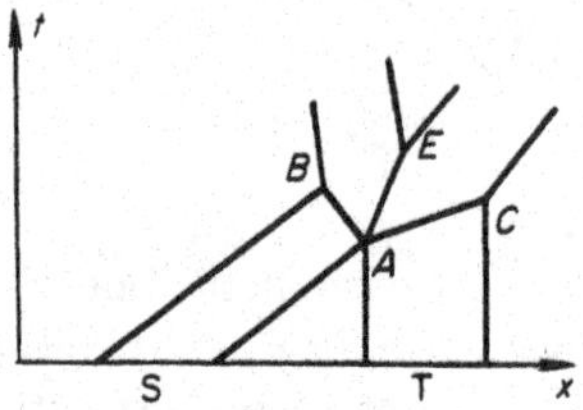

Fig. 11.21. x, t diagram for the "collision" method.

rest and made of the same material. The experimental setup is shown in Fig. 11.20, and an x, t diagram of the process is given in Fig. 11.21. Two shock waves originate at the time of impact and propagate through both bodies

* The pickups were equipped with protective caps to prevent their being closed by the air shock which the metal boundary pushes.

† *Editors' note.* The authors of [1] and the authors used the term "braking" method. The method is sometimes called the "momentum transfer" method (see [60]).

(AB and AC on the x, t diagram). The pressures p and particle velocities u on both sides of the contact surface between the bodies are the same and equal to the values of these quantities behind the shock waves up to the time the shocks reach the opposite boundaries of the specimens*. This velocity u is also the velocity of the contact surface (the line AE). Pressure and velocity distributions after impact are shown in Fig. 11.22.

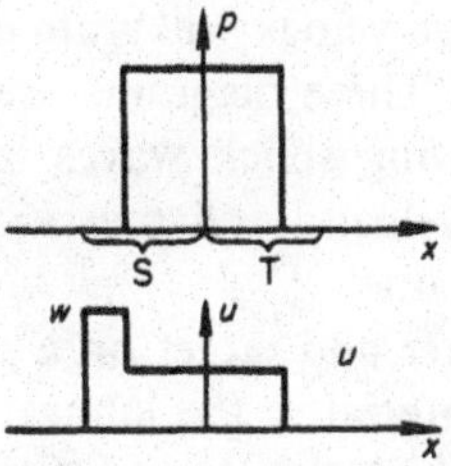

Fig. 11.22. Pressure and velocity distributions after impact in the "collision" method.

Since the materials are identical, the two shock waves are also identical, and the jumps in particle velocity are the same for both waves. The velocity jump for the target is the same as the velocity of the compressed material u, since the target was initially at rest. With respect to the striker, however, the material ahead of the shock moves with the striker velocity w, and behind the wave with the velocity u, so that the absolute value of the velocity jump is equal to $w - u$. Consequently, $w - u = u$ and $u = w/2$. Thus, the problem reduces to measuring the wave velocity D in the target and the striker velocity w. This is carried out experimentally in the same manner as in the free surface method, by using a system of electrical contact pickups.

Using the collision method Hugoniot curves for iron were obtained up to pressures of $p \sim 5 \cdot 10^6$ atm ($D \sim 12$ km/sec, $u \sim 5$ km/sec, and $V_0/V \sim 1.75$) in [1]. Porous iron with a density 1.4 times less than standard was also studied.

The collision method can also be extended to the case when the test target and striker are made from different materials. In this case, however, the striker must be made of a material for which the Hugoniot curve is known. In a number of cases this turns out to be more expedient than making the striker from the test material, since by an appropriate choice of the striker material it is possible from the same explosive charge to obtain a more powerful shock in the test material.

If the striker and target materials are different, then, in spite of the fact that the pressure behind both shock waves is the same, the velocity jumps are no longer the same, and $w - u \neq u$. However, if the Hugoniot curve for the striker is known, then we also know the dependence of the pressure on the particle velocity jump, that is, the function $p = f(w - u)$. On the other hand, the pressure p is related to the jump of the particle velocity in the

*See §24, Chapter I.

target, equal to the velocity of the contact surface u, by (11.32): $p = Du/V_0$. Measuring as before the shock velocity in the target and the striker velocity w, we can find u from

$$f(w-u) = \frac{Du}{V_0}. \tag{11.45}$$

A graphical method, based on the use of a pressure-velocity diagram (see §24, Chapter I), is very convenient for this purpose. These diagrams are widely used when considering various processes involving shock waves in which two contiguous media are involved, since the velocity and pressure at the contact surface between the two media are the same.

Let us consider the collision between the striker and target on a p, u diagram, where u is the particle velocity of the material in the laboratory coordinate system, in this case in that system in which the target is initially at rest. The initial states of the target ($p = 0$, $u = 0$) and of the moving striker $p = 0$, $u = w$, are represented by the points O and A in Fig. 11.23. If the shock

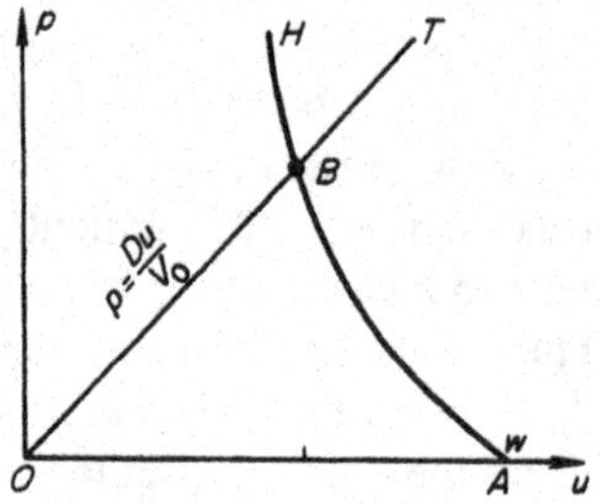

Fig. 11.23. p, u diagram for the "collision" method. HBA is the Hugoniot curve for the striker. OBT is the locus of target states after impact.

velocity measured at the target is D, then the locus of states of the target material behind the shock wave is the straight line $p = Du/V_0$, with the known slope D/V_0. Let us depict the Hugoniot curve for the striker material, considering the dependence of the pressure on the velocity jump instead of on the volume. In this case the velocity jump is equal to $w - u$, so that $p = f(w-u)$. The point of intersection B of the two lines, according to (11.45), determines the state (pressure and particle velocity) behind the two shock waves. If the striker and target are made from the same material, then (as we already know) the point of intersection lies exactly half-way between points O and A ($u = w/2$).

3. *The "calibrated reflection" method**. In this method use is made of the relationships governing the breakup of an arbitrary discontinuity resulting from the reflection of a shock wave from the contact surface between two

* *Editors' note.* Termed "reflection" method by the authors and sometimes referred to as the "impedance match" method (see, e.g., [60]).

media (see §24 of Chapter I). The advantage of this method in comparison with the preceding one is that it does not require the measurement of particle velocities, which are more difficult to measure than the shock front velocity. However, the method requires the use of a calibrated material with a known equation of state. This method was worked out by the authors of [2] together with G. M. Gandel'man.

Let us consider the passage of a strong shock from material A to material B. The wave traveling through material B is always a shock wave, while the wave reflected in A is either a shock wave, in case material B is "harder" than A, or a rarefaction wave, in case B is "softer" than A (this is most easily visualized by considering the two limiting cases, where A is a gas and B is a solid, and where A is a solid and B is a gas). The velocity and pressure distributions in the two cases are given in Fig. 11.24. The corresponding x, t diagrams are also shown there.

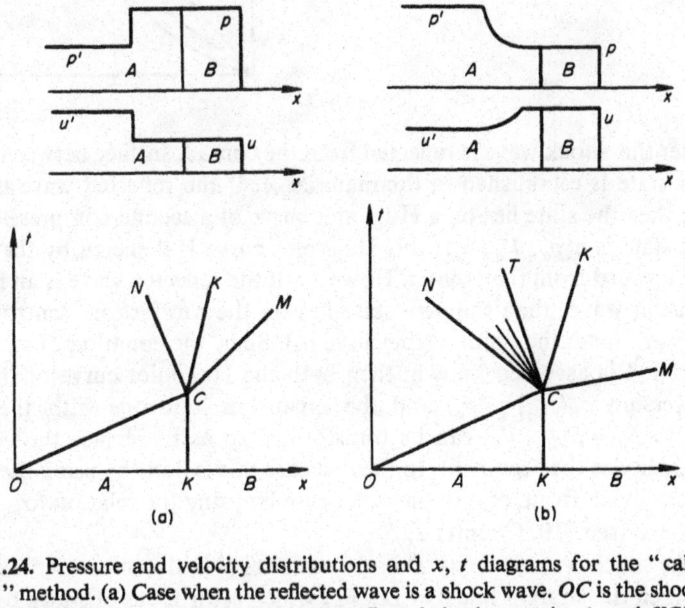

Fig. 11.24. Pressure and velocity distributions and x, t diagrams for the "calibrated reflection" method. (a) Case when the reflected wave is a shock wave. OC is the shock wave in A, CM is the shock wave in B, CN is the reflected shock wave in A, and KCK is the contact surface between A and B. (b) Case when the reflected wave is a rarefaction wave. OC is the shock wave in A, CM is the shock wave in B, CN is the head of the rarefaction wave, CT is the tail of the rarefaction wave, and KCK is the contact surface between A and B.

Let us consider this process on a pressure-velocity diagram (in the initial state both A and B are at rest in the laboratory system of coordinates). We assume that the equation of state of the material A is known. We draw on the

p, u diagram (Fig. 11.25) the Hugoniot curve for the material A, $p_A(u)$ for the first shock wave traveling through the undisturbed material. If we measure experimentally the front velocity of the initial shock D_1, then we can find the state behind it, represented by the point of intersection of the straight line $p = D_1 u/V_{0A}$ with the Hugoniot curve $p_A(u)$ (the point $a(p_a, u_a)$).

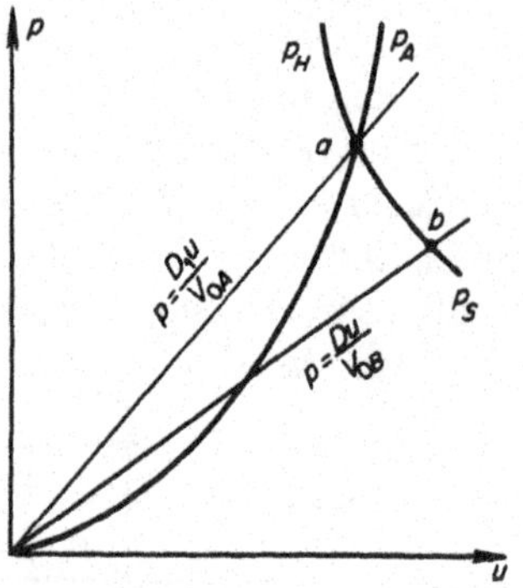

Fig. 11.25. p, u diagram for the "calibrated reflection" method.

After this shock wave is reflected from the contact surface between A and B, a new state is established in the material A. If the reflected wave is a shock wave, then the state lies on a Hugoniot curve of a second compression whose initial state is $a(p_a, V_a, u_a)$; this Hugoniot curve is depicted by the curve p_H going upward from the point a. However, if the reflected wave is an isentropic expansion wave, then the new state lies on the rarefaction isentrope, going downward from the point a (the curve p_S). Since the equation of state for the material A is assumed known, then both the Hugoniot curve of the second compression $p_H(V, V_a, p_a)$ and the expansion isentrope with the entropy equal to $S_a = S(p_a, V_a)$ can be transformed so as to replace the volume by the velocity as the argument. In the first case this is done by using the relations across a shock front, and in the other case by using the relation for an expansion wave (see §10, Chapter I).

If the shock wave velocity D in the material B is also measured experimentally, then the straight line $p = Du/V_{0B}$ which is the locus of states behind this wave can be drawn. The point of intersection b of this line with curve $p_H a p_S$, which is the locus of possible states in the material A after reflection of the shock wave, determines the pressure and velocity behind the shock wave in the material B, equal to the pressure and velocity at the contact surface between A and B (see Fig. 11.24). The p, u diagram of Fig. 11.25 illustrates the second case, in which the reflected wave is a rarefaction wave. In the first case the line $p = Du/V_{0B}$ passes above the line $p = D_1 u/V_{0A}$ and the point of intersection b lies above point a on the Hugoniot curve for the second compression of the material A, described by the curve ap_H*.

Summing up, the calibrated reflection method consists in the following: A shock wave is generated in the plate made of material A with a known equation of state, either directly by an explosive charge, or by impact with another plate previously accelerated by an explosive to a high velocity. This wave passes into specimens of the test material B, with which is also included a sample of material A (a diagram of the experiment is shown in Fig. 11.26).

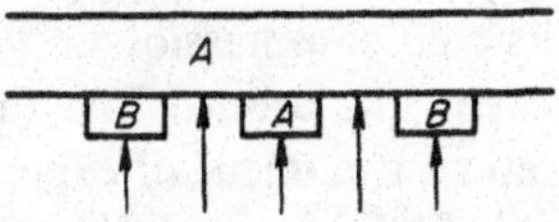

Fig. 11.26. Sketch of the experimental setup in the "calibrated reflection" method.

By recording the closing times of the electrical contact pickups located at the points denoted by arrows in Fig. 11.26, the front velocities D_1 and D are determined. Having constructed the Hugoniot curve $p_A(u)$ on a p, u diagram and having drawn the line $p = D_1 u/V_{0A}$, the point a giving the state behind the shock wave in a material A can be found. Then the Hugoniot curve for the second compression is drawn upward through point a, and the isentrope is drawn downward through point a; the line $p = Du/V_{0B}$ is also drawn. Thereby the sought state point $b(p, u)$ behind the shock wave is determined for the test specimen.

Actually (in the experiments), the pressure changes between states a and b were always small. Under these conditions, as shown by calculations, the curve $p_H a p_S$ can to a high degree of accuracy be represented as a mirror image of the Hugoniot curve of the primary compression at the point a. We note that the slope of the line $p_H a p_S$ at point a is determined by the speed of sound behind the front of the primary shock wave in A. Actually, for either a rarefaction wave or a weak compression wave, $dp = \pm \rho c du$ (see (1.59)); thus the slope of $p_H a p_S$ at point a is $|dp/du| = \rho c = c/V$, where c and V are the speed of sound and the volume in the material A after compression by the first shock wave. Experimental methods for determining the speed of sound behind the shock wave will be considered below.

Al'tshuler, Krupnikov, and Brazhnik have used the calibrated reflection method for obtaining Hugoniot curves for a large number of metals (Cu, Zn,

* In addition, we see here what characterizes the "hardness" of a material. Let us assume that the shock waves are weak and that their speeds are close to the speeds of sound: $D \approx c_B$ and $D_1 \approx c_A$. Material B is harder than A and the reflected wave will be a shock wave if $c_B/V_B > c_A/V_A$ or $\rho_B c_B > \rho_A c_A$. The quantity ρc is sometimes called the acoustic impedance. It determines the relation between pressure and velocity in an acoustic or weak shock wave, with $p = \rho c u$.

Pb, etc.) [2]. This method was used for studying the compressibility of sodium chloride (reference [5]), and was also used in the majority of works of non-Soviet scientists. Iron, aluminum, or brass were most frequently used as the material A.

§13. Determination of cold compression curves from the results of shock compression experiments

One of the most valuable results of experiments on shock compression of solids is the determination for a material of its cold compression curve $p_c(V)$, which characterizes the repulsive forces between the atoms. The functions $p_c(V)$ and $\varepsilon_c(V)$ are determined from a theoretical treatment of experimental data on the Hugoniot curves for the material. In [3] the cold compression curves were determined over a wide range of compressions and pressures by expressing the thermodynamic functions $p(V, T)$ and $\varepsilon(V, T)$ as the sum of three terms in the form of (11.30). The electronic terms p_e and ε_e were written on the basis of purely theoretical considerations (see §§5 and 6), with known experimental values used for the electronic specific heat coefficient β_0.

Using the experimental functional dependences for $p(V)$ and $\varepsilon(V)$ given by the Hugoniot curve, we can regard equations (11.30) as two equations with three unknown functions $p_c(V)$, $\Gamma(V)$, and $T(V)$, where $T(V)$ corresponds to the dependence of the temperature on the volume along the Hugoniot curve. For the third equation we may use the relation between the Grüneisen coefficient $\Gamma(V)$ and the cold compression curve $p_c(V)$, given by the Slater–Landau formula (11.18)*. A numerical solution of this system of equations gives the cold compression curve $p_c(V)$, the function $\Gamma(V)$, and the temperature behind the shock wave T. The data in Table 11.2 and the curves of Fig. 11.8 (§7) were obtained in this manner. Specific results for other metals tested can be found in the tables in [3].

If the Hugoniot curves for both porous and continuous forms of a material are known from experiment, then the relation between the functions $\Gamma(V)$ and $p_c(V)$ is not required. If we consider temperatures that are not too high and neglect the electronic terms p_e and ε_e, we can write

$$\frac{1}{\Gamma} = \frac{\Delta \varepsilon_T}{V \Delta p_T} = \frac{1}{V} \frac{\varepsilon_{\text{por}}(V) - \varepsilon_{\text{cont}}(V)}{p_{\text{por}}(V) - p_{\text{cont}}(V)},$$

where the quantities on the right-hand side are the experimental values on

* In a number of cases a slightly different relation between the functions $\Gamma(V)$ and $p_c(V)$ was used, one given by the Dugdale–McDonald formula [27].

the Hugoniot curves for the continuous and porous forms of the same material compressed to the same volume. The elastic components of pressure and energy are identical in both cases, so that the differences in ε and p are equal to the excess of the purely thermal energy and pressure in the compressed porous material over those of the compressed continuous material. This was the method used in [1] to obtain the cold compression curve for iron (Fig. 11.2).

The cold compression curve for sodium chloride was reported in [5]. Comparison with expressions for the repulsive forces in ionic crystals permitted the determination of the parameters characterizing the interaction forces which enter these equations. The method for calculating the temperature on the Hugoniot curve for comparatively weak waves, when the thermal terms are small in comparison with the elastic terms and the Hugoniot curve follows the cold compression curve closely, is described in [22] (in this case, obviously, the electronic terms were not taken into account).

Frequently, various interpolation equations are employed to characterize the Hugoniot and cold compression curves $p_c(V)$ analytically. This is done by specifying state functions of given forms containing several parameters, which are then determined from experimental data. As an example, we can take the widely used equation $p_c = A[(V_0/V)^n - 1]$, containing the two parameters A and n. In investigating sodium chloride [5] the $p_c(V)$ curve was obtained in analytic form by using a power-law or exponential representation for the repulsive forces in ionic crystals. The constants which enter the equation were determined from data on dynamic compressibility.

Kormer and Urlin [14] constructed an interpolation formula for the cold compression curve of the form

$$p_c = \sum_{n=1}^{6} a_n \left(\frac{V_{0c}}{V}\right)^{\frac{1}{3}n+1}.$$

The coefficients a_n were determined without using experimental data on shock compression, only using the relations between the coefficients and the known parameters of the standard state (the compressibility, the Grüneisen coefficient, etc.) and also the condition that at high pressures the curve should agree with the relation derived from the Thomas–Fermi–Dirac model. Good agreement was obtained with the experimentally determined curves for $p_c(V)$*.

* Subsequently Kormer, Urlin, and Popova [15] refined this method, adding one more term to the series and using one experimental point on the Hugoniot diagram. Very good agreement with the experimental curves was obtained.

3. *Acoustic waves and splitting of waves*

§14. Static deformation of a solid

In the study of shock compression of solids, we have assumed up till now that the pressure in the compressed material is isotropic, that it has a hydrostatic character as in a liquid or gas. The increase in density was then considered as a result of an isotropic compression. Correspondingly, the elastic properties of the material were characterized by a single quantity, the isentropic compressibility $\kappa = -(1/V)(\partial V/\partial p)_S$, which determined the speed of propagation of acoustic compression (and rarefaction) waves, the speed of "sound"

$$c_0 = \left[-V^2\left(\frac{\partial p}{\partial V}\right)_S\right]^{1/2} = \left(\frac{V}{\kappa}\right)^{1/2}$$

This can be done only in the case when the pressures are sufficiently high and the effects connected with the strength of solids and the existence of shear strains and stresses are not important. If the loads are small, then it becomes necessary to take into account the elastic properties of the solid which distinguish it from a liquid. This has an appreciable effect on the character of the dynamic processes and, in particular, on the propagation of elastic compression and rarefaction waves. Thus, it is found that acoustic waves can propagate in a solid with different speeds, depending on the particular conditions. Before considering these dynamic phenomena, let us examine the behavior of a solid under static loads. Here we assume that the deformations and loads are small, so that linear elasticity theory is applicable.

The state of a deformed body is described by two tensors, the strain tensor and the stress tensor. In what follows we shall consider only a few simple cases of homogeneous deformations (where each element of the body is deformed in the same manner), which are characterized by simple and obvious quantities. For this reason we shall not introduce the strain tensor in general form*.

The stress tensor component σ_{ik}, where the subscripts i and k denote the x, y, and z coordinate directions, represents the ith component of a force acting on a unit area of the body whose normal is in the k direction. The components σ_{xx}, σ_{yy}, and σ_{zz} represent normal stresses and the components σ_{xy}, σ_{xz}, and σ_{yz} are tangential or shear stresses (Fig. 11.27). The tensor σ_{ik} is symmetric, so that $\sigma_{xz} = \sigma_{zx}$, $\sigma_{yz} = \sigma_{zy}$, $\sigma_{xy} = \sigma_{yx}$.

Let us consider some examples of deformations.

1. Imagine a cylindrical rod of length L and diameter d, with a compressive

* See, for example, Landau and Lifshitz [28].

force or a pressure p applied to its ends. The z axis is directed along the axis of the rod, as shown in Fig. 11.28. The lateral surface of the rod we assume to be free. Under the action of the load the rod is contracted by a length ΔL

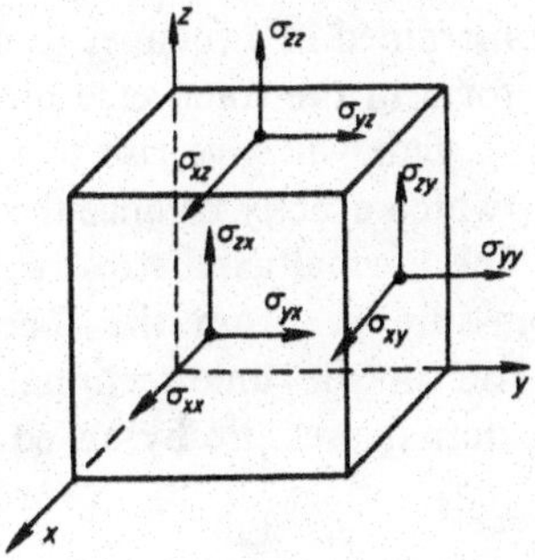

Fig. 11.27. Diagram illustrating the stress tensor components.

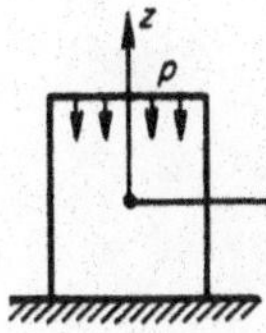

Fig. 11.28. Diagram of a rod in compression.

and it is thickened (the diameter increases by Δd). In this case only the normal stress in the axial direction σ_{zz}, which is minus* the external pressure $\sigma_{zz} = -p$, is different from zero. The normal stresses in the transverse directions, σ_{xx} and σ_{yy}, are absent, since the lateral surface of the rod is free and nothing prevents the rod from expanding in this direction. It is obvious that the tangential or shearing stresses σ_{xy}, σ_{xz}, and σ_{yz} are also equal to zero in the coordinate system chosen.

According to Hooke's law for small deformations the relative contraction of the rod is proportional to the applied force:

$$\frac{\Delta L}{L} = -\frac{p}{E} = \frac{\sigma_{zz}}{E}, \tag{11.46}$$

where E is Young's modulus (this is the definition of Young's modulus). The relative thickening of the rod is proportional to the relative contraction

$$\frac{\Delta d}{d} = -\sigma\frac{\Delta L}{L}, \tag{11.47}$$

* *Editors' note.* It is customary to define the normal stresses as positive when tensile, and we have changed some of the equations for consistency with this practice. Under compression σ_{zz} and $\Delta L/L$ are negative.

where σ is Poisson's ratio. Poisson's ratio is always positive and smaller than $\frac{1}{2}$. This follows from the observations that a compressed rod becomes thicker and that its volume can only be reduced (for constant volume $d^2L = const$, and $\Delta d/d = -\frac{1}{2}\,\Delta L/L$, $\sigma = 1/2$).

2. Let the lateral surface of the rod be constrained in a manner such that for an axial compression the rod cannot deform in the transverse direction (the rod is placed in a shell with rigid walls). This will give rise to normal stresses in the transverse directions $\sigma_{xx} = \sigma_{yy}$ which exactly balance the external lateral forces acting on the shell walls. The normal axial stress σ_{zz} is as before minus the external compressive pressure p. From the theory of elasticity the relative contraction of the rod in the unidirectional axial deformation of this case is related to the external pressure by an equation analogous to (11.46):

$$\frac{\Delta L}{L} = -\frac{p}{E'} = \frac{\sigma_{zz}}{E'}, \tag{11.48}$$

where

$$E' = \frac{E(1-\sigma)}{(1+\sigma)(1-2\sigma)}. \tag{11.49}$$

The quantity E' is always greater than Young's modulus E. In order to decrease the length of a laterally constrained rod by the same amount as that of a free rod it is necessary to apply a larger compressive force. The normal stresses in the transverse directions are

$$\sigma_{xx} = \sigma_{yy} = \frac{\sigma}{1-\sigma}\sigma_{zz} = -\frac{\sigma}{1-\sigma}p. \tag{11.50}$$

Tangential stresses are absent in the coordinate system chosen. All the relationships in the above two examples are equally valid in the case when the rod is extended, is in tension.

3. A body subjected to an isotropic compression (or expansion) changes its volume while retaining its shape, i.e., while remaining similar to itself in shape. An isotropic compression is obtained by applying a constant pressure to the surface of the body. The stress tensor for such a compression is diagonal ($\sigma_{xy} = \sigma_{xz} = \sigma_{yz} = 0$); all three normal components are the same and equal to minus the pressure. This remains true in any coordinate system. The "pressure" in the body is isotropic in this case and is hydrostatic in character, as in a liquid. For small deformations, the relative change in volume* is proportional

* The sum of the diagonal strain tensor components is $u_{xx} + u_{yy} + u_{zz} = \Delta V/V$ and is termed the dilatation. For an isotropic compression $u_{xy} = u_{yz} = u_{xz} = 0$ and $u_{xx} = u_{yy} = u_{zz}$.

to the pressure:

$$\frac{\Delta V}{V} = -\kappa p = -\frac{p}{K}, \tag{11.51}$$

where κ is the compressibility, and its reciprocal $K = 1/\kappa$ is the bulk modulus.

4. Finally, let us consider a pure shear deformation in one direction, as shown in Fig. 11.29. In pure shear the body only changes its shape but not its volume. In the example shown in Fig. 11.29 only the tangential stress

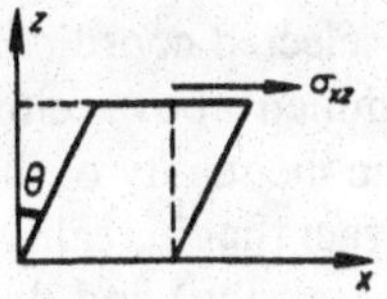

Fig. 11.29. Pure shear strain in one direction.

σ_{xz} is different from zero. All the other components of the stress tensor are equal to zero. According to Hooke's law, the shear strain angle is proportional to the shearing force τ (per unit area), which is equal to the stress σ_{xz}

$$\theta \approx \tan\theta = \frac{\tau}{G} = \frac{\sigma_{xz}}{G}, \tag{11.52}$$

where G is the shear modulus.

As is well known (see [28]), we can represent an arbitrary deformation as the sum of pure shear strains and an isotropic compression (or extension). Because of this interrelationship between the strains in axial compression of a rod and the elementary strains of isotropic compression and shear, the four characteristics of the material E, σ, K, and G are not independent, but are connected by two relations. One can show (see [28], for example) that

$$E = \frac{9KG}{3K+G}, \qquad \sigma = \frac{1}{2}\frac{3K-2G}{3K+G}, \tag{11.53}$$

and, conversely, that

$$G = \frac{E}{2(1+\sigma)}, \qquad K = \frac{E}{3(1-2\sigma)}.^{*} \tag{11.54}$$

Thus, we can rewrite Hooke's law for the axial deformation of a laterally constrained rod (11.48) in terms of the moduli K and G in the form

$$\frac{\Delta L}{L} = -\frac{p}{E'}, \qquad E' = K + \tfrac{4}{3}G. \tag{11.55}$$

* It is evident from this equation that $\sigma \leqslant 1/2$, since $K > 0$ and $E \geqslant 0$.

In order to give some idea of the numerical values of the parameters we note that for iron (with a 1% carbon content)

$$E \approx 2.1 \cdot 10^6 \text{ kg/cm}^2, \qquad G \approx 0.82 \cdot 10^6 \text{ kg/cm}^2,$$

$$K \approx 1.61 \cdot 10^6 \text{ kg/cm}^2, \qquad \sigma \approx 0.28.$$

For a body subjected to an isotropic compression or extension the stress tensor is diagonal in any coordinate system, and all three of its components are the same. For other deformations the stress tensor is diagonal and the tangential stresses vanish only in certain specially selected coordinate systems. The deformations of a rod in compression examined above, either free or constrained laterally, can serve as an example. The inequality of the diagonal elements of the stress tensor is connected with the fact that in reality the deformation is not a pure isotropic compression (or extension) and does contain an element of shear. This is manifested explicitly if we change to another coordinate system, or, equivalently, if we consider forces acting on areas which are inclined with respect to the rod axis. In this case it becomes immediately clear that the inclined areas experience tangential stresses, which shows that shear strains are present.

Let us calculate the tangential stress acting on an area inclined at 45° to the direction of the external pressure (Fig. 11.30). For simplicity we shall not

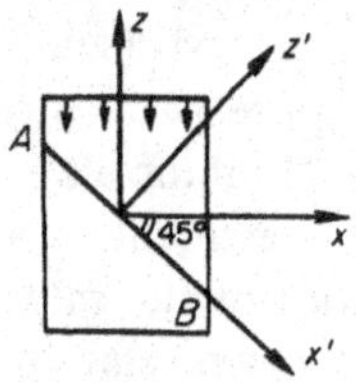

Fig. 11.30. The question of non-diagonality of the stress tensor.

consider a cylindrical rod, but rather a plane layer infinite in the y direction and constrained laterally so that there are no displacements in the x direction. In the x, y, z coordinate system we have stresses σ_{zz} and $\sigma_{xx} = \sigma_{yy}$. In order to find the tangential stress acting on the plane AB, we introduce a new coordinate system, x', y', z', rotated with respect to the old system about the y axis (the axes y and y' coincide). According to the rule for transformation of a tensor on rotating the coordinate system, we find

$$\sigma_{x'z'} = -\sigma_{zz} \cos^2 45^\circ + \sigma_{xx} \cos^2 45^\circ = -\tfrac{1}{2}(\sigma_{zz} - \sigma_{xx}).$$

This is a tangential stress acting in the x' direction on the area AB, whose normal is in the direction of the z' axis.

§15. Transition of a solid medium into the plastic state

One of the characteristic properties of a solid that distinguishes it from a fluid is the stability of its shape, its resistance to shear. A fluid has no resistance to shear and readily assumes any shape as long as this does not require any change in its volume (density). Tangential shear stresses are absent in a fluid in a static state*. A fluid is characterized by the fact that its shear modulus $G = 0$. Formally, for $G = 0$ Poisson's ratio σ is, according to (11.53), equal to $\frac{1}{2}$. The stress tensor in this case is diagonal in any coordinate system, with all its three normal components identical and equal to minus the "hydrostatic" pressure, which is isotropic. The elastic properties of a fluid are characterized only by its compressibility or bulk modulus.

It is well known that, for sufficiently high loads that do not reduce to an isotropic compression, a solid changes its elastic properties and becomes plastic, or flowing, similar in some respects to a fluid. The plastic state of a solid is not characterized by the total absence of tangential stresses, as in the case of a fluid, but by the absence of an increase in the tangential stresses with an increase in shear strains. Starting at certain critical shear strains and stresses, a solid no longer resists any further increase in shear.

Above we have defined the shear modulus G as a coefficient of proportionality between the tangential stress in pure shear and the shearing strain (see (11.52)). As a result of the linearity of the relation between stress and strain, the increments in strain and stress are also proportional

$$\sigma_{xz} = G\theta, \qquad d\sigma_{xz} = G\, d\theta$$

(in pure shear through the angle θ, as shown in Fig. 11.29). When a solid medium is in the plastic state, after the values of the shear strain angle θ and stress σ_{xz} become equal to critical values θ_{cr} and σ_{cr}, there is no longer an increase in stress with increasing strain (or its rate of increase drops sharply). This is illustrated by the $\sigma_{xz}(\theta)$ diagram in Fig. 11.31. If we formally define the shear modulus in this state as the coefficient of proportionality between the increments $d\sigma_{xz}$ and $d\theta$, rather than between the quantities σ_{xz} and θ themselves, then the shear modulus should be set equal to zero.

Let us consider an axial compression of a nonplastic and of a plastic body. Let a cylindrical solid body be placed in a cylindrical container with rigid walls and be compressed by a piston along its axis (Fig. 11.32). A schematic representation of the changes in the positions of the atoms in the body is given in Fig. 11.33. For simplicity we assume a cubic lattice. If the body is nonplastic, then the interatomic distances in the direction of the axis decrease,

* They arise only at the time when the shape changes and do not depend on the strains themselves but on their rate of change.

while they remain unchanged in the transverse directions. In this case the atoms remain "in their places". This is shown in Fig. 11.33b. If, however, the body is plastic, then all interatomic distances are decreased, the lattice is rearranged, and the atoms are redistributed in such a manner that the lattice remains cubic even in the compressed state (Fig. 11.33c). For clarity, the atoms in Fig. 11.33c have been renumbered, without any implication that the atoms in question have been redistributed in the particular manner shown.

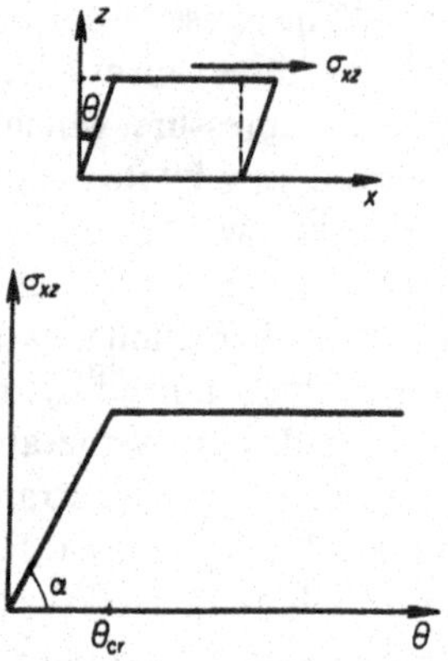

Fig. 11.31. Tangential stress-shear strain angle diagram.

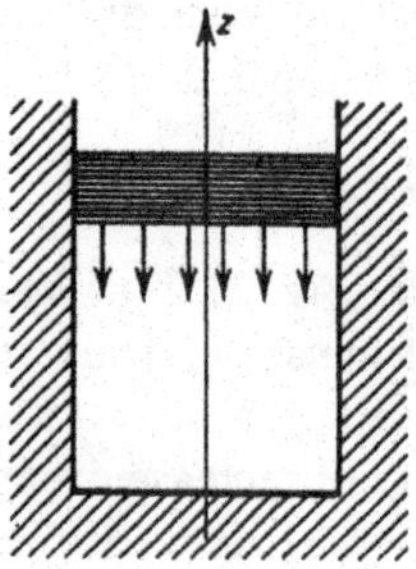

Fig. 11.32. Diagram showing axial compression (constrained) of a rod.

The first case (Fig. 11.33b) contains within it an element of shear. Thus, in the undeformed state (Fig. 11.33a) the projection of atom 2 on the inclined plane *AB* passing through atoms 1–6 of two neighboring horizontal rows falls at point *C* situated at the midpoint of the segment *AB*. In the deformation of a nonplastic body (Fig. 11.33b) point *C* moves closer to point *B*. The inclined rows of atoms are displaced with respect to each other: the upper row 2–7–12 is displaced to the right and downward with respect to the lower row 1–6–11.

However, in the deformation of a plastic body the lattice remains cubic, the projection of atom 5 on the inclined plane *AB* passing through atoms 1–13, labeled point *C* as in the undeformed state, lies at the midpoint of segment AB. The inclined rows of atoms 5–10 and 1–13 are not displaced with respect to each other, as also in the undeformed state.

During deformation a body acquires elastic energy resulting from the work of the external forces producing the deformation. If the body is nonplastic, then this energy is related to the change in volume as well as to the shear. For a given volume the elastic energy is a minimum if the compression is

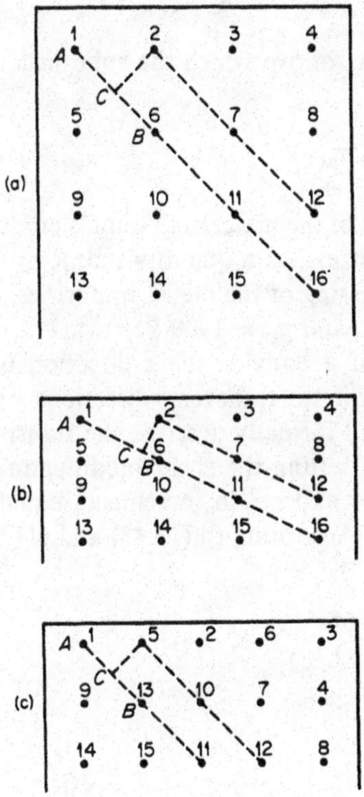

Fig. 11.33. Diagram illustrating the deformation of nonplastic (b) and plastic (c) bodies; (a) shows the undeformed state.

isotropic and there are no shearing strains. Therefore, a nonplastic body subjected to an axial compression to a given volume is in a nonequilibrium state. The equilibrium state for the given volume would correspond to an isotropic compression, that is, to one with a rearranged crystal lattice. The rearrangement of the lattice requires an activation energy, as the atoms must overcome potential barriers*. At small loads rearrangement does not take place and the solid exhibits nonplastic behavior with respect to deformation.

However, when the loads are sufficiently great, the solid loses its firmness or nonplasticity and becomes similar to a fluid, in the sense that it becomes

* It is possible that rearrangement involves a macroscopic breaking up of the particles of the body.

capable of rearranging itself in such a manner that its energy at a given volume is a minimum. In particular, this occurs for axial compression of a body when the tangential stress in a plane inclined at 45° to the direction of the compressive force $\sigma_{x'z'}$ (see the end of the preceding section) exceeds the critical shear stress σ_{cr}. Noting that

$$\sigma_{x'z'} = -\tfrac{1}{2}(\sigma_{zz} - \sigma_{xx}) = -\frac{1}{2}\frac{1-2\sigma}{1-\sigma}\sigma_{zz} = \frac{1}{2}\frac{1-2\sigma}{1-\sigma}p,$$

we find for the critical compressive load p_{cr} above which the solid becomes plastic

$$p_{cr} = \frac{1-\sigma}{1-2\sigma}2\sigma_{cr}. \tag{11.56}$$

In contrast to the thermodynamic constants of the material (Young's modulus or the compressibility), the critical shear stress, as a quantity characterizing the strength, depends strongly on the processing of the metal, impurities, etc. For iron, approximately, $\sigma_{cr} = 600\ \text{kg/cm}^2$, and $p_{cr} = 1900\ \text{kg/cm}^2$.

Let us consider an axial compression of a body in the z direction by a compressive load p. No deformations occur in the transverse directions x and y (the rod is laterally constrained). We shall formally describe the transition from the nonplastic to the plastic state by setting the shear modulus in the proportionality relation between the stress and strain increments equal to zero for loads exceeding the critical value. According to (11.48) and (11.55), for $p = -\sigma_{zz} < p_{cr}$,

$$\sigma_{zz} = (K + \tfrac{4}{3}G)\frac{\Delta L}{L}, \qquad \frac{d\sigma_{zz}}{d(\Delta L/L)} = K + \tfrac{4}{3}G.$$

Then from (11.50) and (11.53)

$$\sigma_{xx} = \sigma_{yy} = (K - \tfrac{2}{3}G)\frac{\Delta L}{L}, \qquad \frac{d\sigma_{xx}}{d(\Delta L/L)} = K - \tfrac{2}{3}G;$$

$$\sigma_{x'z'} = -\tfrac{1}{2}(\sigma_{zz} - \sigma_{xx}) = -G\frac{\Delta L}{L}, \qquad \frac{d\sigma_{x'z'}}{d(\Delta L/L)} = -G.$$

After the load reaches its critical value, in the equations for the derivatives of the stresses (but not in the relations for the stresses themselves) we set $G = 0$. For $p > p_{cr}$ we obtain

$$\frac{d\sigma_{zz}}{d(\Delta L/L)} = \frac{d\sigma_{xx}}{d(\Delta L/L)} = K, \qquad \frac{d\sigma_{x'z'}}{d(\Delta L/L)} = 0. \tag{11.57}$$

The normal stresses $-\sigma_{zz}$, $-\sigma_{xx}$, and $-\sigma_{yy}$ now increase uniformly in correspondence with the bulk modulus (in axial compression $\Delta L/L = \Delta V/V$). The tangential stress in the inclined plane remains constant and given by $\sigma_{x'z'} = \sigma_{cr}$ (the critical strain is $(\Delta L/L)_{cr} = \sigma_{cr}/G$). The stress-strain diagram is shown in Fig. 11.34.

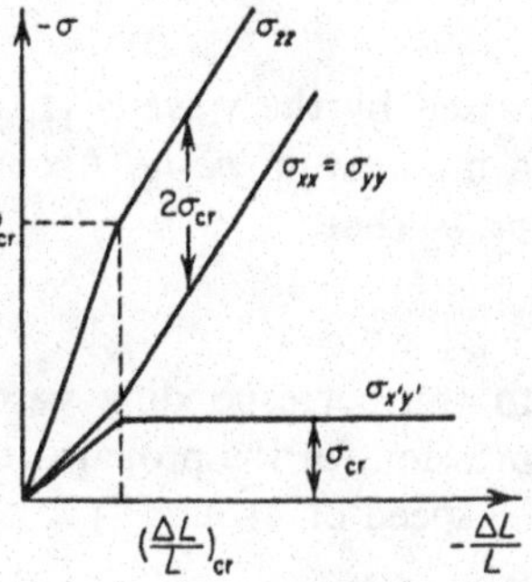

Fig. 11.34. Stress-strain diagram for axial compression of a solid.

For loads less than or of the order of the critical load, σ_{zz} is very different from σ_{xx} and the "pressure" is substantially nonhydrostatic in character. In the limit when the loads are sufficiently large, with $p \gg p_{cr}$, the relative difference $(\sigma_{zz} - \sigma_{xx})/\sigma_{zz} = -2\sigma_{cr}/\sigma_{zz} \to 0$, and all the three normal stresses become almost identical. The tangential stress $\sigma_{x'z'} = \sigma_{cr}$ becomes small in comparison with the normal stresses. It remains constant or increases slowly, much more slowly than before.

§16. Propagation speed of acoustic waves

Let us extend the results of the preceding sections to the case of dynamic loads and find the propagation speeds of acoustic compression (and rarefaction) waves under different conditions. Let a constant compressive force with pressure p be applied at an initial time to the end of a thin rod with a free lateral surface*. A compression wave will travel through the body, with a propagation speed which we denote by c_1. The material between the wave front and the end of the rod deforms as in example 1 of §14, and acquires a constant velocity u in the direction of the axial force. As may be seen from Fig. 11.35 the relative contraction of the rod in the compression region is

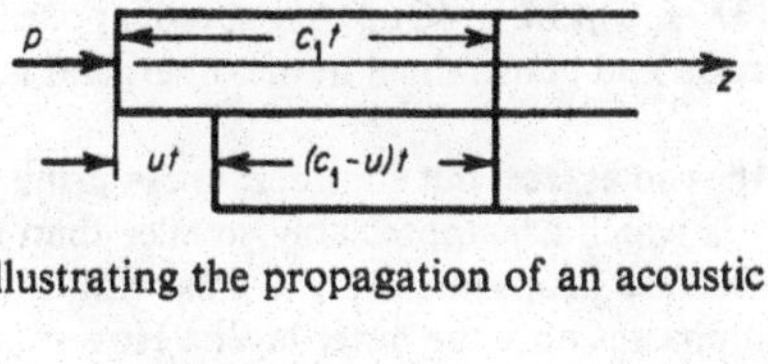

Fig. 11.35. Diagram illustrating the propagation of an acoustic compression wave.

* This statement of the problem is analogous to the piston problem treated in gas-dynamics (see Chapter I).

$[c_1t-(c_1-u)t]/c_1t=u/c_1$. If we consider small loads and deformations, then, according to Hooke's law (11.46),

$$\frac{u}{c_1}=\frac{p}{E}.^{*} \tag{11.58}$$

In a time t the mass of material encompassed by the wave $\rho c_1 t$ (per unit cross sectional area of the rod) acquires a momentum $\rho c_1 tu$. According to Newton's law this momentum is equal to pt, so that

$$p=\rho uc_1. \tag{11.59}$$

This equation is completely analogous to the corresponding gasdynamic equation. From (11.58) and (11.59) the expression for the propagation speed of a compression wave through the rod (the speed of "sound") is

$$c_1=\left(\frac{E}{\rho}\right)^{1/2}.^{\dagger} \tag{11.60}$$

When the compressive load is removed, a tensile or unloading wave is propagated at the same speed.

Let us now imagine a laterally constrained rod, as in example 2 of §14, such that the body is not deformed by the compression wave in the plane perpendicular to the direction of wave propagation‡. Repeating the preceding considerations and using (11.55), we find the speed of "sound" in this case to be

$$c_l=\left(\frac{E'}{\rho}\right)^{1/2}=\left(\frac{K+\frac{4}{3}G}{\rho}\right)^{1/2}. \tag{11.61}$$

The speed c_l is nothing else but the "longitudinal" speed of sound, that is,

* In a dynamic process that takes place adiabatically (isentropically), Young's modulus differs somewhat from that used in statics corresponding to isothermal conditions. This difference is usually negligibly small (see [28]). The same is true of Poisson's ratio and the bulk modulus. The isentropic and isothermal shear moduli do not differ from each other, since shear is not accompanied by a change in volume of the body.

† *Editors' note.* This speed is termed the thin rod wave speed. Wave propagation in a rod of finite width is dispersive, and this wave speed is a limiting speed for large values of the ratio of wavelength to rod width. Analogous is the thin plate wave speed $(E''/\rho)^{1/2}$, where $E''=E/(1-\sigma^2)=4G(3K+G)/(3K+4G)$. This speed applies when a rectangular bar is free in one lateral direction and constrained in the other, and is a limiting speed in the same sense as c_1.

‡ We can also treat the rod as free, but consider times during which the compression wave travels through distances which are appreciably smaller than the diameter. An unloading wave from the lateral surface propagates to the axis with a finite speed, so that over the time considered it encompasses only the outer layers. However, in the central regions close to the axis no transverse displacements have as yet taken place, and the deformation of these layers will be purely axial.

the propagation speed of longitudinal waves in an infinite elastic medium*. Actually, when a compression wave propagates through an infinite medium no displacements take place in the plane perpendicular to the direction of propagation, and the phenomenon takes place in the same manner as for a laterally constrained rod. The speed c_l is always greater than the wave speed in the unconstrained rod, since $E' > E$ (see §14).

The speed c_l is the propagation speed only for sufficiently weak compression (and rarefaction) waves, in which the "pressure" or (more precisely) normal stress perpendicular to the direction of propagation is sufficiently small, less than the critical stress defined by (11.56). If a rarefaction wave propagates through a material previously stressed in compression (let us say an unloading wave), the absolute value of the drop in stress must be less than critical for c_l to be the propagation speed (for more detail on this see §17). However, if the dynamic load is great, greater than critical, then the compressed solid, as shown in the preceding section, goes over to a plastic state, similar to that of a fluid. The wave propagation speed, as we know, is determined by the derivative of the "pressure" with respect to volume, in this case of the normal stress with respect to volume. In the plastic state this derivative is proportional to the bulk modulus, as if the shear modulus were equal to zero. Therefore, the propagation speed of sufficiently strong acoustic compression and rarefaction waves is determined only by the compressibility of the material, and is

$$c_0 = \left(\frac{K}{\rho}\right)^{1/2} = \left(\frac{V}{\kappa}\right)^{1/2}. \tag{11.62}$$

The speed c_l is sometimes called the elastic wave speed, and the speed c_0 the plastic wave speed. The quantity c_0 is always less than c_l; for example, for iron $c_l = 6.8$ km/sec and $c_0 = 5.7$ km/sec. The propagation speed of strong compression waves (shock waves) depends on the wave strength. It is always greater than c_0 or close to this value. The propagation speed of weak disturbances is always equal to c_l, independent of the strength.

* The speed of propagation of transverse waves, in which the particles are displaced perpendicular to the direction of propagation of the wave and in which only shearing strain, without compression and rarefaction takes place, is $c_t = (G/\rho)^{1/2}$; $c_t < c_l$. *Editors' note.* The speed c_l is termed the "dilatation" wave speed, as the dilatation (see (11.51)) obeys a wave equation for which this speed is characteristic. It has also been termed the "irrotational" wave speed, as it is characteristic of any motion which is irrotational, for which the displacement has a scalar potential. The speed c_t is termed the "rotation" (sometimes "distortion" or "shear") wave speed, as the rotation vector (curl of the displacement) obeys a wave equation for which this speed is characteristic. It has also been termed the "equivoluminal" wave speed, as it is characteristic of any motion in which there is no density change. Both c_l and c_t are elastic wave speeds, but only c_l is of importance in the wave propagation treated in this chapter.

Problems of the propagation of rarefaction and compression waves in an elastic-plastic medium with a nonlinear relation between stress and strain, one similar to the dependence $\sigma_{zz}(\Delta L/L)$ shown in Fig. 11.34, were investigated in detail by Rakhmatulin. References to original articles in this field may be found in the review by Rakhmatulin and Shapiro [29]. In the next section we shall consider the simplest case of wave propagation in a material having the properties discussed.

§17. Splitting of compression and unloading waves

Let us see what actually happens when a constant pressure p is applied at an initial time to the surface of a flat body. The pressure is assumed to be sufficiently small that the deformation depends linearly on the pressure, that it follows Hooke's law. Let us draw a p, V diagram for the state of the compressed material behind the wave front. Because of the "anisotropy" of the pressure in the case of small deformations we shall, in place of the pressure, use the normal stress component σ_{zz} acting on an area parallel to the surface of the wave front with the wave taken to propagate in the z direction. The abscissa is taken to be the specific volume of the material. For small deformations and pressures the state is described by Hooke's law in the form (11.55), which, according to (11.61), may be rewritten in the form

$$\sigma_{zz} = \frac{\Delta V}{V} \rho c_l^2, \qquad -\sigma_{zz} < p_{cr}.$$

When the pressure exceeds the critical value p_{cr} and the change in volume exceeds $\Delta V_{cr}/V = p_{cr}/\rho c_l^2$ the body becomes plastic and the slope of the line $\sigma_{zz}(\Delta V)$ changes. According to (11.57) and (11.62), we have in this region

$$\sigma_{zz} = \frac{\Delta V}{V} \rho c_0^2 + const, \qquad -\sigma_{zz} > p_{cr}.$$

The σ_{zz}, V diagram is shown in Fig. 11.36.

If the external pressure $p < p_{cr}$ then one elastic compression wave will

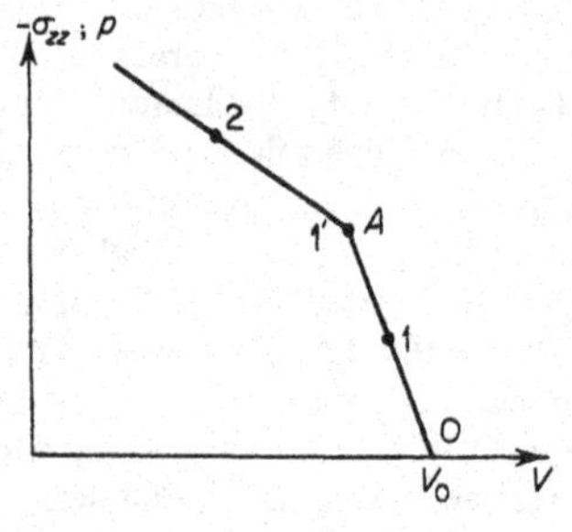

Fig. 11.36. Stress (or pressure)-volume diagram.

propagate through the body with the speed c_l (Fig. 11.37a; state 1 on the σ_{zz}, V diagram of Fig. 11.36). If, however, the applied pressure $p > p_{cr}$, then in the body the final state 2 shown in the σ_{zz}, V diagram is reached. In this case not one but two waves travel through the body: One is an elastic wave

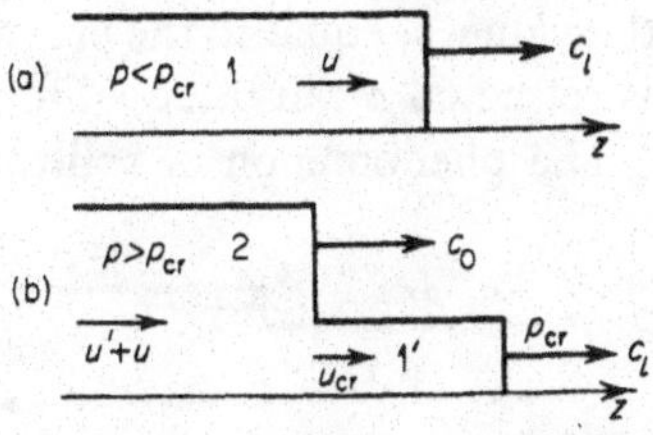

Fig. 11.37. Two cases of propagation of an acoustic compression wave: (a) one elastic wave; (b) a system consisting of a plastic and an elastic wave.

with a strength p_{cr} and with the state 1′ behind the front; this is followed by a plastic wave with the state 2 behind the front (see Fig. 11.37b). Since $c_0 < c_l$, the plastic wave cannot overtake the elastic wave, and the combination of the two waves is stable*. The plastic wave propagates through a slightly compressed medium moving with the velocity $u_{cr} = p_{cr}/\rho c_l$. This velocity is very low; for example, in iron the critical compression by the elastic wave is equal to $\Delta V_{cr}/V = 5 \cdot 10^{-4}$, and the particle velocity $u_{cr} = 3.6$ m/sec. The particle velocity behind the plastic wave is $u' = (p - p_{cr})/\rho c_0$ relative to the moving medium behind the elastic wave, and $u' + u_{cr}$ relative to the undisturbed medium.

If we consider strong compression waves, and even more so shock waves with pressures of hundreds of thousands of atmospheres and higher, then the effects of preliminary compression of the medium by the elastic wave to one or two thousand atmospheres and its acceleration to velocities of the order of several meters per second can be neglected and the plastic wave taken to propagate with respect to the undisturbed medium at rest with the speed c_0 corresponding to the compressibility. Sufficiently strong shock waves propagate at a velocity which appreciably exceeds c_0. If the shock velocity $D > c_l$, then in general no splitting of the wave occurs. The shock wave (plastic) propagates faster than the elastic wave from the very beginning and merges with it into a single wave.

Sufficiently strong unloading waves in an initially compressed medium also split into elastic and plastic waves. Let the medium be unloaded from a pressure p_0 to a pressure p (for example, first a compression wave with a pressure p_0 behind it is generated in the body by compressing it with a piston,

* The existence of the combination of elastic and plastic compression waves was noted by Bancroft *et al.* [30], in a study devoted to phase transitions in iron (for a discussion of this see §19). *Editors' note.* It was also noted by Donnell [61]. The elastic wave is sometimes referred to as a "precursor", particularly when it is much weaker than the plastic wave.

and then after a certain time the pressure at the piston drops to the value p). If $p_0 - p < p_{cr}$, then one elastic unloading wave travels through the compressed medium with the speed c_l. However, if $p_0 - p > p_{cr}$ an elastic unloading wave, in which the pressure drops from p_0 to $p_0 - p_{cr}$, moves ahead followed by a plastic unloading wave propagating at a lower speed. In the plastic wave the pressure drops to the value p, equal to the pressure at the "piston" (in particular, if the piston is retracted p can be equal to zero). These two cases are shown in Fig. 11.38. The phenomenon of splitting the

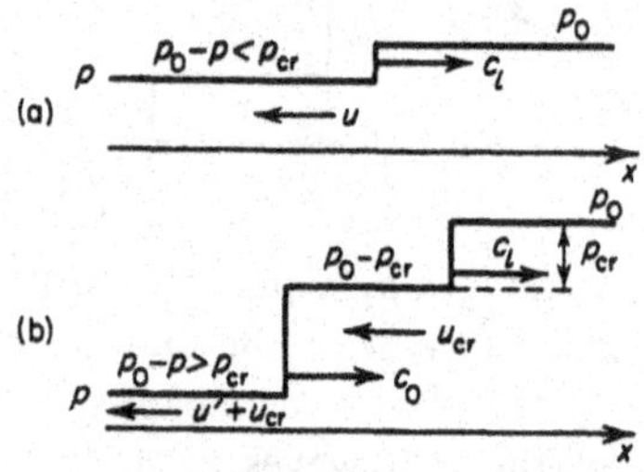

Fig. 11.38. Two cases of propagation of an acoustic unloading wave: (a) elastic wave only; (b) a system consisting of a plastic and an elastic wave.

unloading wave into two waves was observed experimentally by the authors of [4] and will be described in the following section. These authors explained the observed phenomena in the manner discussed above.

§18. Measurement of the speed of sound in a material compressed by a shock wave

The experimental determination of the speed of sound behind a shock front is of great interest. This is the speed of propagation of disturbances which overtake the shock wave and affect its strength*. The speed of sound (or the isentropic compressibility) determines the slope of the isentrope on a p, V diagram that passes through the point representing the state behind the shock front; thus it determines the initial behavior of the compressed material on unloading and its behavior behind the weak secondary shock. A knowledge of the speed of sound is important for establishing the equation of state for the material and for the correct design of shock compression experiments. Finally, values of the speed of sound in a solid at high pressures are also of interest in a number of geophysical problems.

Methods for measuring the speed of sound behind a shock front have been developed by Al'tshuler and Kormer, together with Speranskaya, Vladimirov, Funtikov, and Brazhnik [4]. One of the methods (method of lateral unloading) consists in the following. A cylindrical flanged specimen (Fig. 11.39) is sub-

* We recall that a shock propagates through the medium with the velocity behind the shock subsonic.

jected to a shock compression. Lateral unloading starts after the wave front passes the corner O. Disturbances from the unloading overtake the front and weaken the shock. The front velocity in the weakened outer section of the front surface decreases and the shock curves, as shown in Fig. 11.39, while

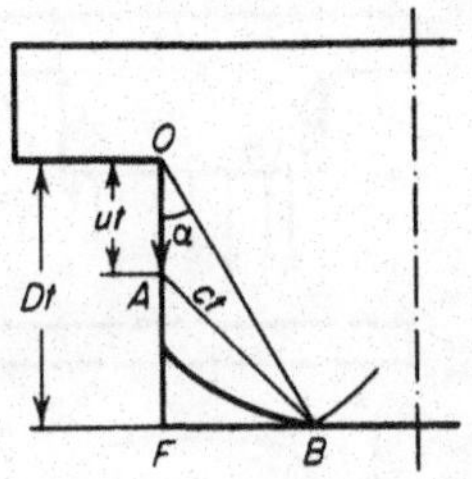

Fig. 11.39. Geometric construction for the lateral unloading experiment.

the central part of the surface not yet reached by the disturbances remains planar and the shock velocity there remains the same as before.

The point at which the weakening of the shock wave begins may be easily found from simple geometric considerations. In the time t which elapses from the time at which the shock passes the corner O, the shock moves through a distance Dt. The material previously situated near the corner is carried ahead through a distance ut to the point A. The earliest disturbances generated at the time the shock passed the corner, which travel through the medium with the speed of sound c, at this time reach a sphere of radius ct about the point A, and the weakening of the shock begins at the point B (see Fig. 11.39). Considering the triangles OBF and ABF, we can relate the speed of sound with the velocities D and u and with the tangent of the unloading angle α:

$$c = D\left[\tan^2 \alpha + \left(\frac{D-u}{D}\right)^2\right]^{1/2}.$$

The problem is reduced to that of the determination of the front velocity D and the angle α (it is assumed that the Hugoniot curve for the medium is known, so that the particle velocity u can be calculated).

The problem is solved experimentally as follows. The shock wave emerges at the free surface with a defined velocity. On the central (unweakened) part of the front surface this velocity is everywhere uniform while on the outer (weakened) part the velocity is less, as shown by the arrows in Fig. 11.40. In the experiment the time at which the free surface reaches a Plexiglas plate P is recorded (this is done by streak or sweeping image photography). The picture obtained on the film is depicted in Fig. 11.40 (the curve on the film results from the luminescence produced when the medium strikes the Plexiglas). Point B is determined from the film and, with the geometry of the experimental setup known, it is then possible to determine the unloading angle α.

It was found that in the case of water the boundary between the weakened and undisturbed regions of the shock surface is very sharp and that the bulk

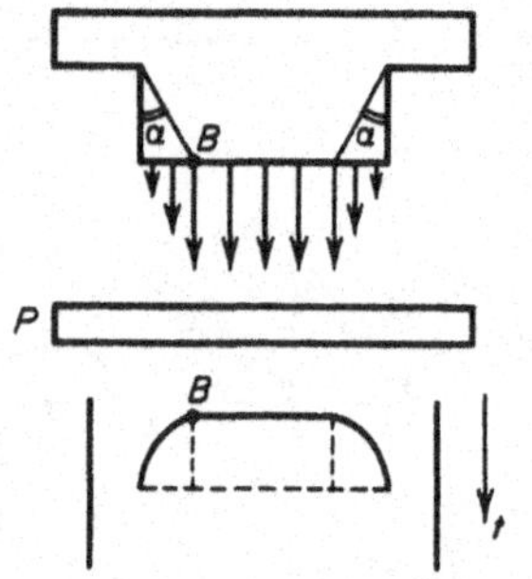

Fig. 11.40. Diagram for the experiment with lateral unloading.

modulus $\rho_0 c^2$ calculated from the speed of sound c is smaller than the slope of the Hugoniot curve (in the variables p, ρ/ρ_0) $\rho_0 dp/d\rho$ at the point corresponding to the state behind the front. This is in complete agreement with the respective locations of the Hugoniot curve and the isentrope as shown in Fig. 11.41.

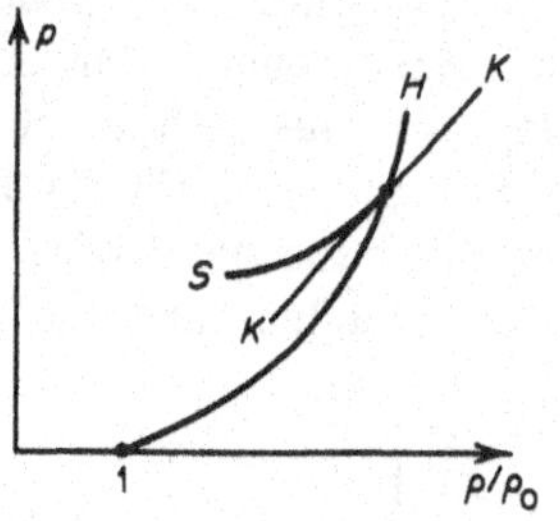

Fig. 11.41. Pressure-density diagram. H is the Hugoniot curve; S is the unloading isentrope; KK is the tangent to the isentrope at a point corresponding to the state behind the shock.

In the case of a metal (iron, copper), however, the shape of the curve on the film is rounded without a sharply defined boundary, as if the outer parts of the front surface were strongly unloaded and those closer to the center (to the axis of the specimen) were unloaded only very weakly. The bulk modulus $\rho_0 c^2$ as calculated using the point where a weak curving of the front surface began was found to be greater than the corresponding slope of the Hugoniot curve $\rho_0 dp/d\rho$ by a factor of approximately 1.5. The experimental data taken from [4] are given in Table 11.3.

This phenomenon was explained by the authors of [4] on the basis of the existence of two speeds of sound in a solid, which we discussed in §§15 and 16. Weak rarefaction disturbances are propagated through the compressed medium with the elastic wave speed c_l (the "pressure" in the medium compressed by the strong shock wave is isotropic). This increased elastic speed of sound corresponds to the beginning of the weak distortion of the front surface; the corresponding modulus $\rho_0 c_l^2$ is found to be too large, larger than the slope of the Hugoniot curve $\rho_0\, dp/d\rho$, because the shock velocity corresponds to the smaller, plastic sound speed. A plastic wave with the decreased plastic sound speed moves through the partially unloaded material. Disturbances of significant strength travel with this speed and appreciably weaken the shock front. The speed of the plastic wave is determined only by the compressibility and it is this speed which must be compared with the slope of the Hugoniot curve. The bulk modulus $\rho_0 c_0^2$, calculated with the plastic sound speed c_0, is found in the case of metals, just as in the case of water, to be smaller than the slope $\rho_0\, dp/d\rho$, in complete agreement with the Hugoniot theory (water, being a fluid, has only one, plastic speed of sound c_0). The existence of two speeds of sound makes extremely difficult the exact determination of the boundary of plastic unloading, which is of major interest because it determines the compressibility of the medium.

In order to eliminate this effect, the authors of [4] developed another method (the "overtaking" unloading method, which in its initial form was proposed by E. I. Zababakhin). This method considers the collision between an accelerated plate and a test sample, made from the same material with a known Hugoniot curve. The x, t diagram for this process is shown in Fig. 11.42. Shock waves OA and OB propagate from the point of collision O

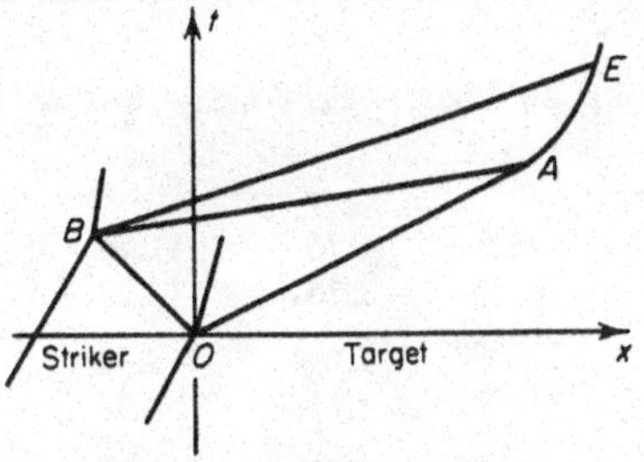

Fig. 11.42. x, t diagram for the "overtaking" unloading method.

through both bodies. After the shock wave in the striker reaches the free surface B, the unloading begins there and a rarefaction wave travels through the medium, overtaking the shock front in the test sample at the point A. From this time onward the shock is attenuated and the trajectory of the front curves upward, as shown in Fig. 11.42. By experimentally determining the trajectory of the shock front AE during the stage of appreciable attenuation,

Table 11.3

DATA FROM LATERAL UNLOADING EXPERIMENT [4]

Material	α, deg	D, km/sec	u, km/sec	Elastic c_l, km/sec	Elastic $\rho_0 c_l^2$, 10^{10} bar	$\rho_0 \frac{dp}{d\rho}$, 10^{10} bar	Plastic $\rho_0 c_0^2$, 10^{10} bar
Water	47.5	4.42	1.52	5.6	31.4	34.2	
Copper	41.0	5.24	0.87	6.33	357.8	288.8	240
Iron	46.5	5.34	0.98	7.15	401.3	298.2	240

with a consideration of the process of propagation of the rarefaction disturbances, we can find the speed of sound in the compressed medium behind the front. We are considering a stage of strong attenuation of the shock wave, which can result only from the plastic wave and not from the weak elastic wave. Hence the speed of sound determined in this experiment is the plastic speed related to the compressibility of the medium (for details of this method see [4]). To illustrate the numerical values we show in Table 11.4 some experimental results. For comparison, the speeds of sound (plastic) c_0 at standard conditions are also given.

Table 11.4

EXPERIMENTALLY DETERMINED SPEEDS OF SOUND AT HIGH PRESSURES

Metal	p, 10^{10} bar	V_0/V	c_0, km/sec	c_0, km/sec (at standard conditions)
Al	195.5	1.76	11.74	5.2
	160.0	1.701	11.23	
Cu	379.6	1.694	9.48	3.9
	311.7	1.638	8.93	
Fe	347.8	1.650	9.48	5.7
	284.9	1.600	9.53	

§19. Phase transitions and splitting of shock waves

Many solids can have different crystalline forms under different conditions. At certain temperatures and pressures, which are related in a defined manner, transitions from one form to another are possible. These transitions, accompanied by changes in volume and the release (or absorption) of latent heat, are called phase transitions of the first kind. Such transitions are often referred to as polymorphic transformations*.

An example of a material capable of undergoing a polymorphic transformation is iron. At atmospheric pressure and a temperature of 910°C iron is transformed from the α phase to the γ phase; the change is accompanied by a 2.5% decrease in volume and the absorption of 203 cal/mole of latent heat. Polymorphic transformations frequently occur at high pressures. In particular, at a pressure of 130,000 atm the above transition in iron takes place at a temperature slightly above standard temperature.

Distinctive phenomena can appear during shock compression of a material capable of undergoing polymorphic transformations at high pressures. These phenomena were treated theoretically (mainly qualitatively) in papers by Bancroft, Peterson, and Minshall [30], Duff and Minshall [31], and Drummond [32]. Experimentally, shock waves in the presence of polymorphic transformations were reported in the first two papers (in the first in iron and in the second in bismuth) and in papers by Dremin and Adadurov [33] (in marble), and Dremin and Karpukhin [34] (in paraffin)†. In a certain range of pressures not one, but two shock waves following one another travel through a material capable of undergoing a polymorphic transformation. This splitting of a shock wave is related to the anomalous behavior of the Hugoniot curve for the material in the region of phase transition. For pressures behind the shock which are not too large the entropy increases only very slightly across the shock wave; hence the Hugoniot curve is close to the isentrope and in examining the above phenomenon we may use the ordinary isentrope.

The isentrope for a material undergoing a polymorphic transformation is depicted schematically in Fig. 11.43. When the material is compressed from its original volume to beyond that of a certain state A, a transition from phase I to phase II begins. The crystal lattice then begins to rearrange itself in such a manner that the new equilibrium positions of the atoms correspond to smaller interatomic distances. Therefore decreasing the volume in the phase transition region requires a much smaller increase in pressure than in the initial phase I. (At absolute zero the phase transition I–II takes place at constant pressure, and the portion AB of the isentrope $S = 0$ is a horizontal line, as shown in Fig. 11.43b.) If there were no rearrangement, the pressure curve would have extended upward from the point A as shown by the dashed lines in Fig. 11.43. In the region AB the material is in a two-phase state. Complete

* For sufficiently strong shock waves the solid will melt, which is also a phase transition of the first kind.

† *Editors' note.* Bethe [62] noted that when isentropic expansion leads to a phase transition from a two-phase state to a one-phase state, compression shocks in general are split and expansion shocks exist.

rearrangement of the lattice and complete transformation of the material from phase I to phase II end at the point B, after which the isentrope of the second phase again goes upward steeply. The compressibility of the material is different in different phases, so that the slopes of the curves corresponding to the single-phase states at points A and B are different in general.

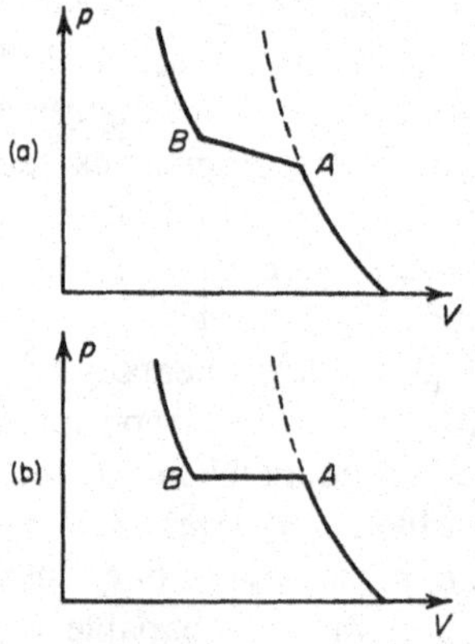

Fig. 11.43. An isentrope for a material undergoing a polymorphic transformation: (a) at a temperature different from zero $T > 0$; (b) at absolute zero.

Let us now consider a solid possessing a Hugoniot curve of the type described above, and assume that at an initial time a constant pressure p is applied at the surface (we shall consider a one-dimensional plane case). We take the pressure to be sufficiently high that we may neglect effects associated with the strength of the material and consider the pressure to be hydrostatic; in other words, we shall disregard the possible existence of an elastic wave (see §17) and consider the shock wave to be plastic.

If the pressure p is lower than the pressure p_A at which the phase transition begins, then an ordinary shock wave propagates through the material, and the state behind the shock corresponds to a point on the Hugoniot curve (say, to point C on Fig. 11.44). The propagation velocity of the shock wave D is determined by the slope of the straight line drawn from the initial state point O to the final state point on the Hugoniot curve

$$D = V_0\left(\frac{p - p_0}{V_0 - V}\right)^{1/2}.$$

If the pressure p is greater than p_E (which corresponds to the intersection with the Hugoniot curve of the line OAE that touches the Hugoniot curve at the point A), say it is equal to p_F, then again only one shock wave will propagate through the body, with the state behind the shock that of point F. However, in this case the material behind the front is in another phase—phase II. The transition from phase I to phase II takes place across the shock front.

Usually the time required for a polymorphic transformation is much larger than that required to establish thermodynamic equilibrium in an ordinary, single-phase material. The situation in this case is quite similar to that which

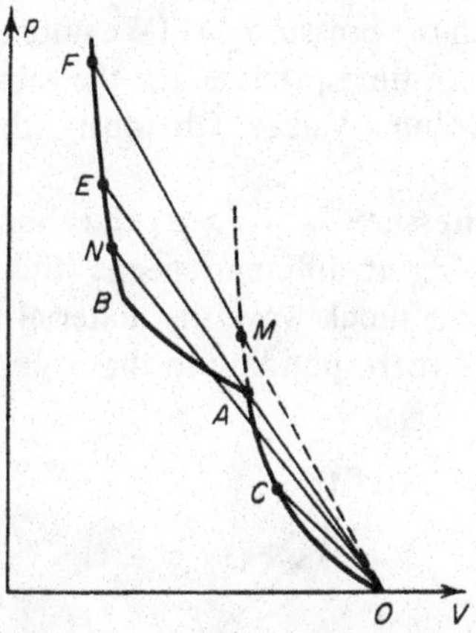

Fig. 11.44. p, V diagram illustrating various cases of shock wave propagation with polymorphic transformation of a material (explanation in text).

takes place in a shock wave propagating through a gas with retarded excitation of some of the degrees of freedom (for example, through a dissociating gas). The direct shock compression leads to the intermediate state M that lies on the extrapolated Hugoniot curve for phase I, corresponding to the absence of a phase transition (this corresponds to the viscous shock in a gas). After that the phase transition begins and the thickness of the front is determined by the relaxation time of the transition, in a manner similar to that in which the shock front thickness in a gas is determined by the dissociation time. The pressure distribution in the shock wave has the form shown in Fig. 11.45, one

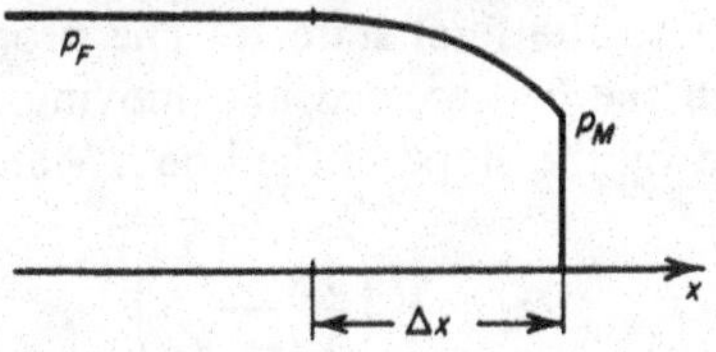

Fig. 11.45. Pressure distribution in a shock wave with phase transition relaxation.

which is completely analogous to the pressure distribution in a shock wave in a dissociating gas. The point characterizing the state in the thickened region of the wave front moves along the straight line segment MF in Fig. 11.44.

Let us now consider the intermediate case in which the pressure applied to the body lies between p_A and p_E, equal to p_N (point N on the Hugoniot curve of Fig. 11.44). The velocity of the shock wave determined by the slope of the line ON is now less than the velocity of the shock with the lower pressure p_A corresponding to the point A. This latter velocity is determined by the slope of the steeper line OA. Therefore, the wave with the pressure p_A moves faster than the shock wave with the pressure p_N. (We note that the line ON intersects the Hugoniot curve three times, that is, for the same wave speed there are three possible pressure-volume values. Obviously, this non-uniqueness is physically meaningless.

At such an intermediate value of the pressure $p_E > p > p_A$ the shock wave is split into two independent waves moving at different speeds (this case is shown separately in Fig. 11.46). In the first shock wave the material is compressed from the initial state O to state A corresponding to the beginning of

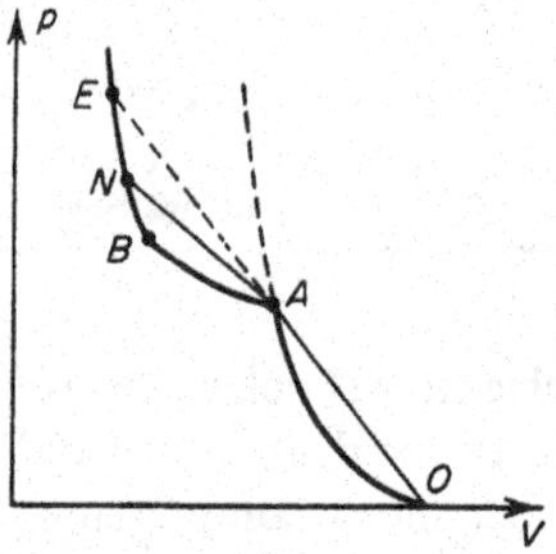

Fig. 11.46. p,V diagram illustrating the splitting of a shock wave.

the phase transition, with the propagation velocity of the first shock through the undisturbed material determined by the slope of the line OA, by

$$D_1 = V_0\left(\frac{p_A - p_0}{V_0 - V_A}\right)^{1/2}.$$

The first wave is followed by another shock in which the material is compressed from state A to the final state N. The propagation velocity of this second wave through the compressed and moving material which is in the state A is determined by the slope of the line AN and is

$$D_2 = V_A\left(\frac{p_N - p_A}{V_A - V_N}\right)^{1/2}.$$

The propagation velocity of the second shock wave relative to the original stationary material is equal to the sum of the velocity D_2 and of the particle velocity of the material behind the first shock u_A

$$D_2' = D_2 + u_A.$$

It is apparent that the second wave does not overtake the first one, so that the combination of the two separate shocks is stable. Indeed, the propagation velocity of the first wave relative to the material behind it is

$$D_1'' = V_A \left(\frac{p_A - p_0}{V_0 - V_A} \right)^{1/2}.$$

Since the slope of the line OA is by definition ($p_N < p_E$) greater than the slope of the line AN, we have $(p_A - p_0)/(V_0 - V_A) > (p_N - p_A)/(V_A - V_N)$ and $D_1'' > D_2$, so that the first wave travels through the material faster than the second one.

A phase transition takes place in the front of the second shock wave. In the initial state A the material is in the first phase, and in the final state N it is either in the second phase (if $p_N > p_B$) or in a two-phase state (if $p_N < p_B$); the transformation in the latter case is incomplete. Since the process of phase transition is retarded, the front of the second shock is found to be greatly thickened, in contrast with the thin front of the first wave. The pressure distribution in the case of the system of two waves is shown schematically in Fig. 11.47. With increasing time the distance between the two wave fronts increases, since their velocities are different, but the pressure profile in the second wave is stationary and propagates unchanged.

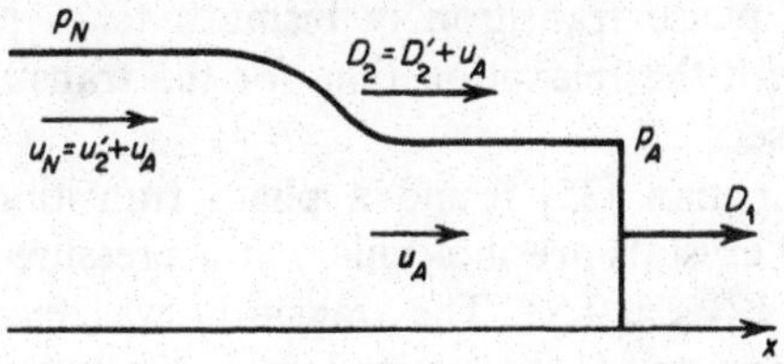

Fig. 11.47. Pressure distribution in the case of splitting of a shock wave into two waves.

The combination of two shock waves in the presence of a phase transition is in many respects analogous to the combination of an elastic and a plastic compression wave considered in §17. The cause for the appearance of the two waves in both cases is the anomalous behavior of the isentrope and of the Hugoniot curve, where by anomalous behavior we mean that there is a region of the p, V curve where it is convex. It was shown in Chapter I that the increase or decrease of entropy across a shock wave depends on the sign of $\partial^2 p/\partial V^2$; this sign determines purely thermodynamic consequences. From this we can conclude that an anomalous Hugoniot curve leads to anomalous kinematic results, in this case to the splitting of a shock wave into two waves. The limiting condition $p > p_E$ for the recombination of the two waves into one corresponds to the situation in the case of the combination of elastic and plastic waves when the plastic wave speed (because of the departure of the Hugoniot curve from Hooke's law) becomes greater than the elastic wave speed, when the second wave overtakes the first one and coalesces with it.

As we pointed out above, the phenomenon of splitting of a shock wave in materials which undergo polymorphic transformations has been observed experimentally. For illustration, we present in Fig. 11.48 the experimental

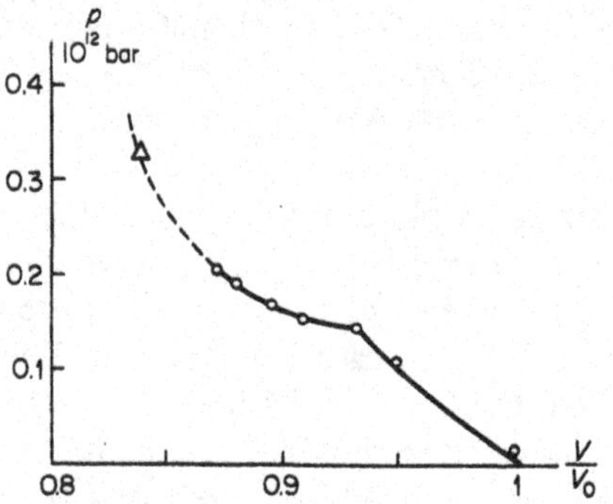

Fig. 11.48. Hugoniot curve for iron in the region of phase transition. △— data from [24], ○— data from [30].

Hugoniot curve for iron in the region of phase transition reported in [30]. We note that a phase transition in bismuth takes place at a pressure of ~28,500 atm, with the relaxation time for the transition at 42°C found to be less than 1 msec.

Alder and Christian [35] found a phase transition of the first kind in iodine I_2 (iodine crystals are molecular) at a pressure $p \approx 7 \cdot 10^5$ atm and a relative volume $V/V_0 \approx 0.53$. The transition was fixed by determining the point at which there was a change in slope in the relation between shock and particle velocities. Calculations show that the temperature behind the wave at the phase transition point is about 1 ev. It is comparable with the dissociation energy of 1.53 ev for iodine molecules. It is assumed that the phase transition is connected with the transformation of a diatomic molecular crystal into a monatomic metallic state.

It is interesting that anomalies in the cold compression curves for metals (and hence also in the Hugoniot curves) similar to those which appear in the presence of polymorphic transformations can also arise even if the atomic lattice is not rearranged. Such anomalies are a result of changes in the structure of electron zones, from overlapping of individual zones under compression. The possibility that metal properties can change with changes in the zone structure has been noted by Lifshitz [36]. The effect of these changes on the cold compression curves for metals and on the appearance of anomalously shaped parts of the curve, with $\partial^2 p/\partial V^2 < 0$, was studied by Gandel'man [37].

For a sufficiently strong shock wave melting of the solid takes place, and this leads to a discontinuous change in slope of the Hugoniot curve. Problems of shock wave melting have been treated in a number of papers [44, 57–59].

§20. Rarefaction shock waves in a medium undergoing a phase transition

According to the general theory presented in §§17, 18, and 19 of Chapter I, an anomalous behavior of an isentrope, one with a segment of the curve which is convex ($\partial^2 p/\partial V^2 < 0$), leads to the possibility of rarefaction shocks. Isentropes for solids undergoing a phase transition represent exactly such a possibility. This was noted in [32]*. Regimes with rarefaction shocks in a metal in the presence of phase transitions were studied by Ivanov and Novikov (and Tarasov) [38], who were the first to give clear experimental evidence of the existence of rarefaction shocks in iron (and steel).

The isentrope of a material undergoing a polymorphic transformation has an anomalous shape in the region of the point of inflection A (Fig. 11.43). Although the second derivative $\partial^2 p/\partial V^2$ at all points where the isentrope behaves normally is positive, nevertheless there exists a segment in the neighborhood of the point A where a chord connecting any two points 1 and 2 lies entirely below the isentrope (Fig. 11.49). This is related to the fact that the mean value of the second derivative on the segment 1–2 is negative†,

$$\left\langle \frac{\partial^2 p}{\partial V^2} \right\rangle_{1\text{-}2} = \frac{(\partial p/\partial V)_2 - (\partial p/\partial V)_1}{V_2 - V_1} < 0.$$

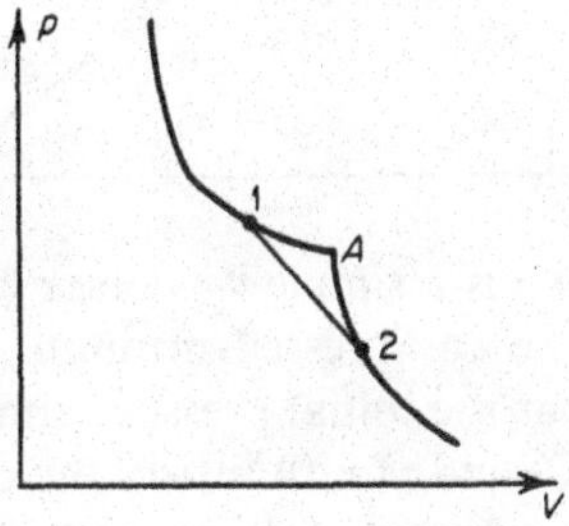

Fig. 11.49. Anomalous segment of an isentrope.

As is known from the general theory, this is the situation which leads to anomalies in the hydrodynamic relationships. The propagation of compression shocks in such a material was considered in the preceding section.

Let us now consider the unloading of a material that has been compressed by a shock wave. We assume that at the time $t = 0$, in a body previously compressed by a shock wave to the state 1 (p_1, V_1), there is a rarefaction wave in which the pressure and volume change smoothly to the values p_2, V_2 (state 2; $p_2 < p_1$, $V_2 > V_1$). The initial pressure distribution as a function of position is shown in Fig. 11.50. We assume that the points of the initial and

* See *Editors' note*, p. 751.

† At all points of the segment 1–2, with the exception of the point of inflection A, $\partial^2 p/\partial V^2 > 0$, but at the point A itself $\partial^2 p/\partial V^2 = -\infty$, so that the mean value on the segment 1–2 is nevertheless negative.

final state, 1 and 2, as well as all of the intermediate points of the smooth distribution lie on the isentrope and that the subsequent process is adiabatic*. Corresponding points in Fig. 11.50 are indicated on the isentrope of Fig. 11.51 by the same letters and numbers.

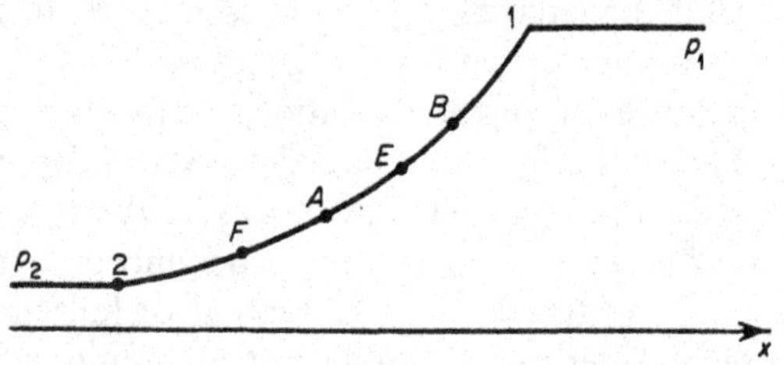

Fig. 11.50. Problem of the evolution of a rarefaction wave; the initial pressure distribution.

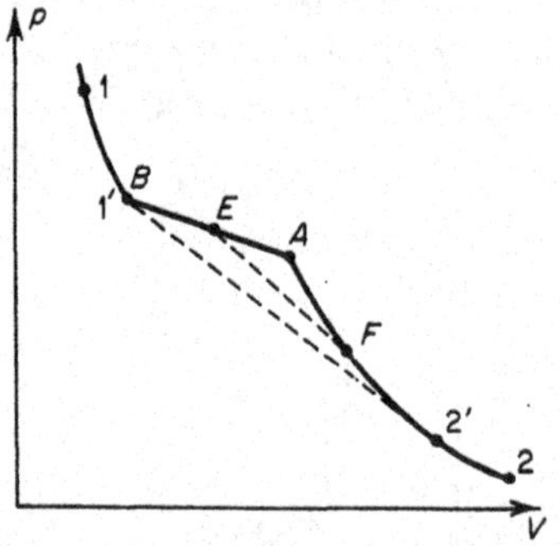

Fig. 11.51. Problem of the evolution of a rarefaction wave; states on the p, V diagram correspond to the distribution shown in Fig. 11.50.

We shall assume that the rarefaction wave is a simple wave (see §8, Chapter I), propagating to the right through the compressed medium. In order for the wave to be simple it is necessary that the initial pressure and velocity distributions with respect to position $p(x, 0)$ and $u(x, 0)$ satisfy the condition that the Riemann invariant $J_-(x, 0)$ be constant. It follows that $J_-(x, t) = const$ at later times. Let us assume that this condition is satisfied. As is known (see §8, Chapter I), the C_+ characteristics in a simple wave propagating to the right are straight lines in the x, t plane; along these lines the pressure and other flow variables are constant.

Let us consider what happens to our initial pressure distribution at later times. To do this we draw on the x, t plane of Fig. 11.52 the C_+ characteristics, which are straight lines of slope $dx/dt = u + c$. The speeds of propa-

* We are considering only low pressures, for which thermal effects are small and the Hugoniot curve practically coincides with the isentrope. In addition, we assume that the phase transitions take place sufficiently rapidly, "instantaneously", so that the states of the material never depart from the thermodynamic equilibrium isentrope.

gation of disturbances (sound speeds) at different points of the initial distribution are determined by the slopes of the tangents to the isentrope at the corresponding points. At the two break points A and B the speed of sound undergoes a jump (the dependence of the speed of sound on volume is shown in Fig. 11.53). The velocity of the material, by virtue of the condition $J_- =$ *const*, is given by $u = -\int c\, d\rho/\rho + const$ and is continuous at the points A and B. Hence the slopes of the characteristics change discontinuously with the jumps in speed of sound.

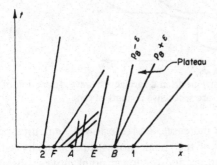

Fig. 11.52. x, t diagram illustrating the evolution of an initial rarefaction wave in a material with an anomalous isentrope.

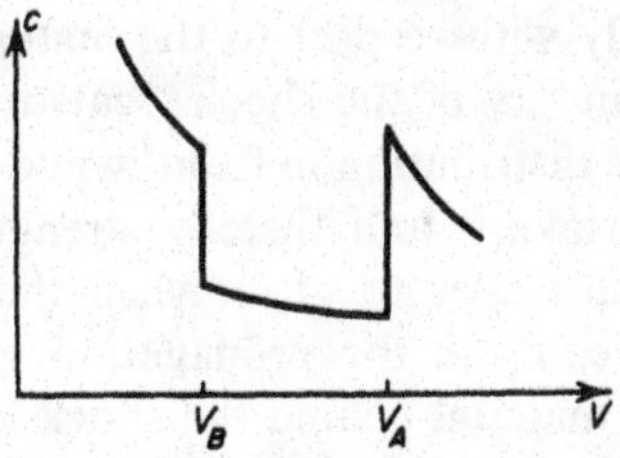

Fig. 11.53. Dependence of the speed of sound on volume, corresponding to the isentrope shown in Fig. 11.51.

Two C_+ characteristics with different slopes, along which the pressure is the same but the speed of sound is different, emanate from the "normal" break point B. These sound speeds correspond to the values immediately before and after the point where the slope of the isentrope changes discontinuously. The propagation speed of the characteristic with the value of the pressure $p_B + \varepsilon$ (ε is infinitesimally small) is greater than that of the characteristic with the slightly lower value $p_B - \varepsilon$. The situation at the "anomalous" break point A is different. Point A also serves as the origin of two characteristics but the characteristic with the higher pressure $p_A + \varepsilon$ propagates slower than the one with lower pressure $p_A - \varepsilon$. Characteristics drawn from points neighboring A tend to intersect (see Fig. 11.52), and the limiting characteristics originating from A behave as if they had intersected at the outset. This means that in the initial pressure distribution a small discontinuity is formed at point A from the very beginning (in the limit $t \to 0$ the discontinuity is infinitesimal) and it grows with time*.

The propagation of the rarefaction wave and the pressure distributions at successive instants of time are shown schematically in Fig. 11.54. The pressure plateau p_B is bounded by the characteristics emanating from the point B and shown in Fig. 11.52. The rarefaction shock wave which originated at point A grows in accordance with its intersection with the characteristics. The jump

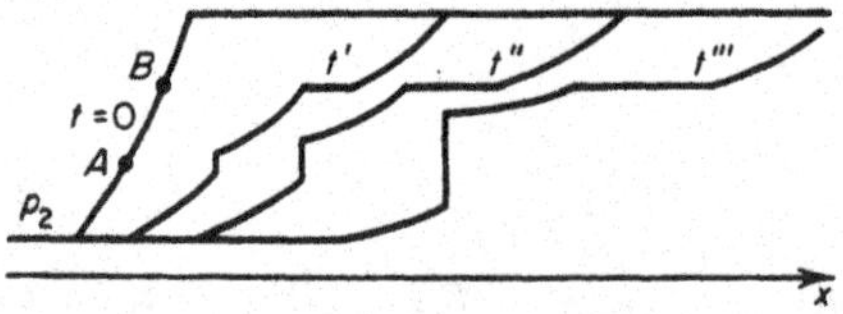

Fig. 11.54. Evolution of the pressure distribution in a rarefaction wave; formation of a rarefaction shock wave. $t = 0$, t', t'', and t''' are successive instants of time.

increases, the upper initial pressure increases, and the lower final pressure decreases, as long as the upper point of the shock moves supersonically with respect to the material ahead of the shock and the lower point moves subsonically with respect to the material behind the shock. In this case the upper boundary of the shock "eats up" portions of the smoothly increasing pressure distribution in front, while the rarefaction disturbances behind the shock overtake it and thereby strengthen it (are "eaten up" by the shock). The jump ceases to grow when the upper pressure reaches the pressure of the plateau and the propagation speed of the lower boundary with respect to the material behind the shock becomes sonic†.

The steady-state position of the discontinuity (the points $1'$–$2'$ on the isentrope of Fig. 11.51) and the pressure distribution in the rarefaction wave are shown in Fig. 11.55. As we know (see §14, Chapter I), the propagation velocities of the discontinuity $1'$–$2'$, u_1 with respect to the material ahead of it and

* The situation for an initially smooth compression wave in a material with normal properties is somewhat different. The characteristics in this case do not immediately intersect (see §12, Chapter I), the pressure distribution steepens gradually, and the compression shock wave does not form immediately. However, for the rarefaction shock the discontinuity is generated at the very beginning and its strength increases proportionally with time. *Editor's note.* With an initially smooth compression wave in the material of Fig. 11.51, a compression shock is generated at the very beginning at point B.

† *Editors' note.* Point $2'$ in the final asymptotic profile is a Chapman–Jouguet point (see p. 346). Thus the determining asymptotic conditions here are analogous to the usual ones in detonations. However, it is possible in principle for $p_2 > p_{2'}$. In this case the Chapman–Jouguet point is never reached and the relative velocity behind the wave remains subsonic in the final profile.

u_2 with respect to the material behind it, are determined by the slope of the line 1′–2′ on the isentrope (actually on the Hugoniot curve)*

$$u_1^2 = V_{1'}^2 \frac{p_{1'} - p_{2'}}{V_{2'} - V_{1'}}, \qquad u_2^2 = V_{2'}^2 \frac{p_{1'} - p_{2'}}{V_{2'} - V_{1'}}.$$

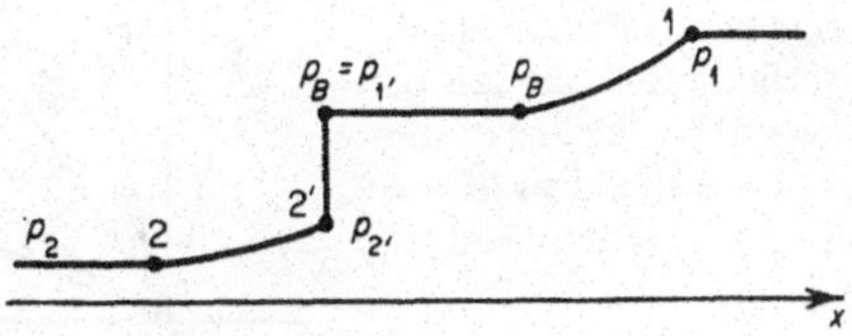

Fig. 11.55. Character of the final pressure distribution in a rarefaction wave. The distribution expands with time, without changing its shape.

It is clear from Fig. 11.51 that the point 2′ is determined by the point of tangency to the isentrope of the line 1′–2′, since at that point $u_2 = c_{2'}$. The propagation velocity of the discontinuity with respect to the material ahead of it u_1 is less than the larger speed of sound at the break point *B*, but greater than the lower speed of sound at that point; the line 1′–2′ has a slope intermediate between the slopes of the two tangents to the isentrope at the point *B*.

In practice a rarefaction wave ordinarily arises when a shock wave emerges at the free surface of a solid. The regime in this case is self-similar; all the C_+ characteristics in the *x*, *t* plane originate from a single point and the entire steady-state pressure distribution shown in Fig. 11.55 is formed at the outset, as in an ordinary self-similar rarefaction wave (see §11, Chapter I). Thus, the rarefaction wave has a complex profile consisting of two parts where the pressure decreases smoothly, a pressure plateau (all of these three parts expand with time in accordance with the self-similarity of the regime), and a rarefaction shock wave (if the surface of the solid is a free surface, point 2 of the final state corresponds to zero pressure). An *x*, *t* diagram for a centered rarefaction wave is shown in Fig. 11.56.

In the experiments described in [38] unusual scabbing phenomena were observed when explosive charges were detonated on the surfaces of iron and steel specimens. The split-off surface was extremely smooth. This phenomenon has been interpreted as resulting from the collision of two rarefaction shock waves, when the pressure on some surface changes in a discontinuous manner from a positive to a negative value. Usually during smooth unloading, the zone of tensile stresses giving rise to the scabbing phenomenon is not

* These equations follow from the conservation laws of mass and momentum across the discontinuity and are equally valid for compression and rarefaction shocks.

sharply defined and the split-off surface is rough, due to the micro-inhomogeneity of the material over the extensive zone of tensile stresses. From an analysis of the complex picture of the motion which takes place under the

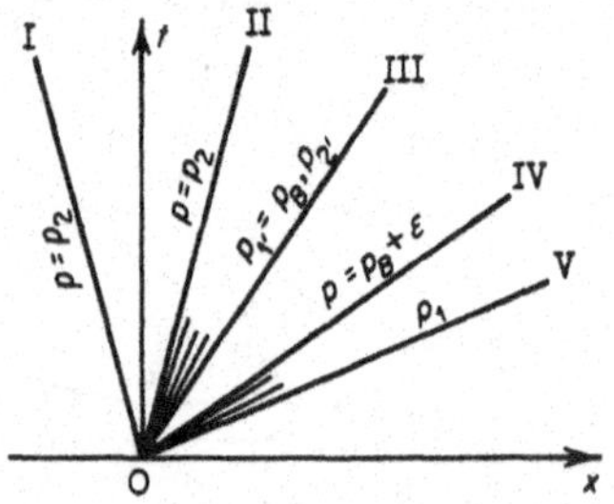

Fig. 11.56. x, t diagram for a self-similar rarefaction wave, formed when a shock wave emerges at a surface. I is the line of the free surface, II is the tail of the rarefaction wave, III is the line of the rarefaction shock, IV is the head of the pressure plateau, and V is the head of the rarefaction wave.

experimental conditions, the authors of [38] were able to conclude that the observed phenomena are attributable to the existence of rarefaction shock waves. This is also substantiated by the fact that in other materials (not iron or steel), in which no phase transitions take place in the pressure range under consideration, no unusual scabbing was observed.

4. *Phenomena associated with the emergence of a very strong shock wave at the free surface of a body*

§21. Limiting cases of the solid and gaseous states of an unloaded material

In §11 we considered the process of unloading of a solid initially compressed by a shock wave, after the shock emerges at the free surface. It was assumed that the shock was not too strong, that the temperature behind the front was relatively low, and that the material when unloaded to zero pressure remained solid. It is clear that if the shock wave is very strong and the internal energy of the heated material ε_1 exceeds by many times the binding energy of the atoms U (equal to the heat of vaporization at zero temperature), then, when the material expands to a low (zero) pressure after the shock wave has emerged from the free surface, the material is completely vaporized and behaves like a gas during the unloading*. In particular, for unloading into a vacuum to

* Sometimes reference is made to the "vaporization" of the material inside the shock wave itself. This statement is incorrect if "vaporization" is understood to denote a phase transition in the ordinary thermodynamic sense. A dense medium can be called a "liquid" or a "gas" only in a conditional sense, depending on the relationship between the kinetic energy of thermal motion of the atoms and the potential energy of their interaction. The transition from a "liquid" to a "gas" in a material heated at constant volume takes place in a continuous manner. In general, we should recall that at pressures and temperatures

strictly zero pressure, the density and temperature at the leading edge of the material are also equal to zero. The density, velocity, and pressure distributions in the unloading wave have the same qualitative character as in a rarefaction wave in a gas (see §§10 and 11 of Chapter I). They are represented in Fig. 11.57.

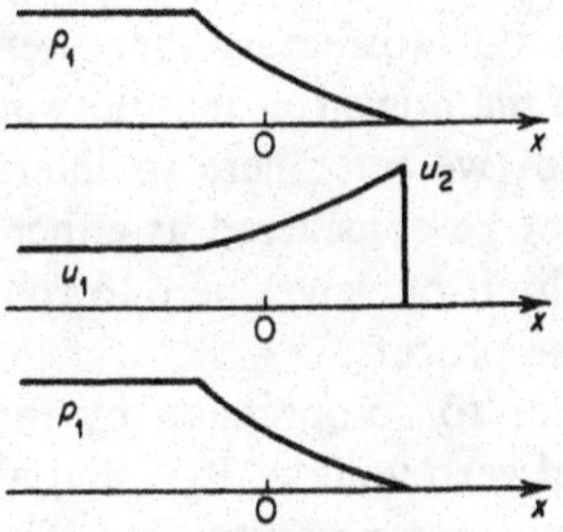

Fig. 11.57. Density, velocity, and pressure distributions after the emergence of a very strong shock wave at a free surface.

The hydrodynamic solution for a self-similar unloading wave can be written in a general form, independent of the thermodynamic properties of the medium. It is given by the equations

$$\frac{x}{t} = u - c, \tag{11.63}$$

$$u + \int \frac{dp}{\rho c} = const \tag{11.64}$$

for a wave moving to the left, and is shown schematically in Fig. 11.57. The integration is carried out at constant entropy *S*, since the unloading process is isentropic. In this case the entropy is equal to the entropy of the material behind the shock front. The constant can be expressed in terms of the properties of the material behind the shock (which are denoted by the subscript "1"). Then (11.64) becomes

$$u = u_1 + \int_p^{p_1} \frac{dp}{\rho c}. \tag{11.65}$$

The velocity of the leading edge of the unloaded material (the velocity of the free surface) is

$$u_2 = u_1 + \int_0^{p_1} \frac{dp}{\rho c}. \tag{11.66}$$

---

above critical the entire material is homogeneous and no phase separation takes place. It should be noted that a statement to the effect that a material in a sufficiently strong shock wave ceases to be solid has a completely physical meaning (the solid material melts).

We have already used (11.66) in §11 in order to obtain the law of velocity doubling. The distribution of hydrodynamic variables in the unloading wave can be found if the thermodynamic properties of the material are known (i.e., the functions $\rho(p, S)$ and $c(p, S)$, with which the integral in (11.65) can be evaluated). The corresponding formulas for a gas with constant specific heats were given in §10 of Chapter I. In the case of interest to us of unloading of a solid this cannot be done as yet, since no satisfactory theory exists for calculating the thermodynamic functions of materials for densities somewhat less than the standard density of the solid (we refer here to intermediate temperatures, for which the material cannot be considered as either a solid or a perfect gas). For this reason we shall limit ourselves here to rough estimates and to a qualitative description of the process.

For simplicity we shall assume that prior to compression by the shock wave the solid was at zero temperature and zero volume V_{0c}, and also that the unloading takes place into a vacuum, to zero pressure. In addition, we shall not make any distinction between the solid and liquid states. The heat of fusion is usually much smaller than the heat of vaporization* (the volume change on melting is also small), and hence when considering phenomena with energies for which the material is completely vaporized we can neglect the effect of melting.

Let us follow the unloading of a given particle of the material on a p, V diagram. Figure 11.58 shows the elastic pressure curve p_c extended into the region of negative pressures, the Hugoniot curve p_H, and the curve OKA separating the single- and two-phase regions. The branch OK up to the critical point K represents the boiling curve (beginning of vaporization), while the branch KA is the saturated vapor curve (beginning of condensation). In addition, the figure also shows several isentropes S, which pass through different possible states behind the shock wave.

Let us consider the simplest limiting cases. Let the shock wave be weak (state 1 on the Hugoniot curve). The compressed material is unloaded along the isentrope S_1, the pressure drops to the point B_1 where the isentrope intersects the boiling curve, after which the solid (or liquid) should, in principle, begin to boil. However, to form nuclei of the new phase, i.e., vapor bubbles, requires a rather large activation energy to destroy the continuity of the material and to form the bubble surfaces. The rate of this process (for metals) at the low temperatures of the order of hundreds and even thousands of degrees is so negligibly small that actually the solid continues to expand and cool down to zero pressure along the "superheated liquid" isentrope shown in Fig. 11.58 by the dashed curve from the point B_1. In its final state

* For example, for lead the heat of fusion is 1/46 times as large, and for aluminum 1/22 times as large as the heat of vaporization.

the volume of the material is $V_2{}^*$, which somewhat exceeds the zero volume V_{0c}, and is heated to a temperature T_2, which is related to the volume difference $V_2 - V_{0c}$ by the thermal expansion relation (see §11). Even if

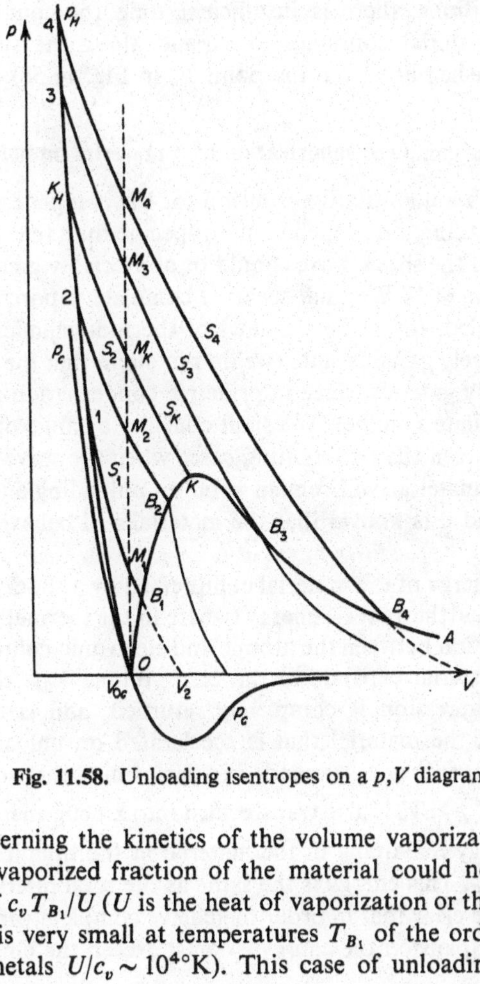

Fig. 11.58. Unloading isentropes on a p,V diagram.

problems concerning the kinetics of the volume vaporization could be disregarded, the vaporized fraction of the material could not exceed a value of the order of $c_v T_{B_1}/U$ (U is the heat of vaporization or the binding energy). This quantity is very small at temperatures T_{B_1} of the order of hundreds of degrees (for metals $U/c_v \sim 10^4$°K). This case of unloading was considered in §11.

In the other limiting case, when the shock wave is very strong (state 4), the unloading isentrope S_4 passes in the purely gaseous region far above the

* In contrast to the notation used in §11, all the quantities in the final unloaded state will be denoted here by the subscript "2", with the subscript "1" referring to quantities behind the shock front.

critical point K, and the material expands as a gas to infinite volume. In general, the isentrope will at some time intersect the saturated vapor curve (the point B_4), after which condensation should begin*. However, if the time for the expansion of the vapor is limited, which is usually the case under laboratory conditions, there is insufficient time for condensation to take place and the material continues to expand along the supercooled vapor isentrope (the dashed line from the point B_4 in Fig. 11.58).

§22. Criterion for complete vaporization of a material on unloading

Let us establish a quantitative criterion for the complete vaporization of a material on unloading, one which is more specific than the obvious condition that the energy of the shock wave should be appreciably greater than the heat of vaporization, $\varepsilon_1 \gg U$. We shall speak of complete vaporization if the material being unloaded, following the laws of thermodynamics, passes through the stage of a purely gaseous state (we do not claim that the final state in this case is also purely gaseous, since in principle condensation must set in when expanding to infinite volume). We shall consider a range of shock strengths intermediate between the two limiting cases when the wave is weak and it is known that the material will remain solid on unloading and when the wave is very strong and it is known that the material will behave as a gas on unloading.

The internal energy of the material compressed by a shock wave is made up of the elastic ε_{c1} and thermal ε_{T1} energy (where in the thermal energy we do not make any distinction between the atomic and electronic contributions). When a compressed material expands to the zero volume V_{0c}, the elastic energy acquired on compression is completely returned, and is transformed into kinetic energy of the material that is accelerated on unloading†. A part of the initial thermal energy ε_{T_1} expended in performing the work of expansion and equal to $\int_{V_1}^{V_{0c}} p_T\, dV$, is also transformed into kinetic energy. Let us denote the thermal energy remaining in the material at the time it reaches the zero volume V_{0c} by ε_T'. This energy is the same as the total internal energy at this instant. It is quite clear that in order to achieve complete vaporization during the subsequent expansion the energy ε_T' must exceed the binding energy U,

$$\varepsilon_T' > U.$$

The question here is what the magnitude of this excess energy should be. In expanding to volumes greater than the zero volume, the excess energy ε_T'

* Condensation in the expansion of vapor into vacuum was considered in detail in Chapter VIII.

† In this case, however, it does not remain concentrated in the same parcel, as for a flow for which Bernoulli's law is applicable; see §11, Chapter I.

is partially expended in performing the work of expansion (this part of the energy is transformed into the kinetic energy of the hydrodynamic motion), and partially into overcoming the binding forces characterized by the negative p_c (this part of the energy is transformed into potential energy).

Let us assume that the energy ε_T' is sufficient to completely vaporize the material, that is, it is sufficient to prevent the pressure $p = p_T + p_c = p_T - |p_c|$ from dropping to zero before the material expands to infinite volume. From the adiabatic relation $d\varepsilon + p\,dV = 0$, it follows from the fact that $d\varepsilon_c + p_c\,dV = 0$ that $d\varepsilon_T + p_T\,dV = 0$. Integrating this equation from the zero volume V_{0c} to infinity, where the thermal energy vanishes, we obtain

$$\varepsilon_T' = \int_{V_{0c}}^{\infty} p\,dV + \int_{V_{0c}}^{\infty} |p_c|\,dV = \int_{V_{0c}}^{\infty} p\,dV + U.$$

The first term represents that part of the excess energy which is expended in the work of expansion, and the second represents the energy expended in breaking the atomic bonds.

Let us represent the pressures p, p_T, p_c on a p, V diagram (Fig. 11.59).

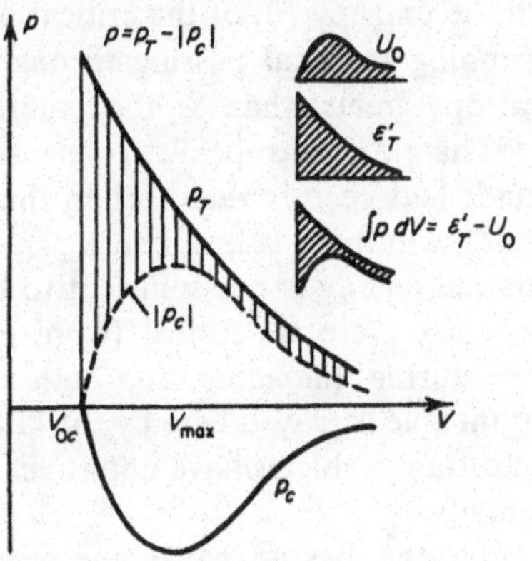

Fig. 11.59. The problem of vaporization of a condensed material on expansion (for explanation see text).

Also shown on the figure are the values of the various energies as represented by the areas defined by the appropriate curves. In the limit of complete vaporization, at that stage of expansion where the binding forces are weakened ($V > V_{max}$), the pressure is close to zero and the thermal pressure is sufficient to overcome the binding forces ($p_T \approx |p_c|$). However, at an earlier stage, when $V_{0c} < V < V_{max}$, the pressure p is high and the thermal pressure is appreciably larger than the elastic pressure, $p_T > |p_c|$. This is clear from the fact that in the state where $V = V_{0c}$, $p' = p_T' = \Gamma \varepsilon_T'/V_{0c} > \Gamma\, U/V_{0c}$, where Γ is the effective Grüneisen coefficient, which is of the order of unity ($|p_c|_{max} \sim U/V_{0c}$). The decrease in thermal pressure on expansion is more or less monotonic (the energy ε_T decreases and the volume V increases). Therefore the curve $p_T(V)$ has the shape shown in Fig. 11.59. From Fig. 11.59 it is clear that the vertically shaded area, equal to the work of expansion $\int_{V_{0c}}^{\infty} p\, dV$, is of the same order of magnitude as the area corresponding to the potential energy U, and that in the limit of complete vaporization the excess energy ε_T' should be roughly twice the binding energy U.

In order to express these rather qualitative considerations quantitatively, the thermodynamic properties of the material must be known in the range of volumes greater than the standard volume of the condensed state, in a range where the binding forces are of importance. Unfortunately, this range of volumes with $V_{0c} < V \lesssim 5V_{0c}$ has been the least investigated, either theoretically or experimentally.

We can approach the evaluation of the shock strength which separates the regions of complete and incomplete vaporization under unloading in a somewhat different manner, by characterizing the vaporization boundary not in terms of the energy ε_T' but in terms of the entropy. It is clear from Fig. 11.58 that the effective boundary between complete and incomplete vaporization on isentropic unloading is defined by a state K_H behind the shock such that the entropy is equal to the entropy S_{cr} of the critical point, the entropy corresponding to the expanding material passing through the critical point K. The fact that for an entropy greater than S_{cr} the material at some instant of time begins to condense (state 3, isentrope S_3, condensation point B_3) means that all the atomic bonds had been broken earlier, that the material had become a gas. Conversely, if the entropy is less than S_{cr} (state 2, isentrope S_2, boiling point B_2), the thermal energy is not sufficient to bring about complete vaporization. For entropies close to critical (from either side) the material is in a two-phase state during unloading, and both vapor and liquid drops are present. An important role is played here by the kinetics of the phase transitions. These very interesting problems have not as yet been studied either theoretically or experimentally.

The entropy criterion, despite its limitations, has an advantage over the energy criterion in that it allows us to approach the estimate of the limiting critical entropy S_{cr} from the "gas side", omitting the poorly investigated range of volumes two to three times greater than the standard volume of the solid. Here, of course, there is also the uncertainty coming from the fact that the critical parameters of liquid metals are unknown as a rule.

Let us illustrate the above qualitative considerations by carrying out an estimate for lead. The entropy of lead at the critical point is calculated by means of the entropy equation (3.18) for a perfect gas which is monatomic, as is lead vapor. We take for the critical temperature $T_{cr} = 4200°K$ and for the volume $V_{cr} = 3V_{0c}$* (ordinarily the critical volume is about three times

* The quantity T_{cr} was estimated in [39]; according to van der Waals equation, $p_{cr} = (3/8)n_{cr}kT_{cr} \approx 2400$ atm.

greater than the standard volume of the liquid). The statistical weight of lead atoms is $g_0 = 9$. Using these parameters gives $S_{cr} = 42.8$ cal/mole · deg*.

The entropy behind the shock can be calculated from the functions $\varepsilon(T, V)$ and $p(T, V)$ described in §6. The simplest procedure is to find the entropy in the state T, V by integrating the equation

$$dS = \frac{d\varepsilon + p\,dV}{T} = \frac{d\varepsilon_T + p_T\,dV}{T}$$

first at constant temperature equal to the standard temperature T_0, from the volume V_0 to V, and then with respect to temperature at $V = const$, from T_0 to T. In the first integration we can neglect the electronic terms, which are negligible at $T_0 \approx 300°$K. For purposes of estimation we take the Grüneisen coefficient to obey $\Gamma(V) \approx \Gamma_0(V/V_0)^m$, where the exponent m for lead, according to the data of Table 11.2, is approximately 1. After integration we obtain

$$S(T, V) = S_0 + c_v \ln\frac{T}{T_0} + \beta_0\left(\frac{V}{V_0}\right)^{1/2}(T - T_0) - \frac{\varepsilon_0}{mT_0}[\Gamma_0 - \Gamma(V)].\dagger \tag{11.67}$$

Here S_0 is the entropy of metallic lead under standard conditions T_0, V_0, and ε_0, which, according to data quoted in [40], is $S_0 = 15.5$ cal/mole · deg. Substituting the shock parameters from Table 11.2 into (11.67), we can find the entropy behind the wave. An entropy close to the critical value S_{cr} is obtained with the following shock wave parameters: $V_0/V_1 = 1.9$, $p_1 = 2.25 \cdot 10^6$ atm, $T_1 = 15{,}000°$K, $\varepsilon_1 = 4.71 \cdot 10^{10}$ erg/g‡ (more exactly, for these parameters $S_1 = 44.5$ cal/mole · deg). The energy ε_T' upon isentropic expansion to the zero volume V_{0c} is found to be equal to $1.9 \cdot 10^{10}$ erg/g, and thus twice the binding energy $U = 0.94 \cdot 10^{10}$ erg/g, in complete agreement with the expected value mentioned above ($T' = 9500°$K, $p_T' = p' \approx 5 \cdot 10^5$ atm). Thus, it is to be expected that for stronger shock waves complete vaporization of lead will occur on unloading. As another example, we present calculations for the strongest shock waves observed experimentally in lead. Namely, for $p_1 = 4 \cdot 10^6$ atm and $V_0/V_1 = 2.2$ the entropy $S_1 = 51.7$ cal/mole · deg, and the energy at the time of expansion to standard volume $\varepsilon_T' = 3.57 \cdot 10^{10}$ erg/g and thus is 3.6 times greater than the binding energy U ($T' = 15{,}000°$K). In

* The use of van der Waals equation to take into account the departure from the perfect gas law leads to a very small correction in the entropy; this correction is $\Delta S_{\text{nonideal}} = \mathscr{R} \ln(2/3) = -0.8$ cal/mole · deg (for the same volume at which the ideal S is calculated). This correction was included in the calculated value of S_{cr}.

† The last term, which depends on Γ, is of little importance, so that the error resulting from the approximation of $\Gamma(V)$ by a power law relation is negligible.

‡ It is interesting to note that the energy behind a shock wave for which complete vaporization just begins is five times greater than the binding energy.

this case, complete vaporization on unloading has apparently already taken place.

In conclusion we wish to emphasize that an unloading wave propagating through a body after a shock wave has emerged at the free surface contains from the outset particles of the material in widely differing states, from those with the pressure p_1 (at the head of the rarefaction wave) down to those with zero pressure (at the free surface). All states through which the given particle passes when going from the pressure p_1 to zero are present in the wave. We also note that the pressure of the particles close to the free surface drops to zero so rapidly that in the case of complete vaporization the vapor in this region is strongly supersaturated, although for thermodynamic equilibrium the medium should be in a two-phase state.

§23. Experimental determination of temperature and entropy behind a very strong shock by investigating the unloaded material in the gas phase

A number of sections in this chapter have been devoted to the study of the thermodynamic properties of solids at high pressures and temperatures and to the description of experimental methods for studying these properties by the measurement of the state and flow variables in the material behind the compression shock. A general feature of these methods is that they allow only the determination of the mechanical state variables of the material, in particular of the pressure, the density, and the total internal energy. Measurement of the kinematic variables of the shock wave, the front velocity and the particle velocity, together with the use of relationships across the shock front, does not make it possible to determine directly such important thermodynamic variables as the temperature and entropy. In order to find the temperature and entropy from mechanical measurements we must adopt some theoretical scheme for characterizing the thermodynamic functions. Earlier we made use of a three-term representation of the pressure and energy in which certain parameters, such as the specific heat of the atomic lattice, the electronic specific heats, and the electron pressure, had to be determined theoretically.

On the other hand, it would be very interesting and important to find some means for the direct experimental determination of the temperature or entropy behind a shock wave to reduce as far as is possible the number of theoretical parameters. Unfortunately, such a procedure is extremely difficult, both in principle and experimentally. Optical methods, one of the major means of measuring high temperatures, can be used only when the body is transparent, while the overwhelming majority of solids are opaque; this particularly includes metals, the solids of greatest interest to us.

The temperature behind the front of a shock wave in Plexiglas was measured optically by Zel'dovich, Kormer, Sinitsyn, and Kuryapin [41]. In these experiments the surface brightness was measured from the front of a very strong shock wave propagating through a transparent material, in this case Plexiglas. The surface brightness was converted to temperature under the assumption that the heated region bounded by the front surface radiates as a perfect black body. The surface brightness was measured in the red and blue regions of the spectrum, and not only the brightness temperature but also the color temperature was determined (see §8, Chapter II). The temperature behind a shock wave with a pressure $p \approx 2 \cdot 10^6$ atm and a density ratio $\rho/\rho_0 \approx 2.7$ was found to be $T \approx$ 10,000–11,000°K. The temperature estimated from the internal energy (known from mechanical measurements) and by making suitable assumptions on the energy balance (the dissociation of Plexiglas molecules is important) shows the measured value of the temperature to be reasonable.

It might have been possible to try to measure the temperature optically at the time when the shock emerges from the free surface. However, in order for the measured temperature to agree with the actual temperature behind the shock, the experiment would have to satisfy almost impossible requirements. Metals are opaque to visible light even in very thin layers $\sim 10^{-5}$ cm. A shock with a velocity of 10 km/sec would travel through such a layer in $\sim 10^{-11}$ sec. Even if it were possible to build a light recording instrument with a resolution time of $\sim 10^{-12}$–10^{-13} sec, in order to record just before the instant at which the wave emerges from the surface, when the wave is separated from the surface by a transparent layer of $\sim 10^{-6}$–10^{-7} cm, it would still be impossible to ensure the simultaneous emergence of the shock at different points of the free surface with the requisite accuracy. In other words, it is impossible to ensure that the front surface is parallel to the free surface to within $\sim 10^{-6}$ cm. If, however, we measure the surface luminosity in the practically measurable time of $\sim 10^{-8}$ sec after the emergence of the wave, then we will record the luminosity not behind the wave front, but behind the unloading wave, since in a time of $\sim 10^{-8}$ sec the unloading wave traverses a layer which is optically very thick, of the order of 10^6 cm/sec × 10^{-8} sec = 10^{-2} cm; such a layer is completely opaque to the light produced in the unloaded region whose temperature we want to determine. (The problem of the luminosity from the surface of an unloading wave will be considered in detail in the next section.)

The possibilities for determining the temperature (and entropy) behind a shock wave experimentally were given in a paper by one of the authors [42]. Let the shock wave be so strong that after it emerges from the free surface the material completely vaporizes on unloading. Then the material at the leading edge of the expanding medium is in the gas phase. If we could somehow measure in the gas phase the mechanical quantities density and pressure, or the temperature, then the entropy could be calculated theoretically, since the thermodynamic properties of gases can be calculated relatively simply (see Chapter III). However, since the unloading process is isentropic the entropy of the material behind the shock wave is the same as the entropy in the gas phase on unloading. Thus, knowing the entropy in the gas phase, we would also know the entropy behind the shock wave.

A method was given in [42] for calculating the temperature along the entire unloading isentrope if the specific internal energy ε is known as a function of pressure and density along the isentrope and if one value of the temperature is known at any point on the isentrope. In this regard, it follows from the thermodynamic identity

$$dS = \frac{d\varepsilon + p\,dV}{T} = \frac{1}{T}\frac{\partial \varepsilon}{\partial p}\,dp + \frac{1}{T}\left(\frac{\partial \varepsilon}{\partial V} + p\right)dV$$

and from the condition that the entropy is a state function and dS a total differential, that

$$\frac{\partial}{\partial V}\left(\frac{1}{T}\frac{\partial \varepsilon}{\partial p}\right) = \frac{\partial}{\partial p}\left[\frac{1}{T}\left(\frac{\partial \varepsilon}{\partial V} + p\right)\right].$$

Differentiating and simplifying we obtain a partial differential equation for the function $T(p, V)$

$$\left(\frac{\partial \varepsilon}{\partial V} + p\right)\frac{\partial T}{\partial p} - \frac{\partial \varepsilon}{\partial p}\frac{\partial T}{\partial V} = T. \tag{11.68}$$

The characteristics of this equation are lines along which

$$\frac{dp}{dV} = -\frac{\partial \varepsilon/\partial V + p}{\partial \varepsilon/\partial p}.$$

But this is the equation for an isentrope. Along the characteristics, then, along an isentrope, according to (11.68)

$$\left(\frac{dT}{dV}\right)_S = -\frac{T}{\partial \varepsilon/\partial p}.$$

From this we obtain

$$T = T_0 \exp\left(-\int_{V_0}^{V} \frac{dV}{\partial \varepsilon/\partial p}\right) = T_0 \exp\left(\int_{p_0}^{p} \frac{dp}{\partial \varepsilon/\partial V + p}\right),$$

where the integrals are taken along the isentrope. The method is simply the application of this formula.

We note that knowing the entropy behind two shock waves whose strengths are close to each other (not necessarily the absolute values of the entropy, but only their difference), we can easily calculate the temperature behind the shock wave from the thermodynamic relation

$$T = \frac{\Delta\varepsilon + p\,\Delta V}{\Delta S},$$

since $\Delta\varepsilon$, p, and ΔV are taken to be known from mechanical measurements. In addition, knowing the temperatures along the Hugoniot curve we can also find the absolute values of the entropy by integrating the thermodynamic relation

$$dS = \frac{d\varepsilon + p\,dV}{T}$$

along the Hugoniot curve, having set the constant of integration equal to the tabulated value of the entropy of the material at standard conditions.

§24. Luminosity of metallic vapors in unloading

It was noted in the preceding section that any attempt to "see" the high-temperature luminosity from the front of a very strong shock propagating through a solid at the time it emerges from the free surface is doomed to failure. Let us consider in more detail what should be observed in this case, what luminosity will be recorded by an instrument directed at the free surface, and how the surface brightness will vary with time. The theory of the phenomenon was given in a paper by the authors [43], and the corresponding experiments were carried out by Kormer, Sinitsyn, and Kuryapin.

Let a very strong shock wave with a front temperature T_1 of the order of several tens of thousands of degrees emerge at the time $t = 0$ from a plane free surface of a metal bounded by a vacuum (the wave front surface is assumed to be strictly parallel to the free surface of the body). The body must be in a vacuum, as otherwise the material which is being unloaded will generate a shock wave ahead of itself in the air; the air temperature can be very high and we would observe the luminosity of the highly heated air instead of the luminosity of the metal, the quantity of interest to us. We consider the shock wave to be so strong that the metal is completely vaporized on unloading and expands in the gas phase. The temperature distributions at $t = 0$ and at a later time are shown in Fig. 11.60. In the time t the rarefaction wave encompasses a layer of the material of thickness $c_1 t$, where c_1 is the speed of sound in the compressed material behind the shock front. Since the material moves with the velocity u_1 in the laboratory system of coordinates, then at the time t the head of the rarefaction wave is at the position $x = (u_1 - c_1)t$ (the initial location of the free surface is taken to be at $x = 0$). The leading edge of the expanding metallic vapor moves ahead with a velocity

u_2 given by (11.66). Since the material in the unloading wave is in the gas phase, the temperature at the vacuum interface is zero, as are the density and pressure.

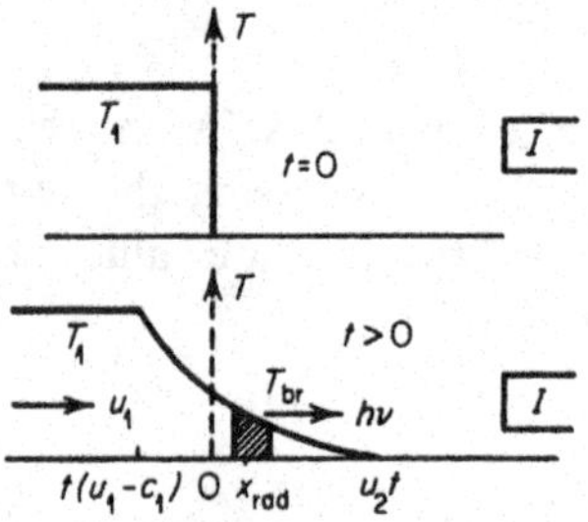

Fig. 11.60. Temperature distributions at the time of emergence of a shock wave from a free surface, at $t = 0$ and at a later time $t > 0$. The radiating layer is shaded. I is a light-recording instrument.

It was stated in the preceding section that metals are opaque in very thin layers of $\sim 10^{-6}$ cm. This means that even as early as $t \sim 10^{-11}$ sec (for a speed $c_1 \sim D \sim 10^6$ cm/sec) the layer of unloaded metal almost completely shields the high-temperature radiation at the temperature T_1, and the metal which was originally heated by the shock wave becomes invisible. Let us examine the nature of the surface luminosity of the material in the continuous spectrum and the type of radiation recorded by an instrument directed toward the planar free surface. The metallic vapor constitutes a monatomic gas, whose radiative properties in the continuous spectrum were studied in detail in Chapter V. The absorption coefficient for visible light is extremely temperature dependent, increasing rapidly with an increase in temperature, with the cold vapor completely transparent in the continuous spectrum. The luminosity of a layer with a temperature distribution similar to that shown in Fig. 11.60 has already been considered in Chapter IX. The phenomenon is completely analogous to that with the luminosity of air in the preheating layer which forms ahead of the compression shock in a strong (supercritical) shock wave. At low temperatures at the vacuum interface the vapor is transparent and radiates only very weakly. On the other hand, in the deeper layers where the temperature is high, the vapor is completely opaque to visible light and does not "release" the photons born in these layers. The photons escaping to infinity from the surface of the material are born in some intermediate radiating layer, removed from the vacuum interface by an optical distance τ_ν of the order of unity (the radiating layer is shown shaded in Fig. 11.60).

Knowing the temperature and density distributions as a function of position and the light absorption coefficient κ_ν for a given frequency ν as a function of temperature and density, we can calculate the brightness temperature for radiation of this frequency using the general relation (2.52). However, we can proceed in a simpler manner by noting that the brightness temperature is the same as the temperature of the radiating layer (the geometric thickness of the radiating layer is small and the temperature in it changes little); we can set up an expression for the optical thickness measured from the vacuum interface and set it equal to unity

$$\tau_\nu = -\int_{u_2 t}^{x(T_{br})} \kappa_\nu(x)\, dx = 1. \tag{11.69}$$

With temperature as the variable of integration in place of x, we write

$$-\int_0^{T_{br}} \kappa_\nu(T) \frac{dx}{dT}\, dT = 1. \tag{11.70}$$

This equation determines the brightness temperature. The derivative of the temperature distribution is calculated using the general solution for a rarefaction wave (11.63) and (11.64).

Since the material at the leading edge, close to the vacuum interface and precisely in that region which contains the radiating layer, is in the gas phase, then by specifying an effective adiabatic exponent γ for the gas we can find the approximate distribution of all quantities in this region in explicit form. To do this (11.64) should be integrated not from the undisturbed compressed material, as was done in deriving (11.65), but from the vacuum interface,

$$u = u_2 - \int_0^p \frac{dp}{\rho c} = u_2 - \int_0^\rho c \frac{d\rho}{\rho}, \tag{11.71}$$

using the isentropic relation $c(\rho, S)$.

The solution will contain the velocity of the interface u_2 and the entropy S as parameters. We shall not write out this solution, but shall find the derivative directly from (11.63) and the differential relation $du = -c\, d\rho/\rho$,

$$\frac{1}{t}\frac{dx}{dT} = \frac{\partial u}{\partial T} - \left(\frac{\partial c}{\partial T}\right)_S = -c\left(\frac{\partial \ln \rho}{\partial T}\right)_S - \left(\frac{\partial c}{\partial T}\right)_S,$$

or

$$\frac{dx}{dT} = -\frac{ct}{T}\left\{\left(\frac{\partial \ln \rho}{\partial \ln T}\right)_S - \left(\frac{\partial \ln c}{\partial \ln T}\right)_S\right\} \approx -\frac{\gamma + 1}{2(\gamma - 1)}\frac{ct}{T}.$$

Here we have used the relation $c \sim \sqrt{T}$, and also the isentropic relation $T \sim \rho^{\gamma - 1}$. Equation (11.70) now takes the form

$$\frac{\gamma + 1}{2(\gamma - 1)}\, t \int_0^{T_{br}} \kappa_\nu(T) \frac{c(T)}{T}\, dT = 1. \tag{11.72}$$

It is evident that the integral and, consequently, the brightness temperature decrease with time. The physical reason for this result is that as the unloading wave encompasses a larger and larger mass of the material with time, the geometric and optical thicknesses of the layer between the vacuum interface and the point with the given temperature increase continuously. Therefore the radiating layer, which is separated from the interface by the specified optical distance of the order of unity, moves into a region of increasingly lower temperatures (Fig. 11.61).

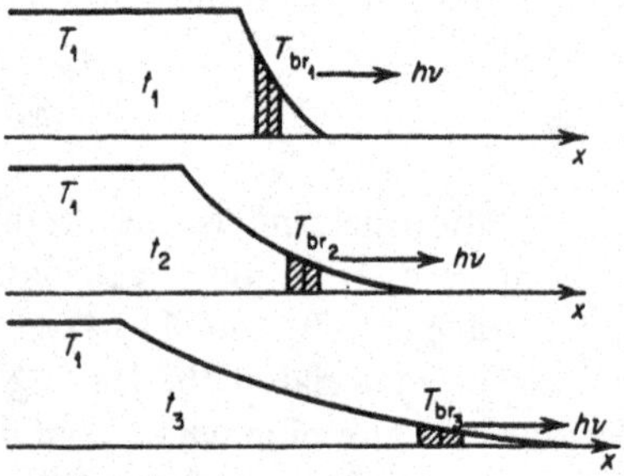

Fig. 11.61. Displacement of the radiating layer (shaded) with time in an unloading wave.

It is remarkable that the interface velocity u_2 has dropped out of the equation (11.72) for $T_{br}(t)$*. The interface velocity is an unknown because it is determined by the thermodynamic properties of the material along the entire unloading isentrope, including the unexplored region where the density is somewhat less than the standard density of the solid. The only parameter contained in (11.72) is the entropy S, which appears because of the dependence of the absorption coefficient κ_ν on the density (the atom number density n), which is related to the temperature by the adiabatic relation

$$n = B(S)T^{1/(\gamma-1)}, \tag{11.73}$$

where $B(S)$ is the entropy constant.

If the basic mechanism of absorption of visible light in metallic vapors is photoelectric absorption by highly excited atoms (and also bremsstrahlung absorption in the fields of the ions), then the absorption coefficient κ_ν can be calculated approximately from (5.44)

$$\kappa_\nu = \frac{a_\nu n}{T^2} e^{-(I-h\nu)/kT}, \tag{11.74}$$

where a_ν is a constant depending on frequency ($a_\nu \sim \nu^{-3}$), and I is the ionization potential.

In dense vapors with very weak ionization an important role may be played by bremsstrahlung absorption in the fields of neutral atoms (see

* Since (11.70) does not explicitly contain the coordinate x but only the derivative dx/dT.

Chapter V). In this case the absorption coefficient κ_ν is proportional to the number of free electrons n_e, that is, to the degree of ionization, and the basic temperature dependence of the absorption coefficient is Boltzmann-like but with a different exponent

$$\kappa_\nu \sim nn_e = b_\nu e^{-I/2kT} n^{3/2}, \tag{11.75}$$

where b_ν depends only weakly on temperature. The general character of the temperature dependence of κ_ν is the same in both cases: $\kappa_\nu \sim e^{-E/kT}$, where $E = I - h\nu$ in the first case and $E = I/2$ in the second case. It should be noted that numerically the two values of E do not differ much for metals (where $I \approx 6$–8 ev and $h\nu \approx 2$–3 ev).

Let us evaluate approximately the integral of (11.72), taking the basic temperature dependence of the integrand to be contained in the exponential factor. Setting all the slowly varying functions which are powers of the temperature constant, we find $te^{-E/kT_{br}} = const$ and obtain a logarithmic decrease of the radiation brightness temperature with time (Fig. 11.62)

$$T_{br} = \frac{const}{\ln t + const'}.$$

Detailed calculations show that the brightness temperature for metals is of the order of 7000–4000°K* at times $t \sim 10^{-7}$–10^{-6} sec, irrespective of the assumed absorption mechanism. In these times the free surface moves away approximately 10^{-1}–1 cm at speeds of $\sim$ 10km/sec.

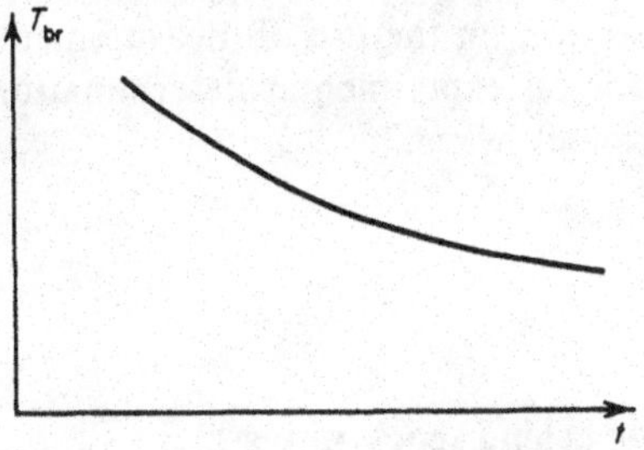

Fig. 11.62. Time dependence of the brightness temperature of the surface of an unloading wave.

§25. Remarks on the basic possibility of measuring the entropy behind a shock wave from the luminosity during unloading

Equation (11.72) contains only one parameter characterizing the shock wave—the entropy S. If the radiative properties of the material, i.e., the function $\kappa_\nu(T, \rho)$, are known, then by plotting the experimental luminosity

* While behind the shock wave the temperature T_1 can reach tens of thousands of degrees.

curve $T_{br}(t)$ we can find the absolute value of the entropy behind a shock wave. On the other hand, by specifying the entropy on the basis of other considerations (calculating it using thermodynamic properties of the compressed solid and measured shock wave parameters), we can extract from the experiment on the luminosity of an unloaded surface data on the radiative properties of metallic vapors, namely to determine the preexponential factor in the expression for the absorption coefficient. It is interesting to note that assuming that only one absorption mechanism exists and that κ_ν is expressed either by (11.74) or (11.75), the final equation for the function $T_{br}(t)$ (11.72) contains only the product of the unknown parameters $a_\nu B(S)$ for the case of (11.74) and $b_\nu B^{3/2}(S)$ for the case of (11.75) (since in (11.74) $\kappa_\nu \sim a_\nu n \sim a_\nu B$, while in (11.75) $\kappa_\nu \sim b_\nu n^{3/2} \sim b_\nu B^{3/2}$). The entropy constant B in the adiabatic equation (11.73) depends on the absolute value of the entropy S as $B \sim e^{-S/\mathscr{R}}$ ($\mathscr{R}$ is the gas constant per mole). This means that by taking the luminosity curves from two experiments with slightly differing shock strengths and determining the parameters, say, the product $a_\nu B$, we find the difference of entropies behind the shock waves even if the optical constant a_ν is not known, from

$$\frac{a_\nu B'}{a_\nu B''} = \exp\left[-\frac{S' - S''}{\mathscr{R}}\right].$$

Here the single and double prime refer to the first and second experiments. As noted in the preceding section, we can also find the temperature behind the shock wave by knowing the entropy difference.

The experiment described above can serve as a specific application of the considerations given in the preceding section on the use of measurements in the gas phase in an unloading wave for an experimental determination of the entropy and temperature behind a shock wave.

5. *Some other phenomena*

§26. Electrical conductivity of nonmetals behind shock waves

Under normal conditions gases are good insulators. Behind sufficiently strong shock waves they become conductors. A similar situation also occurs with solid dielectrics, which conduct electric currents behind strong shock waves. However, while the appearance of conductivity in gases involves simply the thermal ionization that takes place at high temperatures of the order of ten thousand degrees and above, attainable in shock waves, the physical cause of the transformation of solid dielectrics into conductors by shock waves is considerably more complex. It is more likely connected with the compression than with the temperature increase, though in many respects this is still not clear.

The electrical conductivity of condensed materials behind shock waves has been studied by several authors. Brish, Tarasov, and Tsukerman developed a method for the measurement of conductivity and have measured the conductivity of the detonation products of condensed explosives [45], and also of water, Plexiglas, paraffin, and air [46] behind strong shock waves with pressures of up to a million atmospheres. The conductivity of an ionic crystal of sodium chloride at pressures up to a million atmospheres was studied in the previously cited reference [5]. Alder and Christian [47], who measured the electrical conductivity of ionic and molecular crystals of CsI, I_2, CsBr, $LiAlH_4$, and others worked with weaker shock waves (up to 250,000 atm).

The essence of the basic electrical contact method described in [45], used for the measurements of conductivity reported in [45, 46, 5], consists of the following: Electrodes (contacts) E are connected by a shunt resistance R_{sh} and are placed in the body through which the shock wave propagates (Fig. 11.63). As long as the shock wave does not reach the contacts, the resistance of the dielectric is practically infinite. After the shock wave reaches the contacts the dielectric becomes a conductor and the desired resistance R_x is connected in parallel with the resistance R_{sh}.

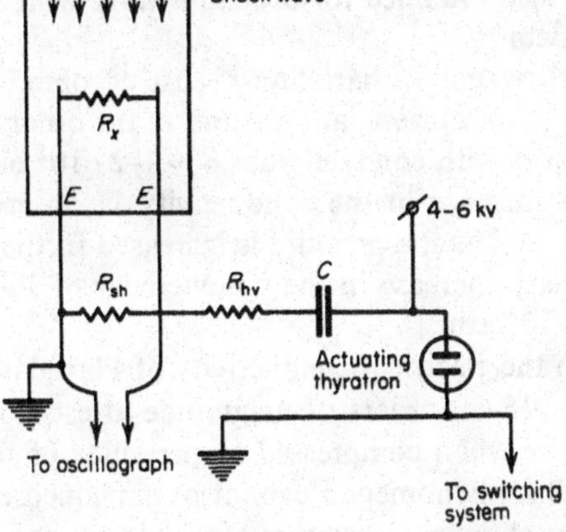

Fig. 11.63. Diagram of the experimental setup for the measurement of electrical conductivity behind a shock wave.

Shortly before the shock wave reaches the contacts the capacitor C, which was previously charged to a voltage of several kilovolts with the help of the actuating thyratron, is discharged through the high-voltage resistance R_{hv}

and the contacts. The resistance $R_{hv} \gg R_{sh}$, so that the current in the circuit is determined only by R_{hv}. The potential difference at the contacts is proportional to the resistance between the contacts. This resistance is equal to R_{sh} before the arrival of the shock wave and $R = R_{sh}R_x/(R_{sh} + R_x)$ after the shock wave reaches the contacts (the resistance R_{sh} is chosen so that it is of the order of R_x). If U_{sh} and U_x correspond to the potential difference at the contacts, then $U_{sh}/U_x = R_{sh}/R = (R_{sh} + R_x)/R_x$. The voltages U_{sh} and U_x are measured by an oscillograph, R_{sh} is known, and the unknown resistance R_x is found from the equation given.

The measured resistance R_x is converted to the specific electrical conductivity of the material by electrolytic simulation. For this purpose the electrodes are immersed into an electrolyte bath maintaining the same geometry as in the experiment. By changing the electrolyte density, a resistance equal to that measured in the experiment is obtained. The unknown conductivity is then equal to the known conductivity of the electrolyte (for other methods of measuring the conductivity of materials behind a shock wave see [45, 5]).

Experiments [46] have shown that the electrical conductivity of dielectrics is increased behind a shock wave by many orders of magnitude. The initial conductivity of distilled water was $\sigma \sim 10^{-5}$ ohm^{-1}cm^{-1}, while at a pressure $p = 10^5$ atm it became $\sigma = 0.2$ ohm^{-1}cm^{-1}. The conductivity behind the shock was completely independent of the initial conductivity of the water, which depends on the purity of the water. The same value of σ behind a shock wave was also obtained for ordinary water with an initial conductivity $\sigma \sim 10^{-3}$ ohm^{-1}cm^{-1}.

Perfect dielectrics such as paraffin ($\sigma \sim 10^{-18}$ ohm^{-1}cm^{-1}) and Plexiglas ($\sigma \sim 10^{-15}$ ohm^{-1}cm^{-1}) were, at pressures of the order of 10^6 atm, converted into fair conductors with conductivities $\sigma \sim 1\text{–}2 \cdot 10^2$ ohm^{-1}cm^{-1}*. In paraffin a significant increase in the conductivity is observed at a pressure of $\sim 6\text{–}7 \cdot 10^5$ atm, and as the pressure is increased further, σ increases rapidly. An extremely sharp increase in the conductivity of Plexiglas takes place at a pressure of $8 \cdot 10^5$ atm.

The change in the electrical conductivity of Plexiglas and paraffin behind a shock wave by 15–20 orders of magnitude attests to the "metallization" of these dielectrics when compressed to pressures of the order of a million atmospheres†. This phenomenon cannot be explained by thermal ionization. It is related to the change in structure of the electron zones of a solid on compression. The zones are brought closer on compression, the distances between

* For comparison with the conductivity of metals we note that for copper $\sigma \sim 10^6$ ohm^{-1}cm^{-1}, for iron $\sigma \sim 10^5$ ohm^{-1}cm^{-1}, and for mercury $\sigma \sim 10^4$ ohm^{-1}cm^{-1}.

† Alder and Christian [47] measured considerably lower electrical conductivities. The "metallization" phenomenon in the comparatively weak waves with which these authors worked manifested itself much more weakly.

them decrease, and this facilitates electron transitions leading to the appearance of free electrons and metallic conductivity in a material which was previously a dielectric*. Qualitative ideas concerning the metallization of any material under sufficiently strong compression were discussed in a paper by Zel'dovich and Landau [48], in which they considered the transformation of metals from the solid to gaseous state. The metallization of hydrogen at high densities was studied by Abrikosov [49].

It must be said that the details of the mechanism of metallization of dielectrics by a shock wave are still not entirely clear, and this phenomenon requires further theoretical and experimental study. In particular, it would be interesting to clarify the separate roles of temperature and compression in increasing the conductivity.

Experiments [5] with sodium chloride, which under normal conditions has a small ionic electrical conductivity, make it possible to assume that the basic role there in the increase of electrical conductivity with increasing shock strength is played by temperature, in contrast to our previous considerations. The curve of $\sigma(T)$ is Boltzmann-like, $\sigma \sim e^{-E/kT}$ with an activation energy $E \approx 1.2$ ev, and this apparently attests to the ionic nature of the conductivity of NaCl behind a shock wave. The limits for the range of shock strengths studied were $p = 10^5$ atm, which gave $T = 440°\text{K}$, $V_0/V = 1.26$, $\sigma = 2 \cdot 10^{-5}$ ohm^{-1}cm^{-1}, and $p = 7.9 \cdot 10^5$ atm, which gave $T = 6150°\text{K}$, $V_0/V = 1.85$, $\sigma = 3.26$ ohm^{-1}cm^{-1}.

§27. Measuring the index of refraction of a material compressed by a shock wave

The thickness of a shock front in solids and liquids is comparable with interatomic distances and much less than the wavelengths of visible light $\lambda \sim 4000$–7300 Å. Therefore light passing through a transparent undisturbed material incident on the surface of a shock front separating the undisturbed from the compressed material is reflected in the same manner as from an ordinary boundary between two different media. The reflection of light from the surface of a shock front in transparent materials, water and Plexiglas, was investigated experimentally by Zel'dovich, Kormer, Sinitsyn, and Yushko [51]. Knowing the index of refraction of the undisturbed material, knowing the angle of incidence, and measuring the reflectivity, it is then possible to

* The effect of pressure on the electrical conductivity of dielectrics had been studied previously (in a region of comparatively low pressures). Thus, Bridgman [50] established that yellow phosphorus, which is a dielectric, is transformed at pressures of 1.2–$1.3 \cdot 10^4$ atm and a temperature of 200°C into a new form, black phosphorus, which has a metallic conductivity. The density of black phosphorus is 1.4 times greater than that of yellow phosphorus.

use the known Fresnel formulas (see [52] for example) to calculate the index of refraction n of a material compressed by a shock wave*. This method is, in general, also applicable to those cases when the material compressed by the shock is opaque. If the absorption mean free path is comparable with the wavelength of light, then, in principle, it is possible to measure both the real and the imaginary parts of the refractive index. To do this it is necessary to determine the degree of polarization of the reflected light and the dependence of the reflectivity on the angle of incidence [54]. A material which is transparent in the undisturbed state becomes opaque behind a sufficiently strong shock wave. The loss of transparency at high pressures can occur for various reasons, from cracking of the material, from phase transitions, or from rearrangement of the electronic levels (in particular, in the "metallization" of dielectrics, mentioned in the preceding section).

The basic experimental arrangement for the study [51] of the reflection of light from a shock front in water is shown in Fig. 11.64. A layer of water is

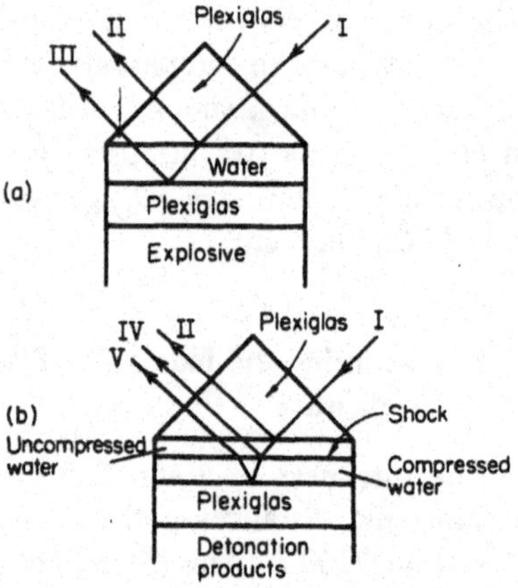

Fig. 11.64. Experimental arrangement for measuring the reflection of light from a shock front: (a) before detonation; (b) during the propagation of the shock through water.

located on top of a Plexiglas plate, which is placed on the flat surface of an explosive charge. A Plexiglas prism is placed on top of the water. The paths of the light rays before the detonation are shown in Fig. 11.64a. Ray I from a light source is incident on the prism, from which emerge rays II and III reflected from the two water surfaces. The paths of the rays after the detonation during the passage of the shock wave through the water are shown in

* The thickness of a shock front in a gas, the thickness of the transition layer between the undisturbed and compressed media, is of the order of a wavelength of light; therefore, the Fresnel formulas are not applicable. However, in gases the index of refraction at different densities is known. A study of the reflection of light under these conditions makes it possible to determine the front thickness. Such measurements were carried out by Cowan and Hornig [53] for weak shock waves (see Chapters IV and VI).

Fig. 11.64b. Ray IV is produced by reflection from the surface of the shock front, while ray V is produced by reflection from the now moving boundary between the compressed Plexiglas plate and the compressed water. Ray V replaces ray III.

The reflected rays are recorded by streak photography. A schematic diagram of a photographic record is shown in Fig. 11.65. Prior to detonation

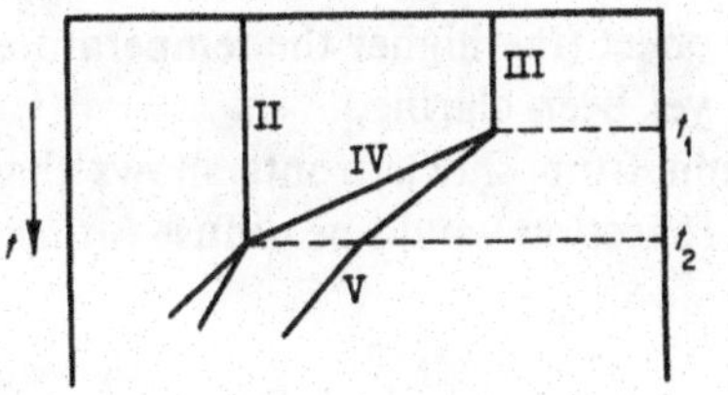

Fig. 11.65. Diagram of a photochronogram.

rays II and III give straight lines on the moving film. At the time t_1 at which the shock wave passes into the water, the two lines produced by rays IV and V appear, with line V now replacing the terminated line III. Line II continues, remaining undisturbed up to the time the shock wave emerges at the upper surface of the water (the time t_2). It is evident from Fig. 11.64b that as the wave front approaches the upper boundary of the water the distance between rays IV and II decreases. At the time of emergence t_2 the rays IV and II come together, and the line IV in Fig. 11.65 reaches the line II. In practice the distance between rays II and III was approximately 20 mm, and the difference in time $t_2 - t_1$ approximately $4 \cdot 10^{-6}$ sec.

The shock front velocity in water was measured by the slope of the line IV. Knowing the Hugoniot curve for water, the density and other parameters behind the front could be determined. The reflectivity was calculated from the ratio of the intensities of the incident and reflected rays; the intensities were determined by photometric methods. The refractive index of compressed water was determined by two methods, one geometric (from the distance between the reflected rays), and the other using the reflectivity. Average values from several experiments, calculated by both methods, were found to be close to each other. As the water density changes from $\rho/\rho_0 = 1.47$ to $\rho/\rho_0 = 1.81$, which corresponds to pressures from 50 to 150 thousand atmospheres, the index of refraction remains almost constant and equal to $n = 1.49 \pm 0.03$ (from the geometric method) or $n = 1.46 \pm 0.03$ (from the reflectivity method). In standard density air $n = n_0 = 1.333$.

Experimental results obtained by other authors on the measurement of the refractive index of water at relatively low pressures are quite well described by the linear relation $n = 1 + 0.334\rho$*, where ρ is the density in g/cm^3.

* The Lorenz-Lorentz formula gives much worse agreement with the experimental data.

This formula also agrees with experimental values for water vapor and with the experimental value of the refractive index for ice at 0°C and $\rho = 0.92$ g/cm^3 equal to 1.311.

The values of the refractive index obtained for water compressed by a shock wave are much lower than the values derived from the above equation. In all probability the difference can be ascribed to a temperature effect (water compressed by a shock wave to a density $\rho = 1.8\rho_0$ was heated to 1100°C). The mechanism of such a temperature effect (the higher the temperature, the lower the index of refraction) has not yet been clarified.

Investigation of the reflection of light from shock fronts shows that the front surface is smooth. Otherwise the reflection would be diffuse rather than specular.

XII. Some self-similar processes in gasdynamics

1. Introduction

§1. Transformation groups admissible by the gasdynamic equations

In Chapter I we have already become familiar with several examples of self-similar motions (the self-similar rarefaction wave, the problem of a strong explosion). In Chapter X we considered self-similar problems in the theory of heat propagation in a stationary medium by thermal conduction. In this chapter we shall consider in detail self-similar motions of one of two basic types. We shall show in the introductory part of the chapter how the gasdynamic equations admit the existence of self-similar solutions and shall present the general characteristics of self-similar motions. It would appear worthwhile at first to become familiar with the general group properties of the gasdynamic equations.

We shall consider one-dimensional adiabatic flows of a perfect gas with constant specific heats, with either planar, cylindrical, or spherical symmetry. Let us write the system of equations for flows of this type. In the continuity equation (1.2) we expand the divergence term and write the equation in a form appropriate to all three types of symmetry. In addition, we divide the equation through by the density ρ. The entropy in the entropy equation (1.13) is expressed by (1.14) (with the density in place of specific volume). The equation of motion (1.6) we leave unchanged. We then obtain the following system of equations for the density, pressure, and velocity as functions of position and time:

$$\frac{\partial \ln \rho}{\partial t} + u\frac{\partial \ln \rho}{\partial r} + \frac{\partial u}{\partial r} + (\nu - 1)\frac{u}{r} = 0,$$

$$\frac{\partial u}{\partial t} + u\frac{\partial u}{\partial r} + \frac{1}{\rho}\frac{\partial p}{\partial r} = 0, \qquad (12.1)$$

$$\frac{\partial}{\partial t}\ln p\rho^{-\gamma} + u\frac{\partial}{\partial r}\ln p\rho^{-\gamma} = 0.$$

In the continuity equation $\nu = 1$, 2, and 3, for the plane, cylindrical, and spherical cases, respectively. The variable r represents the x coordinate in the plane case and the radius in the cylindrical and spherical cases.

Equations (12.1) admit several transformation groups, which we shall presently enumerate. It is assumed that analogous transformations are made at the same time in the initial and boundary conditions of the problem.

(a) The time t enters the equations only in terms of time derivatives. Therefore, a shift in time, carried out by introducing the new variable $t' = t + t_0$, does not alter the equations. The fact that a time shift is possible is related to the arbitrariness in the selection of the initial time.

(b) In the plane case ($\nu = 1$) the coordinate also enters the equations only in terms of distance derivatives. Therefore, a shift in position, related to the arbitrariness in the selection of the coordinate origin, is also possible in the plane case. The introduction of the new variable $x' = x + x_0$ does not alter the equations. This is not possible in the spherical and cylindrical cases, since the radius enters the continuity equation directly and not just in terms of its differential.

The gasdynamic equations contain five dimensional quantities ρ, p, u, r, and t, the dimensions of three of which are independent. For example, if we choose the density, distance, and time as the basic dimensional quantities, then the dimensions of the velocity and pressure are represented in the form $[u] = [r]/[t]$ and $[p] = [\rho][r^2]/[t^2]$. As a result of the existence of three independent dimensional quantities the equations admit three independent similarity transformation groups which are related to the arbitrariness in choice of the units of the basic dimensional quantities.

(1) Let the functions $\rho = f_1(r, t)$, $p = f_2(r, t)$, and $u = f_3(r, t)$ represent solutions of the equations for some specified motion. Let us change the scale of the density without changing the coordinate and time scales by introducing the new variables $\rho' = k\rho$ and $p' = kp$, leaving the remaining variables unchanged. This transformation does not change the equations. If at the same time we similarly transform the boundary and initial conditions by multiplying the density and pressure by k, then the new motion will be described by the functions

$$\rho' = kf_1(r, t), \qquad p' = kf_2(r, t), \qquad u = f_3(r, t).$$

The new motion is similar to the old one, differing only in the density and pressure scales.

(2) Let us change the length scale without changing the density and time scales. The equations are not changed if we transform to the new variables $r' = mr$, $u' = mu$, and $p' = m^2 p$, with the remaining variables ρ and t left unchanged, $\rho' = \rho$ and $t' = t$. This means that if some motion is described by the functions $\rho = f_1(r, t)$, $p = f_2(r, t)$, and $u = f_3(r, t)$, then by a simple change of scales it is possible to describe a new motion in which the distances and velocities are multiplied by a factor of m and the pressure is multiplied

by a factor of m^2 (the density remains the same). The solution for the new motion is expressed by the functions

$$\rho' = f_1(r', t), \qquad p' = m^2 f_2(r', t), \qquad u' = m f_3(r', t).$$

(3) Finally, let us change the time scale, leaving the length and density scales unchanged. The equations admit the following transformation

$$t' = nt, \qquad u' = \frac{u}{n}, \qquad p' = \frac{p}{n^2}, \qquad \rho' = \rho, \qquad r' = r.$$

This means that if in the boundary and initial conditions the velocity is multiplied by the factor n^{-1}, the pressure is multiplied by the factor n^{-2}, and the density is left unchanged, then the new process will be similar to the old one, except that its rate will be multiplied by a factor of n^{-1}.

By the successive application of these three groups of similarity transformations we can obtain solutions for an infinite number of new motions with altered density, length, and time scales. In particular, if the length and time scales are simultaneously increased by the same factor $r' = lr$ and $t' = lt$, then the solution will remain unaltered. Such a transformation is equivalent to the successive application of transformations (2) and (3) with $m = n = l$. In symbolic form this can be expressed

$$u(r, t) \rightarrow lu(lr, t) \rightarrow \frac{1}{l} \cdot lu(lr, lt) = u(lr, lt)$$

with similar rules for the other functions ρ and p*.

§2. Self-similar motions

It was shown in the preceding sections that the gasdynamic equations admit similarity transformations, that there are possible different flows similar to each other which are derivable from each other by changing the basic scales of length, time, and density. The motion itself may be described by the most general functions of the two variables r and t, $\rho(r, t)$, $p(r, t)$, and $u(r, t)$. These functions also contain the parameters entering the initial and boundary conditions of the problem (and the specific heat ratio γ).

* *Editors' note.* The equations of state of the fluid must also follow the transformation. With a general equation of state (an imperfect gas, say) and with the identity of the fluid unchanged under the transformation, the equations admit the time shift and coordinate translation transformations mentioned earlier (also rotation and Galilean transformations), plus the scale transformation $r' = lr$ and $t' = lt$ discussed above. Transformations (1), (2), and (3) separately are not admitted. A perfect gas with γ not constant also admits transformation (1), but not (2) and (3) separately. A free Fermi–Dirac gas (see p. 220) admits transformations for which $m = nk^{1/3}$, and gives an example of a fluid which is not a perfect gas but which admits transformations besides the scale transformation.

However, there exist motions whose distinguishing property is the similarity in the motion itself. These motions are called self-similar. The distribution as a function of position of any of the flow variables, such as the pressure p, evolves with time in a self-similar motion in such a manner that only the scale of the pressure $\Pi(t)$ and the length scale $R(t)$ of the region included in the motion change, but the shape of the pressure distribution remains unaltered. The $p(r)$ curves corresponding to different times t can be made the same by either expanding or contracting the Π and R scales. The function $p(r, t)$ can be written in the form $p(r, t) = \Pi(t)\pi(r/R)$, where the dimensional scales Π and R depend on time in some manner, and the dimensionless ratio $p/\Pi = \pi(r/R)$ is a "universal" (in the sense that it is independent of time) function of the new dimensionless coordinate $\xi = r/R$. Multiplying the variables π and ξ by the scale functions $\Pi(t)$ and $R(t)$, we can obtain from the universal function $\pi(\xi)$ the true pressure distribution curve $p(r)$ as a function of position for any time t. The other flow variables, density and velocity, are expressed similarly.

For self-similar motions the system of partial differential equations of gasdynamics reduces to a system of ordinary differential equations in new unknown functions of the similarity variable $\xi = r/R$. Let us derive these equations. To do this we represent the solution of the partial differential equations (12.1) in terms of products of scale functions and new unknown functions of the similarity variable ξ,

$$\xi = \frac{r}{R}, \qquad R = R(t). \tag{12.2}$$

The pressure, density, velocity, and length scales are not all independent of each other. If we choose R and ρ_0 as the basic scales, then the quantity $dR/dt \equiv \dot{R}$ can serve as the velocity scale and $\rho_0\dot{R}^2$ as the pressure scale. This does not limit the generality of the solution, as the scale is only defined to within a numerical coefficient which can always be included in the new unknown function. We seek a solution of the form

$$p = \rho_0\dot{R}^2\pi(\xi), \qquad \rho = \rho_0 g(\xi), \qquad u = \dot{R}v(\xi), \tag{12.3}$$

where π, g, and v are new dimensionless functions of the similarity variable ξ, in terms of which the differential equations are to be formulated. These functions are here termed the reduced pressure, density, and velocity, respectively, or simply the reduced functions*. The scales R, ρ_0, and $\dot{R}$ are time dependent in some as yet unknown manner.

* *Editors' note.* These functions are termed by the authors the "representatives" of the pressure, density, and velocity, respectively. They are sometimes referred to as the dimensionless pressure, density, and velocity. The similarity variable ξ may be termed the dimensionless or reduced distance.

We now substitute the relations (12.3) into equations (12.1), taking account of the definition of the similarity variable (12.2), and applying relations of the type

$$\frac{\partial \rho}{\partial t} = \frac{d\rho_0}{dt} \cdot g - \rho_0 \frac{dg}{d\xi} \frac{r}{R^2} \frac{dR}{dt} = \dot{\rho}_0 g - \rho_0 g' \xi \frac{\dot{R}}{R},$$

$$\frac{\partial \rho}{\partial r} = \frac{\rho_0 g'}{R},$$

to transform the derivatives. The differentiation of scales with respect to time is denoted by a dot and differentiation of the reduced functions with respect to the similarity variable is denoted by a prime. After some rearrangement we obtain the equations

$$\frac{\dot{\rho}_0}{\rho_0} + \frac{\dot{R}}{R}\left[v' + (v - \xi)(\ln g)' + (v - 1)\frac{v}{\xi}\right] = 0,$$

$$\frac{R\ddot{R}}{\dot{R}^2} v + (v - \xi)v' + \frac{\pi'}{g} = 0, \tag{12.4}$$

$$\frac{R}{\dot{R}} \frac{d}{dt}(\ln \rho_0^{1-\gamma}\dot{R}^2) + (v - \xi)(\ln \pi g^{-\gamma})' = 0.$$

In order that the solution form (12.3) have meaning, so that it is possible to write ordinary differential equations for the new unknown functions $\pi(\xi)$, $g(\xi)$, and $v(\xi)$, it is necessary that the variables t and ξ in equations (12.4) be separated. To do this, in the second equation we must set $R\ddot{R}/\dot{R}^2 = const$, from which (with $const \neq 1$)

$$R = At^\alpha. \tag{12.5}$$

Here A and α are constants (A is dimensional and α is a pure number). In the first equation of (12.4) we must set $\dot{\rho}_0/\rho_0 = const\ \dot{R}/R$, which yields

$$\rho_0 = Bt^\beta, \tag{12.6}$$

where B and β are also constants. The first term in the third equation of (12.4) then automatically becomes a constant. Thus all the scales in the self-similar motion have a power-law dependence on time, and the similarity variable has the form*

$$\xi = \frac{r}{R} = \frac{r}{At^\alpha}. \tag{12.7}$$

* As was noted by Stanyukovich [1], in addition to power-law self-similarity it is also possible to have exponential self-similarity, in which $R = A'e^{mt}$, $\rho_0 = B'e^{nt}$, $\xi = re^{-mt}/A'$, where A', B', m, and n are constants. The exponential solution satisfies the equation $R\ddot{R}/\dot{R}^2 = const$ for $const = 1$. The majority of problems of practical interest have a power-law character.

Equations (12.4) are thus transformed into a system of three ordinary differential equations for the three unknown functions $\pi(\xi)$, $g(\xi)$, and $v(\xi)$. The system contains the constant exponents α and β. In a similar manner, the boundary and initial conditions of the problem are made dimensionless and in turn transformed into conditions on the functions π, g, and v. We shall not write out here the system of equations in general form. Instead, the equations will be presented later in context with their application to specific problems.

In many flows the density scale ρ_0 is constant ($\beta = 0$). This is true, for example, in all cases when a shock (or rarefaction) wave propagates through a gas of uniform density. The exponent β is usually different from zero for those problems in which the spatial distribution of the initial gas density is given by a power law of the type $\rho_{00} = const\ r^\delta$. In these cases the exponent β is defined in terms of the known exponents δ and α (if $\delta = 0$, $\beta = 0$). Thus, the system of equations for the functions π, g, and v (and the boundary conditions) will contain only one new parameter, the similarity exponent α.

The exponents in the scale functions are uniquely related to the exponents α and β (i.e., α and δ). For example, in the case when the density scale is constant ($\beta = 0$, $\rho_0 = const$), $R \sim t^\alpha$, $\dot{R} \sim t^{\alpha-1}$, $\Pi = \rho_0 \dot{R}^2 \sim t^{2(\alpha-1)}$. Since the length scale R is uniquely related to time, we can consider the velocity, density, and pressure scales to be functions of the length scale R, rather than of time. Using the relation $R \sim t^\alpha$, we find

$$\dot{R} \sim t^{\alpha-1} \sim R^{(\alpha-1)/\alpha}, \qquad \rho_0 \sim t^\beta \sim R^{\beta/\alpha},$$
$$\Pi \sim \rho_0 \dot{R}^2 \sim t^{\beta+2(\alpha-1)} \sim R^{[\beta+2(\alpha-1)]/\alpha}.$$

It is evident from the expression for the density scale $\rho_0 \sim t^\beta \sim R^{\beta/\alpha}$ and from the spatial distribution of the initial density $\rho_{00} = const\ r^\delta$, that ρ_0 must have the form $\rho_0 = \rho_{00}(R)$. For example, the initial gas density at the point where the shock wave is located at a time t can serve as the density scale ρ_0 (R is then the coordinate of the shock front). The relationship between the exponents β and δ indicated above then follows: $\beta = \alpha\delta$.

For $\beta = 0$ and $\rho_0 = const$, the functions p, ρ, and u given by (12.3) can be expressed in any of the equivalent forms:

$$\begin{aligned} p &= const\ t^{2(\alpha-1)}\pi(\xi) = const\ R^{2(\alpha-1)/\alpha}\pi(\xi), \\ u &= const\ t^{\alpha-1}v(\xi) = const\ R^{(\alpha-1)/\alpha}v(\xi), \\ \rho &= const\ g(\xi). \end{aligned} \tag{12.8}$$

§3. Conditions for self-similar motion

It is natural to pose the following question: what requirements must be satisfied by the conditions of a problem in order that the motion be self-similar? To answer this question we shall resort to dimensional considerations.

The gasdynamic equations (12.1) do not contain any dimensional parameters other than the dependent variables p, ρ, and u and the independent variables r and t (the only parameter γ is dimensionless). The boundary and initial conditions of the problem do contain dimensional parameters. It is this circumstance which makes it possible to construct the functions $p(r, t)$ and $\rho(r, t)$, since all the five variables p, ρ, u, r, and t have different dimensions, with three of the variables independent. Since the dimensions of pressure and density contain the unit of mass, at least one of the parameters in the problem must also contain a unit of mass. In many cases this is the constant initial density of the gas ρ_0, which has the dimensions* of ML^{-3}. In a number of problems the initial spatial distribution of the density is governed by the power law $\rho_{00} = br^{\delta}$. In this case it is the parameter b whose dimensions are $[b] = ML^{-3-\delta}$.

Let us denote the parameter containing the unit of mass by a. In the most general case it has the dimensions $[a] = ML^kT^s$. Bearing in mind the dimensions of the functions, $[p] = ML^{-1}T^{-2}$, $[\rho] = ML^{-3}$, and $[u] = LT^{-1}$, we can, without any loss in generality, represent them in the form suggested by Sedov [2]

$$p = \frac{a}{r^{k+1}t^{s+2}}P, \qquad \rho = \frac{a}{r^{k+3}t^s}G, \qquad u = \frac{r}{t}V, \tag{12.9}$$

where P, G, and V are dimensionless functions that depend on dimensionless groups containing r, t, and the parameters of the problem. In general there are two dimensionless variables: r/r_0 and t/t_0, where r_0 and t_0 are parameters with dimensions of length and time, which either enter directly in the conditions of the problem, or can be constructed by combining parameters with other dimensions. In this case, the functions P, G, and V will then depend separately on r and t and the problem is not a self-similar one.

We can give a large number of examples of families of similar flows. Let us cite one: the problem of a rarefaction wave generated by withdrawing a piston from a gas with the variable speed $u_1 = U(1 - e^{-t/\tau})$ (see §10, Chapter I). In this example the role of the parameter is played by the constant initial density of the gas ρ_0. In addition, the problem also contains the dimensional parameters $[\tau] = T$, $[U] = LT^{-1}$, and the initial speed of sound $[c_0] = LT^{-1}$ (or the initial pressure p_0, with $c_0^2 = \gamma p_0/\rho_0$). The ratios t/τ and $r/c_0\tau$ or $r/U\tau$ ($r_0 = c_0\tau$ or $U\tau$) can, for example, serve as the dimensionless variables. If length and time scales cannot be constructed from the parameters of the problem, then the variables r and t cannot enter the functions P, G, and V separately; the functions can depend only on a dimensionless combination of

* *Editors' note.* The symbols M, L, and T are introduced to denote mass, length, and time, respectively.

r and t, $\xi = r/At^{\alpha}$, where A is a parameter with the dimensions $[A] = LT^{-\alpha}$. Equations (12.9) then take the form

$$p = \frac{a}{r^{k+1}t^{s+2}} P(\xi), \qquad \rho = \frac{a}{r^{k+3}t^{s}} G(\xi), \qquad u = \frac{r}{t} V(\xi). \qquad (12.10)$$

In this case the problem is self-similar and the expressions (12.10) are equivalent to (12.3), differing from them only in the form of the reduced functions. Let us demonstrate this using as an example self-similar motions with a constant density scale. In this case, $a = \rho_0$, $k = -3$, and $s = 0$, so that the expressions of (12.10) take the form

$$p = \rho_0 \frac{r^2}{t^2} P(\xi), \qquad \rho = \rho_0 G(\xi), \qquad u = \frac{r}{t} V(\xi). \qquad (12.11)$$

Substituting $r = \xi R$ and noting that $\dot{R} = \alpha R/t$, we find that (12.11) and (12.3) are equivalent if

$$P(\xi) = \alpha^2 \frac{\pi(\xi)}{\xi^2}, \qquad G(\xi) = g(\xi), \qquad V(\xi) = \alpha \frac{v(\xi)}{\xi}. \qquad (12.12)$$

The study of self-similar motions is of great interest. The fact that it is possible to reduce a system of partial differential equations to a system of ordinary differential equations for new reduced functions simplifies the problem greatly from the mathematical standpoint and in a number of cases makes it possible to find exact analytic solutions. In addition, the self-similar solutions frequently represent the limits which are approached asymptotically by the solutions of nonself-similar problems. This statement will be clarified later when we consider specific problems.

§4. Two types of self-similar solutions

There exist two rather different types of self-similar solutions. Solutions of the first type possess the property that the similarity exponent α and the exponents of t and R in all scales are determined either by dimensional considerations or from the conservation laws. In this case the exponents are simple rational fractions with integral numerators and denominators. Problems of this type always contain two parameters with independent dimensions*. These parameters are used to construct one parameter whose dimensions contain the unit of mass a (see (12.10)) and another parameter A

* *Editors' note.* There is a type of self-similar solution in which the exponents are determined by the boundary conditions and may be set arbitrarily within certain limits. Although the exponents in such solutions are not simple rational fractions in general, the solutions are to be considered as of the authors' first type, because the two independent parameters exist and the exponents are determinable in advance. See, for example, the solutions of [18] and [19].

that contains only the units of length and time. With the second parameter A it is possible to construct a dimensionless combination, the similarity variable $\xi = r/At^{\alpha}$. The dimensions of the parameter A, $LT^{-\alpha}$, are determined by the similarity exponent α. Two motions of this type were considered in Chapter I, the problem of a self-similar rarefaction wave (§11) and the problem of a strong explosion (§25). In the first case the two independent dimensional parameters are the initial density and pressure of the gas ρ_0 and p_0. They can be combined into a dimensional parameter which does not contain the unit of mass, the initial speed of sound $c_0 = (\gamma p_0/\rho_0)^{1/2}$. The role of the parameter A is played by the speed of sound c_0. Correspondingly

$$\xi = \frac{r}{c_0 t}, \qquad \alpha = 1.$$

The parameters in the problem of a strong explosion are the initial density of the gas $\rho_0 \sim ML^{-3}$ and the explosion energy $E \sim ML^2T^{-2}$. The energy E is always equal to the total energy of the moving gas, and as a result an energy integral appears in the problem. (We recall that the initial pressure and speed of sound p_0 and c_0 in the problem of a strong explosion are assumed to be equal to zero, and hence that these quantities are not parameters of the problem.) The parameters ρ_0 and E are used to construct a parameter which does not contain mass, $A = (E/\rho_0)^{1/5} \sim LT^{-2/5}$, so that the similarity variable is $\xi = r/(E/\rho_0)^{1/5}t^{2/5}$, and $\alpha = \frac{2}{5}$.

For a strong explosion in a medium with variable initial density $\rho_{00} = br^{\delta}$, the explosion energy $E \sim ML^2T^{-2}$ and the coefficient $b \sim ML^{-3-\delta}$ serve as the parameters. They can be used to construct a parameter A not containing the unit of mass

$$A = \left(\frac{E}{b}\right)^{1/(5+\delta)} \sim LT^{-2/(5+\delta)}.$$

The similarity variable has the form

$$\xi = \frac{r}{(E/b)^{1/(5+\delta)}t^{2/(5+\delta)}}, \qquad \alpha = \frac{2}{5+\delta}.$$

(The self-similar problem of an explosion in a medium with variable density was considered by Sedov [2].) The self-similar problem of the propagation of a thermal wave from the point of release of a specified amount of energy is also of this general type (see Chapter X).

As shown in §2, the similarity exponent enters as a parameter in the system of differential equations for the reduced functions. Since in self-similar problems of this type the number α is found immediately from dimensional considerations (or from the conservation laws), the problem is thus reduced to the integration of a system of equations with known boundary conditions and parameters.

In self-similar problems of the second type, the exponent α cannot be found from dimensional considerations or from the conservation laws without solving the equations. In this case the determination of the similarity exponent requires that the ordinary differential equations for the reduced functions be integrated. It turns out that the exponent is found from the condition that the integral curve must pass through a singular point, as otherwise the boundary conditions cannot be satisfied. Examples of self-similar motions of the second type are the problems of an imploding shock wave and of an impulsive load, both of which will be discussed later.

Examination of solutions to specific problems of the second type shows that in all these cases the initial conditions of the problem contain only one dimensional parameter with the unit of mass but lack another parameter which could be used to form the parameter A. This circumstance eliminates the possibility of determining the number α from the dimensions of A. Actually, of course, the problem does have a dimensional parameter $A \sim LT^{-\alpha}$ relevant to it; otherwise it would not be possible to construct the dimensionless combination $\xi = r/At^{\alpha}$. However, the dimensions of this parameter (i.e., the number α) are not dictated by the inital conditions of the problem, but rather are found from the solution of the equations. The numerical value of A cannot be found from self-similar equations alone. It can be determined only by knowing how the given motion arose. Thus, for example, if the self-similar motion originated as a result of some nonself-similar flow that approaches a self-similar regime asymptotically, then the value of A can only be found by a numerical solution of the complete non-self-similar problem in which it is possible to follow the transition of the nonself-similar motion into the self-similar one. These statements will be explained in more detail when we consider specific problems.

Self-similar motions of the first type, in which the similarity exponent is determined by dimensional considerations, were investigated in detail by Sedov. Since Sedov's book [2], which gives an exhaustive treatment of these motions and the solution to a number of specific problems, is available, we shall in this chapter not dwell on self-similar motions of the first type, and shall devote our attention to motions of the second type only.

2. *Implosion of a spherical shock wave and the collapse of bubbles in a liquid*

§5. Statement of the problem of an imploding shock wave

Let us imagine a spherically symmetric flow in which a strong shock wave travels to the center of symmetry through a gas of uniform initial density ρ_0

and zero pressure. We shall not discuss the origin of the wave. The wave could have been generated, for example, by a "spherical piston" which pushed the gas inward, imparting to it a certain amount of energy. As the wave converges to the center the energy becomes concentrated at the front (cumulation), and the wave is strengthened. We shall be interested in the motion of the gas at small distances from the center (say, small in comparison with the initial radius of the "piston"). It is reasonable to assume that at times close to the instant of collapse and at small radii, the motion "forgets" to a considerable extent (which will be defined below) about the initial conditions and reaches some limiting regime which must be determined.

The problem does not contain characteristic parameters of either length or time. The initial radius of the "piston" cannot serve as the scale for the limiting motion in a region whose dimensions are very small in comparison with it. The only length scale is the radius of the shock front R, which itself varies with time. The velocity scale is the time-dependent velocity of the front $dR/dt \equiv \dot{R} \equiv D$. Therefore, it is natural to assume that the limiting motion will be self-similar. We have here no basis for determining in advance the similarity exponent α. Apart from the initial density ρ_0 there are no other evident parameters which could be used for constructing the similarity variable. Of course, the energy of the entire gas, equal to the energy imparted to the gas by the piston, has a definite magnitude. However, in the self-similar region, the dimensions of which are small (of the order of R) and decrease with time as the wave converges to the center, only a small part of the total energy is concentrated and it also decreases with time*. As will be shown below, the energy in the self-similar region, the radius of which is of the order of R and the mass within which is of the order of $\rho_0 R^3$, decreases with time following a power law. However, as $R \to 0$, it decreases slower than R^3 as a result of the strengthening of the shock wave and of the increase in the energy density (pressure). It is evident from what was said that this self-similar motion must be of the second type. The solution will contain some parameter A, whose dimensions are not known in advance, related to the similarity exponent α ($[A] = LT^{-\alpha}$; see §2). If the similarity exponent (or the dimensions of A) is found from the limiting solution, then the numerical value of A remains undetermined. It depends on the initial conditions of the problem, on the motion of the gas as a whole.

As was already stated, the limiting self-similar solution holds only in a region of small dimensions of the order of the radius of the front, and then only close to the instant of collapse of the shock wave, when this radius is small. If we solve numerically the problem of the motion of the gas as a whole

* The assumption that the initial pressure is equal to zero, that we are dealing with a strong wave, also eliminates from the problem the velocity parameter given by the initial speed of sound c_0. This quantity, along with the initial pressure, is equal to zero.

with some initial conditions that ensure that an imploding shock wave will be generated (the problem with a "spherical piston" pushing inward), we shall find that the true solution in a region with a radius which decreases in proportion to the radius of the front will approach the limiting self-similar solution closer and closer with time. The form of the limiting solution does not depend on the initial conditions or on the character of the gas motion at large distances, and in particular does not depend on the manner in which the piston moves. However, the limiting solution does not entirely "forget" the initial conditions. It "forgets" the form of the initial motion, but selects from the entire set of information provided by the initial conditions the single number A which characterizes the intensity of the initial push (a "stronger" push corresponds to a greater value of A).

If the form of the limiting solution does not depend on the initial conditions and on the motion of the gas at a large distance from the center, then the manner in which the true solution approximates the limiting solution will obviously depend on the initial conditions. The closer the initial motion corresponds to the limiting motion, the earlier will the true motion near the front reach the self-similar regime. However, it will reach it sooner or later, regardless of the initial conditions and of the type of motion at large distances. Therefore, we shall seek a self-similar solution to the problem of the implosion of a shock wave. This interesting and important problem was solved independently by Landau and Stanyukovich [1] and by Guderley [3].

§6. Basic equations

The origin for time $t = 0$ is taken to be at the instant of collapse, when $R = 0$. Thus, the time up to the instant of collapse is negative. In this regard, we modify slightly the definition of the similarity variable, setting

$$R = A(-t)^{\alpha}, \qquad \xi = \frac{r}{R} = \frac{r}{A(-t)^{\alpha}}. \tag{12.13}$$

Formally, the solution which we seek includes all of space up to infinity, so that the intervals for the variables are

$$-\infty < t \leqslant 0, \qquad R \leqslant r < \infty, \qquad 1 \leqslant \xi < \infty$$

(actually the self-similar solution holds only in a region with a radius of the order of R, and at large distances it is connected with the solution of the complete nonself-similar problem in some manner). At the shock front $\xi = 1$. The front velocity is directed toward the center and is negative, with $D \equiv \dot{R} = \alpha R/t = -\alpha R/|t| < 0$.

Let us substitute into the gasdynamic equations (12.1) a solution of the self-similar form (12.3). The system reduces to the equations (12.4) in which $\nu = 3$, in accordance with the spherical symmetry of the motion. The problem has a constant density scale $\rho_0 = const$ (we shall satisfy ourselves that this quite obvious statement is valid when we consider the boundary conditions at the shock front). Therefore, the term $\dot{\rho}_0/\rho_0$ in the first equation of (12.4) vanishes and the bracketed terms are equal to zero. The factors, which depend on the scale in (12.4), reduce to the following constants:

$$\frac{R\ddot{R}}{\dot{R}^2} = \frac{\alpha - 1}{\alpha}, \qquad \frac{R}{\dot{R}}\frac{d}{dt}(\ln \rho_0^{1-\gamma}\dot{R}^2) = \frac{2(\alpha - 1)}{\alpha}.$$

We thus obtain the following system of equations for the reduced functions:

$$\begin{aligned} &(v - \xi)(\ln g)' + v' + \frac{2v}{\xi} = 0, \\ &(\alpha - 1)\alpha^{-1}v + (v - \xi)v' + g^{-1}\pi' = 0, \\ &(v - \xi)(\ln \pi g^{-\gamma})' + 2(\alpha - 1)\alpha^{-1} = 0. \end{aligned} \tag{12.14}$$

To simplify the system we make a number of transformations. We use (12.12) to replace the functions π, g, and v by the new reduced functions P, G, and V (of course, it is possible to seek the solution of (12.1) in the form of (12.11) from the very beginning). Further, we also replace the pressure by a new unknown function, the square of the speed of sound*, and correspondingly introduce a new reduced function for the square of the speed of sound. In dimensional variables $c^2 = \gamma p/\rho$. In the form of (12.3) $c^2 = \gamma \dot{R}^2\pi/g = \dot{R}^2 z$, where the reduced function is $z = \gamma\pi/g$. In the new form (12.11), to which we have changed, $c^2 = \gamma(r^2/t^2)P/G = (r^2/t^2)Z$, where the reduced function is $Z = \gamma P/G$. The formulas (12.12) relate the reduced functions z and Z through

$$Z = \alpha^2 \frac{z}{\xi^2}.$$

* The system of gasdynamic equations (12.1) can be also written in terms of the functions ρ, u, and c^2 instead of ρ, u, and p:

$$\begin{aligned} &\frac{\partial \ln \rho}{\partial t} + u\frac{\partial \ln \rho}{\partial r} + \frac{\partial u}{\partial r} + (\nu - 1)\frac{u}{r} = 0, \\ &\frac{\partial u}{\partial t} + u\frac{\partial u}{\partial r} + \frac{c^2}{\gamma}\frac{\partial \ln \rho}{\partial r} + \frac{1}{\gamma}\frac{\partial c^2}{\partial r} = 0, \\ &\frac{\partial}{\partial t}\ln c^2\rho^{1-\gamma} + u\frac{\partial}{\partial r}\ln c^2\rho^{1-\gamma} = 0. \end{aligned} \tag{12.1'}$$

After the introduction of the new variables the system (12.14) takes the form

$$\frac{dV}{d\ln\xi} + (V-\alpha)\frac{d\ln G}{d\ln\xi} = -3V,$$

$$(V-\alpha)\frac{dV}{d\ln\xi} + \frac{Z}{\gamma}\frac{d\ln G}{d\ln\xi} + \frac{1}{\gamma}\frac{dZ}{d\ln\xi} = -\frac{2}{\gamma}Z - V(V-1), \quad (12.15)$$

$$(\gamma-1)Z\frac{d\ln G}{d\ln\xi} - \frac{dZ}{d\ln\xi} = 2\left[\frac{\alpha-1}{\alpha(V-\alpha)} + 1\right]Z.$$

This is a system of three first-order ordinary differential equations in the three unknown functions V, G, and Z of the independent variable ξ.

Let us consider the boundary conditions. At the shock front the conservation laws give the well-known relations between the flow variables behind the front and the speed of the front (see (1.111))

$$\rho_1 = \rho_0\frac{\gamma+1}{\gamma-1}, \qquad p_1 = \frac{2}{\gamma+1}\rho_0 D^2, \qquad u_1 = \frac{2}{\gamma+1}D, \qquad c_1^2 = \frac{2\gamma(\gamma-1)}{(\gamma+1)^2}D^2. \quad (12.16)$$

Substituting the expressions (12.11) for the dimensional quantities in terms of the reduced functions, noting that at the shock front $r = R$ and $\xi = 1$, and also noting that $D \equiv \dot{R} = \alpha R/t$, we obtain the boundary conditions for the reduced functions at $\xi = 1$,

$$V(1) = \frac{2}{\gamma+1}\alpha, \qquad G(1) = \frac{\gamma+1}{\gamma-1}, \qquad Z(1) = \frac{2\gamma(\gamma-1)}{(\gamma+1)^2}\alpha^2. \quad (12.17)$$

Here one should note that it is clear that the density scale depends neither on the time nor on the front radius. Otherwise it would have been impossible to satisfy the condition $\rho_1 = [(\gamma+1)/(\gamma-1)]\rho_0 = const$ at the shock front.

The reduced functions also satisfy boundary conditions at infinity. At the instant of collapse $t = 0$, the velocity, pressure, and speed of sound at any finite radius r are bounded. But with $t = 0$ and finite r, $\xi = \infty$. In order for the quantities $u = (r/t)V$ and $c^2 = (r^2/t^2)Z$ to be bounded when $t = 0$ and r is finite, V and Z must vanish. We thus obtain still another condition which must be satisfied by the solution,

$$V(\infty) = 0, \quad Z(\infty) = 0 \qquad \text{at} \quad \xi = \infty. \quad (12.18)$$

In general, the boundary conditions (12.17) are sufficient to start the integration of equations (12.15) from the point $\xi = 1$ in the direction of increasing ξ, after some value of α has been assigned. However, analysis of these equations, with which we shall be concerned in the following section, show that with an arbitrary value of α it is not possible to obtain a single-valued solution which satisfies (12.17) and arrives at the point (12.18). This is possible only for a certain particular value of α, which is then the desired choice of the similarity exponent.

§7. Analysis of the equations

In this section we shall show how to determine the similarity exponent in the solution of the equations (12.15). In order to do this it is first necessary to analyze the equations. We shall not, however, attempt to present rigorous mathematical proofs or carry out detailed calculations. We shall only consider the most important aspects and present the basic methods for solving the problem. In so doing we shall attempt to emphasize certain features of the problem which are common either to all self-similar solutions, or to solutions of the second type. We shall follow the system of presentation suggested by N. A. Popov, to whom we are thankful for valuable advice.

It becomes immediately evident on inspection of (12.15) that the variable $\ln \xi$, which can be regarded as a new independent variable in place of ξ, enters in the system only as the differential $d \ln \xi$. Similarly, one of the unknown functions, G, appears only as the differential $d \ln G$. This property of equations (12.15), which is characteristic of all self-similar motions, permits the reduction of the system of three differential equations to a single differential equation in V and Z and two quadratures*.

Let us solve the system (12.15) for the derivatives $dV/d \ln \xi$, $d \ln G/d \ln \xi$, $dZ/d \ln \xi$. Instead of writing down the rather lengthy expressions which result, we write the solution of the algebraic system in symbolic form using determinants,

$$\frac{dV}{d \ln \xi} = \frac{\Delta_1}{\Delta}, \qquad \frac{d \ln G}{d \ln \xi} = \frac{\Delta_2}{\Delta}, \qquad \frac{dZ}{d \ln \xi} = \frac{\Delta_3}{\Delta}, \tag{12.19}$$

* This property is not accidental but is a result of the dimensional structure of the gas-dynamic equations, which do not contain any dimensional quantities besides the variables themselves. The fact that a quantity is a logarithmic derivative shows that the choice of units for this quantity is arbitrary. In the case of the density $\rho = \rho_0 G$, this can be seen directly from equations (12.1′) for the quantities ρ, u, and c^2 (see footnote in paragraph following (12.14)). If in the general nonself-similar equations we transform to the new independent variables $\xi = r/At^{\alpha}$ and $\eta = r/r_0$, where A and r_0 are dimensional parameters introduced arbitrarily, then, since no limitations were imposed on the choice of these parameters, they should drop out from the equations. Indeed, the transformation shows that the new variables are contained in the equations only in the form $d \ln \xi$, and $d \ln \eta$ (in the case of self-similar motions all functions depend only on ξ and are independent of η, so that the terms in $d \ln \eta$ vanish).

The dimensionless quantities V and Z are formed from dimensional variables, $V = tu/r$ and $Z = t^2c^2/r^2 = \gamma pt^2/\rho r^2$, without the appearance of any arbitrary parameters; hence, they enter the equations in a free form and not as logarithmic differentials.

where the determinant of the system Δ is given by

$$\Delta = \begin{vmatrix} 1 & V-\alpha & 0 \\ V-\alpha & \dfrac{Z}{\gamma} & \dfrac{1}{\gamma} \\ 0 & (\gamma-1)Z & -1 \end{vmatrix} = -Z + (V-\alpha)^2. \tag{12.20}$$

The determinants Δ_1, Δ_2, and Δ_3 are obtained by replacing the corresponding columns in (12.20) by the right-hand sides of (12.15).

The coefficients of the derivatives and the right-hand sides in equations (12.15) depend only on V and Z and do not depend on G and ξ, so that all the determinants Δ, Δ_1, Δ_2, and Δ_3 are functions of V and Z only. Dividing the third equation of (12.19) by the first, we obtain the first-order ordinary differential equation

$$\frac{dZ}{dV} = \frac{\Delta_3(Z, V)}{\Delta_1(Z, V)}. \tag{12.21}$$

After the solution $Z(V)$ of this equation is found, it can be substituted into the first equation in (12.19) and the function $V(\xi)$ obtained by quadratures. Then, substituting $V(\xi)$ and $Z[V(\xi)]$ into the second equation, the function $G(\xi)$ may be obtained by quadratures.

Actually, only one quadrature is necessary, since the system (12.15) possesses a first integral which has the form of an algebraic relation connecting all the variables. The existence of this integral, termed the adiabatic integral, is related to the law of conservation of entropy on a gas particle path*. In general, the satisfaction of the conservation laws is always accompanied by the existence of corresponding integrals of the self-similar equations. Thus, in the problem of a strong explosion (see §25, Chapter I), the equations admit an energy integral. The main problem, therefore, reduces to

* To derive the adiabatic integral we use the first and third equations of (12.15). The first (continuity) equation is divided by $(V-\alpha)$ and written in the form

$$d \ln G + d \ln(V-\alpha) = -3\, d \ln \xi - \frac{3\alpha\, d \ln \xi}{V-\alpha}.$$

The third (entropy) equation is divided by Z and written in the form

$$d \ln G^{\gamma-1} Z^{-1} = \frac{2(\alpha-1)\alpha^{-1} d \ln \xi}{V-\alpha} + 2\, d \ln \xi.$$

Eliminating from these two equations the differential $d \ln \xi/(V-\alpha)$ and collecting all terms on one side, we obtain an equation of the form $d \ln F\{\xi, G, V, Z\} = 0$. This leads to the integral $F\{\xi, G, V, Z\} = const$, with the constant to be determined from the boundary conditions (12.17).

the solution of (12.21) subject to the boundary conditions (12.17) and (12.18).

Let us consider the behavior of the desired integral curve on the V, Z plane. At the shock front where $\xi = 1$, $V = V(1)$ and $Z = Z(1)$ (see (12.17)). Let us plot this point on the plane, denoting it by the letter A. At infinity, where $\xi = \infty$, $V(\infty) = 0$ and $Z(\infty) = 0$, so that the integral curve $Z(V)$ moves from the point A to the coordinate origin O (Fig. 12.1).

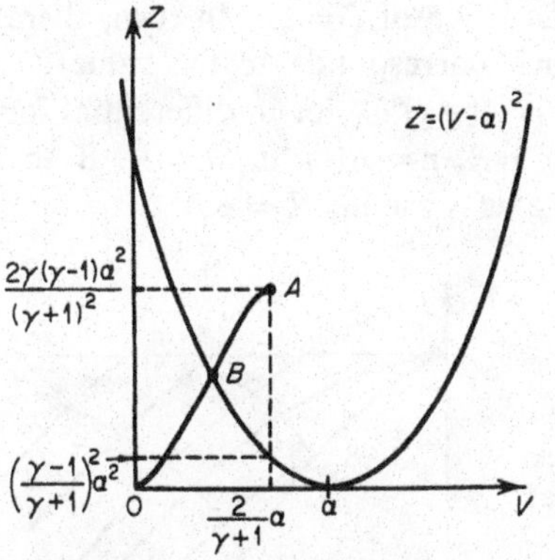

Fig. 12.1. Behavior of the integral curve on the V, Z plane.

In order for the solution of the gasdynamic equations to be physically meaningful, it must be single valued. To each value of the independent variable ξ should correspond unique values of V and Z. This means that ξ as a function of V and ξ as a function of Z or, equivalently, $\ln \xi(V)$ and $\ln \xi(Z)$, should not have extrema. The derivatives $d \ln \xi/dV = \Delta/\Delta_1$ and $d \ln \xi/dZ = \Delta/\Delta_3$ in the correct solution must not become zero in the domain of interest $1 < \xi < \infty$, $0 < \ln \xi < \infty$. But the determinant $\Delta = -Z + (V - \alpha)^2$ is equal to zero on the parabola $Z = (V - \alpha)^2$ in the V, Z plane (Fig. 12.1). It can be easily checked by direct calculation that the point A lies above the parabola, so that the desired integral curve must intersect the parabola on its path from point A to point O. At the point of intersection, in order that the derivatives $d \ln \xi/dV$ and $d \ln \xi/dZ$ not vanish, it is necessary that the determinants Δ_1 and Δ_3 also vanish (it can be checked that if $\Delta = 0$, both Δ_1 and Δ_3 must vanish simultaneously). Thus, the point of intersection of the correct integral curve $Z(V)$ and the parabola is the singular point of (12.21) ($\Delta_1 = 0$, $\Delta_3 = 0$, $dZ/dV = 0/0$).

If we specify some arbitrary value of the similarity exponent α and start the integration of (12.21) from point A, the integral curve will in general have no point of intersection with the parabola or else will intersect it at some ordinary point, and the curve will not correspond to the correct solution. Only for a particular value of α will the integral curve intersect the parabola and pass through the required singular point of (12.21), then moving on to its final point O. This requirement that the correct integral curve must pass through a specific singular point of (12.21) determines the exponent α. The singular point B and a diagram of the correct integral curve are shown in Fig. 12.1 (it can be shown that point B lies on the left branch of the parabola).

At the singular point B, through which the correct integral curve $Z(V)$ passes, the quantities Z and V take on specific values, which also satisfy the equation of the parabola $Z = (V - \alpha)^2$. Since V and Z are functions of ξ, the singular point corresponds to a specific value of $\xi = \xi_0$. In turn, there is a line on the r,t plane, the ξ_0-line, which corresponds to the value $\xi = \xi_0$. The equation of this line is $r = R(t)\xi_0 = A(-t)^\alpha \xi_0$, or in differential form is $dr/dt = \dot{R}\xi_0$. The shock front line is $\xi = 1$, $r = R(t)$, $dr/dt = \dot{R}$. Both lines are shown in Fig. 12.2 (note that the r axis is the line $\xi = \infty$).

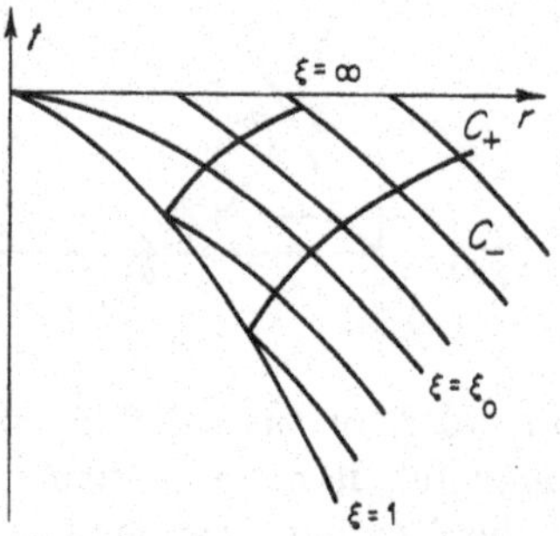

Fig. 12.2. r,t diagram for the implosion of a shock wave. $\xi = 1$ is the shock front line, $\xi = \xi_0$ is the ξ_0-line. Several characteristics of the C_+ and C_- families are also indicated.

The ξ_0-line has the important property that it is one of the C_- characteristics. In order to satisfy ourselves that this is so, we shall transform the dimensional equation for the C_- characteristics $dr/dt = u - c$ to similarity variables. It should be noted here that the speed of sound c is always a positive quantity. The assumed scale for it, $\dot{R}$ or r/t, is negative. Consequently, in taking the root of the expression $c^2 = r^2/t^2 Z$ it is necessary to set $c = -r/t|\sqrt{Z}|$. Thus,

$$\frac{dr}{dt} = u - c = \frac{r}{t} V + \frac{r}{t}|\sqrt{Z}| = \frac{R}{t}\xi(V + |\sqrt{Z}|) = \frac{\dot{R}\xi}{\alpha}(V + |\sqrt{Z}|).$$

We shall consider the C_- characteristics which pass through the ξ_0-line on the r, t plane. To do this, we set $\xi = \xi_0$ in the equation for the characteristics. But for $\xi = \xi_0$

$$Z(\xi_0) = [V(\xi_0) - \alpha]^2, \qquad |\sqrt{Z(\xi_0)}| = \alpha - V(\xi_0)$$

since $V < \alpha$. (We note that for $\xi = 1$, $V(1) = [2/(\gamma + 1)]\alpha < \alpha$, while for $\xi = \infty$, $V = 0$ and the function $V(\xi)$ is monotonic). Therefore, the slope of the C_- characteristics at any point on the ξ_0-line is $dr/dt = (\dot{R}\xi_0/\alpha)[V(\xi_0) +$

$|\sqrt{Z(\xi_0)}| = \dot{R}\xi_0$, which is the same as the slope of the ξ_0-line itself. This means that the ξ_0-line is either the envelope of a family of C_- characteristics, or simply coincides with one of them. It turns out that the second statement is the correct one; the ξ_0-line coincides with a C_- characteristic, and is thus itself a C_- characteristic.

From this result follows an important conclusion on the causality of the phenomena. As we know, the characteristics of the same family never intersect in a continuous flow region. This means that all the C_- characteristics which pass above the ξ_0-line (see Fig. 12.2) never overtake the shock front prior to the instant of collapse. The C_- characteristics passing below the ξ_0-line do overtake the front (the C_+ characteristics originate from the front line). Thus, the ξ_0-line bounds the region of influence. The state of the motion at a given time for points which lie to the right of the ξ_0-line, at distances r greater than $r_0 = R(t)\xi_0$, can in no way affect the motion of the shock wave.

The two special properties of the solution noted above, the passing of the correct integral curve through a singular point, which is possible only for a specific value of the similarity exponent α (determined by this property), and the existence of a ξ_0-line on the r,t plane which corresponds to the singular point and is itself a characteristic bounding the region of influence, are properties peculiar to all self-similar solutions of the second type.

§8. Numerical results for the solutions

In practice the solution and the similarity exponent are found by trial and error. A value of α is assumed, (12.21) is integrated numerically from the initial point A ($\xi = 1$), and the behavior of the integral curve is determined. The value of α is corrected by successive approximations in order to obtain an integral curve which intersects the parabola at the required singular point and then goes to the final point O. Landau and Stanyukovich [1] have given an approximate method which yields a value of α quite close to the correct value. This value was used for the initial guess and then refined. After the exponent α and the function $Z(V)$ are found it is not difficult to determine the functions $V(\xi)$, $Z(\xi)$, and $G(\xi)$.

By such a method [1, 3] the similarity exponent α was found equal to 0.717 for a specific heat ratio $\gamma = 7/5$. In [1] it was also found that $\alpha = 0.638$ for $\gamma = 3$, and it was established that in the limit $\gamma \to 1$, $\alpha \to 1$. The relations governing the radius and velocity of the shock front and the pressure behind the front for $\gamma = 7/5$ are given by

$$R \sim |t|^\alpha \sim |t|^{0.717},$$

$$|\dot{R}| \sim |t|^{\alpha-1} \sim R^{(\alpha-1)/\alpha} \sim |t|^{-0.283} \sim R^{-0.395},$$

$$p_1 \sim |t|^{2(\alpha-1)} \sim R^{2(\alpha-1)/\alpha} \sim |t|^{-0.566} \sim R^{-0.79}.$$

The velocity and pressure distributions u and p as functions of the radius at different times for the case $\gamma = 7/5$ are given in Fig. 12.3, taken from the book of Stanyukovich [1]. The velocity behind the front decreases monotonically with increasing radius, while the pressure first increases slightly and then also drops*. The density behind the front increases monotonically.

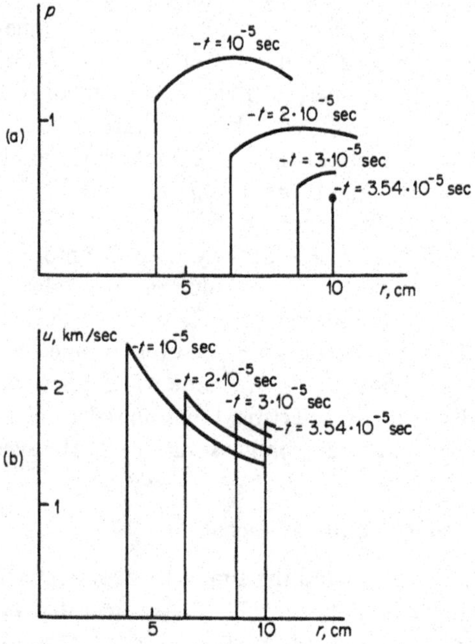

Fig. 12.3. (a) Pressure (in arbitrary units) and (b) velocity distributions at different times during the implosion of a shock wave with $\gamma = 7/5$. The curves are taken from [1].

The shock wave accelerates continuously and is strengthened as it converges to the center. As $t \to 0$ and $R \to 0$ the pressure and temperature behind the front tend to infinity; the density of the gas remains finite, and behind the wave front it is constant and equal to $[(\gamma + 1)/(\gamma - 1)]\rho_0$.

As the shock wave converges, energy becomes concentrated near the shock front as the temperature and pressure there increase without limit. However, the dimensions of the self-similar region decrease with time, and the total energy concentrated within this region also decreases. We consider now a self-similar solution only within some sphere whose radius decreases in proportion to the radius of the front R. The effective boundary of this self-similar region is then considered to be at some constant value of $r/R = \xi = \xi_1$.

* Such behavior of the pressure is not general; for example, when $\gamma = 3$ the pressure as well as the velocity behind the shock front decreases monotonically.

The energy contained in this region, i.e., in a sphere with the variable radius $r_1 = \xi_1 R$, is equal to

$$E_{\text{sim}} = \int_R^{r_1} 4\pi r^2\, dr\, \rho\left(\frac{1}{\gamma - 1}\frac{p}{\rho} + \frac{u^2}{2}\right)$$

$$= 4\pi R^3 \rho_0 \dot{R}^2 \int_1^{\xi_1} g\left(\frac{1}{\gamma - 1}\frac{\pi}{g} + \frac{v^2}{2}\right)\xi^2\, d\xi.$$

The integral with respect to ξ from 1 to ξ_1 is a constant, so that the energy $E_{\text{sim}} \sim R^3\dot{R}^2 \sim R^{5-(2/\alpha)}$. The exponent on R is positive for all real values of the specific heat ratio γ. For example, for $\gamma = 7/5$, $\alpha = 0.717$,

$$E_{\text{sim}} \sim R^{2.21} \to 0 \qquad \text{as} \quad R \to 0.$$

With the integration with respect to ξ extended to infinity ($\xi_1 = \infty$) the integral diverges (this is explained in the following footnote on limiting relationships). Thus, the total energy in all space is infinite within the framework of the self-similar solution. In particular, this conclusion also shows that the self-similar solution cannot be applied to indefinitely large radii r. The energy contained in a sphere of constant radius r can increase (but not indefinitely) with time. If the true solution coincides with (or is a very close approximation to) the self-similar solution at a given time from $r = R$ to $r = r_1 > R\xi_0$, the true solution will continue to agree with (or be very close to) the self-similar solution within a sphere of at least some finite radius smaller than r_1 all the way to and through the instant of collapse. Note that the C_- characteristics to the right of the ξ_0-line in Fig. 12.2 intersect the $t = 0$ axis at finite values of r, and signals from the nonself-similar part of the true solution can only be propagated inward on these characteristics.

The form of the limiting distributions of the flow variables with respect to the radius at the instant of collapse $t = 0$ can be established by using dimensional considerations. We have at our disposal one and only one parameter $A(LT^{-\alpha})$ which can be used to relate the velocity u and the speed of sound c with the radius r. This gives the limiting relationship at $t = 0$

$$u \sim c \sim A^{1/\alpha} r^{1-1/\alpha} = A^{1/\alpha} r^{-(1-\alpha)/\alpha}.$$

Since $\xi = \infty$ at $t = 0$ and $r \neq 0$, the limiting density $\rho_{\text{lim}} = \rho_0 G(\infty)$ is constant with respect to the radius. Consequently, the limiting pressure distribution is given by

$$p = \frac{1}{\gamma}\rho c^2 \sim \rho_0 A^{2/\alpha} r^{-2(1-\alpha)/\alpha}.$$

The limiting relationships $u(r)$, $c(r)$, and $p(r)$ naturally have the same behavior as the relationships at the front during the process $u_1(R)$, $c_1(R)$, and $p_1(R)$ (except for numerical coefficients)*.

The numerical coefficients in the limiting relations for $u(r)$, $c(r)$, and $p(r)$, as well as the limiting value of the density $\rho_{\text{lim}} = \rho_0 G(\infty)$, can only be found by solving the equations of self-similar motion. For $\gamma = 7/5$ the limiting density is $\rho_{\text{lim}} = 21.6\rho_0$ (behind the shock front $\rho_1 = 6\rho_0$). The density at large distances from the front $r \to \infty$ before the instant of collapse is also $\rho = 21.6\rho_0$, since for $R \neq 0$ and $r \to \infty$, $\xi = r/R \to \infty$ and $\rho/\rho_0 = G(\xi) \to G(\infty)$.

The energy concentrated in a sphere of radius r at the instant of collapse is

$$\int_0^r 4\pi r^2\, dr\, \rho\left(\frac{1}{\gamma-1}\frac{p}{\rho} + \frac{u^2}{2}\right) \sim r^{5-2/\alpha}$$

(just as $E_{\text{sim}} \sim R^{5-2/\alpha}$; see above). The energy concentrated in a sphere with a finite radius is finite and tends to zero as $r \to 0$. The larger the sphere, the larger is the energy included in it (within the framework of the self-similar regime).

After the instant of collapse, with $t > 0$, the shock wave reflected from the center propagates outward through the gas which is moving inward toward the center. The motion in this stage is also self-similar, and the similarity exponent does not change. For $t > 0$ the reflected wave propagates following the relation $R \sim t^\alpha$. Calculations show that for $\gamma = 7/5$ the density of the gas behind the reflected shock front is $\rho_{1\text{ref}} = 137.5\rho_0$, and thus the density is 23 times greater than the density behind the front of the incident wave $\rho_1 = 6\rho_0$. The velocity behind the front is positive and the gas is expanding from the center, with the expansion velocity decreasing with time as $\dot{R} \sim t^{-(1-\alpha)}$ starting from infinity†.

* The limiting relationships can also be established analytically starting from the equations for the reduced functions by finding the asymptotic solution in the neighborhood of the point $\xi = \infty$, $V = 0$, and $Z = 0$. We get $V \sim \xi^{-1/\alpha}$, $Z \sim \xi^{2-2/\alpha}$, which, upon transforming to dimensional variables, yields the limiting relationships given in the text. The quantities v^2 and $z = \gamma\pi/g$ are, according to (12.12), proportional to $\xi^2 V^2 \sim \xi^{2-2/\alpha}$ as $\xi \to \infty$; $g(\infty) = G(\infty) = const.$ It is evident from this result that the energy integral diverges as $\xi_1 \to \infty$, since

$$\int_1^{\xi_1} g\left(\frac{1}{\gamma-1}\frac{\pi}{g} + \frac{v^2}{2}\right)\xi^2\, d\xi \sim \int_1^{\xi_1} \xi^{4-2/\alpha}\, d\xi \sim \xi_1^{5-2/\alpha} \to \infty,$$

and the total energy in all space within the framework of the self-similar solution is infinite at any instant of time.

† In a paper by one of the authors [26] a family of self-similar solutions for cylindrical motion is constructed within the acoustic approximation. This family is obtained by means of the superposition of plane waves. The similarity exponent is arbitrary and is chosen in accordance with the initial conditions. For a converging cylindrical shock wave (in the acoustic approximation) the pressure behind the front $p \sim |t|^{-1/2}$, where the front radius $R = c|t|$.

It is interesting that the pressure behind the front of the reflected shock wave is infinite

§9. Collapse of bubbles. The Rayleigh problem

A process which has much in common with the implosion of a shock wave is the collapse of bubbles in a liquid (water). Small bubbles filled with the vapor of the liquid and undissolved gases are frequently formed in a real liquid. The phenomenon of bubble formation is called cavitation. Under steady state conditions the bubble is stable and the internal gas pressure balances the pressure in the liquid. When the liquid moves and goes from a low to a high pressure region, the internal pressure in the bubble (previously formed at the lower pressure) is lower than the new, higher pressure of the liquid. This causes the liquid to move to the center, collapsing the bubble. As in the implosion of a shock wave, the energy of the bubble becomes concentrated as the bubble collapses. As the bubble radius decreases the rate of collapse and the pressure increase and attain very large values. After collapse, a pressure peak is formed in the central region and a shock wave travels away from the center. When such a process takes place near a solid surface, the shock wave can damage the surface material. This is considered to be one of the causes of rapid wear of screw propellers and turbines.

An idealized problem of the liquid motion during the collapse of a bubble was solved by Rayleigh [4]. The liquid was assumed to be ideal (inviscid) and incompressible. The spherically symmetric cavity was regarded as a void, with the pressure within and at the surface of the cavity assumed equal to zero*.

At an initial time let there be a spherical cavity of radius R_0 in the liquid. The pressure in the surrounding liquid is p_0 and the liquid is at rest. After the motion starts the velocity distribution as a function of the radius r is found from the continuity equation with $\rho = const$

$$u = \dot{R}\frac{R^2}{r^2} = \dot{R}\frac{1}{\xi^2}, \qquad \xi = \frac{r}{R}, \tag{12.22}$$

where $R(t)$ is the radius of the cavity and $\dot{R}$ is the velocity of the boundary. Substituting the expression for the velocity into the equation of motion and

---

for $R \neq 0$. The results for the shock wave were obtained earlier by Zababakhin and Nechaev [27]. The pressure behind the front becomes infinite only within the framework of the acoustic approximation, as is explained in [26, 27].

* Apparently the internal vapor pressure in the actual process increases during the last stage of the collapse to such an extent that it withstands the motion thrust of the liquid and forces it to turn back. As a result of the very rapid compression the vapor does not succeed in condensing and its compression at the end is isentropic. However, if we consider the process of collapse of the bubble only up to a radius which is not too small the vapor pressure can be neglected. We may also neglect the surface tension at the boundary with the liquid.

integrating with respect to r from r to ∞ we obtain the pressure distribution

$$p = p_0 + \rho\frac{\ddot{R}R + 2\dot{R}^2}{\xi} - \rho\frac{\dot{R}^2}{2\xi^4}. \qquad (12.23)$$

If we apply this equation at the boundary of the cavity $\xi = 1$, where $p = 0$, we obtain an equation for $R(t)$

$$0 = p_0 + \rho(\ddot{R}R + \tfrac{3}{2}\dot{R}^2). \qquad (12.24)$$

Integrating this equation once under the initial condition $\dot{R} = 0$ when $R = R_0$, we obtain the law for the increase in velocity during collapse

$$\dot{R}^2 = \frac{2p_0}{3\rho}\left(\frac{R_0^3}{R^3} - 1\right).^{*} \qquad (12.25)$$

This equation can also be obtained directly from energy considerations. Let us take the energy of the liquid without the bubble to be zero. The potential energy of the liquid with a bubble of radius R is equal to the work done in overcoming the external pressure forces in the formation of a cavity of volume $4\pi R^3/3$. This work is equal to $p_0 4\pi R^3/3$, independent of the distribution of pressure in the region of the bubble†. The kinetic energy of the liquid is equal to

$$\int_R^\infty 4\pi r^2\frac{\rho u^2}{2}\,dr = \int_R^\infty 4\pi r^2\rho\frac{\dot{R}^2R^4}{2r^4}\,dr = 2\pi\rho\dot{R}^2R^3.$$

The total energy, equal to the sum of the kinetic and potential energies, is conserved, so that

$$2\pi\rho\dot{R}^2R^3 + \frac{p_0 4\pi R^3}{3} = E = \frac{p_0 4\pi R_0^3}{3}. \qquad (12.26)$$

Equation (12.25) follows directly from this relation.

Using (12.24) we can express the pressure distribution (12.23) in the form

$$p = p_0\left(1 - \frac{1}{\xi}\right) + \frac{\rho\dot{R}^2}{2}\left(\frac{1}{\xi} - \frac{1}{\xi^4}\right), \qquad \xi = \frac{r}{R}.$$

* Integration of (12.25) gives the time of collapse of the bubble as $\tau = 0.915\,R_0(\rho/p_0)^{1/2}$. For example, in water with $\rho = 1$ g/cm^3, $p_0 = 1$ atm and $R_0 = 1$ mm, $\tau = 0.915 \cdot 10^{-4}$ sec.

† This statement can be explained as follows. Let us imagine a vessel containing a liquid at a pressure p_0, closed by a movable piston with a surface area S. If a cavity with volume Ω is formed inside, the liquid, due to its incompressibility, will displace the piston through a distance l such that $lS = \Omega$. In doing this the cavity does an amount of work $p_0Sl = p_0\Omega$ on the piston. This work is determined only by the pressure p_0 far from the bubble and is independent of the pressure distribution near the bubble.

(The velocity and pressure distributions are illustrated schematically in Fig. 12.4.)

It is clear from the formula for the pressure that the problem is not self-similar (despite the apparent self-similar form of the velocity (12.22)). This

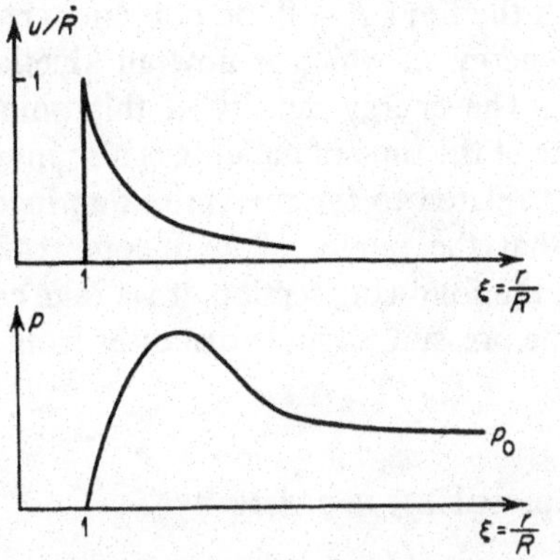

Fig. 12.4. Velocity and pressure distributions in Rayleigh's problem.

is evident from the fact that the problem contains a characteristic length scale R_0 and velocity scale $(p_0/\rho)^{1/2}$. However, in the limit, when the radius of the cavity tends to zero, $R \to 0$, the velocity and pressure increase, tending to infinity, and the solution asymptotically takes on a self-similar character, and

$$p \approx \frac{\rho \dot{R}^2}{2}\left(\frac{1}{\xi} - \frac{1}{\xi^4}\right), \qquad \dot{R}^2 \approx \frac{2p_0}{3\rho}\frac{R_0^3}{R^3}. \tag{12.27}$$

The length scale, the initial radius, becomes too large and the pressure scale p_0 becomes too small to characterize the actual process, whose scales are now the time-dependent cavity radius R and boundary velocity $\dot{R}(R \ll R_0$; $\dot{R} \gg (p_0/\rho)^{1/2}$; $p \sim \rho\dot{R}^2 \gg p_0)$. The motion appears as though it "forgets" the initial conditions. This is demonstrated in particular by the fact that the parameters p_0 and R_0 do not appear separately in the equation of motion of the boundary as before (see (12.25)), but only in the combination $p_0 R_0^3$, proportional to the total energy of the liquid $E = 4\pi R_0^3 p_0/3$ (see (12.27)).

It can be seen that this is a self-similarity of the first type, appearing because the energy is conserved. The dimensional parameters in the self-similar flow are the same as in the problem of a strong explosion, i.e., energy and density. The motion of the boundary is described by (12.27), $\dot{R}^2 \sim E/\rho R^3$ and the pressure by $p \sim E/R^3$. Hence we immediately obtain $R \sim (E/\rho)^{1/5}(-t)^{2/5}$, $\dot{R} \sim (E/\rho)^{1/5}(-t)^{-3/5}$, as in the problem of a strong explosion (the instant of collapse is taken to be zero). The similarity exponent $\alpha = 2/5$.

In the limit $R \to 0$, we obtain from (12.22) and (12.27)

$$u \sim \dot{R}\frac{R^2}{r^2} \sim \frac{R^{1/2}}{r^2}, \qquad p \sim \frac{1}{R^3}\left(\frac{R}{r} - \frac{R^4}{r^4}\right) \sim \frac{1}{R^2 r} - \frac{R}{r^4}.$$

The velocity of the boundary tends to infinity as $\dot{R} \sim R^{-3/2}$, but the velocity at a finite radius $r \neq 0$ tends to zero. In the limit $R \to 0$ the potential energy $4\pi p_0 R^3/3$ tends to zero and the total energy E, which is now all kinetic, is concentrated at the coordinate origin. The energy density at this point is infinite. Unlike the velocity, the pressure at the time of collapse is also infinite at any finite radius $r \neq 0$ (energy is not related to the pressure in an incompressible liquid model). This shows that the model of an incompressible liquid is imperfect. As will be shown in the following section, if we take compressibility into account, we find that the pressure at finite distances from the center is bounded.

§10. Collapse of bubbles. Effect of compressibility and viscosity

The collapse of an empty cavity in water, taking into account compressibility (but not viscosity) was considered by Hunter [5]. The equation of state was assumed to be of the form

$$p = B\left[\left(\frac{\rho}{\rho_0}\right)^\gamma - 1\right]$$

with $\gamma = 7$. Actually, however, in the limit of high pressures the term 1 was dropped, so that the equation of state had a form analogous to that for a perfect gas, $p = B(\rho/\rho_0)^\gamma$. It was assumed that B is a constant independent of the entropy (the flow was assumed isentropic). The value of B was taken equal to 3000 atm.

Numerical solution of the hydrodynamic equations (in terms of the variables u and c) with properly selected initial and boundary conditions showed that, in the limit when the radius of the cavity becomes very small and the velocity of the boundary very large, the solution becomes self-similar. Accordingly, a solution of the equations was sought in the self-similar form $u = \dot{R}v(r/R)$ and $c^2 = \dot{R}^2 z(r/R)$, where the radius of the cavity is $R = A(-t)^\alpha$.* The general properties and behavior of the equations in similarity variables or reduced functions are in many respects analogous to the problem of the imploding shock wave. Numerical integration yielded the value $\alpha = 0.555$ for the similarity exponent (for $\gamma = 7$).

The energy of the entire flow, as in the problem of the implosion of a shock

* For analysis and solution of the equations it was found more convenient to choose $\xi' = -(R/r)^{1/\alpha} = A^{1/\alpha} t r^{-1/\alpha}$ rather than $\xi = r/R = r/A(-t)^\alpha$ as the similarity variable (at the boundary of the cavity $r = R$, $\xi' = -1$; at infinity $r = \infty$, $\xi' = 0$).

wave, is infinite. (The energy contained in a sphere of radius r at the instant of collapse $t = 0$, $R = 0$, is proportional to $r^{1.13}$.) It is the absence of an energy integral which leads to a self-similar problem of the second type. The distributions of velocity, of the square of the speed of sound, and of the density with respect to the radius at the instant of collapse of the cavity, when $R = 0$, are of the form

$$u \sim r^{-(1-\alpha)/\alpha}, \qquad c^2 \sim r^{-2(1-\alpha)/\alpha},$$

$$\rho \sim r^{-2(1-\alpha)/\alpha(\gamma-1)}, \qquad p \sim r^{-2(1-\alpha)\gamma/\alpha(\gamma-1)}.$$

In contrast to the implosion of a shock wave, where the distributions of u and c^2 ($c^2 \sim p/\rho$) have the same form as above, the limiting density is variable in this case. This is a result of the fact that the problem was initially assumed to be isentropic. The sharp increase in c^2 and p is not connected with an increase in entropy, as in a shock wave, but with the increase in density.

To some extent the self-similar solution describes the actual process only in regions of very small radii, when the initial conditions have effectively been "forgotten". A comparison of the self-similar solution with the results of numerical integration of the partial differential equations for initial conditions corresponding to atmospheric pressure in water and to an initial radius of $R_0 = 0.5$ cm showed the following: At the instant of complete collapse $t = 0$, $R = 0$, the self-similar solution is valid in a region whose radius is of the order of 10^{-2} cm. Such a sphere contains about 10–20% of the energy of the liquid, and the pressure at its boundary is of the order of several tens of thousands of atmospheres. Hunter [5] also found the self-similar solution for the shock wave propagating outward from the center after the collapse of the bubble.

The inclusion of the viscosity of the liquid in the calculation leads to interesting laws for the behavior of the flow. The problem of collapse of an empty spherical cavity in an incompressible viscous liquid was solved by Zababakhin [6]. Analysis of the governing equations shows that the character of the flow depends on the Reynolds number $\mathrm{Re} = (R_0/\nu)(p_0/\rho)^{1/2}$, where $\nu = \mu/\rho$ is the kinematic viscosity. When $\mathrm{Re} > \mathrm{Re}^*$ (low viscosity) where Re^* is some critical Reynolds number, the velocity of the cavity boundary $\dot{R}$ becomes infinite as $R \to 0$ in the same manner as in the Rayleigh problem, with $\dot{R} \sim R^{-3/2}$ but with a smaller proportionality coefficient (part of the energy is converted into heat by dissipation). When $\mathrm{Re} < \mathrm{Re}^*$ (high viscosity) viscosity strongly interferes with the acceleration of the liquid and the bubble collapse takes place slowly, taking an infinite time. The cumulation of energy, characteristic of the Rayleigh problem, is absent here. In the intermediate case, when $\mathrm{Re} = \mathrm{Re}^*$, the bubble does collapse in a finite time. The velocity $\dot{R}$ becomes infinite as $R \to 0$, but less rapidly than R^{-1}.

Numerical integration of the equations gives $Re^* = 8.4$ for the critical Reynolds number. For a given liquid under a given pressure, i.e., for given ρ, ν, and p_0, we can define a critical bubble radius R_0^*. For $R_0 < R_0^*$ the cumulation is completely eliminated by the viscosity. Actually, the critical radius is exceedingly small; for example, in water ($\rho = 1$ g/cm^3, $p_0 = 1$ atm, $\nu = 0.01$ cm^2/sec) $R_0^* = 0.8 \cdot 10^{-4}$ cm. Consequently, viscosity has only a weak effect on the collapse of bubbles whose radius exceeds $0.8 \cdot 10^{-4}$ cm.

3. *The emergence of a shock wave at the surface of a star*

§11. Propagation of a shock wave for a power-law decrease in density

It is well known (see [7], for example) that near the surface of a star the density decreases to zero approximately according to the power law

$$\rho_{00} = bx^{\delta}, \tag{12.28}$$

where x is a coordinate measured from the surface into the star and b and δ are constants. This density distribution is a result of the combined action of gravity and thermal pressure. In the establishment of the distribution of temperature, which is proportional to the gas pressure, radiation heat conduction plays an important role (cf. §14, Chapter II). The exponent δ in the density distribution (12.28) is related to the constants appearing in the equation of radiation heat conduction. It is usually of the order of 3.

When internal disturbances accompanied by an increase in pressure take place in the central regions of a star, a shock wave is formed, which travels from the central regions to the periphery and emerges at the surface. The propagation of a shock wave through a gas whose density is decreasing to zero, as occurs near the surface of a star, is accompanied by the concentration (cumulation) of energy. This process is of great interest in astrophysics and relevant to the problem of the origin of cosmic rays (see following section).

There is a certain physical similarity between the cumulation processes in the propagation of a shock wave through a gas whose density decreases to zero, and in the implosion of a shock wave. In both cases energy is imparted to a mass of material that is ever decreasing without limit, in such a manner that the specific energy (per unit mass) increases indefinitely. The difference between the two cases lies in the cause of the decrease in the mass to which the energy is imparted. In the first case the mass decreases as a result of a decrease of gas density, and in the second case as a result of decrease of volume.

We shall be interested in the limiting form of the motion when the shock front is close to the star surface. Under these conditions we can neglect the curvature of the star surface and of the wave front and we can treat the motion as plane. Since the shock is strong, we may neglect gravitational forces. Radiation heat conduction plays an important role in the establishment of the steady-state distributions of gas temperature and density. Over the short period of passage of the very strong shock wave, it does not introduce appreciable changes as a result of the redistribution of heat, so that we may regard the process as approximately adiabatic. Within this formulation the problem of the limiting form of the motion was first solved by Gandel'man and Frank-Kamenetskii [8]. The same problem was later treated by Sakurai [9], who found exactly the same solution, but for other numerical values of the exponent δ in (12.28) and of the specific heat ratio γ. A schematic representation of the shock wave propagation is given in Fig. 12.5.

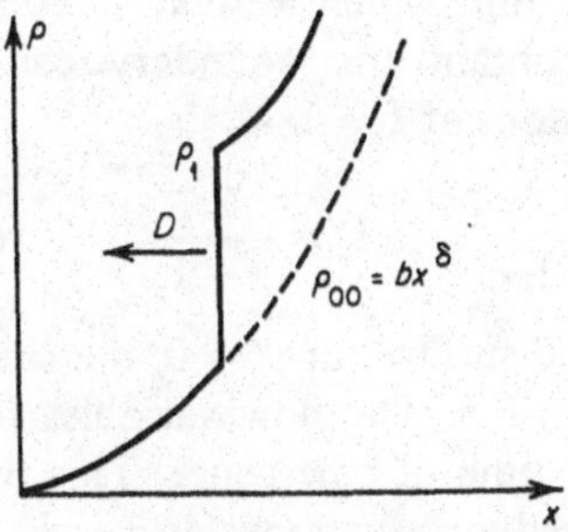

Fig. 12.5. Schematic representation of the emergence of a shock wave at the surface of a star. The density distribution.

The only dimensional parameter for the given conditions of the problem is the constant b, which contains the symbol of mass. There are no other dimensional parameters. It is therefore natural to seek a self-similar solution of the second type. We represent the solution in the form (12.3), (12.5)–(12.7). In accordance with the planar symmetry we denote the coordinate of the wave measured from the star surface $x = 0$ by $X(t)$. As the density scale ρ_0 we take the value of the density of the undisturbed gas ahead of the shock front. Since the wave travels through a gas of variable density, this scale depends on time or, equivalently, on the coordinate of the front X (see the end of §2). The scale ρ_0 is given by

$$\rho_0 = \rho_{00}(X) = bX^{\delta}. \tag{12.29}$$

As in the problem of the imploding shock wave, we take $t = 0$ to be the instant at which the shock wave emerges at the surface, in accordance with which we change the sign of t in the similarity relation

$$X = At^{\alpha} \rightarrow X = A(-t)^{\alpha}.$$

Thus, we seek a solution of the form

$$\rho = \rho_0 g(\xi), \qquad p = \rho_0 \dot{X}^2 \pi(\xi), \qquad u = \dot{X} v(\xi),$$
$$\xi = \frac{x}{X}, \qquad \rho_0 = bX^\delta, \qquad X = A(-t)^\alpha. \tag{12.30}$$

Equations (12.4) in this case ($\nu = 1$) become

$$\begin{gathered} \delta + (v - \xi)(\ln g)' + v' = 0, \\ (\alpha - 1)\alpha^{-1} v + (v - \xi)v' + \frac{\pi'}{g} = 0, \\ (v - \xi)(\ln \pi g^{-\gamma})' + \lambda = 0, \\ \lambda = 2(\alpha - 1)\alpha^{-1} - (\gamma - 1)\delta. \end{gathered} \tag{12.31}$$

The boundary conditions at the shock front, which we take to be strong, are given by (12.16). From these we obtain the boundary conditions analogous to (12.17) for the reduced functions at $\xi = 1$

$$g(1) = \frac{\gamma + 1}{\gamma - 1}, \qquad v(1) = \frac{2}{\gamma + 1}, \qquad \pi(1) = \frac{2}{\gamma + 1}. \tag{12.32}$$

At the time the shock wave emerges at the surface, at $X = 0$, the similarity coordinate $\xi = \infty$ for any nonzero value of x. The flow variables for any finite value of x must be bounded at the time of emergence. This imposes an additional boundary condition on the reduced functions at $\xi = \infty$.

The solution is found in a manner which is completely analogous to the solution of the problem of the implosion of a shock wave. We introduce new reduced functions V, G, and Z, and obtain a system corresponding to (12.15). The system reduces to a single first-order ordinary differential equation in V and Z and to two quadratures. Actually, instead of the two quadratures we have one quadrature and one algebraic relation between the variables, i.e., that from the adiabatic integral. The eigenvalue of the system of equations, the exponent α, is found by a trial and error method in which we numerically integrate the equation for $Z(V)$, and must satisfy the condition that the integral curve pass through the correct singular point. As before, the singular point has a corresponding ξ_0-line on the x, t plane, which is a C_- characteristic and which bounds the region of influence for the motion of the shock front.

In [8] the similarity exponent for $\delta = 13/4 = 3.25$ and $\gamma = 5/3$ was found to be $\alpha = 0.590$. In [9] the exponents α were determined for a number of other values of δ and γ. These results are given in Table 12.1.

The fact that α is always less than one shows that the shock wave is continuously accelerated

$$X \sim |t|^\alpha, \qquad |\dot{X}| \sim |t|^{-(1-\alpha)} \sim X^{-(1-\alpha)/\alpha}, \qquad |\dot{X}| \to \infty \quad \text{as} \quad X \to 0.$$

Correspondingly, the temperature behind the front, which is proportional to the square of the front velocity or to the square of the speed of sound, $T \sim |\dot{X}|^2 \sim X^{-2(1-\alpha)/\alpha}$, also increases without limit. The unbounded increase

Table 12.1

SIMILARITY EXPONENT α

γ	δ = 3.25	2	1	0.5
5/3	0.590	0.696	0.816	0.877
7/5	—	0.718	0.831	0.906
6/5	—	0.752	0.855	0.920

in temperature, as pointed out above, appears because a finite amount of energy is imparted to a mass of gas which decreases to zero. The pressure behind the shock front decreases as the front approaches the surface, despite the increase in the velocity, since the density ahead of the front decreases faster than the temperature (or square of the velocity) increases

$$p_1 \sim \rho_0 \dot{X}^2 \sim X^{\delta - 2(1-\alpha)/\alpha}.$$

It can be easily checked from the results of Table 12.1 that the exponent of X in this equation is always positive, that

$$p_1 \to 0 \quad \text{as} \quad X \to 0.$$

The limiting distributions of the flow variables with respect to the x coordinate at the instant of emergence of the shock wave at the surface $t = 0$, $X = 0$ ($t = 0$, $x \neq 0$ corresponds to $\xi = \infty$), are evidently of exactly the same form as the relations at the shock front. As in the problem of an imploding shock wave, these distributions follow simply from dimensional considerations. At the time $t = 0$ we get

$$u \sim x^{-(1-\alpha)/\alpha}, \qquad T \sim u^2 \sim c^2 \sim x^{-2(1-\alpha)/\alpha},$$

$$\rho \sim x^{\delta}, \qquad p \sim x^{\delta - 2(1-\alpha)/\alpha}.$$

Of course, the same relations follow from the equations in the limit $\xi \to \infty$. The final density distribution is increased by a constant factor with respect to the initial density distribution. The distributions of the flow variables with respect to the x coordinate before emergence and at the instant of emergence of the wave at the surface are shown schematically in Fig. 12.6.

The energy of the gas at $t = 0$ contained in a layer between $x = 0$ and x in a column of unit cross-sectional area is proportional to the quantity

$$\int_0^x \rho u^2 \, dx \sim \int_0^x p \, dx \sim x^{\delta+1-2(1-\alpha)/\alpha}.$$

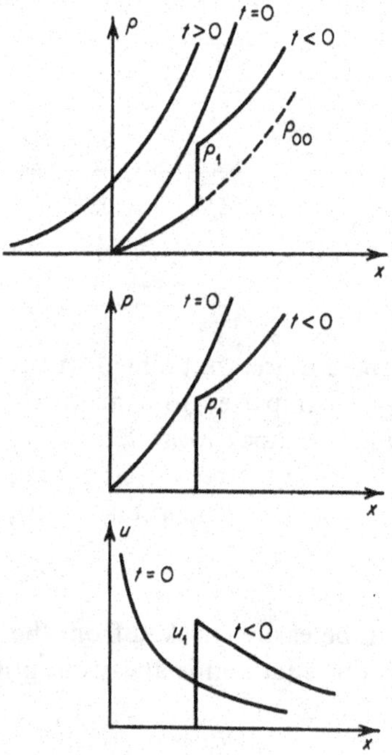

Fig. 12.6. Density, pressure, and velocity distributions for the emergence of a shock wave at the surface of a star. $t < 0$ before emergence, $t = 0$ at the instant of emergence, $t > 0$ after emergence.

As $x \to \infty$ the energy becomes infinite; there is no energy integral. The energy of a layer of finite thickness remains finite and tends to zero, as $x \to 0$. Unlike the case of an imploding shock wave, the energy density, which is proportional to the pressure, also goes to zero at the boundary, as $x \to 0$. Only the temperature or energy per unit mass becomes infinite. An infinite specific energy is imparted to a vanishingly small mass of gas. Of course, the temperature cannot actually become infinite as indicated by the mathematical solution. Thus, for example, when the shock wave comes so close to the surface that the small remaining mass of the layer from $x = 0$ to $x = X$ includes only a small number of gaskinetic mean free paths, gasdynamic considerations are no longer meaningful. The infinite temperature increase can also be limited by physical factors, such as energy lost by radiation from the highly heated gas.

As in the problem of the implosion of a shock wave, the self-similar solution is valid only in a limited region near the boundary $x = 0$. Far from the front the solution is not self-similar and depends on the conditions under which the shock originated. If at a given time the actual solution is very close to the self-similar solution for $1 < \xi < \xi_1 > \xi_0$, it will remain so within some finite distance of the boundary through the instant of emergence.

After the shock wave emerges at the surface, the gas flows into a vacuum, and the initial density, pressure, and velocity distributions are given by the power laws for $t = 0$. As shown in [9], the solution for the outflow stage is also self-similar, but, of course, has a completely different character (the flow is continuous and there are no shock waves). An approximate density distribution for a time $t > 0$ is shown in Fig. 12.6.

§12. On explosions of supernovae and the origin of cosmic rays

It has been suggested that the origin of cosmic rays, of the protons and nuclei with tremendously high energies that are present throughout the universe and that strike the earth, is connected with explosions of supernovae. Such a theory was developed by Ginzburg and I. S. Shklovskii (see the review [10]). The process of the infinite increase in shock strength and of the cumulation of energy in the emergence of a shock wave at the surface of a star from the interior may be the cause of the acceleration of the particles to their tremendously high energies. This idea was used by Colgate and Johnson [11], who considered such a process in detail. They showed by calculations that some of the material ejected from the surface during the explosion of a supernova acquires relativistic velocities and kinetic energies, corresponding to the energies of cosmic rays. (The highest energies of particles presently observed in the cosmic ray spectrum are of the order of 10^8 Bev $= 10^{17}$ ev.) Below we shall present the results obtained by Colgate and Johnson.

Temperatures at the center of supernovae reach the order of 300–500 kev ($\sim 5 \cdot 10^9$ °K). At these temperatures nuclear fusion proceeds up to the formation of the most stable element, iron. The layers further out consist of the lighter elements, carbon, nitrogen, and oxygen. Still closer to the surface the main element is helium, and, finally, the outermost layers consist of hydrogen. Astronomical data show that in the explosion of a supernova a mass of material is ejected that is of the order of one tenth of the entire mass of the star and of the order of the mass of the sun, equal to $M_\odot = 2 \cdot 10^{33}$ g.

Calculations of the mechanical and radiative equilibrium for a star with a mass of $10 M_\odot$ give a behavior for the density and temperature distribution as a function of radius as shown in Fig. 12.7*. The density at the center of the

* Under conditions of radiative equilibrium the density dependence on temperature follows the relation $\rho \sim T^{13/4} = T^{3.25}$. This was the basis for the assumption made in [8] that the density distribution near the surface is given by $\rho \sim x^{3.25}$, although in a layer near the surface the temperature depends only weakly on the coordinate x (the temperature at

star is higher than 10^8 g/cm^3, while at the surface it drops to zero. In any case, propagation of an ordinary shock wave is observable out to layers with $\rho \sim 10^{-5}$ g/cm^3.

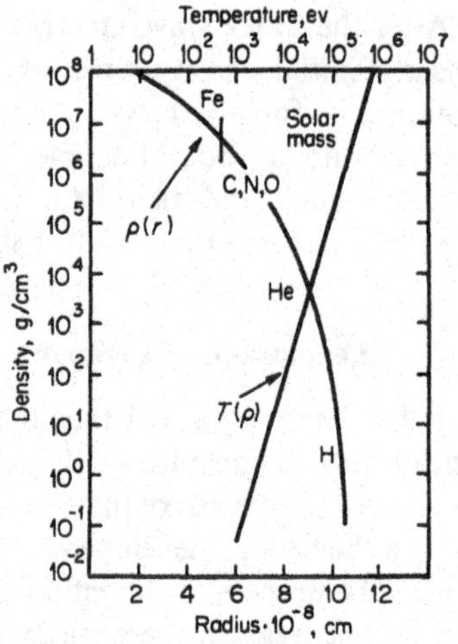

Fig. 12.7. Density and temperature distributions before a star explosion. $\rho \sim T^{3.25}$, corresponding to radiative equilibrium.

It is usually assumed that the energy source for the shock wave is the so-called gravitational instability, which occurs when the isentropic exponent (effective specific heat ratio) in the isentropic equation of state $\gamma < 4/3$. In the central regions of the star, at temperatures ~ 500 kev, the nuclei are highly dissociated. It is well known that the specific heat of a gas markedly increases and the isentropic exponent decreases as a result of dissociation. Small disturbances are amplified as a result of the gravitational instability. The pressure pulse generated grows in strength, and this leads to the formation of a shock wave. The shock moves out from the central region to the surface. The gas behind the shock wave suddenly expands out from the center, and owing to the increase in wave strength the outer layers obtain extremely high velocities.

The material in the peripheral layers, which has acquired the large kinetic energy of the sudden expansion, overcomes the gravitational forces and breaks away from the star after the shock wave emerges at the surface. The star, as it were, sheds a shell. This phenomenon is well known in astrophysics. It is assumed that the Crab nebula was formed in this manner. It has been estimated that an amount of energy of the order of 10^{52} ergs is required to

---

the surface of the star is not equal to zero). On Fig. 12.7 is given the radius of the layer whose mass is equal to the mass of the sun. It must be assumed that this layer is also ejected during explosion. The regions containing the different elements are indicated approximately.

overcome the forces of gravity when a mass equal to the mass of the sun is ejected. Consequently, this is the order of magnitude of the energy which is liberated at the center of the star and goes into the formation of the shock wave.

Hydrodynamic calculations of the propagation of a shock generated by such a source give velocities behind the shock front shown by curve I in Fig. 12.8. The abscissa is the initial density of the material ahead of the front.

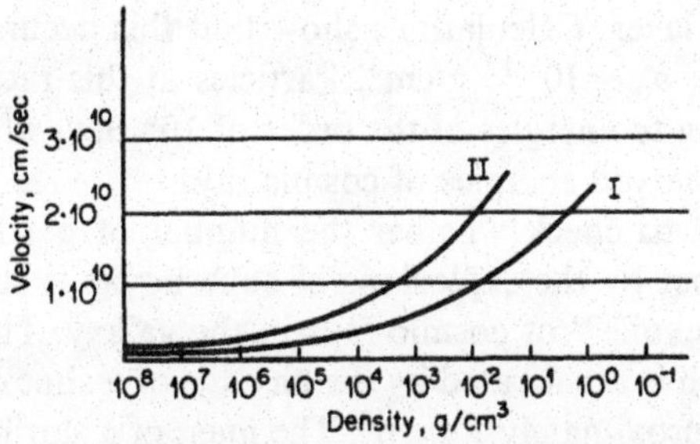

Fig. 12.8. The velocity of the material as a function of its initial density. Curve I is the velocity immediately behind the wave front; curve II is the velocity after expansion.

Curve II indicates the velocity acquired by a layer with the given density after the shock wave emerges at the surface and the material expands. The velocity after expansion is larger approximately by a factor of 2 than the velocity at the time of passage of the shock front. Figure 12.8 shows that the peripheral layers, where the density is less than approximately 30 g/cm^3, acquire velocities behind the strengthened shock wave greater than 10^{10} cm/sec, which is 1/3 the speed of light. Therefore, relativistic effects must be considered in calculating the motion of the shock wave in these peripheral layers. In [11] a numerical calculation was carried out on the basis of relativistic gasdynamics. An approximate analytic solution to the problem was also given, based on the use of characteristic equations and relativistic analogs of the Riemann invariants.

It is interesting to note that the internal energy behind the front of such a powerful shock wave is almost entirely concentrated in equilibrium thermal radiation. An approximate solution shows that the final kinetic energy per unit mass acquired by the material in a layer with an initial density ρ_0 g/cm^3 is of the order of $c^2(30/\rho_0)^{0.64}$ erg/g. If we note that 1 erg/g in hydrogen corresponds to approximately 10^{-12} ev/proton $= 10^{-21}$ Bev/proton, then we find that a kinetic energy of the order of 10^4 Bev is acquired by particles previously contained in a layer with an initial density $\rho_0 \sim 10^{-5}$ g/cm^3. The mass per unit surface area of a layer in a star which surrounds a spherical surface with such an initial density is approximately 1 g/cm^2. Such a thin layer is no longer capable of holding back or "locking in" the thermal radiation, which is out of equilibrium in the outer layers closer to the surface. Therefore, the shock wave can no longer propagate through these outer layers in the same manner as under equilibrium conditions.

As pointed out in [11], the subsequent propagation of the shock wave through a gas of even lower density is connected with the mechanism of plasma oscillations in an essential manner. The shock wave reaches a surface where the Debye length becomes comparable with the length scale of the outer unshocked layer. Calculations show that this occurs at a radius where the initial density $\rho_0 \sim 10^{-12}$ g/cm^3. Particles at this radius are accelerated by the shock wave to energies of the order of 10^8 Bev, which corresponds to the maximum observed energies of cosmic rays.

It is important to check whether the number of particles accelerated to cosmic ray energies by the explosions of supernovae is sufficient to produce the existing "stockpile" of cosmic rays in the galaxy. The initial density of the material which is accelerated by the passage of a shock wave to an energy of ~ 10 Bev is approximately 1 g/cm^3. The mass of a star in a layer surrounding a spherical surface with $\rho_0 \sim 1$ g/cm^3 is of the order of 10^{26} g or $6 \cdot 10^{49}$ protons. We can say that the energy imparted to $6 \cdot 10^{49}$ protons by an explosion exceeds 10 Bev. The lifetime of a high energy proton in the galaxy, with an average particle density of matter in the galaxy of the order of 0.1 particles/cm^3, is $\tau \sim 5 \cdot 10^8$ years. This means that $\sim 5 \cdot 10^8$ years after the "start" of explosions in the galaxy a steady-state number of protons N will be set up. Supernova explosions occur approximately once every 100 years. Consequently, $6 \cdot 10^{49}/100 = 6 \cdot 10^{47}$ protons are born per year, and N/τ protons "die" per year. It follows from the steady-state condition that $N/\tau = 6 \cdot 10^{47}$ protons/year $= const$, that $N = 3 \cdot 10^{56}$. The volume of our galaxy is $V \sim 5 \cdot 10^{68}$ cm^3. The average density of high energy protons is $N/V \sim 6 \cdot 10^{-13}$ cm^{-3}, and their flux is of the order of $Nc/V \sim 2 \cdot 10^{-2}$ cm$^{-2} \cdot$ sec^{-1}. This value is in agreement with observations. According to the calculations given, of the order of $5 \cdot 10^6$ supernova explosions were required to produce the cosmic rays in our galaxy.

4. *Motion of a gas under the action of an impulsive load*

§13. Statement of the problem and general character of the motion

Let us imagine a half-space $x > 0$ occupied by a perfect gas with constant specific heats. Initially, at $t = 0$, the density of the gas is everywhere uniform and equal to ρ_0, and the pressure, temperature, and initial speed of sound are zero. The half-space $x < 0$ is empty. The surface $x = 0$ is the boundary between the gas and the vacuum.

Let a pressure pulse of short duration* be applied to the external surface of the gas (the gas surface is subjected to an impulsive load). Various methods are possible for producing an impulsive load in practice.

(1) In a short time interval τ a plane piston is pushed into the gas with a constant velocity U_1, creating a pressure Π_1 in the gas. To within a numerical coefficient of the order of unity (depending on the specific heat ratio γ) $\Pi_1 \approx \rho_0 U_1^2$. The velocity of the shock D that is created by the action of the piston is close to U_1. After a time interval τ the piston is "instantaneously" withdrawn (the pressure pulse is shown in Fig. 12.9a).

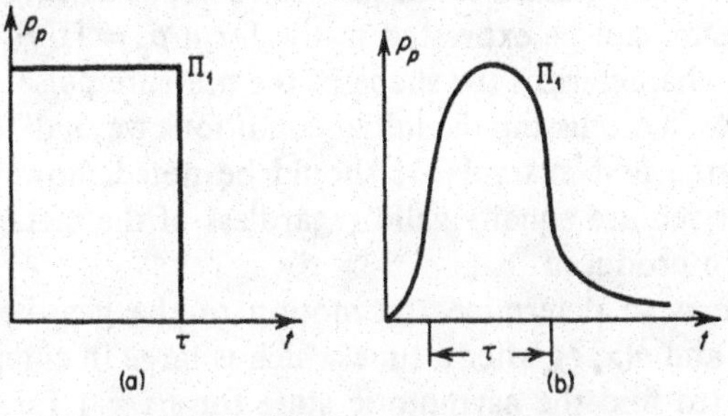

Fig. 12.9. Shapes of initial pressure pulses.

(2) A thin layer of explosive is detonated on the gas surface. If the mass thickness of the layer is m units of mass per unit area and the energy released per unit mass is Q, then the energy released by the explosion per unit area is $E = mQ$. The explosion products suddenly expand with a velocity $U_1 \approx \sqrt{Q}$. Since the products expand suddenly in both directions and since before the detonation the gas was everywhere at rest, the total momentum is equal to zero. However, the momentum of the detonation products moving in one direction is, in order of magnitude, equal to $I \approx mU_1 \approx m\sqrt{Q}$ (per unit surface area). The detonation products generate a shock wave in the gas with a pressure of the order of $\Pi_1 \approx \rho_0 U_1^2$. The time τ over which the pressure acts is determined from the condition that in the time τ the energy and momentum are transferred from the detonation products to the gas,

$$\tau \approx \frac{E}{\Pi_1 U_1} \approx \frac{I}{\Pi_1} \approx \frac{m}{\rho_0 \sqrt{Q}}.$$

During this time the shock wave in the gas will travel through a distance $\sim U_1 \tau \sim (Q\tau)^{1/2}$ and will encompass a mass $\sim \rho_0 (Q\tau)^{1/2} \sim m$, a mass of the order of the mass of the explosive.

* *Editors' note.* The authors' literal term "of short duration" has generally been translated as "impulsive". The term whose literal translation is "blow" is here rendered as "load", or sometimes as "impact". Later in the chapter the term "concentrated" appears, and means concentrated in a spatial sense rather than in a temporal sense.

(3) A thin plate with a small mass m per unit area is made to strike the gas surface with a velocity U_1. The impact of the plate creates a shock wave in the gas, which propagates with the velocity $D \approx U_1$. The pressure in the gas will then be $\Pi_1 \approx \rho_0 U_1^2$. The initial momentum and energy of the plate, $I = mU_1$ and $E = mU_1^2/2$, are transferred to the gas during the time τ in which the plate is decelerated, which is of the order of $\tau \approx E/\Pi_1 U_1 \approx I/\Pi_1 \approx m/\rho_0 U_1$. During this time the shock wave in the gas travels through a distance $U_1\tau$ and encompasses a mass $\rho_0 U_1 \tau \approx m$.

Thus, we shall assume in general that, as shown in Fig. 12.9b, there is a pressure acting on the surface of the gas which drops sufficiently rapidly with time. The pressure can be expressed in the form $p_p = \Pi_1 f(t/\tau)$, where f is a function which characterizes the shape of the pressure pulse. For definiteness and convenience in discussing the initial conditions we shall use the "piston" concept, as in the first example. It should be noted, however, that all the conclusions derived are equally valid regardless of the method by which the impact has been produced.

The problem is to determine the motion of the gas, i.e., the functions $p(x, t)$, $\rho(x, t)$, and $u(x, t)$, after a time which is large in comparison with the impact time τ (to find the asymptotic state for $t/\tau \gg 1$ for a given applied external pressure history). The problem can also be formulated in a slightly different manner. Preserving the shape of the curve $f(t/\tau)$ we let the time τ go to zero, and the pressure Π_1 become infinite, and seek, for finite times, the limiting solutions of the resulting gasdynamic equations. The solution to this problem should, in particular, answer the question of how the pressure Π_1 must increase as $\tau \to 0$, in order to ensure that the pressure in the gas be finite after a finite time t. For example, if the solution contains the combination $\Pi_1 \tau^\beta$, this means that as $\tau \to 0$, Π_1 must increase as $\tau^{-\beta}$

The above problem was formulated and analyzed in a paper by one of the present authors [12] in which the physical features of the resulting motion and of the mathematical solution were explained. The equations were analyzed and integrated numerically by Adamskii [13]; Zhukov and Kazhdan [14], Häfele [15], and von Hoerner [16] found an analytic solution for one particular case ($\gamma = 7/5$). The last two papers are extensions of the work of von Weizsäcker [17], who posed the problem concerning the limits within which the similarity exponent varies for plane motions. It should be noted that in [15–17] the physical meaning was not given for the solution, which was obtained by purely formal means.

The general character of the motion resulting from the action of an impulsive load is illustrated in Fig. 12.10. A shock is propagated through the undisturbed gas with the density ratio across the shock attaining the limiting value $K = (\gamma + 1)/(\gamma - 1)$. Behind the shock the gas expands unhindered into the vacuum; at the vacuum interface the density and pressure drop to zero.

The pressure, density, and velocity decrease with distance behind the shock front. At some point the velocity changes sign, since directly behind the front the gas moves to the right, while at the interface it expands into the vacuum to the left. The strength of the shock decreases with time.

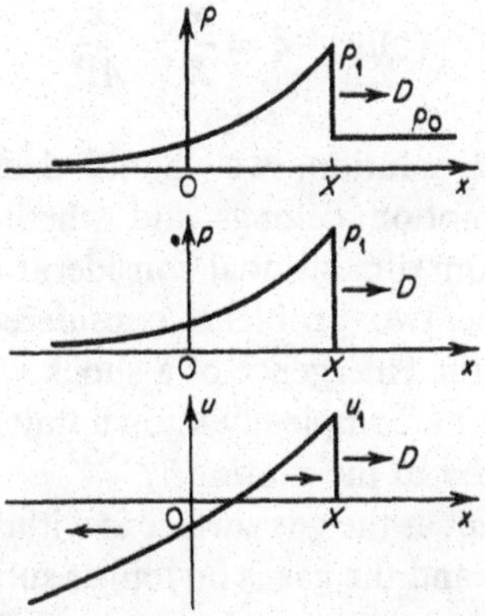

Fig. 12.10. Density, pressure, and velocity distributions for the problem of an impulsive load.

The solution to the problem of an instantaneous pressure pulse should answer the question as to the maximum possible rate of decrease of the strength of a plane shock wave traveling into a gas with a constant initial density. It is clear that if the pressure is applied over a longer time, then the only effect will be to sustain the shock and delay its attenuation. The behavior of the limiting solution is independent of the specific shape of the pressure pulse, i.e., of the form of the function $f(t/\tau)$, as long as it drops sufficiently fast. It was noted above that, under the action of the pressure Π_1, the gas acquires during impact a velocity $U_1 \sim (\Pi_1/\rho_0)^{1/2}$. The gas interface expands into vacuum with a velocity of the same order. In the limit $\tau \to 0$, $\Pi_1 \to \infty$, the velocity of the interface becomes infinite, so that the distributions of p, ρ, and u in the limiting solution, shown in Fig. 12.10, extend to the left to $x = -\infty$.

§14. Self-similar solutions and the energy and momentum conservation laws

The motion which results immediately after applying a pressure pulse is, of course, not self-similar. It is characterized by the time scale τ and the length scale $x_0 = (\Pi_1/\rho_0)^{1/2}\tau$, and depends on the shape of the applied pressure curve $f(t/\tau)$. However, after a sufficiently long time $t \gg \tau$, when the shock front has moved through a distance $X \gg x_0$, the initial scales τ and x_0 are very small in comparison with the natural scales of the motion t and X, and no longer characterize the process. The limiting motion, corresponding to the stage when $t \gg \tau$ and $X \gg x_0$ or, equivalently, to the limit $\tau \to 0$, will be self-similar. The only length scale in this motion is the shock coordinate X, which is variable, while the velocity scale is the front velocity $\dot{X}$. Therefore, a solution should be sought in the self-similar form

$$\rho = \rho_0 g(\xi), \qquad u = \dot{X}v(\xi), \qquad p = \rho_0 \dot{X}^2 \pi(\xi), \qquad \xi = \frac{x}{X} = \frac{x}{At^\alpha}. \quad (12.33)$$

Before proceeding to the mathematical solution, we should decide to which of the two types the self-similar motion belongs and whether the similarity exponent α can be determined from dimensional considerations or from conservation laws. In contrast to the two problems considered previously, the implosion of a shock wave and the emergence of a shock wave at the surface of a star, here at any given time t a completely known finite mass of gas $\rho_0 X$ (per unit surface area) is involved in the motion.

After the piston which produced the impact at the gas surface is withdrawn, the pressure at the vacuum interface is zero, and the gas is no longer subjected to any external forces. Thenceforth the momentum and energy of the gas must be preserved. The momentum of the gas is equal to the momentum produced by the piston pressure

$$I = \int_0^\infty p_p \, dt = \Pi_1 \tau \int_0^\infty f\left(\frac{t}{\tau}\right) d\left(\frac{t}{\tau}\right).$$

To within a numerical factor, this quantity is equal to $\Pi_1 \tau$. The energy of the gas is equal to the work done by the piston over the time during which the pressure acts. In order to calculate this work exactly we would require the full solution of the gasdynamic equations over the period during which the piston acts, since the work is equal to $\int_0^\infty p_p u_p \, dt$, where $u_p(t)$ is the piston speed. The piston speed is unknown in advance if it is the pressure $p_p(t)$ that has been specified. However, to within a numerical factor that depends on the form of the function $f(t/\tau)$, this work is equal to

$$E \approx \Pi_1 U_1 \tau \approx \Pi_1 \left(\frac{\Pi_1}{\rho_0}\right)^{1/2} \tau = \Pi_1^{3/2} \tau \rho_0^{-1/2}$$

(U_1 is the piston velocity scale).

If we substitute into the integral expressions for the momentum and energy of the entire gas the pressure, velocity, and density in the self-similar form (12.33), and take into account the fact that the integrals must vanish in the cold undisturbed region $X < x < \infty$, then the momentum and energy conservation laws (per unit surface area) can be written in the forms

$$I = \int_{-\infty}^{\infty} \rho u \, dx = \rho_0 \dot{X} X \int_{-\infty}^{1} g v \, d\xi = const, \tag{12.34}$$

$$E = \int_{-\infty}^{\infty} \left(\rho \frac{u^2}{2} + \frac{1}{\gamma - 1} p \right) dx = \rho_0 \dot{X}^2 X \int_{-\infty}^{1} \left(\frac{g v^2}{2} + \frac{1}{\gamma - 1} \pi \right) d\xi = const. \tag{12.35}$$

It would seem natural to assume that the dimensionless integrals are constants. Then, each of the two conditions taken separately could serve to determine the similarity exponent α. The conservation of momentum condition would give $\dot{X}X = const$, whence $X \sim t^{1/2}$ and $\alpha = \frac{1}{2}$. On the other hand, the conservation of energy condition gives $\dot{X}^2 X = const$, whence $X \sim t^{2/3}$ and $\alpha = 2/3$. Taken together these conditions contradict each other, since they lead to different values of the exponent α. A paradoxical situation arises in which the conservation laws of momentum and energy, which are the basis of the gasdynamic equations, cannot be satisfied simultaneously. This would seem to indicate that the problem does not have a self-similar solution, since the substitution of such a solution into the conservation laws leads to a contradiction. This contradiction has, however, another resolution. The point is that the self-similar solution, which does exist and which will be found below, is of the second type. The similarity exponent α is found neither from the conservation laws nor from dimensional considerations, but by solving the equations for the reduced functions under the condition that the correct solution pass through a singular point, as in the problems considered in preceding sections.

In order to resolve the above paradox, we note that when the specific heat ratio $\gamma = 7/5$, the similarity exponent is found by means of an analytic solution to be $\alpha = 3/5$*. This exponent lies between the values of α dictated by the conditions of conservation of momentum and energy, $1/2 < 3/5 < 2/3$. It will be shown below that for any value of the specific heat ratio $1 < \gamma < \infty$, the similarity exponent α lies between these limits, $1/2 < \alpha < 2/3$.

In the case where the similarity exponent $\alpha = 3/5$ the dimensions of the parameter A in the relation $X = At^{\alpha}$ are $[A] = LT^{-3/5}$. As we have seen already (see §5), the limiting self-similar motion does not completely "forget" the initial conditions, but from the extensive information included in the initial conditions it "selects" and "remembers" one specific constant A,

* In general, for an arbitrary value of γ the exponent α is not expressible as a rational fraction. However, fortunately, for $\gamma = 7/5$ a solution of the self-similar equations can be found in analytic form, and in this case α is equal to 3/5 (see below).

which somehow characterizes the initial "push". In this case, the limiting solution "selects" one parameter A from the information given by the pressure relation at the piston $p_p = \Pi_1 f(t/\tau)$ and the value of the initial density ρ_0. In order of magnitude A is given by the following combination of characteristic scales:

$$A \approx \left(\frac{\Pi_1}{\rho_0}\right)^{1/2} \tau^{1-\alpha} = \left(\frac{\Pi_1}{\rho_0}\right)^{1/2} \tau^{2/5} \quad (LT^{-3/5}). \tag{12.36}$$

The numerical value of the proportionality coefficient is determined by the form of the pressure relation $f(t/\tau)$.

From the above relation it is easy to determine the manner in which the pressure at the piston Π_1 must become infinite as τ tends to zero in order to ensure a finite (not equal to 0 or ∞) pressure at a finite distance in the limiting motion. In order for the limiting solution to exist, the parameter A must have a finite value, and the product $\Pi_1^{1/2}\tau^{1-\alpha}$, equal to $\Pi_1^{1/2}\tau^{2/5}$ in the case $\gamma = 7/5$, must remain finite as $\tau \to 0$. Therefore, as $\tau \to 0$, Π_1 must increase as $\Pi_1 \sim \tau^{-2(1-\alpha)} \sim \tau^{4/5}$.

We can now clarify the question of the satisfaction of the conservation laws. The momentum imparted by the piston to the gas, or the impulse of the load, is in order of magnitude $I \sim \Pi_1 \tau$, or in proportional form $I \sim \Pi_1 \tau \sim \tau^{2\alpha - 1} \sim \tau^{1/5}$. As $\tau \to 0$, the momentum $I \to 0$. Therefore, the total momentum in the limiting, self-similar motion is zero (the momentum of the gas which is moving with the shock wave to the right is exactly canceled by the momentum of the gas expanding into vacuum to the left; see Fig. 12.10). The momentum conservation law is written in the form

$$I = \rho_0 \dot{X} X \int_{-\infty}^{1} gv\, d\xi \sim t^{1/5} \int_{-\infty}^{1} gv\, d\xi = 0.$$

The only conclusion which follows is the fact that the reduced functions must satisfy the condition $\int_{-\infty}^{1} gv\, d\xi = 0$. It is obvious that we cannot take $\dot{X}X$ to be constant and determine the similarity exponent α in this way.

The energy imparted by the piston to the gas is in order of magnitude $E \sim \Pi_1^{3/2} \tau \rho_0^{-1/2}$, or in proportional form $E \sim \Pi_1^{3/2} \tau \sim \tau^{3\alpha - 2} \sim \tau^{-1/5}$. As $\tau \to 0$, $E \to \infty$. The total energy of the gas in the self-similar motion turns out to be infinite. Conservation of energy

$$E = \rho_0 \dot{X}^2 X \int_{-\infty}^{1} \left(\frac{gv^2}{2} + \frac{1}{\gamma - 1}\pi\right) d\xi \sim t^{-1/5} \int_{-\infty}^{1} \left(\frac{gv^2}{2} + \frac{1}{\gamma - 1}\pi\right) d\xi = \infty$$

shows only that the integral of the dimensionless reduced functions diverges, but says nothing about the value of $\dot{X}^2 X$. The similarity exponent cannot be determined from conservation of energy, either. The fact that the energy is infinite and that the energy integral diverges is connected with the fact that in the exactly self-similar motion that corresponds to the limit $\tau \to 0$ the velocity at which the gas boundary expands into vacuum is infinite (see end of §13). The kinetic energy at the boundary is also infinite since, as $\xi \to -\infty$, the square of the reduced velocity v^2 tends to infinity faster than the reduced density g decreases.

The physical meaning of an infinite energy in the self-similar motion will be discussed below. Here we note only that the actual energy of the gas is of course finite and equal to the work done by the piston. It is simply that the self-similar solution is not valid for a small amount of mass near the gas boundary, and this leads to the divergence of the energy integral.

§15. Solution of the equations

The general method of finding the self-similar solution for the problem of an impulsive load does not differ in principle from the method of solving the problems of an imploding shock wave or the propagation of a shock through a gas whose density decreases with distance following a power law (see §§2 and 3 of this chapter). As before, we shall seek a solution of the gasdynamic equations (12.1) in the self-similar form (12.33) and obtain a system of ordinary differential equations for the reduced functions π, v, and g. These equations become identical with (12.31) when we set δ equal to zero, corresponding to a constant density scale,

$$\begin{gathered}(v-\xi)(\ln g)' + v' = 0,\\ (\alpha-1)\alpha^{-1}v + (v-\xi)v' + g^{-1}\pi' = 0,\\ (v-\xi)(\ln \pi g^{-\gamma})' + 2(\alpha-1)\alpha^{-1} = 0.\end{gathered} \tag{12.37}$$

The boundary conditions (12.32) at the shock front where $\xi = 1$ were given in §11. At the gas-vacuum interface the pressure and density vanish and minus the velocity becomes infinite, so that for $\xi = -\infty$, $\pi(-\infty) = 0$, $g(-\infty) = 0$, $v(-\infty) = -\infty$.

After a number of transformations, the equations again reduce to one first-order ordinary differential equation, one quadrature, and one algebraic relation between all the variables (the adiabatic integral). The similarity exponent is determined by the condition that the desired solution of the differential equation pass through a singular point.

Actually, in [13, 14] the equations were written and solved in Lagrangian rather than in Eulerian coordinates. In the one-dimensional plane case with constant initial density the Lagrangian form leads to simpler and more convenient relations. Of course, the transition from Eulerian to Lagrangian coordinates introduces nothing new in principle. The Lagrangian coordinate is defined as the mass of gas per unit surface area, measured from the vacuum interface,

$$m = \int_{-\infty}^{x} \rho\, dx, \qquad dm = \rho\, dx. \tag{12.38}$$

In place of time in the scale functions it is convenient to introduce the Lagrangian coordinate of the shock front $M = \rho_0 X$, equal to the mass of gas per unit surface area that has been encompassed by the motion at the time t. The Lagrangian similarity variable is given by the ratio

$$\eta = \frac{m}{M}, \tag{12.39}$$

which varies from $\eta = 0$ (at the gas-vacuum interface) to $\eta = 1$ (at the shock front).

The solution is then written in the form

$$p = B\rho_0 M^{-n} f(\eta), \qquad u = \sqrt{B} M^{-n/2} w(\eta), \qquad \rho = \rho_0 q(\eta), \tag{12.40}$$

where B is a parameter of the problem that is related to the parameter A in the relation $X = At^{\alpha}$ and that replaces it in the new formulation. The new reduced functions are now f, w, and q. The new similarity exponent n is uniquely related to the old one α. In fact,

$$M = \rho_0 X \sim t^{\alpha}, \qquad u \sim M^{-n/2} \sim t^{-\alpha n/2}, \qquad u \sim \dot{X} \sim t^{\alpha - 1},$$

whence $\alpha n/2 = \alpha - 1$ and

$$n = \frac{2(1-\alpha)}{\alpha}, \qquad \alpha = \frac{1}{1 + n/2}. \tag{12.41}$$

The mathematical features of the problem, the sequence of transformations of the equations, the analysis, and the specific methods of solution may be found in [13, 14]. Here, we shall discuss in more detail the results for the particular case of $\gamma = 7/5$, for which an exact analytic solution of the equations can be obtained. All the main features of the process are clarified by considering this analytic solution.

The exponents α and n for $\gamma = 7/5$ have the values $\alpha = 3/5$ and $n = 4/3$. The solution in Lagrangian coordinates takes the form

$$f = \eta, \qquad w = -\tfrac{1}{2}(\tfrac{5}{6})^{1/2}(\eta^{-2/3} - 3), \qquad q = 6\eta^{5/3}. \tag{12.42}$$

The pressure, density, and velocity distributions as a function of the Lagrangian similarity variable are shown in Fig. 12.11. We note that by definition $f = p/p_1$, $w(6/5)^{1/2} = u/u_1$, $q/6 = \rho/\rho_1$, where the subscript "1" denotes quantities behind the shock front.

From the definition of the Lagrangian coordinate (12.38) and of the similarity variable (12.39) the solution (12.42) can be easily transformed to the Eulerian variable $\xi = x/X$. In this regard, with the time t and the Lagrangian shock coordinate M fixed,

$$dm = \rho\, dx, \qquad \frac{dm}{M} = \frac{\rho}{\rho_0}\frac{dx}{X}, \qquad \text{and} \qquad d\eta = q\, d\xi.$$

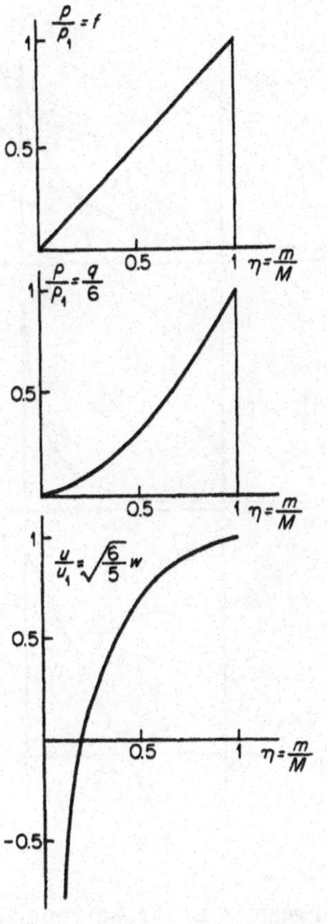

Fig. 12.11. Pressure, density, and velocity distributions in the problem of an impulsive load (in Lagrangian coordinates); $\gamma = 7/5$.

Substituting $q(\eta)$ given by (12.42) into this equation, integrating, and applying the boundary condition $\eta = 1$ at $\xi = 1$ (at the shock front), we obtain

$$\eta = (5 - 4\xi)^{-3/2}, \qquad \xi = \tfrac{1}{4}(5 - \eta^{-2/3}). \tag{12.43}$$

In terms of the Eulerian variable the functions f, w, and q have the form

$$f = (5 - 4\xi)^{-3/2}, \qquad w = -(\tfrac{5}{6})^{1/2}(1 - 2\xi), \qquad q = 6(5 - 4\xi)^{-5/2}. \tag{12.44}$$

The reduced functions f, w, and q are related to the original reduced functions π, v, and g by the relations*

$$\pi = \tfrac{5}{6} f, \qquad v = (\tfrac{5}{6})^{1/2} w, \qquad g = q. \tag{12.45}$$

The pressure, density, and velocity distributions as functions of the Eulerian reduced coordinate are shown in Fig. 12.12. It is interesting to note

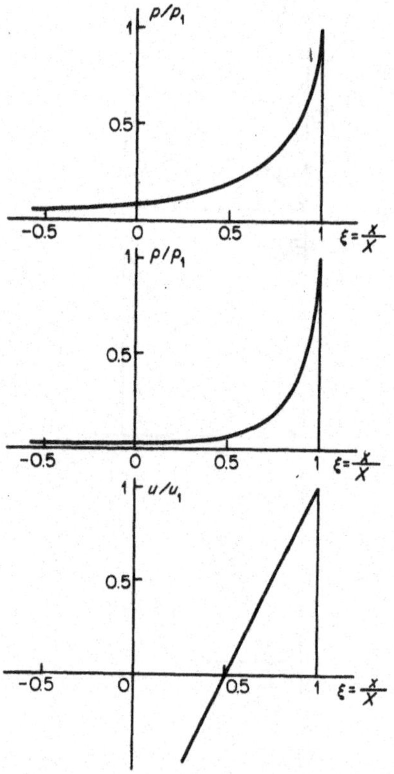

Fig. 12.12. Pressure, density, and velocity distributions in the problem of an impulsive load (in Eulerian coordinates); $\gamma = 7/5$.

that the pressure distribution is linear with respect to the Lagrangian mass coordinate, while the velocity is linear with respect to the spatial coordinate. The velocity becomes zero and changes direction at the point $\xi = \frac{1}{2}$. The mass contained between the initial position of the gas interface $x = 0$ and the wave front at any time constitutes 90 percent of the entire mass set into motion. The remaining 10 percent of the mass is pushed to the left of the initial gas

* We leave it as an exercise for the reader to check, by directly substituting the functions π, v, and g given by (12.45) and (12.44) into (12.37) with $\gamma = 7/5$ and $\alpha = 3/5$, that they indeed satisfy the equations and the boundary conditions (12.32).

interface as a result of the expansion following the shock compression. We note that 78 percent of the mass is moving to the right and 22 percent to the left.

The asymptotic behavior of the solution in the low density region as $\xi \to -\infty$ and $\eta \to 0$ is given by the expressions

$$\begin{aligned} f &\sim (-\xi)^{-3/2}, \quad & w &\sim \xi, \quad & q &\sim (-\xi)^{-5/2}, \\ f &= \eta, \quad & w &\sim \eta^{-2/3}, \quad & q &\sim \eta^{5/3}. \end{aligned} \tag{12.46}$$

The values of the similarity variables corresponding to the singular point through which the solution of the differential equation passes are $\eta_0 = 7^{-3/2} = 0.054$ and $\xi_0 = -\frac{1}{2}$. As in the problem of the imploding shock wave, the ξ_0-line on the x, t plane (the η_0-line on the m, t or the m, M plane) is the characteristic ($dx/dt = u + c$; $dm/dt = \rho c$) that separates the region of influence of the shock from the rest of the space. On the x, t, and m, M diagrams of Figs. 12.13 and 12.14 we have plotted the shock line $\xi = 1$, $\eta = 1$, the singular line $\xi = \xi_0$, $\eta = \eta_0$, and characteristics of both families*. The

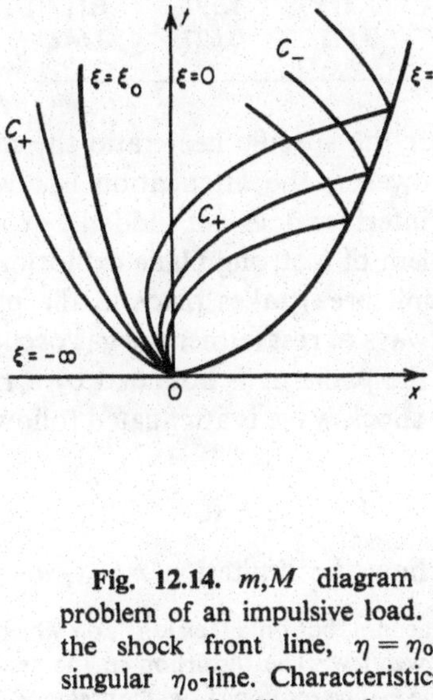

Fig. 12.13. x,t diagram for the problem of an impulsive load. $\xi = 1$ is the shock line, $\xi = \xi_0$ is the singular ξ_0-line. Characteristics of the C_+ and C_- families are shown.

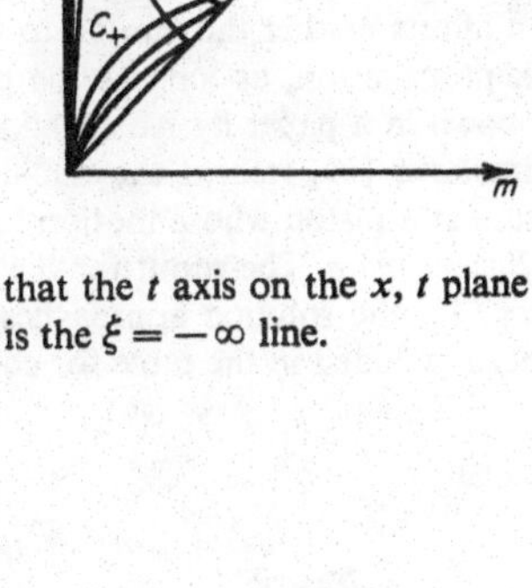

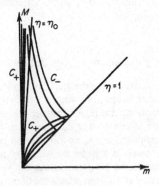

Fig. 12.14. m,M diagram for the problem of an impulsive load. $\eta = 1$ is the shock front line, $\eta = \eta_0$ is the singular η_0-line. Characteristics of the C_+ and C_- families are shown.

* We note that the M axis is the $\eta = 0$ line, and that the t axis on the x, t plane is the $\xi = 0$ line. The negative semiaxis x on the x, t plane is the $\xi = -\infty$ line.

singular line is a C_+ characteristic. All the characteristics of the C_+ family emanate from the origin. Those passing to the right of the singular line overtake the shock front and those passing to the left of it never catch up with it. Thus, the state of the motion in the relatively small mass contained between the vacuum interface and the singular line does not affect the propagation of the shock front*.

In [15] the values of the similarity exponent were obtained by numerical integration for several other values of the specific heat ratio γ. The results are given in Table 12.2†. The pressure, density, and velocity distributions for different values of the specific heat ratio are qualitatively similar to the distributions for the case of $\gamma = 7/5$ (see Figs. 12.11 and 12.12).

Table 12.2

SIMILARITY EXPONENTS IN THE PROBLEM OF AN IMPULSIVE LOAD

γ	1.0	1.1	7/5	5/3	2.8	∞
n	2	1.516	4/3	1.275	1.191	1.117
α	0.5	0.569	3/5	0.611	0.627	0.642

It is clear from the table that the larger the specific heat ratio the more slowly is the shock wave attenuated. However, the attenuation is always faster than in the case when the gas at the interface does not suddenly expand into vacuum but is at rest, as in the problem of a strong plane explosion. If an instantaneous energy release E (per unit area) takes place in the plane $x = 0$ and if the gas in the plane $x = 0$ is always at rest (either the gas occupies the space symmetrically on both sides of the plane or is bounded by a rigid wall), then the energy is conserved and the shock wave is attenuated following the relation

$$p_1 \sim X^{-1} \sim t^{-2/3}; \qquad n = 1, \qquad \alpha = \tfrac{2}{3}.$$

We shall show in the following section how the limitation $n > 1$, $\alpha < 2/3$

* In particular, the gas may border not on a vacuum, but on a "piston", on which the pressure drops following a sufficiently rapid power law. The distortion in the state of the motion in the region between the interface and the ξ_0-line resulting from the presence of the piston affects neither the motion to the right of the ξ_0-line nor the relation governing the shock propagation, as long as the pressure at the piston drops sufficiently rapidly. This was shown in a paper by Adamskii and Popov [18]. This paper and also a paper by Krasheninnikova [19] treated the self-similar problem of the gas motion resulting from the pressure at a piston whose motion follows a power law.

† *Editors' note.* The result for the limit $\gamma \to \infty$ has been added and is from [18]. In the limit $\gamma \to 1$, the solution approaches one with impulse conserved, with $\alpha = 1/2$; this case has been included in the table for completeness.

follows from conservation of energy when the value of γ is arbitrary. We shall also show how conservation of momentum imposes a limitation on the exponents from the other side, $n < 2$, $\alpha > \frac{1}{2}$.

§16. Limitations on the similarity exponent imposed by conservation of momentum and energy

The character of the motion that results from an impulsive load is such that some part of the gas is carried away by the shock wave to the right, while the remaining gas expands to the left into the vacuum. There exists a point which divides these two parts of the gas; we shall denote the coordinate of this point by x^*. The particle velocity of the gas is equal to zero at the point x^*, or $u^* = u(x^*) = 0$. The boundary x^* itself propagates to the right both with respect to the spatial and the mass coordinate. In the self-similar solution the point at which the velocity changes sign corresponds to a specific value of the similarity variable $\xi = \xi^*$; $x^* = \xi^* X$.

Let us consider the volume contained between the shock front surface $x = X$ and the "dividing" surface $x = x^*$. This volume contains a mass (per unit surface area)

$$M^* = \int_{x^*}^{X} \rho\, dx = \rho_0 X \int_{\xi^*}^{1} g\, d\xi = const \cdot \rho_0 X.$$

This mass is a specific fraction of the total mass involved in the motion $M = \rho_0 X$ (for $\gamma = 7/5$, $M^*/M = 0.78$). The remaining mass $M - M^*$ expands to the left. The mass M^* increases with time as $M^* \sim X \sim t^{\alpha}$, as does the total mass M.

The boundary x^* propagates through the mass to the right, and the gas flows out through the surface x^* to the left. The expressions for the momentum and energy of the gas moving with the shock wave to the right are

$$I^* = \int_{x^*}^{X} \rho u\, dx = \rho_0 \dot{X} X \int_{\xi^*}^{1} g v\, d\xi = const \cdot t^{2\alpha - 1}, \tag{12.47}$$

$$E^* = \int_{x^*}^{X} \left(\frac{\rho u^2}{2} + \frac{p}{\gamma - 1}\right) dx = \rho_0 \dot{X}^2 X \int_{\xi^*}^{1} \left(\frac{g v^2}{2} + \frac{\pi}{\gamma - 1}\right) d\xi = const \cdot t^{3\alpha - 2}. \tag{12.48}$$

Undisturbed gas at zero pressure and temperature flows in from the right, through the surface of the shock front into the volume $x^* < x < X$. The undisturbed gas introduces neither momentum nor energy into the volume. Gas with zero velocity, but with a finite pressure p^*, leaves the volume to the left through the surface x^* (the gas leaves the volume not because of its own motion, but because of the propagation of the surface bounding the volume). There is no momentum flow through the surface x^*. The change of momentum in the volume is equal to the pressure applied to its boundary

$$\frac{dI^*}{dt} = p^* > 0. \tag{12.49}$$

The momentum within the volume increases with time. It follows from (12.47) that $2\alpha - 1 > 0$, and $\alpha > \frac{1}{2}$, and (12.41) shows that $n < 2$.

The change of energy in the volume is due only to the loss of internal energy through the surface x^* to the left. Kinetic energy is not lost, since the gas velocity u^* and the kinetic energy are both zero at the boundary x^*. The work of the pressure forces $p^*u^*\,dt$ on the surface x^* is also equal to zero. Therefore,

$$\frac{dE^*}{dt} = -\frac{1}{\gamma - 1}\frac{p^*}{\rho^*}\rho^*\frac{dx^*}{dt} = -\frac{1}{\gamma - 1}p^*\xi^*\dot{X} < 0. \tag{12.50}$$

The energy in the volume decreases with time, it is transferred to the left out of the volume, together with the mass of gas whose velocity changes direction and begins to expand to the left into the vacuum. It follows from (12.48) that $3\alpha - 2 < 0$, $\alpha < 2/3$, and (12.41) then gives $n > 1$. We thus arrive at the following ranges for the similarity exponents

$$\frac{1}{2} < \alpha < \frac{2}{3}, \qquad 2 > n > 1. \tag{12.51}$$

The limit $n = 1$, $\alpha = 2/3$ corresponds to constant energy $E^* = const$, while the limit $n = 2$, $\alpha = \frac{1}{2}$ corresponds to constant momentum $I^* = const$.

§17. Passage of the nonself-similar motion into the limiting regime and the "infinite" energy in the self-similar solution

Strictly speaking, the self-similar solution corresponds to idealized initial conditions, in which the duration of the impact τ is infinitesimally small and the pressure at the piston Π_1 during the impact is infinitely great. In this case, the passage to the limit $\tau \to 0$, $\Pi_1 \to \infty$ is carried out in such a manner that the product $\Pi_1^{1/2}\tau^{1-\alpha}$, which is proportional to the parameter A (see (12.36)), remains finite. Corresponding to the limit $\tau \to 0$, $\Pi_1 \to \infty$, the piston imparts an infinite energy to the gas

$$E \approx \rho_0{}^{-1/2}\Pi_1^{3/2}\tau \sim \tau^{-(2-3\alpha)} \to \infty, \qquad \alpha < \frac{2}{3}, \tag{12.52}$$

and zero momentum

$$I \approx \Pi_1 \tau \sim \tau^{2\alpha-1} \to 0, \qquad \alpha > \frac{1}{2}. \tag{12.53}$$

Let us compare the energy E^* and momentum I^* of that part of the gas which moves to the right in the direction of the shock propagation (see equations (12.47) and (12.48)) with the total gas energy E and momentum I. We obtain

$$\frac{E^*}{E} \sim \left(\frac{\tau}{t}\right)^{2-3\alpha}, \quad \frac{I^*}{I} \sim \left(\frac{t}{\tau}\right)^{2\alpha-1}, \quad \frac{1}{2} < \alpha < \frac{2}{3}. \tag{12.54}$$

The ratio of the energy E^* of the gas moving to the right with the shock wave at a given time t to the initial energy E is the smaller the shorter is the duration of the impact. It is not surprising that in the limit of a vanishingly short duration of impact $\tau \to 0$ the piston is required to do an infinite amount of work (infinite energy of the gas E) in order for the energy of a specific fraction of the mass to remain finite after being decreased by a ratio which approaches zero. All this infinite energy is now concentrated in that part of the mass which expands into the vacuum, or more exactly, at the very edge of the mass of gas. At the edge the gas possesses infinite expansion velocity and infinite kinetic energy. The shorter the duration of impact the greater is the unidirectional momentum I^* at a given time t in comparison with the momentum I imparted by the piston. In the limit $\tau \to 0$, the unidirectional momenta of the gas particles which move to the right and to the left cancel each other out with an accuracy corresponding to the vanishingly small value of I.

Essentially, the idealized limiting solution does not simply correspond to zero duration of the impact τ, but to an infinitely large ratio t/τ. As $t/\tau \to \infty$, $E^*/E \to 0$ and $I^*/I \to \infty$. In interpreting this condition we have considered finite times t, with vanishingly small durations of impact τ, in which limit the work of the piston E was infinite and the momentum I was zero. Closer to reality is another interpretation of this limit, the interpretation in which the duration of the impact does not go to zero but in which the actual duration and energy of the impact E are finite and we consider times t which are large in comparison with τ ($t/\tau \to \infty$ not because $\tau \to 0$, but because $t \to \infty$).

In considering the limiting regime from this point of view we must determine the manner in which the real motion, nonself-similar because τ is finite, asymptotically approaches the limiting regime. How will the fact that the energy of the limiting motion is infinite be reconciled with the fact that the actual work of the piston is finite? The point is that the real solution does not approach the self-similar one uniformly in time. As the time t and the mass of gas involved in the motion $M = \rho_0 X$ increase, the pressure and all other quantities approach values corresponding to the self-similar solution. However, this process does not occur everywhere.

The state of a certain mass m_0 near the vacuum interface that was subjected to the direct action of the piston during impact never approaches the one dictated by the self-similar solution. In order of magnitude this mass is equal to the mass of the gas through which the shock wave travels during the impact, or $m_0 \sim \rho_0 U_1 \tau \sim (\Pi_1 \rho_0)^{1/2}\tau$. The velocity with which this mass expands into vacuum always remains finite and equal in order of magnitude to U_1 ($U_1 \sim (\Pi_1/\rho_0)^{1/2}$, whereas the expansion velocity of the gas interface in the self-similar solution is infinite (as $\tau \to 0$, $\Pi_1 \to \infty$, and $U_1 \to \infty$). Because the motion is isentropic, the entropy of the mass m_0, equal to the initial entropy, is also finite. In fact, $S = c_v \ln p\rho^{-\gamma} + const$, and the value of $p\rho^{-\gamma}$ in the mass m_0 is, in order of magnitude, $\Pi_1 \rho_0^{-\gamma}$. Thus the entropy is bounded for finite τ and Π_1. In the self-similar solution for $\gamma = 7/5$ we have from (12.42)

$$p\rho^{-\gamma} \sim fq^{-\gamma} \sim \eta^{-4/3} \sim m^{-4/3} \to \infty \quad \text{as} \quad m \to 0.$$

Thus, the mass m_0 at the interface always carries the imprint of the initial conditions and its state is not described by the self-similar solution even in the limit $t \to \infty$.

This situation in no way contradicts the general tendency of the actual solution to transform into the self-similar one in the limit $t \to \infty$. The mass m_0 comprises, with time, an increasingly smaller fraction of the entire mass of gas involved in the motion (Fig. 12.15). In the limit $t \to \infty$, this small mass

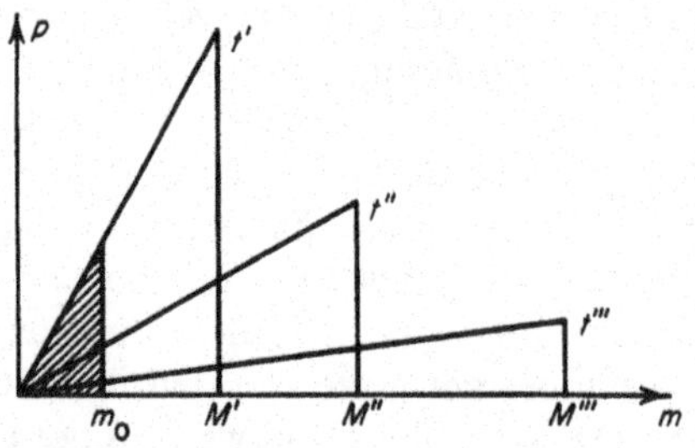

Fig. 12.15. The mass m_0 as a fraction of the total mass M involved in the motion with increasing time ($t''' > t'' > t'$).

need not be considered in the differential equations or in the convergent momentum integral*. However, the replacement, in the small mass m_0, of the actual solution by the self-similar one when evaluating the energy integral

* *Editors' note.* The actual finite momentum I is equal to the difference in the momentum integral over m_0 between the actual solution and the self-similar solution.

causes an essential change in the integral, making it divergent. In the self-similar solution the velocity and kinetic energy become infinite as the vacuum interface is approached ($m \to 0$), whereas actually with a finite pressure at the piston Π_1 and with a finite duration τ, the velocity and kinetic energy of the gas are finite near the interface.

In order to obtain a finite energy of the gas, corresponding to the finite work actually done by the piston, it is necessary when evaluating the energy using the self-similar solution to cut off the integration before the region in which the self-similar solution is inapplicable. We shall evaluate the energy in terms of the Lagrangian coordinate. Then, when integrating the specific energy over the mass of gas encompassed by the motion, we take as the lower limit a mass coordinate equal in order of magnitude to the mass m_0, which is not correctly described by the self-similar solution,

$$E = \int_{m_0}^{M} \left(\frac{u^2}{2} + \frac{1}{\gamma - 1}\frac{p}{\rho}\right) dm = \rho_0 \dot{X}^2 X \int_{m_0/M}^{1} \left(\frac{v^2}{2} + \frac{1}{\gamma - 1}\frac{\pi}{g}\right) d\eta.$$

To this term could be added separately the finite limiting energy of the mass m_0.

Let us carry out the integration in the case $\gamma = 7/5$. The main contribution to the integral is given by the region near the lower limit, where the velocity of the gas and its kinetic energy are very large (in the limit $m_0/M \to 0$, $v \to -\infty$). Therefore, in order to evaluate the integral approximately we use the asymptotic expression for the velocity (12.46) (see also (12.45)). We then obtain

$$E \sim \rho_0 \dot{X}^2 X \int_{m_0/M}^{1} \eta^{-4/3}\, d\eta \sim \rho_0 \dot{X}^2 X \left(\frac{m_0}{M}\right)^{-1/3}.$$

Let us express the variables in this relation in terms of X, using

$$M = \rho_0 X, \qquad \dot{X} = A^{5/3} X^{-2/3} \qquad (\text{since} \quad X = At^{3/5}).$$

Noting the relations $A \approx (\Pi_1/\rho_0)^{1/2}\tau^{2/5}$ (equation (12.36)) and $m_0 \approx (\Pi_1 \rho_0)^{1/2}\tau$, we find that

$$E \sim \rho_0 A^{10/3} X^{-4/3} X m_0^{-1/3} \rho_0^{1/3} X^{1/3} = \rho_0^{4/3} A^{10/3} m_0^{-1/3} \approx \Pi_1^{3/2} \rho_0^{-1/2} \tau.$$

It may be seen that the energy of the entire mass of the gas, with the exclusion of the small mass m_0 to which the self-similar solution does not apply, is constant in time, finite, and equal in order of magnitude to the work done by the piston. The energy contained in the relatively small mass m_0 is of the same order. This mass travels into the vacuum with a velocity of the order of U_1 and with a kinetic energy of the order of $m_0 U_1^2 \approx \rho_0 U_1^3 \tau \approx \rho_0 (\Pi_1/\rho_0)^{3/2}\tau \approx \Pi_1^{3/2} \rho_0^{-1/2} \tau \approx E$. Within the framework of the self-similar solution, however, an infinite amount of energy is concentrated in the mass

m_0, despite the fact that with time it constitutes an ever decreasing fraction of the total mass M of the gas encompassed by the shock.

It is an essential feature that the region of the gas that is not described by the self-similar solution and that produces the divergence in the energy integral if the self-similar solution is extrapolated into it lies beyond the limits of the domain of influence. It is located to the left of the singular line, and thus has no influence on the propagation of the shock wave. Actually, the boundary of the nonsimilar region is described by $m \approx m_0$, and the singular line by $m = \eta_0 M$ ($m = 0.054M$ for $\gamma = 7/5$). As $t \to \infty$ and $M \to \infty$, $m_0 \ll \eta_0 M$.

In order to get some idea how the nonself-similar solution passes to the limiting self-similar regime Zhukov and Kazhdan [14] carried out a numerical integration of the gasdynamic equations with $\gamma = 7/5$ for the rectangular pressure pulse shown in Fig. 12.9a. Figure 12.16 gives the curves of p/p_1,

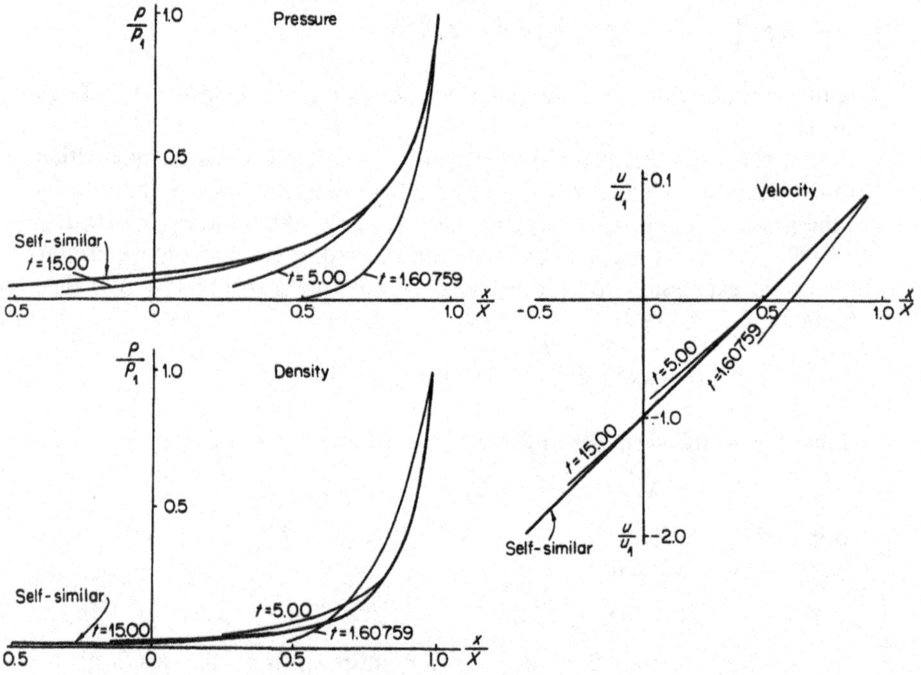

Fig. 12.16. Passage of nonself-similar motion into the self-similar regime. The curves are taken from [14]. The unit of time has been taken as the time of duration of the piston action τ.

u/u_1, and ρ/ρ_1 as functions of the similarity variable x/X for several times (p_1, u_1, and ρ_1 are quantities behind the front). Curves for the exact self-similar solution are also given in this figure. As may be seen from the curves, as early as $t/\tau = 5$ the actual solution is quite close to the self-similar one, and for $t/\tau = 15$ it is almost identical with it. Thus the motion passes into the self-similar regime quite rapidly. From the solution of the nonself-similar problem we can determine the value of the numerical coefficient in equation (12.36) for the parameter A. It turns out to be equal to 1.715, so that $A = 1.715(\Pi_1/\rho_0)^{1/2}\tau^{2/5}$. The value of the numerical coefficient is a characteristic of the shape of the piston pressure pulse. Therefore the value 1.715 is associated with the rectangular pulse with $\gamma = 7/5$.

§18. Concentrated impact on the surface of a gas (explosion at the surface)

Let us consider a "spherical" analogue of the plane motion of a gas subjected to an impulsive load on its surface. A "cylindrical" analogue will be considered at the same time. This problem was treated by one of the present authors [20].

Let the half-space $z > 0$ be occupied by a perfect gas with a specific heat ratio γ. The density of the gas ρ_0 is constant, and the pressure and temperature are zero. The space $z < 0$ on the other side of the plane $z = 0$ is empty. At the initial time $t = 0$ in a mass of gas m surrounding the point 0 at the interface $z = 0$, an energy E is suddenly released. This can occur as the result of an explosion at the surface, or as a result of the concentrated impact at the surface from a high-velocity projectile if the projectile does not penetrate too deep into the fluid but is sharply decelerated near the surface. In this case its kinetic energy is rapidly converted into heat, and something similar to an explosion takes place. A shock wave travels from the point 0 through the gas, while in the other direction the heated gas expands into the vacuum. The initial velocities both in the direction of propagation of the shock wave and in the direction of the vacuum are of the order of $u_0 \sim (E/m)^{1/2}$*.

The surface of the shock front, which is a surface of revolution about the z axis, forms something like a "bowl", as shown in Fig. 12.17. The gas heated by the shock wave flows from the bowl through the round opening (the section of the bowl in the plane $z = 0$) into the vacuum. The outflow of the gas weakens the shock wave in comparison with what would be the case if the opening were closed by a fixed "cover"; this case would correspond to an explosion in an infinite medium.

The shock wave moves most rapidly in the downward direction and most slowly along the surface $z = 0$, where it is strongly attentuated owing to the expansion of the gas into vacuum. Therefore, the front surface is like a hemisphere but elongated in the downward direction. The gas near the front moves

* If the motion is caused by the impact of a projectile then m is of the order of the mass of the projectile, E is of the order of its kinetic energy, and u_0 is of the order of the impact velocity.

in the direction of propagation of the wave. Somewhere within the bowl there is a surface on which the vertical component of the velocity changes direction. Above this surface, which is shown in Fig. 12.17 by the dashed line, the gas

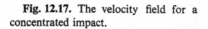

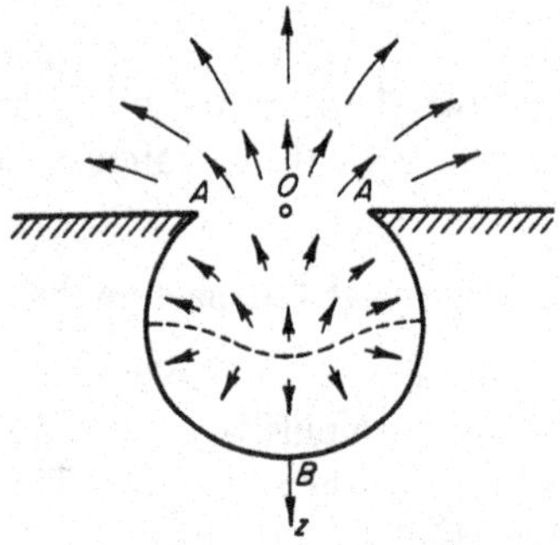

Fig. 12.17. The velocity field for a concentrated impact.

moves in the direction of the vacuum (the approximate directions of the velocity are indicated by the arrows). As the gas moves away from the plane $z = 0$ the expansion velocity in the previously empty space $z < 0$ increases, as shown schematically in the figure by arrows of increasing length*.

It is rather evident that in the limit, when the shock wave encompasses a mass $M \gg m$, the motion is self-similar. In this case the front surface expands in a similar fashion. The coordinate of any point of the front, say point B, increases with time as $z_1 \sim t^\alpha$. The pressure behind the front, say at the same point B, decreases with an increase in the mass M as $p_1 \sim M^{-n}$. Here, the constants n and α are related by the simple expression† $n = 2(1 - \alpha)/3\alpha$.

In the case of a concentrated impact, as in the plane case, the exponent n is limited by the inequality

$$1 < n < 2. \qquad (12.55)$$

In order to check that this is indeed the case, let us consider the stage when $M \gg m$ and $p \sim M^{-n}$ and set up approximate expressions for the energy and vertical z component of the momentum of the gas in the bowl. Taking the dimensional parameters into account in the coefficient of proportionality between p and M but neglecting the numerical coefficients, the average

* Apparently, the gas flowing out of the opening near the plane interface with the undisturbed medium may move along the plane $z = 0$, and the pressure on the plane itself may be equal to zero. It is possible that a "blowout" may occur for certain values of γ, with a conical cavity formed near the plane $z = 0$ outside the hole. It is possible that for some values of γ much more complicated flow patterns may be formed, with, for example, a shock triple point near the point A.

† $M \sim z_1^3 \sim t^{3\alpha}$. The velocity of the gas behind the front u is proportional to $dz_1/dt \sim t^{\alpha-1} \sim p^{1/2} \sim M^{-n/2} \sim t^{-3\alpha n/2}$. From this $\alpha - 1 = -3\alpha n/2$ or $n = 2(1 - \alpha)/3\alpha$.

pressure over the bowl volume can be written as

$$p \sim \frac{E\rho_0}{m}\left(\frac{m}{M}\right)^n \sim p_0\left(\frac{m}{M}\right)^n. \tag{12.56}$$

Here $p_0 \sim E\rho_0/m$ is the initial pressure at the time of impact or explosion. The average velocity of the gas in the bowl is equal, in order of magnitude, to

$$u \sim \left(\frac{p}{\rho_0}\right)^{1/2} \sim u_0\left(\frac{m}{M}\right)^{n/2} \sim \left[\frac{E}{m}\left(\frac{m}{M}\right)^n\right]^{1/2}. \tag{12.57}$$

The energy in the bowl is of the order of

$$E_1 \sim Mu^2 \sim \frac{Mp}{\rho_0} \sim E\left(\frac{m}{M}\right)^{n-1} \sim E_{10}\left(\frac{m}{M}\right)^{n-1}, \tag{12.58}$$

where E_{10} is the initial energy in the bowl, which obviously is of the order of the total energy E. The momentum in the bowl is of the order of

$$I_1 \sim Mu \sim \left[Em\left(\frac{m}{M}\right)^{n-2}\right]^{1/2} \sim I_{10}\left(\frac{m}{M}\right)^{(n-2)/2}, \tag{12.59}$$

where $I_{10} \sim (Em)^{1/2}$ is the initial momentum*.

The energy flows out from the bowl through the opening, since the velocity of the gas in the opening is directed toward the vacuum. Consequently, the energy E_1 contained in the bowl decreases with time (with an increase in the mass M), and according to (12.58), $n > 1$.

We compare the momentum of the gas in the entire bowl to the momentum of that part of the gas that is contained between the surface of the shock front and the dotted surface across which the vertical component of the velocity changes sign and on which it is zero. No momentum flows out through this surface in the vertical direction and the pressure is positive on the surface. Consequently, the momentum increases with time and, according to (12.59), $n < 2$. The vertical momentum of the gas in the bowl is also balanced by the increasing but oppositely directed momentum of the gas which flows out of the bowl and expands into the vacuum. Thus (12.55) can be considered as proved†. The value $n = 1$ would correspond to conservation of energy in the bowl i.e., to an explosion in an infinite medium. The value $n = 2$ would correspond to conservation of momentum.

The same inequality (12.55) is also valid in the cylindrical case, in the case of

* In the case of impact by a projectile, I_{10} is of the order of the momentum of the impacting body.

† We note that the passage to the limit of the self-similar regime corresponds to $m \to 0$. In order for the pressure to be finite, it is necessary that the energy be infinite, $E \sim m^{-(n-1)} \to \infty$, and that the initial momentum be zero, $I_{10} \sim (Em)^{1/2} \sim m^{1-n/2} \to 0$.

a "line" impact. The picture of the motion in a line impact or explosion is qualitatively similar to that given in Fig. 12.17. In this case, however, the explosion takes place not at a single point O, but along a straight line passing through the point O perpendicular to the plane of the figure. The entire motion is symmetric with respect to the plane passing through this line and the z axis. The surface of the front will not form a bowl but an infinitely long "trough", the cross section of which is depicted by the figure. M is the mass per unit length of the trough encompassed by the shock.

We can establish an even narrower interval for the exponent which appears in the relation governing the attenuation of the shock wave. Physically, it is clear that in the case of a concentrated impact the degree of weakening of the shock wave per unit increase of mass is less than in the plane case for the same value of the specific heat ratio. In fact, the relatively smaller the area through which the gas flows out into the vacuum the less pronounced is the attenuating effect of the outflow of gas from the front. In the "spherical" case the area of the opening is much smaller than the surface area of the shock front (see Fig. 12.17). In the plane case both areas are equal. The "cylindrical" case is intermediate in this respect.

Thus denoting the exponents in the relation for the attenuation of the shock wave $p \sim M^n$ by n_1, n_2, and n_3 for the plane, line, and concentrated impacts, respectively, then for the same specific heat ratio

$$1 < n_3 < n_2 < n_1 < 2. \tag{12.60}$$

For example, for $\gamma = 7/5$, $n_1 = 4/3$ and $1 < n_3 < 4/3$. For $\gamma = 5/3$, $n_1 = 1.275$ and $1 < n_3 < 1.275$. The concentrated impact is thus closer to a point explosion in an infinite medium than the plane impact is to a plane explosion.

§19. Results from simplified considerations of the self-similar motions for concentrated and line impacts

As in the plane case, in order to determine the exponent $n(\gamma)$ in the attenuation relation $p \sim M^{-n}$ for the shock wave, it is necessary to solve the equation for the self-similar motion. However, the "spherical" and "cylindrical" problems are incommensurably more complex than the plane problem, since they are essentially two-dimensional and the self-similar motion is described by partial differential equations rather than by ordinary differential equations. The situation is also complicated by the fact that the shape of the shock front surface at which the boundary conditions are specified is not known beforehand and must itself be found from the solution. For this reason even a numerical integration of the self-similar equations can become very difficult.

Some idea of the numerical values of the exponents and of the general characteristics of the motion can be obtained from the simplified considerations presented in [20]. An exact particular solution of the self-similar differential equations was obtained, which is a generalization of the exact solution for the one-dimensional problem (see §15) and which in some respects correctly indicates the features of the two-dimensional process. The solution contains a number of unknown constants. With this rather arbitrary particular solution it is impossible to satisfy the boundary conditions at the shock front. For this reason, the solution, instead of being required to satisfy boundary conditions at the front, was instead required to satisfy general integral relations expressing balances of mass, energy, and the components of the momentum of the gas contained in the bowl or trough. For this purpose, a very simple shape was chosen for the shape of the front surface. The bowl was replaced by a circular cylinder with a flat bottom and the rounded cross section of the trough was replaced by a rectangular cross section (see Fig. 12.18).

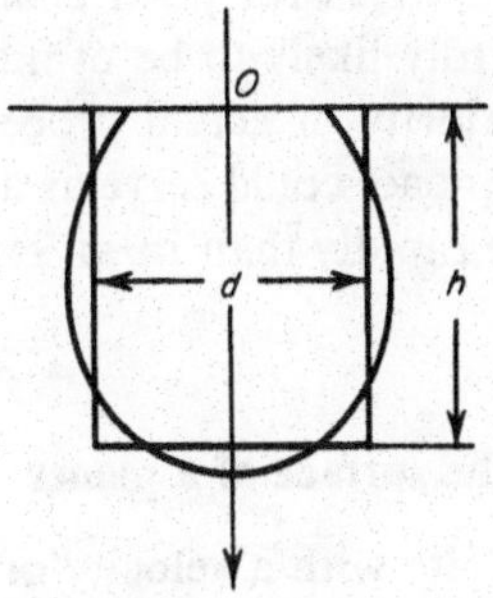

Fig. 12.18. Replacement of the bowl by an equivalent cylinder.

Just as the exact analytic solution in the one-dimensional case exists for only one particular value of the specific heat ratio γ (equal to 7/5), so the approximate solution here, which is a generalization of the exact one-dimensional solution, is applicable with only one particular value of γ. This value, along with the corresponding value of n, is found from the solution.

It was found that in the case of a concentrated impact $n = 1.07$ for $\gamma = 1.205$, the ratio of the height of the cylinder h to its diameter d is equal to 1.05. The density of the gas in the opening $\rho_{\text{open}} = 0.0187\rho_0$, and only 1.6% of the mass encompassed by the shock wave flows out of the opening. The density of the gas at the bottom of the cylinder $\rho_{\text{bot}} = 10.3\rho_0$, which is very close to the actual density behind the shock front $[(\gamma + 1)/(\gamma - 1)]\rho_0 = 10.7\rho_0$. The vertical velocity component changes direction at a depth of $0.846h$ from the opening and at a distance of $0.154h$ from the bottom. In the case of a line impact $n = 1.14$ for $\gamma = 1.266$, $h/d = 1.21$ (h is the height of the trough and d is the width), and 2% of the total mass flows out of the trough.

We see that the exponents n are very close to unity, so that the outflow of gas away from the shock front due to its expansion into vacuum weakens the shock wave only slightly in comparison with an explosion in an infinite medium. This is clearly a result of the fact that only a small fraction of the entire mass flows out from the boundaries of the bowl or trough. The shape of the bowl is evidently appreciably different from hemispherical, the shape which would have corresponded to an explosion in an infinite medium. The height of the bowl, i.e., the cylinder, is approximately equal to the diameter, whereas if a hemisphere is replaced by an equivalent cylinder the height would be approximately half the diameter. The same is also true for the line impact.

Fortunately, the specific heat ratios $\gamma = 1.205$ and $\gamma = 1.266$ for which the approximate solutions are valid are close to the actual values of the effective specific heat ratios for gases at high temperatures with dissociation and ionization important. We note that in the plane case the exponent n decreases monotonically as γ increases. If the situation in the two-dimensional case is the same, which seems quite likely, then for a concentrated impact $1 < n < 1.07$ for $\gamma > 1.205$, and for a line impact $1 < n < 1.14$ for $\gamma > 1.266$. Values appreciably smaller than 1.205 or 1.266 are hardly likely to be of interest in actual processes. As a consequence, in the majority of actual processes for which the problem of a concentrated (or line) impact could serve as a model, the shock wave is attenuated only slightly more rapidly than in an explosion in an infinite medium.

§20. Impact of a very high-speed meteorite on the surface of a planet

The process which takes place when a meteorite with a velocity of several tens or a hundred km/sec (and higher) strikes a planet can serve as a characteristic example of the phenomenon of a concentrated impact. For this problem it is only meaningful to consider either planets lacking an atmosphere, such as the moon, or sufficiently large meteorites. Small meteorites, or rather meteors, are vaporized, "burned up" by atmospheric friction, and never reach the surface of the planet.

When a meteorite impacts it is sharply decelerated and the initial kinetic energy $E = mv^2/2$ (where m is the mass of the meteorite and v is the impact velocity) is largely converted into internal energy or heat. The depth to which the meteorite penetrates the surface is usually of the order of the dimensions of the meteorite itself, so that at the initial time the energy release takes place in a mass of the order of m. A shock wave travels through the ground from the point of energy release*.

We consider only impacts with very high velocities, where the specific energy $v^2/2$ is many times larger than the binding energy (or heat of vaporiza-

* We do not consider low impact velocities, where the deceleration process and the propagation of the shock wave through the meteorite are important, or in other words, where we cannot assume that the energy release is instantaneous.

tion) of the atoms and molecules of the meteorite and of the ground. In this case a stage exists when the shock wave encompasses a ground mass M that appreciably exceeds the initial mass m, but when the material encompassed by the shock wave can be regarded as a dense gas. The ground and the meteorite are totally vaporized during the expansion and expand from the planet surface in a gaseous state. In the stage when the material has not expanded too much, the gas pressure is considerably higher than atmospheric and the existence of an atmosphere (if there is one) can be neglected. The vapor expands as if into a vacuum. As can be seen, we are dealing here with a typical picture of a concentrated impact on the surface of a "gas", as described in the preceding section.

Let us estimate what the required impact velocities are. The heat of vaporization of iron (both iron and stone meteorites occur) is equal to 94 kcal/mole $= 7 \cdot 10^{10}$ erg/g. The heat of vaporization of stone is of the order of 83 kcal/mole $= 5.8 \cdot 10^{10}$ erg/g. This is the value for silica (SiO_2), which is the main component of various soils and rocks. If in addition we also consider the dissociation of the SiO_2 molecules during vaporization, then the total binding energy comes to 203 kcal/mole $= 1.4 \cdot 10^{11}$ erg/g. As an estimate, let us say that complete vaporization requires that the specific energy exceed the heat of vaporization, for which we take the approximate value $U \approx 10^{11}$ erg/g, by a factor of 10. Then the minimum velocity at which a mass of the order of the mass of a meteorite will vaporize will be $v_{\min} \approx (2 \cdot 10 \cdot 10^{11})^{1/2} =$ 14 km/sec. In exactly the same way we can state that the expansion following the passage of the shock wave vaporizes those ground layers which were reached by the shock wave while the specific internal energy ε_1 behind the front is at least of the order of $\varepsilon_k \sim 10U \sim 10^{12}$ erg/g. The assumption that ε_1 must exceed U by a factor of 10 was based on estimates obtained in §22 of Chapter XI, where it was shown that complete vaporization in the unloading of a solid body compressed by a strong shock wave is obtained if the energy behind the wave front is at least five times greater than the binding energy of the material.

In order to estimate the total mass of ground which vaporizes on the impact of a meteorite we must use the relation governing the attenuation of the shock wave. An estimate of this type was first carried out by Stanyukovich [21], who studied the phenomenon of the "explosion" on impact of a meteorite on the surface of a planet as the cause for the formation of lunar craters. He did not take into account the effect of the expansion of the vapor into the vacuum, assuming that the shock wave propagates in the same manner as it would in a strong explosion in an infinite medium, following the relations $p_1 \sim M^{-1}$, $\varepsilon_1 \sim E/M$. The arguments presented in the preceding section substantiate this assumption. The order of magnitude of the vaporized mass M_k is determined by $\varepsilon_k \sim E/M_k$, from which $M_k \sim E/\varepsilon_k \sim m(v^2/\varepsilon_k) =$

$m(v/v_k)^2$, where $v_k = \varepsilon_k^{1/2} \sim 10$ km/sec. For example, if the impact velocity $v \sim 100$ km/sec the vaporized mass of ground exceeds the mass of the meteorite by a factor of 100.

When the energy behind the shock wave drops below $\sim 10^{12}$ erg/g, the ground encompassed by the shock wave no longer vaporizes on unloading. However, the energy in the wave is still sufficient for the mechanical pulverization of the material. The limiting energy required for pulverization is much less than the heat of vaporization. Therefore, the mass of the pulverized material exceeds the mass of the vaporized substance by a large factor. The pulverized material is ejected upward in the form of solid particles, and this results in the formation of the crater. Questions as to the dimensions of craters produced by meteorite impact, as to the role played by the gravitational force which prevents the material from being thrown to large distances, and other questions were considered by Stanyukovich [21].

The explosion-like effects of high-speed meteorite impacts also take place when a body moves through a rarefied atmosphere at hypersonic velocity. The impacts of air molecules with the surface of the body are similar to meteorite impacts at planet surfaces. A "microexplosion" takes place at each impact, and a certain amount of vaporized material is ejected from the surface of the body. The body receives an additional recoil momentum, which increases the drag coefficient and the rate of deceleration of the body in the atmosphere. This phenomenon was considered in a paper by Stanyukovich [22]. The impact of a high-speed body on the surface of a liquid which is assumed to be incompressible has been considered by Lavrent'ev [23].

§21. Strong explosion in an infinite porous medium

Kompaneets [24] solved the problem of a strong point explosion in a plastic compacting medium with constant compaction behind the shock front*. Here we consider the simplified problem (by neglecting plastic shear stress, *eds.*) of the propagation of a shock wave from a point explosion in a porous medium, with the condition that the continuous medium is incompressible (for example, in a sand consisting of incompressible grains). We neglect the strength of the sand grains, and we assume that the adiabatic compression of the material to the density of the continuous medium (total elimination of voids) does not require any expenditure of energy. In other words, the shock is taken to be strong with respect to the strength of the material, but weak with respect to its elasticity (the compressibility of the continuous medium). The initial pressure p_0 is equal to zero.

The average density of the undisturbed medium will be denoted by ρ_0, and the density of the continuous medium (the "sand grains") by ρ_1, $\rho_0 =$

* In [25] the compaction was assumed to depend on the wave strength.

$\rho_1(1-k)$, where k is the porosity, which can vary from zero to one (not the same k as in §10, Chapter XI, *eds.*).

Let a strong explosion take place at some point. An intense initial push is imparted to the material (as for example, if a spherical "piston" were to rapidly expand and then stop). A shock wave will travel through the material, with the material compressed across the shock to the density of the continuous medium and with the voids completely filled. Thereafter the density of the medium no longer changes but remains equal to ρ_1. The material encompassed by the shock wave moves behind the front. A spherical layer with constant density ρ_1 is formed behind the shock front surface, and behind that a cavity, as shown in Fig. 12.19a.

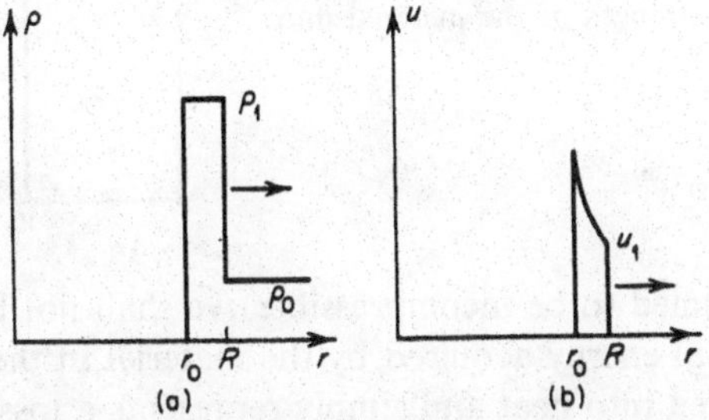

Fig. 12.19. Density (a) and velocity (b) distributions as functions of radius for an explosion in sand with incompressible grains.

If the radius of the wave front is R and the radius of the internal surface of the layer is r_0, then conservation of mass gives

$$M = \frac{4\pi}{3} R^3 \rho_0 = \frac{4\pi}{3}(R^3 - r_0^3)\rho_1$$

or

$$r_0^3 = R^3\left(1 - \frac{\rho_0}{\rho_1}\right) = R^3 k. \tag{12.61}$$

The velocity distribution in the layer is found using the equation of continuity for an incompressible fluid $\nabla \cdot \mathbf{u} = 0$,

$$u = u_1\left(\frac{R}{r}\right)^2, \qquad r_0 < r < R, \tag{12.62}$$

where u_1 is the particle velocity behind the shock front (see Fig. 12.19b). It is related to the front velocity $D = dR/dt$ by

$$u_1 = D\left(1 - \frac{\rho_0}{\rho_1}\right) = Dk.$$

With the assumptions made, the Hugoniot curve for the material has the shape shown in Fig. 12.20*. Let the pressure behind the shock wave be p_1 (point B on the Hugoniot curve). As is known (see §16, Chapter I), a material initially at rest acquires, in a strong shock wave ($p_1 \gg p_0$), the same kinetic and internal energies of $u_1^2/2$ per unit mass. These energies are numerically equal to the area of triangle OAB in Fig. 12.20. Since the solid

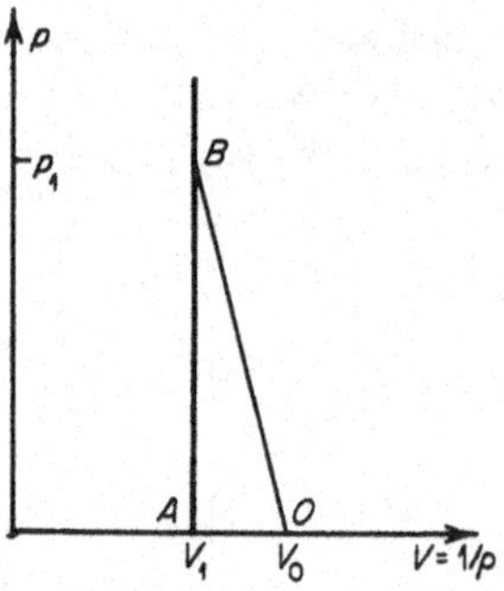

Fig. 12.20. Hugoniot curve for sand with incompressible grains without taking the strength of the material into account.

particles are assumed to be incompressible, we shall not be interested in the fate of the internal energy acquired by the material in the shock wave. This energy is converted into heat and simply represents a loss of the mechanical energy of motion. The decrease in the kinetic energy of the entire mass M in the time dt is thus equal to the increase in the internal energy of a mass dM encompassed by the shock wave in the time dt, and

$$-d\left(\frac{M\overline{u^2}}{2}\right) = -\beta\, d\left(\frac{Mu_1^2}{2}\right) = \frac{u_1^2}{2}\, dM. \qquad (12.63)$$

Here $\overline{u^2} = \beta u_1^2$ denotes the mean square of the velocity in the mass M. The coefficient β is readily calculated from (12.62) and (12.61), and is $\beta = 3/(k + k^{2/3} + k^{1/3})$. Integrating (12.63), we find the relation governing the attenuation of the shock wave

$$u_1^2 = const\ M^{-(1+\beta)/\beta} = const\ M^{-n}.$$

In analogy with preceding sections, we have here denoted the exponent of the mass in the expression for the specific energy by n, with $n = (1 + \beta)/\beta$.

The total kinetic energy is proportional to $E_k = M\overline{u^2}/2 \sim M^{-(n-1)}$, and the momentum is proportional to $I \sim Mu_1 \sim M^{1-n/2}$. Since $\beta > 0$, the exponent n is always bounded between the limits $1 < n < 2$ (cf. the results of §18). In the limiting case of a continuous incompressible medium $k \to 0$, $\beta \to \infty$, $n \to 1$, the energy is conserved, and the momentum increases with

* For a discussion of the shock compression of a porous material when compressibility of the continuous medium is taken into account, see §10, Chapter XI.

time. In the limiting case of an extremely porous substance (a highly "compressible" medium) $k \to 1$, $\beta \to 1$, $n \to 2$, the momentum is conserved, and the energy decreases. In the general case $0 < k < 1$, the energy decreases (it is converted into heat), and the momentum increases. As we see, the situation is the same as that which occurs for an impact on the surface of a gas (see §18). As in the limiting case of a concentrated impact the initial energy is infinite (if $k \neq 0$, $n > 1$ and $E_k \sim M^{-(n-1)} \to \infty$ as $M \to 0$). In a manner analogous to what has been done here, one can also consider an imploding shock wave in a porous medium with incompressible grains.

5. *Propagation of shock waves in an inhomogeneous atmosphere with an exponential density distribution*

§22. Strong point explosion

In Part 4 of Chapter I we considered the problem of a strong explosion in an infinite homogeneous medium. As we know, the earth's atmosphere is not homogeneous and the air density decreases with altitude; the dependence of the density ρ_0 on altitude h can be approximated by the barometric formula $\rho_0 = \rho_{00} e^{-h/\Delta}$. Here ρ_{00} is the density at sea level and Δ is the so-called scale height of the standard atmosphere, which at the earth's surface is approximately equal to 8.5 km*.

Let us examine the propagation of a shock wave induced by a strong point explosion in an inhomogeneous atmosphere. Our interest here is in the stage of the wave motion at which the wave has moved away from the source of the explosion through a distance comparable to the scale height Δ; only in this case does the inhomogeneity of the medium appreciably influence the wave. We shall assume that we are dealing with a strong shock wave (with the pressure behind the front much greater than the pressure ahead of the front). The gasdynamic process is not self-similar, since we have the length scale Δ, and, in addition, the flow is now two-dimensional instead of one-dimensional. In cylindrical coordinates with the vertical axis passing through the point of the explosion, the flow is a function of the z coordinate and of the radius r. The complete solution to this problem can be found only by a numerical integration of the flow equations. It is possible, however, to get some idea of the character of the shock propagation and of the shock shape on the basis of some simple considerations given by Kompaneets [28].

* Actually, since the air temperature also varies with altitude, the density of the earth's atmosphere does not strictly follow the exponential law. The scale height Δ, defined as $\Delta = -(d \ln \rho/dh)^{-1}$, is variable between 6 and 15 km for altitudes below 150 km, and becomes still larger above 150 km.

Let us assume that, as with the strong explosion in a homogeneous medium, the pressure is almost constant throughout the entire volume bounded by the explosion wave, and is thus constant along the back of the surface of the front. We assume also that it is proportional to the average pressure in the volume, to the ratio of the explosion energy to the volume bounded by the wave Ω

$$p_1 = (\gamma - 1)\lambda \frac{E}{\Omega}. \tag{12.64}$$

Here $\lambda(\gamma)$ is a numerical coefficient which can be estimated from the solution to the explosion problem in a homogeneous medium (see Chapter I).

Let the equation of the surface of the shock front in cylindrical coordinates be $f(z, r, t) = 0$. Differentiating, we get

$$\frac{\partial f}{\partial z} dz + \frac{\partial f}{\partial r} dr + \frac{\partial f}{\partial t} dt = 0$$

or

$$\frac{\partial f}{\partial z} D_z + \frac{\partial f}{\partial r} D_r = \mathbf{D} \cdot \nabla f = -\frac{\partial f}{\partial t},$$

where D_z and D_r are the components of the velocity vector $\mathbf{D}$. The normal component of the wave front velocity is given by

$$D_n = -\frac{\partial f/\partial t}{|\nabla f|}.$$

We know, however, that at the front of a strong shock wave

$$D_n = \left(\frac{p_1}{\rho_0}\frac{\gamma + 1}{2}\right)^{1/2},$$

where ρ_0 is the density ahead of the front at a given point on the shock surface. From the last two expressions we obtain

$$\left(\frac{p_1}{\rho_0}\frac{\gamma + 1}{2}\right)^{1/2} = -\frac{\partial f/\partial t}{|\nabla f|}.$$

We now substitute the pressure p_1 from (12.64) and, using the equation for the surface, express the volume Ω bounded by the surface of the wave front in the integral form $\Omega = \int d\Omega$. We assume that the equation $f(z, r, t) = 0$ is solved for the radius $r = r(z, t)$ and consider the atmosphere to be exactly exponential. Combining the above results we obtain for the function $r = r(z, t)$ a partial differential equation, one which is solved exactly in [28].

The evolution of the shock front surface is shown in Fig. 12.21, taken from [28]. This figure shows the wave front cut by a vertical plane passing through the point of explosion (through the z axis), at successive instants of time. The wave is spherical at the beginning, and gradually becomes egg shaped.

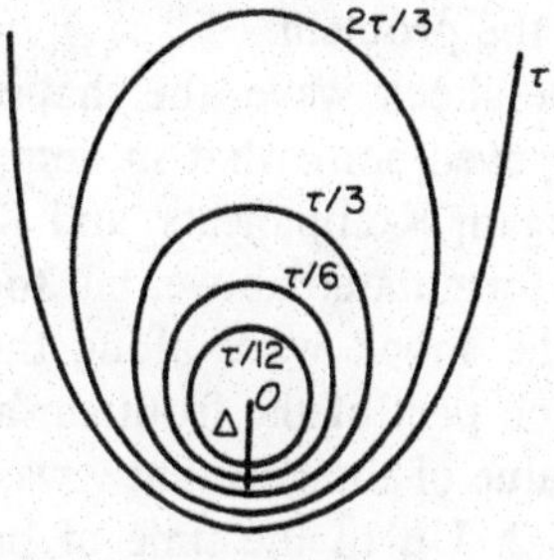

Fig. 12.21. Shock front at successive instants of time for a strong explosion at high altitude. Sections shown are formed by passing a vertical plane through the origin of the explosion. The density of the atmosphere changes by a factor of e over the distance Δ.

The strength of the wave does not change in the same way for its motion in different directions. When moving vertically downward in the direction of the strongest density increase, the shock wave is weakened and decelerates most rapidly. Conversely, when moving vertically upward in the direction of the strongest density decrease, it is even accelerated and, in a finite time τ, it moves to infinity, "breaking through" the atmosphere. The front surface forms something like a "bowl" and, in the huge cavity bounded by this surface, the pressure drops to a very low value (within the framework of the assumptions made, to zero). When moving horizontally the wave is weakened, but more slowly than when moving downward.

The physical reason for the acceleration of the shock wave at large distances when moving upward is easily understood [31]. If R is the distance of the upper point of the shock wave to the explosion center, then the volume of the cavity Ω is proportional to R^3 and the pressure $p_1 \sim E/R^3$. The density ahead of the front is $\rho_0 = \rho_c \exp(-R/\Delta)$, where ρ_c is the density at the level of the explosion. Therefore, as $R \to \infty$ the front velocity

$$D \sim \left(\frac{p_1}{\rho_0}\right)^{1/2} \sim \frac{E^{1/2}}{\rho_c^{1/2} R^{3/2}} e^{R/2\Delta} \tag{12.65}$$

approaches infinity, while the time required by the wave to move upward to infinity,

$$\tau = \int_0^\infty \frac{dR}{D} \sim \frac{\rho_c^{1/2}}{E^{1/2}} \int_0^\infty R^{3/2} e^{-R/2\Delta}\, dR$$

is finite*. It follows from this equation that the time for "breaking through"

* We note that the energy concentrated in a given solid angle whose vertex is located at the point of the explosion does not remain constant as is the case with spherical symmetry. The energy has a tendency to "flow" upward from the bottom and this process also aids in accelerating the shock wave upward. The mass, on the other hand, has a tendency to "flow" downward, and to accumulate near the lower part of the shock.

the atmosphere is $\tau = \nu(\rho_c \Delta^5/E)^{1/2}$, where ν is a numerical constant determined from the complete solution of the problem). It is easy to see that τ is the only quantity with the dimensions of time which can be constructed from the dimensional parameters E, ρ_c, and Δ of the problem.

The relations governing the motion of the shock wave, the shape of the front surface, and the constant ν were improved somewhat in comparison with those given in [28] by Andriankin, Kogan, Kompaneets, and Krainov [29]. These authors analyzed the problem as formulated above, but took into account the pressure distribution behind the shock wave along the front surface. In their treatment the motion of each part of the front surface was determined separately. The more accurate value of the constant ν was found to be $\nu \approx 24$ for a specific heat ratio $\gamma = 1.2$. Up to the time of breaking through the atmosphere the shock wave travels a distance of about 2Δ downward and 3.5Δ horizontally (at the level of the point of the explosion). In accordance with the statement of the problem, the solution given in [28, 29] is no longer valid near the time of breaking through the atmosphere, since then the pressure drops to zero and the shock wave stops moving (under the approximations used).

The inhomogeneity of the atmosphere is felt only in those explosions where the shock wave moves through a distance that exceeds the scale height Δ and still remains sufficiently strong. When this is not so the shock wave will be attenuated before the inhomogeneity is felt, and the explosion takes place in the same manner as in a homogeneous atmosphere. An approximate condition for obtaining a shock front evolution as described above can be defined as in [29], that the ratio of pressures behind the front to the air pressure ahead of the front p_1/p_0 for the propagation of the wave over a distance Δ (as calculated from the relations for an explosion in a homogeneous atmosphere) should exceed $(\gamma + 1)/(\gamma - 1)$ by a factor of, say, 10 (see footnote on page 94 in §25, Chapter I). For example, at an altitude of 100 km, $p_0 \approx 10^{-6}$ atm = 1 bar, and this condition ($p_1/p_0 \approx 100$) is satisfied only for explosions with energies $E > 10^{20}$ erg. At low altitudes, an explosion whose strength is not extremely large takes place in practically the same manner as in a homogeneous atmosphere.

§23. Self-similar motion of a shock wave in the direction of increasing density

Let us consider in more detail the nature of the gasdynamic process in the region of the lowest point of the shock wave produced by a strong explosion which moves vertically downward. In an explosion in a homogeneous atmosphere the pressure in the volume is equalized and is only lower than the pressure behind the shock front by a factor of $\frac{1}{2}$ to $\frac{1}{3}$ (see §§25 and 26 of Chapter I). This internal pressure "maintains" the shock wave, and is responsible for the fact that the wave is attenuated more slowly than it would be in the absence of an internal pressure. The role of the internal pressure becomes particularly understandable if we compare the motion for a plane explosion with the motion for a plane impulsive load (see the preceding part of this chapter). In the latter case, the pressure in the region behind the shock front decreases to zero and there is no internal pressure to maintain the wave. For this reason the shock wave is attenuated faster than in a plane explosion. A similar effect appears in an explosion in an inhomogeneous atmosphere. The pressure in the cavity drops rapidly, owing to the increase in its volume brought about by the upward acceleration of the wave. As a result of this volume increase the internal pressure is no longer large enough to maintain the shock wave which propagates downward, and the gas from the lower layers flows upward away from the front, rushing into the "empty" cavity. The situation to some extent approximates that which takes place in the problem of an impulsive load. Another reason for its similarity to this problem is that, when the shock wave moves away a sufficiently large distance from the point of the explosion, the curvature of the lower part of the front surface becomes small; this part, together with the gas layer behind the front which is adjacent to it, then appears as approximately plane. Thus, to describe the flow of the gas in the lower part of the shock wave approximately it is convenient to consider the idealized problem of the propagation of a plane shock wave in an inhomogeneous atmosphere in the direction of increasing density.

The problem to be considered may be formulated as follows: Let the gas density be distributed in space following the exponential law

$$\rho_0 = \rho^* e^{x/\Delta}, \qquad \Delta = const \tag{12.66}$$

(for convenience the x axis is directed downward). This distribution has the property that the mass of gas concentrated in a column of unit cross-sectional area, from $x = -\infty$, where $\rho_0 = 0$, to $x = X$, is equal to a mass of gas of density $\rho_0(X)$ in a column whose length is Δ,

$$M = \int_{-\infty}^{X} \rho_0(x)\, dx = \rho_0(X)\Delta. \tag{12.67}$$

Let an impulsive load be applied at the initial time $t = 0$, somewhere in the region of very low density, where $x \approx -\infty$. A shock wave will travel through the gas in the direction of increasing density, and the heated gas will expand in the direction of the vacuum.

We shall seek the limiting motion at the stage when the shock wave encompasses a mass of gas M that is much greater than the mass m_0 subjected to the initial impact in the low-density region. The initial pressure and temperature of the gas are taken equal to zero. It is clear that the statement of the problem is entirely analogous to the statement of the problem of an impulsive load applied to the surface of a constant density gas bordering on a vacuum (see §13). The problem of an impulsive load in the case of an inhomogeneous atmosphere was formulated and solved by one of the present authors [30].

It is clear that the limiting motion ($M \gg m_0$) is self-similar. However, this similarity has an unusual character. Unlike the other self-similar motions which have been considered, the conditions of this problem contain a length scale Δ and do not contain a parameter with the units of mass (usually such a parameter is connected with the given initial density of the gas). The quantity ρ^* in (12.66) cannot serve as a parameter, because the x coordinate origin is chosen arbitrarily and thus the choice of ρ^* is arbitrary*.

The x coordinate is defined only to within an additive constant, as a result of which the motion can depend only on a difference of coordinates, but not on the x coordinate itself. The difference in coordinates is the distance measured from the shock front, the coordinate of which will be denoted by X; the motion must depend on the dimensionless distance

$$\xi = \frac{X - x}{\Delta}. \tag{12.68}$$

This quantity is the similarity variable. Here, in contrast to all the other self-similar motions which have been considered, the similarity variable does not contain the units of time. The motion, of course, possesses a certain parameter A that characterizes the strength of the impact†. However, owing to the absence of another parameter with the units of mass it is impossible to construct from the quantities x, A, and Δ a combination with the dimension of time. Consequently, from the independent variables and the parameters x, t, A, and Δ it is impossible to construct a dimensionless variable which would contain the time t.

From the remarks made concerning the dimensional properties of the problem and the arbitrariness of the x coordinate, it is easy to find the relation governing the motion of the shock wave and to write down the general expressions for the unknown functions, the velocity, pressure, and density. The velocity of the shock front is

$$D = \dot{X} = \alpha \frac{\Delta}{t}, \tag{12.69}$$

where the numerical coefficient α depends only on the specific heat ratio γ.

* ρ^* is the density at the point $x = 0$, but we are justified in placing the coordinate origin $x = 0$ at any point at any density.

† It will be related below to the explosion energy.

The coordinate of the front X increases logarithmically with time as

$$X = \alpha \cdot \Delta \ln t + const. \tag{12.70}$$

The expressions for the velocity, density, and pressure of the gas behind the shock front have the form

$$\begin{aligned} u &= u_f \tilde{u} = \frac{2}{\gamma+1} \alpha \frac{\Delta}{t} \tilde{u}, \\ \rho &= \rho_f \tilde{\rho} = \frac{\gamma+1}{\gamma-1} \rho_0(X) \tilde{\rho}, \\ p &= p_f \tilde{p} = \frac{2}{\gamma+1} \alpha^2 \frac{\Delta^2}{t^2} \rho_0(X) \tilde{p}, \end{aligned} \tag{12.71}$$

where the dimensionless reduced functions $\tilde{u}$, $\tilde{\rho}$, and $\tilde{p}$ depend on γ and on the similarity variable ξ, the dimensionless distance measured from the shock front. The reduced functions are defined so that at the shock front ($\xi = 0$) they all become one,

$$\tilde{u}(0) = \tilde{\rho}(0) = \tilde{p}(0) = 1. \tag{12.72}$$

The other boundary condition is that in the vacuum at $x = -\infty$, $\xi = \infty$, $\tilde{p}(\infty) = 0$.

The density $\rho_0(X)$ of the gas directly ahead of the shock front is given in terms of the mass coordinate of the front M by (12.67). In contrast to the geometric coordinate, the mass of the gas encompassed by the shock wave depends on time as usual, following a power law. We have $\dot{M} = \rho_0(X)\dot{X} = M\dot{X}/\Delta = M\alpha t^{-1}$, from which

$$M = At^{\alpha}. \tag{12.73}$$

Here A is the constant of integration, and is also the parameter characterizing the strength of the impact. Its dimensions are $[A] = ML^{-2}T^{-\alpha}$. We can thus substitute into equations (12.71) the explicit dependence of $\rho_0(X)$ on time,

$$\rho_0(X) = \frac{M}{\Delta} = \frac{At^{\alpha}}{\Delta}. \tag{12.74}$$

As was noted before, the self-similarity of the motion is unusual: the velocity, density, and pressure distributions appear as though "tied" to the shock front and move together with the front, without expanding in time (only the amplitudes of these quantities change). However, in Lagrangian coordinates the motion is self-similar in the usual sense. The Lagrangian coordinate m is

$$m = \int_{-\infty}^{x} \rho(x)\, dx = const \cdot M \int_{\xi}^{\infty} \tilde{\rho}(\xi)\, d\xi.$$

Thus ξ and, consequently, also $\tilde{u}$, $\tilde{\rho}$, and $\tilde{p}$ are functions of the similarity variable $\eta = m/M = m/At^\alpha$.

The equations for the self-similar motion are most conveniently solved in terms of Lagrangian coordinates. Substituting (12.71) and (12.74) into the gasdynamic equations

$$\frac{\partial u}{\partial t} + \frac{\partial p}{\partial m} = 0, \qquad \frac{\partial (1/\rho)}{\partial t} - \frac{\partial u}{\partial m} = 0, \qquad p\rho^{-\gamma} = F(m),$$

we obtain equations for the reduced functions $\tilde{u}(\eta)$, $\tilde{\rho}(\eta)$, and $\tilde{p}(\eta)$

$$\begin{aligned} &\tilde{u} + \alpha\eta\tilde{u}' = \alpha\tilde{p}', \\ &\frac{1}{\tilde{\rho}} + \eta\left(\frac{1}{\tilde{\rho}}\right)' = -\frac{2}{\gamma - 1}\tilde{u}', \\ &\tilde{p}\tilde{\rho}^{-\gamma}\eta^{(2/\alpha)+\gamma-1} = 1. \end{aligned} \tag{12.75}$$

Integrating the second of these equations and eliminating $\tilde{\rho}$ and $\tilde{u}$ from the system, we obtain the basic equation for the problem

$$\frac{d\tilde{p}}{d\eta} = \frac{\gamma+1}{2\alpha} \frac{1 - \dfrac{\gamma-1}{\gamma+1}\left(1 - \dfrac{2-\alpha}{\gamma}\right)\tilde{p}^{-1/\gamma}\eta^{-(2-\alpha)/\alpha\gamma}}{1 - \dfrac{\gamma-1}{2\gamma}\tilde{p}^{-(1/\gamma)-1}\eta^{1-(2-\alpha)/\alpha\gamma}}. \tag{12.76}$$

The solution $\tilde{p}(\eta)$ must pass through the two points $\tilde{p}(1) = 1$ and $\tilde{p}(0) = 0$, and this condition determines the exponent α.

An exact solution to the problem may be obtained for the particular case $\gamma = 2$. We have*

$$\begin{aligned} &\alpha = \tfrac{3}{2}, \qquad M \sim t^{3/2}, \qquad D = \frac{3}{2}\frac{\Delta}{t}; \\ &u_f \sim \frac{1}{t}, \qquad \rho_f \sim t^{3/2}, \qquad p_f \sim \frac{1}{t^{1/2}}; \\ &\tilde{p} = \eta, \qquad \tilde{\rho} = \eta^{5/3}, \qquad \tilde{u} = \tfrac{3}{2}(1 - \tfrac{1}{3}\eta^{-2/3}). \end{aligned} \tag{12.77}$$

The solution in Eulerian coordinates has the form

$$\tilde{p} = (1 + 2\xi)^{-3/2}, \qquad \tilde{\rho} = (1 + 2\xi)^{-5/2}, \qquad \tilde{u} = 1 - \xi. \tag{12.78}$$

An analytic solution can also be obtained for the case $\gamma = 1 : \alpha = 1$, $\tilde{p} = \eta$, $\tilde{\rho} = \eta^3$, $\tilde{u} = 1$. This case is of interest only from the point of view of furnishing a limit on the similarity exponent α; it corresponds to infinite compression of

* The solution in Lagrangian coordinates is completely analogous to the analytic solution for the problem of an impulsive load in the case $\gamma = 7/5$. (See (12.42).)

the gas across the wave front, with the result that $\tilde{p}$, $\tilde{\rho}$, and $\tilde{u}$ in Eulerian coordinates become δ functions, proportional to $\delta(\xi)$. Since common values of γ are contained in the interval $1 < \gamma < 2$, we may assume that the corresponding values of the similarity exponent lie in the interval $1 < \alpha < 3/2$*.

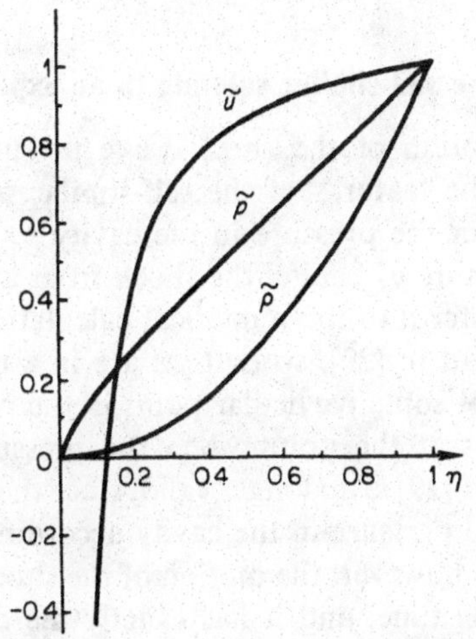

Fig. 12.22. Reduced velocity $\tilde{u}$, pressure $\tilde{p}$, and density $\tilde{\rho}$ as functions of the mass coordinate.

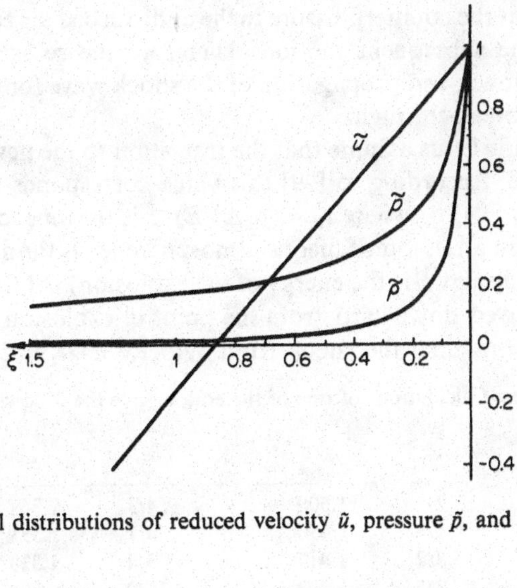

Fig. 12.23. Spatial distributions of reduced velocity $\tilde{u}$, pressure $\tilde{p}$, and density $\tilde{\rho}$ behind the shock wave.

* Consideration of energy and momentum balances, analogous to those given in §16, leads to the general limits $1 < \alpha < 2$.

For arbitrary values of γ the solutions can be found by trial and error, by numerical integration of (12.76). Figures 12.22 and 12.23 present the reduced velocity, density, and pressure distributions as functions of the mass and spatial coordinates which were obtained for $\gamma = 1.25$. In this case the exponent $\alpha = 1.345$*.

§24. Application of the self-similar solution to an explosion

The downward motion of the shock wave produced by a strong point explosion takes on the features of the self-similar motion described in the preceding section after the pressure in the cavity p_c becomes small in comparison with the pressure p_f behind the shock front at the lowest point. Both of these quantities enter into the numerical calculation of the motion of the shock wave carried out in [29], so that we are in a position to "tie-in" the self-similar solution at some particular point in the numerical calculation. It should be noted that at the point when the pressure in the cavity drops sharply the solution of [29] is no longer valid, since the force moving the shock wave vanishes at zero pressure in the cavity according to the approximation made there. Actually, however, the motion of the shock wave continues afterwards for quite a long time, and it has exactly the character of the motion ensuing as a result of an initial "push".

Numerical estimates show that the pressure in the cavity can become much less than the pressure behind the front while the front velocity is still sufficiently large that the counterpressure in the undisturbed air can still be neglected in treating the subsequent motion. Therefore the self-similar solution can describe the downward propagation of the shock wave for some time after the atmospheric break-through.

To be specific let us assume that the transition to the new regime is reached for $p_f/p_c = 10$. According to [29] this value corresponds to a time from the explosion of $t_1 = 19\tau_k$, where $\tau_k = (\rho_c \Delta^5/E)^{1/2}$ is the time scale characteristic of an explosion in an inhomogeneous atmosphere (ρ_c is the air density at the explosion altitude and E is the energy of the explosion). At the time t_1 the shock wave has moved downward from the point of explosion through a distance $z = 1.9\Delta$; at this time the shock front velocity is $D_1 = 2.5 \cdot 10^{-2} \cdot \Delta/\tau_k$.

* *Editors' note.* Calculations of one of the editors give the following values for α in this problem [33]:

γ	α	γ	α
2	1.500	4/3	1.369
5/3	1.450	9/7	1.351
3/2	1.417	5/4	1.338
7/5	1.392	11/9	1.324

Let us extrapolate the limiting shock propagation relations (12.69) and (12.70) to the time of transition to the new regime, with the position and time measured so that the initial condition $D = D_1$ when $X = 0$ is satisfied. Here the process up to the time of transition plays the role of the impulsive load. We obtain the approximate dependence of the front position on its velocity and on time (the coordinate is measured downward from the point of transition to the new regime), with $X = \alpha \cdot \Delta \ln(D_1/D) = \alpha\,\Delta \ln(t/\theta)$. The parameters D_1 and θ are determined in terms of the parameters of the explosion by the expressions $D_1 = 2.5 \cdot 10^{-2}(\Delta/\tau_k) = 2.5 \cdot 10^{-2}(E/\rho_c \Delta^3)^{1/2}$*, $\theta = 40\alpha\tau_k$, and $\alpha = 1.345$ for $\gamma = 1.25$. The impact parameter A is, to the same approximation, given by $A = e^{1.9}\rho_c \Delta\theta^{-\alpha} = 6.7\rho_c \Delta\theta^{-\alpha}$.

It has been estimated using actual numerical values† of the parameters that the distance through which the shock wave moves downward while decelerating from the transition velocity D_1 to the velocity $D \approx 1$ km/sec, a speed a few times the speed of sound in cold air, is approximately 2 to 3 times the scale height Δ. To this is added the distance of about 2Δ downwards from the explosion center, obtained from the theory given in [28, 29]. Thus a shock wave produced by a strong explosion moves downward from the point of the explosion through a distance of about 4 to 5 times Δ in decelerating to a velocity of the order of 1 km/sec.

§25. Self-similar motion of a shock wave in the direction of decreasing density. Application to an explosion

We now consider the self-similar propagation of a shock wave in the exponential atmosphere of (12.66) in the direction from $+\infty$ to $-\infty$. This problem is analogous to that of the emergence of a shock wave at the surface of a star (see Part 3 of this chapter) with the sole difference that there the atmospheric density did not have an exponential but rather a power-law behavior. However, it is the exponential character of the atmosphere which imparts to the motion its special features. The problem as formulated was solved by one of the present authors [31]. It is clear that this solution can also be used to describe the motion of the upper part of the shock wave

* The numerical values of the parameters D_1 and θ depend only weakly on the choice of the transition value of p_f/p_c. Thus, for example, for the last time calculated in [29], $t = 23.4\tau_k$, which was close to the time of the atmospheric break-through, $z \approx 2\Delta$, $D = 2.12 \cdot 10^{-2}\Delta/\tau_k$, $p_f/p_c = 22$.

† *Editors' note.* This estimate appears to be for a high energy, high altitude explosion, say of the order of 10^{24} erg at 100 km altitude. Only a high energy, high altitude explosion will "break-through" or "vent" in the sense of the Kompaneets model. The reader should be warned that at high altitudes the hydrodynamic model of a strong explosion fails, because of the large values of photon mean free paths at very low densities.

resulting from a strong explosion and those neighboring regions which, some time after the explosion, "break loose" from the sphere of influence of the central regions.

We assume that the shock wave emerges at the "boundary" of the atmosphere $x = -\infty$, where $\rho_0 = 0$, at the time $t = 0$; thus we take the time prior to emergence as negative. All our considerations regarding the dimensional properties of the motion in an exponential atmosphere presented at the beginning of §23 hold here as well. Therefore, all of the equations of §23 are applicable. The only feature which must be taken into account is the change in the sign of the time. The front velocity is as before

$$D = \dot{X} = \alpha \frac{\Delta}{t}, \qquad t < 0, \quad D < 0,$$

and the coordinate of the front is now given by

$$X = \alpha \, \Delta \ln(-t) + const.$$

The Lagrangian coordinate of the front is

$$M = A(-t)^{\alpha}, \qquad \alpha > 0.$$

As before, the constant A characterizes the strength of the shock wave source*. Equations (12.71), (12.75), and the basic differential equation (12.76) remain unchanged, and only the region of integration changes. Previously $\eta = m/M$ varied from 1 at the wave front to 0 in the low-density regions. In the present case it varies from 1 at the front to ∞ in the high-density regions. The boundary conditions at the front remain the same as before, while the boundary condition for $\eta = \infty$, which corresponds to the time $t = 0$ $(M = 0)$, is defined so that in the limit $t \to 0$ time disappears from equations (12.71).

Results of numerical calculations for the two specific heat ratios $\gamma = 1.2$ and $\gamma = 5/3$ are shown in Figs. 12.24 and 12.25. The similarity exponents were found to be $\alpha = 6.48$ and $\alpha = 4.90$†, respectively. For sufficiently large η the equations can be solved by means of an approximate analytic solution

* Of course, the similarity exponent α and constant A are unrelated to the corresponding quantities of §23, since the problem is completely different.

† *Editors' note.* Calculations of one of the editors give the following values for α in this problem [33]:

γ	α	γ	α
2	4.57	4/3	5.68
5/3	4.90	9/7	5.90
3/2	5.18	5/4	6.10
7/5	5.45	11/9	6.29
		6/5	6.47

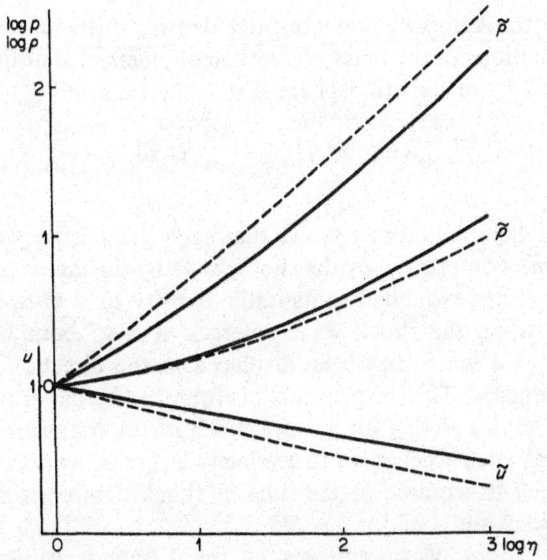

Fig. 12.24. Distributions of reduced pressure $\tilde{p}$, density $\tilde{\rho}$, and velocity $\tilde{u}$ as functions of the mass coordinate for the upward motion of the shock wave. The solid curves are for $\gamma = 1.2$ and the dashed ones for $\gamma = 5/3$.

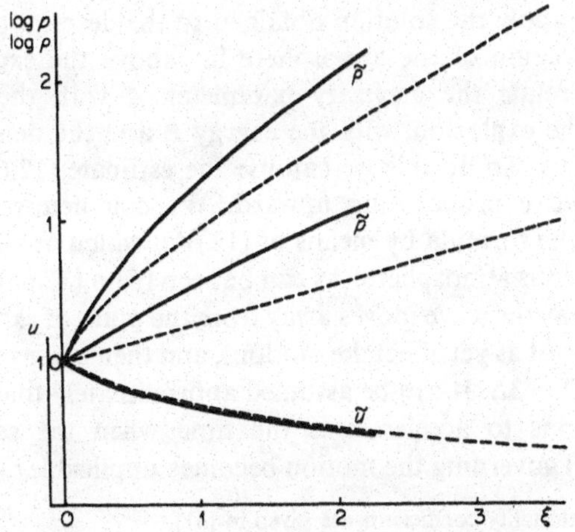

Fig. 12.25. Spatial distributions of reduced pressure $\tilde{p}$, density $\tilde{\rho}$, and velocity $\tilde{u}$ for the upward motion of the shock wave. The solid curves are for $\gamma = 1.2$ and the dashed ones for $\gamma = 5/3$.

from which the velocity, pressure, and density distributions can be determined as functions of the mass at the time of emergence of the shock wave at the "boundary" of the atmosphere. From this solution

$$(u)_{t=0} \sim -m^{-1/2}, \qquad (p)_{t=0} \sim m^{1-2/\alpha}, \qquad (\rho)_{t=0} = \frac{m}{\Delta}^{*}.$$

The third of these equations shows that each gas particle (with Lagrangian coordinate m), compressed by the shock wave by the factor $(\gamma+1)/(\gamma-1)$, at the time $t = 0$ has expanded to its initial density $\rho_0 = m/\Delta$. This means that at the time when the shock wave emerges at the "boundary", the initial atmosphere as a whole has been displaced in the direction of motion by a definite distance d. This displacement is found to be equal to $d/\Delta = 7.50$ for $\gamma = 1.2$ and to $d/\Delta = 4.57$ for $\gamma = 5/3$. Each of the particles of the gas at the time $t = 0$ has been accelerated to a velocity larger by a known factor than the velocity which it acquired at the time of shock compression. This factor is 1.54 for $\gamma = 1.2$ and 1.85 for $\gamma = 5/3$.

After the shock wave emerges at the "boundary" (after "breaking through" the atmosphere) and $t > 0$, the gas continues to expand in the direction of the vacuum. This outflow remains self-similar, since it is characterized by the same pair of dimensional parameters Δ and A, the similarity exponent α remaining unchanged. It was for this motion that the approximate analytic solution of [31] was obtained. It was shown that after "breaking through" the atmosphere the gas particles fly upward practically without further acceleration, by inertia at constant velocity.

In order to apply the solution obtained to the description of the flow field in the upper region of the atmosphere far above the explosion point, it is necessary to relate the arbitrary parameter A with the quantities which characterize the explosion, with the energy E and the density ρ_c at the level of the explosion. To do this we can use the estimate (12.65) for the velocity of a shock wave propagating upward. We can determine the numerical coefficient in the formula by means of (12.64), which applies to an explosion in a homogeneous atmosphere. As can be seen from (12.65), the wave velocity first decreases as the wave moves away from the point of explosion (the density decrease does not as yet affect the motion), and then begins to increase. It has a minimum at $R = 3\Delta$. It can be assumed approximately that the time at which the wave begins to accelerate is the time when the self-similar relation $|D| = \alpha\Delta/(-t)$ governing the motion becomes applicable†. From this relation

* The proportionality coefficients are given in [31].

† *Editors' note.* The effect of the curvature of the rising shock on its propagation in an exponential atmosphere is not negligible. Approximate calculations of one of the editors for $\gamma = 7/5$ give $\alpha = 7.89$ for a rising curved shock wave in place of the 5.45 valid for the plane shock [34].

the time required by the wave to move to infinity is found to be equal to

$$\tau = \frac{\alpha\Delta}{|D|_{\min}} = const\left(\frac{\Delta^5}{\rho_c E}\right)^{1/2}*$$

For $\gamma = 1.2$ the constant is equal to 25, which is almost exactly the value found in [29]. Consequently, the same result could have been obtained by "tying" the self-similar solution to the calculations of [29]. Knowing the time $t = -\tau$ when the wave is at a distance $R = 3\Delta$ above the point of explosion, i.e., at a point with a known Lagrangian coordinate, we can also determine A,

$$A = const\ E^{\alpha/2}\Delta^{1-5\alpha/2}\rho_c^{1-\alpha/2}.$$

In conclusion we note that in principle, the air layers accelerated by the shock wave to high velocities, exceeding the earth's escape velocity, should have broken loose from the earth's gravitational field and "spilled out" into space. However, owing to the ionization of the air that has been highly heated by the strong shock wave, its upward expansion is limited by the decelerating effect of the earth's magnetic field. Some problems relating to the expansion of a highly rarefied plasma into a vacuum containing a magnetic field were considered in [32].

* The time for the wave to move from the point of the explosion to the point $R = 3\Delta$ is much smaller than τ.

Cited references

Editors' note:

In the references cited we have used, as far as possible, the abbreviations for journals and reports used by *Chemical Abstracts*. A list of these abbreviations may be found in the List of Periodicals of the Chemical Abstracts Service published by the American Chemical Society.

Transliteration of Russian names has essentially followed the system adopted by the Library of Congress, but with no distinction between e and ë or between и and й, and with yu used for ю and ya for я. In the case of books translated from Russian into English the transliterated author names are those appearing on the translation. Russian titles have been translated into English, but where a translation is indicated the title given is that appearing on the translated version. A source of an English translation for all cited Russian references has been given whenever known to the editors.

Chapter VII

1. Landau, L. D., and Lifshitz, E. M.
   Fluid Mechanics. Addison-Wesley, Reading, Mass., 1959.
2. Becker, R.
   Stosswelle und Detonation, *Z. Physik* **8**, 321–362 (1922).
3. Morduchow, M., and Libby, P. A.
   On a complete solution of the one-dimensional flow equations of a viscous, heat conducting, compressible gas, *J. Aeron. Sci.* **16**, 674–684, 704 (1949).
4. Meyerhoff, L.
   An extension of the theory of the one-dimensional shock-wave structure, *J. Aeron. Sci.* **17**, 775–786 (1950).
5. Thomas, L. H.
   Note on Becker's theory of the shock front, *J. Chem. Phys.* **12**, 449–453 (1944).
6. Herpin, A.
   La théorie cinétique de l'onde de choc, *Rev. Sci.* **86**, 35–37 (1948).
7. Puckett, A. E., and Stewart, H. J.
   The thickness of a shock wave in air, *Quart. Appl. Math.* **7**, 457–463 (1950).
8. von Mises, R.
   On the thickness of a steady shock wave, *J. Aeron. Sci.* **17**, 551–554, 594 (1950).
9. Lieber, P., Romano, F., and Lew, H.
   Approximate solutions for shock waves in a steady, one-dimensional, viscous and compressible gas, *J. Aeron. Sci.* **18**, 55–60 (1951).
10. Gilbarg, D., and Paolucci, D.
    The structure of shock waves in the continuum theory of fluids, *J. Rational Mech. Anal.* **2**, 617–642 (1953).

11. Bernard, J. J.
Thickness of a steady shock wave, *J. Aeron. Sci.* **18**, 210 (1951).
12. Roy, M.
Sur la structure de l'onde de choc, limite d'une quasi-onde de choc dans un fluide compressible et visqueux, *Compt. Rend.* **218**, 813–816 (1944).
13. Libby, P. A.
The effect of Prandtl number on the theoretical shock-wave thickness, *J. Aeron. Sci.* **18**, 286–287 (1951).
14. Zoller, K.
Zur Struktur des Verdichtungsstosses, *Z. Physik* **130**, 1–38 (1951).
15. Cowan, G. R., and Hornig, D. F.
The experimental determination of the thickness of a shock front in a gas, *J. Chem. Phys.* **18**, 1008–1018 (1950).
Greene, E. F., Cowan, G. R., and Hornig, D. F.
The thickness of shock fronts in argon and nitrogen and rotational heat capacity lags, *J. Chem. Phys.* **19**, 427–434 (1951).
Greene, E. F., and Hornig, D. F.
The shape and thickness of shock fronts in argon, hydrogen, nitrogen, and oxygen, *J. Chem. Phys.* **21**, 617–624 (1953).
16. Mott-Smith, H. M.
The solution of the Boltzmann equation for a shock wave, *Phys. Rev.* **82**, 885–892 (1951).
17. Sakurai, A.
A note on Mott-Smith's solution of the Boltzmann equation for a shock wave, *J. Fluid Mech.* **3**, 255–260 (1957).
Note on Mott-Smith's solution of the Boltzmann equation for a shock wave II, *Research Report, Tokyo Electrical Engineering College*, 1958, No. 6, 49–51.
18. Lord Rayleigh
Aerial plane waves of finite amplitude, *Proc. Roy. Soc.* (*London*), *Ser. A* **84**, 247–284 (1910).
19. Chapman, S., and Cowling, T. G.
The Mathematical Theory of Non-Uniform Gases, Cambridge Univ. Press, New York, 2nd edition, 1958.
Frank-Kamenetskii, D. A.
Diffusion and Heat Exchange in Chemical Kinetics. Izdat. Akad. Nauk SSSR, Moscow, 1947. English transl. by N. Thon, Princeton Univ. Press, Princeton, 1955.
19a. Zhdanov, V., Kagan, Yu., and Sazykin, A.
Effect of viscous transfer of momentum on diffusion in a gas mixture, *Soviet Phys. JETP* (*English Transl.*) **15**, 596–602 (1962).
20. D'yakov, S. P.
Shock waves in binary mixtures, *Zh. Eksperim. i Teor. Fiz.* **27**, 283–287 (1954).
21. Sherman, F. S.
Shock-wave structure in binary mixtures of chemically inert perfect gases, *J. Fluid Mech.* **8**, 465–480 (1960).
22. Cowling, T. G.
The influence of diffusion on the propagation of shock waves, *Phil. Mag.* **33** (7th Series), 61–67 (1942).
23. Zel'dovich, Ya. B.
Propagation of shock waves in a gas in the presence of a reversible chemical reaction, *Zh. Eksperim. i Teor. Fiz.* **16**, 365–368 (1946).
24. Zel'dovich, Ya. B.

Theory of Shock Waves and an Introduction to Gasdynamics. Izdat. Akad. Nauk SSSR, Moscow, 1946.

25. Griffith, W., Brickl, D., and Blackman, V.
Structure of shock waves in polyatomic gases, *Phys. Rev.* **102**, 1209–1216 (1956).
26. Blackman, V.
Vibrational relaxation in oxygen and nitrogen, *J. Fluid Mech.* **1**, 61–85 (1956).
27. Matthews, D. L.
Interferometric measurement in the shock tube of the dissociation rate of oxygen *Phys. Fluids* **2**, 170–178 (1959).
28. Losev, S. A.
Investigation of the dissociation process of oxygen behind a strong shock wave, *Dokl. Akad. Nauk SSSR* **120**, 1291–1293 (1958).
29. Generalov, N. A., and Losev, S. A.
On an investigation of nonequilibrium phenomena behind the front of a shock wave in air. Dissociation of oxygen, *Zh. Prikl. Mekhan. i Tekhn. Fiz.*, 1960, No. 2, 64–73.
30. Britton, D., Davidson, N., Gehman, W., and Schott, G.
Shock waves in chemical kinetics: further studies on the rate of dissociation of molecular iodine, *J. Chem. Phys.* **25**, 804–809 (1956).
Britton, D., and Davidson, N.
Shock waves in chemical kinetics. Rate of dissociation of molecular bromine, *J. Chem. Phys.* **25**, 810–813 (1956).
Palmer, H. B., and Hornig, D. F.
Rate of dissociation of bromine in shock waves, *J. Chem. Phys.* **26**, 98–105 (1957).
31. Losev, S. A., and Osipov, A. I.
The study of nonequilibrium phenomena in shock waves, *Soviet Phys.–Usp.* (*English Transl.*) **4**, 525–552 (1962).
32. Duff, R. E., and Davidson, N.
Calculation of reaction profiles behind steady state shock waves. II. The dissociation of air, *J. Chem. Phys.* **31**, 1018–1027 (1959).
33. Losev, S. A., and Generalov, N. A.
On the nonequilibrium state behind a shock wave in air, *Dokl. Akad. Nauk SSSR* **133**, 872–874 (1960).
34. Biberman, L. M., and Veklenko, B. A.
Radiative processes ahead of a shock-wave front, *Soviet Phys. JETP* (*English Transl.*) **10**, 117–120 (1960).
35. Petschek, H., and Byron, S.
Approach to equilibrium ionization behind strong shock waves in argon, *Ann. Phys.* (*N.Y.*) **1**, 270–315 (1957).
36. Bond, J. W., Jr.
Structure of a shock front in argon, *Phys. Rev.* **105**, 1683–1694 (1957).
37. Rostagni, A.
Ricerche sui raggi positivi e neutrali. V. Ionizzazione per urto di ioni e di atomi, *Nuovo Cimento* **13**, 389–406 (1936).
38. Wayland, H.
The ionization of neon, krypton and xenon by bombardment with accelerated neutral argon atoms, *Phys. Rev.* **52**, 31–37 (1937).
39. Weymann, H. D.
Electron diffusion ahead of shock waves in argon, *Phys. Fluids* **3**, 545–548 (1960).
40. Manheimer-Timnat, Y., and Low, W.
Electron density and ionization rate in thermally ionized gases produced by medium strength shock waves, *J. Fluid Mech.* **6**, 449–461 (1959).

Niblett, B., and Blackman, V. H.
An approximate measurement of the ionization time behind shock waves in air, *J. Fluid Mech.* **4**, 191–194 (1958).

41. Hammerling, P., Teare, J. D., and Kivel, B.
Theory of radiation from luminous shock waves in nitrogen, *Phys. Fluids* **2**, 422–426 (1959).

42. Zel'dovich, Ya. B.
Shock waves of large amplitude in air, *Soviet Phys. JETP* (*English Transl.*) **5**, 919–927 (1957).

43. Shafranov, V. D.
The structure of shock waves in a plasma, *Soviet Phys. JETP* (*English Transl.*) **5**, 1183–1188 (1957).

44. Jukes, J. D.
The structure of a shock wave in a fully ionized gas, *J. Fluid Mech.* **3**, 275–285 (1957).

44a. Tidman, D. A.
Structure of a shock wave in fully ionized hydrogen, *Phys. Rev.* **111**, 1439–1446 (1958).

45. Greenberg, O. W., Sen, H. K., and Trève, Y. M.
Hydrodynamic model of diffusion effects on shock structure in a plasma, *Phys. Fluids* **3**, 379–386 (1960).

46. Krook, M.
Structure of shock fronts in ionized gases, *Ann. Phys.* (*N.Y.*) **6**, 188–207 (1959).
Bond, J. W., Jr.
Plasma physics and hypersonic flight, *Jet Propulsion* **28**, 228–235 (1958).

47. Raizer, Yu. P.
On the structure of the front of strong shock waves in gases, *Soviet Phys. JETP* (*English Transl.*) **5**, 1242–1248 (1957).

48. Raizer, Yu. P.
On the brightness of strong shock waves in air, *Soviet Phys. JETP* (*English Transl.*) **6**, 77–84 (1958).

49. Zel'dovich, Ya. B., and Raizer, Yu. P.
Strong shock waves in gases, *Usp. Fiz. Nauk* **63**, 613–641 (1957).

50. Belokon', V. A.
Disappearance of the isothermal jump at large radiation density, *Soviet Phys. JETP* (*English Transl.*) **9**, 235–236 (1959).

51. Imshennik, V. S.
Shock wave structure in a dense high-temperature plasma, *Soviet Phys. JETP* (*English Transl.*) **15**, 167–174 (1962).
Numerical integration of differential equations of the structure of shockwaves in plasma, *U.S.S.R. Comp. Math. Math. Phys.* (*English Transl.*) **2**, 217–229 (1963).

52. Gustafson, W. A.
On the Boltzmann equation and the structure of shock waves, *Phys. Fluids* **3**, 732–734 (1960).

53. Muckenfuss, C.
Bimodal model for shock wave structure, *Phys. Fluids* **3**, 320–321 (1960).

54. Ziering, S., and Ek, F.
Mean-free-path definition in the Mott-Smith shock wave solution, *Phys. Fluids* **4**, 765–766 (1961).

55. Glansdorff, P.
Solution of the Boltzmann equations for strong shock waves by the two-fluid model, *Phys. Fluids* **5**, 371–379 (1962).

56. Hansen, K., and Hornig, D. F.
Thickness of shock fronts in argon, *J. Chem. Phys.* **33**, 913–916 (1960).
57. Blythe, P. A.
Comparison of exact and approximate methods for analysing vibrational relaxation regions, *J. Fluid Mech.* **10**, 33–47 (1961).
58. Anisimov, S. I.
On the attainment of oscillatory equilibrium behind a shock wave, *Soviet Phys.–Tech. Phys.* (*English Transl.*) **6**, 1089–1090 (1962).
59. Generalov, N. A.
Vibrational relaxation in oxygen at high temperatures. I, *Vestn. Mosk. Univ., Ser. III: Fiz., Astron.*, 1962, No. 3, 51–59.
60. Camac, M.
O_2 vibrational relaxation in oxygen-argon mixtures, *J. Chem. Phys.* **34**, 448–459 (1961).
61. Roth, W.
Shock tube study of vibrational relaxation in the $A\ {}^2\Sigma^+$ state of NO, *J. Chem. Phys.* **34**, 999–1003 (1961).
62. Matthews, D. L.
Vibrational relaxation of carbon monoxide in the shock tube, *J. Chem. Phys.* **34**, 639–642 (1961).
63. Johannesen, N. H., Zienkiewicz, H. K., Blythe, P. A., and Gerrard, J. H.
Experimental and theoretical analysis of vibrational relaxation regions in carbon dioxide, *J. Fluid Mech.* **13**, 213–224 (1962).
64. Hurle, I. R., and Gaydon, A. G.
Vibrational relaxation and dissociation of carbon dioxide behind shock waves, *Nature* **184**, 1858–1859 (1959).
65. Camac, M., and Vaughan, A.
O_2 dissociation rates in O_2-Ar mixtures, *J. Chem. Phys.* **34**, 460–470 (1961).
66. Rink, J. P., Knight, H. T., and Duff, R. E.
Shock tube determination of dissociation rates of oxygen, *J. Chem. Phys.* **34**, 1942–1947 (1961).
67. Patch, R. W.
Shock-tube measurement of dissociation rates of hydrogen, *J. Chem. Phys.* **36**, 1919–1924 (1962).
68. Allen, R. A., Keck, J. C., and Camm, J. C.
Nonequilibrium radiation and the recombination rate of shock-heated nitrogen, *Phys. Fluids* **5**, 284–291 (1962).
69. Wray, K. L., Teare, J. D., Kivel, B., and Hammerling, P.
Relaxation processes and reaction rates behind shock fronts in air and component gases, *Eighth Symposium* (*International*) *on Combustion*, pp. 328–339. Williams & Wilkins, Baltimore, 1962.
70. Lin, S. C.
Low density shock tube studies of reaction rates related to the high altitude hypersonic flight problem, *Rarefied Gas Dynamics* (L. Talbot, ed.), pp. 623–642. Academic Press, New York, 1961.
71. Sayasov, Yu. S.
On the kinetics of oxidation of nitrogen in a normal shock wave, *Zh. Prikl. Mekhan. i Tekhn. Fiz.*, 1962, No. 1, 61–67.
On the structure of an oblique shock wave in a chemically reacting gas, *Zh. Prikl. Mekhan. i Tekhn. Fiz.*, 1961, No. 6, 172–174.
72. Kuznetsov, N. M.
Shock wave structure in air taking into account the kinetics of chemical reactions, *Inzh.-Fiz. Zh., Akad. Nauk Belorussk.*, 1960, No. 9, 17–24.

The kinetics of chemical reactions for expanding air, *Inzh.-Fiz. Zh., Akad. Nauk Belorussk.*, 1962, No. 6, 97–101.

73. Bortner, M. H.
The effect of errors in rate constants on non-equilibrium shock layer electron density calculations, *Planetary Space Sci.* **6**, 74–78 (1961).

74. Lun'kin, Yu. P.
Measurement of entropy in the relaxation of a gas mixture behind a shock wave, *Soviet Phys.–Tech. Phys.* (*English Transl.*) **6**, 810–814 (1962).

75. Dorrance, W. H.
On the approach to chemical and vibrational equilibrium behind a strong normal shock wave, *J. Aerospace Sci.* **28**, 43–50 (1961).

76. Wetzel, L.
Precursor effects and electron diffusion from a shock front, *Phys. Fluids* **5**, 824–830 (1962).

77. Pipkin, A. C.
Diffusion from a slightly ionized region in a uniform flow, *Phys. Fluids* **4**, 1298–1302 (1961).

78. Bortner, M. H.
Shock layer electron densities considering the effects of both chemical reactions and flow field variations, *Planetary Space Sci.* **3**, 99–103 (1961).

79. Blackman, V. H., and Niblett, G. B. F.
Ionization processes in shock waves, *Fundamental Data Obtained from Shock Tube Experiments* (A. Ferri, ed.), pp. 221–241. AGARDograph No. 41, Pergamon Press, New York, 1961.

80. Lin, S. C.
Rate of ionization behind shock waves in air, *Planetary Space Sci.* **6**, 94–99 (1961).

81. Lamb, L., and Lin, S. C.
Electrical conductivity of thermally ionized air produced in a shock tube, *J. Appl. Phys.* **28**, 754–759 (1957).

82. Sisco, W. B., and Fiskin, J. M.
Basic hypersonic plasma data of equilibrium air for electromagnetic and other requirements, *Planetary Space Sci.* **6**, 47–73 (1961).

83. Viegas, J. R., and Peng, T. C.
Electrical conductivity of ionized air in thermodynamic equilibrium, *ARS J.* **31**, 654–657 (1961).

84. Sherman, A.
Calculation of electrical conductivity of ionized gases, *ARS J.* **30**, 559–560 (1960).

85. Pikel'ner, S. B.
Spectrophotometric study of the mechanism of excitation of filamentary nebulae, *Izv. Krymsk. Astrofiz. Observ.* **12**, 93–117 (1954).
Principles of Cosmic Electrodynamics. Fizmatgiz, Moscow, 1961. English transl., Gordon and Breach, New York, 1967.

86. Lin, S. C., and Teare, J. D.
Rate of ionization behind shock waves in air. II. Theoretical interpretations, *Phys. Fluids* **6**, 355–375 (1963).

87. Lin, S. C., Neal, R. A., and Fyfe, W. I.
Rate of ionization behind shock waves in air. I. Experimental results, *Phys. Fluids* **5**, 1633–1648 (1962).

88. Imshennik, V. S., and Morozov, Yu. I.
Shock wave structure taking into account momentum and energy transfer by radiation, *Zh. Prikl. Mekhan. i Tekhn. Fiz.*, 1964, No. 2, 8–21.

89. Jaffrin, M. Y., and Probstein, R. F.
Structure of a plasma shock wave, *Phys. Fluids* **7**, 1658–1674 (1964).
90. Stupochenko, E. V., Losev, S. A., and Osipov, A. I.
Relaxation Processes in Shock Waves. Nauka, Moscow, 1965.
91. Wray, K. L., and Freeman, T. S.
Shock front structure in O_2 at high Mach numbers, *J. Chem. Phys.* **40**, 2785–2789 (1964).
92. Nelson, W. C. (ed.)
The High Temperature Aspects of Hypersonic Flow. AGARDograph No. 68, Pergamon Press, New York, 1964.
93. Biberman, L. M., and Yakubov, I. T.
Approach to ionization equilibrium behind the front of a shock wave in an atomic gas, *Soviet Phys.–Tech. Phys.* (*English Transl.*) **8**, 1001–1007 (1964).
94. Gloersen, P.
Precursor signals from shock waves in xenon, *Bull. Am. Phys. Soc.* **4**, 283 (1959).
95. Kornegay, W. M., and Johnston, H. S.
Kinetics of thermal ionization. II. Xenon and krypton, *J. Chem. Phys.* **38**, 2242–2247 (1963).
96. Kuznetsov, N. M.
The influence of radiation on the ionization structure of the front of a shock wave, *Soviet Phys.–Tech. Phys.* (*English Transl.*) **9**, 483–487 (1965).
97. Biberman, L. M., and Yakubov, I. T.
The state of a gas behind a strong shock-wave front, *High Temp.* (*English Transl.*) **3**, 309–320 (1965).
98. Bronshten, V. A.
Problems on the Motion of Large Meteorites in the Atmosphere. Izdat. Akad. Nauk SSSR, Moscow, 1963.
Bronshten, V. A., and Chigorin, A. N.
Establishment of equilibrium ionization in a strong shock wave in air, *High Temp.* (*English Transl.*) **2**, 774–781 (1964).
99. Biberman, L. M., and Ul'yanov, K. N.
The effect of the emission of radiation on deviation from thermodynamic equilibrium, *Opt. Spectr.* (*USSR*) (*English Transl.*) **16**, 216–220 (1964).
100. Kohler, M.
Reibung in mässig verdünnten Gasen als Folge verzögerter Einstellung der Energie, *Z. Physik* **125**, 715–732 (1949).
101. Tamm, I. E.
On the thickness of a strong shock wave, *Tr. Fiz. Inst. Akad. Nauk SSSR* **29**, 239–249 (1965). (Work completed in 1947.)

Chapter VIII

1. Kantrowitz, A.
Effects of heat capacity lag in gas dynamics, *J. Chem. Phys.* **10**, 145 (1942).
Heat-capacity lag in gas dynamics, *J. Chem. Phys.* **14**, 150–164 (1946).
2. Bloom, M. H., and Steiger, M. H.
Inviscid flow with nonequilibrium molecular dissociation for pressure distributions encountered in hypersonic flight, *J. Aerospace Sci.* **27**, 821–835, 840 (1960).
2a. Li, Ting Y.
Recent advances in nonequilibrium dissociating gasdynamics, *ARS J.* **31**, 170–178 (1961).

3. Kneser, H. O.
Zur Dispersionstheorie des Schalles, *Ann. Physik* **11** (5th Series), 761–776 (1931).
4. Kneser, H. O.
Die Dispersion hochfrequenter Schallwellen in Kohlensäure, *Ann. Physik* **11** (5th Series), 777–801 (1931).
5. Einstein, A.
Schallausbreitung in teilweise dissoziierten Gasen, *Sitzber. Berliner Akad. Wiss.* 1920, 380–385.
6. Nozdrev, V. F.
Application of Ultrasonics in Molecular Physics. Fizmatgiz, Moscow, 1958. English transl., Gordon and Breach, New York, 1963.
7. Gorelik, G. S.
Vibrations and Waves. An Introduction in Acoustics, Radiophysics, and Optics. Fizmatgiz, Moscow, 2nd edition, 1959.
8. Mandel'shtam, L. I., and Leontovich, M. A.
On the theory of sound absorption in fluids, *Zh. Eksperim. i Teor. Fiz.* **7**, 438–449 (1937).
9. Landau, L. D., and Lifshitz, E. M.
Fluid Mechanics. Addison-Wesley, Reading, Mass., 1959.
10. Landau, L. D., and Teller, E.
Zur Theorie der Schalldispersion, *Physik. Z. Sowjetunion* **10**, 34–43 (1936).
11. Ginzburg, V. L.
On a general relation between absorption and dispersion of sound waves, *Akust. Zh.* **1**, 31–39 (1955).
12. U.S. Dept. of Defense
The Effects of Atomic Weapons. McGraw-Hill, New York, 1950.
13. Raizer, Yu. P.
The formation of nitrogen oxides in the shock wave of a strong explosion in air, *Zh. Fiz. Khim.* **33**, 700–709 (1959).
14. Zel'dovich, Ya. B., Sadovnikov, P. Ya., and Frank-Kamenetskii, D. A.
The Oxidation of Nitrogen by Combustion. Izdat. Akad. Nauk SSSR, Moscow, 1947.
15. Tsukerman, V. A., and Manakova, M. A.
Sources of short X-ray pulses for investigating fast processes, *Soviet Phys.–Tech. Phys.* (*English Transl.*) **2**, 353–363 (1957).
15a. Molmud, P.
Expansion of a rarefied gas cloud into a vacuum, *Phys. Fluids* **3**, 362–366 (1960).
16. Belokon', V. A.
Tr. Mosk. Fiz. Tekhn. Inst., 1963, No. 11.
17. Raizer, Yu. P.
Residual ionization of a gas expanding in vacuum, *Soviet Phys. JETP* (*English Transl.*) **10**, 411–412 (1960).
18. Landau, L. D., and Lifshitz, E. M.
Statistical Physics. Addison-Wesley, Reading, Mass., 1958.
19. Raizer, Yu. P.
Condensation of a cloud of vaporized matter expanding in vacuum, *Soviet Phys. JETP* (*English Transl.*) **10**, 1229–1235 (1960).
20. Zel'dovich, Ya. B., and Raizer, Yu. P.
Physical phenomena that occur when bodies compressed by strong shock waves expand in vacuo, *Soviet Phys. JETP* (*English Transl.*) **8**, 980–982 (1959).
21. Frenkel, J.
Kinetic Theory of Liquids. Oxford Univ. Press, London, 1946. Republished, Dover, New York, 1955.

22. Zel'dovich, Ya. B.
Theory of the formation of a new phase. Cavitation, *Zh. Eksperim. i Teor. Fiz.* **12**, 525–538 (1942).
23. Fesenkov, V. G.
Meteoric Material in Interplanetary Space. Izdat. Akad. Nauk SSSR, Moscow, 1947.
24. Narasimha, R.
Collisionless expansion of gases into vacuum, *J. Fluid Mech.* **12**, 294–308 (1962).
25. Pressman, A. Ya.
On the flow of a rarefied gas into a vacuum from a point source, *Soviet Phys. "Doklady"* (*English Transl.*) **6**, 451–453 (1961).
26. Raizer, Yu. P.
Note on the sudden expansion of a gas cloud into vacuum, *Zh. Prikl. Mekhan. i Tekhn. Fiz.*, 1964, No. 3, 162–163.
27. Raizer, Yu. P.
The deceleration and energy conversions of a plasma expanding in a vacuum in the presence of a magnetic field, *Zh. Prikl. Mekhan. i Tekhn. Fiz.*, 1963, No. 6, 19–28. Transl. as *NASA* (*Nat. Aeron. Space Admin.*) *Tech. Transl.* No. TTF-239 (1964).
28. Kuznetsov, N. M., and Raizer, Yu. P.
On the recombination of electrons in a plasma expanding into vacuum, *Zh. Prikl. Mekhan. i Tekhn. Fiz.*, 1965, No. 4, 10–20.

Chapter IX

1. Model', I. Sh.
Measurement of high temperatures in strong shock waves in gases, *Soviet Phys. JETP* (*English Transl.*) **5**, 589–601 (1957).
2. Zel'dovich, Ya. B.
Shock waves of large amplitude in air, *Soviet Phys. JETP* (*English Transl.*) **5**, 919–927 (1957).
Zel'dovich, Ya. B., and Raizer, Yu. P.
Strong shock waves in gases, *Usp. Fiz. Nauk* **63**, 613–641 (1957).
3. Raizer, Yu. P.
On the structure of the front of strong shock waves in gases, *Soviet Phys. JETP* (*English Transl.*) **5**, 1242–1248 (1957).
4. Raizer, Yu. P.
On the brightness of strong shock waves in air, *Soviet Phys. JETP* (*English Transl.*) **6**, 77–84 (1958).
5. Vanyukov, M. P., and Mak, A. A.
High-intensity pulsed light sources, *Soviet Phys.–Usp.* (*English Transl.*) **1**, 137–155 (1958).
6. Schneider, E. G.
An estimate of the absorption of air in the extreme ultraviolet, *J. Opt. Soc. Am.* **30**, 128–132 (1940).
7. Aglintsev, K. K.
Dosimetry of Ionizing Radiation. Gostekhizdat, Moscow, 2nd edition, 1957.
8. Landolt, H. H.
Landolt-Börnstein Zahlenwerte und Funktionen aus Physik, Chemie, Astronomie, Geophysik und Technik. Vol. I. Atom-und Molekularphysik, p. 316. Springer, Berlin, 6th edition, 1950.

9. Messner, R. H.
Der Einfluss der chemischen Bindung auf den Absorptionskoeffizienten leichter Elemente im Gebiete ultraweicher Röntgenstrahlen, *Z. Physik* **85**, 727–740 (1933).
10. Dershem, E., and Schein, M.
The absorption of the $K\alpha$ line of carbon in various gases and its dependence upon atomic number, *Phys. Rev.* **37**, 1238–1245 (1931).
11. U.S. Dept. of Defense
The Effects of Atomic Weapons. McGraw-Hill, New York, 1950.
12. Glasstone, S. (ed.)
The Effects of Nuclear Weapons. U.S. Atomic Energy Comm., Washington, revised edition, 1962 (1st edition, 1957).
13. Sedov, L. I.
Similarity and Dimensional Methods in Mechanics. Gostekhizdat, Moscow, 4th edition, 1957. English transl. (M. Holt, ed.), Academic Press, New York, 1959.
14. Raizer, Yu. P.
The formation of nitrogen oxides in the shock wave of a strong explosion in air, *Zh. Fiz. Khim.* **33**, 700–709 (1959).
15. Raizer, Yu. P.
Glow of air during a strong explosion, and the minimum brightness of a fireball, *Soviet Phys. JETP (English Transl.)* **7**, 331–339 (1958).
16. Zel'dovich, Ya. B., Kompaneets, A. S., and Raizer, Yu. P.
Radiation cooling of air. I. General description of the phenomenon and the weak cooling wave, *Soviet Phys. JETP (English Transl.)* **7**, 882–889 (1958).
17. Zel'dovich, Ya. B., Kompaneets, A. S., and Raizer, Yu. P.
Cooling of air by radiation. II. Strong cooling wave, *Soviet Phys. JETP (English Transl.)* **7**, 1001–1006 (1958).
18. Imshennik, V. S., and Nadezhin, D. K.
Gas dynamical model of a type II supernova outburst, *Soviet Astron.–AJ (English Transl.)* **8**, 664–673 (1965).
19. Abramson, I. S., Gegechkori, N. M., Drabkina, S. I., and Mandel'shtam, S. L.
The passage of a spark discharge, *Zh. Eksperim. i Teor. Fiz.* **17**, 862–867 (1947).
20. Drabkina, S. I.
The theory of the development of a spark discharge column, *Zh. Eksperim. i Teor. Fiz.* **21**, 473–483 (1951).
21. Gegechkori, N. M.
Experimental investigation of a spark discharge column, *Zh. Eksperim. i Teor. Fiz.* **21**, 493–506 (1951).
22. Dolgov, G. G., and Mandel'shtam, S. L.
Density and temperature of a gas in a spark discharge, *Zh. Eksperim. i Teor. Fiz.* **24** 691–700 (1953).
23. Mandel'shtam, S. L., and Sukhodrev, N. K.
Elementary processes in a spark discharge column, *Zh. Eksperim. i Teor. Fiz.* **24**, 701–707 (1953).
24. Sukhodrev, N. K.
On excited spectra in a spark discharge, *Tr. Fiz. Inst., Akad. Nauk SSSR* **15**, 123–177 (1961).
25. Braginskii, S. N.
Theory of the development of a spark channel, *Soviet Phys. JETP (English Transl.)* **7**, 1068–1074 (1958).
26. Zhivlyuk, Yu. N., and Mandel'shtam, S. L.
On the temperature of lightning and the force of thunder, *Soviet Phys. JETP (English Transl.)* **13**, 338–340 (1961).

27. Taylor, G. I.
The formation of a blast wave by a very intense explosion. II. The atomic explosion of 1945, *Proc. Roy. Soc. (London), Ser. A* **201**, 175–186 (1950).

Chapter X

1. Zel'dovich, Ya. B., and Kompaneets, A. S.
On the propagation of heat for nonlinear heat conduction, *Collection Dedicated to the Seventieth Birthday of Academician A. F. Ioffe* (P. I. Lukirskii, ed.), pp. 61–72. Izdat. Akad. Nauk SSSR, Moscow, 1959.
2. Barenblatt, G. I.
On some unsteady motions of a liquid and a gas in a porous medium, *Prikl. Mat. i Mekh.* **16**, 67–78 (1952).
3. Andriankin, E. I., and Ryzhov, O. S.
Propagation of an almost-spherical thermal wave, *Dokl. Akad. Nauk SSSR* **115**, 882–885 (1957).
4. Andriankin, E. I.
Propagation of a non-self-similar thermal wave, *Soviet Phys. JETP (English Transl.)* **8** 295–298 (1959).
5. Barenblatt, G. I.
On the approximate solution of problems of uniform unsteady filtration in a porous medium, *Prikl. Mat. i Mekh.* **18**, 351–370 (1954).
6. Barenblatt, G. I., and Zel'dovich, Ya. B.
On the dipole-type solution in problems of unsteady gas filtration in the polytropic regime, *Prikl. Mat. i Mekh.* **21**, 718–720 (1957).
7. Marshak, R. E.
Effect of radiation on shock wave behavior, *Phys. Fluids* **1**, 24–29 (1958).
8. Nemchinov, I. V.
Some unsteady radiative heat transfer problems, *Zh. Prikl. Mekhan. i Tekhn. Fiz.*, 1960, No. 1, 36–57.
9. Zel'dovich, Ya. B., and Barenblatt, G. I.
The asymptotic properties of self-modelling solutions of the nonstationary gas filtration equations, *Soviet Phys. "Doklady" (English Transl.)* **3**, 44–47 (1958).
10. Kompaneets, A. S., and Lantsburg, E. Ya.
The heating of gas by radiation, *Soviet Phys. JETP (English Transl.)* **14**, 1172–1176 (1962).
11. Kompaneets, A. S., and Lantsburg, E. Ya.
Propagation of a nonequilibrium heat wave with account of the finite velocity of light, *Soviet Phys. JETP (English Transl.)* **16**, 167–171 (1963).

Chapter XI

1. Al'tshuler, L. V., Krupnikov, K. K., Ledenev, B. N., Zhuchikhin, V. I., and Brazhnik, M. I.
Dynamic compressibility and equation of state of iron under high pressure, *Soviet Phys. JETP (English Transl.)* **7**, 606–614 (1958).
2. Al'tshuler, L. V., Krupnikov, K. K., and Brazhnik, M. I.
Dynamic compressibility of metals under pressures from 400,000 to 4,000,000 atmospheres, *Soviet Phys. JETP (English Transl.)* **7**, 614–619 (1958).

3. Al'tshuler, L. V., Kormer, S. B., Bakanova, A. A., and Trunin, R. F.
Equation of state for aluminum, copper, and lead in the high pressure region, *Soviet Phys. JETP (English Transl.)* **11**, 573–579 (1960).
4. Al'tshuler, L. V., Kormer, S. B., Brazhnik, M. I., Vladimirov, L. A., Speranskaya, M. P., and Funtikov, A. I.
The isentropic compressibility of aluminum, copper, lead, and iron at high pressures, *Soviet Phys. JETP (English Transl.)* **11**, 766–775 (1960).
5. Al'tshuler, L. V., Kuleshova, L. V., and Pavlovskii, M. N.
The dynamic compressibility, equation of state, and electrical conductivity of sodium chloride at high pressures, *Soviet Phys. JETP (English Transl.)* **12**, 10–15 (1961).
6. Shnirman, G. L., Dubovik, A. S., and Kevlishvili, P. V.
High-speed photorecording device for SFR (photographic scanning, *eds.*). Izdat. Inst. Tekhn.-Ekonom. Inform. Akad. Nauk SSSR, Moscow, 1957.
7. Dubovik, A. S.
Elements of the theory of mirror scanning, *Zh. Nauchn. i Prikl. Fotogr. i Kinematogr.* **2**, 293–303 (1957).
Mirror compensator of the displacement of photographic film, *Zh. Nauchn. i Prikl Fotogr. i Kinematogr.* **4**, 226–233 (1959).
8. Dubovik (Dubowik), A. S., Kevlishvili, P. V., and Shnirman, G. L.
Zeitlupe mit Mehrfach-Reflexion, *Kurzzeitphotographie Bericht über den IV. Internationalen Kongress für Kurzzeitphotographie und Hochfrequenzkinematographie* (H. Schardin and O. Helwich, eds.), pp. 196–201. Verlag Dr. Othmar Helwich, Darmstadt, 1959.
9. Shnirman, G. L.
Some problems of developing time magnification and photochronographs with a mirror scanner, *Usp. Nauchn. Fotogr., Akad. Nauk SSSR, Otd. Khim. Nauk* **6**, 93–101 (1959).
10. Dubovik, A. S.
Some problems in the theory of mirror scanning, *Usp. Nauchn. Fotògr., Akad. Nauk SSSR, Otd. Khim. Nauk* **6**, 102–112 (1959).
11. Tsukerman (Zuckermann), V. A., and Manakova, M. A.
Röntgen-Blitzquellen zur Untersuchung schnellverlaufender Vorgänge, *Kurzzeitphotographie: Bericht über den IV. Internationalen Kongress für Kurzzeitphotographie und Hochfrequenzkinematographie* (H. Schardin and O. Helwich, eds.), pp. 118–122. Verlag Dr. Othmar Helwich, Darmstadt, 1959.
12. Butslov, M. M., Zavoiskii, E. K., Plakhov, A. G., Smolkin, G. E., and Fanchenko, S. D.
Electron-optical method for studying short-duration phenomena, *Kurzzeitphotographie: Bericht über den IV. Internationalen Kongress für Kurzzeitphotographie und Hochfrequenzkinematographie* (H. Schardin and O. Helwich, eds.), pp. 230–242. Verlag Dr. Othmar Helwich, Darmstadt, 1959.
13. Gombàs, P.
Die statistische Theorie des Atoms und ihre Anwendungen. Springer, Wien, 1949.
14. Kormer, S. B., and Urlin, V. D.
Interpolation equations of state of metals for the region of ultrahigh pressures, *Soviet Phys. "Doklady" (English Transl.)* **5**, 317–320 (1960).
15. Kormer, S. B., Urlin, V. D., and Popova, L. T.
Interpolation equation of state and its application to experimental data on impact compression of metals, *Soviet Phys.–Solid State (English Transl.)* **3**, 1547–1553 (1962).
16. Landau, L. D., and Lifshitz, E. M.
Statistical Physics. Addison-Wesley, Reading, Mass., 1958.
17. Slater, J. C.
Introduction to Chemical Physics. McGraw-Hill, New York, 1st edition, 1939.

18. Landau, L. D., and Stanyukovich, K. P.
On a study of the detonation of condensed explosives, *Compt. Rend.* (*Doklady*) *Acad. Sci. URSS* **46**, 362–364 (1945).
19. Gilvarry, J. J.
Thermodynamics of the Thomas-Fermi atom at low temperature, *Phys. Rev.* **96**, 934–943 (1954).
Solution of the temperature-perturbed Thomas-Fermi equation, *Phys. Rev.* **96**, 944–948 (1954).
Gilvarry, J. J., and Peebles, G. H.
Solutions of the temperature-perturbed Thomas-Fermi equation, *Phys. Rev.* **99**, 550–552 (1955).
20. Latter, R.
Temperature behavior of the Thomas-Fermi statistical model for atoms, *Phys. Rev.* **99**, 1854–1870 (1955).
21. Baum, F. A., Stanyukovich, K. P., and Shekhter, B. I.
Explosion Physics. Fizmatgiz, Moscow, 1959.
22. Walsh, J. M., and Christian, R. H.
Equation of state of metals from shock wave measurements, *Phys. Rev.* **97**, 1544–1556 (1955).
23. Walsh, J. M., Rise, M. H., McQueen, R. G., and Yarger, F. L.
Shock-wave compressions of twenty-seven metals. Equations of state of metals, *Phys. Rev.* **108**, 196–216 (1957).
24. Goranson, R. W., Bancroft, D., Blendin, L. B., Blechar, T., Houston, E. E., Gittings, E. F., and Landeen, S. A.
Dynamic determination of the compressibility of metals, *J. Appl. Phys.* **26**, 1472–1479 (1955).
25. Mallory, H. D.
Propagation of shock waves in aluminum, *J. Appl. Phys.* **26**, 555–559 (1955).
26. McQueen, R. G., and Marsh, S. P.
Equation of state for nineteen metallic elements from shock-wave measurements to two megabars, *J. Appl. Phys.* **31**, 1253–1269 (1960).
27. Dugdale, J. S., and McDonald, D. K. C.
The thermal expansion of solids, *Phys. Rev.* **89**, 832-834 (1953).
28. Landau, L. D., and Lifshitz, E. M.
Theory of Elasticity. Addison-Wesley, Reading, Mass., 1959.
29. Rakhmatulin, Kh. A., and Shapiro, G. S.
Propagation of disturbances in nonlinear-elastic and inelastic media, *Izv. Akad. Nauk SSSR, Otd. Tekhn. Nauk*, 1955, No. 2, 68–89.
30. Bancroft, D., Peterson, E. L., and Minshall, S.
Polymorphism of iron at high pressures, *J. Appl. Phys.* **27**, 291–298 (1956).
31. Duff, R. E., and Minshall, F. S.
Investigation of a shock-induced transition in bismuth, *Phys. Rev.* **108**, 1207–1212 (1957).
32. Drummond, W. E.
Multiple shock production, *J. Appl. Phys.* **28**, 998–1001 (1957).
33. Dremin, A. N., and Adadurov, G. A.
Shock adiabatic for marble, *Soviet Phys.* "*Doklady*" (*English Transl.*) **4**, 970–973 (1960).
34. Dremin, A. N., and Karpukhin, I. A.
Method of determining the Hugoniot curves for dispersive media, *Zh. Prikl. Mekhan. i Tekhn. Fiz.*, 1960, No. 3, 184–187.

35. Alder, B. J., and Christian, R. H.
Destruction of diatomic bonds by pressure, *Phys. Rev. Letters* **4**, 450–452 (1960).
36. Lifshitz, I. M.
Anomalies of electron characteristics of a metal in the high pressure region, *Soviet Phys. JETP* (*English Transl.*) **11**, 1130–1135 (1960).
37. Gandel'man, G. M.
Quantum-mechanical derivation of an equation of state of iron, *Soviet Phys. JETP* (*English Transl.*) **16**, 94–103 (1963).
38. Ivanov, A. G., and Novikov, S. A.
Rarefaction shock waves in iron and steel, *Soviet Phys. JETP* (*English Transl.*) **13**, 1321–1323 (1961).
39. Zadumkin, S. N.
Approximate estimate of the critical temperatures of liquid metals, *Inzh.-Fiz. Zh., Akad. Nauk Belorussk.*, 1960, No. 10, 63–65.
40. *Handbook of Chemistry, Vol. 1.* Goskhimizdat, Moscow, 1951.
41. Zel'dovich, Ya. B., Kormer, S. B., Sinitsyn, M. V., and Kuryapin, A. I.
Temperature and specific heat of Plexiglas under shock wave compression, *Soviet Phys. "Doklady"* (*English Transl.*) **3**, 938–939 (1958).
42. Zel'dovich, Ya. B.
Investigations of the equation of state by mechanical measurements, *Soviet Phys. JETP* (*English Transl.*) **5**, 1287–1288 (1957).
43. Zel'dovich, Ya. B., and Raizer, Yu. P.
Physical phenomena that occur when bodies compressed by strong shock waves expand in vacuo, *Soviet Phys. JETP* (*English Transl.*) **8**, 980–982 (1959).
44. Urlin, V. D., and Ivanov, A. A.
Melting under shock-wave compression, *Soviet Phys. "Doklady"* (*English Transl.*) **8**, 380–382 (1963).
45. Brish, A. A., Tarasov, M. S., and Tsukerman, V. A.
Electric conductivity of the explosion products of condensed explosives, *Soviet Phys. JETP* (*English Transl.*) **10**, 1095–1100 (1960).
46. Brish, A. A., Tarasov, M. S., and Tsukerman, V. A.
Electric conductivity of dielectrics in strong shock waves, *Soviet Phys. JETP* (*English Transl.*) **11**, 15–17 (1960).
47. Alder, B. J., and Christian, R. H.
Metallic transition in ionic and molecular crystals, *Phys. Rev.* **104**, 550–551 (1956).
48. Zel'dovich, Ya. B., and Landau, L. D.
On the relation between the liquid and gaseous states in metals, *Zh. Eksperim. i Teor. Fiz.* **14**, 32–34 (1944).
49. Abrikosov, A. A.
Equation of state of hydrogen at high pressures, *Astron. Zh.* **31**, 112–123 (1954).
50. Bridgman, P. W.
The Physics of High Pressure. Macmillan, New York, 1931.
Recent work in the field of high pressures, *Rev. Mod. Phys.* **18**, 1–93 (1946).
51. Zel'dovich, Ya. B., Kormer, S. B., Sinitsyn, M. V., and Yushko, K. B.
A study of the optical properties of transparent materials under high pressure, *Soviet Phys. "Doklady"* (*English Transl.*) **6**, 494–496 (1961).
52. Landau, L. D., and Lifshitz, E. M.
Electrodynamics of Continuous Media. Addison-Wesley, Reading, Mass., 1960.
53. Cowan, G. R., and Hornig, D. F.
The experimental determination of the thickness of a shock front in a gas, *J. Chem. Phys.* **18**, 1008–1018 (1950).

54. Ginzburg, V. L., and Motulevich, G. P.
Optical properties of metals, *Usp. Fiz. Nauk* **55**, 469–535 (1955).
55. Al'tshuler, L. V.
Use of shock waves in high-pressure physics, *Soviet Phys.–Usp.* (*English Transl.*) **8**, 52–91 (1965).
56. Kormer, S. B., Funtikov, A. I., Urlin, V. D., Kolesnikova, A. N.
Dynamic compression of porous metals and the equation of state with variable specific heat at high temperatures, *Soviet Phys. JETP* (*English Transl.*) **15**, 477–488 (1962).
57. Kuznetsov, N. M.
The break in a Hugoniot curve in a phase transition of the first kind, *Dokl. Akad. Nauk SSSR* **155**, 156–159 (1964).
58. Kuznetsov, N. M.
On the kinetics of shock melting of polycrystals, *Zh. Prikl. Mekhan. i Tekhn. Fiz.*, 1965, No. 1, 112–114. Also *J. Appl. Mech. and Tech. Phys.* (*English Transl.*), 1965, No. 1, 104–106.
59. Urlin, V. D.
Melting at ultra high pressures in a shock wave, *Soviet Phys. JETP* (*English Transl.*) **22** 341–346 (1966).
60. Duvall, G. E.
Some properties and application of shock waves, *Response of Metals to High Velocity Deformation* (P. G. Shewmon and V. F. Zackay, eds.), pp. 165–203. Wiley (Interscience), New York, 1961.
Duvall, G. E., and Fowles, G. R.
Shock waves, *High Pressure Physics and Chemistry*, *Vol. 2* (R. S. Bradley, ed.), pp. 209–291. Academic Press, New York, 1963.
61. Donnell, L. H.
Longitudinal wave transmission and impact, *Trans. ASME* (*Am. Soc. Mech. Eng.*) **52**, APM 153–167 (1930).
62. Bethe, H. A.
Theory of shock waves for an arbitrary equation of state, *Off. Sci. Res. Dev. Rept.* No. 545, 1942.

Chapter XII

1. Stanyukovich, K. P.
Unsteady Motion of Continuous Media. Gostekhizdat, Moscow, 1955. English transl. (M. Holt, ed.), Academic Press, New York, 1960.
2. Sedov, L. I.
Similarity and Dimensional Methods in Mechanics. Gostekhizdat, Moscow, 4th edition, 1957. English transl. (M. Holt, ed.), Academic Press, New York, 1959.
3. Guderley, G.
Starke kugelige und zylindrische Verdichtungstösse in der Nähe des Kugelmittelpunktes bzw. der Zylinderische, *Luftfahrtforschung* **19**, 302–312 (1942).
4. Lord Rayleigh
On the pressure developed in a liquid during the collapse of a spherical cavity, *Phil. Mag.* **34** (6th Series), 94–98 (1917).
5. Hunter, C.
On the collapse of an empty cavity in water, *J. Fluid Mech.* **8**, 241–263 (1960).
6. Zababakhin, E. I.
The collapse of bubbles in a viscous liquid, *Appl. Math. Mech., PMM* (*English Transl.*) **24**, 1714–1717 (1960).

7. Frank-Kamenetskii, D. A.
Nonadiabatic pulsations in stars, *Dokl. Akad. Nauk SSSR* **80**, 185–188 (1951).
8. Gandel'man, G. M., and Frank-Kamenetskii, D. A.
Shock wave emergence at a stellar surface, *Soviet Phys. "Doklady" (English Transl.)* **1**, 223–226 (1956).
9. Sakurai, A.
On the problem of a shock wave arriving at the edge of a gas, *Commun. Pure Appl. Math.* **13**, 353–370 (1960).
10. Ginzburg, V. L., and Syrovatskii, S. I.
Present status of the question of the origin of cosmic rays, *Soviet Phys.–Usp. (English Transl.)* **3**, 504–541 (1961).
11. Colgate, S. A., and Johnson, M. H.
Hydrodynamic origin of cosmic rays, *Phys. Rev. Letters* **5**, 235–238 (1960).
12. Zel'dovich, Ya. B.
Motion of a gas under the action of an impulsive pressure (load), *Akust. Zh.* **2**, 28–38 (1956).
13. Adamskii, V. B.
Integration of the system of self-similar equations in the problem of an impulsive load on a cold gas, *Akust. Zh.* **2**, 3–9 (1956).
14. Zhukov, A. I., and Kazhdan, Ya. M.
On the motion of a gas under the action of a short duration impulse, *Akust. Zh.* **2** 352–357 (1956).
15. Häfele, W.
Zur analytischen Behandlung ebener, starker, instationärer Stosswellen, *Z. Naturforsch.* **10a**, 1006–1016 (1955).
16. von Hoerner, S.
Lösungen der hydrodynamischen Gleichungen mit linearem Verlauf der Geschwindigkeit, *Z. Naturforsch.* **10a**, 687–692 (1955).
17. von Weizsäcker, C. F.
Genäherte Darstellung starker instationärer Stosswellen durch Homologie-Lösungen, *Z. Naturforsch.* **9a**, 269–275 (1954).
18. Adamskii, V. B., and Popov, N. A.
The motion of a gas under the action of a pressure on a piston varying according to a power law, *Appl. Math. Mech., PMM (English Transl.)* **23**, 793–806 (1959).
19. Krasheninnikova, N. L.
On the unsteady motion of a gas displaced by a piston, *Izv. Akad. Nauk SSSR, Otd. Tekhn. Nauk*, 1955, No. 8, 22–36.
20. Raizer, Yu. P.
Motion of a gas under the action of a concentrated impact along its surface (as a result of an explosion on the surface), *Zh. Prikl. Mekhan. i Tekhn. Fiz.*, 1963, No. 1, 57–66.
21. Astapovich, I. S.
Second conference on comet and meteor astronomy, Moscow 29–31 January 1937, *Astron. Zh.* **14**, 248–250 (1937).
Stanyukovich, K. P., and Fedynskii, V. V.
On the destructive effect of meteor impacts, *Dokl. Akad. Nauk SSSR* **57**, 129–132 (1947).
Stanyukovich, K. P.
Elements of the physical theory of meteors and the formation of meteor craters, *Meteoritika*, 1950, No. 7, 39–62.
Elements of the theory of the impact of solid bodies with high (cosmic) velocities, *Iskusstvenyie Sputniki Zemli* **4**, 86–117 (1960). English transl. *Artificial Earth Satellites, Vols. 3–5* (L. V. Kurnosova, ed.), pp. 292–333. Plenum Press, New York, 1961.

22. Stanyukovich, K. P.
On an effect in the area of the aerodynamics of meteors, *Izv. Akad. Nauk SSSR, Otd. Tekh. Nauk, Mekh. i Mashinostr.*, 1960, No. 5, 3–8.
23. Lavrent'ev, M. A.
The problem of piercing at cosmic velocities, *Iskusstvenyie Sputniki Zemli* **3**, 61–65 (1959). English transl. *Artificial Earth Satellites, Vols. 3–5* (L. V. Kurnosova, ed.), pp. 85–91. Plenum Press, New York, 1961.
24. Kompaneets, A. S.
Shock waves in a plastic compacting medium, *Dokl. Akad. Nauk SSSR* **109**, 49–52 (1956).
25. Andriankin, E. I., and Koryavov, B. P.
Shock waves in a variable compacting plastic medium, *Soviet Phys. "Doklady" (English Transl.)* **4**, 966–969 (1960).
26. Zel'dovich, Ya. B.
Cylindrical self-similar acoustical waves, *Soviet Phys. JETP (English Transl.)* **6**, 537–541 (1958).
27. Zababakhin, E. I., and Nechaev, M. N.
Electromagnetic-field shock waves and their cumulation, *Soviet Phys. JETP (English Transl.)* **6**, 345–351 (1958).
28. Kompaneets, A. S.
A point explosion in an inhomogeneous atmosphere, *Soviet Phys. "Doklady" (English Transl.)* **5**, 46–48 (1960).
29. Andriankin, E. I., Kogan, A. M., Kompaneets, A. S., and Krainov, V. P.
Propagation of a strong explosion in an inhomogeneous atmosphere, *Zh. Prikl. Mekhan. i Tekhn. Fiz.*, 1962, No. 6, 3–7.
30. Raizer, Yu. P.
Motion produced in an inhomogeneous atmosphere by a plane shock of short duration, *Soviet Phys. "Doklady" (English Transl.)* **8**, 1056–1058 (1964).
31. Raizer, Yu. P.
Propagation of a shock wave in an inhomogeneous atmosphere in the direction of decreasing density, *Zh. Prikl. Mekhan. i Tekhn. Fiz.*, 1964, No. 4, 49–56.
32. Raizer, Yu. P.
The deceleration and energy conversions of a plasma expanding in a vacuum in the presence of a magnetic field, *Zh. Prikl. Mekhan. i Tekhn. Fiz.*, 1963, No. 6, 19–28. Transl. as *NASA (Nat. Aeron. Space Admin.) Tech. Transl.* No. TTF-239 (1964).
33. Hayes, W. D.
Self-similar strong shocks in an exponential medium, *J. Fluid Mech.*, **32**, 305–315 (1968).
34. Hayes, W. D.
The propagation upward of the shock wave from a strong explosion in the atmosphere, *J. Fluid Mech.*, **32**, 317–331 (1968).

Appendix

Some often used constants, relations between units, and formulas*

Fundamental constants

Speed of light	$c = 2.998 \cdot 10^{10}$ cm/sec
Planck constant	$h = 6.625 \cdot 10^{-27}$ erg · sec
	$\hbar = h/2\pi = 1.054 \cdot 10^{-27}$ erg · sec
Electron charge	$e = 4.803 \cdot 10^{-10}$ esu
Electron mass	$m = 9.108 \cdot 10^{-28}$ g
Proton mass	$m_p = 1.673 \cdot 10^{-24}$ g
Mass of unit atomic weight	$M_0 = 1.660 \cdot 10^{-24}$ g
Boltzmann constant	$k = 1.380 \cdot 10^{-16}$ erg/deg
Universal gas constant	$\mathscr{R} = 8.317 \cdot 10^{7}$ erg/deg · mole
	$= 1.987$ cal/deg · mole
Avogadro number	$N_0 = 6.023 \cdot 10^{23}$ mole^{-1}
Loschmidt number	$n_0 = 2.687 \cdot 10^{19}$ cm^{-3}

Relations between units

Energy $E_0 = 1$ ev $= 1.602 \cdot 10^{-12}$ erg corresponds to:

temperature	$E_0/k = 11{,}605°$K
frequency	$E_0/h = 2.418 \cdot 10^{14}$ sec^{-1}
wavelength	$hc/E_0 = 1.240 \cdot 10^{-4}$ cm $= 12{,}400$ Å
wave number	$E_0/hc = 8066$ cm^{-1}

* Values of the constants are taken from C. W. Allen, *Astrophysical Quantities.* (Athlone Press) Oxford Univ. Press, New York, 2nd edition, 1963. *Editors' note.* Some of the constants and numerical values in this appendix may differ slightly in the last significant figure from those used in the body of the text, which were based on constants given in the first edition (1955) of Allen's book.

In spectroscopy wave number is often used in place of frequency.

Wave number $1/\lambda = 1\ \text{cm}^{-1}$ corresponds to:

wavelength	$\lambda = 10^8$ Å
frequency	$\nu = 2.998 \cdot 10^{10}\ \text{sec}^{-1}$
temperature	$T = h\nu/k = 1.439°\text{K}$
photon energy	$h\nu = 1.240 \cdot 10^{-4}\ \text{ev} = 1.986 \cdot 10^{-16}\ \text{erg}$

$1\ \text{cal} = 4.185 \cdot 10^7\ \text{erg}$, $1\ \text{kcal} = 10^3\ \text{cal}$

1 ev per molecule corresponds to 23.05 kcal/mole

1 volt = 1/300 units of potential in esu

Constants and relations between them

Bohr radius $$a_0 = \frac{h^2}{4\pi^2 me^2} = \frac{\hbar^2}{me^2} = 0.529 \cdot 10^{-8}\ \text{cm}$$

Ionization potential of hydrogen atom $$I_H = \frac{e^2}{2a_0} = \frac{2\pi^2 e^4 m}{h^2} = \frac{e^4 m}{2\hbar^2} = 13.60\ \text{ev}$$

Rydberg constant $$Ry = \frac{I_H}{h} = \frac{2\pi^2 e^4 m}{h^3} = 3.290 \cdot 10^{15}\ \text{sec}^{-1}$$

Electron speed in first Bohr orbit $$v_0 = \frac{2\pi e^2}{h} = \frac{e^2}{\hbar} = 2.188 \cdot 10^8\ \text{cm/sec}$$

Classical electron radius $$r_0 = \frac{e^2}{mc^2} = 2.818 \cdot 10^{-13}\ \text{cm}$$

Compton wavelength $$\lambda_0 = \frac{h}{mc} = 2.426 \cdot 10^{-10}\ \text{cm}$$

$$\bar{\lambda}_0 = \frac{\lambda_0}{2\pi} = \frac{\hbar}{mc} = 3.862 \cdot 10^{-11}\ \text{cm}$$

Rest mass energy of electron $$mc^2 = 511\ \text{kev} = 8.186 \cdot 10^{-7}\ \text{erg}$$

The number "137" = (fine structure constant)$^{-1}$ $\quad \dfrac{\hbar c}{e^2} = \dfrac{hc}{2\pi e^2} = 137.0$

Length ratio $\quad a_0 = \text{“137”}\, \lambda\!\!\!^{-}_0 = \text{“137”}^2 r_0$

Energy ratio $\quad mc^2 = 2I_{\mathrm{H}} \cdot \text{“137”}^2$

Thomson cross section $\quad \varphi_0 = \dfrac{8}{3}\pi r_0^2 = 6.65 \cdot 10^{-25}\ \mathrm{cm}^2$

Mass ratio proton/electron $\quad m_p/m = 1836$

Electric field of a proton at a distance of the first Bohr radius $\quad \dfrac{e^2}{a_0^2} = 1.715 \cdot 10^7\ \mathrm{esu} = 5.145 \cdot 10^9\ \mathrm{volt/cm}$

Area of spectral line with unit oscillator strength $\quad \dfrac{\pi e^2}{mc} = 0.0265\ \mathrm{cm^2/sec}$

Atomic unit cross section $\quad \pi a_0^2 = 0.880 \cdot 10^{-16}\ \mathrm{cm}^2$

Formulas

Radiant energy flux from surface of a perfect black body

$$S = \sigma T^4 = 5.67 \cdot 10^{-5} T_{\mathrm{deg}}^4$$

$$= 1.03 \cdot 10^{12} T_{\mathrm{ev}}^4\ \mathrm{erg/cm^2 \cdot sec}$$

(σ = Stefan-Boltzmann constant)

Equilibrium radiant energy density

$$U_p = \frac{4\sigma T^4}{c} = 7.56 \cdot 10^{-15} T_{\mathrm{deg}}^4$$

$$= 1.37 \cdot 10^2 T_{\mathrm{ev}}^4\ \mathrm{erg/cm^3}$$

Spectral equilibrium radiant energy density

$$U_{\nu p}\, d\nu = \frac{8\pi h\nu^3}{c^3} \frac{1}{e^{h\nu/kT} - 1}\, d\nu \text{ erg/cm}^3$$

(maximum at the frequency $h\nu = 2.822\ kT$)

Spectral equilibrium radiation intensity

$$I_{\nu p}\, d\nu = \frac{cU_{\nu p}\, d\nu}{4}$$

$$= \frac{2h\nu^3}{c^2} \frac{d\nu}{e^{h\nu/kT} - 1} \text{ erg/cm}^2 \cdot \text{sec} \cdot \text{sterad}$$

Saha equation

$$\frac{N_e N_+}{N_a} = A \frac{g_+}{g_a} T^{3/2}\, e^{-I/kT}$$

where

$$A = 2\left(\frac{2\pi mk}{h^2}\right)^{3/2} = 4.83 \cdot 10^{15} \text{ cm}^{-3} \cdot \text{deg}^{-3/2}$$

$$= 6.04 \cdot 10^{21} \text{ cm}^{-3} \cdot \text{ev}^{-3/2}$$

Maxwell distribution function normalized to unity

$$f(v)\, dv = 4\pi \left(\frac{m}{2\pi kT}\right)^{3/2} \exp\left(-\frac{mv^2}{2kT}\right) v^2\, dv$$

$$f(\varepsilon)\, d\varepsilon = \frac{2}{\pi^{1/2}} \frac{\varepsilon^{1/2}}{(kT)^{3/2}}\, e^{-\varepsilon/kT}\, d\varepsilon$$

Electron speed

$$v_e = 5.93 \cdot 10^7 \varepsilon_{\text{ev}}^{1/2} \text{ cm/sec}$$

Speed of particle with atomic weight A

$$v = 1.38 \cdot 10^6 (\varepsilon_{\text{ev}}/A)^{1/2} \text{ cm/sec}$$

Electron mean thermal speed

$$\bar{v}_e = \left(\frac{8kT}{\pi m}\right)^{1/2} = 6.21 \cdot 10^5 T_{\text{deg}}^{1/2}$$

$$= 6.69 \cdot 10^7 T_{\text{ev}}^{1/2} \text{ cm/sec}$$

Particle mean thermal speed

$$\bar{v} = 1.45 \cdot 10^4 \left(\frac{T_{\text{deg}}}{A}\right)^{1/2}$$

$$= 1.56 \cdot 10^6 \left(\frac{T_{\text{ev}}}{A}\right)^{1/2} \text{ cm/sec}$$

Classical damping constant

$$\gamma = \frac{8\pi^2 e^2 \nu^2}{3mc^3} = 2.47 \cdot 10^{-22} \nu^2_{\text{sec}^{-1}}$$

$$= \frac{0.222 \cdot 10^{16}}{\lambda^2_{\text{Å}}} \text{ sec}^{-1}$$

Cross section σ in terms of P_c = average number of collisions per cm at 1 mm Hg and 0°C

$$\sigma = 2.83 \cdot 10^{-17} P_c \text{ cm}^2$$

Specific energy

$$1 \text{ ev/molecule} = \frac{9.65 \cdot 10^{11}}{M} \text{ erg/g}$$

where

M = molecular weight

Author Index

Volumes I and II

Numbers in brackets are reference numbers and are included to assist in locating references in which the authors' names are not mentioned in the text. Italicized numbers indicate the pages on which the references are listed, and those in parentheses the number of citations for the given page.

Subject Index

Volumes I and II